Computational Wind Engineering 2

Computational Wind Engineering 2

Proceedings of the
2nd International Symposium on
Computational Wind Engineering (CWE 96)
Fort Collins, Colorado, USA
August 4-8, 1996

Edited by

Robert Meroney
and
Bogusz Bienkiewicz
Department of Civil Engineering
Colorado State University
Fort Collins, Colorado, USA

1997

ELSEVIER

AMSTERDAM • LAUSANNE • NEW YORK • OXFORD • SHANNON • SINGAPORE • TOKYO

ELSEVIER SCIENCE B.V.
Sara Burgerhartstraat 25
P.O. Box 211, 1000 AE Amsterdam, The Netherlands

Reprinted from Journal of Wind Engineering and Industrial Aerodynamics, 67 & 68 (1997)

ISBN: 0 444 82878 8

This book is printed on acid-free paper.

Transferred to digital printing 2006

Preface

The Second International Symposium on Computational Wind Engineering (CWE-96) was held at Colorado State University in Fort Collins, Colorado, USA, during early August 1996. The objective of this symposium was to examine the accomplishments and challenges posed by the rapid development of Computational Fluid Dynamics (CFD) as applied to the discipline of Wind Engineering (WE).

This symposium follows the lead of the CWE-92 Symposium held at the University of Tokyo under the leadership of Dr. Shuzo Murakami, Institute of Industrial Science. The proceedings of that meeting appeared as Volumes 46&47 of the *Journal of Wind Engineering and Industrial Aerodynamics*, and as a bound volume, *CWE 92: Computational Wind Engineering I* edited by Dr. Murakami, and published by Elsevier Science B.V. in 1993.

In the past four years, the computational foundations identified during the first symposium have grown, been modified, and undergone extensive scrutiny and validation. Although all expectations have not been met, the initial optimism for the impact of CFD on the wind engineering research and design community has been confirmed. Today, CFD calculations are routinely applied to bluff body aerodynamics, bridge aerodynamics, building aerodynamics, atmospheric transport, internal air pollution, pedestrian comfort, and transportation aerodynamics. Significant effort is being spent to verify and validate the accuracy and reliability of numerical methods and codes.

During the Second Symposium, state-of-the-art presentations from the international community were presented, affording researchers and engineers the opportunity to evaluate their most recent findings contributing to the development of this field. Participants representing industry, governments, and academia, world wide, attended this Symposium, which was designed to showcase present CFD activities and identify fruitful areas for further research.

Eighteen technical sessions were organized including five special keynote presentations featuring invited speakers outstanding in the fields of computational fluid dynamics and wind engineering. At the end of the conference, a workshop sponsored by the International Wind Engineering Forum focused on CWE/CFD for Prediction of Wind Effects on Structures. The presentations of this workshop were distributed in a separate proceedings volume.

Selected papers from the Symposium are compiled in this Proceedings volume. All papers included herein have been subjected to strenuous review by several anonymous scientists, and the authors were invited to make minor or major modifications as appropriate. Summaries of the discussions, questions and author responses are also included. The editors anticipate the Proceedings will serve to promote future development in Computational Wind Engineering. I would like to express my sincere gratitude and appreciation of the high technical competence and standards exhibited by all the participants, authors and reviewers.

Finally, I would like to extend my sincere thanks to the many individuals and organizations who assisted during the organization of the CWE-96 Symposium and the subsequent preparation of the Proceedings volume. In particular, I would like to express my special thanks to Professor Bogusz Bienkiewicz, Symposium Co-chairman, Professor Shuzo Murakami, CWE-92 Symposium Chair, Ms. Janet Montera, Symposium Manager, and Ms. Marilee Rowe, Proceedings Secretary, respectively.

The Symposium was organized by the Fluid Mechanics and Wind Engineering Program, Civil Engineering Department, Colorado State University in cooperation with:

AIJ (Architectural Institute of Japan),
Aerodynamics Committee, Aerospace Division, ASCE (American Society of Civil Engineers),
IWEF (International Wind Engineering Forum),
JAWE (Japan Association for Wind Engineering),
AAWE (American Association of Wind Engineering), and
WES (Wind Engineering Society, United Kingdom).

The program was arranged with the assistance of a Scientific Advisory Committee and the Symposium was assisted by members of the Organizing Committee, members of which are listed on the following page. Other individuals assisted as Session Chair and Co-chairs, Session Assistants, and Paper Reviewers.

Robert Meroney
Symposium Chairman

Symposium Organization

Local Organizing Committee

Robert N. Meroney (Chair), Civil Engineering, Colorado State University
Bogusz Bienkiewicz (Co-Chair), Civil Engineering, Colorado State University
Janet L. Montera (Symposium Manager), Civil Engineering, Colorado State University
Patrick J. Burns, Mechanical Engineering, Colorado State University
Jack E. Cermak, Civil Engineering, Colorado State University
John H. Cooley, Academic Computer Networking Services, Colorado State University
Allan T. Kirkpatrick, Mechanical Engineering, Colorado State University
K.C. Mehta, Civil Engineering, Texas Tech University
Melville E. Nicholls, Atmospheric Science, Colorado State University
Roger A. Pielke, Atmospheric Science, Colorado State University
James W. Thomas, Mathematics, Colorado State University
Eric G. Thompson, Civil Engineering, Colorado State University

Scientific Advisory Committee

Joel H. Ferziger, Mechanical Engineering, Stanford University, USA
Ahsan Kareem, Civil Engineering and Geological Sciences, Notre Dame University, USA
Brian Lee, Civil Engineering, University of Portsmouth, United Kingdom
William H. Melbourne, Mechanical Engineering, Monash University, Australia
Shuzo Murakami, Institute of Industrial Science, University of Tokyo, Japan
Michael Schatzmann, Meteorologisches Institut, University of Hamburg, Germany
Naruhito Shiraishi, Civil Engineering, Kyoto University, Japan
Jean Fran Sini, Laboratoire de Mecanique des Fluides, Ecole Centrale de Nantes, France
Theodore Stathopoulos, Centre for Building Studies, Concordia University, Canada
H.W. Tieleman, Engineering Science and Mechanics, Virginia Polytechnique and State University, USA

Reviewers

Keith Ayotte, CSIRO, Australia
S. Belcher, University of Reading, United Kingdom
Bogusz Bienkiewicz, Colorado State University, USA
Darryl Boggs, Cermak, Peterka, and Peterson Engineering, USA
Patrick J. Burns, Colorado State University, USA
Jack E. Cermak, Colorado State University, USA
Chin-Hoh Moeng, National Center for Atmospheric Research, USA
Leighton Cochran, Cermak, Peterka, and Peterson Engineering, USA
Russell Derickson, South Dakota School of Mines and Technology, USA
Joel Ferziger, Stanford University, USA
Roger Gansthrop, BR Research, United Kingdom
J.M.R. Graham, Imperial College, United Kingdom
Paul Heyliger, Colorado State University, USA
I.P. Jones, AEA Technology, Didcot, United Kingdom
Heinke Schluenzen, University of Hamburg, Germany
Ahsan Kareem, Notre Dame University, USA
Allen T. Kirkpatrick, Colorado State University, USA
Martin Korries, University of Hamburg, Germany
Allan Larsen, COWI, Denmark
Brian Lee, University of Portsmouth, United Kingdom
Bernd Leitl, Colorado State University, USA and University of Hamburg, Germany
Claus-Jergen Lenz, University of Hamburg, Germany
M. Leschziner, University of Manchester, United Kingdom
K.C. Mehta, Texas Tech University, USA
William H. Melbourne, Monash University, Australia
Robert N. Meroney, Colorado State University, USA
Akashi Mochida, Niigata Institute of Technology, Japan
Shuzo Murakami, University of Tokyo, Japan
David Neff, Colorado State University, USA
Melville E. Nicholls, Colorado State University, USA
Ulrike Niemeier, University of Hamburg, Germany
A. Nishi, Miyazaki University, Japan
Takashi Nomura, Nihon University, Japan
Y. Ohya, Kyushu University, Japan
Michel Pavageau, University of Nantes, France
Jon Peterka, Colorado State University, USA
Ronald Peterson, Cermak, Peterka, and Peterson Engineering, USA
Roger Pielke, Colorado State University, USA
Mark Powell, National Oceanic and Atmospheric Administration, USA
S. Rafailidis, University of Hamburg, Germany
Michael R. Raupach, CSIRO, Australia
A. Robins, University of Surrey, United Kingdom
Virgil A. Sandborn, Colorado State University, USA

M. Schatzmann, University of Hamburg, Germany
Naruhito Shirashi, Kyoto University, Japan
John Snook, National Oceanic and Atmospheric Administration, USA
W. Snyder, U.S. Environmental Protection Agency, USA
Theodore Stathopoulos, Concordia University, Canada
T. Tamura, Tokyo Institute of Technology, Japan
James W. Thomas, Colorado State University, USA
D. Thomson, Meteorological Office, Bracknell, United Kingdom
O. Tutty, University of Southampton, United Kingdom
P.J. Walklate, Silsoe Research Institute, United Kingdom
Bob Walko, Colorado State University, USA

Contents

Discussions

Discussions

Keynote Presentation

Journal of Wind Engineering
and Industrial Aerodynamics 67&68 (1997) 3–34

Current status and future trends in computational wind engineering

Shuzo Murakami

Institute of Industrial Science, University of Tokyo, 7-22-1, Roppongi, Minato-ku, Tokyo 106, Japan

Abstract

The difficulty of Computational Wind Engineering (hereafter CWE) is described from the viewpoints of Computational Fluid Dynamics (hereafter CFD) technique. The rapid growth of CFD applications to wind engineering is presented. The new trends in turbulence models for applying CWE are noted. The advantages of dynamic subgrid scale (hereafter SGS) models in Large Eddy Simulation (hereafter LES) are clarified.

Keywords: Wind engineering; CFD application; Revised k–ε model; Dynamic LES

Nomenclature

x_i	three components of spatial coordinate ($i = 1, 2, 3$: streamwise, lateral, vertical (or spanwise))
$\langle f \rangle$	time-averaged value of f
$\bar{f}$	filtered value of f
f'	deviation from $\langle f \rangle$, $f' = f - \langle f \rangle$ (in LES, $f' = \bar{f} - \langle \bar{f} \rangle$)
f''	deviation from $\bar{f}$, $f'' = f - \bar{f}$
D	width of square cylinder
H	height of cube
u_i	three components of velocity vector
U_H	time-averaged value of u_1 at the inflow boundary of computational domain at height of H
U_0	time-averaged value of u_1 at the inflow boundary for the case of 2D square cylinder
p	pressure
C_p	instantaneous pressure coefficient ($= (p - \langle p_0 \rangle/(\frac{1}{2}\rho U_0^2)$)
$\langle p_0 \rangle$	reference static pressure
ν_t	eddy viscosity
ν_{SGS}	subgrid eddy viscosity

0167-6105/97/$17.00
PII S0167-6105(97)00230-4

k	turbulence energy ($= \frac{1}{2}\langle u_i' u_i' \rangle$)
P_k	production of k
ε	dissipation rate of k
$\langle u_i' u_j' \rangle$	Reynolds stress
ϕ_{ij}	pressure–strain correlation term (consist of slow term $\phi_{ij(1)}$, rapid term $\phi_{ij(2)}$ and wall reflect term $\phi_{ij(1)}^{w}$, $\phi_{ij(2)}^{w}$)
n	frequency
$S(n)$	power spectrum

When values are made dimensionless, representative length scale (D or H), velocity scale (U_0 or U_H) are used.

1. Introduction

CWE is a youthful field of research. Only 10 years or so have passed since its birth, and its development has been very fast. During the four years from CWE 92 in Tokyo to this meeting in Colorado, much has been achieved. In the last meeting, many papers dealt with basic problems. Since then, new trends of research have treated practical problems or aimed to find new applications. This paper reviews those new achievements and future trends in CWE.

2. Why CWE is difficult – analysis of flowfield around bluff body by CFD

There are two reasons why CWE is difficult. First, flow obstacles, i.e. buildings or structures, always exist in the flowfield within the surface boundary layer. Second, these flow obstacles, the so-called "bluff bodies", have sharp edges at their corners. Very fine grid discretization is required to analyze such flowfields with high precision.

The flowfield around a bluff body is highly complicated, since it is defined by impingement, separation, reattachment, circulation, vortices, etc. as is shown in Fig. 1 1]. Almost all those phenomena which are considered difficult to resolve in the world of fluid dynamics are included here. The most distinctive feature of such a complex flowfield is defined by the distribution of the strain-rate tensor $(\partial\langle u_i\rangle/\partial x_j + \partial\langle u_j\rangle/\partial x_i)$ shown in Fig. 2(2), which is highly anisotropic and changes significantly depending on its position relative to the bluff body [2].

Its complexity becomes clear when compared to the $(\partial\langle u_i\rangle/\partial x_j + \partial\langle u_j\rangle/\partial x_i)$ distribution of the simple boundary layer, as is shown in Fig. 2(1). Compared to the flowfields traditionally treated in the field of CFD, such as channel or pipe flows, the flowfield around a bluff body is much more complicated and difficult to analyze.

Since the production or redistribution of turbulence statistics such as Reynolds stress $\langle u_i' u_j' \rangle$ etc. depend on this strain-rate tensor, the complicated distribution of $(\partial\langle u_i\rangle/\partial x_j + \partial\langle u_j\rangle/\partial x_i)$ results in very complicated turbulence statistics.

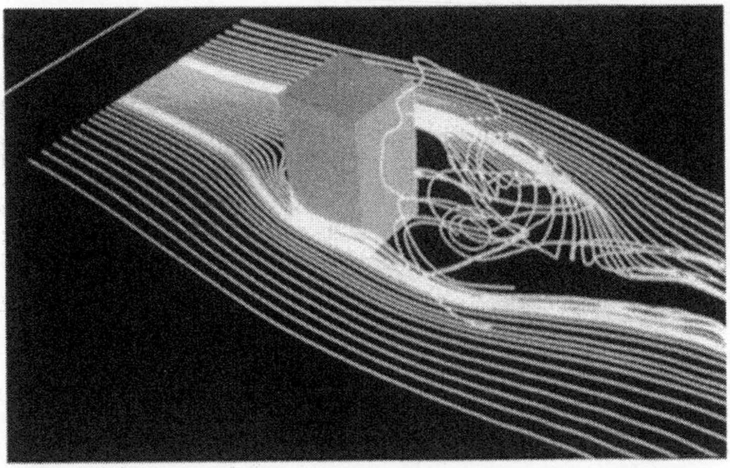

Fig. 1. Instantaneous streamlines past 2D square cylinder (by LES) [1].

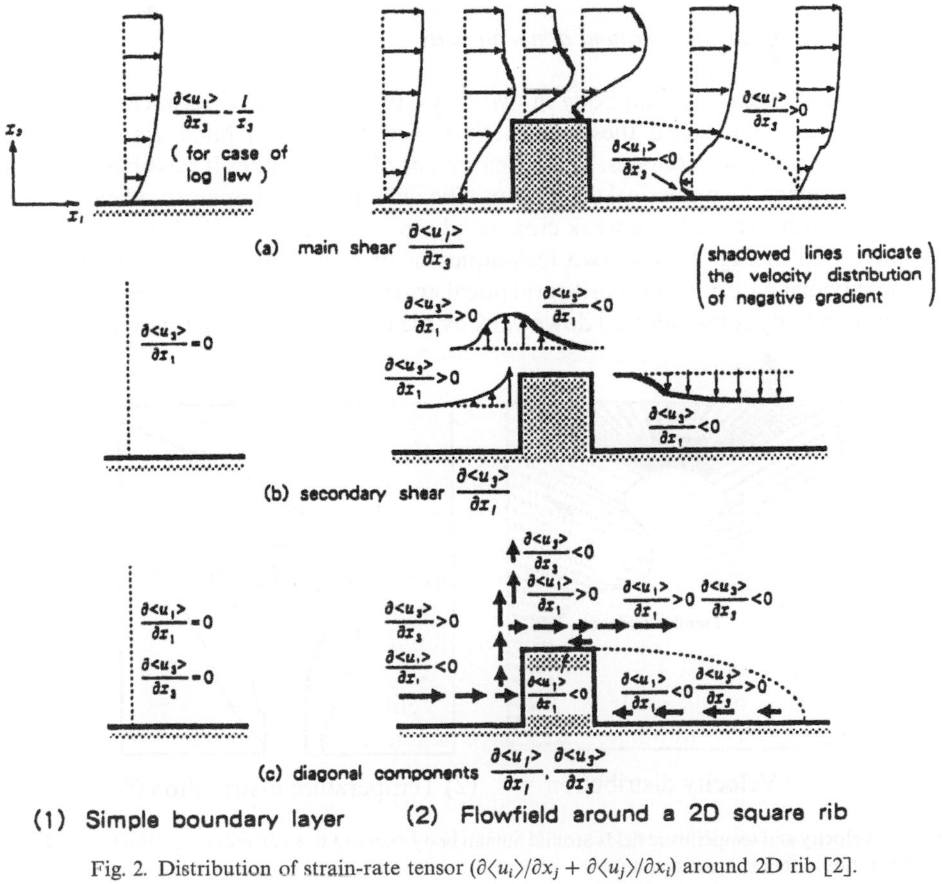

Fig. 2. Distribution of strain-rate tensor ($\partial\langle u_i\rangle/\partial x_j + \partial\langle u_j\rangle/\partial x_i$) around 2D rib [2].

3. Rapid growth of CFD applications in wind engineering

The recent growth of CWE applications can be viewed as follows: (1) from the basic to the application stage; (2) from building to human scale; (3) from building to global scale; (4) from structural engineering problems to environmental engineering problems.

Typical examples can be selected and arranged roughly in terms of scale.

- Analysis of velocity and temperature fields around a human body.
- Analysis of velocity and pressure fields around a bluff body (bridge deck or building structure, etc.).
- Coupled analysis of fluid and structures.
- Analysis of gaseous diffusion around a building or city-block.
- Analysis of pedestrian wind problems near the ground around a high-rise building.
- Analysis of outdoor climate within city-blocks.
- Analysis of city or regional climates.

Current and possible future research on these specific problems are briefly described below.

3.1. Analysis of flowfield around a human body

Wind influences a human body in two ways, thermally and dynamically. It is very important to understand these wind effects on the human body when designing comfortable outdoor climates. The development of CFD makes it possible to analyze this phenomena by numerical simulation. The velocity and temperature fields around a human body exposed to a weak cross wind are shown in Fig. 3 [3]. The grid layout is illustrated in Fig. 4. Here two techniques of grid discretization (generalized curvilinear coordinate and composite grid (local grid)) are utilized. A rising plume around the human body is transported downward by the cross wind. Thus, this approach uses

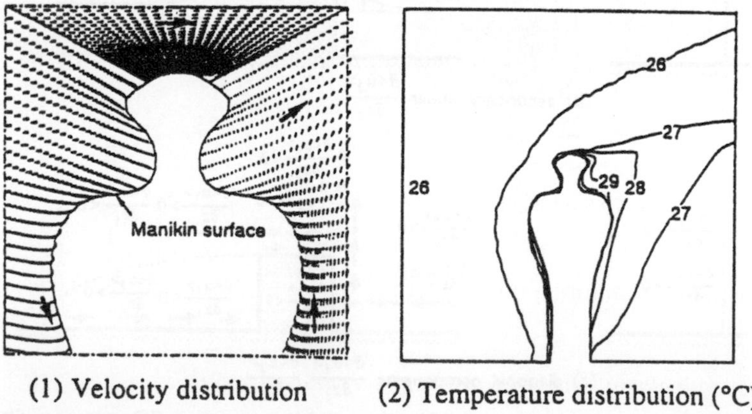

Fig. 3. Velocity and temperature fields around human body exposed to weak cross wind (by k–ε model) [3] (computational thermal mannequin).

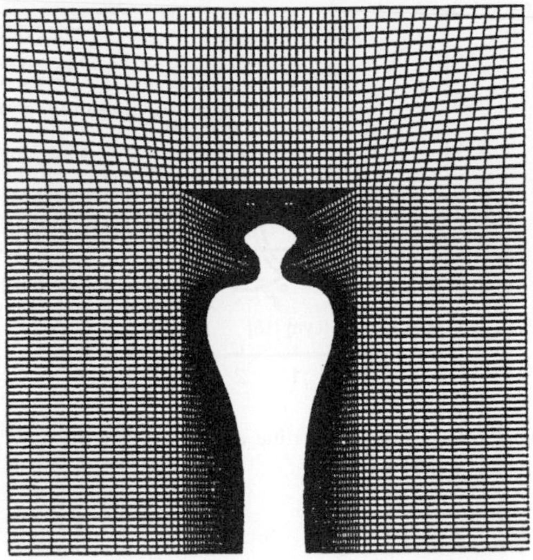

Fig. 4. Grid layout around human body [3] (composite grid based on generalized curvilinear coordinate).

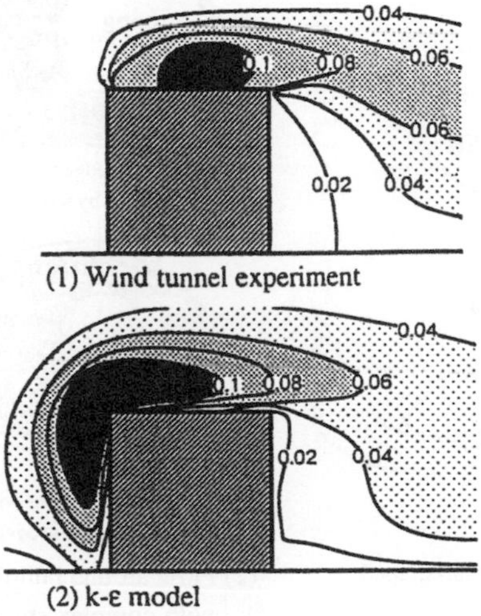

Fig. 5. Over-production of turbulent energy k by k–ε model (center section of cube) [14].

a numerical thermal mannequin in contrast to the traditional experimental thermal mannequin.

3.2. *Analysis of velocity and pressure fields around a bluff body*

CFD analysis of velocity and pressure fields around a bluff body has a long history, see, e.g., Refs. [4–13] among others. Solutions first used the k–ε model, but it is now

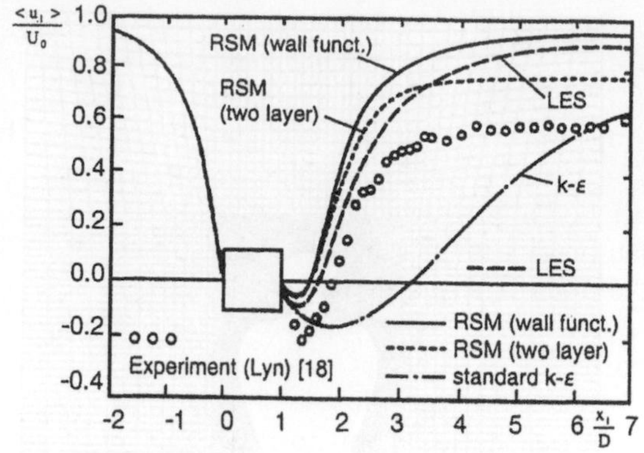

Fig. 6. Time-averaged velocity $\langle u_i \rangle$ along centerline of square cylinder [16] (with various turbulence models). Experimental values from Ref. [18].

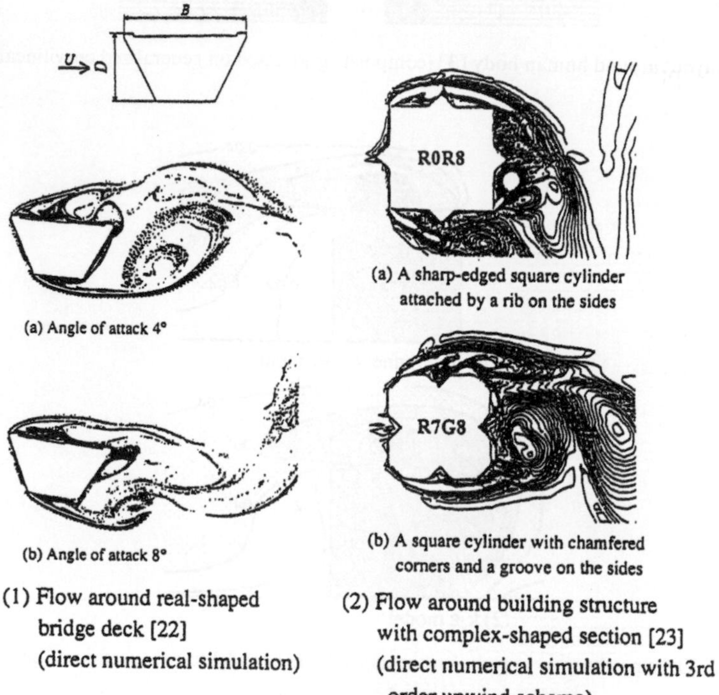

Fig. 7. Practical applications of CFD to bluff body aerodynamics.

well known that analysis of the flowfield around a bluff body with the k–ε model faces many difficulties. This is caused by the fundamental shortcoming of EVM (Eddy Viscosity Modeling) in the k–ε model when applied to a flowfield with impingement. Use of EVM results in the over-production of turbulence energy k around the frontal corners of the bluff body [2,14]. As a result, the predicted values of velocity or pressure differ from the experimental results [2,15]. This over-production of turbulence energy k is illustrated in Fig. 5. Many attempts have been made to revise the k–ε model and several new models have been proposed. The prediction accuracy of these new models has significantly improved. The details of these new k–ε models are described in Section 4 below.

Analysis of bluff body flows by LES (Large Eddy Simulation) has also been completed. LES can predict the flowfield around a bluff body much more accurately than the k–ε model does. The prediction results for the flowfield around a square cylinder by the k–ε model, RSM (Reynolds Stress Model) [12,16], and conventional LES are compared in Fig. 6 [16,17]. LES shows the best reproduction of experimental data, next is the RSM, and the k–ε model gives the poorest result. In recent LES computations the conventional Smagorinsky SGS (subgrid scale) model [19] has been replaced by the dynamic SGS model [20,21]. The development of the dynamic SGS model is one of the most significant improvements in the world of CFD during the past four years. The appearance of dynamic LES makes it possible to predict the velocity and pressure fields around a bluff body with higher accuracy. Details of the dynamic SGS model will be described below.

Practical applications of CFD to bluff body aerodynamics have expanded greatly. Fig. 7 shows applications to real-shaped bridge decks [22] and building structures with complex-shaped cross sections [23].

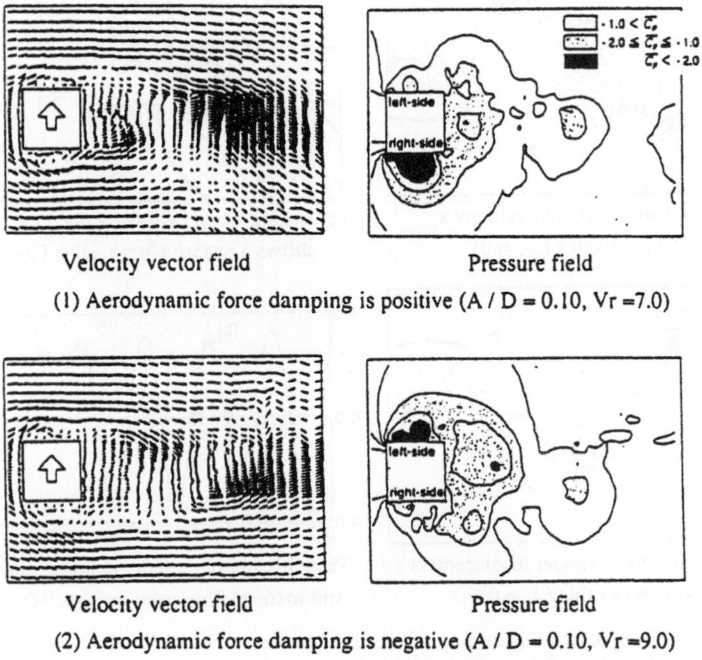

Fig. 8. Interactive analysis of fluid and structure for forced oscillation (by LES) [1].

3.3. *Interactive analysis of fluid and structure*

The interactive phenomenon that occurs between fluid and an elastic structure is very complicated. This is one of the most important research targets of wind engineering. Since CFD is capable of analyzing the pressure and velocity fields of moving fluid, CWE will achieve much in this area. In the world of CFD it is possible to simulate hypothetical situations which cannot be realized in wind tunnel experiments. For example, the 3D disturbances produced by supporting a cylinder cannot be removed in an actual wind tunnel experiment, whereas it is possible in CFD to simulate 2D flow precisely.

In Figs. 8 and 9 [1], analyses of forced oscillation and self-induced oscillation are illustrated. Various characteristics peculiar to the interactive phenomenon between fluid and structure, e.g. positive and negative damping, phase-shift, etc. are well reproduced. Fig. 8(2) shows the situation of negative damping. Analysis of interactive phenomena can be conducted by LES, but currently one difficulty in such an analysis by LES is that too much CPU time is required.

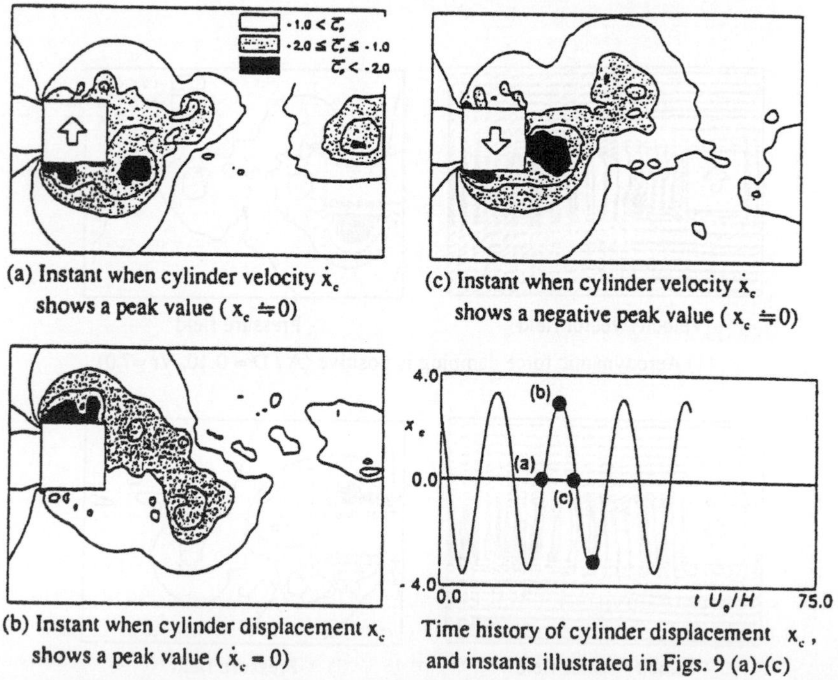

(a) Instant when cylinder velocity $\dot{x}_c$ shows a peak value ($x_c \fallingdotseq 0$)
(c) Instant when cylinder velocity $\dot{x}_c$ shows a negative peak value ($x_c \fallingdotseq 0$)
(b) Instant when cylinder displacement x_c shows a peak value ($\dot{x}_c = 0$)
Time history of cylinder displacement x_c, and instants illustrated in Figs. 9 (a)-(c)
4.0
0.0
- 4.0
x_c
(a)
(b)
(c)
0.0
$t\ U_0/H$
75.0

Fig. 9. Interactive analysis of fluid and structure for free oscillation (by LES) [1]. V_r is the reduced velocity ($= U_0/ND$); N the frequency of the oscillation of the cylinder; A the displacement of the oscillating cylinder; $\dot{x}_c$, x_c the cylinder velocity and displacement in x_2 direction.

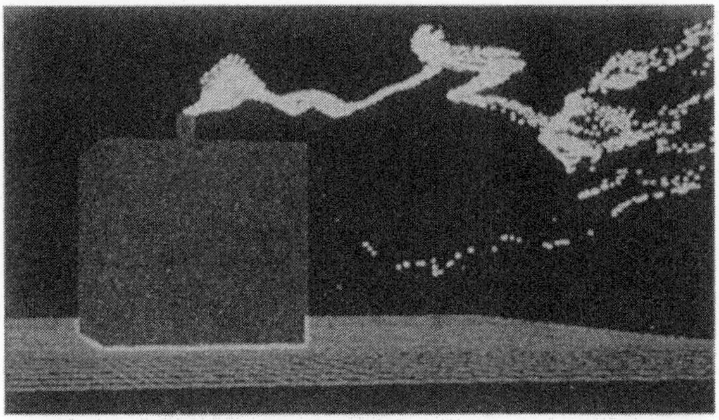

Fig. 10. Gaseous diffusion from the stack on the roof (by LES) [24].

Fig. 11. City blocks in Tokyo reproduced by remote-sensing and CG technique [96].

3.4. *Dispersion around buildings*

Predictions of contaminant dispersion around buildings or in city blocks are important subjects in environmental engineering. The measurement of contaminant concentration in a wind tunnel or in the field can be very tedious. On the other hand, CFD prediction of dispersion phenomenon is very easy if the velocity field is given. From this point of view, CWE is a suitable tool for analyzing contaminant dispersion. Fig. 10 shows the diffusion of smoke discharged from a stack on the roof of an isolated rectangular building as predicted by LES [24].

Fig. 12. High velocity near ground (by k–ε model) [96].

3.5. Analysis of pedestrian wind problems around a high-rise building

Strong wind is often observed around a high-rise building near the ground. Since this strong wind influences pedestrians greatly, architectural design to avoid such strong winds is an important subject for city planning, see, e.g., Refs. [25–30] among others. Prediction of the locations of such strong winds have been made solely on the basis of wind tunnel tests so far, but the development of CFD now makes it possible to predict this phenomenon with fairly good accuracy [31–34]. When conducting CFD analysis of strong wind around real city blocks, the digital data of all building shapes must be provided as input data. A data-base which records the 3D-shapes of all buildings now existing in Tokyo has already been completed, using the technique of remote sensing [96]. Fig. 11 shows an example of city blocks reproduced with Computer Graphics technique using this data-base. Fig. 12 shows the predicted velocity field near the ground using the k–ε turbulence model.

3.6. Analysis of outdoor climate

Prediction of the degree of comfort of an outdoor climate is becoming a new research subject for city planning [35,36], following the conventional concern about indoor climate which has a long history in air-conditioning engineering. In order to evaluate the comfort of an outdoor climate, information about air velocity, temperature, moisture, radiation, etc. is required, as shown in Fig. 13(1). Fig. 13(2) illustrates a computational approach for analyzing the outdoor climate by coupled analysis of CFD and radiation simulation. Here three coupled equations, (1) transport equation of momentum, (2) transport equation of heat, and (3) heat transfer equation by radiation, are analyzed. CFD provides a very powerful tool for coupled analysis to

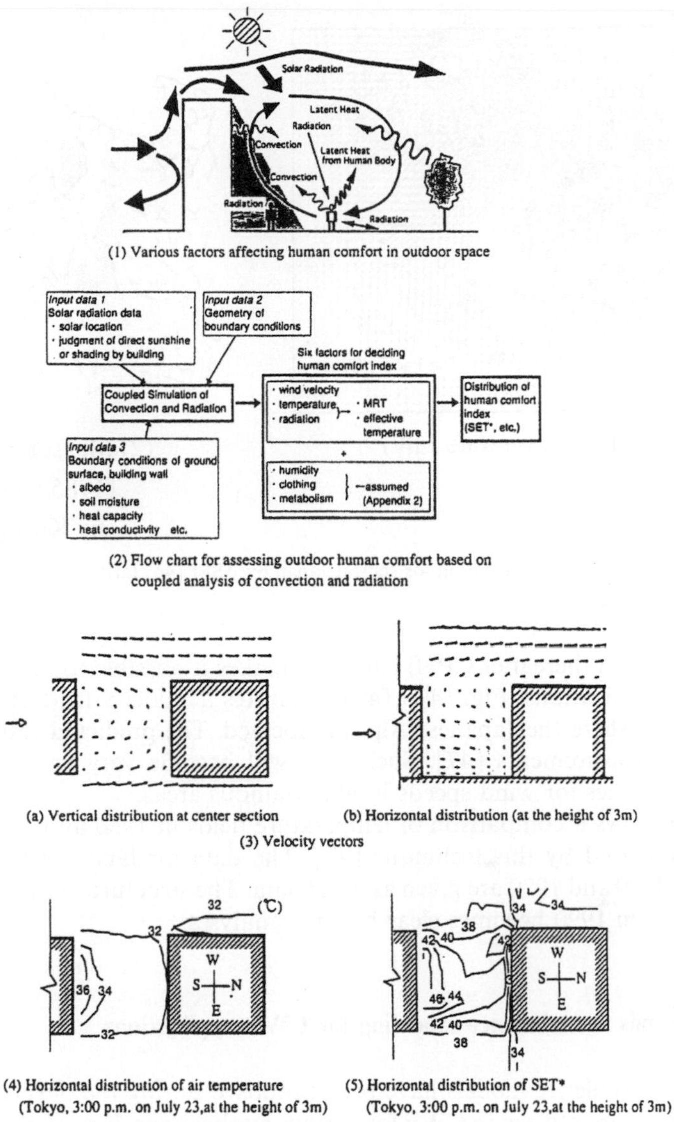

Fig. 13. Analysis of human comfort in outdoor climate [37].

predict the thermal environment. The predicted results of velocity, temperature and SET* (standard effective temperature) are illustrated in Figs. 13(3)–13(5) [37].

3.7. Analysis of regional climate

The prediction of regional climate by CFD has a long history in the field of meteorology, see, e.g., Refs. [38–45] among others. The development of Central

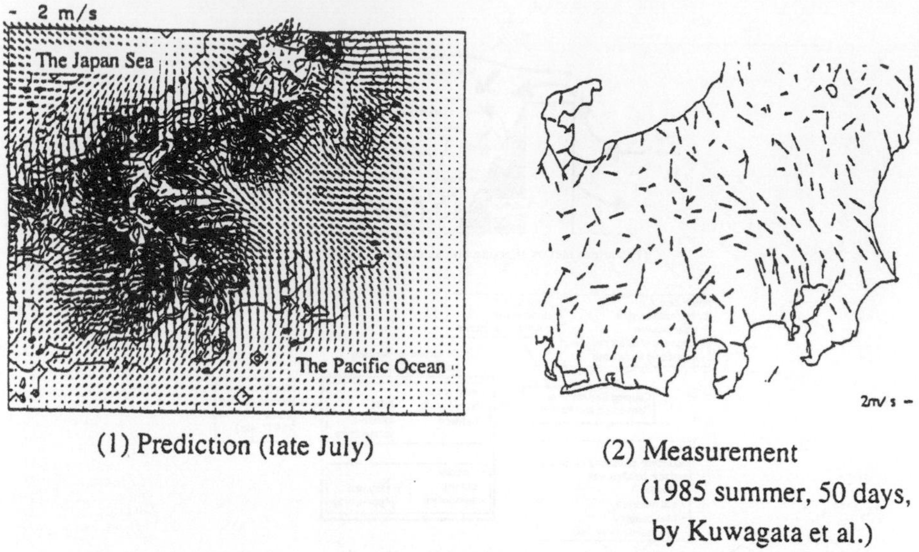

Fig. 14. Wind distribution for central Japan by ASM [46] (at the height of 100 m).

Processing Unit (hereafter CPU) hardware makes it possible to utilize this technique for CWE applications. Fig. 14(1) [46] illustrates a velocity field of the central part over Japan, where the Japanese Alps are located. The predicted results agree rather well with measurements [47]. Such data will become very valuable in providing design guidelines for wind speeds in mountainous areas.

Fig. 15 shows a comparison of temperature fields in 1930 and 1990 in the Tokyo area as analyzed by this technique [48]. The data for land-use and artificial heat release in 1930 and 1990 are given as input data. The structure of the heat island in the Tokyo area in 1990 becomes clear by such analyses.

4. New trends in turbulence modeling for CWE applications

Various turbulence models have been proposed during the development of CFD. These models are utilized in CWE for various applications. Fig. 16 shows the relationship between the types of turbulence modeling and the turbulence scales treated in each model [49]. Three topics can be pointed out in the trends for utilizing turbulence models in CWE; (1) revision of the k–ε model, (2) improvement of RSM, (3) appearance of dynamic LES.

4.1. New revisions of the k–ε model for bluff body aerodynamics

The fundamental shortcoming of the k–ε EVM has been described in the first part of this report. This drawback appears as an over-production of turbulence energy at

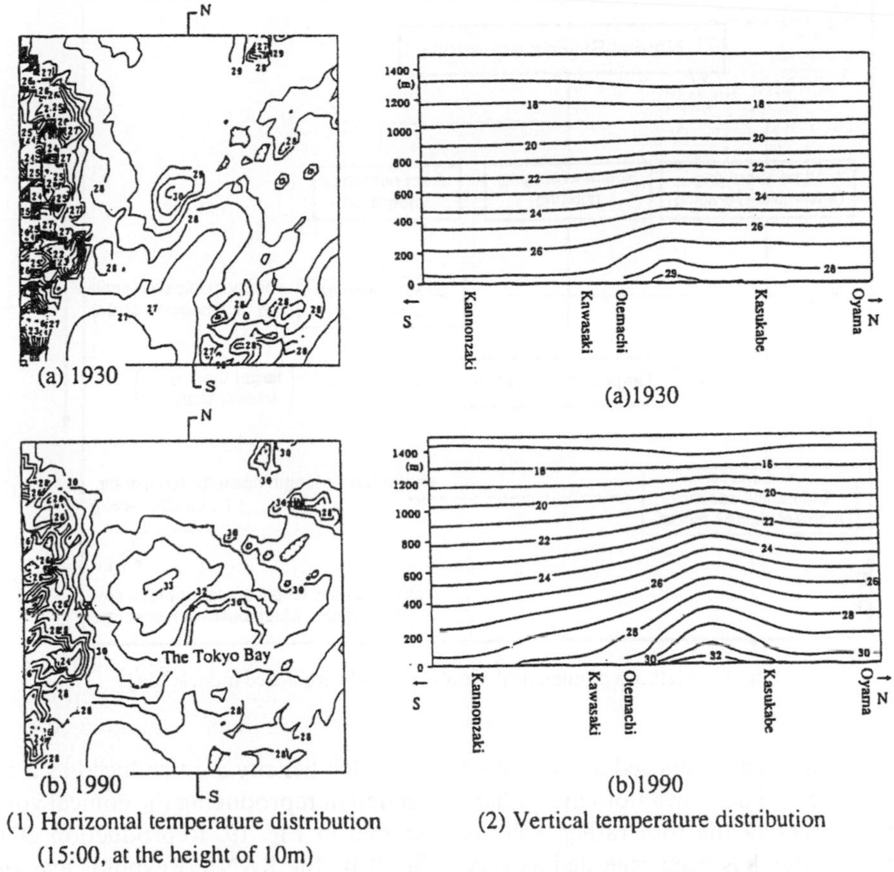

Fig. 15. Comparison of climate of Tokyo in 1930 and 1990 [48].

a region of flow impingement, as is shown in Fig. 5. In recent years various models have been proposed to remove this drawback. The first was the one proposed by Launder and Kato (LK model) [50]. In this model, the production term in the transport equation of turbulence energy k is changed from $P_k = \nu_t S^2$ to $P_k = \nu_t S\Omega$, as noted in Table 1. By this revision, the over-production of k around the frontal corner is much reduced. However, in this model, there exists mathematical inconsistency between the expression of P_k and $-\langle u_i' u_j' \rangle$. Our group, Murakami, Mochida and Kondo [51,52], has proposed a new model (MMK model) which includes removal of the inconsistency observed in the LK model. Basic equations for these models are described in Table 1. The predicted results of turbulence energy k are compared in Fig. 17 (with the standard k–ε, LK and MMK models). Surface pressures are also compared in Fig. 18. The improvement provided by the MMK model is significant.

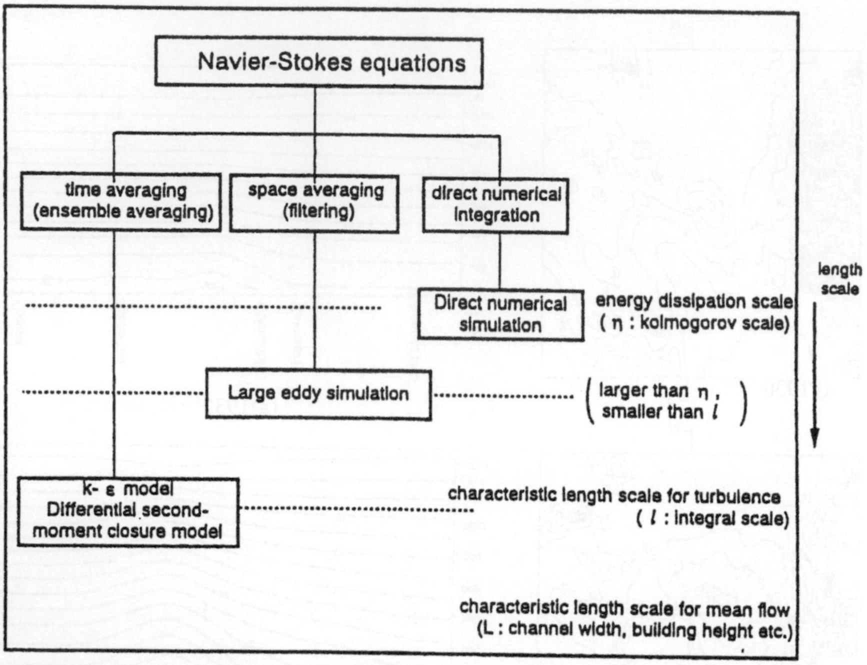

Fig. 16. Methods of numerical simulation and turbulence models [49].

Kawamoto has proposed a k–ε–ϕ model, in which helicity is introduced as a new variable [53]. The Kawamoto model has succeeded in reproducing the conical vortex at the corner of the roof rather well, as is shown in Fig. 19. Reproduction of the conical vortex has been regarded as very difficult by the RANS (Reynolds averaged Navier–Stokes equation) model. This achievement by Kawamoto is an outstanding contribution to research in CFD modeling. Of course, LES can also reproduce it well, as is shown in Fig. 20 [54].

4.2. Improvement in RSM

Efforts to revise the RSM (Reynolds Stress Model) are also continuing. The main focus for RSM modeling in recent years has been to improve the pressure–strain correlation term ϕ_{ij}. Higher-order models of ϕ_{ij} have been proposed by Fu–Launder–Tselepidakis (FLT model) [55] and Speziale–Sarkar–Gatski (SSG model) [56]. SSG and FLT models for ϕ_{ij} are based on quadratic and cubic approximations in $\langle u_i' u_j' \rangle$, in contrast to the linear approximation in the previous ϕ_{ij} modeling. However, only minor improvements can be observed in the prediction results using these sophisticated higher-order models when they are applied to the flowfield around a cube, as shown in Table 2 [57]. Two reattachment lengths, i.e. the reattachment on the roof and that on the floor, are compared here for the three models.

Table 1
Model equations for revised k–ε model [50,52]

1. Standard k–ε model

$$P_k = \nu_t S^2 \quad (\nu_t\text{: Eq. (2)}) \tag{1}$$

$$\nu_t = C_\mu \frac{k^2}{\varepsilon} \tag{2}$$

$$S = \sqrt{\frac{1}{2}\left(\frac{\partial\langle u_i\rangle}{\partial x_j} + \frac{\partial\langle u_j\rangle}{\partial x_i}\right)^2} \tag{3}$$

2. LK model

$$P_k = \nu_t S\Omega \quad (\nu_t\text{: Eq. (2)}) \tag{4}$$

$$\Omega = \sqrt{\frac{1}{2}\left(\frac{\partial\langle u_i\rangle}{\partial x_j} - \frac{\partial\langle u_j\rangle}{\partial x_i}\right)^2} \tag{5}$$

3. MMK model

$$P_k = \nu_t S^2 \quad (\nu_t\text{: Eq. (7) or Eq. (8)}) \tag{6}$$

$$\nu_t = C_\mu^* \frac{k^2}{\varepsilon}, \quad C_\mu^* = C_\mu \frac{\Omega}{S} \quad \left(\frac{\Omega}{S} < 1\right) \tag{7}$$

$$\nu_t = C_\mu^* \frac{k^2}{\varepsilon}, \quad C_\mu^* = C_\mu \quad \left(\frac{\Omega}{S} \geqslant 1\right) \tag{8}$$

x_i: component of spatial coordinate, u_i: velocity component in the x_i direction

A notable contribution to RSM models is the improvement of the wall-reflection term ϕ_{ij}^{w}. The well-known ϕ_{ij}^{w} model proposed by Gibson–Launder [58] produces abnormal turbulence behavior in the impingement region, i.e. the normal stress at the impingement point is amplified near the wall. The normal stress should be reduced due to the limiting behavior of the wall. A new model of ϕ_{ij}^{w} proposed by Craft–Launder [59] has removed this shortcoming for the flowfield around a cube, as shown in Fig. 21. The normal stress $\langle {u'_1}^2\rangle$ is reduced to the levels observed in an experiment [57]. Nonetheless, the application of RSM to wind engineering does not seem very promising when its overly sophisticated modeling is considered.

4.3. New trends in LES

Since the pioneering work by Deardorff [60], only the standard Smagorinsky model [19] has been widely used in the computation of LES, and almost no

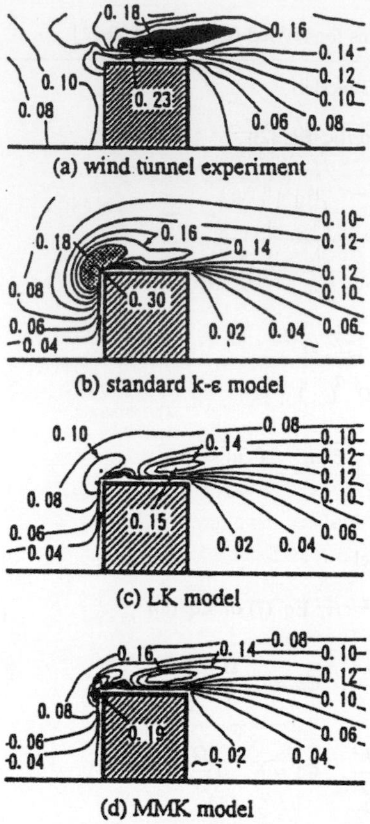

Fig. 17. Comparison of turbulent energy k with various k–ε models (for a 2D square rib) [52].

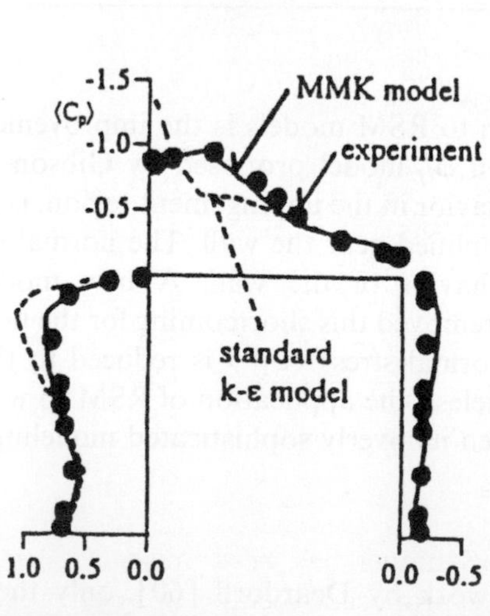

Fig. 18. Comparison of surface pressure with various k–ε models (for cube) [52].

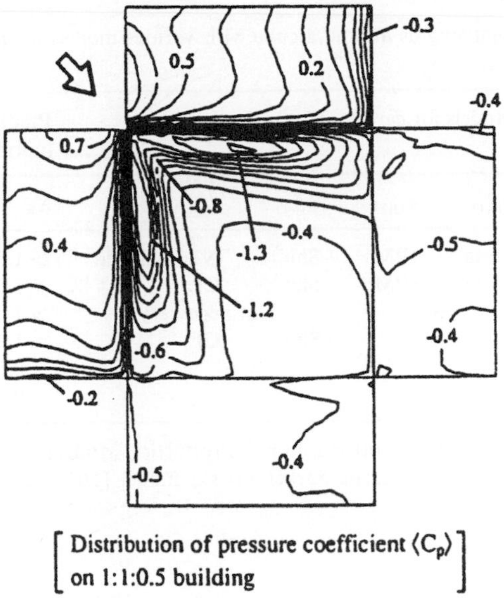

Fig. 19. Conical vortex at the roof corner predicted by revised k–ε model (k–ε–ϕ model by Kawamoto [53]).

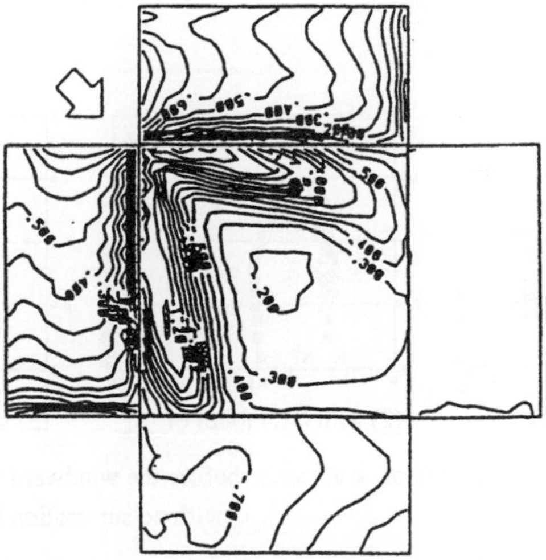

Fig. 20. Conical vortex at the roof corner by LES [54].

improvement to the SGS model has been observed for a long time. However, a dynamic SGS model was proposed recently by Germano et al. [20] and revised by Lilly [88], and its use is spreading explosively.

Table 2
Comparison of reattachment lengths around a cube with various models for pressure–strain correlation ϕ_{ij} in RSM [57]

	Models for ϕ_{ij} and ϕ^{w}_{ij}				Predicted reattachment length (normalized by H)	
	$\phi_{ij(1)}$	$\phi_{ij(2)}$	$\phi^{w}_{ij(1)}$	$\phi^{w}_{ij(2)}$	X_R	X_F
Case 1	Rotta	IPM	Shir	Not included	> 1.0[a]	2.3
Case 2	Rotta	IPM	Shir	CL	> 1.0[a]	2.0
Case 3	SSG		Not necessary		> 1.0[a]	2.1
Case 4	FLT		Shir	CL	> 1.0[a]	2.1
LES (S model)					0.6	1.4
Experiment					0.7	1.2

Note: Rotta: Rotta model [93]; IPM: Isotropization of Production Model [94]; Shir: Shir model [95]; CL: Craft–Launder model [59]; SSG: Speziale–Sarker–Gatski model [56]; FLT: Fu–Launder–Telepidakis model [55].

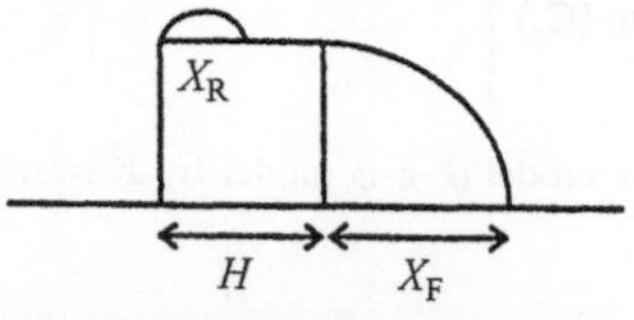

Definition of reattachment lengths X_R and X_F.
[a] $X_R > 1.0$: flow does not reattach on the roof.

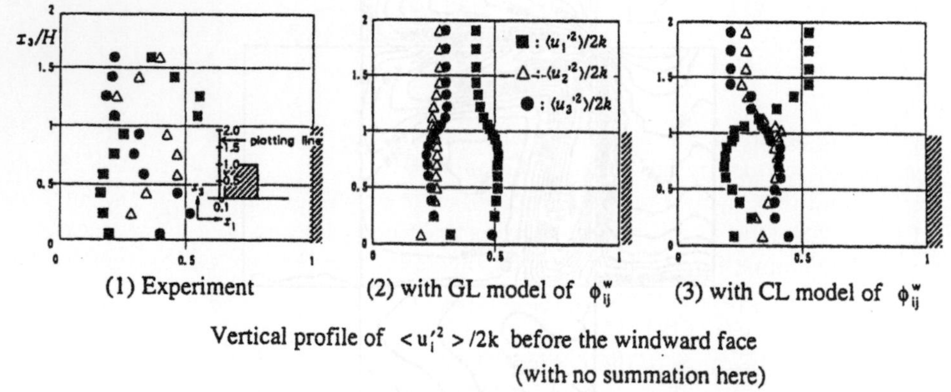

Fig. 21. Improvement of wall-reflection term ϕ^{w}_{ij} in RSM [57]: GL is the Gibson–Launder model [58]; CL is the Craft–Launder model [59].

4.3.1. Improvement of the conventional SGS model

The conventional Smagorinsky model has been used since the start of LES [19]. The Smagorinsky constant C_S has been optimized to values between 0.1 to 0.25 for various flowfields [61–71]. As already pointed out, the flowfield around a bluff body includes various types of flow structures such as impingement, separation, free shear layers, vortices, etc. Thus it is not easy to select one universal value of C_S for analyzing the flowfield around a bluff body. The shortcoming of using a constant value of C_S becomes clear during the application of the conventional Smagorinsky model to CWE problems.

To overcome this drawback of a constant value of C_S, several models have been proposed. The first is the one proposed by Yoshizawa [72], in which C_S is given as a function of space and time using the simplified transport equation of k_{SGS} as discussed in my paper at CWE 92 (cf. Table 3) [2]. Recently, a second model, the now well-known dynamic model was proposed by Germano et al. [20]. Details of that dynamic model are described in the next section.

It is well known that the standard Smagorinsky model (hereafter the S model) is overly dissipative because it does not permit back-scattering (energy cascade from small scale to large scale). In order to overcome this shortcoming, Bardina et al. [73] proposed the scale-similarity model. Recently, a filtered Bardina model [74,75] was proposed by Horiuti. Basic equations for these models are shown in Table 3.

When the standard Smagorinsky model is applied to non-isothermal flowfields, a revision is required. Some attempts to introduce buoyancy effects into the SGS model have been proposed in the field of meteorology by Mason [69], Mason and Derbyshire [70], Moeng [76], Nieuwstadt [77] and others. A model for including such buoyancy effects is shown in Table 4 [49,69,70]. In the field of meteorology a one equation type SGS model is often used [76,77]. Our group applied the SGS model including buoyancy effect proposed by Mason et al. to analyze buoyant gas diffusion around a building model, as shown in Fig. 22 [49]. The advantage of the model including buoyancy effect over the standard S model can be observed in this comparison.

4.3.2. Advantages of the dynamic SGS model in wind engineering applications

The disadvantage of the S model in which a constant C_S is imposed was described in the previous section. In the dynamic SGS model, C ($= C_S^2$) is determined as a variable of space and time using two filters with different characteristic scales, grid filter and test filter [20]. Due to this double filtering procedure, the prediction accuracy of the standard Dynamic Smagorinsky model (hereafter the DS model) for the flowfield around a square cylinder is much improved compared to the result of the S model, as is shown in Fig. 23(1) [78].

The second advantage of the DS model over the S model is the treatment near the wall. The empirical model function f_μ [79] (cf. Table 4) is required in the S model to damp ν_{SGS} in the area near the wall. In the DS model f_μ is not necessary since the value of C automatically goes to zero in the laminar region just near the wall, and the model consequently becomes more elegant.

The third advantage lies in the treatment of the inflow boundary condition. The inflow boundary condition of no velocity fluctuation (laminar flow) can be used in the DS model, a condition which gives some inconsistency in the S model. The inconsistency in the S model becomes clear around the corner of a bluff body where the flow is

Table 3
Models of τ_{ij} and procedure for determining C in LES

$$\frac{\partial \bar{u}_i}{\partial t} + \frac{\partial \bar{u}_i \bar{u}_j}{\partial x_j} = -\frac{\partial \bar{p}}{\partial x_i} + \frac{\partial}{\partial x_j}\left\{ -\tau_{ij} + \nu\left(\frac{\partial \bar{u}_i}{\partial x_j} + \frac{\partial \bar{u}_j}{\partial x_i}\right)\right\} \tag{9}$$

(1) *Base model*

- S model

$$\tau_{ij} - \tfrac{1}{3}\delta_{ij}\tau_{kk} = -2C\bar{\Delta}^2|\bar{S}|\,\bar{S}_{ij} \tag{10}$$

- Scale similarity (Bardina) model

$$\tau_{ij} = B_{ij} \tag{11}$$

$$B_{ij} = \overline{\bar{u}_i\bar{u}_j} - \bar{\bar{u}}_i\bar{\bar{u}}_j \tag{12}$$

- Filter Bardina model

$$\tau_{ij} = L_{ij} + C_{ij} + C_B\,\overline{(\bar{u}_i - \bar{\bar{u}}_i)(\bar{u}_j - \bar{\bar{u}}_j)} \tag{13}$$

$$L_{ij} = \overline{\bar{u}_i\bar{u}_j} - \bar{\bar{u}}_i\bar{\bar{u}}_j \tag{14}$$

$$C_{ij} = \bar{\bar{u}}_i(\bar{u}_j - \bar{\bar{u}}_j) + (\bar{u}_i - \bar{\bar{u}}_i)\bar{\bar{u}}_j \tag{15}$$

- Mixed model

$$\tau_{ij} - \tfrac{1}{3}\delta_{ij}\tau_{kk} = -2C\bar{\Delta}^2|\bar{S}|\bar{S}_{ij} + B_{ij} - \tfrac{1}{3}\delta_{ij}B_{kk} \tag{16}$$

(2) *Procedure for determining* C_S ($C = C_S^2$)

1. Tuning or modeling
   - optimizing C_S according to flow characteristics based on numerical experiment (standard S model, $C_S = 0.1$ (channel flow) – $C_S = 0.25$ (isotropic turbulence))
   - determining from simplified k_{SGS} equation (Yoshizawa model)

$$\frac{C_S}{C_{S0}} = 1 - C_A^{-2}|\bar{S}|^{-2}\frac{D|\bar{S}|}{Dt} + C_B\,\bar{\Delta}^2|\bar{S}|^{-3}\frac{\partial}{\partial x_j}\left(|\bar{S}|^2\frac{\partial|\bar{S}|}{\partial x_j}\right) \tag{17}$$

2. Dynamic procedure with double filtering (Germano identity)
   - Lilly's least square method (optimization of C at each point)

$$C = -\frac{1}{2}\frac{\mathscr{L}_{ij}M_{ij}}{M_{kl}^2} \quad \text{DS model} \tag{18}$$

$$C = -\frac{1}{2}\frac{M_{ij}(\mathscr{L}_{ij} - H_{ij})}{M_{kl}^2} \quad \text{DM model} \tag{19}$$

$$\mathscr{L}_{ij} = \widehat{\bar{u}_i\bar{u}_j} - \hat{\bar{u}}_i\hat{\bar{u}}_j \tag{20}$$

$$M_{ij} = \hat{\bar{\Delta}}^2|\hat{\bar{S}}|\hat{\bar{S}}_{ij} - \widehat{\bar{\Delta}^2|\bar{S}|\bar{S}_{ij}} \tag{21}$$

$$H_{ij} = \widehat{\hat{\bar{u}}_i\hat{\bar{u}}_j} - \hat{\hat{\bar{u}}}_i\hat{\hat{\bar{u}}}_j - \left(\widehat{\overline{\bar{u}_i\bar{u}_j}} - \widehat{\bar{u}_i\bar{u}_j}\right) \tag{22}$$

Table 3 (Continued)

- Meneveau's Lagrangian dynamic model

$$C(\boldsymbol{x}, t) = -\frac{1}{2}\frac{I_{\mathrm{LM}}}{I_{\mathrm{MM}}} \quad \text{LDS model} \tag{23}$$

$$C(\boldsymbol{x}, t) = -\frac{1}{2}\frac{I_{\mathrm{LM}} - I_{\mathrm{HM}}}{I_{\mathrm{MM}}} \quad \text{LDM model} \tag{24}$$

$$I_{\mathrm{LM}} = \int_{-\infty}^{t} \mathscr{L}_{ij}(t')M_{ij}(t')W(t - t')\,\mathrm{d}t' \tag{25}$$

$$I_{\mathrm{MM}} = \int_{-\infty}^{t} M_{ij}(t')M_{ij}(t')W(t - t')\,\mathrm{d}t' \tag{26}$$

$$I_{\mathrm{HM}} = \int_{-\infty}^{t} M_{ij}(t')H_{ij}(t')W(t - t')\,\mathrm{d}t' \tag{27}$$

$W(t - t')$ weighting function

- Ghosal's localization model

$$C(\boldsymbol{x}) = \left[\int K(\boldsymbol{x}, \boldsymbol{y})C(\boldsymbol{y})\,\mathrm{d}y + f(\boldsymbol{x})\right]_{+} \tag{28}$$

where $[\,]_{+}$ denotes the positive part (details of this model are given in Ref. [83]). K and f are defined as functions of $\alpha_{ij} = 2\hat{\bar{\Delta}}^2|\hat{\bar{S}}|\hat{\bar{S}}_{ij}$ and $\beta_{ij} = 2\bar{\Delta}^2|\bar{S}|\bar{S}_{ij}$

still quasi-laminar but large values of the strain-rate tensor exist. The transition from laminar to turbulent near a bluff body is reproduced successfully in the DS model.

Comparison of the distributions of C and velocity near a wall surface is shown in Fig. 24 [80]. In the DS or DM models the values of C are preserved near the wall, in contrast to the S model with damping function. Thus the velocity distribution is improved greatly, as is shown in Fig. 24(2).

4.3.3. New models and new techniques for conducting stable computation of the DS model (cf. Table 3)

(1) *Large fluctuation of C in the DS model.* Since the fluctuation of $C\ (= C_{\mathrm{S}}^2)$ is large in the DS model, it is not easy to carry out a stable computation. This drawback is very serious in the case of bluff body applications. Thus various revisions have been applied to the DS model in order to overcome this shortcoming. These revisions can be grouped into two types; first are improvements to the SGS model, and second are techniques to stabilize the fluctuation of C.

(2) *Introducing scale-similarity model into the DS model.* The DS model requires that the principal axes of the SGS stress term τ_{ij} be aligned with the strain rate tensor.

Table 4
Model equations of LES (S model) with buoyancy effect [69,70]

$$\frac{\partial \bar{u}_i}{\partial x_i} = 0 \tag{29}$$

$$\frac{\partial \bar{u}_i}{\partial t} + \frac{\partial \bar{u}_i \bar{u}_j}{\partial x_j} = -\frac{\partial}{\partial x_i}(\bar{p} + \tfrac{2}{3} k_{\mathrm{SGS}}) + \frac{\partial}{\partial x_j}\left[\left(\frac{1}{\mathrm{Re}} + \nu_{\mathrm{SGS}}\right)\left(\frac{\partial \bar{u}_i}{\partial x_j} + \frac{\partial \bar{u}_j}{\partial x_i}\right)\right] - \mathrm{Frd}\, \bar{c}\delta_{i3} \tag{30}$$

$$\frac{\partial \bar{c}}{\partial t} + \frac{\partial \bar{u}_j \bar{c}}{\partial x_j} = \frac{\partial}{\partial x_j}\left[\left(\frac{1}{\mathrm{Re\ Sc}} + \alpha_{\mathrm{SGS}}\right)\frac{\partial \bar{c}}{\partial x_j}\right] \quad \text{transport equation of buoyant gas} \tag{31}$$

$$\alpha_{\mathrm{SGS}} = \nu_{\mathrm{SGS}} / \mathrm{Sc}_{\mathrm{SGS}} \quad (\mathrm{Sc}_{\mathrm{SGS}} = 0.5) \tag{32}$$

$$\mathrm{Frd} = -\frac{\Delta\rho}{\rho_a}\frac{H_b g}{\langle u_b \rangle^2} \tag{33}$$

(a) Type 1 (the standard type of the Smagorinsky model)

$$\nu_{\mathrm{SGS}} = (C_S f_\mu \bar{\Delta})^2 |\bar{S}| \tag{34}$$

$$|S| = \left[\frac{1}{2}\left(\frac{\partial \bar{u}_i}{\partial x_j} + \frac{\partial \bar{u}_j}{\partial x_i}\right)^2\right]^{1/2} \tag{35}$$

$$C_S = 0.12, \quad f_\mu = 1 - \exp(-x_n^+/25) \tag{36}$$

(b) Type 2 (the Smagorinsky model with buoyancy effect)

$$\nu_{\mathrm{SGS}} = (C_S f_\mu \bar{\Delta})^2 \phi |\bar{S}| \tag{37}$$

- unstable flow (Rf < 0)

$$\phi = (1 - \mathrm{Rf})^{1/2}, \quad C_S = 0.32 \tag{38}$$

- stable flow (Rf > 0, $\mathrm{Rf}_C = 0.33$)

1. $0 < \mathrm{Rf} < \mathrm{Rf}_C \quad \phi = (1 - \mathrm{Rf}/\mathrm{Rf}_C)^2, \ C_S = 0.12$ (39)

2. $\mathrm{Rf}_C < \mathrm{Rf} \quad \phi = 0 \ (\nu_{\mathrm{SGS}} = 0)$ (40)

$$\mathrm{Rf} = -\frac{G_{k\mathrm{SGS}}}{P_{k\mathrm{SGS}}} = -\frac{\mathrm{Frd}}{\mathrm{Sc}_{\mathrm{SGS}}}\frac{\partial \bar{c}}{\partial x_3}\Big/|\bar{S}|^2 \tag{41}$$

$$P_{k\mathrm{SGS}} = -\overline{u_i'' u_j''}\frac{\partial u_i}{\partial x_j} = \nu_{\mathrm{SGS}}|\bar{S}|^2 \tag{42}$$

$$G_{k\mathrm{SGS}} = -\mathrm{Frd}\,\overline{u_3'' c''} = \mathrm{Frd}\,\alpha_{\mathrm{SGS}}\frac{\partial \bar{c}}{\partial x_3} \tag{43}$$

This causes excessive energy back scattering when C becomes negative. The Dynamic Mixed model (DM model) was proposed by Zang et al. [81] and Vreman et al. [82] as a mixture of the DS model and the scale similarity model in order to control the degree of excessive back-scattering (cf. Table 3). When these models are applied to the

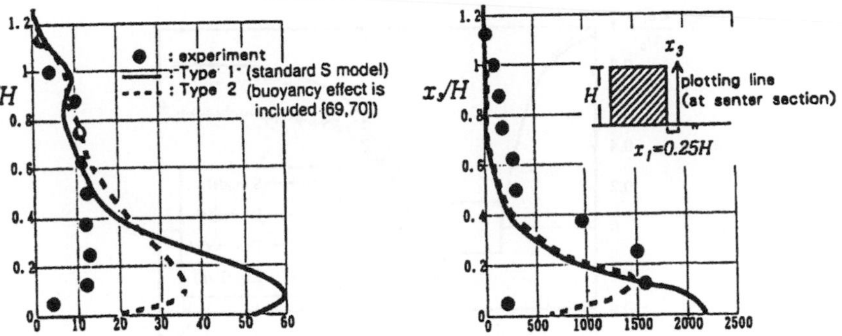

(1) Vertical profiles of $<c>/<c_0>$ (2) Vertical profiles of $<c'^2>/<c_0>^2$

Fig. 22. LES analysis of buoyant gas diffusion [49]. c is the gas concentration; c' the fluctuation of c; $\langle c_0 \rangle$ the standard gas concentration ($= q/(U_H H^2)$); q the gas emission rate.

flowfield around a square cylinder, the DM model yields better results, as is shown in Fig. 23(1).

(3) *Techniques for stabilizing the computation of the DS model.* Various techniques and models have been proposed to stabilize the fluctuation of C by Ghosal et al. [83] (dynamic localization model), Piomelli and Lu [84], Meneaveau et al. [85] (Lagrangian dynamic model), etc.

When the flowfield has a homogeneous direction (e.g. channel flow), C is often calculated using the averaged quantity in the homogeneous direction. The technique of time averaging is also used sometimes [86]. More simply, the technique of clipping is used, i.e. when C becomes negative, it is forced to zero.

In the Lagrangian dynamic model (LD model) proposed by Meneaveau et al. [85], C is calculated using the averaged quantities along the path line (cf. Table 3). The length of averaging is very important. The technique of Lagrangian averaging can be added to the dynamic mixed model. We call it the LDM model.

In the usual formulations of dynamic SGS models, C is taken out of filtering operations as if it were constant during the procedure of determining C dynamically. Obviously, this treatment is not consistent with the main concept of the dynamic SGS model in which C is designed to change in space and time. This inconsistency often causes instability in the calculation. In order to resolve this problem dynamic localization models have been proposed by Ghosal et al. [83] and Piomelli and Lu [84]. In Ghosal's model, a variational formulation is introduced, and C is optimized by minimizing the sum of the squares of the residuals of the equation for determining C integrated over the whole domain. Piomelli and Lu proposed an approximate localization technique [84], where C within the filtering operation is given from a time integration using a first order Euler scheme.

(4) *Comparison of various new dynamic SGS models.* Various models discussed above are compared for the flowfield around a square cylinder in Fig. 23. This was one of the test cases in the workshop on Large Eddy Simulation of flow past bluff

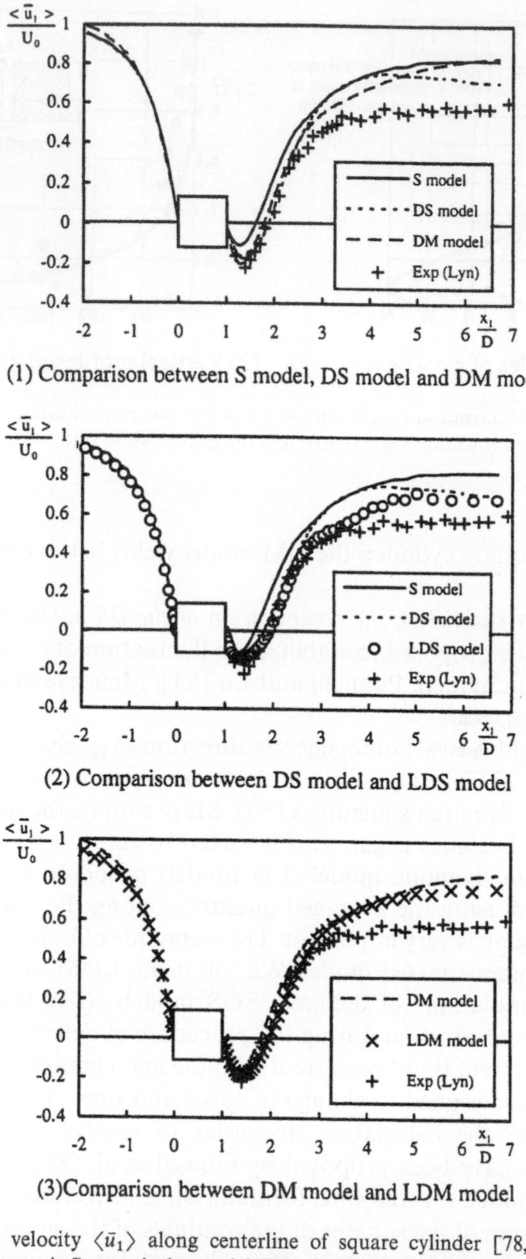

Fig. 23. Time-averaged velocity $\langle \bar{u}_1 \rangle$ along centerline of square cylinder [78,80]: S is the standard Smagorinsky, DS the Dynamic Smagorinsky model, DM the Dynamic Mixed model, LDS the Lagrangian Dynamic Smagorinsky model, LDM the Lagrangian Dynamic Mixed model.

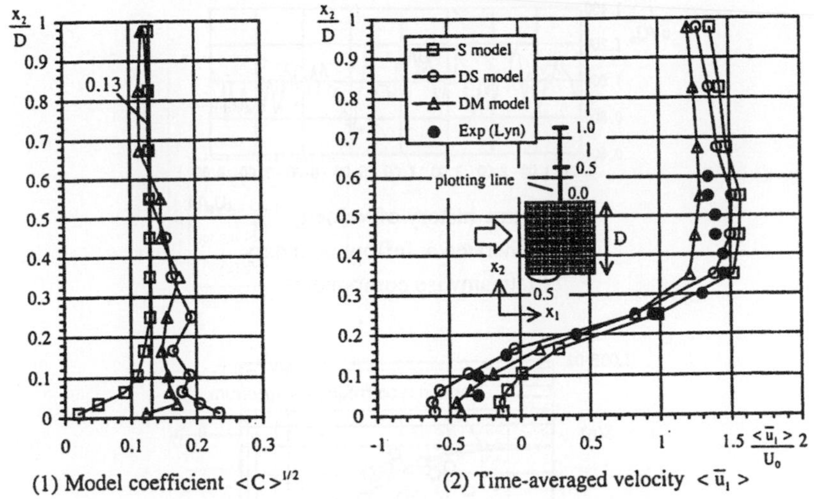

Fig. 24. Comparison of model coefficient C and time-averaged velocity near solid wall [80] ($C = C_S^2$).

Table 5
Advantages and disadvantages of various SGS models

	Consideration to back scatter	Consideration to non-equilibrium effect	Consideration to transition	Stability of computation
S model (static type) (S: Smagorinsky)	× (Not reproduced)	×	×	◎
Scale similarity model (static type)	△ (Too much)	△	○	×
Mixed model (static type)	△,○ (Partly reproduced)	△	△,○	○
Dynamic S model	△ (Too much)	△	○	△
Dynamic mixed model	△,○	△,○	○	△,○
Lagrangian dynamic S model	△	△,○	○	○
Lagrangian dynamic mixed model	△,○	○	○	◎

Note: ◎: functions very well; ○: functions well; △: insufficiently functional; ×: functions poorly.

bodies organized by Prof. Rodi and Prof. Ferziger and held in Germany in June 1995 [87]. As shown in Fig. 23, the dynamic SGS models provide better results than the standard Smagorinsky model. This tendency agrees with the concluding remarks of the workshop [87]. The advantages and disadvantages of various SGS models are compared in Table 5 [19,20,73,81,82,85,88]. Within the experiences of our group, the LDM model seems to give the best result for the flowfield around a cylinder. However,

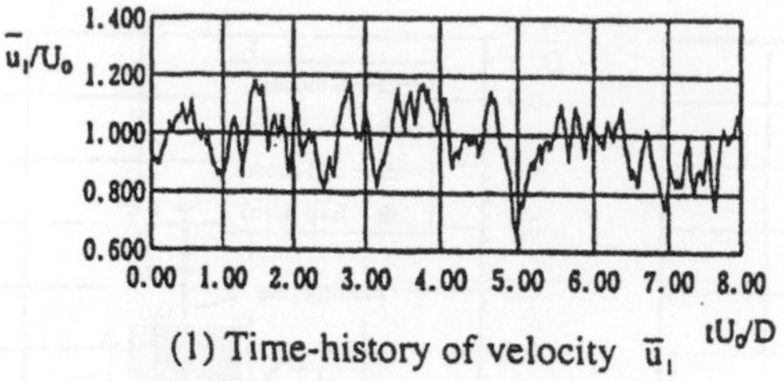

(1) Time-history of velocity $\bar{u}_1$ generated at inflow boundary (streamwise component)

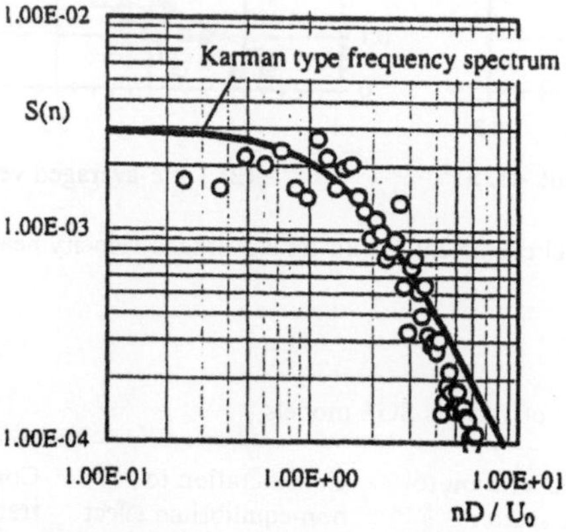

(2) Frequency power spectrum of u_1' generated at inflow boundary

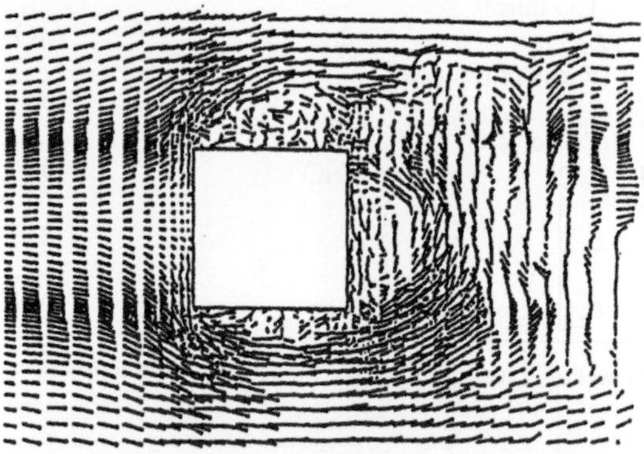

(3) Instantaneous velocity vectors with inflow turbulence

Fig. 25. Generation of velocity fluctuations with prescribed energy spectrum and turbulent intensity [97].

Table 6
Relative performance of various turbulence models for practical applications

Turbulence model	Standard k–ε	Modified k–ε			RSM		LES[a]	
		LK	MMK	Kawamoto k–ε–ϕ	GL	CL	Conventional S	Dynamic SGS
1. Simple flows (channel flow, pipe flow, etc.) (local equilibrium is valid)	○	○	○	○	○	○	○	○
2. Flow around bluff body (with turbulent approaching wind, local equilibrium is not valid)								
(1) Impinging area	×	△,○	△,○	△,○	×	△,○	○	○
(2) Separated area	×	△	○	○	△	△,○	○**	○
(3) With oblique wind angle	×,△	△,○	△,○	○	△	△,○	○**	○
3. Transitional flow (low Re No effects)								
(1) Near wall	○*	○*	○*	○*	○*	○*	○**	○
(2) Non-near-wall	×	×	×	×	×	×	×	○
4. Unsteady flow								
(1) Vortex shedding	×	○	○	○	○	○	○	○
(2) Fluctuation over wide-spectrum range	×	×	×	×	×	×	○	○
5. Stratified flow	×	×	×	×	△	△	△	○

Note: ○: functions well, △: insufficiently functional, ×: functions poorly, ○*: functions well when low Re No type model is employed, ○**: functions well with wall damping function, LK: Launder–Kato [50], GL: Gibson–Launder model for $\phi^{w}_{ij(2)}$ [58], MMK: Murakami–Mochida–Kondo [51, 52], CL: Craft–Launder model for $\phi^{w}_{ij(2)}$ [59], S: Smagorinsky model [19], Dynamic SGS [20, 82, 88].
[a]Detailed comparison is given in Table 5.

various new revisions are proposed continuously, and another good model will surely appear when more trials are added.

(5) *Dynamic SGS model for passive scalar transport.* For improving the prediction of contaminant diffusion, the conventional Smagorinsky model for passive scalar is improved by the dynamic SGS model for passive scalar proposed by Lilly [88] and Cabot and Moin [89].

4.3.4. *Inflow boundary condition for LES*

During the application of LES to wind engineering problems including flow about obstacles, the technique of providing the inflow boundary condition is very important since the periodic boundary condition cannot be used in such cases. Furthermore, inflow is always turbulent in wind engineering, hence, some technique for generating velocity fluctuations is required. Various techniques have been invented to generate such velocity fluctuations [90–92]. The simplest method is to store the time history of velocity fluctuations given from a preliminary LES computation [15]. Second is an artificial generation method in which velocity fluctuations are given by inverse Fourier Transform of prescribed energy spectra and turbulence intensity [90]. An example of velocity fluctuations given by this method is shown in Fig. 25 [97]. Since the treatment of the turbulent inflow boundary condition is a very important subject for wind engineering applications, much attention and effort should be devoted to this technique.

5. Conclusion

Various turbulence models in CFD have varying advantages and disadvantages. A proper choice of turbulence model is necessarily based on an evaluation of the problems peculiar to CWE. The difficulty when CFD is applied to a CWE flowfield around a bluff body, has been described in the first part of this report. The basic measures for the selection and evaluation of turbulence models are (1) prediction accuracy and (2) CPU time required. Various models are evaluated for CWE applications from these viewpoints in Table 6.

We must learn how best to select from these various turbulence models, considering the aim of each application and the advantages of each model. At this stage, it is clear that Dynamic LES gives the best results for many wind engineering applications. One disadvantage of using LES is that too much CPU time is required. However, rapid evolution of CPU hardware will surely overcome this restriction, and wide application of LES to CWE problems will certainly be realized in the near future.

Acknowledgements

Many of the CFD studies discussed here were carried out in cooperation with Professor Mochida, Assistant Professor Tominaga and Mr. Iizuka. The author would like to express his gratitude to each of them for their valuable contributions to these projects.

References

[1] S. Murakami, A. Mochida, S. Sakamoto, CFD analysis of wind-structure interaction for oscillating square cylinder, 9th Int. Conf. on Wind Eng., New Delhi, India, 1995, pp. 671–682.

[2] S. Murakami, Comparison of various turbulence models applied to a bluff body, J. Wind Eng. Ind. Aerodyn. 46/47 (1993) 21–36.

[3] S. Murakami, S. Kato, J. Zeng, CFD analysis of thermal environment around human body, Prep. of Indoor Air'96, July 1996, Nagoya, Japan.

[4] C.W. Hirt, J.L. Cook, Calculating three-dimensional flows around structures and over rough terrain, J. Comput. Phys. 10 (1972) 324–340.

[5] S. Murakami, A. Mochida, K. Hibi, Three-dimensional numerical simulation of airflow around a cubic model by means of large eddy simulation, J. Wind Eng. Ind. Aerodyn. 25 (1987) 291–305.

[6] S. Murakami, A. Mochida, 3D numerical simulation of airflow around a cubic model by means of k–ε model, J. Wind Eng. Ind. Aerodyn. 31 (1988) 283–303.

[7] F. Baetke, H. Werner, H. Wengle, Numerical simulation of turbulent flow over surface-mounted obstacles with sharp edges and corners, J. Wind Eng. Ind. Aerodyn. 35 (1990) 129–147.

[8] D.A. Paterson, C.J. Apelt, Simulation of flow past a cube in a turbulent boundary layer, J. Wind Eng. Ind. Aerodyn. 35 (1990) 149–176.

[9] T. Stathopoulos, A. Baskaran, Boundary treatment for the computational of three-dimensional wind flow conditions around a building, J. Wind Eng. Ind. Aerodyn. 35 (1990) 177–200.

[10] T. Tamura, I. Ohta, K. Kuwahara, On the reliability of two-dimensional simulation for unsteady flows around a cylinder-type structure, J. Wind Eng. Ind. Aerodyn. 35 (1990) 275–298.

[11] S. Murakami, Computational wind engineering, J. Wind Eng. Ind. Aerodyn. 36 (1990) 517–538.

[12] R. Franke, W. Rodi, Calculation of vortex shedding past a square cylinder with various turbulence models, Proc. 8th Symp. on Turbulent Shear Flows, 1993, p. 189.

[13] R.P. Selvam, P.B. Konduru, Computational, experimental roof corner pressures on the Texas Tech building, J. Wind Eng. Ind. Aerodyn. 46/47 (1993) 449–454.

[14] S. Murakami, A. Mochida, Y. Hayashi, Examining the k–ε model by means of a wind tunnel test and large-eddy simulation of the turbulence structure around a cube, J. Wind Eng. Ind. Aerodyn. 35 (1990) 87–100.

[15] A. Mochida, S. Murakami, M. Shoji, Y. Ishida, Numerical simulation of flowfield around Texas Tech building by large eddy simulation, J. Wind Eng. Ind. Aerodyn. 46/47 (1993) 455–460.

[16] W. Rodi, On the simulation of turbulent flow past bluff bodies, J. Wind Eng. Ind. Aerodyn. 46/47 (1993) 3–19.

[17] S. Murakami, A. Mochida, On turbulent vortex shedding flow past 2D square cylinder predicted by CFD, J. Wind Eng. Ind. Aerodyn. 54/55 (1995) 191–211.

[18] D.A. Lyn, S. Einav, W. Rodi, J.H. Park, A laser-Doppler velocimetry study of ensemble-averaged characteristics of the turbulent near wake of a square cylinder, J. Fluid Mech. 304 (1995) 285–319.

[19] J. Smagorinsky, General circulation experiments with the primitive equations, Part 1, Basic experiments, Mon. Weather Rev. 91 (1963) 99–164.

[20] M. Germano, U. Piomelli, P. Moin, W.H. Cabot, A dynamic subgrid scale eddy viscosity model, Phys. Fluids A 3 (7) (1991) 1760–1765.

[21] J. Ferziger, Simulation of complex turbulent flows: recent advances and prospects in wind engineering, J. Wind Eng. Ind. Aerodyn. 46/47 (1993) 195–212.

[22] H. Hirano, A. Maruoka, H. Inoue, Numerical fluid flow analysis and its applications to aerodynamic stability of box girder bridge, 13th Japan National Conf. on Wind Eng., 1994, pp. 561–566 (in Japanese).

[23] T. Tamura, Accuracy of the very large computation for aerodynamic forces acting on a stationary or an oscillating cylinder-type structure, IWEF Workshop on CFD for Prediction of Wind Loading on Buildings and Structures, 1995, 2-1–22.
[24] S. Murakami, A. Mochida, Y. Hayashi, K. Hibi, Numerical simulation of velocity field and diffusion field in an urban area, Energy Building 15–16 (1991) 345–356.
[25] W.H. Melboume, P.N. Joubert, Problems of wind flow at the base of tall buildings, Proc. 3rd Int. Conf. on Wind Effects on Buildings and Structures, 1971.
[26] A.D. Penwarden, Acceptable wind speeds in towns, Building Sci. 18 (1973) 259–267.
[27] N. Isyumov, A.G. Davenport, The ground level wind environment in build-up areas, Proc. 4th Int. Conf. on Wind Effects on Buildings and Structures, 1975, pp. 403–422.
[28] J.C.R. Hunt, E.C. Poulton, J.C. Mumford, The effects of wind on people; new criteria based on wind tunnel experiments, Building Environ. 11 (1976) 15–28.
[29] A.D. Penwarden, P.E. Grigg, R. Raymont, Measurements of wind drag on people standing in a wind tunnel, Building Environ. 13 (2) (1978) 75–84.
[30] S. Murakami, Y. Iwasa, Y. Morikawa, Study on acceptable criteria for assessing wind environment at ground level based on residents' diaries, J. Wind Eng. Ind. Aerodyn. 24 (1986) 1–18.
[31] S. Murakami, A. Mochida, Three-dimensional numerical simulation of turbulent flow around buildings using the k–ε turbulent model, Building Environ. 24 (1) (1989) 51–64.
[32] A. Gadilhe, L. Janvier, G. Barnaud, Numerical and experimental modeling of the three-dimension turbulent wind flow through an urban square, J. Wind Eng. Ind. Aerodyn. 46/47 (1993) 755–763.
[33] S. Takakura, Y. Suyama, M. Aoyama, Numerical simulation of flowfield around buildings in an urban area, J. Wind Eng. Ind. Aerodyn. 46/47 (1993) 765–771.
[34] S. Yamamura, Y. Kondo, Numerical study on relationship between building shape and ground-level wind velocity, J. Wind Eng. Ind. Aerodyn. 46/47 (1993) 773–778.
[35] C.J. Williams, M.J. Soligo, J. Cote, A discussion of the components for a comprehensive pedestrian level comfort criterion, J. Wind Eng. Ind. Aerodyn. 24 (1992) 2389–2390.
[36] M.J. Soligo, P.A. Irwin, C.J. Williams, Pedestrian comfort including wind and thermal effects, Third Asia-Pacific Symp. on Wind Eng., 1993, pp. 13–15.
[37] A. Mochida, S. Murakami, T. Omori, Y. Tominaga, Distribution of SET* in summer season predicted by numerical simulations of convective and radiative heat transports, the 13th Japan National Conference on Wind Eng., 1994, pp. 91–94 (in Japanese).
[38] G.L. Mellor, T. Yamada, A hierarchy of turbulence closure models for planetary boundary layer, J. Appl. Meteorol. 13 (7) (1974) 1791–1806.
[39] G.L. Mellor, T. Yamada, Development of a turbulence closure model for geophysical fluid problem, Rev. Geophys. Space Phys. 20 (4) (1982) 851–875.
[40] R.A. Pielke, Mesoscale Meteorological Modeling, Academic Press, New York, 1984.
[41] T. Yamada, S. Bunker, Development of a nested grid, Second moment turbulence closure model application to the 1982 ASCOT brush creek data simulation, J. Appl. Meteorol. 27 (5) (1988) 562–578.
[42] F. Kimura, S. Takahashi, The effects of land-use and anthropogenic heating on the surface temperature in the Tokyo metropolitan area: a numerical experiment, Atmos. Environ. 25 (2) (1991) 155–164.
[43] H. Yoshikado, Numerical study of the daytime urban effect and its interaction with the sea breeze, J. Appl. Meteorol. 31 (10) (1992) 1145–1164.
[44] R.L. Walko, W.R. Cotton, R.A. Pielke, Large-eddy simulations of the effects of hilly terrain on the convective boundary layer, Boundary-Layer Meteorol. 58 (1992) 133–150.
[45] R.A. Pielke et al., A comprehensive meteorological modeling system – RAMS, Meteorol. Atmos. Phys. 49 (1992) 61–91.
[46] A. Mochida, S. Murakami, S. Kim, R. Ooka, CFD analysis of mesoscale climate in greater Tokyo area, these Proceedings, J. Wind Eng. Ind. Aerodyn. 67&68 (1997) 459–477.
[47] T. Kuwagata, J. Kondo, Estimation of aerodynamic roughness at the regional meteorological stations (AMeDAS) in the central part of Japan, TENKI, March 1990 (in Japanese).
[48] H. Sugiyama, A. Mochida, S. Murakami, T. Ojima, Transport of pollutants convected by heat island circulations, Proc. of Architectural Institute of Japan Annual meeting, 1996 (in Japanese).

[49] S. Murakami, A. Mochida, Y. Tominaga, Numerical simulation of turbulent diffusion in cities, Wind Climate in Cities, Kluwer Academic Publishers, Dordrecht, 1994, pp. 681–701.

[50] M. Kato, B.E. Launder, The modeling of turbulent flow around stationary and vibrating square cylinders, Prep. of 9th Symp. on Turbulent shear flow, 1993, 10-4-1-6.

[51] K. Kondo, S. Murakami, A. Mochida, Numerical study on flow field around square rib using revised k–ε model, 8th Symp. on Numerical Fluid Dyn., 1994, pp. 363–366 (in Japanese).

[52] S. Murakami, A. Mochida, K. Kondo, Y. Ishida, M. Tsuchiya, Development of new k–ε model for flow and pressure fields around bluff body, these Proceedings, J. Wind Eng. Ind. Aerodyn. 67&68 (1997) 169–182.

[53] S. Kawamoto, An improved k–ε–ϕ turbulence model for wind load estimation, Reproduction of conical vortices on the roof of 1:1:0.5 building, 9th Symp. on Numerical Fluid Dyn., 1995, pp. 197–198 (in Japanese).

[54] S. Sakamoto, Flow field around 1:1:0.5 rectangular prism predicted by LES, Proc. of Architectural Institute of Japan Annual meeting, 1995, pp. 169–170 (in Japanese).

[55] S. Fu, B.E. Launder, D.P. Tselepidakis, Accommodating the effects of high strain rates in modelling the pressure-strain correlation, UMIST Mech. Eng. Dept. Rep. TFD/87/5, 1987.

[56] C.G. Speziale, S. Sarkar, T.B. Gatski, Modelling the pressure-strain correlation of turbulence: an invariant dynamical system approach, J. Fluid Mech. 227 (1991) 245–272.

[57] S. Murakami, A. Mochida, R. Ooka, Numerical simulation of flowfield over surface-mounted cube with various second-moment closure models, Prep. of 9th Symp. on Turbulent Shear Flows, 1993, 13-5-1-6.

[58] M.M. Gibson, B.E. Launder, Ground effects on pressure fluctuations in the atmospheric boundary layer, J. Fluid Mech. (1978) 491–511.

[59] T.J. Craft, B.E. Launder, A new model of "Wall-Reflection" effects on the pressure-strain correlation and its application to the turbulent impinging jet, AIAA J. 30 (1992) 2970.

[60] J.W. Deardorff, A three-dimensional numerical study of turbulent channel flow at large Reynolds numbers, J. Fluid Mech. 41 (1970) 453.

[61] J.W. Deardorff, Numerical investigation of neutral and unstable planetary boundary layers, J. Atmos. Sci. 29 (1972) 91.

[62] U. Schumann, Subgrid scale model for finite difference simulation of turbulent flows in plane channels and annuli, J. Comp. Phys. 18 (1975) 376–404.

[63] R.A. Clark, J.H. Ferziger, W.C. Reynolds, Evaluation of subgrid-scale models using an accurately simulated turbulent flow, J. Fluid Mech. 91, part 1 (1979) 1–16.

[64] N.N. Monsour, P. Moin, W.C. Reynolds, J.H. Ferziger, Improved methods for large eddy simulations of turbulence, Turbulent Shear Flows I (1979) 386.

[65] M. Antonopoulos-Domis, Large eddy simulation of a passive scalar in isotropic turbulent, J. Fluid Mech. 104 (1981) 55–79.

[66] S. Biringen, W.C. Reynolds, Large eddy simulation of the shear-free turbulent boundary layer, J. Fluid Mech. 103 (1981) 53–63.

[67] P. Moin, J. Kim, Numerical investigation of turbulent channel flow, J. Fluid Mech. 18 (1982) 341–377.

[68] K. Horiuti, Comparison of conservative and rotational forms in large eddy simulation of turbulent channel flow, J. Comput. Phys. 71 (1987) 343–370.

[69] P.J. Mason, Large eddy simulation of the convective atmospheric boundary layer, J. Atmos. Sci. 46 (1989) 1492–1516.

[70] P.J. Mason, S.H. Derbyshire, Large-eddy simulation of the stable-stratified atmospheric boundary layer, Boundary-Layer Meterol. 53 (1990) 117–162.

[71] K. Mizutani, S. Murakami, S. Kato, A. Mochida, Study on influence of change of Smagorinsky constant, Proc. of Architectural Institute of Japan Annual Meeting, 1991, pp. 483–484 (in Japanese).

[72] A. Yoshizawa, Eddy-viscosity-type subgrid-scale model with a variable Smagorinsky coefficient and its relationship with the one-equation model in large eddy simulation, Phys. Fluid A 3 (8) (1991) 2007–2009.

[73] J. Bardina, J.H. Ferziger, W.C. Reynolds, Improved subgrid-scale models for large-eddy simulation, AIAA paper-80, 1981.
[74] K. Horiuti, Backward cascade of subgrid-scale kinetic energy in wall-bounded and free turbulent flows, Prep. of the 10th Turbulent Shear Flows Symp., vol. 2, 1995, 20.13–18.
[75] K. Horiuti, Subgrid-scale energy production mechanism in large eddy simulation, Proc. Int. Symp. on Mathematical Modeling of Turbulent Flows, 1995, pp. 164–169.
[76] C.H. Moeng, A large eddy simulation model for the study of planetary boundary layer turbulence, J. Atmos. Sci. 41 (1984) 2052–2062.
[77] F.T.M. Nieuwstadt, P.J. Mason, C.H. Moeng, U. Schumann, Large-eddy simulation of the convective boundary layer: A comparison of four computer codes, Proc. 8th Symp. on Turbulent Shear Flows, 1993.
[78] A. Mochida, S. Murakami, Y. Tominaga, H. Kobayashi, Comparison between standard and dynamic type of Smagorinsky SGS model, J. Architecture, Planning and Environmental Eng. Architectural Institute of Japan 479 (1996) 41–47 (in Japanese).
[79] E.R. Van Driest, On turbulent flow near a wall, J. Aeronautical Sci. 23 (1956).
[80] Y. Tominaga, S. Murakami, A. Mochida, Large eddy simulation of turbulent flow past 2D square cylinder using Dynamic Mixed SGS model, 8th Symp. on Numerical Fluid Dyn., 1994, pp. 225–228 (in Japanese).
[81] Y. Zang, R.L. Street, J.R. Koseff, A dynamic mixed subgrid scale model and its application to turbulent recirculating flows, Phys. Fluids A 5 (12) (1993).
[82] B. Vreman, B. Geurts, H. Kuerten, On the formulation of the dynamic mixed subgrid-scale model, Phys. Fluids A 6 (1994) 4057.
[83] S. Ghosal, T.S. Lund, P. Moin, K. Akselvoll, A dynamic localization model for large-eddy simulation of turbulent flows, J. Fluid Mech. 286 (1995) 229–255.
[84] U. Piomelli, J. Lu, Large-eddy simulation of rotating channel flows using a localized dynamic model, Phys. Fluids 7 (1994) 839–848.
[85] C. Meneveau, T.S. Lund, W. Cabot, A Lagrangian dynamic subgrid-scale model of turbulence, J. Fluid Mech., in press.
[86] K. Akselvoll, P. Moin, Large eddy simulation of a backward facing step flow, Eng. Turbulence Modelling and Experiments 2 (1993) 303–313.
[87] LES workshop of Flows past Bluff Bodies, June 26–28 (1995) Rottach–Egern, Tegernsee, Germany, organized by W. Rodi and J.H. Ferziger.
[88] D.K. Lilly, A proposed modification of the Germano subgrid-scale closure method, Phys. Fluids A 4 (3) (1992) 633–635.
[89] W. Cabot, P. Moin, Large eddy simulation of scalar transport with the dynamic subgrid-scale model, Center for Turbulence Research, 1991.
[90] S. Lee, S.K. Lele, P. Moin, Simulation of spatially evolving turbulence and the applicability of Taylor's hypothesis in compressible flow, Phys. Fluids A 4 (7) (1992) 1521–1530.
[91] K. Kondo, S. Murakami, A. Mochida, Study on generation of fluctuating wind velocities for inflow boundary condition of LES (Part 1), 9th Symp. on Numerical Fluid Dyn. 1995, pp. 213–214 (in Japanese).
[92] K. Kondo, S. Murakami, A. Mochida, Generation of velocity fluctuations for inflow boundary condition of LES, submitted to the Proc. of CWE96, 1996.
[93] J.C. Rotta, Statistische theorie nichthomogrner turbulenz, Zeitscr. Phys. 129 (1951) 547.
[94] B.E. Launder, G.J. Reece, W. Rodi, Progress in the development of Reynolds-stress turbulence closure, J. Fluid Mech. 68 (1975) 537.
[95] C.C. Shir, A preliminary numerical study of atmospheric turbulent flow in the idealized planetary boundary layer, J. Atmos. Sci. 30 (1973) 1327.
[96] Y. Morikawa, private communication, Technology Research Center, Taisei Corp.
[97] S. Iizuka, S. Murakami, A. Mochida, S. Lee, Generation of inflow turbulence based on 3D energy spectrum in wave number space, Proc. Architectural Institute of Japan Annual meeting, 1996 (in Japanese).

Bluff Body Aerodynamics

ELSEVIER

Journal of Wind Engineering
and Industrial Aerodynamics 67&68 (1997) 37–49

JOURNAL OF wind engineering AND industrial aerodynamics

Discrete vortex model of flow over a square cylinder

Donald J. Bergstrom *, Jian Wang

Department of Mechanical Engineering, University of Saskatchewan, 57 Campus Drive, Saskatoon, Saskatchewan, Canada S7N 5A9

Abstract

The discrete vortex method (DVM) is used to numerically study the velocity field in the near wake of a 2-D cylinder of square cross-section. Vortex panels are used to model the cylinder surface, at which a no-penetration boundary condition was imposed. The DVM uses a series of individual vortex elements to model the separated shear layers of the cylinder. The discrete-vortex elements interact to form clusters in the wake region, which represent the larger vortical structures present in such flows. By careful choice of numerical parameters, good agreement is obtained with experimental measurements for the mean and fluctuating velocity fields in the near-wake region. The method is especially effective in modelling the unsteady, quasi-periodic motions associated with the large-scale vortex structures in the wake.

Keywords: Discrete vortex model; Vortex shedding; Quasi-periodic flow

1. Introduction

Turbulent flow over a 2-D cylinder represents a classic flow paradigm in bluff-body aerodynamics. The flow is of special interest to wind engineers, because of the unsteady aerodynamic forces which can be produced on structural members exposed to winds. The unsteady forces are associated with the phenomenon of *vortex shedding*, i.e. the quasi-periodic separation of the shear layers from the top and bottom sides of the cylinder. The separated shear layers form large vortical structures in the wake of the cylinder. Given that the flow is quasi-periodic, following Hussain and Reynolds [1] we adopt a triple decomposition of the velocity field as follows:

$$u(t) = \langle u \rangle + \tilde{u} + u', \tag{1}$$

where $\langle u \rangle$ is the time-average value; $\tilde{u}$ is the quasi-periodic fluctuation, or phase component with zero mean; and u' is the stochastic or turbulent fluctuation. For such a flow, the time variation involves both a *coherent* or phase component, as well as an *incoherent* or turbulent component.

*Corresponding author. E-mail: bergstrd@sask.usask.ca.

PII S0167-6105(97)00061-5

Presently, computational fluid dynamics (CFD) has reached a level of maturity that it can be used to predict wind engineering flows such as the cylinder problem described above. Although wind engineers are probably more interested in a prediction of the unsteady pressure field, which is responsible for the aerodynamic forces on the cylinder, we shall view the problem in terms of the velocity field. At the first conference in this series, Rodi [2] reviewed the use of CFD to predict vortex-shedding flow past long cylinders at high Reynolds numbers. One of the geometries he considered was that of a cylinder of square cross-section, for which the separation points are fixed by the sharp upstream corners. He compared the results of predictions based on statistical turbulence models to experimental data where available, as well some large eddy simulation (LES) results. In the statistical approach, the effect of the turbulent fluctuations in the time-average transport equations is modelled as an effective stress called the Reynolds stress. The use of statistical models to predict an unsteady flow pattern presumes that the periodic flow pattern is *quasi-steady* with respect to the turbulence, with a clear separation between the coherent and stochastic scales of motions. The LES on the other hand attempts to model all motions of a size larger than the grid scale. Not surprisingly, Rodi concluded that second-moment models were generally more accurate than k–ε models, and performed almost as well as LES. The use of statistical models for the prediction of vortex-shedding flow over a square cylinder was considered further by Bosch and Rodi [3], where the k–ε model as modified by Kato and Launder [4] was shown to yield much improved results over the standard model. They also suggested that use of a two-layer model to resolve the near-wall viscous region would improve the model performance.

Upon reflection, it would seem that the role of the turbulence model should be relatively less important in a flow like the square cylinder, where the separation points are fixed by geometry. Many of the flows used to test and develop the statistical turbulence models considered in Ref. [2] were dominated by turbulent transport. In contrast, visualization studies of vortex-shedding flows suggests that significant kinetic energy is residing in scales of motion which are larger than the stochastic scales. In this sense, one might look for a more elegant solution technique for solving the velocity field, i.e. one which emphasises the unsteady but non-turbulent structure. In the same conference where Rodi [2] presented his review, a number of other papers, e.g. Ref. [5], used an alternate approach known as the discrete vortex method (DVM) to predict similar bluff body flows. The DVM is a potential flow method which uses a series of individual (discrete) vortex elements released from the separation points of the cylinder to model the separated shear layers. The large vortical structures which form in the wake are produced by the aggregation of these discrete vortex elements into clusters, which interact with each other as they are convected downstream. The DVM would seem to be especially well suited to the simulation of flows with strong vortical structure. The method itself has been extensively developed, and applied to a wide range of flows, see the reviews in Refs. [6,7].

The present study reports a prediction for turbulent flow over a 2-D cylinder of square cross-section using the DVM. Limited experimental data [8,9] are available for the velocity field. More recently, a comprehensive experimental study of the square cylinder has been conducted in Ref. [10]. In our case, the experimental data of Durão et al. [8] was used to calibrate the model in terms of specific values of the numerical parameters. Other papers [11,12] have used the DVM to predict separating flows for similar geometries. However, here attention is focused on the prediction of the velocity field in the near-wake of the cylinder. In terms of the velocity field, the DVM is shown to perform at least as well as the statistical and LES models considered in Ref. [2]. In particular, the extent of the time-averaged recirculation bubble located immediately downstream of the cylinder is in good agreement with the experimental data. The remainder of the paper first describes the numerical method. The predictions for the time-average and fluctuating velocity fields are then discussed, with comparison to experimental results where possible. Finally some conclusions as to the performance and potential of the DVM are presented.

2. Methodology

2.1. Mathematical model

Fig. 1 shows a schematic diagram of the flow geometry. A 2-D cylinder of square cross-section is immersed in a uniform turbulent stream. The cylinder is fixed and the angle of attack is zero. The numerical problem was chosen to closely match the experiment in Ref. [8]. They considered a square cylinder of height $b = 20\,\text{mm}$, placed in a water tunnel with a uniform velocity of $U_0 = 0.68\,\text{m/s}$. The corresponding Reynolds number, $\text{Re} = U_0 b/\nu$, was approximately 14 000.

We assume the flow to be incompressible and irrotational, so that potential theory can be used. A Cartesian coordinate system is used to describe the problem, where x and y denote the streamwise and transverse directions, respectively. We avoid the use of conformal mappings, and instead solve the problem in physical space. Vortex panels, with linear circulation density on each panel, are used to model the surface of the cylinder.

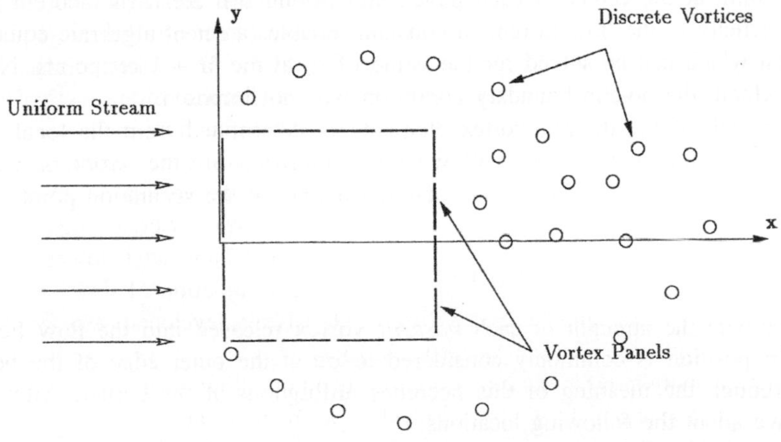

Fig. 1. Schematic diagram of separating flow over a square cylinder.

Unlike a finite difference or finite element calculation, the DVM does not use a numerical grid to discretize the solution domain. For a vortex-shedding flow at high Reynolds numbers, use of a numerical grid would require regions of significantly different levels of refinement. For example, the numerical grid near the solid wall would have to be much finer than that used in the wake region. In our case, we avoid the use of a numerical grid, and instead adopt a Lagrangian approach which follows material elements of the fluid. The separated shear layers are modelled by two series of discrete vortices released from the two upstream corners of the cylinder. The numerical method tracks the vortex elements as they are convected by the velocity field created by the combination of the uniform stream, the vortex panels and the vortex elements.

The mathematical model is developed in terms of the velocity potential function, ϕ. Summing the three contributions, we obtain the following expression:

$$\phi = U_\infty x + \frac{1}{2\pi}\sum_{j=1}^{M}\int_j \gamma(s_j)\tan^{-1}\left(\frac{y-y_j}{x-x_j}\right)\mathrm{d}s_j$$

$$+\frac{1}{2\pi}\sum_{m=1}^{2}\sum_{k=1}^{N}\Gamma_{mk}\tan^{-1}\left(\frac{y-y_{mk}}{x-x_{mk}}\right), \tag{2}$$

where U_∞ is the speed of the uniform flow; M is the total number of vortex panels; γ_j is the circulation density and s_j the length of vortex panel j; Γ_{mk} represents the strength of the kth vortex to be released from the mth separation position; N is the total number of discrete vortices which have been released from each of the two separation positions.

In order to use the expression for the velocity potential given above, we require knowledge of the strengths of the discrete vortices and panels, both of which change in time, as well as the present location of each vortex. The value of the panel density γ_j, at each of the $M+1$ panel endpoints is determined from the boundary condition at the cylinder surface, which requires the normal velocity component to be zero. Applying this constraint at the center of each panel, and noting that Kelvin's theorem requires the net vorticity in the flow to remain constant, enables a set of algebraic equations to be written which can be solved for the value of γ_j at the $M+1$ endpoints. Note that in our method, the no-slip boundary condition was not imposed.

The strength of the discrete-vortex elements is determined from the local velocity field at the separation locations. Following the approach of other workers, e.g. Refs. [5,12], we can use the rate of change of the vorticity at the separation point, i.e.

$$\frac{\mathrm{d}\Gamma}{\mathrm{d}t} = \frac{1}{2}U_\mathrm{s}^2, \tag{3}$$

to approximate the strength of each *nascent* vortex released into the flow field. The separation position is commonly considered to be at the outer edge of the boundary layer, although the meaning of this becomes ambiguous at the corner. After careful testing, we adopt the following locations:

$$x_\mathrm{sep} = 0.05b; \qquad y_\mathrm{sep} = \pm 0.51b$$

for the separation points. Given the local velocity and vortex strength at the separation point, the exact location of the center of the vortex can be inferred. In our calculation, we use *blob*-type vortices, which have a core of finite extent. The radius of the core, σ, which is called the cut-off length is set equal to the displacement of the vortex center from the separation point.

In a real flow, i.e. one in which there is a finite viscosity and (at higher Reynolds numbers) turbulence, the vorticity in the flow is observed to diminish in time. Following Kiya et al. [11], we use a function of the following form to *damp* the initial strength of the discrete vortex with time:

$$\frac{\Gamma(t)}{\Gamma_0} = 1 - \exp\left[-\frac{A_{\rm d}}{t/t_0}\right], \tag{4}$$

where the characteristic time scale $t_0 = b/U_\infty$ is used to non-dimensionalize the time t. After numerical study, the value of the constant, $A_{\rm d}$, was set to 10.

Having explained the method for calculating the strengths of the panels and discrete vortices, it remains to determine the location of the each vortex in the flow field. The convection of the vortices by the flow field was calculated using a Langrangian approach. Based on the velocity field at a given instant of time, the trajectory of a specific vortex element was determined using the following first-order Euler scheme:

$$x_i(t + \delta t) \doteq x_i(t) + v_i \delta t \, . \tag{5}$$

As the vortex elements were convected in the flow, the velocity field responsible for convecting them was itself changing. This in turn causes the circulation density of the vortex panels to change. The manner in which the calculation took into consideration all of these effects is described below.

2.2. Numerical procedure

The calculation procedure involved two major processes: (1) the generation and release of the nascent vortices; and (2) the convection of all of the vortices in the flow field. Notice that the velocity field continually changed in response to both of these effects. The discrete vortex elements were introduced alternately from the two upstream corners of the cylinder. Once a nascent vortex was introduced, it was convected forward in time using the Euler approximation given above. At the end of each convection step, the vortex panel strengths were recalculated based on the new vortex distribution, and the velocity potential function updated. The temporal development of the flow field used a finite time step, δt, which was found to have a significant effect on the evolution of the velocity field. Based on numerical studies, the value of the time step was set to $\delta t/t_0 \doteq 0.16$. Although smaller time steps usually resulted in improved results, finite computational resources set a lower limit on the value. Following the practice of other researchers, we allowed the nascent vortex to be convected forward in time for multiple time steps, specifically three, before introducing the next nascent vortex.

Typically, the development of a flow field required 2000–3000 times steps to reach a fully developed state. Several thousand additional time steps were then required to obtain converged average values. Limited computer resources prevented us from tracking thousands of vortex elements in a single calculation. Furthermore, since the focus of the present study was the velocity field in the near-wake, the influence of the *older* vortex elements as they are convected some considerable distance downstream becomes increasingly small. As a practical compromise, a typical calculation retained only 400 discrete vortex elements in the solution domain. After the first 400 vortices had been created, older vortex elements were systematically removed from the calculation as new vortices were added.

3. Results

As noted above, the various numerical parameters associated with the DVM calculation were calibrated on the basis of comparisons to the experimental measurements of Durão et al. [8] for the velocity field in the near-wake of the cylinder. Using the values identified above, we proceed to discuss the predictions for the velocity field. The numerical results are not strictly *predictive* in the sense that the model has been *tuned* to obtain the best possible agreement for a wide range of flow features.

Fig. 2 shows the instantaneous spatial distribution of discrete vortices in the near-wake of the cylinder. The lines connecting groups of vortex elements shed from the same separation point illustrate the coalescence of individual vortices into groups which represent larger structures. In Figs. 3 and 4 we examine the features of the time-average velocity field. In Fig. 3 we present velocity profiles for the streamwise component, $\langle u \rangle$, at three different transverse sections in the near-wake. The velocity profiles are symmetric, and illustrate the streamwise decay in the momentum deficit. The profile at $x/b = 2.5$ shows a small acceleration of the outer flow due to the obstruction presented by the cylinder. There is very little change in the width of the velocity profile in the

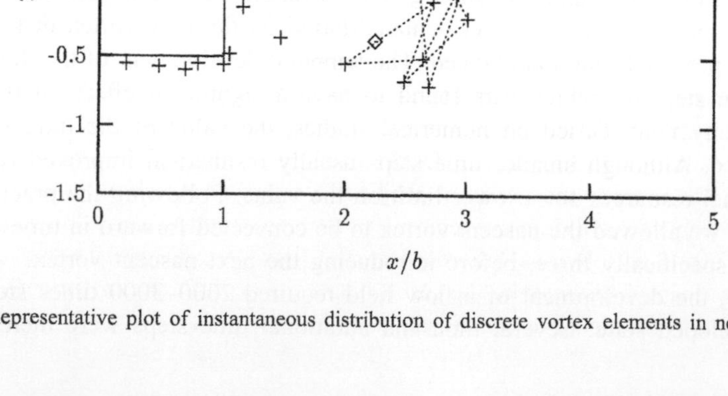

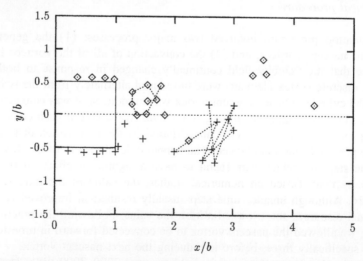

Fig. 2. Representative plot of instantaneous distribution of discrete vortex elements in near-wake.

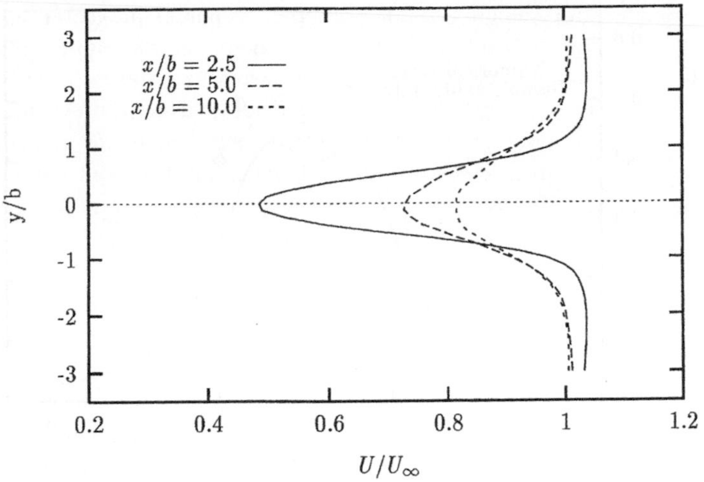

Fig. 3. Time-average velocity profiles downstream of cylinder.

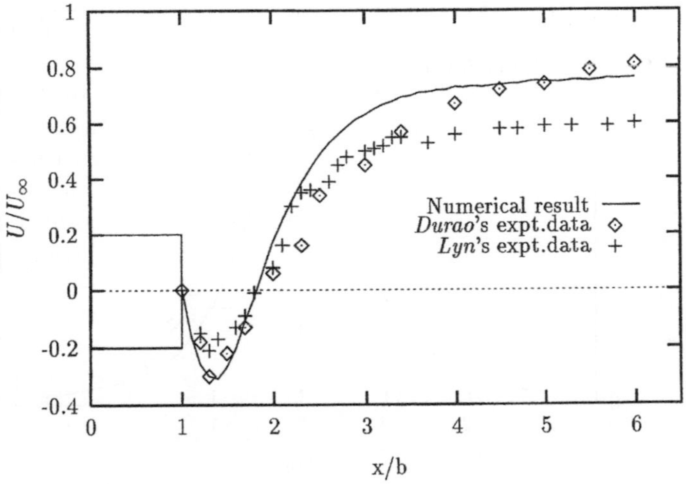

Fig. 4. Time-average velocity along centerline of near-wake.

last two sections, which may warrant further investigation. The decay of the maximum value of the velocity deficit along the wake centerline is considered in greater detail in Fig. 4. In addition to the data of Durão et al. [8], we also show the measurements of Lyn [10] taken from the paper of Rodi [2]. The DVM prediction for the velocity does a good job of resolving the recirculation region immediately downstream of the cylinder. There is an obvious difference in the two sets of experimental data as one moves downstream. This may be the result of differences in the experiments, e.g. aspect ratio, freestream turbulence level, blockage ratio, etc.

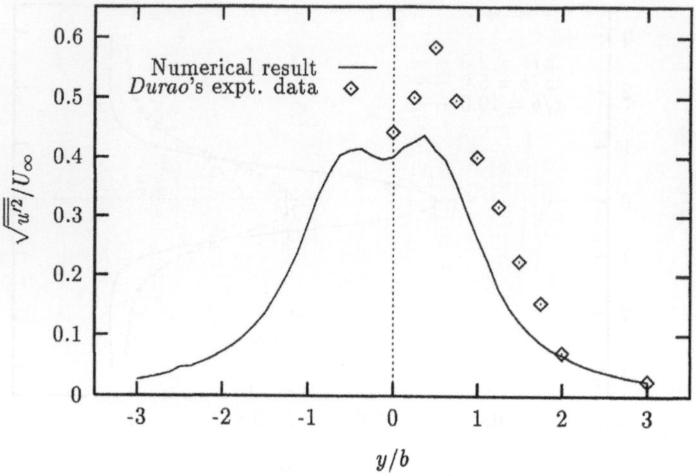

Fig. 5. Profile of streamwise velocity fluctuation at $x/b = 2.5$.

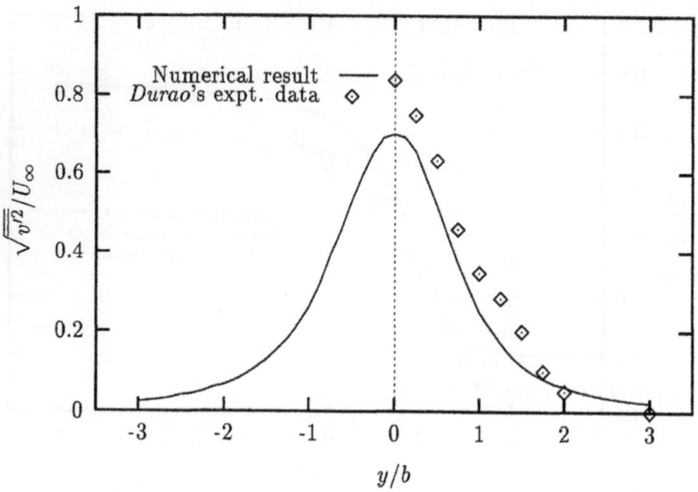

Fig. 6. Profile of transverse velocity fluctuation at $x/b = 2.5$.

In Figs. 5 and 6 we consider the fluctuation of the instantaneous streamwise and transverse velocity components, respectively, at the section $x/b = 2.5$. The experimental data of Durão et al. [8] includes both the effects of the phase and turbulent fluctuations. Since our model does not have a stochastic component, the fluctuation predicted by the numerical result is entirely due to the complex vortex dynamics associated with the temporal development of the flow. The numerical profiles capture the shape and approximate magnitude of the experimental data. The small asymmetries in Fig. 5 may be attributed to the limited size of the ensemble used for averaging.

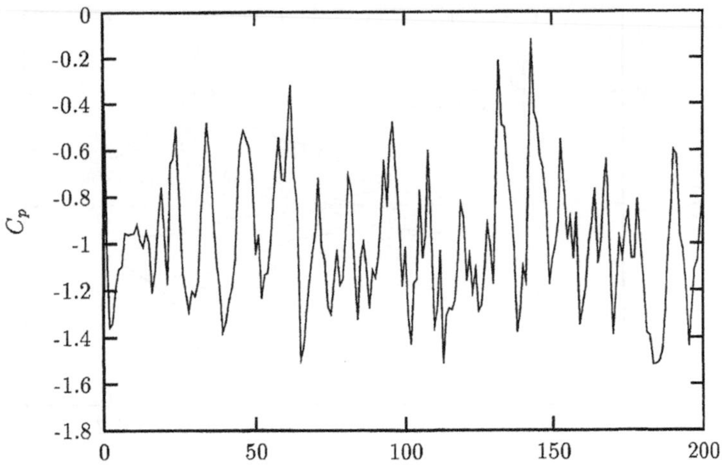

Fig. 7. Time variation of pressure coefficient at midpoint of upper sidewall during the period $3.2 \leqslant T \leqslant 6.4$.

In a wind engineering context, the pressure field is of special interest. The pressure coefficient was calculated using the following relation:

$$C_p = \frac{P - P_\infty}{\rho U_\infty^2/2} = 1 - \left(\frac{U}{U_\infty}\right)^2 - \frac{2}{U_\infty^2}\left(\frac{\partial \phi}{\partial t}\right). \qquad (6)$$

The last term, $\partial\phi/\partial t$, which included contributions from both the vortex panels and discrete vortices presented numerical difficulties, in so far as limitations in the time resolution resulted in some *noise* in our derivative. Fig. 7 shows the time variation of the pressure coefficient at the mid-point of the upper sidewall of the cylinder. The time series evidences a complex behaviour, with a strong quasi-periodic component. The characteristic frequency yielded a Strouhal number of $\mathrm{St} = 0.15$, somewhat larger than the value of 0.12 measured in Ref. [9] and 0.13 measured in Ref. [10]. The mean value of the pressure coefficient at the midpoint of the sidewall was approximately $\langle C_p \rangle \doteq -1.0$, compared to the value of -1.3 measured in Ref. [13] and -2.0 measured in Ref. [14].

Following Ref. [10], we used the pressure signal at the midpoint of the sidewall of the cylinder to examine the phase structure of the velocity field. Ten phase periods were determined for each cycle of the pressure field. The instantaneous velocity field for a given period of time at a specific location in the flow field could then be ensemble averaged for each phase. Using this technique, three representative velocity profiles at $x/b = 4$ are shown in Fig. 8. These phase-averaged profiles clearly show the transverse movement of the maximum velocity deficit in time, following the passage of large-scale vortex structures. The sharp features of the profiles are due to the small ensemble size used for the phase average calculation.

The results above have considered the velocity field. In terms of visualization, a more useful kinematic variable is the stream function. The expression for the stream function can be calculated in an analogous way to that of the velocity potential function, and

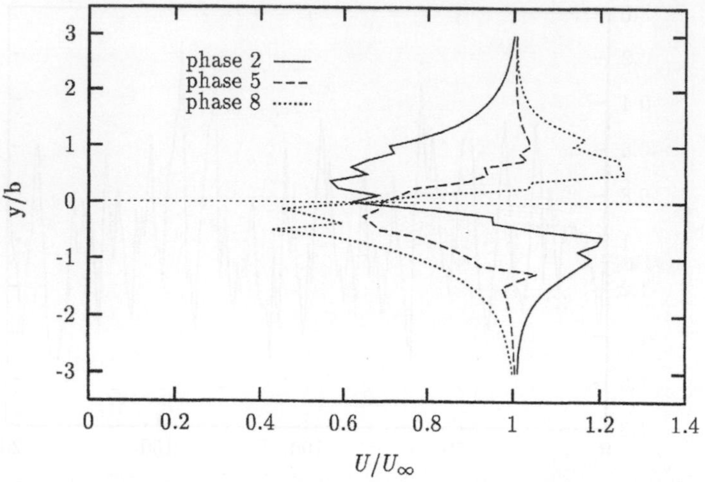

Fig. 8. Phase-average velocity profiles at $x/b = 4$.

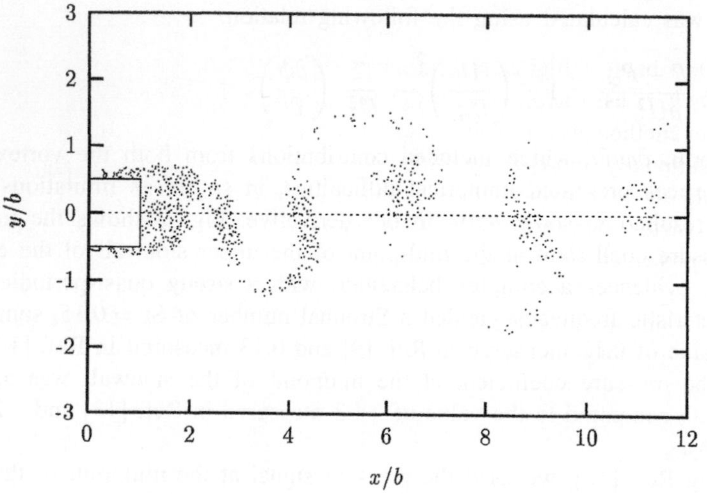

Fig. 9. Instantaneous locations of discrete vortices at $T = 12.8$.

is given by

$$\psi = U_\infty y - \frac{1}{2}\sum_{j=1}^{M}\int_j \gamma(s_j)\ln\{[(x-x_j)^2+(y-y_j)^2]^{1/2}\}\mathrm{d}s_j$$
$$-\frac{1}{2}\sum_{m=1}^{2}\sum_{k=1}^{N}\Gamma_{mk}\ln\{[(x-x_{mk})^2+(y-y_{mk})^2]^{1/2}\}. \qquad (7)$$

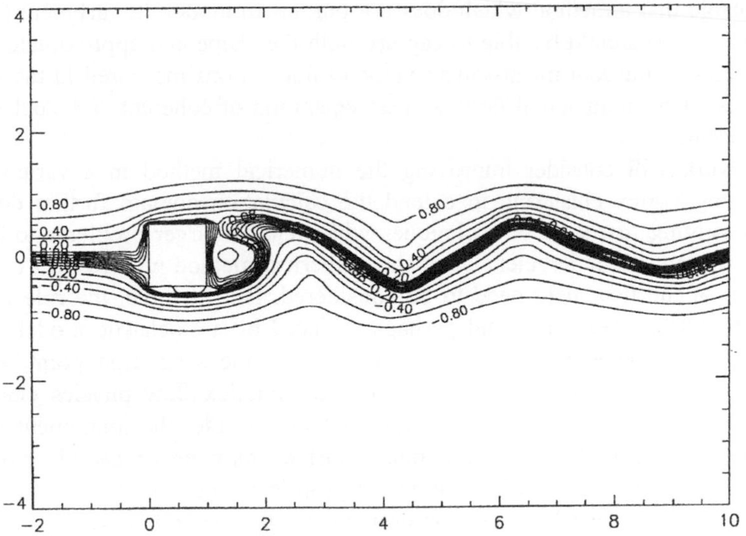

Fig. 10. Streamlines of flow field at $T = 12.8$.

In order to obtain a higher resolution of the flow field, we found it convenient to use a *fine-scale* release approach. In the new approach, each nascent vortex introduced in the former method was replaced by a small group of weak vortices. This approach was computationally more demanding, but enabled a finer resolution of the stream function field to be obtained. Fig. 9 shows the instantaneous spatial distribution of the discrete-vortex elements in the flow field at a given instant of time. Large-scale structures represented by vortex *clusters* are clearly evident in the plot. The associated streamlines are shown in Fig. 10. At this instant in time, a single large vortex can be observed immediately downstream of the cylinder. As the flow develops, it will be shed from the cylinder and move into the wake.

4. Conclusions

In the study reported above, the DVM has been used to investigate vortex-shedding flow over a 2-D cylinder of square cross-section. The DVM is especially well suited for simulation of such a flow, because the primary kinematic variable used to *build* the flow field is vorticity. This enables the DVM to resolve much of the rich and complex vortical structure of the flow field. From the perspective of velocity, we have shown that the method can be *calibrated* to produce results which are in good agreement with many of the experimental measurements available. Visually, the numerical results are especially insightful, since they clearly capture the large-scale vortical structures which dominate the unsteady behavior of the wake. The prediction for the instantaneous pressure field showed some differences from experimental results, and warrants further consideration.

It is notable that a method which does not purport to model the turbulent (stochastic) fluctuations per se should be able to capture both the shape and approximate magnitude of the profiles of the root-mean-square velocity fluctuations measured in the near-wake. This suggests that in an actual flow, a clear separation of coherent and stochastic scales may not occur.

Future work will consider improving the numerical method in a variety of ways, such as using vortex clustering to extend the calculation domain further downstream, and implementing *fast* solution techniques which enable larger problems to be studied. In addition, some of the deficiencies of the numerical method with respect to modelling the physics of the flow field need to be considered. For example, the role of damping needs to be further examined, and perhaps replaced by a stochastic model. Of critical importance is the generation of nascent vortices at the separation point, which also needs to be critically examined in terms of the complex flow physics characterizing such a separation. In this regard, techniques which consider the implementation of the no-slip boundary condition at the cylinder surface may prove useful. Finally, on a cautionary note, the determination of the appropriate value of the various numerical parameters used by the DVM remains problematic for general flows.

Acknowledgements

The support of the Natural Sciences and Engineering Research Council of Canada is gratefully acknowledged.

References

[1] A.K.M.F. Hussain, W.C. Reynolds, The mechanics of an organized wave in turbulent shear flow, J. Fluid Mech. 41 (1970) 241–258.

[2] W. Rodi, On the simulation of turbulent flow past bluff bodies, J. Wind Eng. Ind. Aerodyn. 46&47 (1993) 3–19.

[3] G. Bosch, W. Rodi, Simulation of vortex shedding past a square cylinder near a wall, Proc. 10th Symp. Turbulent Shear Flows, vol. 1, 1995, pp. 4.13–4.18.

[4] M. Kato, B.E. Launder, The modelling of turbulent flow around stationary and vibrating square cylinders, Proc. 9th Symp. Turbulent Shear Flows, 1993, p. 10.4.

[5] T. Inamuro, T. Adachi, H. Sakata, Simulation of aerodynamic instability of bluff body, J. Wind Eng. Ind. Aerodyn. 46&47 (1993) 611–618.

[6] L. Leonard, Vortex methods for flow simulation, J. Comput. Phys. 37 (1980) 289–335.

[7] T. Sarpkaya, Computational methods with vortices – 1988 Freeman Scholar lecture, J. Fluids. Eng. 111 (1989) 5–52.

[8] D.F.G. Durão, M.V. Heitor, J.C.F. Pereira, Measurements of turbulent and periodic flows around a square cross-section cylinder, Exper. Fluids 6 (1988) 298–304.

[9] M.A.Z. Hasan, The near wake structure of a square cylinder, Int. J. Heat Fluid Flow 10 (1989) 339–348.

[10] D.A. Lyn, S. Einav, W. Rodi, J.-H. Park, A laser-Doppler velocimetry study of ensemble-averaged characteristics of the turbulent near wake of a square cylinder, J. Fluid Mech. 304 (1995) 285–319.

[11] M. Kiya, K. Saskai, M. Arie, Discrete vortex simulation of a turbulent separation bubble, J. Fluid Mech. 120 (1982) 219–244.

[12] S. Nagano, M. Naito, H. Takata, A numerical analysis of two-dimensional flow past a rectangular prism by a discrete vortex model, Comput. Fluids 10 (1982) 243–259.

[13] Y. Otsuki, K. Fuji, K. Washizu, A. Ohaya, Wind tunnel experiment on aerodynamic forces and pressure distributions of rectangular cylinders in a uniform stream, Proc. 5th Symp. on Wind Effects on Structures, 1978, pp. 169–175.

[14] P.W. Bearman, E.W. Obasayu, An experimental study of pressure fluctuations on fixed and oscillating square-section cylinders, J. Fluid Mech. 119 (1982) 297–312.

ELSEVIER

Journal of Wind Engineering
and Industrial Aerodynamics 67&68 (1997) 51–64

JOURNAL OF wind engineering AND industrial aerodynamics

Generation of velocity fluctuations for inflow boundary condition of LES

K. Kondo[a,*], S. Murakami[b], A. Mochida[c]

[a] *Kajima Technical Research Institute, 2-19-1, Tobitakyu, Chofu-shi, Tokyo 182, Japan*
[b] *Institute of Industrial Science, University of Tokyo, 7-22-1, Roppongi, Minato-ku, Tokyo 106, Japan*
[c] *Niigata Institute of Technology, 1719, Fujihashi, Kashiwazaki, Niigata 945-11, Japan*

Abstract

In this study, inflow turbulence for LES was generated by the method of Hoshiya [Proc. JSCE 204 (1972) 121–128] based on Monte Carlo simulation considering power spectral density and cross-spectral density as targets. The generated inflow turbulence was modified to satisfy the continuity equation by divergence-free operation based on the method of Shirani et al. [Report TF-15, 1981, Mechanical Engineering Department, Stanford University]. As a result, inflow turbulence satisfied the prescribed spatial correlation and power spectral density and the level of velocity divergence was sufficiently reduced. To confirm the applicability of generated inflow turbulence, it was used as the inflow boundary condition for simulating decaying isotropic turbulence with the same conditions as the experiment by G. Comte-Bellot and S. Corrsin [J. Fluid Mech. 48 (1971) 273–337]. Decay of turbulence kinematic energy in the streamwise direction was reproduced well by LES using generated inflow turbulence.

Keywords: LES; Inflow turbulence; Divergence-free operation; Spatial correlation; Wave number spectrum

Nomenclature

x_i	spatial coordinate in i-direction
$u_i(l,t)$	velocity component at point l at time t
$i = 1$	main flow direction
$i = 2,3$	cross-flow directions
l, p	indexes denoting the two points related to cross-spectral density $H_{lp}(\omega_n)$ ($l = 1, \ldots, M, p = 1, \ldots, l$)

* Corresponding author.

PII S0167-6105(97)00062-7

M	total number of nodal points in the region where inflow turbulence was generated simultaneously
N	total number of frequency intervals
$a_{lp}(\omega_n)$, $b_{lp}(\omega_n)$	Fourier coefficient
$\phi_{lp}(\omega_n)$	phase lag
ω_n	circular frequency
$\Delta\omega_n$	circular frequency interval
$\xi_p(\omega_n)$, $\eta_p(\omega_n)$	independent Gaussian random number with mean value 0 and standard deviation 1
$S(\omega_n)$	cross-spectral density matrix
$H(\omega_n)$	lower triangular matrix obtained from $S(\omega_n)$
$H_{lp}(\omega_n)$	component of $H(\omega_n)$
*	conjugate
Re	real part
Im	imaginary part

1. Introduction

A generation method of velocity fluctuation for an inflow boundary condition (inflow turbulence) of LES with prescribed spatial correlation and turbulence intensity is one of the most important unresolved issues in CFD research. There are several methods of generating inflow turbulence. One is to conduct preliminary computation of turbulent flow fields using LES, such as a channel flow [1,2] or a turbulent flow generated by a turbulence grid settled at the inflow boundary of a computational domain [3], and to store time series of velocity fluctuations for inflow boundary conditions. This method, however, requires a large computational load. Furthermore, turbulence statistics, i.e., spatial correlations, turbulence intensities, etc., from the preliminary computations are not guaranteed to correspond to prescribed target characteristics.

Another method is to artificially generate time series of velocity fluctuations by performing an inverse Fourier transform for prescribed spectral densities. This type of artificial generation method can be classified into two groups. One uses the 3-D energy spectrum in the wave number domain obtained from spatial correlation of velocity as the target [4–6]. This method has the advantage that the condition of continuity can be imposed on the generation procedure. Furthermore, the time series of velocity fluctuations need not be stored, since inflow turbulence is generated at each time step of LES computation. Thus, the computer memory required for this method usually becomes less than that for a later-described method. In boundary layer flow, however, it is hard to prescribe the 3-D energy spectrum as the target. This is a very serious disadvantage of this method from the viewpoint of application to CWE.

The other group uses power spectral density and cross-spectral density in the frequency domain obtained from time series of velocity fluctuations at the same point or two different points [7–10]. Compared with the 3-D energy spectrum, these frequency spectra can be defined as targets with relative ease from measured data of boundary layer flow. This is a very important advantage of the latter method based on frequency spectra over the former method utilizing the 3-D energy spectrum in the wave number domain. However, the condition of continuity cannot be imposed on the generation procedure with this method. Therefore, divergence-free operation is indispensable in making inflow turbulence satisfy the continuity equation after the generation procedure. Furthermore, generated inflow turbulence must be stored before LES computation.

The advantages and disadvantages of these methods were considered, and the latter was employed and tested in this study. Here, inflow turbulence was generated step-by-step based on the method of Hoshiya [7], considering the prescribed power spectral density and cross-spectral density in the frequency domain as targets. Target spectra obtained from the experiment conducted by Comte-Bellot and Corrsin [11] were prescribed at four points spatially distributed. Generated inflow turbulence was modified to satisfy the continuity equation based on the method of Shirani [12]. Finally, it was used as the inflow boundary condition for LES computation for simulating the decaying isotropic turbulence, with the same condition as in the experiment by Comte-Bellot and Corrsin.

2. Method of generating inflow turbulence with prescribed power spectral density and cross-spectral density

The velocity fluctuation $u_i(l,t)$ at point l in the flow field that satisfies the prescribed power spectral density and cross-spectral density can be expressed by Eqs. (1)–(4) using a trigonometric series with Gaussian random coefficients [7]. Here, suffix i means spatial coordinates in the i-direction ($i = 1$ (main flow direction), $i = 2,3$ (cross-flow directions)). Indexes l and p in Eqs. (1)–(4) denote the two nodal points related to cross-spectral density, $H_{lp}(\omega_n)$, which is calculated from time series of velocity fluctuations at points l and p. $H_{lp}(\omega_n)$ is a component of the cross-spectral matrix, $S(\omega_n)$, as shown in Eq. (3),

$$u_i(l,t) = \sum_{n=1}^{N} \sum_{p=1}^{l} [a_{lp}(\omega_n) \cos\{\omega_n t + \phi_{lp}(\omega_n)\} + b_{lp}(\omega_n) \sin\{\omega_n t + \phi_{lp}(\omega_n)\}], \quad (1)$$

$$a_{lp}(\omega_n) = \sqrt{2\Delta\omega_n}|H_{lp}(\omega_n)|\xi_p(\omega_n), \quad b_{lp}(\omega_n) = \sqrt{2\Delta\omega_n}|H_{lp}(\omega_n)|\eta_p(\omega_n), \quad (2)$$

$$S(\omega_n) = H(\omega_n)H^{*\mathrm{T}}(\omega_n)$$

$$= \begin{bmatrix} H_{11}(\omega_n) & & \\ \vdots & \ddots & \\ H_{M1}(\omega_n) & \cdots & H_{MM}(\omega_n) \end{bmatrix} \begin{bmatrix} H^*_{11}(\omega_n) & \cdots & H^*_{M1}(\omega_n) \\ & \ddots & \vdots \\ & & H^*_{MM}(\omega_n) \end{bmatrix}, \quad (3)$$

$$\phi_{lp}(\omega_n) = \tan^{-1}\left\{ -\frac{\mathrm{Im}\, H_{lp}(\omega_n)}{\mathrm{Re}\, H_{lp}(\omega_n)} \right\}. \tag{4}$$

For the meaning of the notation see the Nomenclature.

In the generation procedure, the target spectral matrix $S(\omega_n)$ in Eq. (3) must first be prescribed. The lower triangular matrix $H(\omega_n)$ can be calculated from $S(\omega_n)$. Next, independent Gaussian random numbers $\xi_p(\omega_n)$, $\eta_p(\omega_n)$ in Eq. (2) are simulated by Monte Carlo simulation with the conditions of mean value equal to 0 and standard deviation equal to 1. The Fourier coefficients $a_{lp}(\omega_n)$, $b_{lp}(\omega_n)$ in Eq. (1) can be calculated from Eq. (2) and the phase lag $\phi_{lp}(\omega_n)$ can be calculated from Eq. (4). By substituting these values in Eq. (1), the velocity fluctuation $u_i(l,t)$ can be obtained. If the cross-spectral matrix $S(\omega_n)$ in Eq. (3) is prescribed for all points on the inflow boundary of the computational domain, inflow turbulence can simultaneously satisfy the cross-correlation among all points. However, a huge computer memory is required for this method when a large number of grid points are located in the x_2–x_3 plane. Furthermore, the cross-spectral matrix $S(\omega_n)$ need not be prescribed for all grid points on the inflow boundary of the computational domain, since the cross-correlation value becomes sufficiently small when the distance between two points is large enough. In this study, the following step-by-step method that considers the cross-spectral matrix of four spatially distributed grid points was utilized to generate inflow turbulence based on the method of Hoshiya [7]. Fig. 1 shows the inflow turbulence generation procedure.

(a) First, velocity fluctuations are generated in region ① by considering four-point correlations.

(b) Next, velocity fluctuations are generated in region ② using Fourier coefficients $a_{lp}(\omega_n)$, $b_{lp}(\omega_n)$ at points (2,1),(2,2) of region ①.

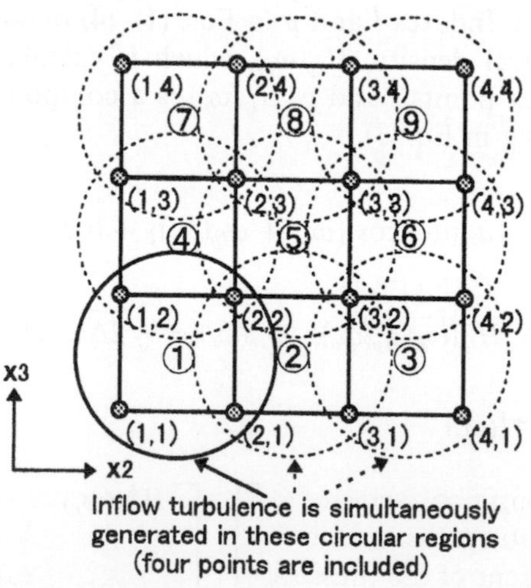

Fig. 1. Inflow turbulence generation procedure.

(c) The same procedure is carried out in region ③.

(d) Velocity fluctuations are generated in region ④ using Fourier coefficients at points (1,2),(2,2) of region ①.

(e) The same procedure is repeated in regions ⑤–⑨.

The target cross-spectral matrix $S(\omega_n)$ was evaluated here from the experimental results of isotropic turbulence measured by Comte-Bellot and Corrsin [11]. In this simulation, we considered that the inflow boundary of LES computation corresponded to the position of normalized distance $U_0t/M = 53.3$ (U_0 is the reference velocity, M the turbulence grid spacing, t the time) from the position of the turbulence grid in the experiment. The decay of turbulence kinematic energies from this position ($U_0t/M = 53.3$) from the results of LES with generated inflow turbulence and from the experiment in Section 5.2 were compared. The target power spectral densities and cross-spectral densities were calculated by Fourier transform from auto-correlation and cross-correlation estimated from the longitudinal correlation $f(r)$, lateral correlation $g(r)$, turbulence intensity and turbulence scale measured by Comte-Bellot and Corrsin. The longitudinal correlation $f(r)$ and lateral correlation $g(r)$ were approximated by Eq. (5) and (6) based on the experimental results,

$$f(r) = \mathrm{e}^{-(r/L_x)^{0.73}}, \tag{5}$$

$$g(r) = \{1 - 0.365(r/L_x)^{0.73}\}\,\mathrm{e}^{-(r/L_x)^{0.73}}. \tag{6}$$

Here, r is the distance and L_x the length scale of longitudinal correlation.

3. Divergence-free operation

The inflow turbulence generated by the method described in the foregoing paragraph does not guarantee fulfilment of the continuity equation. Hence, if the generated inflow turbulence is applied to the inflow boundary condition as it is, convergence of LES computation will be very poor. Shirani et al. [12] proposed a method for obtaining the divergence-free velocity field as the initial condition of temporal simulation. Here, we follow the procedure to satisfy the continuity equation (divergence-free condition) at every time step. The conditions of Eq. (7) and (8) based on the method of Shirani were imposed on the time series of generated inflow turbulence,

$$u_i^{(S)n+1} = u_i^{(R)n+1} - \Delta t \frac{1}{\rho}\frac{\partial \Delta P^n}{\partial x_i}, \tag{7}$$

$$\Delta t \frac{1}{\rho}\frac{\partial^2 \Delta P^n}{\partial x_i^2} = \frac{\partial u_i^{(R)n+1}}{\partial x_i}. \tag{8}$$

Eq. (7) modifies the original velocity $u_i^{(R)n+1}$, which is generated using Eqs. (1)–(4) and does not yet satisfy the continuity equation, to the divergence-free velocity $u_i^{(S)n+1}$ with the gradient of the pressure corrections ΔP^n obtained from the Poisson equation for ΔP^n in Eq. (8) [9]. In order to calculate ΔP^n in Eq. (8), which is expressed

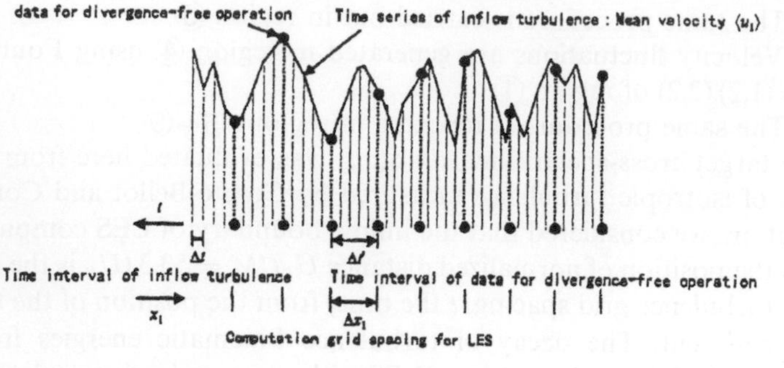

Fig. 2. Treatment of velocity fluctuation data in divergence-free operation.

in spatial coordinates, the time axis of velocity fluctuation must be converted to the spatial coordinate. In this study, the time axis of inflow turbulence was converted to the spatial coordinate in the x_1 direction (main flow direction) based on Taylor's hypothesis of frozen turbulence as shown in Fig. 2. Usually, the grid spacing Δx_1 in the x_1 direction of LES computation is considerably larger than the grid spacing $\Delta x_1'$ corresponding to the time interval Δt of inflow generation, which is evaluated from the relation $\Delta x_1' = \langle u_1 \rangle \Delta t$ based on Taylor's hypothesis. Here, $\langle u_1 \rangle$ is a time averaged value of the u_1 component. If this grid spacing $\Delta x_1'$, which is directly related to the time interval of inflow turbulence, is used as the grid spacing for divergence-free operation, computational grids become too skewed ($\Delta x_1 \ll \Delta x_2, \Delta x_3$), causing some undesirable influences. Hence, data for divergence-free operation were extracted at time interval $\Delta t'$ from the time series of velocity fluctuation with the relationship $\Delta t' = \Delta x_1 / \langle u_1 \rangle$ as shown in Fig. 2. Here, Δx_1 is grid spacing in the x_1 direction for LES computation. In the computations shown in Section 5, $\Delta x_1 = \Delta x_2 = \Delta x_3 = 0.2$. After divergence-free operation, the modified velocity fluctuations were arranged time-serially again.

4. Statistical characteristics of generated inflow turbulence

This section investigates the statistical characteristics of inflow turbulence generated to simulate the decaying isotropic turbulence from the experiment by Comte-Bellot and Corrsin [11].

(1) *Power spectral density*: Fig. 3 compares the power spectral densities of the u_1 and u_2 components of generated inflow turbulence with the target[1]. Power spectral densities of the u_1 and u_2 components of generated inflow turbulence coincide well with the target before divergence-free operation (open circles in Fig. 3a). To

[1] Values presented in this paper were normalized using turbulence grid spacing M and reference velocity U_0 of the wind tunnel test [11].

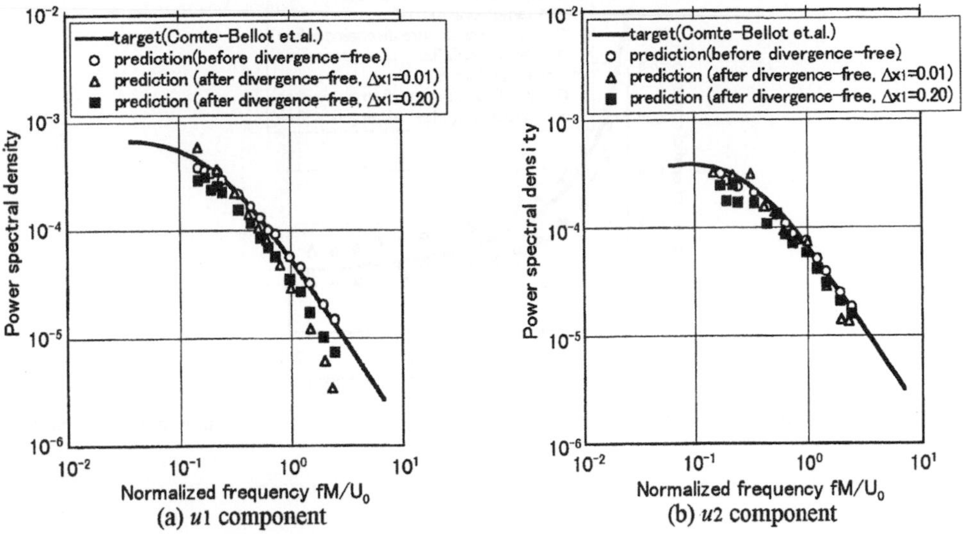

Fig. 3. Comparisons of power spectral density (averaged value on inflow boundary plane): (a) u_1 component; (b) u_2 component.

investigate the influence of the grid skewness on divergence-free operation, two cases with different grid spacings in the x_1 direction are compared in Fig. 3a and Fig. 3b, i.e. $\Delta x_1 = \Delta x_1' = 0.01$ and $\Delta x_1 = 0.2$. The first case employs anisotropic grid spacing $\Delta x_1 = 0.01$, $\Delta x_2 = \Delta x_3 = 0.2$, and the second case employs isotropic grid spacing $\Delta x_1 = \Delta x_2 = \Delta x_3 = 0.2$. When anisotropic grids were used, the gradient of the power spectral density of the u_1 component differed from that of inflow turbulence and its power decreased in the high frequency range $fM/U_0 \geqslant 0.5$ (open triangles in Fig. 3a). This is because divergence-free operation mainly corrected the u_1 component when very thin skewed grids were used in divergence-free operation. When isotropic grids were used, the gradient change, i.e. the power decrease in the high frequency region of the power spectral density of the u_1 component, was improved (full squares in Fig. 3a)[2]. For the power spectral density of the u_2 component, the influence of divergence-free operation was small even when anisotropic grids were used (open triangles in Fig. 3b)[2,3]. As a result of this divergence-free operation, the level of velocity divergence of inflow turbulence decreased from the order of 10^{-1}–10^{-2} to the order of 10^{-5}–10^{-6}, and hence, the convergence of LES computation was significantly improved.

[2] Spectrum power is reduced a little in all frequency ranges by divergence-free operation. This can be overcome by increasing the turbulence intensity level of the inflow turbulence in advance.

[3] The velocity divergence only concerns the longitudinal velocity distribution. However, the time axis of the u_2 (or u_3) component of generated inflow turbulence, which is used for calculation of the power spectrum, corresponds to the lateral velocity distribution. Hence, the effect of divergence-free operation was insignificant in the u_2 and u_3 components in comparison with the u_1 component.

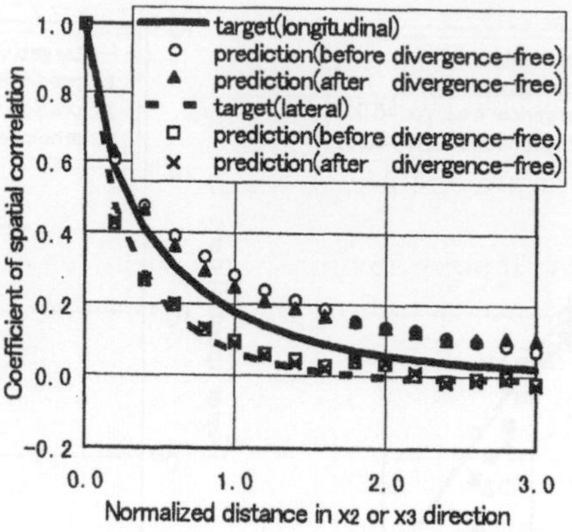

Fig. 4. Comparisons of spatial correlation (u_2 component, averaged value on inflow boundary plane).

(2) *Spatial correlation*: Fig. 4 shows the spatial correlations of the u_2 component. Here, results from divergence-free operation using isotropic grids, $\Delta x_1 = \Delta x_2 = \Delta x_3 = 0.2$, are compared with target values. The differences in spatial correlations between the values before and after divergence-free operation are negligible. For lateral correlation, the predicted value shows very good agreement with the target even though divergence-free operation was applied (open squares and crosses in Fig. 4). However, some difference exists in the longitudinal correlation between the target and predicted values. This is due to the fact that the cross correlations were considered among only four grid points simultaneously in the step-by-step generation procedure utilized in this study. Thus, this can be improved by increasingthe number of grid points for considering the cross correlations simultaneously.

5. Simulation of decaying isotropic turbulence with generated inflow turbulence by LES

5.1. Outline of computations

The Smagorinsky model was used for LES computation. The value of the Smagorinsky constant C_s was 0.23. A second-order centered difference scheme was adopted for spatial derivatives. The Crank–Nicolson scheme was used for the time advancement. The computational domain covered $12(x_1) \times 6(x_2) \times 6(x_3)$. This computational domain was discretized into $60(x_1) \times 30(x_2) \times 30(x_3)$ with uniform grid spacings $\Delta x_1 = \Delta x_2 = \Delta x_3 = 0.2$. A colocated grid arrangement [13] was employed and the inflow turbulence was imposed on the control volume faces at the inflow boundary. Inflow turbulence was generated at a time interval $\Delta t = 0.01$ and was interpolated linearly at a time interval $\Delta t = 0.005$ for LES computation. The convective condition was utilized for the outflow boundary[4]. The periodic boundary condition was applied for four side faces normal to the longitudinal flow direction. Statistical values were determined by time averaging as well as ensemble averaging on the x_2–x_3 plane. Before the inflow turbulence was imposed on the inflow boundary condition of LES, several pre-operations were applied to inflow turbulence as listed below.

Adjustment of mass flux: If mass flux through the inflow boundary plane changes at every time step, the balance of inflow mass flux and outflow mass flux will be disturbed and convergence of LES computation will become poor. In this study, the u_1 component of inflow turbulence was adjusted to satisfy the mass flux balance between inflow and outflow at every time step[5].

Filter treatment: A band pass filter (sharp cut filter) corresponding to the size of the computational domain and computational grid spacing was operated on inflow turbulence.

5.2. Results of LES computation for simulating decaying isotropic turbulence

(1) *Wave number spectrum*: Fig. 5 shows the transition of the wave number spectra of the u_2 component in the x_1 direction. Wave number spectra in Fig. 5a, which are related to the longitudinal correlation of the u_2 component, are calculated from the velocity distribution of the u_2 component in the x_2 direction. Meanwhile, wave number spectra in Fig. 5b, which are related to the lateral correlation of the u_2 component, are calculated from the velocity distribution of the u_2 component in the x_3 direction. These wave number spectra are determined by time averaging as well as ensemble averaging on the x_2–x_3 plane. The gradient change, i.e. the power decrease of wave number spectra are observed evidently for both spectra between the inflow boundary at $x_1 = 0$ and the position of normalized distance $x_1 = 0.5$[6]. The reason for this gradient change in the wave number spectra is examined here. Taking the convection term in the discretized Navier–Stokes (N–S) equation as an example, we must consider that there are two filter effects due to linear interpolation and a centered difference scheme when using the colocated grid [13]. Now, if the filter effect for the function $F(x')$ is defined as $G(x - x')$, the filtered function can be

[4] The convective condition given by the equation below was utilized for the outflow boundary.

$$\frac{\partial u_i}{\partial t} + U_c \frac{\partial u_i}{\partial x_1} = 0.$$

Here, U_c is the convective velocity (value of u_1 component averaged over the x_2–x_3 plane at inflow boundary).

[5] Since the variance of mass flux of inflow turbulence at every time step was small, the effect of this operation on the characteristics of the spectrum, etc. was negligible. Although no adjustments of the u_2 and u_3 components where carried out, averaged values of these components on the inflow plane where almost zero.

[6] The tendencies of the u_1 and u_3 components were almost the same as that of the u_2 component.

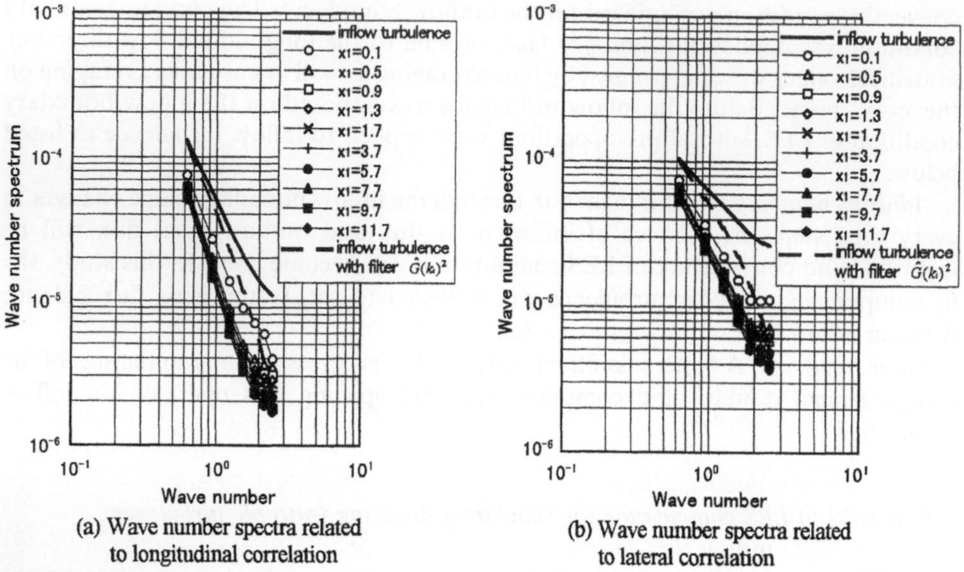

Fig. 5. Transition of wave number spectra in x_1 direction (u_2 component, averaged value on the x_2–x_3 plane): (a) wave number spectra related to longitudinal correlation; (b) wave number spectra related to lateral correlation.

expressed as

$$\overline{F(x)} = \int_{-\infty}^{+\infty} G(x - x')\, F(x')\, \mathrm{d}x'. \tag{9}$$

When $G(x - x')$ is an even function, Eq. (9) can be estimated as Eq. (10) by Taylor expansion using the characteristic of $G(x - x')$ [14–16],

$$\overline{F(x)} = F(x) + \underline{\underline{(\Delta_i^2/24)F''(x)}} + \mathrm{O}(\Delta_i^4). \tag{10}$$

Meanwhile, the error in linear interpolation can be estimated by Taylor expansion as

$$\frac{1}{2}\Big(u_1(i,j,k) + u_1(i+1,j,k)\Big) = u_1\left(i + \frac{1}{2}, j, k\right) + \underline{\underline{\frac{\Delta x_i^2}{8}\frac{\partial^2 u_1(i+\frac{1}{2},j,k)}{\partial x^2}}}. \tag{11}$$

To compare the double underlined term in Eq. (11) with the double underlined term in Eq. (10), linear interpolation can be considered to have a filtering effect with filter width $\Delta_i = \sqrt{3}\Delta x_i$. Furthermore, the filter effect implicitly imposed by the second order centered difference is estimated equivalent to filtering operation with filter width Δx_i. When these two filters act at the same time, their total effect can be approximately evaluated by

$$\Delta_i^2 = (\sqrt{3}\Delta x_i)^2 + (\Delta x_i)^2 = 4\Delta x_i^2, \quad \Delta_i = 2\Delta x_i. \tag{12}$$

When a top-hat filter is used, the total filter effect of linear interpolation and second-order centered difference are approximated by

$$\hat{G}(k_i) = 2\sin(\Delta_i k_i/2)/(\Delta_i k_i), \quad \Delta_i = 2\Delta x_i. \tag{13}$$

By multiplying the wave number spectrum of inflow turbulence (solid line in Fig. 5) by $\hat{G}(k_i)^2$ in Eq. (13), the gradient of the wave number spectrum of inflow turbulence becomes the dotted line (in Fig. 5) and approaches the gradient of computed wave number spectrum in LES. Therefore, the gradient change of wave number spectrum can be attributed to the effects of linear interpolation and centered differences which are implicitly imposed in LES computation. This gradient change of the spectrum may be reduced to a certain degree by making the grid spacing narrower. Furthermore, the power decrease of wave number spectrum in the area $x_1 \leqslant 0.5$ is supposed to reflect the process in which inflow turbulence adapts itself to the discretized N–S equation, since inflow turbulence does not satisfy the N–S equation at the inflow boundary. This can be compensated by increasing the turbulence intensity of the target.

(2) *Spatial correlation*: Fig. 6 indicates the transition of spatial correlation in the x_1 direction obtained from the distribution of the u_2 component on the x_2–x_3 plane. Regarding longitudinal correlation (Fig. 6a), the correlation coefficient at position $x_1 = 0.1$ (open circles in Fig. 6a) just behind the inflow boundary becomes larger than that of the inflow turbulence (solid curve in Fig. 6a). The change in correlation

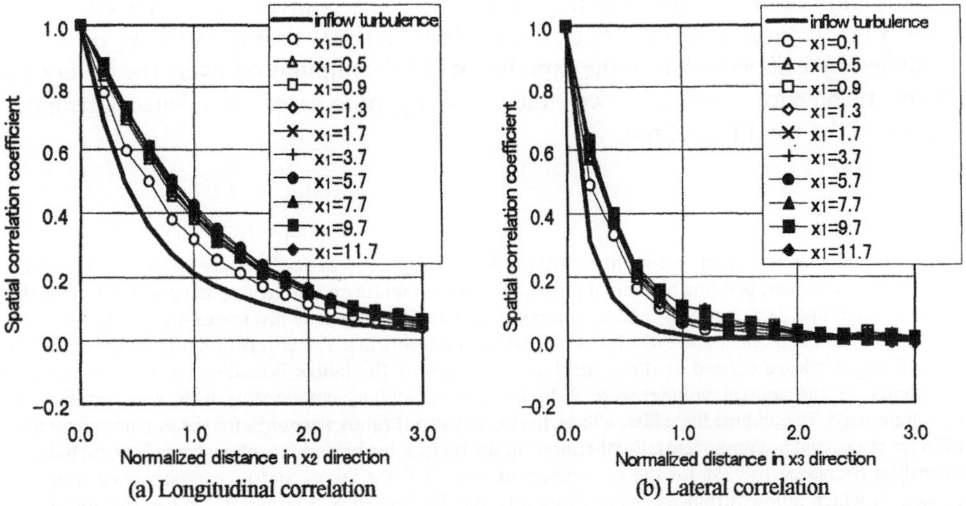

Fig. 6. Transition of spatial correlation in x_1 direction (u_2 component, averaged value on the x_2–x_3 plane): (a) longitudinal correlation, (b) lateral correlation.

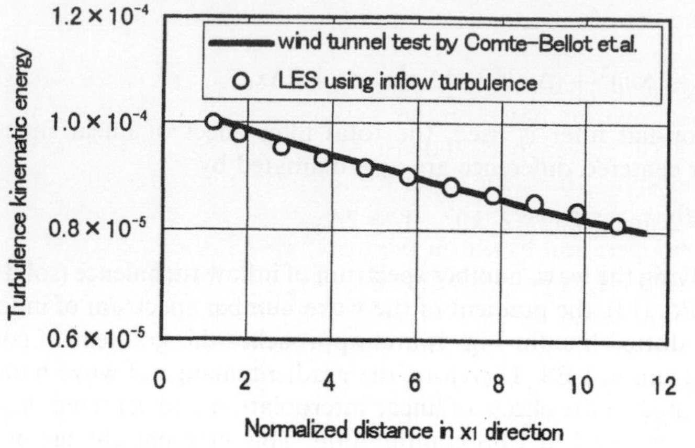

Fig. 7. Longitudinal distributions of turbulence kinematic energy k in decaying isotropic turbulence (averaged value on the x_2–x_3 plane).

coefficients at positions where $x_1 \geqslant 0.5$ is small (open triangles in Fig. 6a). This tendency is similar to that of the lateral correlations (Fig. 6b).

(3) *Turbulence kinematic energy*: As mentioned before, there are several filter effects in the procedure for generating inflow turbulence, such as divergence-free operation, interpolation, etc., and also in the procedure for applying it to LES computation[7]. To compare the results of LES computation using the generated inflow turbulence with the wind tunnel test results, these filter effects must be corrected in advance. In this comparison, the results of LES are aimed to be compared with the results of wind tunnel tests measured by Comte-Bellot and Corrsin [11]. The experimental results are thereby adjusted to compensate the several filter effects involved in LES computations. Fig. 7 compares the decay of kinematic turbulence energy in the x_1 direction predicted by LES with that of the experiment. LES computation using the generated inflow turbulence reproduces well the decaying process of turbulence kinematic energy of the wind tunnel test.

[7] The various filter effects (output/input ratios) on the turbulence intensity level are estimated. The band pass filter effects corresponding to the computational domain width and the grid spacing were 81% for the u_1 component and 74% for the u_2 and u_3 components. The filter effects of divergence-free operation were 84% for the u_1 component and 90% for the u_2 and u_3 components. The effects of linear interpolation of the inflow turbulence defined at the control volume faces at the inflow boundary on the velocities at the center of the control volume were 93% for the u_2 and u_3 components (this was not for the u_1 component). To combine these filter effects, the output/input ratios were 68% for the u_1 component and 62% for the u_2 and u_3 components. Furthermore, in the region just behind the inflow boundary, turbulence intensities decreased to 73% for the u_1 component and 72% for the u_2 and u_3 components during the process in which inflow turbulence adapted itself to the discretized N–S equations. When integrating all these filter effects, the ratios of output/input were 49% for the u_1 component and 45% for the u_2 and u_3 components.

6. Conclusions

(1) Inflow turbulence for LES computation was generated step-by-step using the Monte Carlo simulation based on the method of Hoshiya with the prescribed target spectra. The target power spectral density and cross-spectral density in the frequency domain were obtained from the experimental data of Comte-Bellot and Corrsin. The generated inflow turbulence was modified to satisfy the continuity equation by divergence-free operation based on the method of Shirani.

(2) As a result of these procedures, inflow turbulence succeeded in accurately satisfying the prescribed spatial correlation and power spectral density, and the level of velocity divergence was significantly reduced.

(3) The generated inflow turbulence was used as the inflow boundary condition of LES for simulating the decaying isotropic turbulence given from the experiment by Comte-Bellot and Corrsin [11].

(4) There were several filter effects in the procedure for generating inflow turbulence, i.e. divergence-free operation, interpolation, etc., and also in the procedure for applying it to LES computation. By adjusting the turbulence intensity level corresponding to various filter effects, the decay of turbulence kinematic energy in the longitudinal direction was reproduced very well by the LES computation using the generated inflow turbulence.

References

[1] S. Murakami, Comparison of various turbulence models applied to a bluff body, in: Proc. 1st Int. Symp. on Comp. Wind Eng., Tokyo, August 1992, pp. 164–179.

[2] A. Mochida, S. Murakami, M. Shoji, Y. Ishida, Numerical simulation of flowfied around Texas Tech building by large eddy simulation, in: Proc. 1st Int. Symp. on Comp. Wind Eng., Tokyo, August 1992, pp. 42–47.

[3] S. Sakamoto, S. Murakami, S. Kato, A. Mochida, Numerical study on air flow around 2d square prism (Part 1) Influence of inflow turbulence on surface wind pressure, in: Proc. AIJ Annual Meeting, October 1990, pp. 99–100 (in Japanese)

[4] S. Lee, S.K. Lele, P. Moin, Simulation of spatially evolving turbulence and the applicability of Taylor's hypothesis in compressible flow, Phys. Fluid A 4 (1992) 1521–1530.

[5] M.M. Rai, P. Moin, Direct numerical simulation of transition and turbulence in a spatially evolving boundary layer, J. Comp. Phys. 109 (1993) 169–192.

[6] S. Iizuka, S. Murakami, A. Mochida, S. Lee, Large eddy simulation of turbulent flow past 2D square cylinder using dynamic SGS model (Part 5). Generation of inflow turbulence LES based on 3D energy spectrum in wave number space, in: Proc. AIJ Annual Meeting, September 1996, pp. 537–538 (in Japanese).

[7] M. Hoshiya, Simulation of multi-correlated random processes and application to structural vibration problems, in: Proc. JSCE, No. 204, August 1972, pp. 121–128.

[8] Y. Iwatani, Simulation of multidimensional wind fluctuations having any arbitrary power spectra and cross spectra, J. Wind Eng. 11 (1982) 5–18 (in Japanese).

[9] K. Kondo, A. Mochida, S. Murakami, LES computation of isotropic turbulence based on generated inflow turbulence, in: Proc. 14th Symp. on Wind Eng., December 1996, pp. 227–232 (in Japanese).

[10] T. Maruyama, H. Morikawa, Numerical simulation of wind fluctuation conditioned by experimental data in turbulent boundary layer, in: Proc. 13th Symp. on Wind Eng., December 1994, pp. 573–578 (in Japanese).

[11] G. Comte-Bellot, S. Corrsin, Simple Eulerian time correlation of full- and narrow-band velocity signals in grid-generated, 'isotropic' turbulence, J. Fluid Mech. 48 (2) (1971) 273–337.
[12] E. Shirani, J.H. Ferziger, W.C. Reynolds, Mixing of a passive scalar in isotropic and sheared homogeneous turbulence, report TF-15, 1981, Mech. Eng. Dept., Stanford Univ.
[13] C.M. Rhie, W.L. Chow, Numerical study of the turbulent flow past an airfoil with trailing separation, AIAA J. 21 (November 1983) 1525–1532.
[14] A. Leonard, Energy Cascade in large-eddy simulations of turbulent fluid flows, Adv. Geophys. 18 A (1974) 237.
[15] K. Horiuti, Anisotropic representation of the reynolds stress in large eddy simulation of turbulent channel flow, in: Proc. Int. Symp. on Comp. Fluid Dynamics, Nagoya, August 1989, pp. 233–238.
[16] T. Taniguchi, Large eddy simulation using finite difference method and finite element method, in: Proc. 2nd Asia Workshop on Comp. Fluid Dynamics, Tokyo, December 1996.

ELSEVIER

Journal of Wind Engineering
and Industrial Aerodynamics 67&68 (1997) 65–78

JOURNAL OF wind engineering AND industrial aerodynamics

Calculation of the flow past a surface-mounted cube with two-layer turbulence models

D. Lakehal, W. Rodi*

Institute for Hydromechanics, University of Karlsruhe, Kaiserstrasse 12, D-76128 Karlsruhe, Germany

Abstract

In 3-D steady calculations of the flow around a cube placed in developed-channel flow, various versions of the k–ε model were tested. For the near-wall treatment, standard wall functions were employed, as well as the two-layer approach in which the viscous sublayer is resolved with a one-equation model. Two versions of the one-equation model were tested. In addition, calculations were carried out with the Kato–Launder (1993) modification of the k–ε model which eliminates excessive turbulence production in stagnation regions. The various predictions are compared with the measurements of Martinuzzi and Tropea (1993).

Keywords: Cube; Three-dimensional; Channel-flow; Turbulent; Two-layer model

1. Introduction

The standard k–ε two-equation model using wall functions to bridge the viscous sublayer is the most commonly used turbulence model in practice and has been found to work well in many simpler flow situations, mainly of the shear-layer type. However, in more complex situations involving impingement and separation regions which are always present in the flow around buildings, the use of the isotropic eddy-viscosity concept and of wall functions has revealed deficiencies. Isotropic eddy-viscosity models produce excessive turbulent kinetic energy in impingement regions due to an unrealistic simulation of the normal turbulent stresses which contribute most to the turbulence production in such regions. An ad hoc modification of the k–ε model proposed by Kato and Launder [1] which eliminates the excessive production is tested in this paper. Wall functions are based on the assumption of a logarithmic velocity distribution and of local equilibrium of turbulence at the first grid point placed outside the viscous sublayer. These assumptions are clearly not valid in

* Corresponding author. E-mail: Rodi@ifh.bau-verm.uni-karlsruhe.de.

PII S0167-6105(97)00063-9

separation regions. Therefore, considerable effort has been devoted in the last 10 years to the development and testing of low-Reynolds-number versions of the k–ε model with which the viscous sublayer can be resolved, and many different variants have been proposed [3]. Extensive testing has shown (see, e.g., Ref. [4]) that these models clearly improve the calculations of 2-D separated flows over the use of wall functions, but they have the disadvantage of requiring a high grid resolution in the viscous sublayer (25 to 30 grid points) because of the steep gradients of ε very near the wall. This leads to resolution problems in geometrically complex situations involving a number of walls. Also, basically all variants overpredict turbulence and, consequently, the friction coefficient in adverse pressure gradient boundary layers.

As an alternative, the two-layer approach has recently become popular, in which only the core flow outside the viscosity-affected near-wall region is simulated by the k–ε model. The viscous sublayer is resolved by a simpler model, notably a one-equation model in which the length-scale distribution is prescribed and an ε-equation is not solved. Such models therefore require considerably fewer grid points in the viscous sublayer, of the order of 10–15, and are therefore more suitable for complex situations involving more than one wall for which the near-wall regions have to be resolved. Also, because of the fixed length-scale distribution near the wall, these models have been found to give better predictions for adverse pressure gradient boundary layers than pure k–ε models. A review on the work up to 1990 in this area can be found in Ref. [5].

The objective of this paper is to test various two-layer models, which differ in the near-wall one-equation model used, and also the Kato–Launder modification vis-à-vis the standard k–ε model using wall functions for the flow around a simple building. The flow around a surface-mounted cube placed in a developed channel flow was chosen because (i) it has a simple geometry but has all the important complex features of real building flows as described briefly in Section 2 below, (ii) it was studied experimentally in detail by Martinuzzi and Tropea [2] and (iii) the boundary and inflow conditions are well defined. A further advantage is that LES results are available for comparison [6].

2. Flow around a surface-mounted cube: the experimental study [2]

The geometry of the test case is sketched in Fig. 1. The height of the cube is half of that of the channel. For $\mathrm{Re} = U_B H/\nu = 40\,000$ (U_B is the bulk velocity, H is the cube height) Martinuzzi and Tropea [2] carried out flow visualization studies and detailed LDA measurements from which the mean velocity components and the various Reynolds stresses are available. The entry section of the channel was long enough to have developed channel flow. The on-coming turbulence intensity at roof height is relatively low ($T_u = \sqrt{u'^2}/U_B \approx 0.03$). From their flow visualization studies and the detailed measurements, Martinuzzi and Tropea devised the flow picture given in Fig. 2 which clearly shows the very complex nature of the flow in spite of the simple geometry. The flow separates in front of the cube; on average there is a primary separation vortex but also a secondary one, while instantaneously up to four

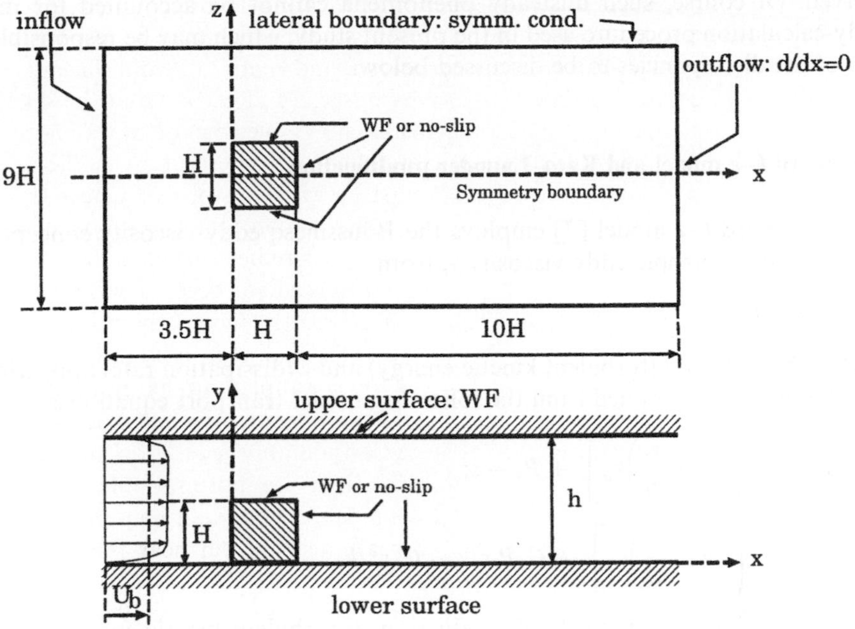

Fig. 1. Geometry of the test case and boundary conditions.

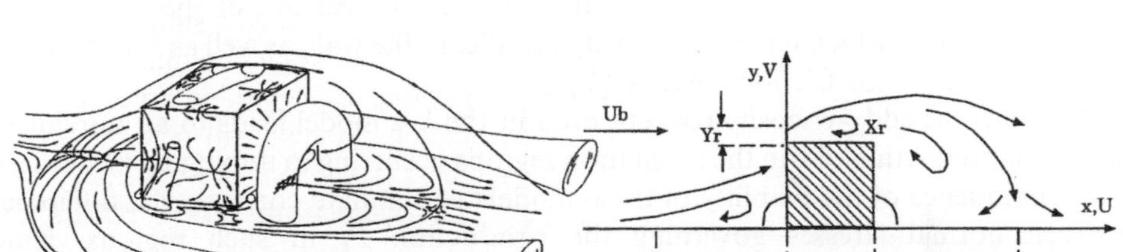

Fig. 2. Schematic representation of the flow around a cube from Ref. [2].

separation vortices were detected. The main vortex wraps as a horse-shoe vortex around the cube into the wake and has a typical converging-diverging behaviour. The flow separates at the front corners of the cube on the roof and the side walls; on average it does not reattach on the roof but there appears to be reattachment on the side walls. A large separation region develops behind the cube which interacts with the horse-shoe vortex. Originating from the ground plate, an arch vortex develops behind the cube. Predominant fluctuation frequencies were detected sideways behind the cube, which were traced to vortex shedding of the flow past the side walls. The Strouhal number was found to be $\mathrm{St} = f/HU_{\mathrm{B}} = 0.15$. Further, bimodal behaviour of the flow separation, and, in particular, of the vortices in front and on the roof was observed. Of course, such unsteady phenomena cannot be accounted for in the steady-calculation procedure used in the present study, which may be responsible for some of the discrepancies to be discussed below.

3. Standard k–ε model and Kato–Launder modification

The standard k–ε model [7] employs the Boussinesq eddy-viscosity concept and determines the isotropic eddy viscosity ν_t from

$$\nu_t = C_\mu k^2/\varepsilon. \tag{1}$$

The distributions of k (turbulent kinetic energy) and ε (dissipation rate) appearing in this relation are determined from the following model transport equations:

$$(U_i k)_{,i} = \left[\left(\nu + \frac{\nu_t}{\sigma_k}\right)k_{,i}\right]_{,i} + P_k - \varepsilon;$$

$$(U_i \varepsilon)_{,i} = \left[\left(\nu + \frac{\nu_t}{\sigma_\varepsilon}\right)\varepsilon_{,i}\right]_{,i} + C_1 P_k \varepsilon/k - C_2 \varepsilon^2/k, \tag{2}$$

$P_k = \overline{u_i u_j} U_{i,j}$ represents the rate of production of turbulent kinetic energy resulting from the interaction of turbulent stresses and mean velocity gradients. The model employs the following standard values of the empirical constants: $C_\mu = 0.09$; $C_1 = 1.44$; $C_2 = 1.92$; $\sigma_k = 1.0$ and $\sigma_\varepsilon = 1.3$. In the standard version of the model, wall functions are used which relate the velocity parallel to the wall, as well as k and ε at the first grid point to the friction velocity [7].

The isotropic eddy-viscosity concept used in the k–ε model leads to an unrealistically high production of k in the stagnation regions occurring in impinging flows. This is a consequence of the inability of these models to simulate correctly the difference between normal stresses governing the production P_k in such regions. Kato and Launder [1] suggested as an ad hoc measure to replace the original production term $P_k = C_\mu \varepsilon S^2$ by $P_k = C_\mu \varepsilon S \Omega$, where $S = k/\varepsilon \sqrt{1/2(U_{i,j} + U_{j,i})}$ and $\Omega = k/\varepsilon \sqrt{1/2(U_{i,j} - U_{j,i})}$ denote, respectively, the strain and vorticity invariants. In simple shear flows the behaviour remains unchanged as $\Omega \approx S$ while in stagnation regions $\Omega \approx 0$ so that the spurious turbulence production is eliminated.

4. Two-layer turbulence models

4.1. Basic concept

The two-layer approach adopted here consists of resolving the viscosity-affected regions close to walls with a one-equation model, while the outer core flow is resolved with the standard k–ε model described above. In the one-equation model, the eddy viscosity is made proportional to a velocity scale and a length scale l_μ. The distribution of l_μ is prescribed algebraically while the velocity scale is determined by solving the k-equation (as in Eq. (2)). The dissipation rate ε appearing as a sink term in the k-equation is related to k and a dissipation length scale l_ε which is also prescribed algebraically. The different two-layer versions available in the literature differ in the use of the velocity scale and the way l_μ and l_ε are prescribed. It should be mentioned that in the fully turbulent region the length scales l_μ and l_ε vary linearly with distance from the wall. However, in the viscous sublayer, l_μ and l_ε deviate from the linear distribution in order to account for the damping of the eddy viscosity and the limiting behaviour of ε at the wall.

4.2. $k^{1/2}$ *velocity-scale-based model* [5]: *TLK*

The approach combines the standard k–ε model in the outer region with a one-equation model due to Norris and Reynolds [8] in the viscous-sublayer employing

$$\nu_t = C_\mu k^{1/2} l_\mu; \quad \varepsilon = k^{3/2}/l_\varepsilon. \tag{3}$$

In this model, the length scale l_μ is damped in a similar way as the Prandtl mixing length by the Van Driest function, so that it involves an exponential reduction governed by the near-wall Reynolds number $R_y = \tilde{U} y_n/\nu$. However, in contrast to the original Van Driest function, R_y uses $k^{1/2}$ as a velocity scale $\tilde{U}$ instead of U_τ which can go to zero for separated flows.

$$l_\mu = C_l y_n f_\mu \quad \text{with } f_\mu = 1 - \exp\left(-\frac{R_y 25}{A_\mu A^+}\right). \tag{4}$$

The constant C_l is set equal to $\kappa C_\mu^{-3/4}$ to conform with the logarithmic law of the wall. The empirical constants appearing in the f_μ-function are assigned the values $A_\mu = 50.5$ and $A^+ = 25$. The reader is referred to Ref. [5] for a review and further details on the choice of the constants. For the dissipation scale the following distribution is used near the wall:

$$l_\varepsilon = \frac{C_l y_n}{1 + C_\varepsilon/(R_y C_l)}, \quad C_\varepsilon = 13.2. \tag{5}$$

The outer (k–ε) and the near-wall model are matched at a location where the damping function f_μ reaches the value 0.95, i.e., where viscous effects become negligible.

The combination of the Kato–Launder correction with the TLK model is hereafter labeled TLKK.

4.3. $(v'^2)^{1/2}$ *velocity-scale-based model* [9]: *TLV*

The development of this model was motivated by the fact that the length scale functions as proposed in Ref. [8], particularly the l_ε-function, are not in agreement with direct numerical simulation (DNS) data, and that the normal fluctuations $(v'^2)^{1/2}$ are a more relevant velocity scale for the turbulent momentum transfer near the wall than $k^{1/2}$. Therefore, the following model using $(v'^2)^{1/2}$ as a velocity scale was proposed in Ref. [9]:

$$\nu_t = \sqrt{\overline{v'^2}}\, l_{\mu,v}, \quad \varepsilon = \sqrt{\overline{v'^2}}\, k/l_{\varepsilon,v} \tag{6}$$

with

$$l_{v,\mu} = 0.33 y_n, \quad \text{and} \quad l_{\varepsilon,v} = 1.3 y_n \Big/ \left[1.0 + 2.12\, \nu \Big/ \sqrt{\overline{v'^2}}\, y_n\right], \tag{7}$$

which is based on DNS data for fully developed-channel flow. As an equation for k is solved, $\overline{v'^2}$ needs to be related to k, which is done through the following DNS-based empirical relation:

$$\overline{v'^2}/k = 4.65 \times 10^{-5}(R_y)^2 + 4.00 \times 10^{-4} R_y, \quad R_y = k^{1/2} y_n/\nu, \tag{8}$$

valid only very near the wall. The matching between the outer and near-wall model is performed at a location where $R_y = 80$.

5. Outline of the computational method

The computer code FAST–3D [10] used for the flow calculations is based on a finite-volume approach for solving the incompressible Navier–Stokes equations. Flows around complex geometries can also be treated since the code is written for curvilinear body-fitted coordinates using Cartesian velocity components. A non-staggered, cell-centred grid arrangement is used. Pressure field oscillations are avoided by means of the momentum interpolation technique due to Rhie and Chow [11]. The pressure–velocity coupling is achieved using the SIMPLE algorithm. The diffusion fluxes are approximated by central differences, while the hybrid linear-parabolic approximation, a second-order low-diffusive and oscillation-free scheme of Zhu [12], is applied for the convection fluxes. The resulting system of difference equations is solved using the strongly implicit solution procedure of Stone [13].

6. Grids and boundary conditions

The computational domain with the various boundaries is shown in Fig. 1. The effect of the location of the inflow and outflow boundaries and of the lateral boundaries was studied and the boundaries were placed at the locations given in Fig. 2 to avoid any influence on the calculations. Because of symmetry conditions, only half of the width of the flow needed to be calculated. Calculations using various grids indicated that with the standard k–ε model using wall functions, grid-independent results could be obtained with 110 × 32 × 32 grid points in the x-, y- and z-directions. A finer mesh consisting of 142 × 84 × 64 grid points was applied in all two-layer computations. The grids were non-uniform, being considerably finer in the near-wall regions. The first cells adjacent to the walls were set with respect to the criteria required for the individual near-wall treatment. Hence, using wall functions the width of the near-wall cell was $0.01H$ and using the two-layer approach $0.001H$, which corresponds to $10 < y^+ < 25$ and $1 < y^+ < 5$, respectively. In the two-layer case, the number of grid points placed in the viscous sublayer was in most regions typically 15–20. The boundary conditions employed are also indicated in Fig. 1: On the ground plate and the cube walls, either wall functions were employed or the no-slip condition and $k = 0$ in the two-layer calculations, while at the upper boundary of the channel wall functions were always used. In a separate calculation, fully developed channel flow was calculated first and the results were then used as inflow conditions. At the exit, zero gradient conditions were applied.

7. Results and discussions

Fig. 3 shows a comparison of streamlines in the plane of symmetry (left) and near the channel floor (right), while Table 1 compares various parameters characterizing the size and location of separation regions as defined in Fig. 2. The experimental results are time-averaged over a long period, and in Table 1 results of an LES calculation [6] are also included. From Fig. 3 it appears that the stagnation point is well simulated by the various models ($Y_s/H = 0.76$). The primary upstream separation location X_f (labelled A in the experimental flow pattern) which is caused by the strong pressure gradient imposed by the obstacle on the oncoming boundary layer is predicted differently by the wall function (WF) and two-layer (TL) models. The WF models predict the separation point (X_f) too close to the obstacle while the TL models give good agreement with experiments, with the exception of the TLV model which predicts somewhat early separation similar to the LES of Ref. [6]. The location of the horse-shoe vortex centre in front of the cube is well captured by the various TL models (at $X \approx -0.35$) while the WF-based models predict the centre further upstream. Also, the TL models are able to reproduce the secondary vortex developing right, in the corner in front of the obstacle (band labelled C) but the resolution of the streamline picture is not good enough to display this. The observations for the adverse pressure-gradient flow region in front of the obstacle allow the conclusion that here the TL models are clearly superior to the ones using wall functions.

Looking now at the separation region on the roof, it can be observed that the k–ε model with wall functions produces a much too small separation zone with unrealistic reattachment on the roof. When the KL modification is switched on, the separation region becomes much longer and in fact there is now no reattachment, but the separation bubble is too thin compared with the experimentally observed one. This improvement is brought about by the significant reduction of the turbulence kinetic energy produced in front of the obstacle, as can be seen from the k-contours given in Fig. 4. As a consequence, in the KL version there is less turbulence swept around the front corner so that the eddy viscosity over the roof is smaller, leading to a longer separation region. A similar effect is brought about by switching from WF to TL models (without KL modification) which also produce a separation bubble without reattachment which is, however, also too thin.

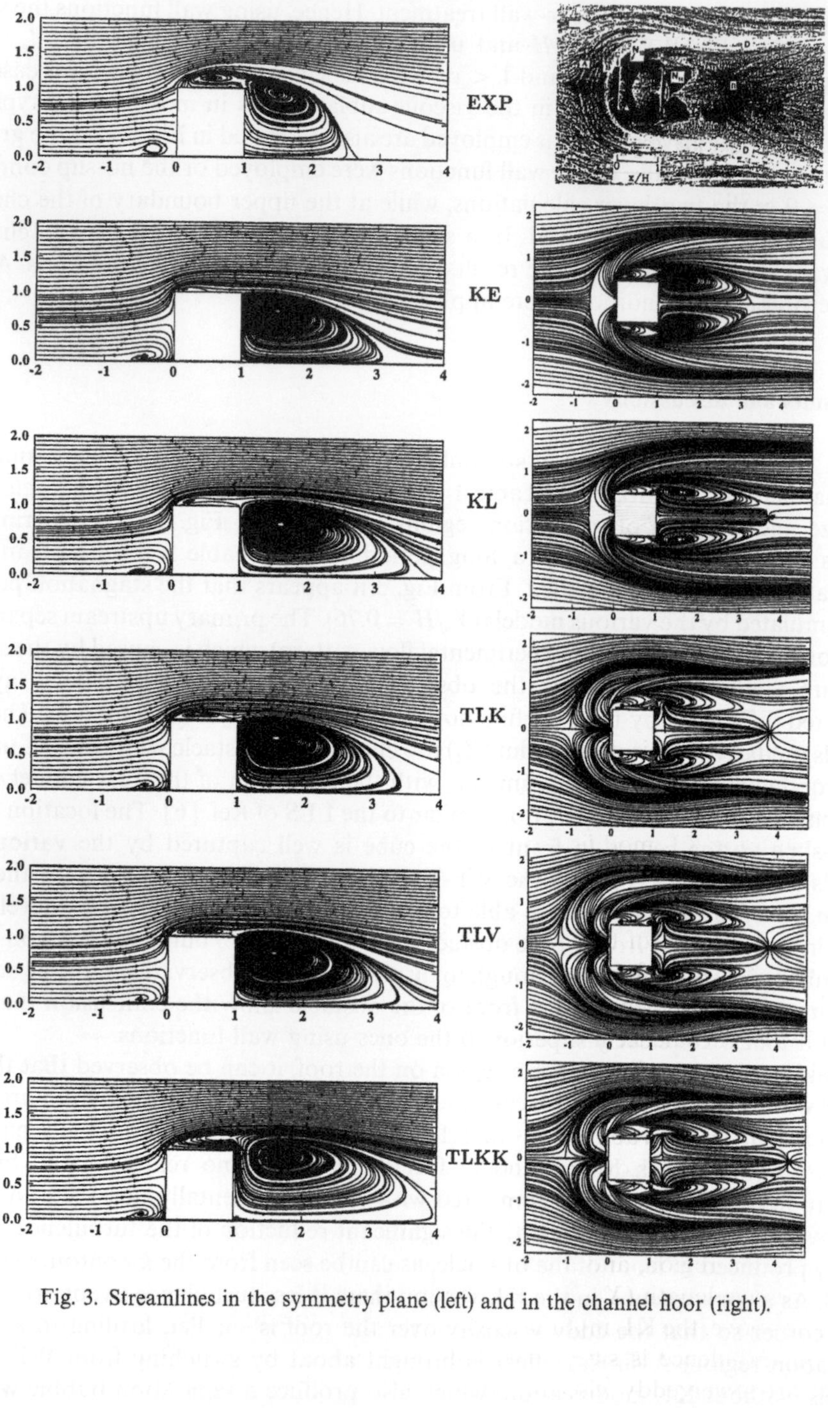

Fig. 3. Streamlines in the symmetry plane (left) and in the channel floor (right).

Table 1
Predicted flow structure parameters over the obstacle

Model	Key	Y_r	X_f	X_r	X_b
Experiment Ref. [2]	Exp	0.17	1.040	—	1.612
k–ε + (WF)	K-E	—	0.651	0.432	2.182
k–ε K–L + (WF)	KL	0.062	0.640	—	2.730
Two-layer k–ε	TLK	0.098	0.950	—	2.680
Two-layer k–ε	TLV	0.102	1.215	—	2.685
Two-layer k–ε	TLKK	0.160	0.950	—	3.405
LES [6]	LES-S	0.162	1.287	—	1.696
LES [6]	LES-D	0.162	0.998	—	1.432

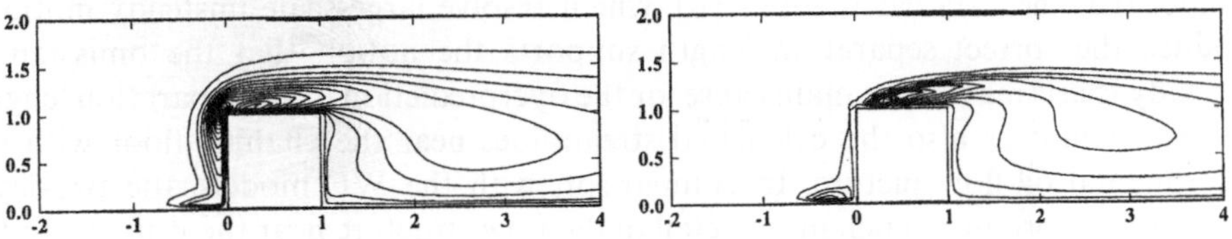

Fig. 4. k-isolines in the symmetry plane, left: KE; right: KL.

This improvement is due to the better resolution of the relatively thin bubble and the more realistic treatment of the near-wall region and is fully in line with previous observations made in 2D calculations (e.g. Ref. [4]). Combining the two-layer approach with the KL modification (TLKK) brings the predicted flow behaviour over the roof closest to the experimental one, i.e., the bubble thickness is now fairly realistic and clearly the separation region on the roof merges with that behind the obstacle. However, the centre of the bubble is predicted somewhat too far downstream. The Kato–Launder modification becomes, of course, more important the higher the approach-flow turbulence is. A recent calculation performed for a similar flow but with a higher turbulence intensity at roof height ($T_u = 0.16$) has shown that using the two-layer approach alone (TLK model) did not prevent unrealistic reattachment of the flow on the roof [14].

From Fig. 3 and Table 1 it is clear that the extension of the separation region behind the cube (X_b) is overpredicted by all models. The standard k–ε model with wall functions gives the smallest value and is closest to the experiments; both the introduction of the KL modification and the two-layer approach increase further the length of the separation region, and a combination of the two approaches (TLKK) gives the most excessive length ($X_b/H = 3.4$ compared with 1.62 in the experiments). As was discussed above, the KL modification reduces the k-production in front of the cube so that less turbulence is swept over the roof and also into the downstream region, leading to lower eddy viscosity there. This explains the longer separation zone vis-à-vis the standard k–ε model calculations. Moving from WF calculations to TL calculations, the resulting larger and also thicker separation zone on the roof of course also increases the separation zone behind the cube, and the lengthening of the separation region when introducing the TL approach is also consistent with a variety of calculations of 2D separated flows. With these, however, the change is in the right direction, whereas here already the basic model produces too long a separation region. As sketched in Fig. 2, the flow field behind the cube is very complex with complex strain fields and curvature effects; the still fairly simple eddy-viscosity k–ε model may not be able to cope with these phenomena and a Reynolds-stress-equation model may do a better job. However, a more serious deficiency of the calculation methods used may be the fact that they are steady and ignore any vortex-shedding effects. In the experiments, vortex shedding from the side walls was observed, and such shedding can contribute greatly to the momentum exchange in the wake and can thereby reduce significantly the length of the separation region behind obstacles. This became very clear in previous calculations of the flow past a square cylinder [15]. Also the fact that the LES calculations [6], which resolve large-scale unsteady motions, produce the correct separation length supports the notion that the omission of unsteady effects may be the main cause for the overprediction of the separation length.

Fig. 3 compares also the calculated streamlines near the channel floor with the experimental oil-flow picture. In comparison with the WF models, the two-layer approaches reproduce much more detail of the flow structure near the wall due to the finer resolution in this region. The figure shows that the horse-shoe vortex is generally predicted quite well by the different approaches. The outer limit of the wake region formed by the lateral arms of the horse-shoe vortex (line D in oil-flow picture) varies however between the different calculation approaches. In the experiment, the width of this wake decreases up to approximately the reattachment point X_b; then it increases again. This feature governed by the rate of rolling is well described by the two-layer models while in contrast the WF model calculations do not produce the converging-diverging behaviour. The two-layer approaches also seem to predict correctly the corner vortices ($N12$) generated downstream of the vertical leading edges of the cube at the channel/obstacle junction. The location of the simulated arch vortices behind the obstacle ($N14$) shows clearly the differences between WF and TL results. In the k–ε WF calculations, basically the whole separation region is occupied by these vortices which contradicts the experimental observation; in the TL calculations these vortices are limited to a smaller area as in the experiments and as predicted similarly by the LES method. Since the large-eddy simulation produces nearly the correct separation length, it yields overall a flow pattern at the channel floor which is in very good agreement with the observed one.

Fig. 5 displays calculated and measured U-velocity profiles at different streamwise locations on the symmetry plane. All streamwise velocity profiles agree well with the measurements at $x/H = -1.0$ upstream of the cube. As was to be expected from the streamlines in Fig. 3, significant differences between the k–ε model results and the other results can already be observed at a location at the middle of the roof ($x/H = 0.5$). Here, and also at the position of the back face of the cube ($x/H = 1.0$), the TL results are in better agreement with the experiments; the best results were obtained with the TLKK model. The KL model performance is similar to that of TLK and

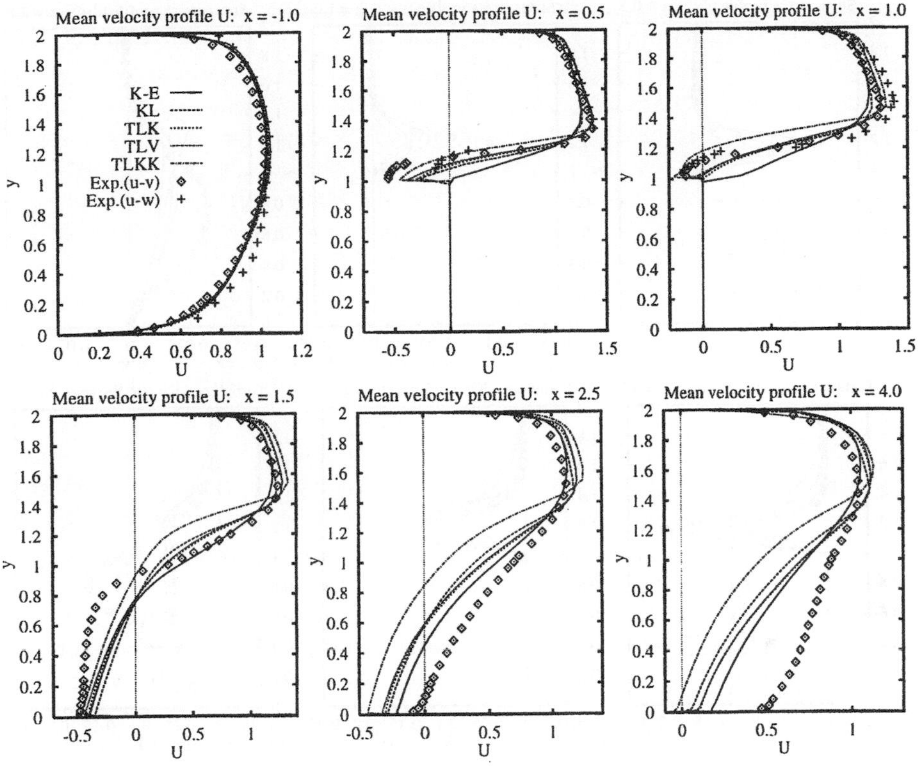

Fig. 5. Comparison of mean velocity profiles U in the symmetry plane.

TLV models, as was to be expected from the streamline pictures. At $x/H = 1.5$, the profiles predicted by the various models are rather similar and agree fairly well with the experiments in the region above the roof height. Below this, the reverse-flow velocity is underpredicted by all models. The further development of the U-profile is of course influenced mainly by the fact that the models predict too large a separation region so that the recovery is underpredicted by various degrees, of course worst by the TLKK model. Here, the LES calculations show much better agreement with the experiments.

Fig. 6 displays three vertical profiles of the shear stress $u'v'$ and turbulent kinetic energy k in the symmetry plane. At $x/H = 0.5$, all models using the TL approach predict fairly well the peak values of $u'v'$ and k except for TLKK which is low, while near the backward edge of the cube, these values are underpredicted by all models. This trend continues further downstream; in the region close to the flow reattachment point ($x/H = 2$ and 2.5) all models behave in a very similar way: the $u'v'$- and k-levels are underpredicted when compared with the experiments. This could partly be due to deficiencies of the eddy-viscosity concept but may again be largely caused by unsteady effects as the LES calculations produce higher levels in significantly better agreement with the measurements.

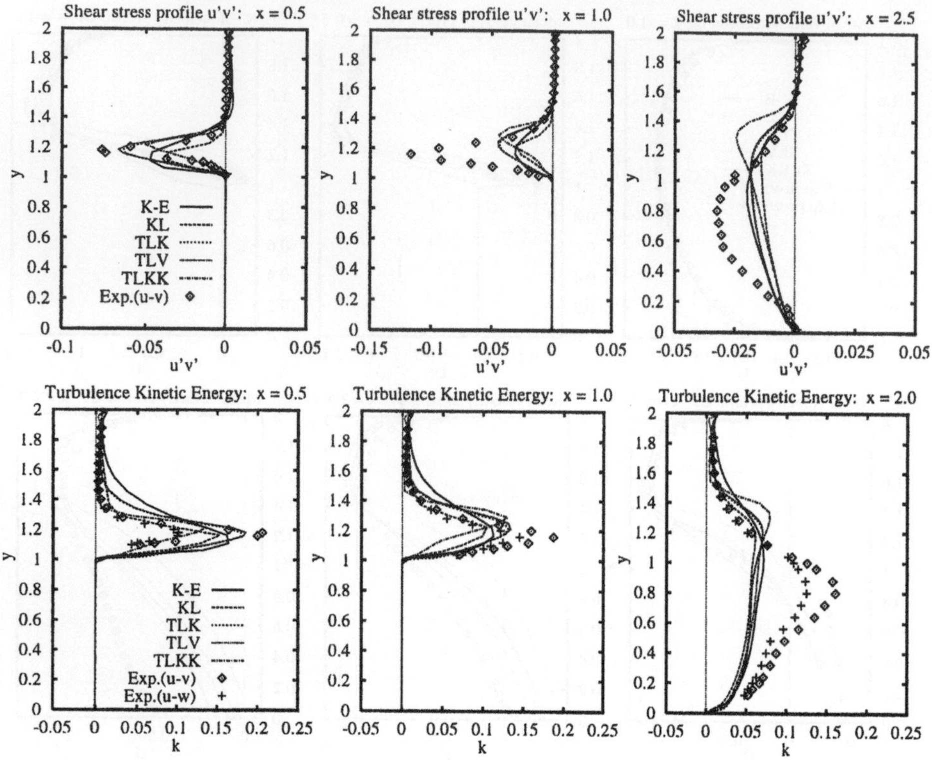

Fig. 6. Comparison of $u'v'$- and k-profiles in the symmetry plane.

8. Conclusions

Turbulent flow past a surface-mounted cubical obstacle placed in developed-channel flow has been investigated with various versions of the k–ε model, including different two-layer approaches, wall function treatment and the Kato–Launder modification. Despite the simple geometry of the obstacle, the flow developing in its vicinity is very complex with multiple, unsteady separation regions, vortices of various kinds, strong curvature and adverse as well as favourable pressure gradients. The models using wall functions cannot reproduce the details of the complex flow structure near the ground, e.g., the converging-diverging behaviour of the horse-shoe vortex, and also produce late separation of the boundary layer ahead of the obstacle. The standard version of the k–ε model further produces a much too small separation region with unrealistic reattachment on the roof. The simulation of this region is improved significantly by including the KL modification which eliminates the excessive kinetic energy production in the stagnation region. Using the two-layer approach, the details of the complex flow structure near the ground wall including the converging-diverging behaviour of the horse-shoe vortex can be resolved much better, the separation location in front of the cube is predicted correctly and also the prediction of the separation region on the roof is improved – when combined with the KL modification the size of the roof separation bubble is predicted fairly well. The price for these improved predictions is, however, quite high since the computing time necessary was a factor of 25 larger than when wall functions were used. The calculations have shown also that, except for the adverse pressure gradient region in front of the obstacle, there is little difference in the results obtained with the TLK and TLV versions of the two-layer model. This indicates that, for the type of flow considered, the calculations are not sensitive to the details of the near-wall one-equation model. This conclusion is supported further by an additional calculation which has been performed using the TL model of Chen and Patel [16] which produced virtually the same results as the TLK model.

All models were found to overpredict the length of the separation region behind the cube, and introducing the Kato–Launder modification and the two-layer approach, which improve the predictions in the front part, both have an adverse effect on the separation length. It would be intriguing to see how Reynolds-stress-equation models perform, but it is likely that the unsteady effects such as the observed vortex shedding from the side walls significantly contribute to the momentum exchange and that the neglect of such effects in the steady calculations has led to the overprediction of the separation length. This notion is supported by the LES results [6] which produced good agreement with the experiments in every respect including the separation length. However, a very high price has to be paid for this as the LES calculations took 160 CPU hours on a SNI S600/20 vector computer while the two-layer model calculations took 6 h and the calculations using wall functions only 15 min.

Acknowledgements

The work reported here was sponsored by the Human Capital and Mobility Programme of the European Union. The calculations were carried out on the SNI S600/20 vector computer of the University of Karlsruhe (Computer Centre).

References

[1] M. Kato, B.E. Launder, The Modeling of turbulent flow around stationary and vibrating square cylinders, Proc. 9th Symp. on Turbulence and Shear Flows, Kyoto, 1993.
[2] R. Martinuzzi, C. Tropea, The flow around surface-mounted prismatic obstacle placed in a fully developed channel flow, J. Fluids Eng. 115 (1993) 85–92.
[3] W. Rodi, N.N. Mansour, Low Reynolds number k–ε modelling with the aid of direct numerical simulation data, J. Fluid Mech. 250 (1993) 509–529.
[4] J.Ch. Bonnin, T. Buchal, W. Rodi, ERCOFTAC workshop on data bases and testing of calculation methods for turbulent flows, University of Karlsruhe, 3–7 April 1995, ERCOFTAC Bull. 28 (1996) 48–54.
[5] W. Rodi, Experience with two-layer models conbining the k–ε model with a one-equation model near the wall, AIAA paper, AIAA-91-0216, 1991.
[6] M. Breuer, D. Lakehal, W. Rodi, Flow around a surface mounted cubical obstacle: comparisons of LES and RANS-Results, in: Computation of Three-Dimensional Complex Flows, M. Deville, S. Gavrilakis, I.L. Rhyming (Eds.), in: Notes of Numerical Fluid Mechanics, vol. 53, Vieweg Verlag, Braunschweig, 1996, pp. 22–30.

[7] B.E. Launder, D.B. Spalding, The numerical computation of turbulent flows, Comput. Meth. Appl. Mech. Eng. 3 (1974) 269–289.

[8] L.H. Norris, W.C. Reynolds, Turbulent channel flow with a moving wavy boundary, Report No. FM-10, Stanford University, Department Mechanical Engineering, 1975.

[9] W. Rodi, N.N. Mansour, V. Michelassi, One-equation near-wall turbulence modeling with the aid of direct simulation data, J. Fluids Eng. 115 (1993) 196–205.

[10] S. Majumdar, W. Rodi, J. Zhu, Three-dimensional finite-volume method for incompressible flows with complex boundaries, J. Fluids Eng. 114 (1992) 496–503.

[11] C.M. Rhie, W.L. Chow, A numerical study of the turbulent flow past an isolated airfoil with trailing edge separation, AIAA J. 21 (1983) 1225–1532.

[12] J. Zhu, A low-diffusive and oscillating-free convective scheme, Commun. Appl. Numer. Meth. 7 (1991) 225–232.

[13] H.L. Stone, Iterative solution of implicit approximations of multidimensional partial differential equations, SIAM J. Numer. Anal. 5 (1968) 530–558.

[14] D. Delaunay, D. Lakehal, C. Barré, C. Sacré, Numerical and wind-tunnel simulation of gas dispersion around a rectangular building, in: Proc. 2nd Int. Symp. on Computational Wind Engineering, Fort Collins, Colorado, 1996.

[15] R. Franke, W. Rodi, Calculation of vortex shedding past a square cylinder with various turbulence models, in: Durst et al. (Eds.), Turbulent and Shear Flows, vol. 8, Springer, Berlin, 1993, pp. 189–204.

[16] H.C. Chen, V.C. Patel, Near-wall turbulence models for complex flows including separation, AIAA J. 26 (1988) 641–648.

ELSEVIER

Journal of Wind Engineering
and Industrial Aerodynamics 67&68 (1997) 79–90

JOURNAL OF
wind engineering
AND
industrial
aerodynamics

Unsteady aerodynamic force prediction on a square cylinder using k–ε turbulence models

Sangsan Lee[1]

Systems Engineering Research Institute, P.O. Box 1, Yuseong, Daejeon 305-600, South Korea

Abstract

In designing structurally safe large-scale civil structures, the dynamic structural behavior in response to unsteady wind loadings should be investigated. Since a typical civil structure is a blunt body, flow separates, reattaches, and forms an unsteady vortex in the wake region behind the structure. Conventional k–ε turbulence models have failed in predicting unsteady turbulent flows around blunt bodies, unless careful modifications suited to the specific cases were made. Systematic investigation of numerical effects on the computation of turbulent flows over a square cylinder has been made. It has been found that some conventional k–ε models may give reasonable predictions when proper numerical parameters are incorporated. Critical numerical aspects are found to be the near-wall grid spacing and the choice of convection schemes, while the temporal accuracy effect is not so important. A numerical prediction is compared in detail not only with experimental measurements but also with the results from a sophisticated turbulence simulation.

Keywords: Turbulent flow; Vortex shedding; Wind loading; Strouhal number; Numerical simulation; k–ε turbulence model; Convection scheme

1. Introduction

Turbulent flow is characterized by apparently random fluctuations in flow quantities, both in space and in time. At the present time, it is impossible to accurately predict engineering flows to the fine details of turbulence, even with the most advanced computers. Only turbulent flows at extremely low Reynolds numbers may be predicted by direct numerical simulation (DNS), where the smallest meaningful length and time scales of turbulence are resolved without introducing any turbulence model [1]. In highly turbulent flows, DNS requires impracticably huge computer resources

[1] E-mail: sslee@seri.re.kr.

PII S0167-6105(97)00064-0

not only in memory but also in CPU time. At high Reynolds numbers, large-eddy simulation (LES) may be performed to compute large scale turbulence motions with a relatively robust turbulence model, namely a subgrid-scale (SGS) model [1]. Recently, active research [2–4] is under way in the field of SGS modeling, so that LES may serve as a research tool with the advent of next-generation supercomputers with TeraFlops (10^{12} floating point operations per second) peak performance. Even the next-generation computers may not have enough capacity to support LES as an engineering design tool. The Reynolds-averaged (or Favre-averaged in compressible flows) Navier–Stokes equations are to be solved for the engineering purpose, where all the turbulence effects on the mean motion are taken into account by eddy viscosity models (EVM: for zero-, one-, and two-equation models) or by Reynolds stress models (RSM: for Reynolds stress models). Mixing-length model [5], a zero-equation model, is widely used to numerically simulate aerodynamic flows, where turbulence effects are not too complex. Even though there has been some success with one-equation model [6], two-equation turbulence model is the most frequently used in engineering predictions of turbulent flows, especially a class of k–ε models. More refined prediction may be generated with RSM by solving Reynolds stress transport equations. Even though RSM, in general, gives better predictions, it is less popular since it tends to require far more computer resources, compared to its k–ε counterparts. Even with the popularity of the k–ε models, they have a fundamental defect due to neglecting rotation, anisotropy, and non-equilibrium effects. Recent review on the conventional turbulence models is given in Bradshaw et al. [7]

Flows over blunt bodies are important in many engineering applications, especially in flows around bridges and buildings. Accurate prediction of such flows using turbulence models in computational fluid dynamics (CFD) is a very demanding issue. The main source of errors in the CFD predictions of such flows is the poor modeling of complex flow characteristics, especially of turbulence. It has been known to be impossible to accurately predict the unsteady turbulent flows in such cases with the conventional two-equation turbulence models. To assess the limit of CFD in such applications, an investigation has been made for canonical flows, such as flows over a circular cylinder [8] and over a square cylinder [9, 10]. In general, unsteady turbulent flows cannot be predicted by two-equation turbulence models except for the case with a special treatment [10]. Recently, it was pointed out that numerical aspects are more critical than the choice of turbulence models in predicting separated turbulent flows by showing that a reasonable wake size behind a triangular cylinder is predicted only by an *unsteady* computation [11]. The unresolved questions from the previous works are: (i) Is it ever possible to accurately predict the unsteady turbulent flow using conventional k–ε turbulence models? (ii) If possible, what are the requirements for the factors affecting the accuracy of such applications?

This paper deals with a generic turbulent flow, namely a turbulent flow over a square cylinder at a relatively low Reynolds number to validate the use of conventional k–ε turbulence models by comparing their predictions with an existing experiment and an elaborate turbulence simulation. The problem is formulated with a brief description of the governing equations and turbulence models. Next, sensitivity of unsteady wind loading prediction to the choice of numerical parameters and turbulence models is discussed in detail. Subsequently, the effects of turbulence models on the detailed flow statistics are described. This paper concludes with a summary and future projections on the application of CFD in wind engineering.

2. Problem formulation

A systematic investigation is made to assess what are the important factors for the accurate prediction of an unsteady turbulent flow. Conventional k–ε turbulence models are tested. Included are the standard k–ε model [12], the RNG k–ε model [13], and the low Reynolds number k–ε model [14]. The effects of numerical parameters – spatial and temporal accuracy, convection scheme – are also investigated. Turbulent flow over a square cylinder is computed numerically as a canonical blunt body flow. Recently, a well-controlled water tunnel experiment was conducted [15], and it is used as a reference case in the present study. All the computational setups are aimed at simulating the experiment, and the computational results are compared with the experimental data.

The experiment described in Ref. [15] is conducted in a water tunnel with $0.56\ \text{m} \times 0.39\ \text{m}$ test section. The square cylinder is $0.04\ \text{m} \times 0.04\ \text{m}$ (7% blockage ratio) with 0.39 m in length, and the free stream flow speed is 0.535 m/s with 2% turbulence intensity. A large-eddy simulation [16] was conducted for the configuration of the experiment. The computational domain is $20D \times 14D \times 2D$ with $104 \times 69 \times 10$ grid points in the streamwise, the lateral, and the axial direction, respectively. The usual LES computation takes about 100 CPU hours on the Fujitsu VP2600 supercomputer (5 GFlops peak performance), for 13 shedding cycles, to get statistically converged unsteady flow patterns.

In the present study, simulation is carried out in two dimensions. The governing equations solved are the conservation of mass and momentum. Transport equations of turbulence kinetic energy (k) and its dissipation rate (ε) are also solved to evaluate the eddy viscosity, ν_T.

To simulate the experimental setup with a minimal computational cost, symmetry boundary condition is imposed at the tunnel walls. Imposition of the symmetry boundary condition (instead of the wall boundary condition) is found to have little effect on the unsteady turbulent flow around the cylinder [9,10]. On the cylinder surface, wall function is used as the boundary condition except for the low Reynolds number model, where the governing equations are integrated to the cylinder surface. The governing equations are approximated through a finite-volume approach with all the flow variables collocated at the cell center [17]. The convection term is treated by upwind, hybrid, and QUICK scheme [18] and all the other spatial derivatives are approximated by the second order central differences. Mass conservation is imposed by solving a pressure correction equation of the SIMPLE type [19], and time advancement is made by an implicit Euler scheme.

3. Effects of turbulence models and numerical parameters

In the following, computational results using the RNG k–ε model are described and compared with the experiments and the LES. The results are validated by investigating the effects of temporal and spatial accuracy. Effects of convection schemes and turbulence models are also discussed.

3.1. The reference simulation

The computation is carried out after the governing equations are normalized by the inflow speed (U) and the cylinder side length (D). The flow Reynolds number ($\mathrm{Re} = UD/\nu$) is chosen as 2.2×10^4 to match the experimental condition [15]. The important numerical parameters of the reference simulation and of the simulations to be discussed later are listed in Table 1 along with those from the experiment [15] and from the LES computation [16]. The reference case (case REF) is simulated using 85×55 grid points in the streamwise and the lateral direction, respectively. The dimensionless grid spacing adjacent to the wall, $\Delta x_i^+ = (\tau_w/\rho)^{1/2}\Delta x_i/\nu$, is ranging from 10 to 40 with the average of about 20. The normalized time step, or a CFL number estimate, $U\Delta t/(\Delta x_i)_{\min}$ is taken as around 0.7, where U is the free stream velocity and $(\Delta x_i)_{\min}$ is the minimum grid spacing. The actual local CFL number exceeds 2 near the corner of the cylinder for the reference case. It takes about 1 CPU second for the reference simulation to advance one time step in CRAY Y-MP C90 (with a processor of 1 GFlops of peak performance) with the absolute cell mass flux residual sum to be less than 10^{-4} of the inflow mass flux at each time step.

Since unsteady regular vortex shedding is the main feature of the flow, the drag (c_D) and lift (c_L) coefficients may represent the overall unsteady behavior of the flow field.

Table 1
Numerical parameters and results with the RNG k–ε turbulence model

Case	Δt^*	Δx^*	Scheme	Model	c_D	c_L	St
REF	0.02	0.035	QUICK	RNG	2.12 ± 0.23	± 2.08	0.133
RT2	0.04	0.035	QUICK	RNG	2.30 ± 0.18	± 2.07	0.135
RT5	0.10	0.035	QUICK	RNG	2.23 ± 0.14	± 2.01	0.138
RXC	0.05	0.074	QUICK	RNG	1.62 ± 0.02	± 0.40	0.141
RXF	0.02	0.012%	QUICK	RNG	2.37 ± 0.44	± 2.04	0.130
fDn	0.02	0.012	QUICK	RNG	2.11 ± 0.21	± 2.04	0.133
RH	0.02	0.035	Hybrid	RNG	1.90 ± 0.05	± 0.77	0.134
RU	0.02	0.035	Upwind	RNG	1.83 ± 0.02	± 0.55	0.130
Exp [15]					2.14 ± 0.09		0.134
LES [16]	0.001	0.022	2nd Central	SGS	2.09 ± 0.13	± 1.60	0.132

Definition: $\Delta t^* = U\Delta t/D$, $\Delta x^* = (\Delta x_i)_{\min}/D$, $c_D = f_D/KD$, $c_L = f_L/KD$ ($K = \rho U^2/2$).
%: Since $(\Delta x_i^+)_{\min} < 10$ in most regions, BC for ε is locally inconsistent.

In evaluating the drag (f_D) and lift (f_L) forces per unit span, viscous stress is also included in addition to pressure. However, the viscous stress contributes less than a percent to the total forces.

Time history of the lift coefficient after an initial transient is taken to estimate the dominant oscillation frequency through FFT. A proper windowing [20] is applied to the signal to remove non-periodicity effects on the Fourier transform. The estimated Strouhal number ($\mathrm{St} = fD/U$) is 0.133 with an uncertainty of 0.003. Other statistics, like the mean and fluctuating force coefficients, are also evaluated using the signal. The mean and the fluctuating amplitude of c_D are 2.12 and 0.23, and the fluctuating amplitude of c_L is 2.08. Since the LES results agree favorably with the experiment, the data not available from the experiment – the amplitude of c_L fluctuation – is taken from the LES as the reference to evaluate the present computations. The statistical results of the REF case compare well with the experiment and also with the LES results, especially for the mean drag coefficient and the Strouhal number. Differences in the unsteady part of force coefficients may be due to inherent three-dimensionality of turbulence which is not accounted for in the current two-dimensional computation. The deterministic unsteady forces in REF tend to be higher than the experiment and the LES results. This may be related to the observations made by Vickery [21] and Wolochuk et al. [22], who concluded that unsteady force coefficients are lower by about 50% for a cylinder in a turbulent stream, when compared with the results obtained in a smooth stream.

3.2. *Effects of temporal and spatial resolutions*

In order to investigate the temporal accuracy effect, the time step has been increased progressively from the reference time step ($\Delta t^* = 0.65$) to $\Delta t^* = 3.25$ (case RT5) with the same grid. Since the unconditionally stable implicit Euler scheme is used for time advancement, numerically stable solutions can be obtained for all cases. The time steps taken for one-period of vortex shedding are decreased from around 700 (case REF) to 140 (case RT5). Even though the force coefficients are found not very sensitive to the time step, the time step used in REF has been used for other cases to accurately evaluate the unsteady loading frequencies.

The effect of mesh refinement is investigated by coarsening (RXC) and refining (case RXF) the meshes in the wall vicinity by more than a factor of 2. The reference resolution is found to be enough for the present study, since no significant difference is noticed from the reference case to the refined case (considering local inconsistency of wall function approach in the case RXF with $(\Delta x_i^+)_{\min} < 10$). Since the use of a coarser grid may affect the computation accuracy, the reference resolution is used for other cases.

In order to validate the properness of the grid used in the wake region behind the cylinder and to investigate the effects of wake resolution to the unsteady forces on the cylinder, only the mesh behind the cylinder is refined by about a factor of 3 from the case REF, which is case fDn in Table 1. Since little differences are observed between the results from case REF and case fDn, it can be concluded that the current wake resolution in case REF is sufficient for the simulation and also that the unsteady forces are rather insensitive to the wake resolution.

3.3. Effects of convection scheme choice

With the lower order convection scheme like hybrid or upwind schemes, unsteady force coefficients are underpredicted. The predicted force coefficients are the largest for the QUICK (case REF), then for the hybrid (case RH), and the smallest for the upwind scheme (case RU) for a given turbulence model. The same trends in the unsteady lift coefficient are also found for other turbulence models. In general, a more dissipative scheme tends to give smaller force coefficients. Added numerical dissipation from the lower order convection schemes erroneously enhances the mixing (or transport), hence reduces the unsteady forces through a similar mechanism found in the standard k–ε model case discussed in Section 3.4.

3.4. Effects of turbulence models

In Table 2, results from different turbulence models – the standard, the RNG, and the low Reynolds number k–ε turbulence model – are described. There is a fundamental defect of the standard k–ε model near the stagnation point, where turbulence kinetic energy (TKE) is produced significantly without any mechanism active in the region to remove or transport the excessively produced TKE. This defect leads to enhanced spurious turbulence motion near the forward stagnation point in the present simulation (case SQ) so that the flow is overmixed to enhance entrainment by the detached shear layers from the forward separation corners, predicting smaller unsteady forces on the cylinder.

The RNG k–ε model has governing equations with similar structures as those of the standard k–ε model with some modifications to model constants: especially the one in the ε transport equation depends on the relative importance of the mean strain rate to a turbulence strain rate measure. In the vicinity of the stagnation region, this modified term enhances the dissipation rate, which results in an increased rate of TKE dissipation. Even though this mechanism does not correspond to the physical

Table 2
Numerical parameters and results with various turbulence models

Case	Δt^*	Δx^*	Scheme	Model	c_D	c_L	St
SQ	0.02	0.035	QUICK	standard	1.75 ± 0.004	0.56	0.138
REF	0.02	0.035	QUICK	RNG	2.12 ± 0.23	2.08	0.133
LX1	0.05	0.015	QUICK	low Re	2.01 ± 0.018	1.13	0.138
LX2	0.02	0.0025	QUICK	low Re	2.08 ± 0.036	1.55	0.134
LX3	0.02	0.0012	QUICK	low Re	2.10 ± 0.04	1.61	0.134
Exp [15]					2.14 ± 0.09		0.134
LES [16]	0.001	0.022	2nd Central	SGS	2.09 ± 0.13	1.60	0.132

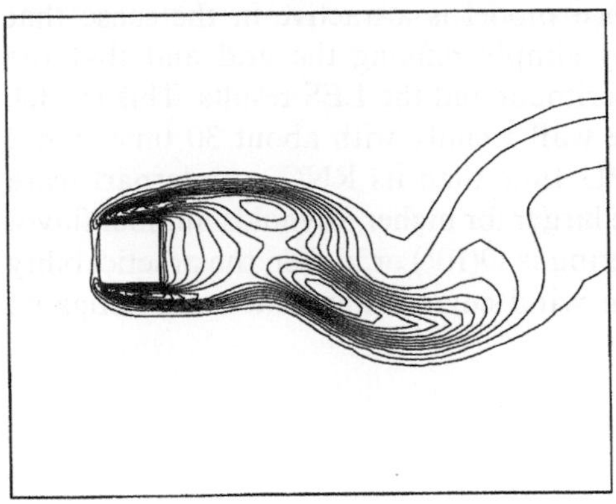

(a)

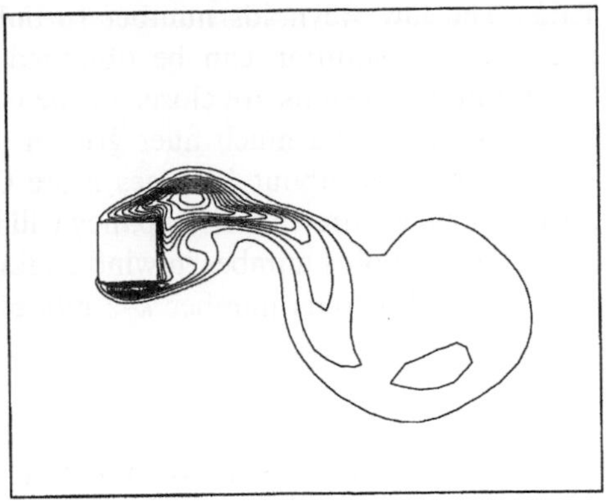

(b)

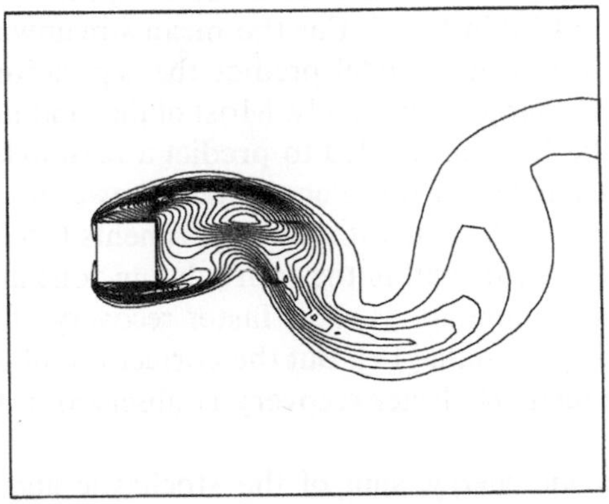

(c)

Fig. 1. Instantaneous stochastic TKE fields from (a) case SQ (standard), from (b) case REF (RNG), and (c) case LX3 (low Reynolds number) at the minimum lift phase of the oscillation ($\Delta k = 0.01$).

phenomena, this ad hoc enhancement of the dissipation rate prevents the augmentation of TKE in the region, which makes the force coefficients from the computation closer to the experiment and the LES results. This enhanced TKE removal in the RNG model is clearly illustrated in Fig. 1 with instantaneous TKE fields from case SQ (with the standard model) and from case REF (with the RNG model). Similar improvement is found with the low Reynolds number k–ε model, which also is shown in Fig. 1.

A low Reynolds number turbulence model is also used to predict unsteady aerodynamic forces on the square cylinder, which generally requires a much more refined grid than the standard and the RNG k–ε models. The grid used is refined progressively as $\Delta x^* = 0.015, 0.0025, 0.0012$, and the final grid is fine enough to give converging results with respect to the spatial resolution, where more than 5 grid points are placed within $y^+ = 10$ with $(\Delta x_i^+)_{\min} < 1$. The total number of grid points are 170 and 110 in the streamwise and the lateral direction respectively, as compared to 85 and 55 in case

REF. The low Reynolds number turbulence model is attractive in the sense that a converging solution can be obtained by simply refining the grid and that the computational results are closer to the experiment and the LES results. This model, however, requires a much finer grid in the wall vicinity with about 30 times more refined grid, and about 10 times more CPU time than its RNG counterpart (case REF). Furthermore, this discrepancy will be larger for higher Reynolds number flows. Since the Reynolds number in wind engineering is $O(10^7)$ or higher, the practicability of the low Reynolds number k–ε model in wind engineering is to be investigated further.

4. Turbulence models and flow statistics

Statistics along the centerline obtained from the computations are compared with the experiments and the LES in Fig. 2. For the mean streamwise velocity, the RNG and the standard k–ε turbulence model predict the separation bubble behind the cylinder too small and too large, respectively. Most of the modified turbulence models [9,10] including the LES [16] have failed to predict a reasonable separation bubble size. The low Reynolds number k–ε turbulence model, however, is found to predict the separation bubble size in agreement with the experiments [15,23]. Prediction of the maximum reverse flow intensity within the recirculating zone also matches well with the experiments. All the computations show faster recovery of the wake behind the cylinder compared with Lyn's data [15]. But the correctness of the experimental data is to be justified, since relatively faster recovery is observed in the other experiment [23].

Total turbulence kinetic energy, sum of the stochastic and the periodic kinetic energy, is shown in Fig. 2b, where all the computations deviate from the experiments. The RNG prediction is closer to the experiments with significant overprediction in the periodic part (refer to Table 1) and underprediction in the stochastic part. The standard model underpredicts both parts, while the low Reynolds number model somewhat underpredicts the periodic part with the largest stochastic part prediction, which is still significantly lower than the experiment [15] (figure not shown).

Pressure statistics around the cylinder are compared with the experiments [15,24,25] and the LES [16] in Fig. 3. The mean pressure around the cylinder is important in terms of structural safety, and the fluctuating pressure intensity is important as a measure of acoustic noise radiated from the cylinder. The standard model predicts the highest side and back pressure, which results from the overprediction of turbulence mixing. All the other computations show reasonable agreement with the experiment [15], but there is some difference in the LES along the side face of the cylinder: The reason for this difference is not yet clarified. All the k–ε turbulence model computations tend to predict higher back pressure, unlike the LES computation. In predicting fluctuating pressure intensity on the cylinder surface, the low Reynolds number model and the LES show good agreement with the available experiments [24,25], while the RNG and the standard model significantly overpredicts and underpredicts the fluctuating pressure, respectively.

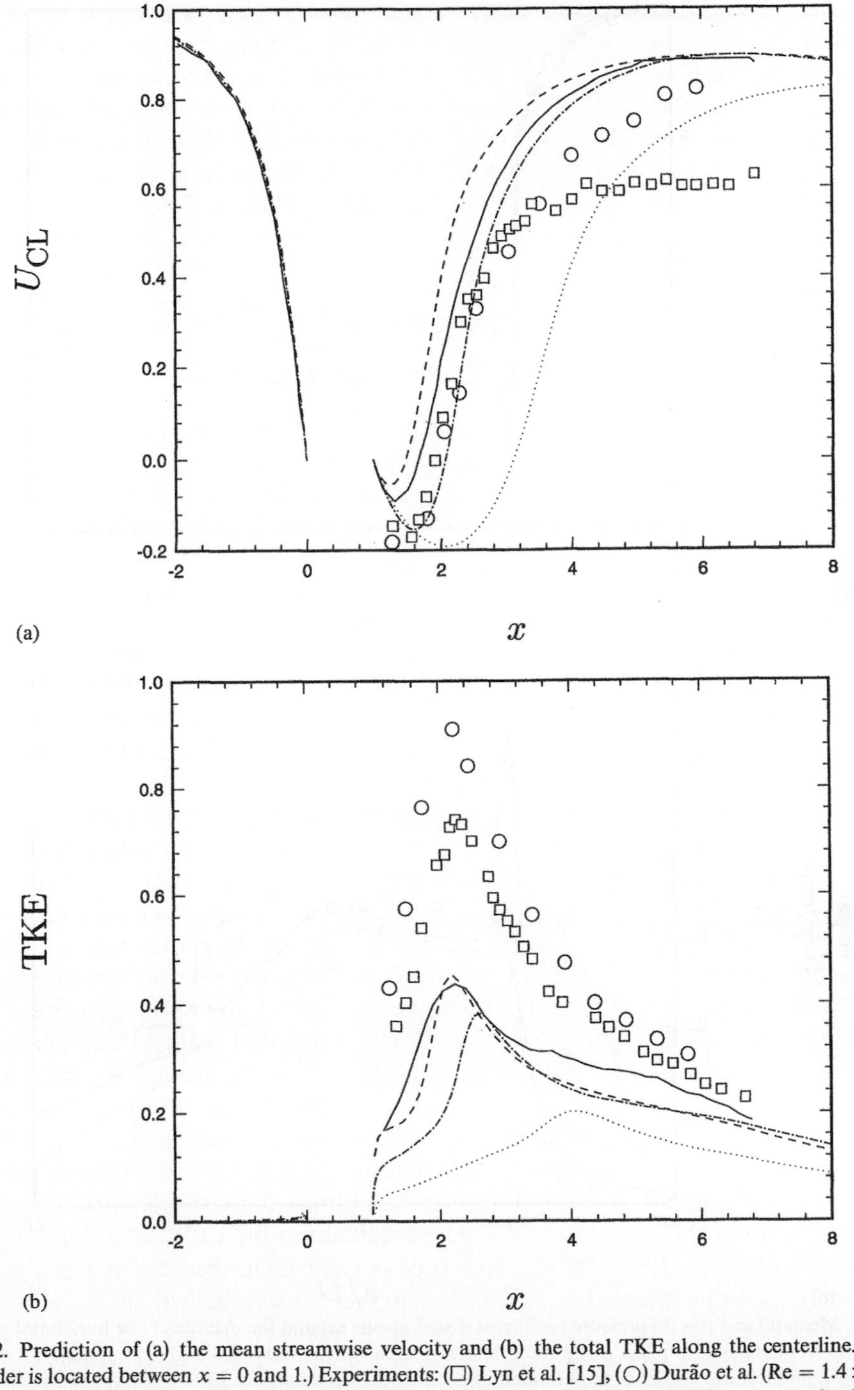

Fig. 2. Prediction of (a) the mean streamwise velocity and (b) the total TKE along the centerline. (The cylinder is located between $x = 0$ and 1.) Experiments: (□) Lyn et al. [15], (○) Durão et al. (Re = 1.4×10^4) [23]. Computations: LES — Murakami et al. [16]; k–ε models - - - - RNG (case REF), · · · · standard (case SQ), —·— low Re (case LX3).

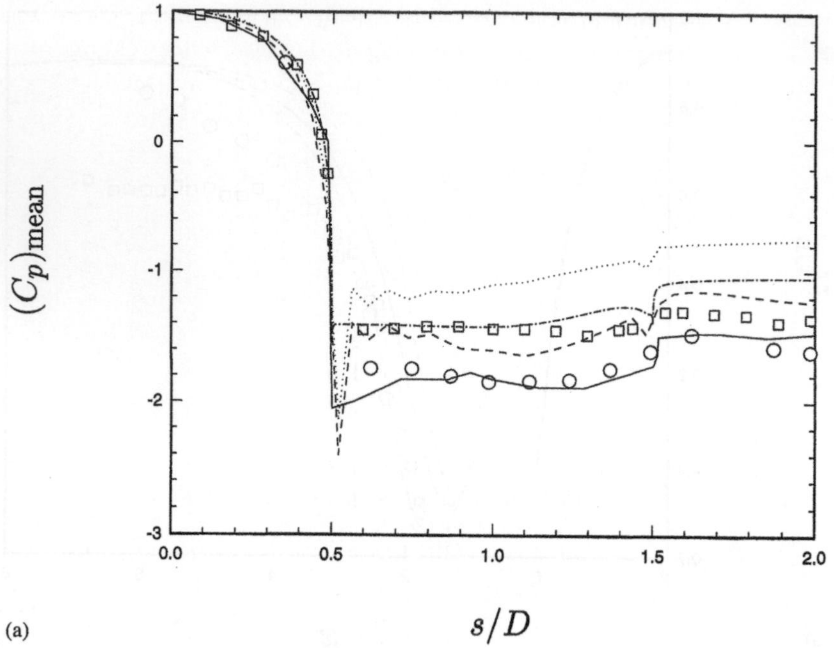

(a)

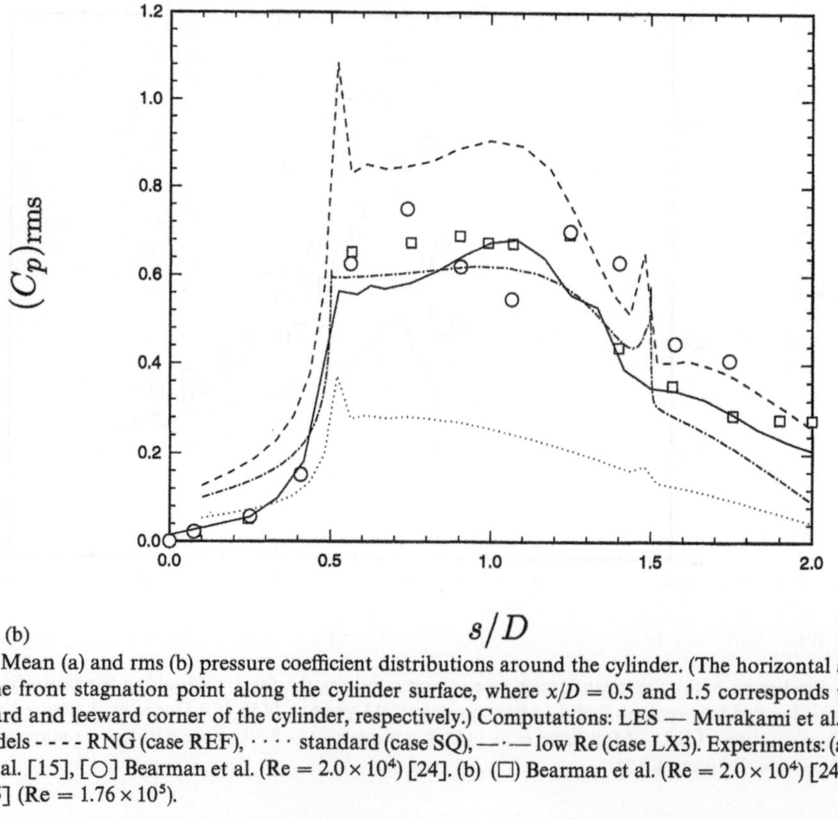

(b)

Fig. 3. Mean (a) and rms (b) pressure coefficient distributions around the cylinder. (The horizontal axis is from the front stagnation point along the cylinder surface, where $x/D = 0.5$ and 1.5 corresponds to the windward and leeward corner of the cylinder, respectively.) Computations: LES — Murakami et al. [16]; k–ε models - - - - RNG (case REF), · · · · standard (case SQ), —·— low Re (case LX3). Experiments: (a) (□) (Lyn et al. [15], [○] Bearman et al. (Re $= 2.0 \times 10^4$) [24]. (b) (□) Bearman et al. (Re $= 2.0 \times 10^4$) [24], (○) Lee [25] (Re $= 1.76 \times 10^5$).

From the investigation of turbulent flow statistics, the low Reynolds number and the RNG turbulence model outperform most turbulence model computations reported so far. And the quality of the low Reynolds number k–ε model with proper resolution is found comparable to that of a large-eddy simulation.

5. Summary and conclusion

Conventional k–ε turbulence models, the RNG and the low Reynolds number k–ε model, are found to successfully reproduce the unsteady force coefficients for turbulent flows over the square cylinder without piling up TKE near the forward stagnation point, which has been regarded as a fundamental deficiency of the standard k–ε model. For proper prediction, however, time accuracy, spatial accuracy and high-order convection scheme are also required. The prediction results are quite sensitive to the spatial resolution, the choice of convection schemes, and the turbulence models adopted. By comparing turbulence statistics with the experiment and the LES results, the low Reynolds number k–ε turbulence model is found to reproduce important flow features around the square cylinder. In summary, the quality of unsteady turbulence simulation is determined not only by the turbulence model, but equally by the numerical parameters.

Severe grid refinement required, however, seems to be a stumbling block against the practical application of these models to computational wind engineering: $(\Delta x_i^+)_{\min} \simeq 20$ in the RNG model and $(\Delta x_i^+)_{\min} \simeq 1$ in the low Reynolds number model are prohibitively stringent for flows with $\mathrm{Re} = O(10^7)$ or higher.

References

[1] R.S. Rogallo, P. Moin, Numerical simulation of turbulent flows, Ann. Rev. Fluid Mech. 16 (1984) 99–137.

[2] P. Moin, K. Squires, W. Cabot, S. Lee, A dynamic subgrid-scale model for compressible turbulence and scalar transport, Phys. Fluids A 3 (1991) 2746–2757.

[3] M. Germano, U. Piomelli, P. Moin, W. Cabot, A dynamic subgrid-scale eddy viscosity model, Phys. Fluids A 3 (1991) 1760–1765.

[4] S. Ghosal, T. Lund, P. Moin, K. Akselvoll, A dynamic localization model for large-eddy simulation of turbulent flows, J. Fluid Mech. 286 (1995) 229–255.

[5] B.S. Baldwin, H. Lomax, Thin-layer approximation and algebraic model for separated turbulent flows, AIAA Paper No. 78–0257, 1978.

[6] P.R. Spalart, S.R. Allmaras, A one-equation turbulence model for aerodynamic flows, AIAA Paper No. 92–0439, 1992.

[7] P. Bradshaw, B.E. Launder, J.L. Lumley, Collaborative testing of turbulence models, in: K.N. Ghia, U. Ghia, D. Goldstein (Eds.), Advances in Computational Methods in Fluid Mechanics, ASME, New York, 1994, pp. 77–82.

[8] R. Mittal, S. Balachandar, Effect of three-dimensionality on the lift and drag of nominally two-dimensional cylinders, Phys. Fluids A 7 (1995) 1841–1865.

[9] R. Franke, W. Rodi, Calculation of vortex shedding past a square cylinder with various turbulence models, in: Proc. 8th Symp. on Turbulent Shear Flow, Munich, Germany, 1991, pp. 20.1.1–20.1.6.

[10] M. Kato, B.E. Launder, The modelling of turbulent flow around stationary and vibrating square cylinders, in: Proc. 9th Symp. on Turbulent Shear Flow, Kyoto, Japan, 1993, pp. 10.4.1–10.4.6.

[11] P.A. Durbin, Separated flow computations with the k–ε–v^2 model, AIAA J. 33 (1995) 659–664.
[12] B.E. Launder, D.B. Spalding, The numerical computation of turbulent flows, Comp. Meth. Appl. Mech. Eng. 3 (1974) 269–289.
[13] V. Yakhot, S.A. Orszag, S. Thangam, T.B. Gatski, C.G. Speziale, Development of turbulence models for shear flows by a double expansion technique, Phys. Fluids A 4 (1992) 1510–1520.
[14] B.E. Launder, B.T. Sharma, Application of the energy dissipation model of turbulence to the calculation of flow near a spinning disc, Lett. Heat Mass Transfer 1 (1974) 131–138.
[15] D.A. Lyn, W. Rodi, The flapping shear layer formed by flow separation from the forward corner of a square cylinder, J. Fluid Mech. 267 (1994) 353–376.
[16] S. Murakami, A. Mochida, On turbulent vortex shedding flow past 2D square cylinder predicted by CFD, J. Wind Eng. Ind. Aerodyn. 54 (1995) 191–211.
[17] C.M. Rhie, W.L. Chow, Numerical study of the turbulent flow past an airfoil with trailing edge separation, AIAA J. 21 (1983) 1525–1532.
[18] B.P. Leonard, A stable and accurate convective modelling procedure based on quadratic upwind interpolation, Comp. Meth. Appl. Mech. Eng. 19 (1979) 59–98.
[19] S.V. Patankar, Numerical Heat Mass Transfer, McGraw-Hill, New York, 1980.
[20] W.H. Press, B.P. Flannery, S.A. Teukolsky, W.T. Vetterling, Numerical Recipes: The Art of Scientific Computing, Cambridge Univ. Press, New York, 1986.
[21] B.J. Vickery, Fluctuating lift and drag on a long cylinder of square cross-section in a smooth and in a turbulent stream, J. Fluid Mech. 25 (1966) 481–494.
[22] M.C. Wolochuk, M.W. Plesniak, J.E. Braun, The effects of turbulence and unsteadiness on vortex shedding from sharp-edged bluff bodies, J. Fluids Eng. Trans. ASME 118 (1996) 18–25.
[23] D.F.G Durão, M.V. Heitor, J.C.F. Pereira, Measurements of turbulent and periodic flows around a square cross-section cylinder, Exp. Fluids (1988) 298–304.
[24] P.W. Bearman, E.D. Obasaju, An experimental study of pressure fluctuations on fixed and oscillating square-section cylinders, J. Fluid Mech. 119 (1982) 297–312.
[25] B.E. Lee, The effect of turbulence on the surface pressure field of a square prism, J. Fluid Mech. 69 (1975) 263–282.

ELSEVIER

Journal of Wind Engineering
and Industrial Aerodynamics 67&68 (1997) 91–102

JOURNAL OF
wind engineering
AND
industrial
aerodynamics

Numerical study of blockage effects on aerodynamic characteristics of an oscillating rectangular cylinder

Atsushi Okajima*, Donglai Yi, Atsushi Sakuda, Tomohito Nakano

Graduate School of Natural Science and Engineering, Kanazawa University, 2-40-20 Kodatsuno, Kanazawa, Ishikawa 920, Japan

Abstract

The flow around a stationary and oscillating square cylinder was numerically simulated at various blockage ratios in order to study the effects of wall confinement on the aerodynamic characteristics of the cylinder. Numerical methods include Direct Simulation (DS) with upwind scheme for laminar flow at Re = 200, 400, 10^3, and k–ε model for turbulence flow at Re = 4×10^3. The simulations were conducted at various blockage ratios of $H/L = 0.04$, 0.067, 0.085, 0.112, 0.166, 0.244, 0.303 and 0.4. For stationary cases, simulations were carried out at Reynolds numbers of Re = 200–10^3. For oscillation cases, simulations were done at Reynolds number of Re = $(1, 4) \times 10^3$ for a square cylinder oscillating at forced oscillatory frequencies of $\mathrm{St_c} = 0.1$, 0.25 and 0.3. Using the SIMPLE algorithm, simulation results show that drag and lift forces and Strouhal numbers all increase with the increase of blockage ratios (i.e. the decrease of the interval of the two parallel confining walls), and the aerodynamic characteristics agree reasonably well with those from experiments. For stationary cases, the lift and drag forces, and vortex-shedding Strouhal numbers all increase with the increase of blockage ratios. The flow streamlines and vorticity contours show conspicuous features of the wake under high blockage ratios. The first-sinking phenomenon due to blockage effect can be successfully captured at Reynolds numbers of only Re = 10^3 and can be confirmed to be caused by the change of non-reattachment of separated shear layers to reattachment flow by the blockage effects. For oscillation cases, the lock-in features are successfully captured. The behavior of phase difference, i.e. the *stable/unstable* features, against the forced oscillatory frequency cannot be almost influenced by the blockage effects.

Keywords: Blockage effect; Oscillating square cylinder; Numerical simulation; Lift force; Drag force; Vortex-shedding; Strouhal number

* Corresponding author. E-mail: okajima@t.kanazawa-u.ac.jp.

PII S0167-6105(97)00065-2

1. Introduction

When channel walls confine the flow past a cylindrical obstruction located between parallel walls, the aerodynamic characteristics of the flow could be different from those for a free-stream type, or the so-called blockage effects arise. Corrections to measured data are required if the initial aim of research is on flow of free-stream type. For the flow past an oscillating rectangular cylinder, the blockage effects can be significant.

The effects of confining walls on measured data have been the subject of many previous investigations. Maskell [1] studied the wall effects on the flow past a bluff body with fixed separation points by doing experiments with a flat-plate normal to the flow. Ranga Raju and Singh [2] studied the blockage effects on the drag on sharp-edged bodies. For two-dimensional stationary rectangular cylinders, Courchesne and Laneville [3] have given a comprehensive comparison of several correction procedures.

While most of the studies concern experimental work, some recent papers employed numerical simulations of confined flow past bluff bodies to investigate the effects of wall confinement. Among them are those by Ota et al. [4], and Li and Humphrey [5]. Ota et al. simulated confined flows past a square cylinder, an elliptic cylinder and an inclined flat plate located between two parallel walls at various blockage ratios, using a discrete vortex method. Calculations included the mean and fluctuating features of surface pressure, lift, drag, and Strouhal number. Li and Humphrey simulated two-dimensional flows of Re = 100, 500, and 10^3 past a stationary square cylinder located asymmetrically, at orientation angles of $\alpha = 0°–20°$, between the parallel walls at the blockage ratio of $H/L = 0.192$.

We also did an experimental investigation of the blockage effects on an oscillating rectangular cylinder with the side ratios of $B/H = 1$, 2, 3 and 4 [6,7] where B is the cross-sectional length in the streamwise direction and H in the cross-flow direction. A very interesting phenomenon has been found during the experimental study. Next, numerical simulations at different blockage ratios were conducted to check the experimental dynamic characteristics and flow patterns around the cylinders. The numerical simulations described in the paper concern the flow past a stationary and oscillatory cylinder of square cross-section with its longitudinal axis aligned parallel to the channel walls and normal to the approaching flow at different blockage ratios.

2. Outline of computation technique

The motion of a continuous medium is governed by the conservation of mass, momentum, and energy. Fundamental equations are the Navier–Stokes equations and the continuity equation based on the assumption of unsteady, viscous, incompressible flow. In the simulations for laminar flows at Re = 200, 400 and 10^3, the Direct Simulation (DS) method was employed. The governing equations are discretized over elementary control volumes on a Cartesian coordinate system; i.e. the finite control volume method is employed. The space derivative terms are discretized by the QUICK scheme for convective terms and by the second-order central difference scheme for all other terms. For time marching, the implicit Crank–Nicolson scheme is used. In the simulations of turbulent flows at $\mathrm{Re} = 4 \times 10^3$, the k–ε model is used. In both cases, the algorithm of computations used is SIMPLE, in which the velocity components are first calculated from the Navier–Stokes equations using a guessed pressure field, and then the pressures and velocities are corrected, to satisfy continuity. The predict-and-correct operation continues until the solution converges. The Poisson equation for the pressure-correction is solved by the Successive-Over-Relaxation (SOR) method. The initial pressure field is from potential calculations in the same computational domain. Corresponding to the forced oscillation of the cylinder, the convective terms $(\boldsymbol{v}\cdot\nabla)\boldsymbol{v}$ in the Navier–Stokes equations are replaced by $((\boldsymbol{v}-\boldsymbol{V})\cdot\nabla)\boldsymbol{v}$, where $\boldsymbol{v}$ is the velocity vector of the incident flow, and $\boldsymbol{V}$ is that of forced oscillation.

The computational domain is a rectangular area in the Cartesian coordinates system, in which rectangular grids are used. In order to change the blockage ratios of the bluff body, we need only to change the number of the mesh grids in cross-flow direction. Fig. 1 shows the computational domains for (a) $H/L = 0.067$ and (b) 0.303. The number of mesh grids in streamwise direction is the same for all blockage ratios.

Boundary conditions are: uniform flow at the upstream inlet and zero-gradient flow at downstream exit, and no-slip conditions at the surfaces of the square cylinder. In experimental studies of the blockage effects [6,7], the square cylinder was towed in a water tank; namely, with respect to the cylinder the confining walls are moving with the same velocity as the flow at the upstream inlet. In the simulation described herein, we also set the boundary conditions at the two confining moving-walls to be uniform flow, for the sake of comparison with the experiments.

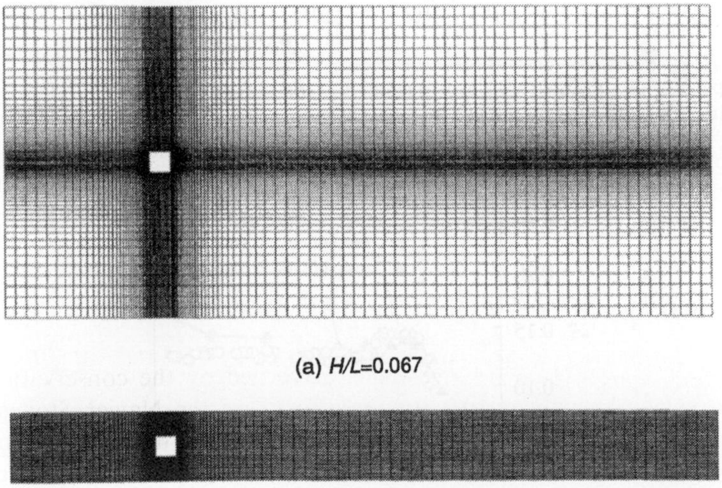

(a) H/L=0.067

(b) H/L=0.303

Fig. 1. Computational domains for a square cylinder at blockage ratios of (a) $H/L = 0.067$ and (b) 0.303.

3. Computational results

The simulations were conducted at the blockage ratios of $H/L = 0.04, 0.067, 0.085, 0.112, 0.166, 0.243, 0.303$ and 0.4. For stationary cases, simulations were carried out for square cylinders at Reynolds numbers of $\mathrm{Re} = 200, 400, 10^3$. For oscillation cases, simulations were only done at Reynolds number of $\mathrm{Re} = (1, 4) \times 10^3$, for a square cylinder at forced (reduced) oscillatory frequencies of $\mathrm{St}_c = 0.1, 0.25$ and 0.3.

3.1. Verification of methodology

The goodness of the numerical procedure was evaluated by testing the program calculating the drag force and the Strouhal number of the flow around a stationary square cylinder and comparing the results with experiments. The Reynolds numbers of simulations were deliberately chosen to be in the range where detailed experimental results are available. The test results are compared with the experiments in Fig. 2 for drag force coefficients and in Fig. 3 for vortex-shedding Strouhal numbers. From

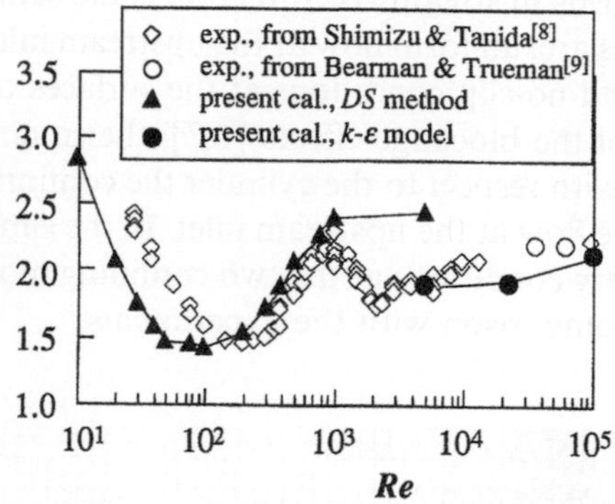

Fig. 2. Comparison of simulated and experimental C_D on a stationary square cylinder at Reynolds numbers of 10–10^5. Experimental values from Refs. [8,9].

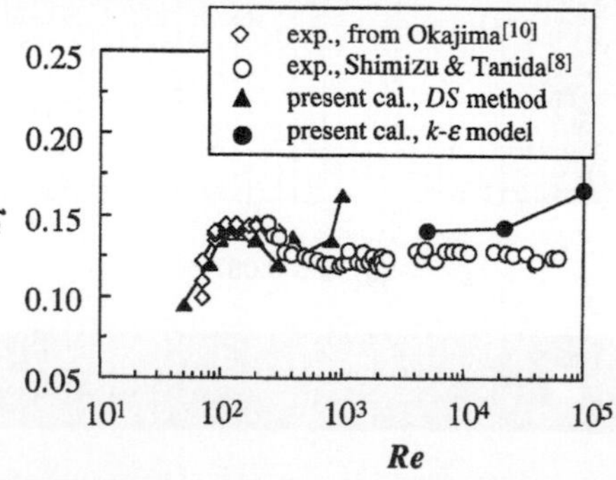

Fig. 3. Comparison of simulated and experimental St of vortex-shedding from a stationary square cylinder at Reynolds numbers of 50–10^5. Experimental values from Refs. [8,10].

the figures, it can be seen that the DS method is well applicable for laminar flow up to Reynolds number of about Re = 10^3 above which the discrepancies between the simulations and experiments are large and the computed drag coefficient has upward tendency. On the other hand, as far as the present test simulations confirmed, the k–ε model is also reasonably suitable for turbulent flow, although the simulated drag force is a little bit larger than that of experiments and the simulated vortex-shedding Strouhal numbers are lower than the experimental ones. Under this classification, the simulation results were regarded to agree well with experiments, and the test calculations illustrated the precision of the simulation algorithm.

3.2. Blockage effects for stationary cylinders

Fig. 4 shows the simulated drag forces on a stationary square cylinder at different blockage ratios and different Reynolds numbers. Experimental measurements are also given for comparison. As shown in the caption of the figure, the full circles represent experimental results [6, 7] at Reynolds number of 4×10^3 for turbulent flow, and the open symbols simulation results.

In our recent experiments [6, 7] at relatively high Reynolds numbers, it was found that, for rectangular cylinders with a side ratio of $B/H = 1$ and 2, the drag forces do not show monotonic increase with the increase of blockage ratio such as the curve with full circles shown in Fig. 4 that one would expect from the knowledge of blockage effects on circular cylinders and thin plates. Instead, the drag forces decrease to a minimum value at first and then increase gradually. The reason for this first-sinking phenomenon is supposed to be that, for a short rectangular cylinder, the base pressure depends mainly on flow patterns, especially the reattachment of shear layers separated from leading edges onto the afterbody of the cylinder. For $B/H = 3$ and 4 rectangular cylinders, such first-sinking phenomenon disappeared because the flow gets reattached even at free-upstream conditions and thus there is no abrupt change of base pressure due to reattachment for rectangular cylinders with afterbody. It is noted in Fig. 4, for simulation of flow around a square cylinder that a first-sinking

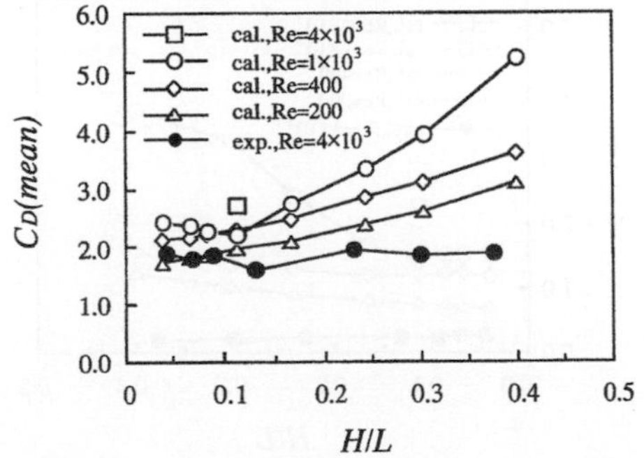

Fig. 4. Mean values of simulated C_D against blockage ratios H/L for a stationary square cylinder at different Reynolds numbers; experimental results [6, 7]: Re = 4×10^3.

phenomenon appears for drag curve of Reynolds number of only 10^3, which can appear in an experimental curve of Reynolds number of 4×10^3, while at low Reynolds numbers such as 200 and 400, the drag force only increases monotonically without the sinking phenomenon, when the blockage ratio increases. Okajima [10] measured the variations of vortex-shedding Strouhal numbers against Reynolds numbers. It is found there that, for a $B/H = 2$ rectangular cylinder, there is an abrupt jump-down in the Strouhal number curve when the Reynolds number increases to $Re = 400$, and the change of flow pattern from the reattached to the non-reattached case is responsible for this jump-down. For the square cylinder, there is a weak decrease in the Strouhal number curve when the Reynolds number increases up to $Re = 400$, and the reason for this weak decrease is the change of wake width from narrow to wide. In the range of low Reynolds numbers of $Re < 400$, the flow has a very narrow wake width. So even for high blockage ratios, there is only a small change of flow patterns with a relatively narrow wake width and thus no first-sinking phenomenon. As discussed in experiments, the sinking phenomenon appearing in the computations at Reynolds number of 10^3 corresponds to the change of flow-pattern from non-reattachment to reattachment when the blockage ratio increases.

In addition, it should be pointed out that the drag forces increase at each blockage ratio when the Reynolds number increases from 200 to 10^3, which agrees well with the experimental measurements in this Reynolds range as shown in Fig. 2.

Fig. 5 shows the variation of simulated and experimental lift coefficients against blockage ratios H/L for a stationary square cylinder at different Reynolds numbers. It can be seen that the experimental values of amplitudes of C_L at Reynolds number of 4×10^3 are almost constant while the simulated values become larger with narrowness of the blockage. Here also, the amplitudes of C_L increase at each blockage ratio when the Reynolds number increases from 200 to 10^3.

It is noted, however, that all the values of C_D and C_L computed under the assumption that the flow is two-dimensional and laminar are very higher than the experimental ones for turbulent flow.

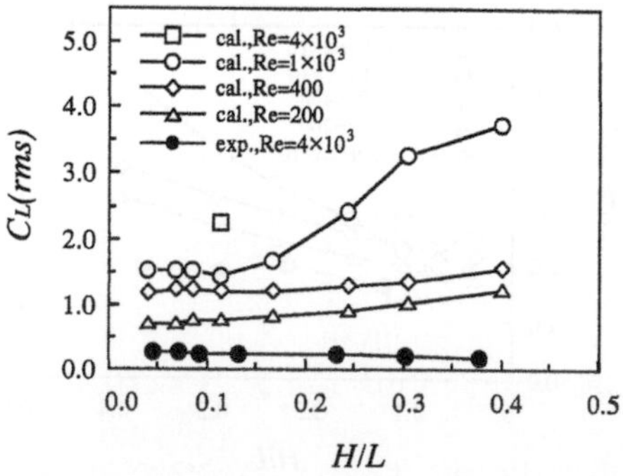

Fig. 5. RMS values of simulated C_L against blockage ratios H/L for a stationary square cylinder at different Reynolds numbers; experimental results [6, 7]: $Re = 4 \times 10^3$.

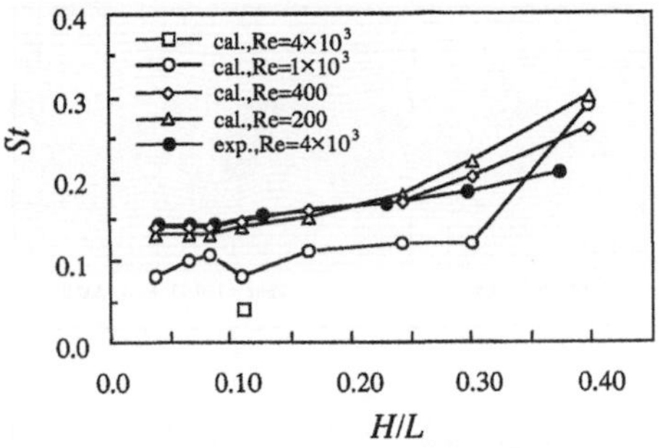

Fig. 6. Values of simulated St against blockage ratios H/L for a stationary square cylinder at different Reynolds numbers; experimental results [6, 7]: $Re = 4 \times 10^3$.

Fig. 6 shows the simulated and experimental variation of Strouhal numbers against blockage ratios for a square cylinder at different Reynolds numbers. The same variation tendency can be found although the increase rates of the curves are different.

3.3. *Flow patterns and vorticity contours at various blockage ratios*

The increase of vortex-shedding Strouhal numbers with the increase of the blockage ratios is clearly shown in flow streamlines and vorticity contours. Figs. 7 and 8 show the instantaneous flow streamlines and vorticity contours at the computational time of $t = 100$ for a stationary square cylinder at Reynolds number of $Re = 200$ and 10^3 and at blockage ratios of (a) $H/L = 0.04$, (b) 0.244, (c) 0.303 and (d) 0.4. As can be seen from Fig. 7a and Fig. 8a, the vortices are loosely formed in the wake both in lateral and in streamwise directions, which represents a smaller value of the vortex-shedding Strouhal number. When the blockage ratio increases to a certain degree as in Fig. 7b and Fig. 8b, the streamlines show a much more regular swing of the wake both in the near and far wakes, and the vortices are forced to form in a narrow space very densely. The very dense arrangement of vortices corresponds to a larger value of the vortex-shedding Strouhal number. When the blockage ratio increases further to $H/L = 0.303$ and 0.4 as in Fig. 7c, Fig 7d, Fig. 8c and Fig. 8d, the vortices are formed in the near wake with further higher vortex-shedding Strouhal numbers. In these cases, as there is not enough space in the lateral direction for the vortices to shed further downstream, the vortices are cracked down by the confining walls and seem to weaken or disappear in the far wake. Fig. 7d clearly shows this disappearance of vortices in the far wake at $Re = 200$.

Furthermore, it is of particular interest that weak vortices are observed on the confining moving-walls at the position just downstream from the cross-section of maximum velocity profile near the cylinder as shown in Figs. 7 and 8.

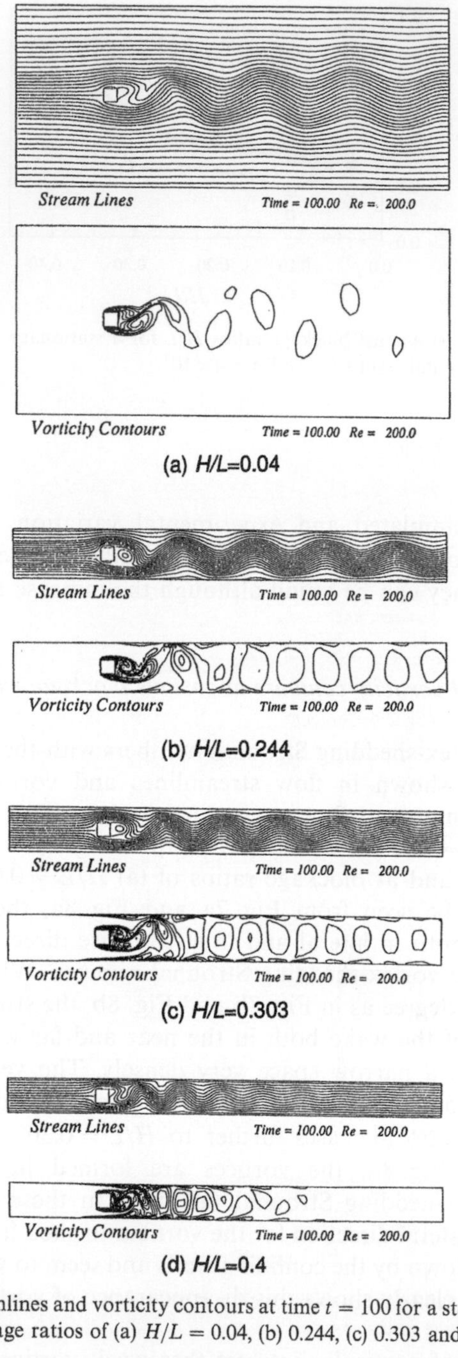

Fig. 7. Instantaneous streamlines and vorticity contours at time $t = 100$ for a stationary square cylinder at Re = 200 at different blockage ratios of (a) $H/L = 0.04$, (b) 0.244, (c) 0.303 and (d) 0.4.

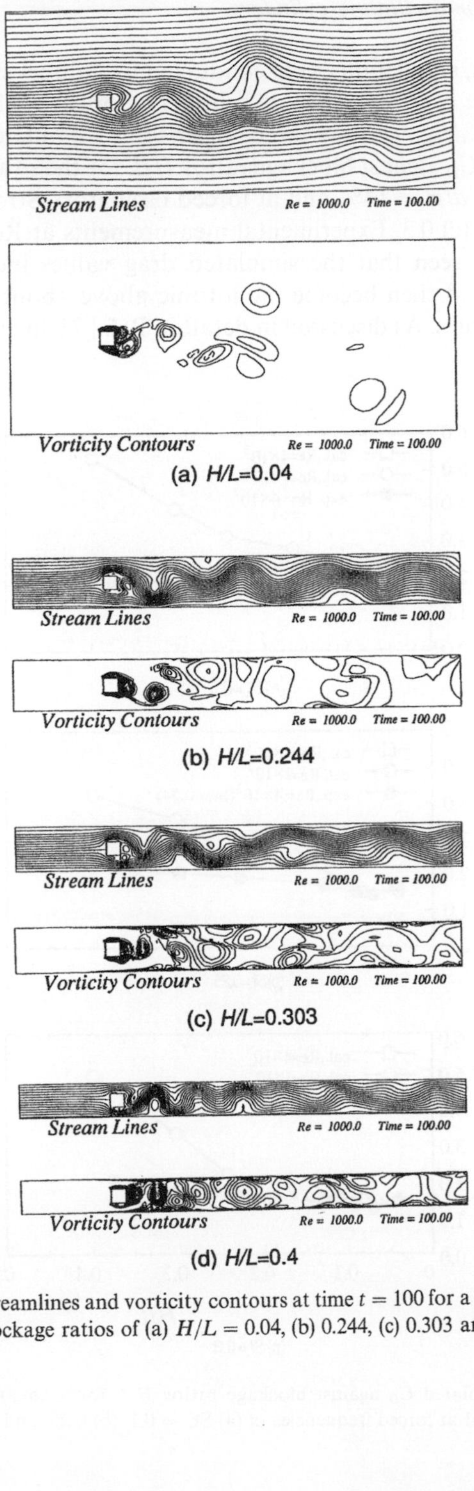

Fig. 8. Instantaneous streamlines and vorticity contours at time $t = 100$ for a stationary square cylinder at $Re = 10^3$ at different blockage ratios of (a) $H/L = 0.04$, (b) 0.244, (c) 0.303 and (d) 0.4.

3.4. Blockage effects for oscillating cylinders

For oscillation cases, simulations were conducted by a direct simulation method at Reynolds number of Re = 10^3 for a square cylinder at forced oscillatory frequencies of $St_c = 0.1$, 0.25 and 0.3. Fig. 9 shows the comparison of simulated and experimental drag force coefficients C_D against blockage ratios H/L for the square cylinder oscillating at an amplitude of $a/H = 14\%$ and at forced oscillatory Strouhal numbers of (a) $St_c = 0.1$, (b) 0.25 and (c) 0.3. Experimental measurements at Re = 4×10^3 are from Refs. [6, 7]. It can be seen that the simulated drag values increase little by little at small H/L ratios, and then become monotonic above about $H/L = 0.1$ with the increase of the H/L values. As discussed in detail in Ref. [7], in experiments the mean

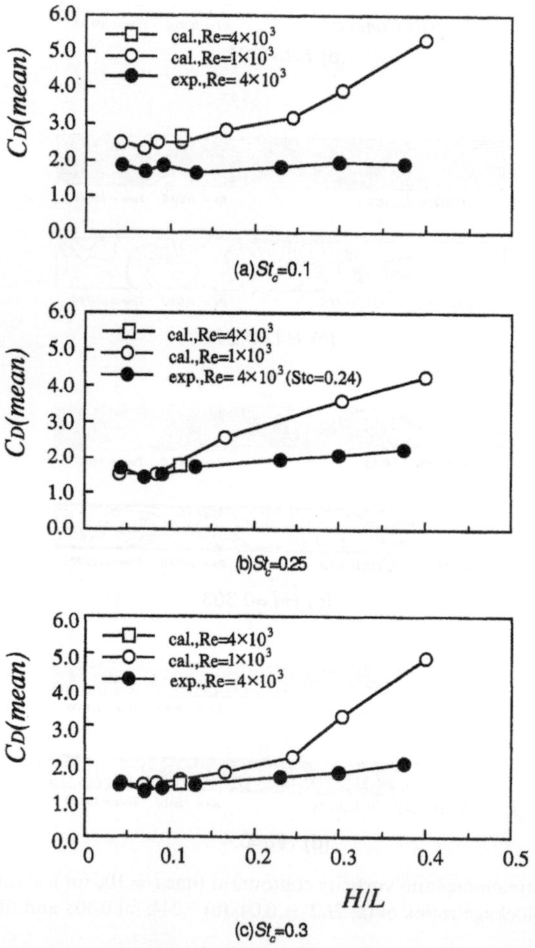

Fig. 9. Mean values of simulated C_D against blockage ratios H/L for a square cylinder oscillating at amplitude of $a/H = 14\%$ and at forced frequencies of (a) $St_c = 0.1$, (b) 0.25, and (c) 0.3.

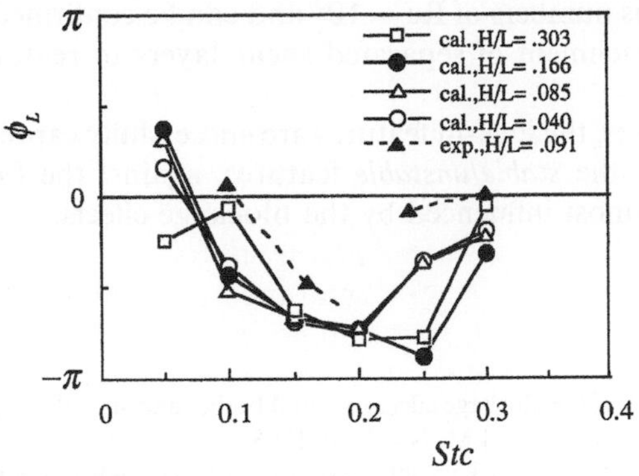

Fig. 10. Phase difference ϕ_L of simulated C_L against forced frequency St_c at different blockage ratios H/L for a square cylinder oscillating at amplitude of $a/H = 14\%$.

values of drag forces also decrease to a minimum value at first and then increase gradually with the increase of H/L ratio. These first-sinking features for oscillation cases may make the problem beyond a simple blockage issue for the case of a stationary cylinder, but a rather combined aerodynamic behavior under other fluid mechanics due to the oscillation of the cylinder. Thus, it is difficult for drag forces to be corrected for that blockage for the case of an oscillating cylinder. Fig. 10 shows the simulated values of phase difference ϕ_L of the lift forces C_L referred to the oscillatory displacement of the cylinder against forced Strouhal number St_c at different blockage ratios H/L and Reynolds number of $Re = 10^3$ for a square cylinder. It can be seen that the phase ϕ_L seems to change slightly except when the ϕ_L values are apt to shift to higher St_c around the blockage ratios of 0.25. It should be noted that the behavior of phase difference, i.e. *stable/unstable* features, against the forced oscillatory frequency cannot be almost influenced by the blockage effects.

4. Concluding remarks

The flow around a stationary and oscillating square cylinder was numerically simulated at various blockage ratios of $H/L = 0.04$, 0.067, 0.085, 0.112, 0.166, 0.243, 0.303, 0.4 and at Reynolds numbers of $Re = 200$ to 4×10^3. For oscillation cases, simulations were done for a square cylinder at forced oscillatory frequencies of $St_c = 0.1$–0.4. The aerodynamic characteristics of the flow around a stationary and oscillating square cylinder under different blockage ratios are simulated and the characteristics agree reasonably well with those from experiments [6, 7].

For stationary cases, the lift and drag forces, and vortex-shedding Strouhal numbers all increase with the increase of blockage ratios. The flow stream-lines and vorticity contours show the conspicuous features of a wake under high blockage ratios. The first-sinking phenomenon due to blockage effect can be successfully captured at Reynolds numbers of Re = 10^3 and can be confirmed to be caused by the change of non-reattachment of separated shear layers to reattachment flow by the blockage effects.

For oscillation cases, the lock-in features are successfully captured. The behavior of phase difference, i.e. the *stable/unstable* features, against the forced oscillatory frequency cannot be almost influenced by the blockage effects.

References

[1] E.C. Maskell, Theory of the blockage effects on bluff bodies and stalled wings in a closed wind tunnel, British Aero. Res. Comm. R. and M. No. 3400, 1963.

[2] K.G. Ranga Raju, V. Singh, Blockage effects on drag of sharp-edged bodies, J. Ind. Aerodyn. 1 (3) (1976) 301–309.

[3] J. Courchesne, A. Laneville, A comparison of correction methods used in the evaluation of drag coefficient measurements for two-dimensional rectangular cylinders, J. Fluids Eng. Trans ASME 101 (1979) 506–510.

[4] T. Ota, Y. Okamoto, H. Yoshikawa, A correction formula for wall effects on unsteady forces of two-dimensional bluff bodies, J. Fluids Eng. Trans. ASME 116 (1994) 414–423.

[5] G. Li, J.A.C. Humphrey, Numerical modelling of confined flow past a cylinder of square cross-section at various orientations, Int. J. Numer. Meth. Fluids 20 (1995) 1215–1236.

[6] D. Yi, A. Okajima, S. Kimura, T. Oyabu, The blockage effects for an oscillating rectangular cylinder at moderate Reynolds number, in: Proc. 3rd Int. Colloq. on Bluff Body Aerodynamics and Applications (BBAAIII), 28 July–1 August, 1996, Blacksburg, Virginia, USA.

[7] D. Yi, Fluid-dynamic forces acting on and flow patterns around an oscillating rectangular cylinder, Ph.D. Thesis, Kanazawa University, Kanazawa, Japan, 1996.

[8] Y. Shimizu, Y. Tanida, Fluid forces acting on cylinders of rectangular cross-section, Trans. JSME B 44 (384) (1978) 2699–2706 (in Japanese).

[9] P.W. Bearman, D.M. Trueman, An investigation of the flow around rectangular cylinders, Aeronaut. Q. 3 XXIII (1972) 229–237.

[10] A. Okajima, Strouhal numbers of rectangular cylinders, J. Fluid Mech. 123 (1982) 379–398.

ELSEVIER

Journal of Wind Engineering
and Industrial Aerodynamics 67&68 (1997) 103–116

JOURNAL OF wind engineering AND industrial aerodynamics

Stably stratified flow around a horizontal rectangular cylinder in a channel of finite depth

Shigehira Ozono*, Noboru Aota, Yuji Ohya

Research Institute for Applied Mechanics, Kyushu University, Kasuga 816, Japan

Abstract

This paper describes a numerical study of the two-dimensional flow of a linearly stably-stratified Boussinesq fluid around a rectangular cylinder in a channel of finite depth. To attain highly stratified flow, a difference scheme combining two grids was applied. Attention was focused upon the interaction between the vortex shedding and the internal gravity waves. A parameter K is used, which is the ratio of the fastest linear internal-wave speed to the mean flow speed. As K increases, the symmetric configuration with the cylinder placed at the mid-depth first stimulates the symmetric vertical dominant mode of the columnar disturbance at $K \sim 2$, but the non-symmetric configuration with the cylinder displaced stimulates the non-symmetric mode at $K \sim 1$. Once a columnar disturbance of a certain mode is created, accompanied by the lee waves, the wave speed of the columnar disturbance is not characterized by cylinder conditions, but by linear theory. A critical change in the Strouhal number is observed, shortly after appearance of the lee wave, which probably influences the dynamics of vortex formation. The likely flow alteration is thought to occur when the lee waves grow in amplitude and the wavelength becomes sufficiently short so that the wake cavity is suppressed/biased by the first troughs of the lee waves.

Keywords: FDM; Stable stratification; Rectangular cylinder; Vortex shedding; Columnar disturbance

1. Introduction

The flow of a uniformly-stratified Boussinesq fluid past a two-dimensional obstacle in a channel of finite depth has been the subject of extensive study. Most of the works were concerned with the flow over surface-mounted obstacles, in which no vortex shedding due to the interaction between the separated shear layers could be introduced, but there are only a few studies of stratified flow in the presence of vortex shedding.

* Corresponding author. E-mail: ozono@riam.kyushu-u.ac.jp.

PII S0167-6105(97)00066-4

Droughton and Chen [1] reported that the flow was observed to be very sensitive to the relative position of a circular cylinder and a slender body in a towing tank for blockage ratio from 0.05 to 0.10. More recently, Boyer et al. [2] observed the flows past a circular cylinder in a stably-stratified fluid under different conditions and depicted a flow regime diagram in a Froude- versus Reynolds-number plane. But the effects of the boundaries were not taken into account, although the blockage ratio ranges from 0.03 to 0.20.

Hanazaki [3] carried out the numerical simulation on the flow past a plate in a channel of finite depth. Although the Reynolds number used is 20 and no vortex shedding is supposed to be created, the results may be even useful for explaining the flow at a high Reynolds number with vortex shedding because they give a clear view of the long internal waves trapped in a channel. According to that work, there are time-dependent oscillations in each vertical mode of the upstream advancing columnar disturbances, which cause unsteadiness in the drag coefficient. On the other hand, Wei et al. [4] developed a linear theory that the columnar disturbance of mode n propagates upstream at the speed of

$$(K/n - 1)U \tag{1}$$

with reference to the obstacle, where K is a parameter defined by $NH/\pi U$, with N being the Brunt Väisälä frequency, H the channel depth and U the basic flow speed. In reality, in the numerical analysis [3], the wave speed in a linearly stratified flow was characterized by parameter K.

By using a density-stratified wind tunnel, we have done an experimental work on the flow around a circular cylinder with a blockage ratio of 0.17 [5]. The Reynolds number used was of the order of 10^3–10^4 and the vortex shedding was expected to be created in the wake throughout. But as stratification is raised, critically from $K \sim 1.0$ the roll-up of separated shear layers was seen to be suppressed and the position of their interaction shifted further downstream in the wake.

The results presented here are part of an ongoing investigation of the mechanism of interaction between the vortex shedding from a bluff body and the internal gravity waves due to stratification, but the immediate goal is to look numerically into the flows around the rectangular cylinder when its placement relative to the channel is varied. Along with the present numerical study, a laboratory experiment [6] was conducted for the symmetric and non-symmetric cases using the rectangular cylinder. Thus, some results are compared with the experimental results.

2. Simulation details

2.1. Flow configuration

The flow configuration is presented in Fig. 1. The rectangular cylinder has a side ratio $d/h = 0.4$, where h is the height of the rectangular cylinder and d is its streamwise

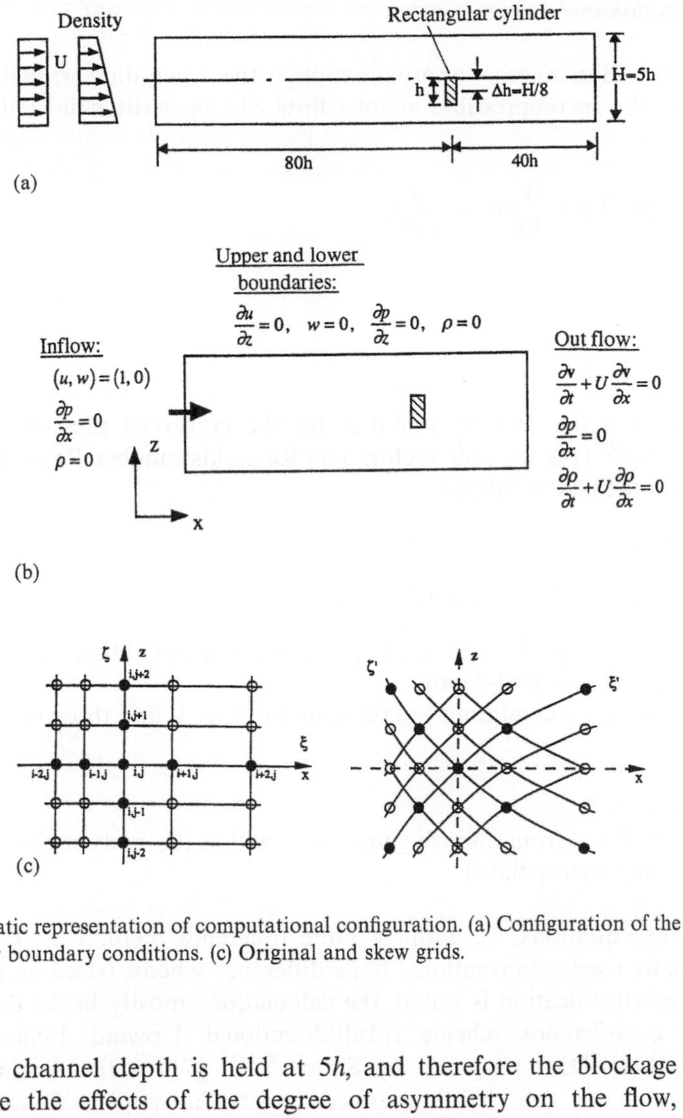

Fig. 1. Schematic representation of computational configuration. (a) Configuration of the non-symmetric case. (b) Outer boundary conditions. (c) Original and skew grids.

length. The channel depth is held at $5h$, and therefore the blockage ratio is 0.2. To examine the effects of the degree of asymmetry on the flow, three cases were chosen. The cases are specified by the displacement Δh from the mid-depth of the channel. As a basic case, the cylinder is settled at the mid-depth of the channel. This is referred to as the *symmetric case* hereafter ($\Delta h = 0$). In the second case, the cylinder is displaced slightly under the mid-depth and this case is called the *non-symmetric case* ($\Delta h = 0.125H$). The last situation is an extreme case in which the cylinder is mounted on the floor, i.e., the *surface-mounted case* ($\Delta h = 0.4H$). The parameter K ranges from 0 to 3.0; the Froude number is from 0.53 to infinity. The Reynolds number based on h was held at 10^3.

2.2. Discretization and solution

Under the Boussinesq approximation, the two-dimensional governing equations for the incompressible viscous flow can be written in a dimensionless form as

$$\frac{\mathrm{D}\boldsymbol{v}}{\mathrm{D}t} = -\operatorname{grad} p + \frac{1}{\mathrm{Re}}\Delta\boldsymbol{v} - \frac{\rho}{\mathrm{Fr}^2}\hat{\imath}, \tag{2}$$

$$\operatorname{div}\boldsymbol{v} = 0, \tag{3}$$

$$\frac{\mathrm{D}\rho}{\mathrm{D}t} = w, \tag{4}$$

where $\boldsymbol{v} = (u, w)$ is the velocity, p and ρ are the perturbed pressure and density, respectively, $\hat{\imath} = (0, 1)$ is the unit vector. The Reynolds number Re and the Froude number Fr are defined as follows:

$$\mathrm{Re} = Uh/\nu,$$

$$\mathrm{Fr} = U/Nh, \quad N^2 = -(g/\rho_0)(\mathrm{d}\rho'/\mathrm{d}z),$$

where g is the gravitational acceleration, ρ' is the dimensional basic density, and ρ_0 is the mean density at the mid-depth.

The outer boundary conditions are presented in Fig. 1, and those on the body are as follows:

velocity: $(u, w) = 0$,
pressure: calculated from imposing $(u, w) = 0$ within the body region,
density: linearly extrapolated.

To solve the equations, we adopt a finite difference method. In a preliminary computation, we used a conventional finite difference scheme based on a single-grid system. But as stratification is raised, the calculations mostly broke down. Hence, we applied a difference scheme (Multidirectional Upwind Finite Difference Method, i.e., MUFDM) proposed by Suito et al. [7] to the spatial term and hence, attained highly stratified flows. According to that paper, this scheme allows the spatial terms to conserve their leading error terms irrespective of the rotation of the coordinate. The reason why this method works well for highly stratified flows is not clear. The more intensified the stratification is, the finer motions are possibly produced away from the obstacle caused by the action of internal waves. The MUFDM is based on a linear combination of two grids with different directions. Thus, comparing the number of points around a specific grid point used for the approximation to spatial derivatives with that in the conventional single grid, we can point out that the scheme in the MUFDM incorporates more information. This feature might be advantageous for the calculation of highly stratified flows.

An original grid is overlaid with a skew one. In the computational plane, the skew grid is rotated with respect to the original one by 45° as presented in Fig. 1. Two coordinate transformations are introduced:

$$x = x(\xi, \zeta), \quad z = z(\xi, \zeta), \tag{5}$$

$$x = x(\xi', \zeta'), \quad z = z(\xi', \zeta'), \tag{6}$$

where (ξ, ζ) and (ξ', ζ') denote the computational plane variables of the original and skew grid, respectively. As long as a rectangular grid is used, a single argument is adequate for the coordinate transformation (i.e., $x = x(\xi)$, $z = z(\zeta)$). But to avoid difficulties in coding, two arguments are commonly used herein. All the spatial terms are discretized in accordance with each transformation rule on each grid. Then the terms from the original and skew grids are weighted with a ratio of 2 : 1, respectively.

The procedure to solve the equations is based on the marker-and-cell (MAC) method [8]. The space derivatives except for the advective terms are approximated with the second-order central differences. The advective terms are approximated with a third-order upwind scheme. A Poisson equation is solved with the successive over relaxation (SOR) method. The time evolution equations are solved with the second-order Runge–Kutta method. Three grids are used, whose size is 201 (horizontal direction) × 71 (vertical). They are clustered towards the body surface to resolve finer-scale fluid motions. The time increment is set to 0.002.

3. Results and discussion

Fig. 2 shows flow patterns for the symmetric case. For the neutral flow ($K = 0$), the well-known von Kármán vortex street is formed in the wake. The vortex shedding survives up to a value of $K = 1.7$, although the wake cavity seems to be suppressed. At $K = 1.8$, however, the vortex shedding almost disappears and coincidentally troughs due to lee waves are discernible downstream at $x \sim 15h$. As K increases further beyond $K = 1.8$, the wavelength of the lee waves gradually shortens in accordance with linear theory. As shown unambiguously for $K = 2.5$, the wake cavity is impeded by the lee waves symmetrically.

Flow patterns for the non-symmetric case are shown in Fig. 3. For the neutral flow ($K = 0$), the vortex street is formed naturally. When K reaches 1.1, a lee wave can be clearly seen with the first trough far downstream at $x \sim 20h$. At $K = 1.4$, the lee wave grows in amplitude and its wavelength becomes sufficiently short so that the first trough of the lee wave blocks the wake cavity, and hence, it is biased towards the lower boundary. Then the vortex shedding seems to be totally inhibited. Note that at $K = 2.5$, the symmetric mode of the lee waves can be seen.

Fig. 4 shows the streamlines for the surface-mounted case. Since such a configuration introduces no interaction between separated shear layers, clear-cut characteristics of the internal waves may be expected. For $K < 1.0$, the separation bubble is shortened with K, but the lee wave is not formed yet. From $K = 1.0$, evidently the lee

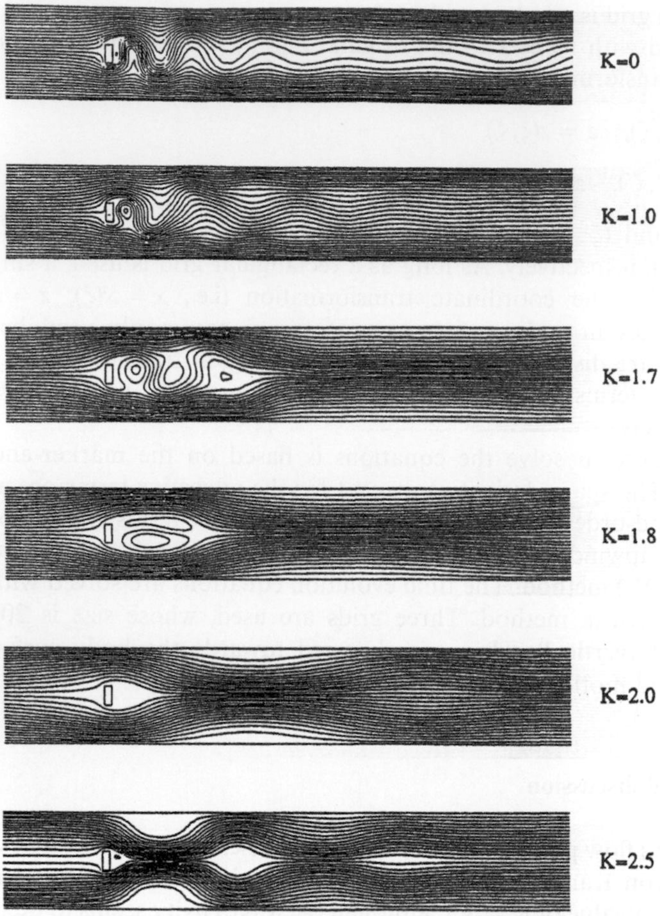

Fig. 2. Streamlines for the symmetric case: $Ut'/h = 100$.

wave is created. Although for larger K somewhat unsteady motions can be seen, the lee waves persist distinguishably as ever.

Figs. 5–7 show the perturbed streamlines at a dimensionless time $t = Ut'/d = 100$, where t' is the dimensional time. The perturbed streamlines can be obtained from the flow field in a frame of reference fixed to the approaching flow. In the symmetric case (Fig. 5), as K becomes greater than 2.0, disturbances with elongated closed streamlines like vortices advance upstream. This is thought to be the columnar disturbances, which is a manifestation of the long internal wave which causes an almost horizontal motion relative to the basic flow. Unlike the symmetric case, it is for K greater than 1.0 that the upstream advancing columnar disturbances are created in the non-symmetric case (Fig. 6). Similarly, in the surface-mounted case (Fig. 7), the upstream columnar disturbance propagation commences immediately after K attains 1.0. These features

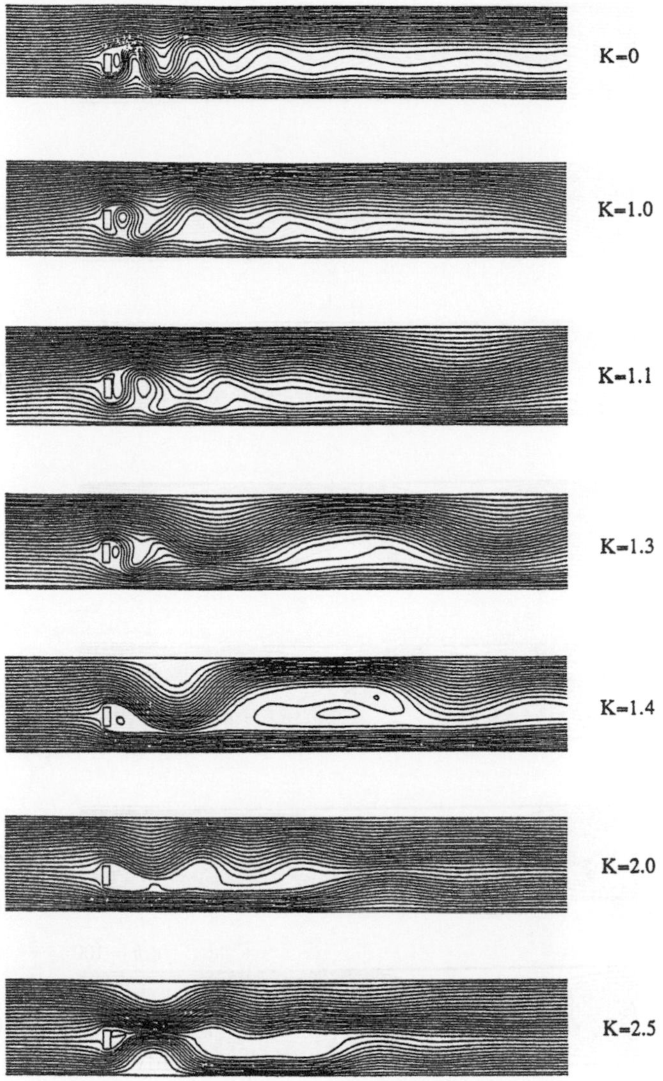

Fig. 3. Streamlines for the non-symmetric case: $Ut'/h = 100$.

indicate that in accordance with the symmetric or non-symmetric configuration, it is determined whether the dominant mode of the waves is symmetric or non-symmetric. In passing, the disturbance amplitude varies in space and the eddy-like disturbances are shed upstream from the body repeatedly as found by Hanazaki [3].

The upstream fronts of columnar disturbances are in good agreement with those predicted by formula (1) which are indicated by arrows in Figs. 5–7. Interestingly, comparing the locations of the fronts for the non-symmetric and surface-mounted

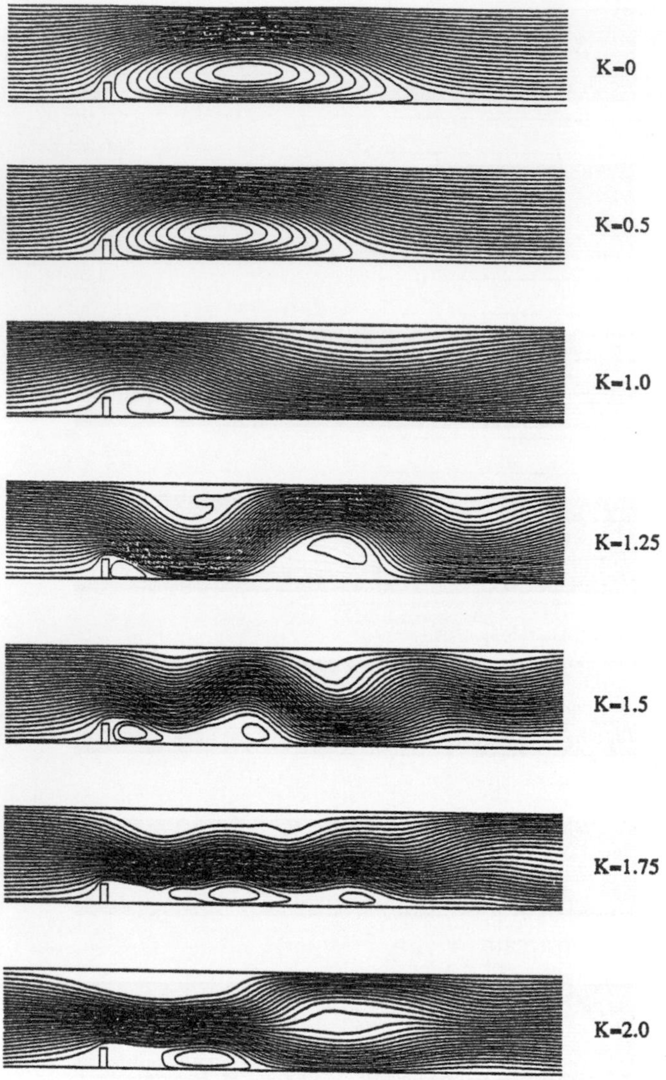

Fig. 4. Streamlines for the surface-mounted case: $Ut'/h = 100$.

cases of mode $n = 1$, we can see that the locations are almost alike for an equal K. These results give us the following understanding. Although the dominant mode of the disturbance is determined by the placement of the obstacle itself (i.e., whether the configuration is symmetric or not), once the disturbances of a certain mode are created, the wave speeds are not characterized by obstacle conditions, but by linear theory.

However, we should mention interesting exceptional cases. According to linear theory, the columnar disturbances of mode $n = 1$ cannot propagate upstream for

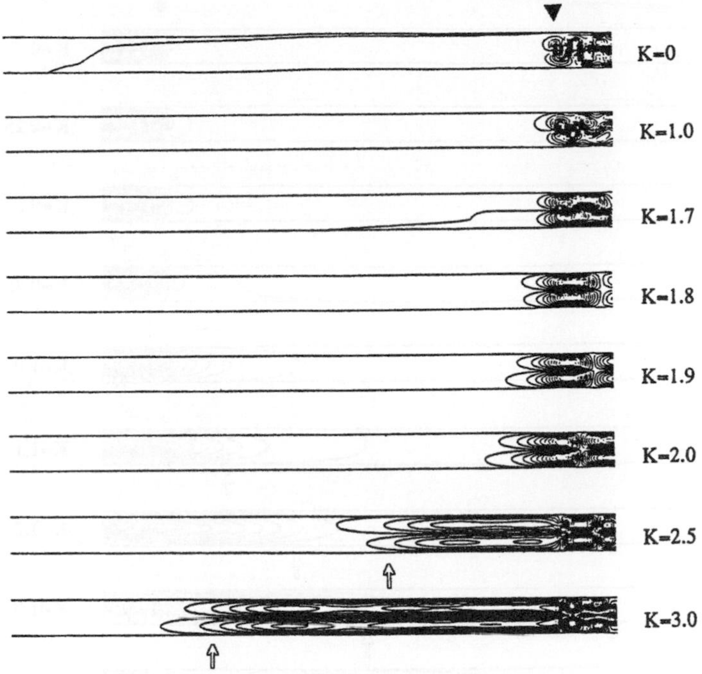

Fig. 5. Perturbation streamlines for the symmetric case: $Ut'/h = 100$. The arrows indicate the locations of the fronts of columnar disturbances of second mode predicted by linear theory. The full triangle shows the location of the cylinder.

$K \leqslant 1$ (see (1)). But in practice, certain disturbances do stretch upstream at $K = 1.0$ (so-called "resonance") in Figs. 6 and 7. Likewise, for the symmetric mode $n = 2$, they cannot propagate upstream for $K \leqslant 2$ in theory, but as shown in Fig. 5, their slight upstream extensions are discernible at $K = 1.8$ and 1.9.

The variation of Strouhal number St with K is plotted in Fig. 8 along with the results from the experiment [6]. The parameter St is defined by fh/U, where f is the dominant frequency of the vortex shedding. For the symmetric case, St decreases with increasing K up to 1.8. However, the vortex shedding is totally inhibited when K exceeds 1.8, so it is unable to define the Strouhal number. For the non-symmetric case, St decreases with K increasing up to 1.1. But with further increase in K, St increases again until the vortex shedding is inhibited at $K = 1.4$. This critical behavior of St is basically in good agreement with that of the laboratory experiment [5], although appreciable discrepancies can be seen for smaller K. Looking into the flow patterns at around $K = 1.1$ in Fig. 3, we can see that the lee wave grows significantly and its first trough approaches the wake cavity, and possibly the lee wave begins to affect the vortex formation. Thus, it should be reasonable to point out that the critical phenomenon in which St takes a minimum at around 1.1 suggests a qualitative change in the dynamics of vortex formation.

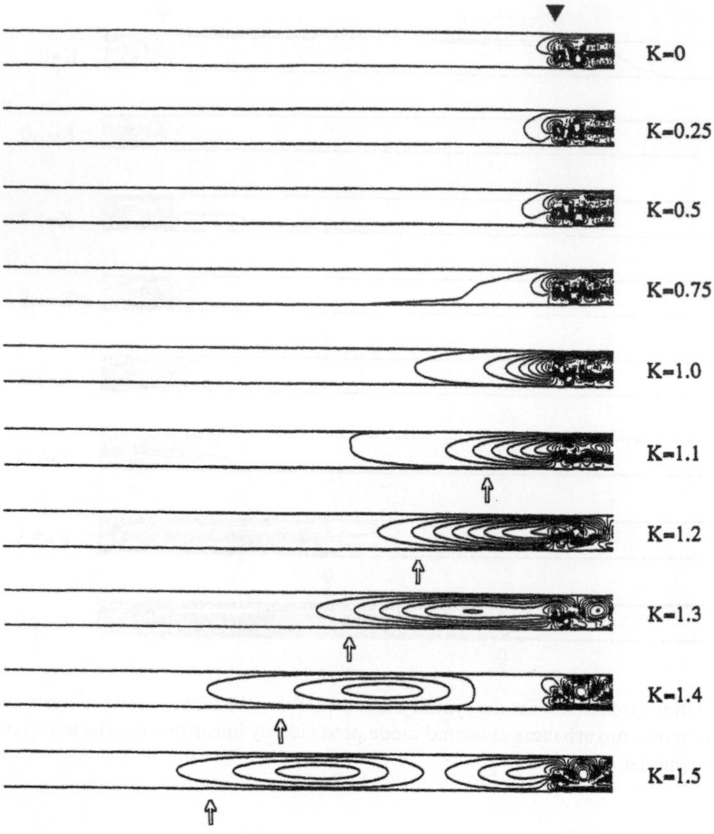

Fig. 6. Perturbation streamlines for the non-symmetric case: $Ut'/h = 100$. The arrows indicate the locations of the fronts of columnar disturbances of first mode predicted by linear theory. The full triangle shows the location of the cylinder.

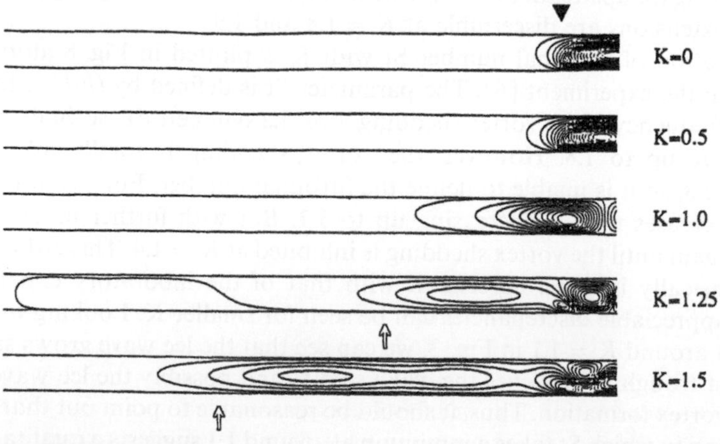

Fig. 7. Perturbation streamlines for the surface-mounted case: $Ut'/h = 100$. See Fig. 6 for legend.

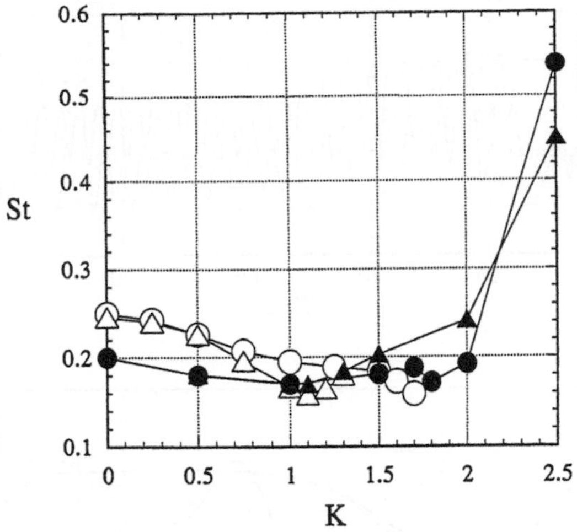

Fig. 8. Variation of Strouhal number with K: ○, symmetric case (numerical); △, non-symmetric case (numerical); ●, symmetric case (experimental); ▲, non-symmetric case (experimental). Experimental data taken from Ref. [5].

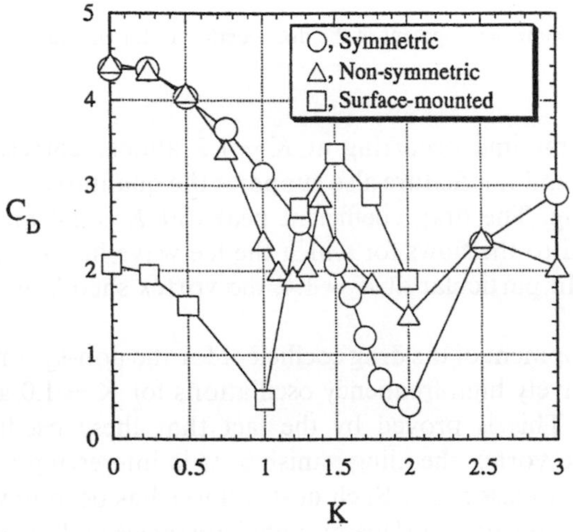

Fig. 9. Variation of the drag coefficient with K: ○, symmetric case; △, non-symmetric case; □, surface-mounted case.

In Fig. 9 the variation in the drag coefficient C_D with K is presented. Here, C_D is defined by $C_D = F_x/((\frac{1}{2})\rho_0 U^2 h)$, where F_x is the drag force. As for the surface-mounted case, C_D is reduced at integral values as predicted by linear theory [9]. Roughly speaking, this trend holds true for both the symmetric and non-symmetric situations. Specifically for the symmetric case, the drag decreases with K increasing up to 2.0. However, with further increase in K, C_D increases rapidly. For the non-symmetric

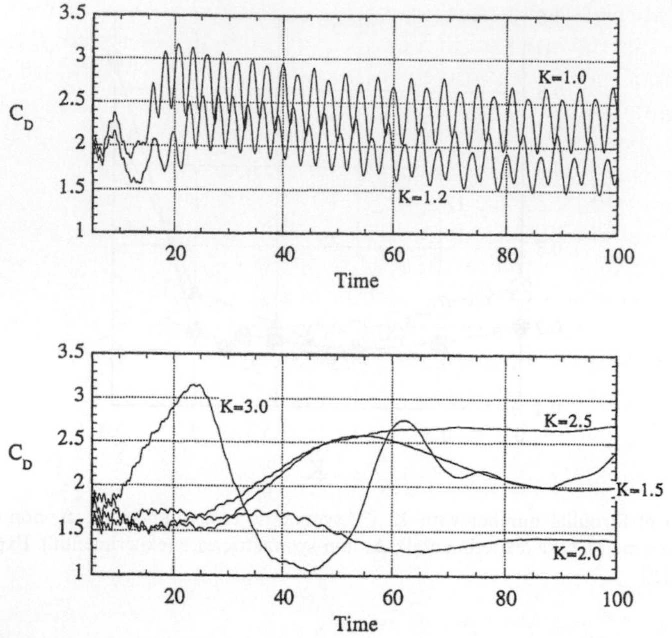

Fig. 10. Time development of the drag coefficient for the non-symmetric case.

case, one of the minima occurring at $K = 1.2$ almost corresponds to that of the Strouhal number and this feature also supports the qualitative change in the dynamics of vortex shedding. The drag coefficient peaks at $K = 1.4$ and $K = 2.5$, and these values correspond to the flows for which the lee wave has a very large amplitude as shown in Fig. 3. In particular at $K = 1.4$, the vortex shedding is totally inhibited by the lee waves.

The time development of the drag coefficient for the non-symmetric case is shown in Fig. 10. The relatively high-frequency oscillations for $K = 1.0$ and 1.2 are caused by vortex shedding. This is proved by the fact that these oscillations disappear for $K = 1.5$, when the vortex shedding vanishes. It is interesting that for $K = 1.5$ some long-period unsteadiness exists. Such unsteadiness has been nowadays a major topic in the study of the flow over surface-mounted obstacles and several explanations were proposed [10,11]. These features of the time development of the drag coefficient hold for the symmetric case.

4. Remarks on a previous experimental result

In our previous laboratory experiment [5], the circular cylinder was supposed to be placed at the mid-depth of the channel. As K increases from zero, the vortex shedding was intensified up to $K \sim 1.0$. From a certain value of $K \sim 1$, however, the roll-up of the separated shear layers was critically suppressed and the position of their interaction shifted farther downstream in the wake. Although this flow change may be tied to the generation of the internal waves, the odd integer of 1.0 is not consistent with the symmetric configuration, because $K = 1$ is a critical value beyond which the first vertical mode appears. For the symmetric case, it is natural that the internal waves of the vertical symmetric (second) mode should occur from $K = 2$.

A plausible explanation for this apparent inconsistency is that the experimental arrangement probably has some imperfections such as the placement of the circular cylinder, and the velocity and/or density in the approaching flow. Then some of them may have driven the non-symmetric first mode in combination, at $K \sim 1.0$. In support of this explanation, Figs. 5 and 6 clearly show that introducing asymmetry into the flow configuration causes the columnar disturbance of first mode at $K \sim 1.0$ unlike the "symmetric case". To confirm it, further studies using a circular cylinder will be necessary.

5. Conclusions

A finite difference simulation was conducted to investigate the two-dimensional flow of a linearly stably-stratified Boussinesq fluid around a rectangular cylinder in a channel of finite depth. To attain a highly stratified flow, a difference scheme combining two grids was applied. The Reynolds number was taken to be sufficiently high so that the vortex shedding from the cylinder can be formed. The main results obtained are as follows:

(1) In the symmetric case, the lee waves occur from $K = 1.8$ close to 2, but in the non-symmetric case, the lee waves appear from K close to 1. The perturbed streamlines evidently reveal that the columnar disturbances evolve upstream from the same values of K. The dominant waves for the symmetric and non-symmetric cases are found to correspond to the second and first vertical mode of internal waves, respectively. Although the dominant mode is determined by the placement of the cylinder itself (i.e., whether the configuration is symmetric or not), once the disturbance of a certain mode is created, the wave speed is not characterized by the cylinder conditions, but by the linear theory.

(2) A critical change in the Strouhal number is observed, shortly after appearance of the lee wave, which probably influences the dynamics of vortex formation. The likely flow alteration is thought to occur when the lee waves grow in amplitude and the wavelength becomes sufficiently short so that the wake cavity is suppressed/biased by the first troughs of the lee waves.

References

[1] J.V. Droughton, C.F. Chen, The channel flow of a density-stratified fluid about immersed bodies, Trans. ASME D 94 (1) (1972) 122–130.

[2] D.L. Boyer, P.A. Davies, H.J.S. Fernando, X.-H. Zhang, Linearly stratified flow past a horizontal circular cylinder, Phil. Trans. Roy. Soc. London A 328 (1989) 501–528.

[3] H. Hanazaki, Upstream advancing columnar disturbances in two-dimensional stratified flow of finite depth, Phys. Fluids A 1–12 (1989) 1976–1987.
[4] S.N. Wei, T.W. Kao, H.P. Pao, Experimental study of upstream influence in the two-dimensional flow of a stratified fluid over an obstacle, Geophys. Fluid Dyn. 6 (1975) 315.
[5] Y. Ohya, Y. Nakamura, Near wakes of a circular cylinder in stratified flows, Phys. Fluids Lett. A 2–4 (1990) 481–483.
[6] Y. Ohya, N. Aota, unpublished.
[7] H. Suito, K. Ishii, K. Kuwahara, Simulation of dynamic stall by multi-directional finite-difference method, 26th AIAA Fluid Dynamic Conf. 19–22 1995, San Diego, CA, 1995.
[8] F.H. Harlow, J.E. Welch, Numerical calculation of time-dependent viscous incompressible flow of fluid with free surface, Phys. Fluids 8 (1965) 2182–2189.
[9] K. Trustrum, An Oseen model of the two-dimensional flow of a stratified fluid over an obstacle, J. Fluid Mech. 50 (1971) 177–188.
[10] I.P. Castro, W.H. Snyder, P.G. Baines, Obstacle drag in stratified flow, Proc. Roy. Soc. London A 429 (1990) 119–140.
[11] T. Uchida, Y. Ohya, S. Ozono, A numerical study of stably stratified flows over a two-dimensional hill – Part I. Free-slip condition on the ground, these Proceedings, J. Wind Eng. Ind. Aerodyn. 67&68 (1997) 493–506.

ELSEVIER

Journal of Wind Engineering
and Industrial Aerodynamics 67&68 (1997) 117–127

JOURNAL OF
wind engineering
AND
industrial
aerodynamics

A critical study on the influence of far field boundary conditions on the pressure distribution around a bluff body

Siva Parameswaran[a,*,1], Ramesh Andra[b,2], Richard Sun[b,3], Mark Gleason[b,4]

[a] *Texas Tech University, Mechanical Engineering Department, Lubbock, TX79409, USA*

[b] *Chrysler Co., Aerodynamics Department, MI, USA*

Abstract

A computational model is developed to help the automotive design engineer to optimize the body shape with minimum wind tunnel testing. Unsteady, Reynolds-averaged, Navier–Stokes equations are solved numerically by a finite-volume method and applied to study the flow around a $\frac{3}{8}$th scale model of 1994 Intrepid. The standard k–ε model is employed to model the turbulence in the flow. The finite volume equations are formulated in a strong conservative form on a three-dimensional, unstructured grid system. The resulting equations are then solved by an implicit, time marching, pressure-correction based algorithm. The steady state solution is obtained by taking sufficient time steps until the flow field ceases to change with time within a prescribed tolerance. For the pressure-correction equation, preconditioned conjugate gradient method is employed to obtain the solution. Numerical predictions were obtained with two different boundary conditions at the far field: (a) no flow across this boundary (b) the gradient of any variable normal to this boundary was set to zero. Drag predictions obtained with boundary condition (b) was in good agreement with the available experimental data.

1. Introduction

Turbulent flow past bluff bodies forms on important class of flow problems which has many practical applications ranging from wind engineering to aerodynamics of automobile. However, numerical simulation for this class of problems becomes difficult due to the geometry of the body and the lack of ability of the standard turbulence models to simulate the complex turbulent structure that exists inside the body-wake.

*Corresponding author.

[1] Associate Professor.

[2] Engineer.

[3] Technical Specialist.

[4] Supervisor.

PII S0167-6105(97)00067-6

In the present study, an attempt is made to predict the salient features of the turbulent flow around a three-dimensional bluff body in ground proximity.

Parameswaran et al. [1] employed k–ε turbulence model to predict the flow around a two dimensional bluff body in ground proximity. Encouraged by the success of the methodology to predict mean quantities, the method is extended to three dimensions. The numerical algorithm remains the same as in Ref. [1] except for few modifications. Similar to earlier work, Cartesian velocity components are solved at the cell center from momentum equations implicitly. At the cell faces, the velocity projection parallel to the line connecting the adjacent grid nodes is solved explicitly from momentum balance to avoid the checker-board type pressure oscillations. The finite-difference equations for the velocity components, pressure and turbulent quantities are solved in a sequential manner at every time step with a modified version of the SIMPLE algorithm, originally developed by Patankar and Spalding [2]. The solution for the pressure-correction equation is obtained with a preconditioned conjugate gradient method described by Meijernik and van der Vorst [3] while all the other variable are solved by Stones' [4] strongly implicit procedure.

In the present study, computational results were obtained for a $\frac{3}{8}$th scale model of 1994 Intrepid, see Ref. [5]. The experimental data were provided by the Aerodynamics and Fluid Dynamics department of Chrysler corporation. In order to resolve the pressure distribution around the body more accurately without employing a huge grid, a smaller computational domain is assumed, i.e., the computational boundaries were not extended up to the slotted-walls of the wind tunnel. Computations were performed with two different boundary conditions at the top and side far-field boundaries.

In the next section, the differential equations governing the mean flow and the turbulence quantities are presented in a general non-orthogonal coordinate system. The numerical algorithm is then described followed by a section on computational results.

2. Mathematical formulation

2.1. Governing equations

The differential equations governing the unsteady, incompressible, turbulent flow in three space dimensions can be expressed in Cartesian tensor form as follows:

$$\text{Continuity}: \quad \frac{\partial u_i}{\partial x_i} = 0. \tag{1}$$

$$\text{Momentum}: \quad \frac{\partial u_i}{\partial t} + \frac{\partial}{\partial x_j}(u_i u_j) = -\frac{1}{\rho}\frac{\partial p}{\partial x_i} + \frac{\partial}{\partial x_j}\left\{\nu_{\text{eff}}\left(\frac{\partial u_i}{\partial x_j} + \frac{\partial u_j}{\partial x_i}\right)\right\}, \tag{2}$$

Here the effective kinematic viscosity, ν_{eff}, is taken as the sum of the molecular and turbulent viscosities as given by

$$\nu_{\text{eff}} = \nu + \nu_{\text{t}}. \tag{3}$$

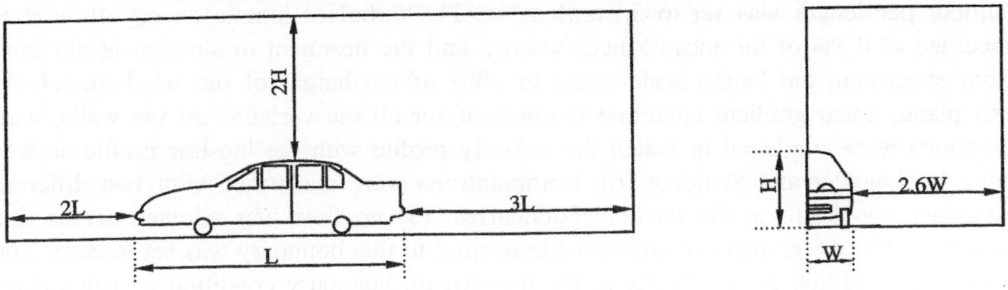

Fig. 1. Computational domain.

2.2. The standard k–ε model of turbulence

The k–ε turbulence model of Launder and Spalding [6] is used in the present study. In this model, the turbulent eddy viscosity is determined from the turbulence kinetic energy k and its dissipation rate ε according to:

$$\nu_t = C_\mu k^2/\varepsilon. \tag{4}$$

The model is composed of two equations, one for k and the other for ε as described below:

$$\frac{\partial k}{\partial t} + \frac{\partial}{\partial x_j}(u_j k) = \frac{\partial}{\partial x_j}\left(\nu + \frac{\nu_t}{\sigma_k}\frac{\partial k}{\partial x_j}\right) + G - \rho\varepsilon, \tag{5}$$

$$\frac{\partial \varepsilon}{\partial t} + \frac{\partial}{\partial x_j}(u_j \varepsilon) = \frac{\partial}{\partial x_j}\left(\nu + \frac{\nu_t}{\sigma_\varepsilon}\frac{\partial \varepsilon}{\partial x_j}\right) + \frac{\varepsilon}{k}(C_1 G - C_2\varepsilon), \tag{6}$$

where

$$G = \nu_t \frac{\partial u_i}{\partial x_j}\left(\frac{\partial u_i}{\partial x_j} + \frac{\partial u_j}{\partial x_i}\right). \tag{7}$$

The empirical constants were assigned their usual values, i.e.,

$$[C_1, C_2, C_\mu, \sigma_k, \sigma_\varepsilon] = [1.44, 1.92, 0.09, 1.0, 1.22].$$

2.3. Boundary conditions

The cross-sectional views of the computational domains are shown in Fig. 1. The inlet plane is located about two body lengths in front of the model and the exit plane is located about three body lengths behind the model. Only one half of the wind tunnel configuration is considered in the computations because of symmetry. The computational domain consists of an inlet boundary, a ground plane, a symmetry plane, an out-flow boundary and two far-field boundaries where the velocities are not known as a priori. At inlet plane, the boundary layer thickness is assumed to be zero and a uniform velocity profile is prescribed. The velocity component parallel to the wind tunnel was set to 1 m/s and the other two components were set to zero. The Reynolds number per length was set to 3.2 million/m. The turbulent kinetic energy at inlet is specified at 0.3% of the mean kinetic energy and the turbulent dissipation is obtained from specifying the length scale equal to 10% of the height of the wind tunnel. At exit plane, linear gradient condition is imposed for all the variable. At the walls, wall functions were employed to match the velocity profile with the log-law profile as was done by Launder and Spalding [6]. Computations were performed with two different boundary conditions at the far-field boundaries: (a) no flow was allowed across this boundary, (b) the gradient of any variable normal to this boundary was set to zero. The boundary condition (b) is similar to the free-stream boundary condition in a boundary layer flow.

2.4. Governing equations in a general coordinate system

The differential equations (1), (2), (5) and (6) have to be expressed in a general, non-orthogonal coordinate system (ξ, η, ζ), so that the boundary conditions at the desired locations could be easily implemented. If ϕ denotes a general scalar variable, the differential equation for the conservation of ϕ in a general ξ–η–ζ coordinate system can be expressed as follows:

$$\frac{1}{J}\frac{\partial}{\partial t}(J\phi) + \frac{1}{J}\frac{\partial}{\partial \xi_j}[(u_k \cdot \beta_k^j) \cdot \phi] = \frac{1}{J}\frac{\partial}{\partial \xi_k}\left[\frac{\Gamma_\phi}{J}(g^{kl})\frac{\partial \phi}{\partial \xi_1}\right] + S_\phi, \tag{8}$$

where $\xi_j (j = 1, \ldots, 3) = \xi, \eta, \zeta$, J is the Jacobian, β_i^j is the area tensor associated with the transformation, and $g^{kl} = \beta_j^k \beta_j^1$ (summation on the j). Γ_ϕ and S_ϕ are the associated diffusivity and source function for the variable ϕ. The detail expressions of the source terms for each variable is given in Appendix A. The area tensor β_i^j and the Jacobian J are given as follows:

$$\beta_i^j = \begin{pmatrix} y_\eta z_\zeta - y_\zeta z_\eta & y_\zeta z_\xi - y_\xi z_\zeta & y_\xi z_\eta - y_\eta z_\xi \\ z_\eta x_\zeta - z_\zeta x_\eta & z_\zeta x_\xi - z_\xi x_\zeta & z_\xi x_\eta - z_\eta x_\xi \\ x_\eta y_\zeta - x_\zeta y_\eta & x_\zeta y_\xi - x_\xi y_\zeta & x_\xi y_\eta - x_\eta y_\xi \end{pmatrix}, \tag{9}$$

$$J = x_\xi y_\eta z_\zeta + y_\xi x_\zeta z_\eta + z_\xi x_\eta y_\zeta - x_\xi y_\zeta z_\eta - x_\eta y_\xi z_\zeta - z_\xi x_\zeta y_\eta, \tag{10}$$

where subscripts ξ, η and ζ denote the derivatives along the coordinates.

3. Method of computation

3.1. General discretized equation

The method of solution is of the finite-volume variety, in which numerical solutions to the transport equation of the preceding section are obtained by dividing the solution domain into a finite number of discrete volumes or 'cells' and determining the numerical values of the dependent variables at the centers of these cells. The steady state solution is obtained by taking sufficient time steps until the solution remains unchanged. For temporal discretization, a time linearized, implicit scheme is selected so that large time steps could be taken without causing numerical instability as described by Issa [7]. Spatially, the standard upwind differencing scheme described in the work of Spalding [8] is employed to treat the nonlinear convective terms, while the diffusive terms are approximated by the central differencing scheme. Eq. (8) is integrated over the control volume for the node P surrounded by nodes E, W, N, S, T and B, to yield the following finite volume equation for ϕ:

$$\begin{aligned}
&\rho \mathrm{Vol}_P \frac{(\phi_P^{n+1}-\phi_P^n)}{\Delta t} + (F_1^n\phi^{n+1})_\mathrm{E} - (F_1^n\phi^{n+1})_\mathrm{W} \\
&\quad + (F_2^n\phi^{n+1})_\mathrm{N} - (F_2^n\phi^{n+1})_\mathrm{S} \qquad + (F_3^n\phi^{n+1})_\mathrm{T} - (F_3^n\phi^{n+1})_\mathrm{B} \\
&= \left(\frac{a^2\Gamma_\phi}{\mathrm{Vol}}\right)_\mathrm{E}(\phi_\mathrm{E}^{n+1}-\phi_P^{n+1}) - \left(\frac{a^2\Gamma_\phi}{\mathrm{Vol}}\right)_\mathrm{W}(\phi_P^{n+1}-\phi_\mathrm{W}^{n+1}) \\
&\quad + \left(\frac{a^2\Gamma_\phi}{\mathrm{Vol}}\right)_\mathrm{N}(\phi_\mathrm{N}^{n+1}-\phi_P^{n+1}) - \left(\frac{a^2\Gamma_\phi}{\mathrm{Vol}}\right)_\mathrm{S}(\phi_P^{n+1}-\phi_\mathrm{S}^{n+1}) \\
&\quad + \left(\frac{a^2\Gamma_\phi}{\mathrm{Vol}}\right)_\mathrm{T}(\phi_\mathrm{T}^{n+1}-\phi_P^{n+1}) - \left(\frac{a^2\Gamma_\phi}{\mathrm{Vol}}\right)_\mathrm{B}(\phi_P^{n+1}-\phi_\mathrm{B}^{n+1}) + S_\phi \mathrm{Vol}_P, \qquad (11)
\end{aligned}$$

where F_1, F_2 and F_3 are the convective fluxes across the east, north and top cell faces. a the area of any of the six cell faces. Vol the volume of a general cell.

$$\phi_\mathrm{E}^{n+1} = \begin{cases} \phi_P^{n+1} & \text{if } F_1^n > 0, \\ \phi_\mathrm{E}^{n+1} & \text{if } F_1^n < 0, \end{cases} \qquad \phi_\mathrm{N}^{n+1} = \begin{cases} \phi_P^{n+1} & \text{if } F_2^n > 0, \\ \phi_\mathrm{n}^{n+1} & \text{if } F_2^n < 0, \end{cases}$$

$$\phi_\mathrm{T}^{n+1} = \begin{cases} \phi_P^{n+1} & \text{if } F_3^n > 0, \\ \phi_\mathrm{T}^{n+1} & \text{if } F_3^n < 0. \end{cases}$$

In Eq. (11), the quantities $\phi_\mathrm{E}^{n+1}, \phi_\mathrm{W}^{n+1}, \phi_\mathrm{N}^{n+1}, \phi_\mathrm{S}^{n+1}, \phi_\mathrm{T}^{n+1}$ and ϕ_B^{n+1} are calculated according to the upwind differencing scheme. The superscripts n and $n+1$ refer to the old and the new time levels. The subscript, in general, refers to the location in the physical plane. F_1, F_2 and F_3 are kept in the old time level. To avoid pressure oscillation, the practice adopted in Ref. [1] is extended to three dimensions. For completeness, this practice is briefly described in the next section.

3.2. Pressure-correction equation

The convective fluxes, F_1, F_2 and F_3 at the cell faces are computed from the velocity projections along the coordinate direction at the cell faces. All velocity projections except the one which is not parallel to the cell face are computed from the Cartesian components stored at the nodes either side of the face. The velocity projection in the direction connecting the adjacent nodes is computed from momentum balance. This procedure is described in detail in Ref. [1]. Here, only the final form of the pressure correction equation is given as:

$$p'_P\left[\frac{a_E^2}{A_E}+\frac{a_N^2}{A_N}+\frac{a_T^2}{A_T}+\frac{a_W^2}{A_W}+\frac{a_S^2}{A_S}+\frac{a_B^2}{A_B}\right]$$

$$= p'_E\frac{a_E^2}{A_E}+p'_N\frac{a_N^2}{A_N}+p'_T\frac{a_T^2}{A_T}+p'_W\frac{a_W^2}{A_W}+p'_S\frac{a_S^2}{A_S}+p'_B\frac{a_B^2}{A_B}+\varepsilon_P, \tag{12}$$

where A_E, A_N, A_T, A_W, A_S, and A_B are the averaged momentum coefficients at the cell faces derived from the momentum coefficients at the cell center and ε_P is the continuity error associated with the cell P is given by

$$\varepsilon_P = F_{1,W} - F_{1,E} + F_{2,N} - F_{2,S} + F_{3,T} - F_{3,B}. \tag{13}$$

3.3. *Solution procedure*

The overall solution is similar to that of Issa [7]. The following steps are carried out at each time step:

(1) Eq. (11) is solved to yield the intermediate velocity field u^*, v^* and w^* with the existing pressure field.

(2) The new fluxes F_1^*, F_2^* and F_3^* are computed from the velocity projections at the cell faces. At each face, the velocity projection along the direction connecting the node is calculated from momentum balance.

(3) The continuity error, ε_P is calculated for each cell from the newly calculated fluxes F_1^*, F_2^*, F_3^* and the pressure correction equation is assembled and solved to yield a new pressure field p^*.

(4) A new velocity field u^{**}, v^{**} and w^{**} is obtained from Eq. (11) by replacing the old pressure field with the newly calculated p^* field.

(5) A new set of fluxes, F_1^{**}, F_2^{**} and F_3^{**} are calculated from the new pressure field p^* and the velocity field u^{**}, v^{**} and w^{**}.

(6) The pressure correction equation is solved to yield an improved pressure field, p^{**}.

(7) Steps (4)–(6) are repeated until the momentum and the continuity equation are satisfied within the preset tolerance.

(8) A new time step is taken.

3.4. *Computational mesh*

A body fitted grid system is generated algebraically to subdivide the computational domain. Unlike in Ref. [9], the computational mesh was generated by ICEM®. Numerical computations were performed with two different grid sizes: (a) 49 812 cells (b) 99 624 cells. The predicted drag coefficient for the coarser grid was 10% higher than the one obtained with the finer grid. The grid size could not be increased beyond 100 000 due to hardware limitations. All the results presented hereafter are obtained with

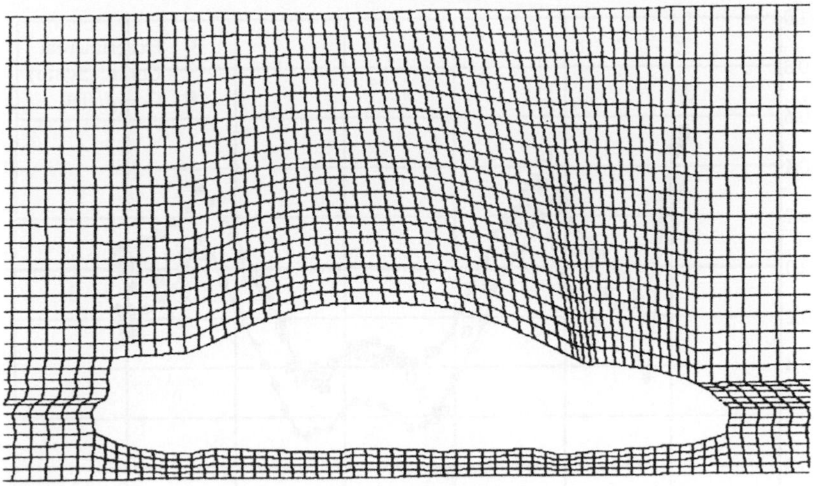

Fig. 2. The grid along the symmetry plane.

the 100 000-cell grid. The grid in the symmetry plane is depicted in Fig. 2 for the computations with 99 624 cells.

The Indigo 2 workstation from Silican Graphics was used as the computational engine. Calculations for 100 000 cell grid required about eight million words of memory, about 300 time steps to reach the steady state and about 20 h of CPU time on the SGI workstation.

4. Results and discussion

The complete geometry description, experimental conditions and the pressure measurements for the Intrepid model was provided by Chrysler Corporation, see Ref. [5]. The IGES description of the model was imported into the workstation and the grid was constructed with the aid of ICEM.

Computed pressure distributions along the symmetry plane of the upper body are compared with the measurements in Fig. 3. It is clear from this figure that the pressure distribution obtained with the gradient boundary condition (hereafter will be referred to as BC2) agreed better with measurements than the one obtained from the "zero flow" boundary condition (hereafter will be referred to as BC1). Pressures obtained using BC2 correctly matches the experimental values near the grill area and also matches the peak pressure near the cowl area. Computed pressures differ significantly from the experimental values near the backlight area. The drag coefficient of 0.29 obtained with boundary condition BC2 agreed better with the experimental value of 0.30. The drag coefficient obtained with boundary condition BC1 was 0.42. Similar high drag coefficients were obtained with BC1 for other bluff bodies also, see Ref. [9]. By imposing

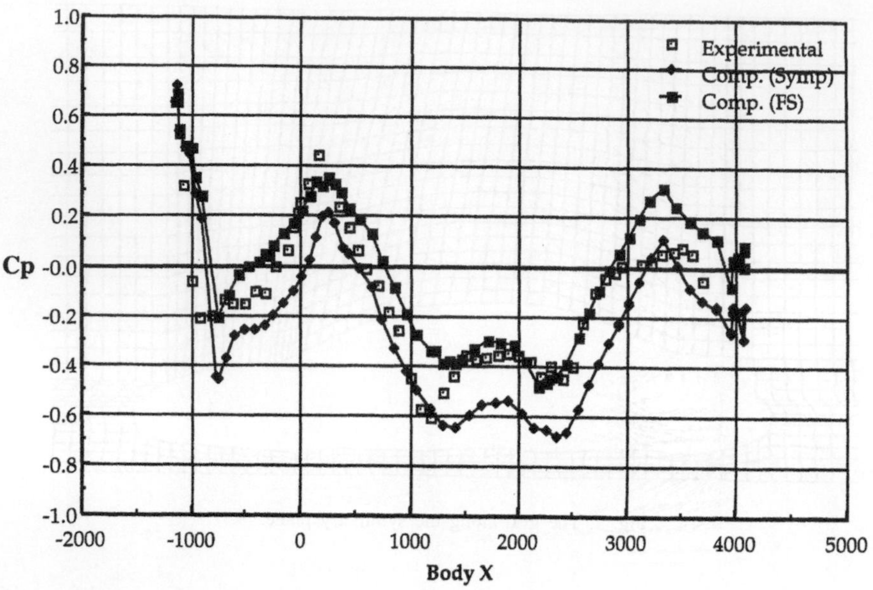

Fig. 3. Pressure distribution along the top surface in the symmetry plane.

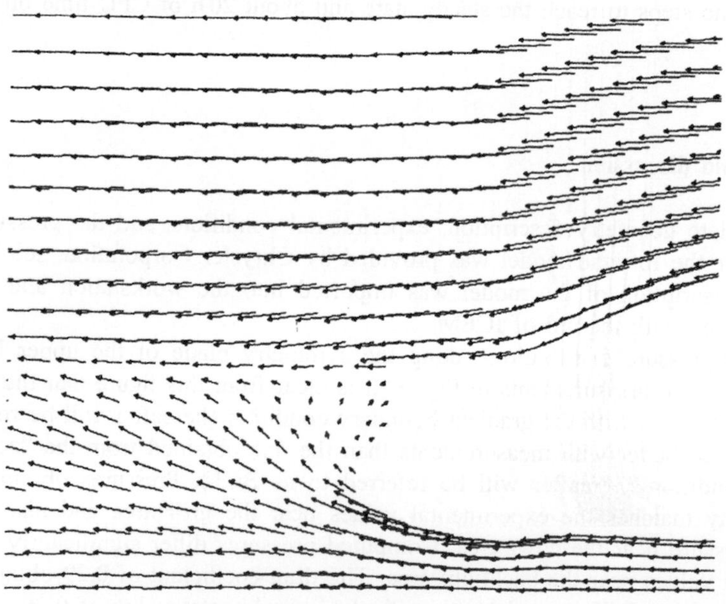

Fig. 4. Velocity vectors near the wake area.

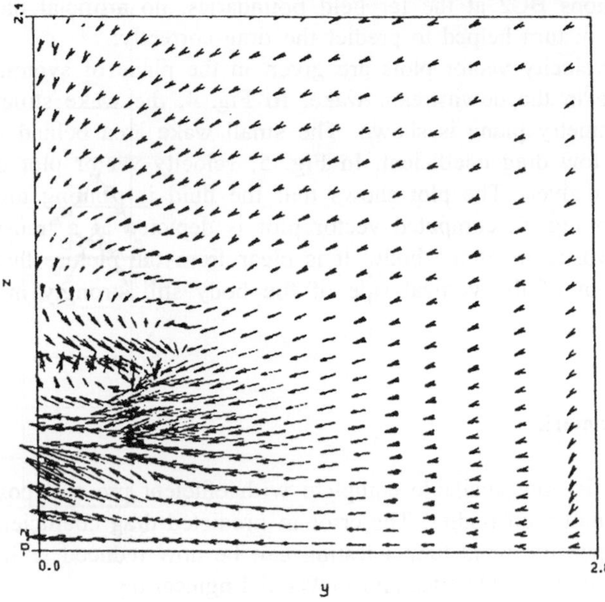

Fig. 5. Velocity vectors in a transverse plane immediately behind the body.

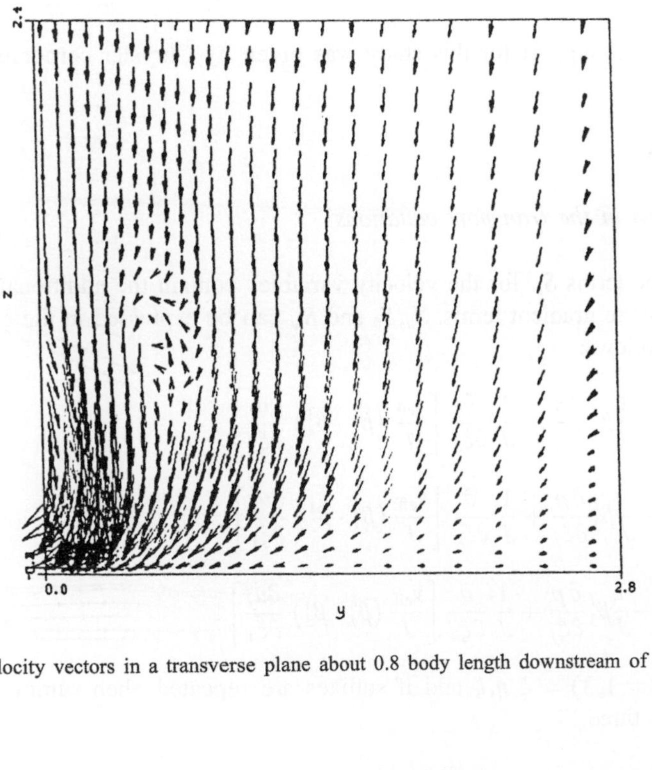

Fig. 6. Velocity vectors in a transverse plane about 0.8 body length downstream of the body.

boundary conditions BC2 at the far field boundaries, no artificial blockage effect is introduced. This in turn helped to predict the drag correctly.

In Figs. 4–6 velocity vector plots are given in the plane of symmetry and in two transverse planes in the downstream wake. In Fig. 4, the wake structure behind the body at the symmetry plane is shown. The small wake area behind the body points correctly to the low drag coefficient. In Fig. 5, velocity vector plot immediately behind the body is given. The plot shows that the fluid is running towards the separation bubble. In Fig. 6, computed vector plot is depicted at a transverse plane 0.8 body length downstream of the body. It is clear from the picture that the horseshoe vortex coming out of the vertical side of the body still strongly interacts with the wake.

5. Concluding remarks

A design tool is now available complete with efficient pre and post processors to analyze flow around bluff bodies. The error in predicted drag coefficient was less than four percent. Expensive wind tunnel testing can be now reduced with the aid of this new tool in Automotive industries and in Wind Engineering.

Acknowledgements

The financial support for this study was given by Chrysler corporation.

Appendix A

Source terms in the transport equations

The source terms S_ϕ for the velocity variables contain the additional diffusion terms and the pressure gradient terms. S_u, S_v and S_w can be expressed in the ξ–η–ζ coordinate system as follows:

$$S_u = -\frac{1}{J}\beta_1^j \frac{\partial p}{\partial \xi_j} + \frac{1}{J}\frac{\partial}{\partial \xi_k}\left[\frac{\nu_{\text{eff}}}{J}\left(\beta_j^k \cdot \beta_1^1\right)\frac{\partial u_j}{\partial \xi_1}\right], \tag{A.1}$$

$$S_v = -\frac{1}{J}\beta_2^j \frac{\partial p}{\partial \xi_j} + \frac{1}{J}\frac{\partial}{\partial \xi_k}\left[\frac{\nu_{\text{eff}}}{J}\left(\beta_j^k \cdot \beta_2^1\right)\frac{\partial u_j}{\partial \xi_1}\right], \tag{A.2}$$

$$S_w = -\frac{1}{J}\beta_3^j \frac{\partial p}{\partial \xi_j} + \frac{1}{J}\frac{\partial}{\partial \xi_k}\left[\frac{\nu_{\text{eff}}}{J}\left(\beta_j^k \cdot \beta_3^1\right)\frac{\partial u_j}{\partial \xi_1}\right], \tag{A.3}$$

where, $\xi_j (j = 1, 3) = \xi, \eta, \zeta$ and if suffixes are repeated, then summation is assumed from one to three.

References

[1] S. Parameswaran, A.B. Srinivasan, R.L. Sun, Numerical aerodynamic simulation of steady and transient flows around two-dimensional bluff bodies using the non-staggered grid system, Numer. Heat Transfer, Part A 21 (1992) 443–461.

[2] S.V. Patankar, D.B. Spalding, A calculation procedure for heat, mass and momentum transfer in three-dimensional parabolic flows, Int. J. Heat Mass Transfer 15 (1972) 1787–1806.

[3] J.A. Meijernik, H.A. van der Vorst, An iterative solution method for linear systems of which the coefficient matrix is a symmetric M-matrix, Math. Comput. 31 (1977) 148–162.

[4] H.L. Stone, Iterative solution of implicit approximations of multi-dimensional partial differential equations, SIAM J. Numer. Anal. 5 (1968) 530–545.

[5] M. Gleason, R. Sun, Private Communications, 1993.

[6] B.E. Launder, D.B. Spalding, The numerical calculation of turbulent flows, Comput. Meth. Appl. Mech. Eng. 8 (1974) 269–289.

[7] R.I. Issa, Solution of the implicitly discretized fluid flow equations by operator splitting, J. Comp. Phys. 62 (1986) 40–65.

[8] D.B. Spalding, A novel finite-difference formulation for differential expressions involving both first and second derivatives, Int. J. Numer. Meth. Eng. 4, pp. 551–562.

[9] S. Parameswaran, I. Kiris, R. Sun, M. Gleason, Flow structure around a 3D bluff body in ground proximity: A computational study, J. Wind Eng. Ind. Aerodyn. 46 & 47 (1993) 791–800.

ELSEVIER

Journal of Wind Engineering
and Industrial Aerodynamics 67&68 (1997) 129–139

JOURNAL OF wind engineering AND industrial aerodynamics

Finite element modelling of flow around a circular cylinder using LES

R. Panneer Selvam[1]

BELL 4190, University of Arkansas, Fayetteville, AR 72701, USA

Abstract

An implicit solution procedure to solve the Navier–Stokes equations using large eddy simulation (LES) and finite element method (FEM) to study flow around a circular cylinder is presented. The drag-crisis phenomena is simulated using a two-dimensional (2D) model. The computed drag and Strouhal numbers for Reynolds numbers (Re) of 10^4, 10^5, 5×10^5 and 10^6 are compared with available experimental and computational results. The drag-crisis trend is noted but the decrease in drag coefficient is not as much as reported in the wind tunnel measurement. The FEM model will be used to study flow around bridges.

Keywords: Computational fluid dynamics; Computational wind engineering; Large eddy simulation; Turbulence; Building loads

1. Introduction

Prediction of wind-induced motion on bridges and other structures using computational fluid dynamics (CFD) is possible due to development in computer hardware and software. In this work a finite element computer model using large eddy simulation is developed and validated with available experimental and computational work. In LES, the fluctuating motions of turbulence can be computed exactly except for eddies that are smaller than the grid size. The smaller eddies are modelled using eddy viscosity models. The LES based on the Smagorinsky's eddy viscosity model used by Murakami and his research group for wind engineering [1] is used in the modelling of drag crisis around a circular cylinder. Kakuda and Toskada [2], Kondo [3] and Tamura et al. [4] used the no-turbulence model to study the flow around a circular cylinder. Deng et al. [5] used Baldwin–Lomax and k–ε turbulence models based on eddy viscosity to study the same phenomena. Nomura and Jiravacharadet [6] and

[1] E-mail: rps@engr.uark.edu.

PII S0167-6105(97)00068-8

Song and Yuan [7] used LES to study the flow around a cylinder. Other than Song and Yuan [7] the other researchers reported their results on drag crisis. Except Tamura's [4] 3D computation the rest of the computed drag in the drag-crisis regime is higher than the wind-tunnel measurement.

Past researchers [2–6] used explicit procedures to solve the Navier–Stokes (NS) equations. Because of stability restrictions on the solution of NS equations; these procedures took enormous computer time. In this work an implicit procedure is investigated to solve the NS equations using LES. Using the implicit procedure one could compute the vortex shedding phenomena around a circular cylinder for 90 s or 9000 time steps in about an hour in SunSparc-20 for 2260 nodes. Hence, it is expected to be an useful tool in the future for the study of flow around bridges.

The finite element method (FEM) is preferred here because of its higher accuracy for the given grid as illustrated by Selvam in Ref. [8]. The solution of the pressure equation usually consumes more than 80% of the computer time. An efficient preconditioned conjugate gradient (PCG) solver is used here. This is possible because the pressure equations using FEM are symmetric.

Flow around a circular cylinder is investigated for Reynolds number Re ranging from 10^4 to 10^6 by solving the 2D NS equations. Here Re is defined as

$$\mathrm{Re} = Vd/\nu ,$$

where d is the diameter of the cylinder, V is the reference velocity and ν is the kinematic viscosity. The computed drag coefficient C_d and lift coefficient C_l are defined as

$$C_\mathrm{d} = F_\mathrm{d}/(0.5\rho V^2) ,$$

$$C_\mathrm{l} = F_\mathrm{l}/(0.5\rho V^2) ,$$

where F_d and F_l are the drag and lift force, and ρ is the density. An attempt is made to study the drag-crisis phenomena using different wall conditions. One is using the Van Driest damping function as tried by Nomura and Jiravachardet [6] and the other is to use the usual law of the wall boundary conditions. The computed C_d and C_l are compared with available experimental and computed results.

2. Computer modelling using large eddy simulation (LES)

2.1. Governing equations

In this work, the LES turbulence model is considered. The 2D and 3D equations for an incompressible fluid using the LES model in general tensor notation are as follows:

Continuity equation: $U_{i,i} = 0 ,$ (1)

Momentum equation: $U_{i,t} + U_j U_{i,j} = -(p/\rho + 2k/3)_{,i} + [(\nu + \nu_t)(U_{i,j} + U_{j,i})]_{,j} ,$ (2)

where $v_t = (C_s h)^2 (S_{ij}^2/2)^{0.5}$, $S_{ij} = U_{i,j} + U_{j,i}$, $h = (h_1 h_2 h_3)^{0.333}$ for 3D and $(h_1 h_2)^{0.5}$ for 2D and $k = (v_t/(C_k h))^2$

Empirical constants: $C_s = 0.15$ for 2D and 0.1 for 3D, and $C_k = 0.094$.

U_i, and p are the mean velocity and pressure respectively, k is the turbulent kinetic energy, v_t is the turbulent eddy viscosity, h_1, h_2, and h_3 are control volume spacing in the x, y, and z directions and ρ is the fluid density. Here area or volume of the element is used for the computation of h. The empirical constants used here are the values suggested by Murakami et al. [1]. Here a comma represents differentiation, t represents time and $i = 1$, 2 and 3 mean variables in the x, y and z directions.

To implement higher-order approximation of the convection term [8] the following expression is used in Eq. (2) instead of $U_j U_{i,j}$:

$$U_j U_{i,j} - \frac{\theta}{2} (U_j U_k U_{i,j})_{,k} \,. \tag{3}$$

Depending on the values of θ different procedures can be implemented. For balance tensor diffusivity (BTD) scheme $\theta = \delta t$ is used; where δt is the time step used in the integration. For the streamline upwind procedure suggested by Brooks and Hughes [9] $\theta = 1/\max(|U_1|/\mathrm{d}x, |U_2|/\mathrm{d}y, |U_3|/\mathrm{d}z)$. Here d$x$, d$y$ and dz are the control volume length in the x, y and z directions. In this computation a value of 0.4 times the expression suggested by Brooks and Hughes for quadrilateral elements is used.

2.2. *Finite element scheme to solve NS equations*

The equations are solved using an implicit method similar to that of Choi and Moin [10] and applied for nonstaggered, node centered, control volume procedure by Selvam in Ref. [11]. The four step advancement scheme for Eqs. (1) and (2) is as follows:

Step 1: Solve for U_i from Eq. (2). The diffusion and higher-order convection terms are considered implicitly to be at the current time and the first-order convection terms are considered explicitly from the previous time step. The pressure is considered in the right-hand side of the equation. This equation leads to a symmetric matrix and the PCG procedure is used for the solution.

Step 2: Get new velocities as $U_i^* = U_i + \delta t(p_{,i})$ where U_i is not specified.

Step 3: Solve for pressure from $(p_{,i})_{,i} = U_{i,i}^*/\delta t$.

Step 4: Correct the velocity for incompressibility: $U_i = U_i^* - \delta t(p_{,i})$ where U_i is not specified.

Step 2 eliminates the checkerboard pressure field when using equal-order interpolation for velocity and pressure in the case of FEM. Implicit treatment of the convective and diffusive terms eliminates the numerical stability restrictions. In this work the time step is kept for CFL (Courant–Frederick–Lewis) number less than one.

The above NS equations are solved by FEM procedure. The velocity and pressure are approximated using equal order interpolation. Hughes and his group [12] and de Sampaio [13,14] used equal-order interpolation and derived the pressure equation from the method of least squares. In this work a different approach which will not produce checkerboard pressure is introduced.

The variables U_i and p are discretized using the shape functions N, i.e.

$$U_i' = N\,U_i \quad \text{and} \quad P' = N\,P\,, \tag{4}$$

where N is a $1 \times n_e$ and $\mathbf{U}_i$ and P are $n_e \times 1$ matrices and n_e is the number of nodes for the element. Substituting Eq. (4) in Eq. (1) and using the regular Galerkine procedure the following matrix expressions are obtained:

$$MU_{i,t} + (D + V_1 + V_2)U_i + Q\,p = F\,, \tag{5}$$

where

$$M = \int_\Omega N^{\mathrm{T}} N \, \mathrm{d}\Omega, \qquad D = \int_\Omega N_{,j}^{\mathrm{T}}(v + v_t)N_{,j} \, \mathrm{d}\Omega,$$

$$V_1 = \int_\Omega N^{\mathrm{T}} u_j N_{,j} \, \mathrm{d}\Omega, \qquad V_2 = \int_\Omega N_{,i}^{\mathrm{T}}(\theta U_i U_j/2)N_{,j} \, \mathrm{d}\Omega, \qquad Q = \int_\Omega N^{\mathrm{T}} N_{,i} \, \mathrm{d}\Omega\,.$$

Here M, D, V_1, V_2, and Q are the mass, diffusion, convection due to first and second order and pressure matrices of size $n_e \times n_e$. F is the $n_e \times 1$ matrix which considers the given Neumann boundary conditions. The element matrices are integrated using 2-point Gauss quadrature initially. Later the geometric terms like the Jacobian and the determinant are computed using one-point integration and assumed constant in the integral. The rest of the terms are integrated exactly using analytical procedures. Because of this the solution can be achieved twice faster for 2D and thrice faster for 3D. The velocities are averaged in each element for integration. The turbulent viscosity is computed at the middle of each element. After approximating in time and assembling all the elements, the equations are solved in the following form:

$$A\,U_i^{n+1} = B\,, \tag{6}$$

where

$$A = (M/\delta t + V_2 + D), \qquad B = (M/\delta t - V_1)u_i^n - QP + F\,.$$

This becomes the step 1 of the implicit procedure. Next the corrected velocity $U_i{}^*$ are calculated using lumped mass matrix M_{L} as follows:

$$U_i^* = U_i + \delta t(QP)/M_{\mathrm{L}}\,.$$

Then the pressures are solved using the following equations:

$$KP = F\,, \tag{7}$$

where

$$K = \int_\Omega N_{,j}^{\mathrm{T}} N_{,j} \, \mathrm{d}\Omega, \quad \text{and} \quad F = -\left(\int_\Omega N^{\mathrm{T}} N_{,i} \, \mathrm{d}\Omega\right) U_i^*/\delta t\,.$$

After the pressures are solved, the velocities are corrected as follows:

$$U_i = U_i^* - \delta t(QP)/M_L.$$

2.3. Solution procedure and convergence criterion

Eqs. (6) and (7) are stored in a compact form as discussed in Refs. [15,16]. The equations are solved by preconditioned conjugate gradient (PCG) procedure. For Eq. (6) an underrelaxation factor of 0.7 is used. The iteration is done until the absolute sum of the residue of the equation reduces to 1×10^{-7} times the number of nodes for each time step. Usually the pressure and momentum equations take about 50 and 10 iterations for PCG solution. To run 2260 nodes for a nondimensional time of 90 takes less than an hour in the Sun Sparcstation 20. For most of the 2D work the time step may be around 0.01 s for CFL less than one.

2.4. Computational grid, boundary and initial conditions

The computational region and boundary conditions are shown in Fig. 1. The cylinder surface is no slip. Close to the wall the damping function tried by Nomura [6] and the law of the wall boundary condition are tried. The law of the wall is implemented by computing the frictional velocity u_* from the following relation:

$$u_s = 2.5u_* \ln(9y^+),$$

where $y^+ = yu_*/\nu$, y is the normal distance from the wall to the next grid point and u_s is the velocity along the wall. Knowing the u_*, the wall shear stress τ_w is computed from the relationship $\tau_w = -\rho u_*^2$ and projected along the x, y and z directions. Here it is assumed that the normal velocity and the normal stresses from the wall are zero. The

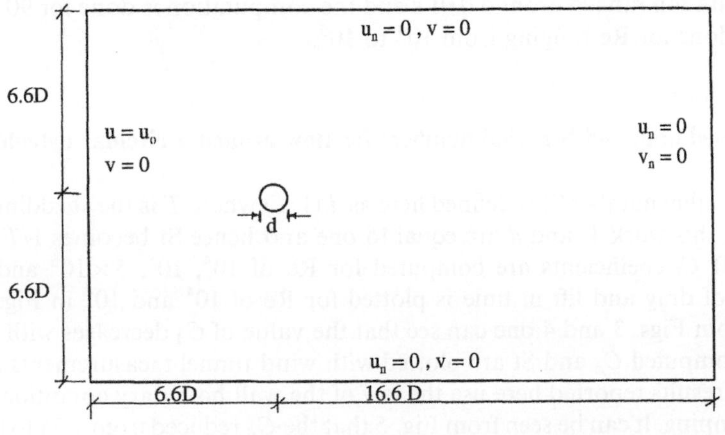

Fig. 1. Solution region and boundary conditions.

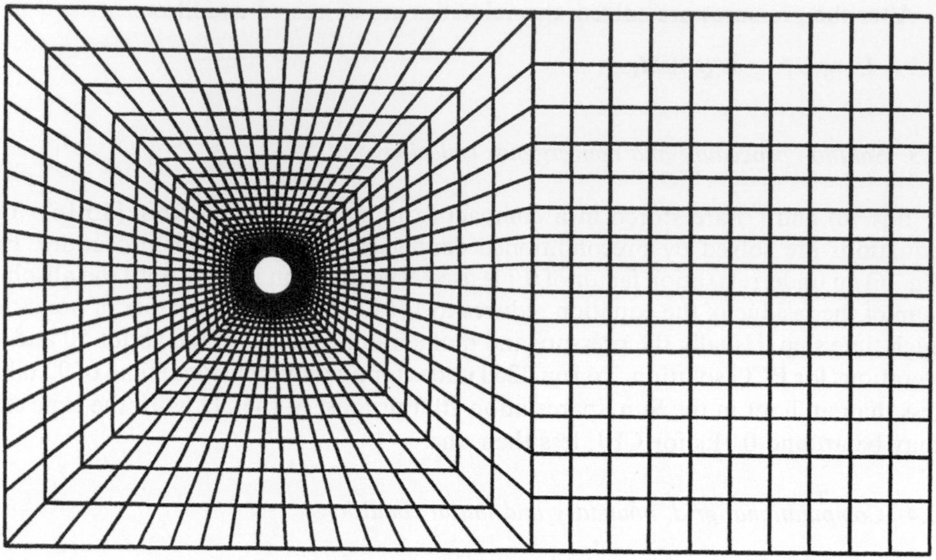

Fig. 2. Finite element mesh.

Van Driest damping function $W = 1 - \exp(-y^+/25)$ is multiplied by the eddy viscosity ν_t for the element close to the wall. If W is one, then ν_t is kept zero.

The upstream boundary has uniform velocities of one and zero in the x and y directions. The outflow boundary is the traction free or normal gradient of the velocities are zero and the side boundaries are slip boundaries. The finite element grid is shown in Fig. 2. The smallest spacing in the radial direction is kept as $0.002d$ where d is the size of the diameter. This spacing is close to $1/(10\sqrt{\text{Re}})$. For this grid 2260 nodes and 2190 four-noded quadrilateral elements are used. To start the solution a uniform velocity of one and zero in the x and y directions are assumed. The time step for computation ranged around 0.01 s and the computation is done for 90 s. Computation is done for Re ranging from 10^4 to 10^6.

3. Computed drag and Strouhal numbers for flow around a circular cylinder

The Strouhal number St is defined here as $d/(VT)$ where T is the shedding period of the lift. In this work V and d are equal to one and hence St becomes $1/T$. The drag C_d and lift C_l coefficients are computed for Re of 10^4, 10^5, 5×10^5 and 10^6. The variation of drag and lift in time is plotted for Re of 10^4 and 10^6 in Figs. 3 and 4. Clearly from Figs. 3 and 4 one can see that the value of C_d decreases with increasing Re. The computed C_d and St are plotted with wind tunnel measurements of Roshko [17]. The results reported here use the law of the wall boundary conditions and Van Driest damping. It can be seen from Fig. 5 that the C_d reduced from 1.34 to 0.77 for Re 10^4 to 10^6. The same trend is also noticed in Figs. 3, 4 and 6. But the computed C_d is

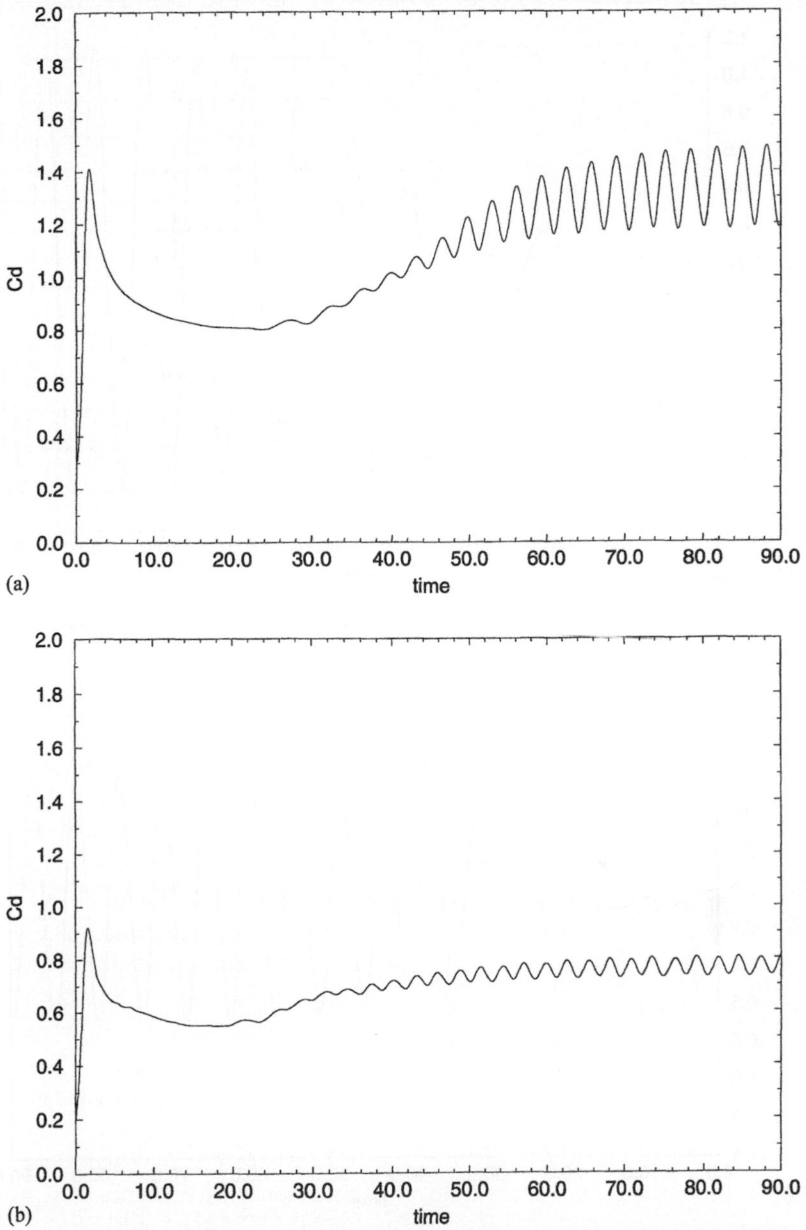

Fig. 3. Time variation of drag coefficient for Re of (a) 1×10^4 and (b) 1×10^6.

higher than the experimental value by about 0.4. In the same way the 1/St decreased from 6.41 s for Re of 10^4 to a value of 5.56 s for Re of 10^6. But the measured value is as much as 3 s as reported by Roshko. Hence, the trend is there but for drag-crisis computation a 3D computation is necessary as reported by Tamura [4]. The Van

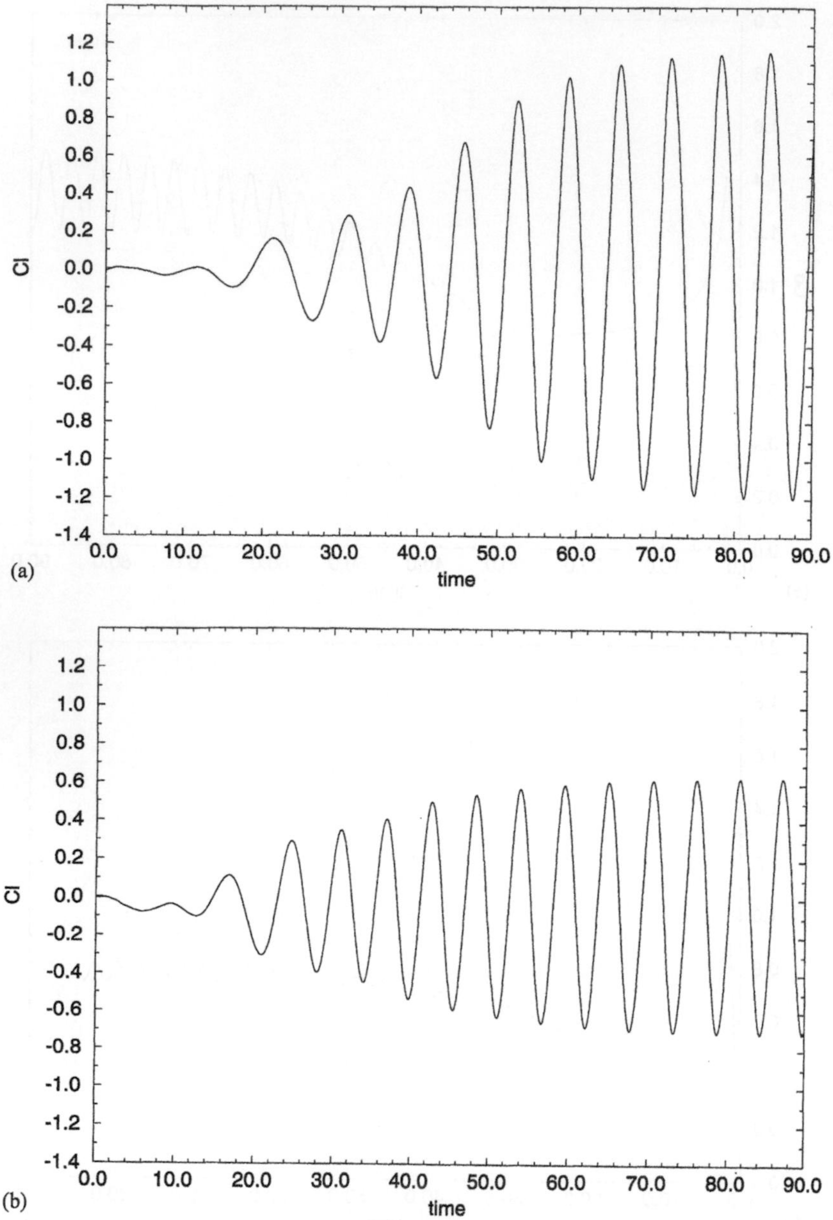

Fig. 4. Time variation of lift coefficient for Re of (a) 1×10^4 and (b) 1×10^6.

Driest wall function and the law of the wall condition on the wall produced similar effects. If special wall conditions are not introduced the drag and St do not have any change for change in Re. Work is under progress for the drag crisis study using 3D LES.

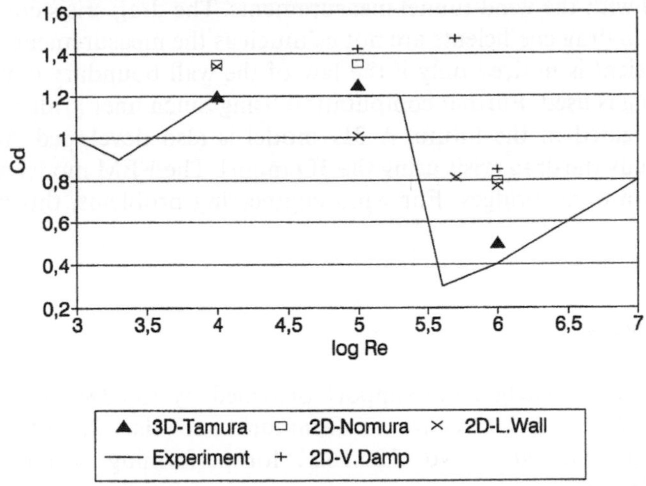

Fig. 5. Comparison of computed drag coefficients with experiment and computed results.

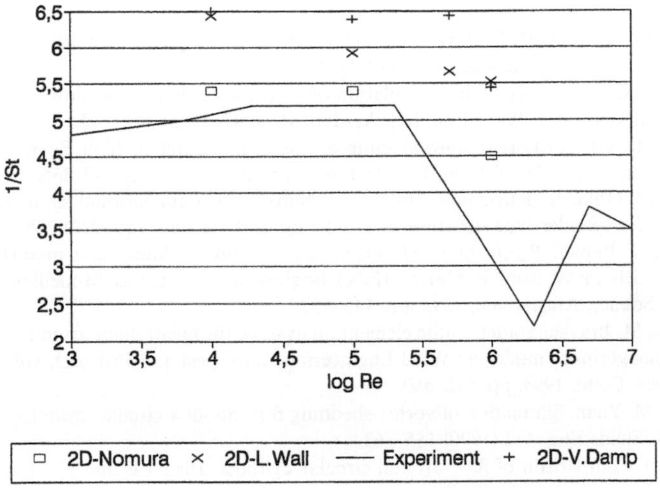

Fig. 6. Comparison of computed Strouhal numbers with experiment and computed results.

4. Conclusions

An efficient large eddy simulation model using FEM is presented. The upwind procedure and implicit solution techniques for LES seem to be a good tool. Currently, the model could solve about 9000 time steps for 2260 nodes in about an hour in SunSparc-20. The 2D LES model is used to study flow over a circular cylinder. The computed drag coefficient and Strouhal numbers for Re of 10^4, 10^5, 5×10^5 and 10^6

are compared with the wind tunnel measurements. The drag crisis could be seen but the reduction in drag coefficients are not as much as the measurement. The reduction in drag coefficient is noticed only if the law of the wall boundary condition or Van Driest damping is used. Further computation using much finer grids and their effects will be investigated in the future. A 3D model is also developed. Work is under progress to study the drag crisis using the 3D model. The FEM model will be used for studying flow around bridges. For wind engineering problems, this model may be a useful tool.

Acknowledgements

The author acknowledges the support provided by the Department of Energy Engineering, Technical University of Denmark and the Danish Technical Research Council under grant No. 5.26.16.31 for performing part of this work as a visiting professor.

References

[1] S. Murakami, A. Mochida, On turbulent vortex-shedding flow past 2D square cylinder predicted by CFD, J. Wind Eng. Ind. Aerdyn. 54/55 (1995) 191–211.

[2] K. Kakuda, N. Tosaka, Numerical simulation of high Reynolds number flows by Petrov–Galerkin finite element method, J. Wind Eng. Ind. Aerdyn. 46 & 47 (1993) 339–347.

[3] N. Kondo, Direct third-order upwind finite element simulation of high Reynolds number flows around a circular cylinder, J. Wind Eng. Ind. Aerdyn. 46 & 47 (1993) 349–356.

[4] T. Tamura, I. Ohta, K. Kuwahara, On the reliability of two-dimensional simulation for unsteady flows around a cylinder-type structure, J. Wind Eng. Ind. Aerdyn. 35 (1990) 275–298.

[5] G.B. Deng, J. Piquet, P. Queutey, M. Visonneau, Vortex-shedding flow predictions with eddy-viscosity models, in: W. Rodi, F. Martelli (Eds.), Engineering Turbulence Modelling and Experiments 2, Elsevier Science, Amsterdam, 1993, pp. 143–152.

[6] T. Nomura, M. Jiravacharadet, Finite element analysis of turbulent flows around a circular cylinder using the Smagorinsky model, in: Wind Engineering Retrospect and Prospect, vol. 2, Wiley Eastern Limited, New Delhi, 1994, pp. 572–580.

[7] C.C. Song, M. Yuan, Simulation of vortex-shedding flow about a circular cylinder at high Reynolds numbers, J. Fluids Eng. 112 (1990) 155–163.

[8] R.P. Selvam, Comparison of flow around circular cylinder using FE and FD procedures, in: S.K. Ghosh, J. Mohammadi (Eds.), Building an International Community of Structural Engineers, vol. 2, ASCE, New York, 1996, pp. 1021–1028.

[9] A. Brooks, T.J.R. Hughes, Streamline upwind/Petrov–Galerkin formulations for convection dominated flow with particular emphasis on the incompressible Navier–Stokes equations, Comput. Meth. Appl. Mech. Eng. 32 (1982) 199–259.

[10] H. Choi, P. Moin, Effects of the computational time step on numerical solutions of turbulent flow, J. Comput. Phys. 113 (1994) 1–4.

[11] R.P. Selvam, Computation of pressures on Texas Tech University building using large eddy simulation, these Proceedings, J. Wind Eng. Ind. Aerodyn. 67 & 68 (1997) 647–657.

[12] T.J.R. Hughes, L.P. Franca, M. Ballestra, A new finite element formulation for computational fluid dynamics: V. Circumventing the Babuska-Brezzi condition: A stable Petrov–Galerkin formulation of the Stokes problem accommodating equal-order interpolation, Comput. Meth. Appl. Mech. Eng. 59 (1986) 85–99.

[13] P.A.B. de Sampaio, A Petrov-Galekin formulation for the incompressible Navier–Stokes equations using equal order interpolation for velocity and pressure, Int. J. Num. Meth. Eng. 31 (1991) 1135–1149.

[14] P.A.B. de Sampaio, P.R.M. Lyra, K. Morgan, N.P. Weatherill, Petrov–Galerkin solutions of the incompressible Navier–Stokes equations in primitive variables with adaptive remeshing, Comput. Meth. Appl. Mech. Eng. 106 (1993) 143–178.

[15] R.P. Selvam, R.P. Elliott, A. Arounpradith, Pavement analysis using ARKPAV, in: S.C. Burns (Ed.), Proc. 45th Highway Geology Symposium, Portland, Oregon, 1994, pp. 241–249.

[16] R.P. Selvam, Implementation of preconditioned conjugate gradient solution procedure for large structural or finite element problems, Report, Department of Civil Engineering, University of Arkansas, Fayetteville, AR 72701, 1994, pp. 1–11.

[17] A. Roshko, Experiments on the flow past a circular cylinder at very high Reynolds number, J. Fluid Mech. 10 (1961) 345–356.

ELSEVIER

Journal of Wind Engineering
and Industrial Aerodynamics 67&68 (1997) 141–154

JOURNAL OF wind engineering AND industrial aerodynamics

Three-dimensional vortical flows around a bluff cylinder in unstable oscillations

T. Tamura*, Y. Itoh

Department of Environmental Physics and Engineering, Tokyo Institute of Technology, 4259 Nagatsuta, Midori-ku, Yokohama 226, Japan

Abstract

In order to investigate the aeroelastic instability of a rectangular cylinder in uniform flows, the three-dimensional unsteady flows around the oscillating cylinder with depth/breadth of 2.0 in heaving mode are computed by means of the direct finite difference method. First we confirm that the computed responses are in good agreement with experimental ones even in nonlinear regimes of vortex-induced oscillation and galloping. In case of unstable oscillations the wake structures are shown by the pressure or vorticity contours around the cylinder and their relations with the aerodynamic forces are examined. Also in view of flow structures and oscillation response we discuss the several stages of the onset of vortex-induced oscillation. As a result we can understand the physical mechanism of unstable phenomena of the rectangular cylinder.

PACS: 47.11.+j; 46.30.My
Keywords: Aeroelastic instability; Vortex-induced oscillation; Galloping; Rectangular cylinder; Three-dimensional flow simulation; Three-dimensional wake structures

1. Introduction

To predict the behavior of aerodynamic unstable oscillations of cylinder-type structures such as vortex-induced oscillation and galloping by means of numerical technique, it is important to deal with the three-dimensional governing equations of flows around an oscillating cylinder. By 3D computation we have succeeded in obtaining good results of the pressure distributions on the side surface of the rectangular cylinder with side ratio of 2.0 which is forced to oscillate transversely to the approaching uniform flows [1].

* Corresponding author. E-mail: tamura@depe.titech.ac.jp.

PII S0167-6105(97)00069-X

In this paper we also set the 3D computational model where the rectangular cylinder with the depth/breadth ratio of 2.0 is elastically mounted for heaving free oscillations in uniform flows. According to previous experiments we can successively find both the vortex-induced oscillation and the galloping due to the self-excited aerodynamic forces as the velocity of the approaching flows increases.

First the computed amplitude of oscillation is compared with the previous experimental one in order to confirm the accuracy of the predicted value by 3D computation. Next we investigate the three-dimensional structures of vortices in the wake of the oscillating rectangular cylinder. We also provide the physical understanding with regard to the unstable phenomena as a result of the interaction between the fluid flows and an elastically mounted rectangular cylinder.

2. Problem formulation

The interaction between the unsteady fluid flow and the dynamic behavior of a body plays an important role in deciding aerodynamic forces. Therefore the governing equations of fluid flows and the equation of motion are alternately solved at each time step.

The governing equations of the three-dimensional incompressible flow are given by the continuity equation and the Navier–Stokes equations as follows:

$$D = \operatorname{div} \boldsymbol{u} = 0, \tag{1}$$

$$\frac{\partial \boldsymbol{u}}{\partial t} + (\boldsymbol{u} \cdot \operatorname{grad})\, \boldsymbol{u} = -\operatorname{grad} p + \frac{1}{\mathrm{Re}} \Delta \boldsymbol{u} \tag{2}$$

where $\boldsymbol{u}$, p, t and Re denote the velocity vector, pressure, time and the Reynolds number, respectively, non-dimensionalized by U_0 (reference velocity), B (reference length), ρ (density) and ν (kinematic viscosity).

On the other hand the free-oscillation of a rectangular cylinder is governed by the equation of motion with one degree of freedom in heaving mode,

$$\frac{\mathrm{d}^2 y}{\mathrm{d}t^2} + \left(\frac{4\pi h}{V_\mathrm{r}}\right)\frac{\mathrm{d}y}{\mathrm{d}t} + \left(\frac{2\pi}{V_\mathrm{r}}\right)^2 y = \frac{\rho B^2 H}{2m} C_\mathrm{L}, \tag{3}$$

where y, t, h, V_r, H, m and C_L denote non-dimensional displacement, non-dimensional time, structural damping ratio, reduced velocity ($V_\mathrm{r} = U_0/Bf_0$, f_0 is the natural frequency of a body), spanwise length of the cylinder, mass of the cylinder and lift coefficient, respectively.

In order to compute the unsteady flows around a moving body, the original governing equations of fluid flow are transformed to the generalized coordinate system including time [2] and approximated by means of the finite difference method. The grid system of the physical domain moves with the motion of the body at each time step, although that of the computational domain is fixed. There is a relation between the computational domain (ξ, η, ζ, τ) and the physical domain (x, y, z, t) as follows:

$$\xi = \xi(x,y,z,t), \qquad \eta = \eta(x,y,z,t), \qquad \zeta = \zeta(x,y,z,t), \qquad \tau = t. \tag{4}$$

Here the time derivative is presented as

$$f_\tau = f_x x_\tau + f_y y_\tau + f_z z_\tau + f_t. \tag{5}$$

The Navier–Stokes equations in the computational domain are rewritten as follows:

$$\frac{\partial \boldsymbol{u}}{\partial \tau} + ((\boldsymbol{u} - \boldsymbol{X}_\tau) \cdot \mathrm{grad})\boldsymbol{u} = -\,\mathrm{grad}\, p + \frac{1}{\mathrm{Re}}\Delta \boldsymbol{u}, \tag{6}$$

where $\boldsymbol{X}_\tau$ stands for the moving velocity (x_τ, y_τ, z_τ) of the grid in the physical domain. Eq. (6) means that the convective velocity in the computational domain is replaced by the relative velocity of the physical domain. The continuity equation is not changed because no time derivative term is included.

The numerical procedures are based on the MAC method [3]. The regular mesh system is employed. All quantities are defined at the same location for a grid. The pressure field is determined by solving the Poisson equation as follows:

$$\Delta p = -\,\mathrm{div}(((\boldsymbol{u} - \boldsymbol{X}_\tau) \cdot \mathrm{grad})\boldsymbol{u}) - \frac{\partial D}{\partial \tau} + \frac{1}{\mathrm{Re}}\Delta D. \tag{7}$$

Providing that the effect of the viscous term is neglected at high Reynolds numbers and the velocity vector of the next time step must satisfy the divergence-free condition, the pressure equation is shown as follows:

$$\Delta p^{n+1} = -\,\mathrm{div}\,((\boldsymbol{u}^n - \boldsymbol{X}_\tau^n) \cdot \mathrm{grad})\,\boldsymbol{u}^n + \frac{D^n}{\delta\tau}, \tag{8}$$

where Δ, n and $\delta\tau$ are the Laplace operator, the time step and the non-dimensionalized time increment. The velocity field is computed by the semi-implicit temporal integration of the Navier–Stokes equations. The scheme is equivalent to the Euler backward scheme except the convection term:

$$\frac{\boldsymbol{u}^{n+1} - \boldsymbol{u}^n}{\delta\tau} + ((\boldsymbol{u}^n - \boldsymbol{X}_\tau^n) \cdot \mathrm{grad})\boldsymbol{u}^{n+1} = -\,\mathrm{grad}\, p^{n+1} + \frac{1}{\mathrm{Re}}\Delta \boldsymbol{u}^{n+1}, \tag{9}$$

where the convection term is linearized using the value at the previous time step for the convective velocity as follows:

$$(u - \boldsymbol{X}_\tau)^{n+1} \cdot \mathrm{grad}\, \boldsymbol{u}^{n+1} \doteqdot (\boldsymbol{u} - \boldsymbol{X}_\tau)^n \cdot \mathrm{grad}\, \boldsymbol{u}^{n+1}.$$

Eqs. (8) and (9) are, respectively, solved at each time step by the SOR method. In order to overcome the numerical instability at high Reynolds numbers, the third-order upwind scheme is incorporated for the convection terms [4]. No turbulence model is adopted. For example, the convection of a quantity f by U, x-component of the convection velocity, at node i is expressed as

$$\left(U\frac{\partial f}{\partial x}\right)_i = U_i \frac{-f_{i+2} + 8(f_{i+1} - f_{i-1}) + f_{i-2}}{12\delta x} + |U_i| \frac{f_{i+2} - 4f_{i+1} + 6f_i - 4f_{i-1} + f_{i-2}}{4\delta x}. \tag{10}$$

Eq. (3) is solved by the linear acceleration method. Considering that the rectangular cylinder has only one degree of freedom in heaving mode, the velocity $\boldsymbol{X}_\tau$ of the grid in the physical domain is given by $(0, y_\tau, 0)$. Therefore, Eq. (3) is expressed as follows:

$$\frac{\mathrm{d}^2 y}{\mathrm{d}\tau^2} + \left(\frac{4\pi h}{V_r}\right)\frac{\mathrm{d}y}{\mathrm{d}\tau} + \left(\frac{2\pi}{V_r}\right)^2 y = \frac{\rho B^2 H}{2m} C_L. \tag{11}$$

3. Computational model

The computational model is shown in Fig. 1. A rectangular cylinder with the depth/breadth ratio (D/B) of 2.0 is elastically mounted in uniform flows (U_0). The Reynolds number is equal to 10^4. The computational domain is a circular cylinder, whose diameter and height are $60B$ and B, respectively. The total number of mesh points is $280 \times 100 \times 10 = 280\,000$. For the boundary condition around the computational region, the Dirichlet condition is given for the velocity of inflow and the Neumann condition is imposed for the velocity of outflow and pressure. On the surface of the cylinder, the velocity of the moving grid at the boundary is given and the Neumann condition is employed for the pressure. The non-dimensionalized time increment is equal to 0.005.

The dynamic properties of a rectangular cylinder are controlled by the structural damping ratio and the mass ratio. The mass damping parameter called "Scruton number" is defined as follows:

$$\mathrm{Sc} = \frac{2m}{\rho BDH} 2\pi h.$$

In this paper the Scruton number (Sc) and the structural damping ratio (h) are equal to 2.0 and 0.005, respectively.

4. Aeroelastic behaviors of a rectangular cylinder

Several types of aeroelastic instabilities are recognized for a rectangular cylinder as a function of the reduced velocity (V_r). The time histories of heaving response at

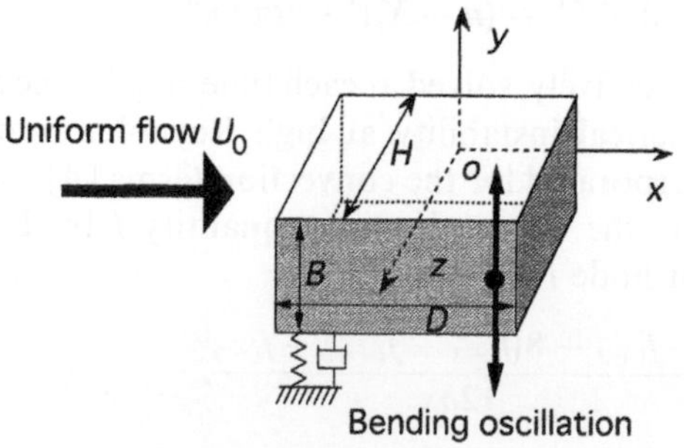

Fig. 1. Model.

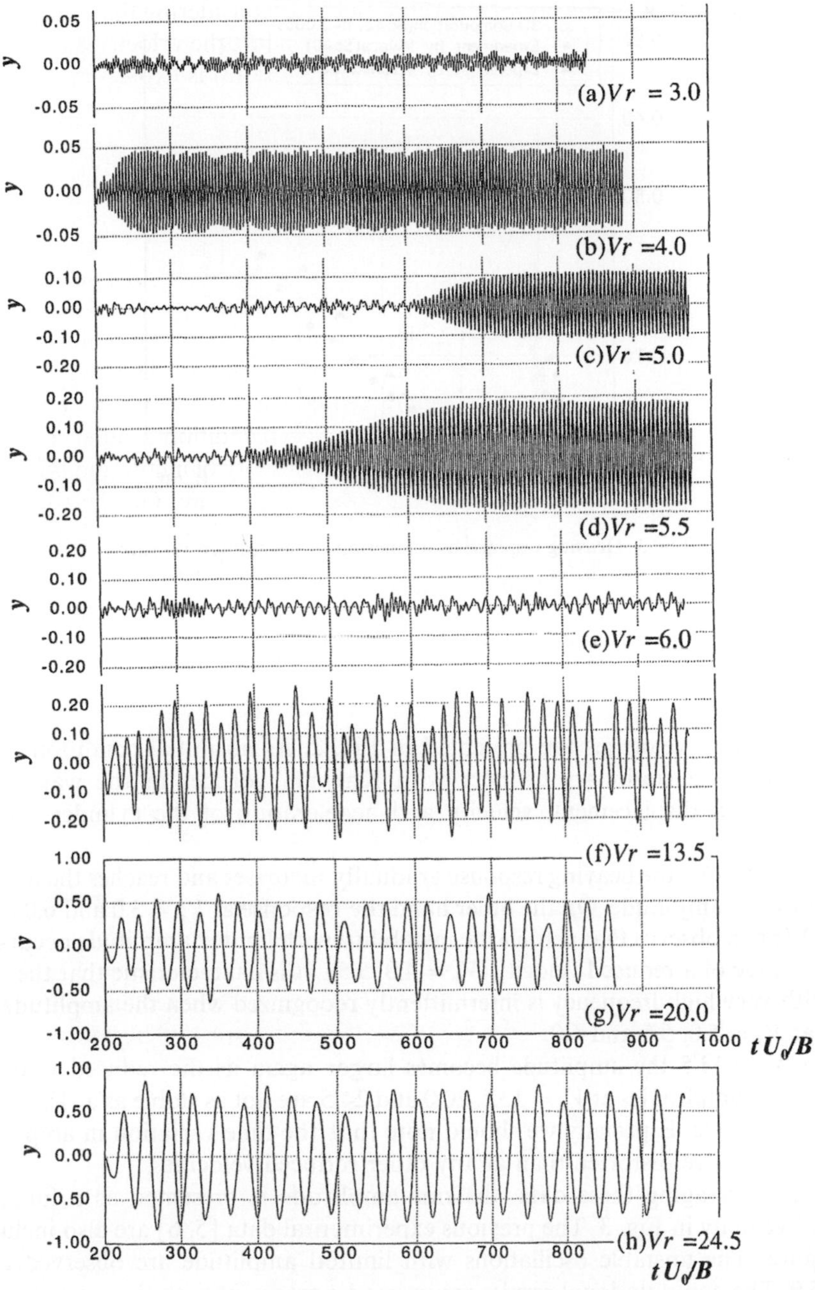

Fig. 2. Time histories of the heaving displacement.

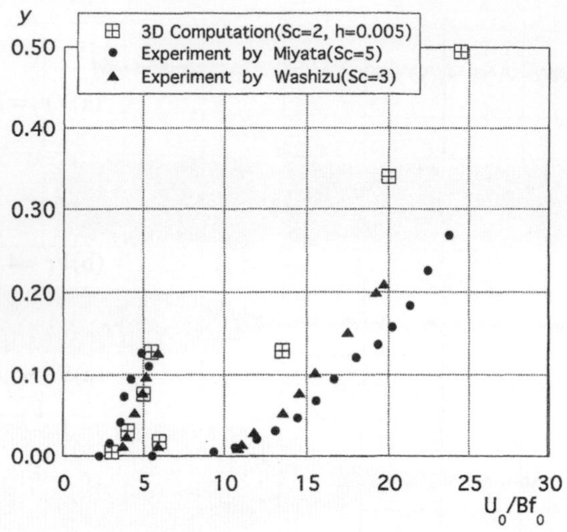

Fig. 3. Heaving response as a function of the reduced velocity.

various reduced velocities are presented in Fig. 2. As an initial condition the flow field around the stationary cylinder with Karman vortices in the wake is used and the initial displacement, velocity and acceleration of the cylinder are equal to 0.0.

At $V_r = 4.0$–5.5 the heaving response gradually increases and reaches the limit cycle with constant amplitude. On the other hand the response at $V_r = 3.0$ and 6.0 remains small. We can observe that aeroelastic instability with limited amplitude occurs in the limited range of a reduced velocity ($V_r = 4.0$–5.5). It is characteristic that the oscillation with very high frequency is intermittently recognized when the amplitude is still small at $V_r = 5.0$, 5.5 and 6.0.

Over $V_r = 13.5$ the amplitude becomes larger again as the reduced velocity increases. The amplitudes at $V_r = 13.5$, 20.0 and 24.5 are not as stable as at $V_r = 4.0$–5.5 in the limit cycle. Especially we should note that the time variation in amplitude at $V_r = 13.5$ is more distorted than at any other reduced velocities.

By time-averaging these data, the rms amplitude is presented as a function of reduced velocity in Fig. 3. The previous experimental data [5, 6] are also included in this figure. The unstable oscillations with limited amplitude are observed around $V_r = 5.0$. The computational results are in good agreement with the previous experimental ones in the vortex-induced oscillations. Over $V_r = 10.0$ the experimental amplitude becomes larger as the reduced velocity increases. In this regime the computed amplitude also varies with qualitatively the same manner as the experimental results in the galloping, but the values are a little larger.

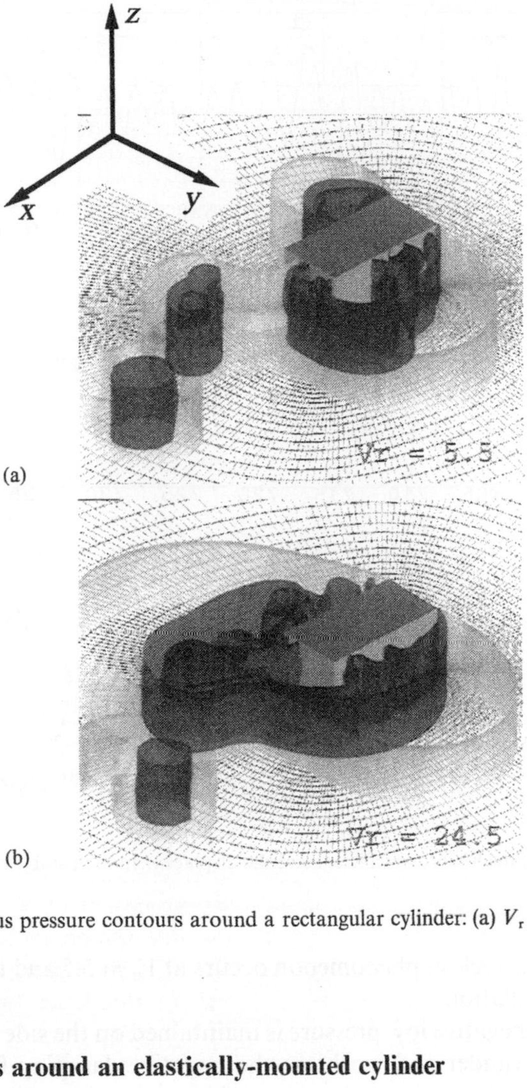

Fig. 4. Instantaneous pressure contours around a rectangular cylinder: (a) $V_r = 5.5$; (b) $V_r = 24.5$.

5. Wake structures around an elastically-mounted cylinder

In this section the physical mechanism of aeroelastic phenomena is discussed through the comparison among the flow structures, oscillation response and aerodynamic force. In Fig. 4 the instantaneous pressure contour surfaces at phase angle 0° ($y = 0.0$, $dy/dt > 0.0$) are displayed. The time histories of response and lift acting on the cylinder and spectral density functions of the lift are displayed in Figs. 5 and 6, respectively.

At $V_r = 5.5$ the characteristic vortices shed from the leading edges of the cylinder can be recognized. The pattern of vortices in the wake is regular as shown in Fig. 4. Then the dominant frequency of lift is coincident with the natural frequency of the cylinder and the spectral density function of fluctuating lift is quite sharp. Thus far it

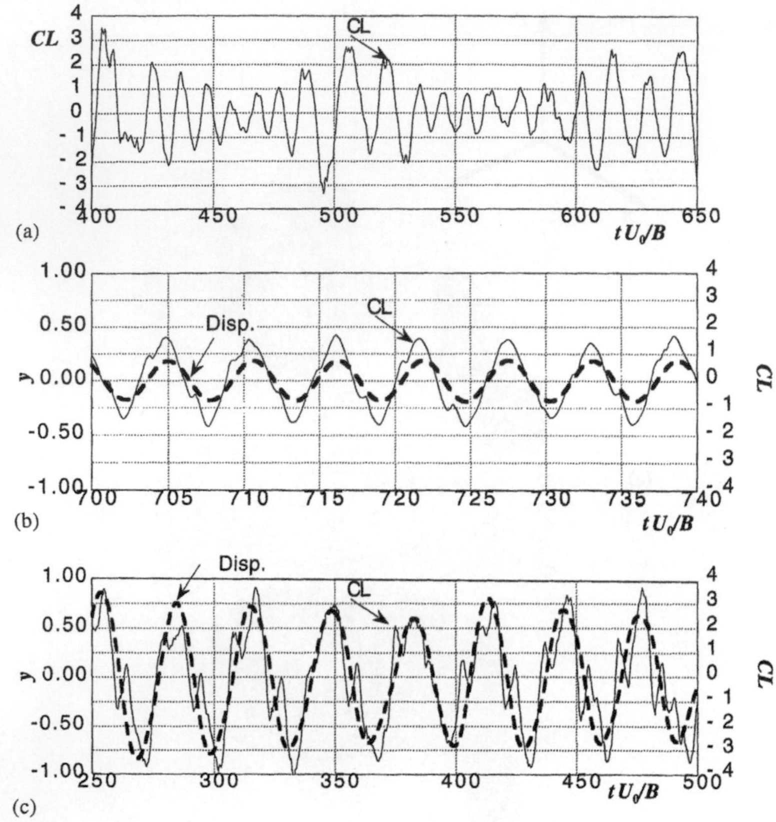

Fig. 5. Time histories of the lift force coefficient and the displacement in unstable oscillations: (a) at rest; (b) $V_r = 5.5$; (c) $V_r = 24.5$.

is confirmed that the lock-in phenomenon occurs at $V_r = 5.5$ and this instability is the vortex-induced oscillation.

At $V_r = 24.5$ the negative low pressure is maintained on the side surface ($y = +0.5$) of the oscillating cylinder and works as the negative damping force. On the other hand, convective vortices are clearly observed near the side surfaces of the cylinder. Because these vortices generate the lift fluctuation with much higher frequency than the natural frequency of the cylinder, these flow structures have no relation to the negative damping force. Due to the convective vortices near the side surfaces the spectral density functions of fluctuating lift at $V_r = 24.5$ are broader than at $V_r = 5.5$. As a result the computed response at $V_r = 24.5$ is more distorted more than at $V_r = 5.5$. Considering that the amplitude at $V_r = 13.5$ near the resonant reduced velocity is unstable and the computed amplitude at $V_r = 13.5$, 24.5 is overestimated, it is supposed that the aerodynamic force, to which a partial force due to vortex-shedding is added, is overestimated by means of this computational model.

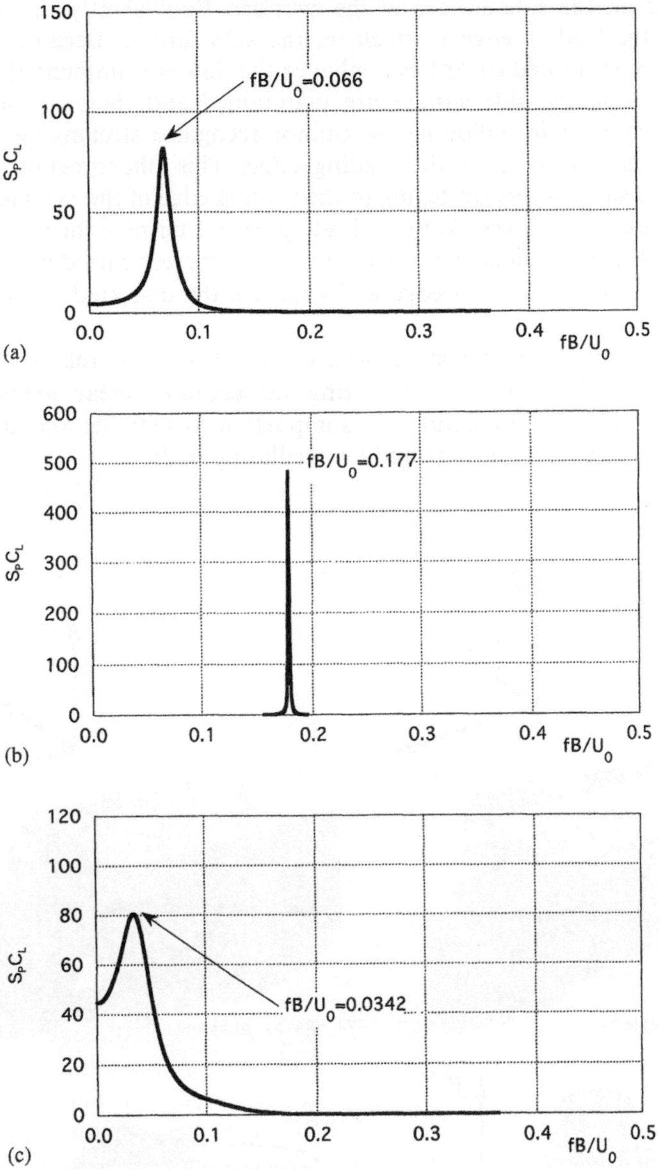

Fig. 6. Spectral density functions of the lift coefficients: (a) at rest; (b) $V_r = 5.5$; (c) $V_r = 24.5$.

The relations between three-dimensional wake structures and the body motion are discussed. Streamwise vortices (ω_x) and spanwise vortices (ω_z) are displayed in Fig. 7. Focusing on the characteristic vortices shed from the leading edge in the vortex-induced oscillation we can see stratified layers of streamwise vortices (ω_x) inside the spanwise vortices (ω_z). At $V_r = 5.5$ the vortices shed from the leading edge appear on the side surface of the cylinder. Equivalently the separated shear flows from the leading edge reattach on the side surface. Because the streamwise vortices in the separated shear layer enhance the flow entrainment, the characteristic large vortices on the side surface are maintained and they are flowing leeward. On the other hand in galloping we cannot recognize streamwise vortices in the separated shear layers near the leading edge. This phenomenon is identical to the separated shear layers stretching to the trailing edge of the cylinder. The flapping of the separated shear layers from the leading edge determine the aerodynamic forces as shown in Fig. 4. Complicated vortex structures are recognized inside the separated shear layers near the trailing edge and generate the distorted response mentioned above.

Accordingly the presence of streamwise vortices has a close relation to the behavior of the separated shear layers. Considering the separated shear layers determine the occurrence of unstable oscillations, it is important to simulate and clarify the three-dimensional wake structures around an oscillating body.

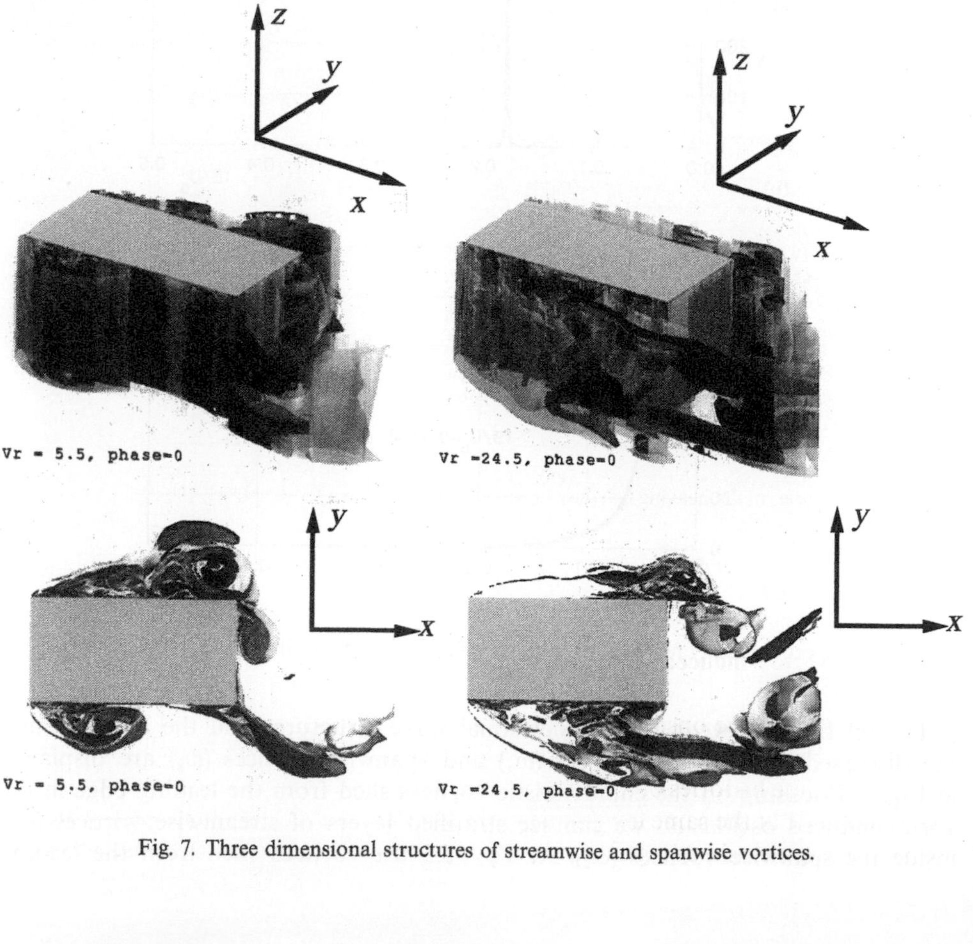

Fig. 7. Three dimensional structures of streamwise and spanwise vortices.

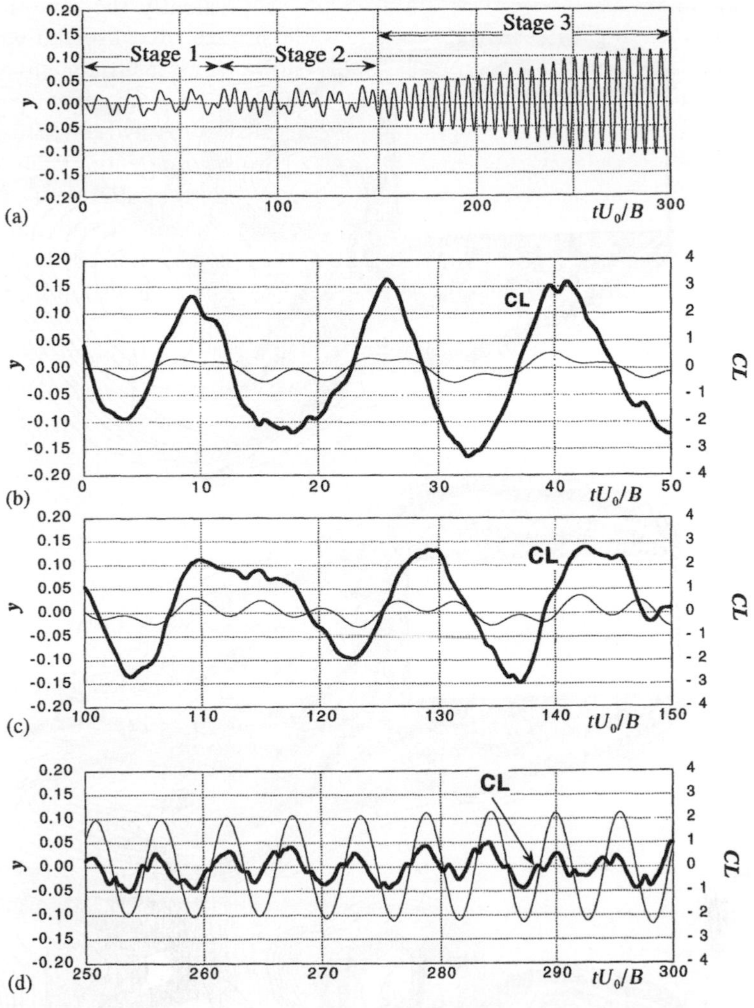

Fig. 8. Time histories of the heaving response in vortex induced oscillations; (a) total, (b) stage 1, (c) stage 2, (d) stage 3.

6. Onset of vortex-induced oscillations

Here we discuss the onset of the vortex-induced oscillation. The time variations in response and lift acting on the cylinder are investigated in Fig. 8. We can find three stages in the time histories of response. In the first stage the cylinder is oscillating with small amplitude at the same low frequency as the frequency of lift fluctuation, which is given by the vortex shedding. In the second stage the frequency of response becomes

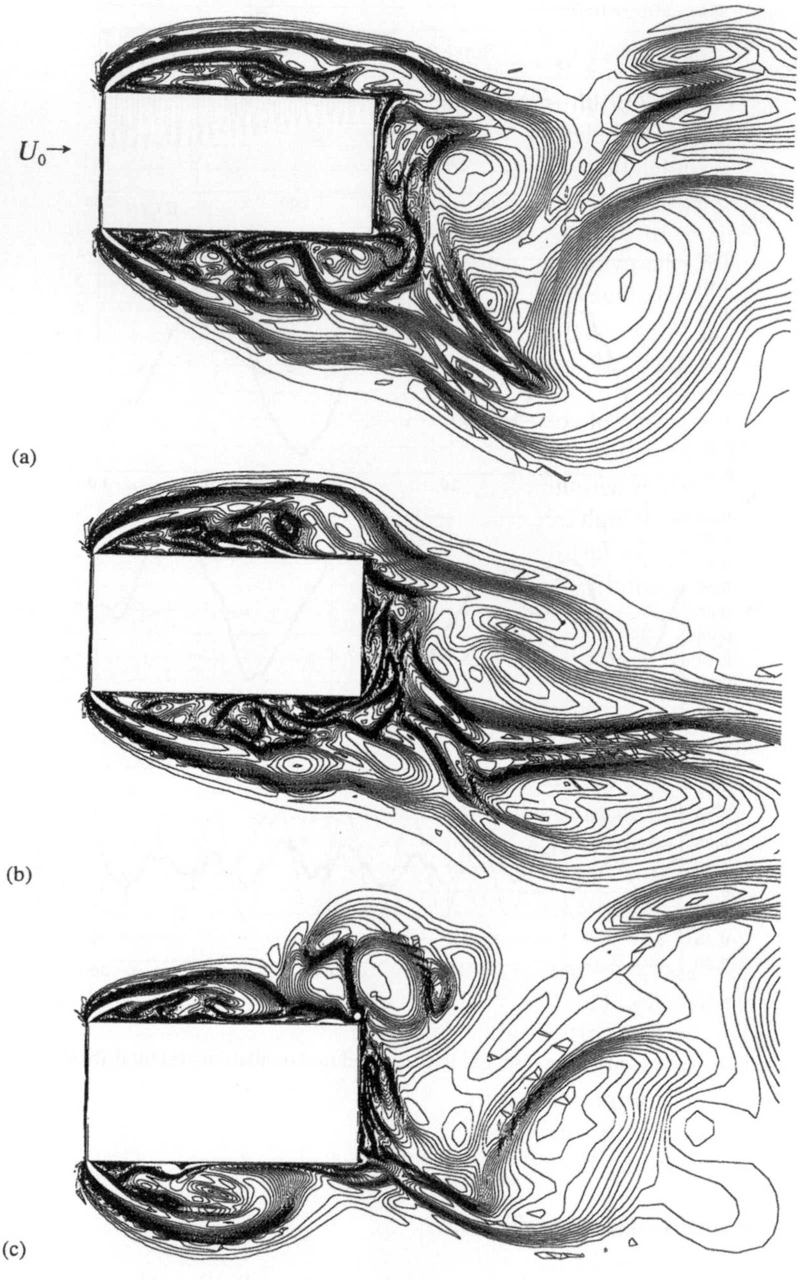

Fig. 9. Change in vortical flows around a rectangular cylinder. (a) $tU_0/B = 40.1$ (stage 1); (b) $tU_0/B = 130.0$ (stage 2); (c) $tU_0/B = 232.9$ (stage 3).

higher however the amplitude is still small. The amplitude and the dominant period of lift fluctuation in the second stage is the same as in the first stage. In the third stage the response becomes larger and fluctuates at higher frequency. The dominant frequency of lift is coincident with that of the response, which means the natural frequency of the cylinder. Finally the lock-in phenomenon occurs.

In order to understand the mechanism of onset of the vortex-induced oscillation, the spanwise vortices (ω_z) around the oscillating cylinder at phase angle 0° are shown in Fig. 9. In the first stage the separated flows from the leading edge form vortices behind the cylinder. In the second stage the vortices move to the trailing edges. In the third stage the vortices are formed near the leading edges on the side surfaces and the vortices are flowing leeward as keeping contact with the side surface of the cylinder. Then the vortex shedding is perfectly locked in the motion of the cylinder.

Taking the position of vortex formation in each stage into consideration, the wake region of the first stage is wider than that of the third stage. As a result the lift in the third stage fluctuates with higher frequency than in the first stage. In the second stage the fluctuations with high frequency are intermittently recognized because the motion of the cylinder is very fast due to the narrow wake. This phenomenon is the sign of onset of the vortex-induced oscillation.

7. Conclusions

The unsteady three-dimensional flows around a freely oscillating rectangular cylinder in heaving mode are simulated by means of the direct finite difference method without any turbulence model.

1. In case of vortex-induced oscillation the computational response is in good agreement with experimental ones. In the galloping the computed amplitude also varies qualitatively the same manner as the experimental results, but the values are a little larger.

2. The relations between the vortex-shedding and the response are discussed as unstable oscillations. The time variation in amplitude of the galloping is distorted more than that of the vortex-induced oscillation. In the galloping regime the aerodynamic force, which is added to a partial force due to vortex-shedding, is overestimated by means of this computational model.

3. It is important to simulate the three-dimensional wake structures which have a close relation to the behavior of the separated shear layers. For example, in the vortex-induced oscillations the characteristic vortices shed from the leading edge have stratified layers of streamwise vortices (ω_x) inside the spanwise vortices (ω_z). In galloping the separated shear layers are flapping in which streamwise vortices are hardly recognized.

4. The mechanism of onset of the vortex-induced oscillation is investigated in detail. We can find three stages in the vortex-induced oscillation. The position of vortex formation determines each stage of the vortex-induced oscillation.

References

[1] T. Tamura, Y. Itoh, A. Wada, Three-dimensional simulations of an oscillating rectangular cylinder, Proc. 6th Int. Conf. on Flow Induced Vibrations, Imperial College, London, 1995, pp. 181–192.

[2] T. Tamura, K. Tsuboi, K. Kuwahara, Numerical simulation of unsteady flow patterns around a vibrating circular cylinder, AIAA Paper 88-0128, 1988.

[3] F.H. Harlow, J.E. Welch, Numerical calculation of time-dependent viscous incompressible flow of fluid with free surface, Phys. Fluids 8 (1965) 2182–2189.

[4] T. Kawamura, K. Kuwahara,Computation of high Reynolds number flow around a circular cylinder with surface roughness, AIAA paper 84-0340, 1984.

[5] T. Miyata, M. Miyazaki, H. Yamada, Pressure distributions measurements for wind induced vibrations of box girder bridges, J. Wind Eng. Ind. Aerodyn. 14 (1983) 223–234.

[6] K. Washizu, A. Ohya, Y. Otsuki, K. Fujii, Aeroelastic instability of rectangular cylinders in a heaving mode, J. Sound Vib. 59 (2) (1978) 195.

Journal of Wind Engineering
and Industrial Aerodynamics 67&68 (1997) 155–167

JOURNAL OF
wind engineering
AND
industrial
aerodynamics

ELSEVIER

Development of a parallel code to simulate skewed flow over a bluff body

T.G. Thomas *, J.J.R. Williams

Department of Engineering, Queen Mary and Westfield College, London E1 4NS, UK

Abstract

This paper outlines a new complex geometry large-eddy simulation code using finite differences and a multigrid Poisson solver written for a parallel computer. The flow domain may be constructed from an arbitrary arrangement of rectangular blocks thus permitting flow in regions with complicated shapes, and may be mapped to any number of processors up to the number of blocks. The code is used to simulate turbulent flow at a Reynolds number of 3000 past a cube placed at ground level in rough terrain with its sides set 45° to mean flow direction. Preliminary results are presented which reproduce the conical vortices on the top of the cube.

1. Introduction

Some of the most destructive wind loads acting on buildings are caused by turbulent gusts at the roof edges and are as associated with flow separation and the production of strong conical vortices; the worst cases are when the wind is approaching the building obliquely and the roof vortex structure has the so called delta-wing vortex pattern. The failure of the roof construction is brought about by the peak pressure suction and is accentuated by the fluctuating nature of the load.

This pattern has been modelled experimentally in environmental wind tunnels. These differ from those commonly used to test aircraft in that it is necessary to reproduce the profile of the approach wind, the turbulent kinetic energy at about the height of the stagnation streamline, and the typical distribution of eddy sizes, in order to get good agreement with full scale measurements; typical tunnels are described by Cook [1], Sykes [2], Wood [3], and Hansen and Sorenen [4], amongst others. The tunnels are able to simulate the mean pressure loads quite accurately, but the fluctuating and peak loads near the roof edges are generally underpredicted, see for example Refs. [5–7]. The distribution of pressure fluctuations, and the peak excursions, are determined to a large extent by the characteristics of the approaching turbulence, as might be expected, and Tieleman [8] has shown that the wind tunnel results can be improved by matching the

*Corresponding author. E-mail: t.g.thomas@qmw.ac.uk.

PII S0167-6105(97)00070-6

turbulent inflow more precisely. Recent measurements on the pressure fluctuations at roof corners are given by Kramer and Gerthardt [9,10], Ginger and Letchford [11,12], and Tieleman et al. [13].

Computational fluid dynamics (CFD) has been extensively applied to predict the wind loads on buildings, see the surveys by Murakami [14] and Laurence and Mattei [15]. Reynolds average turbulence models, such as $k-\varepsilon$ methods are able to predict the mean pressures quite well but are unable to predict the extreme fluctuations – this failure is fundamental because the models are in essence constructed from known average correlations between turbulent quantities. Comparisons between CFD and experimental data have been given by Strathopolous and Zhou [16]. Large-eddy simulation (LES) of turbulence is potentially much better at predicting the fluctuating pressure dynamics because it represents the complete fluid dynamics directly at all scales up to the resolution limit of the mesh and only the subgrid scales (SGS) are modelled. Because the SGS are remote from the energy containing eddies their structure is more universal and hence more easily modelled; in addition, since only a small fraction of the total turbulent energy is represented by a model one might expect that the simulation be relatively insensitive to the details of the model, and this is confirmed by experience. The LES method is well established (see for example the review by Rogallo and Moin [17], and Mason [18]) and has been applied by Murakami and Mochida [19] and Mochida et al. [20] and others to wind loading, and by Werner and Wengle [21] to flow past a cubic obstacle.

In the present paper we outline the development of a LES code which allows the flow domain to be constructed from an arbitrary arrangement of rectangular blocks; the design is naturally parallel and is implemented to run on various massively parallel machines such as the Cray T3D or Intel iPSC960 (or workstation clusters) using either the Parallel Virtual Machine (PVM) or Message passing Interface (MPI) message libraries, as well as on single processor super-scalar machines. The complex geometry permitted allows us to simulate the flow around arrangements of several non-identical rectangular buildings, however the initial application of the code is to turbulent flow over a rough terrain past a cubic building aligned at 45° to the mean flow direction and at a Reynolds number of about 3000. The mesh resolution is just sufficient to resolve the conical vortices on the roof.

2. Governing equations

We consider an incompressible fluid of kinematic viscosity ν in motion with a velocity $u_i=(u,v,w)$ and kinematic pressure p at the points $x_i=(x,y,z)\in\Omega$ within some domain $\Omega\subset\mathbf{R}^3$; here u_i and x_i, $i\in\{1,2,3\}$, are Cartesian coordinates. Following the large-eddy simulation (LES) technique, we seek turbulent solutions of the Navier–Stokes equations on a computational grid of limited resolution Δ and so introduce the spatial filtering operation, indicated by a hat ($\hat{\ }$), so that if $f(x)$ denotes some variable then the filtered variable is given by

$$(\widehat{fx_i}) = \int_\Omega \mathscr{G}(|x_i - x_i'|) f(x_i')\,\mathrm{d}\Omega, \quad x_i, x_i' \in \Omega, \tag{1}$$

where $\mathcal{G} \geqslant 0$ represents a filter kernel of width $O(\Delta)$. The subgrid-scale (SGS) motions are eliminated by filtering the Navier–Stokes equations to give

$$\frac{\partial \hat{u}_i}{\partial t} = -P_{,i} + (-\hat{u}_i\hat{u}_j + r_{ij} - \tfrac{1}{3}\tau_{kk}\delta_{ij} + 2\nu S_{ij})_{,j} + g_i, \tag{2}$$

$$u_{k,k} = 0, \tag{3}$$

$$2S_{ij} = (\hat{u}_{i,j} + \hat{u}_{j,i}), \tag{4}$$

where $P = \hat{p} - \frac{1}{3}\tau_{kk}$ denotes the modified pressure, g_i the kinematic body force per unit volume, S_{ij} the symmetric part of the deformation rate tensor, and t the time. The SGS stress tensor τ_{ij} and SGS kinetic energy are defined (see for example Ref. [22]) as

$$\tau_{ij} \equiv -(\widehat{u_i u_j} - \hat{u}_i\hat{u}_j), \tag{5}$$

$$k \equiv -\tfrac{1}{2}\tau_{kk} \geqslant 0, \tag{6}$$

following the usual sign convention for stress (but opposite of that commonly adopted in SGS modelling). We use a slightly modified form of the Smagorinsky [23] subgrid model to relate the anisotropic part of the SGS stress to the filtered velocity field $\hat{u}_i$,

$$\tau_{ij} - \tfrac{1}{3}\tau_{kk}\delta_{ij} = 2\nu_{sg}S_{ij}, \tag{7}$$

$$\nu_{sg} = \max\{(c_o\Delta)^2 D(\eta)^2 S - \nu, 0\}, \tag{8}$$

where ν_{sg} denotes the subgrid-scale eddy viscosity, c_o the Smagorinsky constant, $S = (2S_{ij}S_{ij})^{1/2}$, $D(\eta)$ a wall damping function defined in terms of $\eta = l_w/\Delta$, and l_w the nearest wall distance. In the present work we use Mason's [18] damping function with (his) $n = 2$, $c_o = 0.1$, and l_w is computed as the shortest distance to any boundary on which a non-zero shear stress is maintained. The SGS kinetic energy is modelled by

$$k = c_e\nu_{sg}(\nu + \nu_{sg})/(c_o\Delta)^2, \tag{9}$$

where $c_e \sim 3.5$ is a constant. In the rest of this paper we drop the hat ($\hat{}$) for convenience of notation unless explicitly required.

3. Wall boundary conditions

On a smooth on-slip wall we use a synthetic boundary condition. Suppose $w_s \in \mathbf{R}^2$, $s \in \{1,2\}$, denotes the projection of the velocity u_i, into the plane of the wall, and $\tau_s \in \mathbf{R}^2$, the traction vector in the plane of the wall. Following Schumann [24], the instantaneous stress is assumed to vary in phase with the instantaneous velocity, so that

$$\tau_s = \frac{|\tau_a|}{|w_a|} w_s. \tag{10}$$

The average stress $|\tau_a| \equiv (\tau_{as}\tau_{as})^{1/2}$ is determined from the average velocity w_{as} by fitting a wall profile

$$w_a/u_* = f(l_w^+), \tag{11}$$

where $u_*^2 = \tau_a$ and $l_w^+ = u_* l_w/\nu$. The potential singular behaviour when $w_a = 0$ is eliminated by ensuring that $f(l_w^+)$ has a limiting form $f \to l_w^+$ hence the ratio $|\tau_a|/|w_a| \to \nu/l_w$, thus recovering the laminar form. The profile function f is chosen as a curve fit to the universally accepted wall profile, see for example Ref. [25]. The average w_a is determined from the time history of w_s using the simple linear filter

$$T_a \frac{d}{dt} w_{as} + w_{as} = w_s(t), \tag{12}$$

in which the characteristic time T_a is chosen to be long compared with the natural time scale of the turbulence $O(h/u_*)$ and short compared with the conditioning time. Initially $w_{as}(0) = w_s(0)$. A rough wall is treated similarly but with $f(l_w^+)$ replaced by $f(l_w^+, z_o^+)$ appropriate for rough terrain.

4. Finite difference discretisation

The filtered Navier–Stokes equations are discretised on a staggered grid by a second-order finite difference method, the arrangement is similar to that used by Schumann [24] and conserves momentum and kinetic energy. The domain Ω is constructed from an arbitrary collection of N_b rectangular blocks with their faces linked to each other or to boundary conditions. The domain is completely specified by the list of block coordinates, which locate the block within a set of global coordinates, and the graph of links, which describe the connections amongst the blocks. The blocks are mapped to $N_p \leqslant N_b$ processors allowing for flexibility in the type and size of parallel machine available. Each block maintains its own partition of the domain plus an overlap region, called a halo, extending for one layer of grid points beyond the block boundary coordinates; as the data within a block is modified the information is propagated into the halo regions of the adjacent blocks via message passing. In order to remove an ambiguity in propagating data between diagonally adjacent blocks the messaging is strictly sequenced so that exchanges along one coordinate direction must complete before starting exchanges along the next direction. The time advancement is by second order Adams–Bashforth method: first the momentum equations are projected over a time step but neglecting the pressure gradient term; enforcing continuity at the next time step requires the solution of a Poisson equation for the pressure which is solved by a parallel multigrid algorithm; the time step is completed by adding the correction velocity due to the pressure gradient term. The method, apart from the pressure solver, is essentially the same as used by Thomas and Williams [26].

The multigrid algorithm uses a sequence of grids constructed by binary subdivision within each block, the lowest level grid is determined by the smallest grid consistent over all the blocks. Clearly it is most efficient when each block permits a degree of binary subdivision but the method operates correctly for any combination of block sizes. Each grid has its own halo region and the messaging operates between grids at the same level of subdivision. The Poisson equation on the bottom grid is solved by a RB-SOR algorithm; because there are only $O(N_p)$ unknowns the CPU time is negligible in comparison with the time to perform the projection and restriction operations on the higher grids provided that the top level grids are chosen to allow sufficient factoring. The RB colouring on the lowest grid must be consistent, so that in multiply connected domains the length of any closed path must be an even number of grid points; this can always be enforced if necessary by not doing the last grid subdivision. The solution algorithm places no additional restrictions on the otherwise arbitrary geometric connections amongst the blocks.

5. Code verification

The code has been tested in its laminar form by predicting the reattachment length of the recirculation region behind a two- dimensional backward facing step as a function of the Reynolds number; the results are in agreement with published experimental data. It has been tested in its turbulent form by predicting the turbulent channel flow between two flat plates and the turbulent secondary circulation in a square duct, both giving good agreement with published experimental data.

6. Flow past a bluff body

The code is at present used to simulate the skewed flow around a cubic ground mounted obstacle. The initial simulation is on a coarse grid and allows us to map out the flow relatively quickly while providing the initial data for a second simulation at a higher resolution. The results presented here are taken from the early stages of a continuing simulation and represent a snapshot of the simulation at its current state of development; it has not yet been run long enough to converge the statistics, nevertheless, the results are representative of the flow structures that are expected to occur in the converged simulation.

The cube has a side h, the reference velocity at the cube height is u_o, and the friction velocity u_τ is equal to γu_o where the constant γ is given by

$$\kappa\gamma^{-1} + \log(z_o/h) = 0, \tag{13}$$

z_o is the roughness scale characterising the inflow profile, and $\kappa = 0.415$ is von Karman's constant. Here, $z_o/h \sim 0.01$ corresponding to $\gamma \sim 0.09$. The Reynolds number R is defined as

$$R \equiv \frac{u_o h}{\nu} = \gamma^{-1} h^+, \tag{14}$$

where $h^+ = u_\tau h/\nu$ is the cube height measured in units of the viscous wall length ν/u_τ, and the flow field around the cube is approximately independent of the Reynolds number provided that $R \geqslant 3 \times 10^3$[27], i.e. $h^+ \geqslant 300$. Experimental measurements [27]

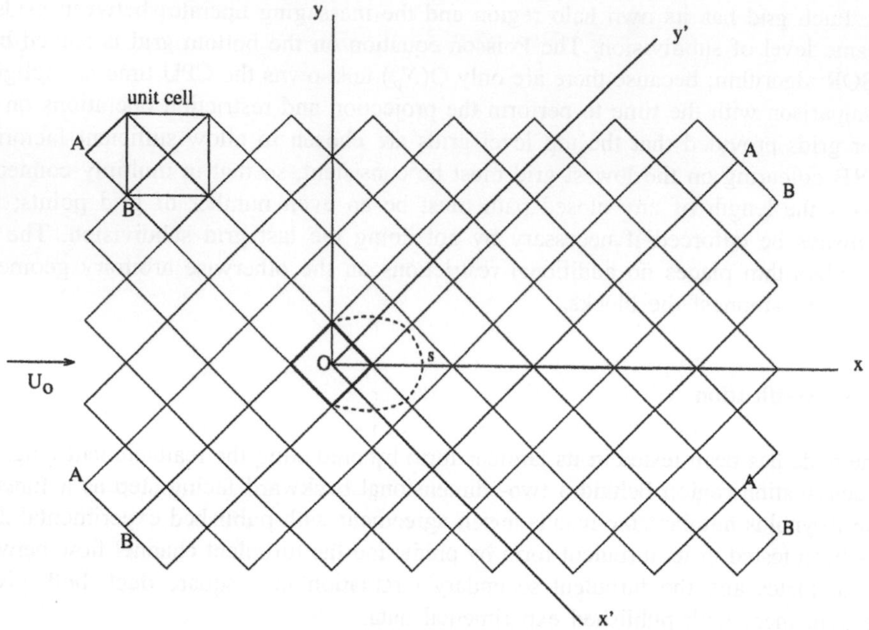

Fig. 1. Coordinates and domain decomposition.

suggest that the turbulence levels induced by the cube above the background turbulence extend to above $6h$ so that provided the computational box is longer than about (say) $8h$ we can generate the inflow by recycling the outflow. The flow is maintained by applying a uniform body force along the streamwise direction so that u_τ is approximately equal to unity. Consequently the inflow parameters are not set a priori but are determined by the simulation. We also use periodic boundary conditions in the lateral direction to eliminate the possibility of unphysical secondary circulation associated with some form of sidewall treatment.

Because the computational mesh is aligned with the cube, the periodicity of the domain must be along axes rotated 45° relative to the mesh: the basic repeat pattern must be a square of side $\sqrt{2}$ and hence consist of two computational blocks. We take a 5 wide by 8 long repeat pattern in the horizontal plane made up from 80 computational blocks and duplicate this pattern vertically 3 times. The cube is created by removing one of the blocks at ground level; the arrangement is shown in Fig. 1, also shown are the mesh based coordinates (x', y', z'), and the unit cell. This skew pattern has the advantage that the streamwise periodic length can be prescribed independently of the lateral periodic length while keeping the mesh direction 45° to the mean flow direction. The resultant computational domain is $8\sqrt{2}h \times 5\sqrt{2}h \times 3h$ using 239 blocks mapped to 128 nodes of a Cray T3D; this is well within the memory limits of the machine but is limited by the CPU time currently available to us. The blockage ratio with this domain is about 6%. An initial coarse grid simulation was performed using blocks with $16 \times 16 \times 16$ grid points, so that the mesh resolution $\Delta^+ \sim 19$. The upper boundary is treated as an impermeable free-slip surface, a rough wall synthetic boundary condition is applied on the floor, and a smooth wall no-slip boundary condition is applied on the faces of the cube.

7. Preliminary results and discussion

The simulation was started from an initial artificial flow constructed from a mean velocity profile, consistent with the value of u_τ over rough terrain, and random fluctuating velocities at about 25% of u_o. This was integrated forward in time, a process called conditioning, for about $10h/u_\tau$ when the present results were taken, the simulation is being continued until successive sets of statistics are no longer varying with time.

The present approach flow conditions are shown in Fig. 2: the velocity profile is roughly constant with u_o equal to $11.2u_\tau$. The boundary layer depth is currently only about 25% of the cube height. The turbulent intensity $(\overline{u}'^2)^{1/2}/u_o$ peaks at 0.23 near the bed and is about 0.1 at the cube height. The shape of the velocity profile is due to a downflow on the centreline which appears to be recycled from the wake; as the simulation proceeds we expect to find more background turbulence developing and a shorter wake, so this effect is probably only temporary and associated with the early development of the flow. Fig. 3 shows gives a good visual impression of the velocity field in the (x, y) plane at half the cube height ($z = h/2$). The regions of slower fluid

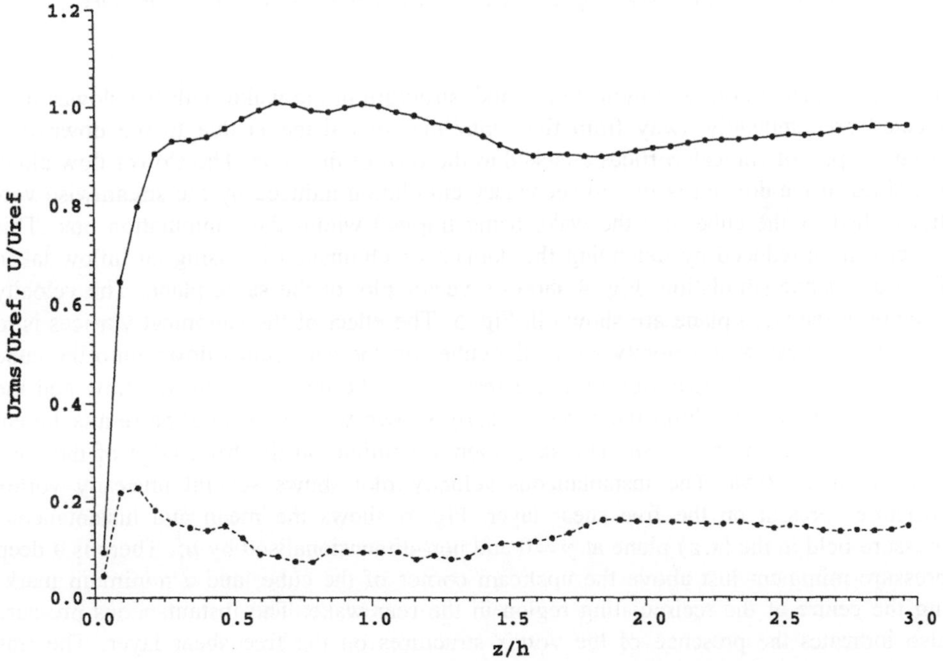

Fig. 2. Approach velocity $\bar{u}/u_o$ and streamwise turbulence intensity $(\overline{u}'^2)^{1/2}/u_o$ profiles upstream of the cube.

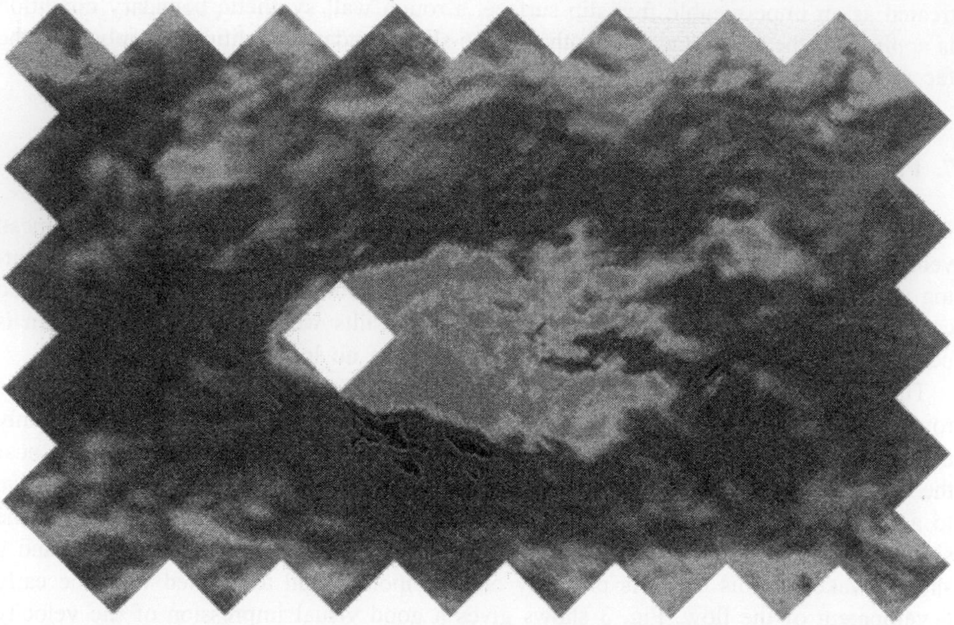

Fig. 3. Velocity magnitude in (x, y) plane at $z = h/2$ (lighter colour corresponds to lower velocity) showing the position of the shear layers separating at the vertical edges and the "fork" shape of the wake.

are clearly marked out and indicate a "fork" structure in the wake with the slower fluid located predominantly away from the centreline, this shape is due to the downwash from the pair of conical vortices attached to the roof of the cube. The slower flow along the sides of the domain is due to secondary circulation induced by the streamwise vorticity shed by the cube into the wake being trapped within the computation box. This effect can be reduced by extending the domain or eliminated by using an inflow taken from a separate simulation. Fig. 4 shows a vector plot of the same plane. The velocity vectors in the (x, z) plane are shown in Fig. 5. The effect of the canonical vortices is to deflect the streamwise velocity above the cube (on the centreline) down onto the cube surface and prevent separation until the rear edge. The region of reverse flow and the recirculation bubble behind the cube is clearly shown with the centreline re-attachment point on the bed near $x = 3h$. The stagnation streamline on the front edge of the cube is at about $z = 0.5h$. The instantaneous velocity plot shows several unsteady vortex structures present on the free shear layer. Fig. 6 shows the mean and instantaneous pressure field in the (x, z) plane at $y = 0$ and non-dimensionalised by u_τ^2. There is a deep pressure minimum just above the upstream corner of the cube, and a minimum marking the centre of the recirculating region in the rear wake. The instantaneous pressure also indicates the presence of the vortex structures on the free shear layer. The rms turbulent intensities in the (x, z) plane on the centreline $y = 0$ non-dimensionalised by u_τ are shown in Fig. 7. The streamwise intensity u_{rms}/u_τ peaks at about 4 in

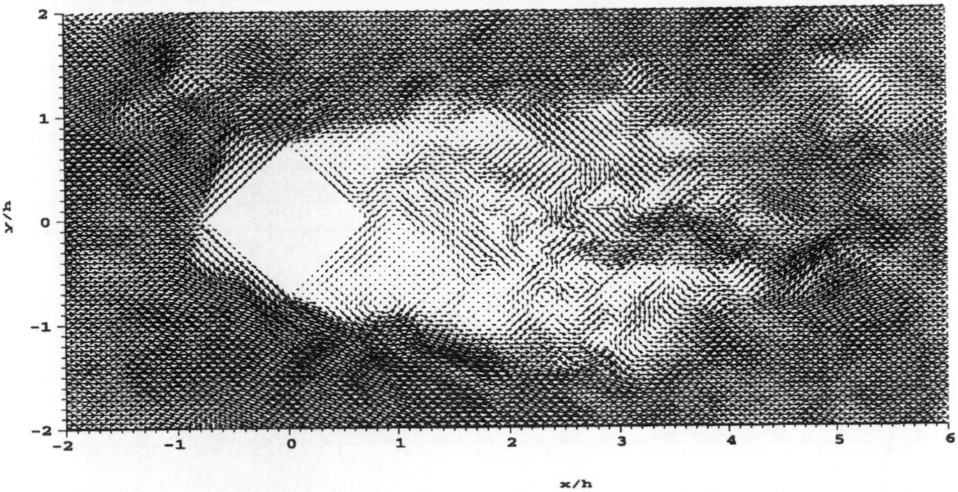

Fig. 4. Velocity vectors in the (x, y) plane at $z = h/2$.

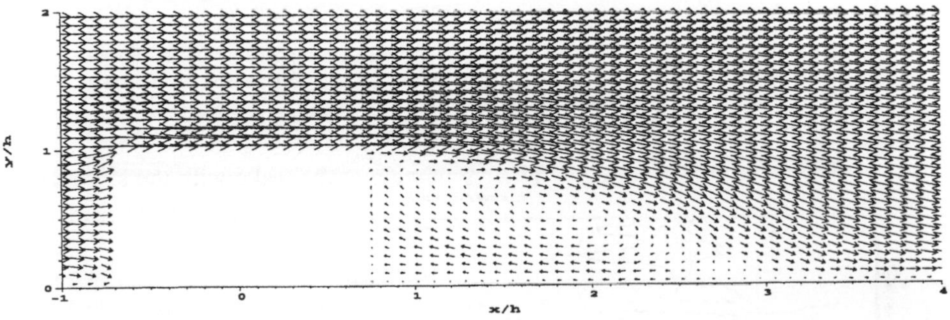

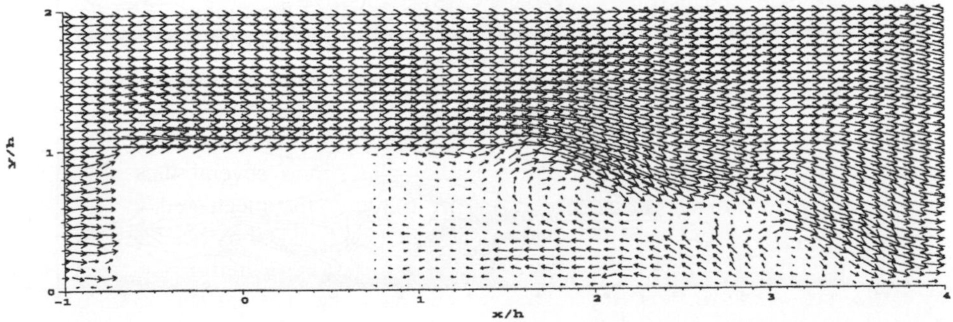

Fig. 5. Velocity vectors in the (x,z) plane at $y = 0$: (a) mean velocity, showing re-attachment point near $x = 3$; and (b) instantaneous velocity.

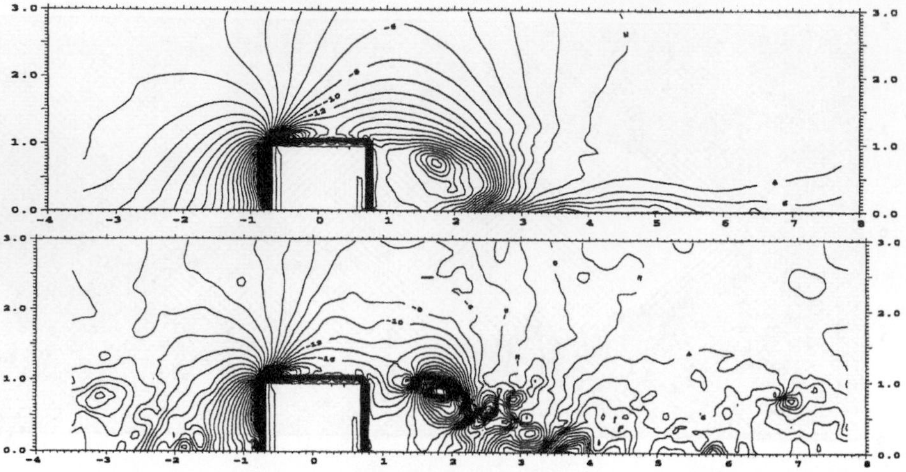

Fig. 6. Pressure in the (x, z) plane at $y = 0$ non-dimensionalized by u_τ^2: (a) mean pressure; (b) instantaneous pressure; contour heights at $-2, 0, 2, \ldots$.

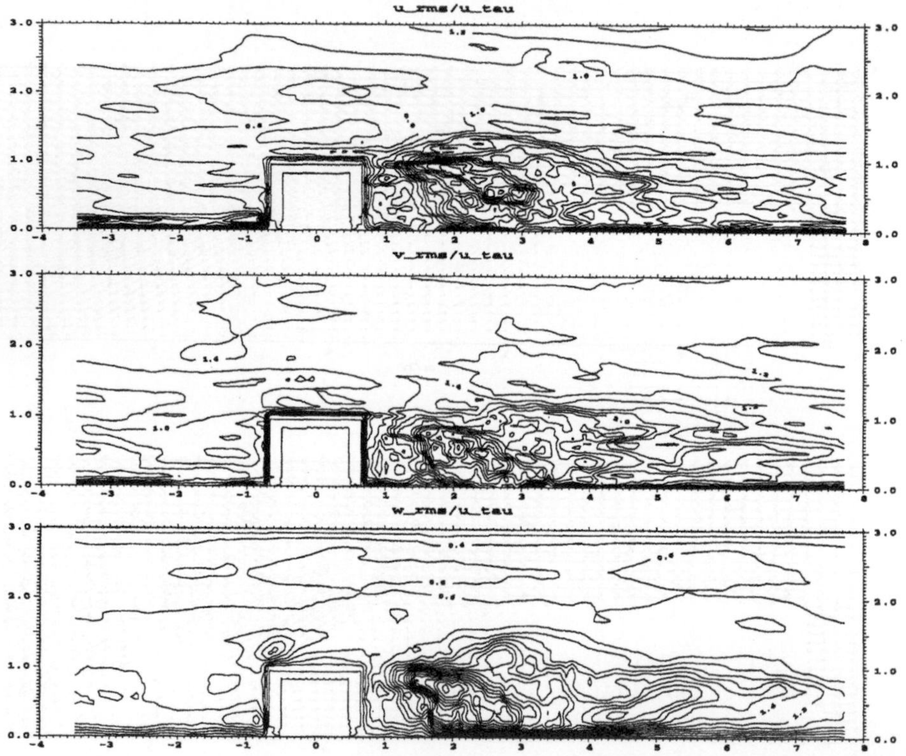

Fig. 7. Turbulent intensities in the (x, z) plane at $y = 0$ non-dimensionalised by u_τ: (a) u_{rms}/u_τ; (b) v_{rms}/u_τ; (c) w_{rms}/u_τ; contours at $0, 0.2, 0.4, \ldots$.

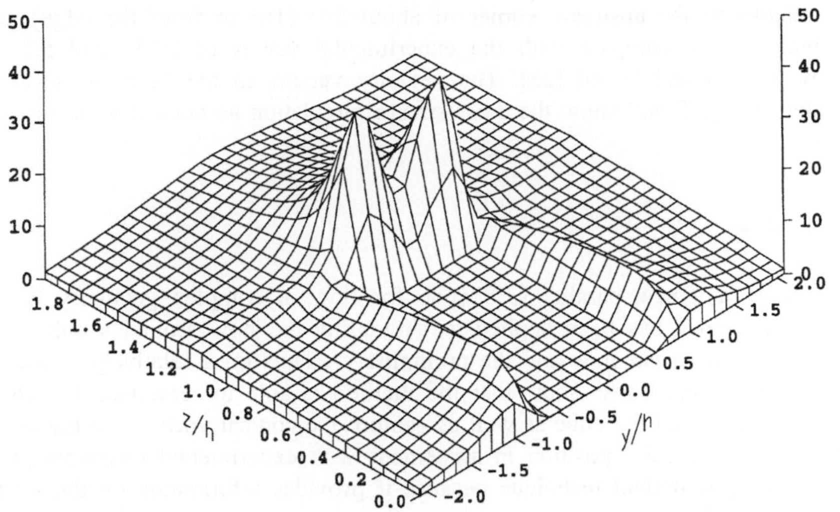

Fig. 8. Pressure (negative) in the (y,z) plane at $x=0$ diagonally bisecting the cube and non-dimensionalised by u_τ^2. The twin peaks at $y=\pm 0.04$, $z=1.074$ mark the cores of the conical roof vortices.

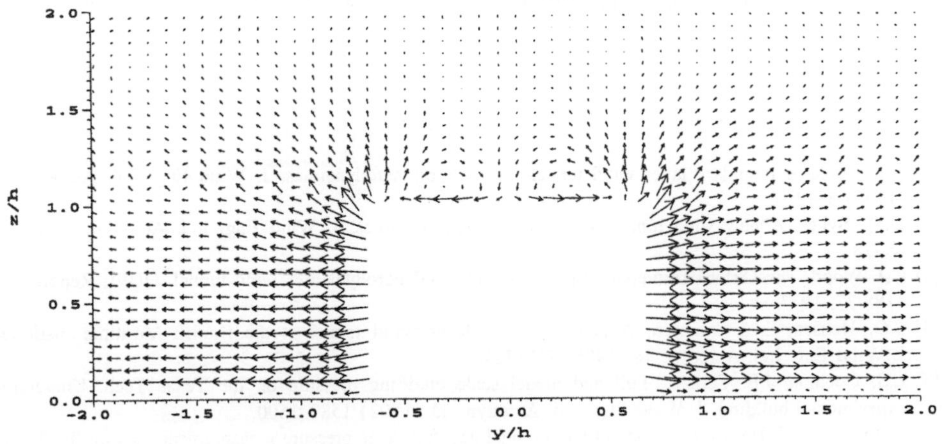

Fig. 9. Velocity vectors in the (y,z) plane at $x=0$ showing a section through the conical roof vortices. The lateral velocity induced on the roof is $\sim 5u_\tau$.

the near wake region behind the cube, the lateral v_{rms}/u_τ and vertical w_{rms}/u_τ intensities have slightly lower peaks of 3 and 3.5 respectively in the same region. There is well defined peak in the vertical intensity just above the upstream corner of the cube.

The location of the conical vortices is clearly mapped by the mean pressure minima shown (but with the opposite sign for clarity) in Fig. 8. At this section, $x=0$, the cores are centred at $y=\pm 0.04$ and $z=1.074$ giving horizontal and vertical angles relative to the upstream corner of about 29° (16° in from the edge) and 5° respectively. These compare with the experimental values of 29.5° and 5.1° measured by Minson and Wood [28]. The velocity vectors in the (y, z) plane at $x = 0$ are plotted in Fig. 9 and show the cross-stream circulation associated with the conical vortices.

8. Conclusions

An outline of a new parallel LES code for flow in complex geometry has been presented. The code has been applied to simulating the flow around a cube aligned 45° relative to the mean flow and the preliminary results on a relatively coarse mesh have been presented. These show that the method is able to reproduce the complex flow behaviour around the cube as well as capture the conical vortices on the cube top and predict their angular position in agreement with experimental observations. LES is a particularly important technique because it provides information on the *unsteady* character of the flowfield.

However, the simulation has run only a short time, and the turbulent statistics will be improved by a longer averaging time and by letting the simulation coverage further towards a statistically steady state. We intent to use the present data as the initial field for a simulation at a higher resolution so that the detailed dynamics of the conical vortices can be investigated.

References

[1] N.J. Cook, A boundary layer wind tunnel for building aerodynamics, J. Wind Eng. Ind. Aerodyn. 1 (1992) 3–12.

[2] D.M. Sykes, A new wind tunnel for industrial aerodynamics, J. Wind Eng. Ind. Aerodyn. 2 (1977) 65–78.

[3] C.J. Wood, The Oxford University 4m × 2m industrial aerodynamics wind tunnel, OUEL Report No. 1188/77, 1977.

[4] S.O. Hansen, E.G. Sorensen, A new boundary layer wind tunnel at the Danish Maritime Institute, J. Wind Eng. Ind. Aerodyn. 18 (1985) 213–224.

[5] L.S. Cochran, J.E. Cermak, Full and model scale cladding pressure on the Texas Tech. University experimental building, J. Wind Eng. Ind. Aerodyn. 43 (1992) 1589–1600.

[6] H. Okada, Y.C. Ha, Comparisons of wind tunnel and full scale pressure measurement tests on the Texas Tech. building, J. Wind Eng. Ind. Aerodyn. 43 (1992) 1601–1612.

[7] R.V. Milford, A.M. Goliger, J.L. Waldeck, Jan Smuts experiment: comparison of full scale and wind tunnel results, J. Wind Eng. Ind. Aerodyn. 43 (1992) 1705–1716.

[8] H.W. Tieleman, Pressures on surface-mounted prisms: the effects of incident turbulence, J. Wind Eng. Ind. Aerodyn. 49 (1993) 289–300.

[9] C. Kramer, H.J. Gerhardt, Wind effects on roofs and roofing systems, J. Wind Eng. Ind. Aerodyn. 36 (1990) 301–308.

[10] C. Kramer, H.J. Gerhardt, Wind pressures on roofs of very low and very large industrial buildings, J. Wind Eng. Ind. Aerodyn. 38 (1991) 285–296.

[11] J.D. Ginger, C.W. Letchford, Peak wind loads under delta wing vortices on canopy roofs, J. Wind Eng. Ind. Aerodyn. 43 (1992) 1739–1750.

[12] J.D. Ginger, C.W. Letchford, Characteristics of large pressures in regions of flow separation, J. Wind Eng. Ind. Aerodyn. 49 (1993) 301–310.

[13] H.W. Tieleman, D. Surrey, J.X. Lin, Characteristics of mean and fluctuating pressure coefficients under delta wing vortices, Proc. of Wind Engineering Society Conference, Cambridge, 1992.

[14] S. Murakami, Computational wind engineering, J. Wind Eng. Ind. Aerodyn. 36 (1993) 517–538.

[15] D. Laurenence, J.D. Mattei, Current state of computational bluff body aerodynmics, J. Wind Eng. Ind. Aerodyn. 49 (1993) 23–44.

[16] T. Strathopolous, Y.S. Zhou, Numerical simulation of wind induced pressures on buildings of various geometries, J. Wind Eng. Ind. Aerodyn. 46-47 419–430.

[17] P.S. Rogallo, P. Moin, Numerical simulation of turbulent flows. Ann. Rev. Fluid Mech, 16 (1984) 99–137.

[18] P.J. Mason, Large-eddy simulation: a critical review of the technique, Q.J.R. Meteorol. Soc. 120 (1994) 1–26.

[19] S. Murakami, A. Mochida, Three-dimensional numerical simulation of air flow around a cubic model by means of large-eddy simulation, J. Wind Eng. Ind. Aerodyn. 25 (1987)291–305.

[20] A. Mochida, S. Murakami, M. Shoji, Y. Ishida, Numerical simulation of flow field around Texas Tech. building by large-eddy simulation, J. Wind Eng. Ind. Aerodyn. 46–47 (1993) 455–460.

[21] H. Werner, H. Wengle, Large-eddy simulation of turbulent flow over and around a cube in a plate channel, Proc. 9th Symp. on Turbulent Shear Flows, Springer, Berlin, 1993.

[22] S. Ghosal, T.S. Lund, P. Moin, K. Akselvoll, A dynamic localization model for large-eddy simulation of turbulent flows, J. Fluid Mech. 286 (1995) 229–255.

[23] J. Smagorinsky, General circulation experiments with the primitive equations: I. The basic experiment, Mon. Weather Rev. 91 (1963) 99–164.

[24] U. Schumann, Subgrid-scale model for finite difference simulations of turbulent flows in plane channels and annuli, J. Comput. Phys. 18 (1975) 376–404.

[25] S. Goldstein, Modern Developments in Fluid Dynamics, Dover, New York, 1965.

[26] T.G. Thomas, J.J.R. Williams, Large-eddy simulation of a symmetric trapezoidal channel at a Reynolds number of 430,000, J. Hydr. Res. 33 (1995) 825–842.

[27] I.P. Castro, A.G. Robins, The flow around a surface-mounted cube in uniform and turbulent streams, J. Fluid Mech. 79 (1977) 307–335.

[28] A.J. Minson, C.J. Wood, Investigation of separation bubbles and inclined edge vortices above model buildings using laser doppler anemometry, Proc. 11th Australasian Fluid Mech. Conf., Hobart, Australia, 1992, pp. 877–880.

ELSEVIER

Journal of Wind Engineering
and Industrial Aerodynamics 67&68 (1997) 169–182

JOURNAL OF wind engineering AND industrial aerodynamics

Development of a new k–ε model for flow and pressure fields around bluff body

M. Tsuchiya[a,*], S. Murakami[b], A. Mochida[c], K. Kondo[a], Y. Ishida[d]

[a] *Kajima Technical Research Institute, 2-19-1, Tobitakyu, Chofu-shi, Tokyo 182, Japan*
[b] *Institute of Industrial Science, University of Tokyo, 7-22-1, Roppongi, Minato-ku, Tokyo 106, Japan*
[c] *Niigata Institute of Technology, 1719, Fujihashi, Kashiwazaki, Niigata 945-11, Japan*
[d] *Kajima Information Processing Center, 1-2-7, Motoakasaka, Minato-ku, Tokyo 107, Japan*

Abstract

It is well known that applications of the standard k–ε model to flowfields around bluff-shaped bodies, often yield serious errors such as overestimation of turbulence kinetic energy k in the impinging region. Murakami, Mochida and Kondo have proposed a new k–ε model which resolves these problems by modifying the expression for eddy viscosity approximation. This paper examines the applicability of this new k–ε model (MMK model) to flowfields around three types of bluff bodies, i.e. a 2D square rib, a cube and a low-rise building model with 1 : 1 : 0.5 shape. The first half of the paper investigates the accuracy of the MMK model in reproducing turbulence characteristics around a bluff body. Results of the MMK model are compared precisely with those of the standard k–ε model, a revised k–ε model proposed by Launder and Kato (LK model) and wind tunnel tests for flow fields around a 2D square rib and a cube. The MMK model is also applied to predicting surface pressures on a low-rise building model with 1 : 1 : 0.5 shape with various wind angles including an oblique one. The accuracy and applicability of the MMK model to wind engineering problems are then discussed by comparing its results with those of the standard k–ε model and of the wind tunnel tests.

Keywords: MMK model; LK model; Standard k–ε model; Turbulence kinematic energy; Flow separation; Mean pressure coefficient; Oblique wind angle

Nomenclature

x_i	three components of spatial coordinate ($i = 1, 2, 3$; streamwise, lateral, vertical)
u_i	velocity component in the x_i direction
$\langle f \rangle$	time averaged value of f

* Corresponding author.

PII S0167-6105(97)00071-8

f'	deviation from $\langle f \rangle, f' = f - \langle f \rangle$
S	shear strain rate scale (cf. Eq. (3) in Table 1)
Ω	vorticity scale (cf. Eq. (5) in Table 1)
P_k	production of k
k	turbulence kinematic energy
ε	turbulence dissipation rate
$-\langle u_i' u_j' \rangle$	Reynold stress
ν_t	eddy viscosity
$\langle u_t \rangle_p$	tangential velocity component at 1st grid adjacent to ground surface or building wall
l	turbulence length scale
H_b	building model height
$\langle u_b \rangle$	time-averaged value of u_1 at the inflow boundary of computational domain at height H_b
k_p	k at 1st grid adjacent to ground surface
h_p	grid spacing in the direction normal to the wall of 1st grid adjacent to ground surface or building surface
τ_w	wall shear stress
ρ	air density
ν	kinematic viscosity
κ	von Karman constant (= 0.4)
C_μ	proportional number (= 0.09)
E	coefficient of generalized log law
C_p	wind pressure coefficient
θ	wind direction

Values presented in the paper are made dimensionless by H_b and $\langle u_b \rangle$.

1. Introduction

There are many difficulties in predicting the complex turbulent flow field around a bluff body such as a building using a standard k–ε model. The standard k–ε model is not suitable for predicting a flowfield that includes impinging, separation, etc. Therefore, this model cannot reproduce well wind pressure distributions acting on a roof surface and wall surfaces of a building [1,2]. This is because the standard k–ε model overestimates production P_k of turbulence kinematic energy k, around the frontal corner of the roof. This problem does not occur in computations using LES (Large Eddy Simulation), ASM (Algebraic Stress Model) and DSM (Differential Stress Model) [2–4]. However, these models have disadvantages of high computational load and computational instability.

Launder and Kato proposed a revised k–ε model (hereafter denoted as LK model [5,6]) which eliminates the excessive production of k around a stagnation point by modifying the expression of P_k. This model succeeds in correcting the overestimation of k around the impinging region. However, the LK model has an inconsistency in modeling of Reynolds stresses $-\langle u_i'u_j'\rangle$ and P_k. In the transport equation of mean flow energy K $(=\frac{1}{2}\langle u_i\rangle\langle u_i\rangle)$, there is a term which has the same form as that of P_k with opposite sign. This term plays a role in transferring the kinetic energy from the mean flow to the turbulence. However, in the case of LK model the term corresponding to P_k in the equation of K does not take the same form as P_k. This is because the LK model only revises the expression of P_k in the equation of k and uses the conventional approximation of $-\langle u_i'u_j'\rangle$ based on the standard k–ε model in the momentum equation. Therefore, P_k in the equation of mean flow energy K and in the equation of k do not appear in the same forms. The authors have proposed a new k–ε model (hereafter denoted as MMK model [7–9]) which resolves this problem by adding the modification not to the expression for P_k, but to the expression for eddy viscosity, ν_t.

The first half of this paper investigates the accuracy of the MMK model in reproducing turbulence characteristics around bluff bodies, i.e. a 2D square rib and a cube. Results of the MMK model are compared in detail with those of the standard k–ε model, the LK model and wind tunnel tests.

Furthermore, the MMK model is applied for predicting surface pressures on a low-rise building model with 1 : 1 : 0.5 shape with various wind angles including an oblique one in the last part of this paper. The accuracy and applicability of the MMK model to wind engineering problems are discussed by comparing the results with those of the standard k–ε model and wind tunnel tests.

2. Outline of revised k–ε models

The LK model (Eq. (4) in Table 1) expresses P_k, as a function of S and Ω (S is the strain rate scale, Ω is the vorticity scale) to eliminate the excessive production of k [5,6]. As described in the previous section, this model has an inconsistency in

Table 1
Model equations (expressions for P_k and ν_t)

1. Standard k–ε model

$P_k = \nu_t S^2$ (1) (ν_t: Eq. (2)) $\quad \nu_t = C_\mu \dfrac{k^2}{\varepsilon}$ (2) $\quad S = \sqrt{\dfrac{1}{2}\left(\dfrac{\partial\langle u_i\rangle}{\partial x_j} + \dfrac{\partial\langle u_j\rangle}{\partial x_i}\right)^2}$ (3)

2. LK model

$P_k = \nu_t S\Omega$ (4) (ν_t: Eq. (2)) $\quad \Omega = \sqrt{\dfrac{1}{2}\left(\dfrac{\partial\langle u_i\rangle}{\partial x_j} - \dfrac{\partial\langle u_j\rangle}{\partial x_i}\right)^2}$ (5)

3. MMK model

$P_k = \nu_t S^2$ (6) (ν_t: Eq. (7) or Eq. (8))

$\nu_t = C_\mu^* \dfrac{k^2}{\varepsilon},\ C_\mu^* = C_\mu \dfrac{\Omega}{S}\ \left(\dfrac{\Omega}{S} < 1\right)$ (7) $\quad \nu_t = C_\mu^* \dfrac{k^2}{\varepsilon},\ C_\mu^* = C_\mu\ \left(\dfrac{\Omega}{S} \geqslant 1\right)$ (8)

x_i: components of spatial coordinate, u_i: velocity components in the x_i direction

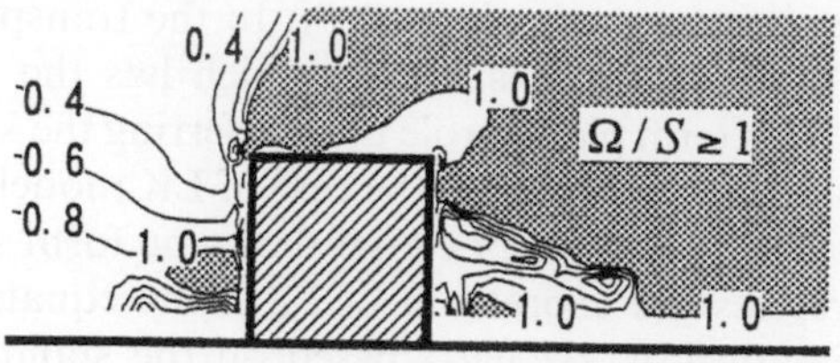

Fig. 1. Distribution of Ω/S (2D square rib, standard k–ε model).

modeling $-\langle u_i'u_j'\rangle$. The MMK model corrects this inconsistency of the LK model by adding the modification to the expression for ν_t [7–9]. In the MMK model, the coefficient C_μ in the definition of ν_t is expressed as a function of Ω/S (Eqs. (7) and (8) in Table 1). Although the expression for P_k for this model (Eq. (6)) is the same as that for the standard k–ε model, the value of P_k becomes smaller than that obtained by the standard k–ε model when $\Omega/S < 1$, because the expression for ν_t in the standard k–ε model is replaced by Eq. (7) in the MMK model. In the flowfield with $\Omega/S \geqslant 1$ (shaded part in Fig. 1), the expression for ν_t in Eq. (7) overestimates P_k. To avoid this overestimation, Eq. (8) is utilized in the MMK model only where $\Omega/S \geqslant 1$.

3. Computation of flowfields around various bluff bodies

3.1. Outline of computation

The applicability of the MMK model is examined for simulating flowfields around the 2D square rib and 3D bluff bodies placed in the surface boundary layer. Here, results obtained from the MMK model are compared with those from the standard k–ε model, the LK model and wind tunnel tests. Computations for the 2D square rib, the cube and the low-rise building model with 1 : 1 : 0.5 shape, are outlined[1,2,3]. Concerning the flow past the 2D square rib, results of the standard

---

[1] For the 2D square rib, the computational domain covers $26(x_1)$ and $24(x_2)$. This domain is discretized into $139(x_1) \times 71(x_2)$ grids. The minimum grid width is 1/20. The length scale is normalized by the height of the model. The boundary condition is shown in Table 2.

[2] For the cube, a composite grid system composed of a fine grid and a coarse grid is utilized [13,14]. The fine grid covers the sub-domain around the model which is very important for an accurate analysis, and the coarse grid covers the whole computational domain. The coarse grid is divided into $30(x_1) \times 30(x_2) \times 30(x_3)$ grids in the area of $17.75(x_1)$, $14(x_2)$ and $7.85(x_3)$. The fine grid is divided into $30(x_1) \times 30(x_2) \times 22(x_3)$ grids. The minimum grid width is 1/40. The length scale is normalized by the height of the model. The boundary condition is shown in Table 3.

[3] For the low-rise building model with 1 : 1 : 0.5 shape, a composite grid system similar to that of the cube is utilized. Wind directions are set by rotating the fine grid (mentioned in footnote 2). The coarse grid is divided into $51(x_1) \times 43(x_2) \times 31(x_3)$ grids in the area of $30(x_1)$, $20(x_2)$ and $10(x_3)$. The fine grid is divided into $35(x_1) \times 35(x_2) \times 22(x_3)$ grids. The minimum grid width is 1/30. The length scale is normalized by the height of the model. The boundary condition is shown in Table 3.

Table 2
Boundary conditions for 2D square rib

Inflow	$\langle u_1(x_2)\rangle$: wind tunnel test result ($\langle u_1(x_2)\rangle \cong x_2^{1/4}$) $\langle u_2(x_2)\rangle = 0$ $k(x_2)$: wind tunnel test result (at model height $k(H_b) = 0.07$) $l(x_2) = (C_\mu k(x_2))^{1/2} (\partial\langle u_1(x_2)\rangle/\partial x_2)^{-1}$ (at inflow boundary $P_k = \varepsilon$) $\varepsilon(x_2) = C_\mu k(x_2)^{3/2}/l(x_2)$ $\nu_t(x_2) = k(x_2)^{1/2} l(x_2)$ 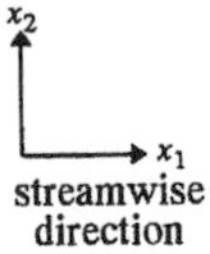
Outflow	$\langle u_1\rangle, \langle u_2\rangle, k, \varepsilon : \partial/\partial x_1 = 0$
Upper face of computational domain	$\langle u_2\rangle = 0, \langle u_1\rangle, k, \varepsilon : \partial/\partial x_2 = 0$
Solid wall	$\frac{\langle u_t\rangle_P}{(\tau_w/\rho)}(C_\mu^{1/2}k_p)^{1/2} = \frac{1}{\kappa}\ln\left[E\frac{1}{2}h_p(C_\mu^{1/2}k_p)^{1/2}/\nu\right]$ $\bar{\varepsilon} = \frac{C_\mu^{3/4}k_p^{3/2}}{\kappa h_p}\ln[Eh_p(C_\mu^{1/2}k_p)^{1/2}/\nu]$ $\varepsilon_p = \frac{C_\mu^{3/4}k_p^{3/2}}{\kappa h_p/2}$ $k : \partial k/\partial x_n = 0, \kappa = 0.4, C_\mu = 0.09$ $E = 9.0$: for roof and wall surface (smooth surface) $E = 0.055$: for ground (rough surface)

k–ε model, the LK model, the MMK model and wind tunnel tests are compared. For flow around the cube and the low-rise building model, results of the MMK model are compared with those of the standard k–ε model and wind tunnel tests (Tables 2 and 3).

3.2. Results of computation of 2D square rib and cube with wind angle normal to windward face

(1) *Turbulence kinematic energy k.*

(a) *2D square rib (Fig. 2)*. Fig. 2 shows the distribution of k around the 2D square rib. In the wind tunnel test, the turbulence kinematic energy k shows its maximum peak above the center of the roof, but k is relatively small around the frontal corner of the roof. On the other hand, the standard k–ε model greatly overestimates k around the frontal corner of the roof due to the overestimation of P_k by this model. Both revised k–ε models, the LK model and the MMK model, improve this overestimation of k. Concerning the peak value of k in the vicinity of the roof, the value for the MMK model is larger than that for the LK model. The MMK model shows closer agreement with the experimental result.

Table 3
Boundary condition for cube and low-wise building model with 1 : 1 : 0.5 shape

Inflow	$\langle u_1(x_3)\rangle$: wind tunnel test result ($\langle u_1(x_3)\rangle \cong x_3^{1/4}$) $\langle u_2(x_3)\rangle = \langle u_3(x_3)\rangle = 0$ $k(x_3)$: for cube = constant value $k(x_3) = 0.025$ for 1 : 1 : 05 model = wind tunnel test result (at model height $k(H_b) = 0.065$) $l(x_3) = (C_\mu k(x_3))^{1/2}\,(\partial\langle u_1(x_3)\rangle/\partial x_3)^{-1}$ (at inflow boundary $P_k = \varepsilon$) $\varepsilon(x_3) = C_\mu k(x_3)^{3/2}/l(x_3)$ $\nu_t(x_3) = k(x_3)^{1/2} l(x_3)$ 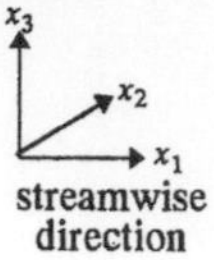
Outflow	$\langle u_i\rangle, k, \varepsilon : \partial/\partial x_1 = 0$
Upper and side faces of computational domain	$\langle u_n\rangle = 0, \langle u_t\rangle, k, \varepsilon : \partial/\partial x_n = 0$ $\langle u_n\rangle$: normal velocity component x_n : spatial coordinate normal to the boundary plane $\langle u_t\rangle$: tangential velocity component
Solid wall	$\dfrac{\langle u_t\rangle_P}{(\tau_w/\rho)}(C_\mu^{1/2}k_p)^{1/2} = \dfrac{1}{\kappa}\ln\left[E\dfrac{1}{2}h_p(C_\mu^{1/2}k_p)^{1/2}/\nu\right]$ $\bar{\varepsilon} = \dfrac{C_\mu^{3/4}k_p^{3/2}}{\kappa h_p}\ln[Eh_p(C_\mu^{1/2}k_p)^{1/2}/\nu]$ $\varepsilon_p = \dfrac{C_\mu^{3/4}k_p^{3/2}}{\kappa h_p/2}, k : \partial k/\partial x_n = 0$ $\kappa = 0.4, C_\mu = 0.09, E = 9.0$

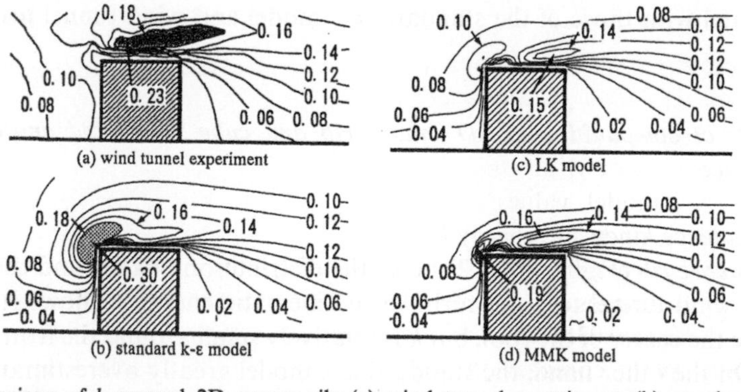

Fig. 2. Comparison of k around 2D square rib: (a) wind tunnel experiment; (b) standard k–ε model; (c) LK model and (d) MMK model.

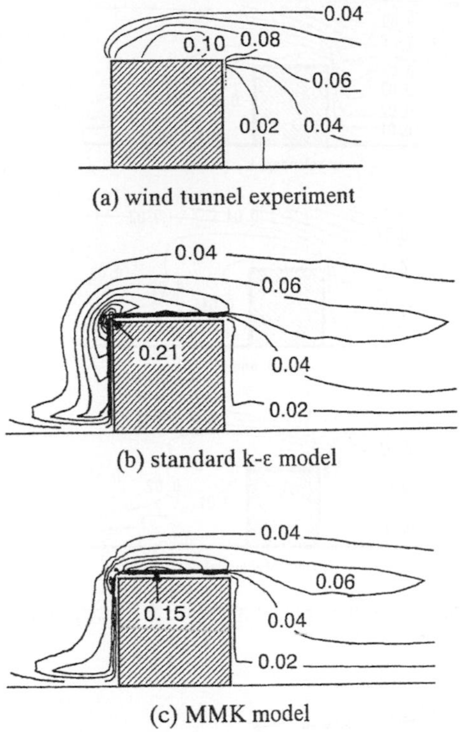

Fig. 3. Comparison of k around cube: (a) wind tunnel experiment; (b) standard k–ε model; (c) MMK model.

(b) *Cube (Fig. 3).* Fig. 3 illustrates the distribution of k around the cube. As mentioned in relation to the 2D square rib, the standard k–ε model greatly overestimates k around the frontal corner of the roof. The MMK model improves this overestimation of k as well as that of the 2D square rib.

(2) *Eddy viscosity ν_t (2D square rib, Fig. 4).*

The distributions of ν_t around the 2D square rib are compared in Fig. 4. For the standard k–ε model, values of ν_t around the 2D square rib range from 0.01 to 0.05. For the LK model, values of ν_t decrease in comparison with the standard k–ε model and range from 0.01 to 0.04 around the rib. Using the MMK model, the distribution of ν_t changes considerably around the windward face of the rib, its values ranging from 0.01 to 0.03. Hence, the maximum peak value of ν_t is reduced further by the MMK model in comparison with that by the LK model. These changes in the distributions of ν_t influence the flowfield, i.e. the flow separation and flow reattachment on the roof, as described below in Section 3.2(3). Because of these changes in the flowfield, surface pressures acting on the roof surface and walls change considerably, as described later in Section 3.2(4). These tendencies are almost the same for the cube (figures are omitted here).

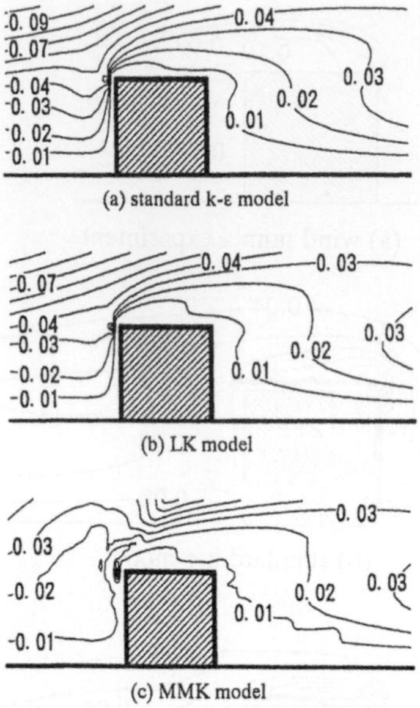

Fig. 4. Comparison of ν_t around 2D square rib: (a) standard k–ε model; (b) LK model; (c) MMK model.

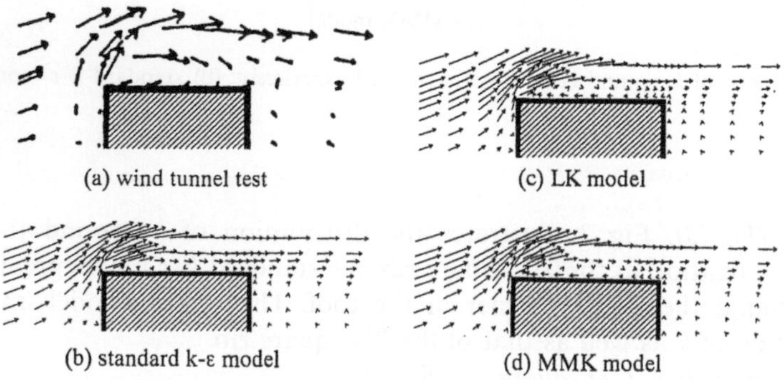

Fig. 5. Comparison of velocity vector around 2D square rib: (a) wind tunnel test; (b) standard k–ε model; (c) LK model; (d) MMK model.

(3) *Mean velocity vector.*

(a) *2D square rib (Fig. 5).* As shown in Fig. 5, flow separation occurs evidently at the frontal corner and the flow does not reattach to the roof in the wind tunnel test under the conditions of approaching wind used here. The reproduction of separation by computation depends on various factors, i.e. grid layout, turbulence model, etc. Since the flow separation can be rather easily reproduced in the computation of the 2D square rib in comparison with that of a 3D bluff body like a cube, flow separation occurs even for the standard k–ε model. However, the flow separation region is

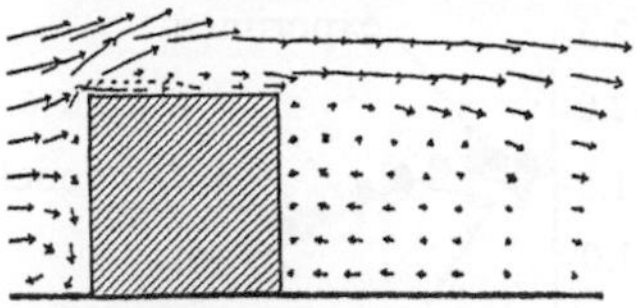

(a) wind tunnel test

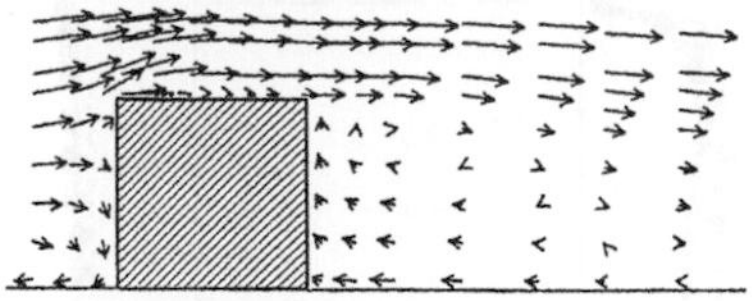

(b) standard k-ε model

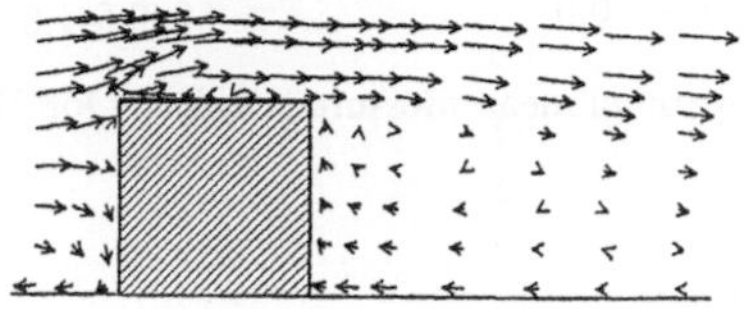

(c) MMK model

Fig. 6. Comparison of velocity vector around cube: (a) wind tunnel test; (b) standard k–ε model; (c) MMK model.

narrower for the standard k–ε model than for the wind tunnel test. Consequently, flow reattachment, which is not observed in the wind tunnel test, occurs at the center of the roof. The reverse flow is weaker for the standard k–ε model than for the wind tunnel test in the vicinity of the frontal corner. Both revised k–ε models indicate improved results in which the separated flows from the frontal corner do not reattach on the roof and the reverse flows become stronger than that of the standard k–ε model. However, the reverse flows are still weaker than that of the wind tunnel test[4].

(b) *Cube (Fig. 6)*. Fig. 6 shows the velocity vector distributions on the vertical section on the center section of the cube. The MMK model can reproduce well the flow separation and reattachment on the roof observed in the wind tunnel test. As pointed out in Refs. [1,2], the standard k–ε model cannot reproduce the flow separation from the frontal corner of the roof. It should be mentioned that the same boundary conditions and the same numerical methods are adopted for both the MMK model and the standard k–ε model.

[4] Though the figure is omitted here, the distance of reattachment in the wake region of the model was about 4.8 in the wind tunnel test, about 4.6 in the standard k–ε model, about 5.2 in the LK model and about 5.3 in the MMK model. Thus, the revised k–ε model tends to estimate a longer distance of reattachment on the floor than that by the standard k–ε model.

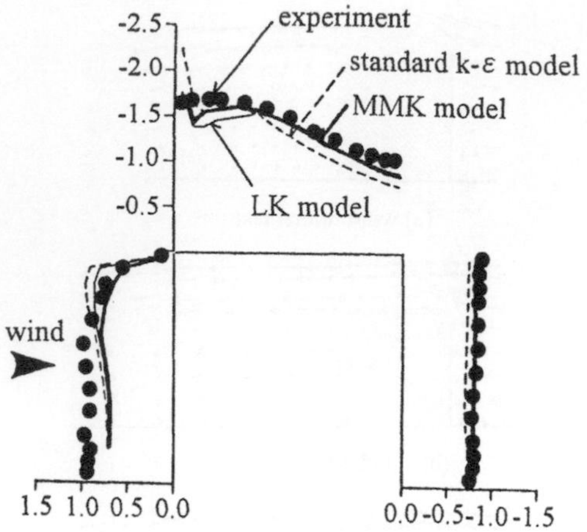

Fig. 7. Comparison of mean pressure coefficient for 2D square rib.

(4) *Mean pressure coefficient.*

(a) *2D square rib (Fig. 7).* Surface pressure distributions on the 2D square rib are compared in Fig. 7 [5]. The results for the standard k–ε model deviate greatly around the frontal corner, and the large negative peak just at the corner decreases rapidly. This distribution is peculiar to the standard k–ε model [1–3]. In the impinging region, the standard k–ε model overestimates the surface pressure on the windward face. These discrepancies observed in the standard k–ε model are closely related to the overestimation of turbulence kinematic energy k, which gives rise to large ν_t around the frontal corner of the roof. Since the MMK model corrects the distribution of k and ν_t, the accuracy in predicting flow separation and surface pressure distribution are improved sufficiently well. The LK model also corrects the discrepancy of the surface pressure, but the correction is insufficient.

(b) *Cube (Fig. 8).* Surface pressure distributions on the cube are compared in Fig. 8. The result of the standard k–ε model is quite different from that of the experiment around the frontal corner of the roof, as for the 2D square rib. In the impinging region, furthermore, the standard k–ε model overestimates the surface pressure on the windward face. However, the surface pressure distribution of the MMK model coincides very well with that of the wind tunnel test not only around the frontal corner of the roof but also in the impinging region on the windward face.

3.3. *Results of computations for oblique wind angles*

In order to confirm the applicability of the MMK model to wind engineering applications, the simulations are carried out for flowfields around a low rise building

[5] Wind pressure coefficients were normalized by the dynamic pressure at a model height H_b without the model both for the computation and the wind tunnel test.

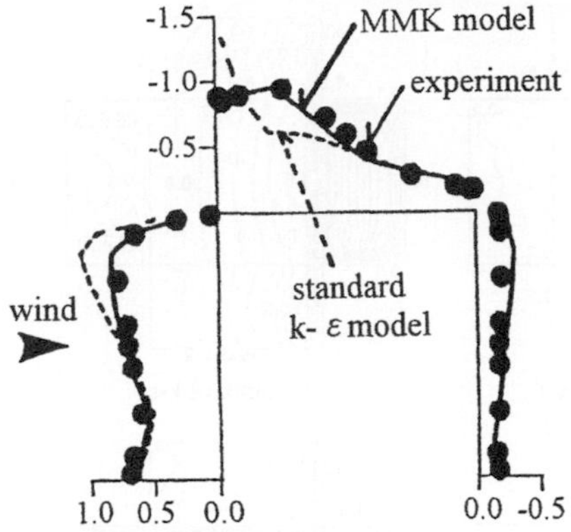

Fig. 8. Comparison of mean pressure coefficient for cube.

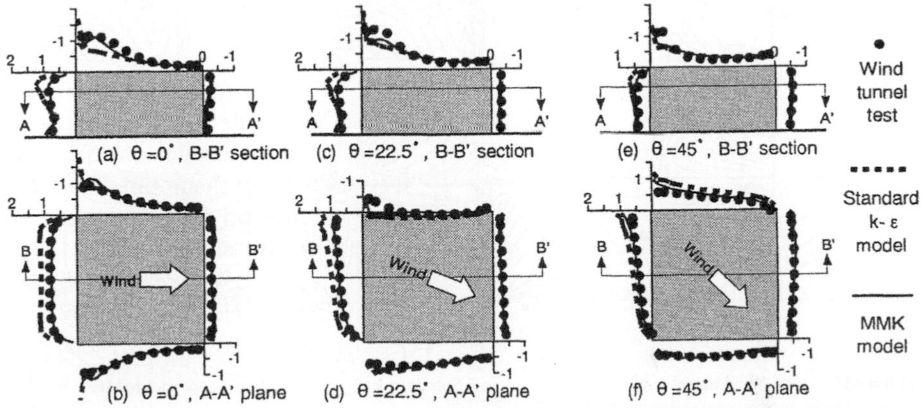

Fig. 9. Comparison of mean wind pressure coefficients for oblique wind angle: (a) $\theta = 0°$, B–B′ section; (b) $\theta = 0°$; A–A′ plane; (c) $\theta = 22.5°$, B–B′ section; (d) $\theta = 22.5°$, A–A′ plane; (e) $\theta = 45°$, B–B′ section;(f) $\theta = 45°$, A–A′ plane.

model with 1 : 1 : 0.5 shape for three wind angles, 0°, 22.5°, 45° [10]. Figs. 9 and 10 compare the distributions of mean wind pressure coefficients acting on the surface of the 1 : 1 : 0.5 model (see footnote 5).

(1) *Comparison of pressure distributions for wind direction 0° (Fig. 9a, Fig. 9b and Fig. 10a).*

(a) *Roof.* It is found from the wind tunnel test that the variation of wind pressure coefficient on the windward side of the roof is small and the absolute value of negative pressure gradually decreases toward the leeward side. The result for the standard k–ε model shows a very different distribution from that of the wind tunnel test, i.e. the negative pressure rapidly decreased near the frontal corner and slowly decreased toward the leeward side (Fig. 9a and Fig. 10a). This is because, although a relatively

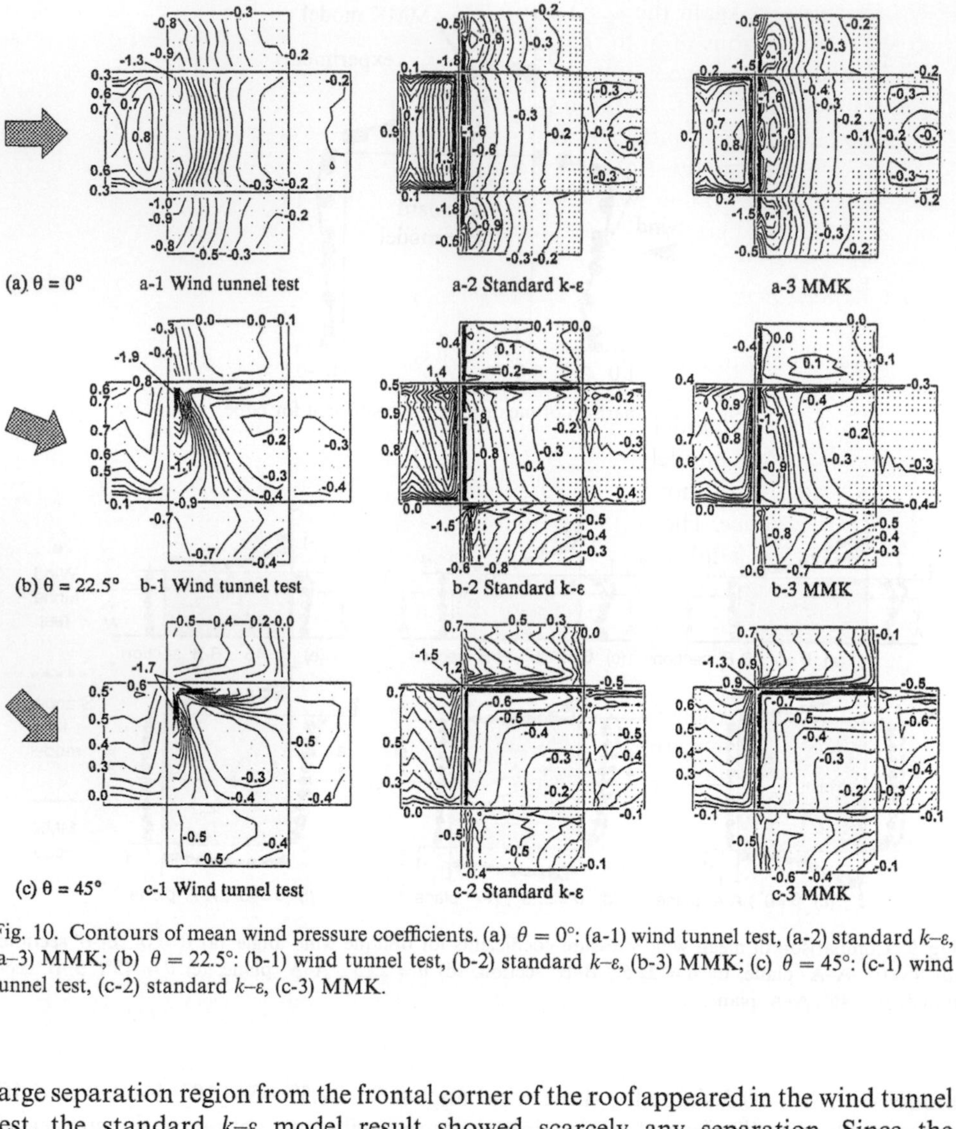

Fig. 10. Contours of mean wind pressure coefficients. (a) $\theta = 0°$: (a-1) wind tunnel test, (a-2) standard k–ε, (a–3) MMK; (b) $\theta = 22.5°$: (b-1) wind tunnel test, (b-2) standard k–ε, (b-3) MMK; (c) $\theta = 45°$: (c-1) wind tunnel test, (c-2) standard k–ε, (c-3) MMK.

large separation region from the frontal corner of the roof appeared in the wind tunnel test, the standard k–ε model result showed scarcely any separation. Since the MMK model can reproduce the flow separation from the frontal corner of the roof, the pressure distribution obtained from the MMK model shows much better agreement with that from the wind tunnel test than that from the standard k–ε model.

(b) *Windward face*. In comparison with the experimental result, the wind pressure coefficient obtained by the standard k–ε model is greatly overestimated in the vicinity of the impinging region, the maximum value exceeding 1.0. However, the result of the MMK model corresponds well to the test result. This is because the MMK model improves the distribution of ν_t near the windward face, and thus the distribution of

Reynolds stresses. Again the same boundary conditions and the same numerics are used for computations of both the MMK model and the standard k–ε model.

(c) *Leeward face.* The computation results for both models correspond pretty well to the test result.

(d) *Side face.* The computed values on the windward corner of the side faces for the MMK model are slightly larger than those obtained from the wind tunnel test. However, they agree rather well with the experimental results. On the other hand, the standard k–ε model greatly overestimates the negative pressure at the windward end of the side faces.

(2) *Comparison of pressure distributions for wind direction 22.5° (Fig. 9c, Fig. 9d and Fig. 10b).*

(a) *Roof.* The wind tunnel test clearly demonstrates the distribution of wind pressure coefficient caused by conical vortices (Fig. 10b-1). However, the distribution of the standard k–ε model is quite different from that of the wind tunnel test. The results of the MMK model are more accurate than those of the standard k–ε model, although they are still not accurate enough (Fig. 9c and Fig. 10b).

(b) *Windward face.* The computational result of the MMK model agrees very well with the test result, while the agreement of the standard k–ε model is very poor as well as for the wind direction 0° (Fig. 9c, Fig. 9d and Fig. 10b).

(c) *Leeward face.* The difference between the results of the standard k–ε model and the MMK model is negligible, and these results agree with the test result.

(d) *Side face.* As can be seen at the upper part of Fig. 9d, the absolute value of negative pressure at the windward corner decreases gradually in the wind tunnel test. However, the values for both models decrease rapidly and only a little improvement is discernible in the result of the MMK model (Fig. 9d and Fig. 10b).

(3) *Comparison of pressure distributions for wind direction 45° (Fig. 9e, Fig. 9f and Fig. 10c).*

(a) *Roof.* In the wind tunnel test, the distribution of wind pressure coefficient caused by conical vortices can be clearly observed, as well as for wind direction of 22.5° (Fig. 10c-1). However, the standard k–ε model cannot reproduce conical vortices at all. The results of the MMK model are more accurate than those of the standard k–ε model, although they are still not accurate enough (Fig. 9e and Fig. 10c). It is thought that to reproduce conical vortices well, further improvements of the model are required, such as the introduction of convective effects etc. [11, 12] into the MMK model.

(b) *Side face.* At both faces of the windward side where the positive pressure appears, the results of the MMK model show better agreement with the test results than those of the standard k–ε model. However, the values for both models correspond well to the test results at both faces of the leeward side showing negative pressure (Fig. 9e, Fig. 9f and Fig. 10c).

4. Conclusions

(1) Flow and pressure fields around three types of bluff bodies such as a 2D square rib, a cube and a low-rise building model with 1 : 1 : 0.5 shape were simulated by three

types of k–ε model, i.e. the standard k–ε model, a revised k–ε model (LK model) and a new k–ε model (MMK model) proposed by the present authors.

(2) The LK model and the MMK model provide better results than the standard k–ε model. The MMK model provides more accurate results than does the LK model in some respects, i.e. better distribution of ν_t, better reproduction of mean surface pressure distributions, etc.

(3) The applicability of the MMK model to various complex turbulent flowfields around buildings seems to be quite good from the viewpoint of engineering applications.

(4) Some points remain to be corrected in the results of the MMK model, as for the roof pressure distribution caused by conical vortices for oblique wind angles.

References

[1] S. Murakami, A. Mochida, Y. Hayashi, Examining the k–ε model by means of a wind tunnel test and large eddy simulation of the turbulence structure around a cube, J. Wind Eng. Ind. Aerodyn. 35 (1990) 87–100.

[2] S. Murakami, Comparison of various turbulence models applied to a bluff body, J. Wind Eng. Ind. Aerodyn. 46&47 (1993) 21–36.

[3] S. Murakami, A. Mochida, Numerical simulation of air flow around surface-mounted square rib by means of ASM and k–ε EVM, in: ASCE 9th Structural Congress, Indianapolis, USA, 1991.

[4] S. Murakami, A. Mochida, R. Ooka, Numerical simulation of flowfield over surface-mounted cube with various second-moment closure models, in: 9th Symp. On Turbulent Shear Flow, 1993, pp. 13-5-1–13-5-6.

[5] B.E. Launder, M. Kato, Modeling flow-induced oscillations in turbulent flow around a square cylinder, in: ASME Fluid Engng. Conf., June 1993, Washington, DC.

[6] M. Kato, B.E. Launder, The modeling of turbulent flow around stationary and vibrating square cylinders, in: 9th Symp. on Turbulent Shear Flows, Kyoto, 1993, pp. 10-4-1–10-4-6.

[7] K. Kondo, S. Murakami, A. Mochida, Numerical study on flow field around square rib using revised k–ε model, in: 8th Symp. on Numerical Fluid Dyn, 1994, pp. 363–366 (in Japanese).

[8] K. Kondo, S. Murakami, A. Mochida, Y. Ishida, Numerical prediction of flow field around 2D square rib using revised k–ε model, J. Wind Eng. (Japan) 63 (1995) 35 (in Japanese).

[9] Y. Ishida, S. Murakami, A. Mochida, K. Kondo, 3-D numerical simulation of turbulent airflow around building using fortified solution algorithm based on composite grid system (Part 4), in: Proc. AIJ Annual Meeting, 1995, pp. 573–574 (in Japanese).

[10] K. Kondo, Y. Ishida, M. Tsuchiya, Numerical simulation of flow field 1 : 1 : 0.5 rectangular prism (Part 1 and 2), in: Proc. AIJ Annual Meeting, 1994, pp. 251–254 (in Japanese).

[11] A. Yoshizawa, Turbulence modeling: construction of eddy-viscosity-type models, in: 7th Symp. on Numer. Fluid Dyn., 1993, pp. 25–33 (in Japanese).

[12] S. Kawamoto, The k–ε–ϕ turbulence model for estimation of wind force, in: 8th Symp. on Numerical Fluid Dyn., 1994, pp. 367–370 (in Japanese).

[13] A. Mochida, Y. Ishida, S. Murakami, Numerical study on flowfield around structures with oblique wind angle based on composite grid system, in: Proc. 7th U.S. National Conf. on Wind Eng., 1993.

[14] S. Murakami, A. Mochida, Y. Tominaga, Numerical simulation of turbulent diffusion in cities, Wind climate in cities, Kluwer Academic Publishers, Dordrecht, 1995, pp. 681–701.

ELSEVIER

Journal of Wind Engineering
and Industrial Aerodynamics 67&68 (1997) 183–193

JOURNAL OF wind engineering AND industrial aerodynamics

Two dimensional discrete vortex method for application to bluff body aerodynamics

Jens Honoré Walther[a,*], Allan Larsen[b]

[a] Danish Maritime Institute, Hjortekærsvej 99, DK-2800 Lyngby, Denmark
[b] COWI A/S, Parallelvej 15, DK-2800 Lyngby, Denmark

Abstract

Two-dimensional viscous incompressible flow past a flat plate of finite thickness and length is simulated using the discrete vortex method. Both a fixed plate and a plate undergoing a harmonic heave and pitch motion are studied. The Reynolds number is 10^4 and the reduced onset flow speed, U/fc is in the range 2–14. The fundamental kinematic relation between the velocity and the vorticity is used in a novel approach to determine the surface vorticity. An efficient influence matrix technique is used in a fast adaptive multipole algorithm context to obtain a mesh-free method. The numerical results are compared with the steady-state Blasius solution, and with the inviscid solution for the flow past an oscillating plate by Theodorsen.

Keywords: Discrete vortex method; Vorticity boundary condition; Conservation of vorticity moments; Flow past a flat plate; Aerodynamic derivatives

1. Introduction

The flow past a flat plate, both fixed and in heave or pitch motion serves as a realistic but simple test case for the numerical study of flow past oscillating bluff bodies. The flow possesses most of the features of bluff body flows, e.g., boundary-layer growth, vortex shedding and interaction between free shear layers. Also the results can be compared with the steady-state solution due to Blasius regarding the boundary-layer characteristics, cf. Ref. [1], and with the inviscid solution by Theodorsen for the oscillating plate [2]. For small amplitudes and relatively high Reynolds numbers the discrete vortex method results are expected to compare well with the inviscid results.

* Corresponding author. E-mail: jhw@danishmaritime.dk.

PII S0167-6105(97)00072-X

In an accompanying paper [3], extensive tests are presented using the present method for aeroelastic analysis of bridge sections.

2. Mathematical formulation

2.1. Kinetic and kinematic relations

The kinetics of a two-dimensional laminar flow with constant kinematic viscosity ν and density ρ in a domain $\mathscr{D}$ bounded by $\partial\mathscr{D} = \mathscr{B}$ is governed by the vorticity transport equation

$$\partial_t \omega + (\boldsymbol{v} \cdot \nabla)\omega = \nu \nabla^2 \omega , \tag{1}$$

where $\boldsymbol{v}(\boldsymbol{x}, t)$ is the velocity and $\boldsymbol{\omega}(\boldsymbol{x}, t) = \omega(\boldsymbol{x}, t)\boldsymbol{e}_z$ the vorticity.

The kinematic relation between the velocity and the vorticity may be formulated as an integral equation, cf. [4]

$$\boldsymbol{v}(\boldsymbol{x}, t) = \boldsymbol{U}(t) - \frac{1}{2\pi} \iint_{\mathscr{D}} \frac{\boldsymbol{\omega}_0 \times (\boldsymbol{x}_0 - \boldsymbol{x})}{|\boldsymbol{x}_0 - \boldsymbol{x}|^2} \, \mathrm{d}\mathscr{D}_0$$

$$+ \frac{1}{2\pi} \oint_{\mathscr{B}} \frac{(\boldsymbol{v}_0 \cdot \boldsymbol{n}_0)(\boldsymbol{x}_0 - \boldsymbol{x}) - (\boldsymbol{v}_0 \times \boldsymbol{n}_0) \times (\boldsymbol{x}_0 - \boldsymbol{x})}{|\boldsymbol{x}_0 - \boldsymbol{x}|^2} \, \mathrm{d}\mathscr{B}_0 , \tag{2}$$

where $\boldsymbol{U}(t)$ is the irrotational onset flow, $\boldsymbol{n}$ is the surface normal, and $\boldsymbol{v}$ is the velocity at the surface. The boundary integral accounts for the vorticity not included in $\mathscr{D}$. Hence, if $\mathscr{D}$ contains all non-zero vorticity, the contribution from the surface integral vanishes.

2.2. Vorticity boundary conditions

Following the work of Wu and Gulcat [5], the value of the vorticity at the solid boundary is found via the kinematic relation (2), where the value of both the volume and surface integrals are known, except the contribution from the surface vorticity, i.e.,

$$\iint_{\mathscr{D}_{\mathscr{B}}} \frac{\omega_0 (\boldsymbol{x}_0 - \boldsymbol{x}_{\mathscr{B}})}{|\boldsymbol{x}_0 - \boldsymbol{x}_{\mathscr{B}}|^2} \, \mathrm{d}\mathscr{D}_0 = \mathscr{T}(\boldsymbol{x}_{\mathscr{B}}) + 2\pi[\boldsymbol{U}(t) - \boldsymbol{v}(\boldsymbol{x}_{\mathscr{B}})] , \tag{3}$$

where $\mathscr{D}_{\mathscr{B}}$ is a "thin" layer adjacent to the surface $\mathscr{B}$, and the vector $\mathscr{T}(\boldsymbol{x}_{\mathscr{B}})$ is the induced velocity from the vorticity in the fluid and solids excluding the surface vorticity

$$\mathcal{T}(\boldsymbol{x}_{\mathcal{B}}) = \oint_{\mathcal{B}} \frac{(\boldsymbol{v}_0 \cdot \boldsymbol{n}_0)(\boldsymbol{x}_0 - \boldsymbol{x}_{\mathcal{B}}) - (\boldsymbol{v}_0 \times \boldsymbol{n}_0) \times (\boldsymbol{x}_0 - \boldsymbol{x}_{\mathcal{B}})}{|\boldsymbol{x}_0 - \boldsymbol{x}_{\mathcal{B}}|^2} \, d\mathcal{B}_0 - \int\int_{\mathcal{D}-\mathcal{D}_{\mathcal{B}}} \frac{\boldsymbol{\omega}_0 \times (\boldsymbol{x}_0 - \boldsymbol{x}_{\mathcal{B}})}{|\boldsymbol{x}_0 - \boldsymbol{x}_{\mathcal{B}}|^2} \, d\mathcal{D}_0 \,. \tag{4}$$

In the vortex method context it is convenient to introduce the surface vortex sheet, γ

$$\frac{\partial \gamma}{\partial n} = \omega \,, \tag{5}$$

where n is the surface normal, thus, using Eq. (5) in (4) gives

$$\int_{\mathcal{B}} \frac{\gamma_0 \boldsymbol{e}_z \times (\boldsymbol{x}_0 - \boldsymbol{x}_{\mathcal{B}})}{|\boldsymbol{x}_0 - \boldsymbol{x}_{\mathcal{B}}|^2} \, d\mathcal{B}_0 = \mathcal{T}(\boldsymbol{x}_{\mathcal{B}}) + 2\pi[\boldsymbol{U}(t) - \boldsymbol{v}(\boldsymbol{x}_{\mathcal{B}})]. \tag{6}$$

The components of the vector equation (6) are Fredholm integral equations in the unknown γ_0. The solution is unique up to a constant, i.e. an infinite number of solutions exist, cf. Ref. [4]. The solution is made unique by imposing the global constraint on the vorticity, that the time rate of change of the total vorticity in both the solid ($\mathcal{S}$) and fluid domain ($\mathcal{F}$) is zero

$$\partial t \int\int_{\mathcal{F} \cup \mathcal{S}} \omega \, d\mathcal{D} = 0. \tag{7}$$

Moreover, if the total vorticity is zero at $t = 0$ it remains zero for $t > 0$. Thus for each solid, Eq. (7) is imposed making the system of equations overdetermined. Eqs. (6) and (7) are solved in the least-squares sense.

In the previous implementations, e.g. Refs. [4, 6–9] Eq. (6) is solved for zero tangential velocity. In the present study Eq. (6) is solved for the normal component. Thus, the no-slip condition is imposed implicitly as a consequence of Eqs. (6) and (7).

3. Vortex method

The Lagrangian solution to Eqs. (1) and (2) involves tracking individual fluid elements ($\boldsymbol{x}_p$, Γ_p) according to

$$\frac{d\boldsymbol{x}_p}{dt} = \boldsymbol{v}(\boldsymbol{x}_p, t) \,, \tag{8}$$

$$\frac{d\omega}{dt} = \nu \nabla^2 \omega \,, \tag{9}$$

where $\boldsymbol{v}(\boldsymbol{x}_\mathrm{p}, t)$ is approximated from Eq. (2)

$$\boldsymbol{v}(\boldsymbol{x}_\mathrm{p}, t) = \boldsymbol{U}(t) - \frac{1}{2\pi}\sum_{j=1}^{n_s} \oint_{\mathscr{B}_j} \frac{(\gamma_0 \boldsymbol{e}_z + \boldsymbol{v}_0 \times \boldsymbol{n}_0) \times (\boldsymbol{x}_0 - \boldsymbol{x}_p) - (\boldsymbol{v}_0 \cdot \boldsymbol{n}_0)(\boldsymbol{x}_0 - \boldsymbol{x}_p)}{|\boldsymbol{x}_0 - \boldsymbol{x}_p|^2} \mathrm{d}\mathscr{B}_0 - \frac{1}{2\pi}\sum_{i=1}^{n_v} \frac{\Gamma_i \boldsymbol{e}_z \times (\boldsymbol{x}_i - \boldsymbol{x}_\mathrm{p})}{|\boldsymbol{x}_i - \boldsymbol{x}_\mathrm{p}|^2} \,. \tag{10}$$

n_s is the number of solids, and n_v is the number of free vortices with vortex strength, Γ_i. The singular vortex–vortex interaction in Eq. (10) is regularised by applying a Gaussian core function. The solids are approximated by polygons and the surface vortex sheet and surface velocity are given a linear variation when evaluating Eq. (10), and when solving Eq. (6).

Eq. (8) is solved numerically by standard ordinary differential equation methods. The diffusion equation (9) is approximated by random walks, thus, the solution to Eqs. (8) and (9) using Euler integration is

$$\boldsymbol{x}_\mathrm{p}^{k+1} = \boldsymbol{x}_\mathrm{p}^k + \boldsymbol{v}(\boldsymbol{x}_\mathrm{p}^k)\Delta t + \boldsymbol{\eta}_\mathrm{p}\,, \tag{11}$$

where $\boldsymbol{x}_\mathrm{p}^k$ is the position of the fluid element at the kth time step, Δt is the time step, and $\boldsymbol{\eta}_\mathrm{p}$ are random numbers with zero mean and variance $2\nu\Delta t$.

The surface vortex sheets are converted into vortex blobs and, subsequently, diffused into the flow by one-sided random walks. Vortices already in the flow entering the solids by a random walk are removed from the calculations. The conservation of the total vorticity (7) is modified accordingly,

$$\sum_{i=1}^{m_j} \frac{\gamma_{ij}^k - \gamma_{ij}^{a,k-1}}{\Delta t} \Delta s_{ij} + 2A_j \frac{\Omega_j^k - \Omega_j^{k-1}}{\Delta t} = 0, \tag{12}$$

for each jth solid. m_j is the number of boundary elements, γ_{ij}^k is the released circulation at the kth time step, $\gamma_{ij}^{a,k-1}$ is the annihilated circulation from the previous time step, Ω_j is the angular velocity, and A_j is the area of the solid.

The costly $\mathcal{O}(n_v^2)$ operations imposed by the kinematic relation (2) are overcome by the $\mathcal{O}(n_v)$ fast adaptive multipole algorithm in Ref. [10], see also Ref. [11] for details regarding the implementation. Also, to limit the number of computational elements, vortices in the wake are amalgamated when sufficiently far downstream the solid bodies, here $|\boldsymbol{x}_p| > 6$. Two vortices are merged to satisfy the zeroth and first-order vorticity moment, provided they satisfy the following expression:

$$\left|\frac{\Gamma_i \Gamma_j}{\Gamma_i + \Gamma_j}\right| |\boldsymbol{x}_i - \boldsymbol{x}_j| < \varepsilon\,, \tag{13}$$

where ε is a small number, cf. Ref. [12].

3.1. Aerodynamic forces

The aerodynamic forces are found by a momentum balance, or by integrating the surface pressure. The derivations are given in Refs. [11,13].

3.1.1. Momentum balance

Considering a momentum balance, the total aerodynamic forces are given by

$$\boldsymbol{F} = -\rho\frac{\mathrm{d}\boldsymbol{\alpha}}{\mathrm{d}t} + \rho\sum_{j=1}^{n_s}\frac{\mathrm{d}}{\mathrm{d}t}\iint_{\mathcal{S}_j}\boldsymbol{v}\,\mathrm{d}\mathcal{D}\,, \tag{14}$$

where $\boldsymbol{\alpha}$ is the first-order moment of the vorticity

$$\boldsymbol{\alpha} = \iint_{\mathcal{D}}\boldsymbol{x}\times\boldsymbol{\omega}\,\mathrm{d}\mathcal{D}\,. \tag{15}$$

For a general translating and rotating body (described by the position of the centre of rotation, $\boldsymbol{x}_{\mathrm{rot}}$ and the angle of attack, θ) it is easy to show that Eq. (14) can be written as

$$F = -\rho\left(\frac{\mathrm{d}\boldsymbol{\alpha}}{\mathrm{d}t}\right)_{\mathcal{F}} + \rho\frac{\mathrm{d}}{\mathrm{d}t}\sum_{j=1}^{n_s}\left[\boldsymbol{v}_{\mathrm{rot}}A_j + 2\Omega_jA_j\boldsymbol{e}_z\times\boldsymbol{x}_{\mathrm{rot}} + 3\Omega_j\boldsymbol{e}_z\times\boldsymbol{M}_j\right], \tag{16}$$

where $\boldsymbol{v}_{\mathrm{rot}}$ is the velocity of the centre of rotation, and $\boldsymbol{M}_j$ is the first-order moment of area of the jth solid

$$\boldsymbol{M}_j = \iint_{\mathcal{S}_j}(\boldsymbol{x} - \boldsymbol{x}_{\mathrm{rot}})\,\mathrm{d}\mathcal{D}\,. \tag{17}$$

The first-order moment of the vorticity is approximated by the sum

$$\boldsymbol{\alpha} \approx \sum_{i=1}^{n_v}\boldsymbol{x}_i\times\boldsymbol{e}_z\Gamma_i\,. \tag{18}$$

3.1.2. Surface pressure

From the integral expression (14) the forces on the individual solid bodies, not given its nature, the local loads can be determined. To this end, the local pressure distribution is needed.

Given the no-slip boundary condition, the Naviér–Stokes equations reduce to

$$\frac{1}{\rho}\frac{\partial p}{\partial s} = -\nu\frac{\partial\omega}{\partial n} - a_s\,, \tag{19}$$

where n and s are the normal and tangential direction, and a_s is the tangential acceleration of the boundary. Now neglecting the streamwise diffusion at the solid boundary, the vorticity transport equation (1) reads

$$\frac{\partial \omega}{\partial t} = \nu \frac{\partial^2 \omega}{\partial n^2}, \tag{20}$$

and using Eq. (5) gives

$$\frac{\partial \gamma}{\partial t} = \nu \frac{\partial \omega}{\partial n}. \tag{21}$$

Inserting Eq. (21) in Eq. (19) gives

$$\frac{1}{\rho}\frac{\partial p}{\partial s} = -\frac{\partial \gamma}{\partial t} - a_s. \tag{22}$$

The flux of circulation, $(\partial\gamma/\partial t)$ at the ith boundary element of the jth solid is given by

$$\left(\frac{\partial \gamma}{\partial t}\right)_{ij} \approx \frac{\gamma_{ij}^{k} - \gamma_{ij}^{a,k-1}}{\Delta t}. \tag{23}$$

Thus Eq. (22) is integrated using Eq. (23) along the solid boundary to compute the pressure distribution. Notice that by virtue of Eq. (12) the pressure is singled-valued since

$$\frac{1}{\rho}\oint_{\mathscr{B}_j} \frac{\partial p}{\partial s}\, \mathrm{d}\mathscr{B} = -\oint_{\mathscr{B}_j} \frac{\partial \gamma}{\partial t}\, \mathrm{d}\mathscr{B} - \frac{\partial}{\partial t}\oint_{\mathscr{B}_j} v_s\, \mathrm{d}\mathscr{B} \tag{24}$$

$$= -\oint_{\mathscr{B}_j} \frac{\partial \gamma}{\partial t}\, \mathrm{d}\mathscr{B} - \frac{\partial}{\partial t}\iint_{\mathscr{S}_j} \omega \cdot e_z\, \mathrm{d}\mathscr{D} \tag{25}$$

$$\approx -\sum_{i=1}^{m_j} \frac{\gamma_{ij}^{k} - \gamma_{ij}^{a,k-1}}{\Delta t}\Delta s_{ij} - 2A_{ij}\frac{\Omega_j^{k} - \Omega_j^{k-1}}{\Delta t} = 0. \tag{26}$$

4. Simulation of unsteady flow past a flat plate

4.1. Fixed plate

The flow past a fixed flat plate at Reynolds number 10^4 is simulated to confirm the ability of the present method to accurately predict boundary-layer growth, and to validate the implicit enforcement of the no-slip condition. The plate is given a finite thickness (H) to obtain a non-trivial potential flow at $t = 0^+$ (the potential flow past an infinite thin plate is $\gamma_i = 0, \forall i$). The necessary thickness of the plate to prevent vortices "diffusing" across the plate is restricted by the Reynolds number and the time-step size, or equivalently, the standard deviation of the random walk, hence $H > 3\sqrt{2\nu\Delta t}$. The chord length (c) and onset flow speed (U) are unity, and a chord-to-thickness ratio of 200 is used throughout. Approximately 400 boundary elements are used to discretize the plate. The vortices are advanced in time using the second-order Adams–Bashforth scheme with a time step of 0.005.

Velocity profiles are computed at 40 x coordinate positions equidistantly spaced along the plate, each profile consisting of 20 velocities. The simulation is carried out to $t = 6$, sampling the velocity profiles from $t = 5$ to 6, with a total of 20 samples.

Fig. 1 shows the position of the vortices close to the plate at time $t = 6$. A boundary layer is clearly visible, and the wake extending six chord lengths downstream (not shown) is unstable showing Kelvin–Helmholtz instability.

The time-average boundary-layer thickness (δ), displacement thickness (δ_1), and momentum thickness (δ_2) are displayed in Fig. 2. An excellent agreement is observed for $x/c < 0.8$. The predicted shape factor is $H_{12} = \delta_1/\delta_2 = 2.5$, -4% off the Blasius value of 2.6. Fig. 3 shows the 10 centre velocity profiles ($0.38 > x/c > 0.68$). Again an excellent agreement is observed, and the no-slip condition is reasonably satisfied.

For lower Reynolds numbers (not presented), the presence of the trailing edge is felt for smaller x co-ordinate values ($x/c < 0.5$). At higher Reynolds numbers the

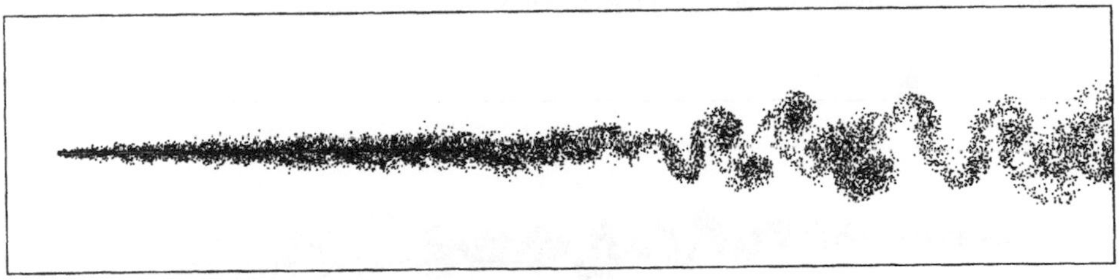

Fig. 1. Position of the vortices for the flow past a fixed flat plate at Reynolds number 10^4. The wake extends approximately six chord lengths downstream (not shown).

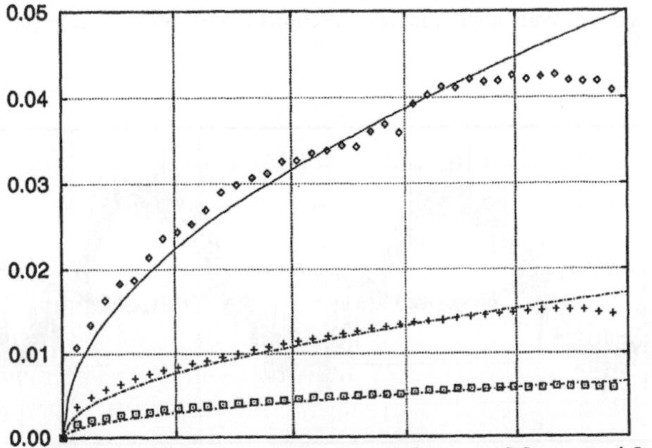

Fig. 2. Time-average boundary-layer characteristics for the viscous flow past a fixed flat plate at Reynolds number 10^4. The profiles are based on 20 velocity samples from $t = 5$–6. Theory: (———) Blasius; present results: (◇) boundary-layer thickness; (+) displacement thickness; (□) momentum thickness.

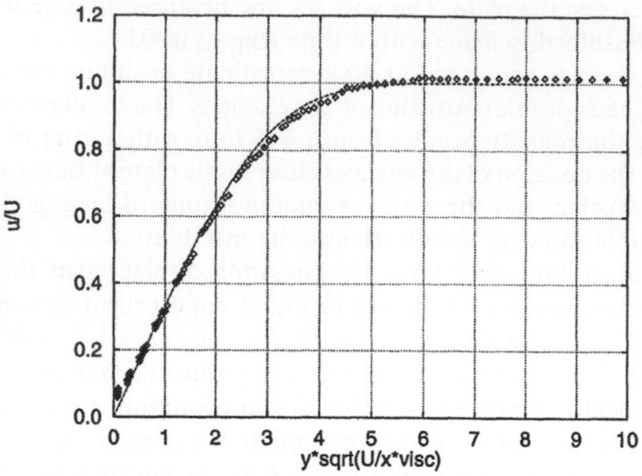

Fig. 3. Time-averaged velocity distribution in the boundary-layer of a fixed flat plate at Reynolds number 10^4. Theory (———), Blasius solution; (◇) present results.

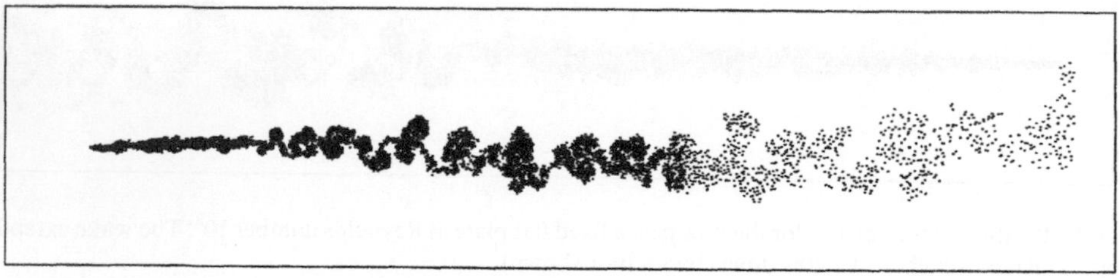

Fig. 4. Position of the vortices for the flow past a flat plate at Reynolds number 10^4 undergoing a harmonic pitch motion. The wake extends approximately 32 chord lengths downstream (not shown). $v_r = 12$ and $t = 32$.

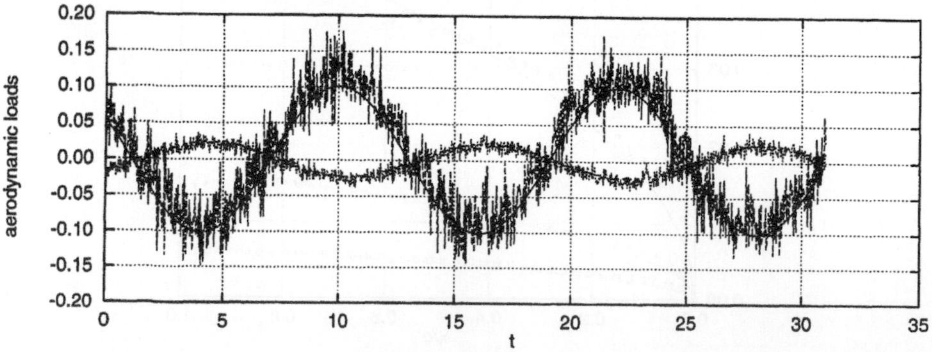

Fig. 5. Time history of the aerodynamic lift and moment for the flow past a flat plate in pitch motion at Reynolds number 10^4. $v_r = 12$. Vortex method: (– –) lift; (– – –) moment. Modelled lift and moment from Eqs. (27) and (28) (—).

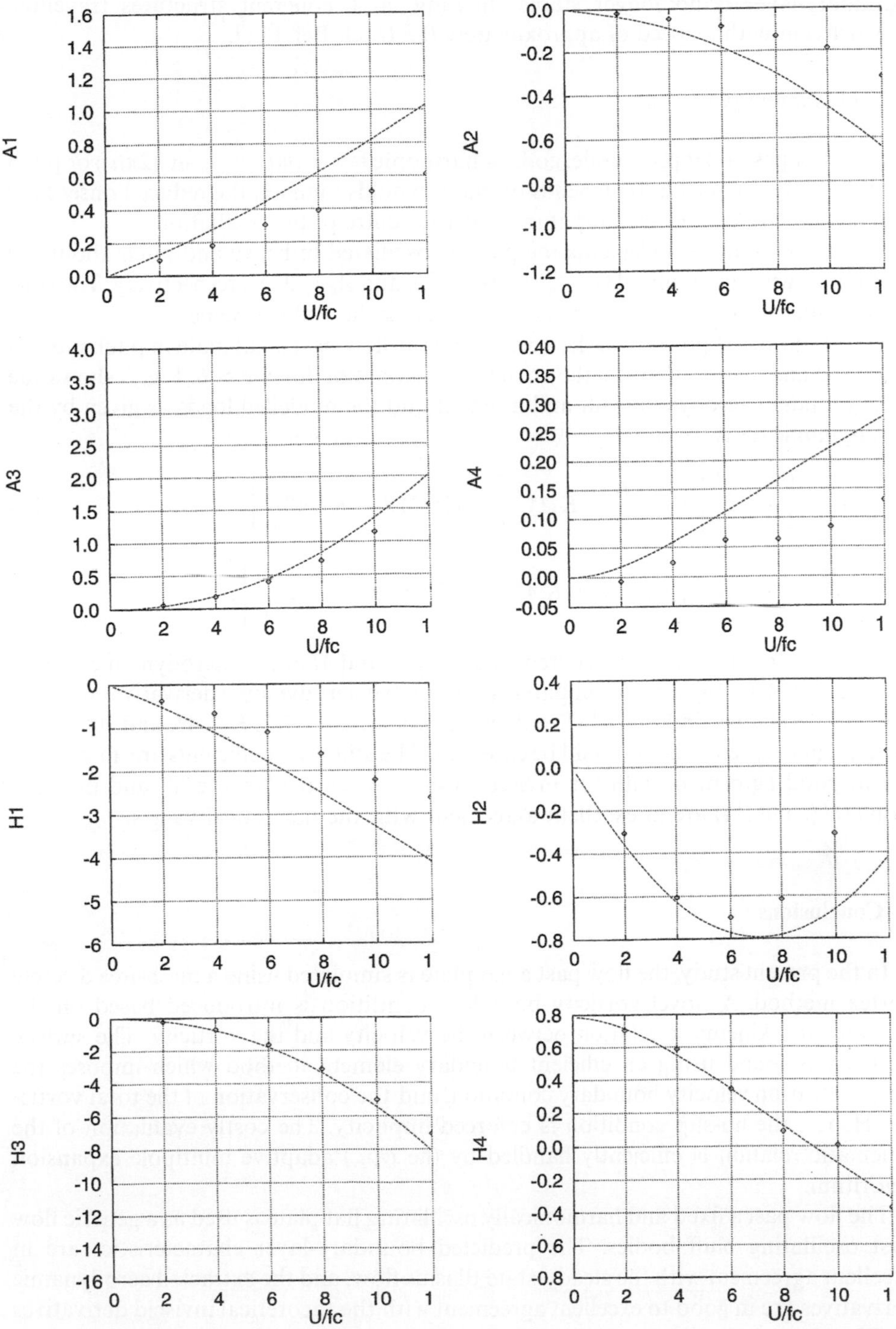

Fig. 6. Aerodynamic derivatives for the flow past a flat plate at Reynolds number 10^4. (◇) present results; (—) inviscid theory [2].

boundary-layer is no longer stable showing large coherent structures travelling downstream with a speed of approximately 0.5 U, cf. Ref. [11].

4.2. Oscillating plate

The flow past a flat plate undergoing a harmonic heave ($h(t) = A_h \sin(2\pi ft)$) or pitch motion ($\theta(t) = A_\theta \sin(2\pi ft)$) depends on the Reynolds number, the reduced onset flow speed, $v_r = U/(fc)$, and the amplitude and the centre of the oscillation.

Using 200 boundary elements the plate is oscillated in heave and pitch about the half-chord with an amplitude of $A_h = 0.02$–0.05, and $A_\theta = 3$–$5°$, respectively. The time step is 0.025 and the vortices are advanced using the Euler scheme.

Fig. 4 shows the position of the vortices for the flow past a flat plate in pitch motion at $t = 32$ and $v_r = 12$. Notice, the vortex amalgamation for $x/c > 6$. Fig. 5 shows the corresponding aerodynamic lift and moment and the modelled loads as given by the aerodynamic derivatives

$$L = \frac{1}{2}\rho U^2 2c\left[KH_1^*\frac{\dot{h}}{U} + KH_2^*\frac{c\dot{\theta}}{U} + K^2H_3^*\theta + K^2H_4^*\frac{h}{c}\right], \tag{27}$$

$$M = \frac{1}{2}\rho U^2 2c^2\left[KA_1^*\frac{\dot{h}}{U} + KA_2^*\frac{c\dot{\theta}}{U} + K^2A_3^*\theta + K^2A_4^*\frac{h}{c}\right], \tag{28}$$

where $K = 2\pi fc/U$ is the reduced frequency, and L and M are the aerodynamic lift and moment, cf. Ref. [14]. Fig. 6 compares the extracted aerodynamic derivatives with the inviscid derivatives in Ref. [2]. The damping coefficients, H_1^*, H_2^*, A_1^*, and A_2^* are, in general, underpredicted as should be expected. The stiffness coefficients are, in general, are in good agreement with the inviscid results, in particular, the lift and moment slope (H_3^*, and A_3^*) are in excellent agreement with the theoretical values.

5. Conclusions

In the present study, the flow past a flat plate is simulated using a mesh-free discrete vortex method. A novel vorticity boundary condition is introduced based on the fundamental kinematic relation between the velocity and the vorticity. The surface vorticity is found using an efficient boundary element method which imposes the no-penetration velocity boundary condition, and the conservation of the total vorticity. Hence, the no-slip condition is enforced implicity. The costly evaluation of the kinematic relation is efficiently handled by the $\mathcal{O}(n_v)$ adaptive multipole expansion algorithm.

The flow past a fixed and harmonically oscillating flat plate is used as a generic flow past oscillating bluff bodies. The predicted boundary-layer characteristics are in excellent agreement with the steady-state Blasius flow, and the extracted aerodynamic derivatives are in good to excellent agreement with the theoretical inviscid derivatives by Theodorsen.

Acknowledgements

This work was financially supported by the COWI foundation.

References

[1] H. Schlichting, Boundary-Layer Theory, 7th ed., McGraw-Hill, New York, 1979.
[2] T. Theodorsen, General theory of aerodynamic instability and the mechanism of flutter, TR 496, NACA, 1935.
[3] A. Larsen, J.H. Walther, Aeroelastic analysis of bridge sections based on discrete vortex simulations, in: 2nd Int. Symp. on Computational Wind Engineering CWE'96, 1996.
[4] J.C. Wu, Numerical boundary conditions for viscous flow problems, AIAA J. 14 (8) (1976) 1042.
[5] J.C. Wu, U. Gulcat, Separate treatment of attached and detached flow regions in general viscous flows, AIAA J. 19 (1) (1981) 20.
[6] R.B. Kinney, Z.M. Cielak, Analysis of unsteady viscous flow past an airfoil: Part I – theoretical development, AIAA J. 15 (12) (1977) 1712.
[7] J.C. Wu, S. Sampath, N.L. Sankar, A numerical study of unsteady viscous flows around airfoils, AGARD Conf. Proc. vol. 227, Ottawa, Canada, 1977, pp. 24-1-24-18.
[8] M.E. Tasim, R.B. Kinney, Analysis of two-dimensional viscous flow over cylinders in unsteady motion, AIAA J. 22 (5) (1984) 586.
[9] I.H. Tuncer, J.C. Wu, C.M. Wang, Theoretical and numerical studies of oscillating airfoils, AIAA J. 28 (9) (1990) 1615.
[10] J. Carrier, L. Greengard, V. Rokhlin, A fast adaptive multipole algorithm for particle simulations, SIAM J. Sci. Statist. Comput. 9 (4) (1988) 669.
[11] J.H. Walther, Discrete Vortex Method for Two-dimensional Flow past Bodies of Arbitrary Shape Undergoing Prescribed Rotary and Translational Motion. Ph.D. Thesis, Department of Fluid Mechanics, Technical University of Denmark, September 1994.
[12] P.R. Spalart, Vortex methods for separated flows, NASA TM, NASA, June 1988.
[13] J.C. Wu, A theory for aerodynamic forces and moments, Technical Report, Georgia Institute of Technology, September 1978.
[14] E. Simiu, R.H. Scanlan, Wind Effects On Structures, 2nd ed., Wiley, New York, 1986.

Journal of Wind Engineering
and Industrial Aerodynamics 67&68 (1997) 195–208

Numerical simulation of flow around rectangular prism

Da-hai Yu*, Ashan Kareem

Department of Civil Engineering and Geological Science, University of Notre Dame, Notre Dame, IN 46556, USA

Abstract

Using the large eddy simulation (LES), both two-dimensional (2D) and three-dimensional (3D) simulations of the velocity and pressure fields surrounding a rigid rectangular prism at a Reynolds number of 10^5 are conducted. A finite difference numerical scheme based on a staggered grid is employed. The convection terms are discretized with either QUICK or central difference schemes, while the Leith method is employed for temporal marching. The computed mean velocity along the symmetry line is compared with experimental results as well as numerical results reported by other investigators. It is in closer agreement with the experimental results than other reported numerical results. The mean, the root mean square (RMS) pressure distribution on the prism surface, and the integral forces (lift and drag) are computed. The correlation coefficients of pressure at different chordwise locations around a square prism are also calculated. These simulations are found to be in very good agreement with available experimental results reported in the literature. Both QUICK and central difference schemes are used to discretize the convection terms, and the upwinding effect of the QUICK scheme is investigated. A grid refinement study is also conducted to evaluate the effects of grid size on the simulation results.

Keywords: Large eddy simulation; Turbulence; Bluff-body flow

1. Introduction

With developments in computational fluid dynamics and the ever-growing capabilities of computers, numerical methods are becoming a promising analysis tool in the modeling of wind-structure interactions. Turbulence models are developed to simulate high Reynolds number flows that are not achievable with direct numerical simulation due to computer limitations. Literature suggests that the large eddy

* Corresponding author. E-mail: dyul@darwin.cc.nd.edu.

PII S0167-6105(97)00073-1

simulation (LES) scheme is more attractive due to its higher accuracy in comparison with techniques based on Reynolds-averaged Navier–Stokes equations. In LES, fluid motions at scales larger than a prescribed filter size are resolved and solved directly, while smaller scales are modeled by a subgrid-scale (SGS) model.

Murakami and his group conducted a series of numerical studies involving two-dimensional square cylinders [1]. Their study has shown that the 3D modeling provides a better representation of the flow characteristics than the 2D case, and the LES modeling provides a better and more realistic simulation of the flow field. Tamura [2] simulated the flow field around a square prism using the dynamic SGS model. He also conducted the same simulation by employing a third-order upwind scheme to discretize the convection terms without a SGS model. He concluded that average and root mean square (RMS) values can be predicted by the computation with reasonable grids, while other statistics, like the correlation of unsteady pressures, did not compare well with the experimental data despite a large number of grid points (4 million). To assess the current state of the art in LES for flow past bluff bodies, a workshop was held in Germany in 1995 [3], where employed methods and obtained results provided by a number of contributors were compared. Besides the numerical simulations, a number of experimental studies are available in the literature. They can be used as a data base for validating numerical results.

2. Governing equations and numerical methods

The Navier–Stokes equations for an incompressible fluid, combined with a sub-grid-scale turbulence model, are used herein in LES of the flow around a rectangular prism. In this study, the Smagorinsky model is used for the subgrid-scale viscosity

$$\frac{\partial \bar{u}_i}{\partial t} + \bar{u}_j \frac{\partial \bar{u}_i}{\partial x_j} = -\frac{1}{\rho}\frac{\partial \bar{P}}{\partial x_i} + (\nu + \nu_{SGS})\frac{\partial \bar{s}_{ij}}{\partial x_j}, \tag{1}$$

$$\frac{\partial \bar{u}_i}{\partial x_i} = 0, \tag{2}$$

$$\nu_{SGS} = (C_s \Delta)^2 \left[\tfrac{1}{2}\bar{s}_{ij}\bar{s}_{ij}\right]^{1/2} \tag{3}$$

$$\bar{s}_{ij} = \frac{\partial \bar{u}_i}{\partial x_j} + \frac{\partial \bar{u}_j}{\partial x_i}, \tag{4}$$

where $i, j = 1, 2, 3$, $\Delta = (Dx_1 Dx_2 Dx_3)^{1/3}$ and $C_s = 0.10$ for the 3D computation. For comparison, representative 2D computational results are also presented in this paper, with $i, j = 1, 2$, $\Delta = (Dx_1 Dx_2)^{1/2}$ and $C_s = 0.15$.

These equations are non-dimensionalized using the length of the front side of the rectangular cross section, L, and the inflow velocity, U_0. The time is non-dimensionalized by L/U_0 and pressure with ρU_0^2, where ρ is the mass density of the fluid.

The Reynolds number is thus defined as LU_0/ν, with ν being the fluid kinematic viscosity.

The boundary conditions on the solid walls are of no-slip and no-penetration type, i.e., all the velocity components of the flow in x_1, x_2 and x_3 directions are set to zero on the surface. Imaginary points inside the solid wall boundaries are specified with a quadratic interpolation in order to be consistent with accuracy of the numerical scheme. The inflow boundary condition is a constant uniform velocity set equal to unity in x_1 direction and zero in x_2 and x_3 directions. The upper and lower sides of the flow domain are set with a condition $\partial/\partial n = 0$. For outflow boundary, a similar free-boundary condition, $\partial/\partial n = 0$, can be used at infinity. However, since the numerical simulation must be conducted on a finite domain, a special treatment of this condition is required. In this simulation, the convective outflow boundary condition is used:

$$\frac{\partial \bar{u}_i}{\partial t} + U_0 \frac{\partial \bar{u}_i}{\partial x_1} = 0, \quad i = 1,2,3. \tag{5}$$

Simulation results confirm this boundary condition to be both stable and accurate for a Reynolds number of 10^5.

In this simulation, the computational domain is discretized with a staggered grid, as the Maker And Cell (MAC) method [4]. In studies involving a subgrid-scale turbulence viscosity model, it is important that the accuracy order of the finite difference scheme be high enough to ensure that the numerical diffusion caused by the discretization does not dwarf the turbulence viscosity. Such being the case, a third-order upwind difference scheme for the convection terms, in conjunction with the Leith-type scheme for the temporal marching, is applied [5–7]. The discretization algorithm has been integrated with the LES model so that the characteristics of high Reynolds number flows are captured.

To study the effect of different schemes for the convection terms in the Navier–Stokes equations, here two schemes are used, namely, the central difference and the QUICK schemes [5]. These two schemes can be expressed in one equation,

$$\begin{aligned}\frac{\partial \phi}{\partial x} \simeq \frac{1}{Dx}\Bigg\{&\left[\frac{1}{2}(\phi_i + \phi_{i+1}) - \frac{q}{8}(\phi_{i-1} - 2\phi_i + \phi_{i+1})\right] \\ &- \left[\frac{1}{2}(\phi_{i-1} + \phi_i) - \frac{q}{8}(\phi_{i-2} - 2\phi_{i-1} + \phi_i)\right]\Bigg\},\end{aligned} \tag{6}$$

where, $q = 1$ corresponds to the QUICK scheme, and $q = 0$ corresponds to the central difference. Detailed derivations of the scheme are omitted for brevity. Interested readers are referred to Ref. [7]. The discretized 3D equation is given by

$$\begin{aligned}\phi_P^{N+1} = \phi_P^N + \{ &- C_e[\tfrac{1}{2}(\phi_P + \phi_E) - \tfrac{1}{2}C_e(\phi_E - \phi_P) \\ &- (\tfrac{1}{8}q + \tfrac{1}{24} - \gamma_1 - \tfrac{1}{6}C_{e^2})(\phi_E - 2\phi_P + \phi_W)] \\ &+ C_w[\tfrac{1}{2}(\phi_P + \phi_W) - \tfrac{1}{2}C_W(\phi_P - \phi_W)\end{aligned}$$

$$
\begin{aligned}
&- (\tfrac{1}{8}q + \tfrac{1}{24} - \gamma_1 - \tfrac{1}{6}C_{\mathrm{w}}^2)(\phi_{\mathrm{WW}} - 2\phi_{\mathrm{W}} + \phi_{\mathrm{P}})] \\
&- C_{\mathrm{n}}[\tfrac{1}{2}(\phi_{\mathrm{P}} + \phi_{\mathrm{N}}) - \tfrac{1}{2}C_{\mathrm{n}}(\phi_{\mathrm{N}} - \phi_{\mathrm{P}}) \\
&- (\tfrac{1}{8}q + \tfrac{1}{24} - \gamma_2 - \tfrac{1}{6}C_{\mathrm{n}}^2)(\phi_{\mathrm{N}} - 2\phi_{\mathrm{P}} + \phi_{\mathrm{S}})] \\
&+ C_{\mathrm{s}}[\tfrac{1}{2}(\phi_{\mathrm{P}} + \phi_{\mathrm{S}}) - \tfrac{1}{2}C_{\mathrm{s}}(\phi_{\mathrm{P}} - \phi_{\mathrm{S}}) \\
&- (\tfrac{1}{8}q + \tfrac{1}{24} - \gamma_2 - \tfrac{1}{6}C_{\mathrm{s}}^2)(\phi_{\mathrm{P}} - 2\phi_{\mathrm{S}} + \phi_{\mathrm{SS}})] \\
&- C_{\mathrm{f}}[\tfrac{1}{2}(\phi_{\mathrm{P}} + \phi_{\mathrm{F}}) - \tfrac{1}{2}C_{\mathrm{f}}(\phi_{\mathrm{F}} - \phi_{\mathrm{P}}) \\
&- (\tfrac{1}{8}q + \tfrac{1}{24} - \gamma_3 - \tfrac{1}{6}C_{\mathrm{f}}^2)(\phi_{\mathrm{F}} - 2\phi_{\mathrm{P}} + \phi_{\mathrm{B}})] \\
&+ C_{\mathrm{b}}[\tfrac{1}{2}(\phi_{\mathrm{P}} + \phi_{\mathrm{B}}) - \tfrac{1}{2}C_{\mathrm{b}}(\phi_{\mathrm{P}} - \phi_{\mathrm{B}}) \\
&- (\tfrac{1}{8}q + \tfrac{1}{24} - \gamma_3 - \tfrac{1}{6}C_{\mathrm{b}}^2)(\phi_{\mathrm{P}} - 2\phi_{\mathrm{B}} + \phi_{\mathrm{BB}})] \\
&+ \gamma_1(\phi_{\mathrm{E}} - 2\phi_{\mathrm{P}} + \phi_{\mathrm{W}}) + \gamma_2(\phi_{\mathrm{N}} - 2\phi_{\mathrm{P}} + \phi_{\mathrm{S}}) \\
&+ \gamma_3(\phi_{\mathrm{F}} - 2\phi_{\mathrm{P}} + \phi_{\mathrm{B}} + s_{\mathrm{p}}Dt\}^N,
\end{aligned}
\tag{7}
$$

where P denotes the present point, W denotes the point west (to the left) of the present point in the x_1 direction, while WW denotes the point farther to the west. Accordingly, the subscripts E (east), N (north), S (south), F (front) and B (back) may be interpreted in a similar fashion. C_{e} is the Courant number on the east side of the present point, i.e., $C_{\mathrm{e}} = u_{\mathrm{e}}Dt/Dx_1$. $\gamma_1 = (\nu + \nu_{\mathrm{SGS}})Dt/Dx_1^2$, $\gamma_2 = (\nu + \nu_{\mathrm{SGS}})Dt/Dx_2^2$, and $\gamma_3 = (\nu + \nu_{\mathrm{SGS}})Dt/Dx_3^2$. Here ν_{SGS} is the subgrid-scale viscosity, calculated using Eq. (3). The pressure field is solved with a successive overrelaxation method, ensuring that the computed flow field is divergence free. The presented equations are for a uniform grid mesh, but both uniform and non-uniform grid meshes were used in the calculations.

The computational domain is $27L$ in the streamwise direction, $18L$ in the cross stream direction, and $2L$ in the spanwise direction. The prism is located at the center in the lateral direction, and its front face is at a distance of 8L from the entrance flow boundary.

The computations were carried out primarily using the computer facilities at the University of Notre Dame and at the National Center for Supercomputing Applications. A flow simulation at a Reynolds number of 10^5 required about 160 h of user time on an SGI Power Challenge computer, to march for a non-dimensional time of 100.

3. Simulation results

3.1. Velocity field

Time histories of the velocity and pressure field are generated using the method described above, with a computed potential flow for the same configuration, or the output from a previous 2D or 3D simulation, serving as the initial flow field. In each case, the vortex shedding process (for $\mathrm{Re} = 10^5$) begins without introducing any initial perturbation. When the 3D simulation is initially started from a 2D flow, it takes about a non-dimensional time of 50 to reach a fully developed 3D flow, and results reported hereafter are for the well-developed period. The statistical values (mean and RMS) are obtained over a time length of 500. Unless otherwise stated, all the reported 3D results in this paper are based on the QUICK scheme, using a grid size of $Dx = Dy = 0.05$ near the bluff body and a uniform $Dz = 0.2$ in the z-direction.

The time-averaged streamwise velocity on the symmetry line is reported in Fig. 1, together with the experimental results by Lyn [8] and Durão et al. [9], and the numerical results by Murakami and Mochida [1] and Franke and Rodi [10, 11]. Upstream of the square, the difference among the compared studies is extremely small. In the wake flow, both the current 2D and 3D simulation results show a good agreement with the experimental results. It should be noted that the three-dimensional results of Murakami and Mochida [1] deviate from the current simulation results, although both are based on LES modeling. The discrepancy may be due to the employed numerical methods and boundary conditions rather than the turbulence model itself. A similar discrepancy is also exhibited by the pressure results shown in the next subsection. Different turbulence models do influence the results, as the differences among the results obtained employing k–ε, RSE and LES are depicted in Fig. 1.

As seen in Fig. 1, the mean values from the 2D simulation are not significantly different from the 3D results, although they are obtained with far less computational

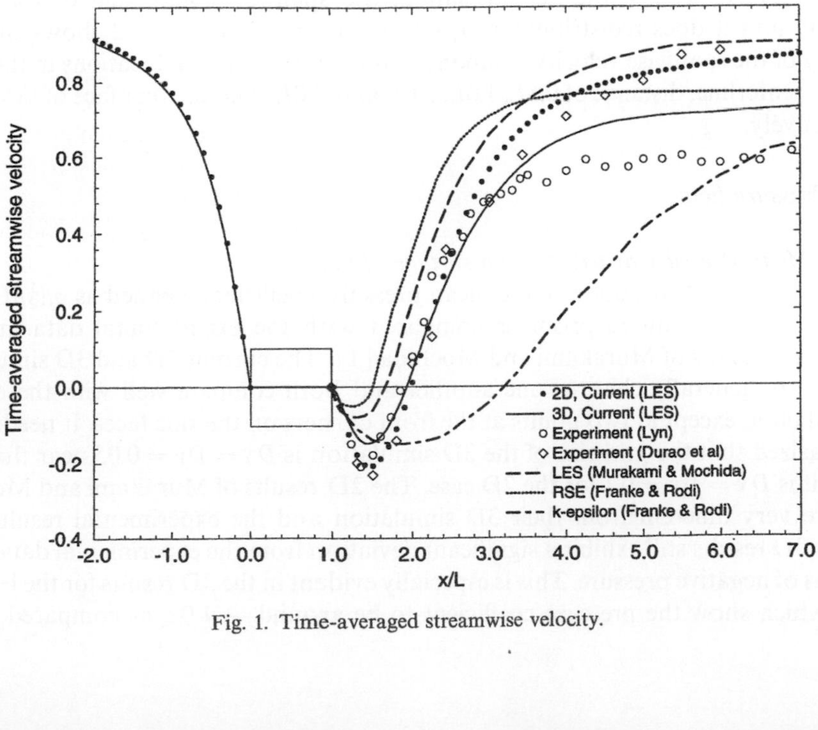

Fig. 1. Time-averaged streamwise velocity.

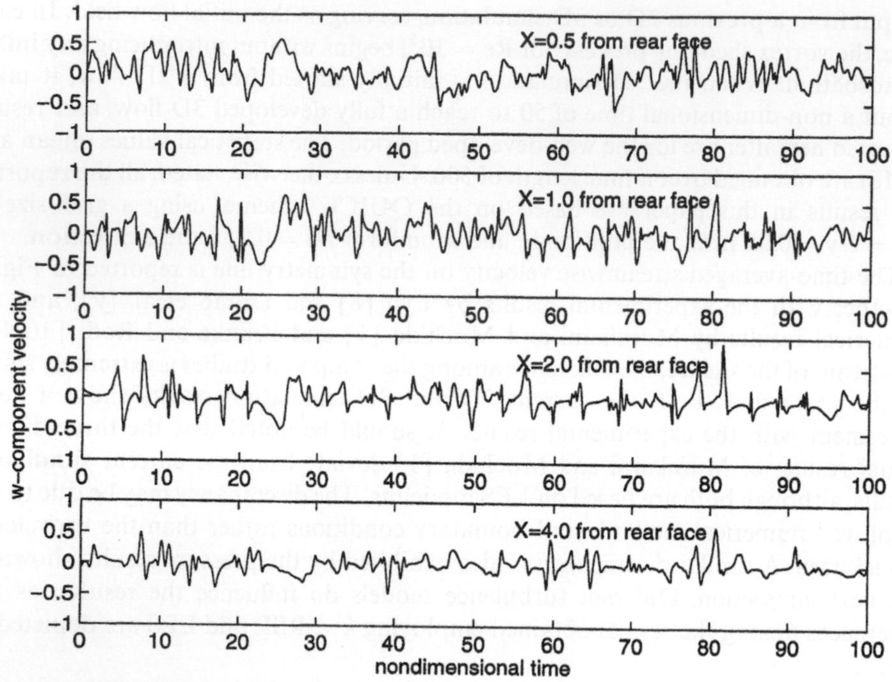

Fig. 2. Time history of *w*-component velocity in the wake.

effort. Nevertheless, in the 3D simulation, the spanwise component of velocity is present, and it does redistribute energy from the *x–y* plane. Fig. 2 shows the time history of the spanwise velocity component at four downstream locations in the wake on the centerline, distance of $0.5L$, $1.0L$, $2.0L$ and $4.0L$ from the rear face of the prism, respectively.

3.2. *Pressure field*

3.2.1. *Mean Pressure distribution on surface of prism*

In Fig. 3, the distribution of the mean pressure coefficient, defined as $\bar{p}/(\frac{1}{2}\rho U_0^2)$, on the surface of a square prism, is compared with the experimental data and the numerical results of Murakami and Mochida [1]. The current 2D and 3D simulation results are generally close to one another and both compare well with the experimental data, except at two points at the front corners on the side faces. It needs to be emphasized that the grid size of the 3D simulation is $Dx = Dy = 0.05$ near the body while it is $Dx = Dy = 0.1$ for the 2D case. The 2D results of Murakami and Mochida [1] are very different from their 3D simulation and the experimental results. The team's 3D results still exhibit a significant deviation from the experimental data in the regions of negative pressure. This is especially evident in the 3D results for the leeward face, which show the pressure coefficient to be around -1.02, as compared to the

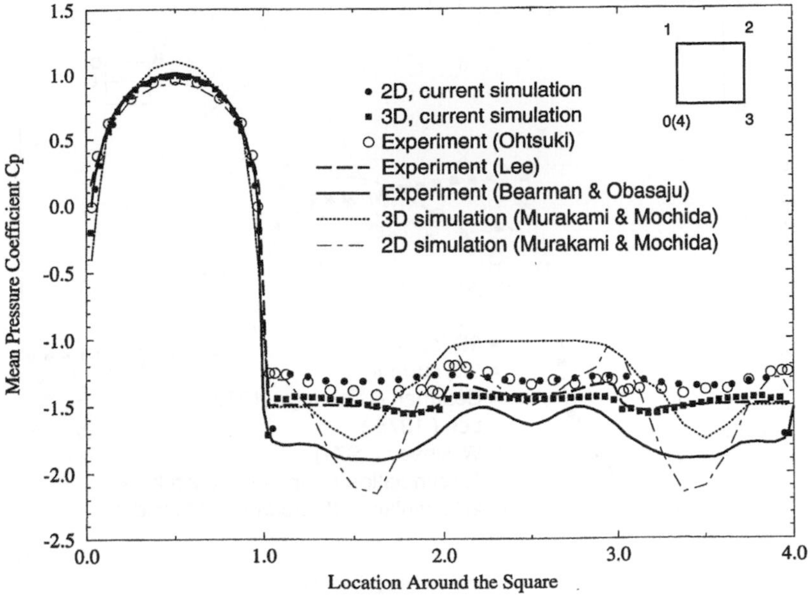

Fig. 3. Distribution of the mean pressure coefficient. Experimental values from Refs. [12–14] and simulations from Ref. [1].

experimental value of near -1.3– -1.5. As mentioned earlier concerning the results for the centerline velocity, the discrepancies between the current simulation and the results by Murakami and Mochida [1] may not be due to the turbulence model, since both used LES, but may instead be a result of the numerical method or boundary condition specifications. For example, the treatment of the outflow boundary by a free boundary condition ($\partial/\partial n = 0$) may introduce errors in the simulated results. As noted earlier, a convective boundary condition is more realistic.

3.2.2. *RMS pressure distribution on surface of prism*

In Fig. 4, RMS values of the pressure fluctuations on the surface of a square prism are presented. The lowest RMS is at the center of the front surface, while higher values of pressure fluctuations appear on the two side faces. On the rear side, the fluctuations decrease as the center line on the back face is approached. Again, numerical results are compared with the available experimental results and are found to be in a good agreement. The range of Reynolds numbers in the cited references in Fig. 4 are: Bearman and Obasaju [13] ($Re = 2 \times 10^4$), Lee [14] ($Re = 1.76 \times 10^5$), and Wilkinson [15] ($Re = 10^4$–10^5). Note that the 3D results of Murakami and Mochida [1] are also very close to the experimental results, while their 2D results differ substantially.

3.2.3. *Integral quantities*

Table 1 lists the integral quantities and their comparison with the experimental results. The 2D and 3D results (with near-body grid size of $\frac{1}{20}$) are close to one another

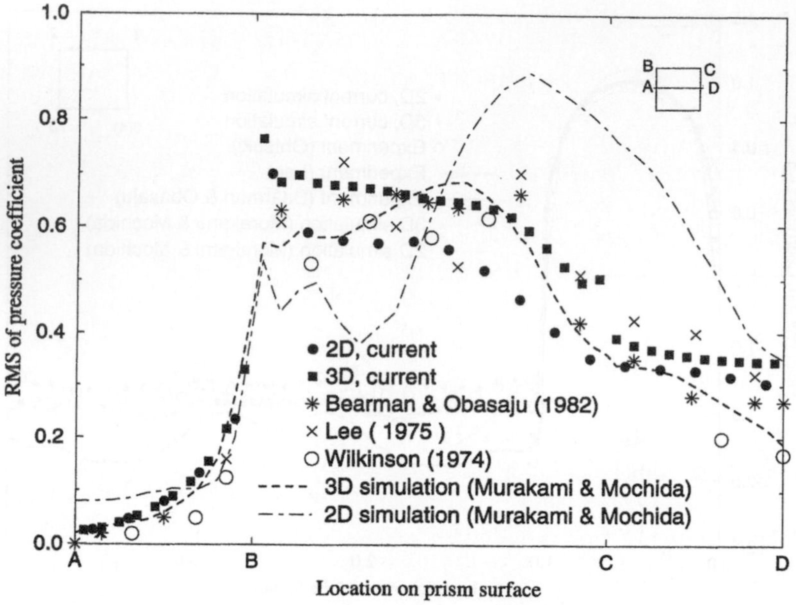

Fig. 4. Distribution of the RMS pressure.

Table 1
Comparison of mean and RMS of C_L and C_D and Strouhal number with experiments

	RMS of C_L	Mean of C_D	RMS of C_D	St. Number
2D, $Dx,Dy = \frac{1}{10}$	1.06	2.01	0.21	0.14
3D, $Dx,Dy = \frac{1}{20}$	1.15	2.14	0.25	0.135
3D, $Dx,Dy = \frac{1}{15}$	1.07	2.19	0.12	0.138
3D, $Dx,Dy = \frac{1}{10}$	0.33	1.78	0.06	0.149
Vickery, 1 [16]	1.32		0.17	0.12
Vickery, 2 [16]	1.27		0.17	
Lee [14]	1.22	2.05	0.22	
Bearman and Obasaju [13]	1.2			0.13
Nakamura and Mizota [17]	1.0			
Okajima (1982) [18]				0.13

and both agree well with experimental findings. At the same time, it is found that 3D results with course grid size ($Dx = Dy = \frac{1}{10}$) give much lower RMS values of both the lift coefficient C_L and the drag coefficient C_D. The results with the grid size of ($Dx = Dy = \frac{1}{15}$) are closer, while the results with the grid size of ($Dx = Dy = \frac{1}{20}$) are the closest to the experimental values. The Strouhal number from simulations with different grid sizes exhibits a similar trend.

3.2.4. *Correlation of pressure fluctuations*

Fig. 5 shows the correlation coefficients of pressure fluctuations on the surface of the prism with reference to a point marked on the upper side face. Both the 2D and 3D results are plotted and compared with the experimental findings by Lee [14]. The results indicate a dominant presence of an antisymmetric correlation pattern associated with vortex shedding, and the numerical results are in very good agreement with experimental data. From Figs. 1, 3–5 and Table 1, it may be observed that the 2D and 3D results are not far apart suggesting dominance of 2D structure in the vortical flow field. This is in contrast with earlier reports by other investigators (e.g., Ref. [1]).

3.3. *Grid refinement study*

To study the accuracy of the numerical scheme and the resolution requirement for the specific flow simulated, a grid refinement study is conducted. Roache [19] suggested a method for reporting numerical convergence uniformly, based on Richardson extrapolation. For the LES, there has been limited experience about reporting of errors associated with grid refinement. One way to refine the grids for an LES is to keep the value of Δ in the Smagorinsky model (Eq. (3)) invariant, while refining Dx, Dy and Dz. The other approach involves refinement of grid uniformly, while allowing Δ vary in each refinement since Δ is defined as $(DxDyDz)^{1/3}$ in the Smagorinsky model. Theoretically, the results should converge to the "true solution" as the grid is refined.

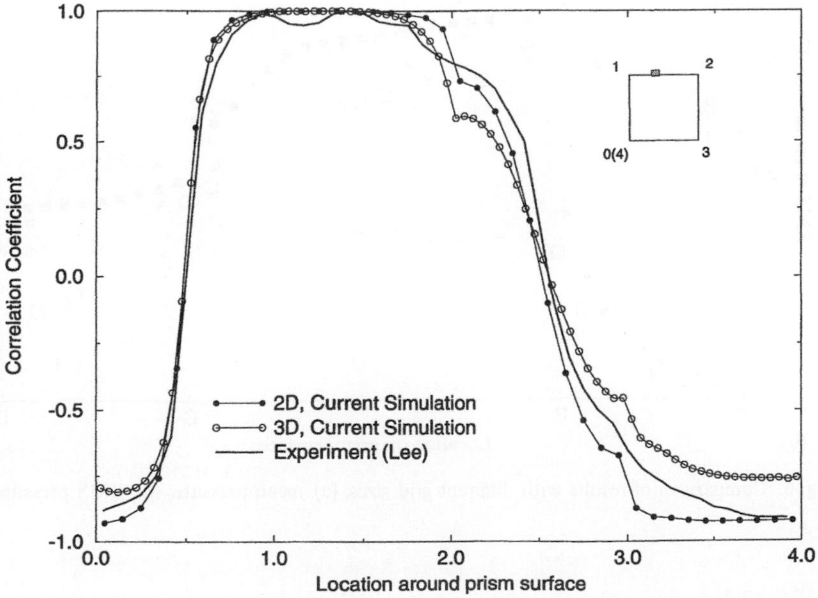

Fig. 5. Chordwise correlation of pressure.

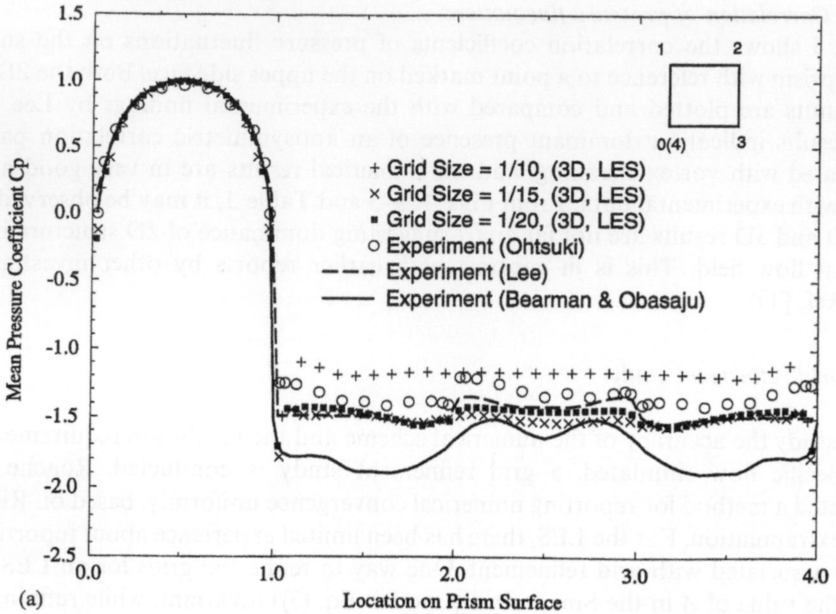

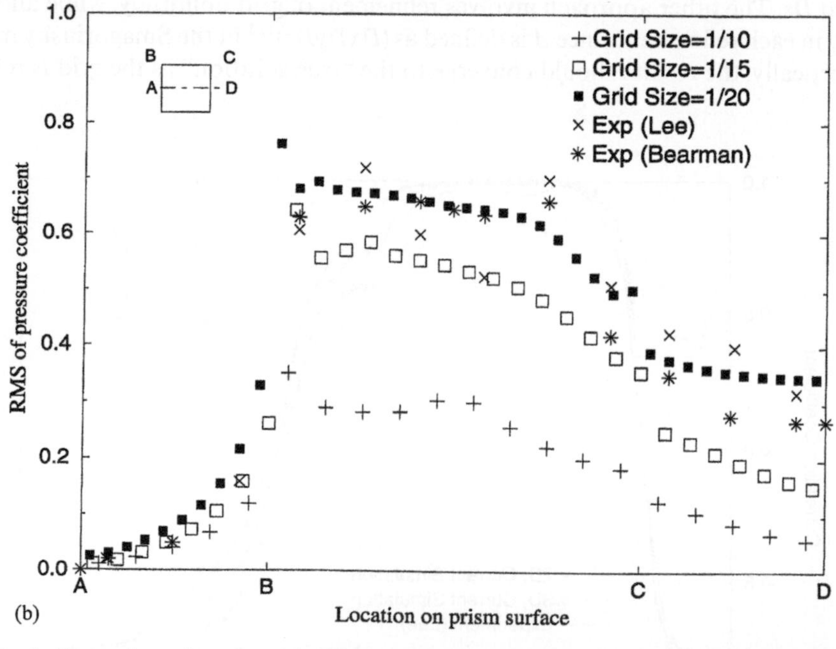

Fig. 6. Comparison of results with different grid sizes: (a) mean pressure; (b) RMS pressure.

These two approaches are not equivalent and may not produce the same results, thus warranting additional investigation through numerical experiments.

In this study, several grid sizes are used, including both uniform and variable-size grids. Δ is also varied as $(DxDyDz)^{1/3}$. For the uniform meshes, three grid sizes are tested, i.e., $Dx = Dy = \frac{1}{10}$ $Dx = Dy = \frac{1}{12}$ (not reported here) and $Dx = Dy = \frac{1}{15}$. To save computational effort, a mesh with variable grid size was used for the case with the near-body grid sizes of $Dx_{\text{body}} = Dy_{\text{body}} = \frac{1}{20}$. For all these cases, the same grid size was used in the spanwise direction, $Dz = 0.2$. The results concerning the mean and RMS pressure distribution on the body surface are presented in Fig. 6, which again confirms the importance of grid resolution. Course grids result in a much lower level of RMS pressure fluctuation, and a higher (less negative) base pressure, which are similar to a flow with excessive dissipation. These deviations cannot be corrected by modifying the C_s value in the SGS model (Eq. (3)), varying the time step Dt, or using a different scheme (QUICK or central difference) for the discretization of the convection terms. There is a clear trend demonstrating that simulations approach realistic values as the grids are refined, with the most refined grids ($Dx_{\text{body}} = Dy_{\text{body}} = \frac{1}{20}$) leading to the closest agreement with the experimental values.

3.4. Comparison of central difference and QUICK schemes

The QUICK and the central difference schemes are used to discretize the convection terms, and the obtained simulation results (with grid size of $Dx_{\text{body}} = Dy_{\text{body}} = \frac{1}{20}$) are compared in Fig. 7. Only insignificant differences are observed and both results are close to experimental data, which indicates the insignificance, at the employed grid resolution, of numerical dissipation in the QUICK scheme due to its upwind-biased nature. The leading order error of the QUICK scheme is proportional to the third power of grid size and the fourth-order derivative of velocity. Thus, the QUICK scheme is slightly dissipative, but does not contain numerical viscosity (if numerical viscosity is defined as error proportional to the second-order derivative of velocity).

3.5. Effect of time step

In a direct numerical simulation of turbulent channel flow, Choi and Moin [20] reported that computational time steps substantially influence the statistical results. In their study, for large time steps ($Dt^+ = 1.6$ and 2, corresponding to CFL = 4 and 5) near or larger than the Kolmogorov time scale (2.4), the simulations resulted in a laminar flow instead of turbulent. Up to now, the authors are not aware of such reports concerning large eddy simulations, and computations are conducted here with two different time steps ($Dt = 0.0125$ and 0.00625, corresponding to CFL = 0.25 and 0.125) to study this effect. Results concerning pressure are reflected in Fig. 7, from which it can be seen that the effect of different Dt are noticeable but not significant. This suggests that a time step of $Dt = 0.0125$ is refined enough to resolve the flow satisfactorily.

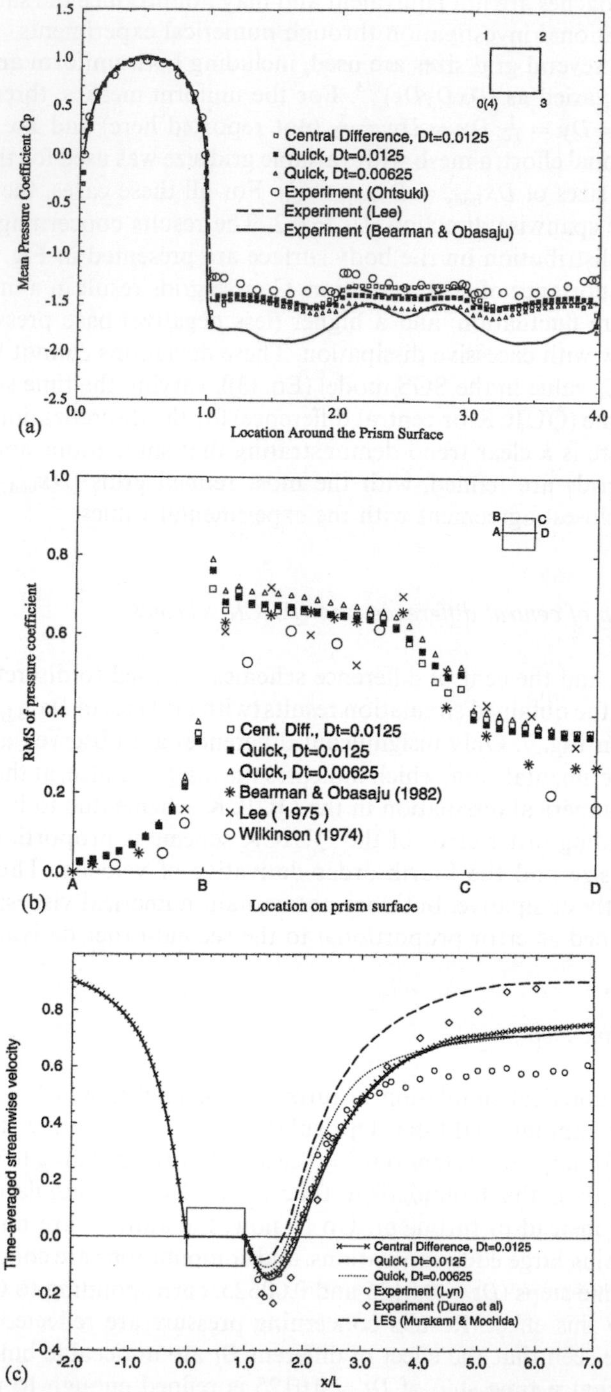

Fig. 7. Comparison of results based on central difference and QUICK methods, and different time steps: (a) mean pressure; (b) RMS pressure; and (c) time-averaged streamwise velocity.

4. Conclusions

Using QUICK and central difference schemes for the convection terms, the Leith method for temporal marching and the Smagorinsky model for subgrid-scale viscosity, 3D numerical algorithm based on a staggered grid gives results in a very good agreement with the available experimental data. A grid refinement study reveals that sufficient resolution is very important in 3D LES simulations. An insufficient resolution leads to much lower RMS values of pressure fluctuations. A comparison of third-order QUICK and central difference schemes for the discretization of the convection terms shows only minor differences, and suggests relative insignificance of the numerical dissipation introduced by the upwind nature of QUICK, as long as the grid resolution is sufficient. The influence of time step is also investigated, and small effects are found for the two tested Dt values ($Dt = 0.0125$ and 0.00625, corresponding to CFL = 0.25 and 0.125).

Acknowledgements

This study was partially supported by NCSA Grant No. BCS960002N, ONR Grant No. 00014-93-1-0761 and NSF Grant No. CMS-9503779.

References

[1] S. Murakami, A. Mochida, On turbulent vortex shedding flow past 2D square cylinder predicted by CFD, J. Wind Eng. Ind. Aerodyn. 54/55 (1995) 191–211.

[2] T. Tamura, Accuracy of the very large computation for aerodynamic forces acting on a stationary or an oscillating cylinder-type structure, IWEF Workshop on CFD for Prediction of Wind Loading on Building and Structures, 9 September, 1995, Tokyo Institute of Technology (Nagatsuta Campus), Yokohama, Japan, pp. 2.1–2.22.

[3] W. Rodi, J.H. Ferziger, M. Breuer, M. Pourquié, 1996, Status of Large Eddy Simulation: Results of a Workshop, personal communication, pp. 1–35.

[4] F.H. Harlow, J.E. Welch, Numerical calculation of time-dependent viscous incompressible flow of fluid with free surface, Phys. Fluids 8 (1965) 2182–2189.

[5] B.P. Leonard, A stable and accurate convective modelling procedure based on quadratic upstream interpolation, Comput. Methods App. Mech. Eng. 19 (1979) 59–98.

[6] R.W. Davis, E.F. Moore, A numerical study of vortex shedding from rectangles, J. Fluid Mech. 116 (1982) 475–506.

[7] D. Yu, A. Kareem, Numerical Simulation of Pressure Field Around Two-Dimensional Rectangular Prism, Technical Report No. NDCE-96-002, Department of Civil Engineering and Geological Sciences, University of Notre Dame, 1996.

[8] D.A. Lyn, Phase-averaged turbulence measurements in the separated shear flow around square cylinder, in: Proc. 23rd Cong. Int. Ass. Hydraulic Research, Ottawa, Canada, 21–25 August 1989, pp. A85–A92.

[9] D.F.G. Durão, M.V. Heitor, J.C.F. Pereira, Measurements of turbulent and periodic flows around a square cross-section cylinder, Exp. Fluids 6 (1988) 298–304.

[10] R. Franke, W. Rodi, Calculation of vortex shedding past a square cylinder with various turbulence models, in: Proc. of the 8th Symp. on Turbulent Shear Flows, Munich, Germany, 1991, p. 189.

[11] W. Rodi, On the simulation of turbulent flow past bluff bodies, J. Wind Eng. Ind. Aerodyn. 46/47 (1993) 3–19.

[12] Y. Ohtsuki, Wind tunnel experiments on aerodynamic forces and pressure distributions of rectangular cylinders in a uniform flow, Proc. 5th Symp. on Wind Effects on Structures, Tokyo, Japan, 1976, pp. 169–175.

[13] P.W. Bearman, E.D. Obasaju, An experimental study of pressure fluctuations on fixed and oscillation square-section cylinders, J. Fluid Mech. 119 (1982) 297–321.

[14] B.E. Lee, The effect of turbulence on the surface pressure field of a square prism, J. Fluid Mech. 69 (1975) 263–282.

[15] R.H. Wilkinson, On the vortex-induced loading on long bluff cylinders. Ph.D. Thesis, Faculty of Engineering, University of Bristol, England, 1974.

[16] B.J. Vickery, Fluctuating lift and drag on a long cylinder of square cross-section in a smooth and in a turbulent stream, J. Fluid Mech. 125 (1966) 481–494.

[17] Y. Nakamura, T. Mizota, Unsteady lifts and wakes of oscillating rectangular prisms, Proc. A.S.C.E.: J. Eng. Mech. Div. 101 (EM6) (1975) 855–871.

[18] A. Okajima, Strouhal numbers of rectangular cylinders, J. Fluid Mech. 123 (1982) 379–398.

[19] P.J. Roache, Perspective: a method for uniform reporting of grid refinement studies, J. Fluids Eng. 116 (1994) 405–413.

[20] H. Choi, P. Moin, Effects of the computational time step on numerical solutions of turbulent flow, J. Comput. Phys. 113 (1994) 1–4.

Keynote Presentation

ELSEVIER

Journal of Wind Engineering
and Industrial Aerodynamics 67&68 (1997) 211–224

JOURNAL OF
wind engineering
AND
industrial
aerodynamics

A fluid mechanicians view of wind engineering: Large eddy simulation of flow past a cubic obstacle

Kishan B. Shah*, Joel H. Ferziger

Department of Mechanical Engineering, Stanford University, Building 500, Stanford, CA 94305, USA

Abstract

We present a fluid mechanicians view of computations of complex 3-D flows in wind engineering. In particular, we focus on large eddy simulation (LES) in which the large scales of motion are computed explicitly while the small or subgrid-scale motions are modeled. This approach can accurately capture the effects of large-scale motions such as wind forces and their fluctuations. Some general issues related to LES in wind engineering are discussed and sample results from simulations of flow past a cubic obstacle are presented.

1. Introduction

Turbulent flows around three-dimensional obstacles are common in nature and occur in many applications including flow around tall buildings, vehicles and computer chips. Understanding and predicting the properties of these flows are necessary for safe, effective and economical engineering designs. Experimental techniques are expensive and often provide data that is not sufficiently detailed. With the advent of supercomputers it has become possible to investigate these flows using numerical simulations.

A number of approaches to computer prediction of flows have been proposed and used. The simplest methods, ones that involve the use of correlations or one- or two-dimensional flow analysis cannot be used to predict complex three-dimensional flows. The next simplest approaches use the Reynolds averaged Navier–Stokes (RANS) equations and a variety of turbulence models. While these are much cheaper than large eddy simulation (discussed below), no single model has proven capable of predicting a wide variety of complex flows, especially when information about the fluctuating part of the flow is required.

* Corresponding author. E-mail: kshah@makalu.stanford.edu.

PII S0167-6105(97)00074-3

This has led to interest in large eddy simulation (LES), in which the large scales of motion are computed explicitly while the small- or subgrid-scale motions are modeled. It is well suited to wind engineering where interest centers on quantities such as forces, moments and their fluctuations which are primarily due to large-scale motions. It should suffice to capture the largest turbulent structures present in the flow.

In this paper, we shall discuss some general issues related to LES in wind engineering and then illustrate them with simulations of flows over cubes as examples. It should be noted from the title that the authors are primarily fluid mechanicians and their view of a different but related field may not be completely accurate.

As already noted, measurement of forces on a bluff body placed in a flow is of considerable importance to wind engineers. It is thus traditional, in experiments on forces on structures, to measure only the pressure distribution around the body and, perhaps, a little data on the velocity field. This, of course, is important, but from a fluid mechanical point of view, it is far from complete. Many people are aware that the forces on the body are due to large coherent structures in the flow but it is rare that this type of experiment is performed to the standards of the fluid mechanics community. Lest this sound too condemning, it should be noted that the bodies studied in wind engineering are much more complex than those considered by fluid mechanicians; there are few measurements performed by anyone that meet the standards applied in simpler flows.

This would hardly be a problem if one could be sure that a simulation which accurately captured the force distribution on the body was accurate in the wider sense. Unfortunately, this is not the case. For some flows, it is possible (indeed, fairly easy) to correctly predict the pressure distribution and miss the velocity distribution by a wide margin. To validate a method solely by its ability to predict the velocity distribution about a single object may be dangerous; relying on the pressure distribution alone is almost surely of no value whatever.

It is always tempting to reduce the difficulty of a problem being simulated. For example, one might try to simulate the flow about a cylindrical body using a two-dimensional technique. One danger is that the result may be a periodic flow when the actual flow is not periodic. It is also well known in fluid mechanics that a flow predicted with a two-dimensional method almost always produces larger drag and side forces than a three-dimensional simulation. Since the flow is always 3-D, the simulation should also be 3-D, even though the cost is much greater.

The average flow may look very different from the instantaneous flow. Even in flow over a symmetric body, the flow is asymmetric at almost every instant while the average flow must be symmetric. Looking solely at the mean flow may give an erroneous impression. We shall see an example of this below.

Finally, one cannot emphasize strongly enough the need for error estimation. One must start by knowing which quantities need to be predicted and with what accuracy. This differs enormously from one field to the next. Fortunately, wind engineers are usually satisfied with lower accuracy than mechanical engineers. Although error estimation is well developed for steady flows, it is in a far from satisfactory state in LES. This is an area that needs considerable effort but the effort is hindered by the very nature of turbulence.

2. Some observations about large eddy simulation

Large eddy simulation (or LES) tries to simulate the largest scales of motion while treating the small scales by a model. This is quite appropriate in wind engineering because it is normally the large scales that create the forces of interest. However, it is important to note that there are flows (the square cylinder flow discussed in Rodi et al. [1] is a good example) in which the pressure distribution depends on large- scale flow structures that originate at relatively small scales. In such flows, it is not known as yet how much resolution is required to accurately predict the pressure distribution on the body. Thus, LES may not be a panacea in wind engineering as, indeed, it is not in other areas of fluid mechanics.

As noted earlier, interest in LES has increased due to the failure of RANS to do an adequate job, especially in cases in which information about the fluctuations is required. LES may be able to do the job but, as just noted, there are cases for which the question is not settled at present.

We shall not go into any detail about LES here as that is adequately covered elsewhere (e.g. Ref. [2]). Rather, we shall give a few observations. The earliest applications of LES in engineering were to relatively simple low-Reynolds number flows in which only a small fraction of the turbulence needs to be modeled. For this type of flow, the quality of the results is insensitive to the quality of the model so long as the numerical method is sufficiently accurate. Excellent and useful results were obtained, giving the impression that success is automatic if LES is used. Unfortunately, this is far from the truth. When the model is called upon to represent more of the turbulent motions, the sensitivity to model quality becomes much more important. A recent workshop [1] demonstrated that success is not guaranteed simply because LES is applied and, especially, that success is very much dependent on the nature of the flow simulated.

Although current models for the small scales, especially those that use the dynamic procedure, are a considerable improvement over those that were used earlier, they are far from perfect and a great deal remains to be done. Interest in subgrid-scale modeling has increased in the last several years, with many interesting proposals having been made, but there is a long way to go. There is not space here to cover these developments but the authors' favorite model is presented below.

Something should also be said about filtering and numerical methods. For complex flows, one is forced to use finite volume or finite element methods. In general, these methods are incapable of resolving any structure whose size is smaller than two grids and attempting to do so will almost always lead to difficulties. The practice of setting the filter width equal to the grid size, used in many recent papers, is one that should be rigorously avoided. Upwind biased convective schemes always introduce some extra dissipation (in many published simulations, the numerical dissipation is larger than the model-produced dissipation) and should be avoided in low-speed flows even though they offer convenience. The results of any numerical simulation, including LES, should not depend on the numerical method used. Indeed, dependence of this kind is a good indicator that the numerical method is not adequate to the task.

In the remainder of this paper we shall give a sample of results from our own simulation of the flow over a cube. In doing so, we shall try to illustrate many of the points that have been made above. It is important to observe that these simulations were expensive and, as a result, only a few of them were made. It follows that, if results of high quality are needed, it is not possible at present to use LES as a design tool.

3. Simulation of flow over a cube

In the remainder of this paper, we shall investigate the flow past a cubic obstacle mounted on one wall of a plane channel using LES. In doing so, we shall present subgrid-scale (SGS) models suitable for complex flows, perform LES of these flows, and show how one can use the results to aid in the identification of dynamically significant large-scale structures. Knowledge of the origins and dynamics of these large-scale structures can greatly assist in devising prediction and control strategies.

Flow around a cube exhibits characteristics common to many wind engineering flows including three dimensionality of the mean flow, separation and large-scale unsteadiness. In spite of its importance, quantitative results for this flow are scarce. Most experiments have examined only the flow in the vicinity of the cube and measured only a few selected quantities such as the overall drag. The most comprehensive experimental data set was reported by Martinuzzi and Tropea [3] for flow past a cube mounted on a wall of a plane channel at high Reynolds number ($\mathrm{Re} = U_{\mathrm{mean}}\delta/\nu = 40\,000$, where δ is the height of the obstacle); we shall compare our results with theirs. We note that this is a flow in a channel, which provides the important advantage that the approach flow is easy to duplicate accurately and is thus an ideal case on which to test and validate computational approaches.

The flow selected differs from the external flows that need to be considered in the study of wind effects on structures in some important ways. In a realistic wind engineering application, the approach flow is often a boundary layer which contains eddies larger than the body and may contain wake effects due to upstream bodies. In these flows, it is very difficult to document the approach flow adequately. This is an important issue that will need to be dealt with in the future but we believe that, for now at least, it is preferable to work with a well-documented flow.

Included in this investigation are the influence of Reynolds number, comparison of several SGS models including a new non-eddy viscosity SGS model [4] and a study of some of the temporal characteristics of the flow. Brevity requires that we concentrate on a few features that illustrate points made above.

3.1. Computational methodology

3.1.1. Numerical method

A second-order accurate-control volume formulation is used on a mesh that is non-uniform in all three directions. The momentum equations are advanced in time with a fractional step method based on the Crank–Nicolson method for the viscous terms and a third-order Runge–Kutta method (RK3) for the convective terms. The resulting velocity field is adjusted to satisfy continuity by solving a Poisson equation for a pressure-like variable; a multigrid procedure solves this problem efficiently. This scheme is stable, accurate, and uses little storage. A staggered variable configuration eliminates the need for a pressure boundary condition. The Poisson equation is solved only on the last substep. A full description of this numerical method can be found in Ref. [5].

The problem was first solved on a relatively coarse grid and the solution exhibited the oscillations found when centered schemes are applied to advective problems. The grid was refined until the oscillations disappeared. This is a necessary but not sufficient test of the accuracy of the method. We continued the grid refinement one step further and found no change in any important quantity; although this indicates that the solution is adequately resolved, it is not a complete test.

3.1.2. Computational domain and boundary conditions

The computational domain consists of a plane channel with a cubical obstacle of dimension (δ), the half-channel height, mounted on one wall (see Fig. 1). The hyperbolic tangent function is used to stretch the grid near the walls with a minimum grid spacing of 0.006δ. The computational mesh is $(192 \times 64 \times 96)$ in the x- , y- , and z-directions, respectively.

The spanwise boundary condition is periodic, implying an infinite periodic array of obstacles. The spanwise width is 7δ, assuring that blockage effects are small. In the streamwise direction, inflow–outflow boundary conditions are used. An instantaneous flow field taken from an LES of fully developed plane channel is used at the inlet and a convective boundary condition [6] is applied at the exit. The streamwise length of the domain is 10δ.

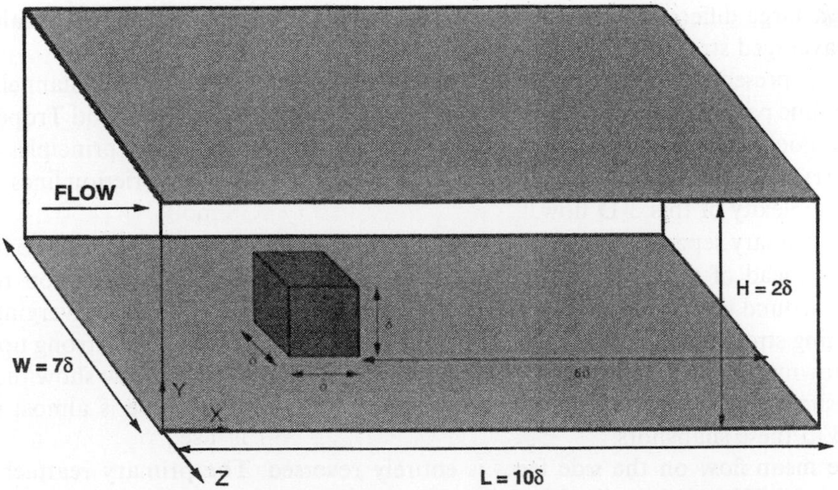

Fig. 1. The solution domain for the flow over a cube mounted on a channel wall.

4. Summary of results

Large eddy simulations were performed at Re = 3000 and 40 000. The low Reynolds number case was chosen to facilitate evaluation of SGS models with adequate wall resolution. Since low Re simulations are relatively inexpensive, we were able to examine various models including several forms of the dynamic model. Previous dynamic models (e.g. Ref. [7]) produced eddy viscosities that had both large positive and large negative values. Investigation revealed that this is due to contributions from the highest wave numbers which are prone to numerical contamination. A new dynamic procedure which removes the high-wave-number portion of the solution prior to application of the dynamic calculation for the model coefficient was used. To further control the oscillation of the coefficient, a mixed-scale-similarity-Smagorinsky version of the model was used; it is described in detail in Ref. [5].

Figs. 2 and 3 show comparisons of the time-averaged vertical profiles of streamwise velocity (U) and its mean-square fluctuation with the experimental results of Martinuzzi and Tropea [3] on the symmetry plane at selected locations. Fig. 4 provides a comparison of the experimental and computational contours of the streamwise velocity on the symmetry plane. The agreement is very good; especially noteworthy is the prediction of the reversed velocity in the recirculation region. The experimental measurements were made using LDA in both the (u–v) and (u–w) planes; the computational results are in better agreement with experimental results taken in the (u–w) plane. The primary separation point ahead of the body and the reattachment point behind the body are predicted quite well; the velocity profiles have a deficit with respect to the experimental results at these points but this does not seem to affect the velocity profiles close to the obstacle or in the recovery-zone downstream of reattachment where the agreements between experiment and computation is again rather good. The quality of computational results deteriorate for the vertical velocity component which has a smaller magnitude (V and $\langle v''^2 \rangle$); the experimental results show a large difference between the results taken in the two planes, especially for time-averaged streamwise turbulence stress $\langle u''^2 \rangle$.

Fig. 5 presents the time-averaged streamlines on the floor of the channel. The streamline patterns are consistent with those observed by Martinuzzi and Tropea [3] and in good agreement with streamline patterns based on kinematic principles given by Perry et al. [8]. These streamlines, which may be viewed as skin friction lines, show the complexity of this 3-D flow.

The primary separation occurs at a saddle point located about one obstacle height (1.05δ) ahead of the obstacle (experimental value = 1.02δ). The separation region wraps around the obstacle and forms a strong horseshoe vortex. The converging and diverging streamlines that mark the extent of this vortex are regions of strong upwash and downwash. Instantaneous pictures (not presented here) of the flow show that the horseshoe vortex is, in fact, highly intermittent; an intact structure is almost never found in these snapshots.

The mean flow on the side faces is entirely reversed. The primary reattachment length of 1.65δ agrees well with the experimental value of 1.61δ; it is much smaller than for 2-D obstacles ($\approx 7.0\delta$). Both the primary separation point ahead of the

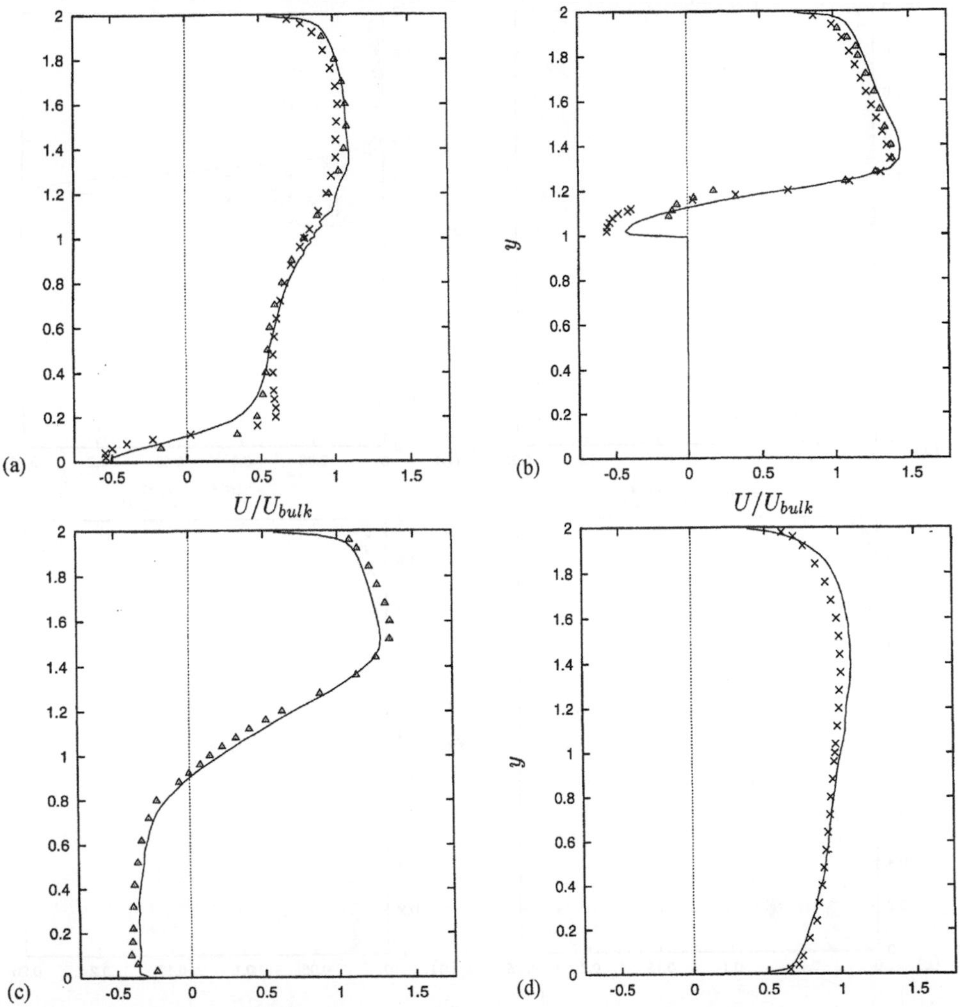

Fig. 2. Time-averaged streamwise velocity at $Re_b = 40\,000$. (a): $x = 2.68$; (b): $x = 3.50$; (c): $x = 4.56$; (d): $x = 9.00$; (———) Computational (LES); ($\times$) and (Δ) are experimental results of Martinuzzi and Tropea [3] corresponding to LDA measurements in u–v and u–w planes, respectively.

obstacle and the rear reattachment points are singular points (zero skin friction) where the so-called separation lines begin and end. The owl-face shaped streamlines in the rear recirculation zone of the obstacle correspond to the base of the arch vortex. The arch vortex is formed by quasi-periodic vortex shedding from the upstream vertical corners that resembles a von Karman street. In fact, the feet of the instantaneous vortices behind the body are staggered and connected by roughly elliptical vortex loops whose tops are downstream of their legs. The intact arch vortex exists only in the mean flow and is an artifact of averaging.

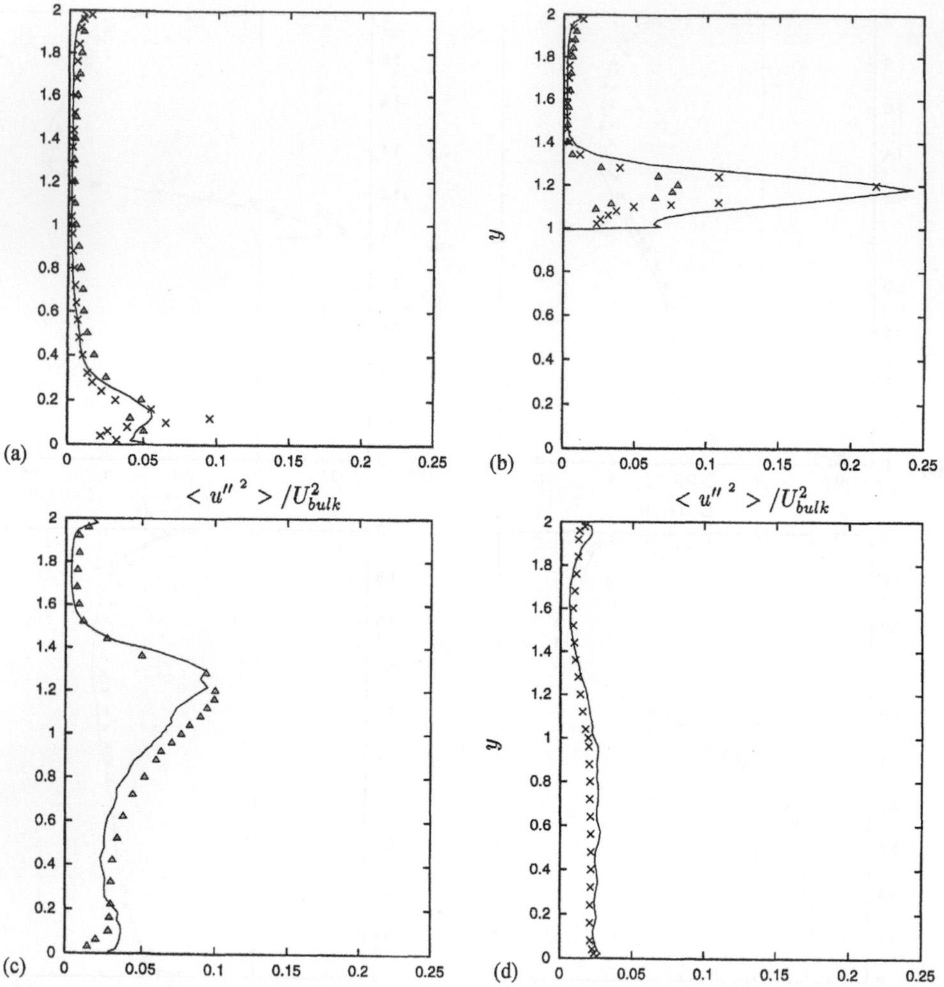

Fig. 3. Time-averaged streamwise turbulence stress $\langle u''^2 \rangle$ at $Re_b = 40\,000$. (a): $x = 2.68$; (b): $x = 3.50$; (c): $x = 4.56$; (d): $x = 9.00$; (———) Computational (LES); ($\times$) and ($\triangle$) are experimental results of Martinuzzi and Tropea [3] corresponding to LDA measurements in u–v and u–w planes, respectively.

Fig. 6 shows a comparison of time-averaged streamlines on the symmetry plane. The overall prediction of the separation region on the roof and behind the obstacle is quite good. The stagnation point is located high on the front face (Fig. 6); fluid striking the body above it goes over the obstacle and does not reattach on the roof. The rear recirculation region is not closed but streamlines originating upstream of the obstacle do not enter this region; fluid enters the rear recirculation region from the sides. Near the top of the recirculation region we find the head of the arch vortex.

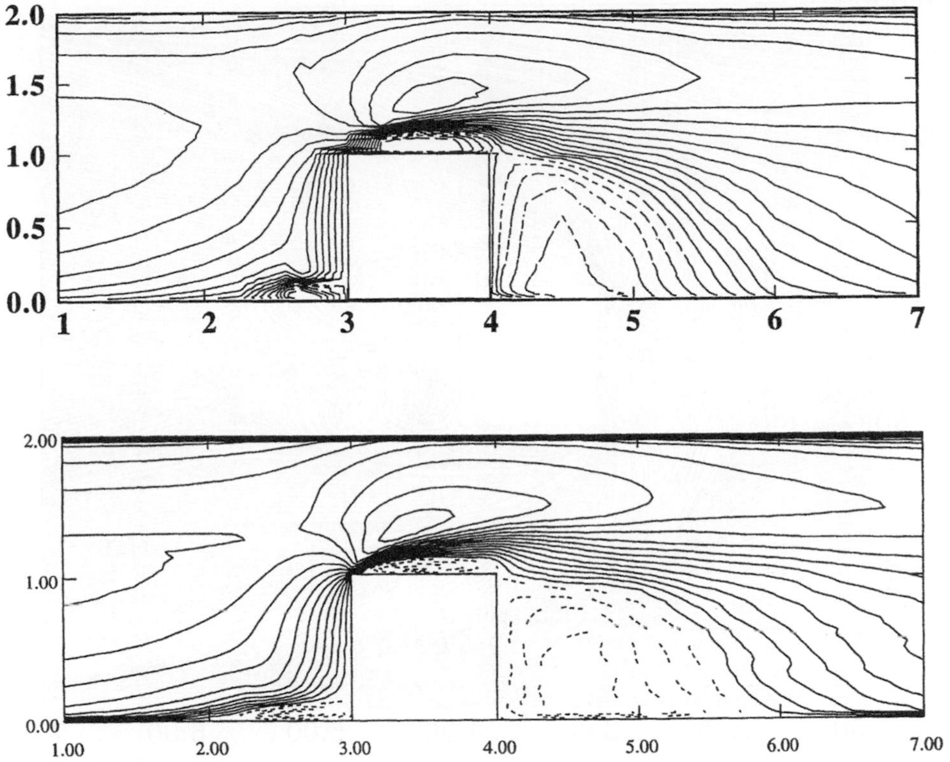

Fig. 4. Time-averaged streamwise velocity on the symmetry plane at $Re_b = 40\,000$. Top: experimental result of Martinuzzi and Tropea [3]; bottom: result from LES. Number of contours = 20, Minimum = − 0.5, Maximum = 1.5.

Streamlines in a cross-flow plane parallel to the rear face (Fig. 7) illustrates the strong vortical motion associated with the horseshoe vortex and its interaction with the corner vortices. The horseshoe vortex widens by diffusion and its center moves upwards from the channel floor in the downstream direction.

These structures are of considerable importance. The horseshoe vortex covers a region of large skin friction and mixing. It also affects the flow on the sides of the obstacle and, depending on its orientation relative to the main flow direction, it can intensify the fluctuating side forces on the obstacle. The arch vortex lies below the strong shear layers shed from the upstream edges of the obstacle. Large fluctuations are observed in this region and contribute considerably to the dissipation.

The flow oscillates roughly between two states; these states are characterized by different vortex locations. The drag on the body is entirely due to the difference of pressure on the front and back faces. Since the flow is reversed on the sides and the roof, the viscous drag is negative and small. A time series of the drag shows a large, quasi-periodic oscillation at a frequency approximately twice that of side forces; the

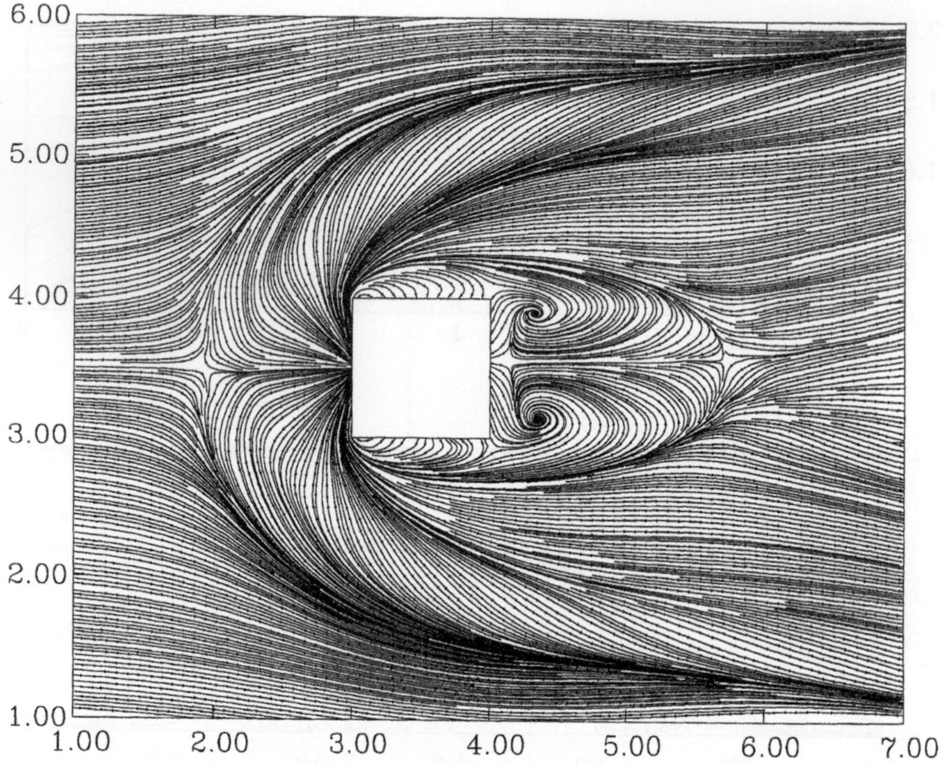

Fig. 5. Time-averaged streamlines on the floor of the channel at $Re_b = 40\,000$.

latter are quite large. The PDF of the drag shows two distinct peaks corresponding to the two flow states. A bimodal PDF of the fluctuating velocity is found in front of the obstacle and seems to be a common feature of flow around bluff bodies. The horseshoe vortex displays quasi-periodic formation, intensification, and transfer of vorticity to the legs of the vortex. The locus of vortex formation varies considerably. In some instances there are two adjacent vortices which coalesce to produce a single strong vortex that interacts strongly with the rest of the flow. The core of the horseshoe vortex moves up from the channel floor with downstream distance. More information on the temporal structures are provided in Ref. [5].

Although the side force (and many other quantities) appear to contain a periodic component (Fig. 8), when the signal is Fourier transformed, no peak is found in the spectrum. This is typical of turbulent flows. Repeatable coherent structures are found but they are not identical in strength or size and do not occur periodically.

Changing the model affects the length of the recirculation zone significantly; this is rather disturbing and the reasons for it are under investigation. In most other respects, the flows at the two Reynolds numbers are quite similar. For more details concerning influence of Reynolds number, subgrid-scale models and the temporal characteristics, the reader is referred to Ref. [5].

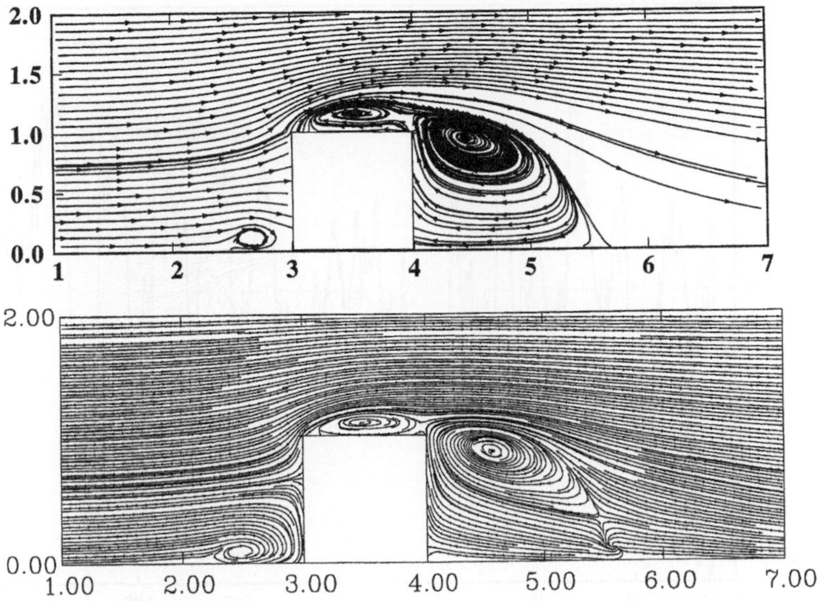

Fig. 6. The streamlines on the symmetry plane at $Re_b = 40\,000$. Top: experimental result of Martinuzzi and Tropea [3]; bottom: result from LES.

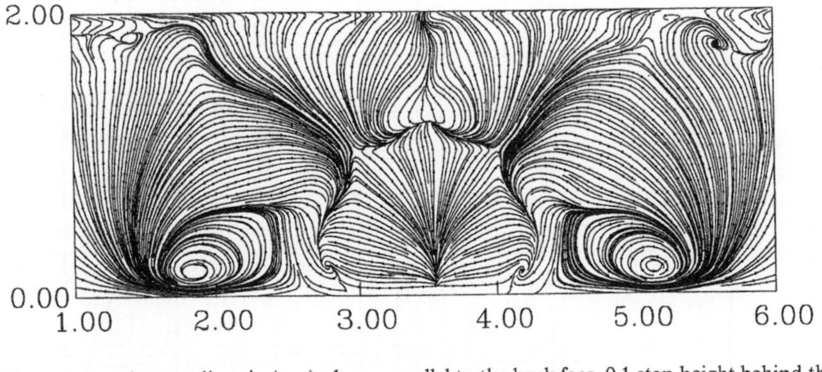

Fig. 7. Time-averaged streamlines in (y–z) plane, parallel to the back face, 0.1 step height behind the cube.

5. Conclusions

We have shown that large eddy simulation (LES) is capable of providing information of importance to wind engineers about flows over bluff bodies. In addition to yielding mean forces and flow parameters, it can provide the fluctuating components of these quantities and structural information about the flow.

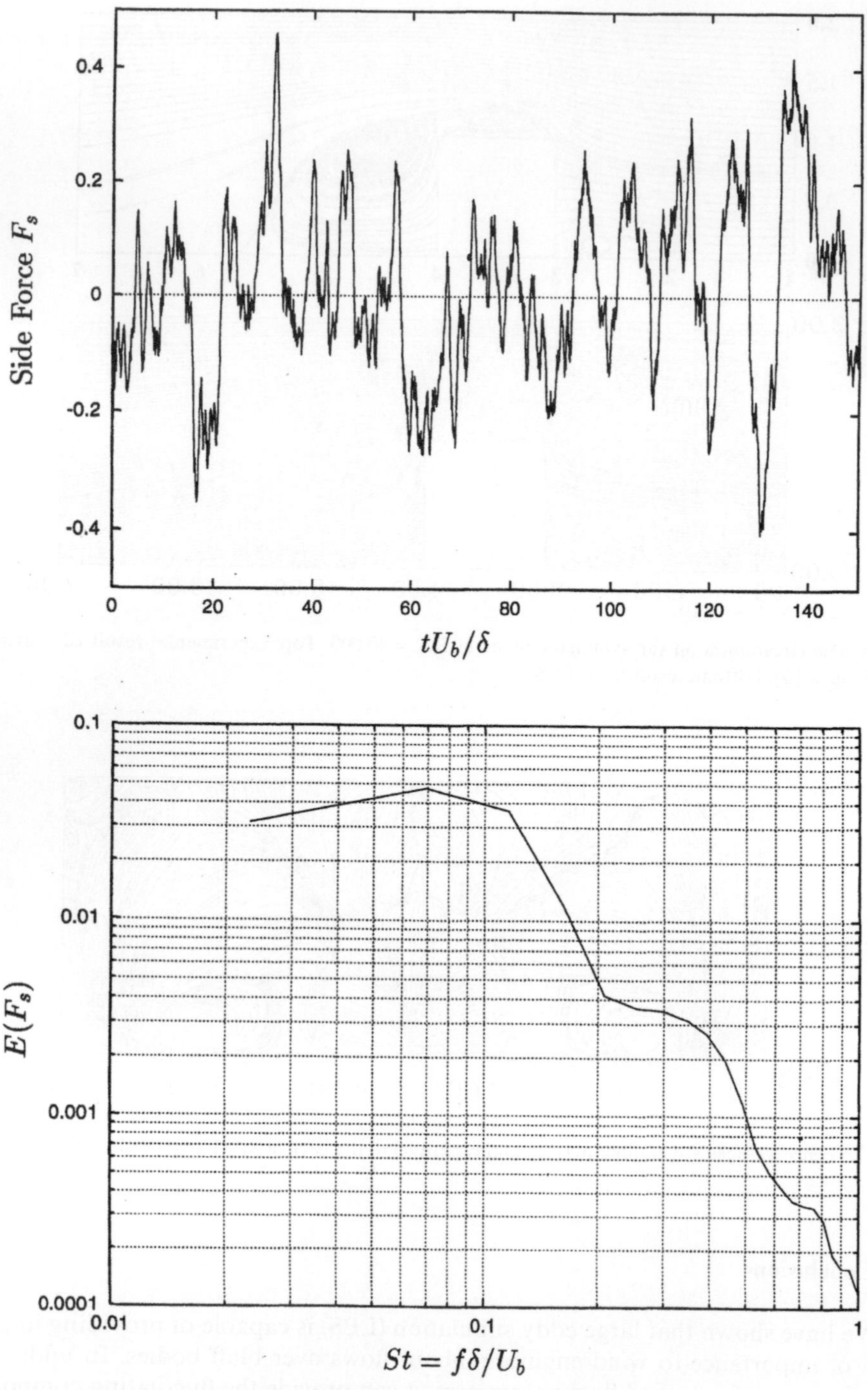

Fig. 8. Time series of the side force (normalized by $\rho U_b^2 \delta^2/2$) and its spectra.

However, the simple fact that one is using LES does not guarantee success. It is possible for LES to yield results that are incorrect qualitatively as well as quantitatively. If LES is to become a useful engineering tool, it will be necessary to establish benchmarks against which numerical methods, subgrid-scale models, and other factors can be tested. These can be obtained from high-resolution LES or high- quality experiments. Experimental testing will remain a necessary adjunct to LES for a long time.

A flow as simple as the one about a cube attached to a wall illustrates many points about turbulent flows over bluff bodies. These flows are only quasi-periodic; for complex flows even that may not be so. The fluctuations can be quite large but there is an important difference between a quasi-periodic flow and a periodic one. (Interaction with vibrations of the body may render the flow truly periodic and one needs to be careful about this.) Coherent structures that are found in the time-averaged flow may not exist in the instantaneous flow, so one needs to be careful in discussing the large vortices found in the averaged flow. Finally, to understand what is happening, it is essential to study the interior of the flow as well as the pressure on the body.

In an earlier paper, one of the authors (Ferziger [9]) suggested a method called coherent structure capturing (CSC) in which only the very largest structures, that are responsible for the intense low- frequency forces on the body, be simulated. If this kind of simulation is performed, there is a danger that a truly periodic flow will result and, as already noted, this can be dangerous. Consequently, we now think that, if CSC is to be developed, it will probably be necessary to include a random force in the simulation.

Acknowledgements

We are thankful to the Karlsruhe group under Prof. Wolfgang Rodi, and, especially, Drs. Michael Breuer and Mathieu Pourquie for providing plots of their results.

References

[1] W. Rodi, J.H. Ferziger, M. Breuer, M. Pourquie, Current status of large eddy simulation: results of a Workshop, ASME J. Fluids Eng., submitted.
[2] J.H. Ferziger, in: T.G. Gatski, M.Y. Hussaini (Eds.), Large eddy simulation, Computation of Turbulent and Transitional Flows, Cambridge University Press, Cambridge, 1996.
[3] R. Martinuzzi, C. Tropea, The flow around surface-mounted, prismatic obstacles placed in a fully developed channel flow, J. Fluids Eng. 115 (1996) 85–92.
[4] K. Shah, J.H. Ferziger, A new non-eddy viscosity subgrid scale model, in: Annual Research Briefs, Center for Turbulence Research, Stanford University, 1995, pp. 73–90.
[5] K. Shah, J.H. Ferziger, Large eddy simulations of flow past a cubic obstacle, Thermosciences Div. Rep. TF-70, Dept. Mech. Eng., Stanford University, 1996.
[6] L.L. Pauley, P. Moin, W.C. Reynolds, The structure of two-dimensional separation, J. Fluid Mech. 220 (1990) 397–411.

[7] M. Germano, U. Piomelli, P. Moin, W.H. Cabot, A dynamic subgrid-scale eddy viscosity model, Phys. Fluids A 3 (7) (1991) 1760–1765.
[8] A.E. Perry, M.S. Chong, A description of eddying motions and flow patterns using critical-point concepts, Ann. Rev. Fluid Mech. 19 (1987) 125–155.
[9] J.H. Ferziger, in: S. Murakami (Ed.), Simulation of complex turbulent flows: recent advances and prospects in wind engineering, Computational Wind Engineering , vol. 1, Elsevier, Amsterdam, 1993.

Bridge Aerodynamics

ELSEVIER

Journal of Wind Engineering
and Industrial Aerodynamics 67&68 (1997) 227–237

JOURNAL OF
wind engineering
AND
industrial
aerodynamics

Fatigue strength design for vortex-induced oscillation and buffeting of a bridge

Masao Hosomi[a,*], Hiroshi Kobayashi[b], Yoshinobu Nitta[a]

[a] *Bridge Research Department, Komai Tekko Inc., 405, Matsuhidai, Matsudo, Chiba 270, Japan*
[b] *Department of Civil Engineering, Ritsumeikan University, 1916, Noji, Kusatsn 525, Japan*

Abstract

Vortex-induced oscillation of bridges appears at relatively low wind speed. Large bridges are sometimes subjected to a large number of oscillation cycles by vortex-induced oscillation and buffeting. Fatigue strength design for these oscillations becomes an important problem for the design of a largc bridge. To execute fatigue strength design reasonably for vortex-induced oscillation and buffeting, we put the characteristics of natural wind (i.e. wind direction, wind speed, and angle of attack) and the characteristics of bridge aerodynamics together. A computer makes fatigue strength design easy under several conditions. In this study, we propose a numerical stochastic method to calculate the number of cycles of vortex-induced oscillation and buffeting, considering that aerodynamic responses appear in a wide range of wind direction.

Keywords: Bridge; Vortex-induced oscillation; Buffeting; Fatigue design

1. Introduction

We usually design a bridge under the assumption that aerodynamic phenomena occur in a narrow range of a yawed wind nearly normal to the bridge axis. The range of yawed wind where an aerodynamic response occurs is not apparent. In the experimental study by the authors [1], vortex-induced oscillation of a rectangular section still appeared at 77.5° of yawed wind. On the girder section, a vortex-induced oscillation may also appear at a large yawed wind. Bridge buffeting in yawed wind was investigated by Tanaka et al. [2] and Kimura et al. [3]. From the above experimental results, we have to design a bridge considering that aerodynamic responses may appear in a wide range of wind directions.

* Corresponding author. E-mail: hosomird@mxm.meshnet.or.jp.

PII S0167-6105(97)00075-5

In this study, we propose a numerical stochastic method to calculate the number of cycles of vortex-induced oscillation and buffeting, considering that aerodynamic responses appear in a wide range of wind directions.

2. Calculation of fatigue indexes

The probability that stress of a bridge member becomes σ_i can be calculated as

$$p(\sigma_i)\,\mathrm{d}\sigma = \mathrm{d}\sigma \iiint p_{\sigma|\alpha,v,\theta}\, p_{\alpha,v,\theta}\, \mathrm{d}\alpha\, \mathrm{d}v\, \mathrm{d}\theta, \tag{1}$$

where, the probability distribution function $p_{\sigma|\alpha,v,\theta}$ gives that a stress σ_i occurs under the conditions of any α (angle of attack), v (wind speed) and θ (wind direction). $p_{\alpha,v,\theta}$ is the joint probability distribution function when the wind conditions α, v and θ occur. $p_{\alpha,v,\theta}$ can be calculated by

$$p_{\alpha,v,\theta} = p_{\alpha|v,\theta}\, p_{v|\theta}\, p_{\theta}, \tag{2}$$

where, $p_{\alpha|v,\theta}$ is the probability that any α occurs under the conditions of any v, θ, $p_{v|\theta}$ is the probability that any v occurs under a condition of any θ and p_θ is the probability that any θ occurs. These pdfs are identified based on the meteorological data. The number of oscillation cycles n_i in which stress σ_i occurs during the lifetime of a bridge T is given by

$$n_i = T f_0 p(\sigma_i)\, \mathrm{d}\sigma, \tag{3}$$

where, f_0 is the frequency of the bridge aerodynamic response.

There are two alternative methods for checking the fatigue strength. One is a conventional method based on cumulative damage. The other method uses Miners' effective stress range. In this study, the above two indices are calculated for comparison. The degree of cumulative damage D can be calculated using

$$D = \sum(n_i/N_i), \tag{4}$$

where, N_i is the fatigue life for stress σ_i. Miners' effective stress range σ_e can be calculated by

$$\sigma_e = \sqrt[m]{\sum \sigma_i^m n_i/N}, \tag{5}$$

where, N is the total number of oscillation cycles, and, m is an exponent constant of the stress-number curve for a given material.

It is difficult to set up the lower limit of the fatigue stress range because we have few sets of data for a fatigue strength of lower stress level. For this problem, two cases are investigated. In the first case, we use the lower limit referred from Guideline of Fatigue Strength Design of steel structure (Japanese Society of Steel Construction) [4]. In the second case, instead of using the lower limit, we use an exponent m for calculation of an allowable fatigue stress range which is smaller than the above mentioned lower limit of fatigue stress range.

3. Check of fatigue strength due to vortex-induced oscillation and buffeting

3.1. Bridge under consideration and its aerodynamic characteristics

A three-span continuous girder bridge which has 120 m center span is studied. The natural frequency of the first vertical bending mode is 0.84 Hz. A 3-D elastic model with a rectangular cross section is used. The ratio of width (B) to depth (D) of the cross section is $B/D = 2$. The structural damping in logarithmic decrement is $\delta_s = 0.009$ (Scruton number Sc = 4.6). Vortex-induced oscillation of the first bending mode appears as shown in Fig. 1, based on the wind tunnel test which has been carried out in a yawed wind. In Fig. 1, β is the yawed angle, V_r is the non-dimensional wind speed, V is the wind speed and f is the natural frequency. The response amplitude in Fig. 1 may be reduced to 40% if we converge the response to that under the condition of $\delta_s = 0.02$ (Sc = 10).

We assume the following two different conditions for investigating the effect of the angle of attack.

Case 1: The same response amplitude occurs for angles of attack between $-5°$ to $5°$.

Case 2: The response does not appear under the condition that the angle of attack is less than or equal to 0°. And the response in Fig. 1 appears under the condition that the angle of attack is 3°. A linear function is used to calculate the response between 0° to 5°.

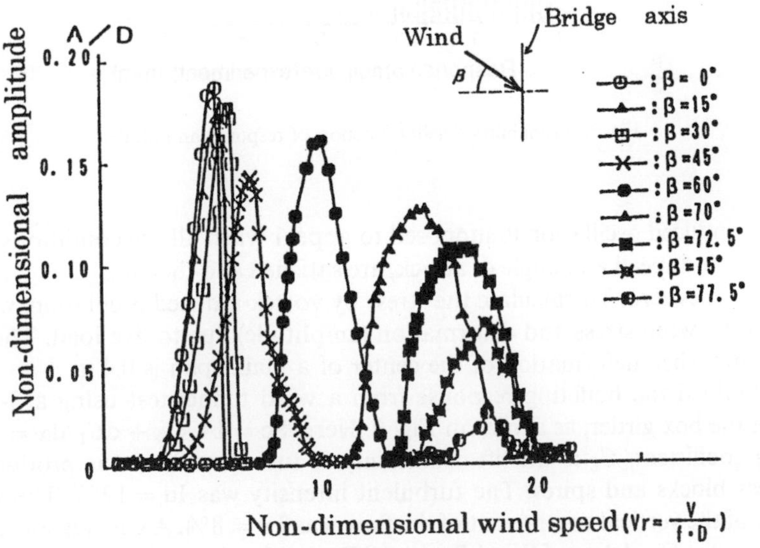

Fig. 1. Vortex-induced oscillation (effect of yawed wind).

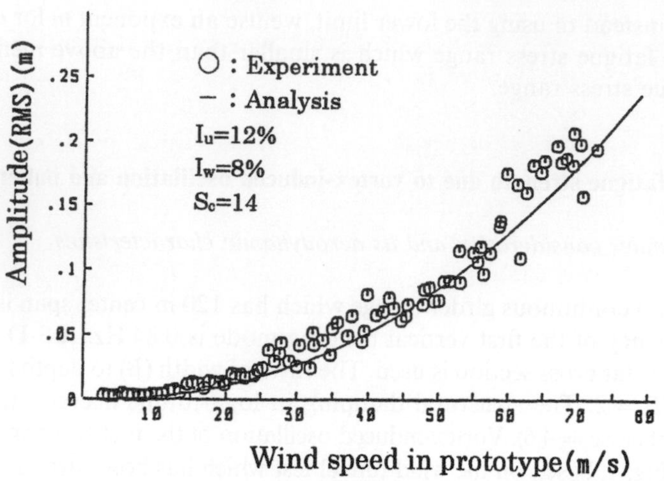

Fig. 2. Buffeting.

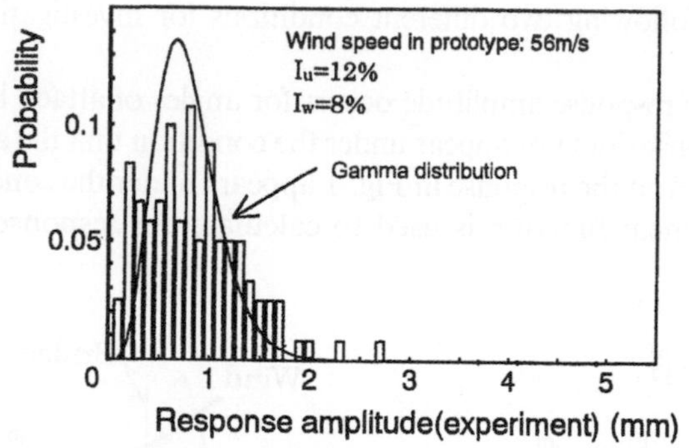

Fig. 3. Probability density function of response amplitude.

Vortex-induced oscillation is supposed to appear when all the conditions, i.e. wind direction, wind speed and angle of attack, are satisfied and then $p_{\sigma|\alpha,v,\theta} = 1$, otherwise, $p_{\sigma|\alpha,v,\theta}$ becomes zero. To calculate the stress by vortex-induced oscillation, we assume a relation between stress and deformation (amplitude) due to live load. The stress is 600 kgf/cm^2 when deformation of the center of a main span is 0.4 m.

We obtained the buffeting response from a wind tunnel test using a 3-D elastic model of the box girder, as shown in Fig. 2. Here, Sc = 14, $C_D + dC_L/d\alpha = 5.3$ (C_D is the drag coefficient, C_L is the lift coefficient). A turbulent flow was produced using roughness blocks and spires. The turbulent intensity was Iu = 12%. The turbulent intensity of the vertical component of the flow was Iw = 8%. A gust response analysis referring to the Guideline of Wind Resistant Design for the Akashi Kaikyo Bridge [5] is also indicated in Fig. 2. In the analysis, the decay factor in the spatial correlation function is 8. The analysis applies Busch and Panafsky's empirical formula as input spectrum of the wind, and simplified Sear's functions as aerodynamic admittance are used. An analytical response of buffeting was employed for this check of fatigue strength. The response of the first bending mode is assumed to be dominant for the prototype. The responses of higher modes are neglected. Fig. 3 shows an example of the probability distribution function of the response amplitude of buffeting in the experiment. We apply a Gamma-distribution function to fit the experimental data, which is indicated as a histogram. We carried out the χ^2-goodness-of-fit test for some cases. The response is assumed to be proportional to the wind speed component normal to the bridge axis in yawed wind referring to Kimura's study [3] (Cosine rule). The maximum wind speed under consideration is 60 m/s.

3.2. *Characteristics of the wind*

The statistical characteristics of the wind were obtained from observations of the natural wind. The probability of wind direction is shown in Fig. 4. The probability distribution function of wind speed on any wind direction is assumed as a Weibull distribution. The Weibull parameters are shown in Table 1. An example of observed values compared with the Weibull distribution applied is shown in Fig. 5. The probability of an angle of attack on any wind direction and any wind speed is shown in Fig. 6. The probability density function is assumed as a normal distribution with mean value of 0°, standard deviation of 1°, for all wind directions and all wind speeds.

3.3. *Characteristics of the material*

The fatigue strength of a bridge member is checked following the design guideline. The member of E rank in the Guideline of Fatigue Strength Design (Japanese Society of Steel Construction) [4] is used. The lower limit of fatigue stress range is 296 kgf/cm^2 corresponding to 44 000 000 oscillation cycles. The exponent constant m of the stress-number curve is 3.

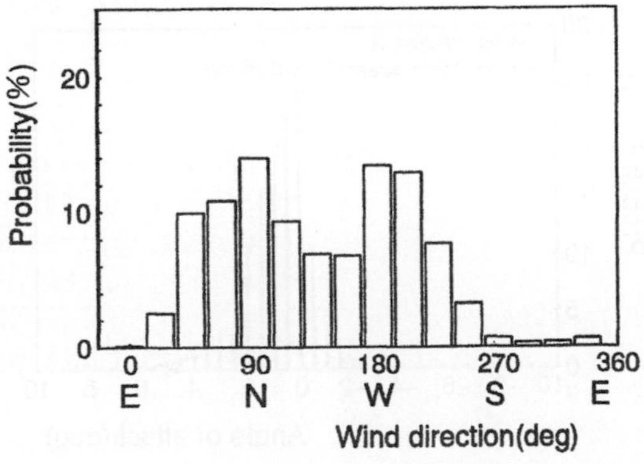

Fig. 4. Probability of wind direction.

Table 1
Weibull parameter

Wind direction	c	k
N	5.16	1.70
NNE	4.78	1.81
NE	4.32	1.65
ENE	3.77	1.38
E	2.74	1.55
ESE	3.27	2.81
SE	1.16	1.01
SSE	2.12	1.45
S	2.20	0.98
SSW	5.49	1.71
SW	5.18	1.77
WSW	7.03	2.00
W	6.27	1.89
WNW	5.32	1.57
NW	1.86	0.91
NNW	5.32	1.87

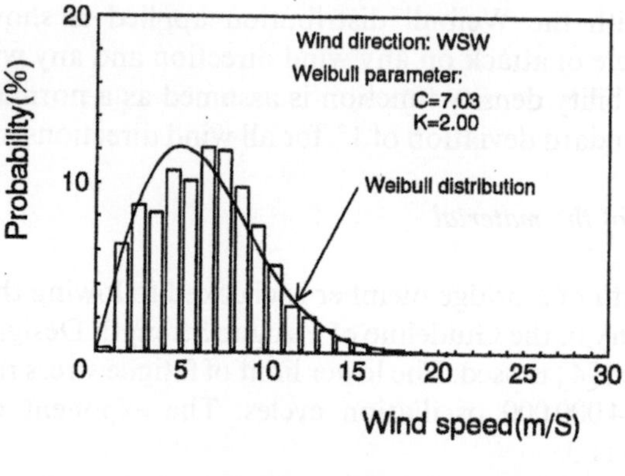

Fig. 5. Probability density function of wind speed.

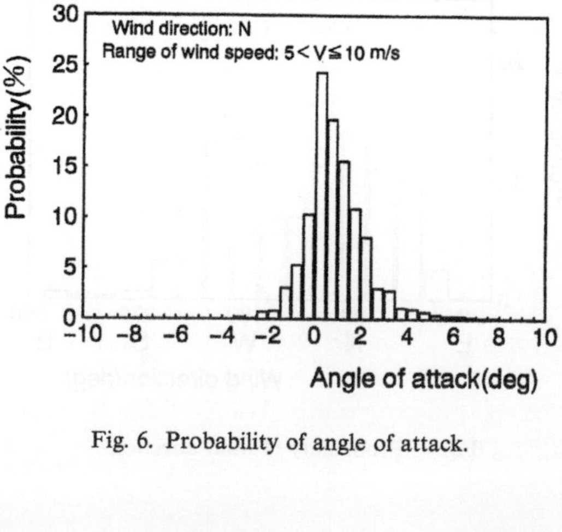

Fig. 6. Probability of angle of attack.

Table 2
Cases of calculation

Cases of calculation	Oscillation	Orientation of bridge	Characteristiics of aerodynamic response	
			β[a]	Angle of attack α
Case 1	Vortex-induced oscillation	NE	Fig. 1	– 5–5°
Case 2		N	Fig. 1	– 5–5°
Case 3		N	– 11.25°–11.25°	– 5–5°
Case 4		N	Fig. 1	0–5°
Case 5	Buffeting	E	Cosine rule[b]	– 5–5°
Case 6		N	Cosine rule	– 5–5°

[a] β is the yawed angle in which aerodynamic response appears.[b] Cosine rule: The response assumed to occur by wind speed component normal to the bridge axis in yawed wind.

4. Result of the calculation

To proceed with a numerical fatigue check, we assume different orientation arrangements of the bridge and aerodynamic characteristics as shown in Table 2. The desired lifetime of a structure is assumed to be 100 years. Cases 1–4 are the cases of vortex-induced oscillations. Cases 5 and 6 are the cases of buffeting. Figs. 7 and 8 show the number of oscillation cycles of the respective response amplitudes. Cases 1 and 2 show the cases of different direction of the bridge axis. The orientation of the bridge axis in case 1 is north–east, and north in case 2. The probability of wind direction affects the number of oscillation cycles on each case significantly. In case 3, the direction of the bridge axis is north, and a response is assumed to appear only for wind directions normal to the bridge axis. The number of oscillation cycles in case 3 is $\frac{1}{2}$ of that in case 2. In case 4, a response is assumed to occur only in the wind of positive angle of attack. The number of oscillation cycles in case 4 decreases as compared to the number in case 3. In cases 5 and 6 for buffeting, oscillation cycles of large response amplitude do not appear.

Table 3 shows the result of the check of fatigue strength for vortex-induced oscillation. In cases 1–3, fatigue check is not safe in both methods of calculation, i.e. the degree of cumulative damage and Miner's effective stress range. Only in case 4, the fatigue check is safe. Case 3 has a small number of oscillation cycles and a small cumulative damage compared with case 2. This difference arises from the fact that the influence of wind direction is taken into account in case 2 and not in case 3.

There is no problem related to a fatigue strength for buffeting as shown in Table 4. The allowable fatigue stress range becomes larger than Miner's effective fatigue stress range. The reason why is that the number of oscillation cycles of the response in question is sufficiently small.

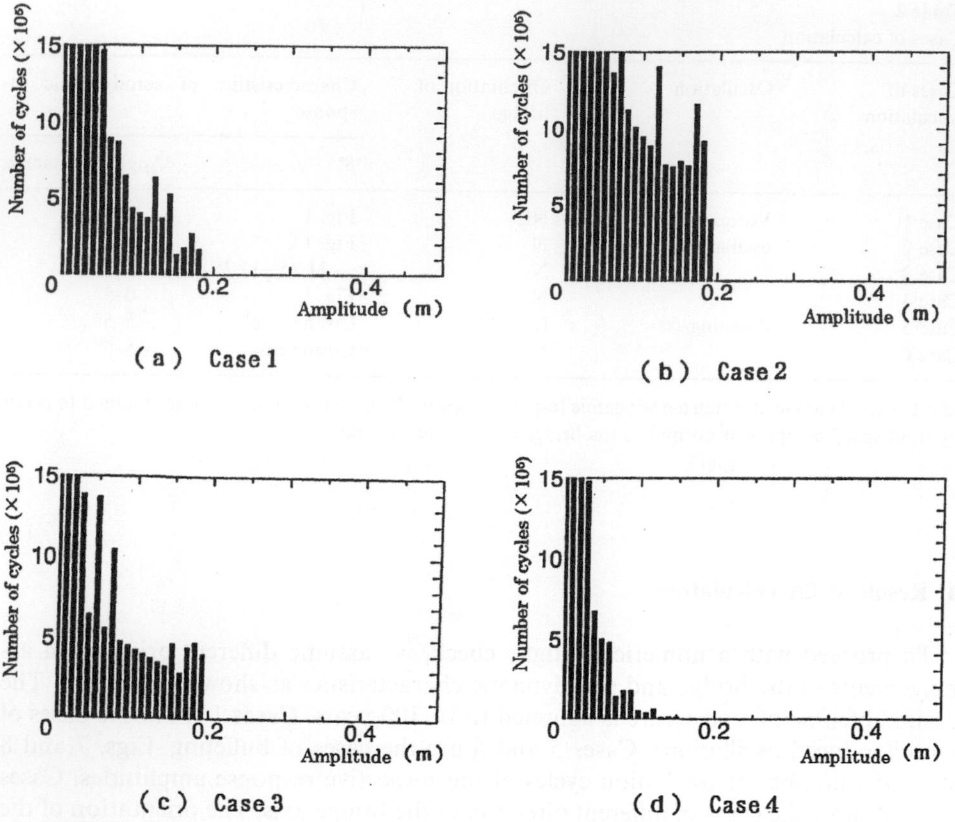

Fig. 7. Number of cycles of vortex-induced oscillation.

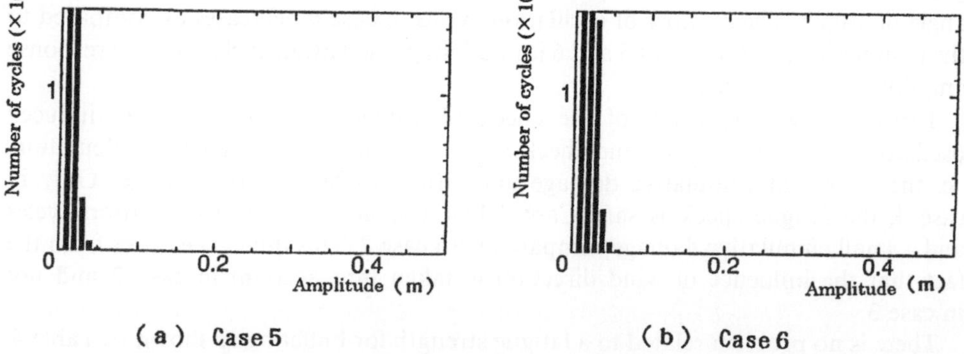

Fig. 8. Number of cycles for buffeting.

Table 3
Fatigue check for vortex-induced oscillation

		Case 1	Case 2	Case 3	Case 4
Using lower limit of stress range	(1) Number of cycles ($\times 10^3$)	26 800	78 600	35 600	1590
	(2) Equivalent stress range (kgf/cm^2)	375	402	415	346
	(3) Allowable stress range (kgf/cm^2)	349	296	317	895
	(2)/(3)	1.07	1.65	1.31	0.39
	Degree of cumulative damage	1.24	4.49	2.23	0.06
Without lower limit of stress range	(1) Number of cycles ($\times 10^3$)	1 370 000	1 600 000	354 000	642 000
	(2) Equivalent stress range (kgf/cm^2)	109	154	201	65.4
	(3) Allowable stress range (kgf/cm^2)	93.9	89.2	148	121
	(2)/(3)	1.16	1.73	1.36	0.54
	Degree of cumulative damage	1.56	5.16	2.52	0.16

Table 4
Fatigue check for buffeting

		Case 5	Case 6
Using lower limit of stress range	(1) Number of cycles	9.99	48.4
	(2) Equivalent stress range (kgf/cm^2)	331	322
	(3) Allowable stress range (kgf/cm^2)	4.85×10^4	2.87×10^4
	(2)/(3)	—	—
	Degree of cumulative damage	3.19×10^{-7}	1.42×10^{-6}
Without lower limit of stress range	(1) Number of cycles ($\times 10^3$)	6820	17 400
	(2) Equivalent stress range (kgf/cm^2)	33.8	35.9
	(3) Allowable stress range (kgf/cm^2)	551	403
	(2)/(3)	0.061	0.089
	Degree of cumulative damage	2.31×10^{-4}	7.09×10^{-4}

Fig. 9 shows the Miner effective fatigue stress range and allowable fatigue stress range for vortex-induced oscillation. The solid line gives the allowable stress range of a typical structural steel of E rank [4], full circles give the results considering stress limit and open circles considering no stress limit. From this figure, we can check the safety of the material subjected to vortex-induced oscillation under various wind environments.

In case of buffeting of the bridge with about 120 m center span, fatigue strength is not a problem at all. But, it is necessary to pay attention that number of vibration is sensitive to the direction of the bridge axis.

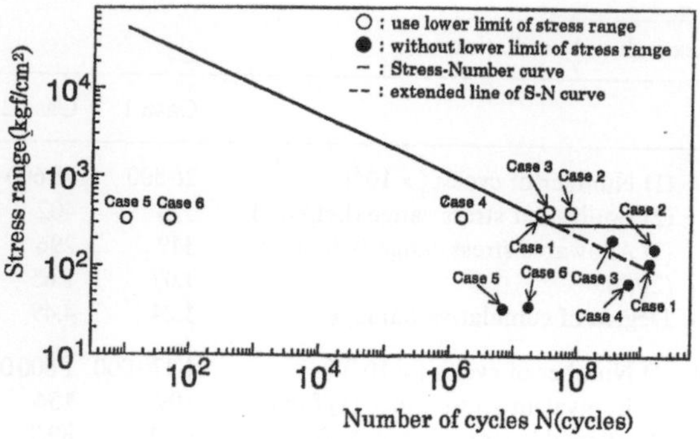

Fig. 9. Miner's effective stress range and S–N curve.

5. Conclusions

In this study, we proposed a numerical stochastic method for calculating the number of cycles of vortex-induced oscillation and buffeting considering that aerodynamic responses appeared in a wide range of wind directions. This method designed in this study will provide a rational fatigue design by considering both the characteristics of responses and the characteristics of wind. We should sum up the numbers of cycles of vortex-induced oscillation, buffeting and other oscillations. The following results are obtained in this study.

The orientation of the bridge axis, the characteristics of aerodynamic responses and the characteristics of the wind affect the number of cycles of vortex-induced oscillation.

The case in which the influence of wind direction is incorporated with the response amplitude of vortex-induced oscillation, and the case in which vortex-induced oscillation appears only in the wind normal to the bridge axis have considerably different number of oscillation cycles and cumulative damage, respectively.

Both degree of cumulative damage and Miner's effective stress range are useful to check fatigue strength.

For buffeting of a bridge with about 120 m center span, fatigue strength is not a problem at all. But, it is necessary to pay attention to the fact that the number of oscillation cycles is sensitive to the orientation of the bridge axis.

References

[1] M. Hosomi, K. Koba, H. Kobayashi, Effect of yawed wind on vortex excited response of bridge girder models, in: Wind Engineering Retrospect and Prospect, vol. 2, Wiley, New York, 1994, pp. 863–870.

[2] H. Tanaka, G.L. Larose, Wind tunnel tests of cable-stayed bridges at their erection stage and effects of wind yawed angles. Proc. Int. Seminar on Utilization of Large Boundary Layer Wind Tunnel, Japan, December 6–7, 1993.

[3] K. Kimura, H. Tanaka, Bridge buffeting due to wind with yawed angles, J. Wind Eng. Ind. Aerodyn. 1992, 41–44.
[4] Japanese Society of Steel Construction, Guideline of Fatigue Strength Design of Steel Structure, Gihodo, 1993.
[5] Honshu-shikoku Bridge Authority, Guideline of Wind Resistant Design for Akashi-kaikyo Bridge, February 1990.

ELSEVIER

Journal of Wind Engineering
and Industrial Aerodynamics 67&68 (1997) 239–252

JOURNAL OF wind engineering AND industrial aerodynamics

Numerical simulation of flow around a box girder of a long span suspension bridge

Shinichi Kuroda[1]

Ishikawajima-Harima Heavy Industries Co., Ltd., 19-10, Mohri 1-Chome, Koto-Ku, Tokyo 135, Japan

Abstract

A numerical simulation of a high Reynolds number flow around a fixed section model with a shallow hexagonal cross-section is presented. The two-dimensional incompressible Navier–Stokes equations are used as the governing equations. The numerical algorithm is based on the method of pseudocompressibility and uses an implicit upwind scheme. The overall characteristics of the measured static force coefficients were well captured by the present computation.

Keywords: Numerical simulation; Navier–Stokes equations; Method of pseudocompressibility; Upwind scheme; Long span bridge; Box girder; Great Belt East Bridge

1. Introduction

At the present, wind tunnel testing is actually the only way to assess the aerodynamic performance and aeroelastic stability of long span bridges. The full bridge model test may be the final phase of the wind resistant design. Section model tests are also used to improve the aerodynamic/aeroelastic performance of the bridge components such as bridge decks or tower shafts. In recent years, specific analytical methods such as the direct flutter FEM analysis [1] have been developed and used to complement the full bridge model test. However, in these analyses, static/dynamic section model tests are mandatory to obtain the wind force data, which are the input for the analyses. Even a section model test is generally very time-consuming and expensive. With the recent rapid improvement of computers, Computational Fluid Dynamics (CFD) has made a remarkable progress. If the section model test can be replaced by the CFD, it will be significant for bridge engineers.

The objective of the present research is to develop a numerical method to simulate the static/dynamic wind tunnel test for section models of long span bridges. The

[1]E-mail: shinichi-kuroda@ihi.co.jp.

PII S0167-6105(97)00076-7

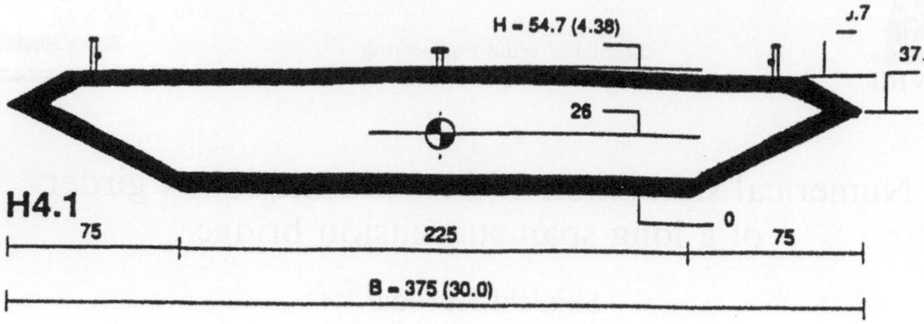

Fig. 1. Section model [2].

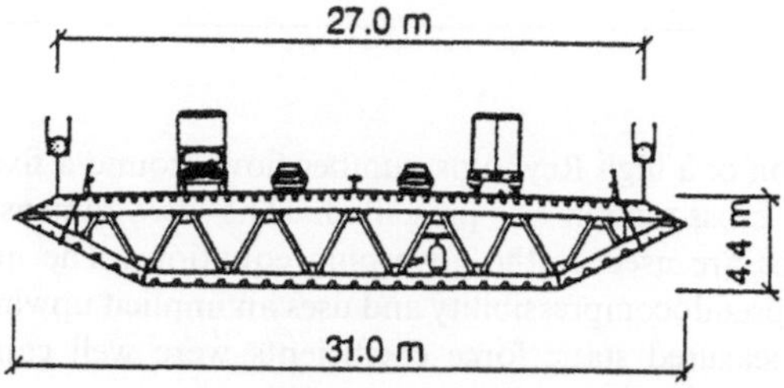

Fig. 2. Box girder cross section of the Great Belt East Bridge [3].

simulation of the static wind tunnel test is the first step of this CFD research. In this paper, a numerical simulation of flow around a fixed section model with a shallow hexagonal cross-section [2] (Fig. 1) is presented. The cross-sectional configuration was a candidate for the girder section of the Great Belt East Bridge [2,3] (Fig. 2). Extensive section model tests, as well as full bridge model tests, for the Great Belt East Bridge were conducted at the Danish Maritime Institute (DMI) and the static force coefficients for the section model of Fig. 1 are found in Ref. [2]. The computed static force coefficients were compared with this data.

2. Numerical procedure

The governing equations are the incompressible two-dimensional Navier–Stokes equations described in a generalized coordinate system. The numerical scheme employs a finite-difference procedure and the equations are solved using the method of pseudocompressibility [4].

In this paper, only the stationary grid was used, however, when considering the translation/deformation of the coordinate system itself, the generalized coordinates are as follows:

$$\xi = \xi(x, y, t), \tag{1a}$$

$$\eta = \eta(x, y, t). \tag{1b}$$

Using these, the governing equations are written in conservation form as follows:

$$\frac{\partial}{\partial \xi}\left(\frac{U}{J}\right) + \frac{\partial}{\partial \eta}\left(\frac{V}{J}\right) = 0, \tag{2a}$$

$$\frac{\partial \hat{u}}{\partial t} = -\frac{\partial}{\partial \xi}(\hat{e} - \hat{e}_v) - \frac{\partial}{\partial \eta}(\hat{f} - \hat{f}_v) = -\hat{r}, \tag{2b}$$

$$\hat{u} = \frac{1}{J}\begin{bmatrix} u \\ v \end{bmatrix}, \tag{2c}$$

$$\hat{e} = \frac{1}{J}\begin{bmatrix} \xi_x p + uU + \xi_t u \\ \xi_y p + vU + \xi_t v \end{bmatrix}, \tag{2d}$$

$$\hat{f} = \frac{1}{J}\begin{bmatrix} \eta_x p + uV + \eta_t u \\ \eta_y p + vV + \eta_t v \end{bmatrix}. \tag{2e}$$

Eq. (2a) represents the conservation of mass and Eq. (2b) the conservation of x and y components of the momentum, respectively. J is the Jacobian of the coordinate transformation, and U and V contravariant components of the velocity vector. ξ_x, etc. are the metrics of the transformation.

$$J^{-1} = x_\xi y_\eta - y_\xi x_\eta, \tag{3a}$$

$$U = \xi_x u + \xi_y v, \tag{3b}$$

$$V = \eta_x u + \eta_y v, \tag{3c}$$

$$\xi_x = \frac{\partial \xi}{\partial x}, \quad \text{etc.} \tag{3d}$$

The viscous fluxes are written as follows with the kinematic viscosity as ν:

$$\hat{e}_v = \frac{\nu}{J}\begin{bmatrix} (\xi_x^2 + \xi_y^2)u_\xi + (\xi_x\eta_x + \xi_y\eta_y)u_\eta \\ (\xi_x^2 + \xi_y^2)v_\xi + (\xi_x\eta_x + \xi_y\eta_y)v_\eta \end{bmatrix}, \tag{4a}$$

$$\hat{f}_v = \frac{\nu}{J}\begin{bmatrix} (\xi_x\eta_x + \xi_y\eta_y)u_\xi + (\eta_x^2 + \eta_y^2)u_\eta \\ (\xi_x\eta_x + \xi_y\eta_y)v_\xi + (\eta_x^2 + \eta_y^2)v_\eta \end{bmatrix}. \tag{4b}$$

In the case of a stationary grid, the generalized coordinates are as follows:

$$\xi = \xi(x, y), \tag{5a}$$

$$\eta = \eta(x, y). \tag{5b}$$

The time derivatives of ξ and η are zero ($\xi_t = \eta_t = 0$). The inviscid fluxes become

$$\hat{e} = \frac{1}{J}\begin{bmatrix} \xi_x p + uU \\ \xi_y p + vU \end{bmatrix}, \tag{6a}$$

$$\hat{f} = \frac{1}{J}\begin{bmatrix} \eta_x p + uV \\ \eta_y p + vV \end{bmatrix}. \tag{6b}$$

Other terms are not changed because the time derivatives of ξ and η do not appear in them. In this paper, Eqs. (6a) and (6b) were used for the inviscid fluxes.

Using the pseudocompressibility method [4], subiterations are required to satisfy the continuity equation at each physical time step. For evaluating the convective terms in the right-hand side, the upwind differencing scheme based on the flux-difference splitting [4] was used. The upwind-biased scheme of third order or fifth order of Refs. [4,5] can be used. For all cases presented in this paper, the fifth order was used. For evaluating the viscous terms, the second-order central differencing was used. To solve the algebraic equation, the implicit line-relaxation scheme was used. Implicit boundary conditions are used at all of the boundaries. At the wall boundary, the no-slip condition was adopted for the velocity, and for the pressure, such a condition was imposed that the component perpendicular to the wall surface of the pressure gradient becomes zero. For the inflow and outflow conditions, characteristic boundary conditions [4] were used.

3. Conditions of computation

In Ref. [2], no description can be found of the wind speed or Reynolds number at which the static force measurements were conducted. However, the following description can be found in Ref. [3]. Steady state wind load coefficients were measured on 1 : 80 and 1 : 300 models for the selected section H9.1 which was identical to the section of Fig. 1 (H4.1) except that the deck width of H4.1 was 1 m narrower than H9.1 (30 m versus 31 m for H9.1). Satisfactory agreement was demonstrated between steady state wind loads obtained from models of different size, indicating that Reynolds number effects are of minor importance in the case of H9.1.

In the present computations, a Reynolds number of 3×10^5 (based on the girder width B) was used, assuming a wind speed of around 12 m/s. The experimental static force measurements were conducted for angles of attack ranging from $-10°$ to $+10°$ [2]. Thus, the computations were carried out at six angles of attack α of $-10°$, $-5°$, $0°$, $5°$, $8°$ and $10°$. The measurements were conducted in turbulent flow with a longitudinal turbulence intensity of 7.5% [2]. However, in the present computations, a uniform smooth flow was assumed as the incoming flow and no turbulence model was used. The turbulent flow simulation together with the three-dimensional extensiton will be presented in a future work.

The O-type grid (221×101) was generated around the girder surface using the hyperbolic grid generation scheme [6]. A close view of this grid is shown in Fig. 3.

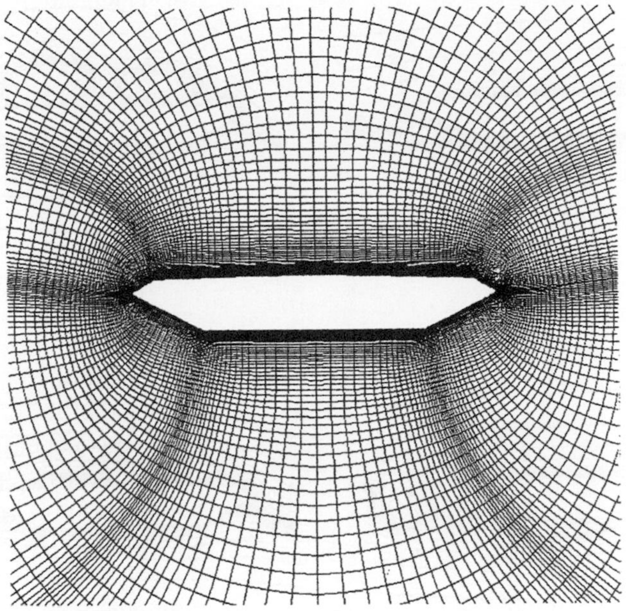

Fig. 3. Grid used in the present computations.

Identical grids were used for all angles of attack. The normal spacing next to the wall was 0.0002 girder widths. The outer boundary of the computational region was placed about 6 times the girder width away from the girder in all directions. The presence of side railings and central crash barrier was not considered in the present computations. The inclusion of such equipments in the computation will be also presented in a future work.

A non-dimensional time step Δt $(= \Delta T\ U/B)$ of 0.01 was used. The sweep for the line-relaxation scheme was 1 time per subiteration step each for ξ and η directions. As the convergence criteria for the iterative calculations, $|p^{m+1} - p^m| < 10^{-3}$ was used for the pressure, and for the momentum equation, the residual of x or y components was not more than 10^{-4}. In the above, the index m indicates the subiteration count at each time step.

4. Computed flow field

The computations were started impulsively from the uniform free-stream conditions and were continued until the non-dimensional time t $(= T\ U/B)$ equal to 40. Only for $\alpha = 10°$, the simulation was continued until $t = 80$, because the period of flow field variation (although not exactly periodic) was long compared with those for other angles of attack. Fig. 4 shows the computed instantaneous streamlines at $t = 40$.

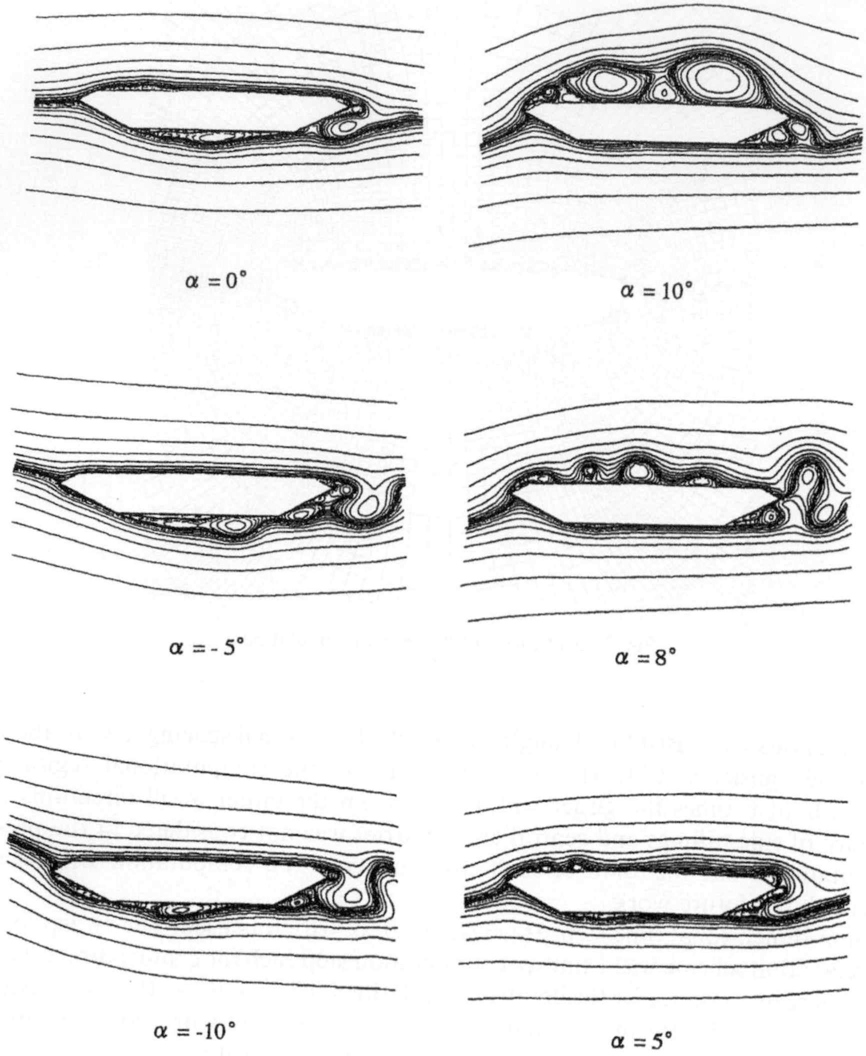

Fig. 4. Instantaneous stream lines.

4.1. *Mean flow pattern*

The mean flow patterns are shown in Fig. 5. The averaging procedure was taken in the range of $t = 20–40$ (for $\alpha = 10°$, the range of $t = 40–80$).

$\alpha = 5°, 8°, 10°$: On the lower surface, the flow is attached to the inclined web surface on the windward side from the stagnation point to the corner. The flow shows a small separation at the corner and reattaches the lower flange soon again and runs to the corner on the leeward side. The small separation region on the windward side does not vary in size in the range of these positive angles of attack.

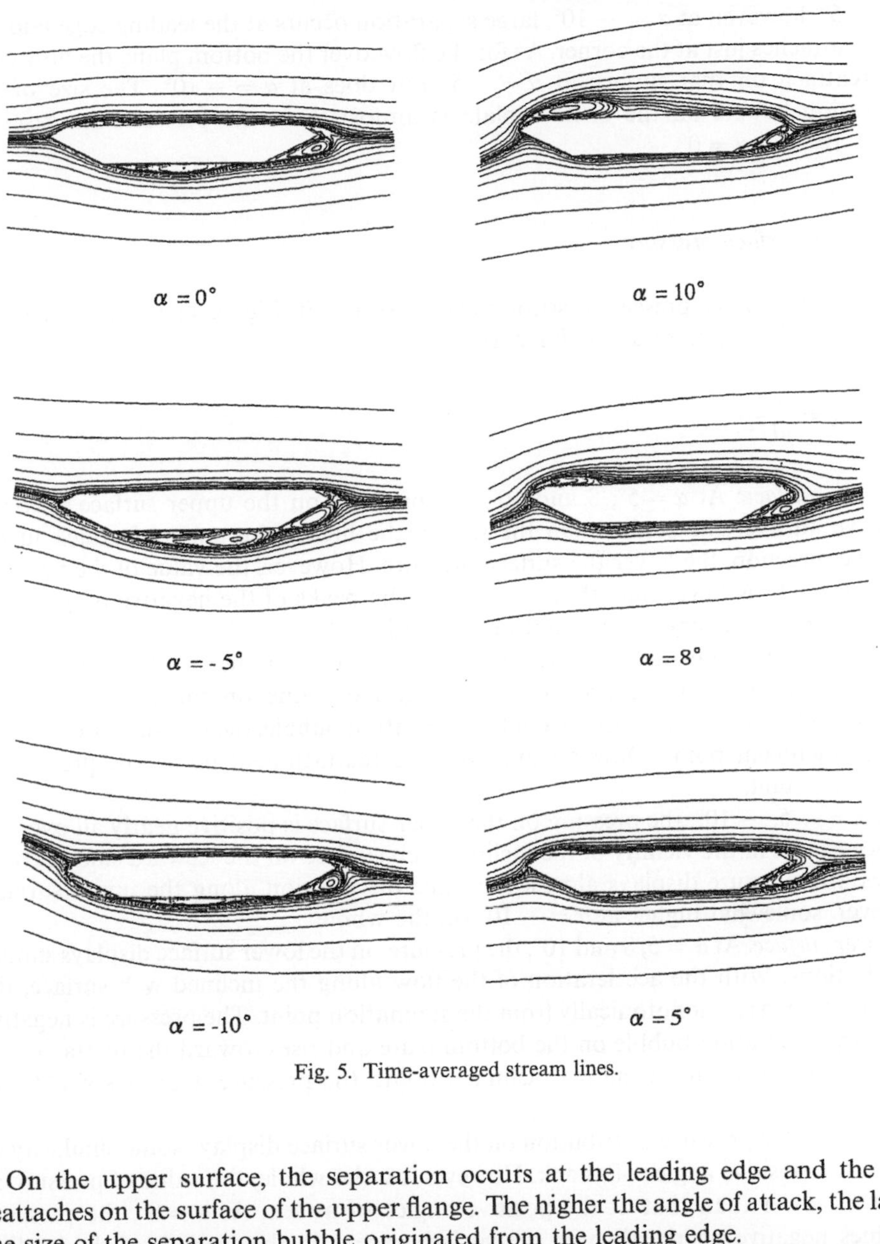

Fig. 5. Time-averaged stream lines.

On the upper surface, the separation occurs at the leading edge and the flow reattaches on the surface of the upper flange. The higher the angle of attack, the larger the size of the separation bubble originated from the leading edge.

$\alpha = 0°$: The separation does not occur at the leading edge and the flow attaches on the upper and lower inclined web surfaces on the windward side. Flow separation occurs at the corners. The size of the separation bubble on the bottom plate is larger than that at the positive angles of attack (5°, 8°, 10°).

$\alpha = -5, -10°$: At both angles, the flow fundamentally attaches on the upper surface. On the lower inclined web surface, the leading edge separation is small at

$\alpha = -5°$, however, at $\alpha = -10°$, large separation occurs at the leading edge and the flow reattaches just at the corner. As for the flow over the bottom plate, the flow does not reattach on that surface at $\alpha = -5°$ but does at $\alpha = -10°$. The size of the separation bubble on the bottom plate is almost the same between the cases of $\alpha = -10°$ and $\alpha = 0°$.

4.2. Mean surface pressure

The mean surface pressure distributions are shown in Fig. 6. The pressure coefficient C_p in Fig. 6 is normally defined as

$$C_p = \frac{2p}{\rho U^2}. \tag{7}$$

Upper surface: At $\alpha = 5°$, 8 and 10°, the pressure on the upper surface becomes negative all over the surface. On the whole, the higher the angle of attack in the positive direction, the lower the surface pressure. However, the value of the negative peak at $\alpha = 8°$ is lower than that at $\alpha = 10°$. The peaks of the negative pressure are located around the corner on the windward side.

At $\alpha = 0°$, the pressure becomes negative from the middle on the inclined web surface, and shows an almost constant (negative) value on the first half of the separation bubble. On the last part of the separation bubble, the pressure rises toward the reattachment point. Downstream from the reattachment point, the pressure is almost constant.

At $\alpha = -5$, $-10°$, the pressure on the upper surface is positive nearly all over the surface except in the vicinity of the inclined web surface on the leeward side. At both angles, the pressure displays almost the same distribution along the upper surface, however, somewhat higher at $\alpha = -10°$ on the whole.

Lower surface: At $\alpha = 5$, 8 and 10°, the pressure on the lower surface displays similar distributions. With the acceleration of the flow along the inclined web surface, the pressure decreases monotonically from the stagnation point. The pressure is negative inside the separation bubble on the bottom plate and rises toward the reattachment point. Downstream from the reattachment point, the pressure decreases slightly in a steady fashion.

At $\alpha = 0°$, the pressure distribution on the lower surface displays some similarity to that at the positive angles of attack. However, on the whole, the value of pressure at $\alpha = 0°$ is lower than that at the positive angles of attack. At $\alpha = 0°$, the pressure becomes negative nearly all over the surface except in the vicinity of the leading edge.

At $\alpha = -5$ and $-10°$, the pressure is negative all over the surface. On the inclined web surface on the windward side, the pressure at $\alpha = -10°$ shows a far lower value than that at $\alpha = -5°$. However, on the last half of the bottom plate, the pressure at $\alpha = -5°$ shows a negative peak and is lower than that at $\alpha = -10°$.

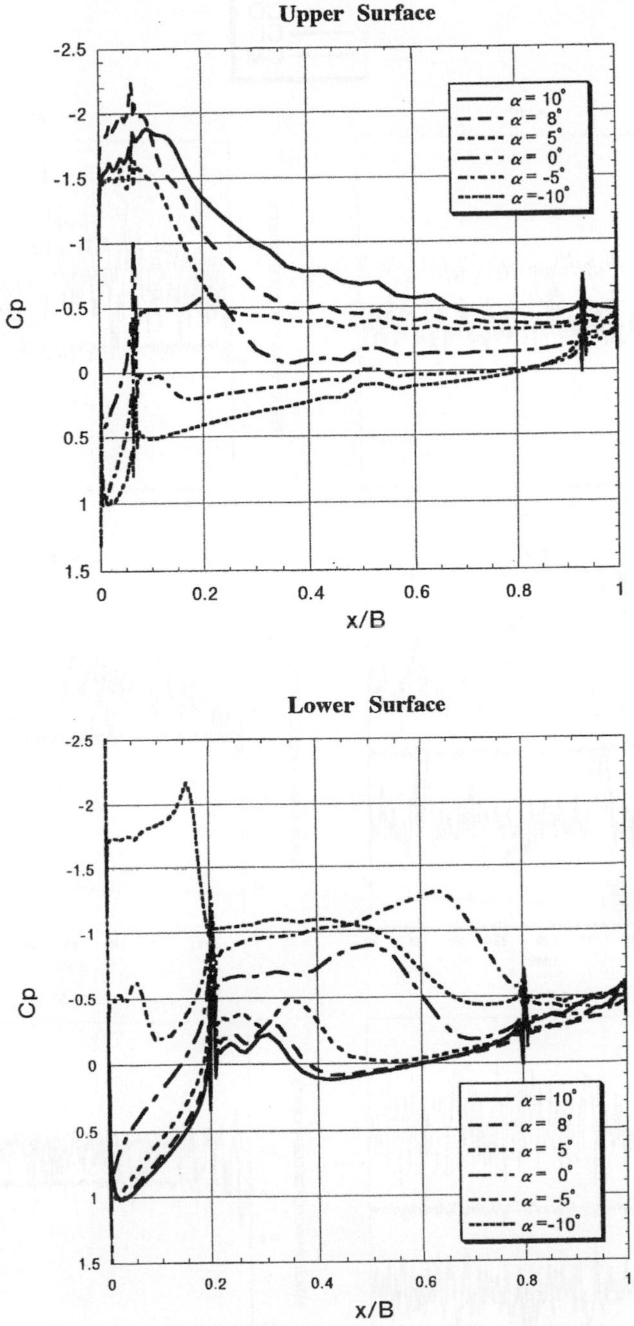

Fig. 6. Time-averaged pressure coefficient distribution.

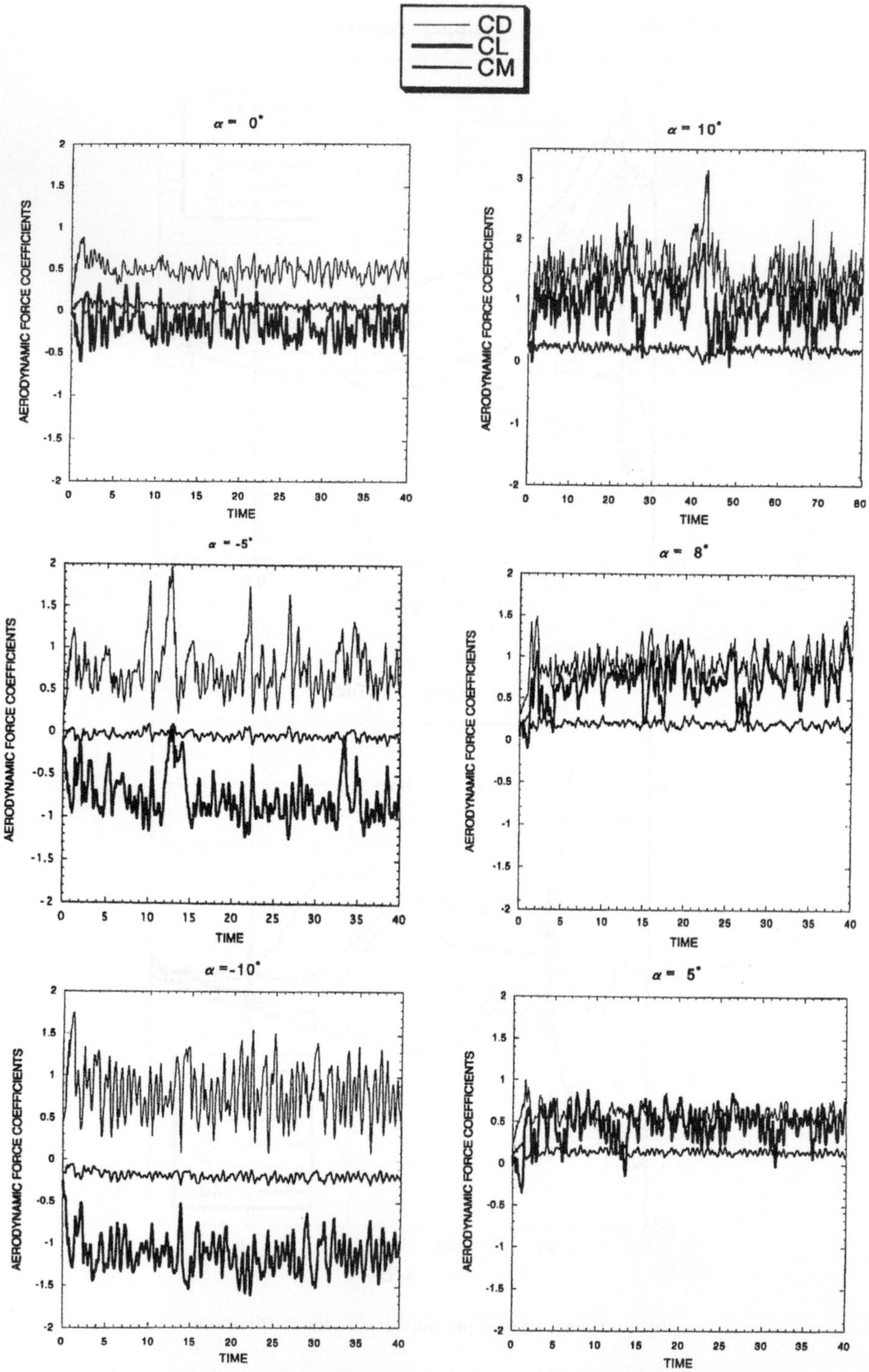

Fig. 7. Time histories of drag, lift and pitching moment.

5. Static force coefficients

In the present computations, the drag coefficient is non-dimensionalized with respect to H, the lift coefficient with respect to B and the pitching moment coefficient with respect to B^2. Here B is the girder width and H the girder depth (see Fig. 1). These aerodynamic coefficients have not been corrected for the side-railings and the central crash barrier. Fig. 7 shows the time series of computed drag, lift and pitching moments. Fig. 8 shows the comparison of static force coefficients between the computation and the wind tunnel test [2]. Computational static force coefficients were calculated by time-averaging the instantaneous values over the interval of $t = 20$–40 (for $\alpha = 10°$, $t = 40$–80).

As a whole, the computed drag and pitching moment agreed well with the measurement from the wind-tunnel experiment. Especially, as for the pitching moment, the agreement between both is fairly well. The computed lifts are also in good agreement with the experiment in the range of positive angles of attack (wind from below). However, in the range of negative angles of attack (wind from above), the computed lift coefficients were somewhat deflected from the experimental data.

The major cause of this discrepancy in the α negative cases can be considered as follows. In the α negative cases, the computed flow was attached to the upper surface of the girder as seen in Figs. 4 and 5. However, in the wind tunnel test, the flow cannot

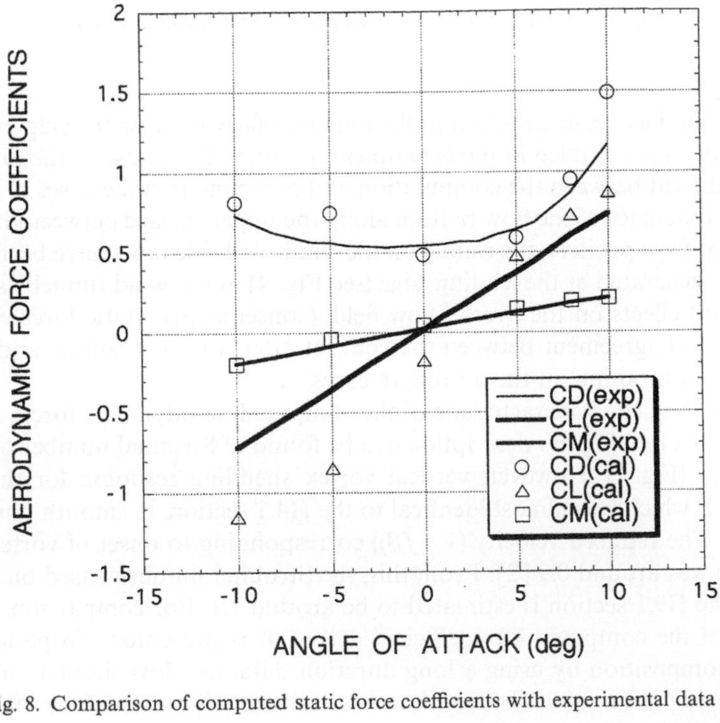

Fig. 8. Comparison of computed static force coefficients with experimental data [2].

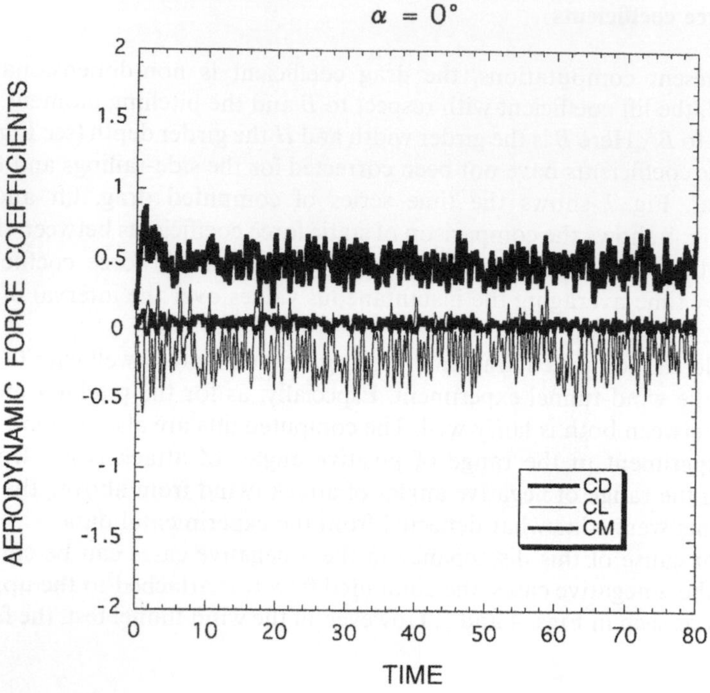

Fig. 9. Time histories of computed aerodynamic forces at $\alpha = 0°$.

be supposed to have been attached to the upper surface, because the side railings and crash barrier were included in the experimental model. The cause of the difference of the lift coefficient between the computation and experiment in the cases of α negative may be the difference of the flow pattern along the upper surface between the two. On the contrary, for α positive, the railing on the windward side must have been buried in the vortices generated at the leading edge (see Fig. 4) in the wind tunnel test and had no significant effects on the overall flow field. Consequently, static force coefficients showed a good agreement between the computation (without railing) and the wind tunnel test (with railing) in the α positive cases.

Finally, a frequency characteristic of the computed aerodynamic forces is referred to cursorily. In Ref. [2], no description can be found of Strouhal number of the fixed H4.1 section (Fig. 1). However, vertical vortex shedding response for the selected section H9.1, which was almost identical to the H4.1 section, in smooth flow is found in Ref. [2]. The reduced velocity ($V/(fB)$) corresponding to onset of vortex induced oscillations was around 0.9 [2]. From this, the Strouhal number based on the girder width for the H9.1 section is estimated to be around 1.1. For comparison, a Fourier transform of the computed lift coefficient at $\alpha = 0°$ is presented. To perform a frequency decomposition by using a long duration data, the flow simulation of $\alpha = 0°$ was continued until $t = 80$. The transform was taken in the range of $t = 20$–80. Fig. 9

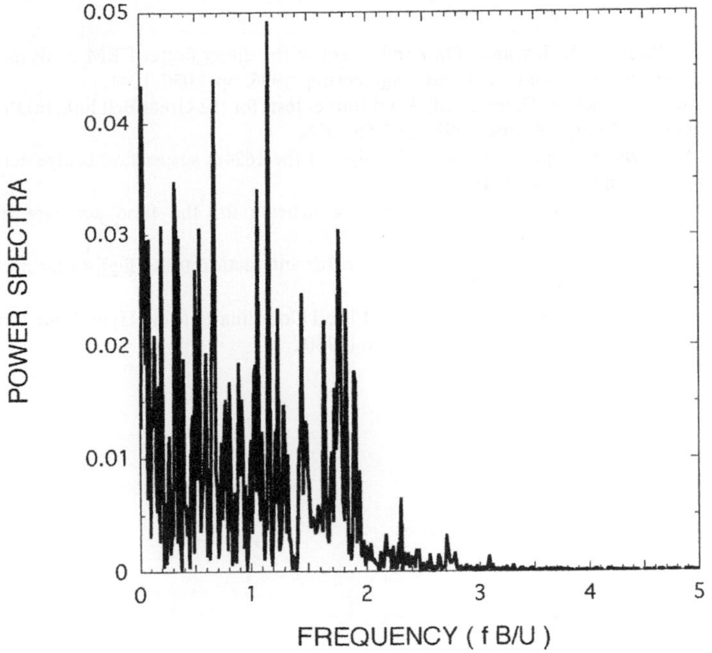

Fig. 10. Power spectrum of computed lift coefficients at $\alpha = 0°$.

shows the time histories of the computed aerodynamic forces and Fig. 10 the power spectrum distribution of fluctuating lift coefficients at $\alpha = 0°$. Although there is a peak at a nondimensional frequency of 1.15, no particular predominant component can be found in the spectrum. It is probable that such a distribution which has no predominant component is physical, because the cross section is shallow. Further investigation is needed to draw a conclusion.

6. Conclusions

The flow over the box girder section model of the Great Belt East Bridge at a high Reynolds number was numerically simulated. Computed static force coefficients were compared with the measurements from the wind-tunnel experiment conducted at DMI. Computed drag and pitching moments agreed well with the experimental data. Computed lift also agreed well in the range of positive angles of attack, but is somewhat larger at negative angles of attack. This discrepancy in the blow-down case is thought to be caused mainly by the absence of equipment such as side railings and crash barriers in the computation. The predicted lift coefficients in the α negative cases may be improved by considering the presence of the bridge equipments.

References

[1] T. Miyata, H. Yamada, K. Kazama, On application of the direct flutter FEM analysis for long span bridges, in: Proc. 9th Int. Conf. on Wind Engineering, 1995, pp. 1030–1041.

[2] T.A. Reinhold, M. Brinch, A. Damsgaard, Wind tunnel tests for the Great Belt link, in: Proc. Int. Symp. on Aerodynamics of Large Bridge, 1992, pp. 255–267.

[3] A. Larsen, Aerodynamic aspects of the final design of the 1624m suspension bridge across the Great Belt, J. Wind Eng. Ind. Aerodyn. 48 (1993) 261.

[4] S.E. Rogers, D. Kwak, An upwind differencing scheme for the time accurate incompressible Navier–Stokes Equations, AIAA J. 28 (2) (1990) 253.

[5] M.M. Rai, Navier–Stokes simulations of blade-vortex interaction using high-order accurate upwind Schemes, AIAA Paper 87-0543, 1987.

[6] J.L. Steger, D.S. Chaussee, Generation of Body-Fitted Coordinates using Hyperbolic Partial Differential Equations, SIAM J. Sci. Stat. Comput. 1 (4) (1980).

Journal of Wind Engineering
and Industrial Aerodynamics 67&68 (1997) 253–265

Aeroelastic analysis of bridge girder sections based on discrete vortex simulations

Allan Larsen[a,*], Jens H. Walther[b]

[a] *COWI, 15 Parallelvej, 2800 Lyngby, Denmark*

[b] *Danish Maritime Institute, Hjortekærvej 99, 2800 Lyngby, Denmark*

Abstract

Two-dimensional viscous incompressible flow past bridge girder cross-sections are simulated using the discrete vortex method. The flow around stationary cross-sections as well as cross-sections undergoing cross-wind vertical (bending) and rotary (torsional) motions are investigated for assessment of drag coefficient, Strouhal number and aerodynamic derivatives for application in aeroelastic analyses. Good to excellent agreement with wind tunnel test results is demonstrated for analyses of forced wind loading, flutter wind speed and vertical vortex-induced response of four practical girder cross-sections. The success of the simulations is attributed to the bluff nature of the cross-sections and to the two-dimensional (2-D) nature of flow around bridge girders.

Keywords: Computational bridge aerodynamics; Aeroelastic instability; Buffeting response; Vortex-induced response; Discrete vortex method

1. Introduction

Structural analysis has seen an explosive development during the past 20 years, moving from linear quasi 3-D modal analysis to full 3-D finite element analysis allowing for non-linear effects and complex boundary conditions. A full dynamic analysis of a bridge structure can now be completed within a week, but acquisition of necessary aerodynamic data from wind tunnel section model testing remains quite time consuming. Typical time consumption for planning, model construction, testing and reporting related to wind tunnel testing of a bridge cross-section is of the order of 6–8 weeks.

* Corresponding author. E-mail: aln@cowi.dk.

PII S0167-6105(97)00077-9

Experience gained from a number of bridge design projects has demonstrated that the time consumption associated with wind tunnel testing of bridge decks puts aeroelastic analyses on the critical path. Reduction of the turn-over time for aeroelastic analyses is thus highly desirable not only from an economical point of view. More important is that a reduction in the turn-over time allows more analyses to be performed within a given budget. The authors have taken up this challenge by developing a relatively fast computer technique based on the discrete vortex method for evaluation of aeroelastic actions on bridges.

The present paper discusses application of the discrete vortex method in aeroelastic analyses of practical bridge girder sections and presents comparisons between numerical simulations and wind tunnel test results. In a companion paper, Walther and Larsen [1] elaborate theoretical aspects of the discrete vortex method (computer code DVMFLOW), developed for bridge aerodynamic applications and discuss the fundamental case of the "flat plate" stationary and in oscillatory motion.

2. Aeroelastic analysis of flexible bridges

Aeroelastic analysis of flexible bridge structures traditionally combines structural dynamic analyses with aerodynamic data derived from wind tunnel model tests. In the analysis 3-D structural properties (mode shapes and structural inertia) are combined with 2-D aerodynamic information assuming conventional "strip theory" to be valid.

An aeroelastic analysis covering the most important design aspects includes horizontal girder response to mean wind and atmospheric turbulence, critical wind speed for onset of flutter and vortex-induced response. Such analyses involve the following 2-D aerodynamic data for the girder cross-section under consideration:

- Drag coefficient C_D: forced horizontal wind response.
- Flutter derivatives $H^*_{1..4}, A^*_{1..4}$: critical wind speed for onset of flutter.
- H^*_1 and Strouhal number St: vertical vortex-induced response.

A detailed discussion of aeroelastic response analyses is offered by Simiu and Scanlan [2].

3. Discrete vortex method for 2-D bridge sections

A distinct feature of flow past bluff bodies, stationary or in motion, is the shedding of vorticity in the wake which balances the change of fluid momentum along the body surface. Similar shedding of vorticity occurs in the wake of streamlined (airfoil-like) bodies in transient motion in a potential flow. The vorticity shed at an instant in time is convected downstream but continues to affect the aerodynamic loads on the body.

Analytical treatment of potential flow past streamlined bodies (e.g. the "flat plate") assumes that the vortical wake is shed from a single point – the trailing edge. The shed vorticity is convected downwind with the speed of the surrounding fluid. This simplified model is not valid for viscous flows past stationary or moving bluff bodies.

As a consequence of viscosity and presence of separation zones the unsteady vortical wake of a bluff body will be shed, not at the trailing edge, but along the entire body contour. Shed vorticity is convected by local mean wind speed and velocity fluctuations associated with viscous diffusion. Shed vorticity will also interact in the wake forming coherent structures – the well-known von Kàrmàn vortex trail. A realistic flow model for bluff bodies was developed within the framework of the "discrete vortex method" and programmed for computer by one of the authors for his Ph.D. Thesis [3]. The resulting numerical code DVMFLOW establishes a 2-D "grid free" time marching simulation of the vorticity field equation. Details of theory and governing equations are presented in a companion paper [1].

The input to DVMFLOW simulations is a boundary panel model of the bridge section contour. The output of DVMFLOW simulations comprises time-progressions of surface pressure distributions and section loads (drag, lift and moment). In addition, maps of the flow field (vector plots) and vortex positions at prescribed time steps are available.

Flutter coefficients (aerodynamic derivatives) to be used as input in flutter routines and for determination of vortex-induced response are obtained by matching calculated time-dependent aerodynamic loads to forced vertical or rotational displacements in accordance with empirical models for self-excited aerodynamic loads. A similar method is used in wind and water tunnel applications [4]. As an alternative to forced motion simulations the cross-sections may be supported by theoretical springs, allowing direct simulation of the aeroelastic flow–structure interaction.

4. Bridge sections analysed

DVMFLOW simulations were carried out for four bridge girder cross-sections for which aerodynamic data were available from wind tunnel model tests. This provided a basis for comparison between numerical simulations and experiment. The bridge sections selected for analysis included:

- "H-shaped" plate section – 1st. Tacoma Narrows Bridge, 850 m main span suspension bridge, USA, 1940.
- "Semi-streamlined" mono-box section – Great Belt East Bridge, suspended spans of a 1624 m main span suspension bridge, currently under construction in Denmark [5].
- "Bluff" mono-box section – Great Belt East Bridge, 193 m multi span approaches leading up to the suspension bridge [6].
- "Semi-streamlined" twin-box section – Chain of 3550 m main span suspension bridges developed during a design study (APP) for a fixed link across the Straits of Gibraltar [10].

Impressions of the bridge structures and geometry of the girder cross-sections are presented in Figs. 1 and 2. Structural properties utilized in the analyses are summarized in Table 1.

Fig. 1. View of bridges analysed. 1st. Tacoma Narrows Bridge (left), Great Belt East Bridge during construction, summer 1996 (middle) and the Gibraltar APP study artist's impression (right).

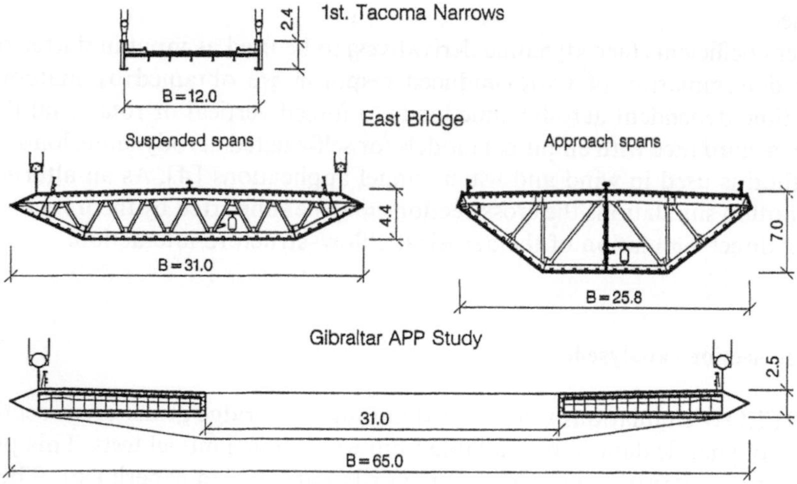

Fig. 2. Bridge girder cross-sections selected for DVMFLOW analysis.

Table 1
Structural properties of bridge cross-sections

Structural property	1st Tacoma Narrows	East Bridge Suspended spans	East Bridge approach spans	Gibraltar APP study
m (kg/m)	4.25×10^3	22.7×10^3	16×10^3	39.5×10^3
I (kgm²/m)	177.73×10^3	2.47×10^6	1.05×10^6	26.7×10^6
f_h (Hz)	0.13	0.099	0.46	0.065
f_α (Hz)	0.20	0.272	2.76	0.093
ζ (rel-to-crit)	0.005	0.002	0.005	0.003

5. Results of flow simulations and comparison with wind tunnel tests

5.1. Drag coefficient and Strouhal number

DVMFLOW simulations were carried out for the four bridge girder cross-sections presented above. Objectives of these simulations were the assessment of the drag coefficient C_D and Strouhal number St at 0° angle of incidence. All simulations were carried out for stationary sections at Reynolds number $Re = 10^5$ based on cross-section width ($Re = UB/\nu$). All simulations on stationary sections were run for a non-dimensional time interval $0 < tU/B < 25$. Input panel models reproduced the large-scale geometry of the cross-sections (deck contour) but finer details such as railings and median dividers were omitted for the present analysis. Modification of DVMFLOW for modelling of such details is in progress.

Typical results of stationary simulations comprise plots of section aerodynamic drag coefficient $C_D = D/\frac{1}{2}\rho U^2 B$ (D is the drag load, B is the section width) and lift coefficient $C_L = L/\frac{1}{2}\rho U^2 B$ (L is the lift force) computed as function of the non-dimensional time tU/B. As an example $C_D(tU/B)$ and $C_L(tU/B)$ time traces obtained for the 1st Tacoma Narrows cross-section are presented in Fig. 3.

The steady-state drag coefficient C_D is obtained as the mean of the drag trace (Fig. 3, top). The Strouhal number, St, is obtained from the non-dimensional period TU/B for one oscillation of the lift coefficient (Fig. 3, bottom) scaled by section depth H: i.e. $St = (H/B)/(TU/B)$.

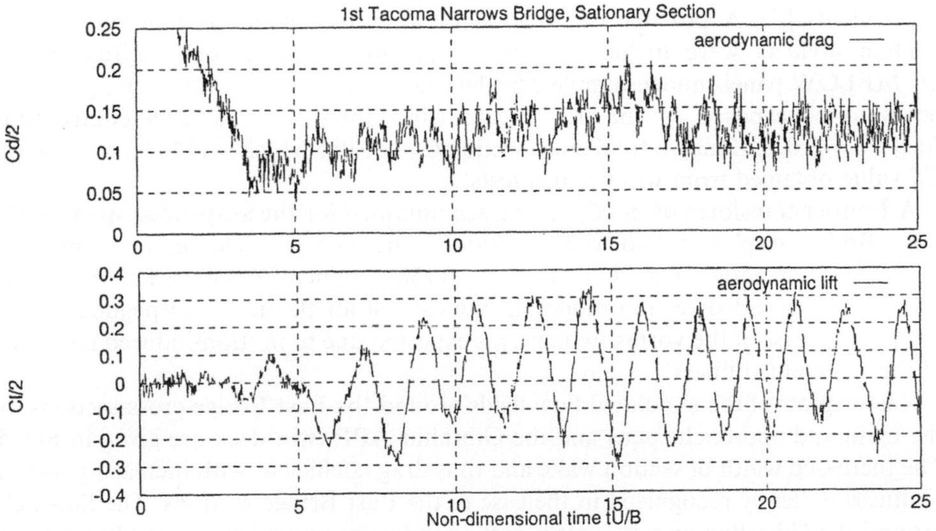

Fig. 3. Simulated time traces of aerodynamic drag (top trace, $C_D/2$ versus time) and lift (bottom trace $C_L/2$ versus time) for the 1st Tacoma Narrows cross-section.

Deck Section Geometry	DVMFLOW		Wind Tunnel	
	C_D	St	C_D	St
East Bridge, Suspension 	0.061	0.168 0.100	0.077	0.158 0.109
East Bridge, Approach 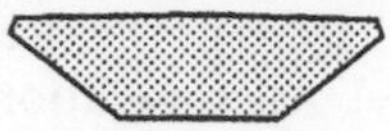	0.179	0.167	0.190	0.170
1st. Tacoma Narrows 	0.28	0.114	0.24 0.30	0.115
Gibraltar APP Study 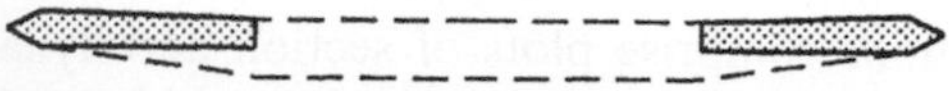	0.060	0.157 0.113	0.059	0.220

Fig. 4. Comparison of drag coefficient and Strouhal number obtained from numerical simulations and wind tunnel tests for four bridge girder cross-sections.

A summary of C_D, St and comparison with wind tunnel test results for the four cross-sections investigated are given in Fig. 4. It is noted that simulated and experimental C_D and St values agree very well for the two bluff sections, the East Bridge approach span [6] and 1st Tacoma Narrows cross-sections [7,8].

In the case of the cross-section of the East Bridge suspended spans, the C_D obtained from DVMFLOW simulations falls 20% below the value obtained from wind tunnel experiments [5]. A possible explanation for this discrepancy is that railings and median dividers were included in the wind tunnel model but omitted in the DVMFLOW panel model. Simple calculations allowing each of the railing components to be exposed to the free stream wind speed yield a drag contribution $\Delta C_D^{\text{railing}} = 0.023$. Hence, a total cross-section $C_D = 0.085$ which is 11% in excess of the C_D value obtained from wind tunnel tests.

A Fourier transform of the C_L time trace obtained for the suspended spans of the East Bridge displayed two vortex shedding frequencies. The main component St = 0.168 is in good agreement with St = 0.158 obtained from experiments with an elastically suspended section model. In the case of an elastically suspended section a slight decrease of the vortex frequency is expected due to motion-induced reduction of the apparent stiffness.

Vector plots of the simulated flow fields around the East Bridge cross-sections for the main and approach spans and the Gibraltar APP twin-deck are given in Fig. 5. The increased width of section wake and thus drag coefficient with increasing section bluffness is clearly recognized in the case of the East Bridge sections. The flow field around the Gibraltar twin-box cross-section indicates impingement and break up of vortices shed from the upwind girder element due to the presence of the down-wind girder element.

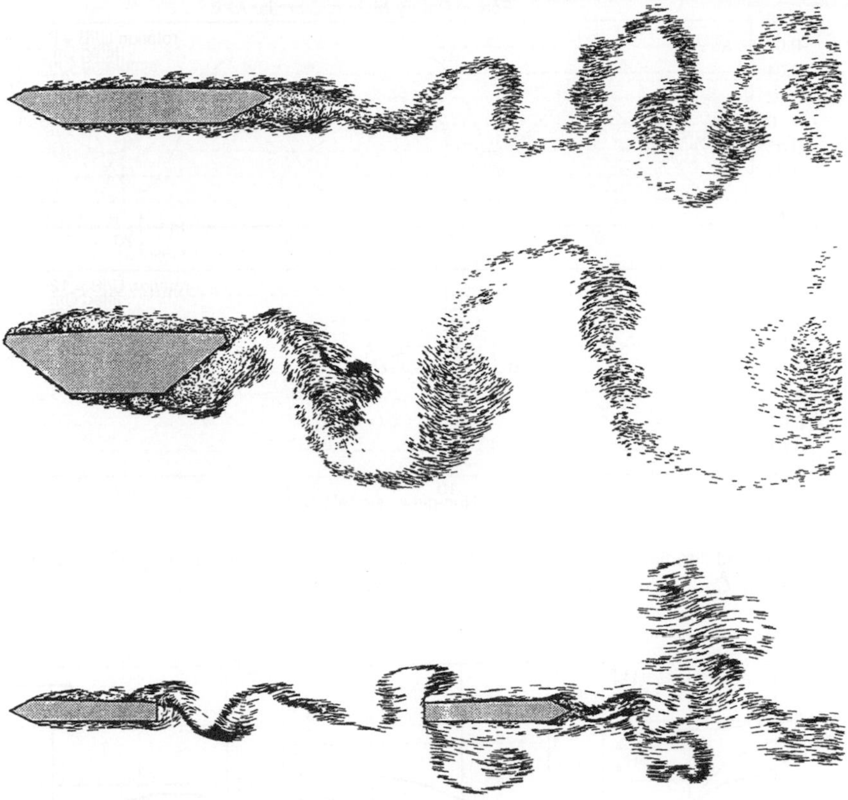

Fig. 5. Simulated flow fields around bridge girder cross-sections. Top: East Bridge suspended spans (shallow mono-box). Middle: East Bridge approach spans (deep mono-box). Bottom: Gibraltar APP twin-deck section.

5.2. 2DOF flutter

2DOF (binary) flutter is encountered for flexible bridges with shallow "semi-streamlined" box girder cross-sections similar to the East Bridge and the Gibraltar APP study. For each of these cross-sections, forced motion simulations were carried out. The simulations involved separate oscillatory vertical and rotational motion about mid-chord for six reduced wind speeds in the interval $U/fB = 4$–14. Vertical amplitudes for the prescribed sinusoidal motion were taken to be $h/B = 0.04$ and $\alpha = 3.0°$–$5.0°$ for rotation about mid-chord. Typical results of forced oscillatory motion simulations comprise plots of section lift and moment coefficients C_L, C_M as function of non-dimensional time. As an example, $C_M(tU/B)$ traces obtained for the Gibraltar APP cross-section at reduced wind speeds $U/fB = 6$ and $U/fB = 12$ are shown in Fig. 6.

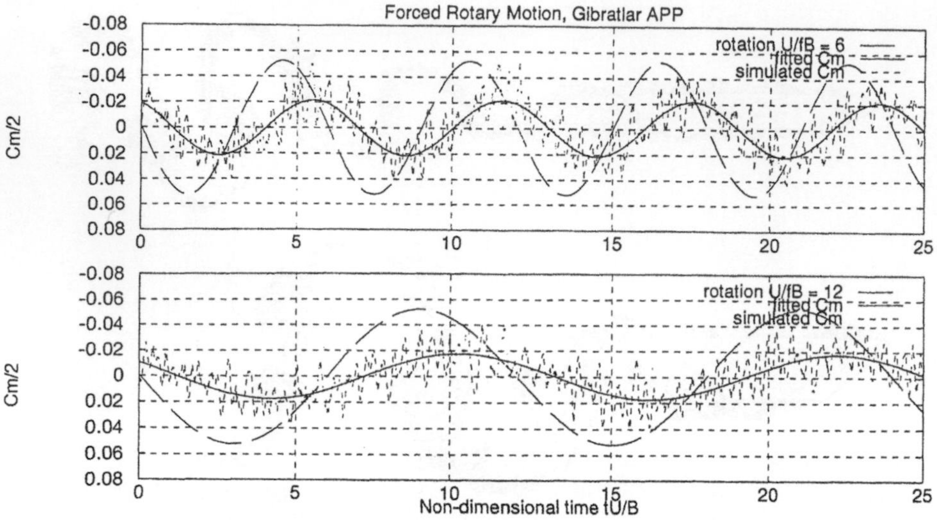

Fig. 6. Rotation and simulated and fitted time traces of aerodynamic moments (top: $C_M/2$ versus time at $U/fB = 6$, bottom: $C_M/2$ versus time for $U/fB = 12$). Gibraltar APP twin-box cross-section.

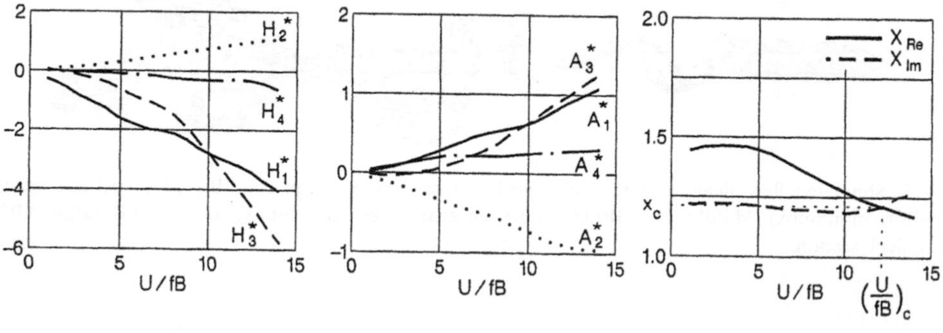

Fig. 7. Aerodynamic derivatives for the Gibraltar APP twin-box ($H^*_{1..4}$ left, $A^*_{1..4}$ middle) and intersection of the real and imaginary parts (X_{Re}, X_{Im}) of the corresponding flutter determinant defining the non-dimensional critical wind speed $(U/fB)_c$.

Simulated time traces of C_L, C_M appear as random narrow band oscillations (due to vortex shedding) modulated by the frequency of forced oscillation. The amplitudes of C_L, C_M at the forced excitation frequency are extracted by least-squares fitting of a sinusoid to the simulated load traces. The aerodynamic derivatives are obtained from the amplitude and phase relationship between the forced motion and the fitted load signals. The critical wind speed for onset of 2DOF flutter is obtained by combining structural properties with $H^*_{1..4}$, $A^*_{1..4}$ in the flutter determinant following the Theodorsen method of Simiu and Scanlan [2]. This procedure is illustrated in Fig. 7.

Deck Section Geometry	DVMFLOW U_C(m/s)	Wind Tunnel U_C(m/s)
East Bridge, Suspension 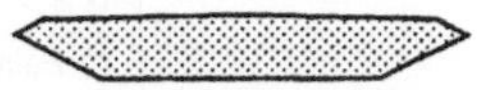	74	73 [5]
Gibraltar APP Study 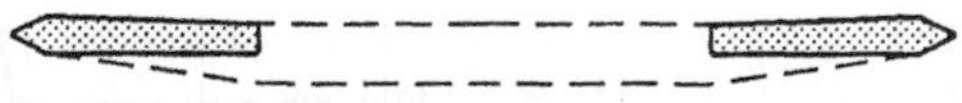	62	66 [10]

Fig. 8. Comparison of critical wind speeds obtained from DVMFLOW simulations and wind tunnel section model tests of the East Bridge suspension and Gibraltar APP girder cross-sections.

A comparison of prototype critical wind speeds U_c obtained from the DVMFLOW simulations and from wind tunnel section model tests of the East Bridge and Gibraltar APP cross sections is given in Fig. 8. Good to excellent agreement is noted indicating a maximum deviation between numerical simulations and wind tunnel tests of about 7%. Indeed satisfactory for design studies of long-span bridges.

5.3. 1DOF flutter

1DOF (torsional) flutter is encountered for flexible bridges with channel type or H-shaped plate girder cross-sections. Forced motion simulations in pure rotation (deck torsion) were carried out for the 1st. Tacoma Narrows Bridge in order to test the ability of DVMFLOW to simulate the strongly non-linear and recirculating flow around this cross-section. The simulations involved three oscillation amplitudes $\alpha = 3°$, $10°$ and $30°$ and were carried out at reduced wind speeds in the range $U/fB = 2$–8. Fig. 9 compares the aerodynamic derivative A_2^* obtained from the DVMFLOW simulations to A_2^* given in Billah and Scanlan (Fig. 4) [9] for incipient torsional motion. The determination of corresponding critical reduced wind speeds $(U/fB)_c$ follows the method given by Simiu and Scanlan [2], detailed in Ref. [9] and further illustrated in Fig. 9 (left). Fig. 9 (right) compares predicted prototype torsional responses to results of the full 3-D aeroelastic wind tunnel tests of the 1st. Tacoma Narrows Bridge carried out by Farquharson [8]. The predictions assume a structural damping $\zeta = 0.005$ (rel-to-crit) as reported for the full bridge model test. Fair agreement is demonstrated for torsional motion amplitudes $\alpha = 3°$ and $10°$ although the DVMFLOW simulations yield higher wind speeds for similar response as compared to the full bridge model tests.

The prototype bridge collapsed in a 19 m/s gale on 7 November 1940, while twisting at amplitudes of approximately $\alpha = 30°$–$35°$. A full-scale wind speed of 19 m/s corresponds to a reduced wind speed $(U/fB)_c = 7.9$ which by virtue of Fig. 9 (left) corresponds to a critical aerodynamic derivative $(A_2^*)_c = 0.2$. Introducing the relevant mass moment of inertia and torsional frequency (Table 1) in the $(A_2^*)_c$-relation ($(A_2^*)_c = 2I\zeta/\rho B^4$) yields a relative structural damping $\zeta = 0.014$ (rel-to-crit) at the

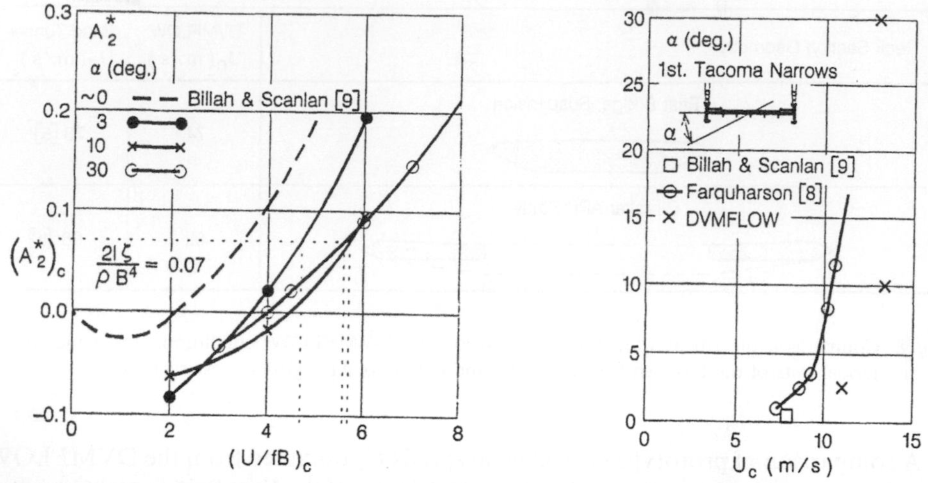

Fig. 9. A_2^* aerodynamic derivative (left) and predicted prototype torsional response (right) obtained from DVMFLOW forced simulations of the 1st. Tacoma Narrows Bridge cross-section.

stage of collapse. This damping level is high compared to $\zeta = 0.005$ but may well be realistic when considering cracking of the concrete deck. Codes of practice yield $\zeta = 0.016$ as typical for reinforced but not pre-stressed concrete structures.

The simulated flow field around the 1st. Tacoma Narrows Bridge cross-section during half a circle of oscillation is shown in Fig. 10. The reduced wind speed is $U/fB = 8$ and the amplitude of twisting motion is $\alpha = 30°$. Heavily recirculating flow below the deck and roll up of a strong leading edge vortex above the deck is evident.

The vortex pattern developing in Fig. 10 resembles the heuristic sketch by Billah and Scanlan [9], Fig. 6.

5.4. Vortex-induced response

Plate- as well as deep box-type cross-sections are known to be prone to vortex-induced oscillations at wind speeds where vortex shedding locks on to a structural eigenfrequency.

In practice, vertical oscillations are usually found to be the most important type of response as vertical eigenfrequencies of bridges tend to be lower than torsional eigenfrequencies. The ability of DVMFLOW to predict vertical vortex-induced oscillations of bridge girders was investigated in two cases, the 1st. Tacoma Narrows Bridge and the East Bridge approach span cross-sections.

Forced vertical motion simulations were run at reduced wind speeds bracketing the Strouhal number obtained from flow simulations on the stationary sections. In the case of the 1st. Tacoma Narrows cross-section a Strouhal number St = 0.114 was obtained from flow simulations reported in Section 5.1. Aerodynamic vortex shedding excitation was thus anticipated at $U/fB = 1.8$. Forced motion simulations were run in

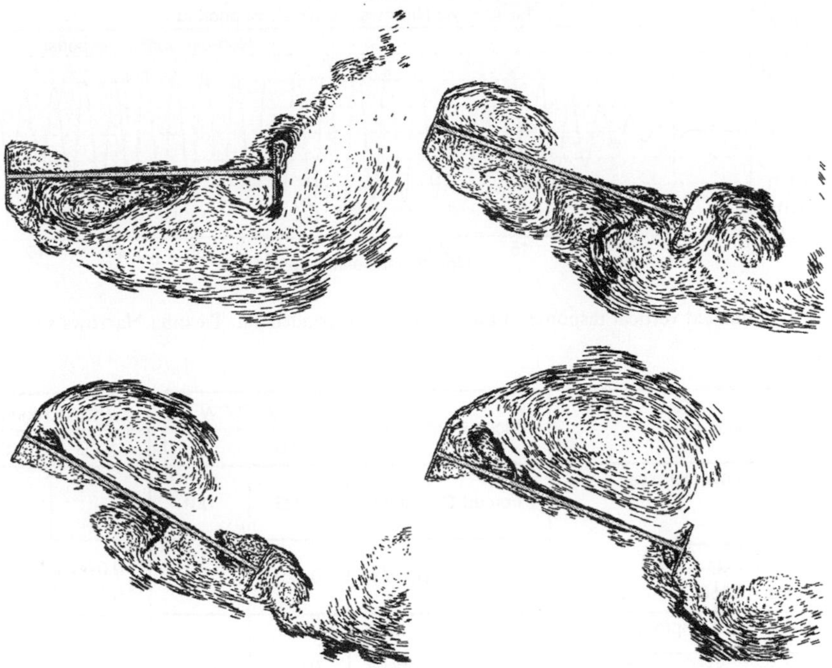

Fig. 10. Simulated flow around the 1st. Tacoma Narrows Bridge twisting at 30° amplitude.

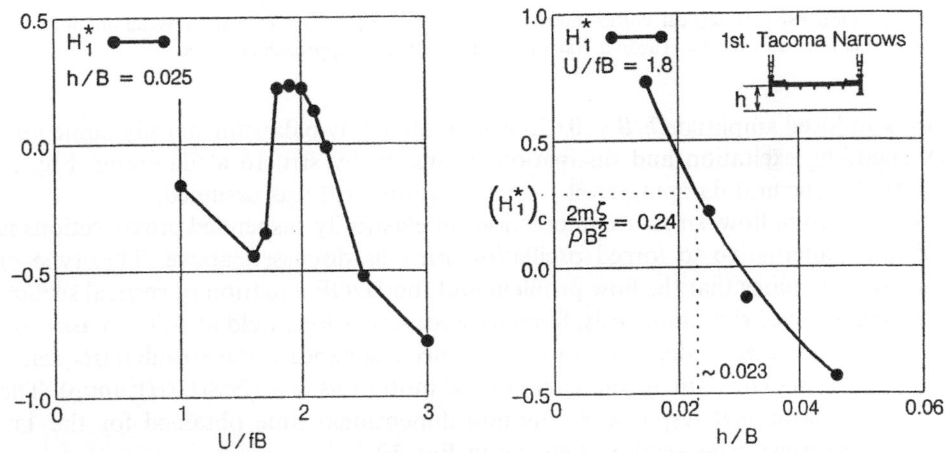

Fig. 11. H_1^* aerodynamic derivative (left) and predicted prototype vertical response (right), 1st. Tacoma Narrows cross-section.

the range $U/fB = 1$–3 and at a constant amplitude $h/B = 0.025$. The results displayed a pronounced shift of the polarity of the H_1^* aerodynamic derivative (from $-$ to $+$) in the vicinity of $U/fB = 1.8$ as expected, Fig. 11 (left). Simulations proceeded at $U/fB = 1.8$ but at varying amplitudes in the range $h/B = 0.015$–0.045. A stationary

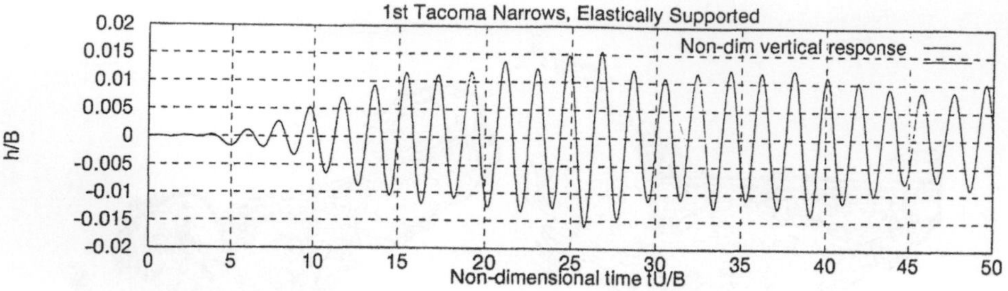

Fig. 12. Simulated vertical response of the elastically suspended 1st. Tacoma Narrows section.

Deck Section Geometry	Mode of Motion	DVMFLOW		Wind Tunnel	
		h/B	St	h/B	St
1st. Tacoma Narrows	Forced Oscillation	0.023	0.11	–	–
 Model scale : *1:50 **1:36	Elastic Suspension	0.020	0.11	* 0.024 [8] ** 0.040	* 0.09 [8] ** 0.12
East Bridge, Approach 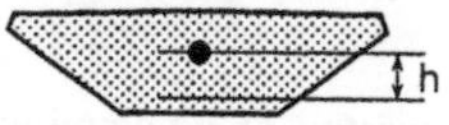	Elastic Suspension	0.014	0.16	0.021 [6]	0.16 [6]

Fig. 13. Comparison of vertical vortex-induced response obtained from DVMFLOW simulations and wind tunnel tests with the 1st. Tacoma Narrows and East Bridge approach sections.

vortex-induced amplitude $h/B = 0.023$ was obtained by balancing aerodynamic vortex shedding excitation and dissipation of energy by structural damping, Fig. 11 (right). A structural damping level $\zeta = 0.005$ (rel-to-crit) was assumed.

Simulation of flow–structure interaction on elastically suspended cross-sections is a distinct alternative to forced oscillation runs as discussed above. This type of simulation requires that the flow problem and the 1DOF equation of vertical section motion are solved simultaneously. Similarity considerations yield identical mass ratio $\rho B^2/m$ for computer model and prototype. Hence resonance at the Strouhal frequency fixes the spring constant in the equation of motion as $k = (2\pi \mathrm{St}U/H)^2/(m/\rho)$. The simulated elastic h/B response versus non-dimensional time obtained for the 1st. Tacoma Narrows cross-section is shown in Fig. 12.

From Fig. 12 it is noted that the stationary vertical response obtained from the simulation of the elastically suspended cross-section is in qualitative agreement with the vertical response obtained from the forced-response simulations.

A comparison of the prototype of vertical vortex-induced response and Strouhal number obtained from DVMFLOW simulations and wind tunnel tests for the 1st. Tacoma Narrows Bridge and the East Bridge approach cross-sections is given in Fig. 13. Satisfactory agreement is demonstrated.

Acknowledgements

The continuing encouragement of colleagues of the COWI bridge department and the financial support of the COWI foundation is highly appreciated.

References

[1] J.H. Walther, A. Larsen, 2D Discrete vortex method for application to bluff body aerodynamics, These Proceedings, J. Wind Eng. Ind. Aerodyn. 67&68 (1997) 183–193.
[2] E. Simiu, R.H. Scanlan, Wind Effects on Stuctures, 2nd ed., Wiley Interscience, New York, 1986.
[3] J.H. Walther, Discrete vortex method for two-dimensional flow past bodies of arbitrary shape undergoing prescribed rotary and translatory motion, AFM-94-11, Ph.D. Thesis, Department of Fluid Mechanics, Technical University of Denmark, 1994.
[4] Q.C. Li, Measuring flutter derivatives for bridge sectional models in water channel, ASCE J. Mech. Eng. 121 (1) (1995) 90–101.
[5] A. Larsen, JWEIA Aerodynamic aspects of the final design of the 1624 m suspension bridge across the great belt, J. Wind Eng. Ind. Aerodyn. 48 (1993) 261–285.
[6] DMI report No. 92194.01: Detailed design, approach bridges, section model tests III, Storebælt East Bridge, March 1993, Restricted.
[7] G. Schewe, Nonlinear flow-induced resonances of an H-shaped section, J. Fluids Struct. 3 (1989) 327–348.
[8] F.B. Farquharson, Aerodynamic stability of suspension bridges, University of Washington Experiment Station, Bull. No. 116, Part I and III, 1952.
[9] K.Y. Billah, R.H. Scanlan, Resonance, Tacoma Narrows bridge failure, and undergraduate physics textbooks, Amer. J. Phys. 59 (2) (1991) 118–124.
[10] A. Larsen, K.H. Ostenfeld, M. Astiz, Aeroelastic stability study for the Gibraltar bridge feasibility phase, IV Int. Coll. Gibraltar Fixed Link, Seville 1995, pp. 273–278.

ELSEVIER

Journal of Wind Engineering
and Industrial Aerodynamics 67&68 (1997) 267–278

JOURNAL OF wind engineering AND industrial aerodynamics

Prediction of vortex-induced wind loading on long-span bridges

Sangsan Lee*, Jae Seok Lee, Jong Dae Kim

Systems Engineering Research Institute, P.O. Box 1, Yuseong, Daejeon 305-600, South Korea

Abstract

Computational methods have been systematically applied for vortex-induced vibrations of long-span bridges by unsteady wind loadings due to vortex-shedding. Two practical bridge cases, the Namehae and the Seohae Bridge in Korea, are analyzed. Wind loading characteristics for turbulent flows over a bridge structure are investigated through computational fluid dynamics for two-dimensional bridge deck section models. With the unsteady wind loadings obtained, three-dimensional dynamic structural analyses are also performed. Reasonable agreements are obtained between the predictions by the computational analyses and the existing experimental measurements. Even though computational wind engineering cannot predict three-dimensional flow characteristics due to current computational and physical modeling constraints, it may serve as a first-stage design tool before serious wind tunnel experiments are performed.

Keywords: Long-span bridge; Vortex shedding; Turbulent flow; Vortex-induced vibration; Loading frequency

1. Introduction

As the design and construction technology of a bridge structure advances, construction of long-span bridges is increasing in numbers. Stiffness and damping of the long-span bridges are relatively low compared with those of short-span bridges. Consequently, the long-span bridges are sensitive to wind, and it is important to consider aeroelastic phenomena such as vortex-excited oscillation, flutter, and galloping in addition to the static structural safety against the static wind loading. In the modern design procedure of these structures, most evaluations of wind loading characteristics and aerodynamic stability analyses have been carried out by wind

* Corresponding author. E-mail: sslee@seri.re.kr.

PII S0167-6105(97)00078-0

tunnel tests. With the rapid growth of the computing power and with the advances in the physical modeling, computational wind engineering has emerged with the hope of supporting or replacing some portion of the expensive and time-consuming wind tunnel tests.

Even with the current high-end supercomputers, application of computational wind engineering is not thought practicable, since the flow Reynolds number involved in wind engineering is so high that accurate turbulent flow prediction is beyond the capacity of present day computers. Most computational wind engineering researches so far have, therefore, focused on the development of computational methods and the refinement of turbulence models. The former efforts are reviewed by Hughes and Jansen [1], including the application of an arbitrary Lagrange–Eulerian (ALE) finite element method, and the latter includes the efforts of Franke and Rodi [2], Kato and Launder [3]. Fujiwara et al. [4] is one of the few groups that have reported on the computation of practical bridge deck structures at low Reynolds numbers. Few works are found in the literature on the computation of unsteady turbulent flows over real bridge structures at the designed Reynolds number range, $Re = O(10^7)$ or higher.

This paper is devoted to the computational method primarily intended to evaluate the steady and unsteady wind loading and to predict a vortex-induced vibration of the bridge structure using readily available commercial softwares. To validate the feasibility of the method, comparative studies between the computational results and wind tunnel tests are carried out for wind loadings and vortex-induced vibrations of typical long-span bridges – the Namhae (suspension) and the Seohae (cable-stayed) Bridge in Korea, respectively.

In the following section, computational methods used are briefly described. In Section 3, computational fluid dynamics analysis results are discussed for the two long-span bridges, where the effects of attack angle and flow speed of incoming free stream are investigated. In Section 4, dynamic structural response of the Seohae Bridge to the unsteady wind loading is investigated. In the concluding section, a summary is given with discussions on the limitations and perspectives of computational wind engineering.

2. Computational methods

In investigating vortex-induced vibrations, it is critically important to accurately predict not only the magnitude and frequency of unsteady wind loadings but also the amplitudes of the resulting structural vibrations. Structural oscillation tends to be violent when the wind loading frequency falls near the structural natural frequencies due to resonance. For the accurate prediction of wind–structure interaction, flow–structure coupled problem should be analyzed, i.e., through the ALE method, where effects of the structural movement are incorporated in the governing equations in the form of mesh movement velocity. Flow around the structure, however, is modified significantly by the structure motion only when the vibration amplitude in the cross-wind direction exceeds 10% of the structure size [5]. Since a structural oscillation in that amount is clearly beyond the design safety limit, the bridge deck section may safely be assumed to be fixed in space throughout the computational fluid dynamics (CFD) analysis in most engineering considerations.

The computational procedure for analysis of the vortex-induced vibration is performed through a two-step process. In the first step, a commercial CFD software, CFDS-FLOW3D [6], is used to analyze turbulent flows around the bridge deck section to predict unsteady wind loadings on the structure. In the second step, a commercial structural analysis software, MSC/NASTRAN [7], is used to compute the structural response under the wind loading predicted in the first step. In the present study, emphasis is put to the CFD analysis, and a brief description of the structural analysis results is followed to validate the overall analysis procedure.

In performing CFD analysis, choice of numerical schemes, grid system, and physical turbulence model is based on the systematic investigations of unsteady turbulent flows over bluff bodies [8, 9]. Numerical methods adopted are the finite-volume formulation for collocated variables with a SIMPLE pressure-correction and a QUICK convection scheme. Use of a two-dimensional model in the CFD analysis is either natural by the two-dimensionality of the physical model or inevitable by unmanageably huge memory and computing resource requirements of a three-dimensional model. Results of the computations presented in this work, however, reasonably validate the use of two-dimensional models. An RNG k–ε turbulence model is used with the grid refined in the vicinity of wall based on the shear layer resolution, rather than on the viscous wall unit [9]. A typical cell Reynolds number near the wall in the CFD analyses is $O(10^6)$. The structural model used for the dynamic structural analysis is validated by comparing the eigenvalues and eigenmodes of the present model with those of the previous analysis [10].

3. Wind loading characteristics

Dimensional specifications of the bridges analyzed are given in Table 1. Since the bridge spans are sufficiently large compared to the bridge deck cross sectional dimensions, two-dimensional models are used for the CFD analyses.

3.1. The Namhae Bridge

The two-dimensional bridge deck section model and grid system around it (with about 11 000 cells) for the CFD analysis are shown in Fig. 1. Incoming flow is uniform at the design speed of $U = 33$ m/s and assumed to be turbulent with turbulence intensity of 3.5%. Since the bridge section is almost streamlined, no appreciable large scale unsteady motion is found. Since the slopes of the force coefficients at zero attack angle, especially that of the lift coefficient, are important in evaluating a linear flutter analysis of the bridge structure, effects of an attack angle on the force coefficients are investigated. The lift, drag and moment coefficients are defined, respectively, as

$$c_L = \frac{L}{\frac{1}{2}\rho U^2 B}, \quad c_D = \frac{D}{\frac{1}{2}\rho U^2 B}, \quad c_M = \frac{M}{\frac{1}{2}\rho U^2 B^2},$$

Table 1
Specifications of the Namhae and the Seohae Bridge

Bridge	Height (H)	Width (B)	Span	Type
Namhae	1.6 m	12 m	404 m	Suspension
Seohae	3.5 m	32 m	470 m	Cable-stayed

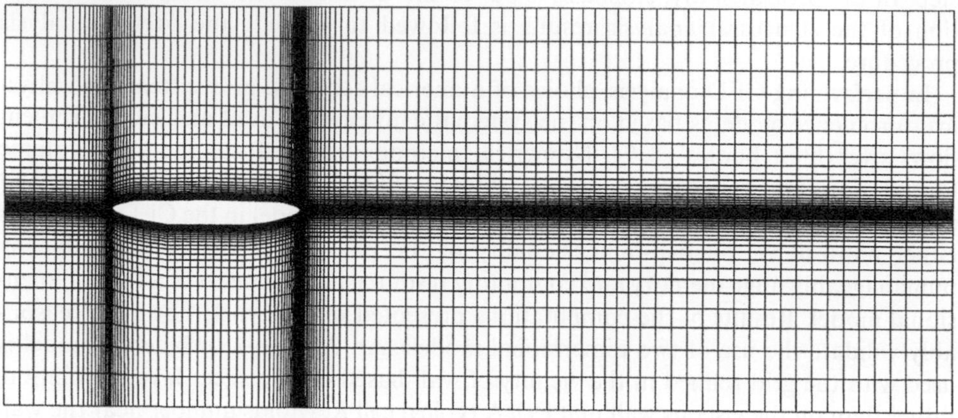

Fig. 1. Finite volume mesh around the Namhae Bridge.

where L, D, M are the lift, drag, and moment per unit span length. The attack angle (α) is changed from $-8°$ to $8°$, where the positive angle is for the clockwise rotation of the deck model, or the nose-up model case.

The results are summarized in Fig. 2, where the slope of the lift coefficient is in good agreement with that from the wind tunnel test [11], considering crude near wall resolution of $\Delta y^+ = \Delta y u_\tau / \nu \simeq 10^3$, where $u_\tau = \sqrt{\tau_w/\rho}$ is a wall-shear speed with τ_w the wall shear stress. Since drag is more sensitive to the quality of near wall flow modeling, discrepancy between the computation and the wind tunnel test is larger for the drag coefficient. On the other hand, lift is dominated by the large-scale pressure field around the model and may be predicted well even without fine near-wall resolution. Now as the lift is larger than the drag force by an order of magnitude in the nearly streamlined body and that it plays the most important role in the linear flutter analysis, current results are not so disappointing even with some disagreements.

3.2. The Seohae Bridge

The representative two-dimensional bridge deck section model and grid system around it (with about 14 000 cells) for the CFD analysis are shown in Fig. 3. Since the

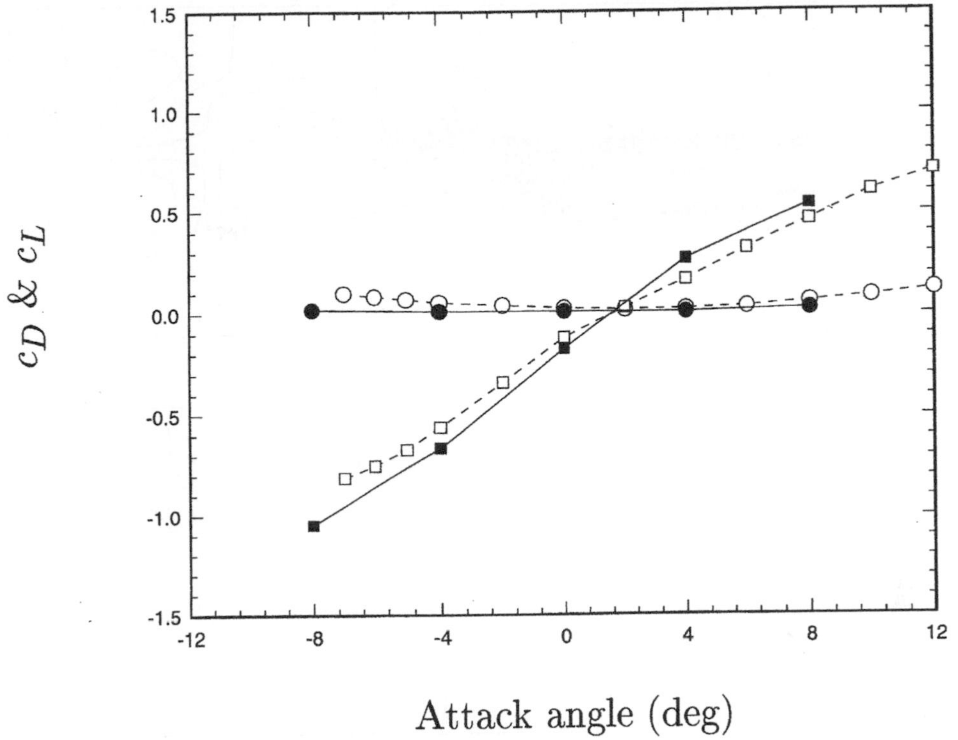

Fig. 2. Drag and lift coefficients for the Namhae Bridge: (□) c_L, (○) c_D, (—) present study, (- - - -) experiment [11].

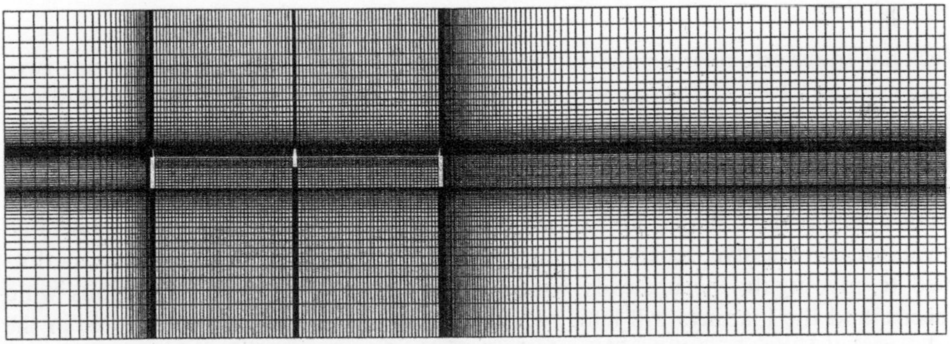

Fig. 3. Finite volume mesh around the Seohae Bridge.

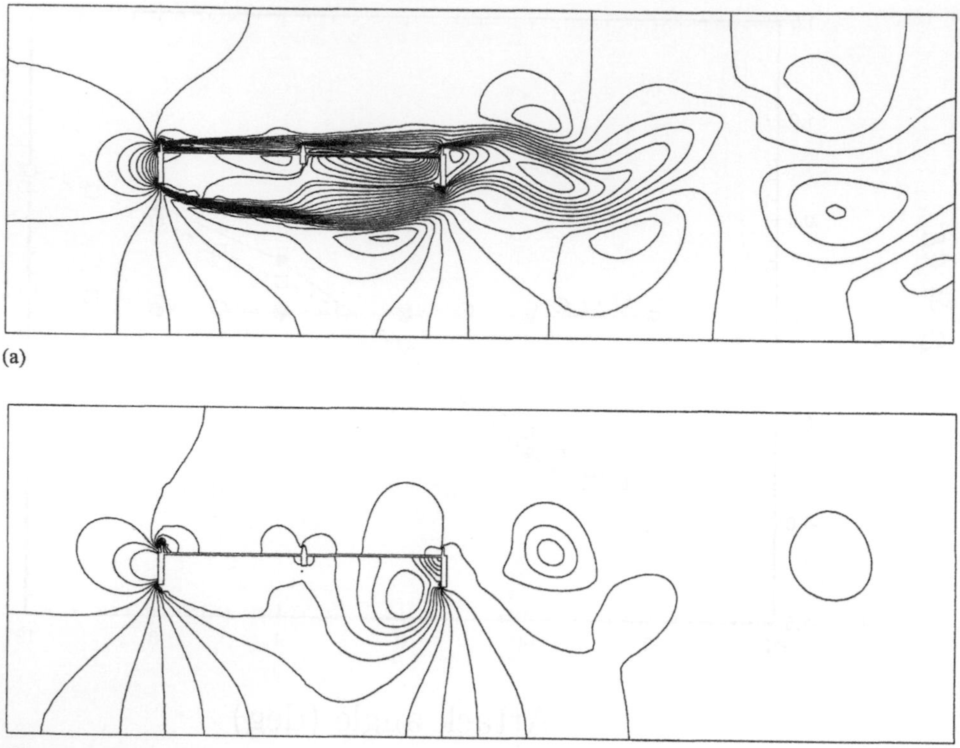

(a)

(b)

Fig. 4. Instantaneous flow fields around the Seohae Bridge at the wind speed of 20 m/s: (a) Streamwise velocity ($\Delta U = 2$ m/s), (b) pressure ($\Delta p = 60$ Pa).

deck has transverse reinforcing structures at every 4 m along the bridge span, any two-dimensional CFD model has some unavoidable deficiencies. Significance of three-dimensional effects not accounted for in the present computation remains to be quantified. Incoming flow is uniform at various speeds between 8 and 30 m/s and assumed to be turbulent with negligible turbulence intensity. Since the bridge deck section is of bluff shape, appreciable large scale unsteady motion is generated by alternate massive separations at the sharp corners. Computation is performed to investigate the unsteady wind loadings on the structure at various wind speeds, and the loadings will later be used as time-dependent external forces in the subsequent dynamic structural analysis. A typical unsteady CFD analysis requires about 10 CPU hours on CRAY C90 which has 1 GFlops peak processor performance.

Instantaneous fields of streamwise velocity and pressure at the wind speed of 20 m/s are shown in Fig. 4, where asymmetric pressure distribution and vortex shedding may be observed. These time-dependent asymmetries lead to unsteady wind loadings on

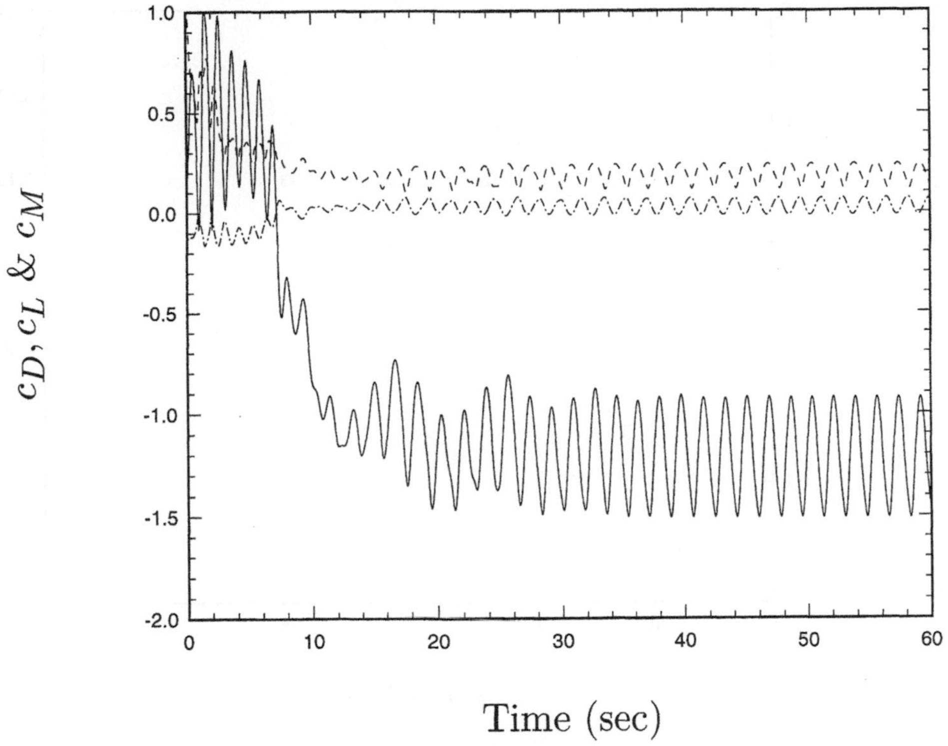

Fig. 5. History of lift, drag and moment coefficients around the Seohae Bridge at the wind speed of 20 m/s. (Moment acting in the clockwise direction is considered to be positive.)

the bridge deck, whose dimensionless measures are shown in Fig. 5. The major contributor to the unsteadiness is alternately generated and shed vortices at the leeward side of the bridge deck, while almost time-independent recirculating regions are sitting at its windward side. Similar flow feature and time dependence are observed at different wind speeds.

The two important factors to be investigated are the dominant frequencies of the wind loadings and their fluctuating amplitudes. Time history of the lift coefficient is taken for the dominant frequency (f) estimation by FFT through a proper windowing [12] to remove the signal non-periodicity effects in the Fourier transform. Dependence of wind loading frequency on the incoming wind speed is shown in Fig. 6, where the frequency is almost linearly proportional to the wind speed. Scaling of the frequency with the wind speed leads to a dimensionless measure, the Strouhal number, $\mathrm{St} = fD/U$. The Strouhal number is almost constant as 0.099 ± 0.0005 beyond $U = 13$ m/s, below which it is about 10% lower as 0.090 ± 0.0005. And magnitudes of the mean and fluctuating force coefficients are found to be fairly insensitive to the wind speed.

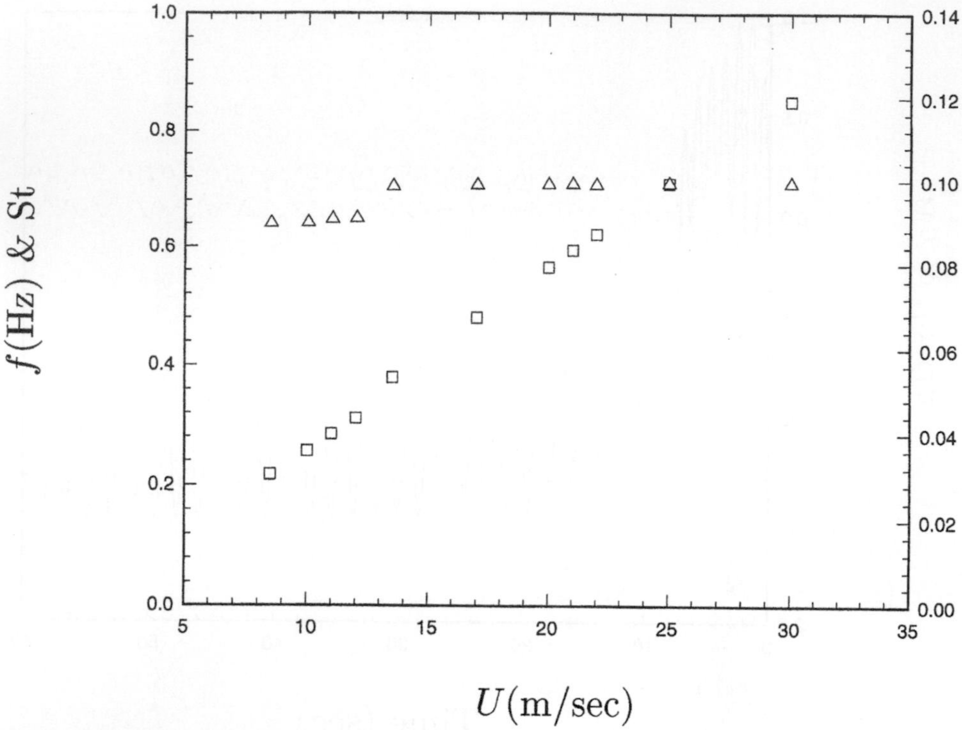

Fig. 6. Change of the Strouhal frequency with the wind speed for the Seohae Bridge: (□) Strouhal frequency and (△) Strouhal number.

Since the Strouhal number and the force coefficients' magnitudes are fairly constant at various wind speeds, wind loading (F_2) at wind speed U_2 may approximately be evaluated by that (F_1) at another wind speed U_1 as

$$F_2(t) \simeq \frac{U_2^2}{U_1^2} F_1\left(\frac{U_2 t}{U_1}\right).$$

Possible errors included in this approximation are due to the variations in the Strouhal number, the fluctuating amplitude and the detailed time history of the force coefficients. If the two speeds are in the constant Strouhal number regime, the total error effects on the structural response are expected to be fairly small. The applicability of this approach is under investigation to check the possibility of an economical application of the computational procedures.

4. Dynamic structural response – Seohae Bridge

Now that the quality of time-dependent CFD analyses performed for the Seohae Bridge cannot directly be checked against a wind tunnel test, a dynamic structural

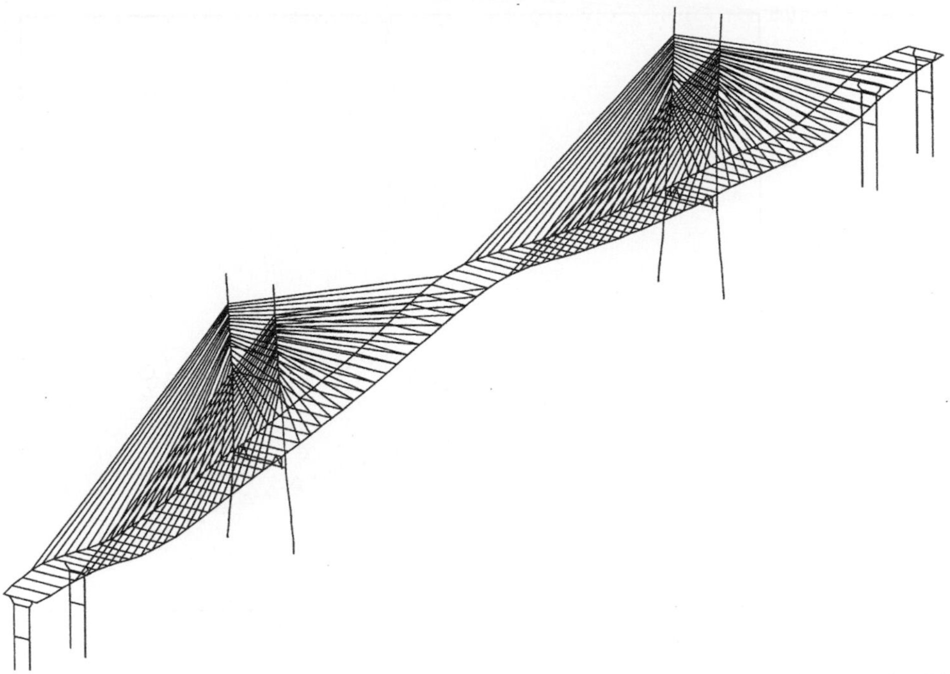

Fig. 7. Structural analysis model for the cable-stayed Seohae Bridge at the fundamental eigenmode with $f = 0.264$ Hz.

analysis for the bridge structure is conducted with the wind loadings obtained from CFD. The dynamic structural response obtained is compared with the wind tunnel experiment [10] not only to indirectly validate the CFD results, but also to justify the effectiveness of the dynamic structural analysis. The structural analysis model for the cable-stayed Seohae Bridge is shown in Fig. 7. Since the wind loadings are computed from a two-dimensional model, complete three-dimensional unsteady forcing information cannot be given for the structural model. Forces are expected to be modified to a certain extent by the presence of towers and cables, which have been neglected in the present CFD analyses. The forces applied to the structure are, therefore, assumed to be uniform in magnitude and to be in-phase along the bridge axis direction. Wind loadings on the tower and the cables are not expected to play an important role in the global dynamic behavior of the bridge deck, and are not included in the present analyses. Since the maximum displacement of the bridge deck is observed at the center of the span in the fundamental eigenmode, the displacement at the midspan is traced to identify the mean and the fluctuating displacement of the bridge.

Numerical parameters of the structural analyses, such as natural frequencies and damping coefficients (expressed in terms of the critical damping coefficient), are matched to simulate the wind tunnel test. Displacements at various wind speeds are computed with the wind loadings obtained from CFD at each speed, which are shown

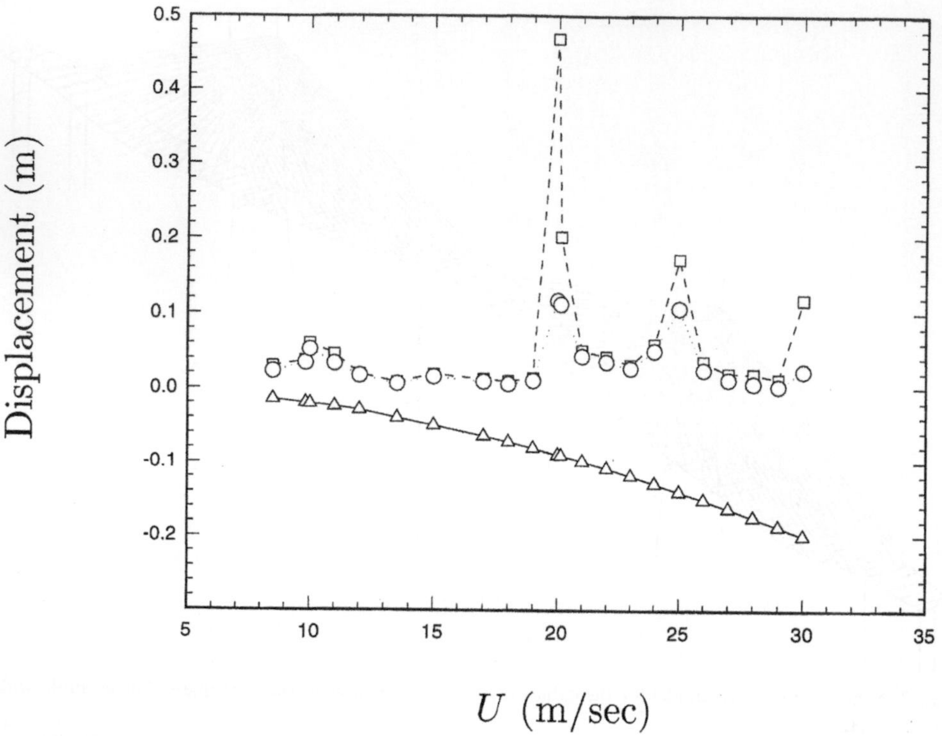

Fig. 8. Mean and fluctuating displacements at the bridge midspan at various wind speeds with 0.6% twisting damping. Mean deflection (△); fluctuating displacement (□) 0.2% bending damping, (○) 1% bending damping.

Table 2
Comparison of the fluctuating displacements at the bridge midspan

	Wind speed (m/s)		Displacement (m)	
	Exp. [10]	Present	Exp. [10]	Present
1st peak	8.0	10.0	0.06	0.059
2nd peak	17.0	20.0	0.42	0.469

in Fig. 8. The mean displacements increase monotonically with almost quadratic dependence on the wind speed, while the fluctuating displacements show some resonant behavior at wind speeds of 10, 20, 25, and 30 m/s. The results for the first two resonant conditions are summarized in Table 2 for comparison with the wind tunnel test. The first two resonant wind speeds compare fairly with the computational results, and the fluctuating displacements under those conditions show similar agreement.

Considering various assumptions introduced in the modeling stage, such agreement is at least promising, although not excellent. Since only a few cases are investigated so far using this procedure, the agreements obtained here should be critically reviewed for other cases.

5. Conclusions

In the hope of applying CFD for practical wind engineering analysis, computational methods have been systematically applied for long-span bridges under unsteady wind loadings due to vortex-shedding. Two practical bridges analyzed are the Namhae suspension and the Seohae cable-stayed bridge in Korea. Wind loading characteristics for turbulent flows over a bridge structure are successfully investigated through CFD for two-dimensional bridge deck section models. Unsteady wind loadings obtained are validated indirectly by performing three-dimensional dynamic structural analyses for the Seohae Bridge. In summary, the computational procedure has successfully been applied for the evaluation of unsteady wind loadings on the bridge deck section at the design flow condition with Re $\simeq O(10^7)$ or higher.

Critical investigation of the effects of various modifications to the bridge deck, such as a guide vane, should be made for more practical applications. Various assumptions introduced in the modeling process should also be justified to systematically validate the simulation technologies.

Even though computational wind engineering cannot predict three-dimensional flow characteristics due to current computational and physical modeling limitations, it may serve as a first-stage design tool before undertaking detailed wind tunnel tests.

Acknowledgements

Financial support of CRAY Research, Inc. for this work is greatly appreciated. Help of Mr. Hong Man Kim and Mr. Young Jin Lee in producing the data presented in the paper has been very useful.

References

[1] T.J.R. Hughes, K. Jansen, Finite element methods in wind engineering, J. Wind Eng. Ind. Aerodyn. 46&47 (1993) 297.

[2] R. Franke, W. Rodi, Calculation of vortex shedding past a square cylinder with various turbulence models, in: Proc. 8th Symp. on Turbulent Shear Flow, Munich, Germany, 1991, pp. 20.1.1–20.1.6.

[3] M. Kato, B.E. Launder, The modelling of turbulent flow around stationary and vibrating square cylinders, in: Proc. 9th Symp. on Turbulent Shear Flow, Kyoto, Japan, 1993, pp. 10.4.1–10.4.6.

[4] A. Fujiwara, H. Kataoka, M. Ito, Numerical simulation of flow field around an oscillating bridge using finite difference method, J. Wind Eng. Ind. Aerodyn. 46&47 (1993) 567.

[5] G.H. Koopman, The vortex wakes of vibrating cylinders at low Reynolds numbers, J. Fluid Mech. 28 (1967) 501.

[6] CFD Services, CFDS-FLOW3D User Manual. Release 3.3, Oxfordshire, UK, 1994.

[7] MSC, MAC/NASTRAN Quick Reference Guide, Version 68, California, USA, 1994.

[8] S. Lee, Unsteady aerodynamic force prediction on a square cylinder using $k-\varepsilon$ turbulence models, these Proceedings, J. Wind Eng. Ind. Aerodyn. 67&68 (1997) 79–90.

[9] S. Lee, Unsteady turbulent flow prediction around rectangular cylinders using two-equation turbulence models, in: Proc. KSME 1996 Spring Annual Meeting B, 1996, pp. 183–188 (in Korean).

[10] Samwoo Institute of Construction Technology report, An experimental study of Seohae Cable-Stayed Bridge, Seoul, Korea, 1993.

[11] Ishikawajima-Harima Heavy Industries Co., Japan, Namhae Bridge final report, 1973.

[12] W.H. Press, B.P. Flannery, S.A. Teukolsky, W.T. Vetterling, Numerical Recipes: The Art of Scientific Computing, Cambridge Univ. Press, New York, 1986.

Vehicle Aerodynamics and Dispersion

ELSEVIER

Journal of Wind Engineering
and Industrial Aerodynamics 67&68 (1997) 281–291

JOURNAL OF wind engineering AND industrial aerodynamics

Numerical simulation of air flow in an urban area with regularly aligned blocks

Ping He[a,*], Tadahisa Katayama[b], Tetsuo Hayashi[b], Jun-ichiro Tsutsumi[c], Jun Tanimoto[b], Izuru Hosooka[b]

[a] *Tsukuba Research Institute, Sanken Setsubi Kogyo Co., Ltd., Kinunodai 4-5-1, Yawara-mura, Ibaraki 300-24, Japan*

[b] *Department of Thermal Energy System, Kyushu University, Kasuga-shi, Fukuoka 816, Japan*

[c] *Department of Civil Engineering and Architecture, University of the Ryukyus, Nishihara-cho, Okinawa 903-01, Japan*

Abstract

Numerical simulation of air flow distribution in a built-up area is an effective way to analyze and predict the urban thermal environment. A cyclic boundary conditions method for the numerical simulation of air flow around a block is used to model the unlimited spread of a built-up area. An equation for the calculation of the pressure difference between the windward and the leeward boundaries is proposed. Another simulation model which has 10 blocks aligned with the wind direction is used for comparison. The inflow boundary conditions are given by a wind tunnel test. The cyclic boundary conditions produced stable calculation results. The simulation results of the cyclic boundary conditions model are similar to those of the 10-block model in the cavity space. There is, however, a little difference between the results of these two models, and between them and the wind tunnel test in the higher area above the cavity and at the crossing point of the streets.

Keywords: Regularly aligned blocks; Air flow in an urban area; Cyclic boundary conditions

1. Introduction

Air flow distribution in a built-up area has a strong influence on the urban climate [1]. It is one of the essential factors to be investigated for a better understanding of the warming of urban areas. The air flow in an urban area is directly influenced by the various factors of urban structure, e.g., scales of blocks and streets, heights of buildings and the whole size of the urban area [2,3].

* Corresponding author.

0167-6105/97/$17.00
PII S0167-6105(97)00079-2

Numerical simulation is one of the most effective methods to predict the air flow distribution. However, the specification of a simulation model is a serious problem, when numerical simulation is applied to the air flow in an urban area. If the simulation area is modeled after an actual urban area, it has to include a large area with an enormous number of calculation points, which makes the actual calculation very difficult. However, central districts in many cities often have a similar style which consists of rectangular blocks separated by straight streets. If they are extremely simplified, they become districts with regularly aligned square blocks of constant size. This simplified urban model can easily be numerically simulated.

Moreover, if the same blocks are aligned regularly, they are regarded as a repeat of a block. If this repeating is expressed by cyclic conditions, the central district of an urban area is represented by one or some block(s) with cyclic boundary conditions. When a numerical simulation model of one block with cyclic boundary conditions can express an unlimited spread of built-up area, it can reduce the space for simulation, computer memories and calculation time. The cyclic boundary conditions of a numerical simulation of air flow over an urban area as expressed by one block are proposed and tested in this paper. It is compared with another simulation model which consists of 10 blocks and a wind tunnel test.

2. Simulation model

A central part of a built-up area is often divided regularly by streets in many cities. A cell surrounded by the streets which is called a block usually consists of several buildings and courtyards. However, it is sometimes regarded as one large building, if the buildings in a block are concentrated densely and their heights are all similar. If such a built-up area is very simply modeled, it may become a series of regularly aligned rectangular cubes.

Fig. 1 shows a model of a built-up area that is a group of simple rectangular cubic blocks of the same shape placed on a two-dimensional regular grid. The gaps between the blocks are thought to be streets. The directions of the streets are supposed to be parallel or normal to the wind direction, and all the streets are assumed to be of the same breadth to simplify the model, which means that all intervals between adjoining blocks are of the same width. The plan of the block is a square of which a side is D that is used as the representative length, and the height is $3D/4$. The interval between two blocks which means the width of the streets is $3D/5$. X, Y and Z-axes are set to the wind direction, the horizontal direction normal to the wind and the vertical upward direction, respectively, as shown in Fig. 1. Such a model was actually used in the wind tunnel test.

It is the aim of this paper to express such a group of the same blocks as a repeat of a block or some blocks by cyclic boundary conditions in numerical simulation. A line of some blocks in the X direction is supposed to be a simulation model. If the ordinary inflow and outflow boundary conditions are used on the boundaries normal to the X direction and the symmetrical boundary conditions are used on the boundaries normal to the Y direction, the simulation model expresses an urban area with a

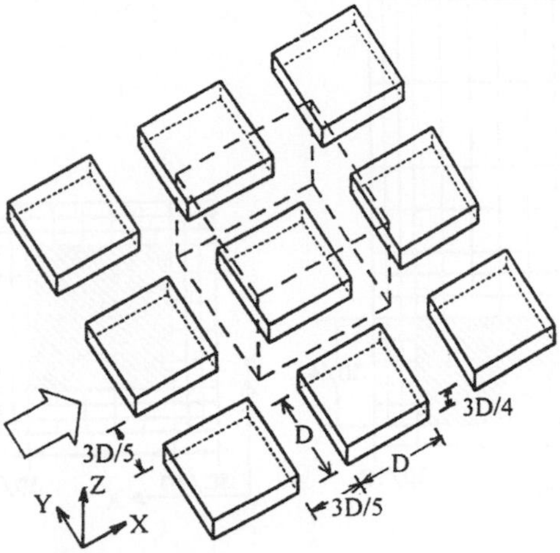

Fig. 1. Simplified model of a built-up area.

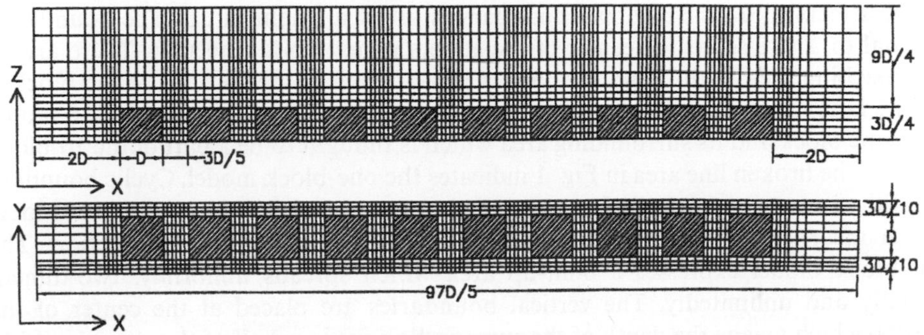

Fig. 2. 10-block model and its calculation grid.

limited length in the X direction and an unlimited width in the Y direction. The symmetrical boundary conditions are thought to be kinds of cyclic boundary conditions. Moreover, if cyclic boundary conditions are used on the boundaries normal to the X direction, the simulation model has an unlimited length in the wind direction. Therefore, one block with cyclic conditions on all the boundaries except the upper and the ground boundaries is thought to be a model of an unlimited spread of a built-up area.

Therefore, two types of simulation models are used in this paper. The former type is a 10-block model shown in Fig. 2. This is a simulation model with 10 blocks which are aligned in a line of the X direction. Cyclic boundary conditions are applied on the boundaries normal to the Y direction, and ordinary inflow and out flow boundary conditions are used on the boundaries normal to the X direction, which means that the simulation area is 10-block length in the X and unlimited width in the Y direction. The plan of the block is a square of which a side is D that is used as the representative

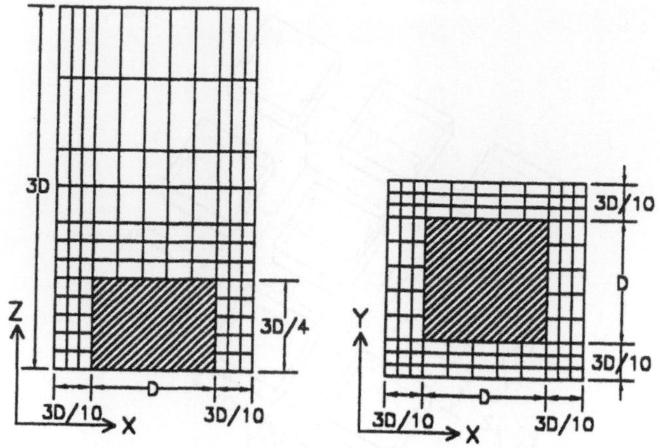

Fig. 3. One-block model and its calculation grid.

length, and the height is $3D/4$. The interval between two blocks which means the width of the streets is $3D/5$. The upper boundary is set at a height of $3D$. The boundaries normal to Y are set at the center of the streets going the X direction. There are an approaching area in the windward and a run-off area in the leeward of the blocks.

The latter type is a one-block model shown in Fig. 3. This is a simulation model with one block and its surrounding area which is thought to be a part of the 10-block model. The broken line area in Fig. 1 indicates the one-block model. Cyclic boundary conditions are adopted on the boundaries not only normal to Y but also normal to X, which means that the simulation area is also unlimited in the X direction. The one-block model expresses a built-up area which spreads uniformly, two-dimensionally and unlimitedly. The vertical boundaries are placed at the center of the streets, which means the depth of the surrounding space is half of the street breadth. Calculation points of the 10-block model and the one-block model are set on structural grids which are shown in Figs. 2 and 3, respectively.

3. Numerical simulation

3.1. Calculation method

The basic equations for the numerical simulation are shown in Table 1. The k–ε two-equation model is used as a turbulence model. Forward differences are applied to the temporal differential terms, and central differences are applied to the spatial terms. However, the advective terms in the k and ε transport equations are changed to up-wind differences. All physical quantities are normalized by the representative values. The representative length is the width of the block, D. The representative velocity is the mean wind speed at a height of D on the inflow boundary of the initial conditions.

Table 1
Basic equations and variables for numerical simulation

$$\frac{\partial U_i}{\partial t} + \frac{\partial (U_i U_j)}{\partial X_j} = -\frac{\partial \Pi}{\partial X_i} + \frac{\partial}{\partial X_j}\left(\nu_t E_{ij} + \frac{1}{\mathrm{Re}}\frac{\partial U_i}{\partial X_j}\right) \quad (1)$$

$$\frac{\partial k}{\partial t} + \frac{\partial (k U_j)}{\partial X_j} = \frac{\partial}{\partial X_i}\left\{\left(\frac{\nu_t}{\sigma_1} + \frac{1}{\mathrm{Re}}\right)\frac{\partial k}{\partial X_i}\right\} + \nu_t E_{ij}\frac{\partial U_i}{\partial X_j} - \varepsilon \quad (2)$$

$$\frac{\partial \varepsilon}{\partial t} + \frac{\partial (\varepsilon U_j)}{\partial X_j} = \frac{\partial}{\partial X_i}\left\{\left(\frac{\nu_t}{\sigma_2} + \frac{1}{\mathrm{Re}}\right)\frac{\partial \varepsilon}{\partial X_i}\right\} + C_1 \nu_t \frac{\varepsilon}{k} E_{ij}\frac{\partial U_i}{\partial X_j} - C_2\frac{k\varepsilon}{\nu_t} \quad (3)$$

$$\frac{\partial U_i}{\partial X_i} = 0 \quad (4) \qquad \nu_t = C_D\frac{k^2}{\varepsilon} \quad (5)$$

$$E_{ij} = \frac{\partial U_i}{\partial X_j} + \frac{\partial U_j}{\partial X_i} \quad (6) \qquad \Pi = P + \frac{2}{3}k \quad (7)$$

X_i = axis, U_i = velocity component of X_i direction, t = time, P = pressure,
ν_t = eddy viscosity coefficient, k = turbulent kinetic energy, ε = energy dissipation rate,
$C_D = 0.09$, $C_1 = 1.59$, $C_2 = 0.18$, $\sigma_1 = 1.0$, $\sigma_2 = 1.3$

The initial conditions of the mean wind speed and the turbulence intensity on the inflow boundary are given by the wind tunnel test which results are shown in Fig. 4. The mean wind speed can be fitted to the power law profile with the exponent $\frac{1}{4}$ which is used as the initial condition of the velocity component normal to the inflow boundary. The tangential velocity components are assumed to be zero on the inflow boundary. The values on the inflow boundary are fixed in the 10-block model, and the free-slip conditions are applied on the outflow boundary. The upper boundary is supposed to be the free-slip condition. The tangential velocity components on the ground and the surfaces of the blocks are supposed to be a power law profile with exponent of $\frac{1}{4}$. The boundary conditions of the other variables on the solid boundaries are free-slip.

3.2. Cyclic boundary conditions

The lateral boundaries normal to Y of both the models are given by symmetrical boundary conditions. The scalar variables and the tangential velocity components at the symmetrical positions on both the boundaries are of the same values. Actually, the averages of these values on both sides are given as the boundary conditions. The normal velocity component is zero on these boundaries.

The variables on the inflow and the outflow boundaries which are normal to X of the one-block model are given by cyclic boundary conditions. The cyclic boundary conditions on these boundaries mean that all the variables on the outflow boundary which are the results of the former calculation step are given to those on the inflow boundary in the next calculation step. The positions of the calculation points in the Y–Z plane are the same positions on these boundaries. However, if the pressure

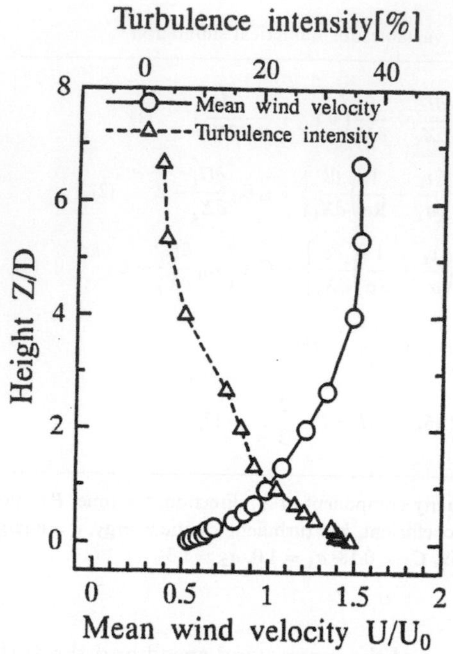

Fig. 4. Mean wind speed and turbulence intensity profiles of the approaching wind of a wind tunnel test.

on the outflow boundary is given to the pressure on the inflow boundary, the pressure difference between the inflow and the outflow boundaries becomes zero, which means that the source power for the wind is lost and the numerical simulation cannot be continued.

There are only few studies [4] on the pressure difference of the cyclic boundary conditions, in which an empirical constant value is given as the pressure difference between the inflow and the outflow boundaries. Therefore, an equation for the calculation of the pressure difference is proposed in this paper. This is made from the momentum equation of the X direction, since the inflow and the outflow boundaries are normal to the X-axis. The pressure gradient on the inflow and the outflow boundaries is calculated as follows:

$$\frac{\Delta \Pi}{\Delta X} = \iint_A \left(V - \frac{\Delta Q}{S \Delta t} \right) \mathrm{d}A \Big/ S, \tag{8}$$

$$V = \frac{\partial}{\partial X_j}\left[v_t \left(\frac{\partial U}{\partial X_j} + \frac{\partial U_j}{\partial X} \right) \right] - \frac{\partial U U_j}{\partial X_j}, \tag{9}$$

where A is the surface of the inflow or the outflow boundary, and dA is the surface element which means each cell on the boundaries actually. S is an area of the inflow or the outflow boundary. Q is the total flow rate on the boundaries. Using Q makes the changes of the pressure gradient on the boundaries more mild than it is calculated at each point. These equations give the average pressure gradient between the outside and the inside cells adjoining the inflow or the outflow boundary. The pressure of the outside cell is calculated from this pressure gradient and the pressure of the inside cell.

The initial values on the inflow boundary of both the models are given by the wind tunnel test. As for the one-block model, preparatory calculations of 200 steps are carried out without using the cyclic boundary conditions at first, and then the cyclic boundary conditions are applied to the inflow and the outflow boundaries.

4. Wind tunnel test

The outline of the experimental model for the wind tunnel test is shown in Fig. 1, which has the same geometry as the numerical simulation models. The actual size of the representative length, D, is 75 mm. The model blocks are aligned in 12 lines parallel to the wind direction, and there are 10 blocks in each line. The intervals between two lines and two blocks in a line are constantly 45 mm that is $3D/5$. Each line of the model blocks, especially the line near the center of these lines is equivalent to the 10-block model.

The wind tunnel used here is a closed circuit type. The test section is 8 m in length, 1.5 m in width and 1 m in height. The model blocks are directly fixed on the floor of the test section. The experimental wind speed is 6 m/s at a height of 600 mm from the floor. The approach mean wind speed and turbulence intensity profiles are shown in Fig. 4. A power law with the exponent of $\frac{1}{4}$ fits the mean wind speed profile up to a height of 300 mm. These profiles are used for the inflow boundary conditions in the numerical simulation of the 10-block model and they are also used for the initial conditions for the one-block model.

5. Results and discussion

5.1. Pressure difference on the cyclic boundary

The fluctuation of the pressure difference of the one-block model after the preparatory calculation is shown in Fig. 5. The pressure difference means the difference between the average pressure on the inflow boundary and that on the outflow boundary. The pressure difference changes violently at the beginning of the calculation. Then, it is stable at the simulation time $t = 20$, and the value then decreases gradually to 0.01 at $t = 80$. The calculation method proposed here creates a stable pressure difference between the inflow and the outflow boundary.

5.2. Mean wind speed profile

The change of mean wind speed profile with simulation time for the one-block model is shown in Fig. 6. These profiles are calculated from the simulation results at

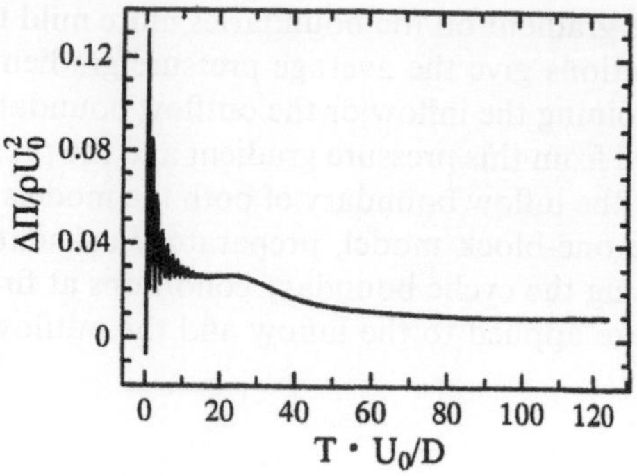

Fig. 5. Fluctuation of pressure difference between the inflow and the outflow boundaries of the one-block model calculated by cyclic boundary condition.

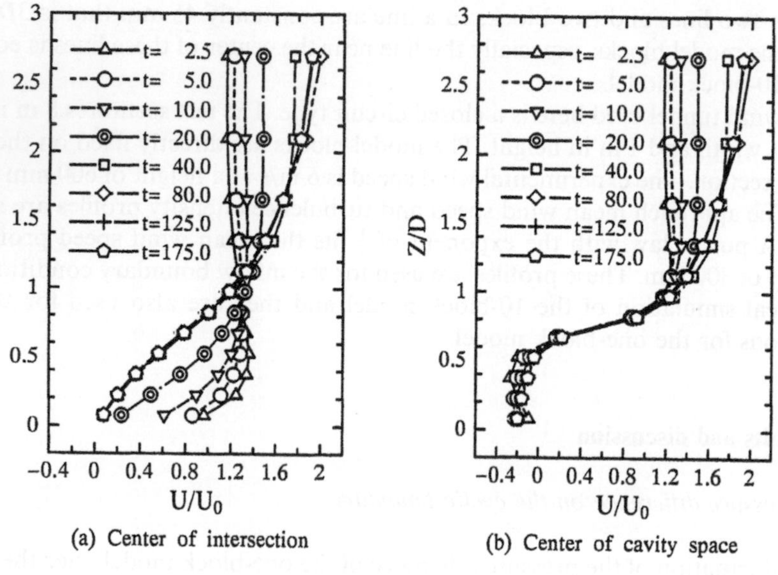

Fig. 6. Change of mean wind speed profile in the one-block model with simulation time.

the central points of the intersection of the streets and the cavity area between the blocks. The center of the intersection is the corner of the calculational area of the one-block model. The cavity area means the street normal to the wind direction.

At the center of the intersection, the mean wind speed decreases below the height of the block and increases over the height of the block as the simulation time progresses. The same tendency is observed in the change of the profiles at the center of the cavity. The wind speed below the height of the block at the center of the cavity space is

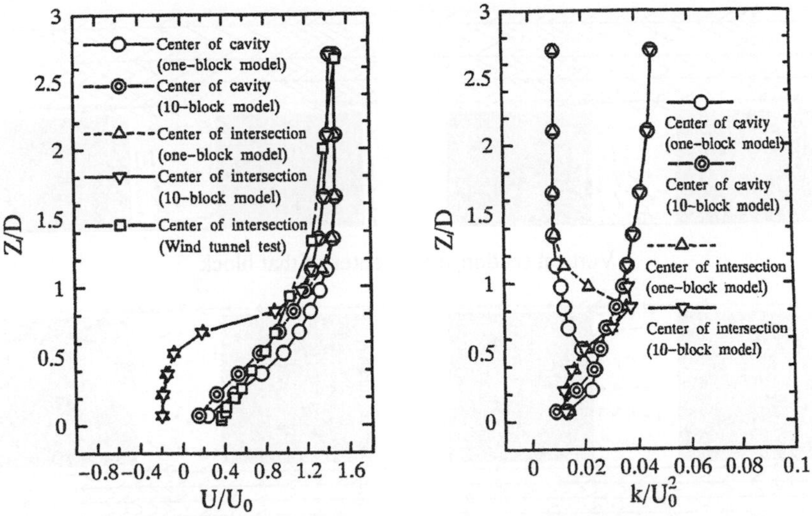

Fig. 7. Comparison of mean wind speed and turbulence energy profiles.

smaller than that at the center of the intersection, and the change of the wind speed at the center of the cavity space with the simulation time is also smaller than that at the center of the intersection.

When the simulation time is 20, the wind speed profiles at these points over the height of the block are similar to the initial conditions on the inflow boundary that is given by the wind tunnel test. Therefore, the wind speed profile at $t = 20$ is similar to the approach wind speed profile of the wind tunnel test. This may be one reason why there is a short flat step in the fluctuation of the pressure difference shown in Fig. 5. The simulation results after $t = 20$ show a different tendency from those before $t = 20$. The air flows after $t = 20$ perhaps indicate the air flow over a built-up area spread unlimitedly. The difference between the wind speed profile at $t = 20$ and that after $t = 20$ is thought to be the effect of the cyclic boundary conditions. The result at $t = 20$ of the one-block model is thought to be suitable for comparison with the results of the 10-block model and the wind tunnel test.

The wind speed profiles at the two points mentioned above by the one-block model, the 10-block model and the wind tunnel test are compared in Fig. 7a. The wind tunnel profile at the center of the cavity space is omitted. The wind direction in the cavity space is very complicated and it is very difficult to measure the correct speed. The results of the 10-block model and the wind tunnel test indicated in Fig. 7a are the data between the seventh and the eighth blocks, since the wind speed profile is stable downwind of this point.

The wind speed profiles in the one-block model, the 10-block model and the wind tunnel test are a little different from each other at the center of the intersection. The distribution patterns of these profiles resemble each other. At the center of the cavity,

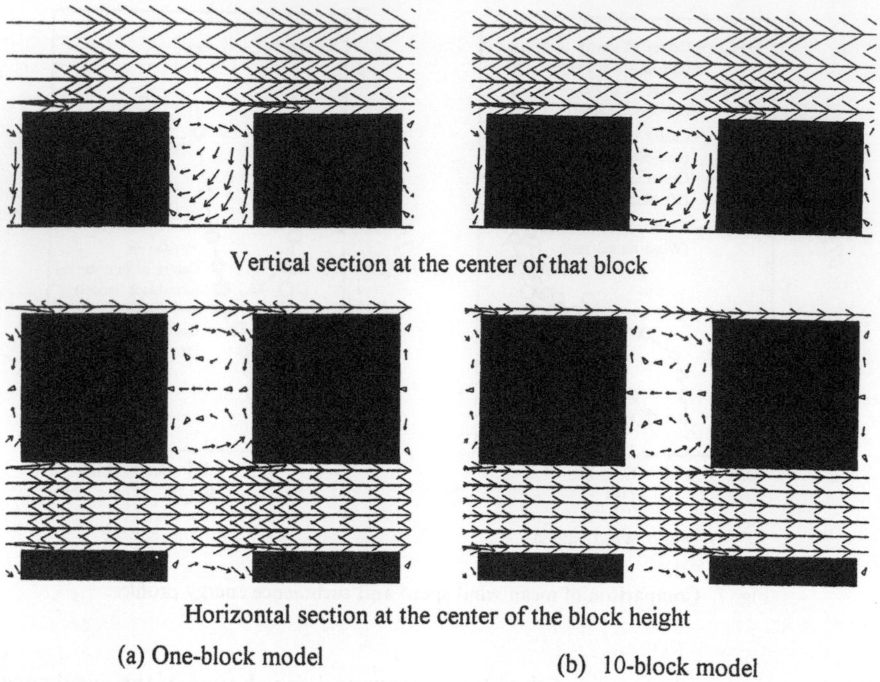

Fig. 8. Distribution of wind velocity vectors in two numerical simulation models.

the wind speed profiles in the one-block model and the 10-block model show a similar curve below the height of the block. However, these profiles over the height of the block are a little different from each other, and each profile is similar to that at the center of the intersection of each model.

5.3. *Turbulence energy profile*

The profiles of turbulence energy for the one-block model and the 10-block model are shown in Fig. 7b. The profiles at the center of the intersection are quite different from each other except in the lower layer. The values of the turbulence energy for the 10-block model are 5 times larger than those for the one-block model. However, these profiles are similar below the height of the block at the center of the cavity space, and each profile is close to that at the center of the intersection of each model over the height of the block.

5.4. *Distributions of mean velocity vector*

The distributions of the mean velocity vectors for the numerical simulation results of the one-block model and the 10-block model are shown in Fig. 8. In the results of both simulation models, circulating secondary flow is observed in the vertical section at the center of the cavity space. In the horizontal section at a half height of the block, the distributions of the mean velocity vectors for both simulation models indicate symmetrical secondary circulations. The distribution patterns of the wind velocity vectors for the one-block model and the 10-block model are similar.

6. Conclusions

Two types of numerical simulation and a wind tunnel test of air flow distributions in a built-up area with regularly aligned blocks were carried out, and the possibility of practical use of the cyclic boundary conditions proposed here was examined. The proposed equation for the pressure difference on the inflow and the outflow boundaries produces a stable result. Mean wind speed profiles in the one-block model at an appropriate simulation time are fairly similar to those in the 10-block model and in the experiment model of the wind tunnel test, which means that the one-block model with the cyclic boundary conditions is useful for a numerical simulation of air flow in an unlimited built-up area. The mean wind speed profile in the one-block model changes with the simulation time, and the profile at the later simulation time is a little different from that in the 10-block model. The reason is that the length in the X direction of the one-block model is unlimited while that of the 10-block model is limited. The turbulence energy in the one-block model decreases with the simulation time, which is the problem in the cyclic boundary conditions to be solved.

References

[1] T. Katayama, A. Ishii, M. Nishida, J. Tsutsumi et al., Cooling effects of a river and sea breeze on the thermal environment in a built-up area, Energy and Buildings 15–16 (1991) 973.

[2] J. Tsutsumi, T. Katayama, M. Nishida, Wind tunnel tests of wind pressure on regularly aligned buildings, J. Wind Eng. Ind. Aerodyn. 41–44 (1992) 1799.

[3] B.E. Lee, M. Hussain, B. Soliman, Prediction natural ventilation forces upon low-rise buildings, ASHRAE J. (1980) 35.

[4] S. Murakami, K. Hibi, A. Mochida, Three dimensional analysis of turbulent flow field around street blocks by means of large eddy simulation (Part 1), J. Archt. Plann. Environ. Eng. AIJ 412 (1990) 1 (in Japanese).

ELSEVIER

Journal of Wind Engineering
and Industrial Aerodynamics 67&68 (1997) 293–304

JOURNAL OF
wind engineering
AND
industrial
aerodynamics

Car exhaust dispersion in a street canyon. Numerical critique of a wind tunnel experiment

Bernd M. Leitl[1], Robert N. Meroney

Fluid Dynamics and Diffusion Laboratory, Civil Engineering Department, Colorado State University, B-227 Engineering Research Center, Fort Collins, CO 80523, USA

Abstract

Due to increasing car traffic in cities, problems related to car induced air pollution in street canyons have become important. Physical modeling in wind tunnels or numerical codes may be used for dispersion simulation when investigating air quality. Rafailidis [in: Annalen der Meteorologie] carried out an extensive set of test runs recently in the BLASIUS wind tunnel at the Meteorological Institute of the University of Hamburg, Germany. In the present study the wind tunnel experiments were simulated numerically using the CFD-code Fluent®. In a first approach, the idealized two-dimensional case was calculated. Several test runs have been carried out to study the effect of emission rate and source design on flow structures and dispersion in the street canyon. It could be shown that alternative emission conditions and the source design might affect the concentration field within a modeled street canyon. A second set of calculations for a simplified three-dimensional simulation of the street canyon setup was performed to investigate the presence of secondary flow patterns found during wind tunnel tests. The lateral flow structure within the street canyon observed during wind tunnel measurements was simulated, and the effect of changing boundary conditions on the secondary flow structure was studied. In the paper the advantages of CFD simulations for planning wind tunnel dispersion tests are discussed.

Keywords: Dispersion; Wind tunnel; Street canyon; Numerical simulation

1. Introduction

Increasing car traffic leads to increasing air quality problems in cities. Since cars are accepted to be the major emission source of air pollutants in urban areas, and

[1] Present address: Meteorology Institute, University of Hamburg, Bundesstrasse 55, D-20148 Hamburg, Germany.

PII S0167-6105(97)00080-9

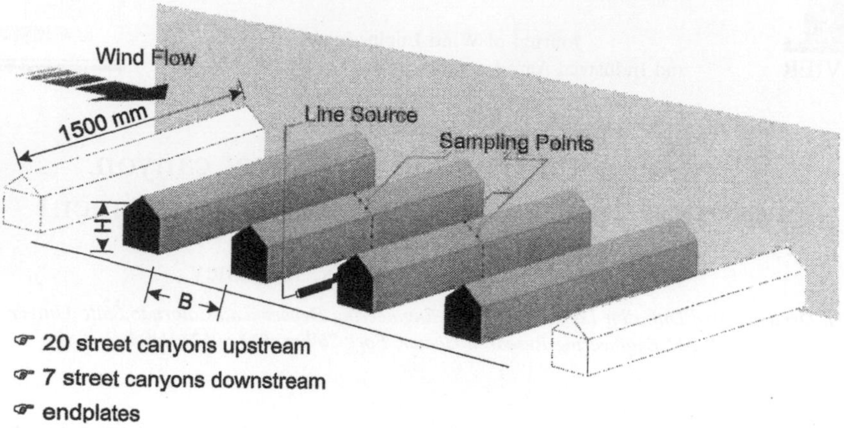

Fig. 1. Wind tunnel setup.

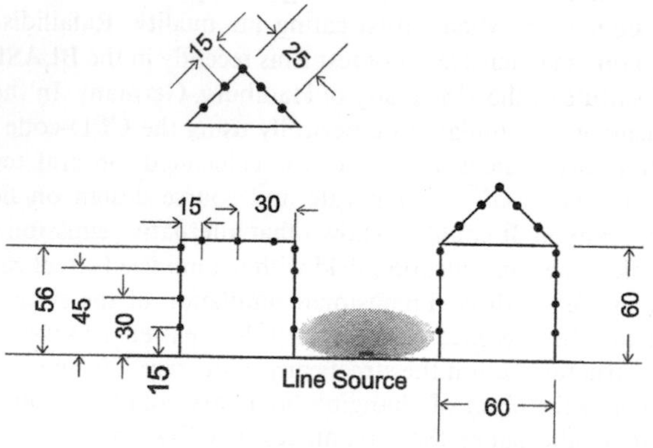

Fig. 2. Configuration of measurement tapping holes at the outline of the test street canyon.

a further increase of city traffic is expected, investigations of dispersion processes in street canyons have become a focal point in environmental research (see Refs. [1, 2]). Recently Rafailidis et al. [3] carried out an extensive set of wind tunnel experiments on gas dispersion in urban street canyons. Using a quasi two-dimensional setup (see Fig. 1) the dispersion of tracer gas emitted by a line source in a closed street canyon was measured for a variety of canyon aspect ratios, B/H, and roof forms. To simulate an urban roughness 20 upstream and 7 downstream street canyons were included. The concentrations were measured in the symmetry plane of the setup at the positions shown in Fig. 2. A more detailed description of the wind tunnel experiments and the physical model is given in Refs. [3, 4]. A significant lateral flow structure (Fig. 3) was detected and measured in the street canyon, where the flow was expected to be quasi two-dimensional. Since the secondary flow structure obviously causes a concentration

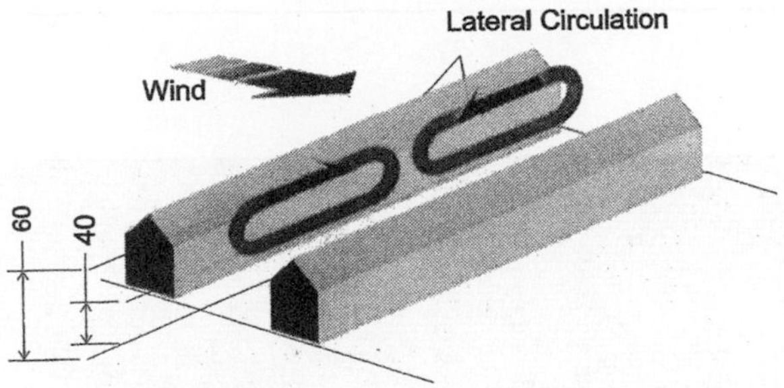

Fig. 3. Lateral flow structure observed during wind tunnel experiments.

gradient in the lateral direction, it was also expected to affect the concentration measurements in the symmetry plane. Extensive flow visualization experiments using a Laser light sheet visualization setup and Laser-Doppler-Velocimeter measurements were performed in order to locate the source, but a final statement about what was causing the 3D flow structure and how to minimize the influence of lateral flow within the street canyon could not be made. Also, it was difficult to predict the magnitude of the influence of the second order flow on mean concentrations measured in the center plane of the street canyon.

A numerical simulation with the CFD-code Fluent® was prepared to give additional information about the mean flow within the street canyon as well as the concentration distribution for a quasi two-dimensional street canyon setup. The Fluent CFD package consists of several tools for defining a discrete flow problem (i.e. grid generation), setting boundary conditions and solving the set of complex equations for conservations of momentum, mass, chemical species and energy. The governing equations are discretized on a curvilinear grid to permit computations over irregular geometries and solved using a control volume based finite difference method. Discrete velocities and pressures are stored in a nonstaggered grid, and interpolation is realized by using a first-order, power-law scheme or optionally by higher-order upwind schemes. The basic solver in Fluent is the SIMPLEC/SIMPLE algorithm with iterative line-by-line matrix solver and multigrid acceleration. For a detailed description of Fluent see Ref. [5].

2. Calculations

2.1. *Two-dimensional calculations – setup*

In a first approach the flow in the street canyon was assumed to be two dimensional. Several discrete 2D setups with different body-fitted grid configurations were tested with respect to their numerical stability before a final grid was selected. An optimum of performance according to memory usage, convergence behavior and computing time could be found for the grid in Fig. 4 which consists of 230×112 cells

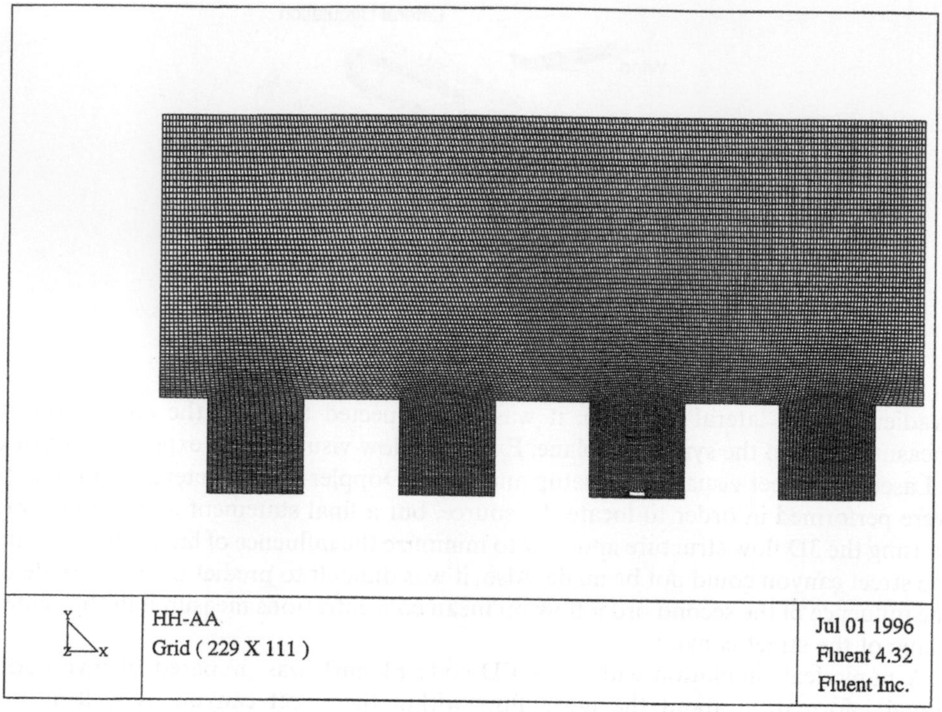

Fig. 4. Computational grid – case A, flat roof configuration.

in a physical domain of 480 mm × 240 mm. The inlet boundary (left side) was defined at the middle of the 3rd up stream building where the wind profile was measured during wind tunnel experiments; hence, the wind tunnel profiles of velocity and turbulence intensity could be used for calculating boundary conditions for the inlet. The wind tunnel data as well as the input profiles of velocity, turbulent kinetic energy and dissipation used for the numerical simulation are shown in Fig. 5. The inlet values of kinetic energy, k, and the dissipation ratio, ε, were calculated from measured velocity profiles and turbulence intensities as well as from a given friction velocity ratio, u_*/u_{ref}, for the wind tunnel setup according to Eqs. (1) and (2) (y is the distance from the wall, κ is the von Karmán constant),

$$\varepsilon = u_*^3/\kappa y, \tag{1}$$

$$k = \tfrac{3}{2}(u')^2. \tag{2}$$

The line source inlet was modeled as a $\mathrm{d}x = 0.2$ mm wide slot (0.25 mm tubing in the wind tunnel) and set as a constant velocity inlet, no turbulence and an inlet velocity of $w_{\text{source}} = 3.81$ m/s according to the source emission rate used during the wind tunnel experiments. When turbulent mixing was simulated, a tracer mass fraction of 1 was applied to the line source inlet. The wind tunnel situation, where 1 l/h ethane was mixed in 100 l/h air was also tested. As expected, the nondimensional

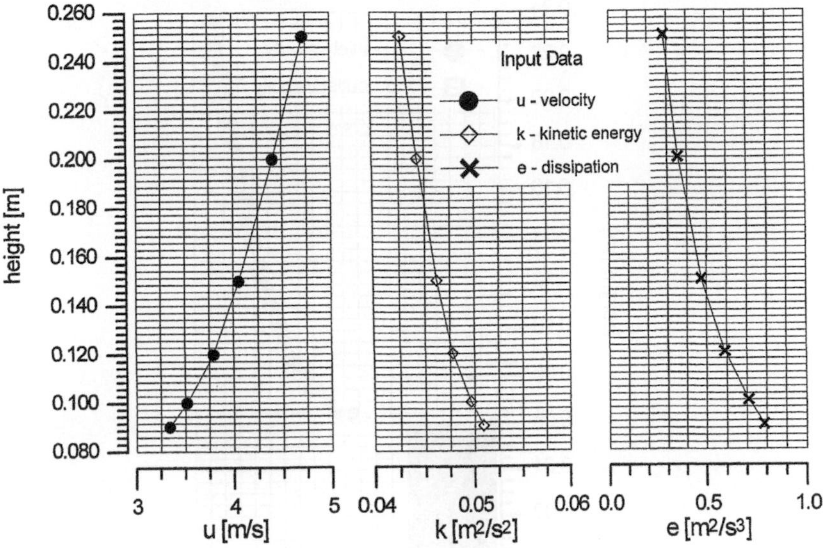

Fig. 5. Inlet profiles of velocity, kinetic energy and dissipation.

concentration values show no dependency on which source concentration is used. The top and right side of the computational domain were defined as outlets to allow for expected blockage effects caused by the models. The model buildings as well as the wind tunnel floor were specified as hydraulic smooth, isothermal walls with $u = w = 0.0$ m/s (u is the horizontal velocity component, w is the vertical velocity component).

To study the influence of applying different turbulence models in Fluent, all configurations were calculated with a standard k–ε model as well as with the newer RNG model (**Re**Normalized **G**roup theory, see Refs. [5, 6] for description). As recommended by Van Oort and Stork [7] the constants of the standard k–ε model were varied to visualize possible improvements of the numerical results compared to the wind tunnel experiments for different constant sets. According to recommendations given in Ref. [5] the turbulence model constants were not changed when the RNG model was used.

Since the wind tunnel results of concentration measurements were provided in a non-dimensional form all calculated concentrations, $\boldsymbol{C}$ were normalized in the same way with source emission rate ($w_{\text{source}}\,\mathrm{d}x_{\text{source}}$, where w_{source} is the exhaust velocity, $\mathrm{d}x_{\text{source}}$ is the width of source), reference velocity, U_{ref}, and the total height of the model building, H (see Eq. (3)),

$$K = CHU_{\text{ref}}/w_{\text{source}}\mathrm{d}x_{\text{source}}. \qquad (3)$$

2.2. *Two-dimensional calculations – results*

Comparisons of the mean velocity fields calculated by a standard k–ε approach versus the velocity field calculated by the RNG approach displayed no major

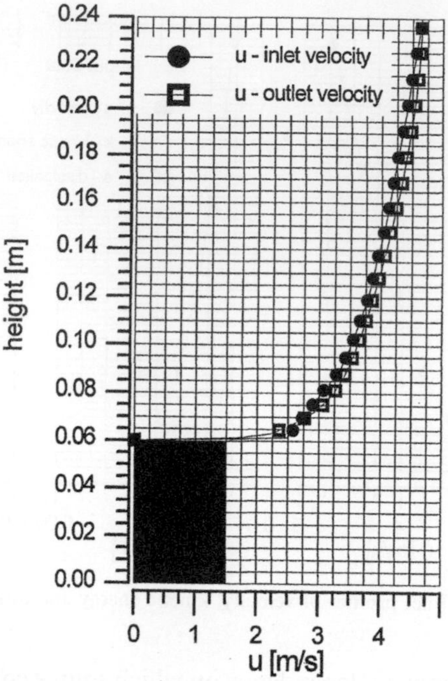

Fig. 6. Comparison between given inlet velocity profile and calculated outlet profile.

differences. In Fig. 6 the given inlet velocity profile (i.e. measured velocity profile) for the flat roof configuration (case A) is compared to the calculated outlet profile. Both profiles are almost identical except for a slight smoothing effect close to the building roofs and the effect of the additional mass source flow. Even if the exhaust velocity of tracer is high compared to the mean velocity of the surrounding canyon flow, the source injects the tracer into the street canyon without major disturbance when operating the source at 101 liter tracer per hour.

Fluent computes similar concentration patterns observed during wind tunnel and field measurements. The main vortex in the canyon dominates the dispersion of car exhaust and generates higher concentration values on the upwind or downwind building for different configurations. Depending on the roof geometry, the flow pattern and the location of the main vortex in the street canyon changes. This causes a change in measured/calculated concentrations at the side walls of the street as well. Fig. 7 shows a direct comparison between measured and calculated concentrations over the upwind and downwind walls for the flat roof configuration (case A). The concentrations calculated for the upwind wall of the street canyon agree well with those measured in the wind tunnel. No major difference can be detected when comparing calculations based on the standard k–ε-model with results from a RNG-simulation except for an area close to the source. At the downwind wall the numerical simulation leads to significantly lower values than measured in the wind tunnel.

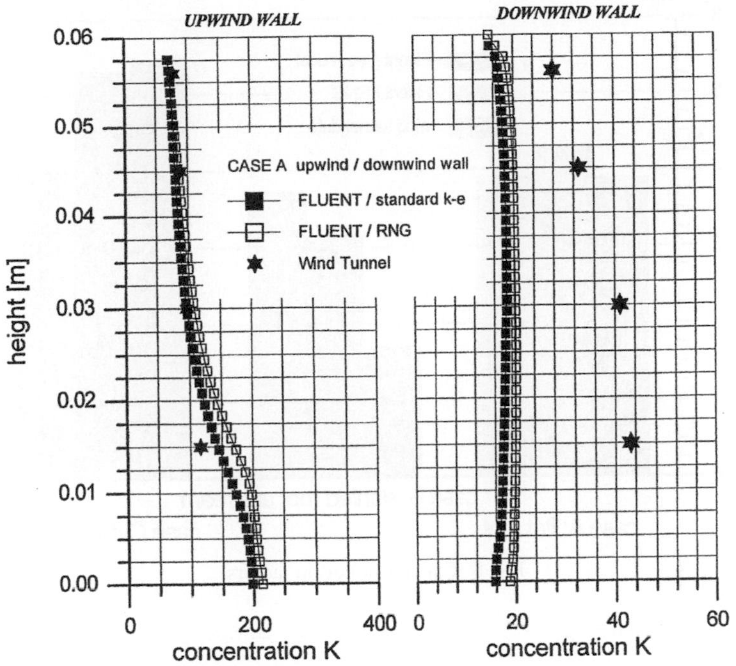

Fig. 7. Comparison between measured and calculated concentrations.

Again the k–ε-model and the RNG-model give approximately the same tendency, but an 'offset' concentration of about 10% for the RNG-model can be detected. Compared with the wind tunnel results the numerical simulations show a difference of up to 58% for the lowest measurement point within the street canyon. For slanted roof configurations (case B: one slanted roof upstream the street canyon, case C: slanted roof on upwind and downwind building) even bigger differences between wind tunnel experiments and numerical simulation were observed. The maximum relative difference based on wind tunnel results was found for case B with values up to 89%. A canyon pollution factor, K^* (K averaged over all sampling points in the canyon), was used to compare the concentration fields for different configurations with the overall K^* values from wind tunnel experiments. As shown in Fig. 8 the averaged concentration value is almost the same in wind tunnel experiment and numerical simulation for the flat roof urban roughness. For slanted roofs the K^* varies up to 50%.

Better agreement between wind tunnel data and numerical simulation could be achieved for slanted roof configuration by changing the constants of the k–ε turbulence model even though there was no general tendency for better results using a modified constant set for all three test cases. Since the calculation of the dispersion within the street canyon is dependent on location and size of the recirculation zones even slightly different calculations of the velocity field had a dramatic effect on the concentrations calculated at the reference points. The results have shown that the

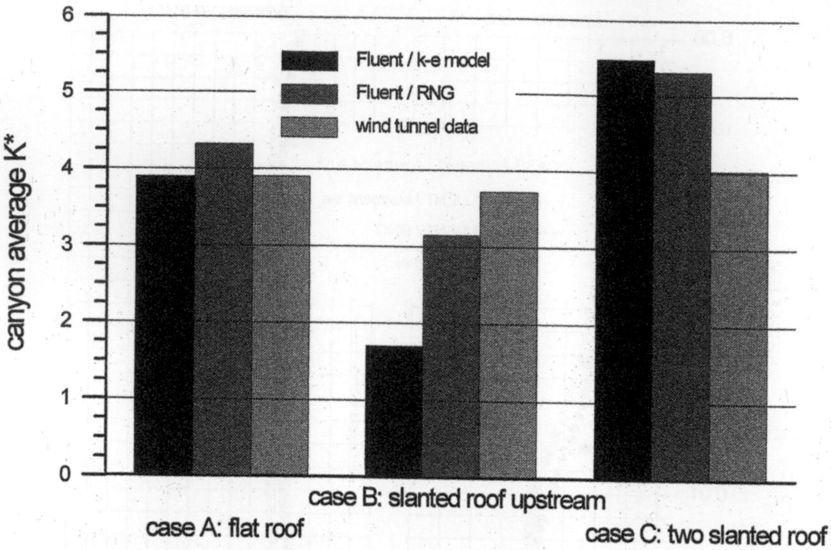

Fig. 8. Canyon pollution factors.

standard constant set cannot be assumed to be universal for complex, high turbulent shear flows with large recirculation areas. Further investigations are required to evaluate constant sets that might lead to a better reproduction of turbulent boundary layer flows and gas dispersion in recirculating areas. However, another reason for discrepancies between the two sets of results might be the lateral flow in the street canyon during wind tunnel experiments studied in the second part of this paper.

2.3. Three-dimensional calculations – setup

To study the three-dimensional flow effects in the street canyon observed in the wind tunnel a simplified 3D setup was generated. A body-fitted mesh with $30 \times 33 \times 61$ (60390) cells were used to define the discrete problem for slanted roof configurations with canyon width ratios $B/H = 1$ and $B/H = 0.5$. To avoid grid skewness problems only a few iterations for grid smoothing were applied. The outline of the geometry as well as two meshed surfaces are shown in Fig. 9. To keep the geometrical resolution of the grid high for a limited total number of cells no upstream or downstream street canyon was included into the discrete model. The flow pattern within the street canyon is dominated by the geometry of the street canyon itself and the separation at sharp edges on the roof. The mean flow field calculation could be assumed to be independent from upstream/downstream street canyons. The inlet profiles for velocity, kinetic energy and dissipation were taken from a 2D simulation with several upstream canyons modeled upwind of the source canyon, and the inlet conditions were derived from wind tunnel measurements (see Section 2.1).

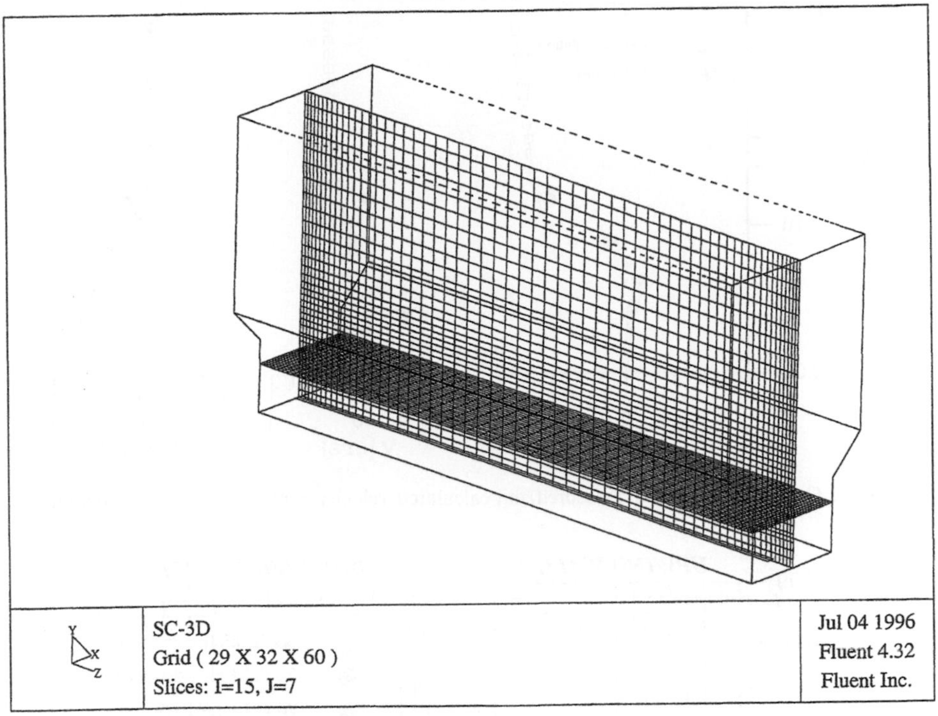

Fig. 9. Simplified 3D street canyon setup.

2.4. *Three-dimensional calculations – results*

A lateral flow structure could be found during the numerical simulation. It could be shown that the major force for the secondary flow is the corner vortex at the end plates of the model canyon. This vortex structure is present for all configurations with end walls or end plates and is caused by the boundary layer at the side walls as well as by the resulting lateral pressure gradient. Similar flow patterns can be observed in a variety of technical flows in rectangular air-conditioning ducts. For a more narrow street canyon ($B/H = 1/2$) the lateral velocity appears to be slightly higher than for a wider canyon ($B/H = 1$). The variation might be caused by the smaller overall velocity in the narrower canyon which is more easily affected by the pressure field as well as the corner vortex at the end plates. A comparison of measured and calculated velocity profiles for the $B/H = 1/2$ setup is given in Fig. 10. The calculated results agree well with Laser Doppler Velocimeter (LDV) measurements in the wind tunnel. Except in an area close to the side walls the resulting concentration field shows no major deviation. The effect of the lateral circulation seems to be small in terms of higher or lower concentrations in the symmetry plane, where the sampling points were located during the wind tunnel experiments. When comparing the calculated concentrations with the results of wind tunnel experiments, the differences are in the same

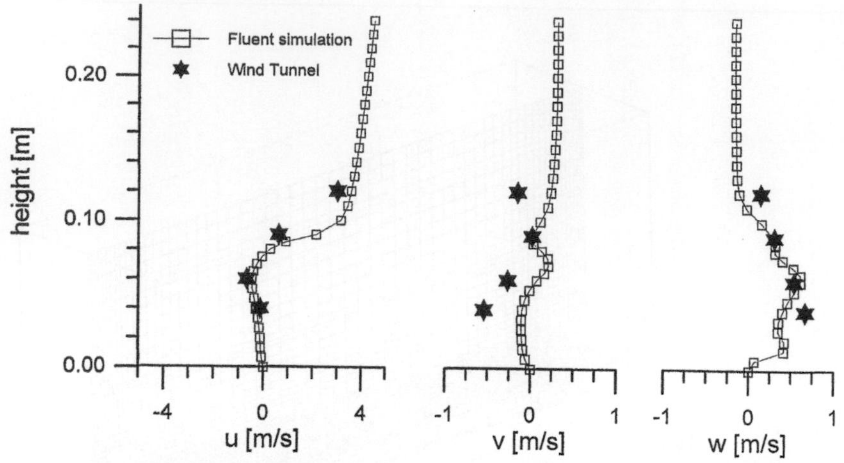

Fig. 10. Comparison between measured and calculated velocity components – 3D simulation.

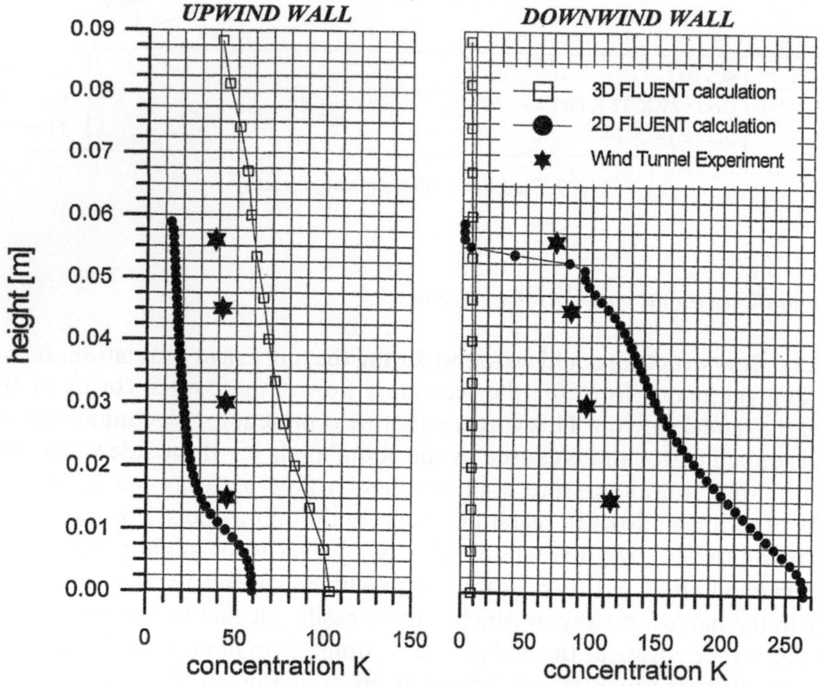

Fig. 11. Calculated and measured concentration profiles – 3D simulation.

order of magnitude as for 2D simulations at the upwind wall (max. 90%). A totally different concentration profile was calculated for the downwind wall. In a 3D simulation an almost constant concentration at a very low level is predicted for the downwind wall where the typical canyon profile (i.e. higher concentrations at the lower elevations) was found during 2D simulations as well as in the wind tunnel (see Fig. 11, case $B/H = 1$).

3. Conclusions

It has been shown that the FLUENT code can simulate the flow field in urban street canyons. Mean flow patterns as well as predicted separation/recirculation areas agree well with the results from physical modeling in a boundary layer wind tunnel. Complex 3D flow structures like secondary vortexes found in the physical model could be simulated, and the calculated velocity components agree reasonably well with LDV measurements from the wind tunnel experiment. The effect of changing source design, source emission rate or wind speed can be simulated, and the resultant changes in the flow field can be predicted.

Standard k–ε model and RNG turbulence modeling give almost the same results for the calculated flow pattern within a street canyon that is bounded by sharp edged buildings. Discrepancies between results from wind tunnel experiments and numerical simulations were found for the predicted concentration fields in a street canyon when simulating car exhaust dispersion. Independent of the turbulence model used the concentration results show differences up to 90%. Using numerical codes like Fluent can help to design and setup wind tunnel experiments; hence reducing the time required to optimize a physical model and expensive pre-runs in a wind tunnel. With a numerical simulation critical points like source design for dispersion simulation can be examined and boundary conditions can be modified.

Acknowledgements

The authors wish to thank Dr. Stilianos Rafailidis (University of Hamburg, Germany) for providing the model drawings and wind tunnel test results for comparisons that made this study possible. Also the support by the Alexander von Humboldt Foundation is gratefully acknowledged.

References

[1] R.N. Meroney, M. Pavageau, S. Rafailidis, M. Schatzmann, Study of line source characteristics for 2-D physical modelling of pollutant dispersion in street canyons, J. Wind Eng. Ind. Aerodyn. 62 (1996) 37–56.

[2] M.C. Murphy, A.E. Davies, Wind tunnel modelling of vehicle emissions on roadways as line sources, in: Proc. Air Pollution Control Association Conf., Dallas, TX, June 1988.

[3] S. Rafailids, M. Pavageau, M. Schatzmann, Wind tunnel simulation of car emission dispersion in urban street canyons, in: Annalen der Meteorologie, Deutsche Meteorologische Gesellschaft, Munich, 1995.

[4] P. Leisen, Windkanaluntersuchungen zur Simulation von Imissionssituationen in verkehrsreichen Straßenschluchten, aus Kolloqiumsbericht Abgasimmissionsbelastungen durch den Kraftfahrzeugverkehr in Ballungsgebieten und im Nahbereich verkehrsreicher Straßen, Verlag TÜV Rheinland, October 1981, pp. 207–234.

[5] Fluent User's Guide, vols. 1–4, Fluent Inc., Centerra Resource Park, Lebanon, NH.
[6] V. Yakhot, S.A. Orzag, Renormalization group analysis of turbulence – I. Basic Theory, J. Sci. Comput. 1 (1) (1986) 1–51.
[7] H. Van Oort, B. Stork, Vergleich von CFD-Berechnungen und Windkanalmessungen der Umströmung einfacher geometrischer Körper, in: E. Plate et al. (Eds.), Windprobleme in dichtbesiedelten Gebieten. D-A-CH '93, Karlsruhe, 1993.

ELSEVIER

Journal of Wind Engineering
and Industrial Aerodynamics 67&68 (1997) 305–311

JOURNAL OF
wind engineering
AND
industrial
aerodynamics

An investigation of the ventilation of a day-old chick transport vehicle

A.D. Quinn[a,*], C.J. Baker[b]

[a] Bio-Engineering Division, Silsoe Research Institute, Wrest Park, Silsoe, Bedfordshire MK45 4HS, UK
[b] Department of Civil Engineering, The University of Nottingham, University Park, Nottingham NG7 2RD, UK

Abstract

The increasing size and complexity of road vehicles used for the transport of day-old chicks has raised concerns about the thermal environment achieved by the ventilation systems, within the load space of such transporters. To address the lack of scientific information available concerning such designs, given the high cost of these vehicles, CFD modelling would seem a viable option. This paper presents the comparison of full scale experimental and CFD model results concerning the air flow with and without heat load in the load space of one particular vehicle. The results show that CFD modelling can predict the mean ventilation flows successfully, but experimental turbulence levels are not reproduced. The numerical results are also sensitive to the boundary conditions assumed for the air conditioning system, making commercial implementation of methods of this type for design purposes potentially problematic at this time.

Keywords: CFD; Validation

1. Introduction

Computational fluid dynamics (CFD) has successfully been applied to the ventilation problems encountered in office buildings and livestock buildings [1–4]. These situations can generally be described as open space problems in which only certain specific zones of the domain are inhabited and air flow blockage is minimal. Studies into more complex situations, such as the ventilation of passenger vehicles, have also been undertaken with some success but have largely been un-validated [5–7]. The purpose of this study was to investigate the potential of using a standard commercial CFD package in a closed and partially blocked situation in which uniformity of conditions is required throughout the domain. The assessment of this model was done

* Corresponding author. E-mail: andrew.quinn@bbsrc.ac.uk.

PII S0167-6105(97)00081-0

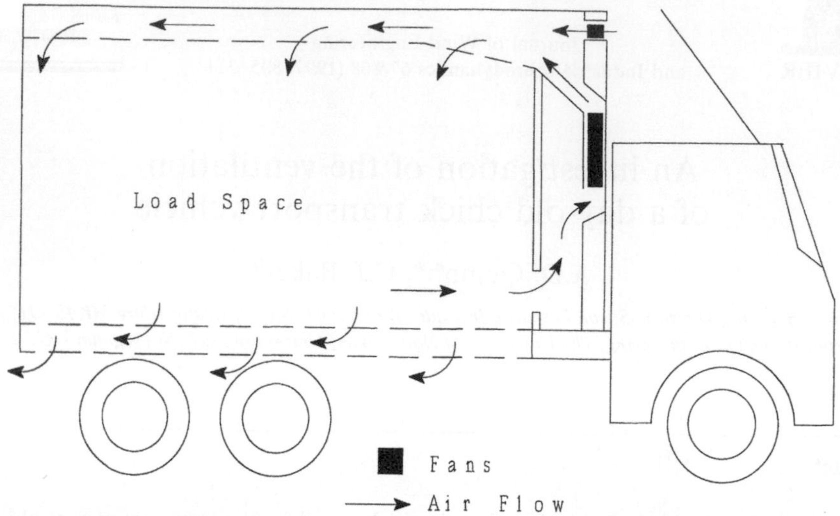

Fig. 1. A diagram of the ventilation system in the vehicle chosen for investigation. The modelled areas consisted of the load space and fan system in the experiments and the load space only in the numerical simulations.

using a comparison method based on statistics from full scale model measurements. The actual level of conditions, for example the dry-bulb temperature, required by neonatal poultry is a matter of physiological debate and therefore not discussed here, it is merely assumed that at whatever level the conditions are required, this level must be uniform throughout the load.

The design of vehicles used for the transport of day-old chicks has not been based on scientific study, rather a small number of designs has been developed by trial and error, over the past thirty years and these are adapted to suit individual situations [8–10]. The problems of providing a suitable ventilation system, capable of maintaining the required uniformity of conditions for neonatal poultry [11], have however been highlighted by the increasing size of vehicles being brought into service. This has emphasized the need for detailed understanding of the ventilation systems in use [12]. For this project a specific current design of transport vehicle was chosen for study, the ventilation system of which is illustrated in Fig. 1. The load space of this vehicle is 7.2 m long 2.35 m high and 2.4 m wide with a capacity of 576 chick boxes each normally holding 100 chicks.

The objectives of this project were to assess the ventilation in the load space of a day-old chick transport vehicle and to use this information in the development of a CFD simulation using PHOENICS software (v1.6.6). In this study the experimental work was carried out with four loading configurations of empty chick boxes to simulate load blockage, in a stationary full scale model of the vehicle load space with ventilation system as in the real vehicle (Fig. 1). The CFD study modelled the isothermal experimental set-ups for consistency.

2. Methodology

The experimental measurements of air velocity were collected in the model load space using a three component ultrasonic anemometer (Gill Solent, 20 Hz sampling rate at $\pm 0.02\ \mathrm{ms}^{-1}$ resolution up to $30\ \mathrm{ms}^{-1}$). These results were then analyzed for the mean and fluctuating components in terms of turbulent kinetic energy (TKE) and spectral composition. The loading configurations of empty chick boxes used were: an empty load space with no chick boxes; a one half load of chick boxes stacked to the front of the load space; a one half load of chick boxes stacked to the sides of the load space and a full load of chick boxes. Approximately 200 experimental measurements were made in each case except for the fully loaded where 75 measurements were made because of the difficulties of placing the instrument among the full load of chick boxes. CFD results of the same situations were obtained from a steady state, isothermal simulation with standard k–ε turbulence model and hybrid differencing scheme (minimum grid $18 \times 17 \times 29$ cells for the load space only). This simulation used a bulk porosity momentum sink to model the chick box load and a simple bird heat model was also included in the final simulations but was not experimentally validated. For the isothermal experimental measurements and simulations a flow rate of $3800\ \mathrm{m}^3\ \mathrm{h}^{-1} \pm 10\%$ was used which was entirely made up of recirculated air from the load space. In the heated simulations this was increased to $5800\ \mathrm{m}^3\ \mathrm{h}^{-1}$ to include the increased flow from the fresh air input fans and the porous floor panels and ducts were included.

These sets of data were used to develop a statistical comparison method for quantitative validation of the CFD model. This method involves the comparison of the predicted value of all measured variables at each experimental location with the experimental data for that location, taking into account the repeatability of the experimental data. In this way an unbiased estimate of acceptability can be calculated which can be used to give a comparative measure of goodness of fit for a series of simulations. For example, consider an experimental measurement of velocity made at position x. The experimental mean being U and the standard deviation of repeated measurements of U being σ. If the numerical simulation predicts velocity U_C for location x then an estimate of the CFD error for this location is $(U_C - U)/\sigma$. This error can then be assessed in terms of the Student's t-test distribution of errors, over the entire domain; i.e. if there exists one repeat of every experimental data point then 80% of errors will lie in the range $\pm 3.08\sigma$ for a normally distributed variable. If therefore such a percentage of the simulated values, for the experimental locations, do lie in the given range of σ then it can be said that such predictions are statistically indistinguishable from the experimental data. This does not remove the need for other qualitative validation techniques to verify flow features but does give a quantitative method for comparisons.

3. Results

The air velocity distribution within the load space was found to be a complex three dimensional composition of recirculating flow, jet impingement and turbulent features

[13]. Figs. 2–5 show the airflow pattern along a cross-section of the load space 0.7 m from the side wall for the experimental and numerical models in the empty and front half loaded cases respectively. The presence of the load, in all the various configurations, had a channelling effect on the air flow through the load space, with significant amounts of air bypassing the chick boxes and being recirculated. The implications of this on the ventilation rate achieved within the boxes was clear in the distribution of temperature in the heated simulations (Fig. 6) where peak temperatures were found in the front central boxes with cooler air bypassing the load. Another significant feature seen in the results was the cooler air entering from beneath the vehicle at the front of the load space in the fully loaded configuration. This undesirable effect reduces the flow through the load and could allow the ingress of exhaust fumes.

Using the comparison technique outlined above it was found that the predictions of mean velocity from the CFD simulations were broadly in accordance with the experimental results, with the large majority within the experimental variability of the observed values (Table 1). The variation in the agreement with the various loading configurations is possibly due to the variation in number and location of the

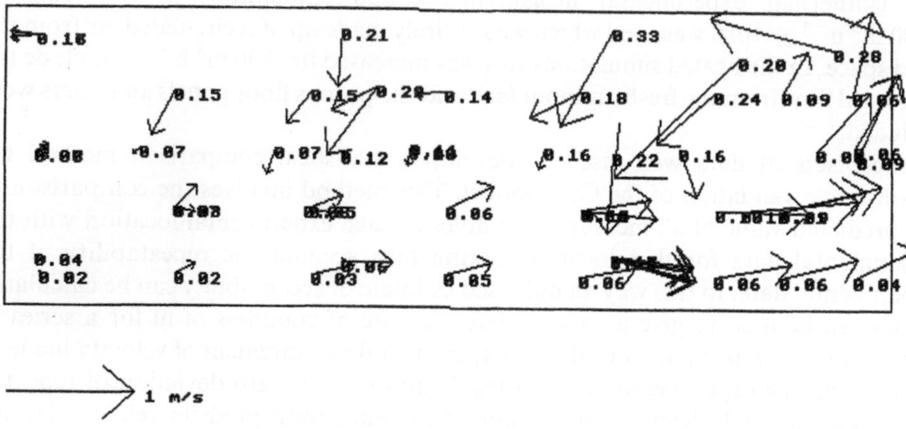

Fig. 2. Experimental results for an empty load space. The vectors indicate air velocity with the TKE value at the base (J kg^{-1}).

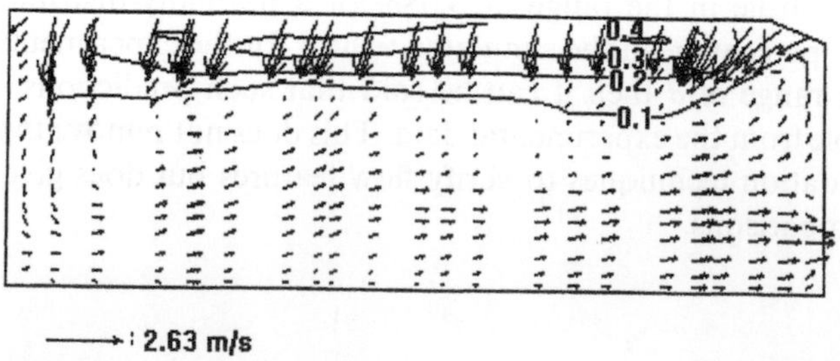

Fig. 3. CFD results for an empty load space.The vectors indicate air velocity with the TKE value as contours (J kg^{-1}).

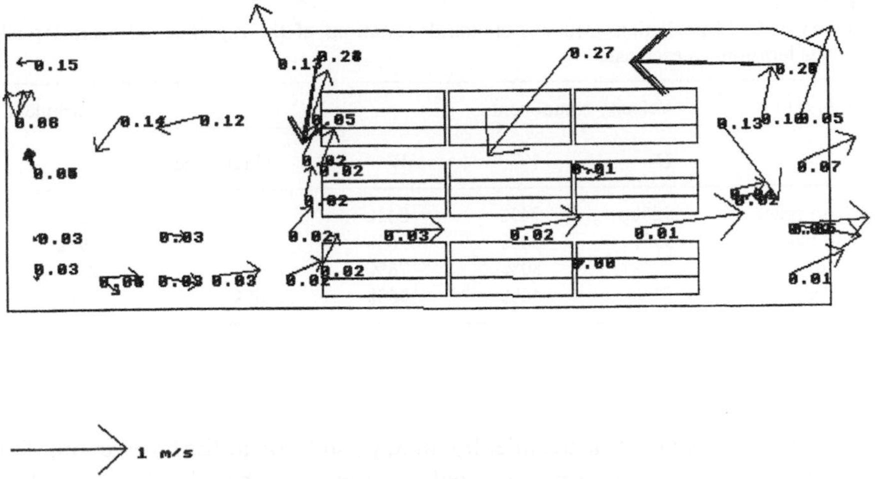

Fig. 4. Experimental results for the front half loaded load space case.

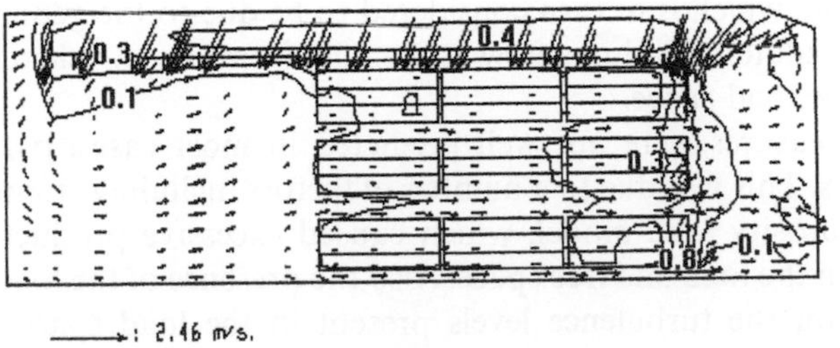

Fig. 5. CFD results for the front half loaded load space case.

Fig. 6. Temperature distribution for the fully loaded, heated model load space simulation.

experimental data collected, which highlights the large amount of data required for meaningful validation of such a complex flow. For each experimental case a number of simulations were undertaken to assess the sensitivity to grid specification, boundary conditions and load description [13]. These showed that the most significant

Table 1
The fraction of CFD predictions within 3 standard deviations of the experimental mean for all the experimental locations in each case

Simulation case	Velocity component				Turbulent KE
	U	*V*	*W*	Magnitude	
Empty load space	92%	74%	84%	76%	66%
Front half load	99%	88%	96%	89%	39%
Side half load	80%	80%	76%	61%	28%
Fully loaded	54%	46%	46%	46%	0%

factor controlling the flow was the inlet boundary condition at the false ceiling. This is illustrated by the flow patterns predicted for the front half loaded case (Fig. 5) compared to the measured pattern in Fig. 4. At the rear of the load space there is a clear difference in the direction of flow and the subsequent recirculation zones. This is considered to be due to the over estimate of the mass flow through the rear section of the false ceiling, which is in turn considered to be due to the geometric simplifications used to calculate the volume flow at the inlet from the jet velocities measured in the model vehicle load space.

Turbulence measures were not well predicted in most cases, particularly in the loaded situations. This was due to a number of factors including streamline curvature and the bulk porosity load model, which caused excessive production of k at the interface between the load and free space. Also the presence of the load had an overall damping effect on the turbulence levels present in the load space which was not reflected by the simulation results.

The simple heat load model included in the heated load space simulations gave a possibly realistic spacial distribution of temperature but further validation would be necessary to determine the accuracy of the absolute temperatures predicted. Further information about the dynamic heat and moisture production of day-old chicks would be necessary to improve these predictions.

4. Conclusions

CFD simulation of this situation has proved to be possible, giving agreement with experimental data, for the velocity components, to within the experimental variability in three of the four cases studied. Particular care was required in the specification of boundary conditions for this type of flow because of the sensitivity of the solution to the description of the inlet jets. Validation of this type of complex flow can be undertaken using the method described to give a statistical measure of goodness of fit to a set of experimental data, although this data must be extensive if the results are to be meaningful.

References

[1] S. Murakami, S. Kato, Numerical and experimental study on room airflow – 3D predictions using the k–ε turbulence model, Building Envir. 24 (1) (1989) 85.

[2] M.G. Yost, R.C. Spear, Measuring indoor airflow patterns by using a sonic vector anemometer, Amer. Ind. Hygiene Assoc. J. 53 (11) (1992) 677.

[3] D. Hope, D. Milholland, A three dimensional ultrasonic anemometer to measure the performance of clean zone air delivery systems, J. IES. 36 (6) (1993) 32.

[4] C.R. Boon, M. Andersen, B.B. Harral, Dynamics of particulate pollutants in an experimental livestock building, ASAE paper no. 94-4586, 1994.

[5] T. Han, Three-dimensional Navier–Stokes simulation for passenger compartment cooling, Int. J. Vehicle Design 10 (2) (1989) 175.

[6] J.W. Wan, J. van der Kooi, Influence of the position of supply and exhaust openings on comfort in a passenger vehicle, Int. J. Vehicle Design 12 (5/6) (1991) 588.

[7] C.H. Lin, T. Han, C.A. Koromilas, Effect of HVAC design parameters on passenger thermal comfort. SAE paper 920264, 1992.

[8] R.H. Hinds, Baby chick transportation – problems and equipment. US Department of Agriculture Report no. 267, 1958.

[9] S. Banks, Anatomy of a chick transporter. World Poultry, September 1983, p. 22.

[10] Anon., Comfort and style for chicks on the move. Poultry World, April 1988, p. 17.

[11] M.A. Michell, The thermal micro-environment experienced by one-day old chicks during road transportation, Unpublished confidential report, Roslin Institute, Edinburgh, 1996.

[12] J. Tamlyn, J.R. Starr, Monitoring environment during transportation of day-olds, Int. Hatchery Practice 1 (6) (1987) 11.

[13] A.D. Quinn, The ventilation of a chick transport vehicle, Ph.D. thesis, The University of Nottingham, 1996.

ELSEVIER

Journal of Wind Engineering
and Industrial Aerodynamics 67&68 (1997) 313–321

JOURNAL OF wind engineering AND industrial aerodynamics

The effect of wind profile and twist on downwind sail performance

P.J. Richards[1]

Department of Mechanical Engineering, The University of Auckland, Private Bag 92019, Auckland, New Zealand

Abstract

The flow over a pair of downwind sails, a mainsail and spinnaker, is successfully modelled using the finite volume package CFDS-FLOW3D. An investigation of the effects of wind twist on the performance of the sails tends to suggest that the major effect is the result of the modified wind profile in the case of twisted flow. The thrust and side force coefficients for logarithmic flow are shown to be similar to published data obtained from wind tunnel tests. Flow visualisation illustrates how the flow is primarily attached to the sails at an apparent wind angle of 100° but is almost completely separated at an angle of 140°.

PACS: 47.11
Keywords: Computational modelling; Downwind sails; Spinnaker

1. Introduction

It is widely recognised by the wind engineering community that in order to obtain realistic results from model tests it is necessary to accurately model both the object being studied and the characteristics of the wind affecting it. For this reason wind tunnel studies of buildings, etc. are almost always conducted in boundary layer wind tunnels which model both the velocity profile and the turbulence of the wind. In studying the wind loads on moving objects, such as yachts, the situation is even more complex since the relative air movement is a combination of the reversed boat velocity and the wind velocity. The vector addition of the components results in a relative wind profile that varies in both magnitude and direction (twist) with height. This phenomenon is not very significant when sailing upwind but can be quite marked when sailing downwind. The need to model both the velocity profile and the wind twist has been

[1] E-mail: pj.richards@auckland.ac.nz.

PII S0167-6105(96)00082-2

recognised by Flay et al. [1] in their development of a twisted flow wind tunnel. This facility was developed as part of a collaborative programme between the University of Auckland, Team New Zealand and North Sails (NZ) Ltd. and was extensively used in the development of downwind sail during Team New Zealand's successful challenge for the America's Cup in 1995.

At the same time that new wind tunnel techniques were being developed at the University of Auckland an investigation was initiated into computational modelling of the flow around downwind sails [2,3]. Although computational models for upwind sails have been established for some centuries the potential flow methods generally applied to these situations cannot be easily used for downwind sails where the twist in the onset flow and the interaction between the sails are important, and large regions of separated flow exist. Earlier research conducted by Richards and Hoxey [4,5] had shown that it is possible to use finite volume codes such as PHOENICS to model both the pressure fields and flow patterns around a low-rise building, the Silsoe Structures Building. Such situations also exhibit large regions of separated flow and complex three-dimensional flow patterns. Many of the lessons learnt in this earlier work have been used while applying another finite volume code, CFDS-FLOW3D, to the downwind sail situation. An example of the resulting computed flow around and between a spinnaker and mainsail combination is shown in Fig. 1. The situation shown is that for an apparent wind angle of 140°, at which point the flow

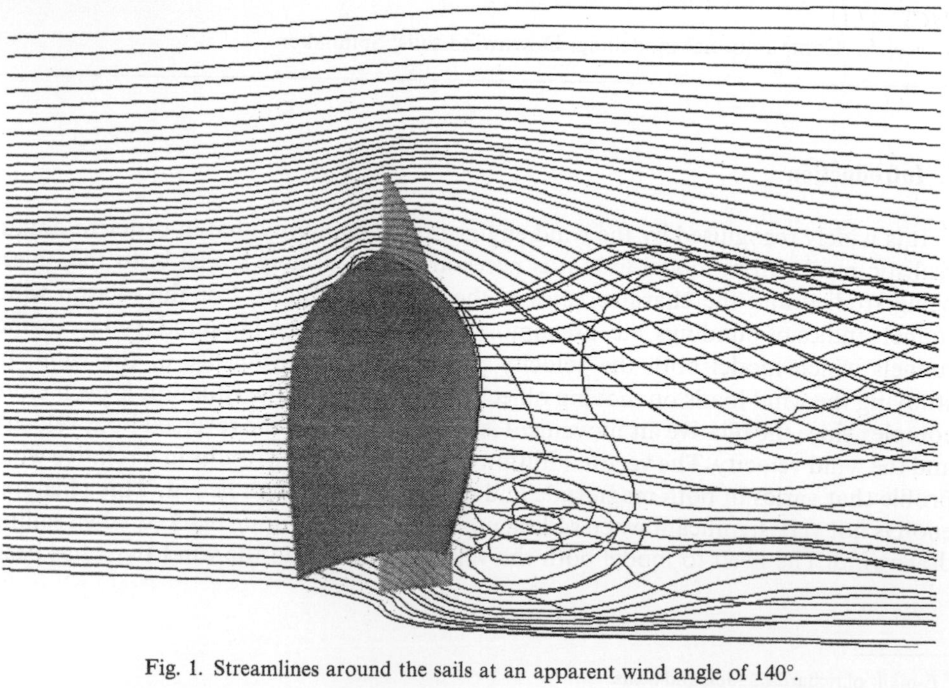

Fig. 1. Streamlines around the sails at an apparent wind angle of 140°.

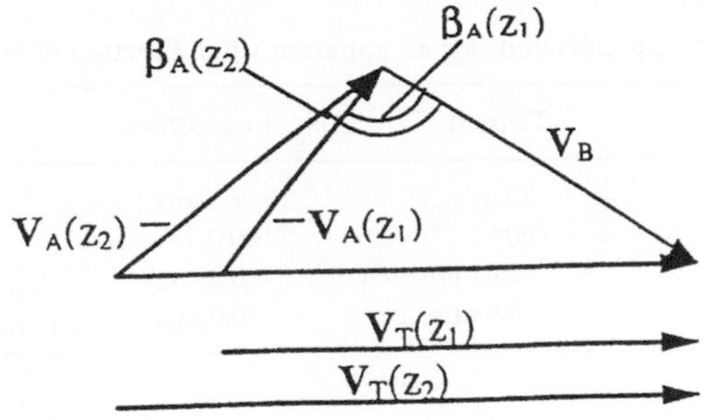

Fig. 2. Apparent wind velocities with wind twist.

is mostly separated from the leeward side of both sails. A strong vortex is seen to be shed from the head of the spinnaker and various recirculating regions exist in the complex wake structure.

2. Twisted flow

In modelling the flow around yacht sails the grid is fixed to the yacht and the flow computed is the flow relative to the yacht, the apparent wind V_A. This relative flow results from the combinations of the yacht's motion V_B and the air's motion, the true wind V_T. These are related by the vector equation (1):

$$V_A = V_T - V_B. \tag{1}$$

Twisted onset flow occurs because the true wind speed varies with height whereas the yacht's velocity is a fixed quantity. Fig. 2 illustrates this situation where the true wind speed at height z_2 is greater than that at height z_1 and as a result the two apparent wind velocities differ in both magnitude and direction.

3. Computational study specifications

In order to investigate the significance of the wind profile and twist on the performance of downwind sails a computational study has been conducted which compares the behaviour of a mainsail – spinnaker combination under different onset flow conditions. The conditions tested were a logarithmic velocity profile with no twist and twisted flow, which was modelled by combining a logarithmic wind profile with the reversed boat speed. In both cases the true wind profile was assumed to fit a simple log law with a surface roughness length $z_0 = 0.001$ m,

$$V_T = u_* \ln(z/z_0 + 1.0)/K, \tag{2}$$

Table 1
Velocity specifications for an apparent wind direction of 140°

	Twisted	Logarithmic
V_T	12 m/s	7.5 m/s
β_T	160°	140°
u_*	0.47 m/s	0.29 m/s
V_B	5.0 m/s	0.0 m/s

where K is the von Karman's constant ($K = 0.4$), u_* the friction velocity, z the height above sea surface and z_0 the surface roughness length.

The particular form of Eq. (2) is used in order to give a zero velocity at the $z = 0$ plane, which represents the mean sea surface. Over open water z_0 varies between 0.1 and 10 mm, and so for this study a value corresponding to moderate wind speeds has been used.

In all tests the apparent wind speed at the top of the mast ($z = h = 27.4$ m) was 7.5 m/s and in the twisted flow case the boat speed was 5 m/s. For each apparent wind direction the spinnaker was set with the pole, which goes between the mast and the bottom windward corner of the spinnaker, perpendicular to the wind at pole height ($z = 3$ m). The apparent wind direction used in this paper is the angle between the apparent wind at spinnaker pole height ($z = 3$ m) and the boat centreline. Table 1 gives the consequent velocity specifications for both twisted and logarithmic onset flow for the case of an apparent wind direction of 140°.

The k–ε turbulence model was used with the boundary conditions specified in the manner recommended by Richards and Hoxey [6]. As a result the inlet turbulence property profiles were:

$$k = 3.33\, u_*^2, \tag{3}$$

$$\varepsilon = u_*^3/K(z + z_0). \tag{4}$$

With twisted flow the sea surface was treated as a moving wall with a velocity $-V_B$. The sails used in this model were based on a pair of sails designed by North Sails (NZ) Ltd. for the Whitbread 60 round the world yacht Winston. The mainsail boom, bottom of the mainsail, was set at an angle of 55° to the boat centreline in all cases.

The finite volume code CFDS-FLOW3D was used in this investigation. The body-fitted grid was generated by a locally written program which takes as its input sail shape data supplied by a sail design program and then constructs a grid around the sails. The grid contained 52 × 33 × 36 cells. The solution domain was rectangular with a length of 90 m in the primary flow direction, 60 m wide and 57.4 m high. Force calculations were achieved by summing over each sail the product of the pressure difference across and the area vector for each cell face that was part of a sail. Graphical post-processing was conducted using the flow visualisation package SeeFD developed by Mallinson [7].

4. Results and discussion

The thrust and side force coefficients obtained for a range of apparent wind angles under the two onset flow conditions are shown in Fig. 3. In this context the thrust force is the horizontal component of the resultant aerodynamic force acting forwards along the boat centreline and the side force is the horizontal component perpendicular to the centreline. In converting these forces into coefficients the force has been divided by the dynamic pressure based on the apparent wind speed at the top of the mast and by the total sail area. It may be noted that the most significant effect of including wind twist is a reduction in thrust coefficient at the higher apparent wind angles. Although these differences result from changes in both the apparent wind's magnitude and direction, examination of these situations suggests that it may be the changes in magnitude that are most significant. Figs. 4 and 5 show the variation in apparent wind direction and apparent wind dynamic pressure ($q_A = 0.61V_A^2$) with height for the case where the apparent wind direction at spinnaker pole height ($z = 3$ m) is 140°. For the twisted flow situation the difference between the apparent wind velocity directions at the top of the mast ($z = 27.4$ m) and pole level ($z = 3$ m) is only 7°, however with twisted flow the apparent wind dynamic pressure is significantly smaller at lower height compared with that for the logarithmic flow. It should be noted that although the amount of wind twist is only moderate with these cases, where it has been assumed that the yacht speed is two thirds of the apparent wind speed at the top of the mast, it is likely to be even more significant with IACC (International America's Cup Class) yachts which can achieve boat speeds much closer to the apparent wind speed.

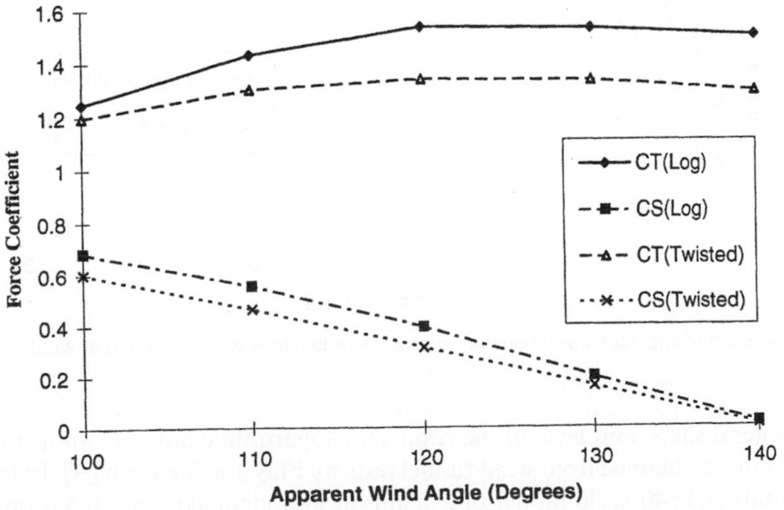

Fig. 3. Thrust and side force coefficients for logarithmic and twisted flow.

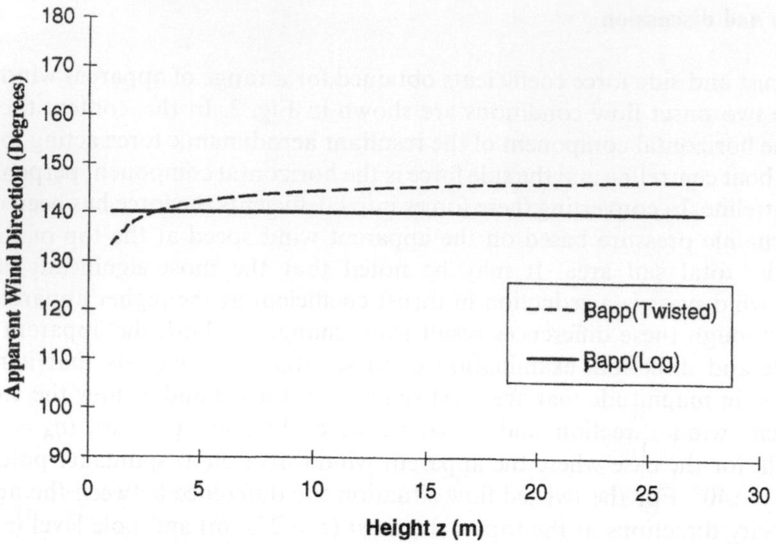

Fig. 4. Variation of apparent wind angle with height – nominal apparent wind angle 140°.

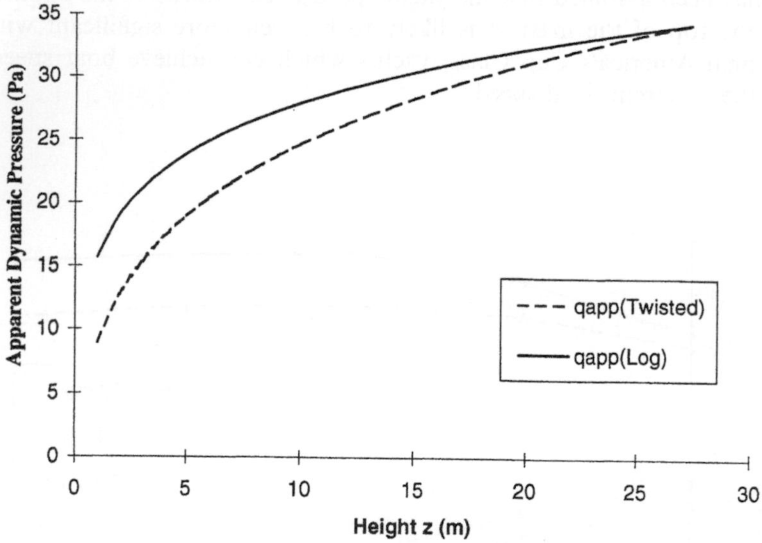

Fig. 5. Variation of apparent wind dynamic pressure with height – nominal apparent wind angle 140°.

The general shape and level of the results for logarithmic flow shown in Fig. 3 are similar to those obtained from wind tunnel tests by Flay and Vuletich [8]. In this wind tunnel study a 1 : 40 scale model of a mainsail and spinnaker combination for the 1986/7 America's Cup 12-Metre Class yacht KZ-7 was tested in both logarithmic and uniform flow at two wind speeds and with two slightly different spinnakers. The logarithmic flow results showed the thrust coefficient to be between 1.05 and 1.25 at 100° and increasing to 1.5–1.6 at 140°. The corresponding computed results from this study are 1.25 and 1.5 at 100° and 140° respectively. Similarly the wind tunnel values for the side force coefficient decrease from 0.6–0.7 at 100° to 0.0–0.15 at 140° while the computed results decrease from 0.68 to 0.01 over the same range. Both the wind tunnel study [8] and the computational study by Hedges et al. [3] showed that testing in uniform flow gave unrealistically high thrust coefficients.

The variations in the forces contributed by the two sails are illustrated in Fig. 6 which gives the lift and drag coefficients for the mainsail and spinnaker separately. In this context the drag force is defined as the component of the resultant aerodynamic force parallel to the apparent wind velocity at spinnaker pole height ($z = 3$ m) and the lift force is the horizontal component perpendicular to the drag. Since, in keeping with normal sailing practice, the spinnaker pole has been set perpendicular to the apparent wind for all apparent wind angles the forces on the spinnaker are relatively constant. However the angle between the apparent wind and the mainsail does change significantly. At an apparent wind angle of 100° the flow over both the mainsail and spinnaker, as illustrated in Fig. 7, is reasonably attached to the sails and significant lift is generated by both sails. At this apparent wind angle it may be noted that the primary contribution to the thrust is from the lift force. In contrast, at an apparent wind angle of 140°, the flow, as illustrated in Fig. 8, is only attached to the leeward side of the spinnaker over a small area and is completely separated from the mainsail. At this wind angle both the lift and the drag

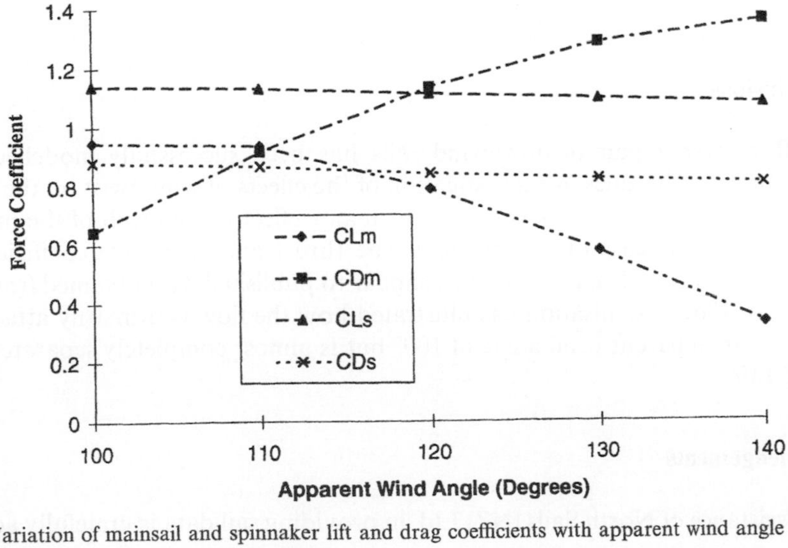

Fig. 6. Variation of mainsail and spinnaker lift and drag coefficients with apparent wind angle – twisted onset flow.

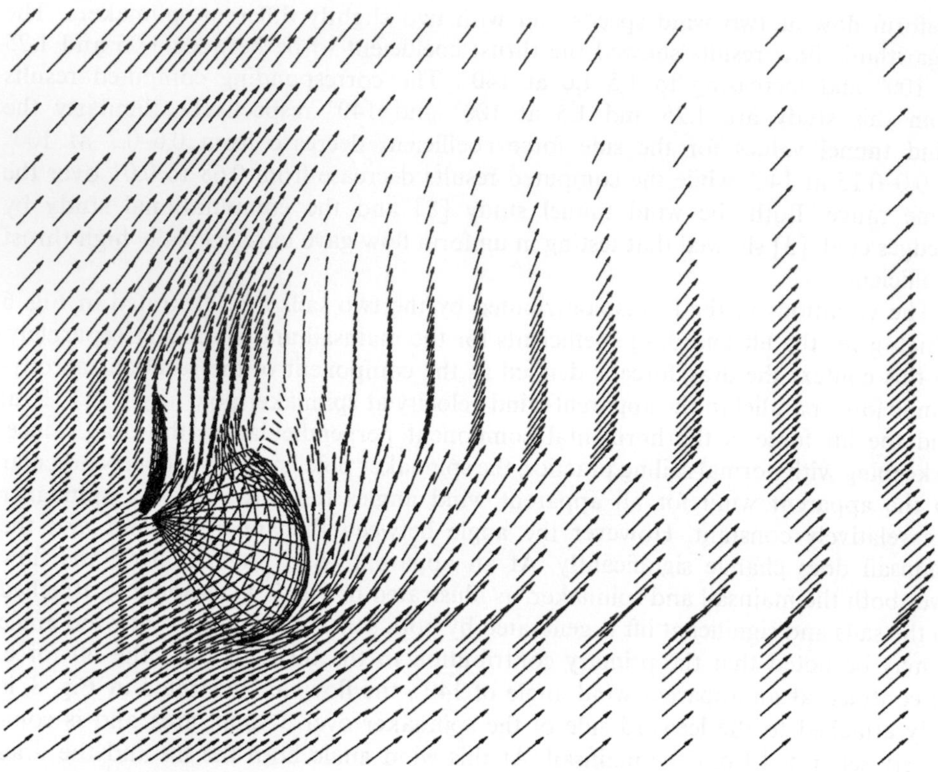

Fig. 7. Visualisation of flow vectors at a height of 6 m and an apparent wind angle of 100°.

forces make contributions to the thrust while their side force contributions tend to cancel.

5. Conclusions

The flow over a pair of downwind sails has been successfully modelled using computational techniques. An investigation of the effects of wind twist on the performance of the sails tends to suggest that the major effect is the result of the modified wind profile in the case of twisted flow. The thrust and side force coefficients for logarithmic flow have been shown to be similar to published data obtained from wind tunnel tests. Flow visualisation has illustrated how the flow is primarily attached to the sails at an apparent wind angle of 100° but is almost completely separated at an angle of 140°.

Acknowledgements

The assistance of North Sails (NZ) Ltd. in providing sail data is gratefully acknowledged.

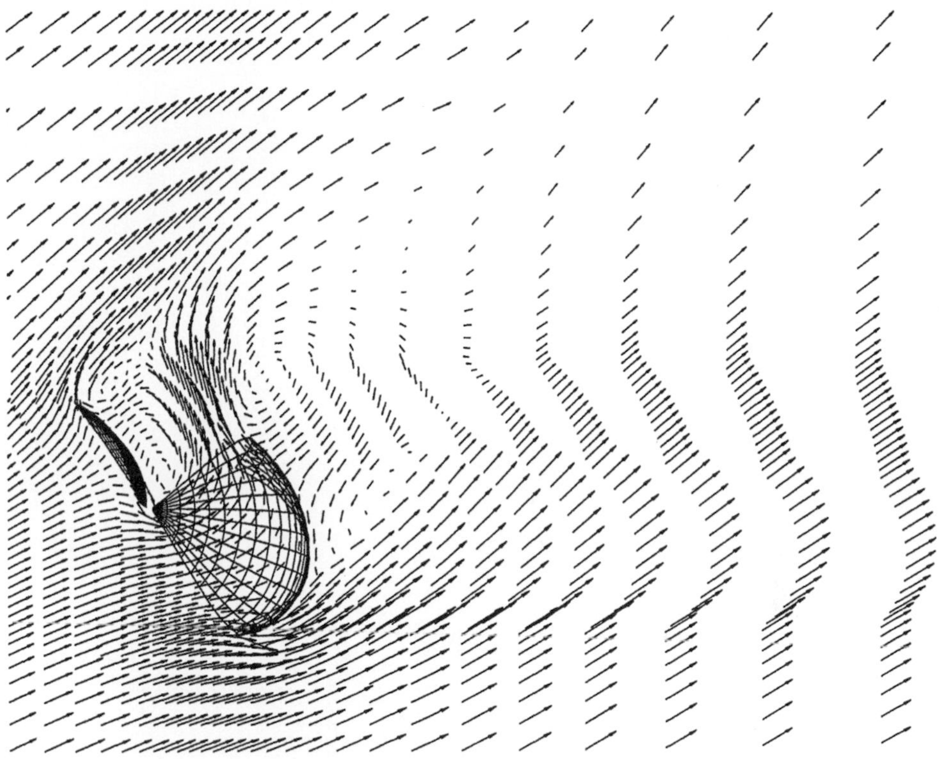

Fig. 8. Visualisation of flow vectors at a height of 6 m and an apparent wind angle of 140°.

References

[1] R.G.J. Flay, N.J. Locke, G.D. Mallinson, New Zealand development of a twisted flow for testing yacht sails, in: Proc. 9th Internat. Conf. on Wind Eng. New Delhi, January 1995.

[2] K.L. Hedges, Computer modelling of downwind sails, ME Thesis, The University of Auckland, NZ, 1993.

[3] K.L. Hedges, P.J. Richards, G.D. Mallinson, Computer modelling of downwind sails, J. Wind Eng. Ind. Aerodyn. 63 (1996) 95–110.

[4] P.J. Richards, R.P. Hoxey, Computational and wind tunnel modelling of mean wind loads on the Silsoe Structures Building, J. Wind Eng. Ind. Aerodyn. 41–44 (1992) 1641–1652.

[5] R.P. Hoxey, P.J. Richards, Flow patterns and pressure field around a full-scale building, J. Wind Eng. Ind. Aerodyn. 50 (1993) 203–212.

[6] P.J. Richards, R.P. Hoxey, Appropriate boundary conditions for computational wind engineering models using the k–ε turbulence model, J. Wind Eng. Ind. Aerodyn. 46 (1993) 145–153.

[7] G.D. Mallinson, SeeFD User Guide, Department of Mechanical Engineering, The University of Auckland, NZ, 1993.

[8] R.G.J. Flay, I.J. Vuletich, Development of a wind tunnel test facility for yacht aerodynamic studies, J. Wind Eng. Ind. Aerodyn. 58 (1995) 231–258.

Structural Response

ELSEVIER

Journal of Wind Engineering
and Industrial Aerodynamics 67&68 (1997) 325–335

JOURNAL OF wind engineering AND industrial aerodynamics

Numerical study on the suppression of the vortex-induced vibration of a circular cylinder by acoustic excitation

S. Hiejima[a,*], T. Nomura[b], K. Kimura[c], Y. Fujino[c]

[a] *Department of Civil Engineering, Kyushu University, 6-10-1, Hakozaki, Higashi-ku, Fukuoka 812, Japan*
[b] *Department of Civil Engineering, Nihon University, 1-8, Kanda-Surugadai, Chiyoda-ku, Tokyo 101, Japan*
[c] *Department of Civil Engineering, University of Tokyo, 7-3-1, Hongo, Bunkyo-ku, Tokyo 113, Japan*

Abstract

In order to investigate the possibility of suppressing flow-induced vibration of structures by applying periodic excitation such as a sound wave to the flow around the structures, numerical simulations of the vortex-induced vibration of a circular cylinder were carried out when a periodic velocity excitation was applied to the flow at two locations on the cylinder surface. A FEM based on the ALE formulation was employed to simulate the vortex-induced vibration, with the time integration being carried out by the predictor–corrector method. The results revealed that the excitation with the transition wave frequency is the most effective in changing the flow characteristics around the cylinder and the characteristics of the vortex-induced vibration. These changes appear to be caused by the promotion of the vortex growth in the early vortex formation behind the cylinder.

Keywords: ALE; Circular cylinder; FEM; Instability; Periodic excitation; Shear layer; Vortex-induced vibration

1. Introduction

It is known that the shear layers which separate from the surface of a bluff body like a circular cylinder form periodic large-scale vortices behind the body. The large-scale vortices cause the vortex-induced vibration when the vortex-shedding frequency coincides with the natural frequency of the elastically supported body. The vortex-induced vibration of a bluff body such as a circular cylinder is well known as a typical flow-induced vibration phenomenon observed in relatively low wind speed range. Although the vortex-induced vibration has limited amplitudes and

* Corresponding author.

PII S0167-6105(97)00083-4

therefore is not catastrophic, it causes fatigue problems of structures or psychological discomfort in a vibrating structure.

The shear flows around a bluff body are sensitive to periodic excitation such as a sound wave with the specific frequency related to the convective instability of the shear layers. And strong fluctuations are easily induced in the separated shear layers when the periodic excitation with the specific frequency is applied to the flow around the body. Nishioka [1] reported that the acoustic excitation applied to the shear flow around an airfoil generates strong fluctuations, which increase the momentum transfer from the outer flow to the boundary layer, thereby suppressing the flow separation from the airfoil surface.

The previous experimental study by the authors [2,3] has showed that the vortex-induced vibration of a circular cylinder can be suppressed using the stimulation of the separated shear layers around the cylinder by an acoustic excitation with the frequency of the transition waves. The transition waves are known as the fluctuation generated by the instability of the separated shear layer around a circular cylinder [4,5], and the transition waves frequency is generally much higher than the vortex-shedding frequency of the cylinder.

In the present numerical simulations, a periodic velocity excitation placed at two locations on the surface of a circular cylinder is employed to stimulate the separated shear layers around the cylinder as an idealization of the acoustic excitation. The purpose of the present study is to investigate the effect of periodic velocity excitation on the flow around the cylinder, and to clarify the relation between the flow characteristics around the cylinder and the characteristics of the vortex-induced vibration when the periodic excitation is applied.

2. Computational method

2.1. ALE description of the Navier–Stokes equations

A finite element method based on the arbitrary Lagrangian–Eulerian formulation (ALE) [6,7] is employed to simulate the interaction between the elastically mounted rigid circular cylinder and the incompressible viscous flow. The ALE formulation is based on a mixed standpoint of the classical Lagrangian and Eulerian description. In the ALE formulation, the motion of the mesh for the fluid analysis is prescribed in accordance with the cylinder motion. The Navier–Stokes equations based on ALE formulation and incompressibility constraint are expressed as follows:

$$\rho \frac{\partial u_i}{\partial t} + \rho(u_j - \hat{u}_j)\frac{\partial u_i}{\partial x_j} = \frac{\partial \tau_{ij}}{\partial x_j} + f_i, \tag{1}$$

$$\partial u_i / \partial x_i = 0, \tag{2}$$

where ρ is the density, u_i is the material velocity, $\hat{u}_i$ is the mesh velocity describing mesh motion, τ_{ij} is the stress tensor, f_i is the body force. Notice that the convection velocity is replaced with the relative velocity $(u_i - \hat{u}_i)$ to the mesh velocity $\hat{u}_i$. The stress tensor τ_{ij} is given as

$$\tau_{ij} = -p\delta_{ij} + \mu\left(\frac{\partial u_i}{\partial x_j} + \frac{\partial u_j}{\partial x_i}\right), \tag{3}$$

where p is the pressure and μ is the dynamic viscosity.

The finite element equations are obtained by applying the streamline upwind/Petrov–Galerkin method [8] to the ALE Navier–Stokes equations with bilinear quadrilaterals for velocities and piecewise constants for pressure. The finite element equations of motion and continuity are written as follows:

$$\boldsymbol{Ma} + \boldsymbol{N}(\boldsymbol{v} - \hat{\boldsymbol{v}})\boldsymbol{v} - \boldsymbol{Gp} = \boldsymbol{f}, \tag{4}$$

$$\boldsymbol{G}^{t}\boldsymbol{v} = 0, \tag{5}$$

where $\boldsymbol{M}$ is the mass matrix, $\boldsymbol{N}(\boldsymbol{v} - \hat{\boldsymbol{v}})$ accounts for convection and viscosity, $\boldsymbol{G}$ is the gradient matrix, $\boldsymbol{a}$ is the acceleration vector, $\boldsymbol{v}$ is the material velocity vector, $\hat{\boldsymbol{v}}$ is the mesh velocity vector, $\boldsymbol{p}$ is the pressure vector, and $\boldsymbol{f}$ is the force vector. Notice that $\boldsymbol{M}$, $\boldsymbol{N}$ and $\boldsymbol{G}$ are time dependent.

2.2. The equation of motion of the rigid body

The circular cylinder mounted on an elastic spring as well as a dashpot is assumed to move only in the across-flow direction. The equation of motion of the cylinder is written as follows:

$$m\ddot{\delta} + c\dot{\delta} + k\delta = X, \tag{6}$$

where $\ddot{\delta}$, $\dot{\delta}$ and δ denote the acceleration, velocity, displacement of the cylinder in the across-flow direction. Then m, c and k are the mass, the damping and the stiffness, respectively. X represents the concentrated force corresponding to δ and is given by

$$X = -\sum_i f_i^b, \tag{7}$$

where f_i^b represents the across-flow direction force at the mesh node i which is located on the cylinder surface. Notice that the cylinder motion is coupled with the flow around the cylinder in terms of X. Then the fluid velocity at the mesh node i is coincident with the cylinder velocity.

2.3. Periodic velocity excitation on the cylinder surface

The periodic velocity excitation [9] to stimulate the separated shear layers is placed at two locations on the cylinder surface (Fig. 1). The angle (ϕ_a) between the upstream stagnation point and the periodic excitation points is fixed at 80°, which is determined

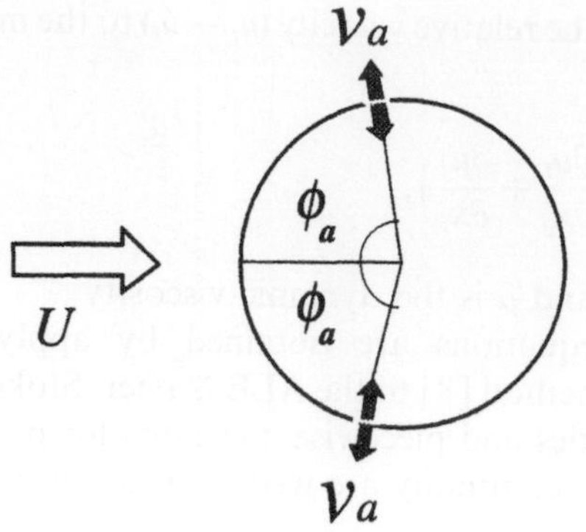

Fig. 1. Periodic velocity excitation on cylinder surface.

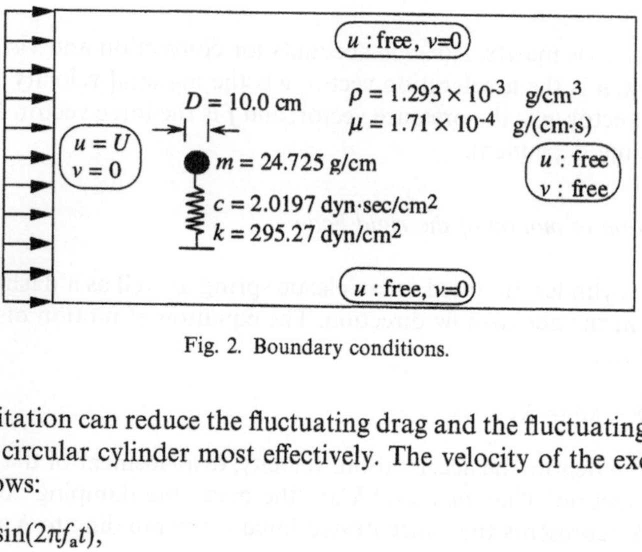

Fig. 2. Boundary conditions.

so that the excitation can reduce the fluctuating drag and the fluctuating lift acting on the stationary circular cylinder most effectively. The velocity of the excitation (v_a) is chosen as follows:

$$v_a = U_a \sin(2\pi f_a t), \tag{8}$$

where U_a and f_a represent the amplitude and the frequency of the periodic excitation, respectively. U_a is set at 10% of the approaching flow velocity (U), and the phases of v_a at two excitation points are the same.

2.4. Boundary conditions and finite element mesh

Fig. 2 shows the boundary conditions, the material properties of the fluid and the parameters of the cylinder–spring system. The approaching flow velocity U is set at 26.4 cm/s and the Reynolds number (Re) based on the cylinder diameter is equal to 2000. A preliminary computation of the flow around the stationary cylinder reveals that the vortex-shedding frequency (f_s) without the periodic velocity excitation is equal to 0.55 Hz at Re = 2000. The parameters of the cylinder–spring system are determined so that the natural frequency of the cylinder–spring system corresponds to

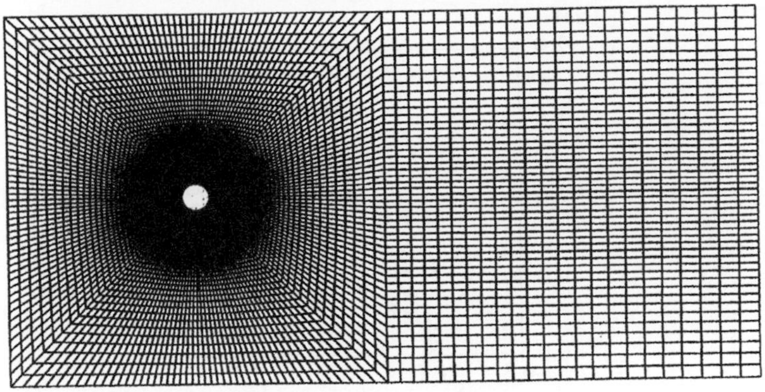

Fig. 3. Finite element mesh.

f_s, and therefore the vortex-induced vibration amplitude becomes the largest at Re = 2000.

Fig. 3 shows the finite element mesh used in the computation. The portion of the ALE mesh is restricted to the square region around the cylinder, and the remaining part is left as Eulerian in order to reduce the computational burden. The mesh velocities are equal to the cylinder velocity on the cylinder surface and zero on the external boundaries, and vary linearly in the region surrounded by the cylinder surface and the external boundaries.

The predictor–corrector method [8,10] is employed for the time integration, and the nondimensional time integration interval is set to 0.01.

3. Results

3.1. The effect of the periodic excitation on the flow around a stationary circular cylinder

The simulations of the flow around a stationary circular cylinder for various frequencies of the periodic velocity excitation revealed that the characteristics of the flow are changed most remarkably when the nondimensional frequency of the periodic excitation (f_a/f_s) is near 4.45, which corresponds to the transition wave frequency obtained by the regression curve of Bloor's experimental data [4]. Similarly, many experimental studies [11–18] reported that an acoustic excitation with the transition wave frequency is effective in changing the characteristics of the flow around a stationary circular cylinder. It is thought that the periodic excitation with the transition wave frequency grows into the strong fluctuation in the shear layer around a cylinder because of the shear layer instability related to the transition wave generation, and the strong fluctuation in the shear layer leads to the change of the flow characteristics around the cylinder. In the present simulation, analysis with respect to the growth rate of the applied periodic excitation has confirmed that the excitation at $f_a/f_s = 4.45$ is enhanced most effectively and grows into the strong fluctuation in the separated shear layer around the cylinder.

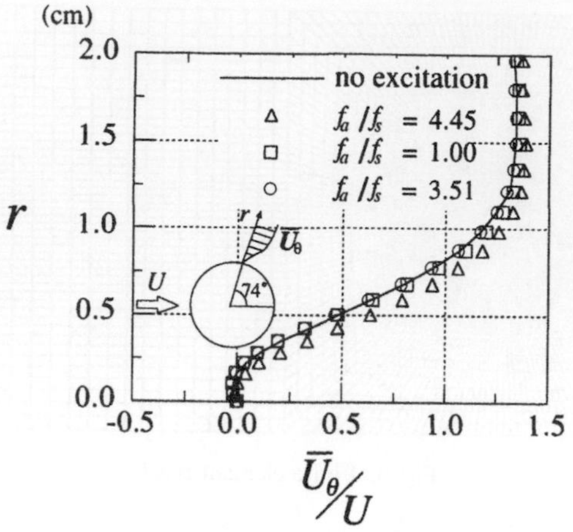

Fig. 4. Mean velocity in the circumferential direction (stationary cylinder).

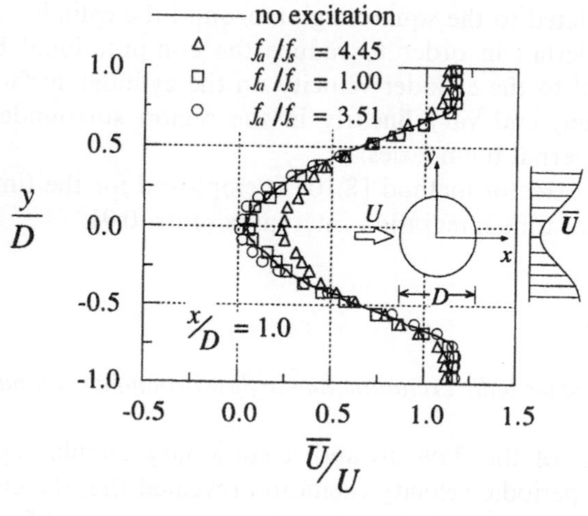

Fig. 5. Mean velocity in the flow direction (stationary cylinder).

Fig. 4 shows the simulation result of the mean velocity ($\overline{U}_\theta$) in the circumferential direction with the periodic excitation at $f_a/f_s = 4.45$, compared to the case of the periodic excitation at $f_a/f_s = 1.00$ and $f_a/f_s = 3.51$, which correspond to the vortex-shedding frequency and the transition wave frequency obtained by the regression curve of Wei's experimental data [5], respectively. As shown in Fig. 4, the periodic excitation at $f_a/f_s = 4.45$ has the effect of increasing $\overline{U}_\theta$ near the cylinder surface and reducing the boundary layer thickness, implying that the separation point of the boundary layer moves downstream. Also the mean velocity ($\overline{U}$) in the flow direction behind the cylinder was altered by the effect of the periodic excitation at $f_a/f_s = 4.45$ (Fig. 5). Fig. 5 shows that the velocity defect behind the cylinder is

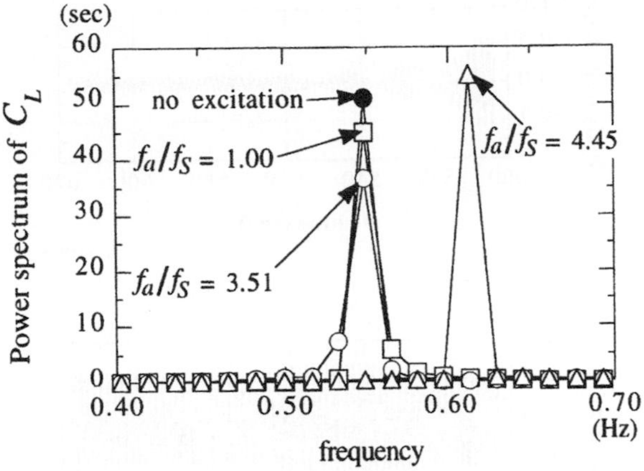

Fig. 6. Power spectrum of lift coefficient (stationary cylinder).

reduced when the periodic excitation at $f_a/f_s = 4.45$ is applied. On the other hand, it is found that the periodic excitation at $f_a/f_s = 1.00$ and $f_a/f_s = 3.51$ are less effective in changing the flow characteristics around the cylinder, as shown in Figs. 4 and 5. Although the simulations with periodic excitation of some other frequencies were conducted, the effects were found to be small. The reason why is that the periodic excitation with frequency such as $f_a/f_s = 1.00$ and $f_a/f_s = 3.51$ cannot be enhanced so effectively in the separated shear layer around the cylinder, compared to the excitation at $f_a/f_s = 4.45$.

Fig. 6 presents the power spectrum of the fluctuating lift coefficient acting on the stationary circular cylinder. The figure tells us that the excitation at $f_a/f_s = 4.45$ has the effect of raising the spectrum peak frequency, that is, the vortex-shedding frequency. The vortex-shedding frequency with excitation at $f_a/f_s = 4.45$ is 10% higher than the vortex-shedding frequency without the excitation.

3.2. *The effect of the periodic excitation on the vortex-induced vibration*

The simulation of the vortex-induced vibration was conducted to examine the effect of the periodic velocity excitation on the characteristics of the cylinder vibration. The simulation result of the vortex-induced vibration with the periodic excitation at $f_a/f_s = 4.45$ is shown in Fig. 7, compared to the excitation at $f_a/f_s = 1.00$, 3.51. The excitations began to be applied to the flow when the vortex-induced vibration amplitude became constant. As presented in the figure, the excitation at $f_a/f_s = 4.45$ is more effective in reducing the vortex-induced vibration amplitude than the cases of $f_a/f_s = 1.00$, 3.51. Although simulations with excitation of some other frequencies were conducted, the effects were found to be very small.

The vibration amplitude reduction in Fig. 7a seems to be caused by the increase of the vortex-shedding frequency in Fig. 6 because the vortex-shedding frequency increased by the excitation at $f_a/f_s = 4.45$ is higher than the natural frequency of the

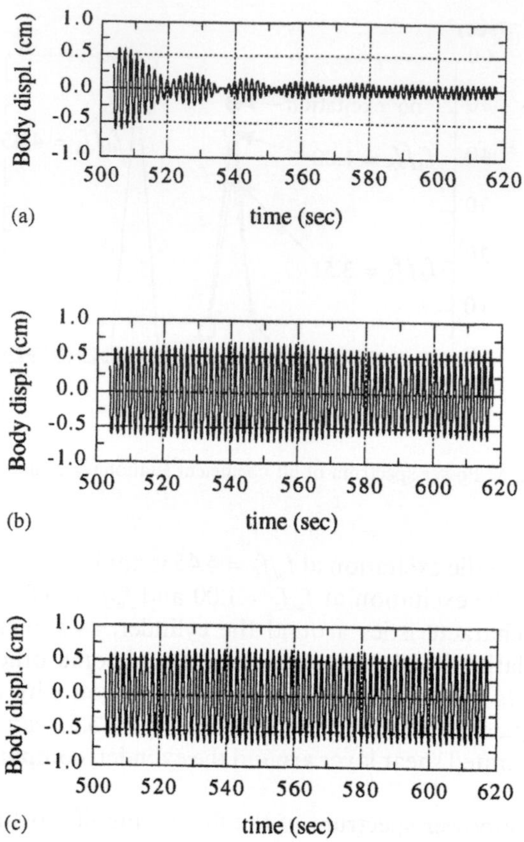

Fig. 7. Time histories of cylinder displacement with periodic excitations: (a) $f_a/f_s = 4.45$, (b) $f_a/f_s = 1.00$, (c) $f_a/f_s = 3.51$.

cylinder, that is, the resonance frequency. Furthermore, in the lower U range, where the vortex-shedding frequency without the excitation is lower than the natural frequency of the cylinder, the vibration amplitude is enhanced by applying the excitation which has the effect of raising the vortex-shedding frequency (Fig. 8). The reason why is that the vortex-shedding frequency increased by the excitation becomes closer to the natural frequency of the cylinder.

Fig. 9 demonstrates a vortex formation in the upper separated shear layer behind the vibrating cylinder when the excitation at $f_a/f_s = 4.45$ is applied at Re = 2000. Also the flow patterns without the excitation are given as a comparison so that the phases are nearly the same as those of the flow patterns with $f_a/f_s = 4.45$. According to the comparison of these figures, it is found that the growth of the vortex in the early vortex formation is promoted by applying the excitation at $f_a/f_s = 4.45$. The promotion of the vortex growth seems to cause the increase of the vortex-shedding frequency in Fig. 6 and some other changes of the flow characteristics.

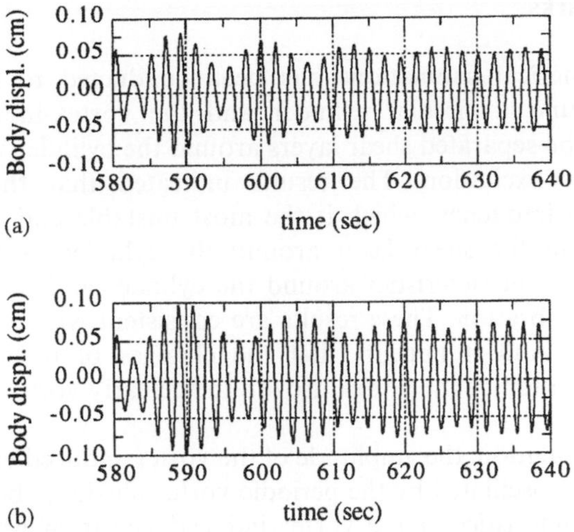

Fig. 8. Effect of the excitation on the vortex-induced vibration in the lower U range (Re = 1700): (a) no excitation, (b) $f_a = 1.93$ Hz.

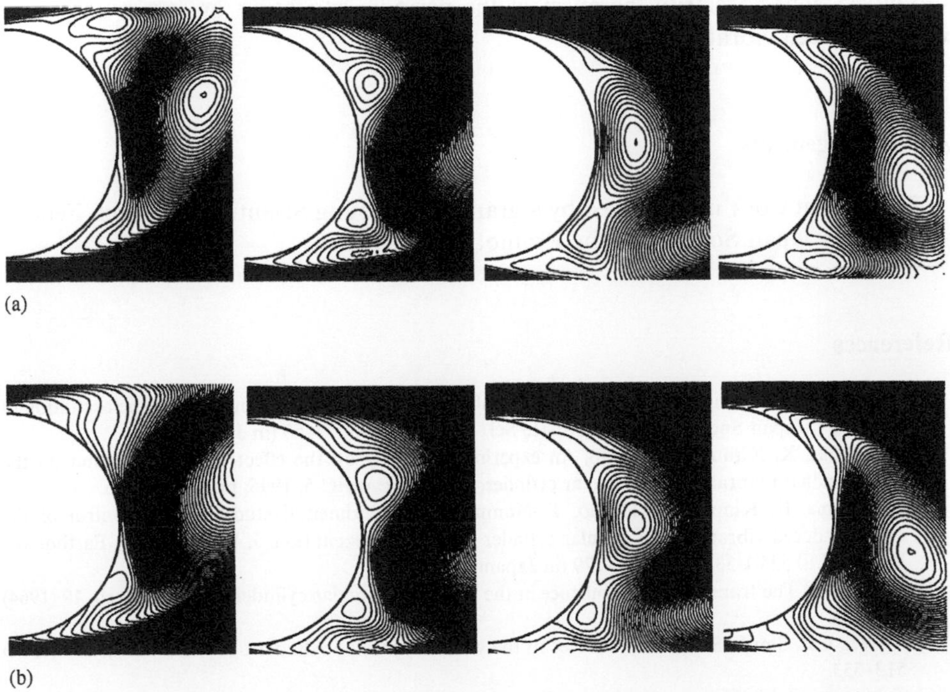

Fig. 9. Comparision of flow patterns behind the cylinder: (a) $f_a/f_s = 4.45$; (b) no excitation.

4. Concluding remarks

A series of numerical simulations has been conducted to examine the flow characteristics around a circular cylinder and the vortex-induced vibration of the cylinder when the separated shear layers around the cylinder were stimulated by a periodic velocity excitation. The results indicated that the excitation with the transition wave frequency, which is the most unstable and can grow into the strong fluctuation in the shear layer around the cylinder, is the most effective in changing the flow characteristics around the cylinder and the characteristics of the vortex-induced vibration. These results are consistent with earlier experimental studies [2,3,11–18]. And it appears that the changes of the characteristics are caused by the promotion of the vortex growth in the early vortex formation behind the cylinder.

In the present simulation, the amplitude of the vortex-induced vibration was small and the cylinder was oscillated by the periodic vortex-shedding behind the cylinder. Therefore the characteristics of the vortex-induced vibration were altered by the change of the vortex-shedding frequency behind the cylinder. However, a larger-amplitude vortex-induced vibration would have the feature of the self-excited vibration, and it is possible that the effect of the periodic excitation on the larger-amplitude vibration was different from that on the smaller-amplitude vibration. So further study should be done to clarify the effect of the periodic excitation on larger-amplitude vortex-induced vibrations.

Acknowledgements

The present work is supported by a grant from Young Scientists Research Fellowships of the Japan Society for the Promotion of Science.

References

[1] M. Nishioka, M. Asai, S. Furumoto, Control of stalled flow over an airfoil by acoustic excitation of vortices, J. Japan Soc. Aeronautical Space Sci. 43 (492) (1995) 53–58 (in Japanese).

[2] S. Hiejima, K. Kimura, Y. Fujino, An experimental study on the effect of applied sound on the vortex-induced vibration of a circular cylinder, Proc. of EASEC-5, 1995, pp. 1231–1236.

[3] S. Hiejima, K. Kimura, Y. Fujino, T. Nomura, An experimental study on the control of the vortex-induced vibration of a circular cylinder by acoustic excitation, J. Struct. Mech. Earthquake Eng. (JSCE) 525/I-33 (1995) 171–179 (in Japanese).

[4] M.S. Bloor, The transition to turbulence in the wake of a circular cylinder, J. Fluid Mech. 19 (1964) 290–304.

[5] T. Wei, C.R. Smith, Secondary vortices in the wake of circular cylinders, J. Fluid Mech. 169 (1986) 513–533.

[6] T. Nomura, T.J.R. Huges, An arbitrary Lagrangian–Eulerian finite element method for interaction of fluid and a rigid body, Comput. Meth. Appl. Mech. Eng. 95 (1992) 115–138.

[7] T. Nomura, Finite element analysis of vortex-induced vibrations of bluff cylinders, J. Wind Eng. Ind. Aerodyn. 46&47 (1993) 587–594.

[8] A.N. Brooks, T.J.R. Hughes, Streamline upwind/Petrov–Galerkin formulations for convection dominated flows with particular emphasis on the incompressible Navier–Stokes equations, Comput. Meth. Appl. Mech. Eng. 32 (1982) 199–259.

[9] O. Inoue, T. Yamazaki, T. Bisaka, Numerical simulation of forced wakes around a cylinder, Int. J. Heat Fluid Flow 16 (5) (1995) 327–332.

[10] T.J.R. Hughes, The finite Element Method, Prentice-Hall, Englewood Cliffs, NJ, 1987.

[11] F.B. Hsiao, J.Y. Shyu, C.F. Liu, R.N. Shyu, Experimental study of an acoustically excited flow over a circular cylinder, in: G.J. Hwang (Ed.), Transport Phenomena in Thermal Control, Hemisphere, New York, 1989, pp. 537–546.

[12] F.B. Hsiao, J.Y. Shyu, Influence of internal acoustic excitation upon flow passing a circular cylinder, J. Fluids Structures 5 (1991) 427–442.

[13] J.A. Peterka, P.D. Richardson, Effect of sound on separated flows, J. Fluid Mech. 37 (2) (1969) 265–287.

[14] J. Sheridan, J. Soria, Wu Jie, M.C. Welsh, The Kelvin–Helmholtz instability of the separated shear layer from a circular cylinder, in: H. Eckelmann et al. (Ed.), Proc. of IUTAM Symp. on Bluff-Body Wakes, Dynamics and Instabilities, Springer, Berlin, 1992, pp. 115–118.

[15] A.B. Zobnin, M.M. Sushchik, Influence of a high-frequency sound field on vortex generation in the wake of a cylinder, Sov. Phys. Acoust. 35 (1) (1989) 37–39.

[16] S. Okamoto, T. Hirose, T. Adachi, The effect of sound on the vortex-shedding from a circular cylinder (Acoustical vibrations directed along axis of cylinder), Trans. JSME, Ser. B 46 (405) (1980) 813–820 (in Japanese).

[17] S. Okamoto, Effect of sound on vortex-shedding from a circular cylinder — Acoustical vibrations directed normal to the axis of cylinder, Trans. Soc. Heating, Air-Conditioning Sanitary Eng. Japan 44 (1990) 1–10 (in Japanese).

[18] G. Yamanaka, T. Adachi, Acoustic influences in vortex-shedding from a circular cylinder, J. Acoust. Soc. Japan 27 (5) (1971) 246–256 (in Japanese).

ELSEVIER

Journal of Wind Engineering
and Industrial Aerodynamics 67&68 (1997) 337–347

JOURNAL OF wind engineering AND industrial aerodynamics

Improvement of pitching moment estimation of an oscillating body by discrete vortex method

F. Nagao[a,*], H. Utsunomiya[a], S. Murata[b]

[a] *Department of Civil Engineering, Faculty of Engineering, The University of Tokushima, 2-1 Minami-Josanjima-cho, Tokushima 770, Japan*

[b] *Ishikawajima-harima Heavy Industries Co. Ltd., Shin-nakahara-cho, Isogo-ku, Yokohama 235, Japan*

Abstract

By using a discrete vortex method which combines the vortex-shedding model with the singularity method, the following subjects are examined in this study: (1) how to decide on a suitable set of parameters to simulate the real flow, (2) a new reattachment model along the body surface and (3) the applicability to an oscillating body of the new calculation formula for the pitching moment proposed by the authors, its validity for bodies at rest had already been confirmed. Furthermore, the physical meaning of the new calculation formula of the pitching moment is also explained. The calculated results show good coincidence with the experimental results obtained from wind tunnel tests.

Keywords: Discrete vortex method; Estimation of unsteady pitching moment; Choice of parameters

1. Introduction

A discrete vortex model which combines the vortex-shedding model with the singularity method is developed for the analysis of an unsteady separated flow past a bluff body. This method is found to be effective to calculate vortical flows; however, there are some problems in it, such as (1) how to decide on the parameters in the discretization of a flow field, (2) the treatment of the reattachment of the separated shear flow on the body surface, (3) the difficulty of the calculation of the pitching moment due to the circular integral along the body, and other problems.

In the study, the following subjects are examined. First, a suitable set of parameters in the calculation scheme is selected from many investigations of

* Corresponding author. E-mail: fumi@ce.tokushima-u.ac.jp.

PII S0167-6105(97)00084-6

systematic arrangement of parameters. Secondly, the applicability of the new proposed formula [1,2] for the pitching moment to an oscillating body is examined. Furthermore, the physical meaning of the new calculation formula of the pitching moment is also explained.

2. Choice of parameters

2.1. Calculation procedure

The calculation procedure was based on the Inamuro model [3]. In this method, the free shear layers are approximated by discrete vortices (wake vortices) and the body is represented by distributed vortices (bound vortices). Fig. 1 shows

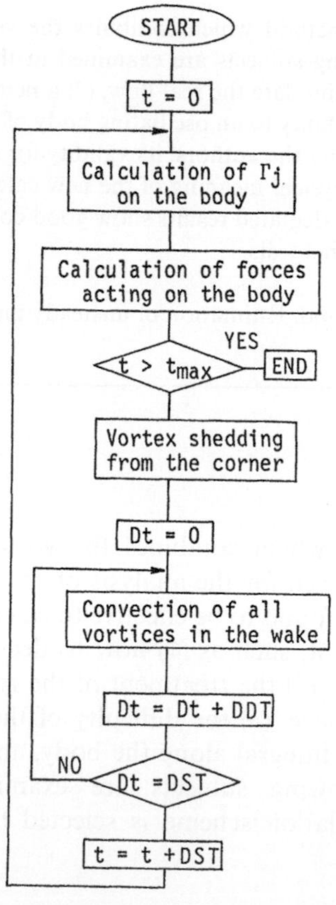

Fig. 1. Flow-chart of calculation procedure.

a flow-chart of the calculation procedure. A complex potential function, f, for a flow field is expressed as follows:

$$f = U \exp(-\mathrm{i}\alpha)z + \frac{1}{2\pi \mathrm{i}}\sum_{n=1}^{N} \Gamma_n \log(z - z_n) + \frac{1}{2\pi \mathrm{i}}\sum_{m=1}^{M} \Gamma_m \log(z - z_m), \tag{1}$$

where, U is the wind velocity, α the angle of attack, Γ_n the circulation of nth wake vortex ($n = 1, 2, \ldots, N$) and Γ_m the circulation of mth bound vortex ($m = 1, 2, \ldots, M$).

The flow field around the body is composed of a uniform flow field and a flow field induced by vortices as shown in Fig. 2. Therefore, Eq. (1) is transformed as:

$$f = U \exp(-\mathrm{i}\alpha)z + \frac{1}{2\pi \mathrm{i}}\sum_{k=1}^{K} \Gamma_k \log(z - z_k) \quad (K = N + M). \tag{2}$$

Aerodynamic drag and lift forces exerted on a bluff body can be calculated from the generalized Blasius theorem, as follows:

$$D - \mathrm{i}L = \mathrm{i}\rho U \exp(-\mathrm{i}\alpha)\sum_{k=1}^{K} \Gamma_k - \mathrm{i}\rho\frac{\mathrm{d}}{\mathrm{d}t}\left(\sum_{k=1}^{K} \Gamma_k \bar{z}_k\right). \tag{3}$$

Here, $\bar{z}_k$ is the complex conjugate of z_k.

2.2. Decision about a suitable set of parameters

For the discrete vortex method, a bluff body in uniform flow is discretized as shown in Fig. 3. The computational parameters treated in the study are given as follows: (1) the distance between bound vortices on the body surface, DS (= ds/d), (2) the radius of the viscous core for the vortex model, CORE(= core/d), which was introduced in order to avoid an extremely large transport velocity when vortices approach too close, (3) the time step for the convection calculation, DDT (= ddt/t_0), and (4) the

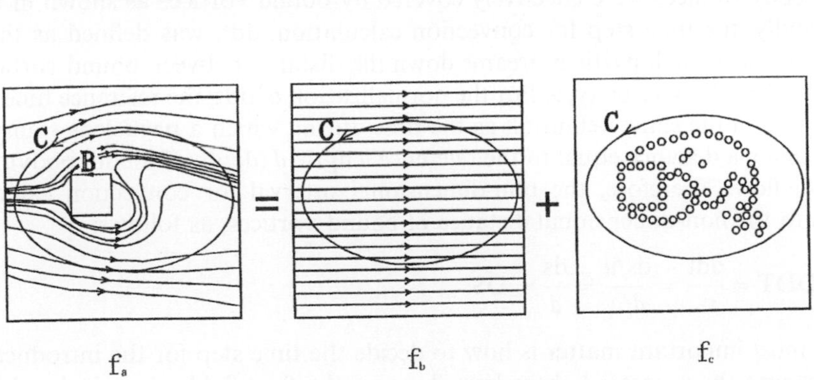

Fig. 2. Flow field in discrete vortex method.

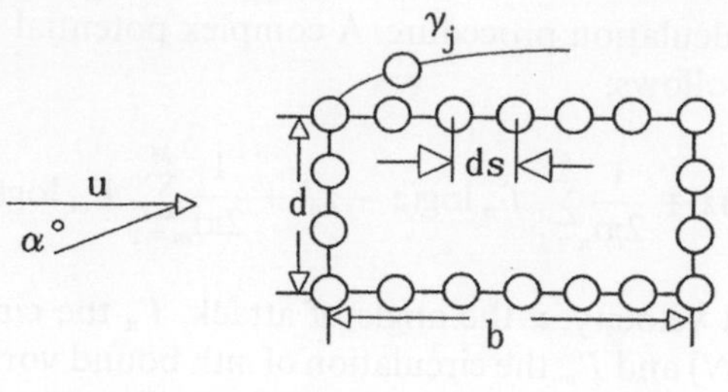

Fig. 3. Parameters around a body in flow field.

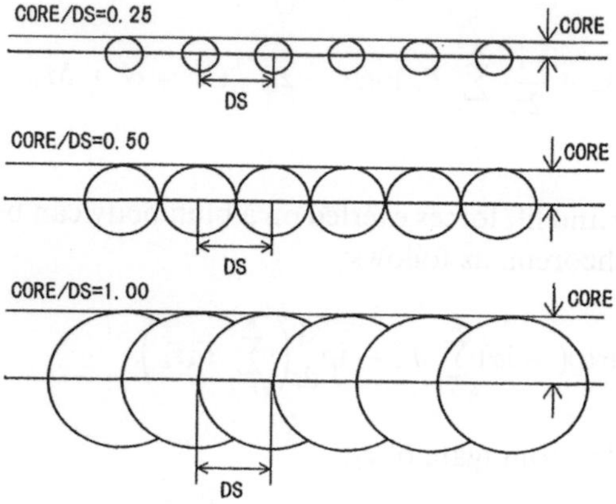

Fig. 4. Relationship between DS and CORE.

time step for generation of nascent vortices, DST (= dst/t_0). For simplicity, the decay of vortices was ignored and the CORE was assumed to be constant.

First, the viscous radius, CORE was fixed as follows:

$$\text{CORE} = \text{DS}/2, \tag{4}$$

where body surfaces were effectively covered by bound vortices as shown in Fig. 4.

Secondly, the time step for convection calculation, ddt, was defined as the time during which a small particle streams down the distance between bound vortices, ds, at the uniform flow velocity, u. For the normalization of ddt, the reference time t_0 was introduced in the same definition as the time during which a particle is transported downstream a distance equal to the reference length, d (depth of a bluff body), by the uniform flow. Therefore, the non-dimensional interval for convection calculation relates to the non-dimensional distance of bound vortices, as follows:

$$\text{DDT} = \frac{\text{ddt}}{t_0} = \frac{\text{ds}/u}{d/u} = \frac{\text{ds}}{d} = \text{DS}. \tag{5}$$

The most important matter is how to decide the time step for the introduction of vortices into the separated shear layer because the flow field was calculated by the movement of wake vortices introduced at the corners of the bluff body. In the study, it

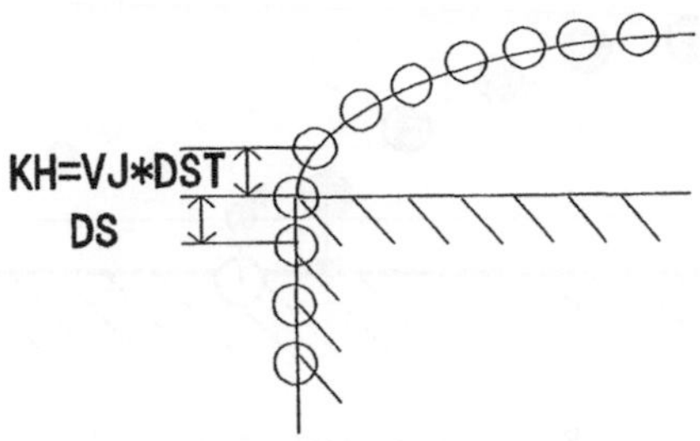

Fig. 5. Details around the separation point.

is assumed that each nascent separated vortex vertically moves a distance almost equal to the distance of bound vortices, ds, during the interval for the generation of vortices as given in Fig. 5:

$$k_h = \mathrm{dst} \cdot \mathrm{vj} \cong \mathrm{ds}, \tag{6}$$

where k_h and vj denote the vertical movement of the introduced vortices and the vertical component of the transport velocity, respectively.

Eq. (6) can be converted into non-dimensional form as follows:

$$K_H = k_h/d = \mathrm{dst} \cdot \mathrm{vj}/d = \frac{\mathrm{dst}}{d/u}\frac{\mathrm{vj}}{u} = \frac{\mathrm{dst}}{t_0}\frac{\mathrm{vj}}{u} = \mathrm{DST} \cdot \mathrm{VJ} \cong \mathrm{ds}/d = \mathrm{DS}. \tag{7}$$

2.3. Reattachment model

In order to reduce the calculation error due to time integration for the movement of wake vortices, many models concerning the reattachment of separated shear flows on the body surface have been considered. The simple and popular reattachment model removes the vortices that have come too close to the body surface from the calculation, taking into account the destruction of the vorticity in the shear layer through its interaction with the body surface. The other reattachment models considering viscous effects on the solid surface were proposed by adding a vertical velocity to each vortex in the vicinity of the edge of the separation bubble [4] or by modifying the approaching transport velocities [5] and other methods.

In this study, based on the same idea of taking into account the viscous effect and the interaction with the body surface, a new reattachment model is proposed which reduces the vorticity close to the body surface, as shown in Eq. (8) and Fig. 6, where the reduction rate of vorticity, RRV, is decided by taking into account the equivalence of decay of vorticity due to reattachment on the body surface and the introduction of vorticity at the separation points.

$$\Gamma_k(t + \mathrm{DDT}) = \mathrm{RRV} \cdot \Gamma_k(t). \tag{8}$$

When the reduction rate of vorticity, RRV, is set to zero, the reattachment model proposed here corresponds to the simplest one that removes the vortices from the calculation.

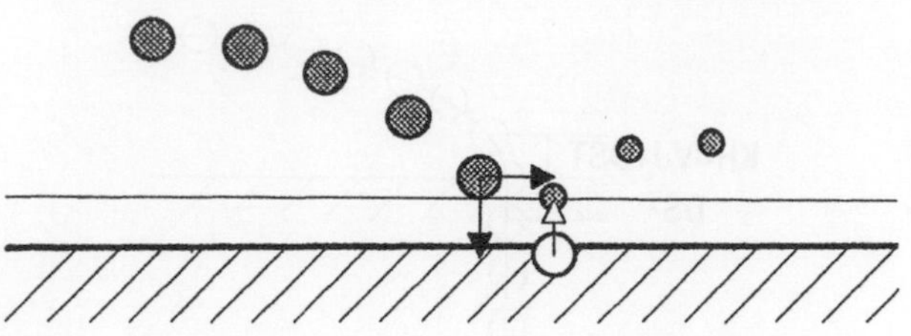

Fig. 6. Reattachment model proposed here.

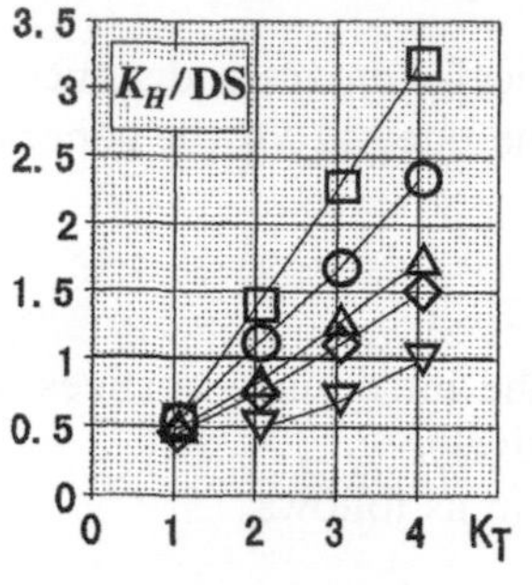

(a) K_H/DS vs. K_T

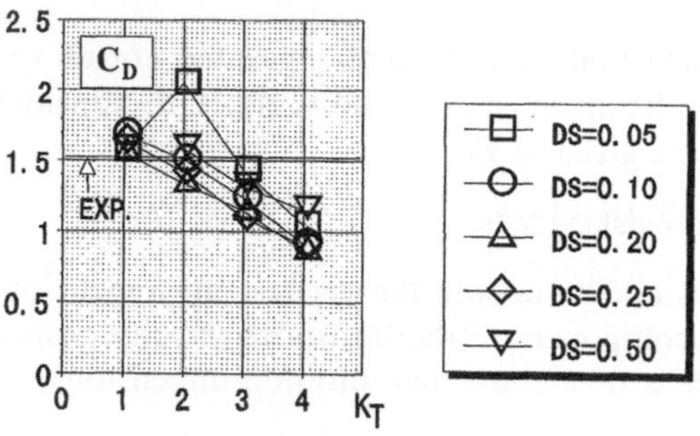

(b) C_D vs. K_T

Fig. 7. (a) K_H/DS versus K_T. (b) C_D versus K_T.

2.4. Drag force exerting on prism at rest

In the previous section, all parameters were connected with the parameter DS. In order to choose the suitable set of parameters, the effect of the change of DS on the drag force exerting on a rectangular prism ($b/d = 2$, b is the width of the rectangular prism) was investigated, the relationships between DS and other parameters being satisfied as mentioned above. The calculation results are shown in Fig. 7, where the horizontal axis variable, K_T, indicates the ratio of DST to DDT. The vertical movement of introduced vortices is given in Fig. 7a and the drag coefficient is shown in Fig. 7b. In the case of DS = 0.1 and DST = 2DDT, the estimated value shows good coincidence with experimental results. Furthermore, the estimated Strouhal number also agrees with the experimental result [6].

2.5. Unsteady lift force in vortex-induced oscillation

Table 1 shows the values of the parameters used here, where all relationships between DS and other parameters were almost satisfied (K_H/DS = 0.96 ≒ 1). Fig. 8 indicates the response of vortex-induced oscillation of a rectangular prism ($b/d = 2$) in the heaving mode [7]. The region of calculation using the forced oscillation method is

Table 1
Parameters used here

DS	0.10
DDT	0.090625
DST	0.181250
CORE	0.05

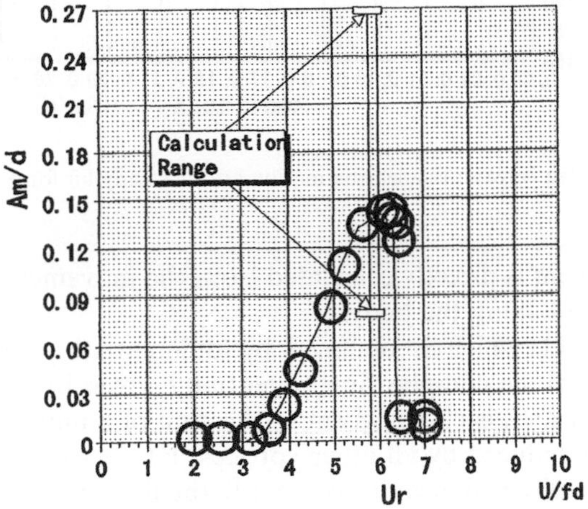

Fig. 8. Heaving response of a rectangular cylinder ($b/d = 2$) [7] and calculation region.

also shown (reduced velocity, $U_r = 5.8$, amplitude normalized by the depth of model, $\mathrm{Am}/d = 0.08$–0.26).

Fig. 9 shows the phase lag between the body motion and the unsteady lift force estimated by fast Fourier transform analysis. In the region of small amplitude ($\mathrm{Am}/d < 0.18$), the estimated results show an exciting force ($\beta > 0$); on the other hand, in the larger amplitude region ($\mathrm{Am}/d > 0.21$), the results show a damping force ($\beta < 0$). These calculated results qualitatively agree with the experimental ones.

Therefore, a suitable set of parameters is obtained for DS = 0.1, and all relationships proposed here between DS and other parameters are satisfied.

3. Pitching moment acting on a bluff body

3.1. New Calculation formula for pitching moment

A new calculation formula for the pitching moment, M, is derived from Blasius' theorem and Eq. (2) [1,2], as follows:

$$M = \rho\,\mathrm{Re}\left(\sum \Gamma_k z_k \frac{\mathrm{d}\bar{z}_k}{\mathrm{d}t} - U\exp(-\mathrm{i}\alpha)\sum \Gamma_k z_k\right). \tag{9}$$

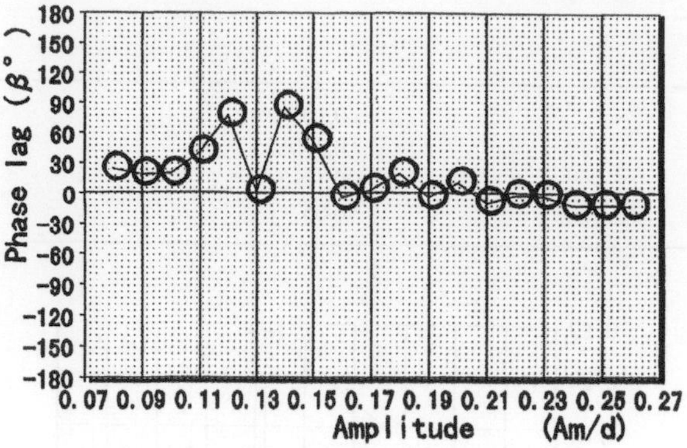

Fig. 9. Phase lag between the heaving motion and unsteady lift force in each Am.

The pitching moment is directly calculated from the movement of vortices instead of the circular integral along the body.

In the velocity field, which is obtained by differentiating (with respect to the complex coordinate, z) the complex flow potential, f, given by Eq. (2), the transport velocity of each vortex, V, is clearly given as the sum of a uniform flow component, $U1$, and the velocity induced by all other vortices, $V1$, as shown in Fig. 10.

To understand the physical meaning of Eq. (9), the first term of the right-hand side is expanded as follows:

$$\rho\,\mathrm{Re}\left(\sum_{k=1}^{K}\Gamma_k z_k\frac{\mathrm{d}\bar{z}_k}{\mathrm{d}t}\right)=\rho\sum_{k=1}^{K}\Gamma_k\left(x_k\frac{\mathrm{d}x_k}{\mathrm{d}t}+y_k\frac{\mathrm{d}y_k}{\mathrm{d}t}\right)=\sum\rho\Gamma_k\frac{1}{2}\frac{\mathrm{d}}{\mathrm{d}t}\left(x_k^2+y_k^2\right)$$

$$=\sum_{k=1}^{K}\frac{\Gamma_k/S_k}{2}(\rho S_k)\frac{\mathrm{d}}{\mathrm{d}t}\left(x_k^2+y_k^2\right)$$

$$=\sum_{k=1}^{K}\frac{\omega_k}{2}\frac{\mathrm{d}}{\mathrm{d}t}\left\{(\rho S_k)\left(x_k^2+y_k^2\right)\right\}=\sum_{k=1}^{K}\Omega_k\frac{\mathrm{d}I_k}{\mathrm{d}t},\tag{10}$$

where, S_k is the area of the vortex ($=\Gamma_k/\omega_k$), ω_k the vorticity, $\rho S_k(x_k^2+y_k^2)$ the inertial moment of the vortex ($=I_k$) and Ω_k ($=\omega_k/2$) the angular velocity of the vortex.

The second term in Eq. (9) is also modified in the same manner, that is, the second term $=\sum\Omega_k\,\mathrm{d}I_{kU}/\mathrm{d}t$, where I_{kU} implies some kind of inertial moment of vortex in the uniform flow field. Therefore, the pitching moment, M, is obtained by

$$M=\sum_{k=1}^{K}\Omega_k\frac{\mathrm{d}}{\mathrm{d}t}(I_k-I_{kU})=\sum_{k=1}^{K}\Omega_k\frac{\mathrm{d}I_{kV}}{\mathrm{d}t}.\tag{11}$$

If vortices were transported only due to the velocities induced by all other vortices as indicated in Fig. 10c, the inertial moments of the vortices, I_{kV}, and the angular momenta of the vortices, $\Omega_k I_{kV}$, should be invariable, respectively. However, in the

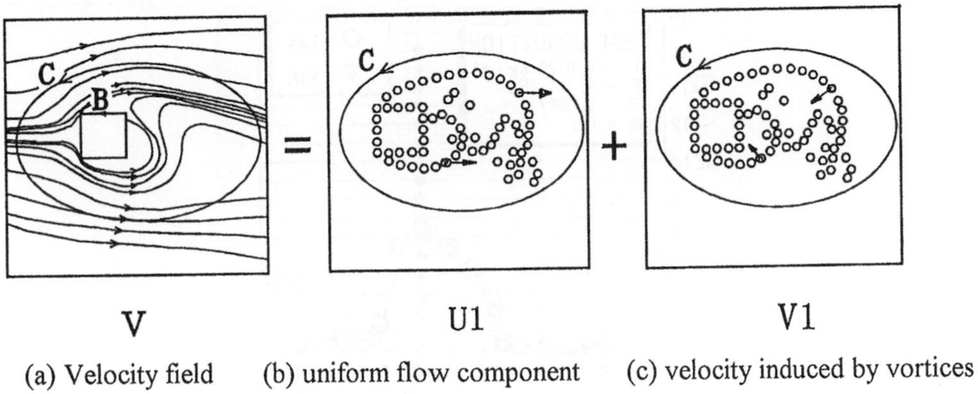

Fig. 10. Velocity field around a body.

flow field, the uniform flow component plays an important role by forcing the vortices downstream as shown in Fig. 10b. Therefore, the change of the inertial moments of the vortices, I_{kV}, should introduce the pitching moment, M.

On the other hand, the pitching moment is defined as the temporal rate of change of the angular momentum of the vortices, as follows:

$$M = \sum_{k=1}^{K} \frac{\mathrm{d}}{\mathrm{d}t}(\Omega_k I_{kV}) = \sum_{k=1}^{K} \frac{\mathrm{d}\Omega_k}{\mathrm{d}t} I_{kV} + \sum_{k=1}^{K} \Omega_k \frac{\mathrm{d}I_{kV}}{\mathrm{d}t} = \sum_{k=1}^{K} \Omega_k \frac{\mathrm{d}I_{kV}}{\mathrm{d}t} \left(\because \frac{\mathrm{d}\Omega_k}{\mathrm{d}t} = 0 \right). \quad (12)$$

Because of the coincidence of Eqs. (11) and (12), it is concluded that the proposed formula for the pitching moment given in Eq. (9) calculates the temporal rate of change of the angular momentum of the vortices.

3.2. Estimation of unsteady pitching moment

As mentioned above, the validity of the new calculation formula for the pitching moment of a square cylinder and box girder bridge section was confirmed for bodies at rest [1,2]. Therefore, the unsteady pitching moment of a rectangular cylinder ($b/d = 2$) during torsional oscillation was investigated.

Fig. 11 shows the response of the torsional-vortex-induced oscillation of a rectangular cylinder with cross section $b/d = 2$ [8], where the solid circles indicate the calculation points for the unsteady pitching moment.

The phase lag, β, between the torsional motion and the unsteady pitching moment is calculated by the spectrum analysis (FFT method) and given in Fig. 12, where the oscillation is excited for positive phase lags ($\beta > 0$); on the other hand, for negative phase lags, a damping force is measured. The phase lag is suddenly changed from positive to negative around the amplitude Am $= 9°$. Therefore, an exciting force is measured for small amplitudes, on the contrary, for larger amplitudes (Am $> 9°$), a damping force is detected.

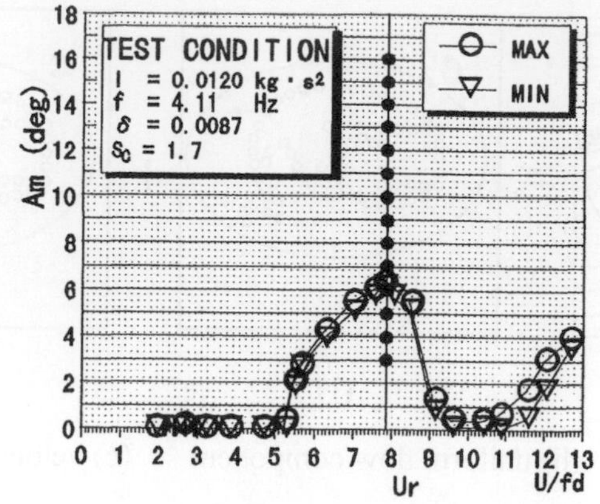

Fig. 11. Torsional response of a rectangular cylinder ($b/d = 2$) [8] and calculation points.

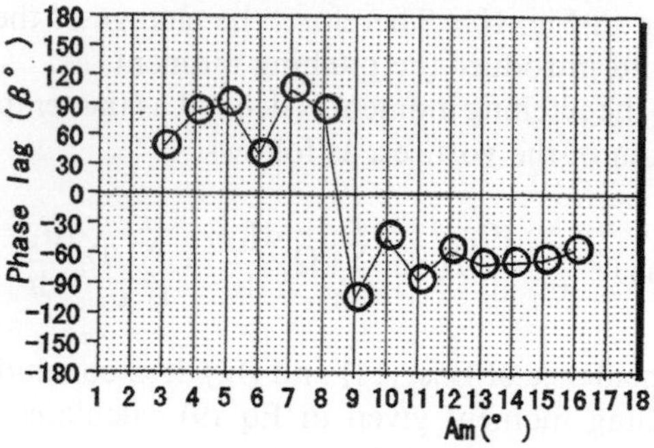

Fig. 12. Phase lag between the torsional motion and unsteady pitching moment in each Am.

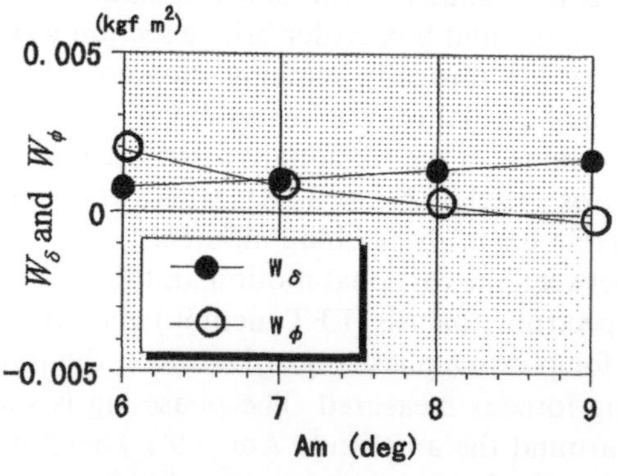

Fig. 13. Relationship between W_δ and W_ϕ.

The response amplitude is estimated by taking into account the structural damping and the unsteady pitching moment, i.e. the work done by the unsteady pitching moment, W_ϕ, which linearly increases with increasing amplitude, is required to equal the damping energy loss per cycle, W_δ, which increases with the second power of the amplitude. The relationship between the two works (W_δ, W_ϕ) is plotted in Fig. 13. From the figure, the response amplitude is evaluated to be about 7°, which coincides with the experimental one.

4. Conclusion

All parameters in the calculation scheme for the discrete vortex method are related by the normalized distance of bound vortices, DS (= ds/d). A suitable set of parameters is achieved by setting DS = 0.1 and by satisfying the relationships between DS and all other parameters, as proposed in this paper.

The new calculation method for the pitching moment, which takes into account the temporal rate of change of the angular momentum of the vortices, was applied to the estimation of the unsteady pitching moment for a rectangular cylinder in torsional oscillation. The calculated results show good agreement with those obtained from wind tunnel experiments.

However, it is desired that the reattachment model should be treated more theoretically.

References

[1] F. Nagao, H. Utsunomiya, T. Oryu, S. Manabe, Aerodynamic efficiency of triangular fairing on box girder bridge, J. Wind Eng. Aerodyn. 49 (1993) 565–574.

[2] H. Utsunomiya, F. Nagao, T. Oryu, Y. Mino, Some improvement of pitching moment calculation in the discrete vortex method, Proc. 12th National Symp. on Wind Engineering, 1992, pp. 177–182 (in Japanese).

[3] T. Adachi, T. Inamuro, Numerical flow simulation of wind engineering, J. Wind Eng. 28 (1986) 29–44, (in Japanese).

[4] J. Kiya, K. Sasaki, M. Arie, Discrete-vortex simulation of a turbulent separation bubble, J. Fluid Mech. 120 (1982) 219–244.

[5] H. Utsunomiya, F. Nagao, H. Uenoyama, Study of flows around rectangular cylinders by finite vortex sheets, J. Wind Eng. 37 (1988) 271–280.

[6] T. Mizota, Experimental studies of separated flows around stationary and oscillating rectangular prisms in a uniform flow, J. Wind Eng. 13 (1982) 15–27 (in Japanese).

[7] N. Shiraishi, M. Matsumoto, On vortex-induced oscillations of bluff cross sections used for bridge structures, Proc. JSCE 322 (1982) 37–50 (in Japanese).

[8] N. Shiraishi, et al., On aerodynamic oscillation and its improvement of structural bluff section, Annuals Disaster Prevention Res. Inst., 1983 (in Japanese).

ELSEVIER

Journal of Wind Engineering
and Industrial Aerodynamics 67 & 68 (1997) 349–359

JOURNAL OF
wind engineering
AND
industrial
aerodynamics

Unsteady pressure evaluation on oscillating bluff body by vortex method

H. Shirato *, M. Matsumoto

Department of Global Environment Engineering, Kyoto University, Kyoto 606-01, Japan

Abstract

This paper describes the unsteady pressure evaluation on an oscillating bluff body using the unsteady Bernoulli's formula. Flow around a body is simulated by the vortex method. A 2-D rectangular cross section is forced to keep heaving oscillation at reduced wind velocity $U/fD = 5$. According to wind tunnel experiments, the movement-induced-type vortex-induced oscillation is observed in this cross section at the focused reduced wind velocity region. And the unsteady pressure on the body side surface is known to show convection nature due to the movement-induced vortices which generate at the leading edge and travel towards the trailing edge in exactly one natural period. In this study, this pressure convection on the side surface is focused and its reproduction using vortex method is investigated. The contribution of each term which is part of the time varying rate of the velocity potential is discussed. It is concluded that the term relating to the shedding vortex movement plays the most important role in giving reliable pressure convection nature.

Keywords: Vortex method; Unsteady pressure; Bernoulli's formula; 2-D bluff body; Vortex-induced oscillation

1. Introduction

The vortex method, in which thin boundary layers on body surfaces as well as the separated shear flow are expressed by a series of vortex elements, is governed by simple potential equations, and shows better performance in the high Reynolds number regime. Therefore, the use of advanced computers is not required. Recently, the vortex method extends its covering fields to separated flow simulation around a bluff body in homogeneous turbulence [1], 3-D problems [2], etc.

On the contrary, when a vortex element is near a body surface, the induced velocity tends to be overestimated when the mirror image method is employed [3]. This fact is also observed when two free vortex elements approach close to each other. In case

* Corresponding author.

PII S0167-6105(97)00085-8

the body surface is expressed by a series of vortex elements, the same problem arises, which spoils the overall precision of the simulation [4]. To improve the reliability of the simulated result, the number of elements should be increased, since, the above undesirable phenomenon occurs more frequently with the increase of the number of vortex elements.

The vortex method requires some special treatment of viscous effect when flow in a relatively low Reynolds number is simulated, such as introducing vortex core growth [5], random-walk method [6], and VIC method to convert the spatial distribution of shedding vortex elements to grid-oriented vorticity field for solving the equation of diffusion process [7–10]. Especially, the velocity evaluation using the VIC method can avoid velocity overestimation due to nearby vortices, and requires a relatively shorter CPU time in comparison to the evaluation using the ordinary Biot–Savart equation when a large number of vortex elements is introduced in the simulation.

In this study, flow containing the separation around an oscillating bluff body is to be simulated by the vortex method. Then, the evaluation of unsteady pressure on body surface is to be discussed in order to get some basic information for the evaluation of wind load on civil engineering structures by CFD. A 2-D rectangular cross section with slenderness ratio $B/D = 2$ (B is the body width parallel to the approaching flow, D is the body depth perpendicular to the approaching flow) is focused. The rectangular section is forced to keep harmonic oscillation in heaving mode, namely, cross-flow oscillation, with a constant amplitude, $y_0/D = 0.1$ (y_0 is the heaving amplitude). The reduced wind velocity, U/fD (U is the approaching wind velocity, f is the frequency of heaving oscillation) is set to 5, at which the vortex-induced oscillation of body-movement-induced-type is likely to occur. Since this reduced wind velocity is much lower than that where the quasi-steady assumption is held, unsteadiness in the aerodynamic properties such as surface pressure, aerodynamic forces, flow pattern, etc. appears significantly [11].

The pressure evaluation on a body surface using vortex method can be done by two ways: one using the unsteady Bernoulli equation [5,12–14] and the other using the Poisson equation [15]. Since the latter requires sufficiently fine resolution near the body surface for vorticity distribution, this study follows the former approach, and will discuss the physical meaning of the equation and the influence of each term on the result. The surface-pressure fluctuation component, which has the same frequency as the body oscillation controls directly the occurrence of the vortex-induced oscillation, shows convecting properties in this wind-velocity region. The phase lag of surface-pressure fluctuation to the body displacement is one of the most important measures of the oscillation [11]. The phase lag is based on the time lag between the most-negative peak of the surface-pressure fluctuation component on the upper-side surface, whose frequency coincides with the body oscillation, to the uppermost heaving body displacement. In this study, the simulation reliability is judged by the reproduction of this pressure-phase lag properties.

2. Computational procedure

The entire flow field is expressed by the complex potential function including potential vortex elements as singularity points,

$$f(z) = \Phi(z) + \mathrm{i}\Psi(z) = Ue^{\mathrm{i}\alpha}\bar{z} + \sum_j \frac{\mathrm{i}\Gamma_j}{2\pi}\ln(z - z_j) + \sum_l \sum_k \frac{\mathrm{i}\Gamma_k^l}{2\pi}\ln(z - z_k^l), \tag{1}$$

where $\Phi(z)$ is the velocity potential, $\Psi(z)$ the stream function, Γ_j the circulation of the jth bound vortex on the body surface, and Γ_k^l the circulation of the kth vortex shed from the lth point on the body surface.

48 finite vortex elements, whose strengths are unknown, are defined as bound vortices along the body surface with a constant distance of $0.125D$. The strength of the bound vortices is decided by the zero-normal relative wind velocity condition as well as the Kelvin's vorticity theorem,

$$\frac{\partial \Phi}{\partial n} - \dot{y} = 0, \tag{2}$$

$$\sum_j \Gamma_j + \sum_l \sum_k \Gamma_k^l = 0, \tag{3}$$

where $\dot{y}$ is the moving velocity of the body in the heaving mode.

All vortices to be shed are set initially in the surrounding flow $0.025D$ apart from the body surface, and shed to the surrounding flow simultaneously. Strength of the newly shed vortices is identical to that of the bound vortices determined by Eq. (2) and Eq. (3) above. Each shed vortex convects with the velocity evaluated at its location

$$\boldsymbol{x}_{t+\Delta t} = \boldsymbol{x}_t + \boldsymbol{u}_{\boldsymbol{x}_t,t}\,\Delta t, \tag{4}$$

where $x(t)$ is the position vector of a vortex element at $t = t$, and Δt is the time step to move a vortex element ($U\Delta t/B = 0.1$).

In the wind velocity calculation process, the finite area around the $B/D = 2$ rectangular cross section is covered with a rectangular mesh with mesh size $\Delta = 0.125D$ as shown in Fig. 1. The surface of the rectangular section coincides with a grid line. Velocity components u, v, vorticity ω, and stream function Ψ are evaluated as in Fig. 2. The entire grid is moving together with the rectangular cross section. The velocity field is determined by solving the Poisson equation of stream function–vorticity distribution

$$\nabla^2\psi = -\omega. \tag{5}$$

To obtain the vorticity, the *modified* VIC method [8] is introduced, which converts the entire circulation of a shed to the *nearest* grid point in the following manner (see Fig. 3):

$$\omega_i = \frac{\Gamma}{\Delta^2}\left\{1 - I\left(x - \frac{\Delta}{2}\right)\right\}\left\{1 - I\left(y - \frac{\Delta}{2}\right)\right\} \quad (i = 1),$$

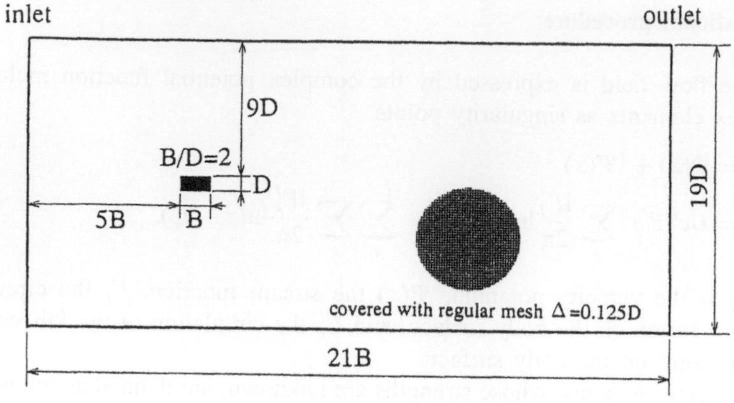

Fig. 1. Computational domain.

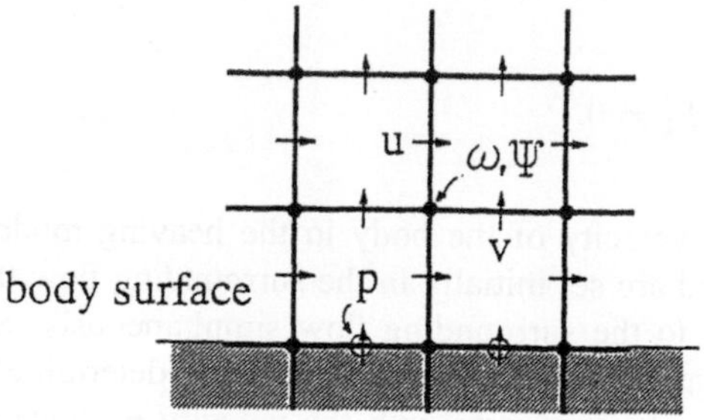

Fig. 2. Location of u, v, ω, Ψ, p in the grid system.

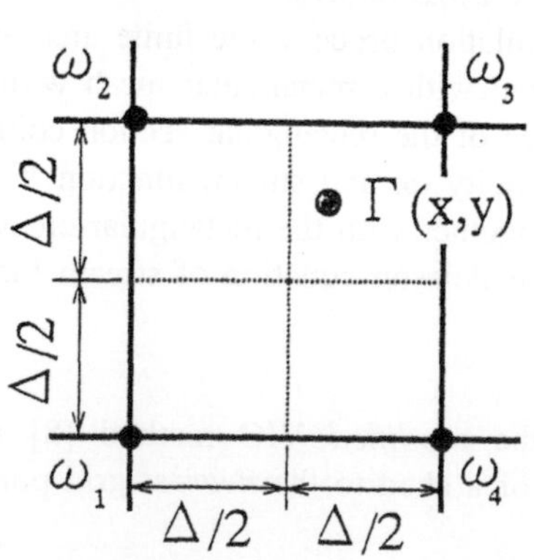

Fig. 3. Vorticity assignment.

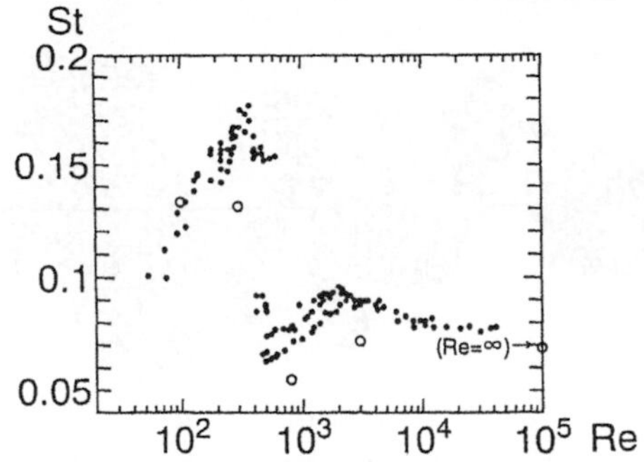

Fig. 4. Strouhal number dependence on Reynolds number (○) present calculation, (•) experiment [16].

$$= \frac{\Gamma}{\Delta^2}\left\{1 - I\left(x - \frac{\Delta}{2}\right)\right\} I\left(y - \frac{\Delta}{2}\right) \quad (i = 2),$$
$$= \frac{\Gamma}{\Delta^2} I\left(x - \frac{\Delta}{2}\right) I\left(y - \frac{\Delta}{2}\right) \quad (i = 3),$$
$$= \frac{\Gamma}{\Delta^2} I\left(x - \frac{\Delta}{2}\right)\left\{1 - I\left(y - \frac{\Delta}{2}\right)\right\} \quad (i = 4), \tag{6}$$

where

$$I(x) = \begin{cases} 1.0 & (x \geqslant 0), \\ 0 & (x < 0). \end{cases} \tag{7}$$

Effect of viscosity is then taken into account by solving the diffusion equation of vorticity

$$\frac{\partial \omega}{\partial t} = \nu \nabla^2 \omega, \tag{8}$$

where ν is the kinematic viscosity.

The simulation can be applied to different Reynolds numbers by adjusting the time duration of the diffusion process, Δt_d, artificially

$$\mathrm{Re}_t = \mathrm{Re}\frac{\Delta t}{\Delta t_d} = \frac{UD}{\nu}\frac{\Delta t}{\Delta t_d}, \tag{9}$$

where Re_t is the target Reynolds number.

As Eqs. (2)–(7) are solved in series, the computation is executed in a time-splitting manner. To confirm the performance of the above algorithm, the effect of Reynolds number on the Strouhal number of a still $B/D = 2$ rectangular cross section is simulated as shown in Fig. 4 [8]. Fig. 5 shows an example of simulated flow pattern in the state of heaving oscillation.

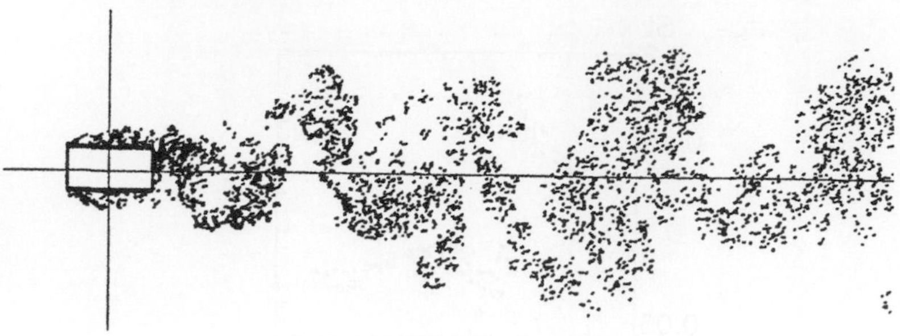

Fig. 5. Simulated result ($Ut/B = 46.0$, Re = 3000).

3. Unsteady pressure on a moving body surface

The unsteady pressure fluctuation on the body surface is evaluated by the unsteady Bernoulli's formula,

$$p = p_0 + \frac{1}{2}\rho\{U^2 - (u^2 + v^2)\} - \rho\frac{\partial \Phi}{\partial t}, \tag{10}$$

where, p is the pressure on the body surface, p_0 the static pressure, and u, v the local wind velocity components on the body surface.

Fig. 6 shows the time history of "row" unsteady pressure fluctuation on the body side surface in heaving motion. Surface pressure is normalized to pressure coefficient, C_p, by static pressure, p_0, and velocity pressure, $\frac{1}{2}\rho U^2$,

$$C_p = \frac{p - p_0}{\frac{1}{2}\rho U^2}. \tag{11}$$

The pressure is evaluated at 16 points on a side surface. Each evaluation point is located between the neighboring bound vortices (see Fig. 2). Although spike noises are seen frequently in the simulated pressure signal, the fluctuation component with the frequency of the heaving motion is somehow recognized. The aerodynamic instability of the body is decided by the heaving frequency component. Fig. 7 shows the RMS distribution of C_p on the side surface. Observed data in wind tunnel experiments [11,17] and numerical simulation by direct simulation [18] are plotted together. The present calculation gives relatively good agreement with the other data. It should be noted that there are two distinct peaks in the RMS distribution of C_p. The upstream side and the other correspond to the development of the movement-induced vortex, and the intermittent expansion of the separation bubble, respectively. The present calculation gives only monotonic increase of C_p along the side surface.

For this rectangular section, the movement-induced-type vortex-induced oscillation can be observed in this wind velocity region, $V_r = U/fD = 5$. Furthermore, the pressure convection towards the trailing edge has been confirmed by the experiments. This nature is caused by the generation of the separated vortex due to body movement and its traveling downstream with the convection velocity. In the movement-induced-type

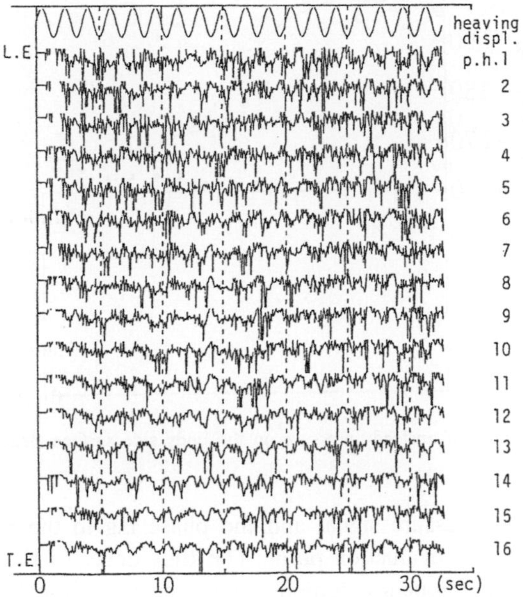

Fig. 6. Time history of "row" unsteady pressure on the body side surface.

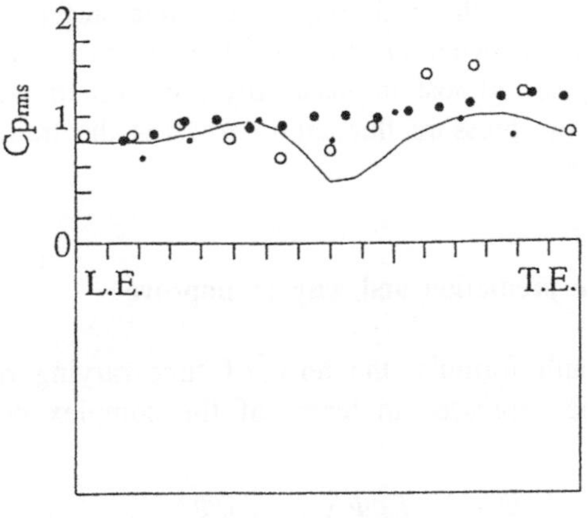

Fig. 7. RMS distribution of C_p on the body side surface: (●) present calculation, $U/fD = 5, y_0/D = 0.1$; (○) experiment, $U/fD = 4.95, y_0/D = 0.05$ [11]; (•) experiment, $U/fD = 5.5, y_0/D = 0.1$ [17]; (—) CFD (DS), $U/fD = 5, y_0/D = 0.114$ [18].

vortex-induced oscillation, the traveling time duration for one separated vortex generated at the leading edge to reach the trailing edge is exactly one natural period of heaving body oscillation. This property is one of the major conditions for the occurrence of movement-induced-type vortex-induced oscillation [11]. To clarify whether the simulated pressure has convection nature, the heaving frequency component is

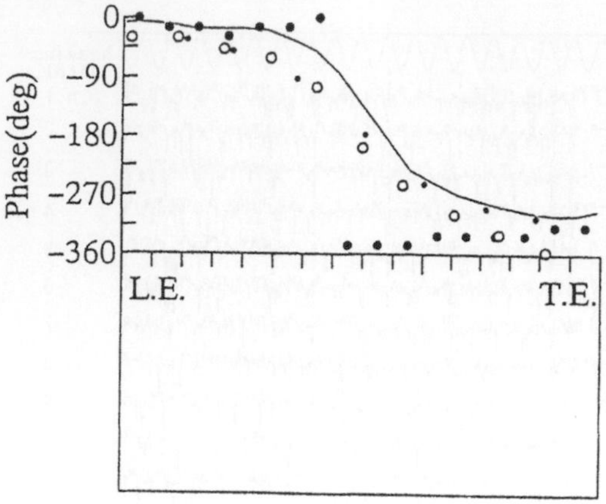

Fig. 8. Phase lag distribution of unsteady pressure on the body side surface (legend as in Fig. 7).

extracted from the "row" pressure signal and the phase lag to the body displacement is evaluated at each pressure evaluation point. The time delay of the negative peak of pressure with the heaving frequency component of the upper-most heaving displacement can be evaluated by the cross-correlation between the "row" pressure signal and heaving displacement.

Fig. 8 plots this time delay as the phase lag. The simulated result has a sudden drop from nearly 0° to 360° at the mid-point of the side surface, which means there is no tendency of pressure convection, but that the pressure fluctuation of heaving frequency component appears almost in-phase anywhere on the side surface. Hence, the convecting nature of the pressure fluctuation towards the trailing edge cannot be caught in this result.

4. Discussion on pressure prediction and way to improve

In the unsteady Bernoulli formula, the term of time-varying rate of the velocity fluctuation, $\partial\Phi/\partial t$, can be expressed in terms of the complex potential function as follows:

$$\frac{\partial \Phi}{\partial t} = \text{Real}\left[\frac{\partial f}{\partial t}\right] = \left(\frac{\partial \Phi}{\partial t}\right)_1 + \left(\frac{\partial \Phi}{\partial t}\right)_2 + \left(\frac{\partial \Phi}{\partial t}\right)_3, \tag{12}$$

where

$$\left(\frac{\partial \Phi}{\partial t}\right)_1 = -\sum_j \frac{\dot{\Gamma}_j}{2\pi}\arg(z - z_j) - \sum_l \frac{\Gamma_0^l}{2\pi\Delta t}\arg(z - z_0^l), \tag{13a}$$

$$\left(\frac{\partial \Phi}{\partial t}\right)_2 = \sum_j \frac{\Gamma_j}{2\pi}\frac{\dot{y}_j(x - x_j) - \dot{x}_j(y - y_j)}{|z - z_j|^2}, \tag{13b}$$

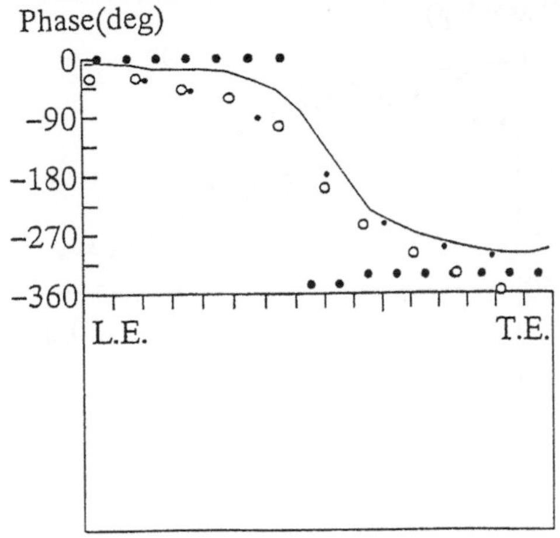

Fig. 9. Phase-lag distribution of the first term, $(\partial\Phi/\partial t)_1$, on the body side surface (legend as in Fig. 7).

$$\left(\frac{\partial \Phi}{\partial t}\right)_3 = \sum_l \sum_k \frac{\Gamma_k^l}{2\pi} \frac{\dot{y}_k^l(x - x_k^l) - \dot{x}_k^l(x - x_k^l)}{|z - z_j|^2}, \tag{13c}$$

The first term in the equation describes the time-varying rate of bound-vortex circulation, while the second term reflects the effect of body movement, and the third term corresponds to the shedding vortices convection. Considering the argument, arg (z), the first term can be reduced further to

$$\left(\frac{\partial \Phi}{\partial t}\right)_1 = \sum_j \frac{\dot{\Gamma}_j}{2\pi} + \sum_l \frac{\Gamma_0^l}{2\pi\Delta t}, \tag{14}$$

where the summation should be along the body surface in the clockwise direction. This term is identical to the Lewis formula for pressure evaluation [3],

$$\frac{\partial \gamma}{\partial t} = -\frac{1}{\rho}\frac{\partial p}{\partial s}. \tag{15}$$

It is known that Eq. (14) consists of time-varying rate of surface-bound vorticity only. Hence, additional factors are necessary in case of flow around an oscillating body, or convection process dominating flow.

Fig. 9 and Fig. 10 show the phase distribution on the side surface evaluated from the first term and third term, respectively. It is confirmed that the second term has minor contribution to the pressure. As for Fig. 9, the plots seem to have almost the same pattern as in Fig. 8. This explains that the predicted unsteady pressure fluctuation is mainly determined by the first term. On the contrary, Fig. 10 gives a fairly realistic result, the pressure fluctuation convects downstream, and consequently, the phase lag is increased towards the trailing edge due to the convecting nature of the separated flow. The fact observed in wind tunnel experiments that the phase lag is caused by the

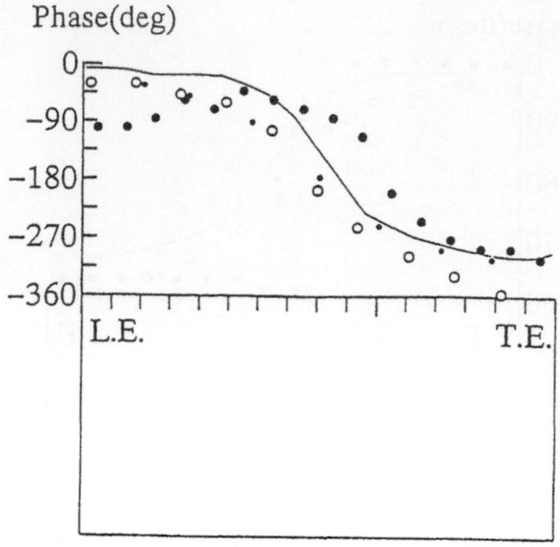

Fig. 10. Phase-lag distribution of the third term, $(\partial\Phi/\partial t)_3$, on the body side surface (legend as in Fig. 7).

convection of the movement-induced vortex generated at the leading edge significantly supports the contribution of the third term.

It may imply, on the improvement of the prediction of pressure convection, that the first term can be reduced artificially to some extent because the magnitude of the first term is strongly related to the time step, Δt, which has no direct relationship with the pressure evaluation but is determined so as to give a reliable flow pattern.

5. Conclusions

The unsteady pressure on an oscillating rectangular 2-D cross section with $B/D = 2$ is simulated using the vortex method. The following remarks can be made as conclusion:

(1) In case of convection-dominating flow, such as focused in this study, the third term of the time-varying rate of velocity-potential function should be most important.

(2) This observation gives the way to tune-up the unsteady pressure prediction by introducing an artificial reduction factor to the first term of the time-varying rate of velocity potential in the unsteady Bernoulli formula.

References

[1] K. Kushioka, T. Saito, A. Honda, T. Inamuro, J. Wind Eng. Ind. Aerodyn. 46 & 47 (1993) 371–379.
[2] Y. Nakanishi, K. Kamemoto, J. Wind Eng. Ind. Aerodyn. 46 & 47 (1993) 363–369.
[3] R.I. Lewis, Vortex Element Methods for Fluid Dynamic Analysis of Engineering Systems, Cambridge University Press, Cambridge, 1991.
[4] H. Shiraro, M. Matsumoto, N. Shiraishi, J. Wind Eng. Ind. Aerodyn. 46 & 47 (1993) 629–637.
[5] R. Inamuro, T. Adachi, H. Sakata, Bull JSME 26 (1993) 2106–2112.
[6] A.J. Chorin, J. Comput. Phys. 27 (1978) 428–442.
[7] J.P. Christiansen, J. Comput. Phys. 13 (1973) 363–379.

[8] H. Shirato, M. Matsumoto, N. Shiraishi, Proc 9th Int. Conf. on Wind Eng., New Delhi, 1995, pp. 261–271.
[9] M. Tsutahara, C. Wang, Y. Kimura, J. Jpn. Soc. Aeron. Space Sci. 42 (481) (1994) 50–58.
[10] J.M.R. Graham, R.H. Arkell, C.-Y. Zhou, J. Wing Eng. Ind. Aerodyn. 50 (1993) 85–96.
[11] N. Shiraishi, M. Matsumoto, Proc. JSCE, 1982, pp. 17–50.
[12] H. Kobayashi, Ph.D. Thesis, 1978.
[13] H. Kawai, Bull Faculty of Engineering Tokyo Denki University, Vol. 11, 1989, pp. 25–35.
[14] M. Murakami, H. Tanaka, Y. Tahara, Y. Himeno, Proc. Str. Eng., JSCE, vol. 42A, 1996.
[15] K. Kamemoto, H. Matsumoto, Y. Yokoi, Flow-Induced Vibration, P.W. Bearman (Ed.), Balkema, 1995, pp. 205–212.
[16] A. Okajima, J. Wing Eng. 17 (1983) 1–20.
[17] H. Yamada, J. Wind Eng. 38 (1989) 115–118.
[18] T. Shimada, Proc. 13th Natl. Symp. on Wind Eng., 1994, pp. 287–292.

Keynote Presentation

ELSEVIER

Journal of Wind Engineering
and Industrial Aerodynamics 67&68 (1997) 363–372

JOURNAL OF
wind engineering
AND
industrial
aerodynamics

Use of meteorological models in computational wind engineering

Roger A. Pielke Sr.*, Melville E. Nicholls

Department of Atmospheric Science, Colorado State University, Fort Collins, CO 80523, USA

Abstract

Computational wind engineering has progressed considerably in recent years. Along with the improved capability to simulate the small-scale flows around architectural structures there is a need to more realistically simulate the larger-scale atmospheric flows in which they are embedded. This would greatly expand the application of numerical models in areas of overlap between the disciplines of atmospheric science and wind engineering. This paper reviews the basic equations used by meteorological models, modifications required to incorporate buildings and how nested grids can be used to span the gap between large separations in scale. Areas requiring future research and model development are discussed and potential applications of a model which can simulate both large-scale meteorological flows as well as small-scale flows around buildings are suggested.

Keywords: Meteorological modeling; Atmospheric modeling

1. Overview of meteorological models used in computational wind engineering

1.1. Basic equations

The development of equations used in meteorological models is discussed in a variety of references beginning with the seminal study of Richardson [1]. Recent summary texts include those of Pielke [2] and Krishnamurti and Bounoua [3].

The physical equations used in atmospheric simulation codes are:

1. A prognostic vector of motion equation defined for a rotating sphere (i.e., the earth). External forces include the pressure gradient force and gravity. The apparent force that results from the coordinate transformation to a rotating system is called the Coriolis force.
2. The ideal gas equation (a diagnostic relation).

* Corresponding author. E-mail: dallas@hercules.atmos.colostate.edu.

PII S0167-6105(97)00086-X

3. A prognostic equation for heat defined in terms of potential temperature. Potential temperature is the temperature that would occur if an air parcel were compressed or expanded to a reference pressure level; usually 10^5 Pa. Potential temperature is directly related to entropy.
   Sources and sinks of heat include those related to phase changes of water, radiative flux divergence, and exchanges from the land, ocean, and ice surfaces.
4. A prognostic equation for water in its three phases: solid, liquid, and vapor. Water is converted among the phases by sublimation, deposition, evaporation, condensation, and transpiration, and is exchanged with the surface. During the conversion heat is released or absorbed.
5. A prognostic equation for other atmospheric trace gases which includes chemical conversions and phase changes. Examples of these trace gases include sulphur dioxide, carbon monoxide, radon, etc.
6. A prognostic equation for mass of air.

Various approximations are made when these equations are applied to specific atmospheric features. For operational weather forecast models, for example, vertical accelerations are neglected and the vertical equation of motion reduces to a diagnostic hydrostatic pressure equation. Moreover, the conservation of mass equation is also often reduced to a diagnostic equation in which either the total or local time tendency of density is neglected. A comprehensive review of the equations and assumptions frequently used in atmospheric models is given in Ref. [4].

For computational wind engineering, the complete three-dimensional equation of motion must be retained since dynamic pressures are essential for the accurate simulation of small-scale flow. Density fluctuations in the conservation of mass equation are ignored, however, for flow that is substantially subsonic.

1.2. Examples

There have been only a limited number of applications of meteorological models to wind engineering problems. Much of this neglect is undoubtedly related to different educational tracks of atmospheric scientists (generally, in a physical science framework) and of engineers. However, since wind engineers and atmospheric scientists work with the same media (i.e., the atmosphere) this void is ripe for substantial scientific achievement.

In the context of physical wind tunnel modeling and numerical modeling, Avissar et al. [5] provided a framework where the capabilities of those two tools overlap. Nicholls et al. [6,7] provided specific examples of improved understanding when a meteorological model is used to simulate airflow over buildings (results from those studies are discussed further in Section 2).

1.3. Procedure of application

Atmospheric numerical models can include large-scale effects such as propagating weather features, rotational effects of the earth, cloud processes, etc. which can be nested down to very small scales of direct relevance to wind engineers. Grasso [8], for example, used six successively finer-scale nests to scale down from regional-scale atmospheric flow to the tornado scale. Ziegler et al. [9] applied four nests to scale down to deep cumulonimbus clouds in which details of the low-level flow were represented. Eastman [10] used two nests to scale down to the eyewall region of Hurricane Andrew in 1992 in which the rapid intensification of the system was realistically modeled. Poulos [11] is using two nests to "telescope down" over the front range of Colorado in order to simulate adequately the local terrain that very substantially influences pollution dispersion from Rocky Flats.

Detailed representations of the atmospheric boundary layer are also a capability of these tools. Such models on that scale are referred to as large eddy simulation (LES) models. Examples include Refs. [12–16].

The procedures to apply atmospheric models to wind engineering studies can be summarized as follows.

- Use grid nesting in a prognostic meteorological model to telescope down in a specific site of interest.
- Use a fine enough spatial scale to represent buildings including roofs and walls.
- Use the wind and turbulence fields provided by the finest model grid as input to physical models of more detailed building structure features.

2. Application to airflow and dispersion around buildings

Models which can simulate both meteorological processes as well as the small-scale flow around buildings have many potential applications. The LES models developed by the wind engineering community to study the flow around buildings and mesoscale meteorological models have many aspects in common. For instance, the numerical techniques used to solve the governing equations and the physical parameterization of subgrid-scale fluxes are very similar. Therefore, it is not necessary to couple a building model and a meteorological model together in some fashion, since to a large extent, the difference between them is the scale of fluid motion to which they are typically applied. A way of bridging the gap between these scales is through the use of multiple nested grids [17]. This enables large-scale meteorological flow patterns to be simulated and by using multiple nested grids to telescope down in scale, the resultant airflow around architectural structures can also be investigated.

The Colorado State University Regional Atmospheric Modeling System (RAMS) is a meteorological model with nested grid capability [6,18]. It employs parameterizations of meteorologically important physical processes, such as surface fluxes, radiation, and microphysics, as well as initialization procedures for incorporating standard weather observations, representation of orography and appropriate boundary conditions. The method for representing buildings in the model is similar to that discussed in Ref. [19]. The normal component of velocity is set to zero on the roofs and walls of the building. A logarithmic function is used to give the relation between the velocity at the grid point adjacent to the surface and the surface shear stress. At the corners of the building, the momentum fluxes are determined by the scheme described in Ref. [6]. When applied to simulate the flow around buildings, this model is an LES model similar to those developed by the wind engineering community. Several investigators have used the LES method to simulate flow around buildings [6,19–22] and it has been found to produce greater accuracy than statistical turbulence models [23]. Most of the testing of LES models has been for the situation of an incident flow perpendicular to a face of the building. However, wind tunnel and field experiments have shown that the maximum wind loads on a roof occur for an oblique incident flow and are associated with the formation of corner conical vortices [24–29]. Therefore, it is important that LES models are tested to see if they are capable of realistically simulating these features. As a test of the RAMS building model, Nicholls et al. [6] attempted to simulate corner conical vortices and had some limited success, although computational limitations precluded the simulation from being run long enough to obtain results which could be statistically analyzed and compared with wind tunnel and field data. Several corner conical vortices formed during the simulation which appear to have some realistic features. Since they are small in scale relative to the height of the building, a large number of grid points were required to resolve them. The fine grid had 80 × 80 × 72 grid points and was embedded in the center of a coarse grid with a similar number of grid points. The grid increment was 1 m for the fine grid and 3 m for the coarse grid. The building was centered within the fine grid and was a cube with 60 × 60 × 60 grid points. Cyclic boundary conditions were prescribed on the coarse grid. As body-induced eddies shed from the building advected through the downstream boundary, they re-emerge at the upstream boundary, eventually producing a turbulent flow field. Fig. 1 shows the pressure perturbation on the roof at

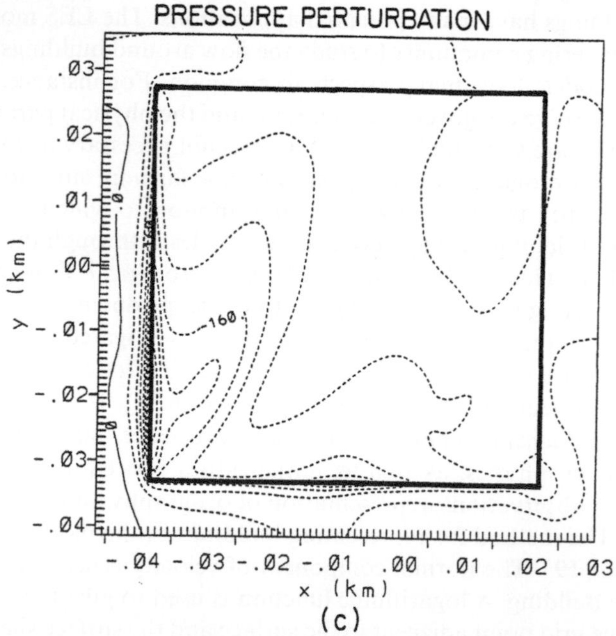

Fig. 1. Pressure perturbations at roof level on a cubic building defined by the thick solid line. The contour interval is 40 Pa.

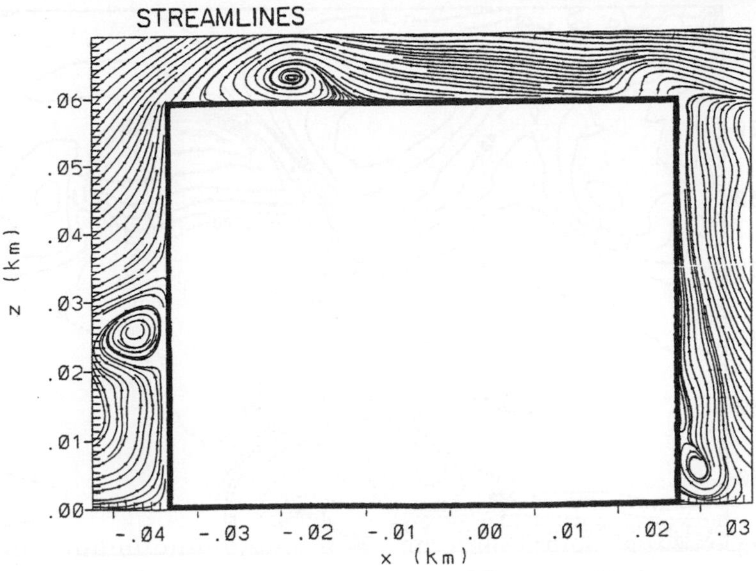

Fig. 2. Streamlines for a vertical cross section through the center of the building showing the roof corner conical vortex.

$t = 80$ s for the fine grid. The wind direction is 225° towards the lower left corner in this figure. At an angle of about 20° to the left edge of the roof is the centerline of a corner conical vortex identifiable in this figure by a relative minima of pressure. The lowest pressure perturbation occurs about a fifth of the way along the left edge of the roof. The corner conical vortices were transient features in this highly turbulent flow field, often being well defined on one side of the roof while virtually disappearing on the other side. Fig. 2 shows an x/z cross section of streamlines through the center of the building which illustrates the well-defined corner conical vortex at $t = 80$ s.

As an example of the application of multiple nested grids to span a wide range of spatial scales, Nicholls et al. [7] conducted a two-dimensional simulation of a microburst-producing thunderstorm and the resultant airflow around a building. Five grids were used with grid increments of 202.5, 67.5, 22.5, 7.5, and 2.5 m. Each successively finer grid was centered within the next coarsest grid with the building centered within the finest grid. The model was initialized using the 2300 UTC 2 August 1985, Dallas-Ft. Worth, Texas microburst sounding used by Proctor [30]. Only the coarsest grid was activated while the thunderstorm developed. When the thunderstorm generated a microburst, the four finer-scale grids were activated. The microburst propagated into the finer grids enabling the small-scale flow around the building to be investigated. Fig. 3 shows the temperature perturbation field at $t = 5400$ s for grid 2. A cold outflow from the thunderstorm can be seen near the surface. The building is just discernible in this figure, centered at $x = 0.3$ km. Fig. 4 shows the streamlines at $t = 5400$ s for grid 3. This shows the microburst vortex centered at a height of 0.8 km and a smaller clockwise rotating vortex which has been shed from the building. There

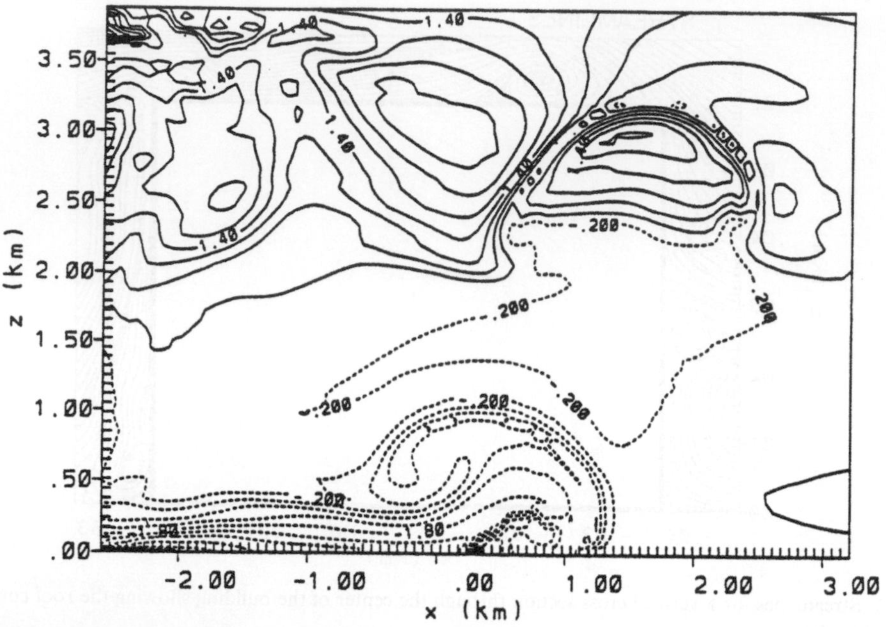

Fig. 3. The temperature perturbation field at $t = 5400$ s for grid 2. The contour interval is 0.4 K.

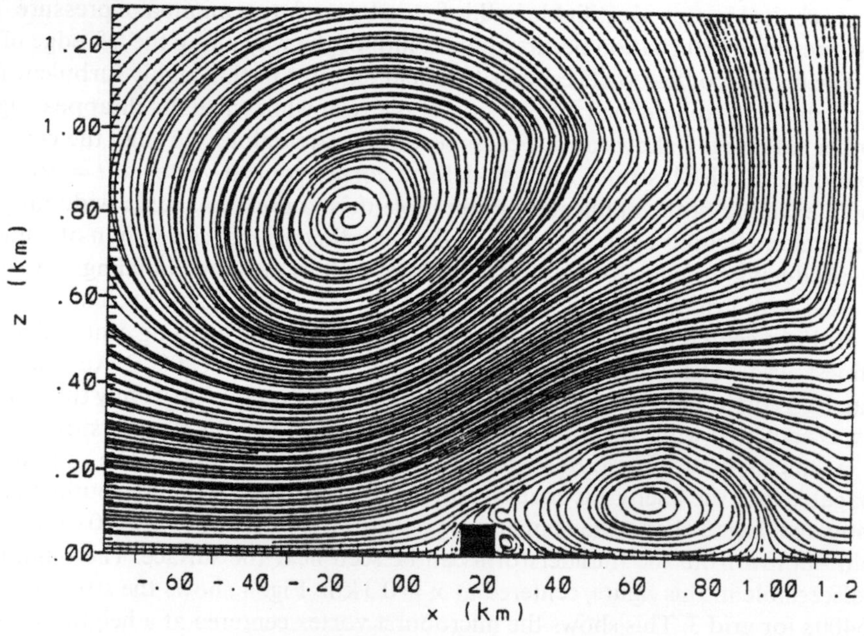

Fig. 4. Streamlines for grid 3 at $t = 5400$ s.

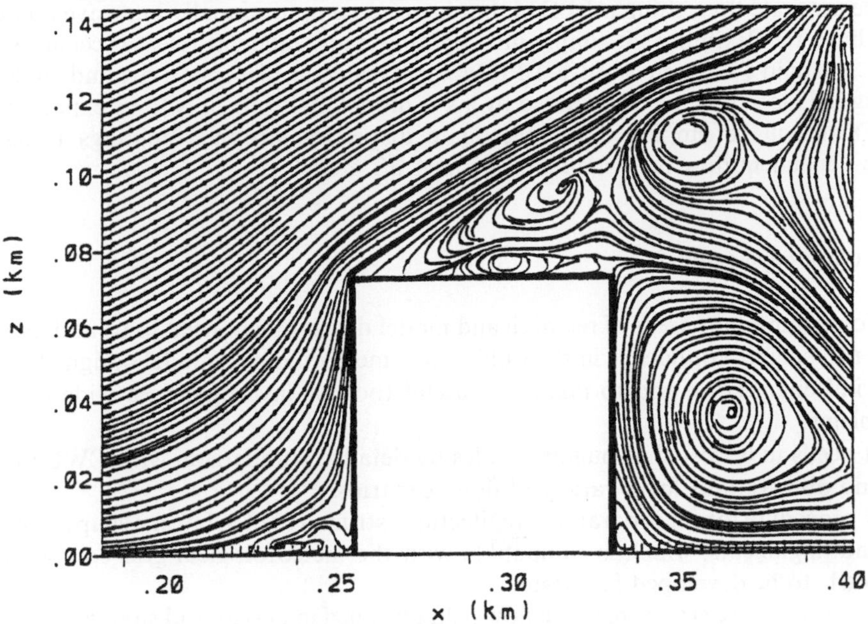

Fig. 5. Streamlines for grid 5 at $t = 5400$ s.

are two finer grids which further "telescope down" in scale. Fig. 5 shows the streamlines at $t = 5400$ s for the finest grid. Separation has occurred at the windward corner of the building. The clockwise rotating eddy which formed behind the building, seen in grid 3, has already advected out of the fine grid. For the simulated microburst, there are actually two wind maxima which occur near the surface. The first strong wind gust is at the leading edge of the outflow, while the second wind maxima is directly beneath the low-pressure center aloft. It is well known that pressure gradients can affect where separation occurs for flow over a flat plate and on the rate at which the boundary layer thickness grows. There was some indication that the large-scale pressure gradient associated with the microburst vortex may have had an effect on the angle the separated flow made with the roof.

Mesoscale meteorological models sometimes utilize dispersion models in order to study how pollutants are transported and their sources and receptors. The Lagrangian particle dispersion model (LPDM) developed in Refs. [31,32] is one such model. It can simulate the release and dispersion of a number of non-buoyant, passive pollutants from multiple sources of various geometries. The LPDM utilizes frequently written out data files created during the mesoscale model simulation. It runs very quickly and can be used to investigate many different release scenarios. Examples of the use of the LPDM can be found in the studies in Refs. [33–35]. Murakami et al. [36] presented results of a LES of the turbulent diffusion of buoyant and heavy gases near a building. In this investigation, equations for the gaseous constituents were added to the LES model and solved during the model run. This approach is inherently more accurate than using a LPDM, but is more computationally intensive and does not have the flexibility for simply investigating many different release scenarios. Lee and Naslund [37] employed a LPDM to investigate dispersion around buildings using the mean flow field simulated by a κ–ε model. In future, it is to be expected that LPDMs will be employed to investigate dispersion around buildings using the time-dependent flow field simulated by an LES model.

3. Future opportunities

Some important areas of research and model development need to be addressed in order to acquire confidence in the validity of a meteorological model designed for use in wind engineering and to make it a useful tool for investigating a wide range of problems:

1. It needs to be tested at building scales by detailed comparison with CWE models and observations for a variety of flow scenarios.
2. A procedure for incorporating architectural structures of arbitrary shape, number, and alignments, which is compatible with the meteorological grid configuration needs to be developed and tested.
3. The turbulence structure of the flow field obtained in nested grid simulations needs to be compared with the observations for various flow scenarios. This knowledge would aid in choosing appropriate nested grid configurations so that realistic turbulence spectra are obtained for a particular problem.
4. The LPDM needs to be tested at building scale.

A model such as this would have applications to a large number of problems where knowledge of wind loading, turbulence, or dispersion of particles are required. It would be particularly useful for situations where accurate prediction of mesoscale flow fields is required. As an example, it could be used to investigate the dispersion of particles around a group of buildings situated in complex terrain for both convectively stable and unstable conditions.

In conclusion, the use of a meteorological model for application in wind engineering studies has been established. Increased computer capabilities in future years will permit additional detail in the representation of building structures in these models and continued research and development could well lead to a powerful tool for investigating a wide range of problems. The main obstacle at present is the continuation of financial research and development support for this marriage of atmospheric science and wind engineering technologies.

Acknowledgements

This work was supported by the Army Research Office under Contract #DAAH04-94-G-0420 and by the Cooperative Program in Wind Engineering, NSF Grant #BCS-8821542. The typing of the paper and editorial review were very ably handled by Dallas McDonald.

References

[1] L.R. Richardson, Weather Prediction by Numerical Process, Cambridge University Press, London, 1922, Reprinted by Dover, 236 pp.
[2] R.A. Pielke, Mesoscale Meteorological Modeling, Academic Press, New York, 1984, 612 pp.
[3] T.N. Krishnamurti, L. Bounoua, An Introduction to Numerical Weather Prediction Techniques, CRC Press Inc., Boca Raton, Florida, 1996, 293 pp.
[4] P. Thunis, R. Bornstein, Hierarchy of mesoscale flow assumptions and equations, J. Atmos. Sci. 53 (1996) 380–397.
[5] R. Avissar, M.D. Moran, R.A. Pielke, G. Wu, R.N. Meroney, Operating ranges of mesoscale numerical models and meteorological wind tunnels for the simulation of sea and land breezes, Bound.-Layer Meteor. (Special Anniversary Issue, Golden Jubilee) 50 (1990) 227–275.
[6] M.E. Nicholls, R.A. Pielke, J.L. Eastman, C.A. Finley, W.A. Lyons, C.J. Tremback, R.L. Walko, W.R. Cotton, Applications of the RAMS numerical model to dispersion over urban areas, in: J.E. Cermak et al. (Eds.), Wind Climate in Cities, Kluwer Academic Publishers, The Netherlands, 1995, pp. 703–732.
[7] M.E. Nicholls, R.A. Pielke, R.N. Meroney, Large eddy simulation of microburst winds flowing around a building, J. Wind Eng. Ind. Aerodyn. 46–47 (1993) 229–237.
[8] L. Grasso, Numerical Simulation of the May 15 and April 26, 1991 tornadic thunderstorms, Atmos. Sci. Paper #596, Colorado State University, Department of Atmospheric Science, Fort Collins, CO 80523, 1996, 151 pp.
[9] C.L. Ziegler, T.J. Lee, R.A. Pielke, Convective initiation at the dryline: a modeling study, Mon. Wea. Rev., 1996, submitted.
[10] J.L. Eastman, Numerical simulation of Hurricane Andrew – Rapid intensification, 21st Conf. on Hurricanes and Tropical Meteorology, 24–28 April 1995, Miami, Florida, AMS, Boston, 1995, pp. 111–113.
[11] G.S. Poulos, The interaction of katabatic winds and mountain waves, Ph.D. Dissertation, Department of Atmospheric Science, Colorado State University, Fort Collins, CO 80523, 1996, in progress.
[12] U. Schumann, Direct and large eddy simulations of stratified homogeneous shear flows, Dyn. Atmos. Oceans 23 (1996) 81–98.
[13] R.L. Walko, W.R. Cotton, R.A. Pielke, Large-eddy simulations of the effects of hilly terrain on the convective boundary layer, Bound.-Layer Meteor. 58 (1992) 133–150.
[14] M.G. Hadfield, W.R. Cotton, R.A. Pielke, Large-eddy simulations of thermally-forced circulations in the convective boundary layer, Part I: a small-scale circulation with zero wind, Bound.-Layer Meteor. 57 (1991) 79–114.
[15] M.G. Hadfield, W.R. Cotton, R.A. Pielke, Large-eddy simulations of thermally forced circulations in the convective boundary layer, Part II: The effect of changes in wavelength and wind speed, Bound.-Layer Meteor. 58 (1992) 307–328.
[16] P.J. Mason, Large-eddy simulation: a critical review of the technique, Quart. J. Roy. Meteor. Soc. 120 (1994) 1–26.
[17] T.L. Clark, R.D. Farley, Severe downslope windstorm calculations in two and three spatial dimensions using anelastic interactive grid nesting: a possible mechanism for gustiness, J. Atmos. Sci. 41 (1984) 329–350.
[18] R.A. Pielke, W.R. Cotton, R.L. Walko, C.J. Tremback, W.A. Lyons, L.D. Grasso, M.E. Nicholls, M.D. Moran, D.A. Wesley, T.J. Lee, J.H. Copeland, A comprehensive meteorological modeling system – RAMS, Meteor. Atmos. Phys. 49 (1992) 69–91.
[19] S. Murakami, A. Mochida, K. Hibi, Three-dimensional numerical simulation of air flow around a cubic model by means of a large eddy simulation, J. Wind Eng. Ind. Aerodyn. 25 (1987) 291–305.
[20] S. Murakami, A. Mochida, Y. Hayashi, Examining the κ–ε model by means of a wind tunnel test and large eddy simulation of the turbulence structure around a cube, J. Wind Eng. Ind. Aerodyn. 35 (1990) 87–100.
[21] T. Tamura, K. Kuwahara, Numerical study of aerodynamic behaviour of a square cylinder, J. Wind Eng. Ind. Aerodyn. 33 (1990) 161–170.
[22] W. Frank, Large-eddy simulation of the three-dimensional flow around buildings, Wind Climate in Cities, Kluwer Academic Publishers, Dordrecht, 1995, pp. 669–679.

[23] S. Murakami, Comparisons of various turbulence models applied to a bluff body, J. Wind Eng. Ind. Aerodyn. 46–47 (1993) 21–36.
[24] J.A. Peterka, J.E. Cermak, Adverse wind loading induced by adjacent buildings, J. Struct. Div. Amer. Soc. Civil Eng. 102 (ST3) (1976) 533–548.
[25] L.S. Cochran, J.E. Cermak, Full and model scale cladding pressures on the Texas Tech Experimental Building, J. Wind Eng. Ind. Aerodyn. 41–44 (1992) 1589–1600.
[26] C. Kramer, H.J. Gerhardt, Wind pressures on roofs of very low and very large industrial buildings, J. Wind Eng. Ind. Aerodyn. 38 (1991) 285–295.
[27] K.C. Mehta, M.L. Levitan, R.E. Iverson, J.R. McDonald, Roof corner pressures measured in the field on a low building, J. Wind Eng. Ind. Aerodyn. 41 (1992) 181–192.
[28] H.W. Tieleman, D. Surry, J.X. Lin, Characteristics of mean and fluctuating pressure coefficients under corner (delta-wing) vortices, J. Wind Eng. Ind. Aerodyn. 52 (1994) 263–275.
[29] J.X. Lin, D. Surry, H.W. Tieleman, Wind tunnel experiments of pressures at roof corners of flat roof low buildings, J. Wind Eng. Ind. Aerodyn. 56 (1995) 235–265.
[30] F.H. Proctor, Numerical simulations of an isolated microburst. Part II: Sensitivity experiments, J. Atmos. Sci. 46 (1989) 2143–2165.
[31] R.T. McNider, Investigation of the impact of topographic circulations on the transport and dispersion of air pollutants, Ph.D. Dissertation, University of Virginia, Charlottesville, Virginia, 1981.
[32] R.T. McNider, M.D. Moran, R.A. Pielke, Influence of diurnal and inertial boundary layer oscillations on long-range dispersion, Atmos. Environ. 22 (1988) 2445–2462.
[33] J.L. Eastman, R.A. Pielke, W.A. Lyons, Comparison of lake-breeze model simulations with tracer data, J. Appl. Meteor. 34 (1995) 1398–1418.
[34] M.D. Moran, R.A. Pielke, Evaluation of a mesoscale atmospheric dispersion modeling system with observations from the 1980 Great Plains mesoscale tracer field experiment, Part I: data sets and meteorological simulations, J. Appl. Meteor. 35 (1996) 281–307.
[35] M.D. Moran, R.A. Pielke, Evaluation of a mesoscale atmospheric dispersion modeling system with observations from the 1980 Great Plains mesoscale tracer field experiment, Part II: dispersion simulations, J. Appl. Meteor. 35 (1996) 308–329.
[36] S. Murakami, A. Mochida, Y. Tominaga, Numerical simulation of turbulent diffusion in cities, Wind Climate in Cities, Kluwer Academic Publishers, 1995, pp. 681–701.
[37] R.L. Lee, E. Naslund, Numerical simulations of turbulent dispersion around buildings via a Lagrangian stochastic particle model, 7th Int. Symp. on Measurement and Modeling of Environmental Flows, San Francisco, CA, 1995.

Terrain Aerodynamics

ELSEVIER

Journal of Wind Engineering
and Industrial Aerodynamics 67&68 (1997) 375–386

JOURNAL OF wind engineering AND industrial aerodynamics

Flow and dispersion over hills: Comparison between numerical predictions and experimental data

David D. Apsley*, Ian P. Castro

University of Surrey, Guildford GU2 5XH, UK

Abstract

A finite-volume solver (SWIFT) for numerical prediction of flow and dispersion over hills is described. Comparison of predictions with experiment are made for (i) adiabatic flow over 2-D hills (RUSHIL experiment) and (ii) stably-stratified flow over real terrain (Cinder Cone Butte). Suitably-modified k–ε turbulence models are shown to yield satisfactory wind speed profiles (windpower) and terrain-amplification factors (dispersion applications).

Keywords: Turbulence modelling; Finite-volume methods; Atmospheric dispersion; k–ε model

1. Introduction

Atmospheric dispersion is greatly complicated by topography. Whilst large areas of orography give rise to a unique climatology, local flow perturbations caused by individual hills are of greater importance for isolated sources. Experimental studies have identified key features of such flows and provide a foundation for the development of predictive tools.

Methods for incorporating complex terrain in dispersion models range from simple plume-height corrections, through flow fields based on linearised equations [1] or potential-flow solutions [2], to computational fluid dynamics. Whilst the last has the *potential* to include all physical processes, its practical application to real terrain [3,4] demands considerable computer resources. Moreover, turbulence modelling in the atmospheric boundary layer (ABL) becomes important: we will address it in this paper.

In Section 2 we describe a finite-volume code for computing flow and dispersion over topography, followed by validation against laboratory and field dispersion data

* Corresponding author. Present address: Department of Mechanical Engineering, UMIST, Manchester M60 1QD, UK.

PII S0167-6105(97)00087-1

in Sections 3 and 4. In Section 5 we summarise the advantages and limitations of this method of prediction.

2. The SWIFT code

SWIFT (Stratified WInd Flow over Topography) is a finite-volume, incompressible, Navier–Stokes solver, using cartesian velocity decomposition on a staggered grid and a pressure-correction method to satisfy continuity. Simple switches control operation in one, two or three dimensions, Cartesian or terrain-fitting curvilinear meshes, transient or steady-state and a variety of (gradient-transport) turbulence models. A full specification can be found in Ref. [5].

2.1. Governing equations

The conservation equations for mass, momentum and potential temperature, with incompressible and Boussinesq approximations, are

$$\frac{\partial U_j}{\partial x_j} = 0,$$

$$\frac{\partial U_i}{\partial t} + \frac{\partial}{\partial x_j}(U_j U_i) = -\frac{1}{\rho_0}\frac{\partial}{\partial x_i}(P - P_a) + \alpha g(\Theta - \Theta_a)\delta_{i3} + \frac{\partial}{\partial x_j}\left[\nu\left(\frac{\partial U_i}{\partial x_j} + \frac{\partial U_j}{\partial x_i}\right) - \overline{u_i u_j}\right],$$

$$\frac{\partial \Theta}{\partial t} + \frac{\partial}{\partial x_j}(U_j \Theta) = \frac{\partial}{\partial x_j}\left(\kappa\frac{\partial \Theta}{\partial x_j} - \overline{u_j \theta}\right),$$

where u_i is velocity, p pressure, ρ density and θ potential temperature. Upper and lower case distinguish mean and turbulent values, respectively. ν and κ are molecular viscosity and diffusivity and α is the coefficient of expansion. Subscript a denotes an ambient value. A comparable scalar equation is solved for concentration C.

Turbulent stresses are derived from the k–ε eddy-viscosity model:

$$-\overline{u_i u_j} = -\tfrac{2}{3}k\delta_{ij} + \nu_t\left(\frac{\partial U_i}{\partial x_j} + \frac{\partial U_j}{\partial x_i}\right), \qquad -\overline{u_j \theta} = \frac{\nu_t}{\sigma_\theta}\frac{\partial \Theta}{\partial x_j}, \qquad \nu_t = C_\mu \frac{k^2}{\varepsilon}.$$

Scalar transport equations are solved for turbulent kinetic energy, k, and dissipation rate, ε:

$$\frac{\partial k}{\partial t} + \frac{\partial}{\partial x_j}(U_j k) = \frac{\partial}{\partial x_j}\left[(\nu + \nu_t/\sigma_k)\frac{\partial k}{\partial x_j}\right] + \Pi_k - \varepsilon,$$

$$\frac{\partial \varepsilon}{\partial t} + \frac{\partial}{\partial x_j}(U_j \varepsilon) = \frac{\partial}{\partial x_j}\left[(\nu + \nu_t/\sigma_\varepsilon)\frac{\partial \varepsilon}{\partial x_j}\right] + (C_{\varepsilon 1}\Pi_k - C_{\varepsilon 2}\varepsilon)\frac{1}{\tau_\varepsilon} \quad (\tau_\varepsilon = k/\varepsilon),$$

where the *total production rate*, Π_k, is the sum of contributions P_k and G_k from mean-shear and buoyancy, respectively. With the exception of σ_ε (chosen for consistency with the log layer), constants take standard values: $C_\mu = 0.09$, $C_{\varepsilon 1} = 1.44$, $C_{\varepsilon 2} = 1.92$, $\sigma_k = 1.0$, $\sigma_\varepsilon = 1.11$, $\sigma_\theta = 0.9$. The following k–ε model modifications were examined.

(i) Streamline curvature modification [6]

$$C_\mu \to \frac{C_\mu}{1 + 0.2\left(\dfrac{k}{\varepsilon}\right)^2 \dfrac{\partial U_s}{\partial n}\dfrac{U_s}{R_c}},$$

where R_c is the radius of curvature and n directed away from the centre of curvature (Ref. [5] discusses a 3-D version). As deduced from the transport equations for $\overline{u_i u_j}$, streamline curvature can diminish or enhance turbulence according to the sign of $(\partial U_s/\partial n)(U_s/R_c)$.

(ii) Dissipation modification (for streamwise strains) [6]: Anomalous increases in dissipation length scale, $l_\varepsilon \equiv C_\mu^{3/4} k^{3/2}/\varepsilon$, predicted in streamwise pressure gradients may be inhibited by rewriting the ε production term as

$$P_\varepsilon \equiv C_{\varepsilon 1}\frac{\Pi_k}{\tau_\varepsilon} \to [C'_{\varepsilon 1}\Pi_k - C''_{\varepsilon 1}\nu_t(2\hat{S}_{12})^2]\frac{1}{\tau_\varepsilon}$$

(where $\hat{S}_{12}$ is the shear strain in a locally streamwise-oriented Cartesian system). The modified constants take values $C'_{\varepsilon 1} = 2.24$, $C''_{\varepsilon 1} = C'_{\varepsilon 1} - C_{\varepsilon 1}$.

(iii) Limited-length-scale k–ε model [5,7]. In near-equilibrium flow l_ε may be limited to some global maximum, l_{max}, by writing

$$P_\varepsilon \to \left[C_{\varepsilon 1} + (C_{\varepsilon 2} - C_{\varepsilon 1})\frac{l_\varepsilon}{l_{max}}\right]\frac{\Pi_k}{\tau_\varepsilon}.$$

In the neutral or stably-stratified ABL a typical maximum eddy size is $l_{max} = \min(h/3, 0.08L_{MO})$ (see Ref. [7]), where h is the ABL height and L_{MO} the Monin–Obukhov length.

2.2. Numerical methods

All calculations to be described in this paper were steady-state. In a curvilinear system $(\xi_i) = (\xi, \eta, \zeta)$ the time-independent conservation equations may be written

$$\frac{\partial}{\partial \xi_i}\left[J\frac{\partial \xi_i}{\partial x_j}(\rho U_j \phi + F_j)\right] = Js, \qquad J \equiv \left|\det\left(\frac{\partial x_i}{\partial \xi_j}\right)\right|,$$

where s is source density and F_j the (Cartesian) components of the flux. Separating F_j into gradient and non-gradient terms, $-\Gamma(\partial\phi/\partial x_j)$ and F'_j, respectively, this can be written in canonical form (no implicit summation over greek indices):

$$\sum_\alpha \frac{\partial}{\partial \xi_\alpha}\left(C_\alpha \phi - \Gamma_\alpha \frac{\partial \phi}{\partial \xi_\alpha}\right) = Js - \sum_\alpha \frac{\partial}{\partial \xi_\alpha}\left(J\frac{\partial \xi_\alpha}{\partial x_j}F''_j\right),$$

where

$$C_\alpha = J\frac{\partial \xi_\alpha}{\partial x_j}U_j\,,\quad \Gamma_\alpha = \Gamma\left(\frac{\partial \xi_\alpha}{\partial x_j}\right)^2,\quad F''_j = F'_j - \Gamma\sum_{\beta\neq\alpha}\frac{\partial \xi_\beta}{\partial x_j}\frac{\partial \phi}{\partial \xi_\beta}\,.$$

Thus, the curvilinear system can be treated as cartesian, but with "off-diagonal" diffusion terms transferred to the rhs. A simple terrain-following curvilinear system is used here:

$$\xi = x\,,\qquad \eta = y\,,\qquad \zeta = \frac{z - z_s(x,y)}{1 - z_s(x,y)/D},$$

where $z_s(x,y)$ and D are the heights of the local terrain and computational domain.

Integrating over control volumes in ξ-space we derive coupled matrix equations

$$\phi_P - \sum_N a_N\phi_N = b_P\,,$$

where the sum is over neighbouring nodes. We used the 2nd-order, upwind-biased, flux-limited *harmonic* advection scheme of Van Leer [8]. For momentum, each Cartesian component is treated as a scalar and (coordinate-wise) pressure gradient separated off:

$$U_P - \sum_N a_N U_N = d_P(P_{P-\xi} - P_P) + b_P\,.$$

(The decomposition relies on the grid being – as here – only mildly distorted from the Cartesian, so that $\partial P/\partial \xi_i$ is primarily responsible for driving U_i.) In the pressure-correction method a relationship between velocity and pressure changes, $U'_P = \delta_P(P'_{P-\xi} - P'_P)$, is used to formulate a mass-conservation equation and steer the iterative solution toward continuity. The SIMPLEX algorithm [9] (equations solved for the δ_P) proves far superior to the original SIMPLE scheme ($\delta_P \equiv d_P$).

Individual matrix equations are solved by line-iteration procedures in conjunction with the tri-diagonal matrix algorithm. Solution of the pressure-correction equation is accelerated by "block-correction" [10] and "anticipated-correction" [11] procedures. The first precedes directional sweeps by adding block corrections ($\phi_{ijk} \to \phi_{ijk} + C_i$) satisfying the integral equations. The second minimises error associated with the directional nature of the sweep by estimating the increment on forward lines not yet updated.

3. The RUSHIL Experiment

The RUSHIL wind-tunnel study [12] made detailed measurements of mean velocity and turbulent stresses and of concentrations from isolated sources near 2-D hills of varying slope, immersed in a deep, rough-wall, adiabatic boundary layer. Smoothed data was made available by Dr. F. Tampieri. Hills H3, H5 and H8 (the number designating the length/height ratio a/H) were examined at a Reynolds number

$U_\infty H/\nu = 3.1 \times 10^4$. Flow separation from the lee slope was observed for H3, but not H8, as confirmed by our simulations.

3.1. Flow-field calculations

Definitive flow calculations were carried out on a variably-spaced 2-D grid of 100×80 nodes, extending $\pm 40H$ (H is hill height) either side of hill centre and to a height of $13.7H$ (at which the wind-tunnel free-stream velocity was maintained constant). Refinement tests confirmed satisfactory grid independence. For concentration calculations, flow solutions were interpolated onto (3-D) grids selectively refined near the source.

Rough-wall, equilibrium-boundary-layer profiles with roughness length $z_0/H = 1.3 \times 10^{-3}$ and friction velocity $u_*/U_\infty = 0.047$ were assumed at inflow, and zero longitudinal gradient was prescribed for all variables at outflow. (Radiation boundary conditions were unnecessary because the calculation was steady-state.) Non-slip boundary conditions were imposed at the lower boundary by wall functions, whilst a constant free-stream velocity was maintained on the upper boundary (with small vertical velocity W so as to satisfy continuity in the topmost line of control volumes).

We first examined turbulence-model modifications for streamwise pressure gradient and curvature. Simulations for H3 (Fig. 1) indicated that the first had a greater impact on the *mean* flow, predicting magnitude and depth of reversed flow and sharp gradients in the separated shear layer better than the standard k–ε model. The distance to reattachment increased from $4.1H$ to $5.0H$ (still short of the experimental value $6.5H$). The curvature correction had little impact on the mean flow because the factor differed significantly from unity only in the outer layer, where mean-flow dynamics were essentially inviscid.

Since the extent of recirculation was important in diffusion calculations we performed all further calculations with the streamwise-pressure-gradient modification. Summit profiles of mean velocity were well predicted (Fig. 2), suggesting that this level of modelling is appropriate for wind-power applications.

3.2. Concentration calculations

Since concentration sources were non-buoyant and isokinetic, concentration calculations were performed after the flow field, using grids selectively refined near the point of emission. In addition we used a locally-analytic imbedding technique [5] to prevent excessive initial diffusion due to grid cells much larger than the plume dimensions.

From a calculation of diffusion over flat terrain we found that an isotropic eddy diffusivity underpredicted lateral spread, but, by rescaling concentrations by the plume centre-line value, that vertical diffusion was well represented. A simple Gaussian-plume analysis shows that *maximum* glc is independent of (a constant) ratio of lateral-to-vertical spread and inversely proportional to the square of the plume height, suggesting that it should be possible to predict *terrain amplification factors*, the

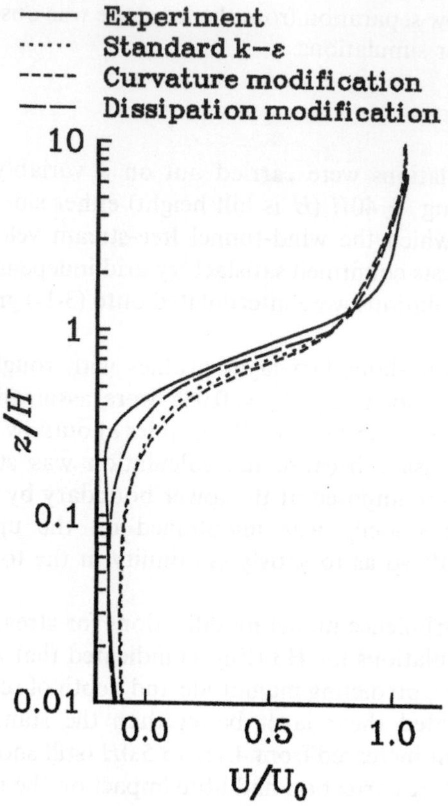

Fig. 1. RUSHIL (H3) – effect of turbulence model on mean flow profiles at downwind base.

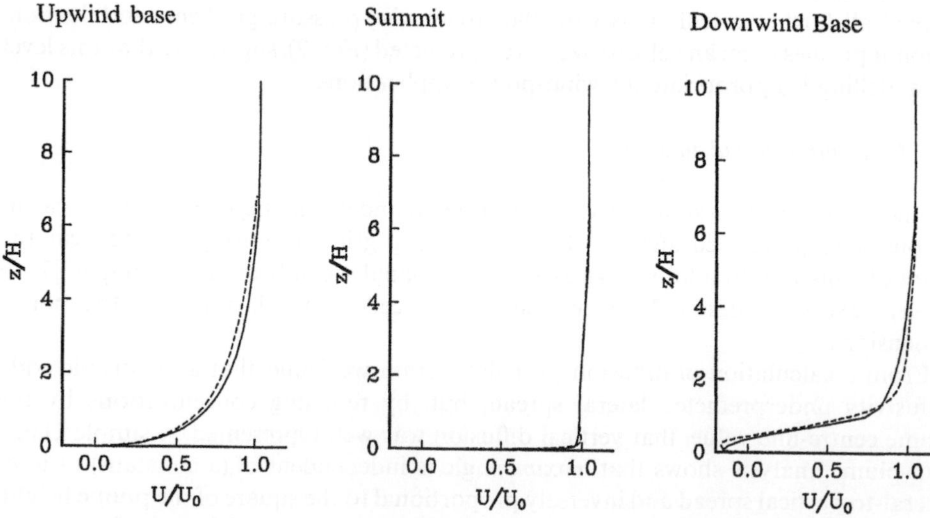

Fig. 2. RUSHIL (H3) – mean flow profiles. (Dashed line – experiment, solid line – predictions.)

relative increase in maximum ground-level concentration (glc) over that for flat terrain.

In Fig. 3 we plot centre-line glc non-dimensionalised by the maximum glc over flat terrain (experimental or computational as appropriate), for sources at height $H/4$ (relative to local terrain height) and various locations over H3. (With the exception of the downwind source for H5 – in which case intermittency of separation cannot be resolved by a time-averaged model – comparable agreement is achieved for the other hills.) Terrain amplification factors are well predicted and reflect both ranking and relative position of maxima for sources in different positions. Highest concentrations arise from a source at the downwind base.

4. Cinder Cone Butte simulation

Tracer releases were carried out by the EPA at Cinder Cone Butte (CCB), a 100 m high isolated hill, to provide data for a complex-terrain model in strongly stratified flow [13]. One particular case-study hour (expt. 206, 0500–0600 local time), which was complemented by a laboratory towing-tank simulation [14], was chosen for numerical calculation.

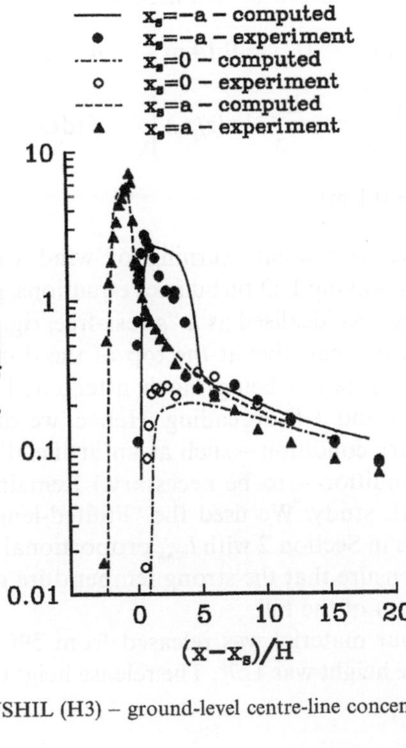

Fig. 3. RUSHIL (H3) – ground-level centre-line concentrations.

A key element of dispersion models for strongly stratified flow around isolated hills is the assumption of a *dividing streamline height*, H_c, given by

$$\tfrac{1}{2}U_a^2(H_c) = \int_{H_c}^{H} (H - z')N_a^2(z')\,dz',$$

where H is hill height and $N \equiv \alpha g(\partial\Theta/\partial z)^{(1/2)}$ buoyancy frequency. In an idealised model, H_c separates a lower layer where fluid is compelled (by energy constraints) to diverge horizontally from an upper layer where the flow passes over the hill. Our case-study hour represented the important case of a source at or near H_c, directly upwind of the hill.

Flow simulations were conducted on a 3-D, terrain-fitting curvilinear mesh of $70 \times 53 \times 40$ nodes with horizontal boundaries at $x = -5000$ m and 8000 m, $y = \pm 2500$ m and domain height 2000 m. (Limited computer resources did not permit detailed grid-dependence tests, but comparable calculations in 2-D suggested that the present grid resolution would be sufficient for a single topographic feature.) The coordinate origin was placed in the draw between CCB's twin peaks and the x-axis aligned with the mean wind (at source height). Surface heights were interpolated from digitised contours. A second grid covering a smaller domain and selectively refined near the source was used to compute concentrations *after* the flow-field calculation.

Power-law and log-linear profiles were used for mean velocity and temperature:

$$U(z) = U_0(z/150\text{ m})^n \qquad (U_0 = 9.14\text{ m s}^{-1}, n = 0.9),$$

$$\Theta = \Theta_0 + \left(\frac{d\Theta}{dz}\right)_\infty \left[z + \frac{L_{MO}}{5}\ln(z/z_0)\right] \qquad ((d\Theta/dz)_\infty = 2.98 \times 10^{-2}\text{ K m}^{-1},$$

$$L_{MO} = 33\text{ m}, z_0 = 0.1\text{ m}).$$

No attempt was made to simulate turning of wind with height. k and ε were determined at inflow by solving 1-D turbulence equations, given the profiles of U and Θ. The upper boundary was idealised as a "stress-free, rigid lid". (Pressure perturbations were observed to be negligible at the top of the domain, any topographically induced internal wave radiation being much attenuated by turbulent dissipation, weakening stratification and 3-D spreading. Hence, we did not consider any more complex upper boundary condition – such as an artificial "viscous sponge" layer or radiation boundary condition – to be necessary.) Remaining boundary conditions were as in the RUSHIL study. We used the "limited-length-scale" k–ε turbulence-model variant described in Section 2 with l_{max} proportional to L_{MO}. The most obvious practical effect was to ensure that the strong temperature gradient at inflow was not severely eroded upstream of the hill.

In the case-study hour material was released from 596 m on bearing 123.6°: the wind direction at source height was 127°. The release height was 35 m, compared with $H_c = 32$ m.

4.1. Mean flow pattern

The calculated streamline flow pattern around CCB is illustrated in Fig. 4. Immediately apparent is a transition from near-horizontal low-level flow to less perturbed

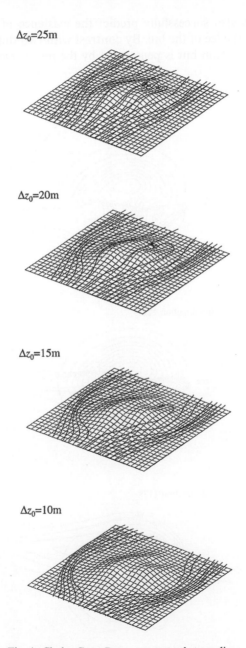

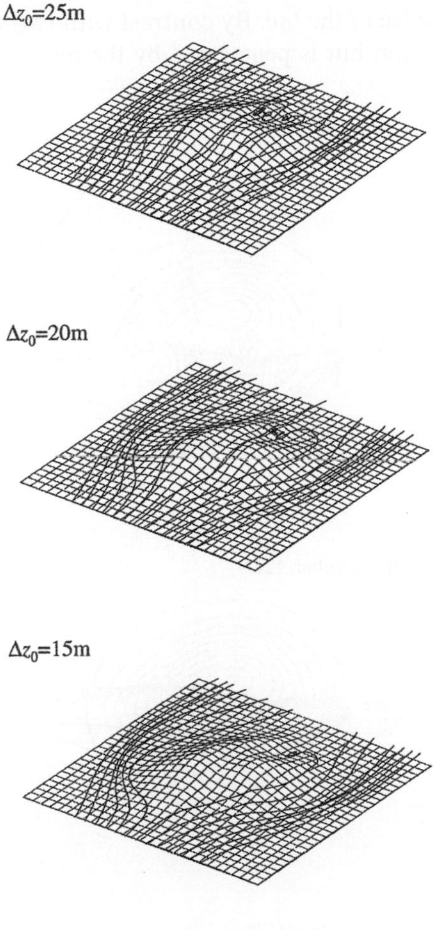

Fig. 4. Cinder Cone Butte – computed streamlines.

flow above. By contrast with the dividing-streamline model the transition is gradual, occurring largely between 15 and 20 m, compared with $H_c = 32$ m. The change in horizontal divergence, however, is responsible for greatly increased lateral plume spread and skewness, since a vertical separation of just 15 m can give rise to lateral separation of 300 m.

The computations also successfully predict the existence of a three-dimensional recirculation zone in the lee of the hill. By contrast with two-dimensional topologies, this is not a "closed" region but is penetrated by the mean streamlines.

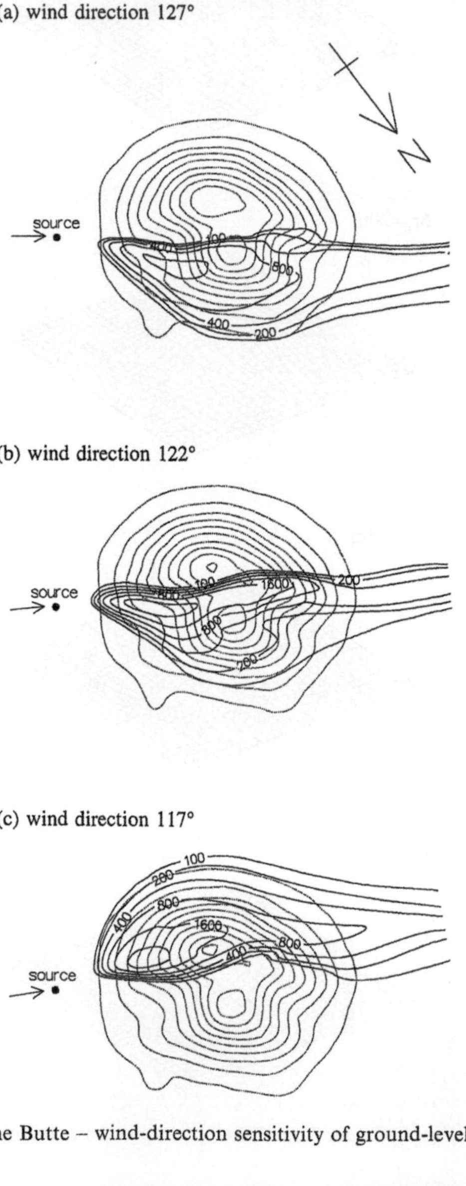

Fig. 5. Cinder Cone Butte – wind-direction sensitivity of ground-level concentrations.

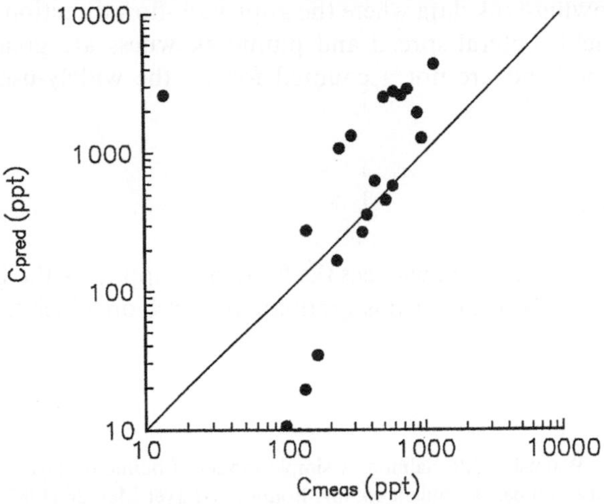

Fig. 6. Cinder Cone Butte – comparison of measured and predicted glc.

4.2. Concentration distributions

Ground-level concentrations were found to be extremely sensitive to wind direction. By conducting a series of flow and dispersion calculations for wind directions 127°, 122° and 117° (rebuilding inflow-oriented grids for each) we obtained the glc distributions illustrated in Fig. 5. These compare favourably with the corresponding towing-tank simulations [14] and demonstrate a plume switching sides of the hill as the approach flow veered by only 10°. Averaged over an hour in the field these switches are responsible for greatly enhanced crosswind spread and lower mean concentrations.

Wind-direction sensitivity meant that the distribution of one-hour average field data was best represented by our computations for 122° and comparison with experimental data is plotted in Fig. 6. Note the tendency to over-predict hourly-average concentrations, for the reasons discussed above.

5. Conclusions

A finite-volume flow solver for computing ABL flow over arbitrary terrain with two-equation turbulence models has been described. Comparison with wind-tunnel data for 2-D hills demonstrate that this level of modelling is capable of predicting summit speed-up and terrain amplification factors provided the basic features of the flow (such as separation zones) are captured: lateral spread is, however, inevitably underestimated with an isotropic eddy-diffusivity model.

Calculations in strongly stratified flow reveal greater complexity than is implied by dividing-streamline methods and a strong sensitivity to wind direction, in good agreement with towing-tank data where the approach-flow direction is better controlled than in the field. Lateral spread and plume skewness are greatly enhanced by low-level divergence and are not accounted for by the widely-used narrow-plume assumption.

Acknowledgements

The first author (DDA) acknowledges the financial support of the Natural Environment Research Council during the postgraduate research on which this paper is based.

References

[1] P.A. Taylor, J.L. Walmsley, J.R. Salmon, A simple model of boundary-layer flow over real terrain incorporating wavenumber-dependent scaling, Boundary-Layer Met. 26 (1983) 169–189.

[2] S.G. Perry, CTDMPLUS: A dispersion model for sources near complex topography. Part I: Technical formulations, J. Appl. Met. 31 (1992) 633–660.

[3] G.D. Raithby, G.D. Stubley, P.A. Taylor, The Askervein hill project: a finite control volume prediction of three-dimensional flows over the hill, Boundary-Layer Met. 39 (1987) 247–267.

[4] P. Dawson, D.E. Stock, B. Lamb, The numerical simulation of airflow and dispersion in three-dimensional atmospheric recirculation zones, J. Appl. Met. 30 (1991) 1005–1024.

[5] D.D. Apsley, Numerical modelling of neutral and stably stratified flow and dispersion in complex terrain, Ph.D. Thesis, University of Surrey, 1995.

[6] M.A. Leschziner, W. Rodi, Calculation of annular and twin parallel jets using various discretisation schemes and turbulence-model variations, J. Fluids Eng. 103 (1981) 352–360.

[7] D.D. Apsley, I.P. Castro, A limited-length-scale k–ε model for the neutral and stably-stratified atmospheric boundary layer, Boundary-Layer Met. 83 (1997) 75.

[8] B.P. Leonard, S. Mokhtari, Beyond first-order upwinding: the ULTRA-SHARP alternative for non-oscillatory steady-state simulation of convection, Int. J. Numer. Methods Eng. 30 (1990) 729–766.

[9] G.D. Raithby, G.E. Schneider, Elliptic systems: finite difference method II, Ch. 7, in: W.J. Mincowycz et al. (Eds.), Handbook of Numerical Heat Transfer, Wiley, New York, 1988.

[10] S.V. Patankar, Elliptic systems: finite difference method I, Ch. 6, in: W.J. Mincowycz et al. (Eds.), Handbook of Numerical Heat Transfer, Wiley, New York, 1988.

[11] J.P. Van Doormaal, G.D. Raithby, Enhancements of the SIMPLE method for predicting incompressible fluid flows, Numer. Heat Transfer 7 (1984) 147–163.

[12] L.H. Khurshudyan, W.H. Snyder, I.V. Nekrasov, Flow and dispersion of pollutants over two-dimensional hills, US EPA Report EPA-600/4-81-067, 1981.

[13] T.F. Lavery, A. Bass, D.G. Strimaitis, A. Venkatram, B.R. Green, P.J. Drivas, EPA complex terrain model development: First milestone report, US EPA Report EPA-600/3-82-036, 1982.

[14] W.H. Snyder, R.E. Lawson, Laboratory simulation of stable plume dispersion over Cinder Cone Butte, Appendix to Ref. [13], 1981.

ELSEVIER

Journal of Wind Engineering
and Industrial Aerodynamics 67&68 (1997) 387–401

JOURNAL OF
wind engineering
AND
industrial
aerodynamics

Cellular convection embedded in the convective planetary boundary layer surface layer

David S. DeCroix, Yuh-Lang Lin*, David G. Schowalter

Department of Marine, Earth and Atmospheric Sciences, North Carolina State University, Raleigh, NC 27695-8208, USA

Abstract

Cellular convection was first studied in the laboratory by Benard [Ann. Chim. Phys. 23 (1901) 62–144] and Rayleigh [Phil. Mag. Ser. 6 (1916) 529–546] investigated these motions from a theoretical perspective. He defined a dimensionless number, now called the Rayleigh number, which is the ratio of convective transport to molecular transport, and found that if a certain critical value is exceeded, cellular convection occurs. Mesoscale cellular convection (MCC) is a common occurrence in the planetary boundary layer. Agee [Dyn. Atmos. Oceans 10 (1987) 317–341] discussed the similarities and differences of MCC and classical Rayleigh–Benard convection. A similar cellular pattern can be seen in the convective boundary layer (CBL) surface layer. It is known that in the CBL, air near the surface converges into thermals producing updrafts. This produces a 'spoke' type pattern similar to the mesoscale cellular or Rayleigh–Benard convection. This paper will focus on applying Rayleigh–Benard convection criteria, using a linearized perturbation method, to the CBL surface layer produced by Large Eddy Simulation (LES). We will investigate the length scales of turbulence in the CBL surface layer and compare them to those predicted from linear theory. Similarities and differences will be discussed between the LES produced surface layer and classical Rayleigh–Benard convection theory.

Keywords: Rayleigh–Benard convection; Cellular convection; Large eddy simulation; Planetary boundary layer

1. Introduction

Cellular convection was first studied in the laboratory by Benard [1]. He used a very thin layer of fluid, about 1 mm deep, which was heated from below at a constant uniform temperature. He noticed that a number of convective hexagonal cells appeared. These convective cells, shown in Fig. 1, were produced by downward motion

* Corresponding author. E-mail: yl_lin@ncsu.edu.

PII S0167-6105(97)00088-3

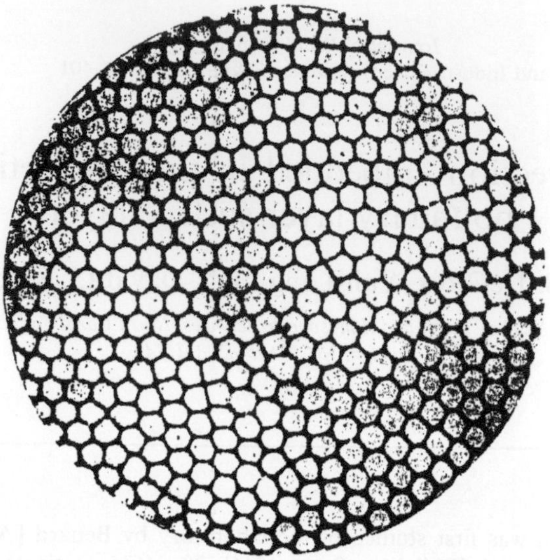

Fig. 1. Benard cells in spermaceti. From Ref. [17] with permission from Dover.

in the cell center and upward motion on the edges shared with adjacent cells. Lord Rayleigh [2] investigated these motions from a theoretical perspective. He defined a dimensionless number, now called the Rayleigh number, which is the ratio of convective transport and molecular transport. He found that if a certain critical value is exceeded, cellular convection occurs.

Classical Rayleigh–Benard (RB) convection was originally thought of as an interesting laboratory phenomenon, but of little meteorological interest. In the 1960's, however, satellites provided meteorologists with their first look at cellular convection in the atmosphere, which often occurs during cold air outbreaks off the eastern coast of North America. In Fig. 2, one sees mesoscale cellular convection (MCC) cells similar to the hexagonal cells Benard discovered in the laboratory. These convection cells may be classified as open or closed. In open (closed) cellular convection, downward motion occurs in the center (edges) of the cell and upward motion on the cell edges (center). Clouds often form in the updraft regions, and in open cellular convection these clouds form hexagonal rings. As discussed by Agee [3] and Stull [4], there are some discrepancies between MCC and RB convection, cell aspect ratio for instance, but the physical mechanism responsible for MCC and RB convection is the same.

A similar cellular pattern can be seen in the convective boundary layer (CBL) surface layer. It is well known that in the CBL, air near the surface converges into thermals producing updrafts. In Fig. 5, one sees a 'spoke' type pattern, which was also observed in Schmidt and Schumann's [5] LES results. This pattern is similar to the mesoscale cellular and Rayleigh–Benard convection. The CBL in Fig. 5 was modeled

Fig. 2. Hexagonal cells north of Cuba. From Ref. [3].

using a large eddy simulation (LES) model developed by Proctor [6,7] and North Carolina State University [8].

This paper will focus on applying Rayleigh Benard convection criteria, using a linearized perturbation method, to the CBL surface layer produced by large eddy simulations. Similarities and differences between the LES produced surface layer and classical Rayleigh–Benard convection theory will be discussed. However, there is an inherent difficulty comparing scales of a laminar transition instability to scales in a fully developed turbulent boundary layer. But, as shown by Brown and Roshko [9], large scale turbulent structures can sometimes be attributed to laminar flow instabilities. We will also investigate the turbulence length scales and structures within the surface layer, and attempt to determine whether the dominant length scales are related to the linear flow instability. We are interested in finding and understanding the length scales in the surface layer, and the planetary boundary layer (PBL), so we can determine what scales interact with aircraft wake vortices. The long term goal of this research, as discussed by Hinton [10], is to quantify the PBL interactions with wake vortices in order to understand and better predict the transport and decay of the vortex.

2. Theoretical model

Emanuel [11] performed a linear stability analysis of the Boussinesq form of the Navier–Stokes equations. He derived a single equation for the perturbation vertical velocity for the flow between two parallel plates and solved it using a normal mode

Table 1
Critical Rayleigh numbers and horizontal wave numbers for flow between parallel plates

Type of boundary	Ra_c	k_c
Both free-slip	657.5	2.22
One no-slip, one free-slip	1100.7	2.68
Both no-slip	1707.8	3.12

approach. This solution was used to determine the critical stability condition for the fluid system. As a result of the nondimensionalization, the Prandtl number σ and the Rayleigh number Ra were defined as

$$\sigma \equiv \frac{\nu}{\kappa}, \qquad \mathrm{Ra} \equiv -\frac{g\beta\Gamma H^4}{\nu\kappa}, \tag{1}$$

where ν is the kinematic viscosity, κ the thermal conductivity, β the coefficient of thermal expansion, Γ the vertical temperature gradient, and H the fluid layer depth.

Emanuel solved the critical flow condition for three vertical (parallel plate) boundary conditions: (1) both upper and lower boundaries are rigid free-slip, (2) the lower boundary is rigid no-slip and the upper is rigid free-slip, and (3) both upper and lower are rigid, no-slip boundaries. Table 1 summarizes these conditions and their corresponding critical Rayleigh and wave numbers. When the Rayleigh number is exceeded, flow instability exists between the two parallel plates.

In this paper we will apply the critical Rayleigh and wave number condition for the second boundary condition type, one no-slip, one free-slip rigid boundaries, to flow in the convective boundary layer surface layer generated by large eddy simulation. The supposition is that the CBL surface layer could be thought of as being restricted in the vertical due to the presence of the ground and the CBL mixed layer, which has no potential temperature gradient, and that the cellular pattern seen is due to this mechanism. This upper boundary condition is not strictly valid for the surface layer flow since the upper boundary is not a rigid free-slip boundary, but a limit to the superadiabatic region.

3. The LES model

The LES model used for the simulations is the TASS model developed by Proctor [6,7]. The model was originally developed for the study of thunderstorms and microbursts, but only required a change in boundary conditions for the simulation of the planetary boundary layer [8]. The equations solved are the three dimensional, fully compressible, non-hydrostatic Navier–Stokes equations. A modified

Smagorinsky first order closure is used in which the eddy viscosity is a function of stability through the local flux Richardson number. Equations for water substances, cloud water, cloud ice, snow, hail and graupel, are present in the model, but were not used in these simulations.

These equations were solved on an Arakawa C type mesh [12]. Periodic boundary conditions have been used in the horizontal directions, while a sponge layer with three grid intervals has been added on the top of the physical domain. At the top boundary, there exists neither heat nor mass transfer.

The lower boundary represents a solid ground plane and employs a no-slip condition. In these simulations, the heat transfer at the surface is computed from the specified ground surface temperature, given as a constant, and the LES computed air temperature at the first grid level above the ground. The reported values of heat flux are the ensemble average of the individual grid point fluxes.

4. Discussion of results

The LES model was run on a domain size of 4 km in the N-S and E-W directions and 2 km in the vertical. The grid resolution used was 50 m laterally, with the vertical resolution varied from 10 m near the surface to 50 m at the domain top enabling higher resolution of the surface layer.

The model was initialized with a vertical profile or sounding of the environmental pressure, temperature, dew point, u and v vector wind components representative of a dry convective boundary layer. It is not an observed sounding, but rather an idealized sounding for numerical experiments. A specified temperature within the lowest level in the model was used to compute a surface heat flux. In order to initiate the convection, a uniform random temperature perturbation of ± 1 K (maximum) was applied to the lowest three vertical levels in the domain. The model was run for 2 h of simulation time to develop fully the convective boundary layer.

To obtain ensemble averages for the properties in the boundary layer, denoted by $\langle\ \rangle$, the variables were averaged in space and time. The variables were averaged at each vertical level in the domain, and then each of those averaged over a time period. After two hours of simulation (spin-up time), the data were output at 2 min intervals for 40 min, producing 20 time-averaging periods. For example, $\langle u'u' \rangle$ was calculated at each vertical level in the model domain by first computing the variance at each grid point in the domain, then horizontally averaging them at each vertical level, and finally averaging each of those horizontal averages over the 20 output times.

Three simulations were performed with progressively lower surface temperatures, in order to create supercritical and subcritical Rayleigh number conditions. These surface temperatures used were 287, 285, and 283.75 K, which produced surface heat fluxes of 237, 61, and 18 W/m^2 and are denoted as the high, medium, and low heat flux cases, respectively. In each case, the mean wind was constant with height and less than 1 m/s to diminish any shear instabilities that could be present due to the environment.

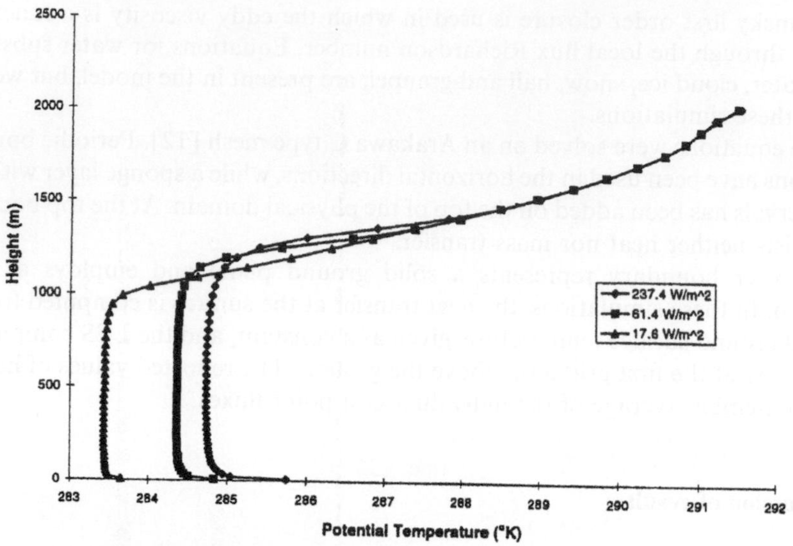

Fig. 3. Ensemble averaged potential temperature profile from LES.

4.1. *Some properties of the convective boundary layer*

Fig. 3 shows the potential temperature profiles of each of the three cases. In each profile, the potential temperature is superadiabatic near the surface, nearly constant within the mixed layer, and has an inversion capping the PBL. The surface layer features of the convective boundary layer constitute a mechanism similar to the laboratory Rayleigh–Benard convection. In the laboratory setup, the lower plate is held at a constant temperature which is greater than the upper plate. In the CBL surface layer, the lower surface is also maintained at a constant temperature greater than the mixed layer. This is analogous to the theoretical results for 'one fixed boundary, one free boundary'.

Figs. 4 and 5 show the horizontal structure of the vertical velocity and potential temperature. Notice the cellular-type appearance of these contours. These cellular patterns, typical of convective surface layers in the PBL, were the motivation of this paper's research.

4.2. *Rayleigh number calculation*

A Rayleigh number, defined in Eq. (1), was calculated from the fields produced in the LES. For this calculation, the turbulent eddy viscosity, ν_e, and conductivity, κ_e, must be used since the flow is turbulent. From the definition of the eddy viscosity,

$$\frac{\tau}{\rho} = \nu_e \frac{\partial \bar{u}}{\partial z} = \langle u'w' \rangle, \qquad (2)$$

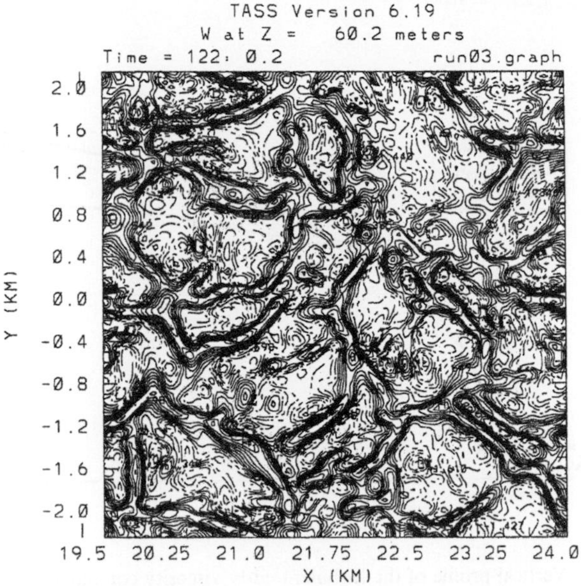

Fig. 4. Vertical velocity contours at 60 m height for the high heat flux case.

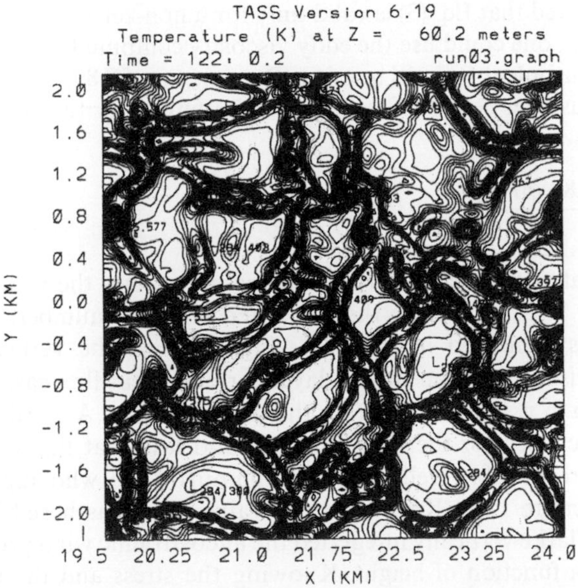

Fig. 5. Potential temperature contours at 60 m height for the high heat flux case.

thus for LES,

$$\nu_e = \frac{\langle u'w' \rangle}{\partial \bar{u}/\partial z}. \tag{3}$$

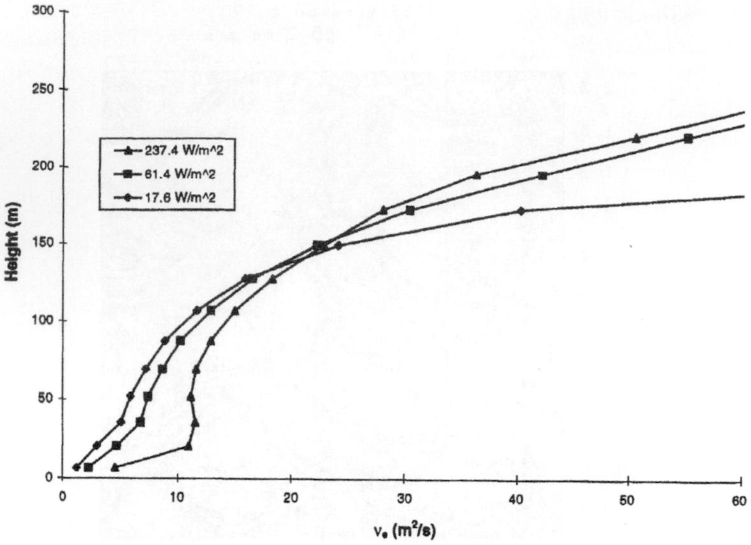

Fig. 6. Vertical profile of the turbulent eddy viscosity computed from LES.

It should be noted that Eq. (3) is valid only for a non-zero mean wind. For the case of free convection, one could use the eddy viscosity computed within the LES closure model. Using a turbulent Prandtl number ($\mathrm{Pr_t} = \nu_e/\kappa_e$) of 0.89, justified for convective conditions by Businger et al. [13], the Rayleigh number for turbulent flow simulated in LES may be written as

$$\mathrm{Ra} = -\frac{0.89 g\beta\Gamma H^4}{\nu_e^2}. \tag{4}$$

What is critical in this definition is how one determines the value of ν_e and H, as they have a very large impact on the computed Rayleigh number.

For the CBLs simulated in the LES, Fig. 6 shows the vertical profile of the turbulent eddy viscosity in the surface layer for each heat flux case. There is a large vertical variation in ν_e and variation for the different cases. A nominal value of 12, 8, and 6 m^2/s was chosen for the high, medium and low heat flux cases, respectively. These compare reasonably well with Krishnamurti [14] who reported a value of 30 m^2/s for the entire PBL. Krishnamurti's value of ν_e was based on Clarke's [15] measurements. Clarke vertically integrated the ageostrophic wind profile to determine the stress, τ, as a function of height. Knowing the stress and the velocity gradient with height, he computed the eddy viscosity, in a similar manner as described above. Given the uncertainties in the measured wind, and the assumptions inherent in numerical modeling, the LES and experimentally derived values of ν_e are quite comparable.

For the region of flow being considered, the surface layer of a convective planetary boundary layer, one must carefully choose values for Γ and H in Eq. (4). Since we are interested in cellular convection within the surface layer, Γ corresponds to the temperature lapse between the surface and the mixed layer, and H to the depth of the surface layer. In order to calculate the potential temperature, θ_0, at the surface roughness height, z_0, Monin–Obukhov similarity theory was used. At each vertical level in the model domain, an average u velocity was computed. Using the average u velocity at the first model level, u_a, the friction velocity, u_*, was computed from:

$$u_* = \frac{k u_a}{\ln(z_a/z_0) - \psi_M(z_a/L)}. \tag{5}$$

Using the fact that the surface heat flux is known, θ_* may be computed from:

$$\langle w'\theta' \rangle = u_* \theta_* . \tag{6}$$

For stability functions, we use the following for a convective boundary layer ($z/L < 0$) [16]

$$\psi_M(z/L) = 2\ln\left(\frac{1+x}{2}\right) + \ln\left(\frac{1+x^2}{2}\right) - 2\arctan(x) + \pi/2, \tag{7}$$

$$\psi_H(z/L) = 2\ln\left(\frac{1+x^2}{2}\right), \tag{8}$$

where

$$x = (1 - 15z/L)^{1/4} . \tag{9}$$

Since the Obukhov length, L, is computed in the LES model at each time step, and with θ_* known, θ_0, the temperature at z_0, can be computed from:

$$\theta_* = \frac{k\{\theta_a - \theta_0\}}{\mathrm{Pr_t}\{\ln(z_a/z_0) - \psi_H(z_a/L) + \psi_H(z_0/L)\}}, \tag{10}$$

where $\mathrm{Pr_t} = 0.89$, the surface turbulent Prandtl number. Using the depth of the surface layer, Δz, and the temperature difference $\Delta\theta = \theta_{ml} - \theta_0$, the Rayleigh number may be computed from:

$$\mathrm{Ra} \approx -\frac{0.89 g \Delta\theta\, \Delta z^3}{\bar{\theta} v_e^2}, \tag{11}$$

where $\bar{\theta}$ is the average temperature in the layer.

In the CBL, the analogy used for the classical RB convection was that the earth's surface represents the 'lower rigid no-slip boundary', and the mixed layer the 'upper free-slip boundary'. Thus the temperature difference, $\Delta\theta$ in Eq. (11) represents the difference between the surface and the mixed layer temperatures divided by the layer depth. With the mixed layer potential temperature, θ_{ml} known from the LES results, and θ_0 computed from Monin–Obukhov similarity theory, the following ratio was used to determine the surface layer depth for a given value of R:

$$R = \frac{\theta_{sl} - \theta_0}{\theta_{ml} - \theta_0}. \tag{12}$$

Table 2
Computed surface layer depth and Rayleigh number of the LES simulations. θ_{ml} is the mixed layer potential temperature (K), θ_0 the temperature (K) at the roughness height z_0, ν_e is the computed eddy viscosity (m^2/s), Δz_{90}, Δz_{95}, and Δz_{99} (m) correspond to $R = 90\%, 95\%, 99\%$, respectively, and Z_i is the mixed layer depth

Heat flux	θ_{ml}	θ_0	ν_e	Δz_{90}	Ra	Δz_{95}	Ra	Δz_{99}	Ra	Z_i	Ra
High	284.73	287.17	12	36	9	69	63	220	2055	1250	1006700
Medium	284.36	285.45	8	36	9	55	32	171	963	1140	762000
Low	283.43	284.62	6	20	3	35	16	127	771	992	981000

By solving Eq. (12) for θ_{sl} and using the temperature profile in Fig. 3 we determine the height above the ground corresponding to this temperature. This is what we are calling the depth of the surface layer, Δz. The values of R considered were 90%, 95%, and 99%. Table 2 summarizes the computed surface layer depth for values of R, θ_{ml}, and θ_0. The chosen values of R are admittedly ambiguous, and unfortunately have a large effect on the computed Rayleigh number.

4.3. Comparisons of linear theory and LES

Table 2 summarizes the Rayleigh numbers computed for the three LES simulations. Using the temperature ratio to determine the surface layer depth, while

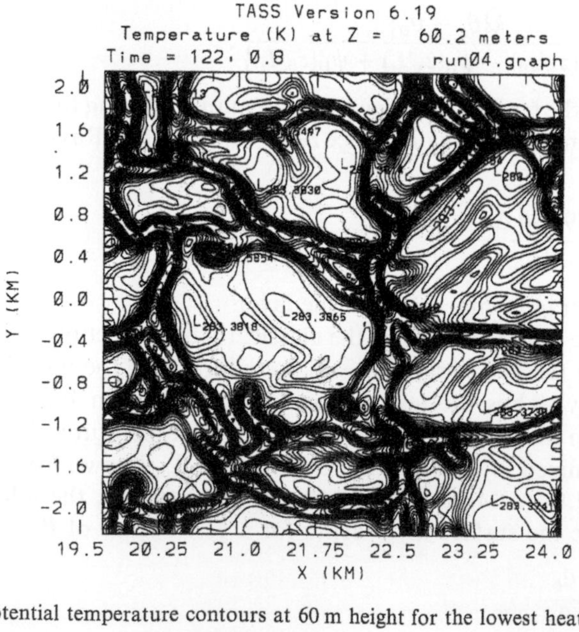

Fig. 7. Potential temperature contours at 60 m height for the lowest heat flux case.

intuitive, yields only three cases for which the computed Rayleigh number exceeds the critical value of $Ra_c = 1100$; the highest heat flux cases. But in all of the cases studied, the cellular structure was still present in the surface layer, as seen in Fig. 7, and the boundary layers are very convective in nature. The parameter Z_i/L is used to quantify the relative importance of convection and shear in the PBL. For the three cases simulated, Z_i/L was -6500, -3500, -700, for the high, medium and low heat flux cases, respectively, indicating all are very convective boundary layers.

Another typical length scale used for the CBL is the mixed layer depth, Z_i. The surface layer is not 'capped' by the mixed layer, but rather the entire PBL is capped by an inversion which limits vertical motion. Perhaps the mixed layer depth would be more consistent with the boundary conditions used to develop the critical Rayleigh number. The mixed layer depth was determined by the height where the vertical profile of the heat flux, $\langle w'\theta' \rangle$, becomes minimum. Using that as the length scale, we computed Rayleigh numbers that greatly exceed Ra_c. But in using this depth, the analogy is inconsistent between the laboratory Rayleigh–Benard flow and the atmosphere. Specifically, the temperature profile of the CBL, shown in Fig. 3, is not consistent with the linear profile used in the laboratory model, nor with that assumed to develop the linear theory estimates of the critical values.

By computing the power spectrum of the potential temperature temperature variance, one may determine dominant length scales present in the flow. An ensemble average spectrum was computed by averaging 1-D spectra horizontally at each model level, then averaging each level over the 20 output times. The power spectrum at a height of 50 m is presented in Fig. 8 for each of the surface heat flux cases, and the peaks in the spectrum represents the more energetic length scales within the surface layer. The dominant peak in the spectrum occurs at $\kappa \approx 1.5 \times 10^{-3}$ cycles/m, or a length scale of 660 m. This roughly corresponds to the cell size seen in Figs. 4 and 5. This corresponds to a cell aspect ratio of roughly 3:1, a 600 m cell size and 220 m surface layer depth. According to RB convection theory, the aspect ratio should be 1:1, so what we are observing is not, strictly speaking, RB convection, but a cellular convection pattern similar to mesoscale cellular convection.

The power spectrum was also computed at 500 m, near the middle of the CBL, and is presented in Fig. 9. Here the dominant peak is near $\kappa \approx 1 \times 10^{-3}$ cycles/m, or 1000 m, which corresponds to the mixed layer depth.

From Table 1, linear theory predicted a dominant wave number of 2.68 for the 'one free-slip, one no-slip' boundary condition. This corresponds to a length scale, L_c,

$$L_c = \Delta z \frac{2\pi}{k_c}. \tag{13}$$

For the highest heat flux case, the surface layer depth using $R = 99\%$ is approximately 200 m. Using a surface layer depth for Δz in the above equation, L_c is 469 m, or $1/L_c$ is 0.002 cycles/m. This nearly corresponds to the highest peak in the power spectrum in Fig. 8. But for all other estimates of the surface layer depth, the critical length scale will be much smaller, corresponding to scales to the right of the dominant

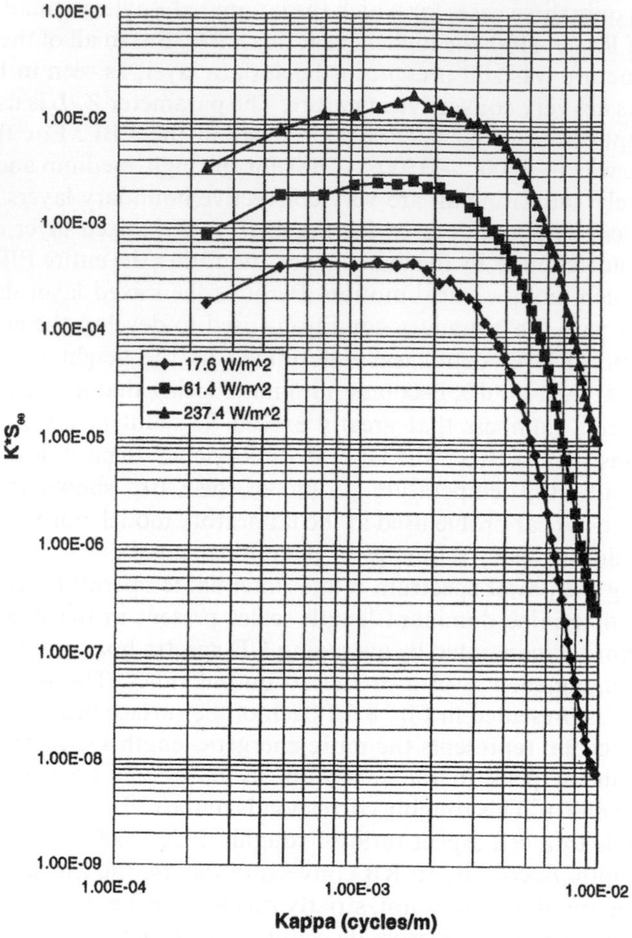

Fig. 8. Potential temperature spectrum at 50 m.

peak in the spectrum, where it is difficult to discern a distinct peak, or length scale, in the spectrum.

While the flow in the surface layer appears to be cellular in nature, a clear comparison to the critical value predicted by the linear stability analysis is not strictly valid. For a clearer comparison, the matter of the boundary condition assumed for the top of the surface layer needs to be addressed. In order to have similar temperature profiles between the CBL and laboratory Rayleigh–Benard flow, we assumed the 'ground to top of the surface layer' within the CBL was analogous to the laboratory flow between two heated parallel plates. But the top of the surface layer is not a rigid, free-slip boundary. We feel this is one possible cause of the discrepancy between our LES computed Rayleigh number and the critical value predicted by the linear stability

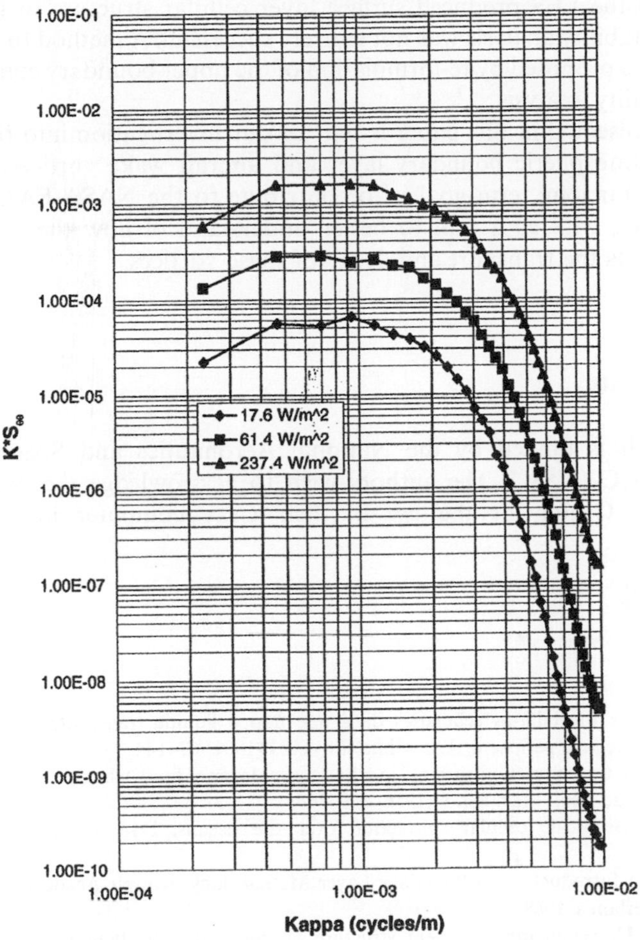

Fig. 9. Potential temperature spectrum at 500 m.

analysis. In order to resolve this, we propose re-evaluating the linear stability analysis and its assumptions. Perhaps a better boundary condition, so as to compare with the CBL surface layer flow, would be to solve the linear equations in Ref. [11] not with a rigid lid, but rather, with flow being bounded at infinity. While we have not pursued this yet, it may yield an analytical result that we could compare to the surface layer flow produced by the LES.

5. Conclusions

This paper presents some preliminary results using large eddy simulation modeling of surface layer turbulence embedded in a convective boundary layer. A direct comparison of the LES produced surface layer cellular structure to that predicted from a linear stability analysis was not entirely successful. A method to allow a better comparison was proposed by re-formulation of the upper boundary condition used in the linear stability analysis.

The work presented in this paper is part of our investigation into the interaction between the atmospheric boundary layer and aircraft wake vortices, discussed by Hinton [10]. Our long term goal is to contribute to the NASA/FAA wake vortex project by using LES as a tool to better understand of how the boundary layer turbulence effects the transport and decay of these vortices.

Acknowledgements

This research is funded by the National Aeronautics and Space Administration, Grant NCC-1-188-5. The authors wish to acknowledge the North Carolina Supercomputer Center for the use of their Cray computer in some of these simulations.

References

[1] H. Benard, Les tourbillions cellulaires dans une nappe liquide transportant de la chaleur par convection en regime permaent, Ann. Chim. Phys. 23 (1901) 62–144.

[2] O.M. Rayleigh, On convection currents in a horizontal layer of fluid, when the higher temperature is on the underside, Phil. Mag. Ser. 6 (1916) 529–546.

[3] E.M. Agee, Mesoscale cellular convection over the oceans, Dyn. Atmos. Oceans 10 (1987) 317–341.

[4] R.B. Stull, An Introduction to Boundary Layer Meteorology, Kluwer Academic Publishers, Dordrecht, Netherlands, 1988.

[5] H. Schmidt, U. Schumann, Coherent structure of the convective boundary layer derived from large-eddy simulation, J. Fluid Mech. 200 (1989) 511–562.

[6] F.H. Proctor, The terminal area simulation system vol I: theoretical formulation, NASA Contractor Report 4046 DOT/FAA/PM-86/50,I, 1987.

[7] F.H. Proctor, Numerical simulations of an isolated microburst. Part I: dynamics and structure, J. Atmos. Sci. 45 (1988) 3137–3159.

[8] D.G. Schowalter, D.S. DeCroix, Y.L. Lin, S.P. Arya, M. Kaplan, Planetary boundary layer simulation using TASS, NASA Contractor Report 198325, 1996.

[9] G.L. Brown, A. Roshko, On density effects and large structure in turbulent mixing layers, J. Fluid Mech. 64 (1974) 775–816.

[10] D.A. Hinton, Aircraft vortex spacing system (AVOS) conceptual design, NASA Technical Memorandum 110184, 1995.

[11] K.A. Emanuel, Atmospheric Convection, Oxford University Press, New York, 1994.

[12] A. Arakawa, Computational design for long term integration of the equations of fluid motion: two-dimensional incompressible flow, Part I, J. Comput. Phys. 1 (1966) 119–143.

[13] J.A. Businger, J.C. Wyngaard, Y. Izumi, E.F. Bradley, Flux-profile relationships in the atmospheric surface layer, J. Atmos. Sci. 28 (1971) 181–189.

[14] R. Krishnamurti, On cellular cloud patterns. Part 1: mathematical model, J. Atmos. Sci. 32 (1975) 1353–1363.

[15] R.H. Clarke, Observational studies in the atmospheric boundary layer, Quart. J. Roy. Met. Soc. 96 (1970) 91–114.
[16] S.P. Arya, Introduction to Micrometeorology, Academic Press, San Diego, 1988.
[17] S. Chandrasekhar, Hydrodynamic and Hydromagnetic Stability, Dover, New York, 1981.

Journal of Wind Engineering
and Industrial Aerodynamics 67&68 (1997) 403–413

JOURNAL OF
wind engineering
AND
industrial
aerodynamics

Flow separation and hydraulic transitions over hills modelled by the Reynolds equations

K.J. Eidsvik*, T. Utnes

NTNU Department of Structural Engineering, N-7034 Trondheim, Norway

Abstract

A model based upon Reynolds equations with a standard (K, ε)-turbulence closure and boundary conditions may predict important features of flows with separation, attachment, hydraulic transitions and internal wave breaking quite robustly and realistically.

Keywords: Mountain flow; Lee waves; Numerical flow simulations

1. Introduction

Flows over hills depend upon many parameters. An idealized hill is characterized by its height, h_m, half-widths along its main horizontal axes, (L_1, L_2), and its surface roughness, z_0. An idealized inflow is characterized by a constant bulk mean velocity, U, with direction β relative to the hill. The depth of the bulk flow and boundary layer are D_0 and D. The Brunt Vaisala frequency is $N^2 = -g/\rho(\partial\rho/\partial x_3)$. The main dimensionless numbers for stratified flows are assumed to be associated with the dispersion equation for internal gravity waves $(c-U)^2 = N^2/(k_1^2 + k_2^2 + k_3^2)$. Here $c = \omega/k_1$ is a phase velocity and k_i are the wavenumber components. The maximum phase velocity occurs for the smallest $(k_1, k_2, k_3) = (0, 0, j\pi/D_0)$, $j = 1, 2, \ldots$ The inverse Froude number, $(c-U)/U \approx ND_0/\pi U$, measures the ratio of this maximum phase velocity relative to the mean flow. The most dominant length scale turns out to be comparable to the stationary wavelength, $\lambda = 2\pi/(k_1^2 + k_2^2 + k_3^2)^{1/2} = 2\pi U/N$ so that the non-dimensional number $L_1/(\lambda/4) \approx N(2L_1)/\pi U$, measures the degree of resonance and wave drag. The non-dimensional number, $h_m/(\lambda/4) \approx (2/\pi)Nh_m/U$, is the typical wave amplitude as related to the dominant horizontal scale and measures maximum wave steepness and degree of non-linearity [1]. The most

* Corresponding author.

PII S0167-6105(97)00089-5

important variability is then supposed to depend upon the following independent dimensionless numbers: L_1/L_2, $h_m/2L_1$, z_0/h_m, β, D/h_m, ND_0/U, $Nh_m/U, \ldots$ Not only is the process dependent upon a many-dimensional parameter vector. Due to its non-linearity, it may also depend strongly upon small variations of the parameter vector and upon how a given parameter vector is approached in time history (bifurcation, hysteresis) [1,2].

There may be several types of flows associated with different foci, saddles and nodes of separation and attachment [1]. In two-dimensional flows, $(L_1/L_2, \beta) \ll (1,1)$, lee-side bluff body separation occurs when, roughly $h_m/(2L_1) > 0.2(1 + Nh_m/U)$. As this curve is approached from above the separation point moves downslope and the recirculation area becomes smaller. Below this curve, for smaller Nh_m/U than about 1, the flow is attached over the whole domain. For larger Nh_m/U there is a low-level jet over the lee slope followed by a hydraulic transition and low-level rotors behind the hill. Significant upstream blocking occurs when the steepness parameter is larger than about 2. In three-dimensions, the flow below the height $x_3/h_m \approx \max(1 - U/Nh_m, 0)$ tends to go around rather than over the hill. Upper-layer wave breaking occur when, very roughly, $Nh_m/U > 0.8 + (0.5 + 3.5(L_1/L_2))(h_m/2L_1)$. Downslope windstorms are characterized by intense turbulence and large-scale fluctuations [1].

For such a complicated process it is difficult to verify if a model is realistic in detail over most of the parameter space. A model with detailed fit to a given set of experimental data at one parameter vector could be inaccurate for other data sets and parameter vectors. The present model has previously been shown to predict robustly and accurately some aspects of separated flows for $Nh_m/U \ll 1$ and different L_1/L_2, β, $h_m/(2L_1)$, z_0/h_m [3]. In the present study this model is extended and applied to stratified flows, and variations with the wave steepness parameter component Nh_m/U are focused. Qualitatively the model also turns out to predict these flows realistically.

2. Model

The Reynolds equations are simplified with the standard Boussinesq density variation and anelastic approximations. Whenever possible the mean-value operator is suppressed in the notations, so that, for instance, the mean density and velocity $\langle\rho\rangle, \langle u_i\rangle$ are denoted by ρ, u_i.

$$\frac{\partial}{\partial t}(\rho u_i) + \frac{\partial}{\partial x_j}(\rho u_j u_i) = -\frac{\partial p}{\partial x_i} - \rho g \delta_{i3} - \frac{\partial}{\partial x_j}\rho\left(\langle u_i' u_j'\rangle - \nu \frac{\partial u_i}{\partial x_j}\right), \tag{1}$$

$$\frac{\partial}{\partial x_i}(\rho u_i) = 0. \tag{2}$$

Assuming no heat supply, the first law of thermodynamics is formulated as conservation of potential temperature: $\theta = T(p/p_0)^{-R/c_p}$. The heat coefficients are

$c_p - c_v = R$ and $c_p/c_v = 1.4$. The density is then given by: $\rho = (p_0/R\theta)(p/p_0)^{c_v/c_p}$, so that variations, say from the hydrostatic state, are given as: $\Delta\rho/\rho = -\Delta\theta/\theta + c_v/c_p \Delta p/p$. Although the justification appears to be weak, the common accepted practice is to disregard the pressure variations in the latter equation. This means that: $\Delta\rho/\rho \approx -\Delta\theta/\theta$ and $\rho'/\rho \approx -\theta'/\theta$. Interpretating θ as $\langle\theta\rangle$ gives

$$\frac{\partial\theta}{\partial t} + \frac{\partial}{\partial x_j}(u_j\theta) = -\frac{\partial}{\partial x_j}\langle\theta' u_j'\rangle . \tag{3}$$

It is obvious that few-equation turbulence closures cannot predict all details of flows with separation, attachment, hydraulic transitions and internal wave breaking [4]. Nevertheless, simple turbulence closures have also been tentatively tried quite successfully. The large Reynolds number (K, ε)-closure is preferred because it appears to be most standard and well documented [3]. The flux terms in the dynamic turbulence equations are approximated with the simplest Boussinesq turbulent viscosity closures like $\langle u_i' u_j' \rangle = \frac{2}{3}K\delta_{ij} - v_t(\partial u_i/\partial x_j + \partial u_j/\partial x_i)$. The turbulent viscosity is $v_t = C_\mu K^2/\varepsilon$ and the Prandtl–Schmidt numbers are $(\sigma_\theta, \sigma_k, \sigma_\varepsilon)$. The coefficients are chosen as standard $(C_\mu, C_{\varepsilon 1}, C_{\varepsilon 2}, C_{\varepsilon 3}, \sigma_\theta, \sigma_k, \sigma_\varepsilon) = (0.09, 1.45, 1.9, 0.8, 1.0, 1.0, 1.3)$. The turbulent intensity is defined as $I = \sqrt{K}/U$, and in geophysical flows K is imagined to be associated only with the smallest-scale fluctuations [3],

$$\frac{\partial K}{\partial t} + \frac{\partial}{\partial x_j}(u_j K) = -\langle u_i' u_j' \rangle \frac{\partial u_i}{\partial x_j} - \frac{g}{\rho}\langle \rho' u_3' \rangle - \varepsilon - \frac{\partial}{\partial x_j}\langle K' u_j' \rangle , \tag{4}$$

$$\frac{\partial \varepsilon}{\partial t} + \frac{\partial}{\partial x_j}(u_j \varepsilon) = \left(-C_{\varepsilon 1}\langle u_i' u_j' \rangle \frac{\partial u_i}{\partial x_j} - C_{\varepsilon 3}\frac{g}{\rho}\langle \rho' u_3' \rangle - C_{\varepsilon 2}\varepsilon\right)\frac{\varepsilon}{K} - \frac{\partial}{\partial x_j}\langle \varepsilon' \mu_j' \rangle . \tag{5}$$

The inflow boundary condition is represented by a stationary fully developed boundary layer over a rough surface where all variables are prescribed. The wall boundary conditions correspond to the same rough surface. At the top frictionless wall conditions are used for the velocity, while the vertical gradient of the dynamic part of the potential temperature is set to be zero. At the outflow boundary the pressure is set equal to its hydrostatic value, and a weak form (in a FEM context) of zero normal derivatives are used for the velocity components, the turbulence quantities and the dynamic part of the potential temperature. These are natural boundary conditions of the form $v_t \partial(.)/\partial n = 0$, implemented in weighted integral form [5,6].

The numerical formulation is based upon a Galerkin finite element method. For the Reynolds equations a fractional step formulation is applied. An intermediate mean velocity is first estimated by neglecting the pressure gradient. The pressure field is estimated from a Poisson equation and used to correct the velocity field. The intermediate velocity, potential temperature and the turbulence equations are solved by an explicit two-step Taylor–Galerkin formulation [3]. It turns out that even features characteristic of separation, hydraulic transitions and wave breaking are simulated robustly, without numerical stability problems.

However, in stratified flows the integrations tend to converge slowly towards quasi-stationary solutions.

3. Separation and hydraulic transition in idealized flows

3.1. Representative predictions

A "Witch of Agnesi" hill is chosen: $h(x_1,x_2) = h_m/(1 + (x_1/L_1)^2 + (x_2/L_2)^2)$. To simplify interpretations, two-dimensional flows are focused. Both the stratification, N, and mean velocity, U, are supposed to be constants down to the boundary layer. The latter is modelled as described in Ref. [3], but significantly thinner to account for stratification [7]. The following non-dimensional numbers are fixed as $(L_1/L_2, h_m/(2L_1), z_0/h_m, \beta, D/h_m, D_0/h_m) = (0, 0.3, 5\times10^{-6}, 0, 0.1, 13)$. The parameters ND_0/U, $N(2L_1)/U$, Nh_m/U are changed by means of N or U so they are all proportional. The inverse Froude number is so much larger than unity that its variations are supposed to be unimportant. Nevertheless, when N and U are considered as local parameters there can be a transition to supercritical conditions. The ratio $h_m/(2L_1)$ is fixed as 0.3 so that the parameter, $N(2L_1)/\pi U$, is approximately equal to Nh_m/U. The steepness parameter is supposed to be the most important independent non-dimensional number.

The integration domain is characterized relative to the hilltop at $(x_1,x_3)/h_m = (0,1)$, by $-12 < x_1/h_m < 35, 0 < x_3/h_m < 14$. Grid variations are tried and the integrations appear to be grid independent. The integrations shown are obtained with 130×40 nodes and a stretched mesh.

It turns out that stratified flows tend to be dynamic. Nevertheless, Figs. 1 and 2 illustrate representative predicted mean streamlines and turbulent intensity for $Nh_m/U = 0, 1, 2, 3$ and constant Reynolds number $Uh_m/\nu = 6.7\times10^8$. There is bluff body separation on the hill only for the smallest Nh_m/U-parameter. For this case model predictions and data have previously been compared quantitatively in flows over cosine-square hills [3]. For parameters as given in Fig. 3, the separation point is then predicted and measured at about $x_1/h_m \approx 0.8$ downstream from the hilltop. Reattachment is measured at $x_1/h_m \approx 5.3$ and predicted insignificantly different, at $x_1/h_m \approx 5.4$. Predicted and measured profiles of mean flow and turbulence intensity show excellent prediction accuracy as illustrated in Fig. 3. It is quite remarkable that the delicate delayed separation behind a smoothly curved hilltop can be predicted as accurately and robustly as this with a model based upon first principles and standard turbulence closures, without coefficient adjustment of any kind [3].

For larger Nh_m/U, there is instead a strong, attached jet over the downslope side of the hill. Although this is a stratified wall jet [8], it is probably so strongly forced by non-turbulence effects that it is reasonably realistically modelled by the (K, ε)-closure. For large enough Nh_m/U the downslope jet is followed by a hydraulic transition with downstream rotors. The strong mean shear near the surface in stratified flows makes the maximal turbulence intensity much larger than for

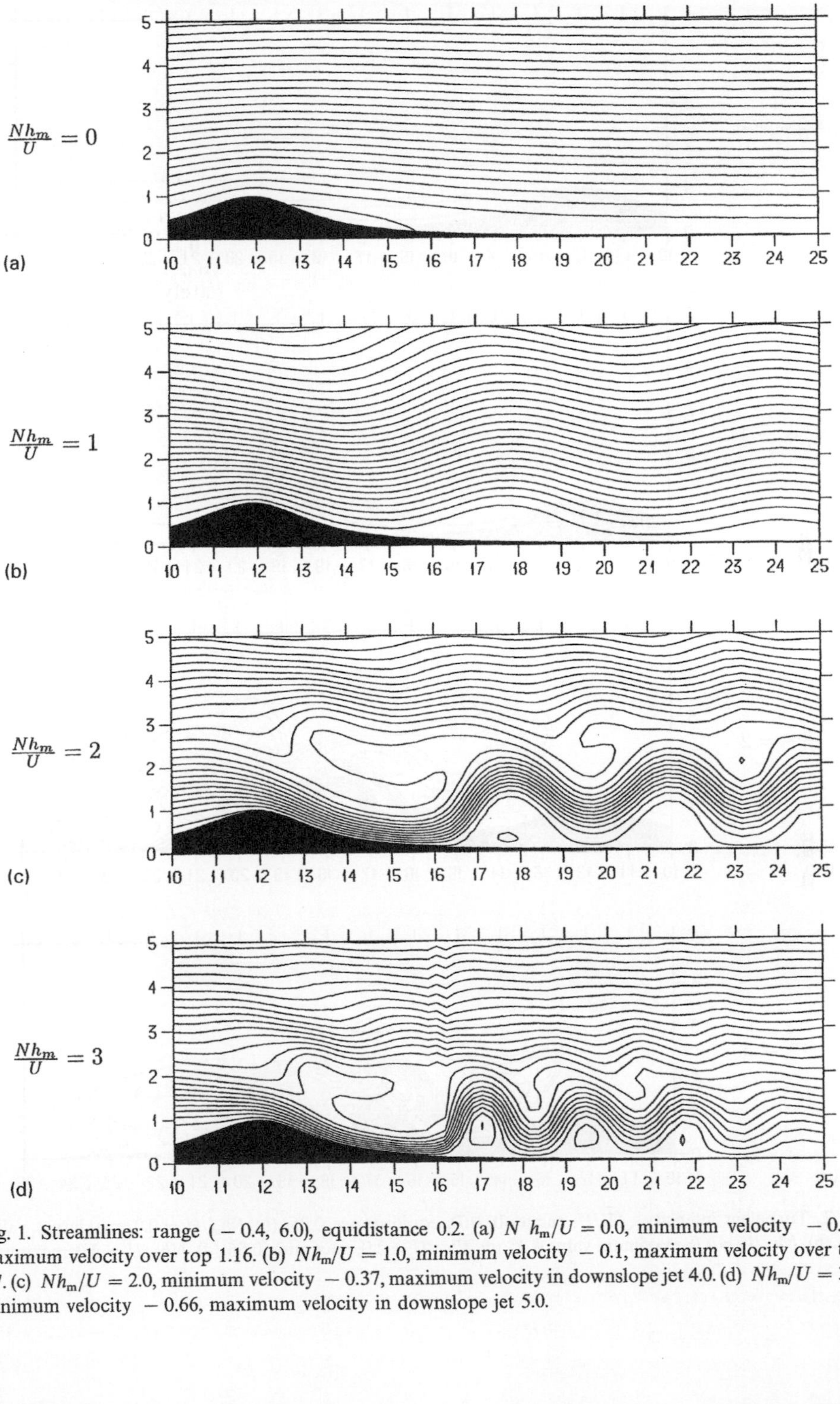

Fig. 1. Streamlines: range (– 0.4, 6.0), equidistance 0.2. (a) $N\ h_m/U = 0.0$, minimum velocity – 0.08, maximum velocity over top 1.16. (b) $Nh_m/U = 1.0$, minimum velocity – 0.1, maximum velocity over top 1.7. (c) $Nh_m/U = 2.0$, minimum velocity – 0.37, maximum velocity in downslope jet 4.0. (d) $Nh_m/U = 3.0$, minimum velocity – 0.66, maximum velocity in downslope jet 5.0.

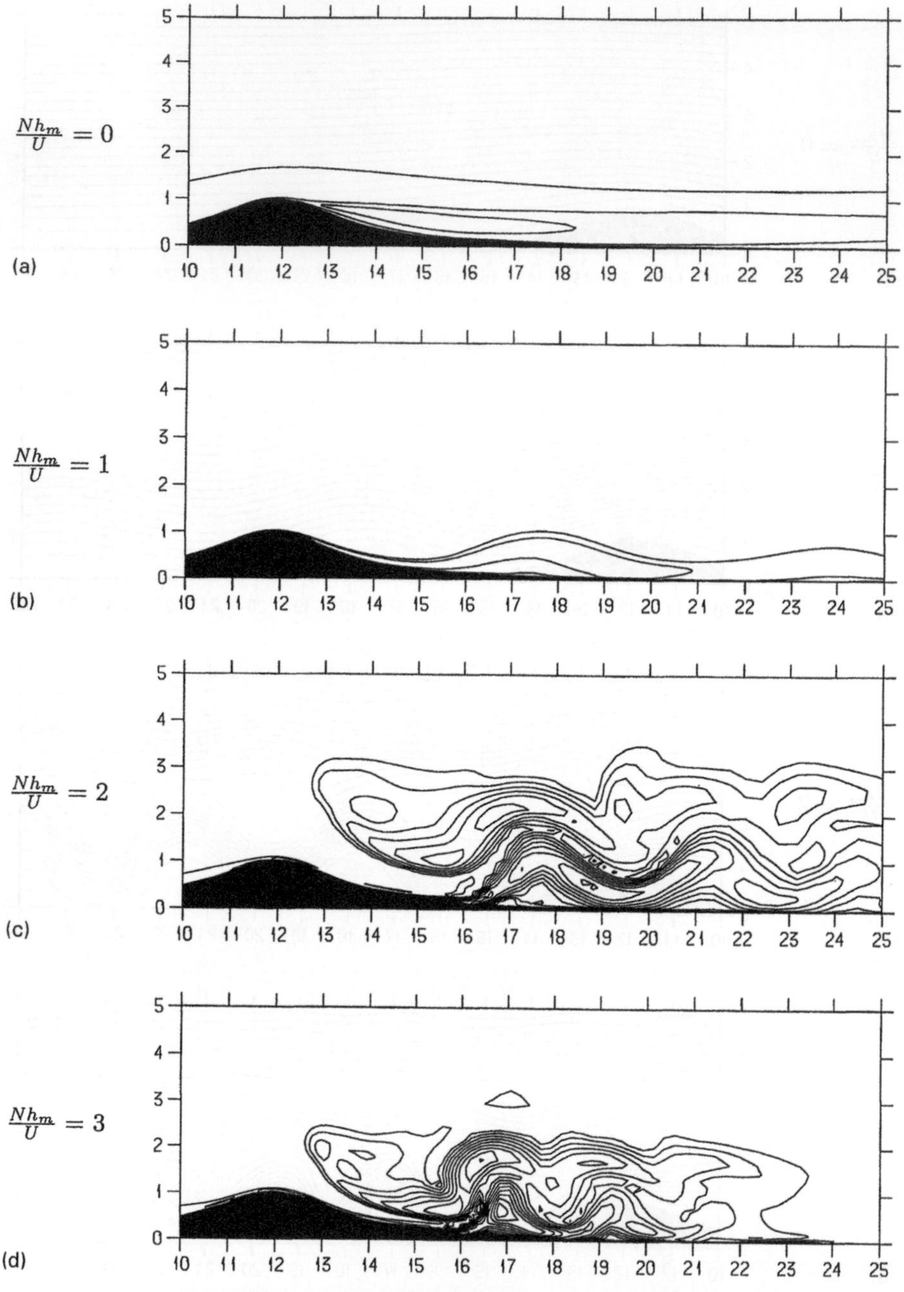

Fig. 2. Turbulent intensity $\sqrt{K}/U$: range (0, 0.7), equidistance 0.05. (a) $Nh_m/U = 0.0$, maximum value 0.18. (b) $Nh_m/U = 1.0$, maximum value 0.25. (c) $Nh_m/U = 2.0$, maximum 0.56. (d) $Nh_m/U = 3.0$, maximum 0.74.

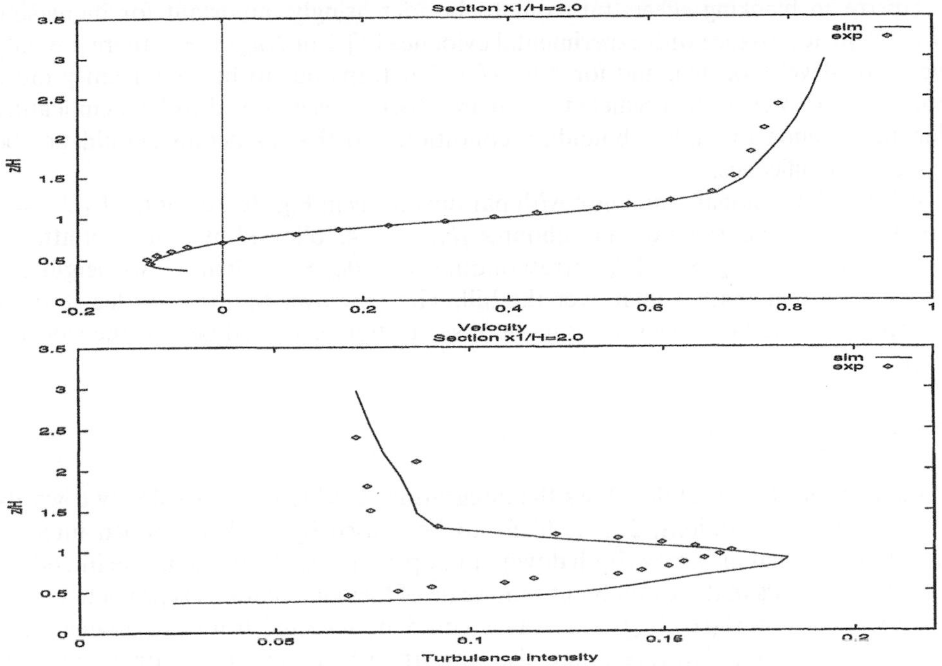

Fig. 3. Typical prediction accuracy for neutrally stratified flows [3]. Profiles of u_1/U and $\sigma_1/U \approx 1.1\sqrt{K}/U$ at $x_1/h_m = 2.0$ downstream from the hilltop. Data points and prediction lines. (L_1/L_2, $h_m/(2L_1)$, z_0/h_m, β, D/h_m, Uh_m/ν) = (0, 0.3, 0.08, 0, 9.0, 3.3×10^4).

a comparable neutrally stratified flow. Roughly, the downslope jet and downstream rotors are associated with extremal relative mean-velocity magnitude, vertical velocity component and turbulent intensity of the order of $(u/U)_{max} \approx 1.0 + 1.5Nh_m/U$, $(u_3/U)_{max} \approx -\min(1, Nh_m/U)(u/U)_{max}$ and $I_{max} \approx 0.2(1 + Nh_m/U)$. Intense turbulence combined with large-scale dynamic internal waves will appear as large-scale fluctuations, commonly observed.

Aloft there are well-defined waves with characteristic wavelength corresponding well to the stationary wavelength, $\lambda/h_m \approx 2\pi(Nh_m/U)^{-1}$. For large enough steepness parameter overturned streamlines with wave breaking is predicted. In reality the upper-layer internal wave breaking may consist of intermittent wave breaking followed by mean-flow readjustments. Although the present model predicts dynamic vortex shedding behind cylinders faithfully (see Ref. [3]), the details of such a intermittent wave-breaking process is likely to be oversimplified. In particular the upper-layer turbulence could be inaccurate. This is because turbulence generation by intermittent wave breaking may be significantly different from turbulence generation in near equilibrium flows where the standard turbulent closures and coefficient fitting have been developed. The downstream decay of wave amplitude must increase with the turbulence, which is confirmed by the simulations illustrated in Figs. 1 and 2.

Upstream blocking effects turn out to be increasingly important for increasing Nh_m/U, in accordance with experimental evidence [1]. For $Nh_m/U = 2$ there are only weak low-level blocking, and for $Nh_m/U = 3$ it turns out to be significantly more dominant. However, the predicted upstream blocking is expected to be significantly dependent upon the inflow boundary conditions, so that its details should not be assigned significance.

A three-dimensional simulation with parameters as in Fig. 1c except for $L_1/L_2 = 1$ predicts a saddle of separation at about $x_1/h_m \approx -4$, followed by a node of attachment at about $x_1/h_m \approx -1.7$, corresponding to $x_3/h_m \approx 0.5$. Below this height the flow goes around rather than over the hill. The flow over the top and lee slope is attached and the first downwind saddle of separation is located behind the steepest slope.

3.2. Dynamic response

As mentioned, in stratified flows the integrations tend to converge slowly towards quasi-stationary solutions. This is likely to be caused by weakly damped internal gravity waves that may move both down- and upstream as illustrated in the introduction. Fig. 4 shows that the dominant time-scale of these fluctuations is predicted to be as slow as about $10(tU/2L_1)$. The waves could be reflected from the downstream boundary condition. However, they also appear to be qualitatively consistent with experimental evidence on large-scale fluctuations and drag variations on the period of long internal waves with near zero group velocity [1]. Stationary solutions may be singular idealizations in stratified flows.

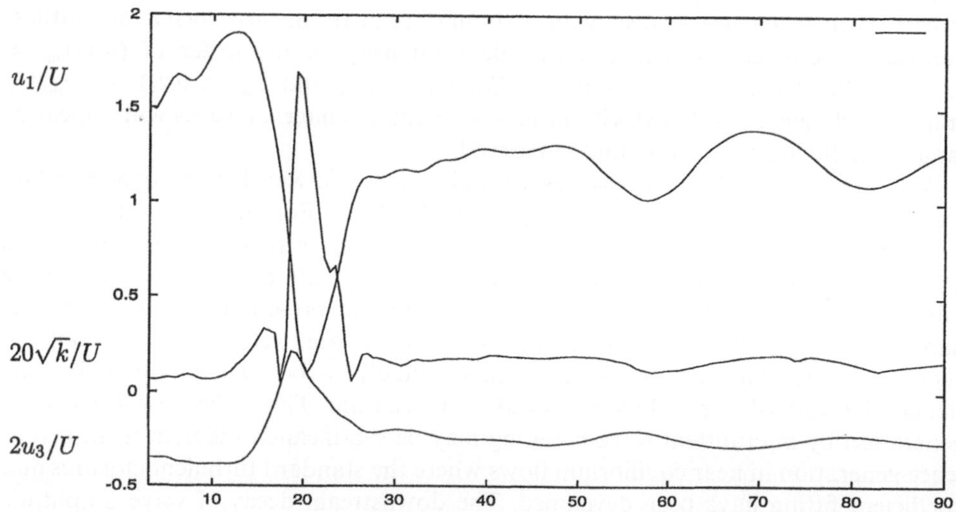

Fig. 4. Transient variations in stratified flow. $(u_1, 2u_3, 20\sqrt{K})/U$ at $(x_1, x_3)/h_m = (3.0, 0.5)$ downstream from a hilltop as functions of non-dimensional time $tU/2L_1$. Parameters as in Fig. 1b and Fig. 2b.

3.3. Bifurcation

The largest transient fluctuation near $tU/2L_1 \approx 20$ in Fig. 4 could to be associated with some "hesitation" about preferring a downslope recirculation flow instead of a downslope jet flow and could indicate a bifurcation-like behaviour. Hysteresis phenomena are readily experienced in association with hydraulic transitions [1]. The possibility of predicted bifurcations is investigated by estimating the response to slow changes of the control parameter Nh_m/U. A bifurcation associated with upper-layer internal wave breaking is particularly kept in mind. A solution is therefore represented very roughly with the minimum alongwind component in the wave breaking area above and behind the hill and the maximum downslope mean-velocity magnitude $[(u_1/U)_{min},(u/U)_{max}]$. Breaking waves are supposed to be associated with negative $(u_1/U)_{min}$. The control variation is obtained by means of a slow change in the dimensionless mean-velocity $dU/d(tU/2L_1) = \alpha U$, corresponding to $d(Nh_m/U)/d(tU/2L_1) = -\alpha(Nh_m/U)$. Although the adjustments to changing parameters are very slow (Fig. 4), integrations with $|\alpha| = 10^{-2}$ should be approximations to successions of stationary states (fixed points [2]). Fig. 5a shows two trajectories of solutions, for increasing and decreasing control variable. Fig. 5b illustrates the relation between the two state variables as the control is varied. The plotting time resolution is $\Delta(tU/2L_1) = 2$.

For decreasing steepness parameter there is an almost constant negative $(u_1/U)_{min}$ and a systematically decreasing $(u/U)_{max}$ down to about $Nh_m/U \approx 1.0$. Here there is a transition to non-breaking conditions associated with a jump in both state variables. For smaller steepness parameter the larger-scale systematic variation is continued together with smaller-scale fluctuations that is probably caused by time rather than parameter variations (Fig. 4). For increasing steepness parameter the development is significantly different. An intermittent transition to breaking occur at about $Nh_m/U \approx 1.4$, but permanent breaking do not take place until $Nh_m/U \approx 1.6$. The fluctuations extends up to a steepness parameter of about 1.8, where the up-going and downgoing trajectories become similar. The fluctuations at larger Nh_m/U than say, 2.2 are least significant because the grid resolution is coarsest here. The simulations suggest a tendency for delayed transition both from non-breaking and breaking conditions. In the range $1 < Nh_m/U < 1.8$ there is a significant hysteresis effect.

As mentioned, the smaller-scale fluctuations in Fig. 5 are probably due to dynamic internal waves rather than steepness parameter variations (Fig. 4). Their typical relative amplitude and timescale are of the order of 0.5 and $10(tU/2L_1)$, respectively. Although significant differences between the fluctuations for the down- and up-going trajectories are suggested by the figure, they may also be considered similar: there are significant dynamic fluctuations in the non-breaking states and only weak fluctuations in the permanently breaking states. It appears reasonable that the wave amplitude is damped when there is breaking with intense turbulence. The difference between the fluctuations of the up- and down-going trajectories may therefore be associated with the hysteresis effect of delayed transition already mentioned.

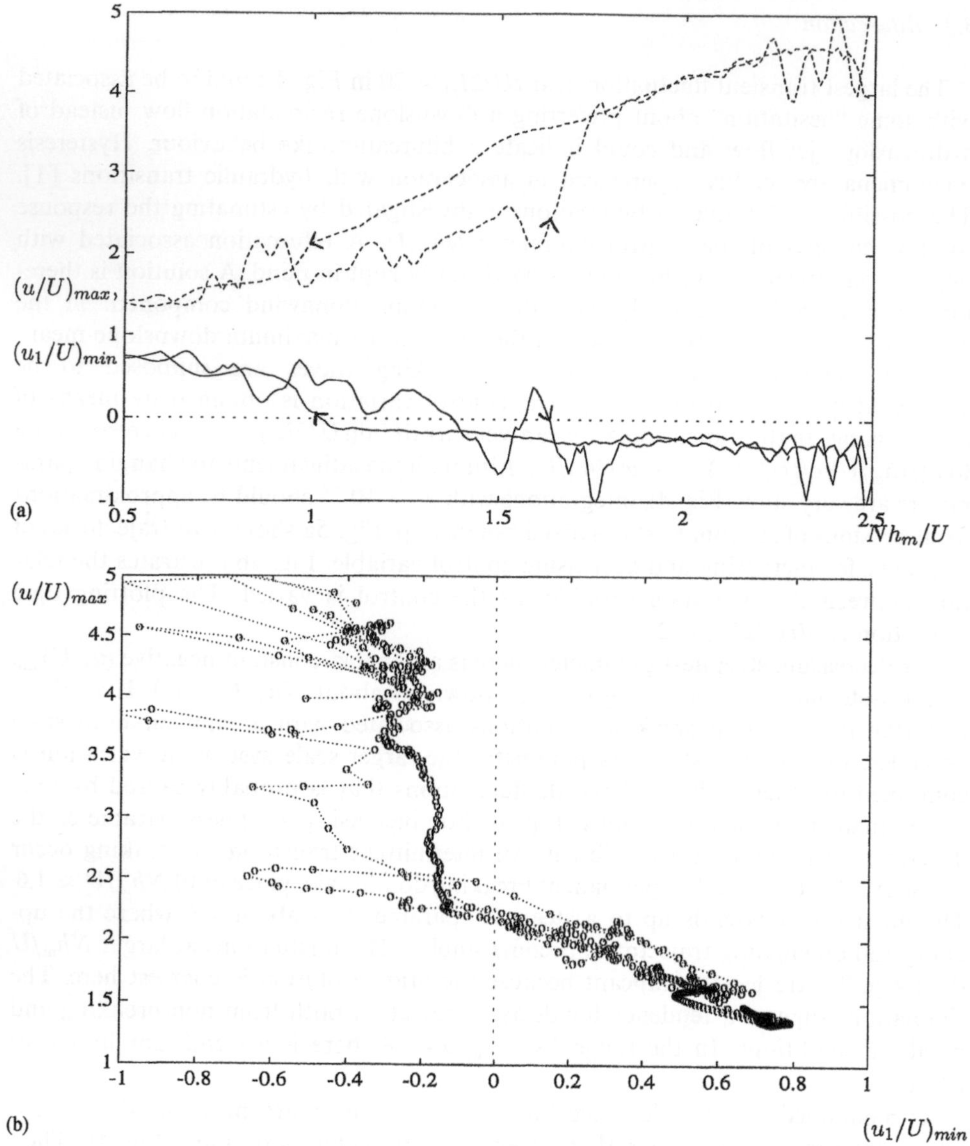

Fig. 5. (a) State-control space representation. The state is characterized by the minimum alongwind component in the breaking wave region and maximum downslope mean velocity magnitude $[(u_1/U)_{min},(u/U)_{max}]$. The control, Nh_m/U, is varied slowly from 0.5 to 2.5, then from 2.5 to 0.5. (b) Relation between the two state variables as the control is varied.

4. Concluding remarks

It is obvious that flows over hills are associated with effects that cannot be predicted in detail by few-equation turbulence closures [4]. Nevertheless, it also appears that important features of such flows can be predicted stably and quite realistically by a numerical model based upon a simple standard turbulence closure and boundary conditions. Separation, hydraulic transitions and internal wave breaking appear to be predicted quite realistically. Even hysteresis effects associated with wave breaking appear to be consistent with general experience.

The boundary and initial conditions (and the time history) of actual stratified flows over hills may not normally be known with sufficient resolution for detailed and accurate forecasting. The main purpose of models like this may therefore rather be to estimate quantitatively and understand how such flows can be.

Acknowledgements

The study is funded by The Norwegian Civil Aviation Administration.

References

[1] P.G. Baines, Topographic Effects in Stratified Flows, Cambridge Monographs on Mechanics, Cambridge Univ. Press, 1995, p. 482.

[2] A.H. Nayfeh, B. Balachandran, Applied nonlinear Dynamics. Analytical, Computational, and Experimental Methods, Wiley Series in Nonlinear Science, New York, 1995, p. 685.

[3] T. Utnes, K.J. Eidsvik, Turbulent flows over mountainous terrain modelled by the Reynolds equations, Bound. Layer Meteorol. 79 (1996) 393–416.

[4] C.G. Speziale, Analytical methods for the development of Reynolds-stress closures in turbulence, Ann. Rev. Fluid Mec. 23 (1991) 107–157.

[5] J.M. Leone, Open boundary condition symposium. Benchmark solution: stratified flow over a backward-facing step, Int. J. Numer. Methods Fluids 11 (1990) 969–984.

[6] R.L. Sani, P.M. Gresko, Resume and remarks on the open boundary condition minisymposium, Int. J. Numer. Methods Fluids 18 (1994) 983–1008.

[7] R.B. Stull, An Introduction to Boundary Layer Meteorology, Kluwer Academic Publishers, Dordrecht, 1988, p. 666.

[8] B. Brørs, K.J. Eidsvik, Dynamic Reynolds stress modeling of turbidity currents, J. Geophys. Res. 97 (1992) 9645–9652.

ELSEVIER

Journal of Wind Engineering
and Industrial Aerodynamics 67 & 68 (1997) 415–424

JOURNAL OF
wind engineering
AND
industrial
aerodynamics

On the flow around a vertical porous fence

Fuh-Min Fang*, D.Y. Wang

*Department of Civil Engineering, National Chung-Hsing University,
250 Kuo-Kuang Road, Taichung, Taiwan, ROC*

Abstract

The turbulent flow around a vertical porous fence was simulated by using a weakly-compressible-flow computational method. A large-eddy-simulation technique together with a subgrid-scale turbulence model is applied to account for the turbulence effects in the flow. By varying the approaching flow condition and the porosity of the fence, the wake characteristics behind the fence are investigated. Results show that as the fence porosity increases, for the mean flow condition the vortex tends to be elongated and results in a reduction of the drag on the fence. In addition, the drag force decreases with an increase of the boundary layer thickness of the approaching flow.

Keywords: Windbreak; Large eddy simulation

1. Introduction

A porous windbreak fence is one of the artificial structures for wind protection. Different from a solid one, the porous fence is made of a series of parallel bars with some porous gaps in between. Since the bars can be easily replaced, it may be an economical selection as a windbreak.

Previous studies [1,2] have shown that the instantaneous and mean flows behind a solid fence have different behavior. The former illustrates a process of flow unsteadiness due to vortex shedding, initiated by the separation at the fence tip. In the mean flow, however, there exists only one vortex immediately downstream. Experimental results [3] reveal that as a stream impinges on a perforated normal plate, the momentum in the wake region is enhanced. On the other hand, for the mean flow case the stationary vortex tends to move further downstream. Accordingly, from the viewpoint of windbreak design, the application of a porous fence may promote the shelter effect of the windbreak.

* Corresponding author.

PII S0167-6105(97)00090-1

2. Goal of study

Physically, the wake characteristics behind the fence depends on the condition of the approaching flow (such as the wind speed, velocity distribution and so on), porosity, and the arrangement of the fence (such as height and orientation). Since the approaching flow are essentially local, a general analysis is then difficult to perform.

In the study, the turbulent flow around a vertical porous fence is simulated numerically. By varying the approaching flow condition and the porosity of the fence, the wake characteristics, corresponding to the mean flow, are analyzed to obtain general guidelines for the design of the windbreak.

The numerical computations are carried out under a simplified condition, as sketched in Fig. 1. The size of the gap is set as one-tenth of the fence height (H). The velocity profile of the approaching flow adopts the shape of a one-to-seven power-law distribution with an edge-velocity of U and a boundary layer thickness of δ. The simulations are performed to monitor the flow under the effects of porosity, the relative boundary layer thickness (δ/H) of the approaching flow at a fixed Reynolds number ($\mathrm{Re} = UH/\nu = 4 \times 10^4$, ν being the kinematic viscosity).

3. Numerical method

A weakly-compressible-flow method [4] is adopted in the flow calculations. For a two-dimensional viscous flow, the continuity and momentum (Euler) equations are

$$\partial\rho/\partial t + \nabla\cdot(\rho \boldsymbol{V}) = 0, \tag{1}$$

$$\partial \boldsymbol{V}/\partial t + \boldsymbol{V}\cdot\nabla \boldsymbol{V} = -(1/\rho)\nabla p + \nu\nabla^2 \boldsymbol{V}, \tag{2}$$

in which ρ, p and $\boldsymbol{V}$ are, respectively, the densities, pressures and the velocity vector. Under a barotropic assumption, the continuity equation can be approximated as

$$\partial p/\partial t + \nabla\cdot(\rho c^2 \boldsymbol{V}) = 0; \tag{3}$$

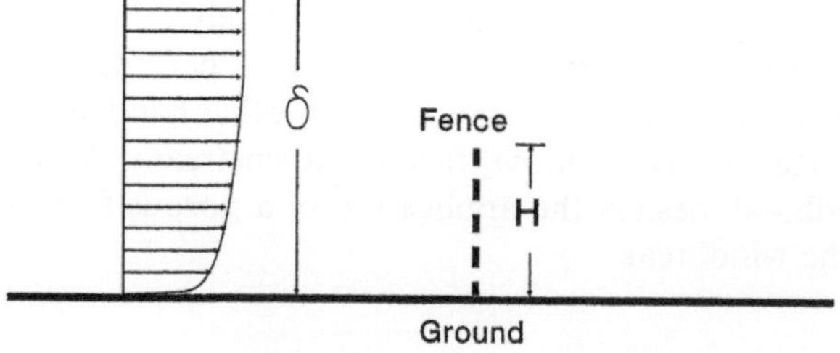

Fig. 1. Schematic of the porous fence.

(c is the sound speed) with an error of the order of the Mach number squared. On the other hand, the momentum equations can also be rewritten in approximate forms as

$$\partial \boldsymbol{V}/\partial t + \nabla \cdot (\boldsymbol{V}\boldsymbol{V}) = -\nabla(p/\rho) + \nu \nabla^2 \boldsymbol{V} \tag{4}$$

with an error of the order of the Mach number [4]. After decomposing the physical quantities into a resolvable-scale and subgrid-scale components, the space-averaged equations are presented in index forms as

$$\frac{\partial \bar{p}}{\partial t} + \frac{\partial (K\bar{u}_j)}{\partial x_j} = 0; \quad K = \rho c^2, \tag{5}$$

$$\frac{\partial \bar{u}_i}{\partial t} + \frac{\partial \bar{u}_i \bar{u}_j}{\partial x_j} = -\frac{\partial \overline{(p/\rho)}}{\partial x_j} + \frac{\partial}{\partial x_j}\left\{ -\overline{u_i' u_j'} - \overline{u_i' \bar{u}_j} - \overline{\bar{u}_i u_j'} - (\overline{\bar{u}_i \bar{u}_j} - \bar{u}_i \bar{u}_j) + \nu \frac{\partial \bar{u}_i}{\partial x_j} \right\}. \tag{6}$$

By adopting Reynolds' averaging assumptions, Eq. (6) becomes

$$\frac{\partial \bar{u}_i}{\partial t} + \frac{\partial \bar{u}_i \bar{u}_j}{\partial x_j} = -\frac{\partial \overline{(p^*/\rho)}}{\partial x_j} + \frac{\partial}{\partial x_j}\left\{ -\left(\overline{u_i' u_j'} - \tfrac{1}{3}\overline{u_i' u_j'}\, \delta_{ij} \right) + \nu \frac{\partial \bar{u}_i}{\partial x_j} \right\}, \tag{7}$$

where δ_{ij} is the Kronecker delta function and $p^* = p + (\rho/3)\overline{u_i' u_i'}$.

As the subgrid-scale stress terms are modeled by

$$-\left(\overline{u_i' u_j'} - \tfrac{1}{3}\overline{u_i' u_j'}\delta_{ij} \right) = \nu_t S_{ij}, \tag{8}$$

where

$$S_{ij} = \left(\frac{\partial \bar{u}_j}{\partial x_i} + \frac{\partial \bar{u}_i}{\partial x_j} \right), \tag{9}$$

Eq. (7) becomes

$$\frac{\partial \bar{u}_i}{\partial t} + \frac{\partial \bar{u}_i \bar{u}_j}{\partial x_j} = -\frac{\partial \overline{(p^*/\rho)}}{\partial x_j} + \frac{\partial}{\partial x_j}[\tau_{ij}/\rho], \tag{10}$$

where τ_{ij} is the combination of the viscous and the subgrid-scale stress terms, and the subgrid-scale diffusion coefficient in Eq. (8) is expressed in the form suggested by Smagorinsky [5] as

$$\nu_t = (C_s \Delta)^2 \left(\frac{S_{ij}^2}{2} \right)^{1/2}, \tag{11}$$

where Δ is the characteristic length of the computational mesh; C_s is the Smagorinsky constant.

After taking out the bars for simplicity, Eqs. (5) and (10) are all in a conservative form as

$$\partial \boldsymbol{G}_i/\partial t + \nabla \cdot \boldsymbol{F}_i = 0, \quad i = 1, 2, 3, \tag{12}$$

where the scalar G_i and the vectors F_i are defined as the rows of the following matrices:

$$G = \begin{bmatrix} p \\ u \\ v \end{bmatrix}, \quad F = \begin{bmatrix} \rho c^2 u & \rho c^2 v \\ u^2 + (p^* - \tau_{xx})/\rho & uv - \tau_{yx}/\rho \\ uv - \tau_{xy}/\rho & v^2 + (p^* - \tau_{yy})/\rho \end{bmatrix}. \tag{13}$$

The computation proceeds by using a volume integration over a specific control volume, $\forall$, as

$$\int_{\forall} \frac{\partial G}{\partial t} \mathrm{d}\forall + \int_{\forall} \nabla \cdot F \, \mathrm{d}\forall = 0. \tag{14}$$

By the divergence theorem, one has

$$\frac{\partial G_m}{\partial t} = -\frac{1}{\forall} \int_S \tilde{n} \cdot F \, \mathrm{d}S, \tag{15}$$

where G_m represents a mean quantity referred to the center of the volume and $\tilde{n}$ is the normal vector. By knowing the G_m quantities of the flow field at a starting time step, the integral form of Eq. (15) can be used to calculate the change of G_m within an elapsed period, Δt, and then the G_m values for the next time step can be updated. The computation uses the MacCormack predictor-corrector scheme [6] to update the flow. The time increment for the unsteady flow calculations is selected at each time step according to the Courant–Friedrich–Lewy criterion [7].

4. Boundary specification

Appropriate values of pressures and velocities are specified at exterior cells (or phantom cells) outside the boundaries to reflect the correct physical nature of the boundaries. For solid boundaries, a no-slip condition is used. At the upstream section, the power-law velocity profile is imposed. On the other hand, the average pressure at the downstream section is chosen as the reference pressure. On the boundary at the top of the domain, the values at the phantom cells are specified according to a zero-gradient assumption in the direction normal to the boundary.

5. Computation domain

To consider both the accuracy and the efficiency of numerical calculations, the computational domain is selected as $26H$ by $10H$. Test results have revealed that the relative error produced by this selection is within 3%. Fig. 2 depicts a typical computation mesh, obtained according to a method based on Poisson equations [8].

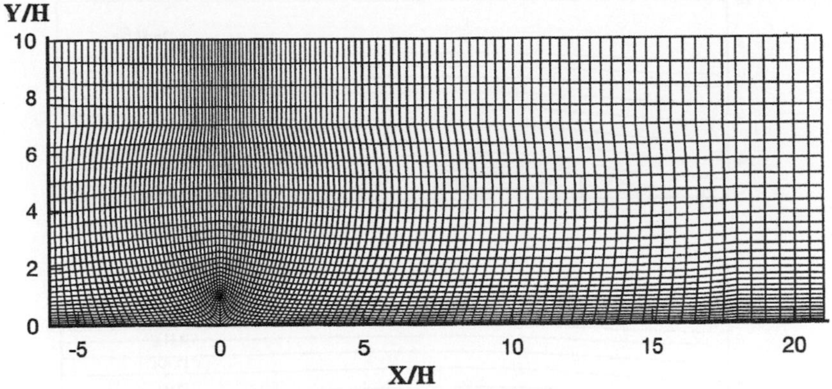

Fig. 2. Typical computational mesh (100 × 30).

6. Results

6.1. *Effect of porosity*

For Re $= 4 \times 10^4$ and $\delta/H = 3$, the flows are calculated with 4 fence porosities ($\eta = 10\%$, 20%, 30% and 40%). The streamline patterns, compared to that of a solid fence ($\eta = 0$), are illustrated in Fig. 3. In all cases, as flow approaches the fence, the streamlines are lifted and produce higher speed above the fence tip. Separation is initiated from the top of the fence and the separation line extends further downstream as the fence becomes porous. Being the upper bound of the vortex area (low-speed region), this line tends to be lower as the fence porosity increases. Consequently, the vortex area is elongated. Meanwhile, when the fence is porous, jet flows are produced from the gaps of the fence. Thus, some portion of the approaching flow momentum is transported through the fence and promote flow unsteadiness in the region immediately downstream of the fence. On the other hand, however, the speed above the fence is reduced (see Fig. 4).

6.2. *Effect of δ/H*

By varying δ/H (1 to 4), a series of calculations are conducted with a fence porosity of 30% at a Reynolds number of 4×10^4 to examine the effect of δ/H. The resulting constant-speed contours of the mean flow (Fig. 5) show that the variation of δ/H has almost no effect on the extension of the separation line. Moreover, as δ/H increases, the magnitude of the speed in the high-speed region decreases.

6.3. *Effect on the drag*

Fig. 6 depicts the relationship between the drag coefficient, C_D (defined by $2D/\rho U^2 H$, D being the drag force on the fence), and δ/H with different porosities.

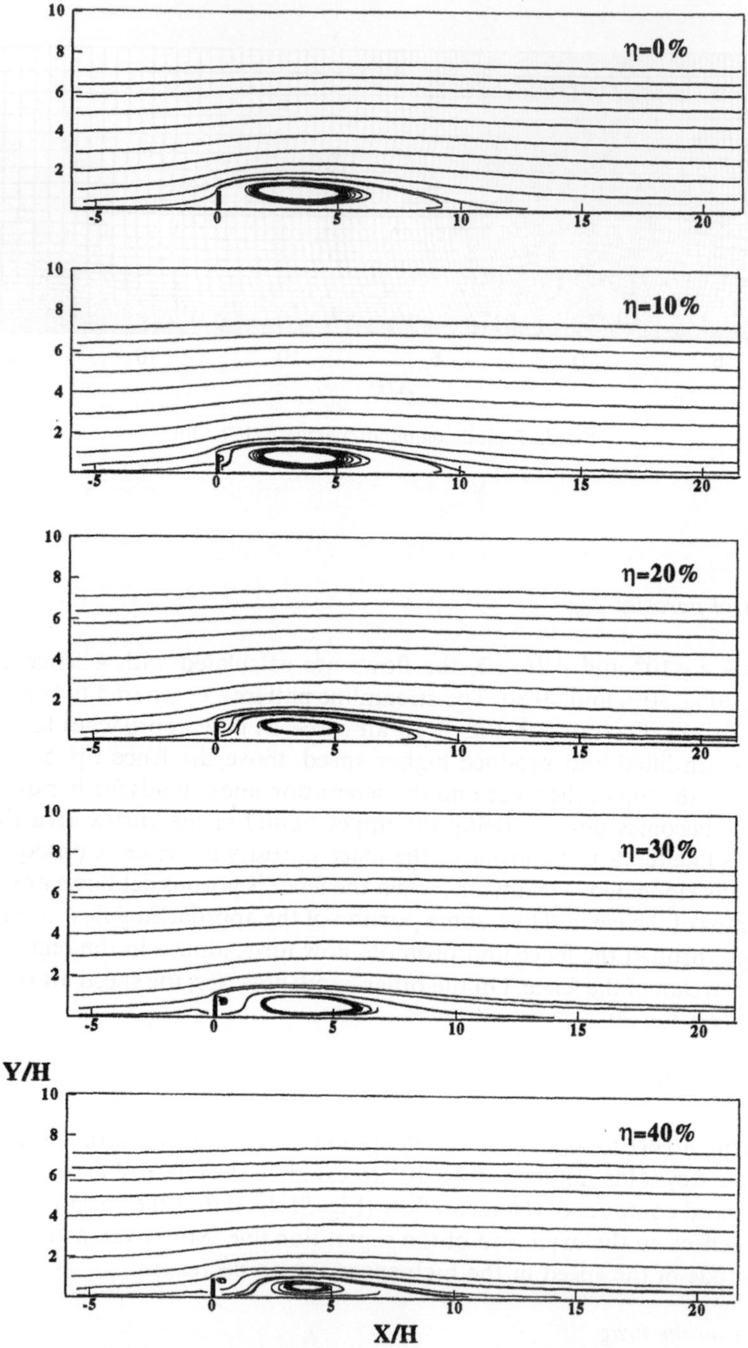

Fig. 3. Streamline patterns for $\delta/H = 3$ at $\mathrm{Re} = 4\times10^4$.

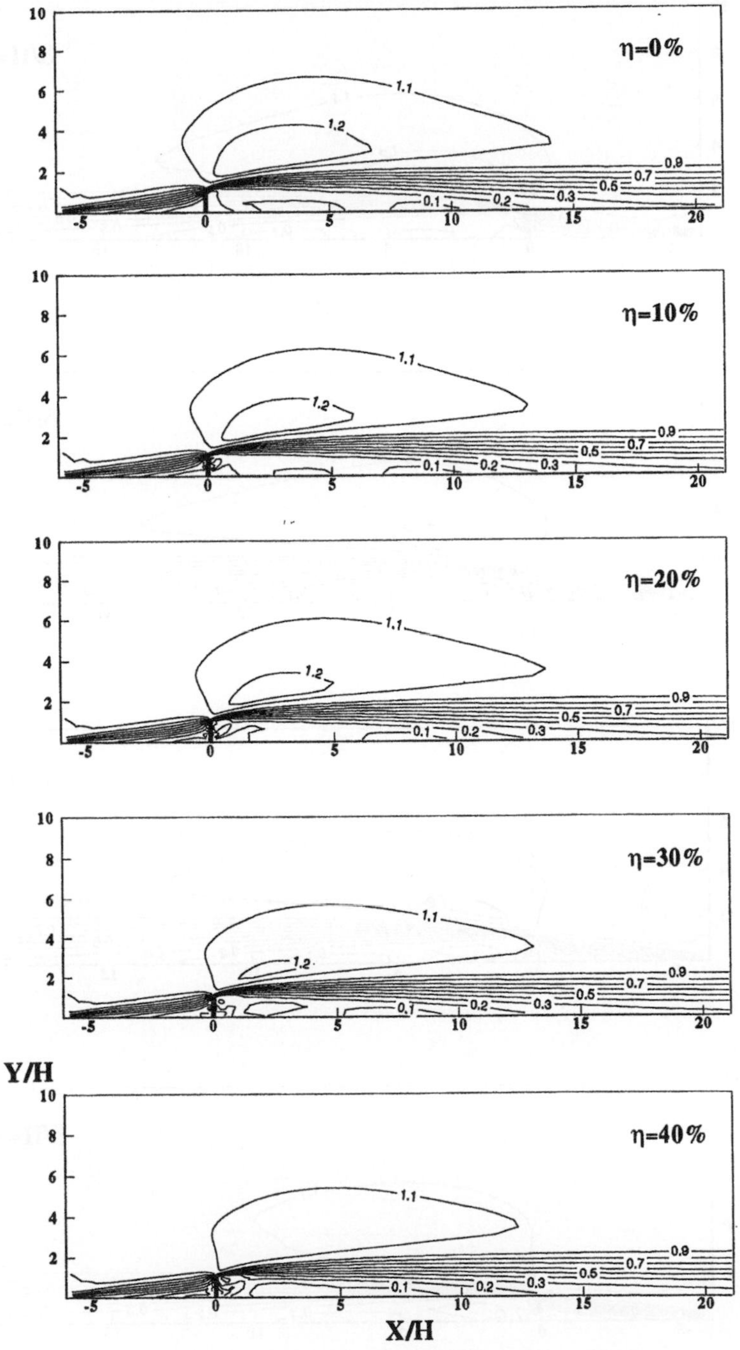

Fig. 4. Constant-speed contours for $\delta/H = 3$ at Re $= 4 \times 10^4$.

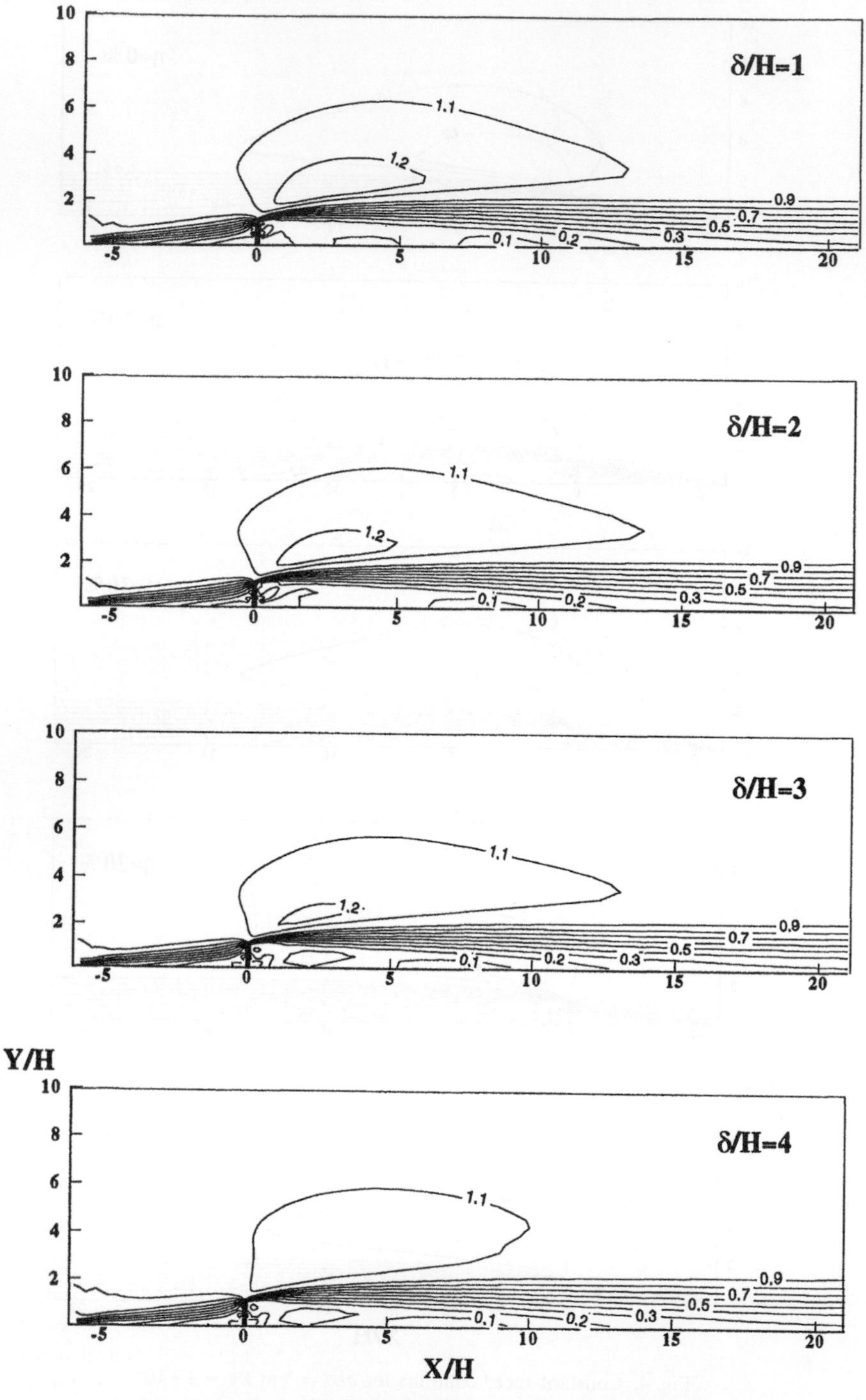

Fig. 5. Constant-speed contours for $\eta = 30\%$ at $Re = 4 \times 10^4$.

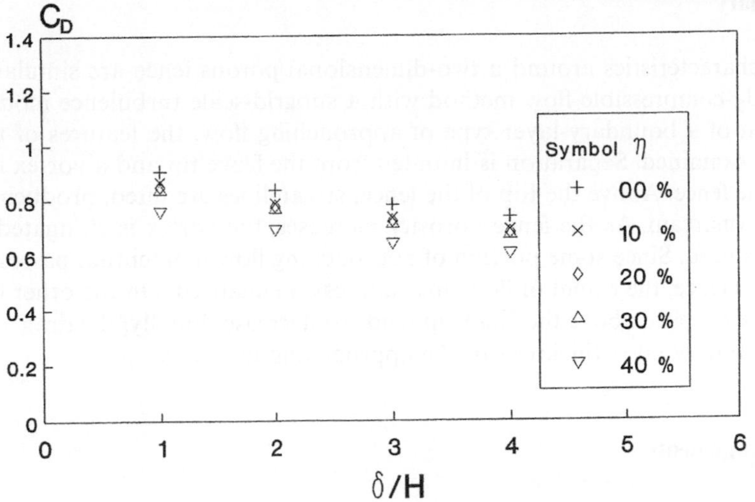

Fig. 6. Relationship between C_D and δ/H for different η.

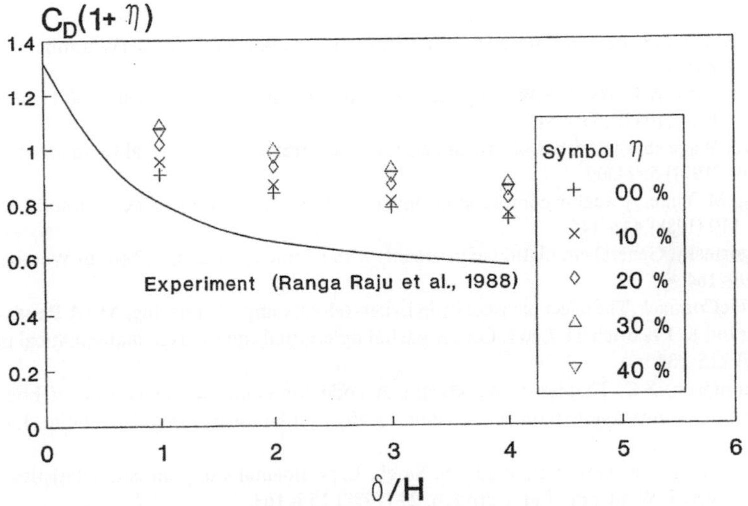

Fig. 7. Comparison of C_D $(1 + \eta)$.

When η or δ/H increases, the value of C_D decreases. Compared to the experimental results of Ranga Raju et al. [9] (Fig. 7), the calculation shows a similar tendency of variation. Without knowing the arrangement of the gaps (size and distribution), the reason why such discrepancy is obtained is difficult to assess and may require more extensive investigations.

7. Summary

Flow characteristics around a two-dimensional porous fence are simulated using the weakly-compressible-flow method with a subgrid-scale turbulence modeling. By the action of a boundary-layer type of approaching flow, the features of the mean flows are examined. Separation is initiated from the fence tip and a vortex is formed behind the fence. Above the top of the fence, streamlines are lifted, producing higher speed downstream. As the fence porosity increases, the vortex is elongated and the drag is reduced. Since some portion of approaching flow momentum passes through the porous fence, the extent of flow unsteadiness is enhanced. On the other hand, the magnitude of speed above the fence tip tends to decrease. Finally, the drag is reduced as the boundary layer thickness of the approaching increases.

Acknowledgements

The work has been supported by the National Science Council in Taiwan under grant No. NSC 83-0410-E-005-022.

References

[1] V. Gupta, K. Ranga Raju, Separated flow in lee of solid and porous fences, J. Hydraulic Eng. ASCE 113 (1987) 1264–1276.
[2] Fuh-Min Fang, W.D. Hsieh, S.W. Jong, J.J. She, Unsteady turbulent flow past a solid fence, J. Hydraul. Eng. ASCE 61 (1997) 131–151.
[3] I. Castro, Wake characteristics of two-dimensional perforated plates normal to an air-stream, J. Fluid Mech. 46 (1971) 599–609.
[4] C. Song, M. Yuan, A weakly compressible flow model and rapid convergence methods, J. Fluids Eng. ASME 110 (1988) 441–445.
[5] J. Smagorinsky, General circulation experiments with primitive equations, Month Weather Rev. 91 (3) (1963) 99–164.
[6] R.W. MacCormack, The effect of viscosity in hyper-velocity impact cratering, AIAA Paper, No. 69-354.
[7] R. Courant, K. Friedrich, H. Lewy, On the partial differential equations of mathematical physics, IBM J. (1967) 215–234.
[8] J.F. Thompson, F.C. Thames, C.W. Mastin, A code for numerical generation of boundary fitted curvilinear coordinate system on field containing any number of arbitrary 2-D bodies, J. Comp. Phys. 24 (1977) 274–302.
[9] K.G. Ranga Raju, R. Garde, S. Singh, N. Singh, Experimental study on characteristics of flow past porous fences, J. Wind Eng. Ind. Aerodyn. 29 (1988) 155–163.

Journal of Wind Engineering
and Industrial Aerodynamics 67 & 68 (1997) 425–436

Application of k–ε model to the stable ABL: Pollution in complex terrain

Asmund Huser*, Pål Jahre Nilsen, Helge Skåtun

Det Norske Veritas, Veritasveien 1, N-1322 Høvik, Norway

Abstract

A procedure for the prediction of stable atmospheric boundary layers (ABL) over complex terrain is developed. The standard k–ε model and the computer programme CFX are used as a basis. The turbulence model is modified for the prediction of stable ABL according to recent published works. The procedure employs databases containing topology, road and traffic data, and is applied to estimate ground pollution concentrations in a stable atmosphere over rough terrain. Examples from road planning in Drammen City show comparison with measurements and effects of planned roads.

Keywords: Air pollution; Environmental aerodynamics; Boundary layers; Computational fluid dynamics

1. Introduction

The aim of this work is to increase knowledge and abilities for prediction of pollution dispersion in a stable atmosphere and over complex terrain. The use of advanced tools, such as 'Computational Fluid Dynamics' (CFD) codes and topography and pollution databases, is essential in this work. The location of Norway, being at a high latitude, causes reduced sun radiation and hence long periods with stable air, especially in the winter. Together with increasing traffic this causes air pollution to become a serious problem, and thus the need for a tool for prediction of pollution dispersion to be developed [1,2].

The two-equation k–ε model is the most popular first order turbulence model. Several variations of this model exist (e.g. k–l model, k–ω model); they have in common that the two turbulence equations produce an *isotropic* eddy-viscosity parameter which is applied to both the momentum equations and the pollution transport equation. The implication of an isotropic eddy-viscosity is that a pollutant

* Corresponding author. E-mail: ahu@dnv.com.

0167-6105/97/$17.00
PII S0167-6105(97)00091-3

will dilute equally fast in all three directions. The buoyancy extended k–ε model applied here is due to Rodi [3]. The work of Duynkerke [4] shows that the isotropic k–ε model compares well with measurements when simulations are performed on a flat atmospheric boundary layer. The two equation q^2–l model of Ref. [5] is also found to predict well the flow over a nearly 2D hill. Based on this, the first order isotropic models perform well for relatively simple geometries, stable atmospheres, and relatively smooth surfaces. The present work applies this turbulence model to a real case on rough terrain and with topology, and reports its performance.

2. Prediction method for the stable ABL

The Reynolds averaged Navier–Stokes equations are solved by the standard programme CFX-F3D (formerly FLOW3D from Harwell). Flow calculations are performed by specifying the wind, temperature and turbulence quantities on the inlet planes, and calculating the air flow and heat transfer in the computational domain until steady state conditions are reached. A quasi steady state is obtained with a constant cooling rate at the surface. This implies that the temperature will decrease downwind, but turbulence quantities and wind are constant. Effluents are released as a source to a transport equation for passive scalar (NO_x). This equation is solved using the constant steady state velocity and temperature field obtained in the flow calculations. The assumption of steady flow conditions is justified by a slow development in the atmospheric conditions under the situations considered. The situations considered are typically a winter day with a stable atmosphere created by low sun radiation with no convective wind generation.

Further assumptions to the flow equations are that no Coriolis force is included, and that the buoyancy term in the vertical momentum equation is neglected. The Coriolis force causes the velocity to turn with height. In the present calculations, the wind is directed by the direction of the valley, and hence the geometry effects are assumed to dominate the Coriolis effect. The buoyancy term in the vertical momentum equation is neglected because it requires special treatment of the pressure outflow boundary conditions. Neglecting the buoyancy effect in the vertical momentum equation causes no wind to occur due to vertical density gradients. These effects may be significant when heavy cold or humid air is draining out from a higher altitude location. In the present model, to compensate for this, drainage flows are forced on the solution when these are known to occur. This is done by specifying a drainage flow as an inlet to the domain at the actual location. The present method is therefore not able to determine where drainage flows will occur. This must be known from observations and field measurements. However, with a mode detailed model, this effect may be incorporated.

2.1. Turbulence model

The k–ε turbulence model is applied in the calculations. The constants used are listed in Table 1. The constants $C_{\varepsilon 1}$ and $C_{\varepsilon 2}$ are standard values from Ref. [6]. Ref. [4]

Table 1
Constants used in the k–ε model

$C_{\varepsilon 1}$	$C_{\varepsilon 2}$	$C_{\varepsilon 3}$	C_{μ}
1.44	1.92 (and 1.83)	1	0.033

uses $C_{\varepsilon 2} = 1.83$ for an atmospheric flow calculation. A reduction of $C_{\varepsilon 2}$ causes the destruction of ε to be reduced, hence ε itself will increase. In the present flat terrain model, the value $C_{\varepsilon 2} = 1.83$ is also used to indicate the effect of this parameter. In the current buoyancy extended version of the code [3], production of ε by buoyancy is zero for stable conditions. Production of k by buoyancy is also zero according to Ref. [3]. The eddy-viscosity is given by

$$\nu_{\mathrm{T}} = C_{\mu}\frac{k^2}{\varepsilon}, \tag{2.1}$$

where k is the turbulent kinetic energy, and ε is the dissipation rate of turbulent kinetic energy. The value of C_{μ} is smaller than the standard value found by wind tunnel tests which is 0.09. In an atmospheric boundary layer C_{μ} is smaller than in a wind tunnel test because production of turbulence is not only due to wall shear stress, but also gravity waves, topography and other large-scale effects.

2.2. Inlet conditions

Inflow profiles are given for the wind speed (u), potential temperature (θ), turbulent kinetic energy (k) and dissipation rate of turbulent kinetic energy (ε). By enforcing these profiles on the inlet planes, the wind speed and the atmospheric stability are defined. The surface roughness, stability class and the surface heat transfer rate are parameters that are set to specify the state of the atmosphere. These parameters are both set on the surface and on the inlet profiles. Formulas for the inflow profiles are given in Ref. [7] and read:

$$u = u_{*}\frac{1}{\kappa}[\ln(z/z_0) - \psi_{\mathrm{M}}(z/L) + \psi_{\mathrm{M}}(z_0/L)], \tag{2.2}$$

where $\psi_{\mathrm{M}} = -17(1 - \mathrm{e}^{-0.29z/L})$, and $\kappa = 0.41$ is the von Karman constant;

$$\Delta\theta = \theta_{*}\frac{1}{\kappa}[\ln(z/z_2) - \psi_{\mathrm{H}}(z/L) + \psi_{\mathrm{H}}(z_2/L)], \tag{2.3}$$

where $\psi_{\mathrm{H}} = -5z/L$ and z_0 is used as a reference length of temperature. The difference between the surface temperature and the air temperature at z is given by $\Delta T = \Delta\theta - gz/C_p$. The surface roughness z_0 may vary from 0.0002 m for open sea to 1 m for suburbs and forests [8]. In the present cases, a constant value of 1 m is used at

Table 2
Mean Obukov length for stable atmospheres, Ref. [7]

Class parameter	d	e	f	g
Mean Obukov length, L (m)	10 000	350	130	60

the inlet and at the surface. The friction velocity (u_*) is found by inserting a known reference wind speed into Eq. (2.2). The reference wind speed at 10 m above the surface is applied. The temperature scale (θ_*) is found by the definition of the Obukov length scale:

$$L = \frac{u_*^2}{\kappa g \theta_* / T}. \tag{2.4}$$

The Obukov length is a measure of stability and defines the stability classes in Table 2, where d is neutral and g is very stable.

Another important parameter is the boundary layer depth, h, which is the height where the velocity shear vanishes. It is given in Ref. [4]:

$$h = 0.4\sqrt{u_* L / f}, \tag{2.5}$$

where f is the Coriolis parameter. ($f = 0.000125$/s in Drammen.) The heat flux at the ground is given by $H = \theta_* u_* \rho C_p$ (W/m^2). In a stable boundary layer the effects caused by velocity shear and density stratification have opposite effects on dilution. The shear layer produces turbulence and hence enhances mixing, whereas the stable air reduces mixing. The correct combinations of the stability parameters (Obukov length) and velocity shear (roughness height) may be estimated when comparing measurements and model predictions.

The inlet profiles for the turbulence parameters k and ε are found in Ref. [4] and read:

$$k = \frac{u_*^2}{\sqrt{C_\mu}}\left(1 - \frac{z}{h}\right)^2, \tag{2.6}$$

$$\varepsilon = \frac{u_*^3}{\kappa}\left(\frac{1}{z} + \frac{4}{L}\right). \tag{2.7}$$

Combining Eqs. (2.1), (2.6) and (2.7), the normalised eddy-viscosity at the inlet is

$$\frac{\nu_T}{\kappa u^* h} = \frac{(1 - z/h)^4}{h/z + 4h/L}. \tag{2.8}$$

2.3. Wall boundary conditions

Wall boundary conditions for the velocity, temperature and scalar are given by the law of the wall which reads for a general variable ϕ:

$$\frac{\phi}{\phi^*} = \frac{\sigma_\phi}{\kappa}\ln\left(\frac{z}{z_0}\right), \tag{2.9}$$

where σ_ϕ is the turbulent Prandtl number (= 0.9 for temperature and scalar).

2.4. Outlet, sides and upper boundary

Outlet boundary conditions are given by specifying the pressure, and setting no gradients in the flow direction on the other variables. On the side planes of the domain and the upper boundary, symmetry conditions are imposed. That is to say no air is entering or leaving these boundaries.

2.5. Equation of state

The equation of state applies the weakly compressible assumption:

$$\rho = \frac{p_{\mathrm{ref}} w}{R\theta}, \tag{2.10}$$

where $p_{\mathrm{ref}} = 1.013105$ Pa is constant, $w = 29$ is the molecular weight of air and $R = 8314$ J/(K kmol) is the universal gas constant. The potential temperature (θ) is used in the calculations in order to compensate for the weakly compressible assumption.

2.6. Estimation of sources

The major NO_x sources used for this study are emissions from vehicles. In addition emissions from a few major industry stack sources are included, as well as diffuse sources, mainly originating from domestic heating.

The emissions from vehicles are implemented as quasi-line-sources. The road network is divided into line segments. For each segment the release per road length and time unit are computed, based on parameters such as the amount of vehicles, average speed, fraction of heavy trucks and the slope of the road. Minor roads, having less than 5000 vehicles passing per day are excluded.

Each linear road segment contributes to the source of the cells corresponding to the length of the segment within each cell. When the contributions to each horizontal grid cell are added for all line segments, the total source for each cell is applied as a point source in the grid cell closest to the ground. Emissions from vehicles contribute to approximately 80% of the total sources.

Diffuse sources are also added as a point source in the grid cell closest to the ground. For stack sources, standard plumerise formulas are applied to find the equilibrium height of the plume, and the stack sources are entered into the model in the grid cell corresponding to the position of the stack and a height above the ground which equals the equilibrium height.

3. Results and discussion

Results from two simulation models are presented here. First, flow over a flat terrain and second, flow over topology. For both models the development of the boundary layer is presented in terms of eddy-viscosity profiles. The eddy-viscosity is defined in Eq. (2.1) and is important for pollution dilution because it enters the diffusion term in the passive scalar equation.

3.1. Flow over flat terrain

A simple geometrical model measuring 10 by 5000 m is used. Velocity, temperature and turbulence profiles, given in Section 2.2, are set on the inlet. The parameters for the two cases presented here are given in Table 3. The normalised eddy-viscosity profiles are presented in Fig. 1 indicating that the eddy-viscosity increases when going downwind. The measured maximum value of normalised eddy-viscosity [4] is 0.04, indicating that the present calculations are over-predicting the eddy-viscosity. Measurements that make the basis for the normalised eddy-viscosity in Ref. [4] are from flat terrain with little topology and low roughness $z_0 = 0.01$ m. In the high roughness case, a roughness of $z_0 = 1$ m is used. The over-prediction of normalised eddy-viscosity is smaller for smooth terrain than for rough terrain. Measurements of eddy-viscosity in rough terrain are lacking, and hence the correct value of the maximum eddy-viscosity is unknown. Reducing the $C_{\varepsilon 2}$ parameter from 0.92 to 0.83 causes the maximum eddy-viscosity to decrease toward the measured value (see

Table 3
Parameters used in flow over flat terrain and topology

	Case 1, Rough surface	Case 2, Smooth surface	Run 3, Topology model
Wind speed at 10 m (m/s)	1.5	4	1
Stability class	e	f	f
Roughness z_0 (m)	1	0.01	1
u^* (m/s)	0.26	0.23	0.16
θ^* (K)	0.013	0.029	0.014
h (m)	340	196	163
L (m)	350	130	130
Ground heat flux (W/m^2)	– 3.87	– 13.4	– 2.4

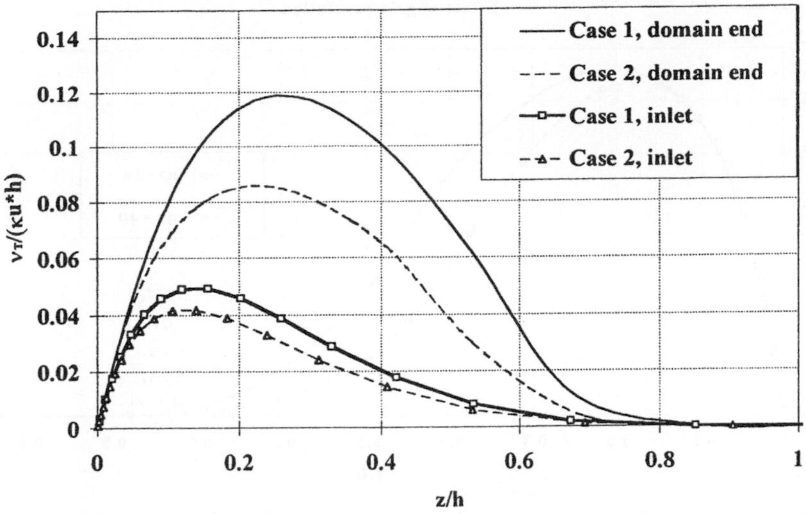

Fig. 1. Normalised eddy-viscosity at the inlet and 5 km downwind the inlet. The inlet profiles are given by Eq. (2.8) and compared with measurements for smooth surfaces. Comparing smooth and rough surface model. $C_2 = 0.92$.

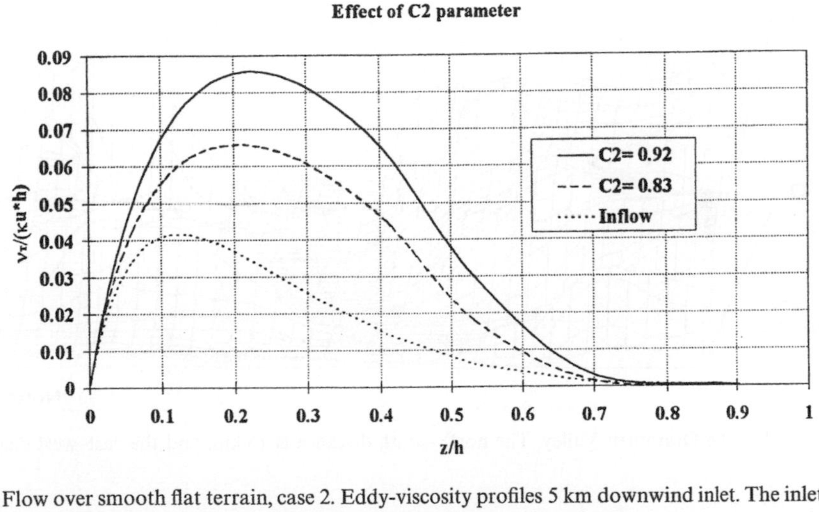

Fig. 2. Flow over smooth flat terrain, case 2. Eddy-viscosity profiles 5 km downwind inlet. The inlet profile is given by Eq. (2.8) and compared with measurements for smooth surfaces.

Fig. 2). This indicates that the value of 0.83 is more appropriate for atmospheric flows. Still, the value of 0.83 does not give the correct behaviour, indicating the uncertainty of this parameter. The grid resolution has a small impact on the solution in the outer part of the boundary layer as indicated in Fig. 3.

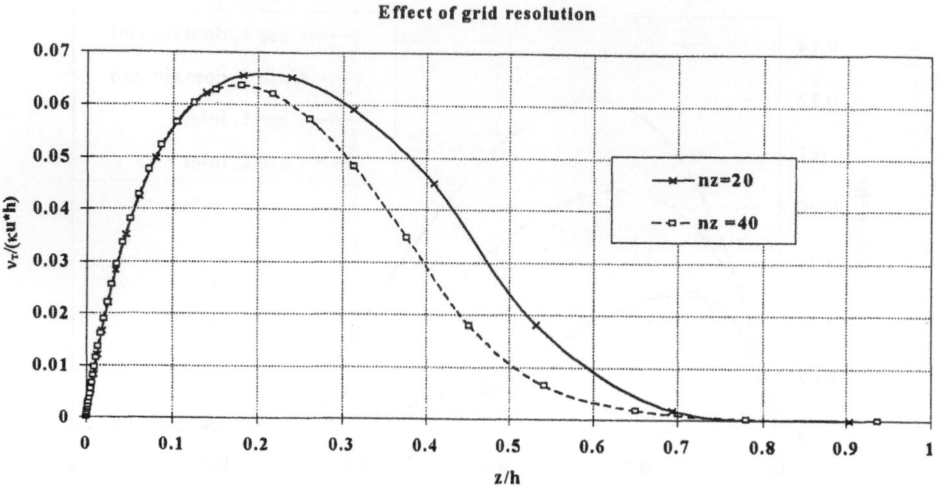

Fig. 3. Flow over smooth flat terrain, case 2. Eddy-viscosity profiles 5 km downwind inlet. $C_2 = 0.83$. Grid resolution: $n_x = 20$ and $n_z = 20$ (crosses); and $n_x = 50$ and $n_z = 40$ (squares), where n_x and n_z are the number of nodes horizontally and vertically, respectively.

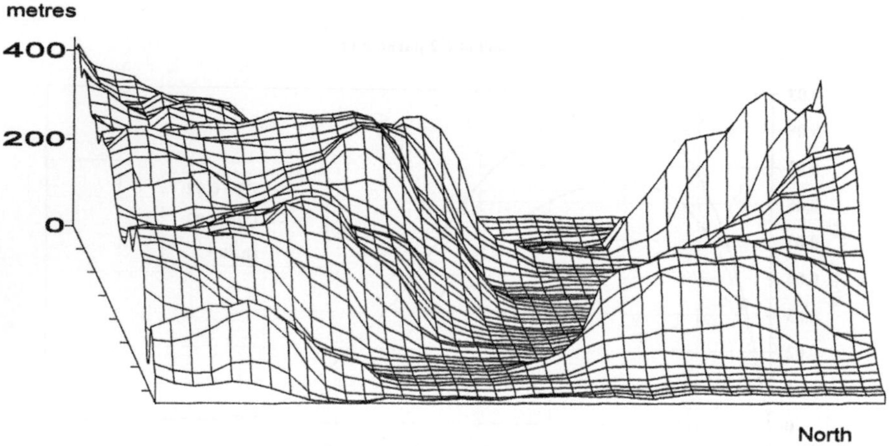

Fig. 4. Model of the Drammen Valley. The north–south distance is 16 km, and the east–west distance is 20 km.

3.2. *Flow over topology*

The city model contains an area of 20 by 16 km including the Drammen valley and the surrounding hills, which are up to 400 m high (see Fig. 4). In this model, the finest grid cells are down to 100 by 100 m.

Results show the normalised eddy-viscosity for one of the runs. The parameters are presented in Table 3. The normalised eddy-viscosity is shown to increase above the

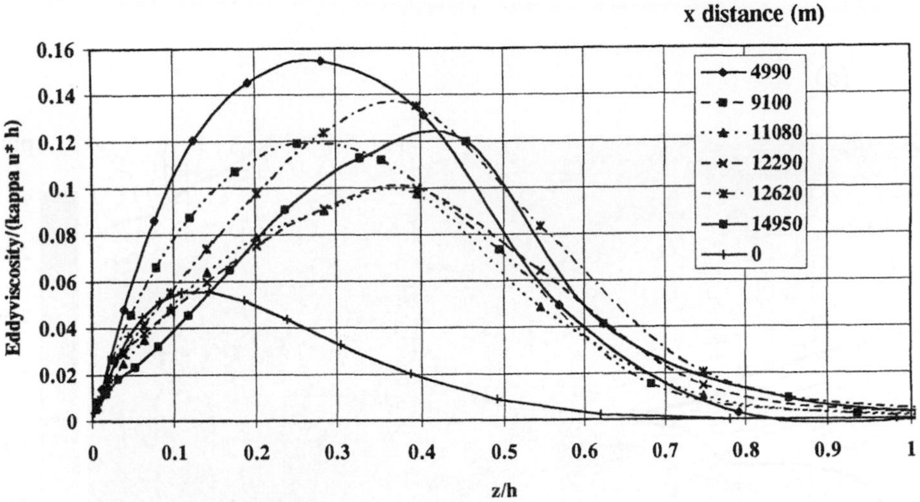

Fig. 5. Normalised eddy-viscosity at six locations downwind the inlet. The inlet profile is given by Eq. (2.8) and compared with measurements for smooth surfaces.

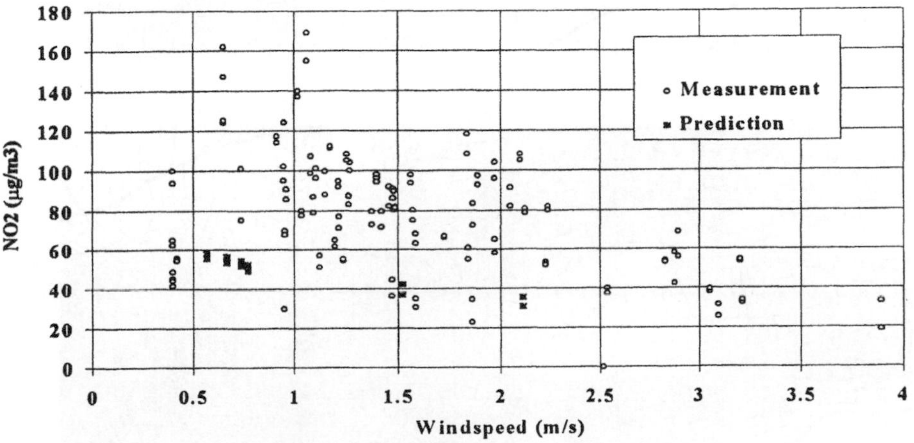

Fig. 6. Comparison of predicted concentration of NO_2 with measured values.

inlet value of 0.055 (see Fig. 5). After 5000 m the maximal value of the normalised eddy-viscosity stays below 0.14 indicating that the boundary layer is not developing further. A surface roughness of 1 m is above the roughness of 0.01 which is used in Ref. [4], explaining some of the discrepancies.

3.3. Evaluation of the model

The model was found to give the right trend of pollution concentration with increasing wind speed (see Fig. 6). Both the surface roughness and the stability

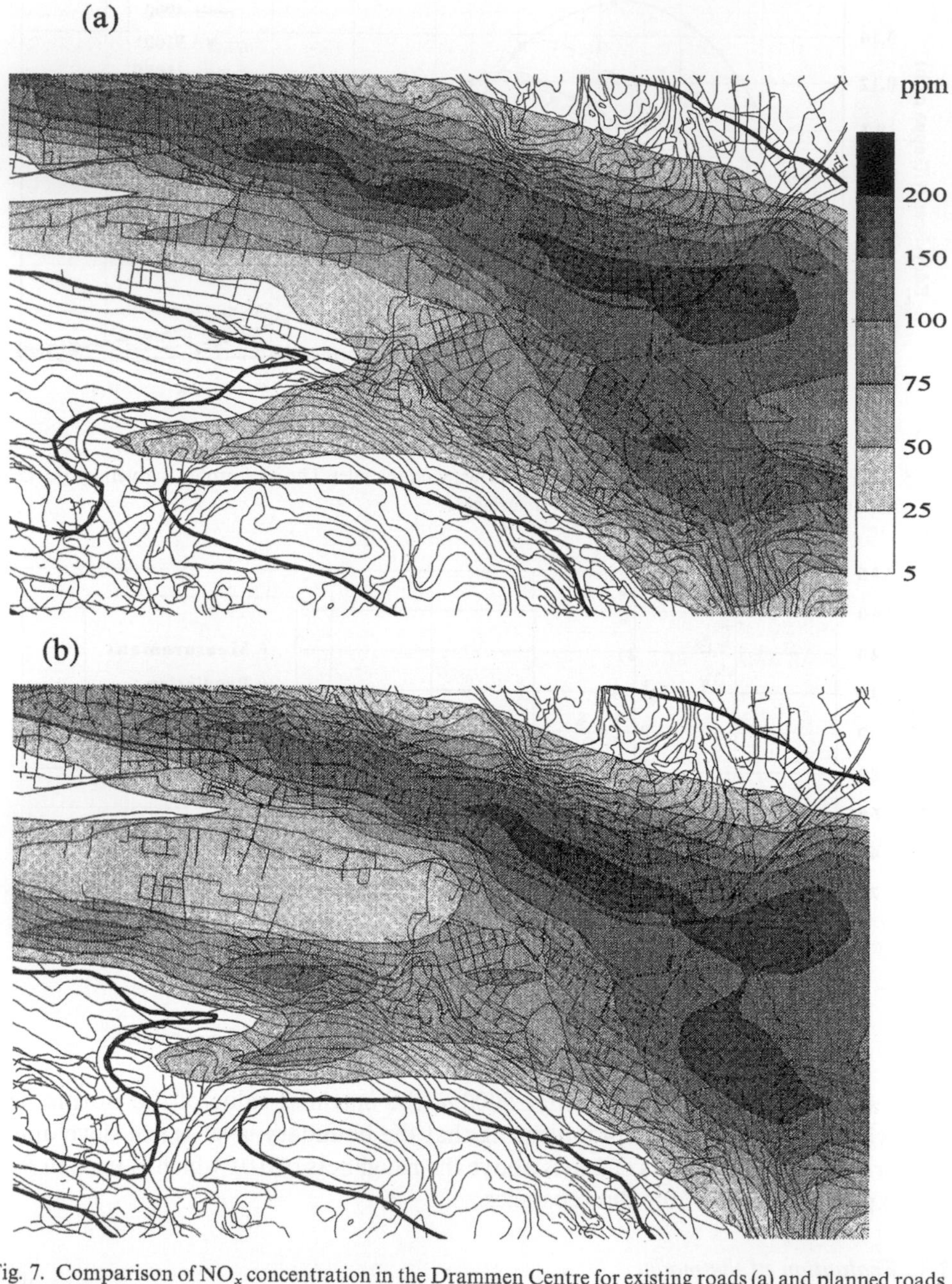

Fig. 7. Comparison of NO_x concentration in the Drammen Centre for existing roads (a) and planned roads in 2005 (b).

parameters were fixed, based on qualitative judgements, prior to the calculations. The present results indicate that too high roughness and too weak stability has been used, both factors causing too much mixing and hence lower concentrations than measured on the worst days in Drammen. By specifying the stability and surface roughness parameters, the corresponding atmospheric conditions may be simulated. The present results may be used as a basis for selecting other values for the roughness and stability parameters.

The model is found useful for predicting airpollution in e.g. road planning or planning of location of other releases to the atmosphere. As an example, Fig. 7 shows the effect of planned roads in Drammen on NO_x concentration.

Results show that the *normalised* eddy-viscosity is over-predicted compared to measured values [4]. This may be caused by three factors: Too high surface roughness, adjustments to the $C_{\varepsilon 2}$ parameter and the isotropic model.

The surface roughness length of forests and suburbs (1 m) is used in the whole domain in the Drammen model. Measurements of the eddy-viscosity (or other turbulence parameters) found in the literature are uncertain for such high roughness, therefore the value of the calculated eddy-viscosity cannot be directly verified. Also, the surface roughness does vary in the Drammen area from $z_0 = 0.01$ m over flat areas and water surfaces to the order of 1 m over forests and urban areas. A lower surface roughness may therefore be justified. In a more detailed model, a variable surface roughness should be given over the area.

The turbulence constant $C_{\varepsilon 2}$ takes different values in the literature, and could be adjusted further to calibrate the model against measured pollutant concentrations.

Atmospheric stability is specified on the flow inlets by the temperature profile, and on the surface by the heat flux from the air. A logarithmic temperature profile is used, which has the steepest temperature gradient near the surface. A more detailed model will have to include the possibility to model inversion layers and anisotropy turbulence models. An inversion layer acts as a 'lid' on the valley and will trap the pollutants inside. The logarithmic temperature profile is an assumption which is difficult to check because no profile measurements have been performed in Drammen. Observations of temperature profiles elsewhere [7] show both profiles with the strongest gradient at the ground and profiles with a gradient both at the ground and in a layer above the ground. Implementation of an anisotropic turbulence model will cause the vertical dilution to reduce. This will enable us to predict more correctly the inversion which is trapping the pollutant inside the valley.

The grid resolution may also be a small source of inaccuracies. It is important to concentrate the grid-lines where the gradients are highest. Therefore, when reducing the surface roughness, the grid-lines in the z direction must be refined near the ground.

Acknowledgements

This study has been sponsored by Statens Vegvesen Buskerud (the Road Authorities of Drammen). The support of the contact person, Mr. Bjørn Haram, is much appreciated. The authors would also like to acknowledge the valuable comments of Dr. Ivar Nestaas and Mr. Sigmund Olavsen. Funding provided by Det Norske Veritas is appreciated.

References

[1] S. Olafsen, Forstudie luftmodell – Drammen og Nedre Eiker, DNV Industry rapport no. 94-3634, 1994.

[2] A. Huser, P.J. Nilsen, H. Skåtun, Beregning av luftforurensing i Drammen og omegn – Utvikling av beregningsmodell, DNV Industry rapport no. 95-3287, 1995.

[3] W. Rodi, Calculation of stably stratified shear layer flows with a buoyancy – extended K–ε turbulence model, in: J.C.R. Hunt (Ed.), Turbulence and Diffusion in Stable Environments, Oxford University press, Oxford, 1985, pp. 111–143.

[4] P.G. Duynkerke, Application of the E–ε turbulence closure model to the neutral and stable atmospheric boundary layer, J. Atmos. Sci. 45 (5) (1988).

[5] T. Yamada, Simulation of nocturnal drainage flows by a q^2l turbulence closure model, J. Atmos. Sci. 40 (1983) 91.

[6] CFDS-FLOW3D Release 3.3, Users manual.

[7] A.P. van Ulden, A.A.M. Holtslag, Estimation of atmospheric boundary layer parameters for diffusion applications, J. Clim. Appl. Met. 24 (1985).

[8] J. Counihan, Adiabatic atmospheric boundary layers: A review and analysis of data from the period 1880–1972, Atmospheric Environment 9 (1975) 871–905.

ELSEVIER

Journal of Wind Engineering
and Industrial Aerodynamics 67&68 (1997) 437–448

JOURNAL OF wind engineering AND industrial aerodynamics

A numerical study of the wind field in a typhoon boundary layer

Yan Meng*, Masahiro Matsui, Kazuki Hibi

Environmental Engineering Department, Institute of Technology, Shimizu Corporation, No. 4-17, Etchujima, 3-chome, Koto-ku, Tokyo 135, Japan

Abstract

The wind field in a typhoon boundary layer (TBL) has been investigated by a numerical model. The results show that vertical profiles of wind speed in the TBL can be satisfactorily stated by conventional power-law expressions. To describe the structure of strong wind in the TBL, two parameters have been suggested: one is a dimensional parameter, f_λ, approximately representing the absolute vorticity in the wind field, and the other is a non-dimensional parameter, ξ, characterizing the heterogeneity of vorticity in the radial direction of a typhoon. Substituting the parameter f_λ by Coriolis parameter f, the gradient height z_g during typhoons can be predicted by the same formula as that used during non-typhoon climates. The ratio of surface to gradient wind speeds $G(r)$ and the inflow angle γ_s in the TBL are also examined using the present numerical results, and the formulae for predicting them are presented.

Keywords: Typhoon boundary layer; New external parameters; Prediction of the vertical wind profile; Ratio of surface to gradient wind speeds; Inflow angle

Nomenclature

v_θ, v_r	horizontal components of wind
$v_{\theta g}$, v_{rg}	horizontal components of gradient wind
v'_θ, v'_r	differences between wind velocities and gradient winds
v''_θ, v''_r, v''_z	velocity fluctuations
c	translation velocity of typhoon
θ	angle, counterclockwise positive from east
r	radial distance from typhoon center
z	elevation

* Corresponding author. E-mail: meng@sit.shimz.co.jp.

PII S0167-6105(97)00092-5

β	approach angle of typhoon, counterclockwise positive from east
γ	inflow angle or geostrophic angle
r_m	radius of maximum wind
f	Coriolis parameter
u_*	friction velocity at the surface
z_0	roughness length
κ	von Karman constant, 0.4
U	wind speed
K_m	turbulent exchange coefficient
q^2	twice the turbulent energy
σ_u	standard deviation of the fluctuation of wind speed
l	master turbulence length scale
l_0	master turbulence length scale as z approaches infinity
B_1	closure parameter, 21.3, see Eq. (7)
S_q	closure parameter, 0.2, see Eq. (8)
α	empirical parameter, 0.1, see Eq. (11)
γ_1	closure parameter, 0.222, see Eq. (12)
f_λ, ξ	see Eqs. (13) and (14)
Ro_λ	modified surface Rossby number
$G(r)$	ratio of surface to gradient wind speeds
U_g	gradient wind speed
z_g	gradient height used in power law expression
α_u	power law exponent for the wind speed profile

1. Introduction

It is well known that in most areas on the southeastern coasts of China and Japan, strong winds are associated with two types of weather systems. The first of these is called the non-typhoon wind climate and represents all winds not related to tropical storms or typhoons. The second type of wind climate is referred to as the typhoon wind climate and represents tropical cyclones originating in the Pacific Ocean. To exactly evaluate wind loads on structures in these regions, a detailed understanding of wind structure in both types of weather systems is required.

During the last few decades, extensive research on wind structure during strong wind conditions has been carried out by meteorologists and engineers. As a result, valuable data for various types of terrain have been gathered from sites, and several empirical models for expression of the vertical profiles of wind speed have been proposed. However, most of them are designed for strong winds in non-typhoon conditions. Information on wind structure during severe tropical cyclones is still insufficient, although Choi [1] has reported some upper-level wind data during typhoons from radar sounding records.

In the present study, a series of numerical simulations are conducted to provide an adequate description of the vertical wind profile during a typhoon. New parameters are suggested to describe the structure of strong wind in the typhoon boundary layer (TBL). The relation between basic parameters of the power law and external parameters of the TBL are then examined using the numerical results. Finally, the ratio of surface to gradient wind speeds $G(r)$ and the inflow angle γ_s in the TBL are also investigated and formulae for predicting them are presented.

2. Numerical model

2.1. Description of model

A model for describing wind field in the typhoon boundary layer has been proposed by the authors [2]. Fig. 1 is a brief illustration of the model, in which the typhoon-induced mean wind velocity $\boldsymbol{v}$ is expressed by the addition of the gradient wind $\boldsymbol{v}_g$ in the free atmosphere, and the component $\boldsymbol{v}'$, caused by friction on the ground surface. Assuming that the gradient wind pattern moves at the translation velocity of typhoon $\boldsymbol{c}$ in the free atmosphere, gradient wind velocity $\boldsymbol{v}_g$ was analytically obtained by means of perturbation analysis. Furthermore, the equation for the component $\boldsymbol{v}'$ was linearized, considering that the unsteady term in this equation is smaller than the

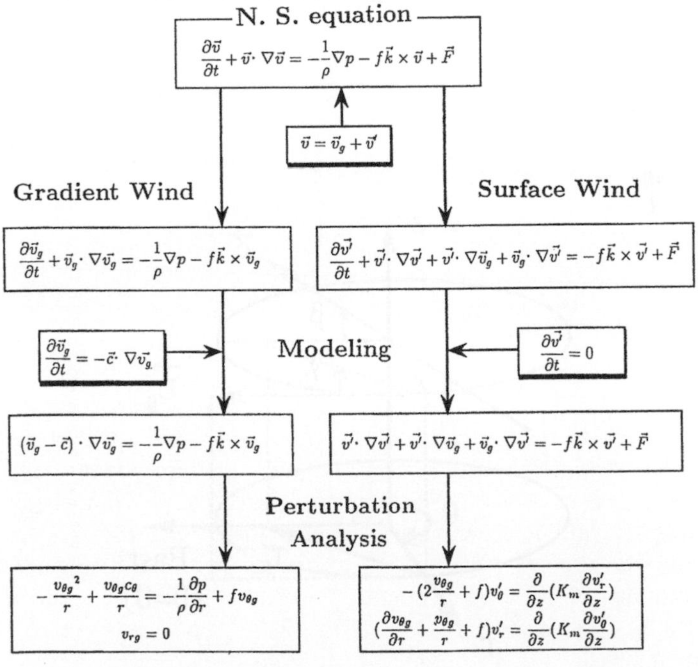

Fig. 1. Summary of the model describing wind field during a typhoon.

turbulent viscosity term and the inertia term, and the component $\boldsymbol{v}'$ is smaller than the gradient wind velocity $\boldsymbol{v}_g$. Detailed explanations of the typhoon model are described in Ref. [2].

Consequently, gradient wind velocity $\boldsymbol{v}_g$ is expressed as

$$v_{\theta g} = \frac{c_\theta - fr}{2} + \sqrt{\left(\frac{c_\theta - fr}{2}\right)^2 + \frac{r}{\rho}\frac{\partial p}{\partial r}}, \tag{1}$$

$$v_{rg} = 0, \tag{2}$$

and the equation of motion for $\boldsymbol{v}'$ is written as

$$-\left(2\frac{v_{\theta g}}{r} + f\right)v'_\theta = \frac{\partial}{\partial z}\left(K_m \frac{\partial v'_r}{\partial z}\right), \tag{3}$$

$$\left(\frac{\partial v_{\theta g}}{\partial r} + \frac{v_{\theta g}}{r} + f\right)v'_r = \frac{\partial}{\partial z}\left(K_m \frac{\partial v'_\theta}{\partial z}\right), \tag{4}$$

where $c_\theta = -c\sin(\theta - \beta)$. Fig. 2 shows the coordinate system used in this study. Approach angle β and angle θ are defined as counterclockwise positive from the east. The boundary condition at the upper atmosphere is

$$\boldsymbol{v}'|_{z\to\infty} = 0 \tag{5}$$

and the boundary condition above the ground surface is obtained from the logarithmic formula

$$U = \frac{u_*}{\kappa}\ln\frac{z}{z_0}, \tag{6}$$

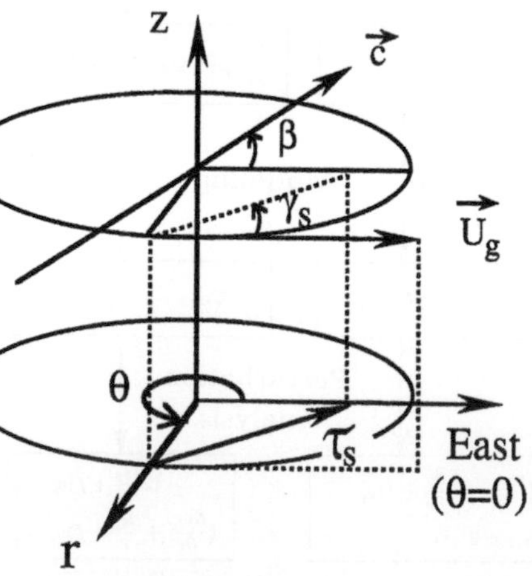

Fig. 2. Coordinate system.

in which $U\ (=\sqrt{v_\theta^2 + v_r^2})$ is zero wind speed at $z = z_0$, κ is the von Karman constant, and u_* is the von Karman friction velocity at the surface.

The expression of the vertical distribution of K_m is an important factor for successful simulations of the atmospheric boundary layer (ABL). The present study used a turbulence closure model proposed by Mellor and Yamada [3] in which K_m is related to turbulence kinetic energy $q^2/2$ and turbulence length l through the expression

$$K_m = B_1^{-1/3} q l, \tag{7}$$

and $q^2/2$ is obtained by solving the turbulence kinetic energy equation as follows:

$$\frac{\partial}{\partial z}\left(qlS_q\frac{\partial q^2/2}{\partial z}\right) - \overline{v_\theta'' v_z''}\frac{\partial v_\theta}{\partial z} - \overline{v_r'' v_z''}\frac{\partial v_r}{\partial z} - \frac{q^3}{B_1 l} = 0, \tag{8}$$

in which $-\overline{v_\theta'' v_z''}$ and $-\overline{v_r'' v_z''}$ are turbulence second moments obtained from a set of diagnostic equations and are expressed as

$$(-\overline{v_\theta'' v_z''}, -\overline{v_r'' v_z''}) = B_1^{-1/3} q l \left(\frac{\partial v_\theta}{\partial z}, \frac{\partial v_r}{\partial z}\right). \tag{9}$$

For turbulence length scale l, Blackadar's interpolation formula is adapted as

$$l = \frac{\kappa z}{1 + \kappa z / l_0}, \tag{10}$$

which interpolates between two limits $l \to \kappa z$ as $z \to 0$ and $l \to l_0$ as $z \to \infty$. l_0 is assumed to be proportional to the ratio of the first to the zeroth moment of the profile $q(z)$. Thus,

$$l_0 = \alpha \frac{\int_0^\infty zq\, dz}{\int_0^\infty q\, dz}, \tag{11}$$

where α is an empirical constant and is set as 0.1. This expression works well for the boundary layer, as pointed out by Mellor and Yamada [3].

The standard deviation of the fluctuation of wind speed is obtained as

$$\sigma_u^2 = (1 - 2\gamma_1) q^2. \tag{12}$$

The values of closure parameters $(B_1, S_q, \gamma_1) = (21.3, 0.2, 0.222)$ have been determined from neutral turbulence data.

Finite-difference methods are used to obtain the numerical solution. The space derivative terms are approximated with central differences, and all equations are iteratively solved by the successive over relaxation (SOR) method.

2.2. External parameters in the TBL

In order to describe the structures of strong wind in the TBL, the following two parameters are introduced,

$$f_\lambda = \left(\frac{\partial v_{\theta g}}{\partial r} + \frac{v_{\theta g}}{r} + f\right)^{1/2}\left(2\frac{v_{\theta g}}{r} + f\right)^{1/2}, \tag{13}$$

$$\xi = \left(2\frac{v_{\theta g}}{r} + f\right)^{1/2} \Big/ \left(\frac{\partial v_{\theta g}}{\partial r} + \frac{v_{\theta g}}{r} + f\right)^{1/2}, \tag{14}$$

and rearranging Eqs. (3) and (4) gives

$$-\xi f_\lambda v'_\theta = \frac{\partial}{\partial z}\left(K_m \frac{\partial v'_r}{\partial z}\right), \tag{15}$$

$$\frac{1}{\xi} f_\lambda v'_r = \frac{\partial}{\partial z}\left(K_m \frac{\partial v'_\theta}{\partial z}\right), \tag{16}$$

where f_λ has the same dimension as vorticity and ξ is a non-dimensional parameter.

To investigate the radial dependence of f_λ and ξ, the pressure data of typhoon Mireille, observed at 1600 JST on 27 September 1991 [2], are used to calculate the gradient wind speed. Table 1 summarizes the parameters of the typhoon as obtained from the Japanese Meteorological Agency.

Fig. 3a illustrates a variation of f_λ with r/r_m in the $\theta = \beta$ direction. The dashed line represents a mean value of the Coriolis parameter f in the area dominated by typhoon Mireille. As expected, f_λ shows a large value for small r/r_m and gradually approaches the Coriolis parameter f for large r/r_m. Note from Eq. (13) that f_λ approximately expresses absolute vorticity in the ABL during a typhoon. Since relative vorticity $(\partial v_{\theta g}/\partial r + v_{\theta g}/r)$ associated with the typhoon gradually decreases as the distance away from the typhoon center increases, f_λ reaches a value close to the Coriolis parameter f for large r/r_m. This fact implies that, under typhoon condition, f_λ must be adopted as an external parameter to substitute the Coriolis parameter f. Rapid increase of the vorticity associated with a typhoon at places located near the typhoon center has been reported by Mitsuta et al. [4], and the basic shape of the vorticity is similar to f_λ.

Fig. 3b presents the variation of ξ with r/r_m in the $\theta = \beta$ direction. Although the parameter ξ shows a value close to 1 for both small and large r/r_m, it gives a broad peak in the region $2 < r/r_m < 3$. It is obvious from Eq. (14) that the condition for

Table 1
Summary of typhoon parameters used in this study

Date (JST)	Time (JST)	Latitude (deg)	Longitude (deg)	β (deg)	C (m/s)	P_c (hPa)	ΔP (hPa)	r_m (km)
91.09.27	16	32.8	129.7	50.1	17.1	940.0	73.0	85.4

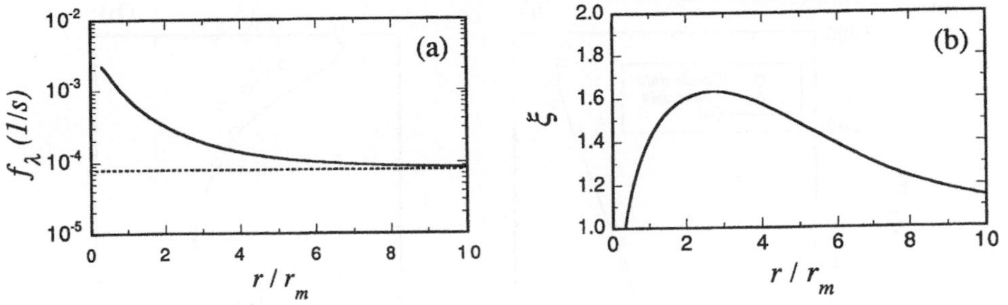

Fig. 3. Variation of f_λ and ξ with r/r_m.

$\xi = 1$ is that the distribution of vorticity is uniform in the radial direction. This fact indicates that ξ is a parameter characterizing heterogeneity of vorticity in the radial direction. Since the vorticity field is similar to that induced by solid body motion for small r/r_m and is close to uniform for large r/r_m, this condition is approximately satisfied in both regions. In the region where ξ is larger than 1, the coefficient multiplying v'_θ in Eq. (15) is larger than that multiplying v'_r in Eq. (16). As a result, the inflow angles on the surface do not only depend on the roughness length z_0, but also on the distance away from the typhoon center, as discussed in Section 3.3.

It is well known that, under non-typhoon condition, characteristics of the vertical wind profile in the ABL with neutral stability and horizontal homogeneity depend only on the external parameters, U_g, f, z_0, while, under typhoon condition, the external parameters of the ABL are U_g, f_λ, z_0, ξ. In the next section, our task is to show the wind structure in the TBL and to formulate the relationship between the basic parameters of the power law and these external parameters.

2.3. Code validation

Before moving to the main task, it was necessary to simulate the strong wind records resulting from a developed extratropical depression that occurred at Hokkaido on 4 September 1981, in order to confirm the validity and accuracy of the numerical model. The wind data were obtained from a 213 m high meteorological observation tower located at the center of the Tsukuba Science City. This area is a part of the Kanto Plain, where there is no distinct elevated terrain within a radius of about 20 km. The area within a radius of 150 of the tower is covered with short grass and, farther out, irregular patches of ground are dotted with young pine plantations, no more than 10 m in height.

The wind velocities at six levels (10, 25, 50, 100, 150 and 200 m) were measured by three-dimensional sonic anemometers mounted at the end of 6 m booms extending from the edge of the tower. All signals from the anemometers were sampled at 20 Hz and analyzed by statistical techniques for a run of 100 min in duration. Since the developed extratropical depression was far from the observation site, there were very

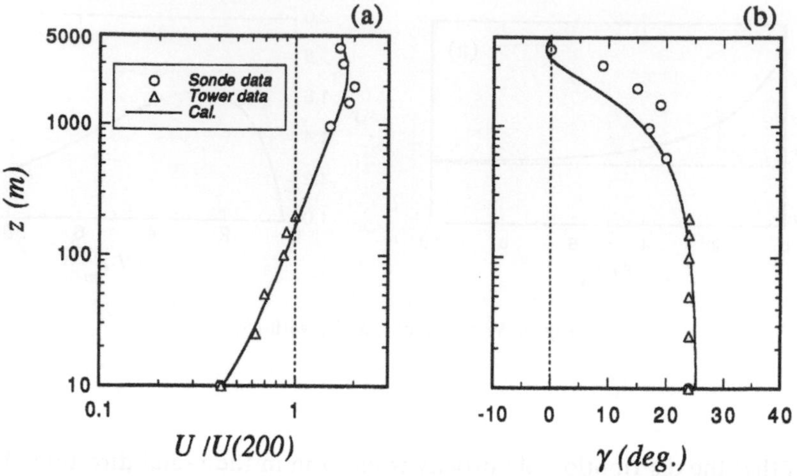

Fig. 4. Comparison of wind speed and direction computed using the numerical model and observed data; symbols: observed; solid line: simulated.

few variations in mean wind speed and direction from 08 : 00 to 17 : 00 h. External parameters of the ABL used in this simulation are set at the following values: $U_g = 27.5$ m/s, $f = 0.857 \times 10^{-4}$ s^{-1}, $z_0 = 0.9$ m. The magnitude of the roughness length z_0 was derived using the log-law to fit the wind speeds obtained at the lower three levels (10, 25 and 50 m).

Fig. 4 shows a comparison of wind speed and wind direction using the numerical results and observed data. Wind speed profiles are normalized by the wind speed at a height of 200 m (top of tower). Agreement between the computed values (solid lines) and observed data (symbols) obtained from the tower is almost total. The validity and accuracy of the numerical model is therefore confirmed.

3. Results and discussion

3.1. An application for simulating typhoon Mireille

The wind field resulting from typhoon Mireille is simulated to examine the radial dependence of the vertical wind profile. The parameters of the typhoon, as shown in Table 1, are used to calculate the gradient wind speed. The roughness length z_0 is set at 0.001, 0.01 and 0.1 m, covering flat coastal areas to rural terrain. Simulations are performed at several typical positions in the radial direction.

Fig. 5a shows the variation of the gradient wind speed and pressure difference with the radius r/r_m. Vertical profiles of wind speed and direction for the case of $z_0 = 0.01$ m are plotted in Fig. 5b and Fig. 5c. Simulated wind speeds (solid lines) are normalized by the gradient wind speed U_g. The dashed lines shown in Fig. 5b and

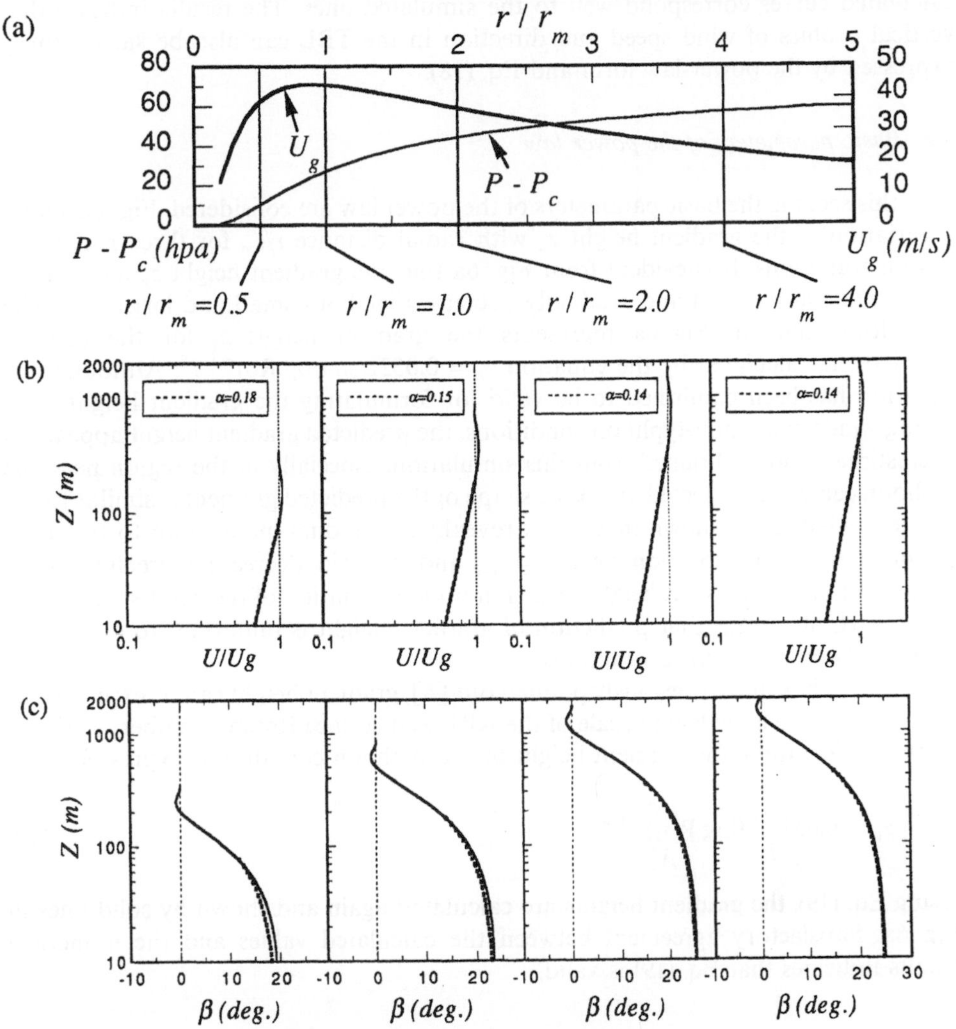

Fig. 5. Vertical profiles of wind speed and direction at several radial locations.

Fig. 5c represent curves calculated using Eqs. (17) and (18),

$$\frac{U(z)}{U_g} = \left(\frac{z}{z_g}\right)^{\alpha_u}, \tag{17}$$

$$\gamma(z) = \gamma_s \left(1.0 - 0.4\frac{z}{z_g}\right)^{1.1}, \tag{18}$$

in which z_g denotes the gradient height where the wind speed attains the gradient value. Eq. (17) presents a conventional power-law expression proposed by Davenport [5], and Eq. (18) was suggested by the authors [6,7]. Under the gradient height, the calculated curves correspond well to the simulated ones. The results indicate that vertical profiles of wind speed and direction in the TBL can also be satisfactorily expressed by the power-law form and Eq. (18).

3.2. Basic parameters of the power law

In this section, the basic parameters of the power law are considered. Fig. 6a shows a variation of the gradient height z_g with radial distance r/r_m for three roughness terrain conditions. It is evident from Fig. 6a that the gradient height z_g has a maximum value at $r/r_m = 4$ and gradually decreases for both small and large r/r_m. The dash-dotted line in Fig. 6a represents the gradient height z_g for the case of $z_0 = 0.01$ m, calculated by the equation $z_g = 0.052U_g/f(\log \mathrm{Ro})^{-1.45}$. Although this equation has been confirmed to be valid for formulating the gradient height z_g of strong winds under non-typhoon conditions, the predicted gradient height appears to overestimate those obtained from this simulation, especially in the region near the typhoon center. As expected, the basic shape of the predicted gradient is similar to the gradient wind speed shown in Fig. 5a, revealing a gradual increase up to the maximum value at the location near $r = r_m$, and a rapid decrease thereafter. Since overestimation of gradient height results in underestimation of the wind speed in the lower ABL, it is clear that prediction of gradient height is important to evaluating wind loads on engineering structures.

As the authors have previously pointed out [6], gradient height can be expressed as a function of the large length scale of the ABL and surface Rossby number. With the same consideration, the gradient height in the typhoon condition is expressed as

$$z_g = 0.052\frac{U_g}{f_\lambda}(\log \mathrm{Ro}_\lambda)^{-1.45}. \tag{19}$$

Using Eq. (19), the gradient heights are calculated again and shown by solid lines in Fig. 6a. Satisfactory agreement between the calculated values and the numerical results indicates that Eq. (19) is valid.

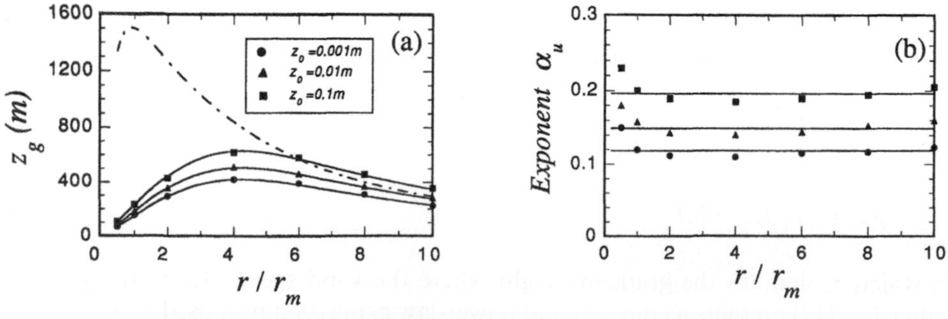

Fig. 6. Radial dependence of the gradient height z_g and the power exponent α_u.

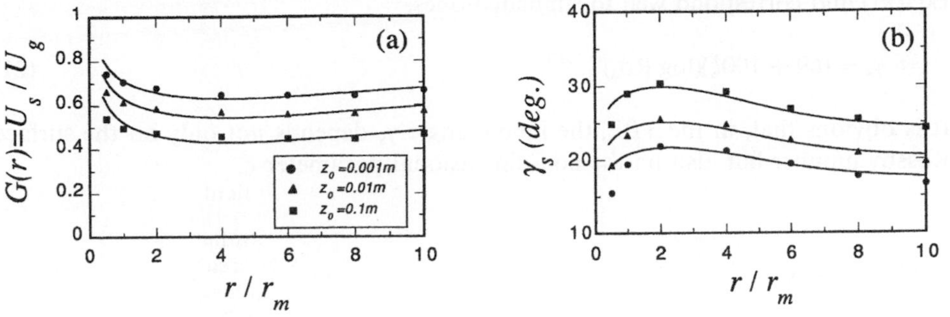

Fig. 7. Radial dependence of the wind speed ratio $G(r)$ and the inflow angle γ_s.

The effect of terrain on the shape of the vertical wind profile is represented by the exponent α_u in the power law. Fig. 6b illustrates a variation of the power exponent α_u with r/r_m for three roughness terrain conditions. With the exception of the core region of the typhoon, the power exponents for the three cases are practically steady and can be satisfactorily predicted by Eq. (20) [6],

$$\alpha_u = 0.27 + 0.09 \log z_0 + 0.018(\log z_0)^2 + 0.0016(\log z_0)^3. \tag{20}$$

Although the value of gradient height strongly depends on the large length scale of the ABL, the power exponent can approximately be expressed as a function of the small length scale of the ABL, z_0. If the roughness length z_0 does not change during a typhoon, the power exponents can be expected to be constant and can be predicted from roughness terrain conditions using Eq. (20). This means that if the pressure data associated with a typhoon or hurricane are given, a vertical profile of wind speed and direction at an object site can be readily obtained using the equations given above.

3.3. Wind speed ratio $G(r)$ and inflow angle γ_s

It is well known that wind speed ratio $G(r)$ rapidly increases in the region near the typhoon or hurricane center, as reported by Georgiou et al. [8]. To explain this phenomenon, wind speed ratio $G(r)$ at a height of $z = 10$ m is calculated using the present numerical results, and shown in Fig. 7a. It can be seen that the wind speed ratio $G(r)$ (symbols) obtained from the simulation displays the same tendency as the observation data [8]. Solid lines shown in Fig. 7a are calculated by the power law with fixed exponents obtained by Eq. (20) and gradient heights by Eq. (19), and satisfactorily correspond to the simulated ones. This indicates that rapid increases in wind speed ratio can be explained by abrupt decreases in gradient height in the region near the typhoon or hurricane center.

Fig. 7b presents a variation of the inflow angle γ_s at a height of $z = 10$ m with r/r_m. A broad peak appears at the location $r/r_m = 2$. The shape of the inflow angle γ_s is similar to that of the non-dimensional parameter ξ. Solid lines are calculated by

Eq. (21) and correspond well to simulated ones,

$$\gamma_s = (69 + 100\xi)(\log \mathrm{Ro}_\lambda)^{-1.13}. \tag{21}$$

It is obvious that, in the TBL, the inflow angle γ_s depends not only on the surface Rossby number but also on the non-dimensional parameter ξ.

4. Conclusions

The wind field in a typhoon boundary layer has been investigated by a numerical model. The following summarizes the major findings and conclusions of this study:

1. Two important parameters, f_λ and ξ, have been suggested to describe the structure of strong wind in the TBL. The parameter f_λ approximately represents the absolute vorticity in the ABL, and influences the gradient height. The parameter ξ characterizes the heterogeneity of vorticity in the radial direction and causes an increase in the inflow angle.
2. Vertical profiles of wind speed in the TBL can be satisfactorily stated by the conventional power-law expression. The gradient height is a function of the length scale U_g/f_λ and modified surface Rossby number Ro_λ, and the power exponent can be approximately expressed as a function of the small length scale of the ABL, z_0.
3. The ratio of surface to gradient wind speeds $G(r)$ and the inflow angle γ_s in the TBL are also examined using the present numerical results, and are satisfactorily predicted by the formulas proposed in this study.

References

[1] E.C.C. Choi, Gradient height and velocity profile during typhoons, J. Wind Eng. Ind. Aerodyn. 13 (1983) 31–41.

[2] Y. Meng, M. Matsui, K. Hibi, An analytical model for simulation of the wind field in a typhoon boundary layer, J. Wind Eng. Ind. Aerodyn. 56 (1995) 291–310.

[3] G.L. Mellor, T. Yamada, A hierarchy of turbulence closure models for planetary boundary layer, J. Atmos. Sci. 31 (1974) 1791–1804.

[4] Y. Mitsuta, N. Monji, O. Tsukamoto, H. Asai, Meteorological study of typhoon 7705, Disaster Prevention Research Institute Annuals of Kyoto University, No. 21B, 1978, pp. 405–415 (in Japanese).

[5] A.G. Davenport, The relationship of wind structure to wind loading, Proc. Symp. on Wind Effects on Buildings and Structures, vol. 1, Nalt. Phys. Lab., H.M.S.O., 1965, pp. 53–102.

[6] Y. Meng, M. Matsui, K. Hibi, Characteristics of the vertical wind profile in the neutrally atmospheric boundary layers, Part 1 Strong wind during non-typhoon climates, J. Wind Eng. 65 (1995) 1–15 (in Japanese).

[7] Y. Meng, M. Matsui, K. Hibi, Characteristics of the vertical wind profile in the neutrally atmospheric boundary layers, Part 2 Strong winds during typhoon climates, J. Wind Eng. 66 (1996) 3–14 (in Japanese).

[8] P.N. Georgiou, A.G. Davenport, B.J. Vickery, Design wind speeds in regions dominated by tropical cyclones, J. Wind Eng. Ind. Aerodyn. 13 (1983) 139–152.

ELSEVIER

Journal of Wind Engineering
and Industrial Aerodynamics 67&68 (1997) 449–457

JOURNAL OF
wind engineering
AND
industrial
aerodynamics

Statistical–dynamical downscaling of wind climatologies

Heinz-Theo Mengelkamp[a,*], Hartmut Kapitza[a], Ulrich Pflüger[b]

[a] *GKSS-Forschungszentrum, D-21494 Geesthacht, Germany*
[b] *German Weather Service, D-63004 Offenbach, Germany*

Abstract

A statistical–dynamical downscaling procedure is applied for an investigation into the availability of wind power over a region of 80 × 87 km which covers flat and hilly terrain. The approach is based on the statistical coupling of a regionally representative wind climate with a numerical atmospheric mesoscale model. The large-scale wind climatology is calculated by a cluster-analysis of a time series of radiosonde data over 12 years. Any of the resulting 143 clusters represents a particular combination of geostrophic wind components and vertical temperature gradient. For each cluster, a highly resolved steady-state wind field is simulated with a non-hydrostatic mesoscale model. These wind fields are statistically evaluated by weighting them with the corresponding cluster frequency. The resulting three-dimensional wind field and the frequency distributions of windspeed and direction are compared with observations at synoptic stations.

Keywords: Wind field simulation; Downscaling; Wind climate

1. Introduction

Information about regional or local wind conditions is often asked for when considering environmental aspects in planning of urban development, studying emission scenarios of power plants or siting of wind turbines. Usually wind climatologies are based on windspeed data from long-term near-surface observations at synoptic stations. However, the spatial density of such data is usually inadequate to deduce high-resolution windspeed maps on regional scales, and observations at the standard height of 10 m may be disturbed by close obstacles and vegetation. In particular, over complex and heterogeneous terrain, the data may also be influenced by dynamically or thermally induced local circulations which probably cannot be corrected for. We aim at estimating the wind climate on regional and local scales by the use of long-term

* Corresponding author. E-mail: mengelkamp@gkss.de.

PII S0167-6105(97)00093-7

upper-air data and a numerical model. Near-surface observations are used for comparison reasons only.

Wippermann and Gross [1] applied a statistical–dynamical downscaling procedure to construct surface wind roses in the upper Rhine valley knowing the frequency distribution of the geostrophic wind. Restricting their calculations to stably stratified situations and using the two-dimensional version of a non-hydrostatic mesoscale model they showed that over orographically structured terrain surface-wind distributions can realistically be simulated by coupling statistics of upper-air data and a numerical model. A three-dimensional, three-layer hydrostatic model was used by Heimann [2] to calculate annual frequency distributions of surface winds over the Main-Taunus-area. To account for stability effects, cycles of typical days for each of twelve wind-direction sectors were simulated assuming an annual mean value for the geostrophic wind speed. Frey-Buness et al. [3] extended this approach to the regionalization of climate change scenarios by classifying large-scale weather situations derived from multi-year episodes of global climate simulations. Segal et al. [4] used a two-dimensional version of a hydrostatic mesoscale model to study wind energy characteristics over central Israel. The weather classification consisted of three typical cases for which full daily cycles were simulated due to the strong influence of thermal circulation patterns.

2. The downscaling procedure

The statistical–dynamical downscaling procedure is based on the assumption that the climate for a given region can be described by a classification of an appropriate set of atmospheric parameters which characterize typical weather situations. For each class or weather situation, a simulation is performed with a high-resolution model giving three-dimensional fields of relevant atmospheric variables. Weighting these fields with the class-frequency results in mean distributions. The downscaling procedure is sketched in Fig. 1.

Because we concentrate on wind climate, the geostrophic wind and the vertical temperature gradient are chosen as the basic data set. These data can be taken from large-scale simulations or from upper-air observations. Here a 12-year time series (1979–1991) of radiosonde data at 850 hPa (wind) and between 100 m and 1.5 km (temperature gradient) was classified using a cluster analysis. A total of 143 clusters was found to represent the time series appropriately under the constraints of three fixed stability classes, a maximum number of 150 clusters due to limited computer resources, and reasonable standard deviations for wind speed and direction in each cluster. A steady-state simulation was performed for each cluster with the non-hydrostatic mesoscale model GESIMA (Geesthacht Simulation Model of the Atmosphere). The resulting wind fields are weighted with the corresponding cluster frequency giving the average three-dimensional wind field and the frequency distributions of wind speed and direction.

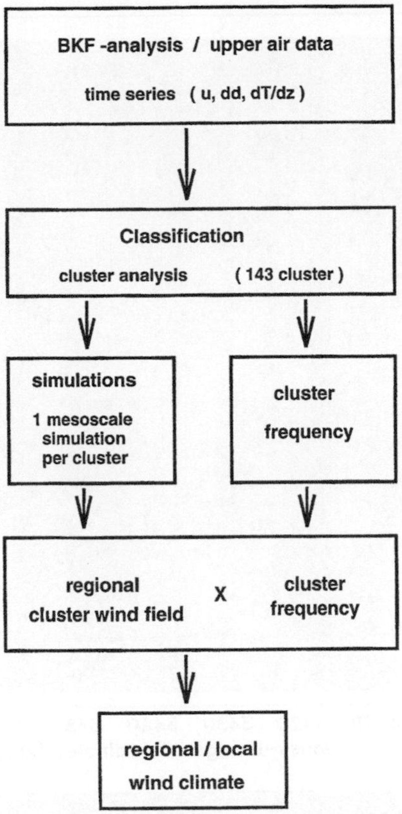

Fig. 1. Sketch of the regionalization approach.

3. The mesoscale model and simulation domain

The mesoscale model GESIMA and some verification studies are described in Refs. [5–7]. Because we are interested in a steady-state solution for a particular geostrophic forcing and fixed background temperature profile, only the dynamic part of the model is of concern. The governing equations are the Boussinesq approximated, anelastically filtered equations of motion in flux-conserving form. The turbulence parameterization is based on the prognostic equation for turbulent kinetic energy with boundary-layer approximations. Mahrer et al. [8] calls this the dynamic mode of a prognostic model.

The solution domain (Fig. 2) covers a region of 80×87 km. The grid spacing in the horizontal is set to 1×1 km, and the vertical grid spacing increases from 20 m at the surface to 400 m at the upper boundary which is located at 3480 m. A total number of $80 \times 87 \times 19$ grid points is used. The steady-state solution was assumed to have been reached when the mean of the absolute values of the differences in

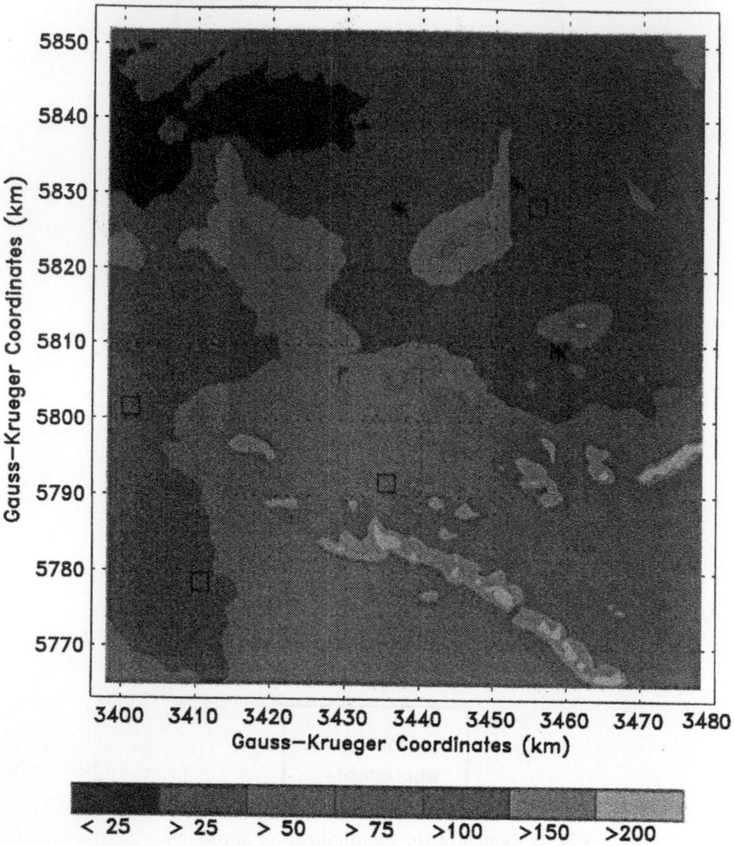

Fig. 2. Orography of the simulation area with locations of wind observations (squares denote synoptic weather stations, stars wmep stations). The height is given in meter a.m.s.l.

wind-speed components between two consecutive time steps asymtotically reached a lower limit.

The simulation area (Fig. 2) covers flat terrain as well as the "Teutoburger Wald" area with hills up to 330 m a.m.s.l. Locations of wind observations that are used for verification purposes are marked. The surface roughness distribution was described by 10 land-use classes with roughness lengths ranging from 2 mm for lakes (class 1) to 2 m for the center of big cities (class 10) according to Wieringa [9]. Forested areas were characterized by a roughness length of 1 m and a displacement height of 8 m (class 8). The hilly regions are almost completely forested (Fig. 3) whereas the flat terrain is characterized by farmland with bushes and smaller groups of trees (classes 4 and 5). Large areas of moorland (classes 2 and 3) can be found in the northeastern part of the domain.

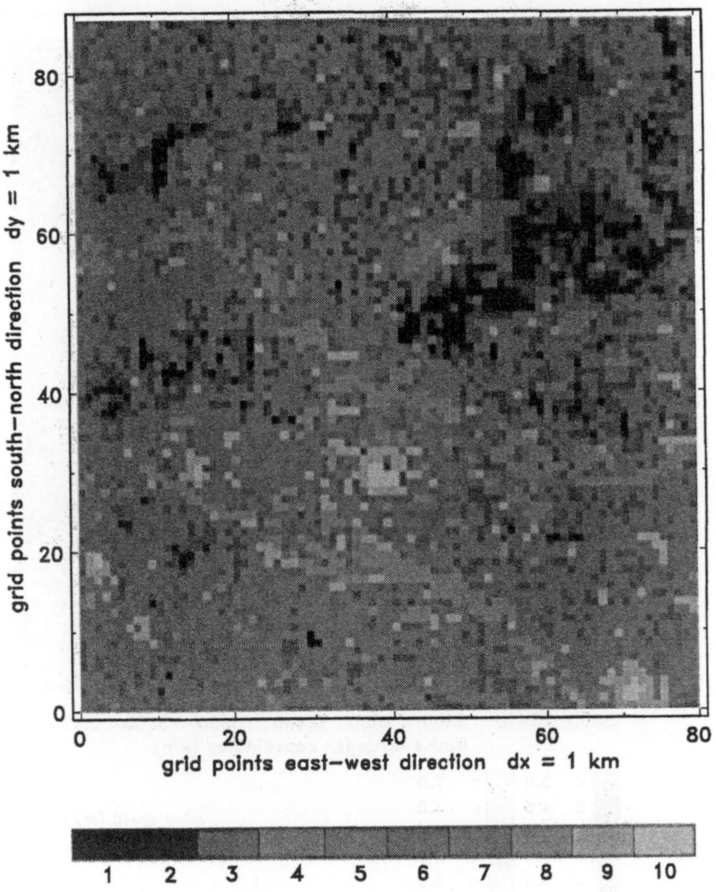

Fig. 3. Roughness characteristics of the simulation domain. See text for roughness class description.

4. Wind climate and verification

The mean wind field at 50 m height is shown in Fig. 4. Windspeed values between 5 and 6.5 m/s can be found over most of the flat and smooth terrain. Regions with values below 5 m/s coincide almost all with forested areas and bigger cities. The retardation effect implied by roughness and displacement height dominates the orographically induced speed-up over the relatively low and gentle hills. To separate these effects, three test-runs were made for the first cluster only which represents the most frequent weather situation with southwesterly winds and a moderate geostrophic forcing. In test-run 1, the roughness length was identical over the whole region and no displacement height was introduced. Variations in the wind field are then due to orographic effects only. For the second test-run, the roughness distribution as shown in Fig. 3 was applied but no displacement height was accounted for while

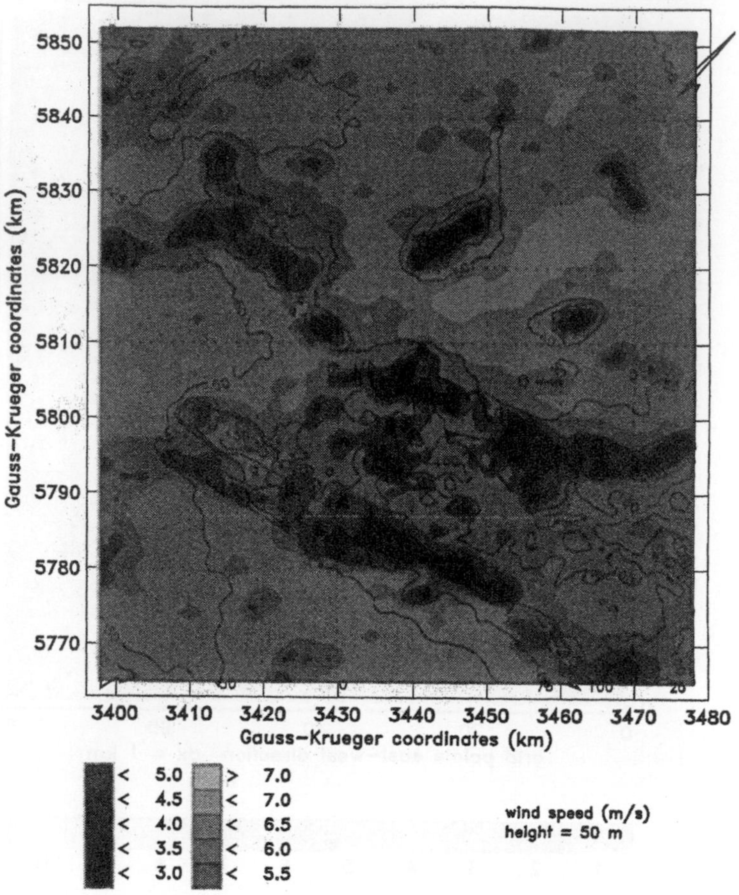

Fig. 4. Regional distribution of mean wind speed at 50 m height.

test-run 3 includes the displacement height in addition. Fig. 5 shows vertical profiles of wind speed at a location over flat, smooth terrain (grid point $x = 15$, $y = 20$ in Fig. 3) and a forested hilly site (grid point $x = 38$, $y = 20$). The profile over flat terrain is the same for all three test-runs (solid line) while the profiles at the hilly site clearly show the orographically induced acceleration in shifting the whole wind profile to the right (dashed line). In test-run 2, the roughness effect at the hilly site dominates below 70 m height, while in test-run 3, the displacement height leads to a further wind speed reduction and the orography effect is evident only above 100 m.

A comparison of the simulated and observed frequency distributions of wind speed and direction is shown in Fig. 6 for the weather station Diepholz (coordinates 3455/5828 in Fig. 2) which is representative of all stations. There are relatively large differences at low wind speeds and an overrepresentation of southwesterly wind directions.

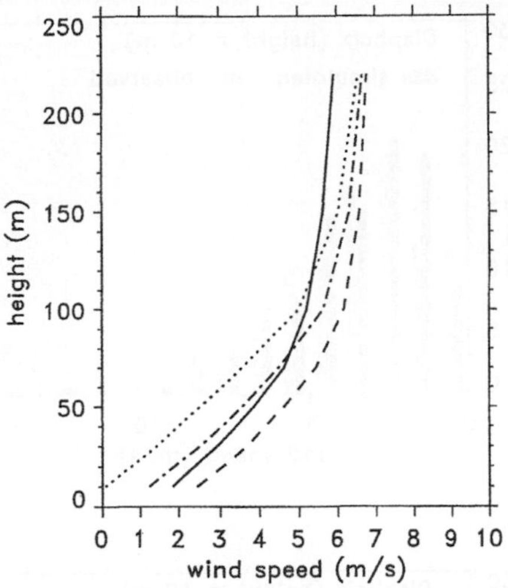

Fig. 5. Vertical wind profiles over flat smooth terrain (solid line) and a forested hilly site (dashed line: same roughness length at both locations and no displacement height; dash-dotted line: different roughness length but no displacement height; dotted line: different roughness length and displacement height over the forested hill).

Table 1
Mean wind speeds and energy flux densities at weather stations and WMEP observations (obser.) compared to simulated data (simul.). The coordinates refer to Fig. 2

	Coordinates x/y	u (m/s) obser.	u (m/s) simul.	E (W/qm) obser.	E (W/qm) simul.
Diepholz	3455/5828	3.8	3.6	89	70
Hopsten	3401/5801	4.1	4.1	99	100
Greven	3410/5778	3.7	3.6	74	76
Osnabr.	3435/5791	3.2	3.0	42	63
WMEP 1	3452/5831	3.5	4.0	70	100
WMEP 2	3436/5828	3.4	3.6	74	75
WMEP 3	3458/5809	2.9	3.5	45	75

The average wind speed and energy flux density observed at the synoptic stations and WMEP stations (a special German Wind Measuring and Evaluation Program for wind energy assessment) at 10 m height and calculated for the same height and the respective model grid are listed in Table 1. The agreement between observed and simulated wind speed values at the synoptic stations is good whereas the energy flux density shows differences up to about 20%. The WMEP stations reported from 1991 to 1993. This different time period, and to some extent sheltering effects, may have caused less agreement with the simulated data than is the case for the long-term synoptic data. The energy flux density was calculated for each cluster wind field

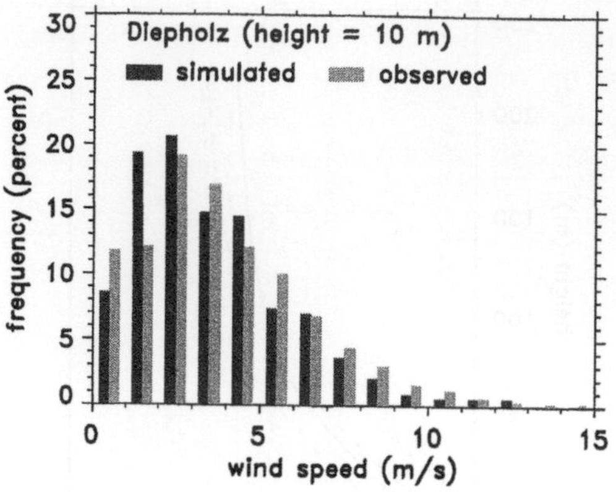

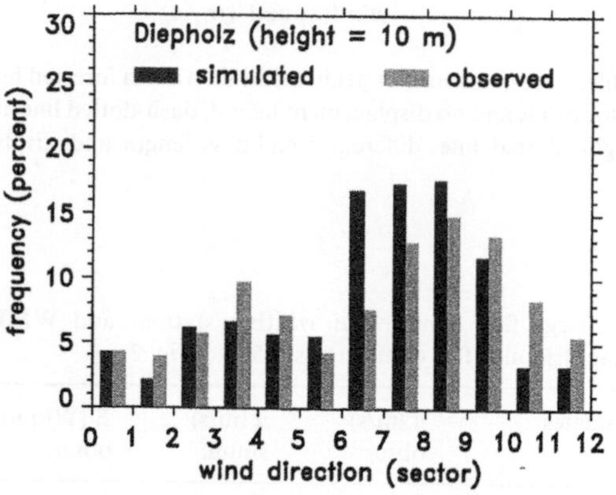

Fig. 6. Frequency distribution of wind speed and direction for the observation point "Diepholz". Wind direction sector 0 refers to the north direction, sector 3 to east and sector 6 to the south direction.

because this parameter is more suitable for comparing the power of the wind at different locations than the mean wind speed as it takes into account the nonlinear relationship between wind speed and wind energy.

5. Conclusions

A statistical–dynamical downscaling procedure was applied for estimating mean wind fields and the frequency distributions of wind speed and direction over complex heterogeneous terrain. The comparison with statistics of observed data sets shows a fairly good agreement concerning the mean wind speed and energy flux density at weather stations which reported during the same time period as the upper-air data were observed. Less agreement was found with data taken over a much shorter time period. Considering the difficulty of comparing point observations with simulated volume averages one should aim at comparing simulated data with observations at greater heights where disturbances by close obstacles may be much less important. Also, the area of representativeness of such observations might come close to the resolution which today is possible to achieve with mesoscale models.

Acknowledgements

This study was supported by Deutsche Bundesstiftung Umwelt.

References

[1] F. Wippermann, G. Gross, On the construction of orographically influenced wind roses for given distributions of the large-scale wind, Beitr. Phys. Atmos. 54 (4) (1981) 492–501.

[2] D. Heimann, Estimation of regional surface layer wind field characteristics using a three-layer mesoscale model, Beitr. Phys. Atmos. 59 (4) (1986) 518–537.

[3] F. Frey-Buness, D. Heimann, R. Sausen, A statistical-dynamical downscaling procedure for global climate simulations, Theoret. Appl. Climatol. 50 (1995) 117–131.

[4] M. Segal, Y. Mahrer, R. Pielke, Numerical study of wind energy characteristics over heterogeneous terrain – central Israel study, Bound.-Layer Meteor. 22 (1982) 373–392.

[5] H. Kapitza, D. Eppel, The non-hydrostatic mesoscale model GESIMA. Part I: Dynamical equations and tests, Beitr. Phys. Atmos. 65 (2) (1992) 129–146.

[6] D.P. Eppel, H. Kapitza, M. Claussen, D. Jacob, W. Koch, L. Levkov, H.-T. Mengelkamp, N. Werrmann, The non-hydrostatic mesoscale model GESIMA. Part II: Parameterizations and Applications, Beitr. Phys. Atmos. 68 (1) (1995) 15–42.

[7] H.-T. Mengelkamp, Boundary-layer structure over an inhomogeneous surface: simulation with a non-hydrostatic mesoscale model, Bound.-Layer Meteor. 57 (1991) 323–341.

[8] Y. Mahrer, M. Segal, R. Pielke, Mesoscale modelling of wind energy over non-homogeneous terrain, Bound.-Layer Meteor. 31 (1985) 13–32.

[9] J. Wieringa, Roughness-dependent geographical interpolation of surface wind speed averages, Quart. J. Roy. Meteor. Soc. 112 (1986) 867–889.

ELSEVIER

Journal of Wind Engineering
and Industrial Aerodynamics 67&68 (1997) 459–477

JOURNAL OF wind engineering AND industrial aerodynamics

CFD analysis of mesoscale climate in the Greater Tokyo area

Akashi Mochida[a,*], Shuzo Murakami[b], Toshio Ojima[c], Sangjin Kim[b], Ryozo Ooka[b], Hirokatsu Sugiyama[d]

[a] *Niigata Institute of Technology, 1719, Fujihashi, Kashiwazaki, Niigata 945-11, Japan*
[b] *Institute of Industrial Science, University of Tokyo, 7-22-1, Roppongi, Minato-ku, Tokyo 106, Japan*
[c] *Waseda University, 3-4-1, Okubo, Shinjuku-ku, Tokyo 169, Japan*
[d] *Tokyo Electric Power Company, 1-1-3, Uchisaiwaicho, Chiyoda-ku, Tokyo 100, Japan*

Abstract

The results of CFD analyses of mesoscale climate in the Tokyo area are presented. Here, the model for geophysical flow problems developed by Mellor and Yamada is used for turbulence closure. In the first half of the paper, the accuracy of CFD analyses is examined by comparing their results with the measured data. The topography and present land-use situation in Japan are incorporated into the predictions by using the numerical data-base provided by the National Land Agency of Japan. For comparison, a computation which does not consider the effects of the distribution of land-use conditions at the present is also carried out. In the latter part, urban climates during the 1930s and the 1990s are analysed by the CFD method developed here. By comparing the results of these analyses, the effects of urbanization on heat island circulations over the Tokyo area are investigated.

Keywords: Urban climate; Heat island; Land-use; Artificial heat release

Notation

x,y,z	coordinates (x: east–west direction, y: north–south direction, z: vertical direction)
U,V,W	mean velocities for x, y, and z directions, respectively
u,v,w	velocity fluctuations
q	turbulence energy ($q = (\langle u^2\rangle + \langle v^2\rangle + \langle w^2\rangle)^{1/2}$)
T	absolute temperature
Θ	potential temperature ($\Theta = (P_0/P)^{R/C_p}T$)
Θ_v	virtual potential temperature

* Corresponding author. E-mail: mochida@abe.niit.ac.jp.

PII S0167-6105(97)00060-3

P_0 a reference pressure, 1000 mb
Q_w mixing ratio of total water (vapor + liquid)
β moisture availability
l master length scale
β_v thermal expansion coefficient ($\beta_v = 1/\langle\Theta_v\rangle$)
g gravity acceleration (9.8 m/s^2)
z_g ground elevation
z^* terrain-following vertical coordinate
C_p specific heat of air at constant pressure [J/(kg K)]
ρ_a air density (kg/m^3)

1. Introduction

The results of CFD analyses of mesoscale climate in the Tokyo metropolitan area are described in this paper. Rapid urban development accompanied by increase of artificial heat release has significantly changed the regional climate in the Tokyo area and caused various environmental problems [1–3].

Urban climate is highly influenced by the different land-use conditions, e.g. asphalt road, green vegetation area, river, pond, etc. and also by artificial heat release. Historically, the effects of land-use and artificial heat release on the urban environment have been investigated from different viewpoints by many authors in the fields of meteorology, city planning and building engineering using a one-dimensional (1D) boundary layer model [4,5]. Obviously, this type of 1D analysis is seriously limited by 1D simplification, since various flow phenomena peculiar to urban climate, such as heat island circulation, etc., which are highly three-dimensional (3D), cannot be reproduced by 1D analysis. Thus, 3D analysis is required for investigating the effects of urbanization on local climate. Methods of 3D CFD analysis of regional climate have been developed and tested by many researchers in the field of meteorology, e.g. by Pielke et al., Yamada et al., etc., where attention has mainly been paid to meteorological and geophysical phenomena, such as cloud cluster, orographic effects, tornadoes, etc. [6–15]. There have been only a few studies based on 3D analysis in which the effects of land-use and artificial heat release have been considered [12,13].

In this study, the urban climate in the Tokyo area is analyzed by 3D numerical predictions. The turbulence closure model for geophysical flow problems developed by Mellor and Yamada is used for turbulence modelling [8–11]. The present situation of land-use in Japan is incorporated into the predictions by utilizing the numerical data-base provided by the National Land Agency of Japan, etc. [16].

In the first half of the paper, the accuracy of 3D CFD analyses is examined by comparing their results with measured data. For comparison, a computation which does not consider the effects of the distribution of land-use conditions at present is also carried out here in order to clarify the effects of urbanization on local climate. In the latter part, the climatic change in Tokyo between the 1930s and the 1990s is numerically investigated using the CFD method examined in the first part of the paper.

2. Outline of CFD analyses

2.1. Model equations

Mellor and Yamada proposed a hierarchy of turbulence closure models for geophysical flow problems ranging from the zero-equation model (level 1) to the Differential Second-moment Closure Model (DSM, level 4) [8,9]. This hierarchy of models was obtained by systematically simplifying the second-moment closure model based on estimation of the degree of importance of each term of the model equations. Their 2.5 model level was employed in this study [8,9]. This model is regarded as a simplified Algebraic Second-moment Closure Model (ASM) used in engineering. All computations presented here were carried out by using HOTMAC (Higher Order Turbulence Model for Atmospheric Circulation) developed by Yamada and Bunker [10,11].

2.2. Boundary conditions

(1) *Surface heat energy balance*: The heat energy balance at ground surface expressed as Eq. (1) should be satisfied in computations. As shown below, Eq. (1) can be transformed into Eq. (8) using relations in Eqs. (2), (3), (5)–(7). Surface temperature, T_G is estimated based on Eq. (8) in this study. The process of derivation of Eq. (8) is described below.

First, the heat energy balance at ground surface is written as follows:

$$R_{SOG} + R_{LG\downarrow} - R_{LG\uparrow} - (H_G + LE_G + C_G) + A_G = 0, \tag{1}$$

where, R_{SOG} is the incoming direct solar radiation absorbed by the surface, $R_{LG\downarrow}$ is the incoming long-wave radiation, $R_{LG\uparrow}$ the outgoing long-wave radiation, H_G the sensible heat flux, LE_G the latent heat flux, C_G the ground heat flux by heat conduction in a downward direction and A_G is the artificial heat release. Positive and negative signs of each term in Eq. (1) mean incoming and outgoing fluxes onto the ground surface, respectively.

Here, R_{SOG} and $R_{LG\uparrow}$ can be expressed as

$$R_{SOG} = (1 - \alpha_G)SO, \tag{2}$$

$$R_{LG\uparrow} = \varepsilon_G \sigma T_G^4 + (1 - \varepsilon_G) R_{LG\downarrow}, \tag{3}$$

where, ε_G is the emissivity of the surface and σ is the Stefan–Boltzman constant. The subscript G denotes the value at ground surface. By Eq. (3), the net long-wave radiation at the ground surface, R_{NG} can be written as

$$R_{NG} = R_{LG\downarrow} - R_{LG\uparrow} = \varepsilon_G R_{LG\downarrow} - \varepsilon_G \sigma T_G^4. \tag{4}$$

H_G, LE_G and C_G are given by

$$H_G = -\rho_a C_p u_* T_*, \tag{5}$$

$$LE_G = -\rho_a L u_* Q_*, \tag{6}$$

$$C_G = K_S(\partial T_S/\partial z_S)|_G, \tag{7}$$

where, u_*, T_*, and Q_*, defined as $u_* \equiv \sqrt{\tau/\rho_a}$, $T_* \equiv -H_G/\rho_a C_p u_*$, and $Q_* \equiv -E_G/\rho_a u_*$, are the friction velocity, temperature scale and water vapor scale, respectively. ρ_a is the air density.

By substituting Eqs. (2), (3), (5)–(7) into Eq. (1), we obtain

$$(1-\alpha_G)\mathrm{SO} + \varepsilon_G R_{LG\downarrow} - \varepsilon_G \sigma T_G^4 + \rho_a C_p u_* T_* + \rho_a L u_* Q_* - K_S(\partial T_S/\partial z_S)|_G + A_G = 0. \tag{8}$$

In order to estimate H_G and LE_G defined by Eqs. (5) and (6), values of u_*, T_*, and Q_* should be given. These values are evaluated by the following formulas given by the similarity theory of Monin–Obukhov (cf. Appendix A):

$$V_s(z) = \frac{u_*}{\kappa}[\ln\{(z+z_0)/z_0\} - \Psi_m(\zeta)], \tag{9}$$

$$\Theta(z) - \Theta_G = \frac{\mathrm{Pr}_1}{\kappa} T_*[\ln\{(z+z_{0t})/z_0\} + \ln(z_0/z_{0t}) - \Psi_h(\zeta)], \tag{10}$$

$$Q_w(z) - Q_{wG} = \frac{S_{ct}}{\kappa} Q_*[\ln\{(z+z_{0v})/z_{0v}\} - \Psi_v(\zeta)], \tag{11}$$

where, $V_s(z)$ in Eq. (9) is the horizontal wind speed $(U^2+V^2)^{1/2}$.

In order to estimate Q_{wG} in Eq. (11), two types of parameterizations, i.e. β method and α method are used in Sections 3 and 4, respectively. Details of these two methods are described in Appendix B.

Using the soil temperature at 1 m below the ground surface as the boundary condition, the C_G (ground heat flux) term in Eq. (1) is given by the following heat conduction equation:

$$\partial T_S/\partial t = \partial/\partial z_S[K_S(\partial T_S/\partial z_S)]. \tag{12}$$

As is already shown in Eq. (7), C_G is estimated by

$$C_G = K_S(\partial T_S/\partial z_S)|_G.$$

SO and $R_{LG\downarrow}$ in Eq. (8) are given following Kondrat'yev [22] and Sasamori [17], respectively.

(2) *Top and lateral boundary conditions*: At the top of the computational domain, $U = 0$ m/s (velocity component in east–west direction), $V = 0.5$ m/s (velocity component in the north–south direction), $\Delta\Theta_v = 0$, $Q_w \cong 0$, $q^2 = 0$, and $l = 0$. The lateral boundary values for U, V, Θ, Q, q^2 and $q^2 l$ are obtained by integrating the 1D equations corresponding to the 3D transport equations for the variables, by imposing the condition that variations in the horizontal directions are zero. The soil temperature 1 m below the ground surface is set at 22°C [23] and the water surface temperature is set at 25°C [24].

2.3. Initial condition

The initial wind direction is set southerly in the whole computational domain. Horizontal wind component V is assumed to obey Eq. (9) from the ground up to the height where the wind speed reaches 0.5 m/s. Above this height, V is set to be constant ($V = 0.5$ m/s, cf. Fig. 1). The vertical profile of potential temperature Θ_v is initially assumed to increase linearly with height (Fig. 2). Initial potential temperatures are set homogeneous in horizontal directions. The initial relative humidity, except at the ground surface, is set at 50%. Initial values for the mixing ratio of total water vapor at the ground are given by using the β method in Section 3 and the α method in Section 4. The initial conditions for turbulence kinetic energy $q^2/2$ and turbulence length scale l are given by solving the model level 2 proposed by Mellor and Yamada [8,9], in which distributions of $q^2/2$ and l are calculated by algebraic expressions using the initial wind and temperature profiles.

3. Effects of land-use conditions on mesoscale climate in Central Japan

3.1. Computed cases and surface parameters

In this section, two cases of numerical experiments are carried out. Computed cases are summarized in Table 1. In case 1, the distribution of land-use at present is

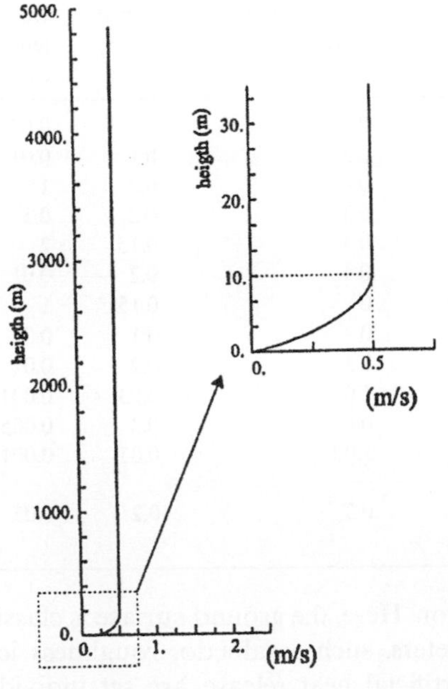

Fig. 1. Initial vertical profile of velocity V.

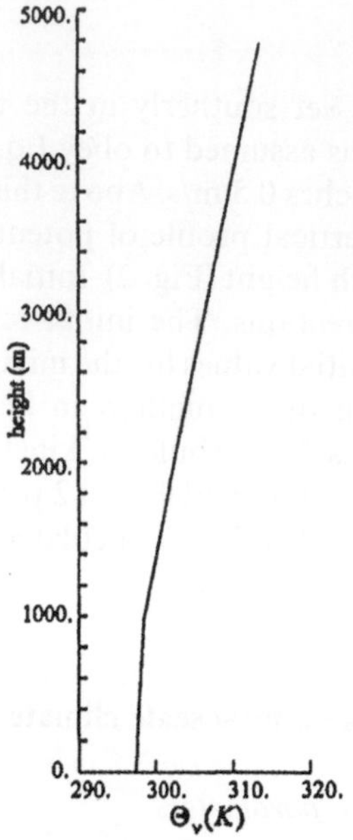

Fig. 2. Initial vertical profile of Θ_v.

Table 1
Computed cases for Section 3 and surface parameters [6,20,21]

Case	Land-use conditions	Soil moisture availability β	Albedo	Roughness length z_0 (m)	Artificial heat release (W/m^2)
Case 1	Rice paddy	0.6	0.2	0.05	0
	Farming	0.3	0.1	0.01	0
	Orchards 1	0.4	0.2	1	0
	Orchards 2	0.3	0.2	0.5	0
	Forest	0.3	0.15	2	0
	Vacant land	0.4	0.2	0.01	0
	Buildings	0	0.15	1	50
	Paved road	0	0.1	0.01	4
	Other land	0.3	0.2	0.01	0
	River site	1.0	0.03	0.001	0
	Coast	0.6	0.3	0.005	0
	Ocean	1.0	0.03	0.001	0
Case 2	Assumed to be uniform grassy plain (cf. footnote [1])	0.3	0.2	0.05	0

incorporated into the prediction. Here, the ground surface is classified into 12 types of land-use, and surface parameters, such as albedo, roughness length, soil moisture availability β [20,21] and artificial heat release are set individually following the boundary conditions (Table 1). The present situation of land-use in Japan is given by

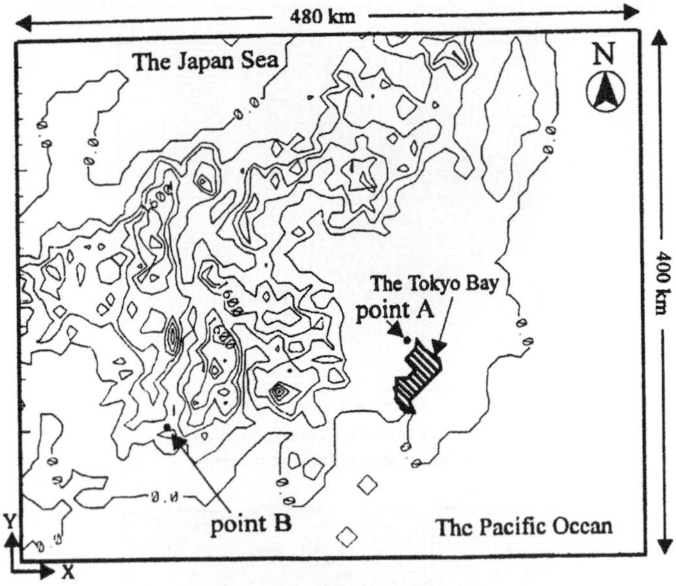

Fig. 3. Computational domain.

utilizing the numerical data-base for land-use compiled by the National Land Agency of Japan [16]. In case 2, the entire ground surface is assumed to be a grassy plain (Table 1)[1]. The topography is reproduced in both cases using the numerical data-base provided by the National Land Agency. The daytime flowfield variation for the two cases are predicted imposing the same initial condition and the same boundary condition at the top of the computational domain, which are the typical meteorological conditions during late July in Japan.

3.2. Computational domain and grid arrangements

Fig. 3 illustrates the computational domain, which covers 480 km (east–west, x) × 400 km (north–south, y) × 5 km (vertical direction, z). This domain is discretized into 60 (x) × 50 (y) × 20 (z) grids. Grid spacings in x and y directions are 8 km and grid spacing in z direction adjacent to the ground is set at 4 m[2].

3.3. Comparison of predicted results with measurements

Fig. 4 shows wind velocity vectors at 100 m height at 3:00 p.m., late July. Fig. 4(1) and Fig. 4(2) are the computed results after 33 h from the initial condition. Here, the

[1] In case 2, all parts of the ground surface are assumed to be a plain covered with high grass. Ground surface is assumed to be naked soil, such as farming. Based on these assumptions, surface parameters, i.e. β, albedo, roughness length are determined for case 2 following Watanabe [21], etc.

[2] Grid spacings in z direction are 4, 4, 4, 4, 23, 62, 100, 139, 176, 216, 254, 292, 331, 369, 407, 447, 484, 523, 561, and 600 m.

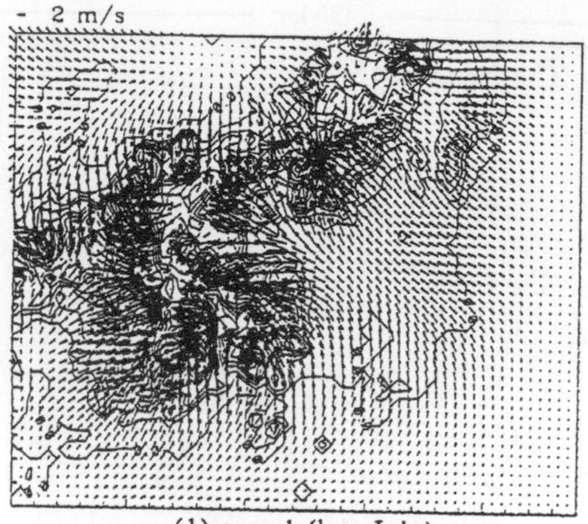

(1) case 1 (late July)
(Surface boundary condition is given by land-use condition at present)

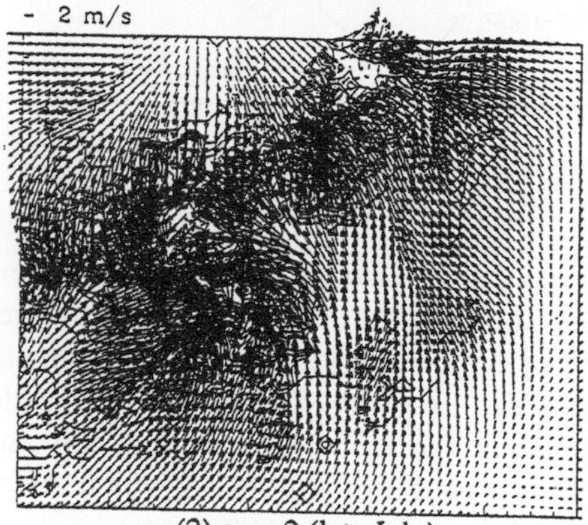

(2) case 2 (late July)
(All parts of surface boundary are assumed to be a grassy plain)

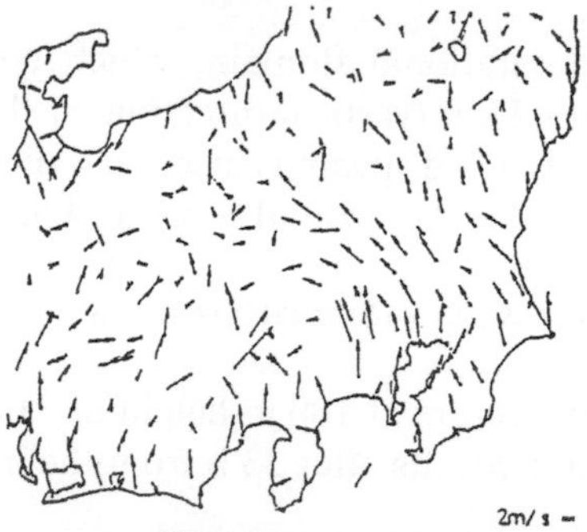

(3) meteorological data
(1985 summer 50 days, by Kuwagata et al. [25])

Fig. 4. Horizontal distribution of velocity vectors (at 3:00 p.m. at the height of 100 m). (1) Case 1 (late July) (surface boundary condition is given by land-use condition at present); (2) case 2 (late July) (all parts of surface boundary are assumed to be a grassy plain); (3) meteorological data (1985 summer 50 days, by Kuwagata et al. [25]).

results of cases 1 and 2 are compared with the measured data [25]. It is seen in the measured data that the prevailing wind direction is southwest in the Kanto plain at this time (Fig. 4(3)). In the result of case 1, in which the present situations of land-use conditions are given as input data for the surface boundary condition, the flow pattern peculiar to sea breeze is well reproduced. The agreement with the measured data is fairly good in this case. On the other hand, wind velocities are generally predicted too high and wind directions are quite different from the measured data in some areas in the result of case 2, in which all parts of the ground surface are assumed to be a grassy plain; consequently, the agreement with the measured data becomes much poorer.

Fig. 5 shows the surface temperature distributions for the same time as Fig. 4. In case 1, the highest temperature is observed in the centre of Tokyo, reflecting the present situation of land-use. This tendency agrees well with the results observed in measurements [26,27]. On the other hand, the result of case 2 is different from those of case 1 and of the measurements. The differences in velocity vectors shown in Fig. 4 are closely related to the differences in the surface temperature distributions in Fig. 5. The mechanism which causes these differences is discussed in the next section by examining the heat energy balance at the ground surface.

Fig. 6 shows the diurnal variations of air temperature at 2 m height in Otemachi (downtown Tokyo) and in Hachioji (a typical suburban area). The results of case 1 agree rather well with the measured data [28], while the nocturnal temperature is predicted too low in case 2.

3.4. *Effects of land-use on predicted results*

Figs. 7 and 8 show comparisons of the surface heat energy balance in an urban area (point A in Fig. 3) and a forest area (point B in Fig. 3) for cases 1 and 2, respectively. The values of parameters imposed as the surface boundary conditions are summarized in Table 2. The numerical data-base of the National Land Agency provides land-use information for each 100 m × 100 m mesh [16]. Since horizontal grid spacing in cases 1 and 2 is 8 km (x) × 8 km (y), the value of each parameter in Table 2 is obtained by

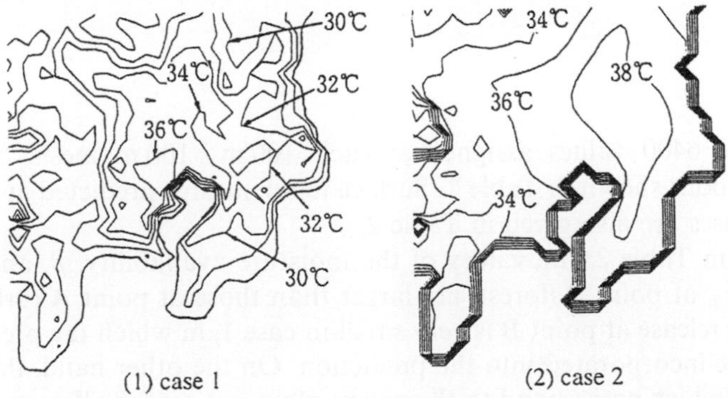

Fig. 5. The surface temperature distributions at 3:00 p.m. (late July). (1) Case 1 and (2) case 2.

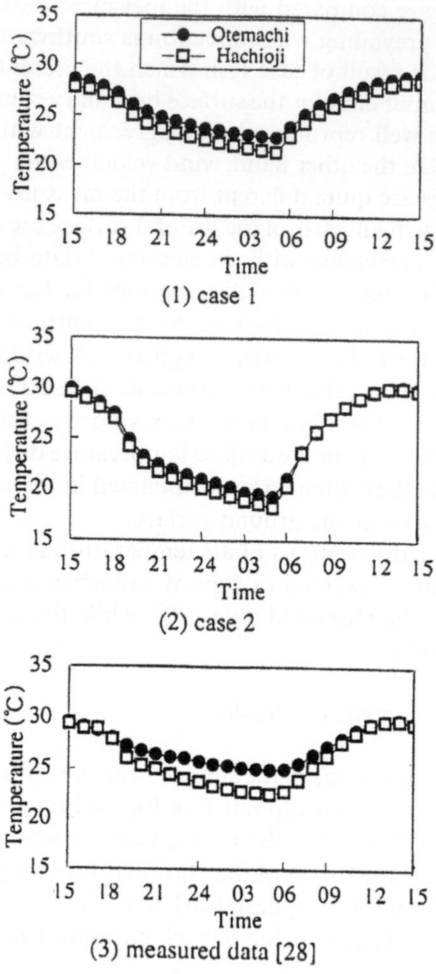

Fig. 6. Diurnal variation of air temperature at the height of 2 m. (1) Case 1; (2) case 2; (3) measured data [28].

averaging the 6400 values assigned at each 100 m × 100 m mesh, based on the parameterizations shown in Table 1. Surface temperatures predicted at points A and B for both cases are also given in Table 2.

As shown in Table 2, the values of the moisture availability, β, and the roughness length, z_0, at point B (forest) are larger than those at point A (urban), and the artificial heat release at point B is very small in case 1, in which the present land-use conditions are incorporated into the prediction. On the other hand, the same parameter values, which correspond to the grassy plain, are set for all parts of ground in case 2.

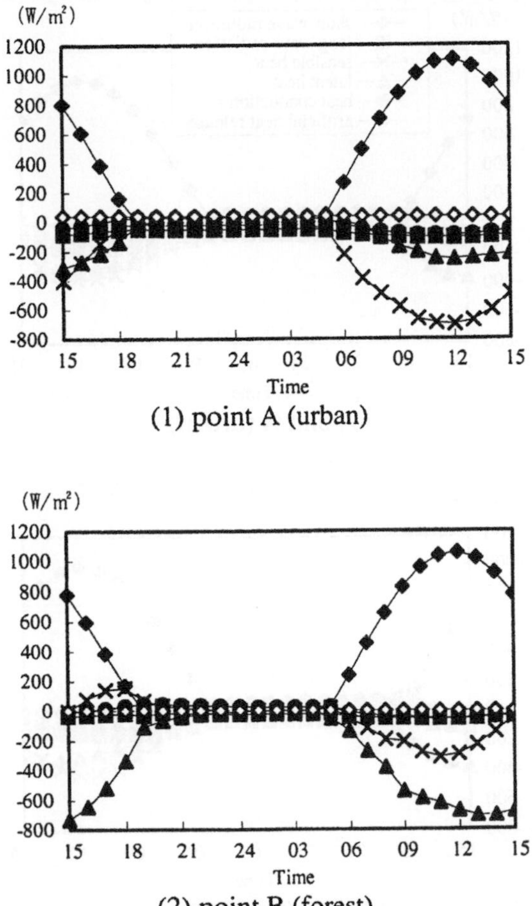

Fig. 7. Diurnal variation of heat energy balance in case 1: (1) point A (urban) and (2) point B (forest). Positive and negative signs indicate incoming and outgoing flux to and from the surface, respectively.

Table 2
Surface parameters imposed at points A and B and predicted surface temperatures at points A and B

Case	Point	Soil moisture availability β	Albedo	Roughness length z_0 (m)	Artificial heat elease (W/m^2)	Surface temperature (prediction at 3:00 p.m.)
Case 1	Point A	0.076	0.147	0.891	40.234	37.0°C
	Point B	0.411	0.148	1.725	1.291	28.3°C
Case 2	Point A	0.3	0.2	0.05	0	36.8°C
	Point B	0.3	0.2	0.05	0	34.9°C

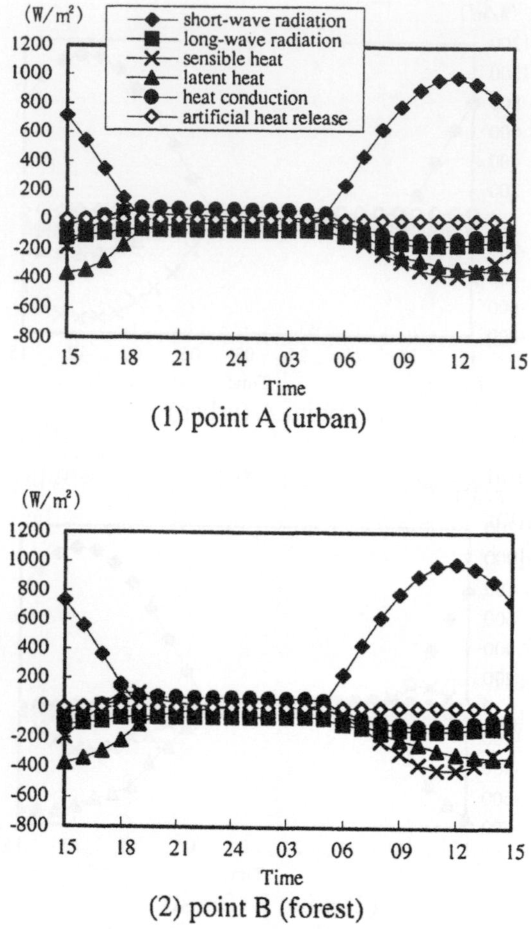

Fig. 8. Diurnal variation of heat energy balance in case 2: (1) point A (urban) and (2) point B (forest). Positive and negative signs indicate incoming and outgoing flux to and from the surface, respectively.

In the result of case 1, the absolute value of daytime latent heat flux at point B is significantly larger than that at point A, while the absolute value of daytime sensible heat flux becomes smaller at point B compared with that at point A. As shown in Table 2, the value of moisture availability, β, imposed at point B is much larger than that at point A in case 1. This is the reason why the estimated absolute value of latent heat flux at point B becomes large in the result of case 1 as shown in Fig. 7. This large negative value of latent heat flux gives rise to a decrease of ground surface temperature at point B, compared with that at point A (Table 2). Consequently the difference between the surface and the air temperatures becomes small at point B. This makes the absolute value of sensible heat flux at point B much smaller than that at point A. On the other hand, as is shown in Fig. 8, the difference in the heat energy balance between points A and B is naturally quite small in case 2, in which all parts of the ground surface are assumed to be a grassy plain. Thus, the difference in surface temperatures between these two points is also quite small in case 2 (Table 2). From the comparison between the results of cases 1 and 2, it can be concluded that the effects of the surface parameters, such as β, z_0, etc. on the prediction accuracy are significant. Therefore, it is very important to select adequate values for the surface parameters in order to predict accurately the distribution of surface temperature in an urban area, which significantly influences the distributions of air temperature and wind velocity. Effects of urbanization on the heat island circulation over the Tokyo metropolitan area are numerically investigated in the next section.

4. Effects of urbanization on heat island circulation in Tokyo: comparison between urban climates during 1930s and 1990s

4.1. Computed cases and surface parameters

Two cases of CFD analysis are carried out here in order to examine the effects of the change in land-use conditions and the increase of artificial heat release based on the values during the 1930s and the 1990s in Tokyo [29]. The surface parameters of these two cases are summarized in Table 3. Fig. 9 illustrates the distribution of artificial

Table 3
Computed cases for Section 4 and surface parameters

Case	Land-use conditions	Roughness length z_0 (m)	Albedo	Surface relative humidity α	Thermal conductivity (W/m K)	Thermal diffusivity (m^2/s)
Case 3 (1930s)	Forest	2	0.15	0.5	1.05	5.7×10^{-7}
	Buildings	1	0.1	0	1.7	8.1×10^{-7}
	Paved road	0.01	0.1	0	0.7	5.0×10^{-7}
	River site	0.001	0.1	1	0.63	1.5×10^{-7}
	Ocean	0.001	0.1	1	0.63	1.5×10^{-7}
Case 4 (1990s)	Rice paddy	0.05	0.2	1	0.63	1.5×10^{-7}
	Farming	0.01	0.1	0.5	1.05	5.7×10^{-7}
	Orchards 1	1	0.2	0.4	1.05	5.7×10^{-7}
	Orchards 2	0.5	0.2	0.4	1.05	5.7×10^{-7}
	Forest	2	0.15	0.5	1.05	5.7×10^{-7}
	Vacant land	0.01	0.2	0.2	1.05	5.7×10^{-7}
	Buildings	1	0.1	0	1.7	8.1×10^{-7}
	Paved road	0.01	0.1	0	0.7	5.0×10^{-7}
	Other land	0.01	0.2	0.2	1.05	5.7×10^{-7}
	River site	0.001	0.1	1	0.63	1.5×10^{-7}
	Coast	0.005	0.3	0.9	2.2	7.4×10^{-7}
	Ocean	0.001	0.1	1	0.63	1.5×10^{-7}

heating imposed in case 4 (1990s). The computational domain covers 136 km (east–west, x) × 136 km (north–south, y) × 5 km (vertical direction, z) for the cases in Table 3. This domain is discretized into 34 (x) × 34 (y) × 16 (z) grids. Grid spacings in x and y directions are 4 km and the grid spacing in z direction adjacent to the ground is 4 m. Distribution of initial velocity is set to be zero in the whole computational domain in these analyses.

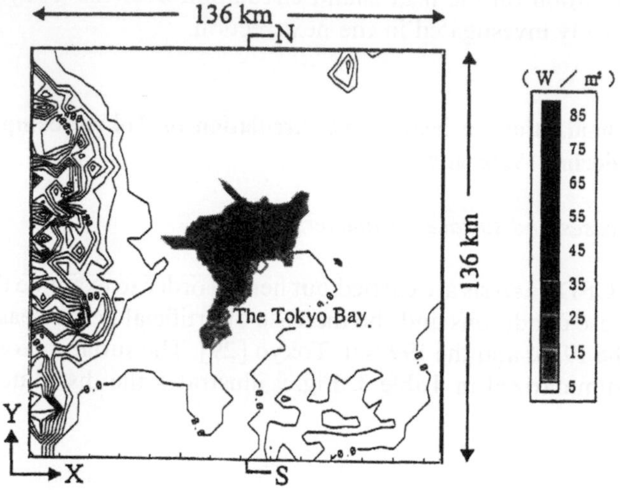

Fig. 9. Distribution of artificial heating imposed in case 4 (1990s).

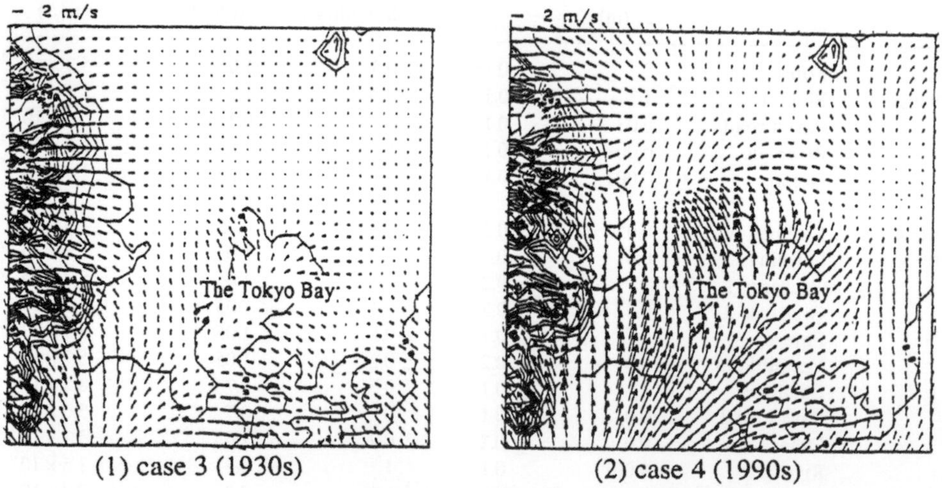

Fig. 10. Horizontal distribution of velocity vectors (at 3:00 p.m. at the height of 10 m). (1) Case 3 (1930s); (2) case 4 (1990s).

4.2. Comparison between urban climates during 1930s and 1990s

Fig. 10 illustrates wind velocity vectors at 10 m height at 3:00 p.m. It is observed that the velocity of the sea-breeze increases significantly in the 1990s in comparison with that in the 1930s. The vertical distribution of temperature on a north–south section (cf. Fig. 9) are compared in Fig. 11. The temperature distributions show maximum peaks in central Tokyo for both cases. The region indicating high temperature is expanded greatly in case 4 compared with case 3. The temperature increase at the peak from the 1930s to the 1990s is about 3°. The temperature difference between urban and rural areas also becomes larger in case 4 in comparison with that in case 3. Fig. 12 shows the vertical velocity vector at the same section as for Fig. 11. The effects of urbanization on wind climate are clearly shown in this figure. As is illustrated here, the heat island circulation is intensified in case 4, and the sea-breeze reaches far more inland in case 4 in comparison with case 3.

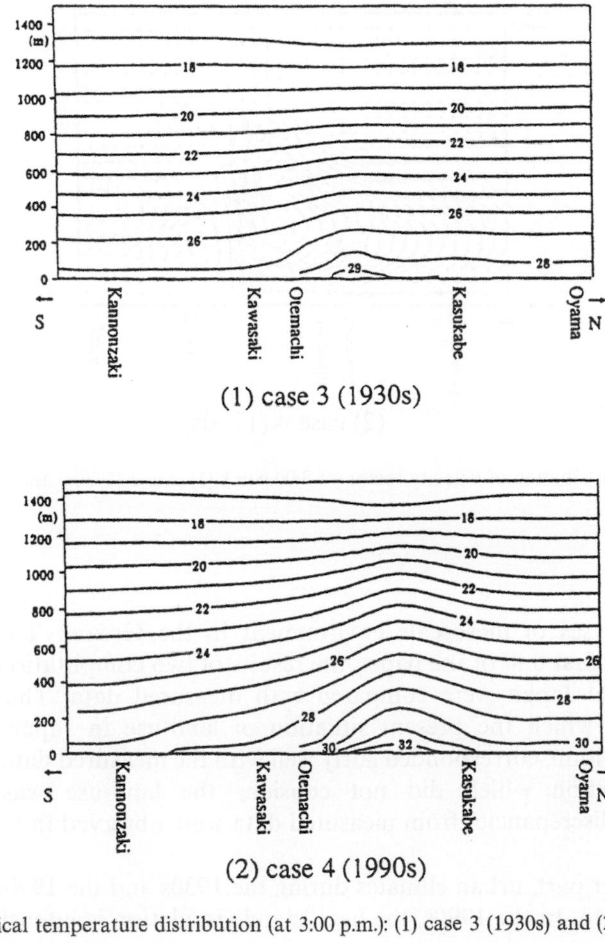

Fig. 11. Vertical temperature distribution (at 3:00 p.m.): (1) case 3 (1930s) and (2) case 4 (1990s).

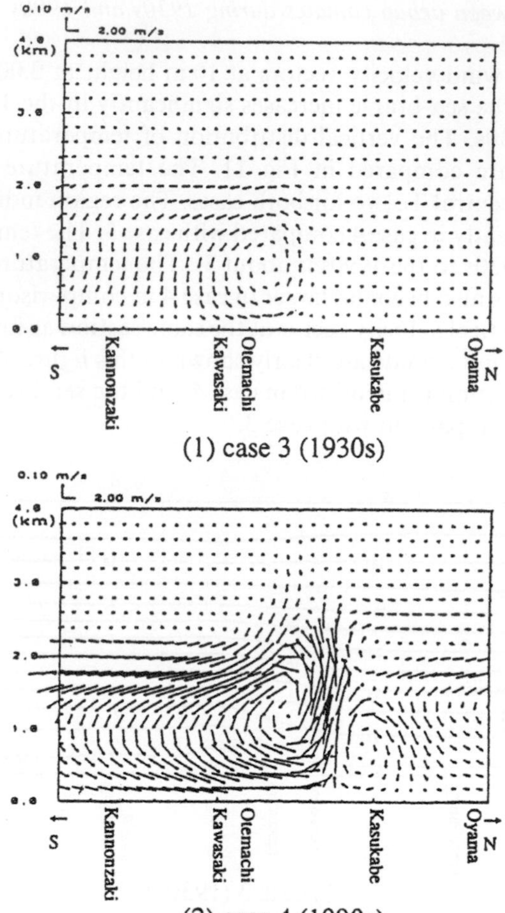

Fig. 12. Vertical distribution of velocity vector (at 3:00 p.m.): (1) case 3 (1930s) and (2) case 4 (1990s).

5. Conclusions

(1) CFD analyses of mesoscale environment in the Greater Tokyo area were presented. In the first half of the paper, the results of two computations of mesoscale climate in central Japan were compared with measured data. The results of the computation, in which the present situation of land-use in Japan was given as a boundary condition, corresponded fairly well with the measured data. For comparison, a computation which did not consider the land-use was also carried out. Significant discrepancies from measured data were observed in the result of this case.

(2) In the latter part, urban climates during the 1930s and the 1990s in the Tokyo area were analysed. In the 1990s, the heat island circulation is intensified by a large

temperature difference between urban and rural areas and sea breeze reaches far more inland in comparison with the results for 1930s.

(3) As is demonstrated here, CFD analysis has great potential in comprehending the very complicated climatic system in an urban area.

(4) However, more effort should be devoted to examining and improving the prediction accuracy, so CFD analysis can be used for practical applications. We believe that the CFD method will become a very powerful tool for city planning in the future.

Acknowledgements

The authors would like to express their gratitude to Dr. Tetsuji Yamada for his useful comments and assistance.

Appendix A

The parameter ζ in Eqs. (9)–(11) is the nondimensional height z/L, where L is the Monin–Obukhov length $-u_*^3/\kappa\beta_V g H_t$; κ is the von Karman constant; z_0, z_{0t}, and z_{0v} are the roughness lengths for wind, temperature, and water vapor, respectively [18], and Pr_t and S_{ct} are the turbulence Prandtl and Schmidt numbers. The terms Ψ_m, Ψ_h, and Ψ_v are correction terms for the atmospheric stability given by [19]

$$\Psi_m(\zeta) = \int_0^\zeta (1 - \Phi_m(\zeta'))/\zeta' \, d\zeta', \tag{A.1}$$

$$\Psi_h(\zeta) = \int_0^\zeta (1 - \Phi_h(\zeta))/\zeta' \, d\zeta', \tag{A.2}$$

$$\Psi_v(\zeta) = \int_0^\zeta (1 - \Phi_v(\zeta'))/\zeta' \, d\zeta'. \tag{A.3}$$

The following formulations are used for Φ_m, Φ_h, and Φ_v for unstable conditions,

$$\Phi_m(\zeta) = (1 - 15\zeta)^{-1/4}, \tag{A.4}$$

$$\Phi_h(\zeta) = (1 - 15\zeta)^{-1/2}, \tag{A.5}$$

$$\Phi_v(\zeta) = (1 - 15\zeta)^{-1/2}. \tag{A.6}$$

For stable conditions,

$$\Phi_m(\zeta) = \Phi_h(\zeta) = \Phi_v(\zeta) = 1 + 5\zeta. \tag{A.7}$$

Appendix B

In the analyses in Section 3, Q_{wG} in Eq. (11) is given by using q_G computed by the following equation:

$$q_G = q(z) + \beta(q_{sat} - q(z)). \tag{B.1}$$

Here, Q_{wG} is set equal to q_G.

In Eq. (B.1), β is the surface moisture availability [20,21], q_G and $q(z)$ are the specific humidity at the ground surface and at height of z. q_{sat} is the saturation specific humidity at the surface temperature.

On the other hand, q_G is estimated by the following equation in Section 4:

$$q_G = \alpha q_{sat}, \tag{B.2}$$

where, α is the relative humidity of the air in the vicinity of the ground and q_{sat} is the saturated water vapor at surface temperature. The definition of q is different for the β method (Eq. (B.1)) and the α method (Eq. (B.2)). In the β method, q denotes the specific humidity, while q is the total water vapor in the α method. If there is no liquid water in the air, such as mist, both become equal.

References

[1] Environmental Agency of Japan, Metropolitan area: toward its conservation and creation, 1990.
[2] H. Yoshikado, M. Tsuchida, Winter air pollution structures as related with the sea breeze and the urban heat island in the Tokyo Bay area, 7th IUAPPA Regional Conference, 1994.
[3] Toshio Ojima, Changing Tokyo metropolitan area and its heat island model, Urban Climate, Planning and Buildings 1 (1991) 191–203.
[4] M.A. Atwater, Thermal effects of urbanization and industrialization in the boundary layer, Bound.-Layer Met. 3 (1972) 229.
[5] T. Ojima, M. Moriyama, Earth surface heat balance chances caused by urbanization, Energy Build. 4 (1982) 99–114.
[6] R.A. Pielke, Mesoscale Meteorological Modeling, Academic Press, London, 1984.
[7] R.A. Pielke, W.R. Cotton, R.L. Walko, C.J. Tremback, W.A. Lyons, L.D. Grasso, M.E. Nicholls, M.D. Moran, D.A. Wesley, T.J. Lee, J.H. Copeland, Comprehensive meteorological modeling system RAMS, Meteor. Atmosph. Phys. 49 (1992) 69–91.
[8] G.L. Mellor, T. Yamada, A hierarchy of turbulence closure models for planetary boundary layer, J. Appl. Met. 13 (7) (1974) 1791.
[9] G.L. Mellor, T. Yamada, Development of a turbulence closure model for geophysical fluid problem, Rev. Geophys. Space Phys. 20 (4) (1982) 851–875.
[10] T. Yamada, S. Bunker, Development of a nested grid, second moment turbulence closure model and data simulation, J. Appl. Met. 27 (5) (1988) 562.
[11] T. Yamada, S. Bunker, A numerical model study of nocturnal drainage flows with strong wind and temperature gradients, J. Appl. Met. 28 (1989) 545.
[12] M.R. Hjelmfelt, Numerical Simulation of the Effects of St. Louis on Mesoscale Boundary-Layer Airflow and Non-Urban Effects, American Meteorological Society, Providence, RI, 1982, pp. 1239–1257.
[13] F. Kimura, S. Takahashi, The effects of land-use and anthropogenic heating on the surface temperature in the Tokyo metropolitan area: A numerical experiment, Atmos. Environ. 25 (2) (1991) 155.

[14] H. Yoshikado, Numerical study of the daytime urban effect and its interaction with the sea breeze, J. Appl. Met. 31 (10) (1992) 1145.
[15] I. Uno, Quantitative evaluation of a mesoscale numerical model simulation using four-dimensional data assimilation of complex airflow over the Kanto Region, Atmos. Environ. (1995) 351.
[16] National Land Numerical Information, Planing and Coordination bureau, National Land Information office, Ministry of Construction Geographical Survey Institute, 1992.
[17] T. Sasamori, The radiative cooling calculation for application to general circulation experiments, J. Appl. Meteor. 7 (5) (1968) 720.
[18] J.R. Garratt, B.B. Hicks, Momentum, heat and water vapour transfer to and from natural and artificial surfaces, Quart J. Roy. Met Soc. 99 (1973) 680.
[19] H.A. Panofsky, Determination of stress from wind and temperature measurements, Quart. J. Roy. Meteor. Soc. 23 (1963) 495.
[20] J. Kondo, T. Watanabe, Studies on the bulk transfer coefficients over a vegetated surface with a multilayer energy budget model, J. Atmos. Sci. 49 (1992) 2183.
[21] W. Tsutomu, J. Japan Soc. Hydraulic & Water Resources 5 (1992) 39.
[22] Kondrat'yev, Radiation regime of inclined surfaces, WMO Technical Note No. 15, Secretariat of the World Meteorological Organization, Geneva, Switzerland, 1977.
[23] M. Miura, T. Ojima, The soil temperature difference between urban and suburban areas by measurement at 95 spots, J. Archit. Plann. Environ. Eng. AIJ 454 (1993) 35.
[24] Marine Climatological Charts of the North Pacific Ocean for 1987, 1988, 1989, 1990 by Marine Department, JMA, Japan Meteorological Agency, 1992.
[25] T. Kuwagata, J. Kondo, Estimation of aerodynamic roughness at the regional meteorological stations (AMeDAS) in the Central Part of Japan, TENKI 37 (3) (1990) 55.
[26] W. Gao, S. Miura, T. Ojima, Site survey on formation of river in Koto-ku, Tokyo. Thermal effects of the open space with green area on the urban environment, Part II, J. Archit. Plann. Environ. Eng. AIJ 456 (1994) 75.
[27] I. Ken, G. Weijin, O. Ryuzo, T. Yoshihide, M. Akashi, M. Shuzo, O. Toshio, Micro climate of ecological city project in Shitamachi, Tokyo by numerical study, in: Summaries of Technical Papers of Annual Meeting Architectural Institute of Japan, 1995, pp. 951–952.
[28] Data from AMeDAS (Automated Meteorological Data Acquisition System, 1990–1994), Japan Meteorological Agency.
[29] H. Sugiyama, A. Mochida, S. Murakami, T. Ojima, Numerical analysis of urban heat island over Tokyo metropolitan area, (part 2), Transport of pollutants convected by heat island circulations, in: Summaries of Technical Papers of Annual Meeting Architectural Institute of Japan, 1996.

ELSEVIER

Journal of Wind Engineering
and Industrial Aerodynamics 67&68 (1997) 479–492

JOURNAL OF wind engineering AND industrial aerodynamics

On the validation of high-resolution atmospheric mesoscale models

K. Heinke Schlünzen[1]

Meteorological Institute, University of Hamburg, Bundesstrasse 55, 20146 Hamburg, Germany

Abstract

The central ideas of a validation concept are outlined. The concept is applicable to high-resolution atmospheric mesoscale models, which calculate wind, temperature and humidity fields as well as pollutant concentrations and which might be used as regulatory models. The validation concept takes into account model characteristics, the model realization and results from selected case studies. It includes five criteria: *completeness*, *comprehensibility*, *code quality*, *result quality* and *result control*. The concept may be used by model developers and users to determine the validity of their models and to evaluate its applicability for the simulation of atmospheric phenomena, developing in an area of horizontal extension between 10 and 300 km in the lower troposphere in mid-latitudes.

Keywords: Validation; Verification; Evaluation; Mesoscale model; Validation criteria

1. Introduction

High-resolution models, which are applicable for the simulation of atmospheric phenomena in the mesoscale-β and mesoscale-γ range, have become more and more sophisticated in recent years. They are three-dimensional, include prognostic equations for wind, temperature and humidity, but also for cloud and rain water content and tracer concentrations. Recently, some first attempts have been made to apply these models in environmental impact assessment studies. In some countries it is discussed whether and how highly sophisticated atmospheric models can be used as regulatory models. However, before the models can be applied in support of economically and environmentally important decisions, they have to be evaluated with respect to their accuracy for typical applications.

[1] E-mail: schluenzen@dkrz.de.

PII S0167-6105(97)00095-0

In the present paper a validation concept is presented that might be used to evaluate high-resolution mesoscale models. Before the validation concept is outlined (Section 3), the meaning of some words used in this paper should be defined. They are often used in a less specific sense when discussing the validation of models. The definitions given here are not new but summarize the ideas of other authors [1,2] and reference books [3–5].

1.1. *Verification (checking the correctness of a model)*

To *verify* a model completely, it has to be proved that the model is able to simulate all atmospheric phenomena of the model application area with the correct solution [1]. Such a proof could only be constructed if we knew all atmospheric phenomena of the application area, could simulate them all, and could successfully compare model results with measurements. Apart from the fact that it is not at all a simple task to compare model results and measurements, a proof is impractical for complex models, because we neither know all the atmospheric phenomena nor all the initial data. The impossibility of a proof becomes even more evident if one considers that all possible variations of input parameters have to give realistic results. Therefore, complex atmospheric models cannot in general be verified completely. However, model results can be verified for single-case studies, e.g., by comparing them with measurements. In addition, a verification of single aspects of a model or of a simple model might be possible.

1.2. *Validation (checking the applicability of a model)*

To consider a model as *valid* it has to fulfil a set of criteria, which are specific for the application area of the model. Tests of the model have to be performed against these criteria (including: completeness; comprehensibility; code quality; comparisons with experimental data, other model results, analytic solutions; and result control). Fulfilling the validation criteria does not necessarily imply that the results agree with reality. The accuracy required of the model results depends on the validation criteria. However, no one would denote a model as valid, which gives results that are in contrast to reality. If there is at least one phenomenon which cannot be modelled within prescribed tolerances, then the model can no longer be considered 'validated'. To ensure that the model results are close to reality, the validation criteria have to be formulated in a way which depends on the purpose of the model [6].

It is worth noting that some sources [7] assign some of the criteria put here under the heading 'validation' under other headings. Nevertheless, the importance of the criteria is also emphasized.

1.3. *Evaluation (assessment of the validity of a model and its results)*

A model is evaluated for simulating atmospheric phenomena of a certain scale, if the model is validated and the full application area of the model is specified. To *evaluate* a model, it has to be assessed with respect to its limit of application.

2. Application area of the model and consequences for model structure

The validation concept is developed for regulatory models applicable to the simulation of atmospheric phenomena of the mesoscale-γ and mesoscale-β. We focus on phenomena that can be simulated without explicit inclusion of obstacles (e.g. buildings, trees) in the model. The effects of surface inhomogeneities on flow fields are incorporated by using the corresponding roughness length. This means that a model validated by applying the proposed concept is not validated for flows close to obstacles. To avoid misuse, we restrict the *horizontal grid resolution to above* 500 *m and below* 5 *km*. The *horizontal extent of the model should be at least* 10 *km and not more than* 300 *km*. For larger model areas we would have to consider horizontally inhomogeneous large-scale situations. Models with these characteristics outlined before can be used to simulate flow fields and pollution transport for source distances from about four times the grid spacing at a minimum, to some hundreds of kilometres at a maximum, depending on the numerical and filtering schemes applied.

Regulatory models are, in general, used to derive concentration data close to the surface, e.g. at a height of 2 or 10 m. The data might be calculated by linear or logarithmic interpolations between the model grid points close to the surface. To reduce errors, the *lowest grid point should not be more than* 10 *m above the ground with a vertical grid resolution of* 20 *m or less*. Since a validation concept should not only check the validity of the surface layer data but also provide justification for the calculated values, we need to simulate processes in the whole planetary boundary layer (PBL) and, in order to model entrainment and detrainment processes accurately, some thousands of metres above the PBL. Therefore, the *model should extend to heights well above* 5000 *m* in order to calculate surface data. The *vertical resolution might be as large as* 500 *m at the highest levels*. With the characteristics of the grid spacing outlined above we are able to simulate phenomena like sea-breeze circulations including the return current, or lee waves above mountains, and can consider their influence on wind and temperature fields when calculating values in the surface layer. However, *deep tropical convection*, which might reach up to the tropopause level, cannot be simulated realistically with the described grid structure and *is thus excluded*.

The regulatory models applied up to now are used to calculate, for example, annual means or 98 percentile concentration values [8]. For this purpose relatively simple models are applied, mostly based on Gaussian plume dispersion models, assuming horizontal homogeneity of terrain and e.g. of the horizontal wind fields. For elevated sources this assumption might be fulfilled. However, sources in mountainous regions need a more complex calculation of wind and temperature fields, and the assumption of horizontal homogeneity is no longer valid. High-resolution mesoscale models can be applied to *simulate wind, temperature and humidity as well as concentrations* on the scale outlined above over complex terrain. The models *consider the effects of terrain, surface temperature and surface humidity* on meteorological variables and tracer concentrations. However, they are still quite expensive to run, and thus full statistics needed to derive 98 percentile values or annual means cannot be calculated without excessive effort. Instead, the models can very well be applied for single-case studies, for

Table 1
Validation areas distinguished in the validation concept. The *extended validation* (extended version of the model, including condensation, deposition, radiation budget in the atmosphere) includes all sub-criteria of the *basic validation* (basic version of the model; no condensation, no deposition, no radiation budget in the atmosphere), the validation of *thermodynamics* is part of the *tracer-dynamics*. The complexity of the validation increases from left to right and top to bottom

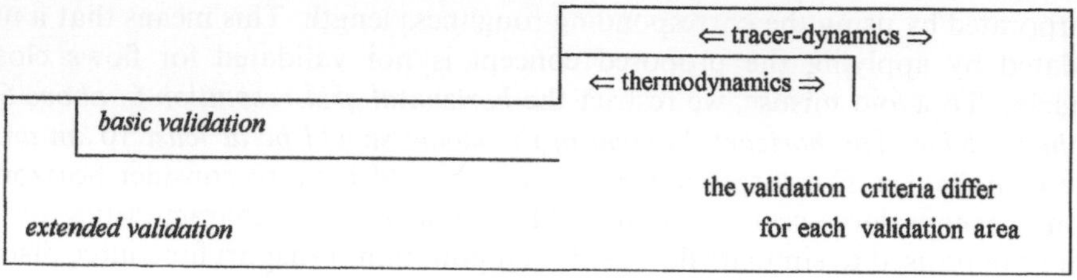

example, investigations of worse case scenarios. These were less easy to investigate with the regulatory models applied hitherto. For this reason we focus on *single-case studies* in the validation concept. The simulation period lies *between several hours and some tens of hours. Temporal changes of the large-scale situation are not considered.*

The models might be applied as meteorological pre-processors for atmospheric transport models or they might contain the simulation of tracer transport. Due to the different applications, two validation areas are distinguished: the meteorological and the tracer-specific part of the model (Table 1). The validation of the meteorological part will be denoted *thermodynamics* throughout this text, and corresponds to a validation concept for a high-resolution mesoscale model used as a pre-processor for air-quality models. The validation of the tracer-specific part of the model is not restricted to tracer specific components of the model alone but includes the meteorological part too. Therefore the concept cannot be applied to transport model results alone but is always a test of the meteorological pre-processor as well. To make this clear, this validation field will be denoted *tracer-dynamics* throughout this text. A model cannot be validated for tracer-dynamics without first validating it for thermodynamics.

Besides distinguishing between the validation areas of thermodynamics and tracer-dynamics, an additional separation is proposed with respect to the intended model applications. For idealized situations (e.g. flow fields in mountainous terrain, tracer transport in coastal regions) the *basic validation* may be sufficient, which is restricted to a moist atmosphere without condensation and does not include deposition processes or a radiation budget in the atmosphere (Table 1). The basic validation includes most of the atmospheric phenomena typically taken into account under regulatory aspects. However, for environmental impact assessment studies wet deposition and thus precipitation might also be important. This broader range of applications is covered by the *extended validation.* Chemical transformations are not accounted for; the validation scheme is *restricted to non-reacting species.*

3. Validation concept

Several ideas can be found in the literature for the validation of high-resolution mesoscale models concerning model construction, numerical and physical parameters [7,9–12] and transportive quantities [13–15]. Many authors validate their model by comparing model results with measurements, analytic solutions or other model results for some or several case studies ([16–23] and many others). They infer the ability of the model to simulate the investigated case from qualitative and quantitative comparisons (case-by-case verification). However, a general conclusion on the validity of the model cannot be drawn, because not all atmospheric phenomena are simulated (see Section 1) and the reasons behind the right model results cannot be demonstrated by data comparisons alone. The applicability of the model to other cases can only be assessed if the model physics and model realization are evaluated too.

3.1. General validation criteria

The validation concept includes five general criteria:

Completeness: the model physics have to allow the simulation of the atmospheric phenomena.

Comprehensibility: the model and its validation must be transparent to others.

Code quality: the model realization (e.g. the computer code) must be clear and easy to use.

Result quality: the model results in specified applications have to be consistent with analytic solutions, measurements or other model results, and should only differ within prescribed tolerances.

Result control: during the model runs all data have to be checked on-line for plausibility; further off-line checks have to be performed.

Each of the five validation criteria consists of several sub-criteria that have to be specified with respect to the application area of the model. The sub-criteria given in the following section are formulated for the application area outlined in Section 2.

3.2. Sub-criteria of the validation concept

3.2.1. Completeness

An outline of the model structure has already been derived (in Section 2). Here, some details on equations, parameterizations and approximations are given.

The prognostic equations for wind, temperature, humidity and pollutants (basic validation) and for liquid water and ice content (extended validation) to be used in the model can be found in Refs. [6,9]. To obtain a sufficient accuracy in the vertical wind component it must be calculated using a prognostic equation [24]. For the phenomena considered the continuity equation might be used in its complete or in the anelastic form [24]. For simulating convection as realistically as possible, the density deviations from the mean should not only include temperature deviations but also pressure and humidity deviations [25]. Furthermore, since we are interested in modelling surface layer data correctly, Coriolis force effects have to be accounted for in the model [24].

The determination of the parameterizations to be used in high-resolution mesoscale models is more difficult than the specification of equations and approximations, because there is still not enough knowledge to decide for one specific scheme. The best parameterizations can hopefully be identified much more precisely, after currently running world-wide projects have compared parameterizations used in models (cloud microphysics, turbulence, land surface parameterization). For the moment we have to remain more general. The turbulence parameterization should allow for different atmospheric stabilities to ensure a realistic simulation of the vertical profiles. The stress tensor should be symmetric [26] and the calculated fluxes should be continuous functions of height and stability [25]. Some turbulence parameterizations seem to be sufficient to simulate a convective PBL [27,28]. In other cases an extra convection parameterization should be included [25].

A cloud microphysics parameterization is essential for the extended validation. However, selecting one particular scheme is impossible at present. The same is true for a radiation flux parameterization in the atmosphere. Both should at least consider cloud and rain water and ice if necessitated by sufficiently low temperatures.

The choice of a specific surface energy or humidity budget scheme can also not be made at present. However, to facilitate scenario studies (essential for environmental impact assessment) surface temperature and humidity should be predicted and not simply be prescribed or kept constant. For this purpose at least a 'force restore' method as presented in Ref. [29] should be used in the basic version of a model.

Steep terrain influences might be important in the model studies. Therefore, the equations which might be solved in a boundary-following coordinate system [9], have to be transformed without simplification of the transformation terms. The effects of shadows of mountains have to be included for terrain slopes of more than 10° if the surface energy budget is solved.

For the tracer-dynamics, point, line, area and volume sources, including both hot and cold emissions, have to be introduced into the basic version of the model. In the extended version dry and wet deposition processes should be included too, since both are important in the atmosphere. As mentioned for the other parameterizations, current scientific knowledge does not indicate that one specific scheme is better than all others. For dry deposition a resistance model concept is quite common [30–33], which considers atmospheric stability, tracer and surface characteristics. Surface resistance data can be found, for example, in Ref. [34].

Wet deposition can be simulated by including wash-out factors, which can be used effectively in models with a coarse resolution. This is not sufficient for high-resolution models. More complicated schemes where wet deposition is better simulated [35–37], on the other hand, require excessive computer resources. Despite this they should be applied if wet deposition is significant for the simulated concentrations.

The model characteristics described above are based on current scientific knowledge. Newer findings need to be introduced into the validation criteria in order to keep them up-to-date.

3.2.2. Comprehensibility

To be properly understood a model has to be well documented and the documentation must be available. The model code should be accessible too, because more model users and applications decrease the probability of errors in the model and of wrong model results. Model structure and qualities, including the basic theory as well as the validation performed have to be published in peer reviewed journals. A more detailed paper on the model validation should be accessible in a comprehensive report.

3.2.3. Code quality

Complex atmospheric models are solved on computers and thus a program code is needed. Since different computers have different accuracy, the dependence of model results on the computer accuracy should be checked and reported by the model developers when they validate and evaluate the model. Further theoretical aspects of code development could be listed here, which mostly concern the theory of writing verifiable programs. However, only those sub-criteria are suggested here that seem necessary for an atmospheric model.

To ensure the transfer of the model code to different computer programming language standards should be adhered to. Programming conventions should be prescribed and observed when writing and editing the program. This is essential to make a program readable by a third party. A list of variables, data flow diagrams and call-trees complete the documentation of the model code. Most models allow case-dependent parameter and initial data settings (e.g. number of grid-points, grid spacing, profiles of meteorological variables). These should be checked automatically by the program for consistency to avoid misuse of the model. Any resulting abnormal termination should be accompanied by informative error descriptions.

3.2.4. Result quality

To check the result quality it would be best to perform as many case studies as possible, but such testing would often cost too much time. Since all tests should be carried out after each model change, to ensure that improvements in one part do not cause a worsening in another part of the model, it is essential to reduce the number of validation test cases to a necessary minimum. The test cases should detect shortcomings with respect to the application area of the model described in Section 2. They should be simple in view of the time needed to do the model runs. For these reasons comparisons that need statistics of several model runs are not recommended. Basic tests for checking numerical schemes, the influence of grid spacing, parameterizations or boundary conditions are also not considered, because these tests need often idealized model structures and are mostly very time consuming. However, they should be done during model development in order to decide on a specific scheme. The validation suggested here has to check a three-dimensional model as a whole. Some case studies are quite simple with regard to the number of interacting processes and the model physics. This allows one to detect basic errors and to use restrictive error measures. For comparisons with measurements the full model physics is needed. This allows one to detect errors in the non-linear processes.

3.2.4.1. Basic validation of thermodynamics.. Most test cases are only briefly described here. More details can be found in Ref. [12]. For the basic validation of thermodynamics four case-studies may be sufficient (flow over flat terrain, flow over a mountain, dry sea-breeze system, comparison with a PBL data set). For the case 'flow over flat terrain' very idealized runs have to be performed (input data in Table 2), which could help to detect shortcomings in programming and numerics. The model is verified for the specific application if the wind veering in southern and northern hemisphere are opposite within a deviation of 0.1°. This is also the maximum deviation for the direction normalized wind profiles, resulting from different directions of the geostrophic wind. The wind velocities should differ less than 0.01 m/s between the different cases. Since the runs are performed for an unstratified atmosphere, the temperature differences have to be very small. They should be lower than 100 times the computer accuracy times at a temperature scale of 100 K.

In the case of 'flow over a mountain', model results and analytic solution used by Lilly and Klemp [38] are compared for different wind velocities and stratifications. In this test the coordinate transformation and stratification influences are checked. To avoid having to change the model code in order to do this comparison, this case should be performed with a very low roughness length, with a no-slip boundary condition at the surface and with the usual turbulence parameterization. With these model features, analytic solutions and computer model results differ considerably in the PBL. Therefore, the results should only be compared and evaluated above the PBL. The input data and evaluation measures for this test are given in Tables 3 and 4.

The simulation of a dry sea-breeze system is an idealized two-dimensional case, used to check stratification influences and the surface energy budget. The inland penetration of a sea-breeze front is compared with data, calculated as mean values from several observed sea breeze fronts [39]. As a PBL data set traditionally the Wangara data of day 33/34 [40] are used [41–43]. This data set is suggested here despite the unstationarity of the observed large-scale situation, since the data are

Table 2
Input data for the 'flat terrain' test case

Input variable	Value
Geostrophic wind	5 m/s from east, south, west, north
Stratification	Neutral
Surface temperature at sea level	288.16 K, no temporal changes
Relative humidity	None
Surface pressure at sea level	1013.25 hPa
Roughness length	0.10 m
Latitude	50° (northern or southern hemisphere)
Horizontal grid size	2000 m
Model area (horizontal)	− 4 to 4 km, − 4 to 4 km (4 × 4 grid points)
Vertical resolution	20 to 1000 m, non-uniform
Model top	8 km
Integration time	6 h
Control data	Each 3 h

Table 3
Input data for the 'flow over a mountain' test case

Input variable	Value
Geostrophic wind	10 m/s from west
Stratification	Stable, 0.005 K/m
Surface temperature at sea level	290 K, no temporal changes
Relative humidity	None
Surface pressure at sea level	1000 hPa
Roughness length	0.0001 m
Latitude	30° northern hemisphere
Horizontal grid size	1000 m
Model area (horizontal)	− 80 to 80 km, − 2 to 2 km (160 × 4 grid points)
Vertical resolution	20 to 1000 m, non-uniform
Model top	11 km
Orography height z_s	$z_s(x) = 300\frac{3000^2}{3000^2 + x^2}$
Integration time	9 h
Control data	Every 2 h

Table 4
Evaluation measures for the 'flow over a mountain' test case. The results should be evaluated between 1000 and 5000 m, except for the wave length

Case	Error measure	Allowed value
Stable stratification, Geostrophic wind 10 m/s	Wavelength	4830 ± 1210 m
	Correlation coefficient (u-component)	⩾ 0.9
	Correlation coefficient (vertical wind)	⩾ 0.9
	Correlation coefficient (temperature)	⩾ 0.9
	Maximum deviation (v-component)	< 0.5 m/s
Stable stratification, Geostrophic wind 5 m/s	Wavelength	2415 ± 605 m
	Correlation coefficient (u-component)	⩾ 0.9
	Correlation coefficient (vertical wind)	⩾ 0.9
	Correlation coefficient (temperature)	⩾ 0.9
	Maximum deviation (v-component)	< 0.5 m/s
Neutral stratification, Geostrophic wind 10 m/s	Correlation coefficient (u-component)	⩾ 0.9
	Correlation coefficient (vertical wind)	⩾ 0.9
	Maximum deviation (v-component)	< 0.5 m/s

widely distributed. However, if the model developer has an independent data set this might be used as an alternative, since the Wangara data have already been used during the development of most turbulence parameterizations.

3.2.4.2. Basic validation of tracer-dynamics. Two additional tests should be performed for the basic validation of tracer-dynamics (changes in emission rates and stratification). The 'emission rate' test is used to check numerical accuracy and the correct implementation of the transport equation. For this test the concentrations, normalized with the emission rates, should differ by less than 100 times computer accuracy times normalized concentration value, because the concentration of a passive tracer is linearly dependent on the emission rate. The test case 'stratification' is a comparison with an analytic solution again (Gaussian dispersion model; see e.g. Ref. [44]), used to check the influence of stratification on dispersion. Similar to the 'flow over a mountain' test case, the simplifications with respect to terrain and meteorology do not have to be assumed in the more complex and more realistic high-resolution numerical model. Therefore, the comparison between the results can only be qualitative; the maximum concentration should be further away for stable stratification than for neutral or unstable stratification. In addition, the concentration patterns should be well correlated ($\geqslant 0.9$) with the analytic solution in source height. Above and below the source height, the correlation is lower due to the veering of the wind.

3.2.4.3. Extended validation.. In an extended validation additional model characteristics like precipitation and deposition have to be validated. For these characteristics two further tests should be performed for the thermodynamics (wet sea-breeze system, comparison with a 3-D meteorological data set) and another two tests for the tracer-dynamics (deposition influence, comparison with a 3-D tracer-dynamics data set). The 'wet sea-breeze' case should use the same input data and configuration as the 'dry sea-breeze' simulation, but the formation of clouds should now be allowed. In this case the inland penetration of the sea-breeze front should be accelerated [39,45,46]. The test is used to check the correct influence of condensation on dynamical systems. However, the test does not check the correctness of the microphysics parameterization scheme.

The 'deposition influence' can be checked by performing model runs with and without dry and/or wet deposition. Wet deposition and precipitation should be well correlated ($\geqslant 0.9$). The reduction of total mass in the atmosphere should not differ by more than 1% from the deposited amount. Each tracer has to be considered in order to validate the deposition scheme. The test checks the mass budget of the model. To examine physical and chemical accuracy, comparisons with measurements should be performed for different vegetation types.

For the case studies that need comparisons with experimental data, the selected data sets should be well documented, accessible to other users, evaluated with respect to data quality and, last but not least, they should be maintained to ensure the availability of the data set for a longer time. Since comparisons of model results and measurements are very time consuming and not many data-sets are available to other users, it is suggested in the first instance that a data set available to the model developer should be used.

3.2.5. Result control

Even after a successful validation of a model by applying the first four validation criteria, wrong results cannot be ruled out (impossibility of complete verification [1]).

Table 5
Evaluation measures for on-line and off-line model control; bias $|D|$, correlation coefficient R

	⇐ Tracer-dynamics ⇒	
	⇐ Thermodynamics ⇒	
On-line control data, to be used at each grid point (below 6 km) and each time step[a]	$-60 \leqslant \bar{u}, \bar{v} \leqslant 60$ (m/s) $-6 \leqslant \bar{w} \leqslant 6$ (m/s) $200 \leqslant \bar{\theta} \leqslant 400$ (K) $0 \leqslant \bar{q}_1^1 \leqslant 1.1 q_{1\,\mathrm{sat}}^{1\,\mathrm{water}}$ $0 \leqslant \bar{q}_1^1 \leqslant 1.1 q_{1\,\mathrm{sat}}^{1\,\mathrm{ice}}$, $T < 253$ K $0 \leqslant \bar{q}_1^{2C}, \bar{q}_1^{2R}, \bar{q}_1^3 \leqslant 4$ (g/kg)	$0 \leqslant \bar{C} \leqslant QtV_Q/\rho_Q$ (kg/kg) $0 \leqslant v_D \leqslant 0.1$ (m/s)
Off-line control data for test runs with double horizontal resolution, reduced integration time; comparison of filtered and aggregated data well above highest topography	$\lvert D\rvert(\bar{u},\bar{v}) \leqslant \max\begin{cases} 1\ (\mathrm{m/s}) \\ 0.1(\bar{u},\bar{v},\bar{w}) \end{cases}$ $\lvert D\rvert(\bar{w}) \leqslant \max\begin{cases} 0.1(\mathrm{m/s}) \\ 0.1(\bar{w}) \end{cases}$ $\lvert D\rvert(\bar{\theta}) \leqslant 1$ K $D_{\max}(\bar{q}_1^1) \leqslant 1$ (g/kg) $D_{\max}(\bar{q}_1^{2C}, \bar{q}_1^{2R}, \bar{q}_1^3) \leqslant 0.4$ (g/kg) $R(\bar{u},\bar{v},\bar{w},\bar{\theta}) \geqslant 0.9$ $R(\bar{q}_1^1, \bar{q}_1^{2C}, \bar{q}_1^{2R}, \bar{q}_1^3) \geqslant 0.9$	$R(\bar{C}) \geqslant 0.9$

[a]Values for water based on measured data (J. Jensen, 1994, personal communication); threshold values for humidity, concentrations and fluxes for plausibility reasons; threshold values for deposition velocities from measured data; all other values US standard atmosphere [6]. Attention: wind velocities might differ considerably without being erroneous for intense vortices, which can have values of more than 100 m/s.

By application of the *result control* criterion, the probability is reduced that a new case study might give unreasonable results. With this objective two groups of sub-criteria may be applied: several *on-line result controls* can be performed during the model run (time series of variables at a control grid point and of integral values; plausibility limits for each variable as summarized in Table 5). This allows a continuous check of the model results. By performing *off-line result controls* (repeating the run for a shorter integration time with a double resolution to check the sufficiency of parameterizations; evaluation measures are summarized in Table 5) further control steps can be formalized. Finally, expert knowledge is always essential: the model results should be compared with measurements or other model results; they should be evaluated taking all available information into account. If the results are wrong for one single case, the model is falsified. In this case the model should be improved and fully validated in its new configuration. If only the unsuccessful single-case study were repeated, the improvement might be restricted to this case and it might make all other cases worse. Only a model validated by application of all five criteria can be considered valid – and only as long as no falsification has occurred.

4. Conclusions

In this paper the central ideas of a validation concept for mesoscale atmospheric models are presented. Some examples of sub-criteria are given; more details can be found in Ref. [12]. The suggested criteria and sub-criteria are presented here to define a validation concept which might also stimulate further discussion on model validation. Future knowledge will hopefully lead to clearer choices of parameterizations to be used in high-resolution mesoscale models. However, current knowledge already allows one to define some characteristics of the models and to identify case studies and measures of errors.

The validation concept might be used to define the validity of a (regulatory) model and to evaluate it with regard to its applicability for simulating atmospheric phenomena of the scale specified in Section 2. In addition, it might help model developers and users to detect model shortcomings.

It should be emphasized here that the quality of the initial data is crucial for the quality of the results of a model. If the initial data are poor or incomplete, the results of a model can be unrealistic. The resulting falsification cannot be distinguished from a falsification induced by an inaccurate model. Therefore, the initial data have to be chosen carefully in order to receive realistic results from validated models.

Acknowledgements

The comments of David Webber and the referees on this paper are appreciated.

References

[1] K.R. Popper, Logik der Forschung, Verlag J.C.B. Mohr (Paul Siebeck), Tübingen, 1982, pp. 450.
[2] N. Oreskes, K. Shrader-Frechette, K. Belitz, Verification, validation, and confirmation of numerical models in earth sciences, Science 263 (1994) 641–646.
[3] McGraw-Hill, Dictionary of Scientific and Technical Terms, McGraw-Hill, New York, 1978.
[4] IEEE, Standard Dictionary of Electrical and Electronics Terms, Wiley-Interscience, New York, 1977.
[5] Webster, New International Dictionary of the English Language, Encyclopaedia Britannica, Inc., 1976.
[6] J.A. Dutton, Dynamics of Atmospheric Motion, Dover, New York, 1995, pp. 617.
[7] Model Evaluation Group, Model Evaluation Protocol, European Communities Directorate-General XII, Science Research and Development, Version 5, 1994, pp. 14.
[8] M. Schatzmann, Atmospheric dispersion models for regulatory purposes in the Federal Republic of Germany. Part II: the current situation, Int. J. Environ. Pollution 5 (1995) 431–440.
[9] R.A. Pielke, Mesoscale Meteorological Modeling, Academic Press, London, 1984, pp. 612.
[10] P.J. Roache, Computational Fluid Dynamics, Hermosa Publishers, Albuquerque, 1982, pp. 446.
[11] S.R. Hanna, Mesoscale meteorological model evaluation techniques with emphasis on needs of air quality models, in: R.A. Pielke, R.P. Pears, (Eds.), Mesoscale Modelling of the Atmosphere, Meteorol. Monographs, American Meteorol. Soc. 25/47 (1994) 47–58.
[12] K.H. Schlünzen, Validierung hochauflösender Regionalmodelle, Ber. aus dem Zentrum f. Meeres- und Klimaforschung, Meteorologisches Institut, Universität Hamburg, A23, 1996, pp. 184.

[13] W. Klug, G. Graziani, G. Grippa, D. Pierce, C. Tassone (Eds.), Evaluation of long range atmospheric transport models using environmental radioactivity data from the Chernobyl accident, The ATMES Report, Elsevier Applied Science, London, New York, 1991, pp. 347–351.

[14] H. Hass, A. Ebel, H. Feldmann, H.J. Jakobs, M. Memmesheimer, Evaluation studies with a regional chemical transport model (EURAD) using air quality data from the EMEP monitoring network, Atmos. Environ. 27A (1993) 867–887.

[15] J.C. Weil, R.I. Sykes, A. Venkatram, Evaluating air-quality models: review and outlook, J. Appl. Meteorol. 31 (1992) 1121–1145.

[16] T.L. Clark, R. Gall, Three-dimensional numerical model simulations of airflow over mountainous terrain: a comparison with observations, Mon. Weather Rev. 110 (1982) 766–791.

[17] U. Schumann, T. Hauf, H. Höller, H. Schmidt, H. Volkert, A mesoscale model for the simulation of turbulence, clouds and flow over mountains: formulation and validation examples, Beitr. Phys. Atmos. 60 (1987) 413–446.

[18] K.H. Schlünzen, Numerical studies on the inland penetration of sea breeze fronts at a coastline with tidally flooded mudflats, Beitr. Phys. Atmos. 63 (1990) 243–256.

[19] A.M. MacDonald, C.M. Banic, W.R. Leaitch, K.J. Puckett, Evaluation of the Eulerian acid deposition and oxidant model (ADOM) with summer 1988 aircraft data, Atmos. Environ. 27A (1993) 1019–1034.

[20] G. Groß, Statistical evaluation of the mesoscale model results, in: R.A. Pielke, R.P. Pears (Eds.), Mesoscale Modelling of the Atmosphere, Meteorol. Monographs, American. Meteorol. Soc. 25/47 (1994) 137–154.

[21] T.L. Clark, W.D. Hall, Multi-domain simulations of the time dependent Navier–Stokes equations: benchmark error analysis of some nesting procedures, Comput. Phys. 92 (1991) 456–480.

[22] D.M. Mocko, W.R. Cotton, Evaluation of fractional cloudiness parameterizations for use in a mesoscale model, J. Atmos. Sci. 52 (1995) 2884–2901.

[23] M. Xue, K.K. Droegemeier, V. Wong, A. Shapiro, K. Brewster, Advanced regional prediction system. ARPS Version 4.0 User's Guide. Centre for Analysis and Prediction of Storms, The University of Oklahoma, http://wwwcaps.uoknor.edu, 1995, pp. 295–328.

[24] F.K. Wippermann, The applicability of several approximations in mesoscale modelling – a linear approach, Beitr. Phys. Atmos. 54 (1981) 298–308.

[25] K.H. Schlünzen, Mesoscale modelling in complex terrain – an overview on the German nonhydrostatic models, Beitr. Phys. Atmos. 67 (1994) 243–253.

[26] H. Tennekes, J.L. Lumley, A First Course in Turbulence, The MIT Press, Cambridge, 1972, pp. 300.

[27] A.A.M. Holtslag, van E. Meijgaard, W.C. de Rooy, A comparison of boundary layer diffusion schemes in unstable conditions over land, Bound.-Layer Meteorol. 76 (1995) 69–95.

[28] C. Lüpkes, K.H. Schlünzen, Modelling the Arctic convective boundary-layer with different turbulence parameterizations, Bound.-Layer Meteorol. 79 (1996) 107–130.

[29] J.W. Deardorff, Efficient prediction of ground surface temperature and moisture, with inclusion of a layer of vegetation, J. Geophys. Res. 83 (1978) 1889–1903.

[30] J.S. Chang, R.A. Brost, I.S.A. Isaksen, S. Madronich, P. Middelton, W.R. Stockwell, C.J. Walcek, A Three-dimensional Eulerian acid deposition model: physical concepts and formulation, J. Geophys. Res. 92 (1987) 14681–14700.

[31] S.M. Joffre, Modeling the dry deposition velocity of highly soluble gases to the sea surface, Atmos. Environ. 22 (1988) 1137–1146.

[32] M.L. Wesely, Parameterization of surface resistance to gaseous dry deposition in regional scale numerical models, Atmos. Environ. 23 (1989) 1293–1304.

[33] K.H. Schlünzen, S. Pahl, Modification of dry deposition in a developing sea-breeze circulation – a numerical case study, Atmos. Environ. 26A (1992) 51–61.

[34] C.J. Walcek, R.A. Brost, J.S. Chang, SO_2, sulfate and HNO_3 deposition velocities computed using regional landuse and meteorological data, Atmos. Environ. 20 (1986) 949–964.

[35] N. Chaumerliac, E. Richard, J.-P. Pinty, E.C. Nickerson, Sulfur scavenging in a mesoscale model with quasi-spectral microphysics: two-dimensional results for continental and maritime clouds, J. Geophys. Res. 92 (1987) 3114–3126.

[36] A.I. Flossmann, H.R. Pruppacher, A theoretical study of the wet removal of atmospheric pollutants. Part III: the uptake, redistribution, and deposition of $(NH_4)_2SO_4$ particles by a convective cloud using a two-dimensional cloud dynamics model, J. Atmos. Sci. 45 (1988) 1857–1871.
[37] S. Wurzler, P. Respondek, A.I. Flossmann, H.R. Pruppacher, Simulation of the dynamics, microstructure and cloud chemistry of a precipitating and a non-precipitating cloud by means of a detailed 2-D cloud model, Beitr. Phys. Atmos. 67 (1994) 313–320.
[38] D.K. Lilly, J.B. Klemp, The effects of terrain shape on nonlinear hydrostatic mountain waves, J. Fluid Mech. 95 (1979) 241–261.
[39] J.E. Simpson, D.A. Mansfield, J.R. Milford, Inland penetration of sea-breeze fronts, Quart. J. Roy. Meteorol. Soc. 103 (1977) 47–76.
[40] R.H. Clark, A.J. Dyer, P.R. Brook, D.G. Reid, A.J. Troup, The wangara experiment: boundary layer data. Technical Paper 19, CSIRO Division of Atmospheric Research, Mordialloc, 1971, pp. 340.
[41] T. Yamada, G. Mellor, A simulation of the Wangara atmospheric boundary layer data, J. Atmos. Sci. 32 (1975) 2309–2329.
[42] J.C. André, G. DeMoor, P. Lacarrére, G. Therry, R. DuVachat, Modelling the 24-hour evolution of the planetary boundary layer, J. Atmos. Sci. 35 (1978) 1861–1883.
[43] J.R. Mahfouf, E. Richard, P. Mascard, E.C. Nickerson, R. Rosset, A comparative study of various parameterizations of the Planetary Boundary Layer in a numerical mesoscale model, J. Climate 26 (1987) 1671–1695.
[44] F. Pasquill, F.B. Smith, Atmospheric Diffusion, Ellis Horwood, Chichester, 1983, pp. 437.
[45] L.C.J. Van de Berg, J. Oerlemans, Simulation of the sea-breeze front with a model of moist convection, Tellus 37A (1985) 30–40.
[46] K.H. Schlünzen, U. Krell, Mean and local transport in air, in: J. Sündermann (Ed.), Circulation and Contaminant Fluxes in the North Sea, Springer, Berlin, 1994, pp. 317–344.

ELSEVIER

Journal of Wind Engineering
and Industrial Aerodynamics 67 & 68 (1997) 493–506

JOURNAL OF wind engineering AND industrial aerodynamics

A numerical study of stably stratified flows over a two-dimensional hill – Part I. Free-slip condition on the ground

T. Uchida[a,*], Y. Ohya[b]

[a] *Interdisciplinary Graduate School of Engineering Sciences, Kyushu University, Kasuga 816, Japan*
[b] *Research Institute for Applied Mechanics, Kyushu University, Kasuga 816, Japan*

Abstract

Stably stratified flows over a two-dimensional hill in a channel of finite depth are analyzed numerically by using a newly-developed multi-directional finite-difference method at a Reynolds number Re = 2000. To simplify the phenomena occurring in the flow around the hill, the free-slip condition for the velocity is assumed on the ground, and the nonslip condition is imposed only on the hill surface. Attention is focused on the unsteadiness in the flow around the hill for the cases of $K(= NH/\pi U) > 1$ where N and U are the buoyancy frequency and free-stream velocity and H is the domain depth. The flow unsteadiness is discussed, being associated with shedding of the upstream advancing columnar disturbance.

Keywords: Direct numerical simulation; Finite-difference method; Stably stratified flow; Two-dimensional hill

1. Introduction

In the atmospheric boundary layer, buoyancy force associated with density stratification plays a key role. It damps or enhances vertical motion directly in the stratified flows. When a stably stratified fluid flows over topography, vertically disturbed air parcels can lead to generation of downstream lee waves, which cause drastic changes in the flow pattern.

It is well known from laboratory experiments (see, e.g. Refs. [1,2]) that, in the linearly stratified flow of finite depth, columnar disturbances, which have discrete vertical modes, propagating upstream of an obstacle always appear when lee

* Corresponding author.

PII S0167-6105(97)00096-2

waves exist. The columnar disturbance is a manifestation of the long internal wave which causes an almost horizontal motion relative to the basic flow. According to the linear theory, for stratified fluid of finite depth H, the flow is characterized by the parameter $K\ (= NH/\pi U)$ where N and U are the buoyancy frequency and free-stream velocity. For weak stratification (i.e. $K \leqslant 1.0$), all modes have velocities less than free-stream velocity and are swept downstream with no lee waves, while for strong stratification (i.e. $K > 1.0$), stationary lee waves can form and columnar modes propagate upstream.

For stably stratified flows over a two-dimensional obstacle in a channel of finite depth, there are many findings from both experiments [1,2] and numerical simulations [3–7]. The most striking feature of those findings is the unsteadiness of the flow around an obstacle for strong stratification. Castro et al. [2] found in their experiments that the obstacle drag in linearly stratified flows shows a persistent periodical oscillation under a certain stratification. Similar phenomena have been confirmed in some numerical calculations under similar situations [4–7]. As to the mechanism of the flow unsteadiness, various discussions have been given [4–7], however, it still remains unclear.

The numerical simulations reported above have been made at low Reynolds numbers only as high as Re = 100. Therefore, it is difficult to compare those results with experimental ones (Re = 10^3–10^4) directly. We have tried to clarify the effect of stable stratification on unsteady, separated and reattaching flows behind an obstacle at higher Reynolds numbers both experimentally [8] and numerically [8,9]. In general, for higher Re, the flow over an obstacle becomes more complex owing to the unsteady separated and reattaching flow behind the obstacle. In the present numerical study, we have investigated the linearly stratified flows over a two-dimensional hill at Re = 2000 by using a newly-developed multi-directional finite-difference method. Particular emphasis is given to the mechanism of flow unsteadiness around a hill for strong stratification (i.e. $K > 1.0$). Therefore, to simplify the phenomena occurring in the flow around a hill, the free-slip condition for the velocity is assumed on the ground, and the nonslip condition is imposed only on the hill surface.

2. Numerical model

2.1. Physical domain and computational grid

We consider a linearly stratified flow of incompressible and nondiffusive fluid past a two-dimensional hill in a channel of finite depth. A typical physical domain is shown in Fig. 1. The inflow boundary is set sufficiently far away to delay the arrival of upstream reflection ($x = -420h$ for all cases). A cosine function is chosen for the geometry of the hill with its profile given by $h(x) = 0.5\{1 + \cos(\pi x/a)\}$, where the width parameter a is set equal to 1. In order to simulate the flow around the hill with high accuracy, a body-fitted coordinate system is employed. The number of grid points in the x- and z-directions are 421×101 for all cases. Moreover, we have studied

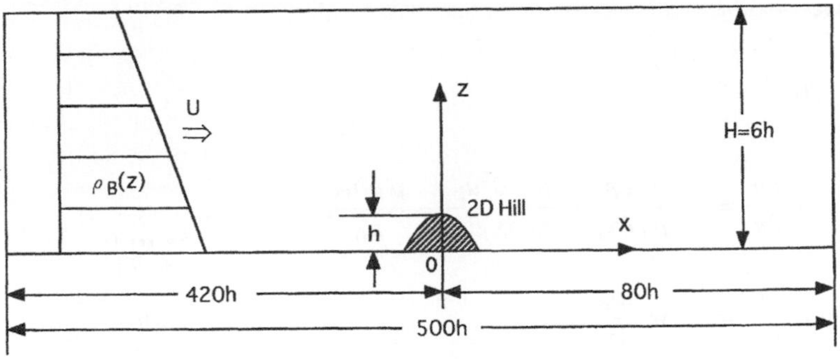

Fig. 1. Physical domain.

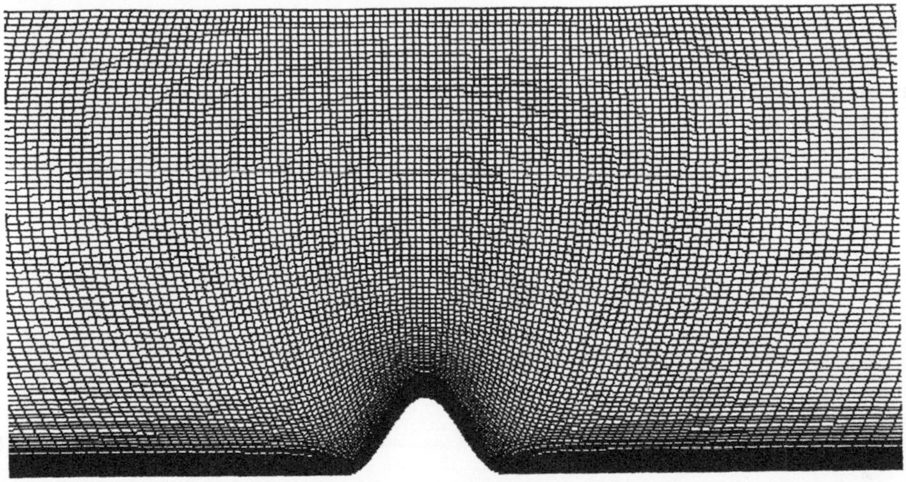

Fig. 2. Computational grid near the hill.

the other two grid systems which have 351 × 51 and 421 × 76 points, respectively. The changes in the grid resolution were found to have negligible effect on the numerical results, therefore, we expect that the present calculations are free from grid dependence. The computational grid near the hill is shown in Fig. 2. The grid points in the physical domain are concentrated toward the hill surface and the ground. The vertical smallest grid spacing is $3.0 \times 10^{-3}h$ at the hill top.

2.2. Governing equations

Under the Boussinesq approximations, the dimensional governing equations consist of the continuity equation, the Navier–Stokes equations, and the density equation as follows,

$$\frac{\partial u_i}{\partial x_i} = 0, \tag{1}$$

$$\frac{\partial u_i}{\partial t} + u_j \frac{\partial u_i}{\partial x_j} = -\frac{1}{\rho_0}\frac{\partial p'}{\partial x_i} + \frac{\mu}{\rho_0}\frac{\partial^2 u_i}{\partial x_j \partial x_j} - \frac{\rho' g \delta_{i3}}{\rho_0}, \tag{2}$$

$$\frac{\partial \rho'}{\partial t} + u_j \frac{\partial \rho'}{\partial x_j} = -w\frac{\mathrm{d}\rho_\mathrm{B}}{\mathrm{d}z}, \tag{3}$$

ρ', p' are the perturbation density and pressure defined as

$$\rho = \rho_\mathrm{B}(z) + \rho', \qquad p = p_\mathrm{B}(z) + p', \tag{4}$$

where $\rho_\mathrm{B}(z)$ and $p_\mathrm{B}(z)$ are the undisturbed distributions at the inflow boundary, and $\rho_\mathrm{B}(z)$ decreases upward linearly, $u_i = (u, w)$ is the velocity, ρ_0 is the reference density, g is the acceleration due to gravity, and μ is the viscosity coefficient.

Nondimensionalizing the variables using the free-stream velocity U, the hill height h, and the reference density ρ_0, we have three dimensionless equations as follows:

$$\frac{\partial u_i}{\partial x_i} = 0, \tag{5}$$

$$\frac{\partial u_i}{\partial t} + u_j \frac{\partial u_i}{\partial x_j} = -\frac{\partial p}{\partial x_i} + \frac{1}{\mathrm{Re}}\frac{\partial^2 u_i}{\partial x_j \partial x_j} - \frac{\rho}{\mathrm{Fr}^2}\delta_{i3}, \tag{6}$$

$$\frac{\partial \rho}{\partial t} + u_j \frac{\partial \rho}{\partial x_j} = w. \tag{7}$$

These equations include two dimensionless parameters, i.e., the Reynolds number Re ($= \rho_0 Uh/\mu$) and the Froude number Fr ($= U/Nh$), where N is the buoyancy frequency defined as $N^2 = -(g/\rho_0)(\mathrm{d}\rho/\mathrm{d}z)$. It should be noted that for the finite depth flow not only Re and Fr but also K ($= NH/\pi U$), which contains the domain depth H, is a parameter necessary to determine the character of the internal gravity wave. Since K is rewritten as $K = H/\pi h\mathrm{Fr}$, it is determined by H/h when Fr is a known number. In other words, K reflects the effect of boundary conditions though it does not appear explicitly in the governing equations. We use the parameter K as a stability parameter in this study.

2.3. *Finite-difference method*

Through a coordinate transformation ($x = x(\xi, \zeta)$, $z = z(\xi, \zeta)$), the governing equations (Eqs. (5)–(7)) are solved in the computational domain using a finite-difference

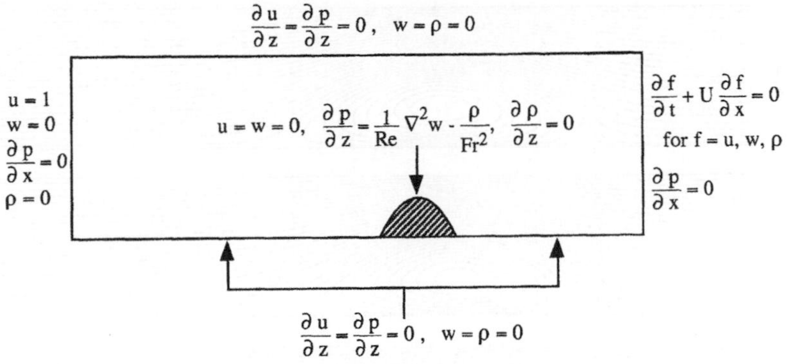

Fig. 3. Boundary conditions.

method. The numerical method used is an extension of the MAC method to the incompressible stratified flow, and the Poisson equation for pressure derived by taking divergence of Eq. (6) is solved by the SOR method. The regular grid arrangement for variables is employed. The Euler explicit method with first-order accuracy is used for time advance in Eqs. (6) and (7). All of the spatial derivatives except for the convective terms are approximated by a second-order central scheme. For the convective terms, a third-order upwind scheme is employed to minimize the numerical dissipation of velocities and density. Moreover, for all the spatial derivatives, a newly-developed multi-directional finite-difference method [10], which leads to a calculation of much higher accuracy, is used.

The boundary conditions are shown in Fig. 3. The condition $\partial u/\partial z = 0$ on the upper boundary and the ground except for the hill in Fig. 3 is the free-slip condition. In general, understanding the evolving flow field is complicated by a number of factors, particularly in boundary layer phenomena. To avoid such a troublesome boundary layer effect, which usually appears in laboratory experiments, we chose the free-slip condition on the ground both upstream and downstream ($|x| > a$) of the hill. The nonslip condition is imposed only on the hill surface ($|x| \leqslant a$). Impulsive-start ($u = 1, w = 0, p = 0, \rho = 0$) is employed for an initial condition. We calculate the flows at a Reynolds number Re = 2000 for a wide range of K ($0 \leqslant K \leqslant 3.0$). The nondimensional time step is 2×10^{-3}.

3. Results and discussions

3.1. Weak stratification ($0 \leqslant K \leqslant 1.0$)

Fig. 4 shows the instantaneous streamlines around the hill for weak stratification cases of $0 \leqslant K \leqslant 1.0$ at a nondimensional time $t = 200$. A stationary vortex behind the hill is observed and its length is shortened as K increases. It should be noted that for the cases of $K = 0$–0.8, a stationary vortex becomes longer gradually as time

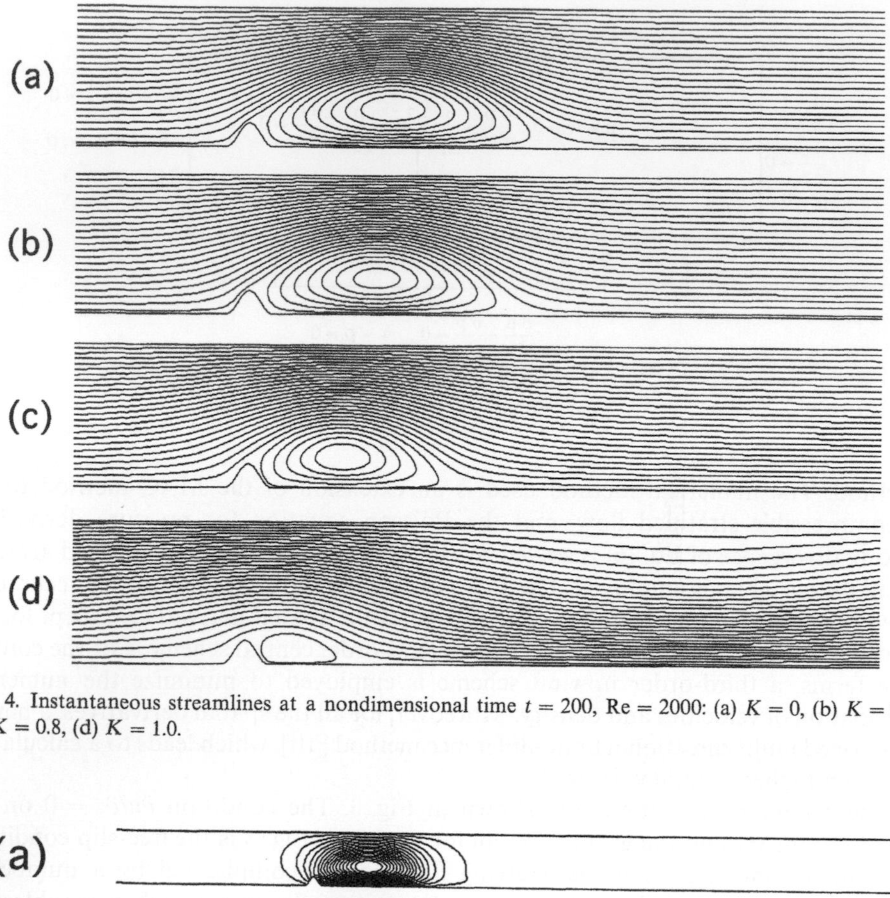

Fig. 4. Instantaneous streamlines at a nondimensional time $t = 200$, Re = 2000: (a) $K = 0$, (b) $K = 0.5$, (c) $K = 0.8$, (d) $K = 1.0$.

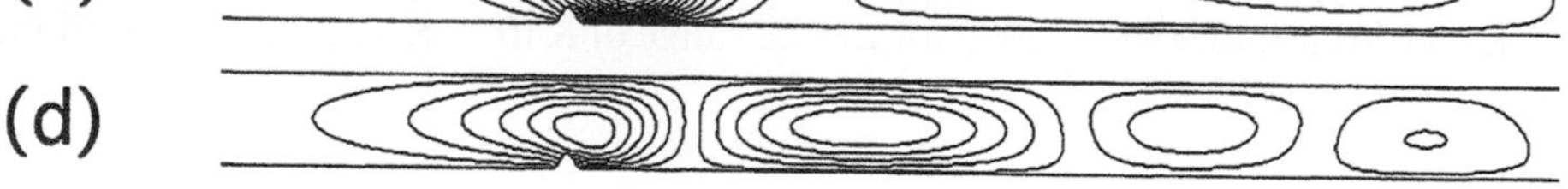

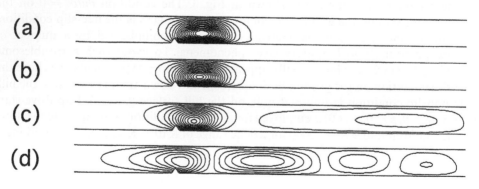

Fig. 5. Perturbation streamlines $\Delta\psi$ $(-4 \leqslant \Delta\psi \leqslant 2)$ at $t = 200$, Re = 2000: (a) $K = 0$, (b) $K = 0.5$, (c) $K = 0.8$, (d) $K = 1.0$.

proceeds. For $K \geqslant 0.8$, stationary lee waves with long wavelength appear (these can be observed more clearly in Fig. 5c and Fig. 5d), but for the case of $K = 0.8$, this lee wave disappears at $t = 500$. Thus, the stratification acts largely to modify the separated wake due to the generation of lee waves.

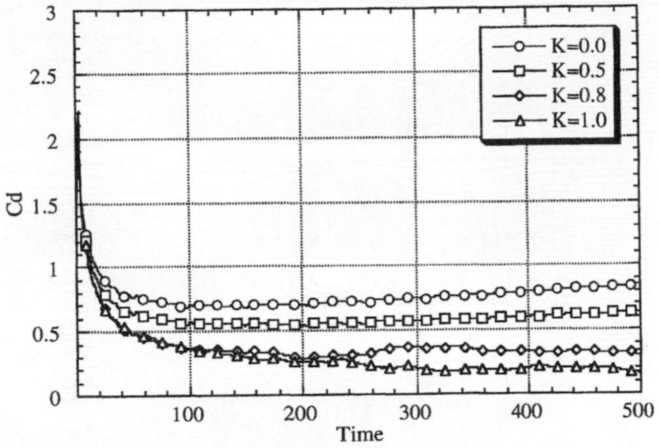

Fig. 6. Time development of drag coefficient C_d for the cases of $0 \leqslant K \leqslant 1.0$, Re = 2000.

Fig. 5 shows the perturbation streamlines ($\Delta\psi$), which are depicted by the perturbation velocities u' and w', i.e. $u' = U - u$, $w' = w$, at $t = 200$. These perturbation streamlines make clear the existence of lee waves and columnar disturbances propagating upstream of the hill. In Fig. 5c and Fig. 5d, closed perturbation streamlines, which mean the existence of lee waves, are observed in the downstream of the hill. In Fig. 5a–5d, systematic upstream propagations of columnar disturbances are not yet observed, although extended upstream disturbances are observed for the case of $K = 1.0$ in Fig. 5d.

Time development of drag coefficient C_d over a period of integration $0 \leqslant t \leqslant 500$ is shown in Fig. 6. The behavior of C_d for all cases suggests that the flow around the hill for weak stratification cases of $0 \leqslant K \leqslant 1.0$ reaches an almost steady condition. But exactly speaking, these C_d values never reach a constant value, reflecting gradual elongation of the size of the stationary vortices.

The stable stratification effects on the flow over a hill for weak stratification cases of $0 \leqslant K \leqslant 1.0$ at a Reynolds number of 2000 are almost similar to those at low Reynolds numbers (Re $\leqslant$ 100) of Castro [3], Hanazaki [4,5] and Paisley et al. [7].

3.2. *Strong stratification ($1.0 < K \leqslant 3.0$)*

Fig. 7 shows the instantaneous streamlines around the hill for strong stratification cases of $1.0 < K \leqslant 3.0$. The lee wavelength is gradually shortened as K increases. The first upward flow in the lee wave motion induces a rotor (secondary separation) on the downstream ground. We can see another rotor on the upper boundary, the location of which gradually approaches $x = 0$ as the lee wavelength is shortened, as seen in Fig. 7a–7d. These rotors seem to behave as a downstream obstacle. Moreover, for the cases of $K = 2.5$ and 3.0 in Fig. 7e and Fig. 7f, "lee wave-breaking" can be observed.

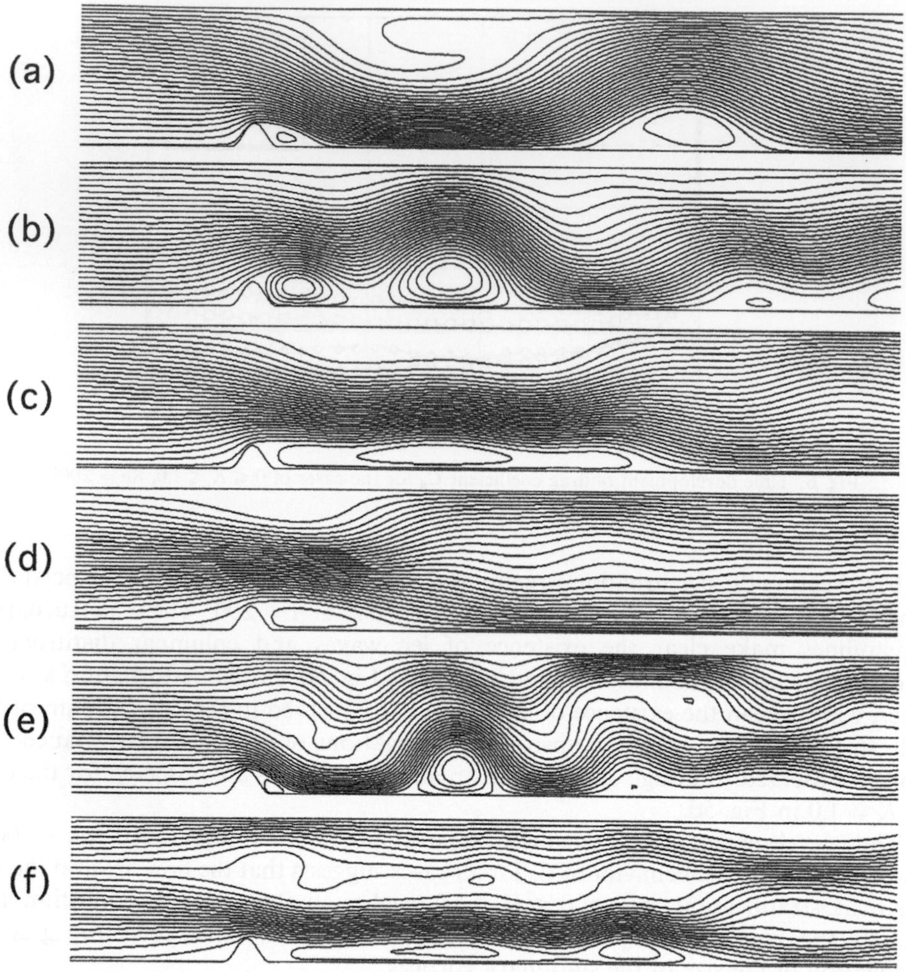

Fig. 7. Instantaneous streamlines at $t = 200$, Re = 2000: (a) $K = 1.25$, (b) $K = 1.5$, (c) $K = 1.75$, (d) $K = 2.0$, (e) $K = 2.5$, (f) $K = 3.0$.

Fig. 8 shows the perturbation streamlines ($\Delta\psi$). The wave of mode $n = 1$ begins to propagate upstream in the form of a columnar disturbance. The "detaching" of the eddy is seen since the strong and weak part appear, in turn, in each mode of upstream advancing columnar disturbances. As K becomes larger, the mode $n = 1$ eddy is detached successively at a shorter period. When $K \geqslant 1.75$ in Fig. 8c–8e, the columnar disturbances with mode $n = 2$ also propagate upstream of the hill and when $K = 3.0$ in Fig. 8f, those with mode $n = 3$ are seen clearly. It should be noted that the columnar disturbances with mode $n = 1$ generated as "detaching eddies" have only the clockwise circulation for the case of $K = 1.25$, while for the cases of $K \geqslant 1.5$, they have both clockwise and counter-clockwise circulations. A detaching eddy (columnar

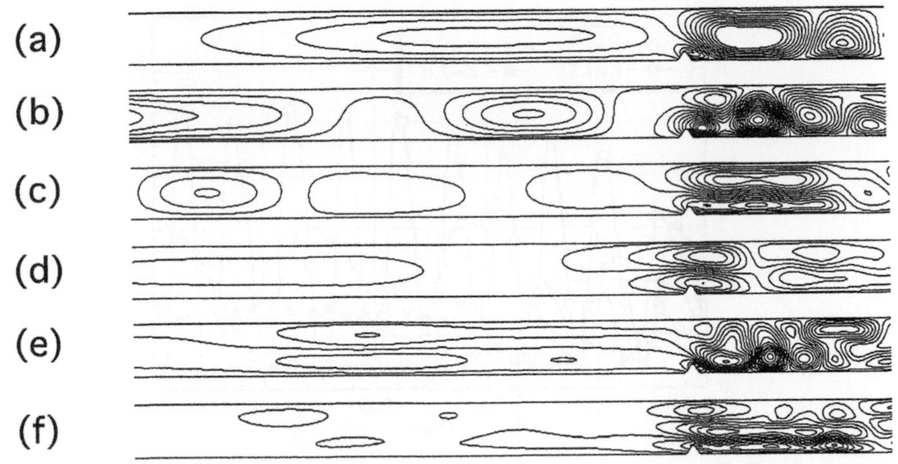

Fig. 8. Perturbation streamlines $\Delta\psi$ $(-4 \leqslant \Delta\psi \leqslant 2)$ at $t = 200$, Re = 2000: (a) $K = 1.25$, (b) $K = 1.5$, (c) $K = 1.75$, (d) $K = 2.0$, (e) $K = 2.5$, (f) $K = 3.0$.

disturbance) with clockwise circulation is generated from below around the hill, on the other hand, the one with counter-clockwise circulation is generated from the above near the upper boundary. The origin of the detaching eddy (columnar disturbance) from the above seems to be a rotor on the upper boundary when it is generated very close to $x = 0$ by a short lee wavelength.

According to the linear theory, the propagation speed of the columnar disturbance (eddy) with mode $n = 1$ is given by $(K - 1)U$. The value obtained by the present calculation is consistent with the prediction of the linear theory for the cases of $K \geqslant 1.25$.

Fig. 9 shows the time development of C_d for the cases of $1.25 \leqslant K \leqslant 2.0$ over a period of integration $0 \leqslant t \leqslant 500$. The most striking feature in Fig. 9 is the persistent periodical C_d oscillations for the cases of $K = 1.25$ and 1.5, while for the cases of $K = 1.75$ and 2.0, these oscillations rapidly decay after $t = 200$. These C_d oscillations suggest that the flow around the hill under strong stratification is intrinsically unsteady. Especially for the case of $K = 1.5$, the amplitude of C_d oscillations is very large. These features of C_d variations in time for the cases of $K = 1.25$–2.0 are very similar to the results of Paisley et al. [7]. We will discuss a possible mechanism of the flow unsteadiness in more detail in the next section.

3.3. *Discussions of the flow unsteadiness ($1.0 < K \leqslant 2.0$)*

Hanazaki [4,5] asserted that the C_d oscillations were the result of the columnar disturbance (eddy) detaching from the obstacle. Castro et al. [2] suggested that the C_d oscillations were driven by the changes in the effective value of K upstream of the obstacle as a consequence of the changes in the strength of the columnar disturbances. Lamb [6] stated that no upstream columnar disturbances which permanently modify

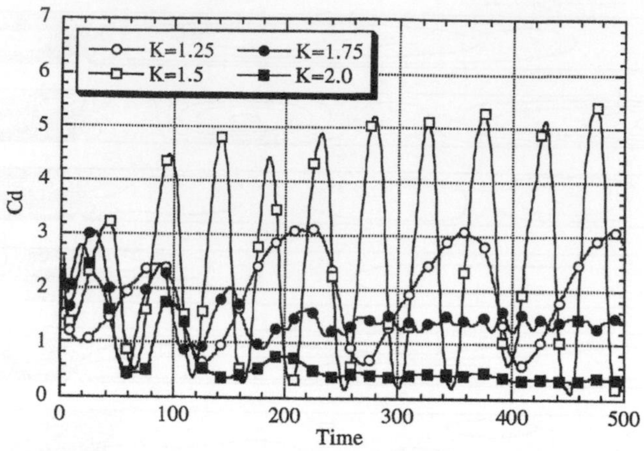

Fig. 9. Time development of C_d for the cases of $1.25 \leqslant K \leqslant 2.0$, Re = 2000.

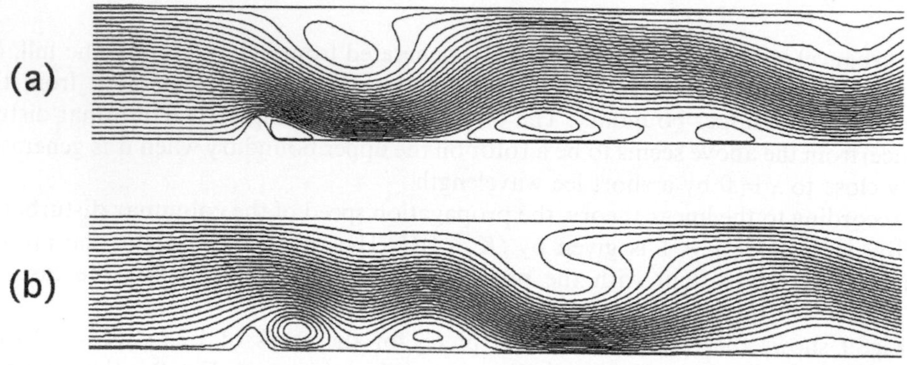

Fig. 10. Instantaneous streamlines for the case of $K = 1.5$, Re = 2000: (a) high-C_d state ($t = 425$), (b) low-C_d state ($t = 450$).

the upstream flow conditions were observed even though the C_d oscillations were seen. Paisley et al. [7] suggested that the unsteady behavior was the result of nonlinear processes.

We consider the flow unsteadiness for the cases of $1.25 \leqslant K \leqslant 2.0$ as follows. Fig. 10a and Fig. 10b show the instantaneous streamlines at high- and low-C_d states, respectively, for the case of $K = 1.5$. There is a large difference in the flow pattern between the two states. In a high-C_d state, we can see large-amplitude lee waves over the hill, while in a low-C_d state, these are small. Correspondingly, a small recirculating eddy behind the hill is induced in a high-C_d state, while a large one in a low-C_d state. Fig. 11a and Fig. 11b show the $\Delta\psi$ patterns corresponding to high- and low-C_d states, respectively. We should note that the columnar disturbance (eddy) with mode $n = 1$

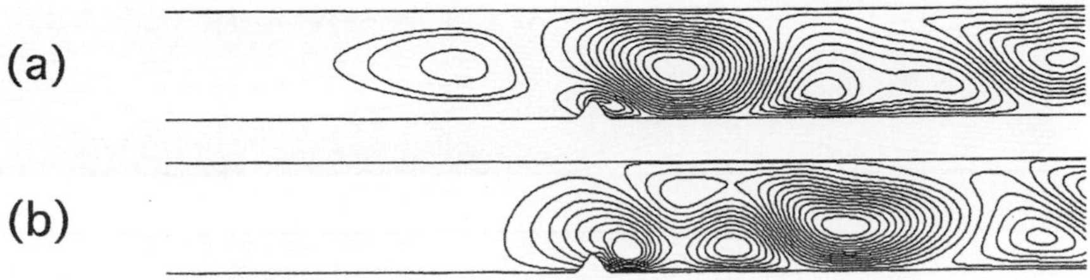

Fig. 11. Perturbation streamlines $\Delta\psi$ ($-4 \leqslant \Delta\psi \leqslant 2$) for the case of $K = 1.5$, Re = 2000: (a) high-C_d state ($t = 425$), (b) low-C_d state ($t = 450$).

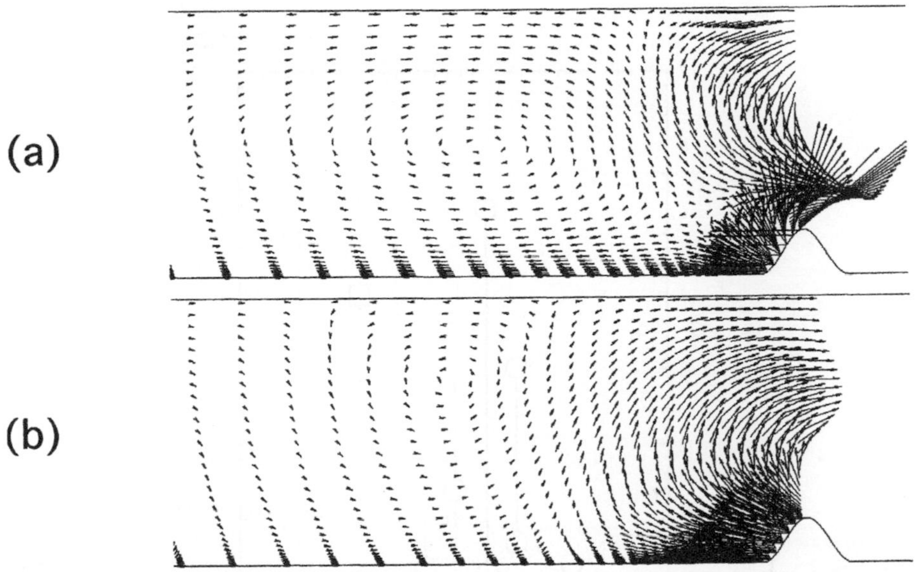

Fig. 12. Perturbation velocity vectors for the case of $K = 1.5$, Re = 2000: (a) high-C_d state ($t = 425$), (b) low-C_d state ($t = 450$).

has just detached from the hill in a high-C_d state, while the one is just about to detach from the hill in a low-C_d state. To examine the perturbation-flow field in more detail, the corresponding perturbation-velocity vectors near the hill are displayed in Fig. 12a and Fig. 12b. First, it can be observed that the columnar disturbance (eddy) with mode $n = 1$ has a clockwise circulation. Therefore, when the columnar disturbance (eddy) has just detached from the hill, it induces a downward flow in front of the hill as seen in Fig. 12a, on the other hand, when the columnar disturbance (eddy) is just about to detach from the hill, it induces an upward flow in front of the hill as seen in Fig. 12b. These downward and upward flows generated by the columnar disturbance (eddy) affect the approaching flow directly; the former tends to press down the approaching flow and the latter tends to lift it up. As a result, the curvature of the

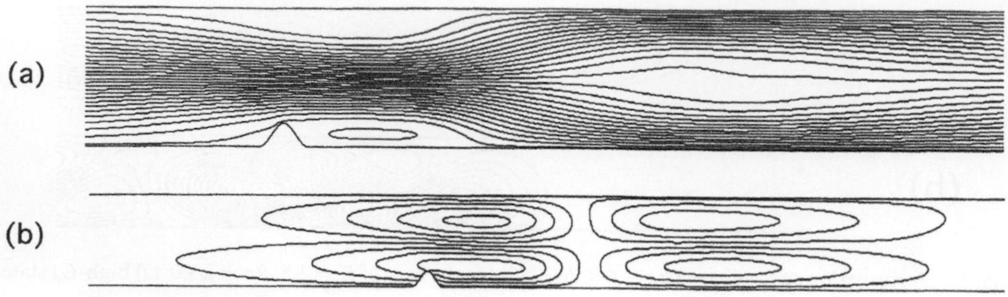

Fig. 13. Flow around the hill reached a steady condition for the case of $K = 2.0$ at $t = 500$, Re = 2000: (a) instantaneous streamlines, (b) perturbations streamlines $\Delta\psi$ ($-4 \leqslant \Delta\psi \leqslant 2$).

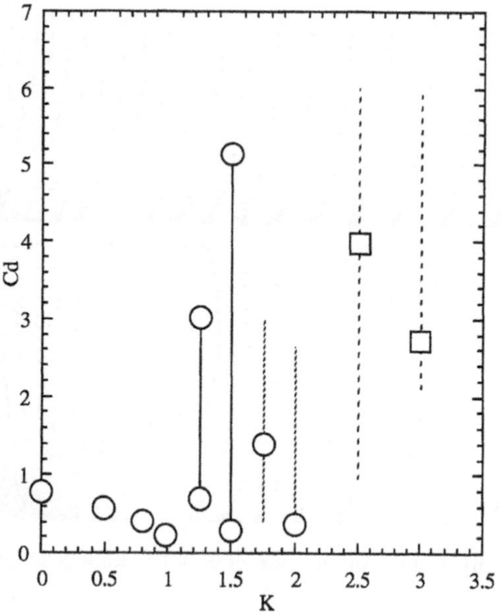

Fig. 14. Variations of C_d with K, Re = 2000. Full and dotted lines indicate persistent and decaying C_d oscillations, respectively. The behavior of C_d for $K = 2.5$ and 3.0 are still unclear.

separated and reattaching shear layer behind the hill becomes large as seen in Fig. 10a, while it becomes small as seen in Fig. 10b. Since the base pressure of the hill can be directly influenced by the curvature of the separated and reattaching shear layer, C_d becomes a high value due to the large curvature and a low value due to the small curvature. Thus, the alternating states of high- and low-C_d are caused by the shedding of the columnar disturbance (eddy) in the upstream of the hill. Accordingly, the period of C_d oscillations in Fig. 9 is consistent with that of the shedding of the columnar disturbances.

Next, we consider the reason of the decay of C_d oscillations for the cases of $K = 1.75$ and 2.0. For these cases, the columnar disturbance with mode $n = 2$ becomes dominant, for example, as shown in Fig. 13a and Fig. 13b for the case of $K = 2.0$. This leads to the flow around the hill to be a symmetric one with the horizontal center axis of finite domain. Therefore, as time proceeds, the flow around the hill tends to reach a steady condition for the cases of $K = 1.75$ and 2.0.

3.4. *Variation of C_d with $K (0 \leqslant K \leqslant 3.0)$*

Fig. 14 shows the variation of C_d with K for the cases of $0 \leqslant K \leqslant 3.0$. We can see that C_d decreases locally at integral values of $K = 1.0$, 2.0 and 3.0. The overall trend of C_d is similar to the previous numerical studies [4, 5, 7–9].

4. Conclusions

We have performed numerical calculations on the stratified flows over a two-dimensional hill in a channel of finite depth at $Re = 2000$ for a wide range of K $(0 \leqslant K \leqslant 3.0)$. To simplify the phenomenon occurring in the flow around the hill, the free-slip condition for the velocity is assumed on the ground, and the nonslip condition is imposed only on the hill surface. The major conclusions of the present study can be summarized as follows:

(1) As K increases, the recirculating eddy behind the hill is suppressed and its length is shortened. For the cases of $K \geqslant 0.8$, lee waves appear and affect strongly the flow behind the hill. The lee wavelength is shortened as K increases. For the cases of $K \geqslant 1.25$, the lee wave motion induces a rotor (secondary separation) on the downstream ground and another rotor on the upper boundary.

(2) From the perturbation streamlines ($\Delta\psi$), which are depicted by the perturbation velocities, the upstream advancing columnar disturbances (eddy) with mode $n = 1$ are observed when lee waves exist. When $K \geqslant 1.75$, the columnar disturbances with mode $n = 2$ also propagate upstream of the hill and when $K = 3.0$, those with mode $n = 3$ are seen clearly. The propagation speeds obtained by the present calculations are consistent with the prediction of the linear theory for the cases of $K \geqslant 1.25$.

(3) For the cases of $0 \leqslant K \leqslant 1.0$, the time series of the drag coefficient C_d suggest that the flow around the hill under weak stratification reaches an almost steady condition. On the other hand, the persistent periodical C_d oscillations for the cases of $K = 1.25$ and 1.5 are observed, and for the cases of $K = 1.75$ and 2.0 the C_d oscillations rapidly decay after $t = 200$.

(4) For the cases of $K = 1.25$ and 1.5, the flow unsteadiness associated with the periodical variation of C_d is caused by the "shedding" of the columnar disturbance (eddy) from the hill. A high-C_d state corresponds to the situation where the columnar disturbance (eddy) has just detached from the hill, while a low-C_d state corresponds to the situation where the columnar disturbance (eddy) is just about to detach from the hill.

(5) For the cases of $K = 1.75$ and 2.0, the flow around the hill reaches a steady condition because of the appearance of the columnar disturbance with mode $n = 2$.

Acknowledgements

We would like to thank the referees for their helpful suggestions which have been incorporated in the final manuscript. Thanks are also due to Dr. Ozono of Research Institute of Applied Mechanics of Kyushu University for his variable discussions in numerical simulations.

References

[1] P.G. Baines, Topographic Effects in Stratified Flows, Cambridge University Press, Cambridge, 1994, pp. 224–336.

[2] I.P. Castro, W.H. Snyder, P.G. Baines, Obstacle drag in stratified flow, Proc. Roy. Soc. London A 429 (1990) 119–140.

[3] I.P. Castro, Effects of stratification on separated wakes: part I. Weak static stability, Proc. 3rd IMA Meetings on Stably Stratified flows, Leeds, 1989.

[4] H. Hanazaki, Upstream advancing columnar disturbances in two-dimensional stratified flow of finite depth, Phys. Fluids A 1 (1989) 1976–1987.

[5] H. Hanazaki, Drag coefficient and upstream influence in three-dimensional stratified flow of finite depth, Fluid Dynamics Res. 4 (1989) 317–332.

[6] K.G. Lamb, Numerical simulations of stratified inviscid flow over a smooth obstacle, J. Fluid Mech. 260 (1994) 1–22.

[7] M.F. Paisley, I.P. Castro, N.J. Rockliff, Steady and unsteady computations of strongly stratified flows over a vertical barrier, in: Stably Stratified Flows: Flow and Dispersion over Topography, Clarendon, Oxford University Press, Oxford, 1994, pp. 39–59.

[8] Y. Ohya et al., Stratified flow over a 2-D semicircular cylinder in fluid of finite depth, Proc. 12th Japan Wind Engng Symp., 1992, pp. 13–18 (in Japanese).

[9] T. Uchida et al., A numerical study of stably stratified flows over a two-dimensional hill – Part I. Free-slip condition on the ground, Proc. 9th Japan CFD Symp., 1995, pp. 465–466 (in Japanese).

[10] H. Suito, K. Ishii, Simulation of dynamic stall by multi-directional finite- difference method, AIAA 95-2264, 1995.

Keynote Presentation

ELSEVIER

Journal of Wind Engineering
and Industrial Aerodynamics 67&68 (1997) 509–532

JOURNAL OF
wind engineering
AND
industrial
aerodynamics

Computational wind engineering: Past achievements and future challenges

Theodore Stathopoulos[1]

Centre for Building Studies, Concordia University, Montreal, Que., Canada H3G 1M8

Abstract

The paper reviews the current state of the art in computational wind engineering, particularly as it relates to applications of numerical flow modelling for the evaluation of wind effects on buildings and their environment. The variability of computational results is presented and compared with that of wind tunnel measurements. Concerns are expressed regarding the current application of the numerical approach in the design practice in cases for which the computational results may not be adequate. Future challenges regarding the improvement of computational wind engineering methodologies are discussed and the importance of identifying resolution and numerical errors is emphasized.

1. Introduction

Current practice for determining wind velocities, building cladding pressures and overall wind-induced loads or dispersion of pollutants around building configurations is mainly by measurements in boundary layer wind tunnels simulating the main features of atmospheric flow conditions. For instance, much of the wind-load information contained in wind standards and building codes of practice has been obtained from research carried out by physical modelling. Nevertheless, research in boundary layer wind tunnels has its own limitations. These include but are not limited to the inability to fully model swirling flow impacts on structures (tornadoes) and some types of gust fronts (downbursts); the Reynolds number limitations in accurate modelling wind flows over curved surfaces or flows inside buildings such as natural ventilation or diffusion of pollutants within a partially open enclosure; and the inability to model wind flows in atmospheric boundary layers with various stability conditions.

In the framework of Computational Wind Engineering (CWE), which deals with the application of Computational Fluid Dynamics (CFD) methodologies in the

[1] E-mail: statho@cbs-engr.concordia.ca.

0167-6105/97/$17.00
PII S0167-6105(97)00097-4

classical wind engineering and building aerodynamics problems, numerical solutions have been developed with the potential to overcome these limitations. However, such solutions have not yet fully demonstrated their ability even for simple building geometries exposed to strong horizontal winds. On the other hand, numerical prediction of fluid flows and their effects on solid bodies has developed to the point where commercial programs are available for use in many cases. These programs have had limited application to wind engineering because of the difficulties in the modelling of flow conditions, the complex building configurations and the large area of model domain required to define the geometry of the surroundings. Such factors translate into fine computational grids and large computer memory and time requirements.

The area of CWE should not be confused with the area of Computer-Aided Wind Engineering (CAWE), which refers to the development and utilization of software to be used either directly or indirectly for wind engineering applications (design, measurements, wind condition assessment, simulation, etc.). CAWE applications

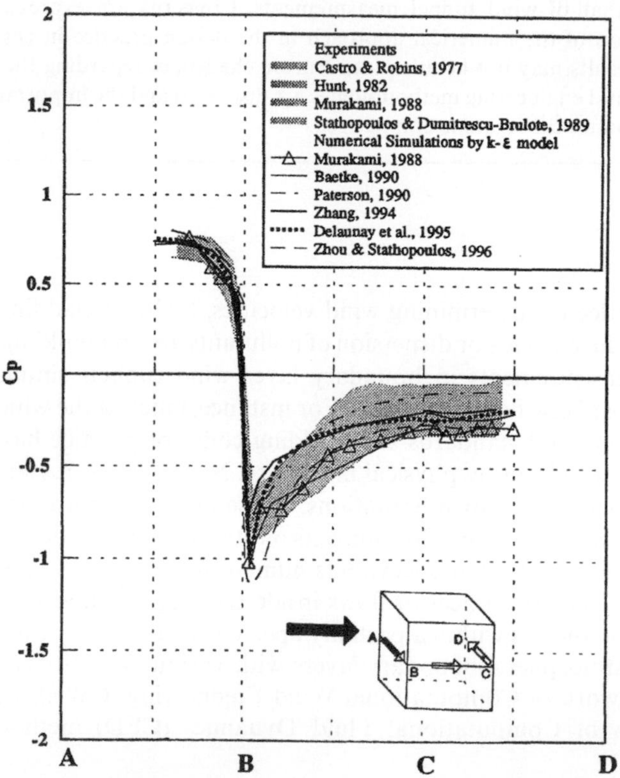

Fig. 1. Comparison of mean wind pressure coefficients on a cubic building for suburban exposure: experiments and numerical simulations by using the k–ε model – horizontal plane.

include the development of computerized tools, e.g. knowledge-based expert systems and neural networks for the assessment of pedestrian-level wind conditions or the loads on buildings under the influence of adjacent buildings; the software recently developed for the utilization of the ASCE-7/95 wind load standard [1]; the software for the development of time series to simulate wind speed or pressure signals for applications to extreme value analysis, fatigue design or codification of wind effects; the software for the automation of wind tunnel or full scale measurements of wind effects, etc.

The paper reviews some of the accomplishments in the numerical evaluation of wind effects on buildings and their environment and presents them in the context of variability of wind tunnel measurement results. CWE applications in specific areas are presented and analyzed whereas future challenges regarding the improvement of current approaches are discussed. The importance of turbulence modelling for wind engineering applications is emphasized.

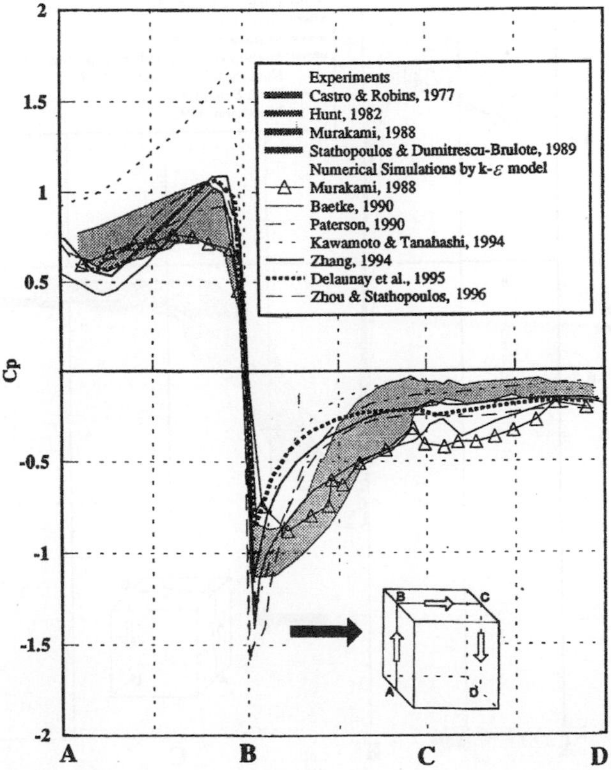

Fig. 2. Comparison of mean wind pressure coefficients on a cubic building for suburban exposure: experiments and numerical simulations by using the k–ε model – vertical plane.

2. Computational evaluation of wind pressures on buildings

The great majority of applications regarding the numerical evaluation of wind pressures on buildings refers to the basic *cubic* shape exposed to wind perpendicular to its face. This is easily explained since, in addition to the simplicity of this shape, admittedly there are several experimental studies and results available for cubes. In the present work it has been decided to examine experimental data available from four different studies [2–5]. All these studies refer to the so-called suburban exposure, i.e. exposure B in the North American wind codes and standards; the exponents of the power law velocity profiles range in these studies from 0.21 to 0.27. The experimental results appear as a range of values rather than as individual numbers associated with various points of the building envelope. Presenting the results in this format, it is feasible to compare experimental and numerical scatters and get an appropriate perspective regarding the comparisons from the design point of view.

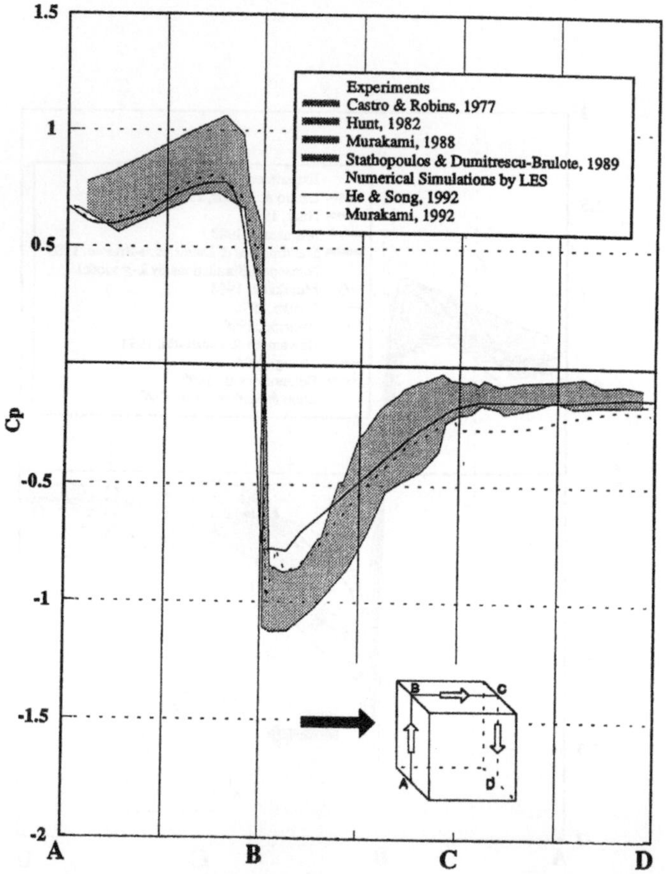

Fig. 3. Comparison of mean wind pressure coefficients on a cubic building for suburban exposure: experiments and numerical simulations by LES model.

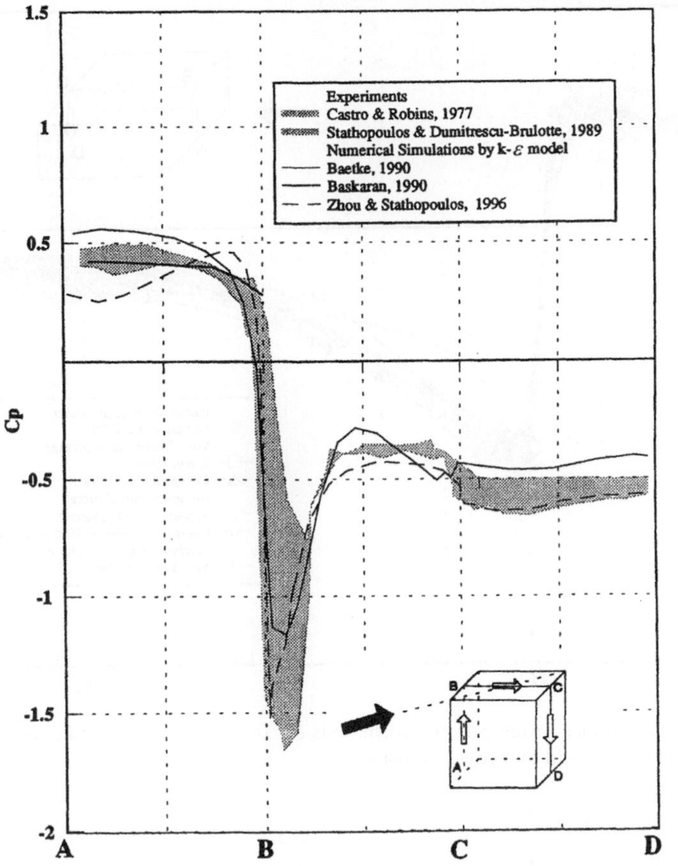

Fig. 4. Comparison of mean wind pressure coefficients on a cubic building for suburban exposure: experiments and numerical simulations by using the k–ε model – oblique wind direction.

Fig. 1 compares mean wind pressure coefficients on the middle horizontal plane of a cube with relevant results from numerical simulations carried out by using the k–ε model. The experimental values are indicated by the shaded zone in the diagram. Numerical results have been obtained by studies carried out in Japan [4], Germany [6], Australia [7], USA [8], France [9] and Canada [10]. The scatter of the numerical results appears similar with that of the experimental data with the exception of the front edge B of the cube, at flow separation in which the numerical results indicate higher suctions. Results for the vertical middle plane of the cube are presented in the same format in Fig. 2. Clearly, the roof and the leeward wall areas appear problematic regarding the numerical results, in which the odd results of Ref. [11] (showing on the windward wall mean pressure coefficient values above 1) have also been included. Recent modifications of the k–ε model [12,13] have improved the discrepancies

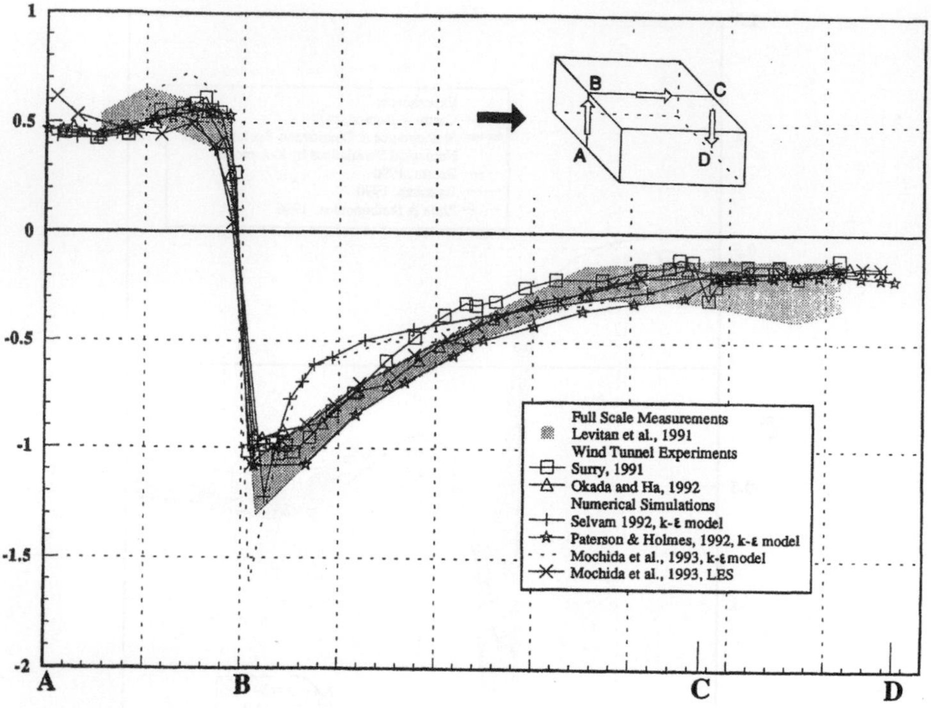

Fig. 5. Comparison of mean wind pressure coefficients on Texas Tech building: full scale measurements, wind tunnel experiments and numerical simulations.

between the experimental and numerical values on the roof, although such adjustments may be of an ad hoc nature and improve the situation only for some particular cases.

Fig. 3 compares experimental and numerical data, the latter obtained by using the large eddy simulation (LES) model which, although it is expected to provide better results, does not show a very significant improvement [14,15]. Limited data for wind blowing at an angle, say 45° to the building face, are presented in Fig. 4 in which the numerical simulations have been carried out by using the k–ε model [6,10,16]. Larger discrepancies in the numerical values appear on the windward wall and the roof areas of the model, while the situation in terms of agreement with the experimental values becomes worse if the data on a vertical plane near the edge of the cube is considered.

It is worth considering the case of the full scale Texas Tech experimental building, since field pressure coefficients are available and have been compared with wind tunnel values. Such comparisons shed a new light in the degree of success of numerical simulations to represent the wind pressure field on the building envelope. Fig. 5 shows full scale measurement results [17], wind tunnel experimental data [18,19], some of which were obtained prior to the full scale measurements, as well as numerical data

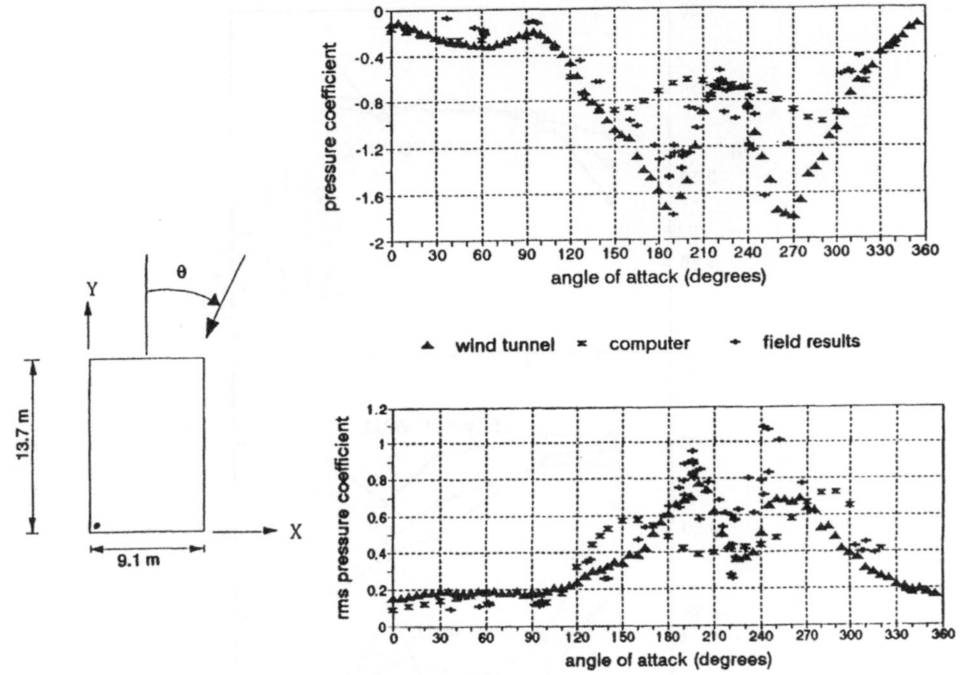

Fig. 6. Comparison of mean and rms pressure coefficients computed and measured on the roof corner of the Texas Tech building, after Ref. [23].

obtained by using either the k–ε model [20–22] or the LES model [22]. Results are always for wind perpendicular to the building wall, i.e. the so-called textbook wind case. The comparison is rather promising, since with the exception of one set of numerical results, the rest show agreement with the full-scale data similar to that between the wind-tunnel/full-scale comparisons. Numerical simulations by the LES model are also superior to those by the k–ε model. However, when results for different wind directions were obtained, the optimism already developed diminishes quickly. Fig. 6 shows both mean and rms pressure coefficients measured both in full scale and in wind tunnel for a roof corner point of the Texas Tech building [23]. Numerical results obtained with the k–ε model for the same point show drastic differences from both the full scale and the wind tunnel values, particularly for the critical wind azimuths ranging from 170° to 280°.

As far as taller prismatic buildings is concerned, numerical data are limited. Fig. 7 shows comparisons of average pressure coefficients for a square building with different H/B ratios. The average pressure coefficients have been derived by computing the arithmetic mean of the most critical values of pressure coefficients obtained for each building surface. The results computed using the zonal treatment method [16] are mostly in good agreement with the measured data. Clearly, encouraging improvements are obtained for the building side wall irrespective of the building height. The

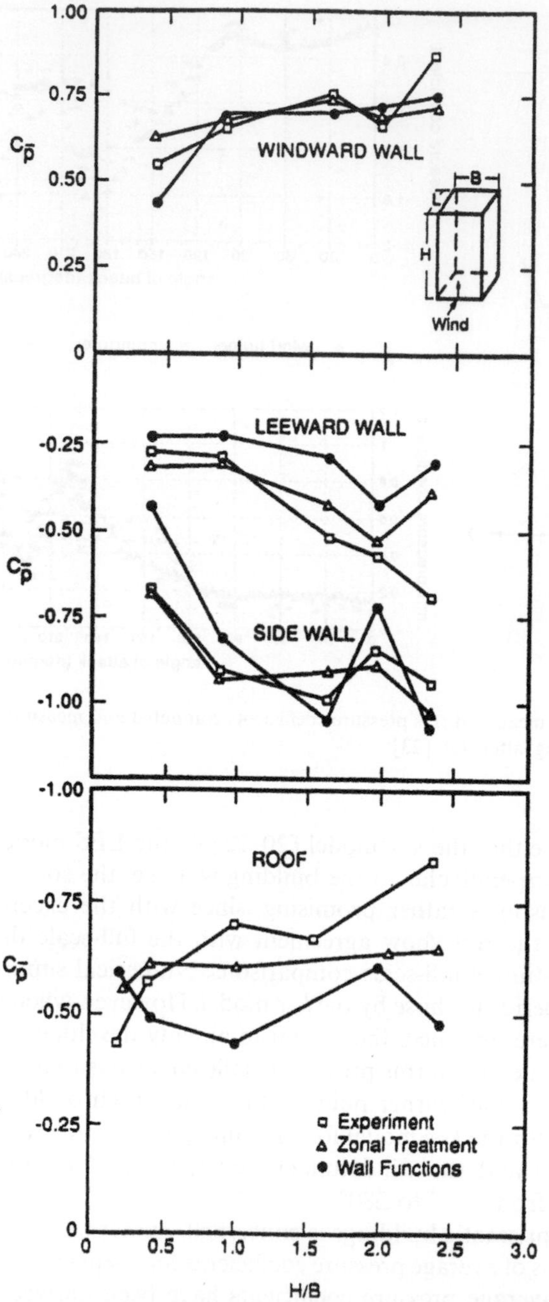

Fig. 7. Computed and measured averaged pressure coefficients on the building envelopes, after Ref. [16].

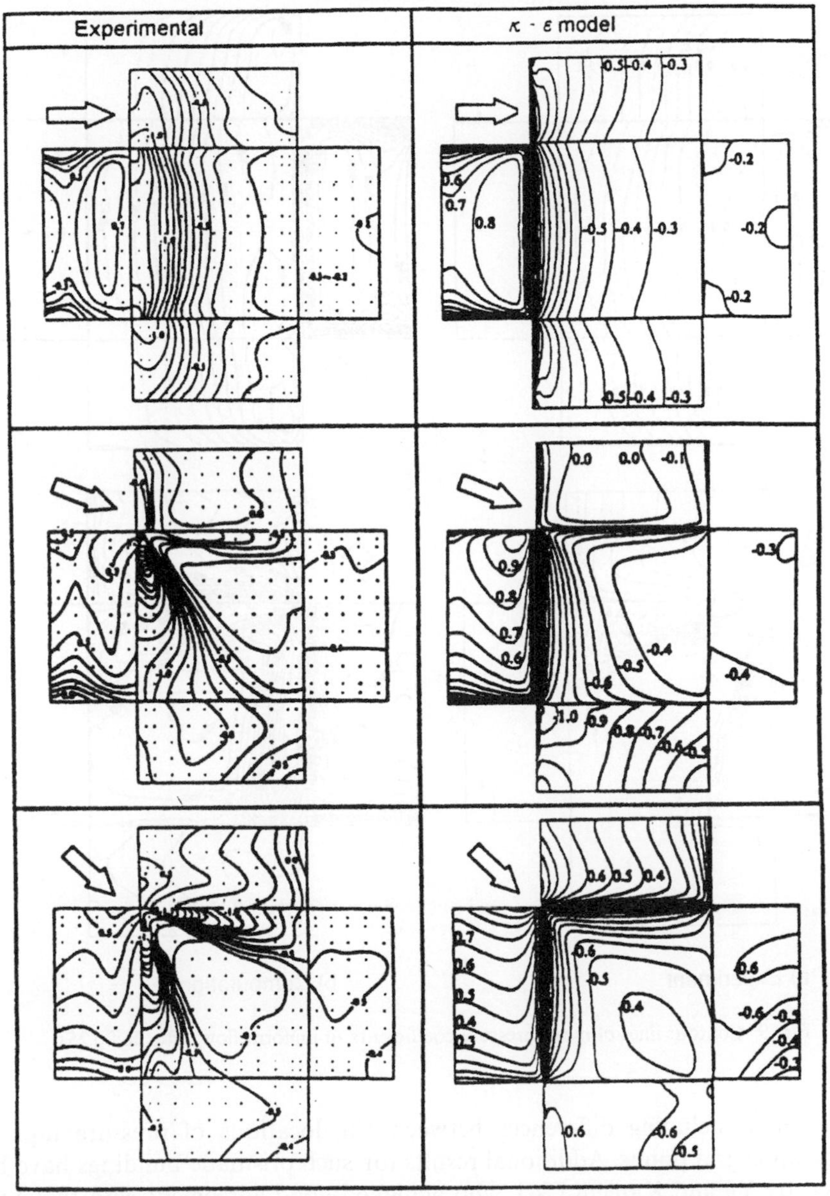

Fig. 8. Contour lines of mean pressure coefficients in 1/4 shear flow, after Ref. [25].

suction values measured for the leeward wall increase as H/B increases. In the computational approaches, this trend breaks down for $H/B > 2$; at $H/B = 2.4$, a significant difference can be noticed between the computed and measured values. This discrepancy may also be due to experimental uncertainties involved in the

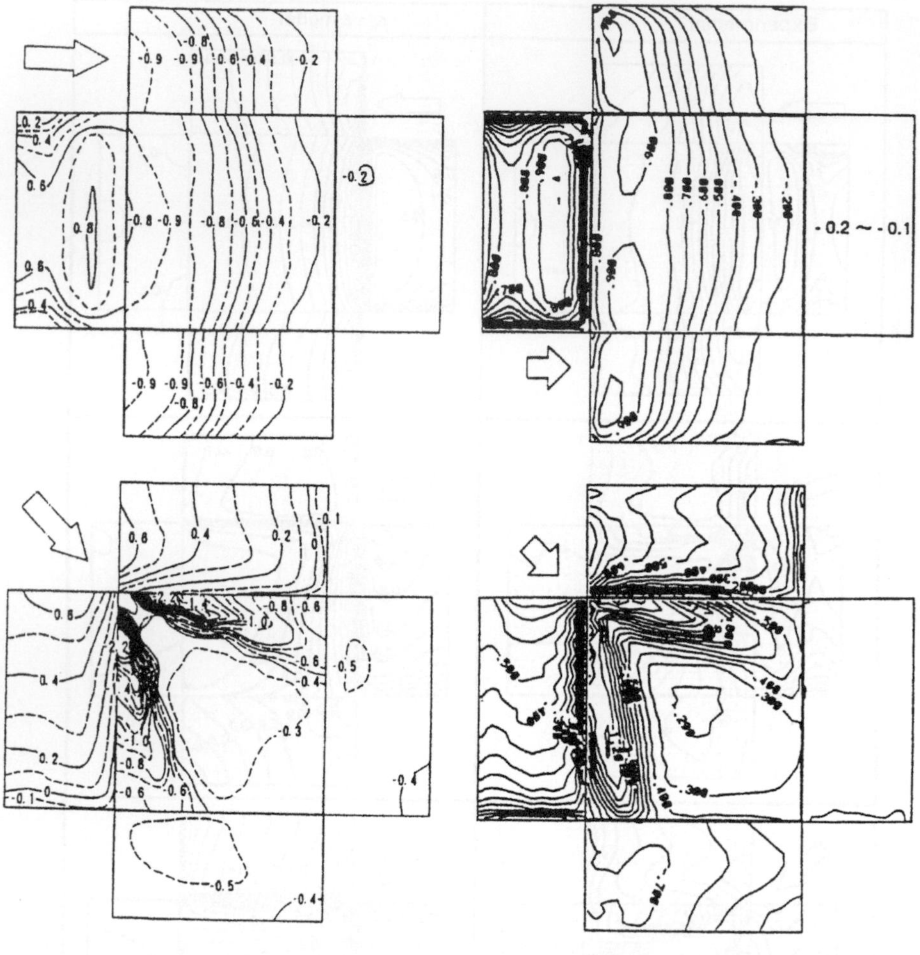

a) Experiment b) Computation

Fig. 9. Contour lines of mean pressure coefficients in uniform flow, after Ref. [25].

measurements, including differences between the locations of pressure taps and computational grid points. Additional results for such prismatic buildings have been presented by Yu and Kareem [24], who achieved good agreement with the experimental data consisting of mean pressure coefficients when they applied the large eddy simulation. However, the agreement with the RMS values is much less satisfactory.

Kawai presented the report of the Architectural Institute of Japan concerning computational wind engineering activities [25] in 1995. Fig. 8 shows contour maps of mean pressure coefficients measured and computed by the k–ε model on the envelope of a low square building whose height was equal to half its width. Data are shown for three wind directions, i.e. 0°, 22.5° and 45°. In accordance with Ref. [25]

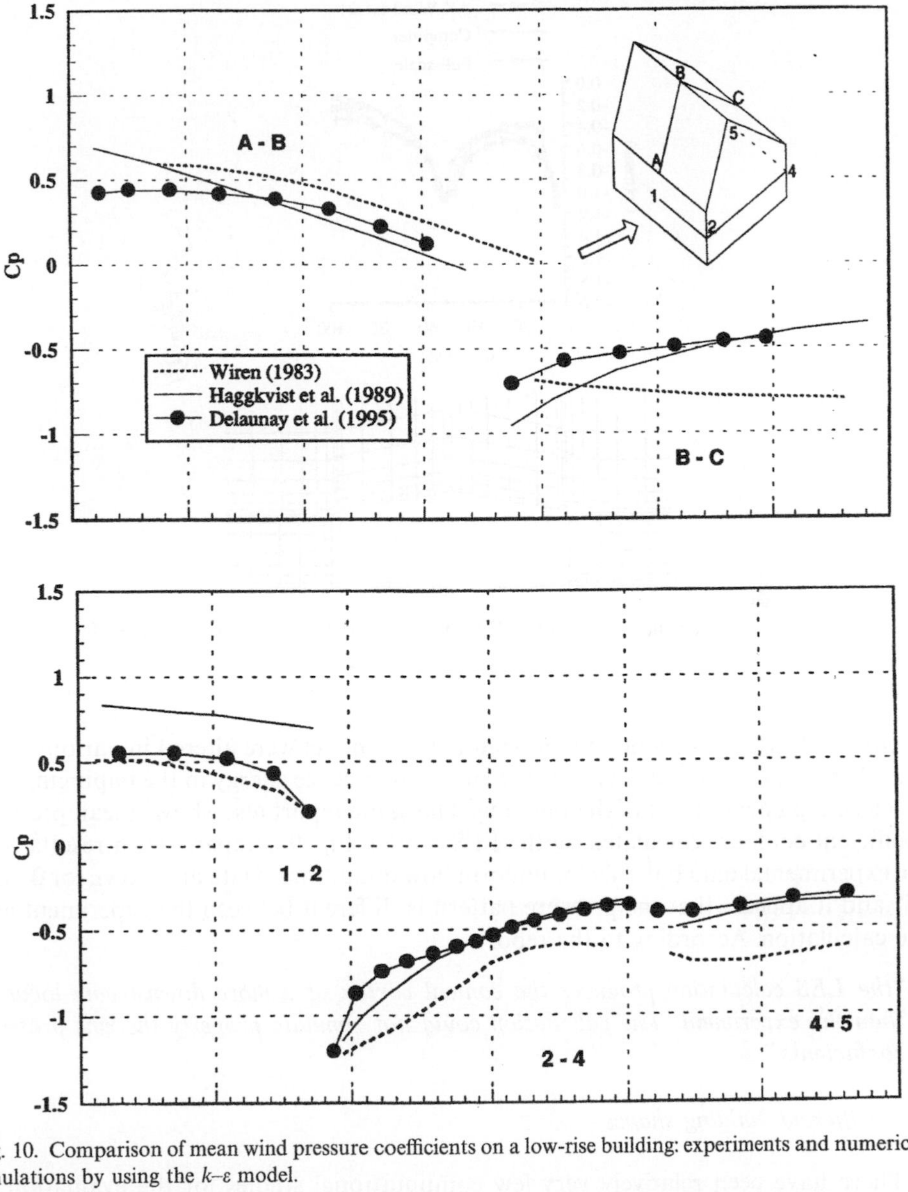

Fig. 10. Comparison of mean wind pressure coefficients on a low-rise building: experiments and numerical simulations by using the k–ε model.

"The maps obtained by the calculation seem to correspond with the experimental map at the first sight but the large local suctions are observed only in the area very close to the windward edgelines on the roof in the calculation, whereas the large suction zones spread from the edge to the roof center induced by the separation bubble and the conical vortex in the experiment."

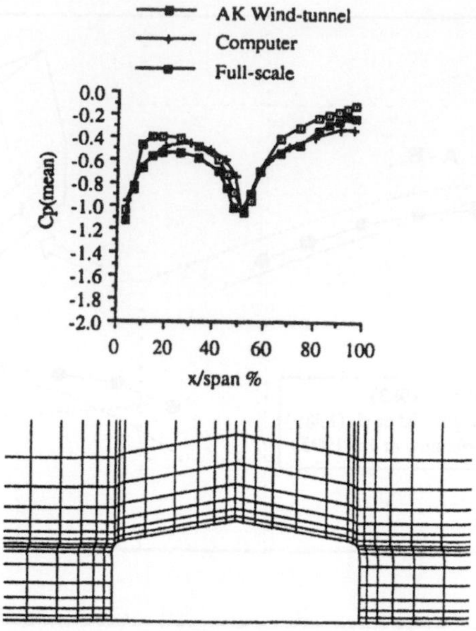

Fig. 11. Mean pressure coefficients on the SILSOE gable roof building: wind-tunnel, full-scale and computational results.

On this basis, the production terms in the k–ε model were altered in various ways [12,13] to supress the extra production of the turbulence energy in the impinging and accelerating flow in front of the building. The same report also shows mean pressure coefficient contours calculated by the LES model. Fig. 9 compares such results with the experimental data but only for uniform flow conditions. Data are shown for 0° and 45° and it appears that the pressure pattern is different between the experiment and the calculation. According to the report,

"*the LES calculation produces the conical vortex at a more downstream location than the experiment. The calculation could not simulate properly the rms pressure coefficients.*"

2.1. Different building shapes

There have been relatively very few computational studies for the evaluation of wind pressures on buildings with shapes other than rectangular. Fig. 10 shows experimental and numerical data consisting of mean pressure coefficients measured on the envelope of a low building with a gabled roof of high slope. The numerical data presented in Refs. [26,9] have been derived by utilizing the PHOENICS code. Ref. [27] has used a stepped diagram to model the roof slope, whereas Ref. [9] has used body-fitted coordinates for this purpose. There are discrepancies between the two numerical simulations particularly for the roof and the windward wall surface and

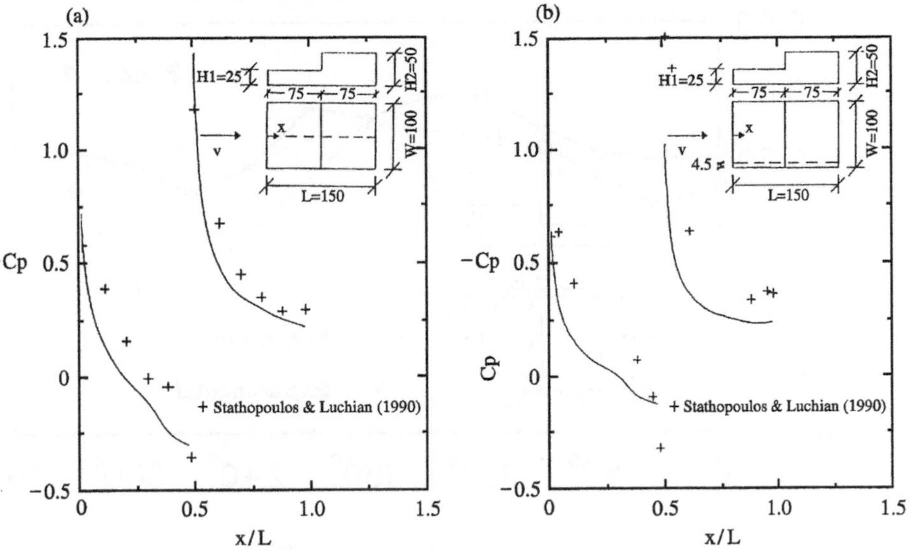

Fig. 12. Mean pressure coefficients along the centreline and edgeline of a stepped roof (0° azimuth), after Ref. [30].

also differences between the numerical and the experimental values – the latter taken from Ref. [26]. In general, there is a better agreement between the results obtained from the body-fitted coordinate solution and the experimental values.

Fig. 11 shows results for the Silsoe full-scale building which has a gabled roof of lower slope [28]. Three sets of mean pressure coefficients have been presented for the central line of the roof for wind perpendicular to the building wall: full-scale, wind-tunnel data obtained in the University of Auckland and computational results. In spite of the relatively coarse grid also shown in the figure, the comparison between computational and full-scale data is rather satisfactory, in fact better than that between the full-scale and wind-tunnel results. However, more recent computational results for free-standing walls [29] have not been as successful in representing the full-scale data.

A computational evaluation of wind pressures on the envelope of an L-shaped building has been carried out by Stathopoulos and Zhou [30]. Typical results for wind perpendicular to the building wall are presented in terms of mean pressure coefficients in Fig. 12 for the central and edge line of the roof. The comparison with the experimental data for the same geometry obtained by Stathopoulos and Luchian [31] is fair, particularly for the central line of both roof levels. Fig. 13 shows mean pressure coefficients for two particular points on the lower and upper level of the roof as obtained numerically and experimentally for all different wind directions. Again, a weak agreement is apparent for the point near the roof corner whereas a much better agreement appears for the point on the middle of the edge of the roof. Finally, Fig. 14 shows mean pressure coefficients for the central and edge line of the roof for an

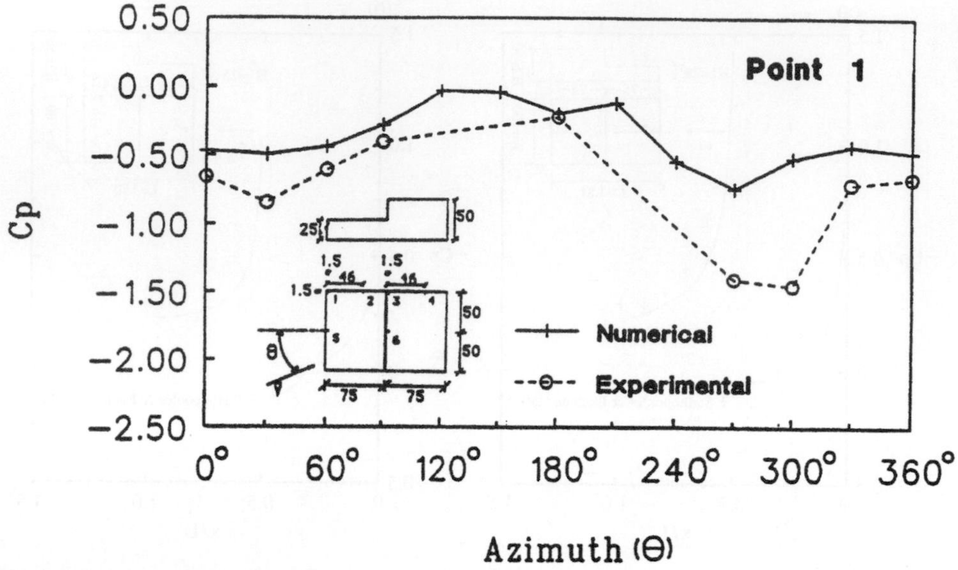

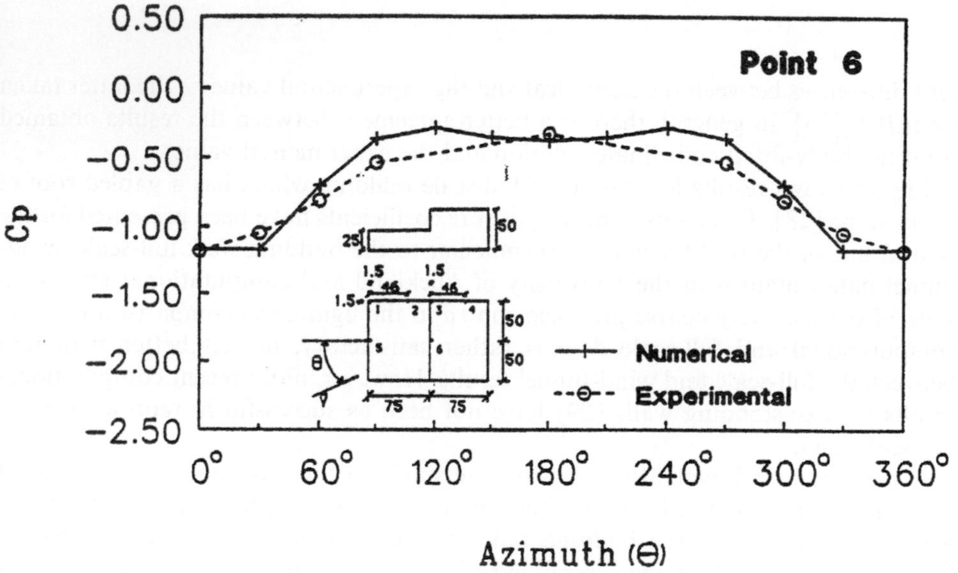

Fig. 13. Mean pressure coefficients computed and measured on a stepped roof, after Ref. [30].

azimuth of 60°. The agreement between numerical and experimental results is tolerable for the central line but it breaks down completely for the edge line of the roof. Clearly the k–ε model cannot represent the vortices generated along the edge of the roof.

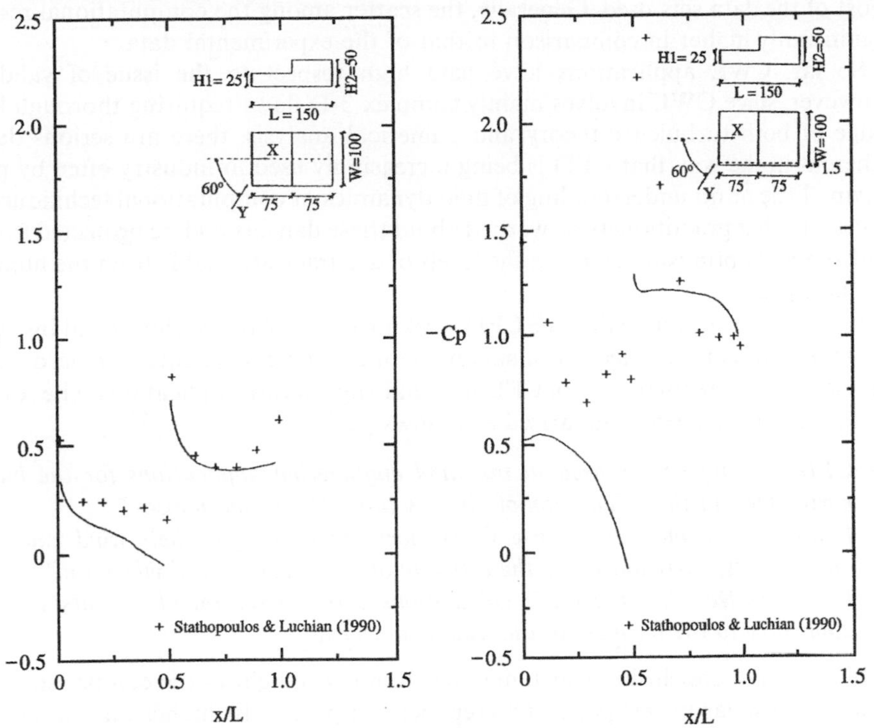

Fig. 14. Mean pressure coefficients along the centreline and edgeline of a stepped roof (30° azimuth), after Ref. [30].

3. Computational wind engineering applications

The validation of CFD results by reference to appropriate experimental data is extremely important and exercises such as this of the Aylesbury collaborative experiment for wind tunnel data [32] are very useful in order to place the computational results into the right perspective. Current experience has shown that numerical results do not compare as well among themselves as those from physical modeling.

Ref. [25] compares mean pressure coefficients obtained for the central section of a prismatic building for a wind direction of 0° by a number of experimental and numerical studies. Data have been obtained in different laboratories for turbulent flow represented by a wind velocity profile with a power-law exponent equal to 0.25 but turbulence at roof height may not have been the same for all cases. Fig. 15 shows all the experimental data and the numerical results obtained by various computational approaches. The pressure coefficients on the windward face scatter very much both in the experiments and the numerical calculations, perhaps due to different modelling conditions. As far as the roof and the leeward face is concerned, scatter is somewhat reduced in spite of the large differences of the approaching turbulence intensity for most of the data sets used. Generally, the scatter among the computational results is significantly higher in comparison to that of the experimental data.

So far, CWE applications have paid high respect to the issue of validation. However, since CWE involves mainly complex 3-D flows requiring thorough knowledge of both turbulence theory and numerical analysis, there are serious dangers inherent in the way that CFD is being increasingly used in industry often by people having little or no understanding of fluid dynamics or computational techniques. It is necessary that practitioners be warned about these dangers and be protected from the wide-spread optimism regarding the levels of accuracy attainable from the numerical evaluations.

It is worth mentioning that the CFD working group of the Architectural Institute of Japan has recently carried out a survey among practicing engineers regarding their opinion about the usefulness of CFD for wind engineering applications. The results of this survey have been summarized as follows [25]:

"– *CFD is being widely used in the wind engineering applications for the business particularly in the calculation of air flow in buildings and houses.*"
"– *Engineers are planning to use CFD more frequently for their wind engineering applications, particularly for the estimate of the wind-induced vibration.*"
"– *Engineers feel that the CFD calculations produce reasonable results which are applicable to the business in the wind engineering.*"

It is this last conclusion which may be alarming in light of the comparisons made and shown so far in this paper. Discrepancies appear to be higher than what design would tolerate, particularly for cases that require complex building shapes, surroundings and results other than simple, mean pressure coefficients.

It is the author's opinion that areas of applications of computational wind engineering may initially include the assessment of wind environmental conditions around building complexes as well as, partly, the evaluation of pollutant concentrations in the building environment. This is because these are cases for which mean values of the variables involved, i.e. wind speed, concentration, etc. may be usable in an approximate way to provide some insight, valuable for preliminary design purposes.

The feasibility of evaluating wind environmental conditions around multiple building configurations has been examined for an actual downtown location in Montreal [16]. Fig. 16 shows the cluster of buildings A, B, C, D, E and F around the building of interest X. These have been modelled for both numerical computation by using the k–ε model as well as for the experimental measurements. The non-uniform grid distribution used for this arrangement contains nearly 300 000 nodes. In spite of inadequacies in the vector plot of computed velocity field around this group of buildings, the comparison of computed and measured velocity ratios at 2 m above the ground level in the presence and absence of building X is generally satisfactory with a maximum discrepancy of the order of 30%. It is noteworthy that the maximum discrepancies between the experimental and numerical data appear in highly complex recirculating flow regions for which neither the measured nor the computed values can be considered very accurate. Regarding concentrations in problems of pollutant dispersion, Fig. 17 shows good comparisons between computational and experimental

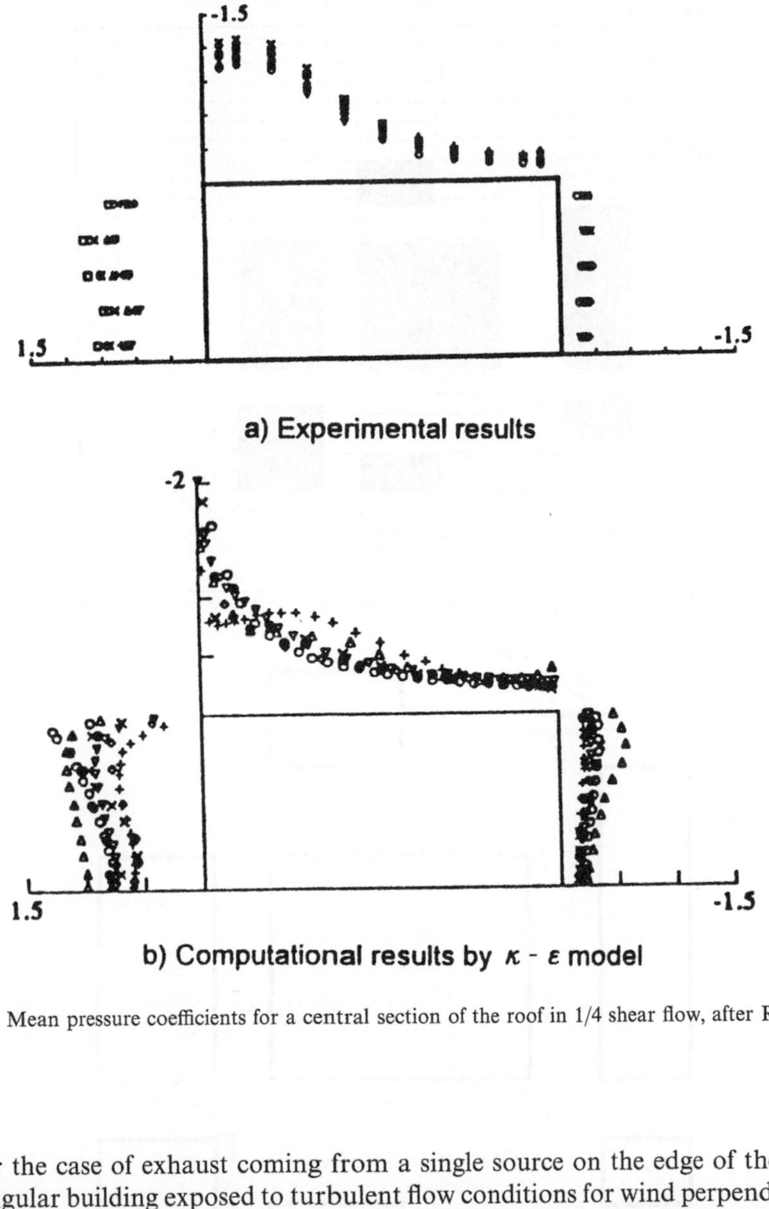

Fig. 15. Mean pressure coefficients for a central section of the roof in 1/4 shear flow, after Ref. [25].

data for the case of exhaust coming from a single source on the edge of the roof of a rectangular building exposed to turbulent flow conditions for wind perpendicular to the building wall [33]. Clearly, not only the maximum ground level concentrations are predicted well by the numerical solution, but also the location of the maximum ground level concentration has only been slightly underestimated by the numerical results. On the other hand, computational evaluations carried out by using the FLUENT code for some simple geometries [34] have been much less successful. It appears that although CFD is definitely a good friend of wind engineering, it has not yet become a true ally.

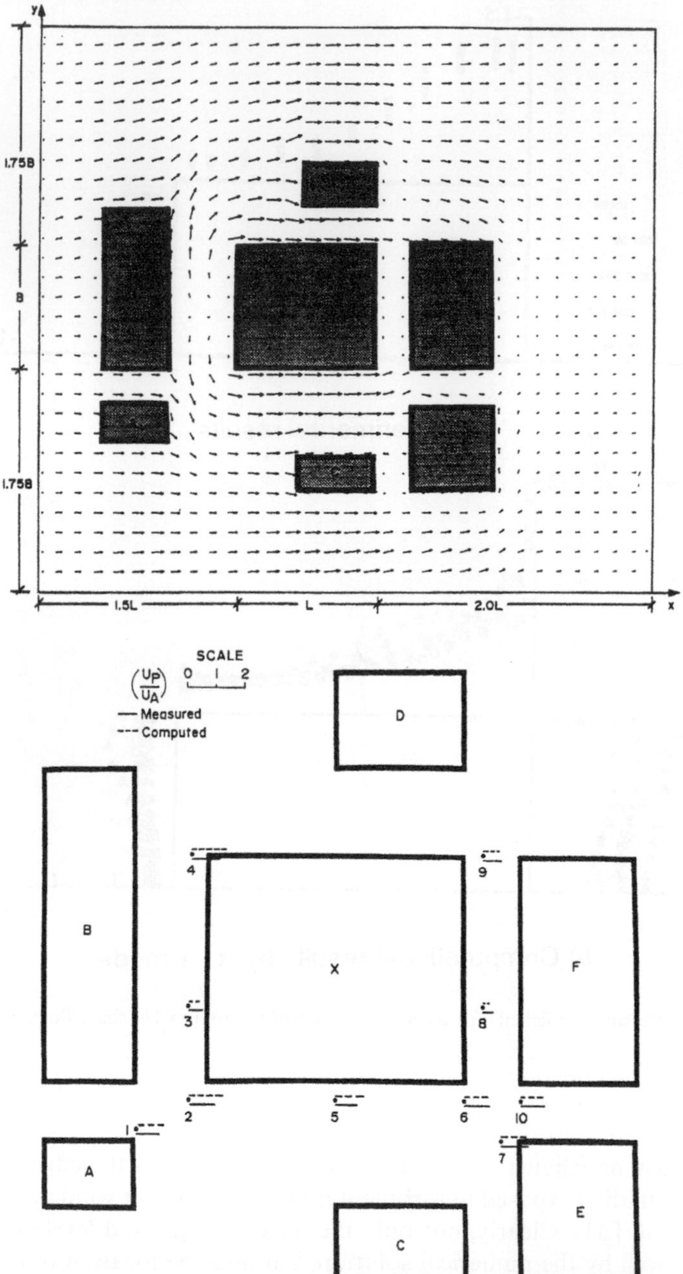

Fig. 16. Vector plot of computed velocity field around a group of buildings and comparison of computed and measured velocity ratios at 2 m from the ground level, after Ref. [16].

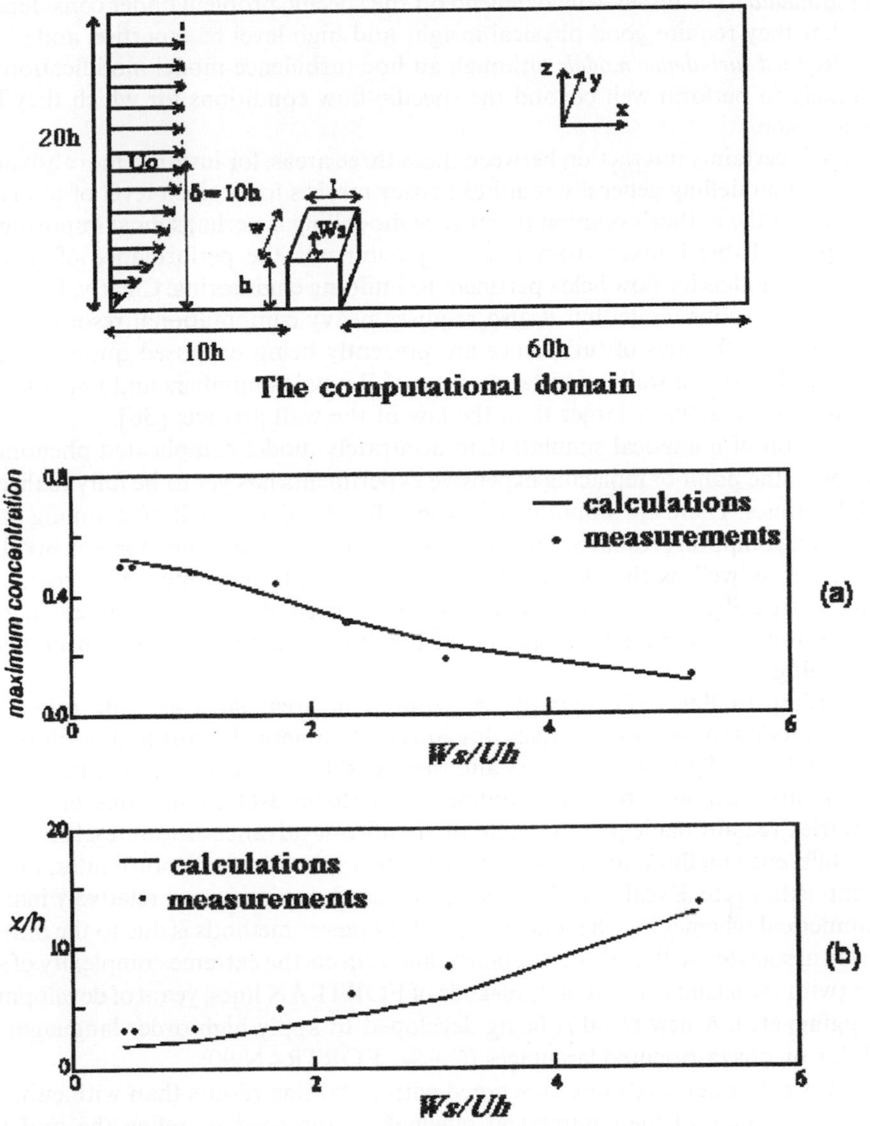

Fig. 17. Computational results versus experimental data (a) maximum ground level dimensionless concentration, (b) position of the maximum ground level concentration, after Ref. [33].

4. Future challenges

There are several areas of improvement in CWE. The most significant of which are:

(a) *Numerical accuracy* by using higher-order approximations coupled with grid-independence checks;

(b) *Boundary conditions*, which depend on the specific problem under consideration so that they require good physical insight and high level of expertise; and
(c) *Refined turbulence models* although ad hoc turbulence-model modifications are unlikely to perform well beyond the specific flow conditions for which they have been made.

There is certainly interaction between these three areas, for instance more advanced turbulence modelling generally requires denser meshes for a given level of numerical accuracy. In the author's opinion turbulence modelling is perhaps most important. In this regard, Table 1 taken from Ref. [35] compares the performance of different turbulence models for flow fields pertinent to building engineering. Clearly, LES is the most advantageous model but it also requires heavy computational resources. Furthermore, new theories of turbulence are presently being proposed questioning the validity of law of the wall for certain ranges of Reynolds numbers and implying that shear forces are actually larger than the law of the wall predicts [36].

The vision of numerical simulation to accurately model complicated phenomena, perhaps to the point of replacing expensive experiments has yet to be fully realized in fluid dynamics. There are a number of reasons for this delay, including among others the limited computing resources, the inability to model complex phenomena involving turbulence, as well as the challenge of multiple length and time scales co-existing within a single problem. Although, there have been dramatic advances in computing power, there have been significant but less dramatic advances in numerical methodologies.

According to Ref. [37], research is required in areas such as code robustness, physical models for flows in complex domains and numerical errors and resolution. In general, CFD codes lack robustness and they tend to be touchy and temperamental, which limits their user base. Capabilities to perform 3-D simulations in complex geometries remain inadequate in spite of significant advances in finite element and finite difference methodologies, generalized meshes, body-fitted coordinates, domain decomposition, etc. Existing codes tend to be based on simple and relatively inaccurate numerical schemes but the reticence to adopt newer methods is due to the inherent time lag associated with code development and reflects the extreme complexity of such codes (with thousands or tens of thousands of FORTRAN lines, years of development, debugging, etc.). A new trend is being developed to apply high order languages, say modular or object-oriented languages (C+ +, FORTRAN 90).

Most practitioners are more concerned with obtaining results than with either the order of accuracy of their numerical schemes or the need to refine the grid until converged grid-independent solutions are obtained. Some journals now require a "systematic discussion of numerical errors". However, these concepts may be largely irrelevant for problems which typically involve such wide ranges of length and time scales that the use of cells and time steps small enough to fully resolve them is prohibitive. New methodologies are required for estimating, bounding and minimizing discretization errors in situations of very coarse resolutions. Adaptive strategies can be developed to minimize errors subject to the constraints of the available resolution and computational resources. Spatial resolution can be enhanced by refining the grid (h-refinement), increasing the order of approximation (p-refinement)

Table 1
Relative performance of various turbulence models for practical applications, after Ref. [35]

Turbulence model	Standard k–ε	Modified k–ε			RSM		LES	
		LK	MMK	Kawamoto (k–ε–ϕ)	GL	CL	Conventional S	Dynamic SGS
1. Simple flows (channel flow, pipe flow, etc.) (local equilibrium is valid)	○	○	○	○	○	○	○	○
2. Flow around bluff body (with turbulent approaching wind, local equilibrium is not valid)								
(1) Impinging area	×	△, ○	△,○	△,○	×	△,○	○	○
(2) Separated area	×	△	○	○	△	△,○	○**	○
(3) With oblique wind angle	×,△	△,○	△,○	○	△	△,○	○**	○
3. Transitional Flow (low Re number effects)								
(1) near-wall	○*	○*	○*	○*	○*	○*	○**	○
(2) non-near-wall	×	×	×	×	×	×	×	○
4. Unsteady flow								
(1) vortex shedding	×	○	○	○	○	○	○	○
(2) fluctuation over wide-spectrum range	×	×	×	×	×	×	○	○
5. Stratified flow	×	×	×	×	△	△	△	○

Note: ○: functions well; △: insufficiently functional; ×: functions poorly; ○*: functions well when low Re number type model is employed; ○**: functions well with wall damping function; LK: Launder–Kato; MMK: Murakami–Mochida–Kondo; S: Smagorinsky model; GL: Gibson–Launder model; CL: Craft–Launder model; Dynamic SGS.

or grid clustering. Note that these can all be done adaptively as the computations progress [37].

5. Concluding remarks

Computational Wind Engineering is in its infancy and has a long way to go to become truly useful to the design practitioner; although CFD has tremendous potential, it also has significant limitations for several wind engineering flows. It is unlikely that CFD for general turbulent flow will evolve to a computational wind tunnel, at least not in the foreseeable future. Work on turbulence modelling will continue but a shift of emphasis towards LES has already started and is anticipated to be more intense in the future. Experiments and computer modelling have to be carried out in parallel to complement each other; at present, computer modelling can be used as a predictive tool to obtain a wider range of design alternatives at reduced cost – laboratory experiments may follow on finalized/reduced options.

Acknowledgements

The author would like to express his appreciation to Dr. Robert Meroney, Chair of the 2nd International Symposium on Computational Wind Engineering for his kind invitation to deliver this keynote lecture. The assistance of Mr. Ye Li in organizing some of the data presented in this paper is gratefully acknowledged. Appreciation is also expressed to Mr. George Stathopoulos, whose expert skills have been utilized in the preparation of the manuscript.

References

[1] ASCE 7-95, Minimum design loads for buildings and other structures, ASCE Standard, American Society of Civil Engineers, 1995.
[2] I.P. Castro, A.G. Robins, The flow around a surface-mounted cube in uniform and turbulent streams, J. Fluid Mech. 79 (2) (1977) 307–335.
[3] A. Hunt, Wind-tunnel measurement of surface pressure on cubic building models at several scales, J. Wind Eng. Ind. Aerodyn. 10 (1982) 137–163.
[4] S. Murakami, A. Mochida, 3-D numerical simulation of airflow around a cubic model by means of the k–ε model, J. Wind Eng. Ind. Aerodyn. 31 (1988) 283–303.
[5] T. Stathopoulos, M. Dumitrescu-Brulotte, Design recommndations for wind loading on buildings of intermediate height, Canad. J. Civil Eng. 16 (6) (1989) 910–916.
[6] F. Baetke, H. Werner, Numerical simulation of turbulent flow over surface-mounted obstacles with sharp edges and corners, J. Wind Eng. Ind. Aerodyn. 35 (1990) 129–147.
[7] D. Paterson, C. Apelt, Simulation of flow past a cube in turbulent boundary layer, J. Wind Eng. Ind. Aerodyn. 35 (1990) 149–176.
[8] C.X. Zhang, Numerical predictions of turbulent recirculating flows with a k–ε model, J. Wind Eng. Ind. Aerodyn. 51 (1994) 177–201.
[9] D. Delaunay, D. Lakehal, D. Pierrat, Numerical approach for wind loads prediction on buildings and structures, J. Wind Eng. Ind. Aerodyn. 57 (1995) 307–321.

[10] Y.S. Zhou, T. Stathopoulos, Application of two-layer methods for the evaluation of wind effects on a cubic building, ASHRAE Trans. 102(1) (1996).
[11] S. Kawamoto, T. Tanahashi, High-speed GSMAC-FEM for wind engineering, Comput. Meth. Appl. Mech. Eng. 112 (1994) 219–226.
[12] M. Tsuchiya, S. Murakami, A. Mochida, K. Kondo, Y. Ishida, Numerical study on surface pressures of low-rise building uisng revised k–ε model, Proc. Third International Colloquium on Bluff Body Aerodynamics & Applications, Virginia Polytechnic Institute and State University, Blacksburg, Virginia, USA, 1996.
[13] S. Kawamoto, Improved turbulence models for estimation of wind loading, these Proceedings, J. Wind Eng. Ind. Aerodyn. 67&68 (1997) 589–599.
[14] J. He, C.C.S. Song, Computation of turbulent shear flow over surface-mounted obstacle, ASCE J. Eng. Mech. 118 (11) (1992) 2282–2297.
[15] S. Murakami, A. Mochida, Y. Hayashi, S. Sakamoto, Numerical study on velocity-pressure field and wind forces for bluff bodies by k–ε, ASM and LES, J. Wind Eng. Ind. Aerodyn. 41–44 (1992) 2841–2852.
[16] A. Baskaran, Computer simulation of 3D turbulent wind effects on buildings, Ph.D. Thesis, Centre for Building Studies, Concordia University, Canada, 1990.
[17] M.L. Levitan, K.C. Mehta, W.P. Vann, J.D. Holmes, Field measurements of pressures on the Texas Tech building, J. Wind Eng. Ind. Aerodyn. 38 (1991) 227–234.
[18] D. Surry, Pressure measurements on the Texas Tech building: wind tunnel measurements and comparisons with full scale, J. Wind Eng. Ind. Aerodyn. 38 (1991) 235–247.
[19] H. Okada, Y.C. Ha, Comparison of wind tunnel and full-scale pressure measurement tests on the Texas Tech building, J. Wind Eng. Ind. Aerodyn. 41–44 (1992) 1601–1612.
[20] R.P. Selvam, Computation of pressures on Texas Tech building, J. Wind Eng. Ind. Aerodyn. 41–44 (1992) 1619–1627.
[21] D.A. Paterson, J.D. Holmes, Computation of wind pressures on low-rise structures, J. Wind Eng. Ind. Aerodyn. 41–44 (1992) 1629–1640.
[22] A. Mochida, S. Murakami, M. Shoji, Y. Ishida, Numerical simulation of flowfield around Texas Tech building by Large Eddy Simulation, J. Wind Eng. Ind. Aerodyn. 46–47 (1993) 455–460.
[23] R.P. Selvam, Roof corner pressures on the Texas Tech building: numerical and field results, Structures Congress '92, San Antonio, Texas, USA, 1992.
[24] D. Yu, A. Kareem, Numerical simulation of flow around rectangular prisms, these Proceedings, J. Wind Eng. Ind. Aerodyn. 67&68 (1997) 195–208.
[25] H. Kawai, Activities in AIJ concerning CFD for wind-resistant structural design, IWEF Workshop on CFD for prediction of wind loading on buildings and structures, Tokyo Institute of Technology, Yokohama, Japan, September 9, 1995.
[26] K. Haggkvist, U. Svensson, R. Taesler, Numerical simulations of pressure fields around buildings, Building Envir. 24 (1) (1989) 65–72.
[27] B.G. Wiren, Effects of surrounding buildings on wind pressure distributions and ventilative heat losses for a single-family house, J. Wind Eng. Ind. Aerodyn. 15 (1983) 15–26.
[28] P.J. Richards, R.P. Hoxey, Computational and wind tunnel modelling of mean wind loads on the Silsoe structures building, J. Wind Eng. Ind. Aerodyn. 41–44 (1992) 1641–1652.
[29] P.J. Richards, A.P. Robertson, R.P. Hoxey, Full-scale measurements and computational predictions of wind loads on free-standing walls, these Proceedings, J. Wind Eng. Ind. Aerodyn. 67&68 (1997) 639–646.
[30] T. Stathopoulos, Y.S. Zhou, Numerical simulation of wind-induced pressures on buildings of various geometries, J. Wind Eng. Ind. Aerodyn. 46–67 (1993) 419–430.
[31] T. Stathopoulos, H.D. Luchian, Wind pressures on buildings with stepped roofs, Canad. J. Civil Eng. 17 (1990) 569–577.
[32] B.L. Sill, N.J. Cook, C. Fang, The Aylesbury comparative experiment: a final report, J. Wind Eng. Ind. Aerodyn. 41–44 (1992) 1553–1564.
[33] A.P. Kazantzidou, M.N. Christolis, C.A. Christidou, N.C. Markatos, Mathematical simulation of flow and concentration fields around prisms, in: D.A. Sotiropoulos, D.E. Beskos, (Eds.), Proc. Second

National Congress on Computational Mechanics, Technical University of Crete, Chania, Crete, Greece, June 26–28, 1996, pp. 698–705.
[34] B. Leitl, P. Klein, M. Rau, R.N. Meroney, Concentration and flow distributions in the vicinity of U-shaped buildings: Wind-tunnel and computational data, these Proceedings, J. Wind Eng. Ind. Aerodyn. 67&68 (1997) 745–755.
[35] S. Murakami, A. Mochida, S. Iizuka, New trends in turbulence models for prediction of wind effects on structures, IWEF Workshop on CWE/CFD for Predcition of Wind Effects on Structures, Colorado State University, Fort Collins, Colorado, USA, August 9, 1996.
[36] B. Cipra, A new theory of turbulence causes a stir among experts, Science Mag. 272 (May 17, 1996).
[37] R.W. Douglas, J.D. Ramshaw, Perspective: future research directions in computational fluid dynamics, Trans. ASME 116 (1994) 212–215.

Building Aerodynamics

ELSEVIER

Journal of Wind Engineering
and Industrial Aerodynamics 67&68 (1997) 535–545

JOURNAL OF
wind engineering
AND
industrial
aerodynamics

Numerical considerations for simulations of flow and dispersion around buildings

Ian R. Cowan, Ian P. Castro*, Alan G. Robins

EnFlo, Department of Mechanical Engineering, University of Surrey, Guildford, Surrey GU2 5XH, UK

Abstract

This paper presents some aspects of computational work undertaken as part of a multi-partner European Union project on flow and dispersion around buildings. Attention is concentrated on those features of the numerical methods which need particular care if adequate predictions are to be obtained. Some results from a few of the 15 or so test cases being computed by all the partners using a commercially available CFD code are used to illustrate the dangers that attend such calculations. It is shown that typical numerical solutions obtainable in an industrial context are likely to be strongly dependent not only on the turbulence model but also, and often more importantly, on mesh design and the numerical method. For example, a solution obtained with, say, the standard k–ε turbulence model on a course grid can give results closer to experimental laboratory data than would be obtained with improved gridding and/or numerical schemes. Statements concerning apparent accuracy can therefore be misleading.

Keywords: Pollutant dispersion; Numerical modelling; Wind tunnel experiments

1. Introduction

The very existence of the second CWE Conference illustrates the exploding use of CFD for Wind Engineering applications. This use is not confined to the research environment but is already common in industry, where both computing power and available manpower are often very limited. Industrially-based CFD users are commonly faced not only with geometrically complex problems but also with severe restrictions on the resources available for implementation of their particular CFD code. The latter is often one of the increasingly sophisticated, commercially available codes and, although User Manuals are always supplied with such codes, it is often far from straightforward to obtain credible predictions; considerable expertise is required to make the most appropriate use of the code's capabilities for specific problems.

* Corresponding author. E-mail: i.castro@surrey.ac.uk.

PII S0167-6105(97)00098-6

Because of the wide range of user-inputs and decisions required for any particular problem (e.g. meshing details, boundary conditions, etc.) it seems quite likely that different user-groups might produce rather different solutions, even if the same code were used by each group. There is already evidence of this in fields where, arguably, user-expertise is more refined than it generally is in the Wind Engineering context (see, e.g. Ref. [1]).

In order to assess the likely extent of solution variability arising in this way, 'Project EMU' was conceived. It involved four partners – a University Research group, a Government Laboratory and two engineering Consultancies (one large and one small) – and required each partner to use the same code in predicting a wide range of hazardous gas releases in typical topographical situations. These range from steady releases from a single building in a neutrally stable atmosphere over homogeneous terrain, to unsteady releases with multiple obstacles in stable conditions and releases from within a complete process plant on a coastal site. A selection of the cases have also been modelled in a wind tunnel. Results for each case and from each Group are currently being compared and assessed and will be reported in due course – some initial details are reported by Cowan [2] and Hall and Cowan [3]. To assist in this final assessment, however, we have undertaken (at Surrey) a number of additional computations for a small subset of the total of 15 or so cases, in which alternative choices (of mesh arrangements, differencing schemes, etc.) have been used. In this paper we concentrate on various features of our solutions in order to illustrate some of the uncertainties inherent in CFD approaches to industrial hazard assessment. The intention is *not* to suggest that such approaches should be avoided, nor to discuss in detail the (undoubtedly serious) limitations of the (k–ε) turbulence model used for many of these cases; rather, it is to provide, by illustration, appropriate caution for industrially-based CFD users and to emphasise the questions that must be asked before realistic reliability statements can be made about CFD solutions for complex problems.

Selected results for just three of the test cases (Stage A of the EMU project) will be presented here. Space constraints do not allow us full descriptions of the modelling procedures; these can be found in Cowan [4] and here we mention only the more salient points. The test cases themselves, summarised in Section 2, were 'imposed' on each of the four partners. Section 3 outlines our own decisions concerning the basic details of the modelling procedures and the different variants used. The more crucial computational results, along with some comparisons with experimental data, are presented in Section 4, with conclusions are given in the final section.

2. Release case specifications

In each of the Stage A scenarios, gas releases occurred from or near a single L-shaped building, as shown in Fig. 1. Case A1 was a passive, low-momentum, large-area, continuous release from a 'doorway' in the longer 'courtyard' wall of the building which was parallel to the upstream flow; case A2 was a transient release over 90 s of a buoyant, high momentum, small area jet from a similar position; case A3 was

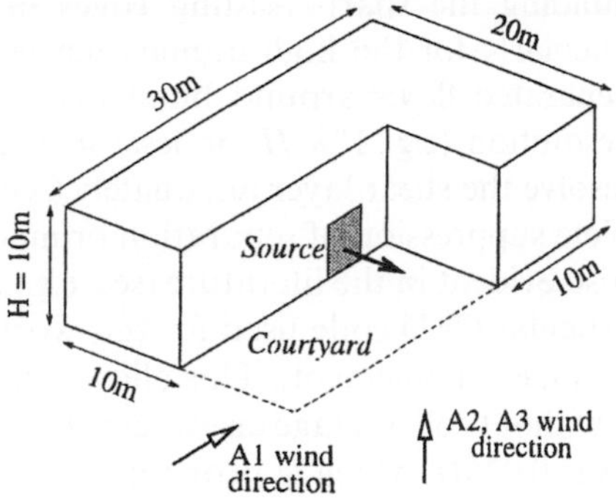

Fig. 1. Building geometry.

an instantaneous release of a large volume, heavy gas cloud which initially filled the building's courtyard. In cases A2 and A3 the building was at 45° to the approach flow which, in all three cases, was a neutrally stable atmospheric boundary layer with a thickness (δ) of 20 times the building height (H). For case A3 it was assumed that at time $t = 0$ the gas cloud was fully mixed into the open courtyard and that for $t < 0$ the courtyard was enclosed, i.e. the initial flow field was that computed for flow past a cuboid building. Other specific details are included in Fig. 1 and are given in terms of full-scale conditions. All three cases were intended to represent simplified versions of typical industrial accidental release scenarios although all three were isothermal; more realistic releases were studied in later stages but are not discussed here.

The approach boundary layer was the same (apart from its direction) in all cases and was specified (for all partners) by a surface roughness length of $z_0 = 0.12$ m, velocity at $z = 10$ m of $U_{10} = 5$ m/s and thickness $\delta = 200$ m. The latter is rather small for a neutral atmospheric boundary layer but was chosen in view of the maximum δ/H (20) that was readily achievable in the wind tunnel experiments (for A1 and A2). The exact forms of the inlet velocity and turbulence distributions were left up to the individual groups – our choices are outlined in Section 3.

In all cases mean concentration data were extracted at a series of downwind points (the 'sensor positions'). For the transient cases the data were extracted at 0.5 s intervals for a 2 min period. These (and other) data were used for comparisons between the four groups, who had complete freedom (apart from the choice of CFD code) over all other aspects of the computations – as would normally be the case for a real industrial problem.

3. Computational decisions

We discuss first the design of the meshes. Our general approach was to maximise the mesh density (and minimise mesh spacing) in regions where high gradients in the flow variables could be expected, but without allowing mesh expansion ratios to exceed about 1.2 as higher values can lead to significant errors. The obvious critical regions are those surrounding the sharp leading edges of the building and the near-source regions (particularly for the high-momentum jet case, A2). Turbulence closure calculations of separated flows around blunt obstacles have demonstrated the need for very fine resolution (e.g. 1% H or less) in regions where separation is expected, in order to resolve the shear layer adequately (see, e.g., Ref. [5]). Coarser grids lead either to complete suppression of separation or much too early a reattachment and this feature is also evident in the literature (see, e.g., Refs. [6–8]). One of the helpful features of the particular CFD code used for this work is its ability to handle embedded meshes (i.e. local mesh refinement). This allows one to employ small mesh sizes around the building without cell wastage elsewhere. The minimum mesh spacing (Δ_{min}) used in this work was $0.025H$, which is probably too large to resolve properly the flow around the building edges. However, finer meshes would have led either to unacceptably large grids (i.e. computationally too expensive) or to unacceptable expansion ratios and/or too small domain sizes. Of course, finer meshes are always possible given sufficient resources, but one of the intentional constraints in the present work was to keep mesh sizes within the limits often forced in the industrial context.

Survey of the published literature (and potential flow calculations around simple 2–D objects) suggest that the computational domain should not have upstream, downstream and cross-stream dimensions smaller than $5H$, $15H$ and $\pm 4H$, respectively. (Larger domains may be necessary for some dispersion problems). Increases in these dimensions for cases A1 and A2 (obtained by maintaining the same mesh in the smaller domain and adding further grid nodes beyond it, rather than simply stretching expansion ratios) were found to have only a small influence on the velocity and concentration fields downwind of the building, at least compared with the effects of other changes (see below). For all three cases, computations were performed on two meshes, here termed (for case A1) the fine and coarse grids. The former had $\Delta_{min}/H = 0.025$ and about 182 000 cells whereas the latter had $\Delta_{min}/H = 0.075$ and about 80 000 cells. It was important to ensure that there were no significant numerical errors around the boundaries of the embedded mesh regions; errors would certainly occur if these boundaries were in regions of large gradients in the flow variables, but tests confirmed that in this respect our mesh design was adequate. In case A2, it was also crucial to ensure good resolution around the high-momentum jet issuing from the wall. The effects of inadequate resolution are illustrated later.

Secondly, we mention the spatial discretisation scheme used for the convective terms. The CFD code contained a number of options, ranging from 'standard' (first-order) upwind differencing (UD, actually zero-order on a non-uniform mesh), through second-order upwind (LUD), to self-filtered central differencing (SFCD), which is based on the more common CONDIF scheme in which central differencing is blended with just sufficient upwinding to remove non-physical overshoots. The four partners were free to choose whatever scheme they wished. For the case A1, the LUD scheme was used on the momentum and the turbulence equations with SFCD for the scalar concentration equation, whereas SFCD was used for all equations in the two transient cases (A2 and A3). It is well-known that this aspect of the numerical procedure can be crucial in terms of accuracy, with errors being very dependent on the particular scheme used (see, e.g. Ref. [9]). In the present context we demonstrate later the effects of changing the scheme.

Thirdly, we comment on the boundary conditions. Inlet profiles were compatible with the simulated neutral boundary layer developed in the EnFlo wind tunnel for the experimental phases of the work. For all these neutral flow cases the flow was characterised by $u^*/U_0 = 0.05$, $z/\delta = 0.0006$ (with $\kappa = 0.4$) and $U = 5$ m/s at $z/\delta = 0.05$. The turbulence energy profile was estimated via the measured longitudinal energy as $k = (29/32)u'^2$, using the typical relations $v'/u' = 0.75$ and $w'/u' = 0.5$ in the usual notation, while the dissipation rate (ε) was taken as $u^{*2}(1 - z/\delta)\partial U/\partial z$ (i.e. assuming local equilibrium and a linear profile for the shear stress). For the 45° case, inflow occurred through two of the four sides of the domain and outflow through the other two; the inflow profiles were given by an appropriate rotation of the 0° inflow velocity. Outflow boundary conditions were of the usual zero-gradient type on all variables and for the side boundaries (in the 0° cases) it was assumed that the domain width was sufficiently large that symmetry conditions could be safely imposed. Symmetry conditions were also imposed at the top of the domain.

Finally, although for all test cases all partners used the standard k–ε turbulence model, we undertook some additional computations with a modified version. It is well-known that standard k–ε fails to model appropriately the response to high-streamwise strain rates. These always occur in wind flow over buildings and lead to unphysically high levels of turbulence energy upwind of the building's leading edges unless specific steps are taken to improve the model. It is, in fact, relatively simple to effect a significant improvement (see, e.g. Ref. [10]), using a modification similar to that originally proposed by Leschziner and Rodi [11]. Space does not allow discussion of the details, but the effects are shown later.

4. Some typical results

4.1. Effects of the mesh resolution

Fig. 2 shows velocity vectors on a horizontal plane for the coarse mesh simulation of case A1, demonstrating the complex flow field that exists around this asymmetrical

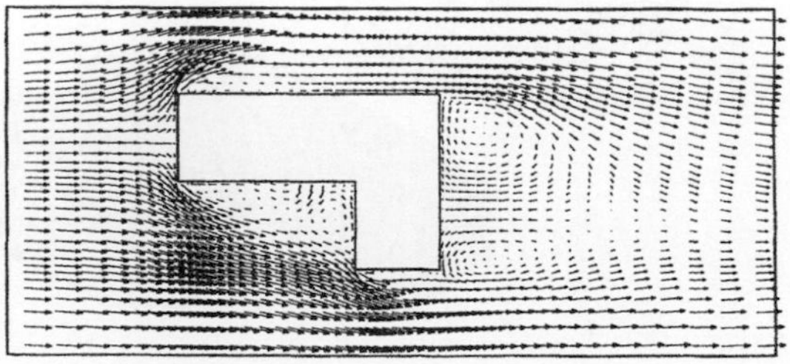

Fig. 2. Velocity vectors at $z/H = 0.33$. Case A1.

building. Separation occurs at the rear edges and on side surfaces, with subsequent reattachment in the latter case. These recirculation regions are resolved better by the finer grid but do not change in extent significantly. However, the turbulence energy distribution *does* change noticeably and, since the near-source dispersion of the source fluid is dominated by a combination of the turbulent diffusion and the recirculating flow around the building, the concentration field is also changed by increased mesh resolution. Fig. 3 shows typical concentration isosurfaces. The plume splits with the most of the scalar being swept around and the remainder passing over the building. With the fine mesh, the recirculating zone draws the plume further in behind the building, resulting in higher concentrations there. This is associated with a reduction in the vertical spread of the plume so that over the majority of the domain concentrations are increased near the ground and reduced for $z/H < 1$.

4.2. Effects of differencing scheme

More significant changes in the results occur if alternative differencing schemes are used. An investigation of the velocity field for case A1 showed that the strength of the separation at the leading edges of the building, and the associated backflow, was highly dependent upon the discretisation scheme. The highly diffusive UD scheme completely suppressed all separation, except from the rear of the building, whereas the second-order upwind scheme yielded strong separation on both sides of the building. The SFCD scheme gave results between the two, but note that no separation was observed on the roof of the building for any of the schemes, even for the fine mesh simulations. It was also found that the different schemes led to significant differences in the turbulence kinetic energy field near the doorway and at the building's corners and these (along with the mean flow changes around the building) led to significant differences in the downstream concentration field. Fig. 4, for example, shows ground-level concentration levels for one building height downstream; note that here the general spread in the results with the various differencing schemes is not reduced

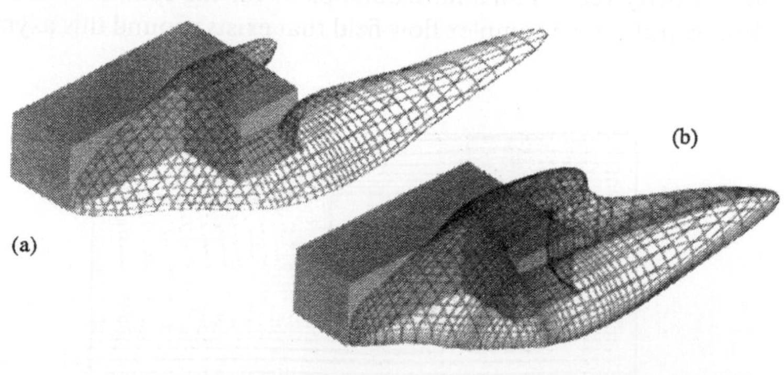

Fig. 3. Normalised concentration isosurface ($C/C^* = CU_hH^2/C_sQ_s = 1$), case A1. (a) Coarse grid; (b) fine grid.

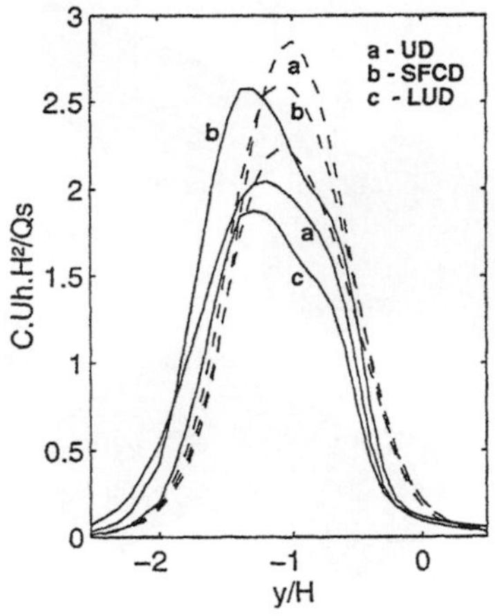

Fig. 4. Spanwise surface C/C^* profiles at $x/H = 1$ for case A1. (———) Coarse; (- - -) fine grid.

much by using the finer grid. Elsewhere, however, the mesh effects can dominate differencing scheme effects.

4.3. *Effects of turbulence model*

As expected, in all cases the standard k–ε model produced excessive turbulence upstream of the building's forward-facing walls. This unphysical result was largely removed by an appropriate modification to the modelling of the term representing production of turbulence-energy dissipation [10]. The resulting change is illustrated by the turbulence isosurface shown in Fig. 5, where the large reduction in turbulence levels around the building is clear. It was also found that turbulence levels were generally lower for the coarse-mesh computations, which has implications discussed in the following section.

For this project the major issue was the concentration levels downstream of the building. Fig. 6 shows some cross-stream profiles for coarse (C)- and fine (F)-mesh computations of case A1 with the standard (Std) and modified (PDM) k–ε model, compared with experimental data obtained in the EnFlo tunnel. The largest changes occur near ground level close to the building. Note, in particular, that the coarse mesh results are in better agreement with the experimental data. This might seem surprising – finer resolution gives a worse solution! However, the coarse mesh agreement is undoubtedly coincidental, with the numerical diffusion compensating somewhat for the overproduction of turbulence around the building. It is worth emphasising that increasing mesh resolution will generally yield a more accurate solution of the *modelled* equations, but this may well be further from the true Navier–Stokes solution (i.e. the laboratory experiment) than the result obtained on a less adequate mesh. It is consequently possible to be seriously misled by the apparent agreement between a computation and a physical experiment.

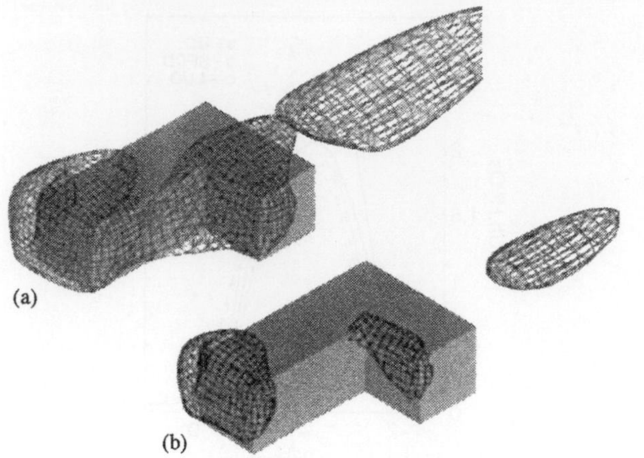

Fig. 5. Turbulence energy isosurface ($k/U_0^2 = 0.02$), for case A1. (a) Standard k–ε; (b) k–ε/PDM.

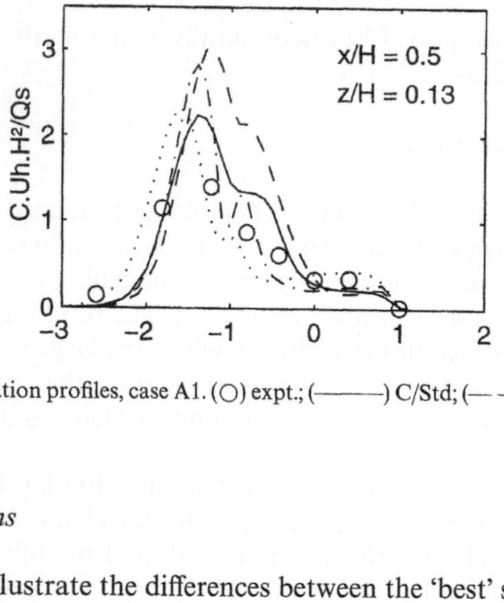

Fig. 6. Spanwise concentration profiles, case A1. (○) expt.; (———) C/Std; (— — —) F/Std; (- - -) C/PDM; (-- -) F/PDM.

4.4. Other comparisons

In this section we illustrate the differences between the 'best' solutions obtained by all partners, for all three test cases. Fig. 7 shows a set of typical cross-stream concentration profiles near the ground, obtained by all four groups for case A1. Although one solution is closer to the laboratory data than others, none of them agree very well with the latter and the differences between each computation are generally of the same order as the differences between any one of them and the experiments. This feature is also evident for the two transient cases (although case A3 was not studied in the laboratory). Fig. 8 shows a concentration isosurface for case A2 after 90 s of release (just before the jet is turned off). In this case, resolution of the high-momentum buoyant jet is particularly crucial and inadequate gridding or differencing schemes lead to serious misrepresentation of the jet path near the source and subsequent entrainment by the wake. The resulting large variations in the results (not shown here for reasons of space) can be mostly explained on this basis. Fig. 9a shows the slumping of the heavy gas cloud in case A3 at two times after release and Fig. 9b shows concentration profiles at 40 s from release. Again, there are significant disparities in

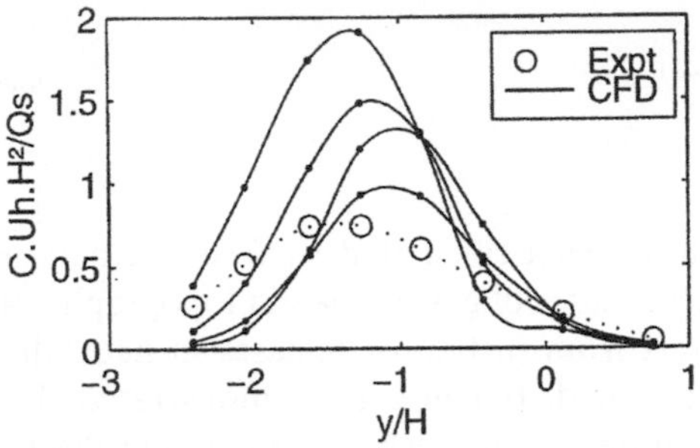

Fig. 7. Concentration profiles obtained by all partners; (○) expt. data; case A1.

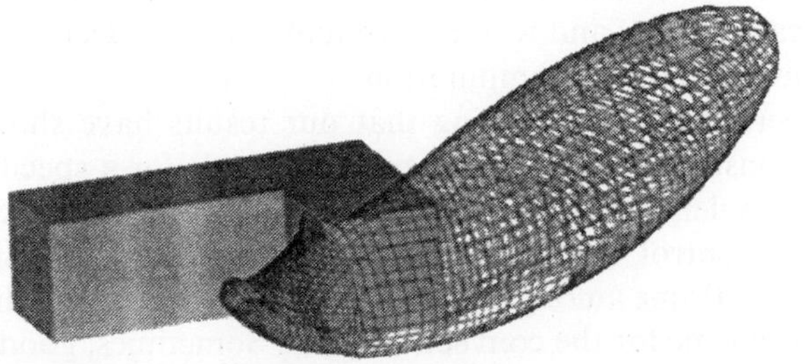

Fig. 8. Concentration isosurface ($C/C_s = 0.01$), case A2.

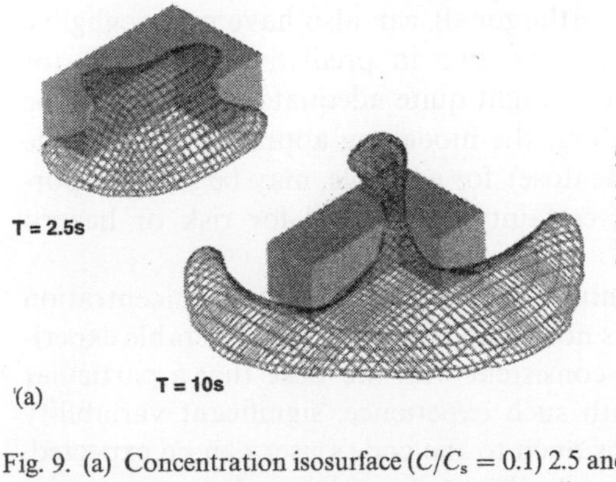

(a)

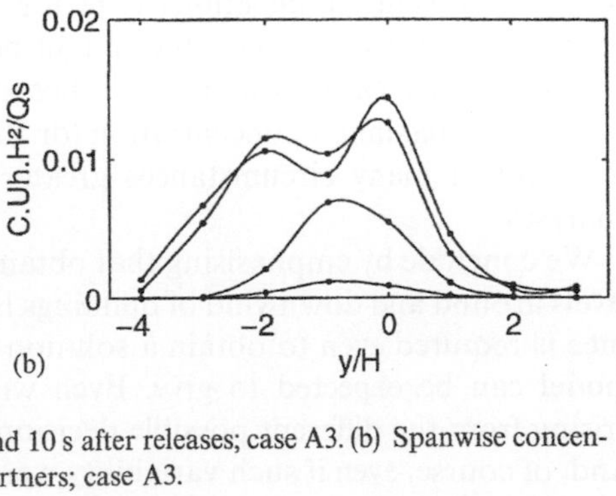

(b)

Fig. 9. (a) Concentration isosurface ($C/C_s = 0.1$) 2.5 and 10 s after releases; case A3. (b) Spanwise concentration profiles at $x/H = 0.5$, $z/H = 1$ after 40 s, all partners; case A3.

the prediction of concentration level and distribution. Since the same CFD code and the same turbulence model were used by all partners for cases A2 and A3, these large differences must be explained on the basis of different choices of mesh and differencing schemes. The effects of other parameters (i.e. different choices of inlet flow profile values, domain sizes, etc.) are probably relatively minor, although detailed analyses of these results is still being undertaken so other 'user-factors' may turn out to have been important.

5. Conclusions

Computed solutions have been calculated by four partners using the same CFD code for a number of specific cases of gas releases in topography of varying complexity. In this paper, some results from just three releases in neutrally stable atmospheric conditions have been presented. It has been demonstrated that the results can be significantly dependent on details of mesh design, spatial discretisation scheme and turbulence model. This is not a surprising conclusion but emphasises the need for great care in undertaking such Wind Engineering problems with standard CFD packages, which have usually been designed for a very wide range of applications. Different CFD users will almost invariably use somewhat different approaches in undertaking such calculations and will have to make specific decisions about a number of the crucial features of the computation.

It is therefore, perhaps, not surprising that our results have shown that even if different users are constrained to use the *same* CFD code for a specific problem and even if the *same* turbulence model is used, other factors over which the user has variable degrees of control can lead to large differences in predicted flow and dispersion parameters. Prime amongst these other factors are the mesh design and the spatial differencing scheme for the convective terms. Sometimes, good but coincidental agreement with laboratory experimental data can be achieved, when more accurate numerical procedures would lead to solutions further from the data. Other factors which due to space constraints have not been discussed, like details of the surface boundary conditions, the inlet turbulence profiles, the domain size or the particular type of mesh used (e.g. orthogonal or non-orthogonal), can also have non-negligible effects. For some applications a factor of, say, two in predictive variability (or accuracy) of *concentration* values might be thought quite adequate, but it would be unwise on that basis to be too cavalier about the modelling approaches used. The location of maximum concentration (or the dose), for example, may be more important and in many circumstances greater certainty is required for risk or hazard analyses.

We conclude by emphasising that obtaining adequate predictions of concentration levels around and downwind of buildings is not straightforward. Considerable experience is required even to obtain a solution consistent with the 'best' that a particular model can be expected to give. Even with such experience, significant variability arising from the different possible decisions open to the code's user can be expected. And, of course, even if such variability were somehow removed entirely, uncertainties about the adequacy of the turbulence (or other physical) modelling procedures embodied in the code would remain. However, our results indicate that in practice these uncertainties are, in some circumstances at least, likely to be no greater than those arising from 'user-variability' and may, in fact, be less important.

References

[1] M.V. Casey, Simulation of turbulent flows for industrial applications: a Fluid Engineer's view, in: S. Gavrilakis, L. Machiels, P.A. Monkewitz (Eds.), Advances in Turbulence, Kluwer, Dordrecht, 1996, pp. 157–162.

[2] I.R. Cowan, A comparison of wind tunnel experiments and computational simulations of dispersion in the environs of buildings, 4th Workshop on Harmonisation within Atmospheric Dispersion Modelling for Regulatory Purposes, Oostende, May 1996.

[3] I.R. Cowan, R.C. Hall, Uncertainty in CFD- modelling of Wind Engineering problems, 3rd UK Conf. Wind Engineering, Oxford, September 1996.

[4] I.R. Cowan, Project EMU: Stage A Summary, EnFlo, Mech. Eng. Department Report, June 1996.

[5] N. Djilali, I.S. Gartshore, M. Salcudean, Turbulent flow around a rectangular plate. Part II: numerical predictions, J. Fluids Eng. 113 (1991) 60–67.

[6] D.A. Paterson, C.J. Apelt, Computation of wind flows over three-dimensional buildings, J. Wind Eng. Ind. Aerodyn. (1986) 193–213.

[7] F. Baetke, H. Werner, H. Wengle, Numerical simulation of turbulent flow over surface-mounted obstacles with sharp edges and corners, J. Wind Eng. Ind. Aerodyn. 35 (1990) 129–147.

[8] S. Murakami, Comparison of various turbulence models applied to a bluff body, J. Wind Eng. Ind. Aerodyn. 46/47 (1993) 21–36.

[9] M.E. Jones, I.P. Castro, Studies in numerical computations of recirculating flows, Num. Methods Fluids 7 (1987) 793–823.

[10] I.P. Castro, D.D. Apsley, Flow and dispersion over topography; a comparison between numerical and laboratory data for two-dimensional flows, Atmos. Environ. 31 (1997) 839–850, submitted.

[11] M.A. Leschziner, W. Rodi, Calculation of annular and twin parallel jets using various discretisation schemes and turbulence model variations, J. Fluids Eng. 103 (1981) 352–360.

ELSEVIER

Journal of Wind Engineering
and Industrial Aerodynamics 67&68 (1997) 547–558

JOURNAL OF wind engineering AND industrial aerodynamics

A numerical study of wind flow around the TTU building and the roof corner vortex

Jianming He*, Charles C.S. Song

St. Anthony Falls Laboratory, Department of Civil Engineering and Minnesota Supercomputer Institute, University of Minnesota, Minneapolis, MN 55414, USA

Abstract

Comprehensive studies on wind flow around the TTU (Texas Tech University) building were conducted in both field test and wind-tunnel modeling. The results are very useful benchmark database for computational wind engineering studies. Therefore, a number of numerical studies have been carried out since then to explore a deeper insight into the physical nature of the flow characteristics. However, the previous numerical studies have not reported much about the results on the roof corner vortex, which is believed to be very important in engineering applications. This paper presents a recent numerical study of wind flow around the TTU building using an advanced CFD method with large eddy simulation approach. The main focus of this paper is the roof corner vortex. Two approaching wind directions (215° and 225°) to the building were investigated with three mesh systems. The three-dimensional roof corner vortex pattern is successfully simulated. The mean values are generally in good agreement with the wind tunnel modeling and field test, but the rms values are not comparable since the numerical results only include the resolvable part due to the large-scale eddies while the small-scale eddies are modeled by SGS model and not included in the comparison. However, as the mesh system is made finer, the rms values are closer to the measured data since more smaller scale eddies are directly resolved.

Keywords: CFD; Wind; Vortex

1. Introduction

The United States National Science Foundation sponsored Colorado State University/Texas Tech. University Cooperative Program on Wind Engineering provided

* Corresponding author. E-mail: mf13001@sk.msc.edu.

0167-6105/97/$17.00
PII S0167-6105(97)00099-8

a detailed study on wind effects on the Texas Tech. University Experimental Building, as a typical example of low-rise buildings and structures. The study includes the field test in Lubbock on the high plains of Texas [1] and 1 : 100 model test in the Meteorological Wind Tunnel (MWT) at Colorado State University [2]. It has been reported that the wind-tunnel modeling data generally agree well with the field test, except the peak and rms pressure induced by the roof corner vortices, which have very complicated turbulence structures. The wind-tunnel modeling investigators believed that the discrepancies are due to (1) the geometric scaling of the pressure taps under the corner vortices in the wind tunnel test; (2) the viscous stresses in the small vortices on the 1 : 100 wind-tunnel model; and (3) mismatch of the gusts and direction change between the field test and wind-tunnel modeling. The highly turbulent vortex flow is also a great challenge for numerical study. It is reported [2] that a number of researchers have conducted numerical studies on the TTU building using different CFD models. The computed results also indicated some discrepancies with field test and wind tunnel test in the roof corner vortex region. Since this study is focused on the investigation of the roof corner vortex features using an advanced CFD method with large eddy simulation approach, only two wind directions were tested. In order to demonstrate the effects of the mesh system on the resolvable structures of turbulence eddies, three mesh systems were used for each wind direction.

2. Governing equations

In wind engineering, most flows can be considered to be at low Mach number state, for which the time evolution can be adequately described by a weakly compressible flow model [3]:

$$\frac{\partial p}{\partial t} + K\frac{\partial u_i}{\partial x_j} = 0, \tag{1}$$

$$\frac{\partial u_i}{\partial t} + \frac{\partial (u_j u_i)}{\partial x_j} + \frac{1}{\rho_0}\frac{\partial p}{\partial x_i} = \nu\frac{\partial^2 u_i}{\partial x_j \partial x_j}, \tag{2}$$

where $K = \rho_0 a_0^2$ is the bulk modulus of fluid elasticity, and a_0 is the sound speed.

For turbulent flow, the large eddy simulation approach is adopted. By means of the box filter operation over a small control volume in space, the momentum equation can be written as

$$\frac{\partial \bar{u}_i}{\partial t} + \frac{\partial (\bar{u}_j \bar{u}_i)}{\partial x_j} + \frac{1}{\rho_0}\frac{\partial P}{\partial x_i} = \nu\frac{\partial^2 \bar{u}_i}{\partial x_j \partial x_j} - \frac{\partial \tau_{ij}}{\partial x_j}, \tag{3}$$

where $P = p + \frac{1}{3}\overline{u'_k u'_k}$ is a modified pressure, and $\tau_{ij} = \overline{u'_i u'_j} - \frac{1}{3}\delta_{ij}\overline{u'_i u'_j}$ is the sub-grid scale turbulent shear stress.

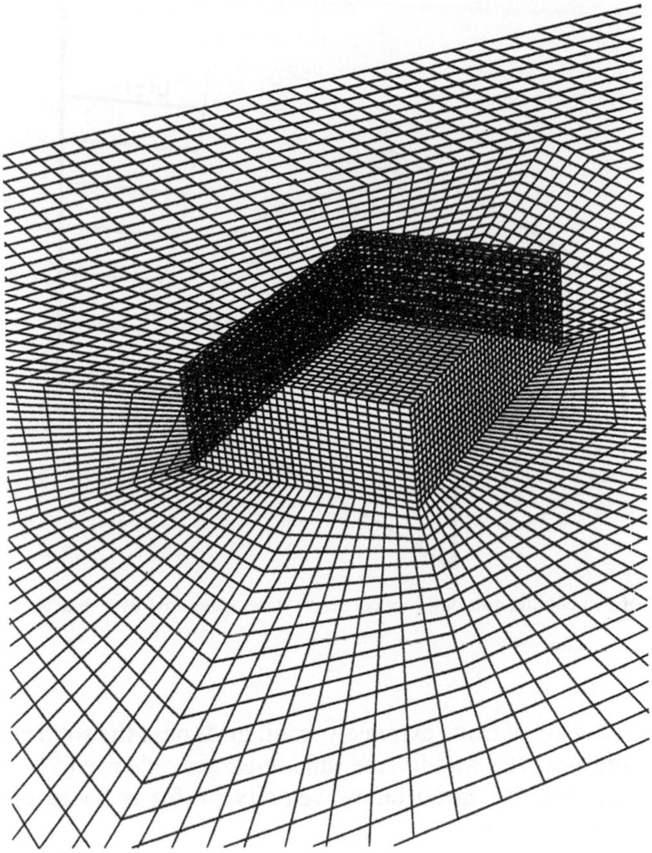

Fig. 1. The coarse mesh system around the TTU building.

The simplest and most widely used sub-grid model of the small-scale turbulence is due to Smagorinsky [4]:

$$\tau_{ij} = -\nu_t\left(\frac{\partial \bar{u}_i}{\partial X_j} + \frac{\partial \bar{u}_j}{\partial X_j}\right) \tag{4}$$

and

$$\nu_t = (C_s\Delta)^2(\bar{S}_{ij}\bar{S}_{ij})^{1/2}, \tag{5}$$

where C_s is a model coefficient ($C_s = 0.1$ was used here),

$$\bar{S}_{ij} = \frac{\partial \bar{u}_i}{\partial X_j} + \frac{\partial \bar{u}_j}{\partial x_i},$$

and Δ is a length associated with the filter ($\Delta = V^{1/3}$, where V is the volume of control volumes).

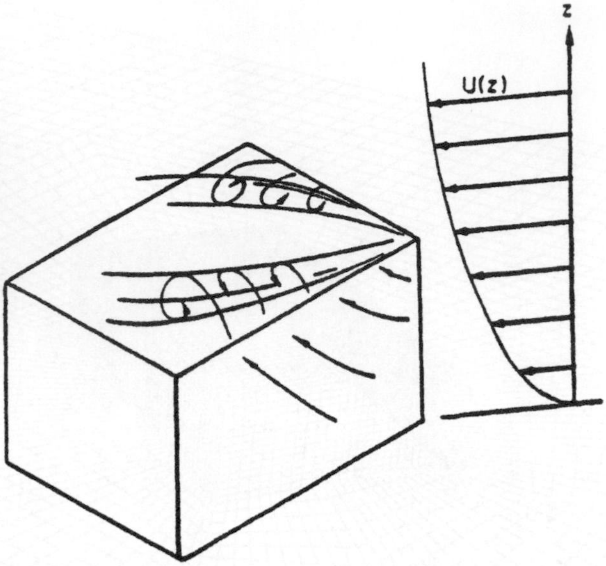

Fig. 2. Schematic of two counter rotating roof corner vortices generated by a quartering wind flow (after Ref. [8]).

To avoid the necessity of using extremely small grids near the solid wall, the log law is assumed to hold near the wall for the time-averaged velocity [5].

At the upstream end, the mean inflow velocity profile is based on a power-law distribution:

$$\frac{\bar{U}(z)}{\bar{U}(10)} = \left(\frac{z}{10}\right)^{0.15}. \tag{6}$$

In addition, the turbulence intensity of the velocity profile is assumed to follow the Gaussian distribution with a standard deviation of $\acute{O} = 0.15$.

3. Model applications

The method described above was employed to study the wind flow characteristics around a surface-mounted cube [6], and a high-rise building [7]. In this study, wind flow around the low-rise TTU was investigated, and main focus was placed on the corner roof vortex. It is known that when the wind flow does not approach in the direction normal to the face of the building, such as a quartering wind, two counter rotating vortices along two side corners on the roof could be generated. This highly turbulent vortex flow always produces large suction pressure, which is the key factor resulting in house roof damage.

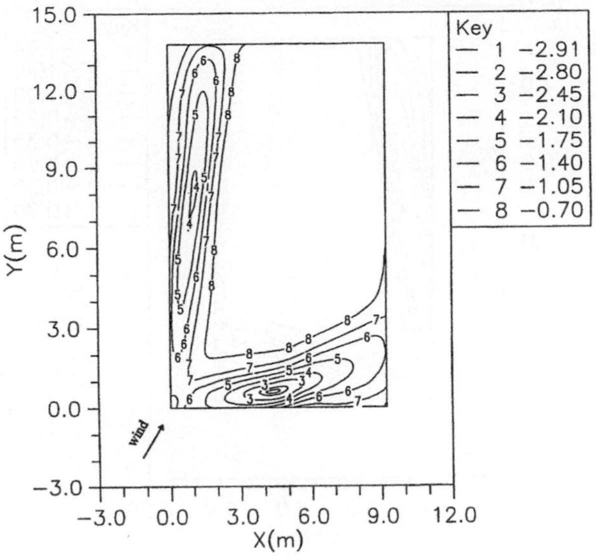

Fig. 3. Mean pressure coefficient on the roof surface, computed using the fine mesh system, wind azimuth of 215°.

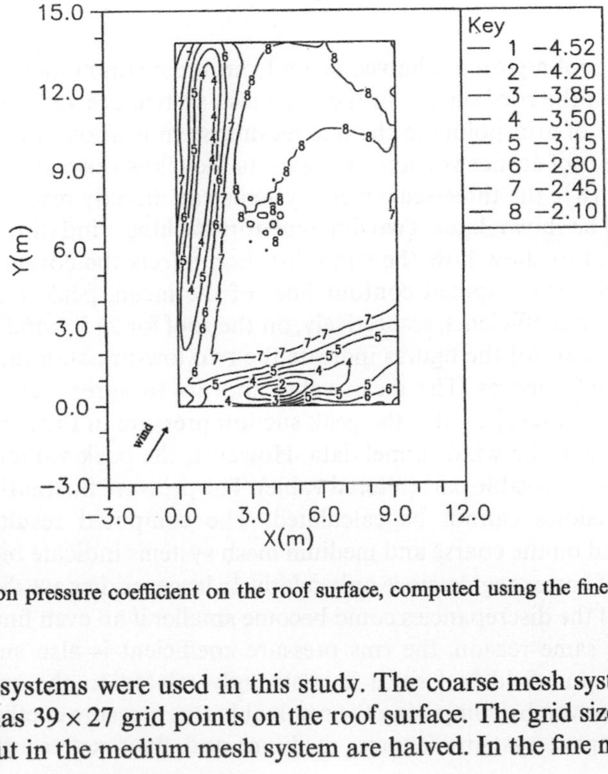

Fig. 4. Peak suction pressure coefficient on the roof surface, computed using the fine mesh system, wind azimuth of 215°.

Three mesh systems were used in this study. The coarse mesh system is shown in Fig. 1, which has 39 × 27 grid points on the roof surface. The grid sizes Δ on the roof and roundabout in the medium mesh system are halved. In the fine mesh system, the

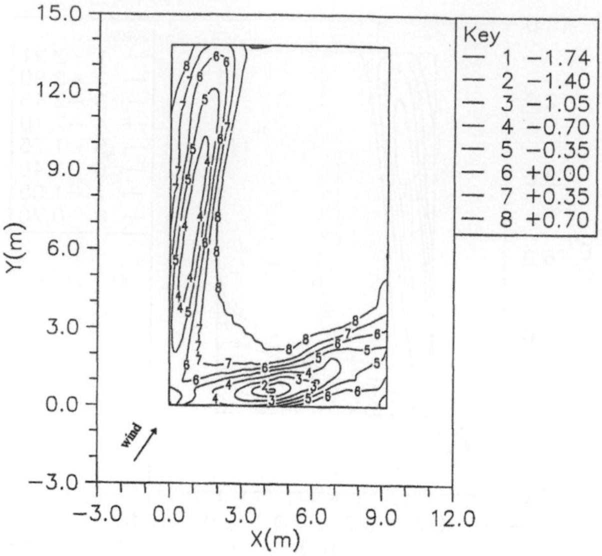

Fig. 5. Peak pressure coefficient on the roof surface, computed using the fine mesh system, wind azimuth of 215°.

grid sizes in the roof region are halved again from the medium mesh system. Hence, the grid points on the roof surface in the fine mesh system are 153×105. As a result, the total number of grid points in the fine mesh system is about 1.2×10^6. Since the eddy scale of the roof corner vortices increases as they flow downstream, as shown in Fig. 2 (after Ref. [8]), the three-scale mesh systems can directly resolve different eddy levels, which will be shown later. Two different approaching wind directions (215° and 225°) are modeled to show how the wind direction affects the corner roof vortices.

Figs. 3–5 show the computed contour lines of the mean, peak suction and peak maximum pressure coefficients, respectively, on the roof for 215° wind direction using the fine mesh system. All the figures indicate the two low-pressure regions generated by the corner roof vortices. The mean value appears to agree well with the wind-tunnel data and field test [2]. But the peak suction pressure in the corner roof center region is higher than the wind-tunnel data. However, the peak values in Figs. 4 and 5 include only the resolvable part (filtered value). The pressure fluctuations induced by the small-scale eddies cannot be calculated. The computed results of the eddy fluctuations based on the coarse and medium mesh systems indicate big discrepancies in the corner roof vortex region since only relatively larger eddies are directly resolved. It is believed that the discrepancies could become smaller if an even finer mesh system is used. For the same reason, the rms pressure coefficient is also smaller than the wind-tunnel test and field test since it is strongly related to the small-scale eddy fluctuations. However, by comparing the resolvable rms pressure coefficients based on the three scale mesh systems (coarse, medium, and fine) representing three level

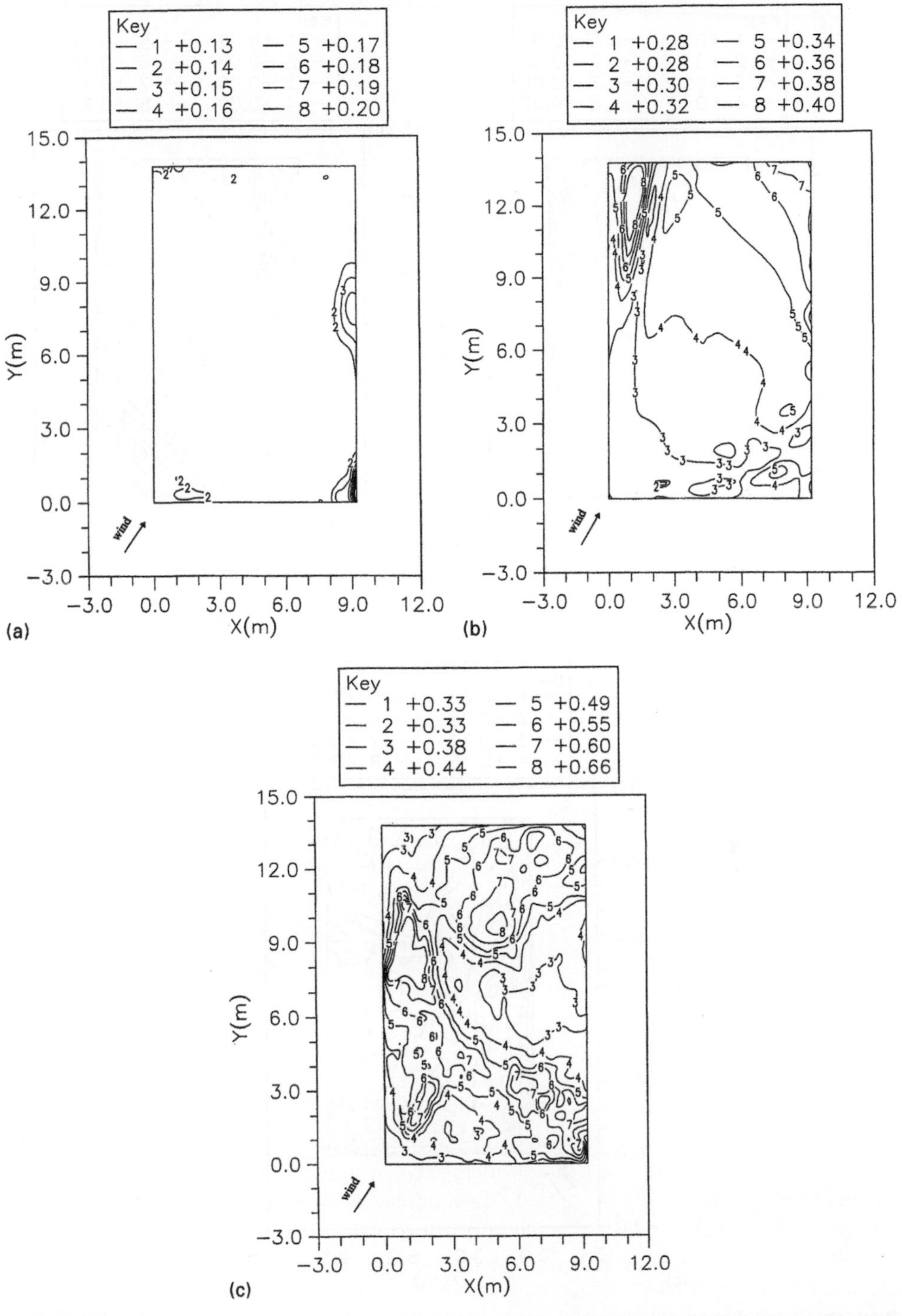

Fig. 6. Computed rms pressure coefficient on the roof surface, wind azimuth of 215°, (a) coarse mesh, (b) medium mesh, (c) fine mesh.

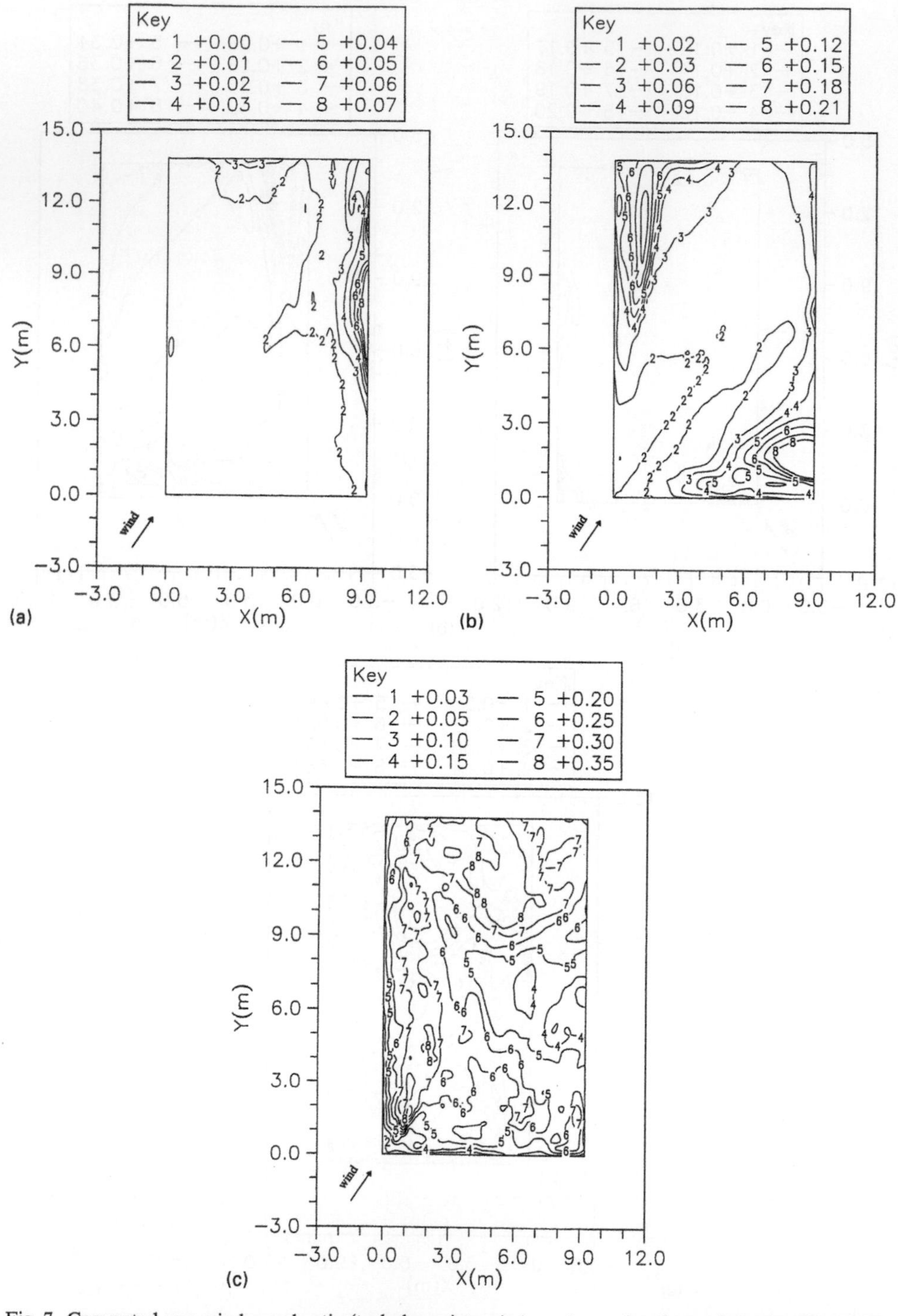

Fig. 7. Computed rms wind speed ratio (turbulence intensity) on the roof surface, wind azimuth of 215°, (a) coarse mesh, (b) medium mesh, (c) fine mesh.

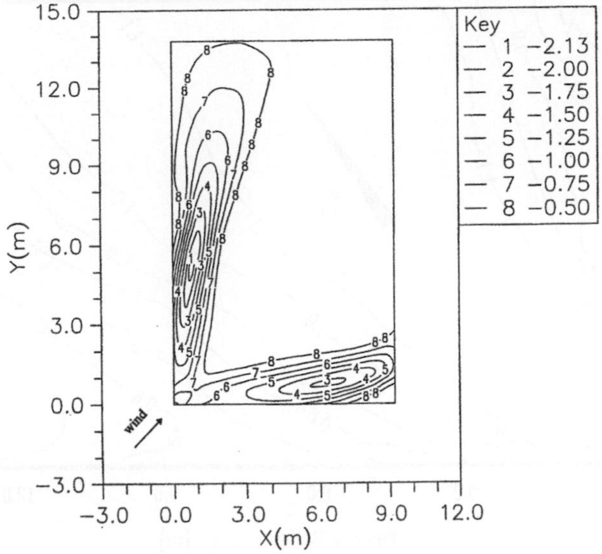

Fig. 8. Mean pressure coefficient on the roof surface, computed using the fine mesh system, wind azimuth of 225°.

filter sizes, as shown in Fig. 6a–6c, respectively, it is found that the fluctuations due to small-scale eddies modeled by the fine mesh system behave more randomly. The resolvable rms wind speed ratios (turbulence intensity) on the roof based on the three mesh systems show a similar pattern, as indicated in Fig. 7a–7c. Apparently, the coarse mesh system does not directly resolve the fluctuations in the corner vortex regions.

For the 225° wind direction case, the wind direction turns 10° in clock-wise direction. The main features of the corner vortices based on the computed results are quite similar to those in the above case. The mean pressure coefficient on the roof in this case is shown in Fig. 8. Compared with Fig. 3, the notable difference due to different wind direction is the location switch of the lowest pressure corner vortex region.

Comparison of the computed mean velocity on the roof with the wind tunnel test (after Ref. [2]) is shown in Fig. 9 for the approaching wind of 315°. In the computation model, this wind direction can be interpreted from the case of 225° wind direction due to geometrical similarity. Generally, the computed results agree very well with the wind-tunnel data. The low wind speed regions due to the roof corner vortices are also successfully resolved. However, at the lower right corner, the computation does not show the high speed spot with the mesh system since the mesh system is still not fine enough in this small area. The resultant vorticity of X and Y components ($\sqrt{\omega_x^2 + \omega_y^2}$) at the middle of the roof corner vortices is shown in Fig. 10. This figure clearly indicates the existence of the two roof corner vortices.

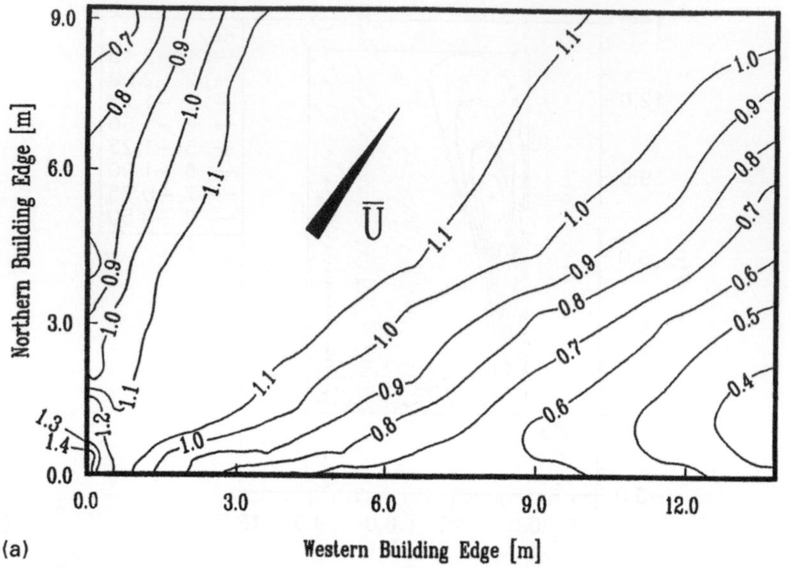

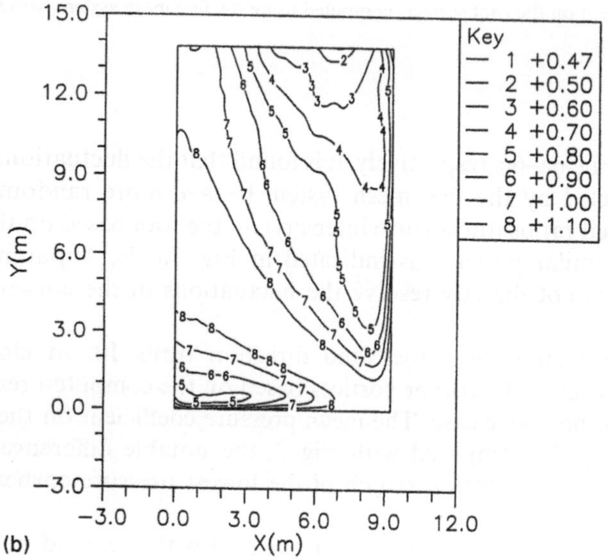

Fig. 9. Comparison of the computed mean wind speed ratio with the wind-tunnel test data at 1 mm above the roof surface of a 1 : 25 model [2], wind azimuth of 315°, (a) wind-tunnel data, (b) computed results.

4. Conclusions

The large-scale structures of the roof corner vortex generated by a quartering wind over the TTU building are resolved successfully. The mean values generally agree well with the wind-tunnel data and field test. By applying three different mesh sizes, which

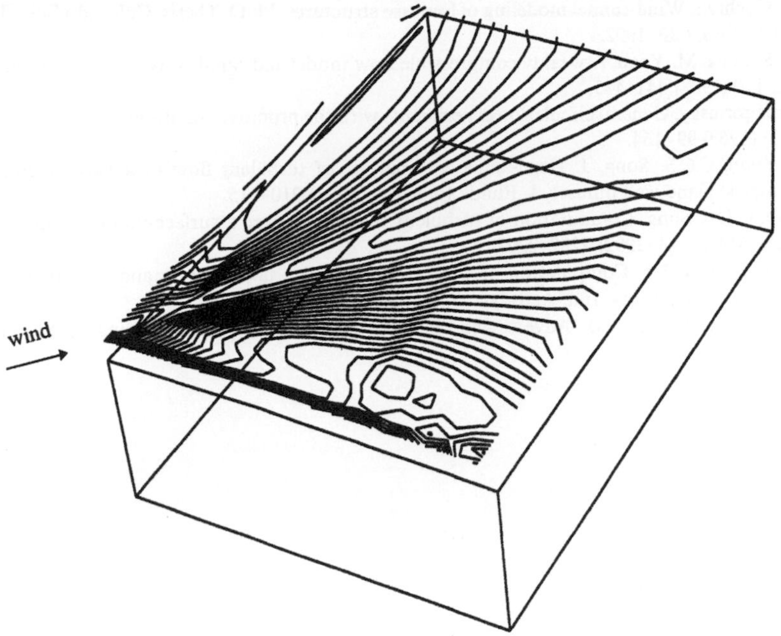

Fig. 10. Computed resultant vorticity of X and Y components ($\sqrt{\omega_x^2 + \omega_y^2}$) at the center of the roof corner vortices, wind azimuth of 215°.

represent three-level filter scales, it is found that finer fluctuations can be resolved with a finer mesh system. Since only the resolvable values can be obtained in the computation, it is difficult to compare the computed pressure fluctuations with wind tunnel and field data. However, the computed results show that if the mesh is fine enough, the resolvable solution could catch all scale levels of the fluctuating eddies and the computation will be closer to the wind-tunnel and field test data. It would be interesting to try to use a grid size comparable to the size of measuring instruments and compare the results on equal size basis.

Acknowledgements

This study is collaborated with CPP, Inc. The initiation of the interesting research topic by Dr. J.A. Peterka in CPP, Inc. is appreciated. The computer time has been provided by the Minnesota Supercomputer Institute, University of Minnesota through its grant program.

References

[1] C.V. Chok, Wind parameters of Texas Tech University field site, M.Sc. Thesis, Texas Tech. University, Lubbock, TX, August, 1988.

[2] L.S. Cochran, Wind-tunnel modeling of low-rise structures, Ph.D. Thesis, Colorado State University, Fort Collins, CO, 1992.
[3] C.C.S. Song, M. Yuan, A weakly compressible flow model and rapid convergence methods, J. Fluids Eng. 110 (1988) 441–445.
[4] J. Smagorinsky, General circulation experiments with the primitive equations, Monthly Weather Rev. 91 (3) (1963) 99–164.
[5] M. Yuan, C.C.S. Song, J. He, Numerical analysis of turbulent flow in a two-dimensional non-symmetric plane wall diffuser, J. Fluids Eng. 113 (1991) 210–215.
[6] J. He, C.C.S. Song, Computation of turbulent shear flow over a surface-mounted obstacle, J. Eng. Mech. ASCE 118 (1992) 2282–2297.
[7] C.C.S. Song, J. He, Computation of wind flow around a tall building and the large-scale vortex structure, J. Wind Eng. Ind. Aerodyn. 46 (1993) 219–228.
[8] J.A. Peterka, J.E. Cermak, Adverse wind loading induced by adjacent buildings, J. Struct. Div. 102 (1976) 533–548.

Journal of Wind Engineering
and Industrial Aerodynamics 67&68 (1997) 559–572

Wind-driven rain distributions on two buildings

Achilles Karagiozis[a,*], George Hadjisophocleous[b], Shu Cao[b]

[a] *National Research Council Canada, Institute for Research in Construction, Building Performance Laboratory, Ottawa, Canada K1A-0R6*

[b] *National Research Council Canada, Institute for Research in Construction, National Fire Laboratory, Ottawa, Canada K1A-0R6*

Abstract

Wind-driven rain is an important consideration in the hygrothermal performance of building envelope parts. Wind-driven rain (in liquid form) can increase the amount of moisture present in the structure by more than 100 times that due to vapor diffusion. To date, very little work that provides field or laboratory wind driven rain data to moisture transport models is available. This information is a definite requirement as a boundary condition by the more sophisticated hygrothermal models such as LATENITE and WUFIZ which consider both vapor and liquid moisture flows. In this paper, the wind driven rain striking the exterior facade of two buildings (one twice the size of the other) is generated using a three-dimensional computational fluid dynamics (CFD) model that solves the air flow and particle tracking of the rain droplets around these two buildings. These simulations were carried out for a city center region. Four factors which govern wind-driven rain are investigated in this work: (a) upstream unobstructed wind conditions, (b) the rainfall intensity, (c) the probability distribution of raindrop sizes, and (d) the local flow patterns around the building. All four of these governing factors make wind-driven rain on a building facade very distinct. Simulations were carried out for three wind speeds of 5, 10 and 25 m/s, three rainfall intensities of 10, 25 and 50 mm/h and three wind directions 0°, 30° and 45° from the west face of the buildings. In this paper, only the results of the 0° wind direction are discussed. The results show distinct wetting patterns on the top of the building of both the two buildings which is most concentrated at the corners when the wind was normal to the facade surface. For the tallest building a distinct wetting pattern is displayed in the mid-height of the building. This information from wind engineering is directly employed for the design of building envelope moisture control. Results on a series of simulations are presented to demonstrate the effect of wind conditions, rain intensities, the interaction between the two buildings, and the droplet sizes on the wetting patterns on the faces of the short and tall building.

Keywords: Wind flow; Rain; Moisture transport; High-rise buildings; Multiple-building wind simulations; Wind driven rain; Numerical modeling

* Corresponding author.

0167-6105/97/$17.00 Published by Elsevier Science B.V.
PII S0167-6105(97)00100-1

1. Introduction

Hygrothermal (combined heat-air and moisture) performance in building envelopes dictates to large extent the durability and service life of the building envelope. The dominant role of hygrothermal performance on the resistance to deterioration (durability) is mainly because hygrothermal processes can occur in all three states namely vapor, liquid and ice. Each of these three states contributes differently to the deterioration mechanisms created by the response of the building envelope system to both interior and exterior environment excitations. Deterioration can exist in various forms, i.e. surface damage (discolourization by efflorescence), ageing processes (chemical damages) (moisture induced salt migration), structural cracking (due to thermal and moisture gradients), corrosion of steel, and mold or bacteria growth. The exact description of these deterioration processes is still not well understood, however, as the amount of available moisture increases so does the severity of degradation of the construction materials.

The exterior surface of a building envelope (facade) is constantly interacting with the ambient temperature, solar radiation, wind pressure, relative humidity, and wind-driven rain. Wind-driven rain is defined here as rain droplets carried along by wind having a characteristic angle with respect to the vertical. Recently, a few sophisticated heat-air-moisture transport models LATENITE [1], TRATMO2 [2], MATCH [3], WUFIZ [4] have appeared mostly within the research communities to predict the long-term hygrothermal behavior of building envelope systems. These models, to varying degrees of complexity, can handle vapor and liquid transport, crack flows, latent heat effects and wind-driven rain. In the recent paper by Karagiozis and Salonvaara [5] wind-driven rain was determined to be an important contributor to the total amount of moisture entering the structure. Indeed for masonry wall systems, wind-driven rain has been documented as the single most important source of moisture for certain brick-veneer exterior facade walls [1]. The amount of available water striking the surface of a wall can exceed the precipitation rate by a factor of 20. As an example, an east-facing wall in Vancouver can receive up to 400 kg/m^2 in one year period. Karagiozis and Salonvaara [5] showed that a brick cavity wall which included the effects of wind-driven rain retained approximately 36 times the amount of moisture than a wall considering only vapor transfer at the boundaries for the city of Vancouver. This is clearly depicted in Fig. 1 for a typical brick cavity wall.

In situations where moisture accumulates at rather high levels, structural deterioration due to freeze-thaw cycles may occur. Wind-driven rain entering the facade of a structure accelerates the normal deterioration process and therefore reduces the service life of the building. Therefore, any short term or long term, hygrothermal study on building envelopes requires the accurate prescription of wind-driven rain information as boundary condition inputs, especially when these studies lead to design guidelines.

Wind-driven rain is a complex phenomenon itself, relatively unresearched and still not fully understood. Rain droplets with a wide range of sizes are transported by wind that has a distinct 3-D behavior near buildings. The droplets trajectories are time

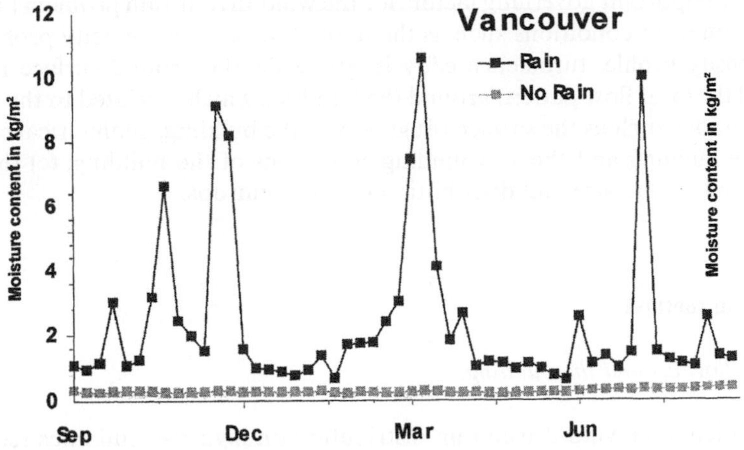

Fig. 1. The effect of wind-driven rain on the amount of moisture present in a brick masonry wall.

dependent due to the effect of turbulence. Furthermore, rain droplet size distributions vary randomly with respect to time and space. For these reasons, the amount of rain striking the exterior surfaces of a building is unique to that building as it depends on the local geometry of the building, topography around the building, wind speed, wind direction, rain intensity and rain droplet distribution.

Knowledge available on wind-driven rain, albeit limited, has been predominately determined by field experiments [6–9]. Recently however, investigations employing Computational Fluid Dynamics (CFD) methods [10–12] have appeared. Lacy [6], the British Code of Practice (BCP) [13], Wisse [12], Schwarz and Franc [8], Hens and Mohamed [9], Choi [10,11], Surry et al. [14], and Karagiozis and Hadjisophocleous [15] are some of the authors who worked in this area. The present study numerically examines wind-driven rain hitting the exterior surfaces of two buildings, one being half the height of the tall one. As stand-alone buildings are a rarity, this study determines the effects of the interference of a short building on the wetting patterns of a tall building. The wetting patterns for the short building are also examined in this paper.

2. Wind-driven rain

Wind-driven rain, as discussed above, is strongly affected by the local wind around a building. Rain droplets, in the absence of wind, fall down vertically, under the influence of gravity. This would imply that the amount of rain hitting the walls of a building in the absence of wind will be theoretically zero. Near the vicinity of a building, raindrop trajectories are affected not only by the unobstructed wind flow, but also strongly affected by the particular flow distribution surrounding the building.

Some of the important governing factors for the wind-driven-rain problem [11] are (a) the upstream wind conditions such as the unobstructed wind velocity profile, turbulence intensity profile, turbulence eddy length scale, the ground surface roughness height, (b) the local flow pattern around the building, which is related to the upstream wind conditions such as the surface roughness of the building, geometrical configuration of the building and the surrounding conditions of the building, (c) the rainfall intensity and (d) the size and distribution of the raindrops.

3. Modeling method

3.1. *Wind flow around the buildings*

The prediction of wind-driven rain distribution on high-rise buildings requires the determination of the local 3-D time-averaged velocities, velocity fluctuations and the time-averaged pressures. The flow field around a bluff body, even for a simple cube, is very complicated. It includes a turbulent boundary layer, stagnation region, separation, reattachment, circulation and von Karman's vortex streets. The equations governing wind flow around buildings are the incompressible Navier–Stokes equations and the continuity equation. Closure to these equations is provided by various turbulence models, such as the k–ε two equation model, the algebraic second-moment closure model (ASM) and the Large Eddy Simulation. Specific details about these models can be found in the works of Rodi [16], Launder and Ying [17], and Murakami [18]. The k–ε model is one of the most commonly used model in the field of wind engineering and it has been adopted for this study. For this study the 3-D CFD code TASCflow [19] was used. The exact mathematical formulation of the model employed in this study is given by Karagiozis and Hadjisophocleous [15]. The flow was considered as time-averaged flow, and the output of the code included the kinetic, dissipation energy, velocity vector field and pressure. The additional development of a Langrangian model for water spray injection and tracking enabled the authors to investigated wind-driven rain.

3.2. *Rain droplet modeling*

In this section, information is presented regarding the size distribution of rain droplets, followed by a discussion on the 3-D Lagrangian Turbulent Particle Tracking method employed.

3.2.1. *The size distribution of raindrops*

Calculation of the amount of rainfall impinging on the building facades requires knowledge on the size distribution of raindrops in a particular storm. Best [20] carried out an extensive experimental study of rain droplet size distributions and their relationship with rainfall intensity.

3.2.2. Rain trajectories using the Lagrangian particle tracking method

Many engineering problems, such as wind-driven rain, involve the study of mixtures containing a continuous phase which exhibits fluid properties and a dispersed phase which is discretely distributed in the fluid. A Lagrangian tracking model can be used to predict the behavior of the dispersed phase. The Lagrangian tracking method tracks several individual rain droplets through the flow field, and involves the integration of particle paths through the discretized domain. Individual rain droplets are tracked from their injection point until they escape the domain or some integration limit criterion is met. The rain droplets are assumed to have a spherical shape and their density much greater than the air fluid density. In the present Lagrangian particle method, no influence of rain droplet on turbulent flow is assumed. However, the authors developed the model to keep track of the location and droplet characteristics and inject new particles every time step into the solution domain based on the rain characteristics.

4. Building structures and computational domain

In this study, two buildings located in an area sparsely populated by low-rise buildings are considered. The geometric configuration of the two buildings is shown in Fig. 2. For simplicity, the two building will be referred to as tall and short building, respectively. The width of the buildings, defined as h ($h = 15$ m), is used as the measuring unit for the geometry of the domain. The simulation domain, with a downstream length of $16h$, an upstream length of $16.0h$, a lateral width of $4h$ on each side,

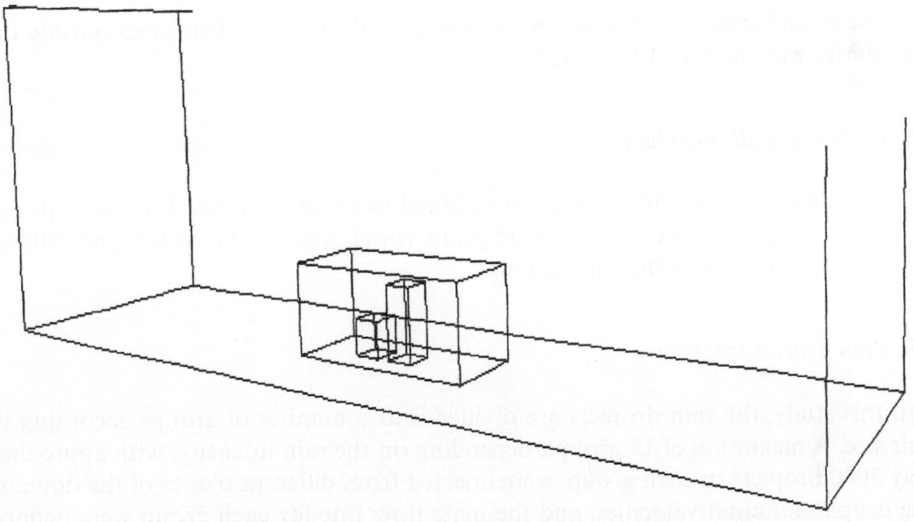

Fig. 2. Geometry for the two buildings ($h = 15$ m (width)).

and a vertical height of $10h$ was discretized. The distance between the tall and short building is $1h$. The computational domain was extended far before and after the building structure to accommodate far field effects. To avoid using a highly non-uniform spacing grid system, grid embedding close to the two buildings was employed. The short building was represented by a cube with sides equal to h. The tall building represented by a building model of $1 \times 2 \times 1h$ was used with $81 \times 31 \times 31$ control volumes in the grid embedding zone and $36 \times 11 \times 10$ control volumes outside the embedding zone in the x, y and z directions respectively. Fig. 3a and Fig. 3b, display the grid distribution, in both the x–y and x–z planes.

5. Boundary conditions

5.1. Upstream boundary

Wind velocity, turbulence intensity profiles for the upstream inlet were assigned values as given by Baskaran [21], suburban open country conditions.

5.2. Upper faces of the computational domain

The upper faces of the computational domain are defined as a slip-wall with zero shear stresses.

5.3. Downstream boundary and side faces of the computational domain

These boundaries are set to be pressure-specified openings. Pressures outside the boundaries were assumed to be zero.

5.4. High-rise wall boundary

The building's walls and ground are assumed as rough surfaces. For the high-rise walls and ground, an equivalent sand-grain roughness height of 0.1 and 0.03 m, respectively, are used in the simulations.

5.5. Rain droplet tracking

In this study, the rain droplets are divided into a number of groups according to their size. A maximum of 18 groups, depending on the rain intensity, with approximately 3000 droplets in each group, were injected from different planes of the domain. The droplet terminal velocities, and the mass flow rate for each group were defined prior to the simulations at the injection joints.

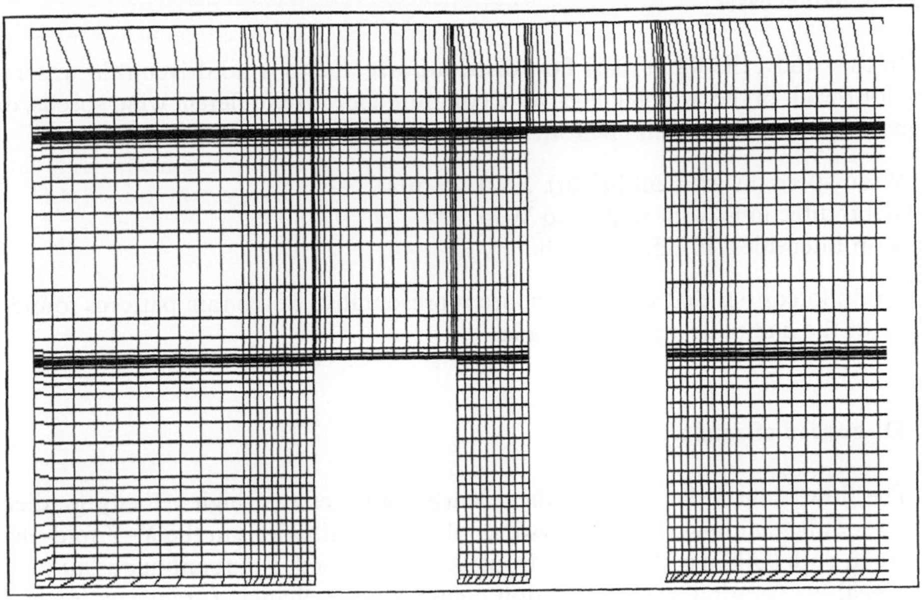

(a)

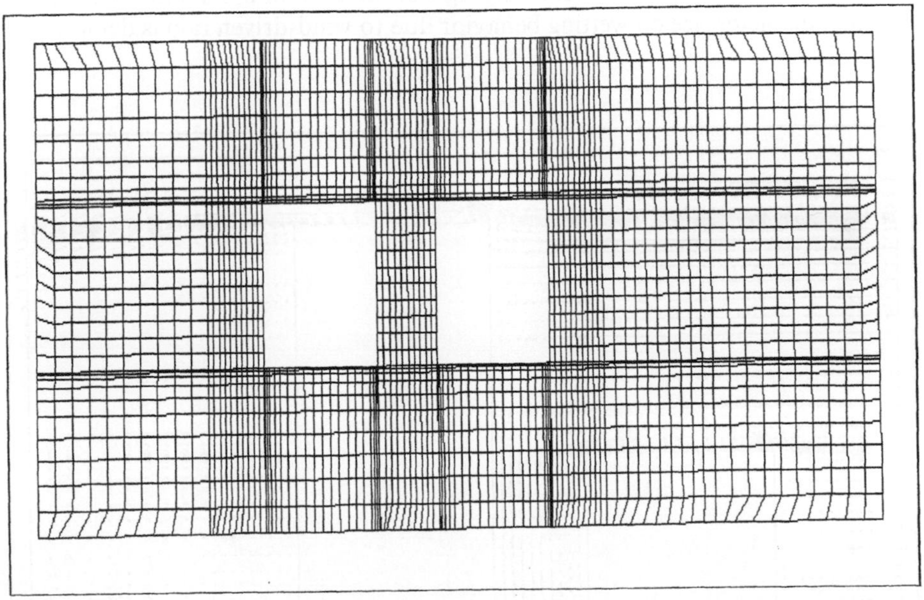

(b)

Fig. 3. (a) Grid distribution for the two high-rise buildings in X–Y plane. (b) Grid distribution for the two high-rise buildings in Y–Z plane.

6. Simulation cases

In this study, the water mass flow distributions, due to wind-driven rain water on the four walls of the two buildings, were computed for three different wind speeds, one wind angle and three rain intensities, as follows:

- Wind speed (at gradient height): 5, 10, 25 m/s.
- Wind directions (with respect to West): 0°
- Rain intensities: 10, 25, 50 mm/h.

The rain intensities that were chosen are representative of the rain patterns found in most of the habitable areas of Canada.

7. Discussion of results

The flow around the two buildings clearly indicates that high velocity gradients surround the buildings. Fig. 4 shows the velocity distributions in the $X–Y$ mid-plane through both buildings. As the rain droplets are dispersed in the continuous air phase, the local acceleration and deceleration create rapidly changing transport forces on each rain droplet. Fig. 5a–5d show a limited number of wind-driven rain trajectories for various rain droplet diameters ranging between 0.5 and 5.0 mm. The influence of droplet diameter size on wetting behavior due to wind-driven rain is demonstrated

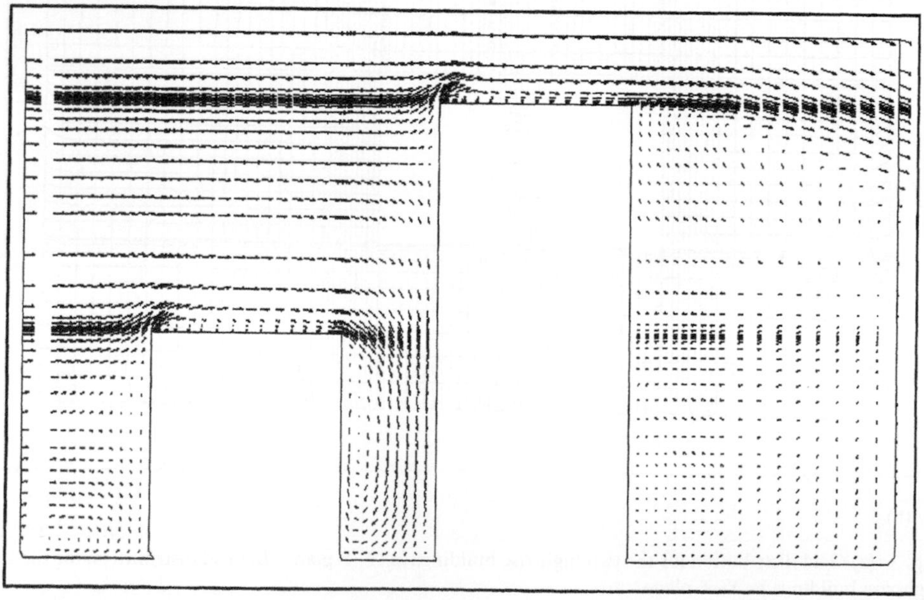

Fig. 4. Velocity distribution surrounding the two buildings in centerline $X–Y$ plane.

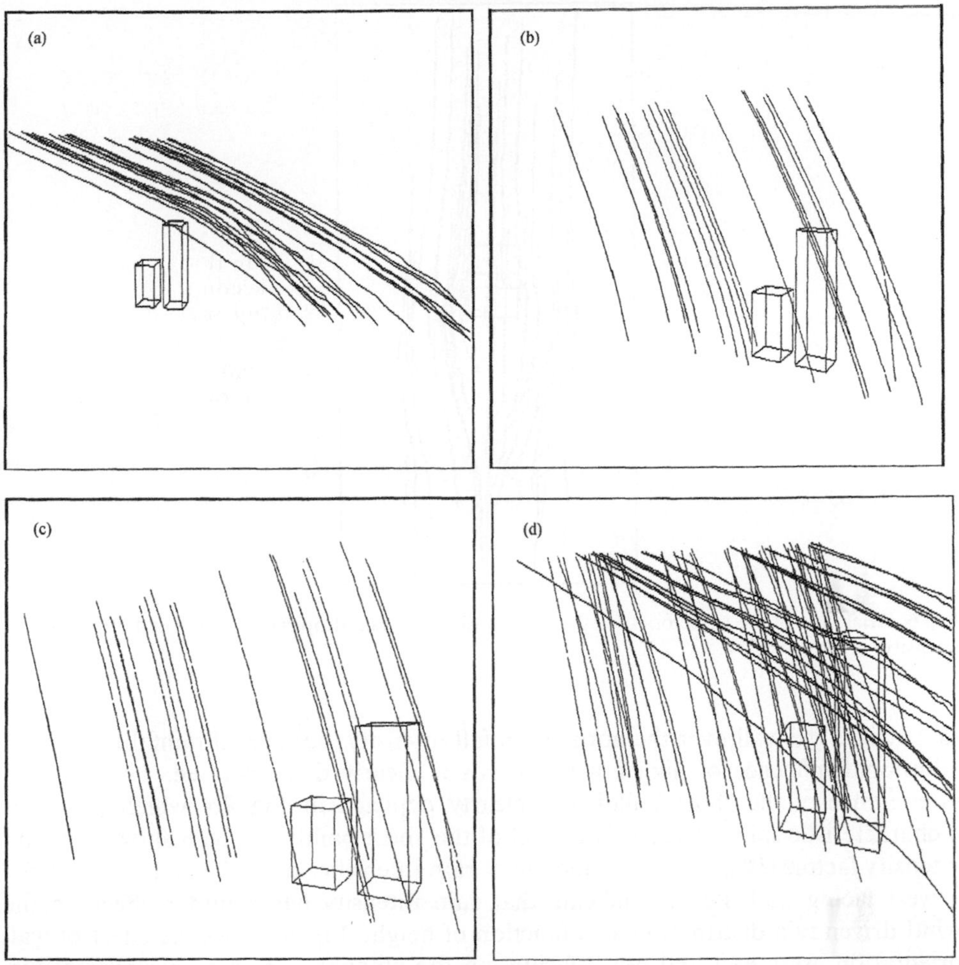

Fig. 5. Trajectories formed at 25 m/s rain intensity of 50 mm/h: (a) Rain-droplets diameter of 0.5 mm, (b) droplet diameter of 3.0 mm, (c) droplet diameter of 5.0 mm and (d) all sizes between (0.5 and 5.0 mm) droplet diameter.

to be very important. These figures, at the gradient height velocity of 25 m/s, show that the velocity field has a greater influence on the droplet trajectories for small diameter particles than for the much larger (5.0 mm) diameter droplet. For the 5.0 mm rain droplet diameter sizes, the droplet trajectories become more parallel to each other. While the figures depict trajectories with fairly straight lines, in 3-D presentation this is far from true. Fig. 6 shows trajectory results for the 1.0 mm rain droplet diameter as one could view from the top of the buildings. Fig. 7 displays the effect of wind speed on the rain intensity striking the windward (west) face of the short and tall building. The intensity ratio, I_{wall}/I_{rain} factor, is defined here as the intensity of rain

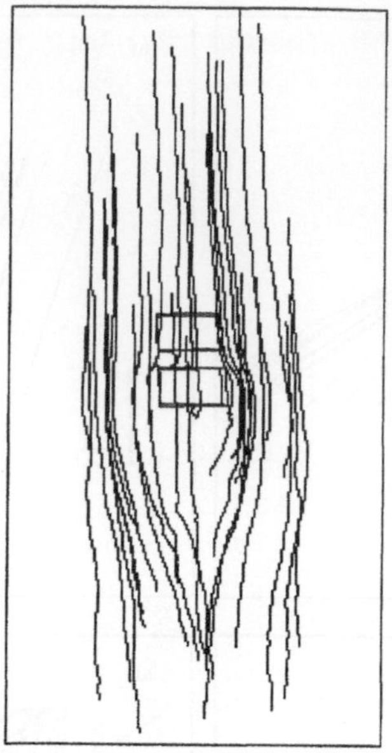

Fig. 6. Arial view of the trajectories for the 1.0 mm diameter rain droplets at 25 m/s and 50 mm/h rain intensity.

that strikes the wall over the normal rainfall intensity (horizontal rainfall). Here the additional influence of higher wind speeds is clearly demonstrated. The effect of interference of the short building is clearly depicted in Fig. 7b, which produces a characteristic spike at the height level of the short building. Fig. 8 shows the rain intensity factors (I_{wall}/I_{rain}) as a function of rain intensities of 10, 25 and 50 mm/h for a west facing wall. Results indicate that rain intensity has a minor effect on the wind-driven rain distribution as a function of height. Fig. 9 shows the effect of wall orientation, west, east and north facing, on the amount of rain each wall receives, when employing a rain intensity of 50 mm/h and a wind speed of 25 m/s. The south wall receives a similar amount of rain as the north wall, but is not shown due to illustration reasons. A distinct vertical wetting pattern distribution from the bottom to the top of the building is present. The west face, which is normal to the wind speed, receives more wind-driven rain, and the upper top area of the building receives the highest amount of rain for all faces. Irregularities in the trend, as we move from bottom of the building to the top are due to the randomness of the turbulent Lagrangian particle tracking method used.

8. Conclusions

A CFD method for predicting the wetting patterns on high-rise structures during wind-driven rain is presented in this paper. For both the short and tall buildings, the

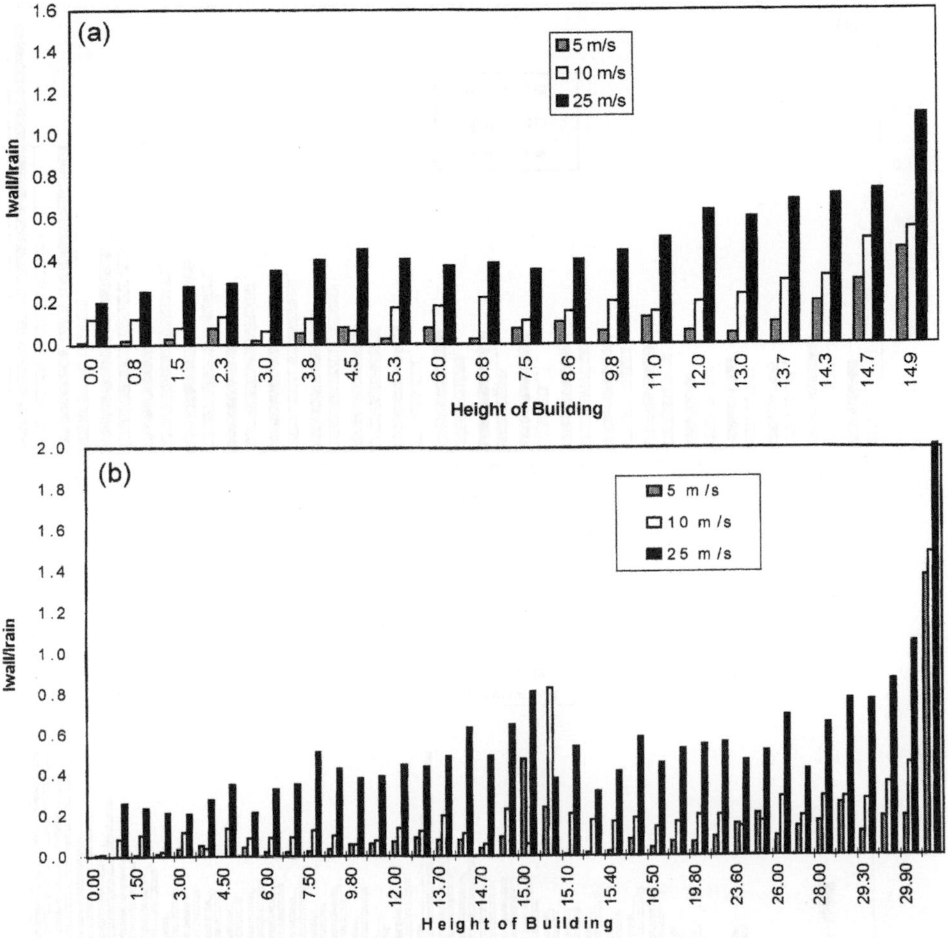

Fig. 7. Wind-driven rain intensity factors for the west facing wall as a function of wind speed, using $I_{rain} = 25$ mm/h for the (a) short and (b) tall buildings.

amount of wind-driven rain striking the buildings increase from bottom to top. The rain wetting patterns observed on the tall building were different, than in the case when the short building was not present. An overall 20% reduction of the total rain mass flow striking the exterior surfaces was observed for the tall building. However, much higher rain intensities are observed at mid-height regions of the tall building, which coincide with the height of the short building. The results also show that the downstream side (East is the case considered) of the wall receives little rain except near the top due to the wind induced recirculation in that region. This investigation found that, at a 25 m/s wind speed, the upper top areas of the tall high-rise building received slightly more than twice the horizontal rain intensities. The results show that the

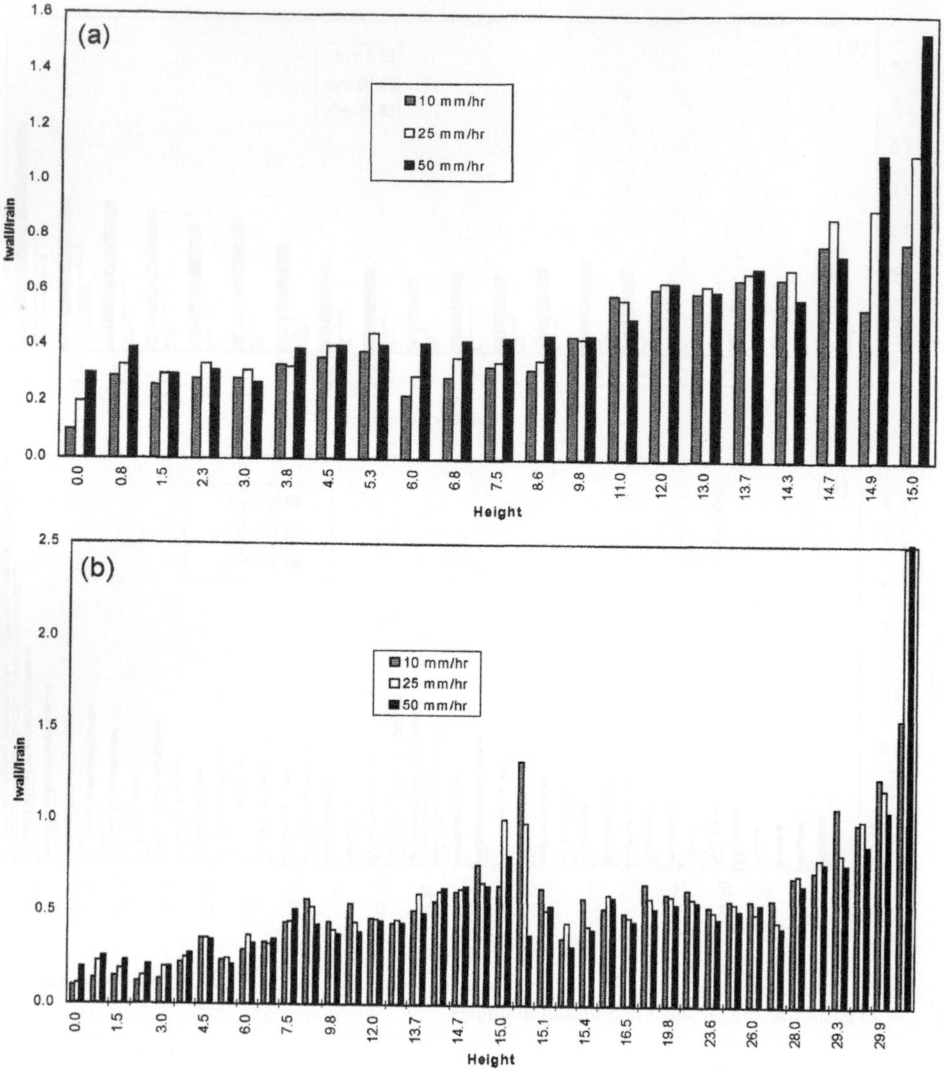

Fig. 8. Wind-driven rain intensity factors for the west facing wall as a function of rain intensities, using $U = 25$ m/s, for the (a) short and (b) tall building.

rainfall intensity does not significantly affect the wetting pattern of the building walls. Higher wind speeds, however, were found to significantly increase the amount of water received on the building facade. The higher the wind speed, the higher the amount of water received.

Information generated using this complex 3-D flow and turbulent particle tracking approach, is essential for accurate hygrothermal modeling. The amount of rain hitting the exterior facades of a building must be incorporated into hygrothermal models, so

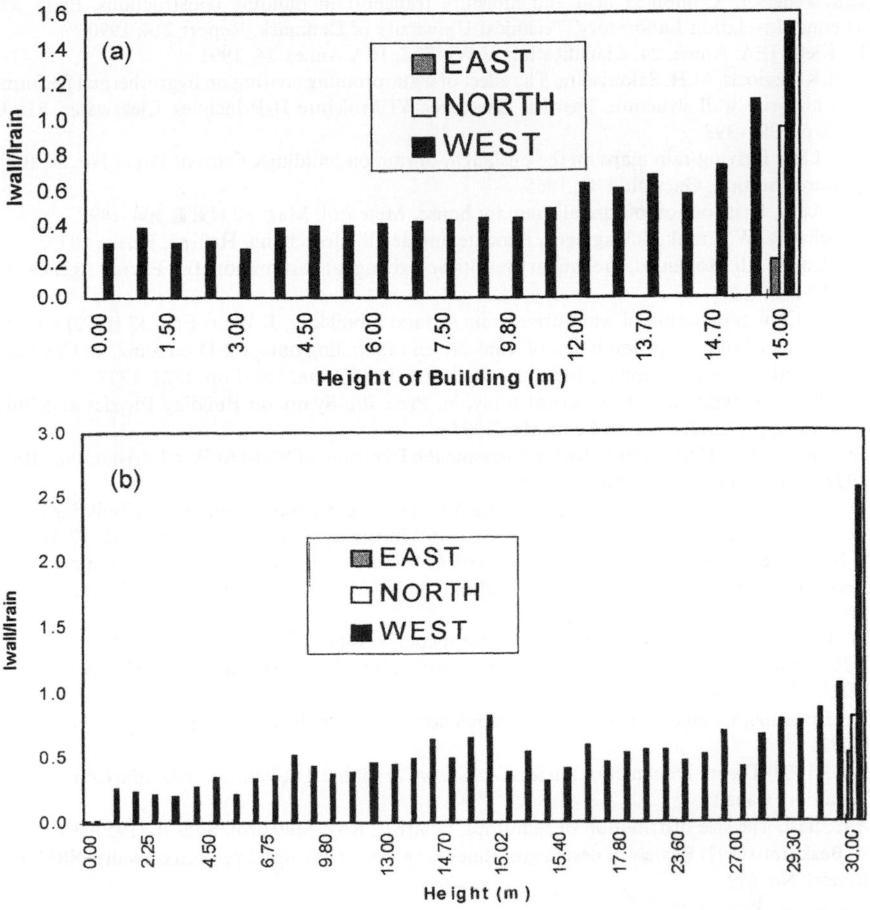

Fig. 9. Wind-driven rain intensity factors for the west, east, and north faces of the high-rise building, using $U = 25$ m/s, and $I_{rain} = 50$ mm/h for (a) the short and (b) the tall building.

as to account for liquid transport at the surfaces. The analysis of durability issues with respect to the hygrothermal performance can only start when models are equipped with methods of incorporating driving rain, that is a function of building geometry, wind speed and orientation, rain intensity and interference objects such as the presence of other building and structures.

References

[1] A. Karagiozis, Overview of the 2-D hygrothermal heat-moisture transport model LATENITE, Internal IRC/BPL Report, 1993.

[2] H.M. Salonvaara H.M., "TRAMTO2" VTT Finland, 1994.

[3] C.R. Pedersen, Combined heat and moisture transport in building constructions, Ph.D. Thesis, Thermal Insulation Laboratory, Technical University of Denmark, Report 214, 1990.
[4] H. Kieβl, IEA, Annex 24, Classification of WUFIZ, IEA Annex 24, 1991.
[5] A.N. Karagiozis, M.H. Salonvaara, The effect of waterproofing coating on hygrothermal performance of a high-rise wall structure, Thermal Envelopes VI/Moisture II-Principles, Clearwater, FL, USA, 1995, pp. 391–398.
[6] R.E. Lacy, Driving-rain maps on the onslaught of rain on buildings, Current Paper No. 54, Building Research Station, Garston, UK, 1965.
[7] R.E. Lacy, Distribution of rainfall round a house, Meteorol. Mag. 80 (1951) 184–189.
[8] B. Schwarz, W. Frank, Schlagregen, Berichte aus der Bauforschung, Heft 86, Berlin, 1973.
[9] H. Hens, F. Ali Mohamed, Preliminary results on driving rain estimation, Int. Energy Agency, Annex 24, T2-B-94/02
[10] E.C.C. Choi, Simulation of wind-driven-rain around a building, J. Wind Eng. 52 (1992) 60–65.
[11] E.C.C. Choi, Numerical simulation of wind-driven rain falling onto a 2-D building, in: Cheung, Lee, Leung (Eds.), Computational Mechanics, Balkema, Rotterdam, 1991, pp. 1721–1727.
[12] J.A. Wisse, Driving rain, a numerical study, in: Proc. 9th Symp. on Building Physics and Building Climatology, Dresden, 14–16 September 1994.
[13] British Standard Code of Practice for Assessing the Exposure of Walls to Wind-driven Rain. BS 8104: 1992, British Standards Institution, 1992.
[14] D. Surry, D.R. Inculet, P.F. Skerlj, J.X. Lin, A.G. Davenport, Wind, rain and the building envelope, Wind, Rain and the Building Envelope Seminar, University of Western Ontario, 16–17 May, 1994.
[15] A.N. Karagiozis, G.V. Hadjisophocleous, (1996) Wind-driven rain on high-rise buildings, Thermal Performance of the Exterior Envelopes of Buildings VI/Moisture II, Clearwater, FL, December 1995, pp. 399–405.
[16] W. Rodi, Examples of turbulence models for incompressible flows, AIAA J. 20 (1981) 267.
[17] B.E. Launder, D.B. Spalding, The numerical computation of turbulent flows, Comp. Meth. Appl. Mech. Eng. 3 (1974) 269.
[18] S. Murakami, Comparison of various turbulence models applied to a bluff body, J. Wind Eng. 52 (1992).
[19] ASC (1993) Theory Documentation for TASCflow, Version 2.2, Advanced Scientific Computing Ltd., Waterloo, Canada.
[20] A.C. Best, The size distribution of raindrops, Quart. J. Roy. Meteorol. Soc. 76 (1950) 16–36.
[21] A. Baskaran (1992) Review of design guidelines for pressure equalized rainscreen walls, NRC Internal Report No. 629.

ELSEVIER

Journal of Wind Engineering
and Industrial Aerodynamics 67&68 (1997) 573–587

JOURNAL OF wind engineering AND industrial aerodynamics

Chained analysis of wind tunnel test and CFD on cross ventilation of large-scale market building

Shinsuke Kato*, Shuzo Murakami, Takeo Takahashi, Tomochika Gyobu

Institute of Industrial Science, University of Tokyo, 7-22-1, Roppongi, Minato-ku, Tokyo 106, Japan

Abstract

The "chained analysis of wind tunnel test and CFD (computational fluid dynamics)" for cross ventilation of a large indoor space is proposed and an analysis example for a large-scale wholesale market building is demonstrated. The procedure of the "chained analysis" is as follows. With a wind tunnel model experiment, the overall airflow rate of cross ventilation and the wind pressure distribution at the building openings are measured. CFD simulation for the indoor flow field is then carried out with the boundary conditions of the measured wind pressure distribution. The simulated overall airflow rate of cross ventilation is then compared with that obtained from the wind tunnel test and the reliability of the CFD is confirmed. Detailed analysis of contaminant distribution in space is conducted with the obtained flow field from CFD simulation. With this procedure, reliable and detailed analyses of cross-ventilation characteristics of large indoor spaces become possible. To demonstrate the above-mentioned procedure, cross ventilation of a large-scale wholesale market building was analyzed and ventilation efficiency in the indoor space was evaluated.

Keywords: Numerical simulation; Wind tunnel test; Turbulent flow; Cross ventilation; Indoor air quality; Ventilation efficiency

1. Introduction

To comprehend the characteristics of cross ventilation, a detailed analysis of flow fields both in and around the building concerned is required. An accurate airflow rate prediction of cross ventilation is only obtained with an accurate CFD simulation of the energy dissipation (total pressure loss) of the flow fields [1]. That is, the airflow rate depends on both the drag (total pressure loss) characteristics of the internal flow and the total (or static) pressure difference between the windward and leeward

* Corresponding author. E-mail: kato@iis.u-tokyo.ac.jp.

PII S0167-6105(97)00101-3

openings of the buildings. An accurate energy dissipation simulation of flow field (i.e., accurate turbulent flow simulation) is not so easy to carry out since both flow fields require an enormous amount of CFD calculation with a computer.

Breaking down the above-mentioned situation, wind tunnel testing should be introduced for the basic part of the analysis. The basic part ensures reliability and practical usefulness by evaluating the overall airflow rate of cross ventilation [2]. In other words, the fundamental parameters of the phenomenon such as the total pressure difference between windward and leeward openings, the overall airflow rate of cross ventilation, etc., are evaluated by wind tunnel testing. Using these parameters, CFD is then carried out to analyze the space distribution properties. The authors call this procedure "chained analysis of wind tunnel test and CFD".

2. Chained analysis of wind tunnel test and CFD

2.1. Analysis of cross ventilation

Analysis of cross ventilation is divided into two steps. The first step consists in analyzing the basic properties. Overall cross-ventilation airflow rate should be analyzed first and is one of the most important factors in cross-ventilation analysis. With the overall airflow rate, the space-averaged features of contaminant concentration can be roughly estimated from the contaminant generation rate.

The second step consists in analyzing the air distribution in the indoor space. Contaminant and heat (temperature) have their own distributions in space and are not dealt with as uniform. When the zone occupied by people is limited to a large indoor space, the contaminant concentration and temperature are the most important, and those of other zones need not be strictly controlled. Analysis of their distributions in the room is indispensable to their efficient control.

2.2. Airflow rate analysis

2.2.1. Total pressure loss evaluation

It is usually explained that the airflow around a building creates a driving force of cross ventilation and that the airflow rate of the cross ventilation is governed by this driving force (total pressure difference) and the drag (total pressure loss) of the cross-ventilation flow. In this context, the wind pressure difference between the windward and leeward openings is assumed to be the driving force and expression of the ability of cross ventilation. To evaluate the wind pressure with a certain degree of accuracy, the airflow around the building should be predicted with a certain degree of accuracy.

2.2.2. Difficulty of applying CFD

For analyzing the airflow around the complicated geometry usual in an actual building case study, accurate CFD simulation remains a difficulty at the present. Such flow includes separations, reattachments, circulation, etc. It is characterized by unisotropic and inhomogeneous turbulent properties. The turbulence models used for simulation should be deliberately selected to reproduce such complicated flow features [1]. The analysis grid should also be sufficiently fine to avoid any grid resolution dependence of the simulation results. These situations lead to enormous amounts of calculation and sometimes make it impossible to carry out the simulation.

2.2.3. Established wind tunnel test

A wind tunnel experiment, on the contrary, can reproduce turbulent flow far easier than CFD since it utilizes real physical phenomena. It has been proven that it usually provides useful and reliable predictions of the airflow rate and pressure distributions around building openings [2]. It is reasonable to use wind tunnel testing to analyze the overall airflow rate of cross ventilation and wind pressure distributions around a building instead of CFD at the present.

2.3. Room air and contaminant distribution

2.3.1. CFD utilization

Indoor airflow may be analyzed either by CFD or a model experiment using a larger scale room model by imposing measured boundary conditions at the opening. Since a model experiment with a rather large scale ($\frac{1}{2}$–$\frac{1}{10}$) model is quite expensive, CFD simulation becomes more useful. Mean velocity (or mean momentum), mean concentration and heat are mainly transported by mean convection and not so much by turbulent diffusion in enclosed space. In this context, their distributions in space are mainly determined by mean convection and are easily predicted with a rather simple turbulence model such as the standard k–ε model, even if the energy dissipation process (total pressure loss) is not so accurately predicted. This evidence is quite useful for utilizing CFD for predicting the internal flow and contaminant diffusion fields of cross ventilation.

2.3.2. Tuning CFD boundary conditions

As mentioned before, if a simple turbulence model and a rather coarse grid system are used for CFD simulation, the obtained airflow rate is not so reliable since the energy dissipation (total pressure loss) process is not always simulated accurately. In this case, the airflow can be simulated by changing the pressure conditions a little bit, making it possible to obtain the airflow rate values corresponding to wind tunnel testing.

In this manner, the velocity distribution of the simulation can be controlled to correspond to the wind tunnel test. With this flow field, contaminant and heat distributions in the indoor space are analyzed. In Fig. 1, a "chained analysis of wind tunnel test and CFD" flow chart is shown.

2.4. Ventilation efficiency analysis

Contaminant distribution analysis based on CFD enables us to easily analyze the ventilation efficiency distribution in a space. In an indoor space, the ventilation ability

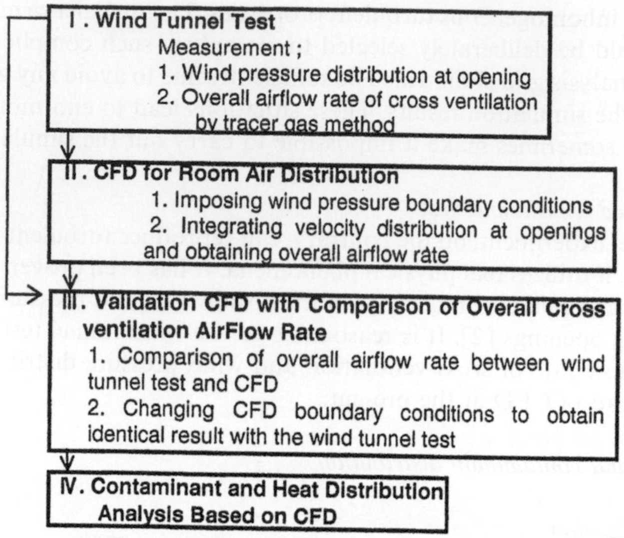

Fig. 1. Chained analysis of wind tunnel test and CFD.

of exhausting contaminants and supplying fresh outside air differs from point to point [3]. The age of inflow air and the contribution ratio of the exhaust opening, etc., are estimated on the basis of the CFD result. Details of ventilation efficiency indices and calculation method are described by Kato et al. [3].

Using the internal flow field of cross ventilation, the ventilation efficiency analysis is easily executed.

3. Case study of chained analysis

3.1. Naturally ventilated wholesale market building

3.1.1. Large-scale wholesale fish market space analyzed

A two-story, large-scale wholesale market building (220 m by 320 m by 45 m in size) is under planning. It is to be constructed in Tokyo, Japan. Here, ventilation by natural force is to be utilized to save energy and also to reduce maintenance requirements. The indoor climate was analyzed by chained analysis of the wind tunnel test and CFD [4,5].

The analyzed wholesale market building is shown in Fig. 2. The first floor is for a fish market and the second for a vegetable and fruit market. In the study, behaviors of cross ventilation in both markets were analyzed; however, only that for the fish market is reported.

3.1.2. Heat generation in the fish market

A large amount of heat generation is expected in the fish market space from artificial lighting, motor vehicles, and exhausted heat from refrigerators and so on.

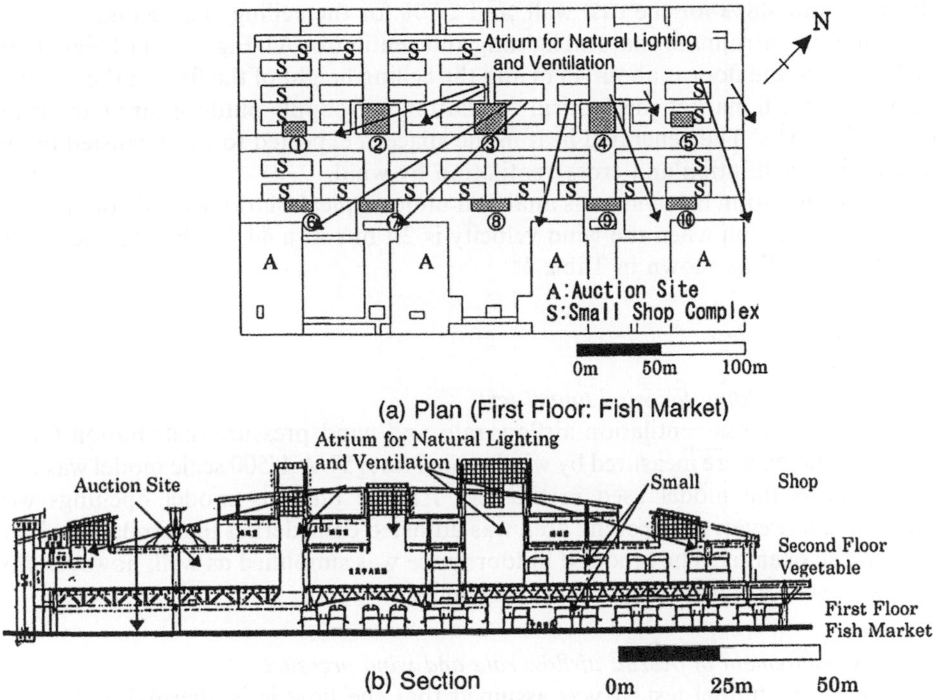

Fig. 2. Wholesale market building with wind forced cross ventilation.

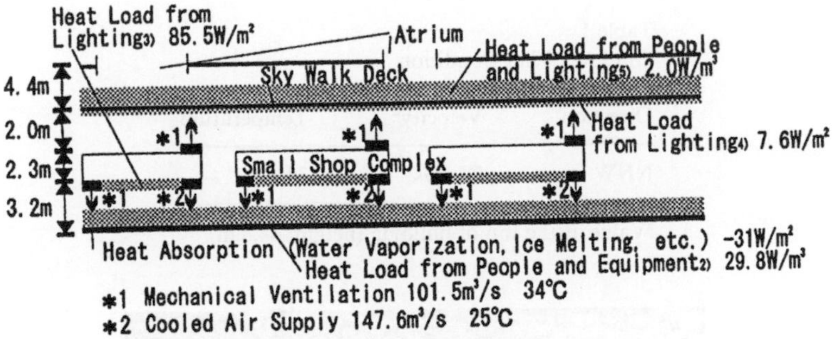

Fig. 3. Outline of heat load condition for the fish market.

The total heat generation rate amount is up to 4400 kW. Near the floor, a heat absorption of − 2800 kW by ice used for cooling fish, by chilled fish, and by water vaporization from the wet floor is also expected. A schematic view of the heat load in the fish market is shown in Fig. 3.

3.1.3. Cross ventilation conditions

Wind-forced natural ventilation is expected through wall openings. The ratio of the opening area to the wall is 24% for the north wall, 46% for the south wall, 47% for the west wall, 48% for the east wall, and 1.7% for the ceiling. The ceiling openings have high-rise atriums for natural lighting and ventilation (cf. Fig. 2). The height of the atriums from the floor is about 45 m and the ceiling height of the fish market is about 12 m. In the fish market, mechanical ventilation will supply outdoor air at an airflow rate of 250 m^3/s. The generated heat in the space is expected to be exhausted by this mechanical ventilation and cross ventilation by wind.

Cross-ventilation behavior was analyzed under typical outdoor wind conditions in the summer season when the wind velocity is 2.1 m/s at a 44.8 m height and a wind direction NNW as shown in Table 1.

3.2. Wind tunnel test

3.2.1. Model building for wind tunnel test

The overall cross-ventilation airflow rate and wind pressure distribution for the market building were measured by wind tunnel test [5]. A 1/500 scale model was used. Fig. 4 shows the model used in the test. All the building model openings were simplified; however, the opening area was adjusted considering the similarity of flow drag at the opening. The modeled indoor space was simplified as well; however, flow obstacles such as small shops were reproduced.

3.2.2. Measurement of overall airflow rate and wind pressure

In the wind tunnel test, it was assumed that the flow is isothermal and has no buoyancy effect. The overall cross-ventilation airflow rate was measured by tracer gas

Table 1
Outdoor Wind Condition

Direction	Velocity[a]	Temperature
NNW	2.1 m/s	33°C

[a]Value at the top of building (height: 44.8 m).

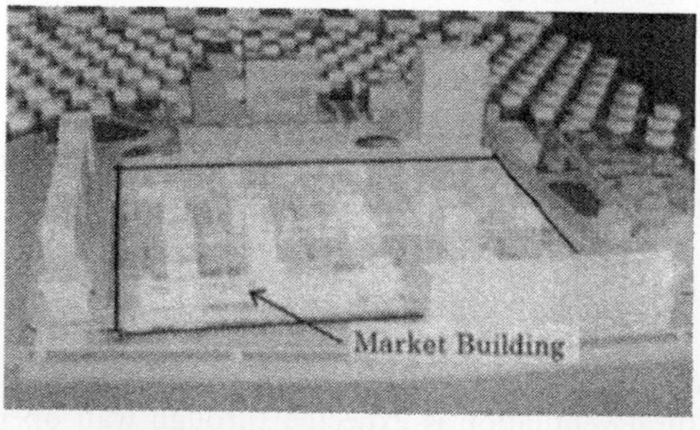

Fig. 4. Building model for the wind tunnel experiment (1/500).

method. In the measurement, instead of the averaged tracer gas concentration at exhaust openings, space-averaged tracer gas concentration (average concentration of about 50 points in the model building space) was used to estimate the overall airflow rate since, in wind tunnel testing, it is hard to distinguish exhaust openings. The difference between average concentration at the exhaust openings and that in the space is 0% when room air is perfectly mixed; however, it is usually estimated up to 15% under well-mixed room air and under uniform tracer gas generation throughout the space [3]. In the wind tunnel test, uniform tracer gas generation throughout the space could not be realized and a further 10% was added for the difference between them. In other words, the measured overall airflow rate in the wind tunnel test showed a variance of 25%.

The static pressure near the opening was measured and used for the CFD boundary condition. Variance in measured pressures was estimated to be less than 1%.

3.2.3. Temperature distribution analysis

Using a larger scale building model of 1/200, temperature distribution in the fish market under cross ventilation was also analyzed by the wind tunnel test. The correspondence of the Archimedes number (or Froude number) between the model test and the actual scale was considered. To reproduce buoyancy flows from the heat source, light (helium) gas was utilized. The temperature was estimated from the helium gas concentration. In the wind tunnel test, it was difficult to simulate both hot (light) and cold (heavy) air simultaneously; only hot (light) air was reproduced and its concentration was measured. The details of the wind tunnel test have been reported by Gyobu et al. [5].

3.3. CFD

3.3.1. Modeling of analyzing space

In CFD, flow fields in the building were calculated with pressure boundary conditions [4]. The velocity field of the outer area was not included. All obstacles inside the building were modeled with simplified configurations. Since buoyant flow by an interior heat source is likely to be affected by the location of the heat source, the positions of the heat sources were modeled precisely as shown in Fig. 3.

The flow fields of the first and the second floor will not be connected in the building; thus, each flow field was separately simulated. In Table 1, and Fig. 3, the simulation condition is shown.

3.3.2. Turbulence model and discretization method

The turbulence model used here is the standard k–ε model (Viollet type). The control volume method and a collocated grid system were used. The indoor space was divided into a $57 \times 103 \times 20 = 117\,420$ grid system.

3.3.3. Boundary conditions

At the openings, the given static pressures were set for the boundary condition of the mean momentum equations. The boundary conditions of the tangential velocity component at the openings were given as the free slip condition. Inflow turbulence intensity was assumed to be 10% and 1/7 times the height of the openings was given as the turbulence length scale. At the solid walls, the generalized log-law relation was assumed to give the wall boundary conditions. The given heat fluxes were set for the heat boundary conditions.

3.4. Results

The total CPU time required for internal flow field simulation of the fish market space was about 100 h by 8 GFLOPS computer.

3.4.1. Airflow rate

In Table 2, the airflow rate obtained by the simulation is tabulated. The overall airflow rate is 1 231 m^3/s. It is about 125% of the measured value by the wind tunnel test. In this simulation, the windward static pressures were given as rather overestimated values identical to the measured static pressures. Around the openings where the static pressure decreases toward the opening by transforming itself into dynamic pressure, it was possible that the measured static pressures were higher than those at the center of the openings. Considering this situation, the given pressures at inflow openings may reduce significantly in the simulation. Total indoor flow pressure loss had a tendency to be underestimated in the CFD simulation. Given these effects, the simulation showed a somewhat higher value of the overall cross-ventilation airflow rate. However, the discrepancy was about 25% and within the usual variance of the wind tunnel test measurement. Therefore, in this case no modifications for pressure boundary conditions were made.

3.4.2. Velocity distribution

In Figs. 5–9, the simulation results of the velocity fields for the fish market are shown. In the room, air velocity ranges from 0.2 to 0.5 m/s (cf. Fig. 5) under conditions where the outer air velocity is 2.1 m/s at a height of 44.8 m. The outside air is flowing into the space from the east side of the building and the inside air is flowing out through the west side. Since the wind pressure distributions of the north and south side are both negative, the inside air is also flowing out through the north and south side

Table 2
Airflow rate for each opening

Openings for outflow	m^3/s	Opening for inflow	m^3/s
North wall opening	405.7	East wall opening	691.5
West wall opening	408.2	Southeast side wall opening	273.4
Southwest side wall opening	222.9	Atrium 2,4	17.1
Atrium 1,3,5,6–10	194.2	Mechanical ventilation	249
Total	1231	Total	1231

Note: Outdoor wind condition: wind direction: NNW, wind velocity: 2.1 m/s.

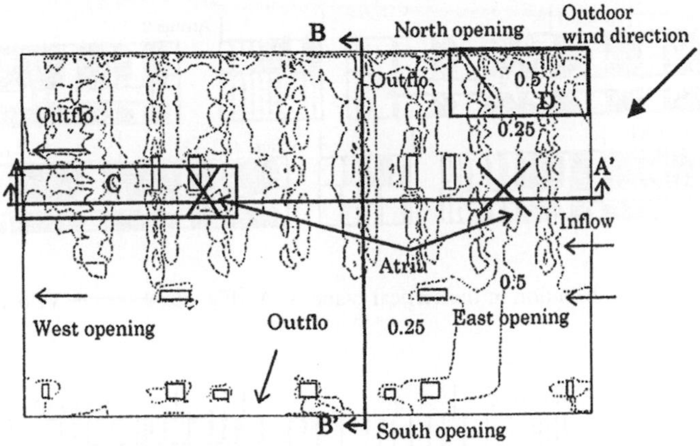

Fig. 5. Contour lines of absolute velocity in the horizontal plane (height: 1 m, fish market, unit: m/s).

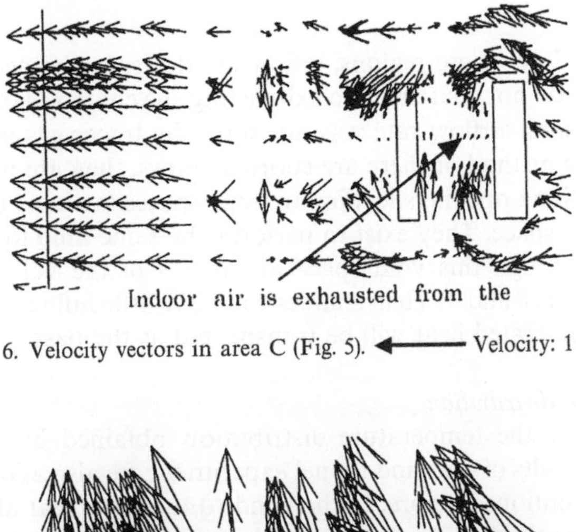

Fig. 6. Velocity vectors in area C (Fig. 5). ⟵ Velocity: 1 m/s.

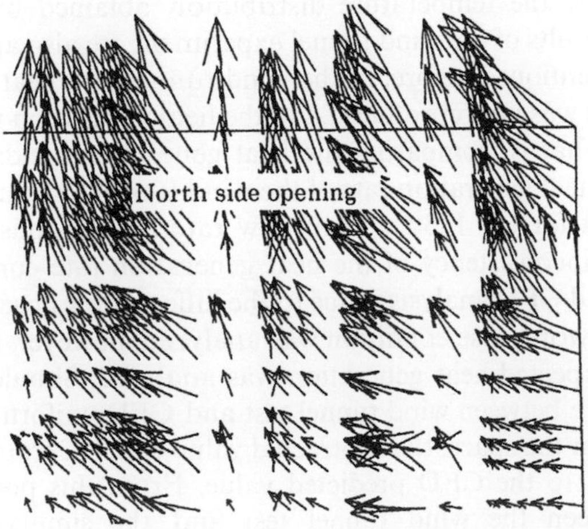

Fig. 7. Velocity vectors in area D (Fig. 5). ⟵ Velocity: 1 m/s.

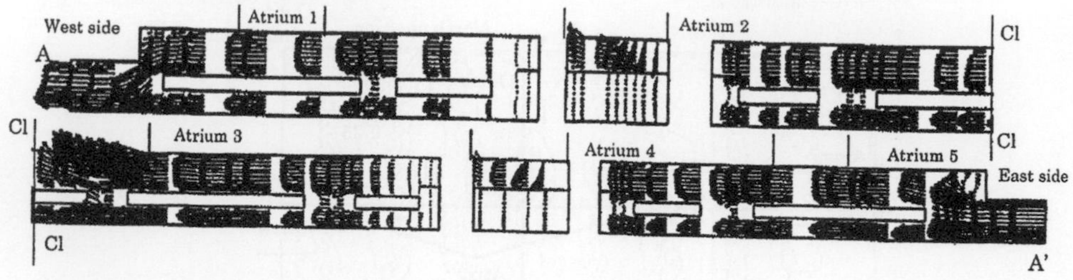

Fig. 8. Velocity distribution in the vertical plane A–A' (Fig. 5). ⟵ Velocity: 1 m/s.

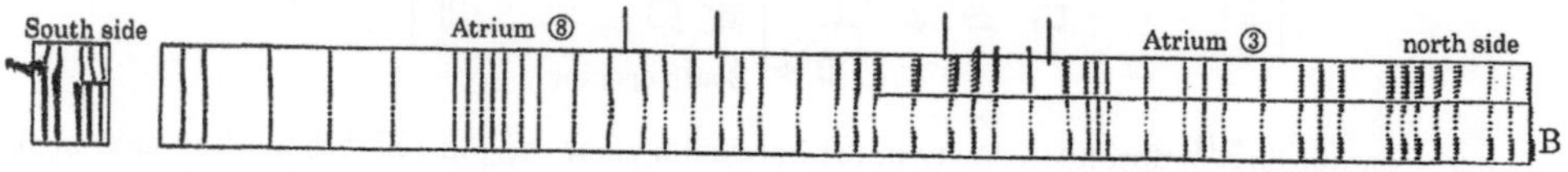

Fig. 9. Velocity distribution in the vertical plane B–B' (Fig. 5). ⟵ Velocity: 1 m/s.

openings (cf. Fig. 7). In these regions, inflow air flows out immediately, in shortcut flows. These flows do not contribute to exhausting the contaminants generated inside the space. The overall airflow rate was measured by tracer gas method in the wind tunnel test. In this method, if there are shortcut flows, they are underestimated.

There are 10 ceiling openings leading to void spaces (atriums for natural ventilation) in the indoor space. They exist to perform the same kind for ventilation as the wall openings. Through this void space about 1/3 of the volume of air is to be exhausted (cf. Figs. 6, 8 and 9). Heat sources will have little influence on the flow fields. In this case, the generated heat will be transported as the passive contaminant.

3.4.3. Temperature distribution

Figs. 10–12 show the temperature distribution obtained by simulation. In the illustrations, the results of the wind tunnel experiment are also added to the figures in parentheses. As mentioned before, in the wind tunnel test, heat absorption near the floor (− 2800 kW) was not reproduced and the heat generation rate was estimated as over 51 000 kW. That is, comparing the heat generation conditions used for CFD analysis, the total heat generation rate of the wind tunnel test was larger by 3500 kW. The 3500 kW heat for the 1230 m^3/s airflow rate corresponds to 2.4°C higher air temperature. The inconsistency of the heat generation rate conditions between the wind tunnel test and CFD analysis is due to the different purposes of analysis. In wind tunnel testing, the worst case condition was analyzed; however, in CFD analysis, the average state of expected heat generation was analyzed. Should the discrepancy in heat generation rate between wind tunnel test and CFD uniformly affect the temperature distribution in the space, the subtracted value of 2.4°C from wind tunnel testing should correspond to the CFD predicted value. From this point of view, the correspondence between the wind tunnel test and the simulation is not so bad (cf. Figs. 10–12).

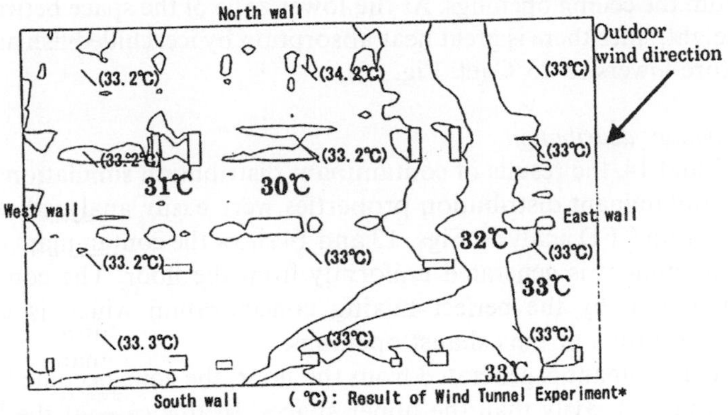

Fig. 10. Temperature distribution in the horizontal plane (height: 1 m, fish market). In the wind tunnel experiment, heat absorption by ice, etc., near the floor was not reproduced. Supposing uniform diffusion of this heat absorption throughout the space, the temperature falls 2.4°C.

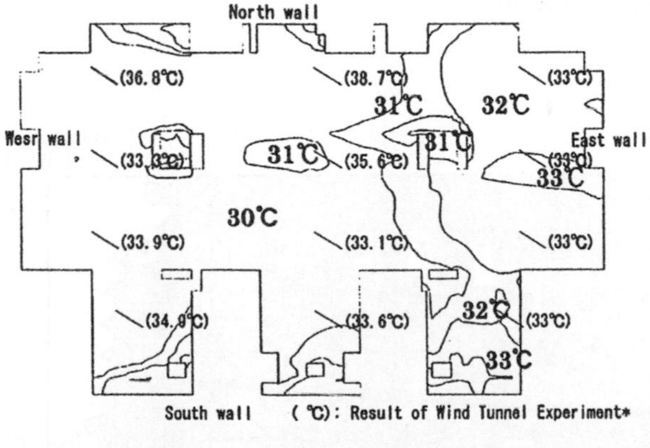

Fig. 11. Temperature distribution in the horizontal plane (height: 8.75 m). *cf. Fig. 10.

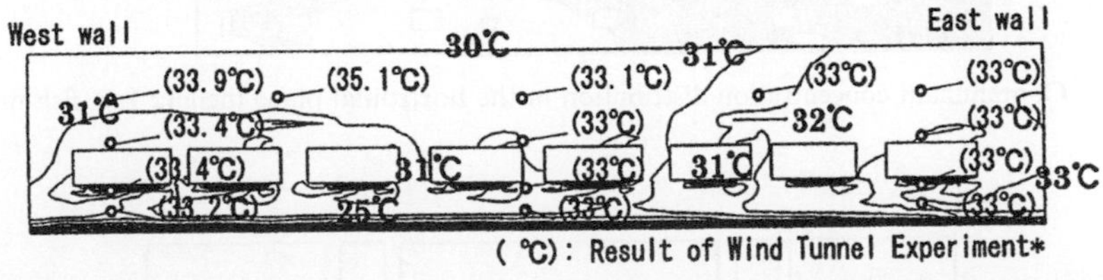

Fig. 12. Temperature distribution in the vertical plane A–A (Fig. 5). *cf. note to Fig. 10.

From the east side wall, outer hot air at 33°C flows into the room (cf. Figs. 10 and 11). The inflow air mixes with the cool air of the lower zone of the space and the temperature decreases to 30°C to 31°C. It seems that heated air from the heat load from lighting and so on does not mix with the room air and is efficiently exhausted from the ceiling openings. At the lower zone of the space between the floor and 80 cm height, since there is great heat absorption by ice, chilled fish and so on, the air temperature lowers to 25°C (cf. Fig. 12).

3.4.4. Contaminant distribution

In Figs. 13 and 14, the results of contaminant distribution simulation in the space are shown. Contaminant distribution properties were easily analyzed with the flow field obtained from CFD analysis. Figs. 13 and 14 show the contaminant distributions when the contaminant is generated uniformly from the floor. The concentration is non-dimensionalized by the perfect mixing concentration which is equal to the averaged concentration at the exhaust openings.

When the contaminant is generated from the floor, the zone occupied by people is contaminated more heavily than the upper spaces. In this case, at the leeward side (west side region) occupied zone, the contaminant concentration becomes 7–8 times the perfect mixing concentration. Comparing the higher concentration of the occupied zone, the concentration is not so high in the upper space. Above the sky walk deck level (7.5 m above the floor), the contaminant concentration becomes less than the perfect mixing concentration.

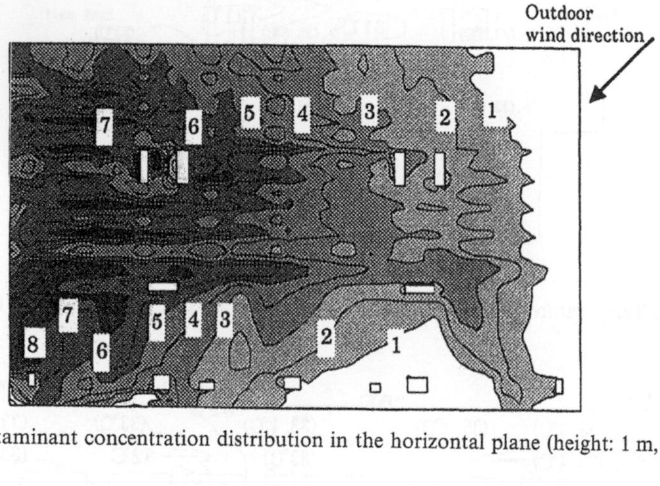

Fig. 13. Contaminant concentration distribution in the horizontal plane (height: 1 m, fish market).

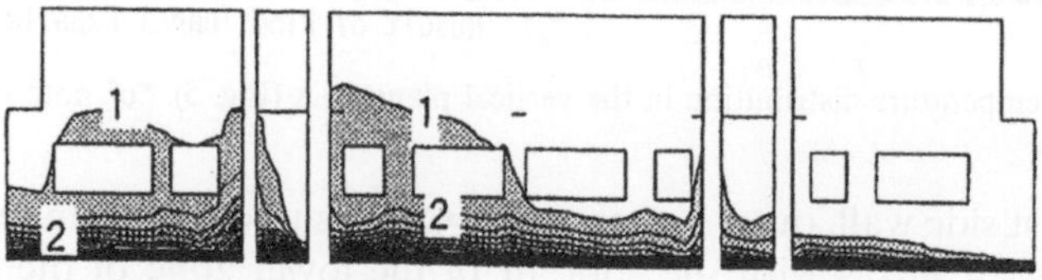

Fig. 14. Contaminant concentration distribution in the vertical B–B plane (Fig. 5) (contaminant is uniformly generated near the floor).

3.4.5. Ventilation efficiency

Fig. 15 shows the space distribution of inflow air age. The values show the required mean time for inflow air to reach a concerned point. The values are non-dimensionalized to the nominal air exchange time which is the quotient of the indoor space volume to the airflow rate. It shows a value of unity at the center of the space and almost two at the leeward side wall (west side wall). It is natural that at the leeward side zone the ventilation efficiency greatly deteriorates. Ventilation efficiency analysis shows the quantitative degree of ventilation efficiency and, in this case, ventilation efficiency near the west wall is two times worse than that of the center zone.

Age distribution analysis is done with contaminant concentration analysis of the condition of uniform contaminant generation throughout the space. In this case, the non-dimensionalized space-averaged concentration is 1.25. This means that the space-averaged concentration is 1.25 times the averaged concentration at the exhaust openings under this condition. In wind tunnel testing, the overall airflow rate is estimated from the space-averaged tracer gas concentration. If the same ratio could be assumed for the wind tunnel test, it means that the estimated overall airflow rate would be 80% of the real airflow rate. In such a case, CFD simulation corresponds well to wind tunnel testing.

Figs. 16 and 17 show the distribution of the contribution ratio of the ceiling opening (3) (atrium exhaust opening (3)). The contribution ratio at one point for the exhaust opening shows which ratio of air is exhausted from the exhaust opening concerned. In Fig. 16, the value of 0.1 at one point means that 10% of the air at that point is to be exhausted from the ceiling exhaust (3). Fig. 17 shows the territory of the exhaust opening (3). The air in the territory of the exhaust opening is to be exhausted from the exhaust opening. This is of course true for generated contaminant. Generated contaminant in the territory of the exhaust opening is to be exhausted from the opening and does not diffuse into the space.

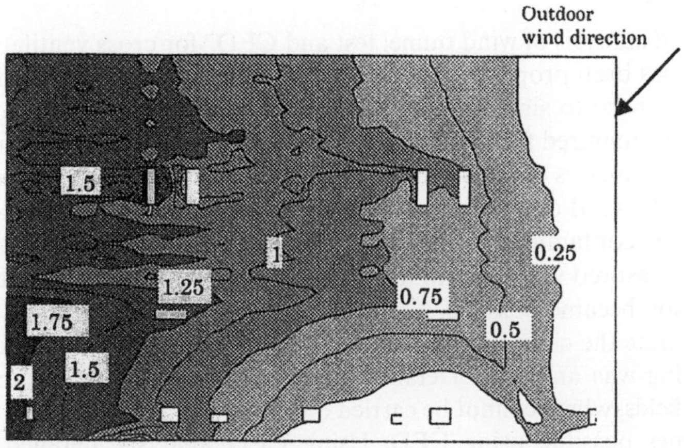

Fig. 15. Age distribution of inflow air height: 1 m (contaminant concentration distribution where contaminant is uniformly generated throughout the space).

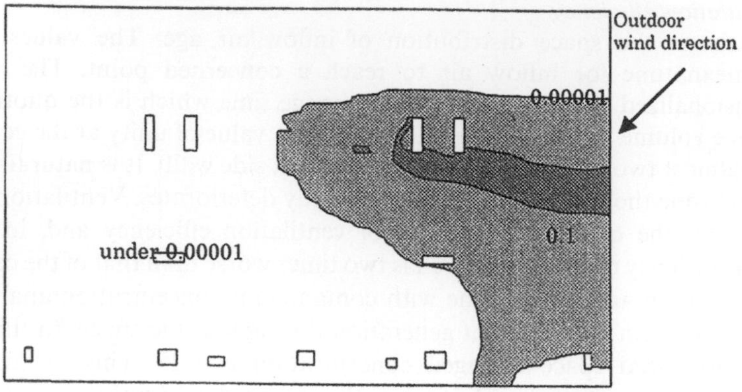

Fig. 16. Distribution of contribution ratio of atrium exhaust opening (3) in the horizontal plane (height: 1 m, fish market).

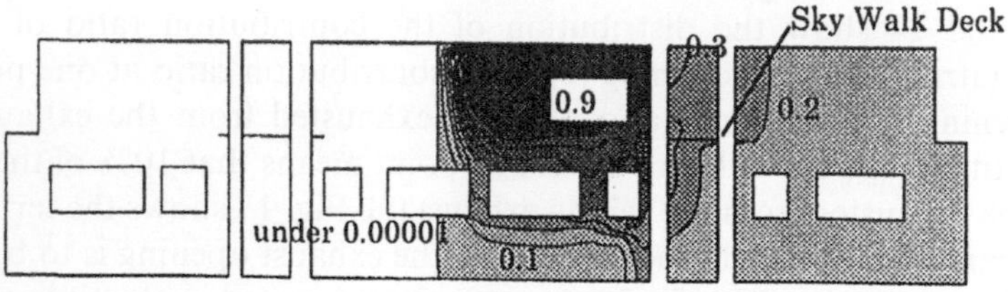

Fig. 17. Distribution of contribution ratio of atrium exhaust opening (3) in the vertical plane B–B (Fig. 5).

4. Conclusions

The "chained analysis of wind tunnel test and CFD" for cross ventilation of a large indoor space has been proposed. The method was devised on the recognition that at the present it is hard to simulate flow fields both in and around a building with the certain accuracy required for practical use. In this method, using wind tunnel testing, fundamental parameters of cross ventilation, such as total pressure difference between windward and leeward openings and overall cross-ventilation airflow rate are measured. The air and contaminant distributions in the room are then simulated with CFD based on the measured values. With this procedure, a reliable and detailed analysis of cross ventilation became possible.

To demonstrate the chained analysis, cross ventilation of the large-scale wholesale market building was analyzed. Detailed analysis of indoor flow, temperature, and contaminant fields, which cannot be carried out by wind tunnel test with a small-scale model, becomes possible using CFD. Using CFD analysis, detailed analysis for ventilation efficiency was also done. Analysis of ventilation efficiency was carried out quite easily by CFD only, this is one of the great merits of CFD analysis.

Acknowledgements

The authors would like to express their gratitude to the concerned central wholesale market, Tokyo Metropolitan Government, Nihon Sekkei Inc. and Azusa Sekkei Co. Ltd. for their cooperation in this study.

References

[1] S. Kato, S. Murakami, A. Mochida, S. Akabayashi, Y. Tominaga, Velocity pressure field of cross ventilation with open windows analyzed by wind tunnel and numerical simulation, J. Wind Eng. Ind. Aerodyn. 41–44 (1992) 2575–2586.

[2] S. Murakami, S. Kato, S. Akabayashi, K. Mizutani, Y.D. Kim, Wind tunnel test on velocity pressure field of cross ventilation with open winds, ASHRAE Trans. 97 (1991).

[3] S. Kato, S. Murakami, H. Kobayashi, New scales for evaluating ventilation efficiency as affected by supply and exhaust openings based on spatial distribution of contaminant, In: S. Murakami (Ed.), Room Air Convection and Ventilation Effectiveness, ASHRAE, pp. 177-186.

[4] T. Gyobu, S. Murakami, S. Kato, T. Takahashi, Study on indoor climate of large-scale space. Wind tunnel test on air and temperature fields of wholesale market building, Annual Meeting of Architectural Institute of Japan, 1995 (in Japanese).

[5] T. Gyobu, S. Murakami, S. Kato, T. Takahashi, Study on indoor climate of large-scale space. CFD analysis on velocity and temperature fields of wholesale market building, Annual Meeting of SHASE of Japan, 1996 (in Japanese).

ELSEVIER

Journal of Wind Engineering
and Industrial Aerodynamics 67&68 (1997) 589–599

JOURNAL OF wind engineering AND industrial aerodynamics

Improved turbulence models for estimation of wind loading

Shinji Kawamoto[1]

Flat Glass Division, Nippon Sheet Glass Co., Ltd., Anesaki-kaigan 6, Ichihara-city, Chiba 299-01, Japan

Abstract

The objective of this work is to develop a cost efficient and accurate turbulence model for the estimation of wind loading on buildings. This paper presents two types of k–ε–ϕ turbulence model, which are stable, simple, cost efficient and high-accuracy turbulence models, for estimation of wind loading on a bluff body. The variable ϕ, an anisotropy parameter of the Reynolds stresses, decreases due to the deceleration effect through an adverse pressure gradient, and increases due to the acceleration effect through a favorable pressure gradient. The coefficient of the kinematic eddy viscosity, C_μ, is defined by a function of ϕ and the helicity H, and the coefficient of production term of turbulence energy dissipation rate in the conventional standard k–ε turbulence model, $C_{\varepsilon 1}$, is defined by a function of ϕ. The favorable pressure gradient flow and the flows around a square prism are used for adjusting each coefficient, and the flows around a low-rise building in a $\frac{1}{4}$ shear flow are used for adjusting the coefficient of the helicity effect and for the verification of these models. The distributions of the mean pressure coefficient using these two types of k–ε–ϕ turbulence model are improved dramatically compared with those using the standard k–ε turbulence model.

Keywords: Turbulence modelling; Improved k–ε model; k–ε–ϕ model; Wind loading

Nomenclature

H	normalized helicity
k	turbulence kinetic energy
P	time-averaged pressure
P_k	production term of k
Re	Reynolds number
t	time

[1]E-mail: nsg10392@taxp2.nsg.co.jp.

PII S0167-6105(97)00102-5

U_i	time-averaged velocity vector
x_i	spatial coordinate vector
ε	dissipation rate of k
ν_t	kinematic eddy viscosity
ϕ	anisotropy parameter
Ω_i	time-averaged vorticity vector

1. Introduction

The standard k–ε turbulence model [1] has been applied to various practical problems, and the accuracy of this model has been proved. Then various improved turbulence models have been developed for high-accuracy prediction of turbulent flows, and these turbulence models have been verified by computing simple flow fields such as the channel flow and the backward-facing step flow [2,3]. However, most of the improved turbulence models are not effective for the complicated flow problems in the engineering fields.

An improved k–ε turbulence model for the impinging flow field was presented by Kato and Launder [4]. The model was improved further for the prediction of wind loading on buildings [5], and the accuracy of the pressure distribution is improved compared with the standard k–ε model. But the model does not always reproduce accurate pressure distributions for various angles of incidence [6], because the improvement is effective only for the impinging flow field.

On the other hand, an improved k–ε model was presented for predicting high-accuracy pressure distributions [7], in which the adverse pressure gradient effect and the impinging effect are introduced as the effects of wall-reflection, and favorable pressure gradient effect is introduced as a correction term for the turbulence kinetic energy production of the standard k–ε model. The improved k–ε model was developed into a k–ε–ϕ turbulence model [8], in which the advection-diffusion equation of an anisotropy information, ϕ, of the Reynolds stresses is introduced, and the pressure distributions on a square prism for some angles of incidence, α, are improved. Moreover, the effect of helicity was introduced to the k–ε–ϕ turbulence model [9], and the conical vortices on the roof of a low-rise building for angle of incidence $\alpha = 45°$ were reproduced.

In this paper, the latest two types of k–ε–ϕ turbulence model are presented. In the k–ε–ϕ model A, the adverse pressure gradient effect and the impinging effect are considered as the wall-reflection effects of pressure–strain correlations. The over-production of turbulence kinetic energy must be controlled at the impinging flow field [10]. The impinging effect as the wall-reflection is effective for the normal impingement to a face, e.g., the flow around a square prism for the angle of incidence $\alpha = 0°$. However, the impinging effect as the wall-reflection effect has no effect near the upstream edge of the face for oblique impingement, e.g., near the upstream edge of the flow around a square prism for the angle of incidence $\alpha = 45°$. In the k–ε–ϕ turbulence model B, the impinging effect is included in the adverse pressure gradient effect, which is not connected with wall-reflection effects to improve the drawback of the model A.

2. Basic equations

Table 1 shows the basic equations of the k–ε–ϕ turbulence model A. In this mode, the deceleration effect through impingement and adverse pressure gradient, the acceleration effect through favorable pressure gradient, and the helicity effect, as shown in Fig. 1, are introduced. The normal stress contribution to the turbulence kinetic energy production term is completely replaced for the favorable pressure gradient area, as shown in Eq. (6), because the corresponding normal stress contribution of the standard k–ε model, which is always positive, cannot reproduce the relaminarization. Eq. (8) is the advection–diffusion equation of the anisotropy information ϕ, $0 \leqslant \phi \leqslant 1$, of the Reynolds stresses, and the equation contains the source term and the sink term. The source term means increment of ϕ as the acceleration effect through favorable pressure gradient, and the sink term means decrement of ϕ as the deceleration effect through impingement and adverse pressure gradient. In the model A, source and sink terms are considered as effects of the wall-reflection of the pressure–strain correlation, i.e., these terms are represented by using the normal distance from the lth wall $h_n^{(l)}$, tangential velocity to the lth wall $U_{Si}^{(l)}$, and unit inner normal vector of the lth wall $n_i^{(l)}$. Though the variable ϕ is called the anisotropy parameter, this variable corrects not only the anisotropy effect of the Reynolds stresses but also other effects such as the imperfection of the definition of the kinematic eddy viscosity.

Yoshizawa and Yokoi [11] found that the helicity is the parameter which represents the weakness of the turbulence kinetic energy cascade process. This helicity effect is necessary to reproduce helical flow exactly, e.g., the conical vortices on the roof of building for $\alpha = 45°$, and the effect is introduced in the coefficient of the vortex viscosity C_μ, shown in Eq. (15) in the latest k–ε–ϕ turbulence model. Though model A does not satisfy the Galilean indifference, we can reproduce good solutions in practical problems with this model.

Table 2 shows the difference of the k–ε–ϕ turbulence model B from model A. The model B is much simpler than model A because the impinging effect is included in the adverse pressure gradient effect, which is not connected with the wall-reflection but the total time derivative of pressure. Therefore the impinging effect, i.e. the adverse pressure effect, is much larger near the upstream edge of the face for oblique

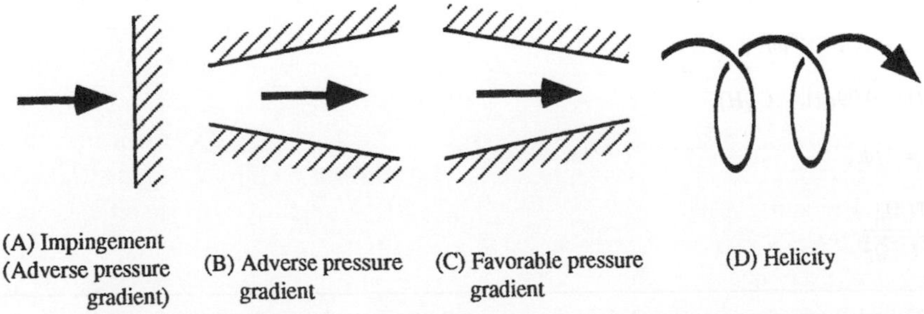

Fig. 1. Flow field model for improvement of the k–ε–ϕ turbulence models.

Table 1
Basic equations of the k–ε–ϕ turbulence model A

$$\frac{\partial U_i}{\partial t} + U_i \frac{\partial U_i}{\partial x_j} = -\frac{\partial P}{\partial x_i} + \frac{\partial}{\partial x_j}\left\{\left(\frac{1}{\mathrm{Re}} + \nu_t\right)\left(\frac{\partial U_i}{\partial x_j} + \frac{\partial U_j}{\partial x_i}\right)\right\}, \tag{1}$$

$$\frac{\partial U_i}{\partial x_i} = 0, \tag{2}$$

$$\nu_t = C_\mu(\phi, H)\frac{k^2}{\varepsilon}, \tag{3}$$

$$\frac{\partial k}{\partial t} + U_j \frac{\partial k}{\partial x_j} = \frac{\partial}{\partial x_j}\left\{\left(\frac{1}{Re} + \frac{\nu_t}{\sigma_k}\right)\frac{\partial k}{\partial x_j}\right\} + P_k - \varepsilon, \tag{4}$$

$$\frac{\partial \varepsilon}{\partial t} + U_j \frac{\partial \varepsilon}{\partial x_j} = \frac{\partial}{\partial x_j}\left\{\left(\frac{1}{\mathrm{Re}} + \frac{\nu_t}{\sigma_\varepsilon}\right)\frac{\partial \varepsilon}{\partial x_j}\right\} + \frac{\varepsilon}{k}(C_{\varepsilon 1}(\phi)P_k - C_{\varepsilon 2}\varepsilon), \tag{5}$$

$$P_k = \begin{cases} C_{F1}\left(k\dfrac{\partial P}{\partial x_i}\dfrac{U_i}{U_j U_j}\right) + P_{ks} & \text{if } \dfrac{\partial P}{\partial x_i}\dfrac{U_i}{|U_j|} < 0, \\ P_{kn} + P_{ks} & \text{elsewhere,} \end{cases} \tag{6}$$

$$P_{kn} + P_{ks} = \nu_t\left(\frac{\partial U_i}{\partial x_j} + \frac{\partial U_j}{\partial x_i}\right)\frac{\partial U_i}{\partial x_j}, \tag{7}$$

$$\frac{\partial \phi}{\partial t} + U_j \frac{\partial \phi}{\partial x_j} = P_\phi - S_\phi + \frac{\partial}{\partial x_j}\left\{\nu_t \frac{\partial \phi}{\partial x_j}\right\}, \tag{8}$$

$$P_\phi = \alpha_{F2}(1 - \phi), \tag{9}$$

$$S_\phi = (\alpha_I + \alpha_A)\phi, \tag{10}$$

$$\alpha_{F2} = \sum_l \begin{cases} -C_{F2}\left(\dfrac{k^{1/2}}{\varepsilon}\dfrac{\partial P}{\partial x_i}\dfrac{U_{Si}^{(l)}}{\sqrt{U_j^2}}\right)\dfrac{k^{3/2}}{h_n^{(l)}\varepsilon} & \text{if } \dfrac{\partial P}{\partial x_i}\dfrac{U_{Si}^{(l)}}{\sqrt{U_j^2}} < 0, \\ 0 & \text{elsewhere,} \end{cases} \tag{11}$$

$$\alpha_I = \sum_l \begin{cases} -C_I\left(\dfrac{k^{1/2}}{\varepsilon}\dfrac{\partial P}{\partial x_i}n_i^{(l)}\right)\dfrac{k^{3/2}}{h_n^{(l)}\varepsilon} & \text{if } \dfrac{\partial P}{\partial x_i}n_i^{(l)} < 0, \\ 0 & \text{elsewhere,} \end{cases} \tag{12}$$

$$\alpha_A = \sum_l \begin{cases} -C_A\left(\dfrac{k^{1/2}}{\varepsilon}\dfrac{\partial P}{\partial x_i}\dfrac{U_{Si}^{(l)}}{\sqrt{U_j^2}}\right)\dfrac{k^{3/2}}{h_n^{(l)}\varepsilon} & \text{if } \dfrac{\partial P}{\partial x_i}\dfrac{U_{Si}^{(l)}}{\sqrt{U_j^2}} > 0, \\ 0 & \text{elsewhere,} \end{cases} \tag{13}$$

$$U_{Si}^{(l)} = U_i - (U_i \cdot n_i^{(l)})n_i^{(l)}, \tag{14}$$

$$C_\mu(\phi, H) = 0.09\phi(1 - C_H H), \tag{15}$$

$$C_{\varepsilon 1}(\phi) = 1/\phi, \tag{16}$$

$$H = \frac{|U_i \Omega_i|}{\sqrt{U_j^2 \Omega_k^2}}. \tag{17}$$

Note: P_{kn} and P_{ks} are the normal and shear stress contributions to the production term of the standard k–ε turbulence model, respectively.

Table 2
Basic equations of the k–ε–ϕ turbulence model B (only the part different from model A)

$$P_k = \begin{cases} C_{F1}\left(\frac{DP}{Dt}\right) + P_{ks} & \text{if } \frac{DP}{Dt} < 0, \\ P_{kn} + P_{ks} & \text{elsewhere}, \end{cases} \tag{18}$$

$$S_\phi = \alpha_A \phi, \tag{19}$$

$$\alpha_{F2} = \begin{cases} C_{F2}\left(-\frac{1}{\varepsilon}\frac{DP}{Dt}\right)^{1/2} & \text{if } \frac{DP}{Dt} < 0, \\ 0 & \text{elsewhere}, \end{cases} \tag{20}$$

$$\alpha_A = \begin{cases} C_A R_t \left(\frac{1}{\varepsilon}\frac{DP}{Dt}\right)^{1/3} & \text{if } \frac{DP}{Dt} > 0, \\ 0 & \text{elsewhere}, \end{cases} \tag{21}$$

$$C_{\varepsilon 1}(\phi) = 1/\phi^2. \tag{22}$$

impingement as well. More over, the model B is precise because this model satisfies the Galilean indifference.

3. Adjustment of coefficients

3.1. Adjustment of relaminarization

The relaminarization in a strongly favorable pressure gradient flow is used to adjust the coefficient of the normal stress contribution to the turbulence kinetic energy production term, C_{F1}. Fig. 2 shows the outline of this experimental equipment [12]. A fully developed turbulent boundary layer is subjected to strongly favorable pressure gradient. Fig. 2 presents also the change of the turbulence kinetic energy along the stream during the favorable pressure gradient. Though the standard k–ε turbulence model does not reproduce the relaminarization in the favorable pressure gradient, both types of k–ε–ϕ turbulence model recover the accuracy in the relaminarization phenomenon. The boundary conditions of ϕ are $\phi = 1$ on the inflow boundary and $\partial\phi/\partial n_i = 0$ on the wall and outflow boundary in this computation.

The reproduction of the relaminarization effect is very important for high-accuracy prediction of the mean pressure coefficient on a bluff body. The fluid around the upstream edge is accelerated rapidly because of the strongly favorable pressure gradient in the flows around a square prism for $\alpha = 0°$ and $\alpha = 12.5°$. Therefore, we cannot reproduce the separation from the upstream corner correctly by using the standard k–ε model or the improved k–ε models which employ only impinging effect.

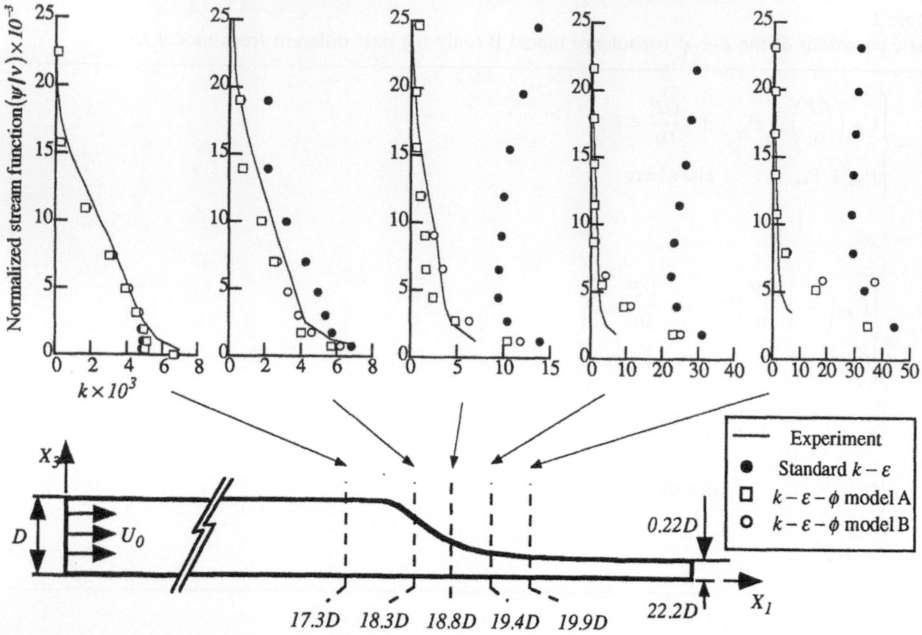

Fig. 2. Turbulence kinetic energy in favorable pressure gradient flow.

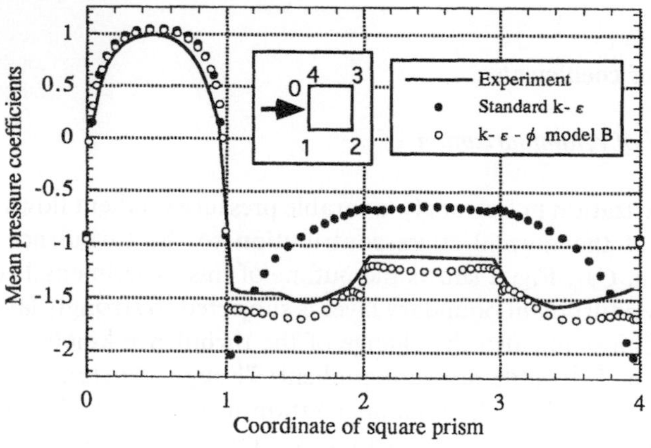

Fig. 3. Distributions of mean pressure coefficient on a square prism in a turbulent stream ($\alpha = 0°$, turbulent intensity of 8%).

3.2. Adjustment of ϕ-equation

The three coefficients of the advection–diffusion equation in the k–ε–ϕ model A and the two coefficients in the model B must be adjusted. In this stage, the coefficient C_{F1} is already fixed, as shown in Section 3.1, and the coefficient C_{H} is not related to these

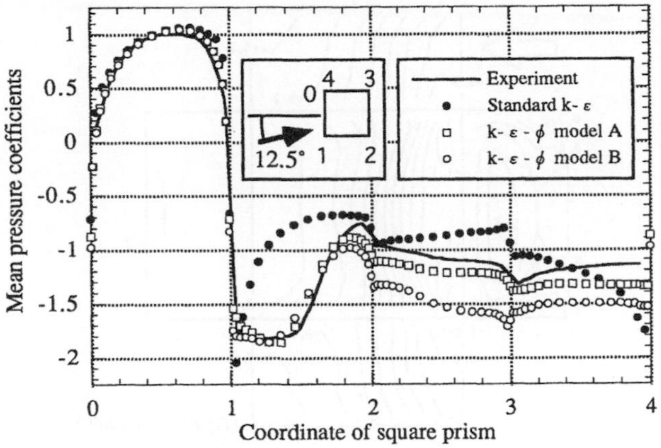

Fig. 4. Distribution of mean-pressure coefficient on a square prism in a turbulent stream ($\alpha = 12.5°$, turbulent intensity of 8%).

adjustments because two-dimensional problems such as the flow around a square prism are applied. Figs. 3 and 4 present the distributions of the mean pressure coefficient on a square prism in a turbulent stream for two angles of incidence, which is used for the adjustment of the coefficients. The model A yields better solutions than model B for these problems in the present state, because the model A has been tuned including the equation system for a long time. The vortices at the back of the square prism of the model B for $\alpha = 12.5°$ are much closer to the wall than those of model A. In consequence, the absolute value of the back-pressure coefficient becomes larger as shown in Fig. 4. The error is caused by the overproduction of the dissipation ratio of the turbulence kinetic energy, and as a result the rapid decrease of the turbulence kinetic energy at the back of the square prism. The equation system of the model B, especially Eq. (22), must be improved further for high-accuracy prediction of the back-pressure coefficient of the square prism.

3.3. *Adjustment of helicity effect and verification*

Figs. 5 and 6 show the distributions of mean-pressure coefficient on the $1:1:0.5$ building in a $\frac{1}{4}$ shear flow. The coefficient of helicity effect C_H is adjusted so that the mean-pressure distribution on the roof of the building is reproduced for $\alpha = 45°$, shown in Fig. 6C. Though the peak values of the negative pressure in the conical vortices are not reproduced exactly, the conical vortices themselves are reproduced, and the pressure distribution on the roof is improved dramatically compared with that of the standard k–ε model.

The standard k–ε model causes fatal errors for the prediction of the mean-pressure coefficient on a building, i.e., large stagnation pressure on the front face of the building, sharp peak and large gradient of the negative pressure behind the upstream

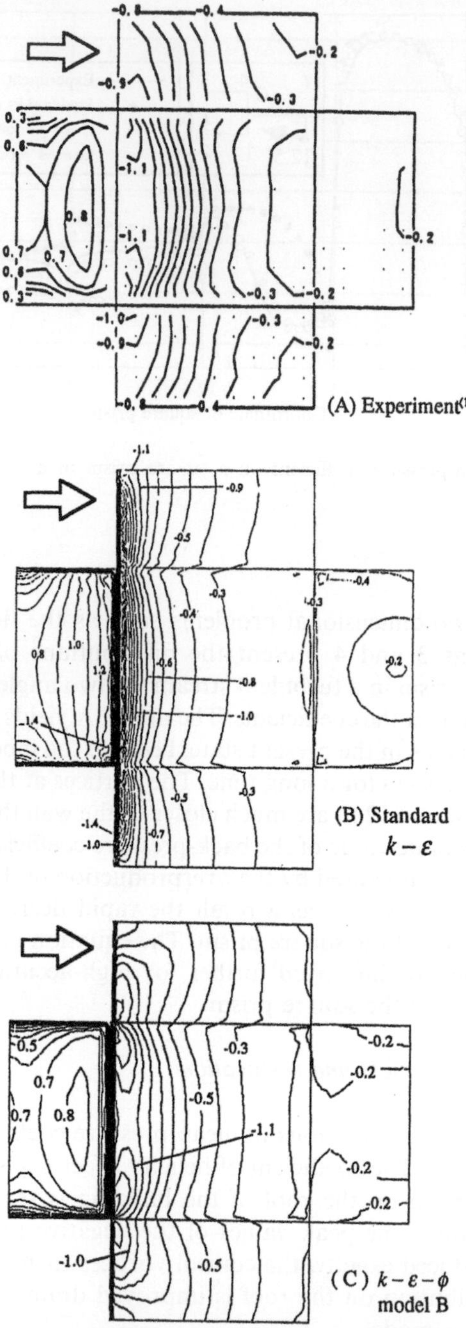

Fig. 5. Distributions of mean pressure coefficient on the 1 : 1 : 0.5 building in a $\frac{1}{4}$ shear flow ($\alpha = 0°$): (A) experiment from Ref. [13]; (B) standard k–ε model; (C) k–ε–ϕ model B.

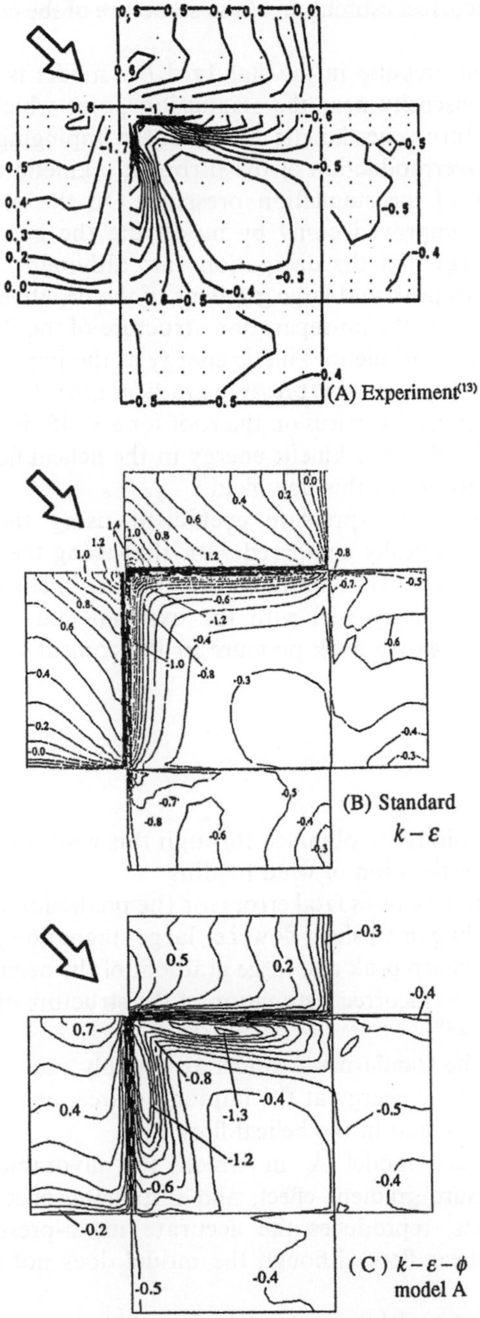

Fig. 6. Distributions of mean pressure coefficient on the 1 : 1 : 0.5 building in a $\frac{1}{4}$ shear flow ($\alpha = 45°$). Conditions as in Fig. 5.

edge for $\alpha = 0°$, and incorrect estimation of the structure of the conical vortices on the roof for $\alpha = 45°$.

The large stagnation pressure in the standard k–ε model is caused by the large gradient of the eddy viscosity near the stagnation point which is brought by the overproduction of the turbulence kinetic energy at the impinging area. Therefore, the improvement on the overproduction of the turbulence kinetic energy is effective to improve the accuracy of the stagnation pressure. On the other hand, the large stagnation pressure is improved easily by modifying the boundary conditions of turbulence kinetic energy and the dissipation rate although it is not an essential improvement. The sharp peak and large gradient of the negative pressure behing the upstream edge is caused by the no-separation structure of the flow which is brought by overproduction of the turbulence kinetic energy at the impinging area and at the upstream edge, the strongly favorable pressure gradient area. The incorrect estimation of the structure of the conical vortices on the roof for $\alpha = 45°$ is essentially caused by the overproduction of turbulence kinetic energy in the helical flow field. The helicity effect is the most important for this flow field.

The accuracy of the mean-pressure coefficient using the k–ε–ϕ turbulence model is improved dramatically compared with that using the standard k–ε model for both $\alpha = 0°$ and $\alpha = 45°$. Though the mean-pressure coefficient using the k–ε–ϕ turbulence model agrees fairly well with the experimental result for $\alpha = 0°$, the absolute values of the negative peak pressure in the conical vortices are somewhat small for $\alpha = 45°$.

4. Conclusions

The following conclusions are obtained through this work on the improvement of turbulence models for estimation of wind loading.

1. The standard k–ε model causes fatal errors for the prediction of the mean-pressure coefficient on a building in a $\frac{1}{4}$ shear flow, i.e., large stagnation pressure on the front face of the building, sharp peak and large gradient of the negative pressure behind the upstream edge, and incorrect estimation of the structure of the conical vortices on the roof for $\alpha = 45°$.
2. The fatal errors in the standard k–ε model are mainly caused by overproduction of the turbulence kinetic energy at the impinging area, at the strongly favorable pressure gradient area, and in the helical flow field.
3. The k–ε–ϕ turbulence model A, in which the favorable pressure gradient effect, adverse pressure gradient effect, and impinging effect are introduced as wall-reflection effects, reproduces the accurate mean-pressure coefficients on a building in a $\frac{1}{4}$ shear flow although the model does not satisfy the Galilean indifference.
4. The k–ε–ϕ turbulence model B, in which the favorable pressure-gradient effect and adverse pressure-gradient effect are introduced irrelevantly to the wall-reflection effect, also reproduces the accurate mean-pressure coefficients on a building in a $\frac{1}{4}$ shear flow.

Though the k–ε–ϕ turbulence model A and model B need further improvements, these models will become powerful tools for practical estimation of the wind loading on buildings including slow fluctuations based on the Kármán vortex.

Acknowledgements

The author would like to thank Mrs. S. Matsunaga, Mrs. R. Toda and Miss M. Miyashita for carrying out part of these computations and making fair figures.

References

[1] B.E. Launder, D.B. Spalding, The numerical computations of turbulent flows, Comp. Meth. Appl. Mech. 3 (1974) 269–289.

[2] S.A. Orszag et al., in: Renormalization Group Modeling and Turbulence Simulations, Near-Wall Turbulent Flows, Elsevier, Amsterdam, 1993, pp. 1031–1046.

[3] S. Kawamoto, S. Kawabata, T. Tanahashi, Numerical analysis of wind around building using high-speed GSMAC-FEM, J. Wind Eng. Ind. Aerodyn. 46&47 (1993) 115–120.

[4] M. Kato, B.E. Launder, The modeling of turbulent flow around stationary and vibrating square cylinder, Proc. 9th Symp. on Turbulent Shear Flows, 10-4, 1-6, 1993.

[5] K. Kondo, S. Murakami, A. Mochida, Numerical study on flow field around square rib using revised k–ε model, 8th CFD Symp., 1994, pp. 363–366 (in Japanese).

[6] M. Tsuchiya, S. Murakami, A. Mochida, K. Kondo, Y. Ishida, Numerical simulation of wind pressures acting using revised k–ε model, 9th CFD Symp., 1995, pp. 199–200 (in Japanese).

[7] S. Kawamoto, An improved k–ε turbulence model for accurate prediction of pressure distribution, 7th CFD Symp., 1993, pp. 323–326 (in Japanese).

[8] S. Kawamoto, The k–ε–ϕ turbulence model for estimation of wind force, 8th CFD Symp., 1994, pp. 367–370 (in Japanese).

[9] S. Kawamoto, An improved k–ε–ϕ turbulence model for wind load estimation, 9th CFD Symp., 1995, pp. 197–198 (in Japanese).

[10] S. Murakami, Comparison of various turbulence models applied to a bluff body, J. Wind Eng. Ind. Aerodyn. 46&47 (1993) 21–36.

[11] A. Yoshizawa, N. Yokoi, Coherent structures in turbulence and their relationship with breaking of spatial symmetry, Seisan-Kenkyu, Monthly J. Inst. Ind. Sci. University of Tokyo 4 (2) (1994) 50–55 (in Japanese).

[12] R.F. Blackwelder, L.S.G. Kovasznay, Large-scale motion of a turbulent boundary layer during relaminarization, J. Fluid Mech. 53(1) (1972) 61–83.

[13] T. Tamura et al. (AIJ Working group), Numerical prediction of wind loading on buildings and structures, IWEF Workshop, 1996.

ELSEVIER

Journal of Wind Engineering
and Industrial Aerodynamics 67&68 (1997) 601–609

JOURNAL OF wind engineering AND industrial aerodynamics

Large-eddy simulation of wind effects on bluff bodies using the finite element method

Sungsu Lee*, Bogusz Bienkiewicz

Department of Civil Engineering, Colorado State University, Fort Collins, CO 80521, USA

Abstract

In this paper, a finite element formulation of the large-eddy simulation (LES) is applied to calculate two-dimensional, turbulent flow past a square rib mounted on the floor of a channel at Reynolds number of 40 000. The Adams–Bashforth and Crank–Nicholson schemes are employed for time integration. The Smagorinsky model for the subgrid scale and upwind scheme for the advection are used. The fully developed turbulent flow is imposed as inflow. No-slip condition is enforced on the surface of solid walls. The simulated results are compared with the numerical and experimental studies reported by other researchers.

Keywords: LES; FEM; Bluff bodies; High Reynolds number

1. Introduction

Numerical simulations of turbulent flows past bluff bodies have been motivated by the emerging and projected developments in computational technology and an increasing cost of experimental studies. Among a number of existing numerical techniques, the large-eddy simulation (LES) appears to be the most promising and of potential for future practical applications. At present, its main drawback is a demand for large computational resources.

There have been a number of preliminary wind engineering investigations employing LES, initiated by a study reported by Murakami et al. [1]. Early simulations were based on Smagorinsky model for modeling unresolved subgrid scales (SGS) [2]. Potential of this scheme has been clearly demonstrated by Deardorff [3], in a numerical study of a turbulent channel flow. Despite its limitations, it remains the most popular SGS model, mainly due to its simplicity. Various implementations in the SGS

* Corresponding author. E-mail: sl128035@lamar.colostate.edu.

0167-6105/97/$17.00
PII S0167-6105(97)00103-7

model have been proposed. They include artificial damping function, the scale-similarity model [4], and the dynamic subgrid-scale model [5].

Wind engineering problems involve interaction of turbulent flow with bluff bodies of complex geometry. Applied aspects of many problems lead to focus on overall understanding of this interaction and its effects on wind-induced loading and structural response. Details of turbulent flow structure are not of main interest in such analysis.

In this paper, a finite element formulation of LES is described. The developed three-dimensional computer code is employed in simulation of a test case, a two-dimensional turbulent flow past a square rib in a channel, at Reynolds number of 40 000. The conventional Smagorinsky SGS model and the first-order upwind scheme are employed. Preliminary results for a moderate spatial resolution are presented and compared with numerical and experimental data published by other researches.

2. Background

The governing equations of LES for incompressible flow of viscous fluid with constant properties are the Navier–Stokes equations and the equation of continuity. For a two-dimensional case, they read as follows:

$$\frac{\partial u_i}{\partial t} + \frac{\partial u_i u_j}{\partial x_j} = -\frac{1}{\rho}\frac{\partial p}{\partial x_i} + \nu\frac{\partial^2 u_i}{\partial x_j^2} \quad (i, j = 1,2),$$

$$\frac{\partial u_j}{\partial x_j} = 0, \tag{1}$$

where ρ is the mass density and ν is the kinematic viscosity of the fluid. The instantaneous velocity and pressure can be expressed in terms of their spatially filtered values, $\bar{u}_i$ and $\bar{p}$,

$$u_i = \bar{u}_i + u_i' \quad \text{and} \quad p = \bar{p} + p', \tag{2}$$

where u_i' and p' are the subgrid-scale (SGS) components. The SGS are modeled in this paper by representation proposed by Smagorinsky [2]. Substitution of Eq. (2) in Eq. (1), filtering the governing equations and implementation of the Smagorinsky model lead to the following set of equations:

$$\frac{\partial \bar{u}_j}{\partial x_j} = 0, \qquad \frac{\partial \bar{u}_i}{\partial t} + \bar{u}_j\frac{\partial \bar{u}_i}{\partial x_j} = -\frac{1}{\rho}\frac{\partial P}{\partial x_i} + (\nu + \nu_{\mathrm{SGS}})\bar{S}_{ij},$$

$$\nu_{\mathrm{SGS}} = (C_{\mathrm{S}}\,\Delta)^2|\bar{S}|, \qquad \bar{S}_{ij} = \frac{1}{2}\left(\frac{\partial u_i}{\partial x_j} + \frac{\partial u_j}{\partial x_i}\right),$$

$$|\bar{S}| = \sqrt{2\bar{S}_{ij}\bar{S}_{ij}}, \qquad P = \bar{p} + \frac{\rho}{3}\,\delta_{ij}\overline{u'_k u'_k}, \tag{3}$$

where C_S is the Smagorinsky constant and the filter width, Δ, is the square root of the finite element area. In the present study, C_S is set to be equal to 0.15.

Eq. (3) are written in a non-dimensional form resulting from using L_0 and U_0 as the characteristic length and velocity scales, respectively. The Reynolds number is thus defined as $L_0 U_0/\nu$, where time and pressure are non-dimensionalized using the time scale $T_0 = L_0/U_0$ and the pressure scale $P_0 = \frac{1}{2}\rho U_0^2$, respectively.

3. Numerical schemes

The spatial discretization of Eq. (3) is performed using the finite element method (FEM). Quadrilateral elements with constant pressure and linear interpolation functions for velocity field are employed. The Galerkin central difference scheme is employed for diffusion terms while upwind representation [6] is used for the (nonlinear) advection. The resulting algebraic equations used in computation are

$$\boldsymbol{M}\frac{\mathrm{d}\boldsymbol{V}}{\mathrm{d}t} + [\boldsymbol{L} + \boldsymbol{N}(\boldsymbol{V})]\boldsymbol{V} + \boldsymbol{G}\boldsymbol{P} = \boldsymbol{Q}, \quad \boldsymbol{G}^{\mathrm{T}}\boldsymbol{V} = 0, \tag{4}$$

where $\boldsymbol{M}, \boldsymbol{L}, \boldsymbol{N}, \boldsymbol{G}$, and $\boldsymbol{G}^{\mathrm{T}}$ represent mass, diffusion, advection, gradient and divergence operator, respectively, while $\boldsymbol{V}$ and $\boldsymbol{P}$ denote velocity and pressure vector, respectively. The diffusion operator, $\boldsymbol{L}$ consists of terms dependent on fluid viscosity (ν) and eddy viscosity (ν_{SGS}). The eddy viscosity is evaluated at the center of each element, for each time step.

Time integration is performed using Adams–Bashforth algorithm of the second-order for the (nonlinear) advection and Crank–Nicholson for the linear terms, respectively. In such an approach, the matrix decomposition is performed only once unless time increment is changed. Since the FEM with linear velocity-constant pressure leads to *checkerboard* modes in computed results, the pressure field is smoothed by the least-square method [6].

4. Specifications of simulated case

Fig. 1 shows the geometry of the tested case, which was also investigated by other researchers. In a study reported by Yang and Ferziger [7], periodic upstream/downstream boundary conditions and a relatively large computational domain were employed. The periodicity constraint was relaxed by Werner and Wengle [8], who used fully developed turbulent channel flow as inflow. This approach is adopted in this paper. First, a turbulent channel flow with periodic conditions is generated, and the velocity time series in a channel cross-section normal to the main flow is stored. Second, flow past the square rib is simulated using the stored time series of the velocity as the inflow. Free-slip boundary condition is imposed at the surface of outflow. On the solid surface, no-slip condition is employed. Reynolds number, $\mathrm{Re} = U_0 H/\nu$, is set to be 40 000.

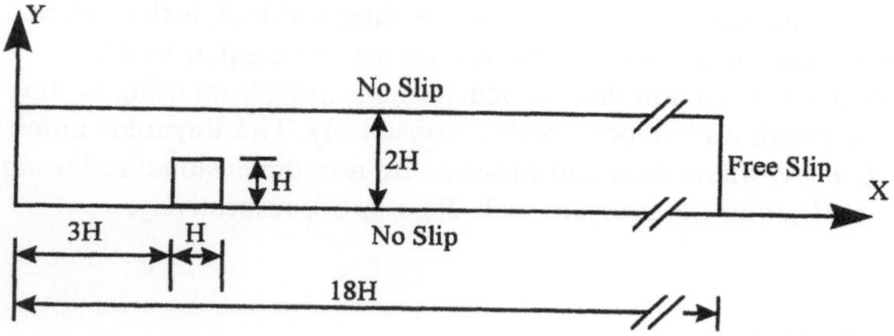

Fig. 1. Geometry and boundary conditions.

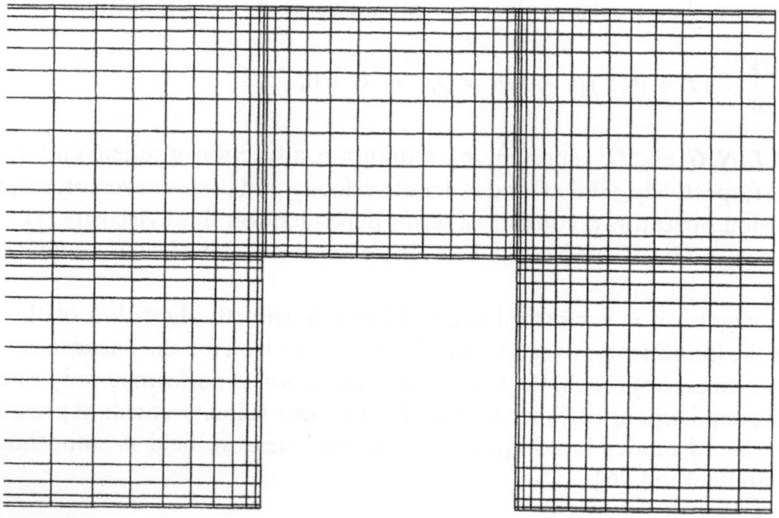

Fig. 2. Mesh distribution near rib, Case II.

The computations were performed for two mesh sizes. Case I had 1100 elements while 1708 elements were employed in Case II. In both the cases, non-uniform discretization was used to achieve high mesh density near the solid wall. The minimum grid spacing (next to the wall) was $0.01818H$ for Case I, and $0.01212H$ for Case II. Taking into account the importance of grid spacing near the surface, both the cases represent a relatively low grid resolution. Testing of a more refined (grid) resolution was impossible due to limitations of computation resources available for this study. The close-up view of the mesh near the rib for the Case II, is shown in Fig. 2. In simulations, the (non-dimensional) time increment was chosen to be 0.002 and the computations were performed for 40 000 time steps.

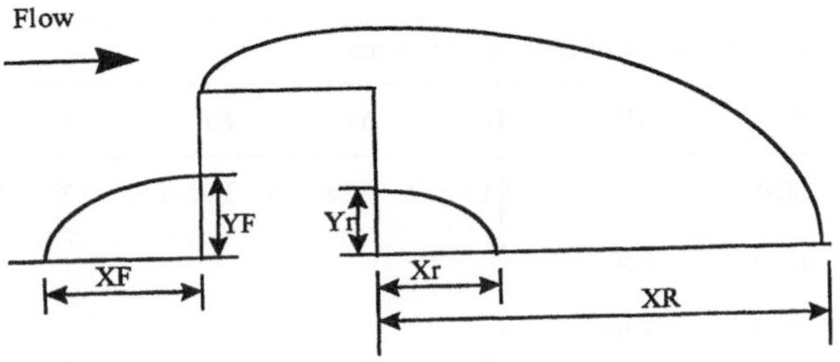

Fig. 3. Schematic diagram of separation and reattachment zones.

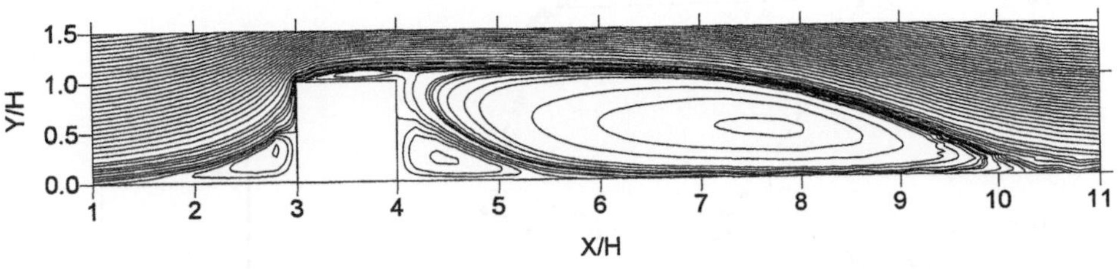

Fig. 4. Time-averaged streamlines.

5. Results and discussion

The flow past a two-dimensional rib develops separation and reattachment zones depicted schematically in Fig. 3. There are primary recirculation zones in upstream and downstream regions and several secondary ones, of much smaller size. The reattachment length of the primary downstream recirculation zone is denoted by XR in Fig. 3. The separation and the reattachment lengths of the primary upstream recirculation are represented, respectively, by YF and XF. The corresponding parameters for the secondary recirculation in the downstream direction are Yr and Xr, respectively. The time-averaged streamlines of the computed flow are depicted in Fig. 4. The primary and the secondary recirculation zones are clearly shown. The flow separated at the leading edge of the rib does not reattach to the top of the rib. This is consistent with the results of other reported numerical simulations [7,8]. Table 1 compares characteristic lengths of the recirculation zones. The comparison of the results for different Reynolds numbers is made in view of the experimental findings of Armaly et al. [9] and Tropea et al. [10], who showed that in turbulent flow the recirculation characteristics are nearly independent of Reynolds number. All the lengths are normalized by the rib height, H. The results of the present study are in a good agreement with the data reported by other researchers.

The computed mean velocity near the windward corner on the top of the rib is shown in Fig. 5. The present results are compared with those of three-dimensional LES by Werner and Wengle [8], and the experimental data reported by Dimaczek

Table 1
Comparison of characteristic lengths of recirculation zones

	Re	XR	Xr	Yr	XF	YF	Grids
Three-dimensional LES [7]	3210–3330	6.42–7.01	1.13–1.76	0.28–0.36	1.35–1.51	0.28–0.40	112 × 48 × 40
Three-dimensional LES [8]	42 500	6.50					160 × 64 × 64
Experiment [11]	42 500	7.10					—
Present (2D)	40 000	6.42	1.72	0.64	1.97	0.49	Case II

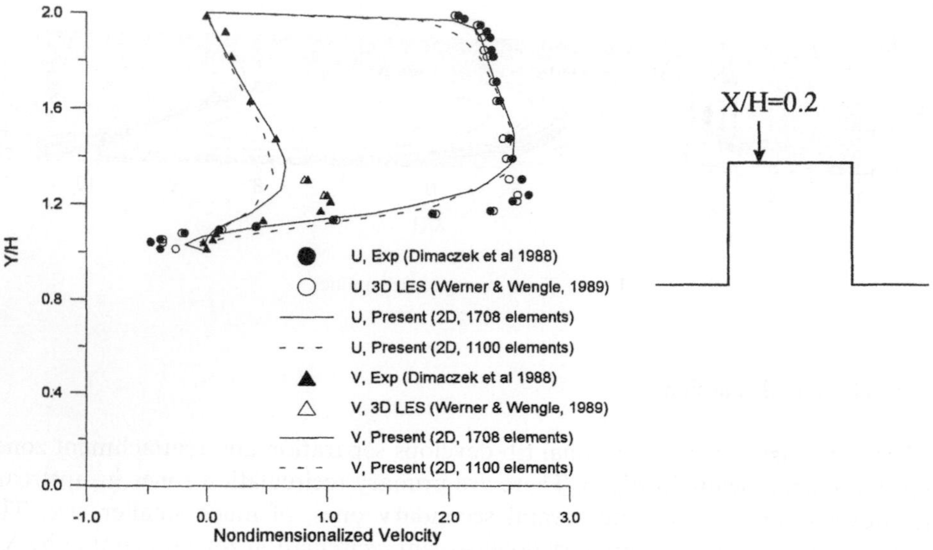

Fig. 5. Comparison of mean velocity at $X/H = 0.2$.

et al. [11]. The horizontal U and vertical V velocity components are in a good agreement with those reported by other researchers, except for the region adjacent to the rib. The discrepancy between the present and compared data is larger for V component. Taking into account the fact that the grid resolutions near the upper wall and the top of the rib are identical, and the flow field near the upper wall is in better agreement, the region in the separated zone needs a higher grid resolution. Increasing the number of elements from 1100 to 1708 leads to a slightly better agreement with the compared data.

The root-mean square (RMS) of the fluctuating horizontal velocity component of the resolvable scale is compared in Fig. 6. The result shows large departure from the experimental data. This is also attributed to the rough grid system used in this study. A trend exhibited by the results near the top of the rib, confirms this conclusion. As the number of elements is increased, the magnitude of the velocity RMS in this region,

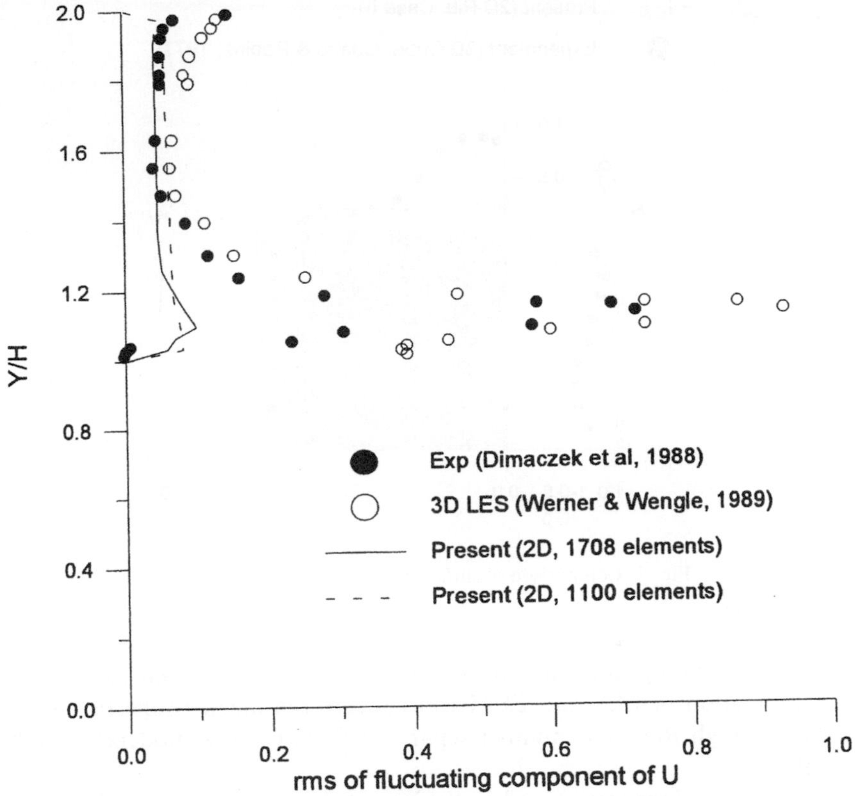

Fig. 6. Comparison of RMS of fluctuating component of horizontal velocity at $X/H = 0.2$.

especially peak value near the surface, increases and gets closer to the experimental values.

The time-averaged surface pressure coefficient, computed for the square rib, is compared in Fig. 7 with the experimental results for the flow past a cube [12]. The computed pressure on the windward and the leeward surface is in a better agreement with the experimental results than the pressure on the top surface. The discrepancy between the compared values is attributed to three-dimensional effects and the insufficient grid resolution employed in the present study.

6. Concluding remarks

Turbulent flow past a square rib mounted in a channel was simulated at Reynolds number of 40 000, in two-dimensional large-eddy simulation, using the finite element method. Due to the limitations in the available computational resources, a relatively small number of elements (1100 and 1708) was employed.

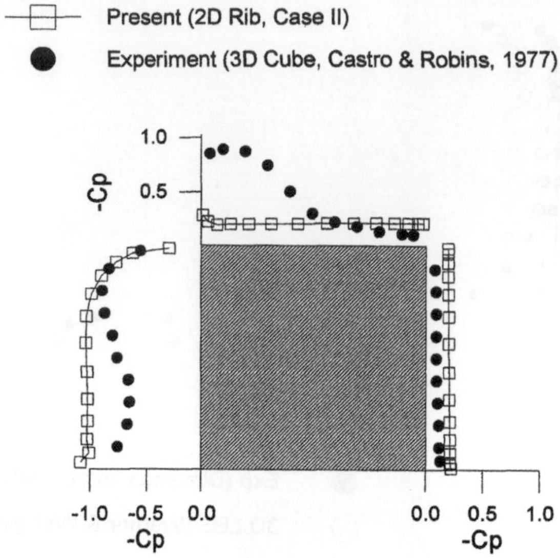

Fig. 7. Comparison of surface mean-pressure coefficient.

Conclusions resulting from this study can be summarized as follows:

- The combination of LES and FEM can capture the mean properties of a large structure of high Reynolds number separated flow, using a moderate number of elements.
- The time-averaged separation and reattachment characteristics are well reproduced using the present approach.
- The computed results near the top of the rib exhibit large departure from the experimental data. To improve computational results, significantly refined grids are needed, especially near the solid wall.

Simulations with a significantly larger number of elements are planned to be performed next. Since the developed code is capable of handling three-dimensional problems, near future plans include of three-dimensional computations of flow for cases with two- and three-dimensional geometry.

References

[1] S. Murakami, A. Mochida, K. Hibi, Three-dimensional numerical simulation of air flow around a cubic model by means of large eddy simulation, J. Wind Eng. Ind. Aerodyn. 25 (1987) 291–305.

[2] T.S. Smagorinsky, General circulation experiment with primitive equations: Part I, Basic experiments, Monthly Weather Rev. 91 (1963) 99–164.

[3] J.W. Deardorff, A numerical study of three-dimensional turbulent channel flow at large Reynolds numbers, J. Fluid Mech. 41 (1970) 453–480.

[4] J. Bardina, J.H. Ferziger, W.C. Reynolds, Improved subgrid scale models for large eddy simulation, AIAA papers, 1980, pp. 80–1357.

[5] M. Germano, U. Piomelli, P. Moin, W.H. Cabot, A dynamic subgrid-scale eddy viscosity model, Phys. Fluids A 3 (7) (1991) 1760–1765.

[6] T.J.R. Hughes, W.K. Liu, A. Brooks, Finite element analysis of incompressible viscous flows by the penalty function formulation, J. Comput. Phys. 30 (1979) 1–60.

[7] K.S. Yang, J.H. Ferziger, Large-eddy simulation of turbulent obstacle flow using a dynamic subgrid-scale model, AIAA J. 31 (8) (1993) 1406–1413.

[8] H. Werner, H. Wengle, Large-eddy simulation of turbulent flows, over a square rib in a channel, Proc. 7th Conf. Turb. Shear Flows, 1989, pp. 418–423.

[9] B.F. Armarly, B.F. Armaly, F. Durst, J.C.F. Pereira, B. Schonung, Experimental and theoretical investigation of backward facing step flow, J. Fluid Mech. 127 (1983) 473–496.

[10] C.D. Tropea, R. Gackstatter, The flow over two-dimensional surface-mounted obstacles at low Reynolds number, J. Fluid Eng. 107 (1985) 489–495.

[11] G. Dimaczek, C. Tropea, A.B. Wang, Turbulent flow over two-dimensional, surface-mounted obstacles: plane and axisymmetric geometries, in: H.-H. Fernholz, H.E. Fieldler (Eds.), Advances in Turbulence 2, Springer, Heidelberg, 1989, pp. 114–121.

[12] I.P. Castro, A.G. Robins, The flow around a surface-mounted cube in uniform and turbulent streams, J. Fluids Mech. 79 (1977) 307–335.

ELSEVIER

Journal of Wind Engineering
and Industrial Aerodynamics 67&68 (1997) 611–625

JOURNAL OF wind engineering AND industrial aerodynamics

Suppression of flow-induced oscillations using sloshing liquid dampers: analysis and experiments

V.J. Modi[1,*], M.L. Seto[2]

Department of Mechanical Engineering, University of British Columbia, Vancouver, BC, Canada V6T 1Z4

Abstract

Control of vortex resonance and galloping type of wind-induced instabilities using passive rectangular nutation dampers, involving dissipation of energy through sloshing liquid, is studied numerically accounting for nonlinear effects. It includes the effects of wave dispersion as well as boundary-layers at the walls, floating particle interactions at the free surface, and wave-breaking. Predictions of the analysis are substantiated through extensive free surface dynamics, wind tunnel and flow visualization experiments. Results suggest that the numerical approach, based on the shallow water waves in conjunction with a nonlinear force excitation model near resonance, is able to capture the system dynamics rather well.

Keywords: Nutation damper; Wind induced instabilities; Vortex resonance and galloping control

Nomenclature

A_1	linear coefficient in the lift-oscillator model; Eq. (11)
A_3	coefficient of the cubic term in the lift-oscillator model; Eq. (11)
a	height of a small amplitude wave
B_x, B_y, B_z	body forces per unit mass in x, y, z directions, respectively
b	coefficient used in the lift-oscillator model; Eq. (11)
C_A	constant associated with excitation force; $\rho H_m \times 2L_m/8\pi^2 \mathrm{St}^2 m_s$
C_L	lift coefficient
c	wave speed
c_s	damping of the wind tunnel model

* Corresponding author.
[1] Professor Emeritus; Fellow ASME, AIAA.
[2] Graduate Research Assistant.

PII S0167-6105(97)00104-9

E	energy density; energy per unit volume of the damper fluid
$\dot{E}$	rate of energy dissipation per unit volume; dE/dt
F_s^*	nondimensional sloshing force; Eq. (10)
F_e^*	nondimensional wind induced excitation force, $C_A\omega_{nv}^2 C_L$; Eq. (10)
H_m	width of the wind tunnel model
h	quiescent liquid height
k_s	stiffness of the wind tunnel model
L_m	length of the square and circular cross-section wind tunnel models
l	half the length of the rectangular damper
M_l	mass of the liquid
m_s	mass of the wind tunnel model
n	positive integer
p	pressure
St	Strouhal number; $f_v H_m/V$
U	reduced wind velocity; $V/\omega_n H_m$
$\bar{U}$	damper liquid velocity vector; $u_i + v_{i,}j + w_k$
u, v, w	velocity components in the x, y, and z directions, respectively
V	wind speed
V_l	volume of the liquid
W	width of rectangular damper
X	displacement of the damper in the x-direction
x_{st}	nondimensional structural displacement
α	damping parameter
γ	$z + h/\delta$
δ	boundary-layer thickness; $\sqrt{2\nu/\omega}$
ε_e	amplitude of the damper motion
κ	wave number
ζ_s	structural damping ratio
ρ	liquid density
ν	kinematic viscosity;
σ	$(\tanh \kappa h) / \kappa h$
τ	nondimensional period
τ_1	parameter used to nondimensionalize time; $l\sqrt{gh\sigma_1}$
ϕ	velocity potential
ω	circular frequency for a wave, rad/s; $c\kappa$
ω_e	frequency of the damper motion
$\hat{\omega}_e$	nondimensional forcing frequency; ω_e/ω_l
ω_l	liquid natural frequency
ω_n	undamped natural frequency; $\sqrt{k_s/m_s} = 2\pi f_n$
ω_{nv}	dimensionless vortex shedding frequency
λ	wavelength; $2\pi/\kappa$
χ	$\tanh \kappa(1 + \eta/h)/\tanh \kappa h$

1. Introduction

Wind effects on large aspect ratio structures such as bridges, towers, masts, antennae, smokestacks, transmission line conductors and similar bluff bodies have been of interest to scientists and engineers for a long time. Flow induced vibration problems are more relevant today because of the tendency to build higher and longer structures [1]. Furthermore, advances in the metallurgical sciences and computer aided design have tended to reduce the safety margin towards a lower structural stiffness. Obviously for a given strength, lighter building materials are preferred for reduced cost and lower weight induced stresses. On the other hand, the increased flexibility makes them prone to flow induced vibrations. Hence the control of fluid–structure interaction instabilities has become a subject of considerable importance. Of particular interest is the effectiveness of energy dissipation to damp the motions. Several methods have been developed over years and used in practice, however, the nutation damper is particularly attractive due to its efficiency, simplicity in design, ease of construction and maintenance [2–5]. A recent paper by Modi et al. [6] reviews the literature in the area of nutation damping of wind-induced instabilities. It suggests that numerical approach to fluid dynamics of a nutation damper and fluid–structure interaction response in presence of such energy dissipation mechanism have received, relatively, less attention [7–11].

The classical term ‘nutation’ represents a component of the general three-dimensional rotation in the Eulerian description. Pitch and roll librations of satellites as well as the fundamental oscillations of cantilever structures such as somestacks, buildings, etc. closely represent the Eulerian nutation, hence the designation ‘nutation damper’. Lately, several authors have referred to it as the ‘sloshing liquid damper’, ‘tuned liquid damper’, ‘tuned sloshing damper’, etc.

With this as background, the paper presents three distinct aspects associated with a comprehensive study aimed at damping of wind induced instabilities as indicated below:

1. numerical simulation of sloshing liquid in a rectangular nutation damper accounting for nonlinearities and large scale motions;
2. numerical prediction of the structural response during vortex resonance, in presence of a rectangular nutation damper, considering conjugate character of the problem where the sloshing liquid excites the structural vibration and the structural response induces the sloshing motion;
3. validation of the numerical studies through extensive test-programs involving:
   - evaluation of energy dissipation through measurement of the reduced damping and added mass parameters using a specially constructed Scotch-Yoke harmonic excitation facility operating in the low frequency range (0.1–1.0 Hz);
   - visualization of sloshing modes and wave-breaking as affected by the excitation frequency to get better physical appreciation of the free surface dynamics;
   - a comprehensive set of wind tunnel experiments near vortex resonance using two-dimensional circular and square prisms.

Experimental results confirm predictions of the numerical analysis concerning character of the free surface dynamics and effectiveness of the rectangular nutation damper in controlling the vortex resonance. As can be expected, details of the numerical and experimental procedures as well as the amount of information obtained through a planned variation of the system parameters are extensive. For conciseness, the focus here is on the method of approach and a few typical results useful in establishing trends. More details are given by Seto [12].

2. Numerical approach

From a practising engineer's point of view, undertaking of the test-program would be demanding in terms of time, effort and cost. With the advent of computers, a simple, numerical approach, if available, can prove to be quite attractive. However, energy dissipation through sloshing liquid is a complex process involving a large number of parameters and their intricate interactions. Hence to develop a relatively simple model that can capture the physics of the problem is a challenging task.

The experimental results obtained by Welt and Modi [4] as well as Seto and Modi [13], using a variety of nutation dampers, suggest the following:

1. The nutation damper appears to be quite effective at low liquid heights. This indicates that the relatively simple shallow water wave theory [14], suitably modified to account for viscous dissipation and dispersion, may present one avenue to follow.
2. Viscosity and wave breaking seem to be the major factors contributing to dissipation.

Based on these observations, it was decided to approach the problem of sloshing in a rectangular damper as a shallow water wave phenomenon (Fig. 1). Characteristic features of the analysis are summarized below:

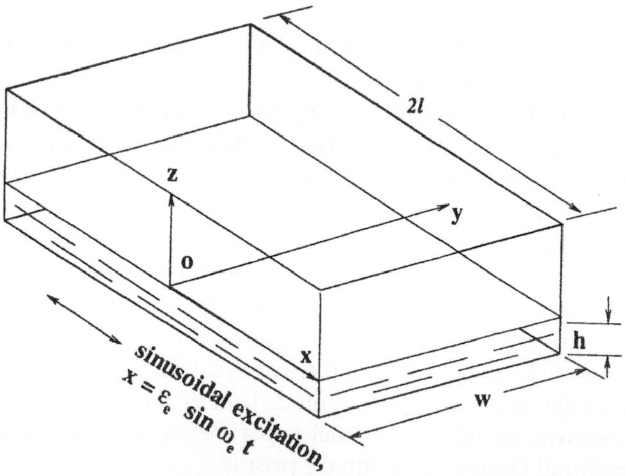

Fig. 1. Geometry of the rectangular nutation damper, subjected to a sinusoidal excitation, considered for the study.

1. The liquid is considered homogeneous, irrotational and incompressible.
2. The walls of the damper are treated as rigid.
3. The analysis accounts for nonlinearities in the governing equations as well as boundary conditions. Thus it is applicable at resonance when a large amplitude sloshing motion is encountered.
4. The analysis considers dispersion as well as dissipation in the damper. Effect of floating particles is also accounted for.
5. It should be emphasized that the structural response acts as an excitation for the damper thus affecting the sloshing motion of the liquid and its dissipation. On the other hand, the nutation damping affects the structural response. The analysis accounts for this conjugate character.
6. Such a comprehensive treatment of the subject accounting for nonlinearities, resonance conditions, and conjugate interactions between the structure and the damper is indeed rare.

To begin with equations governing sloshing motion of a liquid in a rectangular cross-section container, subjected to harmonic excitation, were derived and their potential flow solution obtained consistent with boundary conditions. The solution was subsequently modified to account for viscous dissipation and dispersion of the waves. The solution obtained through a finite difference scheme provides information about the free surface dynamics. Finally, dynamics of a vertical cantilever beam-type structure with the damper at its tip was studied, accounting for the conjugate character of the problem, and the numerically predicted results were compared with the experimental data.

2.1. Potential flow analysis

For nonviscous, incompressible flows the governing continuity and momentum equations can be written as:

$$\nabla \cdot \boldsymbol{U} = 0, \qquad \text{i.e. } \nabla^2 \phi = 0\,, \tag{1}$$

$$\frac{\mathrm{d}u}{\mathrm{d}t} = -\frac{1}{\rho}\frac{\partial p}{\partial x} + B_x, \qquad \frac{\mathrm{d}v}{\mathrm{d}t} = -\frac{1}{\rho}\frac{\partial p}{\partial y} + B_y, \qquad \frac{\mathrm{d}w}{\mathrm{d}t} = -\frac{1}{\rho}\frac{\partial p}{\partial z} + B_z. \tag{2}$$

Here $\boldsymbol{U}$ is the velocity vector, $u\boldsymbol{i} + v\boldsymbol{j} + w\boldsymbol{k}$; ϕ the velocity potential; p the pressure; B_x, B_y, B_z the body forces per unit mass in x, y, z directions, respectively; ρ the density; and t the time. The velocity potential is taken in the form

$$\phi = F(x, y, t)G(z), \tag{3}$$

with the kinematic and dynamic boundary conditions as

$$\frac{\partial \eta}{\partial t} + \frac{\partial \phi}{\partial x}\frac{\partial \eta}{\partial x} + \frac{\partial \phi}{\partial y}\frac{\partial \eta}{\partial y} - \frac{\partial \phi}{\partial z} = 0, \qquad w_s = \frac{\partial \eta}{\partial t} + u_s\frac{\partial \eta}{\partial x} + v_s\frac{\partial \eta}{\partial y}, \tag{4a}$$

$$\frac{\partial \phi}{\partial t} + g\eta = \frac{1}{2}(\nabla \phi)^2. \tag{4b}$$

Here η corresponds to the free surface elevation, subscript s represents fluid properties at the surface, and g is the acceleration due to gravity. Now, integrating along the liquid depth to average variation in the z-direction gives the equations of motion governing the surface dynamics as

$$\frac{\partial}{\partial x}\left[u_s\frac{\tanh\kappa(h+\eta)}{\kappa}\right]+\frac{\partial}{\partial y}\left[v_s\frac{\tanh\kappa(h+\eta)}{\kappa}\right]+\frac{\partial\eta}{\partial t}+u_s\frac{\partial\eta}{\partial x}+v_s\frac{\partial\eta}{\partial y}=0\,, \quad \text{(5a)}$$

$$\frac{\partial u_s}{\partial t}+g\frac{\partial\eta}{\partial x}+\left[\frac{1-\tanh^2\kappa(h+\eta)}{2}\right]\frac{\partial}{\partial x}\left(\frac{u_s^2}{2}\right)$$
$$+\left[\frac{1+\tanh\kappa(h+\eta)}{2}\right]\frac{\partial}{\partial x}\left(\frac{v_s^2}{2}\right)-\frac{\partial}{\partial y}(u_s v_s)\tanh^2\kappa(h+z)$$
$$+\left(\frac{g}{\kappa}\right)\frac{\partial\eta}{\partial x}\left[\frac{\partial^2\eta}{\partial x^2}+\frac{\partial^2\eta}{\partial y^2}\right]\tanh\kappa(h+\eta)=\ddot{X}\,, \quad \text{(5b)}$$

$$\frac{\partial v_s}{\partial t}+g\frac{\partial\eta}{\partial y}+\left[\frac{1-\tanh^2\kappa(h+\eta)}{2}\right]\frac{\partial}{\partial y}\left(\frac{u_s^2}{2}\right)$$
$$+\left[\frac{1+\tanh\kappa(h+\eta)}{2}\right]\frac{\partial}{\partial y}\left(\frac{v_s^2}{2}\right)-\frac{\partial}{\partial x}(u_s v_s)\tanh^2\kappa(h+z)$$
$$+\left(\frac{g}{\kappa}\right)\frac{\partial\eta}{\partial y}\left[\frac{\partial^2\eta}{\partial x^2}+\frac{\partial^2\eta}{\partial y^2}\right]\tanh\kappa(h+\eta)=0\,, \quad \text{(5c)}$$

where κ is the wave number, h the quiescent liquid height, and g the gravitational acceleration.

2.2. Energy dissipation and modified equations of motion

Waves at the free surface experience attenuation due to three different energy dissipation mechanisms: (i) damping at the bottom and side walls where the viscosity effects are dominant; (ii) dissipation within the body of the fluid where the viscous effects are not significant and the flow may be considered as irrotational; and (iii) dissipation at the free surface due to wave interactions. Ultimately, the objective is to modify the original nonlinear momentum Eqs. (5b) and (5c) through incorporation of the energy dissipation and dispersion terms. To that end, a general expression for energy dissipation in a sloshing liquid is obtained first. It is then applied to assess contributions from the three sources of dissipation mentioned above.

The velocity potential for a small amplitude shallow water wave can be easily shown to be [15]:

$$\phi=\frac{a\omega}{\kappa}\frac{\cosh\kappa(h+z)}{\sinh\kappa h}\cos(\omega t-\kappa x)\,.$$

Here a is the amplitude of the propagating wave and ω the circular frequency. The mean kinetic energy per unit volume, T, for a damper filled to the height h with the shallow water approximation and small amplitude waves becomes

$$T \approx \frac{1}{2}\rho \int_{-h}^{0} \overline{(\nabla\phi)^2}\,\mathrm{d}z = \frac{1}{4}\rho\omega^2 a^2(\kappa \tanh \kappa h)^{-1},$$

where the bar indicates average over the period. Recognizing that in a conservative dynamic system of small wave motion, the mean kinetic energy equals the mean potential energy, the total energy (E) can be written as

$$E = \tfrac{1}{2}\rho(\omega a)^2(\kappa \tanh \kappa h)^{-1}.$$

Hence

$$\frac{\mathrm{d}E}{\mathrm{d}t} = \frac{1}{2}\rho\omega^2 2a\frac{\mathrm{d}a}{\mathrm{d}t}(\kappa \tanh \kappa h)^{-1}. \tag{6}$$

As $E \propto a^2$ for small waves,

$$\frac{\mathrm{d}E}{\mathrm{d}t} \propto 2a\frac{\mathrm{d}a}{\mathrm{d}t},$$

where $a = a_0 \mathrm{e}^{-\alpha t}$; a_0 is the initial amplitude; and α is the damping parameter. The corresponding general expression for energy dissipation due to fluid viscosity can be written in tensor notation as [16]

$$\frac{\mathrm{d}E}{\mathrm{d}t} \approx -\frac{1}{2}\rho\nu \overline{\int_{-h}^{0}\left(\frac{\partial u_i}{\partial x_j} + \frac{\partial u_j}{\partial x_i}\right)^2 \mathrm{d}z}, \tag{7}$$

where ν is the kinematic viscosity. Comparing Eqs. (6) and (7) gives the damping parameter α. Applying this procedure to evaluate energy dissipation through the above mentioned sources gives

$$\alpha = (\alpha_\mathrm{b} + \alpha_\mathrm{s}) + \alpha_\mathrm{v} + \alpha_\mathrm{f} = \left[\left(1 + 2\frac{h}{w}\right) + \cosh \kappa h\right]\frac{\kappa}{2}\sqrt{\frac{\omega\nu}{2}}\frac{1}{\sinh 2\kappa h} + 2\nu\kappa^2,$$

where α_b is the boundary-layer contribution at the damper bottom, $(\kappa/2)\sqrt{\omega\nu/2}\,(1/\sinh 2\kappa h)$, α_s the boundary-layer contribution at the damper side wall, $2(h/w)(\kappa/2)\sqrt{\omega\nu/2}\,(1/\sinh 2\kappa h)$, α_v the contribution from the central volume of the liquid, $2\nu\kappa^2$, α_f the contribution from the free surface boundary-layer, $\nu\sqrt{2\nu/\omega}\,\kappa^3 \tanh \kappa h$.

Several researchers have attempted to account for the free surface 'cleanliness' by incorporating an empirical parameter referred to as the 'contamination factor' (S). One can readily accomodate such an approach by introducing the parameter as indicated below:

$$\alpha = \left[1 + 2\frac{h}{w} + S \cosh \kappa h\right]\frac{\kappa}{2}\sqrt{\sqrt{\frac{\sigma}{\sigma_1}}\frac{\kappa\nu}{2}}\frac{1}{\sinh 2\kappa h} + 2\nu\kappa^2, \tag{8}$$

where σ is the wave speed $(=(\tanh \kappa h)/\kappa h)$; and σ_1 the resonant wave speed; σ for $\kappa\kappa_1$; $n\pi/2$.

Thus, the present study is also applicable to a nutation damper with floating particles by assigning a suitable value to S. The next logical step would be to incorporate the damping parameter in momentum equation (5b) obtained earlier. Furthermore, it would be desirable to nondimensionalize the set of governing Eqs. (5a), (5b) and (5c) to make the analysis applicable to a large class of rectangular dampers operating under a wide variety of conditions. The length and time scales were selected to emphasize shallow water consideration of the analysis. Putting:

$$x^* = \frac{x}{l};\quad y^* = \frac{y}{l};\quad z^* = \frac{z}{h};\quad t^* = t/\tau_1;\quad \eta^* = \frac{\eta}{h};\quad u^* = \frac{u}{c};\quad v^* = \frac{v}{c};$$

$$X^* = X/l;\quad Y^* = Y/l;\quad \kappa^* = \kappa l;\quad \sigma = \frac{\tanh \kappa h}{\kappa h};\quad \chi = \frac{\tanh kh(1+\eta^*)}{\tanh \kappa h};\quad \beta = \frac{h}{l};$$

$$l = (\text{damper length})/2;\quad c = \text{wave speed};\quad \tau_1 = l/\sqrt{gh\sigma_1};$$

the nondimensionalized equations of motion can be written as (dropping the asterisk):

$$\frac{\partial \eta}{\partial t} + \left(\frac{\partial u_s}{\partial x} + \frac{\partial v_s}{\partial y}\right) + \left(\frac{\sigma}{\sigma_1} - 1\right)\left(\frac{\partial u_s}{\partial x} + \frac{\partial v_s}{\partial y}\right)$$
$$+ \frac{\sigma}{\sigma_1}\left[\frac{\partial}{\partial x}(\chi - 1)u_s + \frac{\partial}{\partial y}(\chi - 1)v_s\right]\frac{1}{\sigma_1}\left(u_s\frac{\partial \eta}{\partial x} + v_s\frac{\partial \eta}{\partial y}\right) = 0; \tag{9a}$$

$$\frac{\partial u_s}{\partial t} + \frac{\partial \eta}{\partial x} + \left[\frac{1-(\chi \tanh \kappa h)^2}{2\sigma_1}\right]\frac{\partial}{\partial x}\left(\frac{u_s^2}{2}\right)$$
$$+ \left[\frac{1+(\chi \tanh \kappa h)^2}{2\sigma_1}\right]\frac{\partial}{\partial x}\left(\frac{v_s^2}{2}\right) - \frac{\partial}{\partial y}(u_s v_s)\frac{(\chi \tanh \kappa h)^2}{\sigma_1}$$
$$+ \sigma\chi\beta^2\frac{\partial \eta}{\partial x}\left[\frac{\partial^2 \eta}{\partial x^2} + \frac{\partial^2 \eta}{\partial y^2}\right] + \alpha\tau_1 u_s = -\sigma_1 \dot{X}; \tag{9b}$$

$$\frac{\partial v_s}{\partial t} + \frac{\partial \eta}{\partial y} + \left[\frac{1-(\chi \tanh \kappa h)^2}{2\sigma_1}\right]\frac{\partial}{\partial y}\left(\frac{u_s^2}{2}\right)$$
$$+ \left[\frac{1+(\chi \tanh \kappa h)^2}{2\sigma_1}\right]\frac{\partial}{\partial y}\left(\frac{v_s^2}{2}\right) - \frac{\partial}{\partial x}(u_s v_s)\frac{(\chi \tanh \kappa h)^2}{\sigma_1}$$
$$+ \sigma\chi\beta^2\frac{\partial \eta}{\partial y}\left[\frac{\partial^2 \eta}{\partial x^2} + \frac{\partial^2 \eta}{\partial y^2}\right] = 0. \tag{9c}$$

Here σ_1 represents the wave speed corresponding to the wave number κ_1. It identifies the resonant liquid sloshing condition. Note, for $\chi \neq 1$, the equations retain non-linear contributions consistent with the shallow water analysis. $\alpha\tau_1$ is the dissipation parameter while β, associated with the second order partial derivatives of η, is the measure of dispersion.

2.3. Fluid-structure interaction dynamics

The rectangular nutation damper was mounted on a cantilevered beam-type structure undergoing vortex-induced motion normal to the flow (Fig. 2) represented by

$$m_s\ddot{x} + c_s\dot{x} + k_s x = F_s + F_e,$$

where m_s, c_s, k_s correspond to the mass, viscous damping and stiffness of the system, respectively, F_s is the sloshing force, and F_e the aerodynamic excitation. The aerodynamic force on the structure is represented as

$$F_e = C_L(\tfrac{1}{2}\rho V^2)(H_m L_m),$$

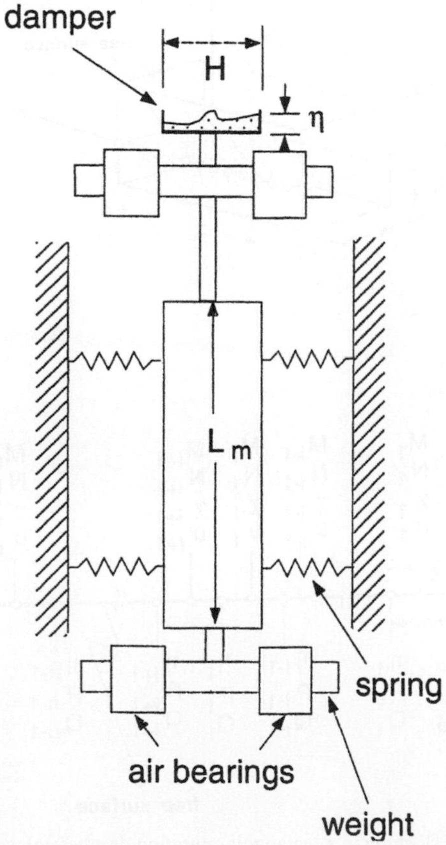

Fig. 2. Schematic diagram showing the test-arrangement for the fluid–structure interaction model.

where C_L is the lift coefficient; V the freestream velocity; H_m/L_m the height and length of the model, respectively. It is convenient to write the equations of motion in a nondimensional form using the definitions as follows:

$$x_{st} = \frac{x}{H_m}; \quad \tau = t\sqrt{\frac{k_s}{m_s}} = t\omega_n; \quad \zeta_s = \frac{c_s}{2m_s\omega_n};$$

$$\omega_{nv} = \frac{\omega_v}{\omega_n} = \mathrm{St}\left(\frac{V}{f_n H_m}\right); \quad C_A = \frac{\rho H_m^2 L_m}{8\pi^2 m_s \mathrm{St}^2},$$

where ω_n is the natural frequency of the structure; ω_v the vortex shedding frequency; St the Strouhal number, $f_v H_m/V$; giving

$$\ddot{x}_{st} + 2\zeta_s\dot{x}_{st} + x_{st} = F_s^* + F_e^*. \tag{10}$$

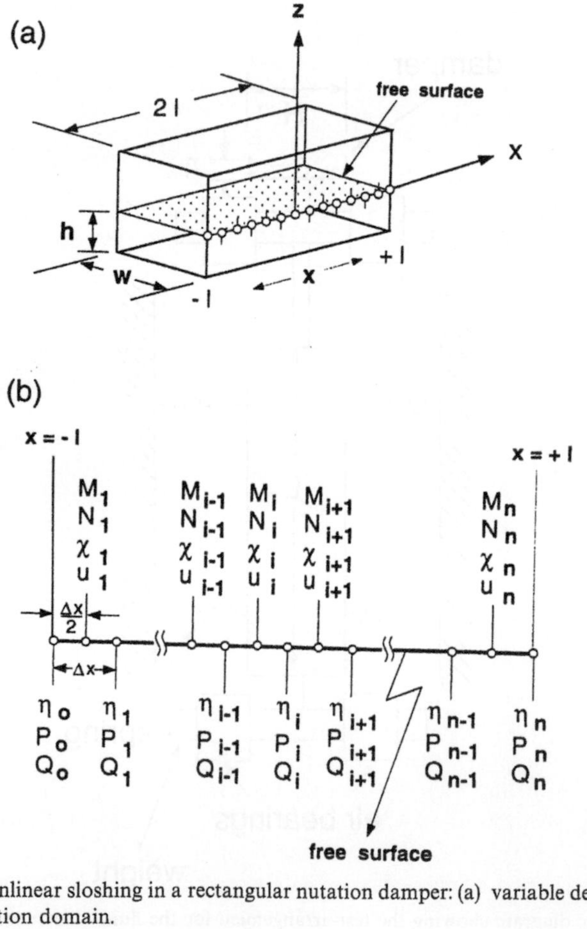

Fig. 3. Modelling nonlinear sloshing in a rectangular nutation damper: (a) variable definition; (b) discretization over the solution domain.

The sloshing force F_s^* was obtained through integration of the pressure at the side walls while the F_e^* was taken as proposed by Hartlen and Currie [17]. Accounting for the fact that

1. the vortex shedding frequency is proportional to the wind speed, i.e. the Strouhal number (St) remains essentially constant;
2. the system has self-exciting and self-limiting (limit cycle) type of response;

the lift variation can be characterized through the equation

$$\ddot{C}_L - A_1\omega_{nv}\dot{C}_L + \frac{A_3}{\omega_{nv}}(\dot{C}_L)^3 + \omega_{nv}^2 C_L = b\dot{x}_{st}, \tag{11}$$

where A_1, A_3 and b are constants with values based on information compiled over years [17].

2.4. Numerical solution for the fluid–structure interaction dynamics

One is now faced with the solution of a highly nonlinear and coupled set of Eqs. (9a), (9b), (9c)–(11). A finite difference scheme was selected for solution of the above set of equations. The damper width $2l$ was discretized using the nodes at the free surface as indicated in Fig. 3. The above equations were integrated simultaneously to yield the time dependent surface displacements η_i and velocities u_i. With the time history of the surface established, added mass and reduced liquid damping can be calculated quite readily. The iteration process, in general, converged rapidly. A typical case required around 0.3 s (≈ 15 iteration cycles) to give the transient solution at a given instant of time using a Sun Sparc II workstation. In general, around 1.5 hours were required to obtain the steady state solution.

3. Results and discussion

Fig. 4 shows the time history of the propagating wave over one cycle, obtained numerically, with an excitation frequency $\hat{\omega}_e = 1.05$. This represents the near resonance condition leading to a large amplitude free surface motion. The resulting reduced damping was also predicted to be large.

Fig. 5 presents variation of the reduced damping with liquid height as obtained using the numerical procedure. Experimental data are also presented to assess accuracy of the numerical analysis. Considering the complex character of the sloshing dynamics and relatively simple character of the numerical model, the agreement is surprisingly good, particularly for $h/W > 0.2$. This suggests that wave dynamics and viscosity, which are accounted for in the numerical analysis, are the major parameters contributing to the energy dissipation. Note, the numerical results are conservative compared to the experimental data at lower liquid heights suggesting a change in relative contributions from different sources. It appears that at a smaller h/W, the energy dissipation is essentially due to wavebreaking, i.e. viscous dissipation at the walls and from the irrotational body of the liquid are

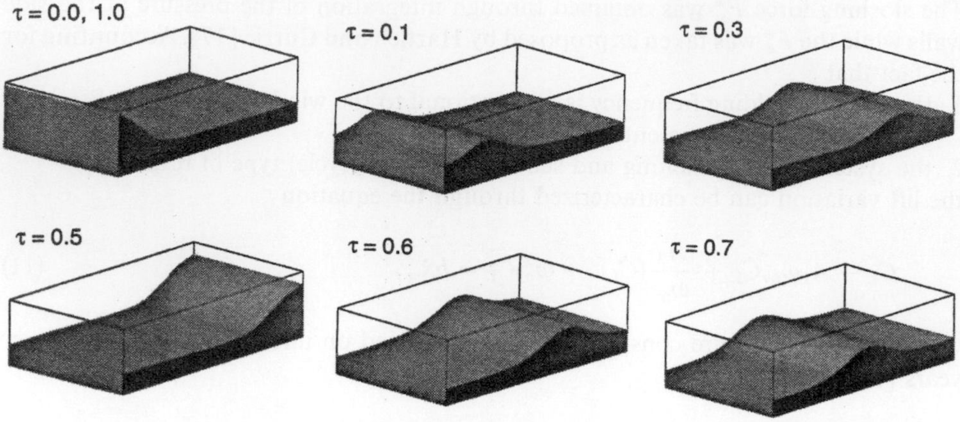

Fig. 4. Animation of the numerical data at a frequency $\hat{\omega}_e = 1.05$. Note the train with a single wave and large amplitude at the wall. The sloshing force was found to be higher.

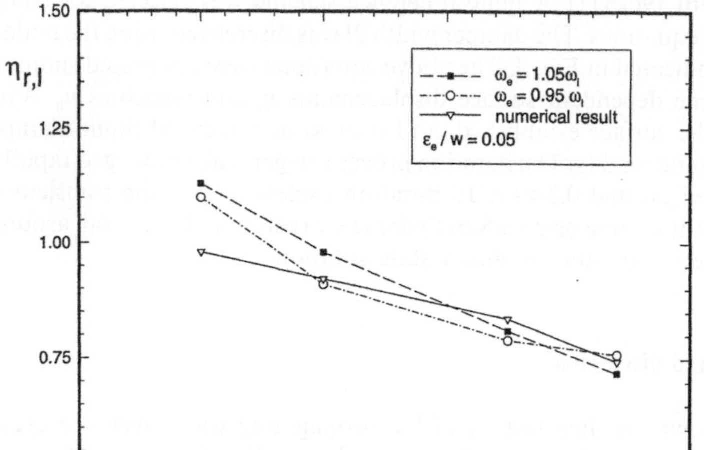

Fig. 5. Variation of the reduced damping with liquid height and excitation frequency as predicted by the numerical analysis. (∇) refers to numerical results corresponding to resonance. Correlation with the experimental results is good.

relatively small. On the other hand, the present analysis does not account for the impact dynamics of the wave striking the damper wall. However, during the experiment, the strain gauge transducer is able to record the sloshing force rather accurately. Furthermore, at lower liquid heights, part of the damper bottom becomes exposed. As a consequence, the effective wetted area is signifigantly reduced. However, the numerical analysis does not account for this and hence the discrepancy for $h/w < 0.2$.

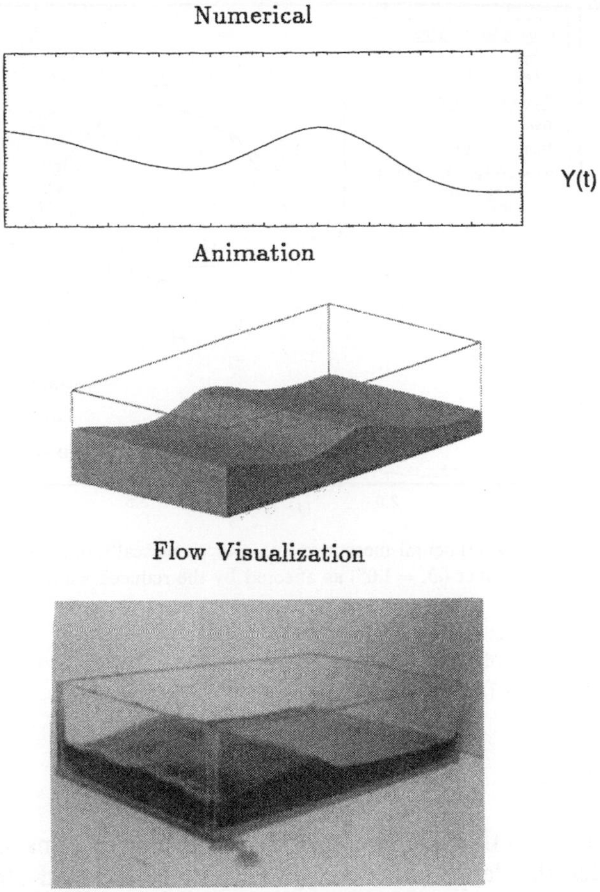

Fig. 6. Comparison between the numerical model, its animation and flow visualization results for a rectangular damper free surface showing a single propagating wave.

It would be of interest to show a typical case of surface waves as obtained numerically, as well as their animation, and compare them with the free surface character observed during the flow visualization study. In a sense, this would help in assessing the accuracy of the numerical model. This is shown in Fig. 6. The correlation is indeed quite good. The comparison clearly shows effectiveness of the numerical analysis in capturing essential features of the free surface dynamics even under such demanding situation of resonance.

Numerically calculated response of the structure with a systematic increase in the wind speed is shown in Fig. 7, for both the conditions of with and without a rectangular damper. The damper to structure mass ratio is 4% with the structural frequency $f_n = 1.05$ Hz. The numerical model is able to capture the classical resonance phenomena quite well. The reduction in amplitude by over 85% at resonance attest to the remarkable effectiveness of the nutation damper.

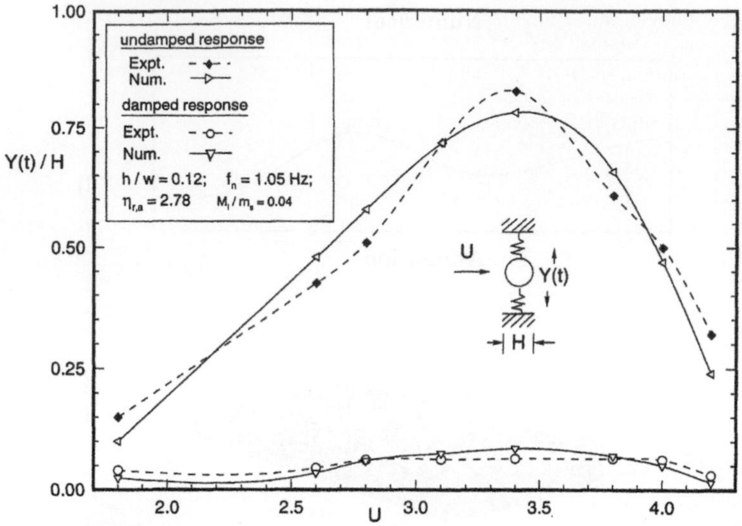

Fig. 7. Comparison between experimental measurements and numerically obtained structural displacement, x_{st}/H_m, near vortex resonance ($\hat{\omega}_e = 1.05$) as affected by the reduced wind speed $U(V/\omega_n H_m)$. The agreement is quite good.

This versatile and powerful numerical tool can now be used with confidance to design nutation dampers for real-life applications.

4. Concluding remarks

This study pertains to the development of numerical algorithms for prediction of the sloshing modes, their animation, energy dissipation, and fluid–structure interaction dynamics in the presence of nutation damping.

The numerical simulation, based on the nonlinear dissipative as well as dispersive shallow water model, is able to predict the free surface dynamics and reduced damping quite well considering the complex character of the problem. Furthermore, it proved to be successful in predicting vortex resonance response, a challenging conjugate problem.

Acknowledgements

The investigation reported here was supported by the Natural Sciences and Engineering Research Council of Canada, Grant No. A-2181.

References

[1] J.E. Cermak, Application of fluid mechanics to wind engineering, Freeman Scholar Lecture, J. Fluids Eng. Trans. ASME 97 (1) (1975) 9–38.

[2] F. Welt, A study of nutation dampers with application to wind induced oscillations, Ph.D. Thesis, University of British Columbia, 1988.
[3] F. Welt, V.J. Modi, Vibration damping through liquid sloshing. Part 1: A nonlinear analysis, Trans. ASME J. Vib. Acoust. 114 (1992) 10–16.
[4] F. Welt, V.J. Modi, Vibration damping through liquid sloshing. Part 2: Experimental results, Trans. ASME J. Vib. Acoust. 114 (1992) 17–23.
[5] Y. Tamura, R. Kousaka, V.J. Modi, Practical application of nutation damper for suppressing wind-induced vibrations of airport towers, J. Wind Eng. Ind. Aerodyn. 43 (1992) 1919–1935.
[6] V.J. Modi, F. Welt, M.L. Seto, Control of wind-induced instabilities through application of nutation dampers: A brief review, Eng. Struct. 17 (9) (1995) 626–638.
[7] J. Miles, Resonantly forced surface waves in a circular cylinder, J. Fluid Mech. 149 (1984) 15–31.
[8] K. Amano, M. Koizumi, M. Yamakawa, Three-dimensional analysis method for potential flow with a moving liquid surface using a boundary element method, in: D.C. Ma, J. Tani, S.S. Chen, W.K. Lin (Eds.), ASME Trans. Sloshing and Fluid–Structure Vibration, PVP-157, 1989, pp. 127–132.
[9] T.C. Su, Y. Wang, Numerical simulation of three-dimensional large amplitude sloshing in cylindrical tanks subjected to arbitrary excitations, in: D.C. Ma, J. Tani, S.S. Chen (Eds.), ASME Trans. Flow–Structure Vibration and Sloshing, PVP-191, 1990, pp. 127–148.
[10] L.M. Sun, Y. Fujino, M. Pacheco, P. Chaiseri, Modelling of tuned liquid damper (TLD), J. Wind Eng. Ind. Aerodyn. 41–44 (1992) 1883–1894.
[11] W.D. Iwan, R.D. Blevins, A model for vortex induced oscillation of structures, Trans. ASME, J. Appl. Mech. 41 (1974) 581–586.
[12] M.L. Seto, An investigation on the suppression of flow induced vibrations of bluff bodies, Ph.D. Thesis, University of British Columbia, 1996.
[13] M.L. Seto, V.J. Modi, Nutation damping of wind induced instabilities: Experimental and numerical studies, Proc. 1st Int. Symp. on Advances in Vibration Issues, Active and Passive Vibration Mitigation, Damping and Seismic Isolation, in: ASME/JSME PVP Conf., PVP-309, 1995, pp. 31–41.
[14] J.J. Stoker, Long waves in shallow water, in: Water Waves, Interscience, New York, 1957, pp. 291–326.
[15] T. Weiyan, Stability analysis and boundary procedures, in: Shallow Water Hydrodynamics – Mathematical Theory and Numerical Solution for a Two-Dimensional System of Shallow Water Equations, Elsevier, New York, 1992, pp. 388–413.
[16] H. Lamb, Hydrodynamics, 6th Ed., Dover, New York, 1945, pp. 562–571, 579–581.
[17] R.T. Hartlen, I.G. Currie, Lift-oscillator model of vortex-induced vibration, ASCE J. Eng. Mech. Div. 96 (1970) 577–591.

ELSEVIER

Journal of Wind Engineering
and Industrial Aerodynamics 67&68 (1997) 627–638

JOURNAL OF
wind engineering
AND
industrial
aerodynamics

On the application of Thomson's random flight model to the prediction of particle dispersion within a ventilated airspace

A.M. Reynolds[1]

Silsoe Research Institute, Wrest Park, Silsoe, Bedford MK45 4HS, UK

Abstract

The correct prediction of mean particle concentrations within ventilated airspaces is an important practical problem. Recently, much progress has been made in understanding how to formulate correctly random flight models for particle trajectories in such inhomogeneous turbulent flows. Thomson showed that random flight models which satisfy the well-mixed condition (i.e. give the correct steady-state distribution of particles in phase space) are essentially correct. Thomson subsequently formulated a random flight model which satisfies the well-mixed condition for inhomogeneous Gaussian turbulence. Prompted by the success of this model in predicting particle dispersion in one-dimensional and two-dimensional inhomogeneous turbulent flows, an assessment is made of the ability of the model to predict correctly mean particle concentrations in a three-dimensional inhomogeneous turbulent flow within a mechanically ventilated airspace. When used in conjunction with the statistical properties of the air flow predicted by the k–ε model, Thomson's model is shown to predict mean particle concentrations in close accord with experimental findings. It is suggested that skewness of the turbulent velocity fluctuations is of secondary importance in determining particle dispersion in highly inhomogeneous turbulent flows, compared to the effects of strong mean streamline straining and large gradients in Reynolds stress. This view is supported by numerical studies undertaken using an extension of Thomson's model, which takes partial account of the skewness of velocity fluctuations. It is further suggested that when predicting particle dispersion in strongly inhomogeneous turbulent flows, it may not be necessary to devise random flight models which satisfy the well-mixed condition for more realistic (i.e. non-Gaussian) probability distribution functions of turbulent velocity fluctuations. This would allow difficulties associated with determining such probability distribution functions and the increased complexity of such models to be circumvented.

Keywords: Turbulence; Dispersion; Ventilation; Model

[1] E-mail: areynolds@bbsrc.ac.uk.

PII S0167-6105(97)00105-0

1. Introduction

The correct prediction of particle dispersion within ventilated airspaces is an important practical problem in many biology- and electronic-based industries. For example, in the food processing and micro-electronic manufacturing industries, the dispersion and subsequent deposition of micro-organisms and particulates can result in a deterioration in the quality of food products and micro-electronic components.

Recently, much progress has been made in understanding how to correctly formulate random flight models for dispersion in inhomogeneous turbulent flows [1], typical of those found within ventilated airspaces. Thomson [1] showed that random flight models which satisfy the well-mixed condition (i.e. the model gives the correct steady-state distribution of particles in phase space) are essentially "correct" in the sense that (i) they give the correct small time behaviour of the velocity distribution of particles from a point source, (ii) the Eulerian equations derived from the model are compatible with the true Eulerian equations and (iii) their forward and reverse time formulations of dispersion are consistent.

Thomson [1,2] subsequently formulated a random flight model which satisfied the well-mixed condition for inhomogeneous Gaussian turbulence. This model has been successful in predicting particle dispersion in one-dimensional [3] and two-dimensional [4] inhomogeneous turbulent flows. Prompted by this success, in this paper an assessment is made of the ability of the model to predict correctly particle dispersion in a three-dimensional inhomogeneous flow within a ventilated airspace. The model is used in conjunction with the statistical properties of the air flow predicted by the standard k–ε model. Predicted mean particle concentrations are shown to be in close accord with experimental findings. Predictions are compared with those obtained using an extension of Thomson's model, which approximately satisfies the well-mixed condition for skewed turbulence (i.e. where only the first three moments of the velocity distribution are taken into account). An assessment is made of the importance of skewness of turbulent velocity fluctuations in determining particle dispersion in strongly inhomogeneous turbulent flows.

2. Thomson's model

Thomson [2] considered a very general class of model for particle trajectories in turbulent flows which takes the form,

$$\boldsymbol{u}^i_{n+1} = \boldsymbol{u}^i_n - \Delta t \boldsymbol{T}^{ij}(x_{n+1})\boldsymbol{u}^j_n + \Delta t[\beta_{ij}(x_{n+1})\boldsymbol{u}^j_n + \gamma^{ijk}(\boldsymbol{x}_{n+1})\boldsymbol{u}^j_n\boldsymbol{u}^k_n] + \boldsymbol{\mu}^i_{n+1},$$

$$\boldsymbol{x}_{n+1} = \boldsymbol{x}_n + \boldsymbol{u}_n \Delta t, \tag{1}$$

where $\boldsymbol{x}_n$ and $\boldsymbol{u}_n$ are the position and velocity of the particle after the nth time step, of size Δt. Superscripts indicate Cartesian components and the Einstein summation convention applies. This class of model is defined in terms of a timescale tensor, $\boldsymbol{T}$, two tensors, β and γ, and the random vector, $\boldsymbol{\mu}$. The joint probability density function of the components of the random vector $\boldsymbol{\mu}$ is determined by requiring that the model satisfy the well-mixed condition. Thomson [2] considered the case of inhomogeneous Gaussian turbulence. That is, the one-point joint Eulerian probability distribution function was taken to be joint normal everywhere even though moments vary with position. Thomson's analysis is readily extended to the case of inhomogeneous skewed turbulence. For skewed inhomogeneous turbulence the first five moments of the random vector, $\boldsymbol{\mu}$, are non-zero to order Δt. The first three moments of the random vector, $\boldsymbol{\mu}$, are given by

$$\overline{\mu^i} = \Delta t\left[\frac{\partial \sigma^{il}}{\partial x^l} + U^l\frac{\partial U^i}{\partial x^l} + \boldsymbol{T}^{il}U^l - \beta^{il}U^l - \gamma^{ilm}(U^lU^m + \sigma^{lm})\right] + \vartheta(\Delta t^2),$$

$$\overline{\mu^i\mu^j} = \Delta t\left[\sigma^{il}\frac{\partial U^j}{\partial x^l} + \frac{U^l}{2}\frac{\partial \sigma^{ij}}{\partial x^l} + \boldsymbol{T}^{il}\sigma^{jl} - \beta^{il}\sigma^{jl} - \gamma^{ilm}(U^l\sigma^{jm} - U^m\sigma^{jl}) + \tfrac{1}{2}\frac{\partial S_{ijl}}{\partial x_l} - \gamma^{ilm}S^{jlm}\right] + \Delta t[i\leftrightarrow j] + \vartheta(\Delta t^2), \quad (2)$$

$$\overline{\mu^i\mu^j\mu^k} = \Delta t\left[\tfrac{1}{2}\sigma^{il}\frac{\partial \sigma^{jk}}{\partial x^l} - \gamma^{ilm}\sigma^{jl}\sigma^{km} + \frac{1}{2}S^{jkl}\frac{\partial U^i}{\partial x^l} + \frac{1}{6}U^l\frac{\partial S^{ijk}}{\partial x^l} + \boldsymbol{T}^{il}S^{jkl} - \beta^{il}S^{jkl} - \gamma^{ilm}U^lS^{jkm}\right]$$
$$+ \Delta t\,[i\leftrightarrow j + i\leftrightarrow k + j\leftrightarrow k + i\rightarrow j, j\rightarrow k, k\rightarrow i + i\rightarrow k, j\rightarrow i, k\rightarrow j] + \vartheta(\Delta t^2),$$

where $\boldsymbol{U}$ is the average velocity, $\boldsymbol{\sigma}$ is the tensor containing the covariances of the turbulent velocity fluctuations and S is the triple correlation of turbulent velocity fluctuations at the particle position.

Currently there is insufficient evidence to decide which values of the tensors β and γ will yield the most accurate particle trajectories, so their values are chosen to simplify the implementation of the model. Thomson [2] choose

$$\beta^{ij} = \frac{\partial U^i}{\partial x_j} - \frac{1}{2}\left(\frac{\partial \sigma^{il}}{\partial x_j}\right)(\sigma^{-1})^{lm}U^m, \quad \gamma^{ijk} = \frac{1}{2}\left(\frac{\partial \sigma^{il}}{\partial x_k}\right)(\sigma^{-1})^{lj}, \quad (3)$$

so that for Gaussian turbulence, the distribution of the random vector, $\boldsymbol{\mu}$, can be taken to be Gaussian.

For the Lagrangian velocity structure function to be consistent with inertial sub-range scaling, the covariance of the random vector, $\boldsymbol{\mu}$, must depend only upon the mean dissipation rate of turbulent kinetic energy, ε, and the Lagrangian velocity structure constant, C_0. For Gaussian turbulence, this requires that the time-scale tensor, $\boldsymbol{T}$, take the form,

$$T^{ij} = \tfrac{1}{2} C_0 \varepsilon (\sigma^{-1})^{ij}. \tag{4}$$

For Gaussian turbulence, the first two moments of the random vector, $\boldsymbol{\mu}$, are then given by

$$\overline{\mu^i} = \Delta t \left[\frac{1}{2} \frac{\partial \sigma^{il}}{\partial x^l} + T^{il} U^l \right] + \vartheta(\Delta t^2), \quad \overline{\mu^i \mu^j} = \Delta t C_0 \varepsilon \delta^{ij} + \vartheta(\Delta t^2), \tag{5}$$

This model is identical to the model proposed later, in 1987, by Thomson [1] which is based upon the Ito and Fokker–Planck equations.

For skewed turbulence, there is no choice for the tensors β and γ for which the distribution of $\boldsymbol{\mu}$ can be taken to be Gaussian. In this paper, the tensors β and γ will be taken to have the form given in Eq. (3) even when turbulence is skewed. Then, for weakly non-Gaussian turbulence, the distribution of the random vector, $\boldsymbol{\mu}$, will be weakly non-Guassian and it is appropriate to take account of only the lower-order moments of the random vector, $\boldsymbol{\mu}$.

For skewed turbulence, there is no choice for the timescale tensor, $\boldsymbol{T}$, for which the Lagrangian velocity structure function derived from the model can be made to be consistent with inertial sub-range scaling. However, in the present application to turbulent dispersion in a strongly inhomogeneous flow this inconsistency is not considered to be important. This is because inertial sub-range scaling is not exact and is likely to be violated in strongly inhomogeneous flows where gradients in mean velocity and turbulent kinetic energy are large, and where "large-scales" of turbulence are anisotropic [5]. This is especially the case for low Reynolds-numbers ($\mathrm{Re} \approx 10^4$) flows at which an inertial sub-range would just become established under ideal circumstances. Indeed, spectra obtained for the ventilating flow under condition, which has an inlet Reynolds number $\mathrm{Re} \approx 5 \times 10^4$, showed little evidence of an inertial sub-range. The extension of Thomson's model to the case of skewed turbulence will not, however, be appropriate for simulating particle trajectories in atmospheric or oceanic flows where Reynolds-numbers are sufficiently high ($\mathrm{Re} \approx 10^5$–10^7) for an inertial sub-range to be established. The timescale tensor, $\boldsymbol{T}$, will be taken as given in Eq. (4) even for skewed turbulence.

3. Experimental facility

Mean particle concentrations were measured in a full-scale section of a building that was designed to study air flow patterns under controlled climatic conditions. The facility is a cross-section of a typical UK intensive livestock building of one bay, 3 m wide. The dimensions and geometry of the building are shown in Fig. 1. The section is surrounded by an outer temperature-controlled shell allowing precise ambient conditions to be established. The inlet air and the air surrounding the building section were set at 20°C.

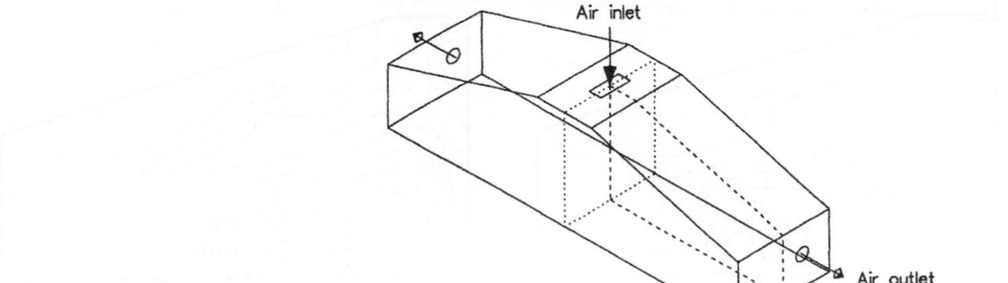

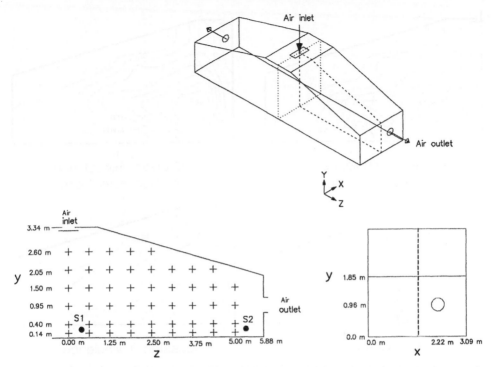

Fig. 1. Dimensions and geometry of the building section. Location of the sampling points (+). In the coordinate system shown, the locations, S1 and S2, of the dust generator are (0.5, 0.2, 0.5 m) and (0.5, 0.2, 5.38 m). The central y–z plane containing the locations at which particle concentrations were measured is indicated by a dashed-line. Also shown are the locations (+), on the central plane, at which particle concentrations were measured.

The room is ventilated by a jet ventilation system with air entering horizontally in each direction from an adjustable baffled ceiling inlet. Air is exhausted through two fans located in the side walls of the building. The inlet velocity of the jet was 6.0 m s^{-1}, the inlet Reynolds number 5×10^4 and the total ventilation rate was 1.22 m^3 s^{-1}, which is equivalent to 44 air changes per hour.

Dust was generated using a TSI Inc. fluidized bed aerosol generator. The dust material was Arizona road dust, a standard test dust, having density $\rho = 2650$ kg m^{-3} and with 99% of the particles having diameters between 0 and 2.5×10^{-6} m. Mean concentrations of particles with diameters between 1 and 2×10^{-6} m were measured at 49 points (see Fig. 1) on a vertical plane through the building using a RION optical particle counter. The measured mean particle concentrations corresponding to two locations S1 and S2 (see Fig. 1) of the dust generator are shown in Figs. 2 and 3.

A more complete description of the experimental facility and measurements of mean particle concentrations can be found in Ref. [6].

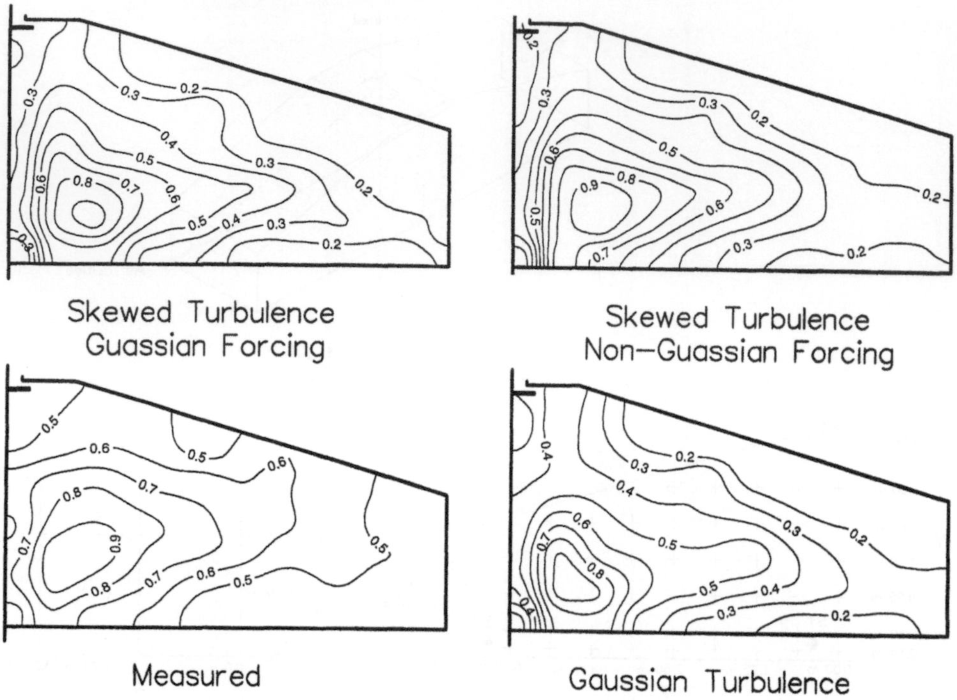

Fig. 2. Measured and predicted mean particle concentrations on the central y–z plane of the building section for particle dispersion from source S1. Predictions were obtained using Thomson's model. Predictions are shown for the case of Gaussian turbulence and skewed turbulence. For Gaussian turbulence the well-mixed condition is exactly satisfied. For skewed turbulence the well-mixed condition is approximately satisfied. Predictions are shown for two approximation schemes. First, where the distribution of the random vector, $\boldsymbol{\mu}$, is approximated by a Guassian distribution. Second, where the partial account (see text) is taken of the non-Gaussian distribution of the random vector, $\boldsymbol{\mu}$. Predictions were obtained for the Lagrangian structure constant, $C_0 = 7$. Measured and predicted particle concentrations have been normalized by the maximum particle concentrations.

4. Numerical implementation

In calculating particle dispersion it was assumed that the presence of the particles did not affect the statistical properties of the air flow. The statistical properties of the air flow can then be predicted independently of the particles.

The statistical properties of the air flow within the building section, under isothermal conditions, were predicted using the standard k–ε model. The flow were treated as being stationary. It was then sufficient, because of spatial symmetry (see Fig. 1), to predict properties of the flow in just one half of the building section. Predictions were made using a body-fitted grid containing $45 \times 48 \times 50$ cells. Further computations using a grid containing 330 000 cells indicated that grid-independent predictions had been obtained. To account for viscous effects, near the walls of the building section, log-law wall functions were used. A detailed discussion of the predictions

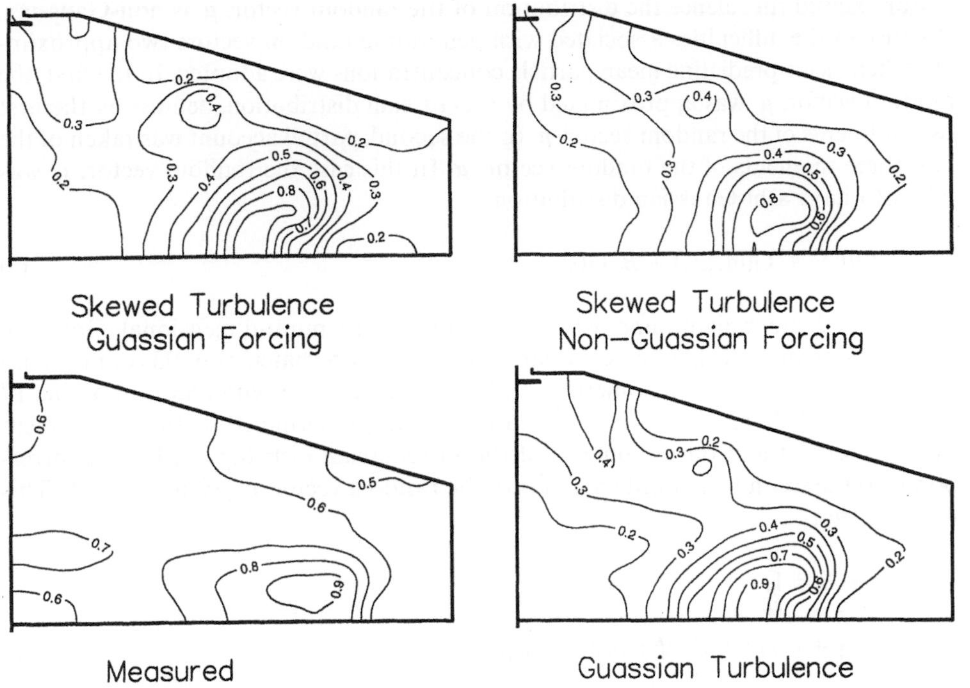

Fig. 3. Measured and predicted mean particle concentrations on the central y–z plane of the building section for particle dispersion from source S2. Predictions were obtained using Thomson's model. Predictions are shown for the case of Gaussian turbulence and skewed turbulence. For Gaussian turbulence the well-mixed condition is exactly satisfied. For skewed turbulence the well-mixed condition is approximately satisfied. Predictions are shown for two approximation schemes. First, where the distribution of the random vector, $\boldsymbol{\mu}$, is approximated by a Guassian distribution. Second, where partial account (see text) is taken of the non-Gaussian distribution of the random vector, $\boldsymbol{\mu}$. Predictions were obtained for the Lagrangian structure constant, $C_0 = 7$. Measured and predicted particle concentrations have been normalized by the maximum particle concentrations.

obtained using a $26 \times 24 \times 34$ grid and their comparison with experimental findings can be found in Ref. [7]. The components of predicted mean air velocity were found to be within 20% of experimental values. Predicted values of mean turbulent kinetic energy were found to be within a factor of two of experimental findings.

Predictions for the velocity covariances, σ_{ij}, were obtained, from the predictions for the mean velocity field, the mean turbulent kinetic energy and the mean dissipation rate of turbulent kinetic energy, using the Boussinesq approximation. Predictions for the triple velocity correlations, S_{ijk}, were obtained using the model proposed by Launder et al. [8],

$$S^{ijk} = -C_s \frac{k}{\varepsilon}\left(\sigma^{il}\frac{\partial \sigma^{jk}}{\partial x_l} + \sigma^{jl}\frac{\partial \sigma^{ik}}{\partial x_l} + \sigma^{kl}\frac{\partial \sigma^{ij}}{\partial x_l}\right), \tag{6}$$

where C_s is a constant, $C_s = 0.11$.

For skewed turbulence the distribution of the random vector, $\boldsymbol{\mu}$, is non-Gaussian. Because of the difficulties associated with generating random vectors two approximate schemes for predicting mean particle concentrations were adopted. In the first, the random vector, $\boldsymbol{\mu}$, was approximated by a Gaussian distribution, defined by the first two moments of the random vector, $\boldsymbol{\mu}$. In the second, partial account was taken of the first three moments of the random vector, $\boldsymbol{\mu}$. In this case the random vector, $\boldsymbol{\mu}$, was taken to have a bi-Gaussian distribution,

$$P(\mu) = A \cdot \mathrm{G}(m_A, \sigma_A) + B \cdot \mathrm{G}(m_B, \sigma_B), \tag{7}$$

where A, B are weights and where the $\mathrm{G}(m,\sigma)$ are multi-dimensional Gaussian distributions with mean m and covariances σ. The appropriateness of this approximation has been examined in numerical studies which are reported in Section 5. The 17 parameters, A, B, $m^i = m^i_A = m^i_B$, σ^{ij}_A and σ^{ij}_B are fully determined by the requirement that the second and third moments of the bi-Gaussian distribution, Eq. (7) corresponds to the second and third moments of the random vector, $\boldsymbol{\mu}$, given in Eq. (2). This requires that

$$\begin{aligned} &A + B = 1, \\ &A(\sigma^{ij}_A + m^i m^j) + B(\sigma^{ij}_B + m^i m^j) = \overline{\mu^i \mu^j}, \\ &A(\sigma^{ij}_A m^k + \sigma^{ik}_A m^j + \sigma^{jk}_A m^i) + B(\sigma^{ij}_B m^k + \sigma^{ik}_B m^j + \sigma^{jk}_B m^i) = \overline{\mu^i \mu^j \mu^k}. \end{aligned} \tag{8}$$

This distribution produces a random vector, $\boldsymbol{\mu}$, those components have mean, m_i rather than the required mean, $\overline{\mu^i}$, given in Eq. (2). This can be corrected for by adding a term $-m^i + \overline{\mu^i_{n+1}}$ to the deterministic part of Thomson's model, Eq. (1). This scheme is preferable to one using a more general bi-Gaussian distribution, with $m^i_A \neq m^i_B$, where the 20 parameters, A, B, m^i_A, m^i_B, σ^{ij}_A and σ^{ij}_B would then be fully determined by the requirement that the first three moments of the bi-Guassian correspond to the first three moments of the random vector, $\boldsymbol{\mu}$.

It was not always possible, however, to find a bi-Gaussian distribution with the required moments. This is because the first three moments of the random vector, $\boldsymbol{\mu}$, are not necessarily consistently specified. They are not necessarily consistently specified because: for multi-dimensional flows the choice for β, γ and $\boldsymbol{T}$ is somewhat arbitrary (non-unique); the first and second moments of the velocity distribution are not necessarily consistent with the third-order moments of velocity predicted by the model of Launder et al. [8]; the interpolation scheme adopted does not enforce statistically consistent interpolated quantities. When an appropriate bi-Gaussian distribution could not be found, the third-order moments of the random vector, $\boldsymbol{\mu}$, were neglected and a Gaussian distribution, defined by the first two moments of the random vector, $\boldsymbol{\mu}$, was adopted. The appropriate bi-Gaussian distribution was found to exist for about 10% of a typical particle trajectory. Similar difficulties have been encountered when Thomson's model is used in conjunction with meteorological data to predict dispersion in atmospheric boundary layers. It is found that the meteorlogical data cannot be used directly but first require an "analysis" (in a sense similar to the current meterological practice) in order to produce internally consistent sets of moments of random vector, $\boldsymbol{\mu}$ [9].

Thomson's model, Eq. (1), was integrated using a fourth-order Runge–Kutta method. The moments of the random vector, $\boldsymbol{\mu}$, are given in Eq. (2) to order Δt. For this to be valid the size of time-step Δt, used in the numerical integration of Eq. (1) must be small compared to the time scales of the turbulence (i.e. the reciprocals of the eigenvalues of $\boldsymbol{T}$) and must be small enough to ensure that the particle motions adequately resolve any inhomogeneities in mean flow, $\boldsymbol{U}$, velocity covariance $\boldsymbol{\sigma}$ and triple velocity correlation, $\boldsymbol{S}$. The sensitivity of predicted mean particle concentrations to the size of the time-step, Δt, was investigated in numerical studies. All predicted mean particles concentrations presented within this paper are insensitive to a decrease in the value of the time step, Δt, that was used.

Particles were treated as "marked fluid particles" (i.e. they are assumed to travel at the local velocity of the air) until they were transported toward a surface within one "stopping distance", characterized by the particle relaxation time and the wall-ward velocity [10]. At this point, particles are assumed to deviate from the local turbulent motion, arriving at the surface through "free-flight" by virtue of their inertia. Free-flight deposition is widely believed [11] to be the main contributor to particle deposition. The other mechanisms of particle deposition include Brownian diffusion and turbulent diffusion. In this paper free-flight deposition is taken to be the only mechanism of particle deposition and once deposited particles are assumed not be to re-entrained.

Nodes in the computational grid, used in predicting the statistical properties of the air flow, formed 50 parallel x–y planes. Piece-wise linear interpolation was used to interpolate, in the z-direction, the statistical properties of the air flow from the nodes of the computational grid, surrounding the particle, to the particle position. Nodes in the x–y planes did not form an orthogonal system, so piece-wise linear interpolation could not be used. The Rinka and Cline method [12] was used to interpolate, in the x- and y-directions, the predicted values of the statistical properties of the air flow from the nodes of the computation grid, surrounding the particle, to the particle position. Predicted mean particle concentrations obtained using this scheme were found not to be statistically different from those obtained using a more sophisticated, but slower, weighted least-squares interpolation scheme. In this scheme the values of the statistical properties of the air flow were interpolated from both the nearest-neighbouring and next-nearest-neighbouring nodes of the computational grid, surrounding the particle, to the particle location.

Ensemble mean particle concentration fields were calculated from the simulated independent trajectories of 25 000 particles. All predicted mean particle concentration fields presented in this paper are statistically stationary; that is independent of increasing the number of particles within the ensemble. Multiple contributions to the ensemble, averages due to particles revisiting points in the flow domain, were removed. This enables direct comparisons to be made between predicted mean particle concentrations and measured particle concentrations, obtained by extracting particles at given points within the flow domain.

5. Predictions for mean particle concentrations

Before Thomson's model can be implemented an appropriate value of the Lagrangian structure constant C_0 must be determined. Sawford [13] showed that for random flight models which assume that the collective evolution of $(\boldsymbol{x}, \boldsymbol{u})$ is Markovian, e.g. Thomson's model, the supposedly universal constant C_0 is not universal but rather depends on Reynolds number. He subsequently attributed the variability in the "best" estimates for C_0 obtained using such models to variations in the Reynolds number across the various dispersion experiments. The dependence of predicted mean particle concentrations on the value of C_0 has been investigated in numerical studies. In accord with theoretical [13] and experimental [14,15] estimates for C_0, the "best" correspondence between predicted and measured mean-particle concentrations was obtained with $C_0 = 7$.

Predicted mean-particle concentrations obtained using the Thomson's model for Gaussian turbulence are shown in Figs. 2 and 3. Predicted locations of maximum mean particle concentrations and the shape of the contours of constant particle concentration, for both source S1 and S2, are seen to be in accord with experimental findings. Less well predicted, however, are the gradients in mean particle concentration.

The over-prediction of gradients in mean-particle concentrations can be attributed, at least in part, to the neglect of non-stationarity. Measurements of air velocity obtained using sonic anemometers and a non-stationary analysis (time-dependent k–ε model, $26 \times 24 \times 68$) reveals time-periodic oscillations of the statistical properties of the air flow. The predicted and measured period of this oscillation is approximately 28 s. These oscillations are expected to influence particle dispersion because the residence times of particles is predicted to be comparable to or longer than the period of oscillation. Particles released from sources S1 and S2 are predicted to remain in the air flow for an average of 59 and 32 s, respectively, [10].

Predicted mean particle concentrations obtained using the model for skewed turbulence are shown in Figs. 2 and 3. Predictions were obtained taking account of just the first two moments and taking partial account of first three moments of the random vector, $\boldsymbol{\mu}$. Consequently, because the distribution of the random vector, $\boldsymbol{\mu}$, is approximated, the well-mixed condition is not satisfied exactly to order Δt. However, predicted mean particle concentrations obtained using the two approximation schemes were not found to be statistically different. This indicates that the neglect of third and higher moments of the random vector, $\boldsymbol{\mu}$, and the subsequent violation of the well-mixed condition does not significantly affect the predicted particle dispersion in the air flow under consideration. Furthermore, a comparison of Figs. 2 and 3 shows that these predicted mean particle concentrations are not significantly different from those predicted by the Thomson model for Gaussian turbulence.

6. Conclusions

It is concluded that Thomson's model for Gaussian turbulence, when used in conjunction with the statistical properties of air flows by the k–ε model, can be successful in predicting mean particle concentrations of particles dispersing in highly inhomogeneous turbulent flows typical of those found within ventilated airspaces. This suggests that skewness, of turbulent velocity statistics, is of secondary importance in determining particle dispersion, in highly inhomogeneous turbulent flows, compared to the effects of strong mean streamline straining [16,17] and large gradients in Reynolds stress. This view was supported by numerical studies undertaken using an extension of Thomson's model, which approximately satisfied the well-mixed condition for skewed turbulence. However, the sensitivity of particle dispersion on skewness predicted by this model will, at least, in part, be dependent upon the particular truncation of the moments of the random vector, $\boldsymbol{\mu}$, that was adopted and upon the ability of the model of Launder et al. [8] to represent correctly the skewness of turbulent velocity statistics.

The success of Thomson's model for Gaussian turbulence can also be understood in terms of the "thermodynamic constraint", which requires that an initially uniform distribution of material is maintained [18]. Satisfaction of the thermodynamic constraint becomes increasingly more important with increasing inhomogeneity of the flow. Thomson's model for Gaussian turbulence, necessarily satisfies the "thermodynamic constraint" because it satisfies, to second-order, the well-mixed condition for the true probability distribution function of turbulent velocity fluctuations. Indeed Thomson's model for Gaussian turbulence is the simplest random flight model which satisfies the thermodynamic constraint. It is not necessary for satisfaction of the "thermodynamic constraint" to satisfy the well-mixed condition, for the true probability distribution function of turbulent velocity fluctuations, to third- or higher-order. With the consequence that when predicting turbulent dispersion in strongly inhomogeneous flows it is appropriate to approximate the true probability distribution function for turbulent velocity fluctuations by a Gaussian distribution.

It is suggested that when predicting particle dispersion in highly inhomogeneous flows, it is not necessary to devise random flight models satisfying the well-mixed condition for more realistic (i.e. non-Gaussian) probability distribution functions of turbulent velocity fluctuations. This would extend the range of applicability of Thomson's model from being restricted to weakly inhomogeneous turbulent flows to include strongly inhomogeneous turbulent flows.

Acknowledgements

I am indebted to B.B. Harral for making available his unpublished CFD predictions for the air flow within the ventilated building. I would like to thank M. Andersen and C.R. Boon for helpful discussions regarding their experimental results and for making available unpublished findings. Numerical studies of particle dispersion using a weighted least-squares interpolation scheme were performed by M. Wilson.

References

[1] D.J. Thomson, Criteria for the selection of stochastic models of particle trajectories in turbulent flows, J. Fluid Mech. 180 (1987) 529–556.

[2] D.J. Thomson, A random walk model of dispersion in turbulent flows and its application to dispersion in a valley, Quart. J. Roy. Met. Soc. 12 (1986) 511–530.
[3] J.D. Wilson, B.L. Legg, D.J. Thomson, Calculation of particle trajectories in the presence of a gradient in turbulent-velocity variance, Boundary-Layer Meteorol. 27 (1983) 163–169.
[4] T.K. Flesch, J.D. Wilson, A two-dimensional trajectory-simulation model for non-Gaussian, inhomogeneous turbulence within plant canopies, Boundary-Layer Meteorol. 61 (1992) 349–374.
[5] J.C.R. Hunt, J.C. Vassilicos, Kolmogorov's contributions to the physical and geometrical understanding of small-scale turbulence and recent developments, Proc. Roy. Soc. London A 434 (1991) 183–210.
[6] C.R. Boon, M. Andersen, Particle concentrations in an experimental livestock building, J. Agric. Eng. Res., in preparation.
[7] B.B. Harral, C.R. Boon, A comparison of predicted and measured air flow patterns in a mechanically ventilated livestock building, Building and Environ., in preparation.
[8] B.E. Launder, G.J. Leece, W. Rodi, Progress in the development of a Reynolds-stress turbulence closure, J. Fluid Mech. 68 (1975) 537.
[9] F. Tampieri, C. Scarami, U. Giostra, G. Brusasca, G. Tinarelli, D. Anfossi, E. Ferrero, On the application of random flight models in inhomogeneous turbulent flows, Ann. Geophys. 10 (1992) 749–758.
[10] S.K. Friedlander, H.F. Johnstone, Deposition of suspended particles from turbulent gas streams, Ind. Eng. Chem. 49 (1957) 1151–1156.
[11] J.W. Brooke, T.J. Hanratty, J.B. McLaughlin, Free-flight mixing and deposition of aerosols, Phys. Fluids 6 (10) (1994) 3404–3415.
[12] R.L. Rinka, A.K. Cline, A triangle-based C^1 interpolation scheme, Rocky-Mountain J. Maths. 14 (1984) 223–237.
[13] B.L. Sawford, Reynolds number effects in Lagrangian stochastic models of turbulent dispersion, Phys. Fluids A 3 (1991) 1577–1586.
[14] S.R. Hanna, Lagrangian and Eulerian time-scale relations in the day-time boundary layer, J. Appl. Meteorol. 20 (1981) 242–249.
[15] B.L. Sawford, F.M. Guest, Uniqueness and Universality of Lagrangian stochastic models for turbulent dispersion, Proc. 8th Symp. Turbulent Diffussion AMS, San Diego, CA, 1988, pp. 96–99.
[16] J.C.R. Hunt, Turbulent diffusion from sources in complex flows, Ann. Rev. Fluid Mech. 17 (1985) 447–485.
[17] A.M. Reynolds, Modelling particle dispersion within a ventilated airspace, Fluid Dyn. Res. 1996, submitted.
[18] B.L. Sawford, Generalized random forcing in random-walk turbulent diffusion models, Phys. Fluids 29 (1986) 3582–3585.

ELSEVIER

Journal of Wind Engineering
and Industrial Aerodynamics 67&68 (1997) 639–646

JOURNAL OF wind engineering AND industrial aerodynamics

Full-scale measurements and computational predictions of wind loads on free-standing walls

A.P. Robertson[a], R.P. Hoxey[a], P.J. Richards[b], W.A. Ferguson[c]

[a] Silsoe Research Institute, Wrest Park, Silsoe, Bedford MK45 4HS, UK
[b] University of Auckland, Private Bag 92019, Auckland, New Zealand
[c] Building Research Establishment, Garston, Watford WD2 7JR, UK

Abstract

Recent developments in the wind loading codes for the UK, Australia and Europe have introduced new, more onerous, pressure coefficient data for the design of free-standing walls. These data, derived from wind-tunnel studies conducted in the mid-1980s in the UK and Australia, have been called into question by various interested parties. In 1993, a research programme was initiated to undertake an independent, full-scale study of the wind pressures on free-standing walls in order to critically appraise the new data and to determine reliable design data. The full-scale, variable-geometry experimental facility which includes automatic, rapid, data-logging instrumentation is described. To supplement this full-scale work, CFD investigations in 2 and 3 dimensions have been undertaken at the University of Auckland using the PHOENICS finite volume code, Version 2.1, with a k–ε turbulence model. Comparisons are presented which reveal that despite the simplicity of its structural form, the free-standing wall exhibits surprising aerodynamic effects which render it an excellent and highly challenging test case to model computationally.

Keywords: Wind loads; Wind pressures; Pressure coefficients; Walls; Measurements; Modelling; Design

1. Introduction

Considerable research interest has developed recently over the determination of the wind loads on one of the simplest of structures – free-standing walls. Following detailed wind-tunnel measurements in the UK and Australia some 10 years ago [1], new wind loading data for free-standing walls appeared in major design Standards and other documents published over recent years, see, e.g., Ref. [2]. Concerns over inconsistencies between these data sets, and over comparisons with recent full-scale measurements by Silsoe Research Institute (SRI) and computational analyses made at Auckland University [3], led to new full-scale parametric research being initiated at

PII S0167-6105(96)00106-2

SRI by the Buiding Research Establishment in 1993 [4–6]. This on-going research has been augmented by further CFD studies, and by further collaborative wind-tunnel studies at Oxford University [7,8]. Selected results of mean pressure and force coefficients are presented which show this simple case to be a very challenging task to model computationally.

2. Full-scale test facility and data

This unique test facility has been described in detail previously [4–6]. The experimental wall (Fig. 1) is situated on the flat and exposed site at SRI and measures 2 m high by 0.215 m thickness (representing typical masonry walls). For experimental convenience, it was constructed of 2 m square modular panels which enabled its length to be varied between 2 m and 18 m (i.e. $1h$ and $9h$ where h is wall height). Perpendicular return corners and gaps could also be introduced (Fig. 2). One of the panels contained 15 pressure tappings on each face which were located at corresponding positions in 3 rows of 5 columns. All the pressure transducers, switching solenoids and power supplies were mounted inside the panel. The instrumented panel also contained a precision load-cell which enabled the wind-induced over-turning moment on the panel to be monitored independently of pressure measurements. The panel could be installed at any position along the wall to measure the pressures and moment loading at that position. A reference sonic anemometer, static probe and directional pitot tube were mounted at 2 m height on a mast positioned some 15 m from one end of the wall.

During three seasons of recording, 12 different wall configurations have been monitored and some 1300 h (over 5.5 Gb) of 5 Hz data have been recorded.

Fig. 1. General view of the experimental wall and the reference mast instrumentation.

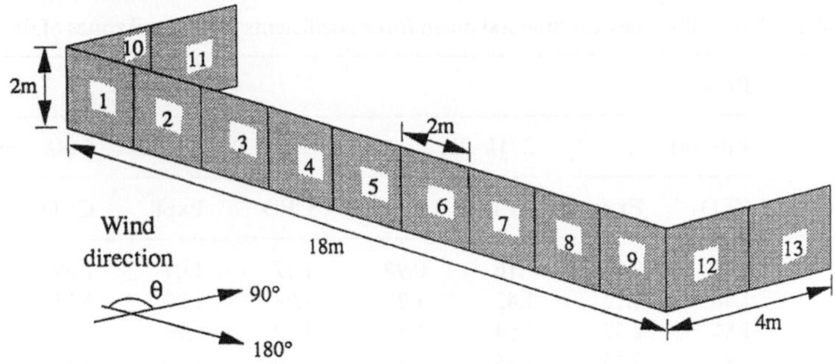

Fig. 2. Schematic of the modular panel layout of the experimental wall including optional return corner panels.

3. CFD studies

Computations in 2 and 3 dimensions were undertaken at Auckland University using the PHOENICS (Version 2.1) finite volume code with a k–ε turbulence model. A rectangular grid with domain boundaries extending $15h$ from the edges of the wall was used with the boundary conditions described in Ref. [9]. Wall pressures were non-dimensionalised by the reference pressure at wall height $z = h$ at the inlet $15h$ upstream, and the solution domain extended $15h$ downstream of the wall. Minimum cell size close to the wall was $0.066h$. For a $9h$ long wall, the grid contained 34 cells perpendicular to the wall, 62 cells parallel to the wall, and 28 cells vertically.

A log-law boundary layer velocity profile extending over the total depth was used. The boundary conditions were such that in the absence of the free-standing wall the velocity profile was in equilibrium and would propagate through the solution domain without change.

The effect of Jensen number (h/z_0) was investigated as part of the study, but otherwise the value used was the full-scale value of 200 (full-scale surface roughness parameter $z_0 = 0.01$ m), for which turbulence intensity at wall height h was 13.7%. The thickness/height ratio of the wall was 0.1, which was very close to the full-scale value of 0.108.

4. Selected results

4.1. Comparisons of force coefficients for a 9h long wall

Mean pressure coefficients, C_{pe}, were evaluated at each of the 30 pressure tappings by non-dimensionalising the measured mean pressures by the free-stream mean wind dynamic pressure at the height of the wall (2 m). Net pressure coefficients were then evaluated by differencing the corresponding front and back pairs of coefficient values.

Table 1
Comparison of CFD and full-scale experimental mean force coefficients for 4 panel zones of the 9*h* long wall

	Panel zone							
	1 (0–1*h*)		2 (1*h*–2*h*)		3 (2*h*–3*h*)		5 (4*h*–5*h*)	
Wind dir. θ (deg)	CFD	Expt.	CFD	Expt.	CFD	Expt.	CFD	Expt.
90	1.10	0.98	1.16	0.98	1.12	1.07	1.08	0.91
100	1.48	1.16	1.42	1.25	1.28	1.20	1.13	1.01
110	1.81	1.57	1.54	1.53	1.32	1.28	1.10	1.02
120	1.87	2.15	1.45	1.79	1.20	1.28	0.96	0.97
130	1.87	2.54	1.33	1.74	1.07	1.17	0.82	0.86
140	1.72	2.40	1.15	1.43	0.91	0.99	0.68	0.70
150	1.36	1.81	0.84	0.98	0.67	0.75	0.50	0.53
160	0.96	1.07	0.54	0.54	0.42	0.50	0.33	0.35
170	0.50	0.45	0.25	0.22	0.19	0.25	0.14	0.17
180	0.00	0.00	0.00	0.00	0.00	0.00	0.00	0.00

Mean force coefficients, C_f, were then determined by integrating the net coefficients according their tributary surface areas (the distribution of tappings on the experimental panel was such as to give an equal surface area to each tap). This also gave the heights to the centres of pressure which were found generally to be $(0.5 \pm 0.05)h$, i.e. close to the mid height of the wall panel. Thus the force coefficients relate the mean net loading on the panel to the mean wind dynamic pressure at wall height, they relate to the face area of the loaded zone, and can be taken to represent a uniformly distributed load.

A three-dimensional CFD study was undertaken to predict the mean force coefficients on a wall of length 9*h* with no return corners. Force coefficients were evaluated for each of the 9 panel zones of length *h* along the wall, and for wind directions ranging in 10° increments from perpendicular to parallel (90° to 180°, Fig. 2) to the wall. Full-scale results are available for panel positions 1, 2, 3, and 5 (Fig. 2) and these are compared with the CFD solutions in Table 1.

The CFD and experimental results are in reasonable agreement for the perpendicular (90°) wind direction (the CFD results overestimate by 15% on average). The results are in good agreement for the centre panel 5 (4*h*–5*h*), and are reasonably good for panel 3 (2*h*–3*h*) over the full range of wind directions. However, for the windward end panel 1 where the full-scale results show the highest loadings to arise for 130° glancing winds, and for the adjacent panel 2, the CFD results significantly underestimate loadings. This is shown in Fig. 3 which compares the CFD and experimental C_f values for the the first 3 panel positions of the 9*h* long wall. The CFD results show a similar pattern but give maximum loads occurring at slightly nearer to normal winds, and significantly underestimate the high loads on the first two panels.

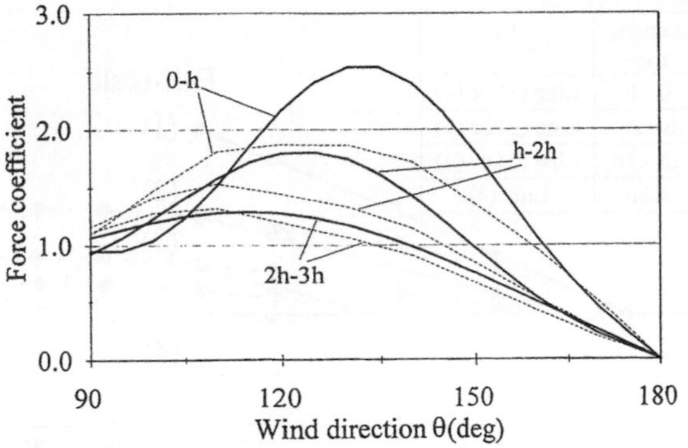

Fig. 3. Comparisons of mean force coefficients for first 3 zones at windward end of 9*h* long wall (solid lines, full-scale; dashed lines, CFD).

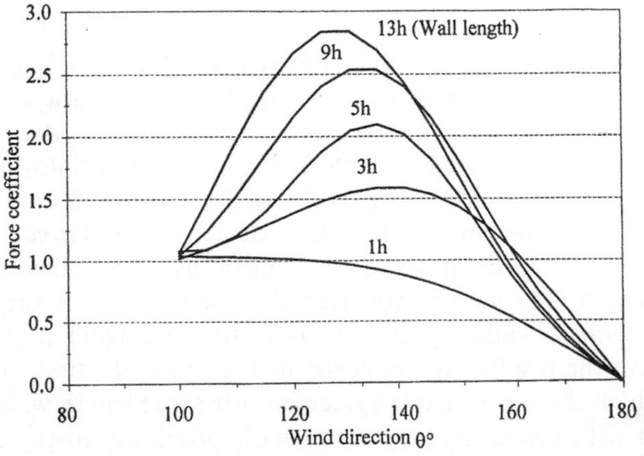

Fig. 4. Mean force coefficients for the end panel of the full-scale experimental wall for the wall at each of 5 different lengths.

4.2. *Effect of wall length*

The full-scale results have shown clearly the strong dependence wind loads have on wind direction and on position along the wall (Fig. 3). Greatest loads arise at the windward end of a wall for a wind at 40°–45° from normal to the wall, and the loads steadily decrease with distance from the windward end. It became evident from the full-scale measurements on walls of different length (from 1*h* to 9*h*) that the magnitudes of the high loads on the end panel also increased surprisingly with wall length.

To investigate this effect further, the wall was temporarily extended to 13*h* and measurements were repeated. The results for the end panel are shown in Fig. 4 from

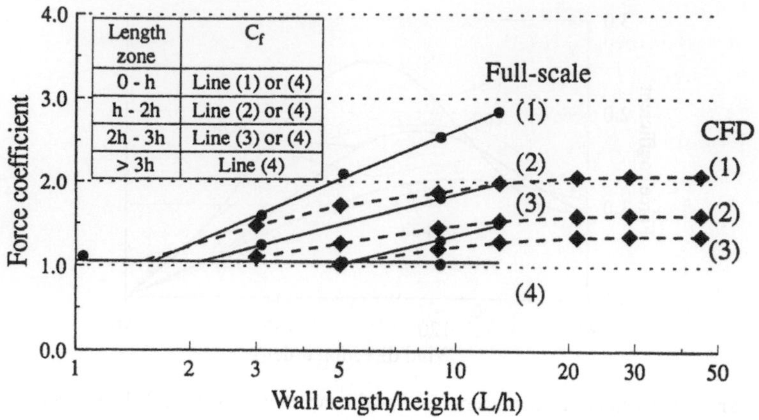

Fig. 5. Comparisons of maximum mean force coefficients for zones of walls of different lengths (solid lines, full-scale; dashed lines, CFD).

which it can be seen that the 13h wall was still insufficiently long to identify the asymptotic maximum loads for increasing length. Similar data were obtained for other panel zones, from which the maximum mean force coefficients, irrespective of wind direction, were obtained and plotted (Fig. 5, solid lines) as a function of wall length, L, non-dimensionalised by wall height, h (plotted on a log scale). The solid lines in Fig. 5 thus present tentative design data derived from the full-scale measurements on walls up to 13h in length, as is described more fully elsewhere [6]. Owing to the uncertainty over the load levels on the windward end panel zones of longer walls, CFD was used to investigate a greater range of lengths. This gave the results superimposed in Fig. 5 for the first three panel zones (dashed lines) which show reasonable agreement for short length walls but increasing underestimation with increasing length and with proximity to the end of the wall. The CFD results suggest, however, that asymptotic values are attained for a length of approximately 20h which, if confirmed, is a valuable indication in formulating design data.

4.3. *Effect of Jensen number* (h/z_0)

To date, investigations of the effect of Jensen number have not been possible in the full-scale study, although information on the existence of any dependency would be valuable. An initial investigation was therefore made using a 2-dimensional CFD analysis. Computed mean pressure coefficients, C_{pe}, for the front and rear face, and the mean net, or force coefficient, C_f, are summarised in Table 2.

These results show loads to increase with decreasing h/z_0, and show that this is due very predominantly to suctions increasing on the rear face (pressures on the front face remaining nearly constant). Physically, this may be attributed to the higher onset velocities that occur above a height h at lower h/z_0 values. The CFD results show

Table 2
Mean pressure coefficient dependence on Jensen number from 2-dimensional CFD study

Pressure coefficient	h/z_0: 20	66	200	666	2000
C_{pe} (front)	0.51	0.47	0.46	0.47	0.49
C_{pe} (rear)	− 1.21	− 1.05	− 0.97	− 0.92	− 0.89
C_f (net)	1.72	1.52	1.43	1.39	1.38

there to be higher velocities immediately above the wall at lower h/z_0 values, which is consistent with the higher approach velocities above height h. A check was made of blockage effect by doubling the height of the solution domain but this made no significant difference to the solution.

However, this trend of increasing loads with decreasing h/z_0 is at variance with the wind-tunnel results of Letchford [10] who found increased loads on a wall twice the height of another (due again almost entirely to changes in suctions on the rear face), although this was attributed to a blockage effect (the maximium blockage being 6.7%). Letchford and Holmes [1] also cite earlier work by Baines [11] who found that near-uniform flow produced more negative leeward suctions than did boundary layer flow on infinitely long walls. Holmes [12], and Letchford and Holmes [1], have concluded from earlier measurements and their own that, provided pressure coefficients are based on wind velocity at wall height, h, the coefficients are, for practical purposes, insensitive to Jensen number over the range 40–1000.

There are thus conflicting indications of the effect of Jensen number on wind loads on free-standing walls, and the dependency question remains unanswered, although it seems likely that for practical cases the dependency will be small.

5. Conclusions

Detailed and careful full-scale wind pressure measurements such as those described here provide definitive data for design and for the validation of modelling techniques. Earlier full-scale/CFD comparisons on two-dimensional free-standing walls [3] provided encouraging and useful results. Subsequent three-dimensional comparisons reveal that, despite its inherently simple structural form, the free-standing wall creates a complex three-dimensional flow field which provides a very challenging test case for the computational modeller, and further effort is needed to develop a reliable CFD research and design tool. Over 1300 h, or 5.5 Gb, of 5 Hz full-scale data covering 12 different wall arrangements are already available and may be exploited for this purpose.

One interesting discovery from the full-scale data acquired so far are the high loads that are generated at the windward end of a wall for winds at 40°–45° from normal, and more intriguingly that these loads increase markedly with increasing wall length, even for walls up to $13h$ in length which was the maximum wall length possible in the full-scale test arrangement. In the present study, the CFD results have provided a useful prediction that these high end loads reach their asymptotic values at a length of approximately $20h$.

Data recommended for use in design have been presented recently [6], as have loading data for walls containing gaps [13]. Further studies will continue to investigate the effect of perpendicular return corners at the end of walls of different length, and the sheltering and interaction effect of a parallel wall at different spacings from the test wall.

Acknowledgements

The full-scale work reported here was commissioned by the UK Building Research Established and funded by the Department of the Environment. Grateful acknowledgement is extended to them for supporting the continuation of full-scale wind load testing.

References

[1] C.W. Letchford, J.D. Holmes, Wind loads on free-standing walls in turbulent boundary layers, J. Wind Eng. Ind. Aerodyn. 51 (1994) 1–27.

[2] British Standards Institution, BS 6399: Loading for buildings: Part 2: Code of practice for wind loading, BSI, London, 1995.

[3] A.P. Robertson, R.P. Hoxey, P.J. Richards, Design code, full-scale and numerical data for wind loads on free-standing walls, J. Wind Eng. Ind. Aerodyn. 57 (1995) 203–214.

[4] A.P. Robertson, R.P. Hoxey, J.L. Short, W.A. Ferguson, S. Osmond, Wind loads on free-standing walls: a full-scale study, in: Proc. 9th Int. Conf. on Wind Engineering, New Delhi, India, 9–13 January, 1995, pp. 457–468.

[5] A.P. Robertson, R.P. Hoxey, J.L. Short, W.A. Ferguson, S. Osmond, Full-scale testing to determine the wind loads on free-standing walls, J. Wind Eng. Ind. Aerodyn. 60 (1996) 123–137.

[6] A.P. Robertson, R.P. Hoxey, J.L. Short, W.A. Ferguson, S. Osmond, Wind loads on boundary walls: full-scale studies, in: Proc 3rd Int. Colloq. on Bluff Body Aerodynamics & Applications, BBAA III, Blacksburg, Virginia, USA, 28 July–1 August, 1996, pp. AV13–AV16, J. Wind Eng. Ind. Aerodyn., to be published.

[7] C.W. Letchford, A.P. Robertson, Measures to reduce wind loading at the leading ends of free-standing walls, in: Proc. 5th National Australian Wind Engineering Society Workshop, Tanunda, South Australia, 22–23 February, 1996.

[8] C.W. Letchford, A.P. Robertson, Mean wind loading at the leading ends of free-standing walls, J. Wind Eng. Ind. Aerodyn., submitted.

[9] P.J. Richards, R.P. Hoxey, Appropriate boundary conditions for computational wind engineering models using the k–ε turbulence model, J. Wind Eng. Ind. Aerodyn. 46/47 (1993) 145–153.

[10] C.W. Letchford, Wind loads on free-standing walls, University of Oxford, Department of Engineering Science, Report OUEL 1599/85, 1985.

[11] W.D. Baines, Effects of velocity distribution on wind loads and flow patterns on buildings, in: Proc. 1st Int. Conf. on Wind Engineering, Teddington, UK, 1963.

[12] J.D. Holmes, Pressure and drag on surface-mounted rectangular plates and walls, in: Proc. 9th Australasian Fluid Mechanics Conf., Auckland, New Zealand, 8–12 December, 1986, pp. 383-386.

[13] A.P. Robertson, R.P. Hoxey, J.L. Short, W.A. Ferguson, P.A. Blackmore, Full-scale measurements of wind loads on boundary walls with and without gaps, in: Proc. 3rd UK Conf. on Wind Engineering, University of Oxford, 16–18 September, 1996, pp. 149–152.

ELSEVIER

Journal of Wind Engineering
and Industrial Aerodynamics 67&68 (1997) 647–657

JOURNAL OF wind engineering AND industrial aerodynamics

Computation of pressures on Texas Tech University building using large eddy simulation

R. Panneer Selvam[1]

BELL 4190, University of Arkansas, Fayetteville, AR 72701, USA

Abstract

An implicit solution procedure to solve the Navier–Stokes equations using large eddy simulation (LES) for flow around building is presented. Using this model the pressures around the Texas Tech University (TTU) building are computed using different inflow turbulence conditions and compared with available field mean and peak pressure coefficients. The turbulence is generated by keeping periodic boundary conditions, by using Gaussian distribution and random number generator and actual TTU field wind data. The computed mean pressures are in good agreement with field measurement. The peak pressures computed using Gaussian distribution are much higher than the field measurement. The peak pressures using the TTU wind data are much closer to the field measurements than with the three methods.

Keywords: Computational fluid dynamics; Computational wind engineering; Large eddy simulation; Turbulence; Building loads

1. Introduction

Methods to compute turbulent flow around buildings are described by the author in Ref. [1]. For time-dependent flow phenomena large eddy simulation (LES) and direct simulation (DS) are used. Of the two, DS consumes more computer time and storage space than LES. Hence LES is preferred for practical applications rather than DS. In LES, the fluctuating motions of turbulence can be computed exactly except for eddies that are smaller than the grid size. The smaller eddies are modelled using eddy viscosity models. Because of this, LES can compute fluctuating pressures on buildings due to turbulence. This suggests that LES will be used more and more as computing hardware and software technology improve. Past researchers [2–6] used explicit procedures to solve the Navier–Stokes (NS) equations. Because of stability restrictions

[1] E-mail: rps@engr.uark.edu.

PII S0167-6105(97)00107-4

on the solution of NS equations; these procedures took enormous computer time. In this work an implicit procedure is investigated to solve the NS equations using LES.

He and Song [3] computed the pressures on a cube without using any turbulence for inflow and kept the other boundaries as open boundaries. They reported mean pressures. Murakami and his research group [6] generated initial horizontal velocities by means of random numbers. Their initial turbulence intensities had approximately the same magnitude as those from wind-tunnel experiments. Using periodic lateral boundaries they tried to show the time development of the flow over a building as classified by Piomelli [7]. Later, Baetke et al. [2] and Mochida et al. [5] used the turbulence generated for a channel flow using periodic boundary condition as the inflow. The other boundaries are kept as outflow or open boundaries. In this approach the spatial variation of the turbulence is considered. The mean and RMS pressures are reported [5], but they did not report peak values. In another work of He and Song [4], the inflow turbulence is generated as a Gaussian distribution using random numbers. In later works [8,9], the inflow turbulence is generated using spectral density and random numbers. In this work previous approaches used in Refs. [2–6,8,9] are evaluated by applying these techniques to compute the pressures on the Texas Tech University (TTU) building and comparing the results with available field measurements. In addition, a new approach takes actual turbulence generated in the field and uses it to compute pressures. The relative merits of each approach and their relevance to wind engineering problems are discussed.

2. Computer modelling using large eddy simulation (LES)

2.1. Governing equations

In this work, the LES turbulence model is considered. The three-dimensional equations for an incompressible fluid using the LES model in general tensor notation are as follows:

$$\text{Continuity equation:} \quad U_{i,i} = 0. \tag{1}$$

$$\text{Momentum equation:} \quad U_{i,t} + U_j U_{i,j} = -(p/\rho + 2k/3)_{,i} + [(\nu + \nu_t)(U_{i,j} + U_{j,i})]_{,j}, \tag{2}$$

where $\nu_t = (C_s h)^2 (S_{ij}^2/2)^{0.5}$, $S_{ij} = U_{i,j} + U_{j,i}$, $h = (h_1 h_2 h_3)^{0.333}$ for 3D and $(h_1 h_2)^{0.5}$ for 2D and $k = (\nu_t/(C_k h))^2$.

Empirical constants: $C_s = 0.15$ for 2D and 0.1 for 3D, and $C_k = 0.094$.

U_i, and p are the mean velocity and pressure respectively, k is the turbulent kinetic energy, ν_t is the turbulent eddy viscosity, h_1, h_2, and h_3 are control volume spacing in the x, y, and z directions and ρ is the fluid density. The empirical constants used here are the values suggested by Murakami et al. [10]. Here a comma represents differentiation, t represents time and $i = 1, 2$ and 3 mean variables in the x, y and z directions.

To implement a higher order approximation of the convection term [11] the following expression is used in Eq. (2) instead of $U_jU_{i,j}$:

$$U_jU_{i,j} - \frac{\theta}{2}(U_jU_kU_{i,j})_{,k}. \tag{3}$$

Depending upon the values of θ different procedures can be implemented. For balance tensor diffusivity (BTD) scheme [11,12] $\theta = \delta t$ is used, where δt is the time step used in the integration. For the streamline upwind procedure suggested by Brooks and Hughes [13], $\theta = 1/\max(|U_1|/\mathrm{d}x, |U_2|/\mathrm{d}y, |U_3|/\mathrm{d}z)$. Here d$x$, d$y$ and dz are the control volume length in the x, y and z directions.

2.2. Finite difference scheme and computational grid

A nonstaggered grid system is used. Velocity components were defined at the vortex or nodes. All spatial derivatives were approximated by centered differences. For convection terms, the scheme becomes a regular center difference if $\theta = 0$ and the scheme becomes some form of upwinding if θ is not equal to zero. The equations are solved using an implicit method similar to that of Choi and Moin [14]. The four step advancement scheme for Eqs. (1) and (2) is as follows:

Step 1: Solve for U_i from Eq. (2) using backward Euler method for convection and diffusion.

Step 2: Get new velocities as $U_i^* = U_i + \delta t(p_{,i})$ where U_i is not specified.

Step 3: Solve for pressure from $(p_{,i})_{,i} = U_{i,i}^*/\delta t$.

Step 4: Correct the velocity for incompressibility: $U_i = U_i^* - \delta t(p_{,i})$ where U_i is not specified.

Step 2 eliminates the checkerboard pressure field when using a nonstaggered grid. Implicit treatment of the convective and diffusive terms eliminates the numerical stability restrictions. In this work the time step is kept for the CFL (Courant–Frederick–Lewis) number less than one.

The TTU building has dimensions of 9.1 m × 13.7 m × 4 m as reported by Levitan et al. [15]. The building is equally divided into six divisions or seven points in each direction. The region is divided into 43 × 35 × 20 grid points. The computational domain has a downstream length of $18.2H_b$, an upstream length of $9.1H_b$, and a lateral width of $13.7H_b$ on both sides of the building. A vertical height of $5H_b$ is considered from the ground. H_b is the height of the building. The smallest mesh interval near the building is $H_b/6$. The time interval δt for time-marching is 0.1 on a time scale nondimensionalized by H_b and U_b for most of the work. Here U_b is the time-averaged approach wind velocity at the building height.

2.3. Solution procedure and convergence criterion

Eq. (2) is solved using point iteration with an underrelaxation factor of 0.7 for each time step. Usually about 10–20 iterations are sufficient to reduce the absolute sum of the residue to be less than 0.01. The pressure equations take considerable computer time. In this work a preconditioned conjugate gradient (PCG) procedure is used. The iteration is done until the absolute sum of the residue of the pressure equation reduces to 0.01 for each time step. The algorithm for the implementation of PCG on a finite difference grid was presented by the author [16] and the advantages of the procedure are discussed in Ref. [17]. For a $43 \times 35 \times 20$ grid to reach steady state about 75 time steps are needed using a time step of 0.1. In Sun-Sparcstation 20 the total computer time is about 45 min for the 75 time steps.

2.4. Boundary conditions and initial conditions

Two types of problems are considered for analysis. For Type 1, all the lateral boundaries are considered to be periodic. The top boundary is considered to be free slip, i.e., the velocity normal to the plane is considered to be zero and the normal gradient of the other velocities to be zero. The initial condition for velocity profiles is kept as a logarithmic profile. The mean wind speed is considered as 8.6 m/s at 4 m from the ground, and the roughness length of the ground is considered to be 0.024 m as reported from field measurement [15]. It is assumed that the wind flow is parallel to the 9.1 m side of the building. A mean wind speed of 1 m/s at unit building height and the corresponding roughness length of 0.006 m are used for nondimensional solution of the equations. The initial velocity fluctuations for x and y velocities are considered by means of random numbers with standard deviation of 0.2 and 0.1, respectively. This boundary conditions are same as used in Ref. [6]. For Type 1, when streamline upwinding ($\theta = 0.9/\max(|U_1|/dx, |U_2|/dy, |U_3|/dz)$) is used for the convection term, the turbulence disappeared slowly. Hence the usual central difference procedure, i.e., $\theta = 0$, is considered in this work for comparison. It seems that the dispersive error in approximating the convection term using the usual central difference for the convection term acts as turbulence generator as reported by the author [18].

The lateral boundary normal to the 9.1 m side of the building is considered as the inflow boundary for Type 2. On the opposite lateral side the normal gradient of all the velocities kept as zero (outflow boundary) is used. For the other two lateral and top boundaries free slip boundary conditions are used. The mean wind speed is considered to be the same as the one for Type 1. The following cases of inflow velocities are considered for Type 2:

Case 1: Mean velocity as the inflow and $\theta = 0$ (no upwinding). This is similar to the approach of He and Song [3].

Case 2: Mean velocity as the inflow and $\theta = 0.9/\max(|U_1|/dx, |U_2|/dy, |U_3|/dz)$.

Case 3: Turbulent velocity as the inflow. Here the turbulent part is generated as a Gaussian distribution using random numbers. The standard deviation is kept as 0.2 for U_1 and 0.1 for U_2. At each time step, at each inflow grid point, the turbulent velocity is generated using random numbers having Gaussian distribution. This is similar to the work of He and Song [4].

Case 4: Turbulent velocity as the inflow. Here the turbulent part is generated using the actual Texas Tech measurement data reported by Levitan et al., [15]. The data is available over a 15 min period and the data is recorded at every 0.1 s. The data is started for each grid point randomly within the first 500 data points to consider the cross variation at the inflow.

For cases 3 and 4 the spatial correlations of velocity fluctuations are not the same in the boundary layer. For case 3, the variation is random and for case 4, randomly picked from the first 500 data points of the field measurement. Further work is necessary to implement the proper variation in the lateral direction. For cases 3 and 4, $\theta = 0$ and $\theta = 0.9/\max(|U_1/\mathrm{d}x, |U_2|/\mathrm{d}y, |U_3|/\mathrm{d}z)$ were tried. Both gave similar results and $\theta = 0$ is used for comparison here. Here $\theta = 0$ means central differencing.

The usual law of the wall boundary condition is introduced on the wall. The time step used is 0.1 s for all cases and types. For cases 3 and 4 turbulence is added, only after 50 time steps. Totally 3000 time steps were considered for cases 3 and 4 in type 2 and for type 1 at this time and 75 time steps were sufficient to reach steady state for cases 1 and 2 in type 2. The procedure suggested by Ogawa et al., [8] to develop inflow turbulence using spectral density and random numbers is also tried. The effect is similar to case 3.

3. Comparison of numerical solution with field measurement of TTU building

To study the structure of the turbulence at the inflow, the nondimensionalized turbulent velocity at the building height for type 1 and case 3 and 4 of type 2 are plotted in Fig. 1. In this plot only 40 s are considered. It can be seen that case 3 of

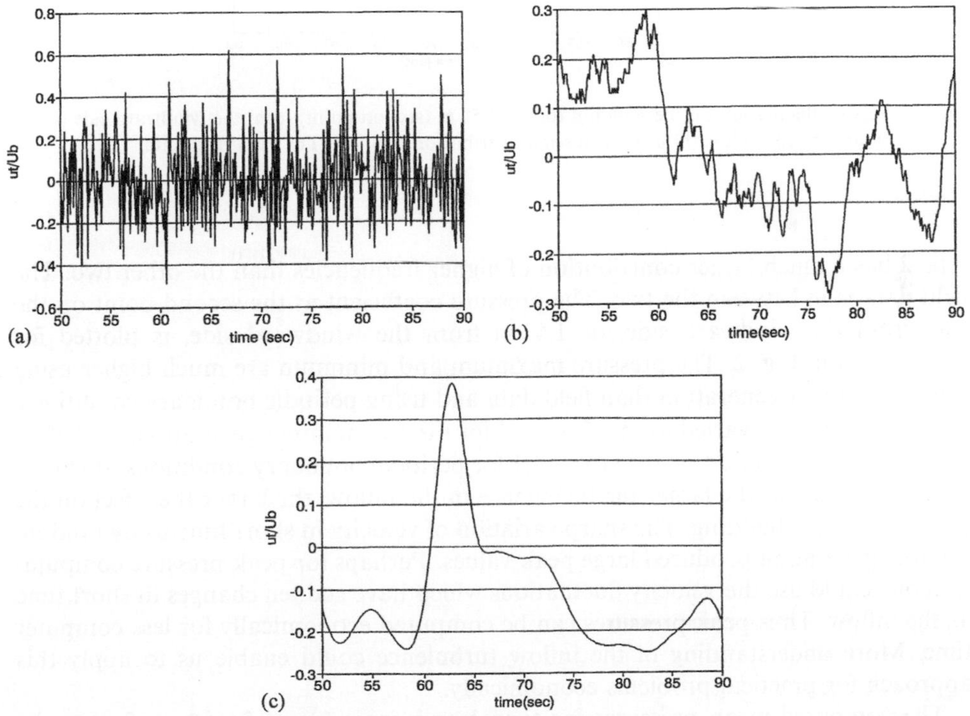

Fig. 1. Velocity fluctuation of inflow for the x component at the building height (a) for periodic boundary condition, (b) for using Gaussian distribution and (c) TTU wind field data.

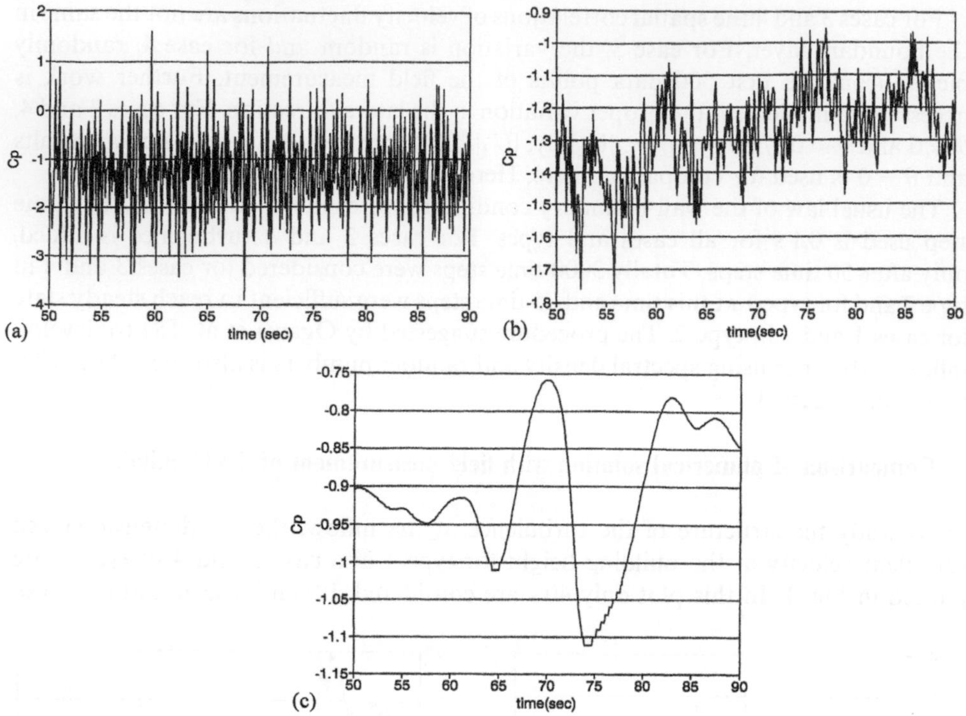

Fig. 2. Pressure fluctuation on the building roof at 1.52 m (second point) from the windward side (a) for periodic boundary condition, (b) using Gaussian distribution and (c) TTU wind field data.

type 2 has a much larger contribution of higher frequencies than the other two. The field data is in between the two. The pressure coefficient at the second point on the roof from the windward side, or 1.52 m from the windward side, is plotted for comparison in Fig. 2. The pressure maximum and minimum are much higher using random number generation than field data and using periodic boundary conditions. The peak pressures varied from 1.5 to − 4 for random number generation, − 0.95 to − 1.8 for field data and − 0.75 to − 1.1 for periodic boundary conditions. It can be seen from this that the larger the turbulence in the inflow, the larger the effect on the pressures on the building. The sharp variation of velocity in short time using random number generation produced large peak values. Perhaps for peak pressure computation one could use the velocity fluctuations which have sudden changes in short time as the inflow. Thus peak pressures can be computed economically for less computer time. More understanding of the inflow turbulence could enable us to apply this approach for practical problems economically.

The computed mean pressures for type 1 and cases 1 and 2 of type 2 over the building at the center plane are plotted in Fig. 1. The field measurements and the previous work of the author [19] using the k–ε turbulence model are also reported in

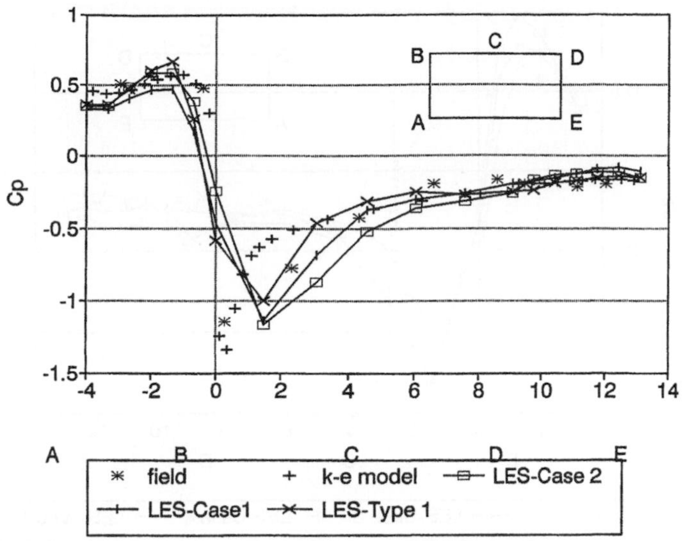

Fig. 3. Mean pressure coefficient at the centerline of the building along the 9.1 m side from field, k–ε model, Cases 1 and 2 of Type 2 and Type 1.

Fig. 1 for comparison. Here the pressure coefficients are calculated from:

$$C_p = 2(p - p_0)/(\rho V^2),$$

where p_0 is the reference pressure at freeflow, considered to be the pressure at the inflow along the centerline and V is the approach wind velocity at the building height. It can be seen from Fig. 3 that the different LES procedures have the same pattern of pressures. The cases 1 and 2 in type 2 took about 45 min of CPU time in Sparc station 20 whereas type 1 took more than 30 times of that. Hence for mean flow one could use cases 1 and 2 advantageously. The pressure coefficient at the windward roof from case 2 is little higher than case 1. The pressure coefficients using the k–ε turbulence model [19] are also reported for comparison. The sudden drop using the k–ε model on the roof from the windward side is not present when using LES modelling. When finer points are used on the windward side [17] the pressure increases on the windward roof edge. This may be due to the discontinuity in the pressure calculation using control volumes for cells. Here the pressures are solved on the walls directly. Hence this problem may be avoided using the new grid arrangement. Further work is underway to run many cases using finer grids to validate the procedure.

The mean pressure coefficients using cases 3 and 4 of type 2 are compared with type 1 and field measurements in Fig. 4. Even though 15 min of averaging time is considered for the field measurement, an averaging time of 5 min (3000 time steps) is considered at this time for type 1, cases 3 and 4 of type 2 to compute the mean and peak pressures due to lack of computer time. The computed pressure coefficients using cases 3 and 4 on the roof and windward side are higher than the field measurement by

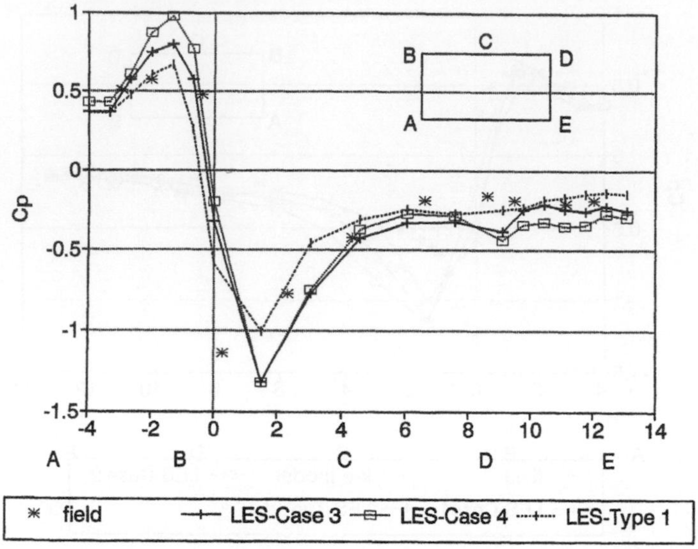

Fig. 4. Mean pressure coefficient at the centerline of the building along the 9.1 m side (a) field data, (b) case 3 of Type 2, (c) Case 4 of Type 2, and (d) Type 1.

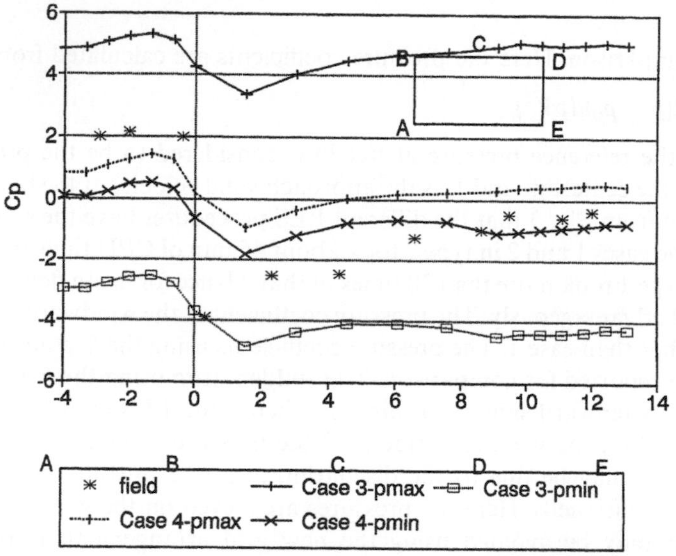

Fig. 5. Peak pressure coefficients at the centerline of the building along the 9.1 m side (a) field data, (b) Case 3 of Type 2 and (c) Case 4 of Type 2.

as much as 30%. In Fig. 5 the peak pressures from cases 3 and 4 of type 2 are compared with the field measurement. The computed peak pressures using case 3 are overpredicted by as much as 150% on the windward side and about 100% on the roof. On the leeward side the values are overpredicted by 8 times the field measurement.

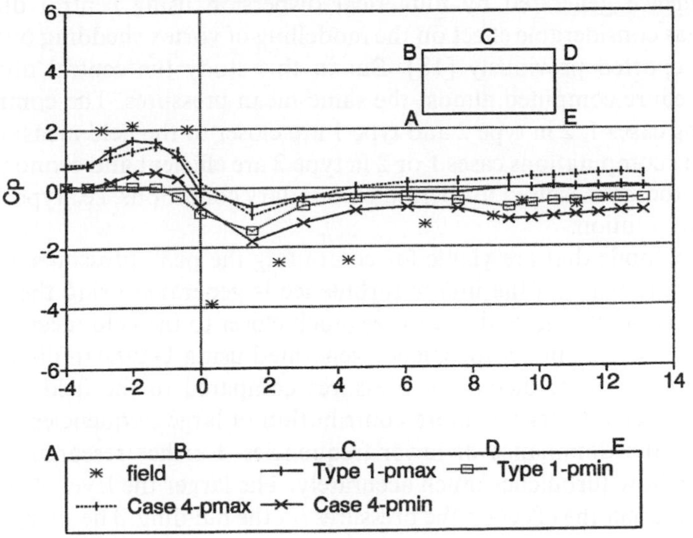

Fig. 6. Peak pressure coefficients at the centerline of the building along the 9.1 m side (a) field data, (b) Case 3 of Type 2 and (c) Type 1.

The computed case 4 pressure coefficients are less than the field value on the windward side and roof. On the leeward side they are almost the same. On the windward side C_p are underpredicted by 30% for case 4. Close to the middle of the roof C_p is underpredicted by 60% for case 4. On the windward side of the roof the error is as low as 10%. The peak C_p from periodic boundary conditions (type 1) is compared with the field measurement and case 4 of type 2 in Fig. 6. The C_p from type 1 are lower than that of case 4 on the roof. On the windward and leeward side they are almost the same. Over all it can be concluded that the use of inflow turbulence using the field data computes the peak pressure coefficient much closer to the field measurement than the other procedures to generate turbulence. Further work is underway to investigate the effect of a finer grid on the peak pressures and to run for a longer time duration of the flow.

4. Conclusions

An efficient implicit solution procedure to solve the Navier–Stokes (NS) equations using large eddy simulation (LES) is presented. The procedure is capable of solving the NS equations in about 45 min for 75 time steps in a Sun-Sparc station 20 for about 30 000 grid points. Usually the pressures are stored at the center of the cell. The pressures computed using this approach produced more pressures when the grid is refined on the roof as reported previously [16]. This discontinuity in pressure may be eliminated by solving the pressures on the wall at the grid vortex as used in this work. Validation is underway by applying the method to a finer grid.

The turbulence generated by numerical dispersion using central difference for convection has considerable effect on the modelling of vortex shedding over a circular cylinder as reported previously [18]. But in this study the central difference and upwind procedure computed almost the same mean pressures. The computed mean pressure using cases 1, 2 in type 2 and type 1 are closer to the field measurement. For mean pressure computations cases 1 or 2 in type 2 are efficient and economical. When the upwind procedure is used for periodic boundary conditions, i.e., type 1, it became a steady state solution.

Different methods that are viable for computing the peak pressures are surveyed. A new procedure wherein the inflow turbulence is generated using the actual wind data seems to compute the peak pressures much closer to the field measurement than other procedures. The inflow turbulence generated using Gaussian distribution and random numbers overpredicted the pressures compared to the field measurement. Also the turbulent velocity has more contribution of large frequencies. Whereas the field data has the dominance of lower frequencies. Further research is needed to simulate the inflow turbulence much accurately. The larger the level of turbulence at the inflow, the larger the effect on the pressures on the building. The sharp variation of velocity in a short time using the random number generation produced large peak values. Maybe for peak pressure computation one could use the velocity fluctuations which have sudden changes in short time as the inflow and compute peak pressures economically for less computer time than running them for long time as is done in wind-tunnel tests. More understanding of the inflow turbulence and its effect on the building could enable us to compute peak pressures using computer modelling in the coming days.

The variation of turbulence in the lateral direction at the inflow, the variation of flow direction with time and proper modelling of inflow turbulence can greatly improve the results. Further research is underway to investigate the impact of grid resolution on peak pressures.

The proposed new approach of using actual turbulence provides a reasonable pressure coefficient around buildings. The implicit procedure to solve the NS equations and the preconditioned conjugate procedure to solve the pressure equations makes the LES procedures more viable for use in practical problems.

Acknowledgements

The author acknowledges the support provided by Dr. D.A. Smith, Texas Tech University by providing TTU wind data and the valuable discussions with D.A. Smith and Dr. L. Cochran, CPP Wind Engineering consultants, Fort Collins to understand the boundary layer turbulence.

References

[1] R.P. Selvam, Numerical methodologies and turbulent flow simulation in computational wind engineering, in: M. Sanayei (Ed.), Restructuring: America and Beyond, vol. 2, ASCE, New York, 1995, pp. 1269–1272.

[2] F. Baetke, H. Werner, H. Wengle, Numerical simulation of turbulent flow over surface-mounted obstacles with sharp edges and corners, J. Wind Eng. Ind. Aerdyn. 35 (1990) 129–147.
[3] J. He, C.C. Song, Computation of turbulent shear flow over a surface-mounted obstacle, J. Eng. Mech. ASCE 118 (1992) 2282–2297.
[4] J. He, C.C. Song, Wind load computation of building complex with large eddy simulation, in: M. Sanayei (Ed.), Restructuring: America and Beyond, vol. 2, ASCE, New York, 1995, pp. 1431–1434.
[5] A. Mochida, S. Murakami, M. Shoji, Y. Ishida, Numerical simulation of flowfield around Texas Tech building by large eddy simulation, J. Wind Eng. Ind. Aerdyn. 46 & 47 (1993) 455–460.
[6] S. Murakami, A. Mochida, K. Hibi, Three-dimensional numerical simulation of air flow around a cubic model by means of large eddy simulation, J. Wind Eng. Ind. Aerdyn. 25 (1987) 291–305.
[7] U. Piomelli, Large-eddy simulation of turbulent flows, TAM report no. 767, University of Illinois at Urbana-Champaign, 1994.
[8] T. Ogawa, T. Suzuki, Y. Fukuoka, Large eddy simulation of wind flow around dome structures by the finite element method, J. Wind Eng. Ind. Aerdyn. 46 & 47 (1993) 461–470.
[9] G. Turkiyyah, D. Reed, J. Yang, Fast vortex methods for predicting wind induced pressures on buildings, J. Wind Eng. Ind. Aerdyn. 58 (1995) 51–79.
[10] S. Murakami, A. Mochida, On turbulent vortex-shedding flow past 2D square cylinder predicted by CFD, J. Wind Eng. Ind. Aerdyn. 54/55 (1995) 191–211.
[11] R.P. Selvam, K.S. Rao, A.H. Huber, Numerical simulation of pollutant dispersion around a building, Proc. 9th Joint Conf. on Applications of Air Pollution Meteorology with A&WMA, American Meteorological Society, 1996, pp. 329–332.
[12] J. Dukowicz, J. Ramshaw, Tensor viscosity method for convection in numerical fluid dynamics, J. Comput. Phys. 32 (1979) 71.
[13] A. Brooks, T.J.R. Hughes, Streamline upwind/Petrov-Galerkin formulations for convection dominated flow with particular emphasis on the incompressible Navier-Stokes equations, Comput. Meth. Appl. Mech. Eng. 32 (1982) 199–259.
[14] H. Choi, P. Moin, Effects of the computational time step on numerical solutions of turbulent flow, J. Comput. Phys. 113 (1994) 1–4.
[15] M.L. Levitan, J.D. Holmes, K.C. Mehta, W.P. Vann, Field measured pressures on the Texas Tech building, J. Wind Eng. Ind. Aerdyn. 38 (1991) 227–234.
[16] R.P. Selvam, Computation of flow around Texas Tech building using k–ε and Kato–Launder k–ε turbulence model, Eng. Struct. 18 (1996) 856–860.
[17] R.P. Selvam, Numerical simulation of flow and pressure around a building, ASHRAE Trans. 102 (1996) 765–772.
[18] R.P. Selvam, Comparison of flow around circular cylinder using FE and FD procedures, in: S.K. Ghosh, J. Mohammadi (Eds.), Building an International Community of Structural Engineers, vol. 2, ASCE, New York, 1996, pp. 1021–1028.
[19] R.P. Selvam, Computation of pressures on Texas Tech building, J. Wind Eng. Ind. Aerdyn. 43 (1992) 1619–1627.

ELSEVIER

Journal of Wind Engineering
and Industrial Aerodynamics 67&68 (1997) 659–670

JOURNAL OF wind engineering AND industrial aerodynamics

Application of computational techniques for studies of wind pressure coefficients around an odd-geometrical building

Sang-Ho Suh[a,*], Hyung-Woon Roh[a], Ha-Rim Kim[b], Kwang-Yerl Lee[c], Kyu-Suk Kim[d]

[a] *Department of Architectural Engineering, Woosung University, San 1-6, Jugang-dong, Dong-ku, Taejon 300-100, South Korea*

[b] *Kisan Corp., 361-1, Hangang-ro 2-ka, Yongsan-ku, Seoul 140-132, South Korea*

[c] *Department of Architecture, Daelim College of Technology, Bisan-dong, Dongahn-ku, Anyang-shi, Kyonggi-do 430-715, South Korea*

[d] *Department of Architectural Engineering, Dongguk University, Pil-dong, Chung-ku, Seoul 100-715, South Korea*

Abstract

Most studies on wind flows around buildings are focused on the flow fields of buildings with prismatic shapes. Only a few works on wind flows around odd-geometrical buildings are reported. Comprehensive studies for the flow phenomena in odd-geometrical buildings are of significance for practical building construction and heating–ventilating–air conditioning applications. Characteristics of approaching wind are determined by the climate data based on the history of typhoons that have passed through Korea. The wind characteristics obtained by the statistical data are used for numerical simulations. Three-dimensional wind flows around the model of odd-geometrical building are simulated using the finite volume method. Velocity vectors and average-wind-pressure coefficients for equivalent static load are calculated by numerical results. The magnitudes of the wind pressure coefficients for the model building are relatively small compared with those of prismatic-shaped buildings. The largest average-wind-pressure coefficient for southwestern wind of 1.04 can be found near the top on the south wall for an odd-geometrical building with wind openings, the average-wind-pressure coefficients are decreased by approximately 19–27% on the windward surfaces and 32–53% on the leeward surfaces compared with those for buildings with prismatic shapes.

Keywords: Odd-geometrical building; Flow phenomena; Average wind pressure coefficients; Wind openings; Numerical simulation

* Corresponding author. E-mail: suh@scssh.soongsil.ac.kr.

PII S0167-6105(97)00108-6

1. Introduction

Buildings in large cities come under the impact of the typhoons which pass through Korea several times a year. Proper assessments of the wind loadings and the wind environment are necessary to ensure the safety of buildings and to provide pleasant living conditions for residents [1–3].

Construction of residential and commercial-combination-type high-rise buildings have increased recently. The shapes of these types of buildings are often irregular because of restrictions of construction sites and building regulations. Information on the wind loadings of such odd-geometrical buildings are not available in or out of Korea. According to Korean building regulations, it is permitted to determine wind loadings using the wind tunnel test.

Thus, studies on wind flows in odd-geometrical buildings are of importance to evaluate structural safety and occupant comfort in practical applications. Due to the limitations on wind tunnel experiments, there is a general tendency to simulate numerically using computational techniques.

The objective of this study is to determine the wind pressure coefficients on the odd-geometrical building which is to be constructed in the downtown area of a south costal city in Korea. This study also aims at investigating the effects of wind openings on the wind flows.

2. Analyses of wind climate

In order to evaluate flow phenomena and wind loadings around the odd-geometrical building, wind speed and direction are determined by the climate data based on the history of typhoons that have passed through Korea.

Fig. 1 is the path of typhoons and extra-tropical storms that have passed through Korea in the last 50 years. From the wind climate analyses for the southern coastal area in which the building is to be constructed, winds coming from southwestern directions are dominant from April to September.

The predicted mean-hourly wind speed for a 100-year return period is 41.4 m/s [4]. This wind speed, used for numerical simulation, includes the effects of both the typhoons and extratropical storms that have passed through Korea.

3. Model description

In order to obtain the information on the wind environment and the wind loadings, the odd-geometrical building to be constructed in the downtown area has been selected as a model for numerical simulation. The perspective of odd-geometrical building is shown in Fig. 2.

The height of the building is 98.1 m at its roof level with an additional plan shape resembling an elongated rhombus. The plan dimensions are 153.25 and 59.8 m. The most important characteristic of this building is its central opening. This study

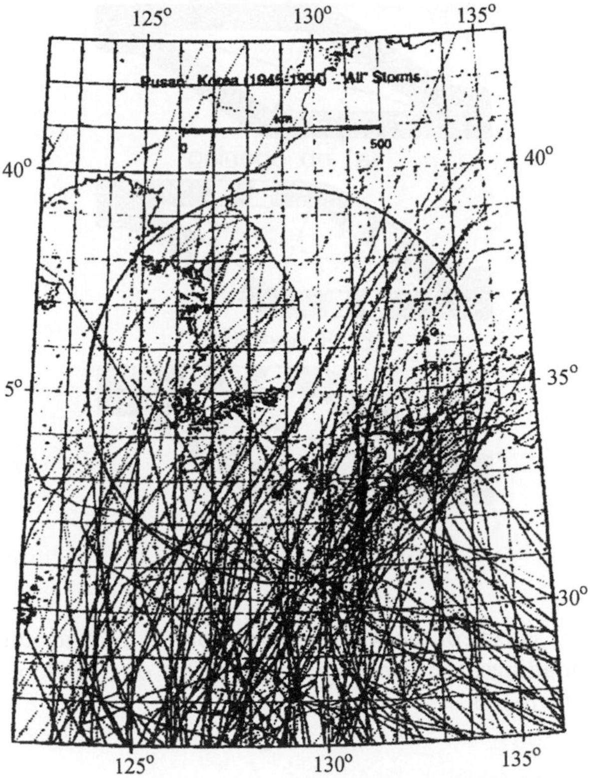

Fig. 1. Path of typhoons and extratropical stroms in the last 50 years.

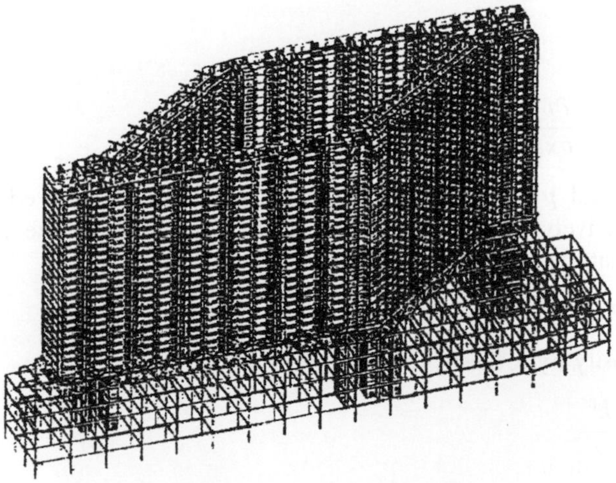

Fig. 2. Perspective of the model building.

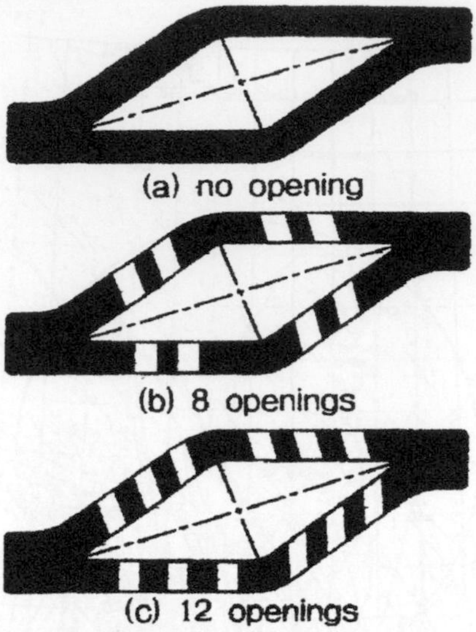

Fig. 3. Types of different openings.

also investigates the effects of the wind openings on the flow around a building. To reduce the wind loadings, the wind openings are set on the floor 25 m above the ground as shown in Fig. 3.

4. Numerical simulation

The following continuity and momentum equations in tensor form are used to solve the flow characteristics of numerical simulation [5–8].

$$\frac{\partial u_j}{\partial x_j} = 0\,, \tag{1}$$

$$\rho u_j \frac{\partial u_i}{\partial x_j} = -\frac{\partial p}{\partial x_i} + \frac{\partial}{\partial x_j}\left[\mu_e\left(\frac{\partial u_i}{\partial x_j} + \frac{\partial u_j}{\partial x_i}\right)\right] - \frac{2}{3}k\delta_{ij}\,, \tag{2}$$

where u_j, ρ, μ_e and p are the velocity vector, density, effective eddy viscosity and pressure, respectively. The effective viscosity is the sum of the laminar and the turbulent viscosity as follows:

$$\mu_e = \mu_\ell + \mu_t\,. \tag{3}$$

From the viscosity hypothesis, the effective viscosity is given by

$$\mu_e = \rho C_\mu \frac{k^2}{\varepsilon}\,, \tag{4}$$

where the value of C_μ is 0.085.

In this study, for high Reynolds number flow the RNG k–ε model is used instead of the standard k–ε model. The model, which is derived from a renormalization group analysis of the Navier–Stokes equations, differs from the standard model only through a modification to the equation for ε.

The equations describing the turbulence model are as follows [9],

$$\rho u_i \frac{\partial k}{\partial x_j} = \frac{\partial}{\partial x_j}\left[\frac{\mu_e}{\sigma_{k,e}}\frac{\partial k}{\partial x_j}\right] + P - \rho\varepsilon\,, \tag{5}$$

$$\rho u_i \frac{\partial \varepsilon}{\partial x_j} = \frac{\partial}{\partial x_j}\left[\frac{\mu_e}{\sigma_{\varepsilon,e}}\frac{\partial \varepsilon}{\partial x_j}\right] + \frac{\varepsilon}{k}(C_1 P - C_2\rho\varepsilon - R)\,, \tag{6}$$

where C_1 and C_2 are the turbulence model constants and P is the rate of production of turbulent kinetic energy,

$$P = \mu_t \frac{\partial u_i}{\partial x_j}\left[\frac{\partial u_i}{\partial x_j} + \frac{\partial u_j}{\partial x_i}\right]. \tag{7}$$

A new term, R in Eq. (6) comes from the RNG theory and vanishes in weakly strained turbulence to give a standard form of the k–ε model. It has been proposed to have the following form:

$$R = \frac{\eta(1 - \eta/\eta_0)}{(1 + \beta\eta^3)}\,, \tag{8}$$

$$\eta = \frac{k}{\varepsilon}(P/\mu)^{0.5}\,, \tag{9}$$

where, the values of η_0 and β are 0.015 and 4.38, respectively. The model constants used in the RNG k–ε model are tabulated in Table 1.

The oscillating problem is removed by adapting the Rhie–Chow algorithm. The QUICK scheme is adapted for discretization of convective terms and the SIMPLE algorithm for treating the pressure term in the momentum equations. The algebraic multi-grid (AMG) method is used to obtain the iterative solution of the finite-volume discretization equations. For simulation, SUN SPARCstation 20 is used taking 172 800 s to compute the wind flows and wind pressure on each model. Fig. 4 represents the computational domain of the model. The three-dimensional mesh for the odd-geometrical building model is presented in Fig. 5. The simulations have been performed using the fluid dynamic general-purpose code CFDS-FLOW3D.

Table 1
Constant values for the RNG k–ε model

Constant	$\sigma_{k,e}$	$\sigma_{\varepsilon,e}$	C_1	C_2
Value	0.7179	0.7179	1.42	1.68

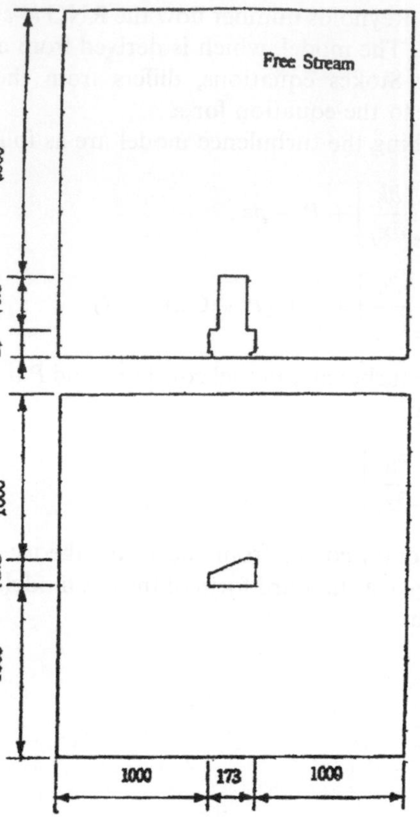

Fig. 4. Computational domain of the model.

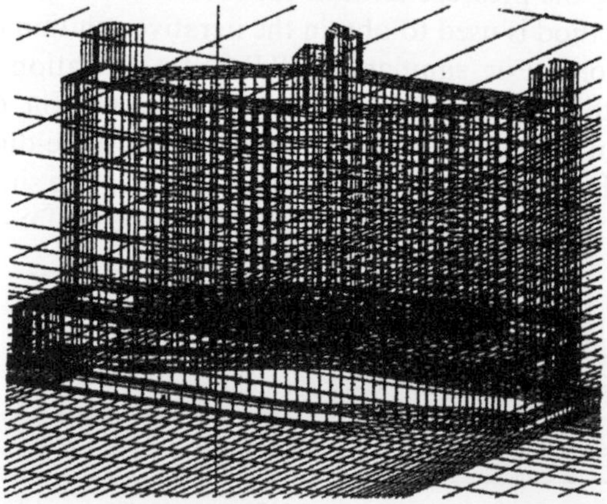

Fig. 5. Three-dimensional mesh for the model.

5. Results and discussion

5.1. Verification of numerical simulation

Numerical and experimental results [6] around a prismatic shape model are compared to verify the validity of the numerical simulation. This validity can be analyzed with various turbulence models as shown in Fig. 6. As turbulence model the standard k–ε model, RNG k–ε model, and ASM model were applied. The calculated results are quantitatively in good agreement with experimental data. The RNG k–ε model, above all turbulence models, is in good agreement with the experimental results. The average wind pressure coefficient of the front surface is 0.48 and that of the roof and rear surface is − 0.44 and − 0.17, respectively. In this case the wind pressure coefficients on the roof and side wall surface show rapid decrease at the back.

5.2. Flow phenomena around the model building

To observe the flow phenomena around the model building, velocity vectors are simulated by the finite volume method. In this paper flow phenomena around the model building are focused on the winds coming from the southwestern directions often passing through in summer.

For winds from the southwest the velocity vectors at "PIT" floor and at the middle of the building are presented in Fig. 7. Flow patterns at the "PIT" floor are very complicated. At the windward surface some flows are skewed to the roof floor and then passed along the walls, the others are flow in the central opening. In the leeward side two recirculating zones are formed at the upper part and the lower part of the "PIT" floor. Such phenomena does not occur in buildings with prismatic shapes.

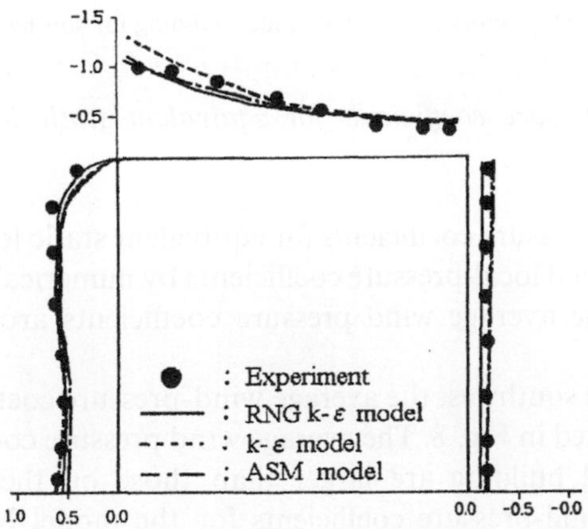

Fig. 6. Comparison of wind pressure coefficients obtained experimentally and numerically.

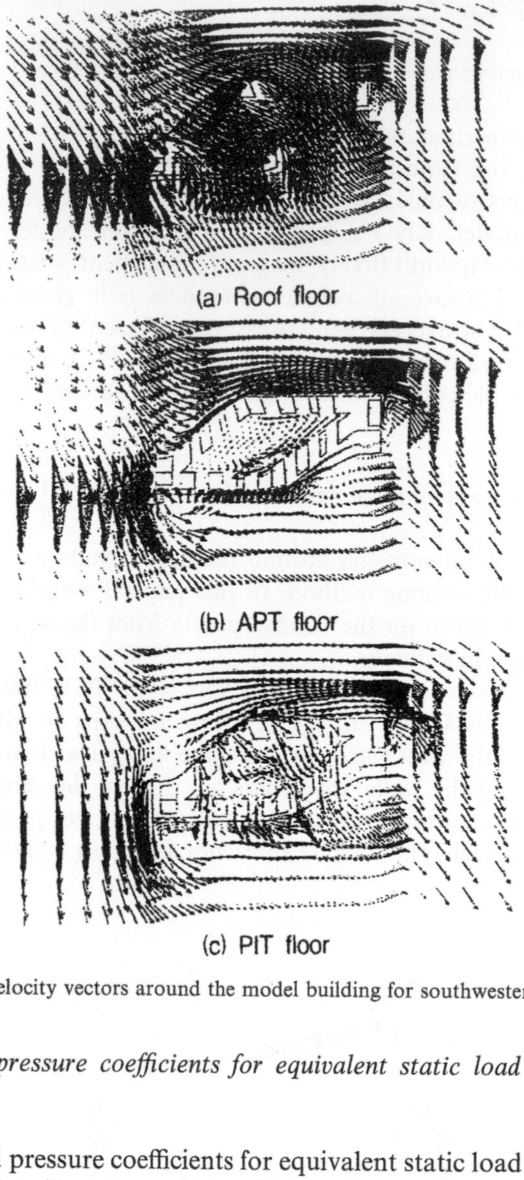

Fig. 7. Velocity vectors around the model building for southwestern wind.

5.3. *Average wind pressure coefficients for equivalent static load around the model building*

The average wind pressure coefficients for equivalent static load are determined by integrating the predicted local-pressure coefficients by numerical simulation. Effects of wind direction on the average wind pressure coefficients around the building are investigated.

For winds from the southwest the average wind-pressure coefficients for equivalent static load are presented in Fig. 8. The average wind pressure coefficients on the upper surface of the model building are larger than those on the lower surfaces. The magnitudes of the wind-pressure coefficients for the model building are relatively small compared with those of prismatic-shaped buildings. The largest average wind

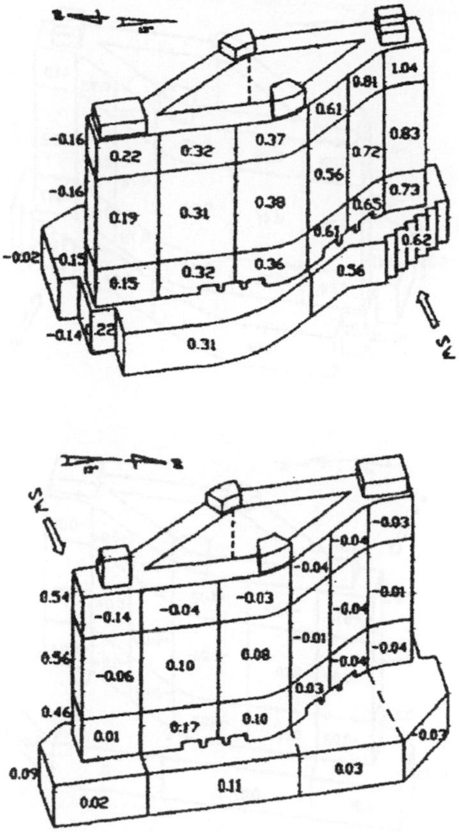

Fig. 8. Average pressure coefficients on the building surfaces for southwestern direction wind.

pressure coefficient for the southwestern wind of 1.04 can be found near the top of the south wall. For the odd-geometrical building with wind openings, the average wind pressure coefficients decrease by approximately 19–27% on the windward surfaces and 32–53% on the leeward surfaces compared with those for the building with prismatic shapes [6,10]. This is mainly attributed to the geometry of the model building. It is a relatively wide building with H/W aspect ratios of about 0.69 and 1.8. Due to its small aspect ratio, the "tip" effect is expected to substantially reduce the average pressure coefficients.

To investigate the effects of wind openings, the average-wind-pressure coefficients on the model building are simulated by numerical technique. The average-wind-pressure coefficients in the model building without openings and with 8 openings are shown in Figs. 9 and 10, respectively.

The wind pressure coefficients for the model building with openings decrease by approximately 20–25% on surfaces. The average wind pressure coefficients on the leeward surface decrease remarkably than those on the leeward surface.

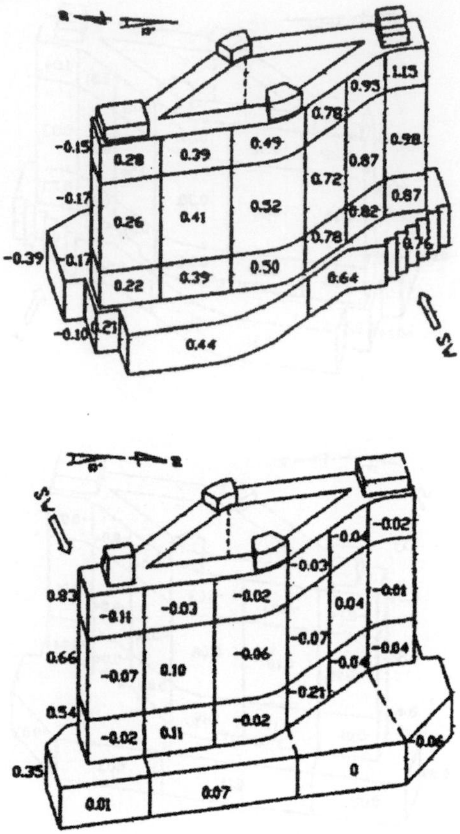

Fig. 9. Average pressure coefficients on the building surfaces without openings.

The pressures within the central opening are relatively uniform and the contribution of wind loads from the walls within the central opening are small.

6. Conclusions

Flow phenomena and wind pressure coefficients around the odd-geometrical building are investigated by using numerical simulations. From the wind climate analyses, winds coming from the southwestern directions are dominant from April to September.

Three-dimensional wind flows around the model of an odd-geometrical building are simulated using the finite volume method. Velocity vectors and wind pressure coefficients for equivalent static load are calculated using numerical results. The magnitudes of the wind pressure coefficients for the model building are relatively small compared with those of prismatic-shaped buildings. The largest average wind

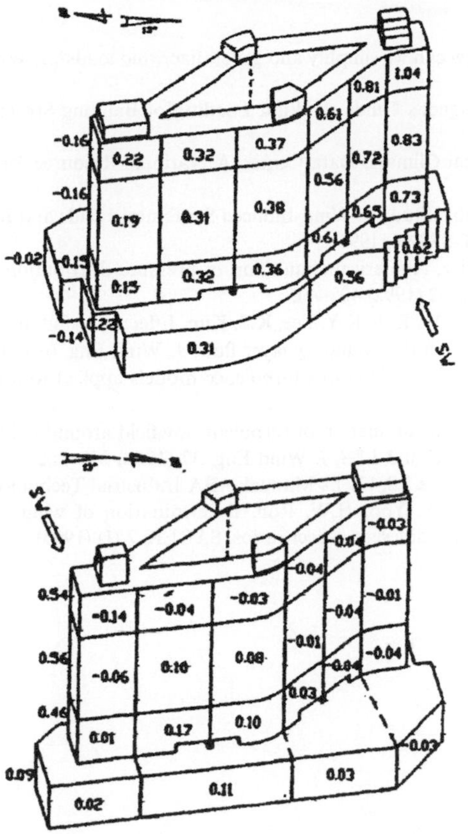

Fig. 10. Average pressure coefficients on the building surfaces with 8 openings.

pressure coefficient for southwestern wind of 1.04 can be found near the top on the south edge wall.

For the odd-geometrical building with wind openings the average wind pressure coefficients decrease by approximately 19–27% on the windward surfaces and 32–53% on the leeward surfaces compared with those for the building with prismatic shapes.

Wind pressure coefficients for the model building with 12 openings decrease by approximately 20–25% on all surfaces. Especially, the average wind pressure coefficients on the leeward surface decrease remarkably than those on the leeward surface.

Acknowledgements

The authors would like to express their appreciation to the Boundary Layer Wind Tunnel Laboratory, University of Western Ontario for the wind climate data.

References

[1] A.G. Davenport, How can we simplify and generalize wind loads?, J. Wind Eng. Ind. Aerodyn. 54/55 (1995) 657–669.
[2] N.J. Cook, The Designer's Guide to Wind Loading of Building Structures – Part 1, B.R.E., 1985, p. 156.
[3] U.S. NOAA, National Climatic Data Center, A National Resource for Climate Information, April 1995.
[4] T.C.E. Ho, N. Isyumov, Study of Wind-Induced Structural Loads and Responses for the Dong Nam World Plaza, BLWT-SS4-96, 1996.
[5] Y. Sun, B. Bienkiewiez, Numerical simulation of pressure distribution Underneath Roofing Paver System, J. Wind Eng. 52 (1992) 400–405.
[6] H.R. Kim, S.H. Suh, H.W. Roh, K.Y. Lee, K.S. Kim, Effects of wind direction changes on the surface pressure of 3-D model in the boundary layer flow, J. Wind Eng. 63 (1995) 69–70.
[7] S. Murakami, Comparison of various turbulence models applied to a bluff body, J. Wind Eng. 52 (1992) 162–179.
[8] S. Murakami, Numerical simulation of turbulent flowfield around cubic model: current status and application of k–ε model and LES, J. Wind Eng. 37 (1988) 239–252.
[9] CFDS-FLOW3D Release 3.3, User's Manual, AEA Industrial Technology Harwell Lab., UK, 1994.
[10] S.H. Suh, K.Y. Lee, S.S. Yoo, H.W. Roh, Determination of wind pressure coefficients around prismatic structure with different aspect ratios, SAREK 7 (1) (1995) 52–62.

ELSEVIER

Journal of Wind Engineering
and Industrial Aerodynamics 67&68 (1997) 671–685

JOURNAL OF
wind engineering
AND
industrial
aerodynamics

Numerical prediction of wind loading on buildings and structures – Activities of AIJ cooperative project on CFD

Tetsuro Tamura[a,*], Hiromasa Kawai[b], Shinji Kawamoto[c], Kojiro Nozawa[d], Shigehiro Sakamoto[e], Takeshi Ohkuma[f]

[a] *Department of Environmental Physics and Engineering, Tokyo Institute of Technology, 4259 Nagatsuta, Midori-ku, Yokohama 226, Japan*
[b] *Tokyo Denki University, Hatoyama, Saitama 350-03, Japan*
[c] *Nippon Sheet Glass Co., Ichihara, Chiba 299-01, Japan*
[d] *Shimizu Corporation, Chiyada, Tokyo 100, Japan*
[e] *Taisei Corporation, Shinjuku, Tokyo 169, Japan*
[f] *Kanagawa University, Rokkakubashi, Yokohama 221, Japan*

Abstract

This study presents the activities of the Architectural Institute of Japan (AIJ) concerning the Computational Fluid Dynamics (CFD) for a numerical prediction of wind loading on buildings and structures. In the AIJ project, the flows and the pressures around a low-rise building (breadth : depth : height = 1 : 1 : 0.5) have been computed by 10 members of the working group, who mainly employed the k–ε model or the large eddy simulation for turbulent flows. Here, on the basis of the results of the AIJ project, future subjects for further development of the CFD technique are discussed. Also we present how to find the way to realize the practical use of the CFD technique on prediction of wind loading.

Keywords: Computational fluid dynamics; Turbulence simulation; Modified k–ε model; Large eddy simulation; Wind loading; Low-rise building; Pressure distribution

---

1. Introduction – AIJ project and subjects for further development of CFD

AIJ started a cooperative project on CFD in 1992 by establishing the working group, in order to investigate the current status and the future possibility of the CFD for practical use. Accordingly, a total of 27 members[1] including structural

* Corresponding author. E-mail: tamura@depe.titech.ac.jp.

[1] The working group members are: T. Tamura (Chairman), K. Kondo, S. Kawamoto, M. Hachiya, T. Ishikawa, N. Kato, S. Kawabata, H. Kawai, M. Kawamura, T. Maruyama, K. Miyashita, H. Mukai, H. Noda, K. Nozawa, S. Ohgaki, T. Ohkuma, K. Ohtake, Y. Okuda, T. Saito, S. Sakamoto, K. Shimada, M. Shimura, Y. Suyama, Y. Tamura, N. Tsuchiya, T. Yamada, K. Yoshie.

PII S0167-6105(97)00109-8

designers joined the project from several universities, research institutes and private companies.

At the first stage of the project, we bring into focus an evaluation of the CFD technology in view of numerical accuracy and computational requirements for the prediction of wind loading. Previous research works are surveyed and an applicability of present techniques to wind loading problems is discussed.

Also it was planned to solve the common specified problem in wind engineering by various numerical methods. Accordingly, a half cube on a flat plate was adopted as a computational model of a low-rise building (Fig. 1). Many members solved the flow around a building using a numerical scheme and the turbulent model selected by themselves (Table 1). Experiments were also carried out for this problem by several groups at the same time (Table 2). In order to examine the results of the CFD project of AIJ, the national symposium for the numerical prediction of the wind loading on buildings and structures was held in Tokyo, Japan on 10 November 1994 [1]. The report was written in Japanese and published at that time. In 1995, a one-day international workshop of the related theme was coordinated by IWEF [2] and part of the results were provided internationally. Now the report of the CFD project is to be rewritten in English and to be presented at the IWEF Workshop in Colorado, 9 August 1996 [3].

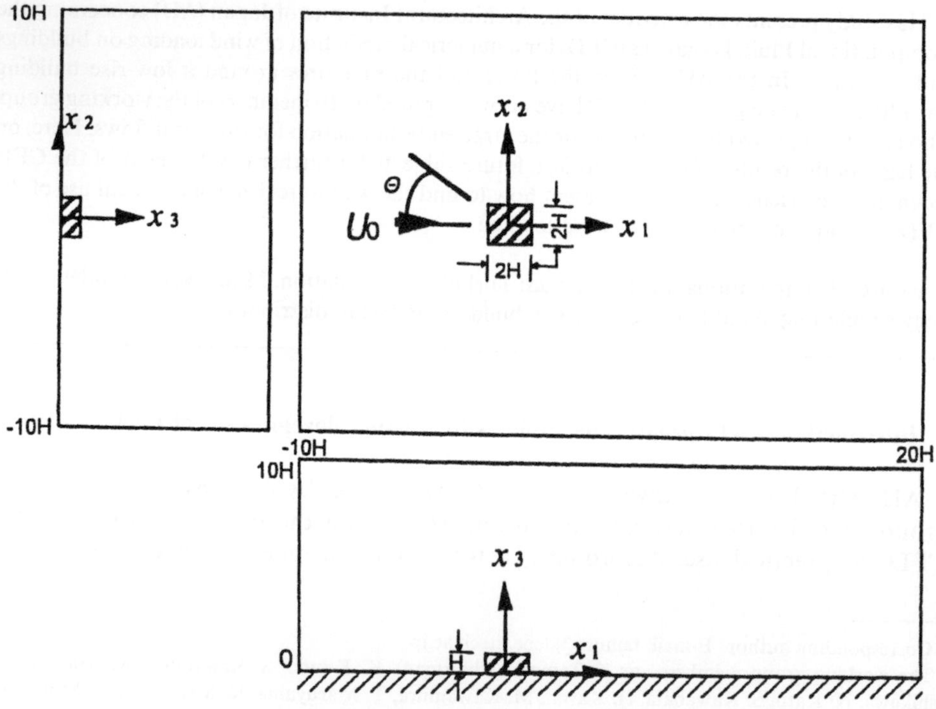

Fig. 1. Computational model of the AIJ project.

Table 1
Computations in the AIJ project (51 cases)

⟨CASE⟩				Boundary condition (inflow)	Software	Discritization	Turbulence model	Scheme for convection terms	Computational method and time integral scheme
Without model	Wind direction								
	0°	22.5°	45°						
CSNNAS1	CS00AS1	CS25AS1	CS45AS1	k constant	ORIGINAL	FEM	Improved k–ε	SU/PG	GSMAC, implicit
CSNNAS2	CS00AS2	CS25AS2	CS45AS2	ESNNVLKJ	ORIGINAL	FEM	Improved k–ε	SU/PG	GSMAC, implicit
CSNNAS3	CS00AS3			ESNNVLKJ	STAR-CD	FVM	Standard k–ε	Mixed upwind	SIMPLE, steady solution
CSNNCN1	CS00CN1	CS25CN1	CS45CN1	ESNNVLCN	STREAM	FVM	Standard k–ε	First-order upwind	SIMPLE, steady solution
CSNNCN2	CS00CN2	CS25CN2	CS45CN2	ESNNVLCN	STREAM	FVM	Improved k–ε	First-order upwind	SIMPLE, steady solution
CSNNHZ1	CS00HZ1	CS25HZ1	CS45HZ1	ESNNVLHZ	STREAM	FVM	Standard k–ε	First-order upwind	SIMPLE, implicit
CSNNKJ1	CS00KJ1	CS25KJ1	CS45KJ1	ESNNVLKJ	ORIGINAL	FVM	Standard k–ε	QUICK	MAC, implicit
CSNNNH1	CS00NH1	CS25NH1	CS45NH1	ESNNVLKJ	ORIGINAL	FEM	Standard k–ε	Galerkin	GSMAC + BTD, explicit
CSNNNH2	CS00NH2	CS25NH2	CS45NH2	ESNNVLKJ	ORIGINAL	FEM	Improved k–ε	Galerkin	GSMAC + BTD, explicit
CSNNOB1	CS00OB1			k constant	STREAM	FVM	Standard k–ε	QUICK	SIMPLE, implicit
CSNNTK1	CS00TK1	CS25TK1	CS45TK1	ESNNVLTK	α-FLOW	FDM	Standard k–ε	QUICK	SMAC, explicit
	CS00HZ1	CU25HZ1	CU45HZ1	EUNNVLHZ	STREAM	FVM	Standard k–ε	First-order upwind	SIMPLE, implicit
	CU00KY1			Very low turbulence	ORIGINAL	FDM	Standard k–ε	First-order upwind	MAC, explicit
CUNNKY2	CU00KY2			EUNNVLKY	ORIGINAL	FDM	Standard k–ε	First-order upwind	MAC, explicit
	CU00NH1			Very low turbulence	ORIGINAL	FEM	Standard k–ε	Galerkin	GSMAC + BTD, explicit
	CU00NH2			Very low turbulence	ORIGINAL	FEM	Improved k–ε	Galerkin	GSMAC + BTD, explicit
	CU00SM1			Smooth flow	ORIGINAL	FDM	LES,DSGS	Third-order upwind	MAC, explicit
	CU00TS1		CU45TS1	Smooth flow	ORIGINAL	FDM	LES,SGS	Second-order	HSMAC, mixed

Table 2
Wind tunnel experiments in the AIJ project (65 cases)

Velocity measurement		Pressure measurement			Approaching flow	Turblence intensity (%)	Turbulence scale (cm)	Model size $B \times D \times H$ (cm)	Wind tunnel size (Measurement section) $B \times H$ (cm)
Without model	0°	Wind direction							
		0°	22.5°	45°					
ESNNVLKN	ES00VLKN	ES00CPKN	ES25CPKN	ES45CPKN	1/4 shear flow	17	21	20 × 20 × 10	120 × 90
ESNNVLTD	ES00VLTD	ES00CPTD	ES25CPTD	ES45CPTD	1/4 shear flow	26	18	12 × 12 × 6	120 × 120
ESNNVLKY	ES00VLKY	ES00CPKY	ES25CPKY	ES45CPKY	1/4 shear flow	22	–	12 × 12 × 6	250 × 200
ESNNVLTK	ES00VLTK	ES00CPTK	ES25CPTK	ES45CPTK	1/4 shear flow	19	30	20 × 20 × 10	200 × 180
ESNNVLKJ	ES00VLKJ	ES00CPKJ	ES25CPKJ	ES45CPKJ	1/4 shear flow	27	55	20 × 20 × 10	250 × 200
ESNNVLSM	ES00VLSM	ES00CPSM	ES25CPSM	ES45CPSM	1/4 shear flow	20	35	20 × 20 × 10	210 × 260
ESNNVLNH	ES00VLNH	ES00CPNH	ES25CPNH	ES45CPNH	1/4 shear flow	16	30	20 × 20 × 10	200 × 200
ESNNVLCN	ES00VLCN	ES00CPCN	ES25CPCN	ES45CPCN	1/4 shear flow	22.7	23	12 × 12 × 6	180 × 180
EUNNVLKN	EU00VLKN	EU00CPKN	EU25CPKN	EU45CPKN	Smooth flow	2.3	–	20 × 20 × 10	120 × 90
EUNNVLTD	EU00VLTD	EU00CPTD	EU25CPTD	EU45CPTD	Smooth flow	0.2	–	12 × 12 × 6	120 × 120
EUNNVLKY	EU00VLKY	EU00CPKY	EU25CPKY	EU45CPKY	Smooth flow	1.2	–	12 × 12 × 6	250 × 200
EUNNVLKJ	EU00VLKJ	EU00CPKJ	EU25CPKJ	EU45CPKJ	Smooth flow	2.0	–	20 × 20 × 10	250 × 200
EUNNVLSM	EU00VLSM	EU00CPSM	EU25CPSM	EU45CPSM	Smooth flow	0.15	–	20 × 20 × 10	210 × 260

The results of the AIJ cooperative project consist of the following two parts:

– Part 1. The current status of CFD technology in wind engineering.
– Part 2. Challenge of wind loading estimation on a low-rise building.

With regard to the current status of CFD technology of Part 1, we also submitted a questionnaire to the wind and structural engineers in research institutes and private companies. The examples for practical use of CFD were investigated. The following conclusions are obtained:

(1) The technique of CFD is widely used for applications in wind engineering, especially environmental problems such as wind flow around buildings.
(2) Structural engineers are planning to use the CFD technique for wind-load estimations, especially for structural problems such as wind-induced vibrations.

Concerning the analysis of flows around a low-rise building in Part 2, the following computational models are employed:

(1) k–ε turbulence model in $\frac{1}{4}$ shear flow,
(2) large eddy simulations in a uniform flow.

As important conclusions, we emphasize the limitation of the standard k–ε model and the necessity of an improved turbulence model for wind-force estimation on buildings and structures. We also show the numerical accuracy of the large-eddy simulation for the fluctuating statistics of wind flows and pressures. Accordingly, it is concluded that the future possibilities of the large eddy simulations are promising, but the current results are scattered, depending on the usage of LES or a numerical model.

On the basis of these conclusions, this paper concretely discusses how to develop the numerical technique and turbulence modeling for practical use in the estimation of wind loading. Especially we bring into focus the improvement of the k–ε model and the appropriate methodology for LES.

2. CFD technique on prediction of wind loading for practical use

Thus far, the techniques of CFD have been widely used in various kinds of engineering fields. Even in wind engineering, this technique was much developed and at present CFD is usually employed to some problems as often as the wind tunnel technique. For example, the flow field of environmental wind can be numerically predicted in the limit of simple geometry such as a single cubic model for a building. However, it is well known that the flow around a simple cube is still very much complicated in view of fluid mechanics, because the flow is separated and generates various kinds of vortices around it. Accordingly, we should note that much time and effort for the development of CFD technology is required to realize the application to prediction of flows with complex structures.

Due to accumulation of a great number of researches, the CFD technique has recently reached use for prediction of wind loading. Wind forces on buildings are determined by the integration of the pressures acting on the model. However, the pressure field is very sensitive to the flow patterns, which is usually very complicated around a bluff body like a building. It is not easy to estimate the pressure field due to the complex flow structures and their unsteadiness, because only a highly accurate but not costly numerical scheme can produce sufficient and reasonable results, where the numerical dissipation and the phase error should be removed as much as possible. In order to realize the practical use of CFD techniques to these problems, we have concluded to adopt two methodologies.

For complex flow structures around a bluff body, the turbulent energy transfer among various flows is also very complex. Therefore, the improvement of turbulence modeling is required, but too much complicated formulation of turbulence modeling is not appropriate for the practical use in wind engineering. That is to say, we employ the improvement of the k–ε model as a first way and do not employ the Reynolds stress equation models at the present stage. For the unsteadiness of flows and pressures, taking into account recent development of subgrid-scale modeling for the fine structures of turbulence, we employ the LES technique as a second way. This means it is concluded that the LES technique can be used even for unfavorable flow fields such as transitional flows, separated flows or complicated behaviors of vortices.

In the following sections, we discuss the scope of above two ways with regard to numerical techniques.

3. Improvement of k–ε model

The original k–ε model was proposed about 20 years ago [4] and has been applied to various kinds of problems in engineering. The applicability and the limitation of this model becomes clear now. Especially, the flows in wind engineering often have complicated structures such as separations, impinging and strong pressure gradient. Fig. 2 shows the mean pressure coefficients in the central section in the $\frac{1}{4}$ shear flow at the wind direction of 0°. The pressure coefficients on the windward surface scatter very much both in experiments and computations. On the other hand, the pressure coefficients on the roof and the leeward surface do not scatter so much in spite of the large difference of the approaching turbulence intensity. The significant difference is observed for the profile of the mean pressure on the roof between the results of the experiments and the computations. The mean suction computed by the k–ε model shows a rapid increase near the windward corner of the roof. On the other hand, the experimental suction is constant near the corner and decreases gradually to the downward end. This difference comes from the over production of the turbulence energy in the accelerating flow in front of the model. As a result an improved version of the k–ε model is required.

Kato and Launder proposed an improved k–ε model for the impinging flow field [5]. This model led to the improvement of pressure distributions on the frontal surface on buildings. This basic idea for turbulence modeling is very promising and effective in view of both accuracy and cost, because the model is very simple as before and modified only due to the focus on the essence of physical behaviors of various flow

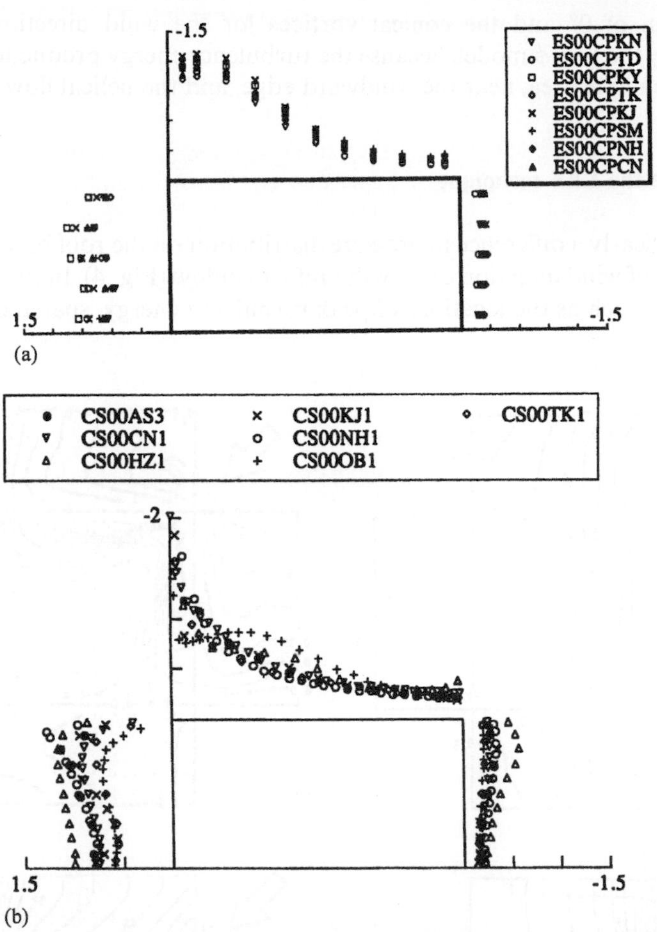

Fig. 2. Mean pressure coefficients on vertical section ($x_2 = 0.0$) in the $\frac{1}{4}$ shear flow. (a) Experiments [1]. (b) Computations by the standard k–ε model.

structures. Kawamoto proposed a different type of improved k–ε model [6], where the adverse pressure gradient effect and the impinging effects are introduced as the effects of wall-reflection, and the favorable pressure gradient effect is introduced as a correction term for the turbulence kinetic energy production of the standard k–ε model. Due to this model, the pressure distributions on a low-rise building are much improved, especially on the roof as well as on the frontal surface (Fig. 3).

Kawamoto has been developing the k–ε–ϕ model for prediction of wind loading with much higher accuracy [7]. The variable ϕ, an anisotropy parameter of the Reynolds stresses which varies through adverse pressure gradient and favorable pressure gradient, is introduced to the standard k–ε model, and then the kinematic eddy viscosity is affected by the variable ϕ. Moreover, the kinematic eddy viscosity is affected by the helicity. In consequence, the separation from windward edge for the wind direction of 0° and the conical vortices for the wind direction of 45° are reproduced by the k–ε–ϕ model, because the turbulence energy production is controlled at the impinging area, near the windward edge, and the helical flow field (Fig. 3).

4. How to use the LES technique

There was clearly a difference in pressure distribution on the roof between two LES computations of wind direction of 0° with uniform inflow (Fig. 4). In the results, other characteristics, such as the locations of peak turbulence energy, spatial correlation of

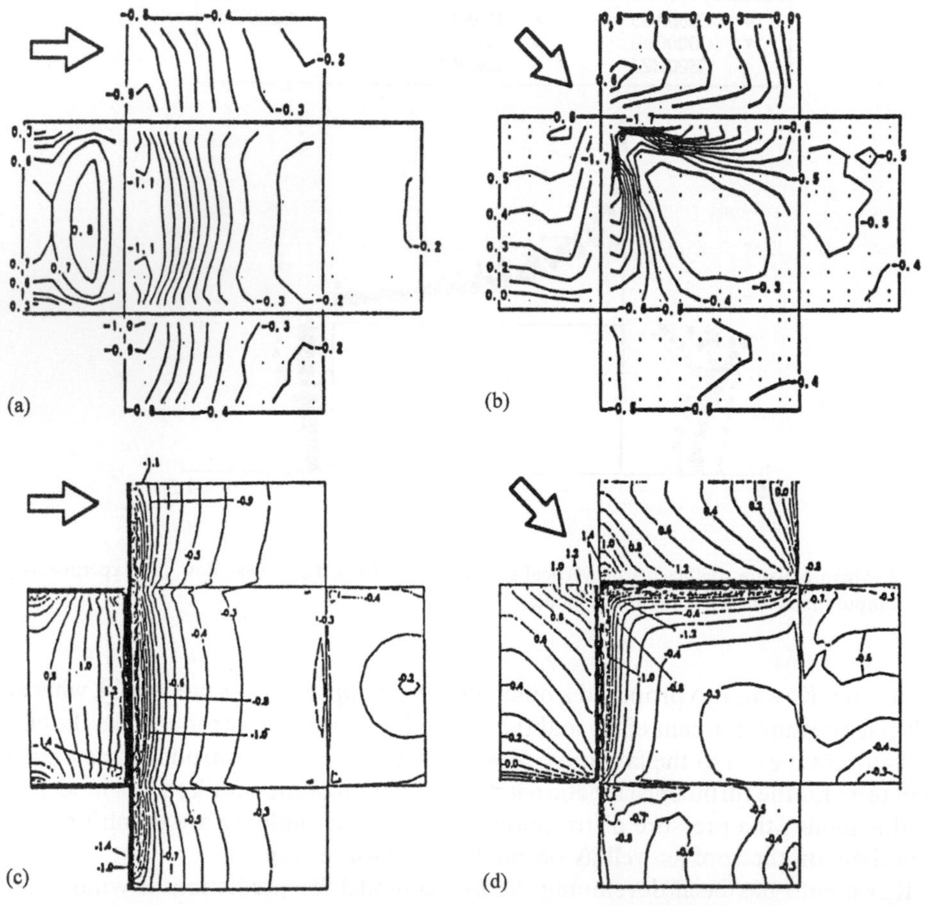

Fig. 3. Distributions of mean pressure coefficients on a low-rise building (breadth : depth : height = 1 : 1 : 0.5) in $\frac{1}{4}$ shear flow: (a) experiment [1], wind direction 0°; (b) experiment [1], wind direction 45°; (c) standard k–ε model, wind direction 0°; (d) standard k–ε model, wind direction 45°; (e) improved k–ε model, wind direction 0°; (f) improved k–ε model, wind direction 45°; (g) improved k–ε–ϕ model, wind direction 0°; (h) improved k–ε–ϕ model, wind direction 45°.

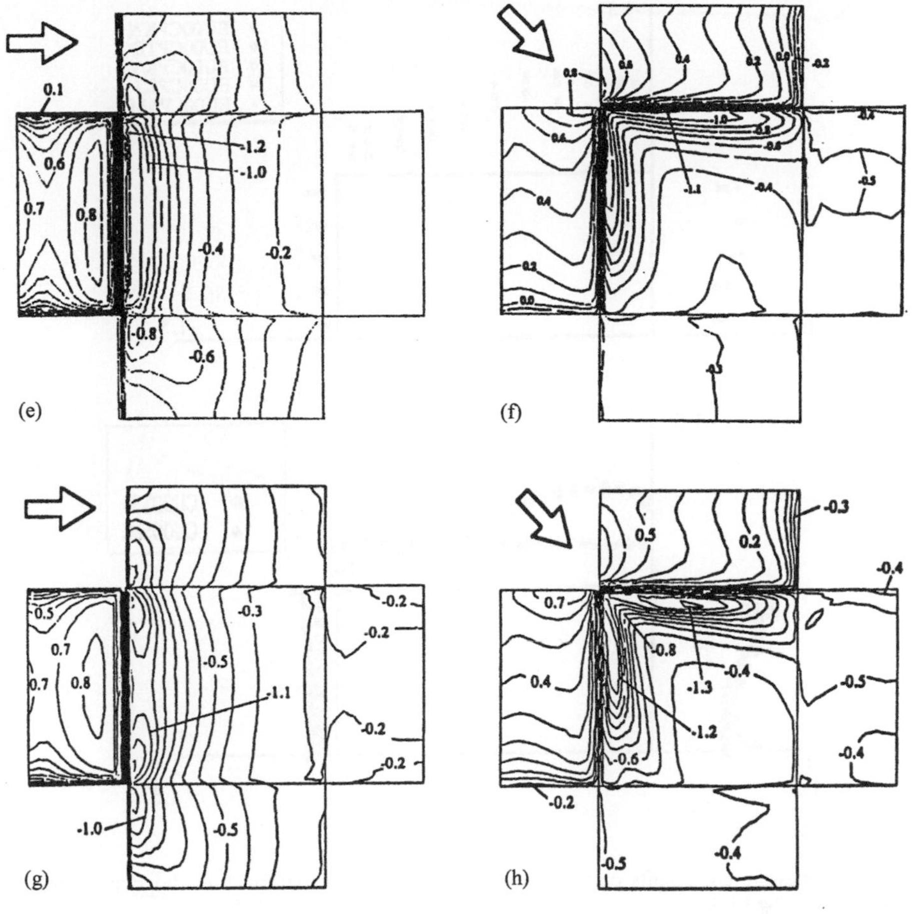

Fig. 3. Continued.

fluctuating pressure on the roof, etc., were different from each other. These differences were also observed in the wind tunnel tests and we could find similar results with each LES computation. However, the computational conditions are not so variable as the characteristic of approaching flows of the wind tunnel tests. We could expect similar results in computations if the numerical methods are properly applied. We attempted to identify the main cause of the difference of the two LES calculation results and find a way to improve LES computations.

4.1. Numerical method of LES computations

Two LES computations use some different conditions, such as subgrid-scale model of LES and boundary condition on the floor [1]. The case (the code in Ref. [1] is CU00TS1) which used the Smagorinsky model [8] for the subgrid-scale model was

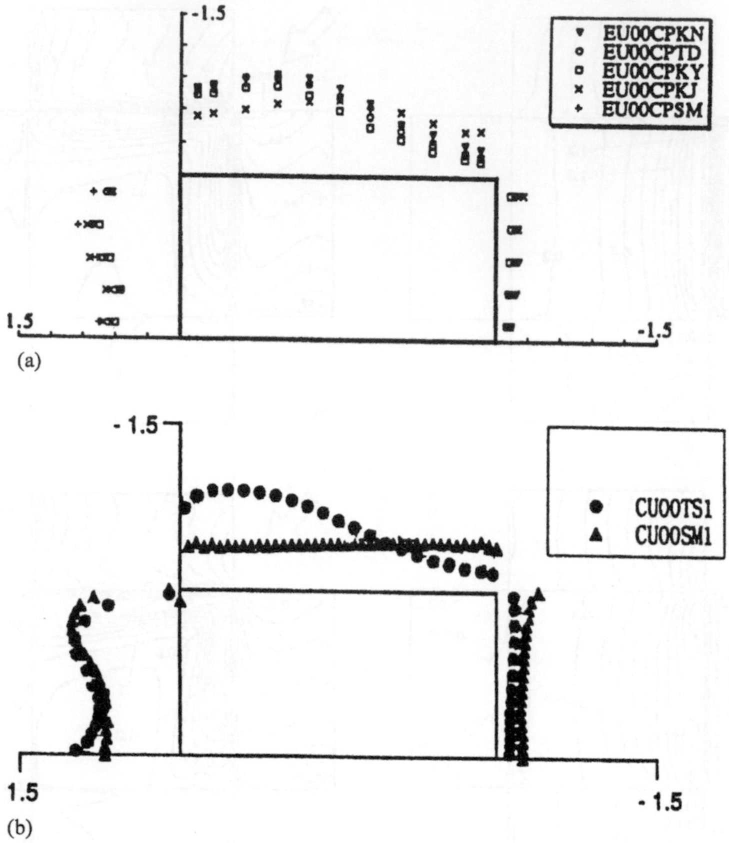

Fig. 4. Mean pressure coefficients on vertical section ($x_2 = 0.0$) in uniform flow. (a) Experiments [1]. (b) Computations by LES.

made of artificial boundary conditions for the floor and the surface of the model [9]. For discretization in time, the Adams–Bashforth scheme was applied for convection terms, and the Crank–Nicolson scheme for viscous terms with staggered mesh. In other case (the code in Ref. [1] is CU00SM1) a dynamic subgrid-scale model [10] is used with non-slip boundary condition on the floor and the surface of the model. For discretization in space, a second-order central differencing was used with collocation grid system except for the convection terms. For the convection terms, a third order upwind scheme was applied. The Reynolds number of the latter case was 5000 that was an order smaller than 50 000 of the former case.

4.2. *Testing based on the variant conditions*

The variable results by two LES computations come from the behavior of the separated flows. The slow development of the separated flow, in which flow fluctuates

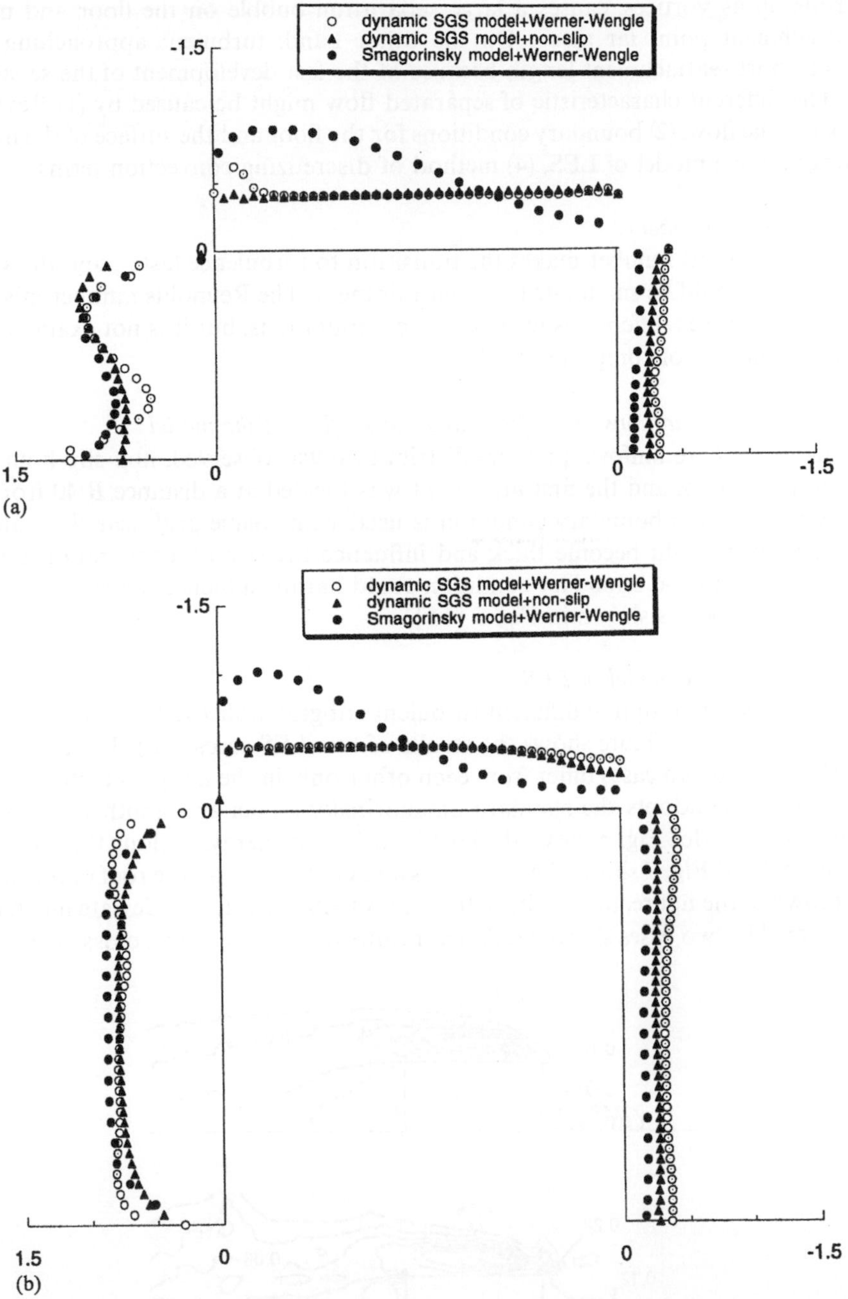

Fig. 5. Effects of boundary conditions on mean pressure coefficients on central section. (a) Vertical section ($x_2 = 0.0$). (b) Horizontal section ($x_3 = 0.5H$).

and rolls up as vortices, causes a large separation bubble on the floor and makes a reattachment point far away. On the other hand, turbulent approaching flow makes a short reattachment length because of the fast development of the separated flow. The different characteristic of separated flow might be caused by (1) Reynolds number of the flow, (2) boundary conditions for the floor and the surface of the model, (3) subgrid-scale model of LES, (4) method of discretizing convection terms.

4.2.1. Reynolds number

Higher Reynolds number makes the transition to turbulence faster, and the separation bubble would be small due to the entrainment. The Reynolds number might be a large factor of the differences in two LES computations, but it is not examined yet due to a high cost of computation.

4.2.2. Boundary conditions of the floor and the surface of the model

In the case where uniform pressure distribution was observed, non-slip boundary condition was used, and the first grid point was located at a distance $B/40$ from the wall. When non-slip boundary condition is used with coarse grid near the wall, the boundary layer might become thick and influence the size of a separation bubble. However, the artificial boundary condition could hardly influence the pressure distribution on the roof as shown in Fig. 5 [11].

4.2.3. Subgrid-scale model of LES

The influence of adopting different turbulent subgrid-scale (SGS) models is shown in Figs. 6 and 7. The figure shows the results of two LES cases [12]. The calculation conditions of the two cases differ from each other only in the adopted turbulent SGS model. One case adopts the standard Smagorinsky model and another adopts the dynamic SGS model. Fig. 6 shows the kinetic turbulent energy, k, along the center line of the model and Fig. 7 shows the mean pressure coefficients on the roof of the model. As is shown in the figure, the results of these two LES cases do not deviate much from each other. The two cases showed different results in the space–time cross-correlation

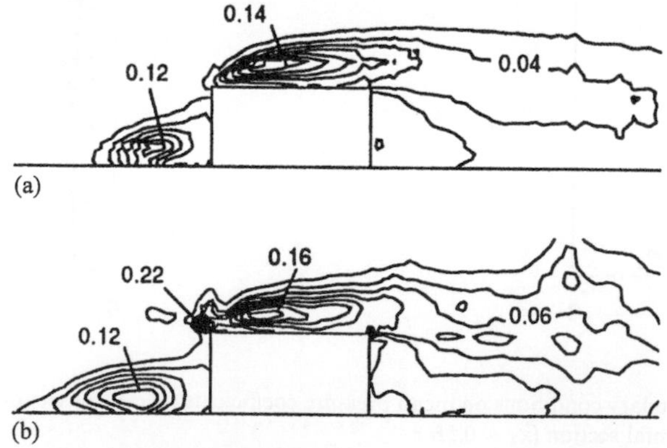

Fig. 6. Kinetic turbulent energy. (a) Smagorinsky model. (b) Dynamic SGS model.

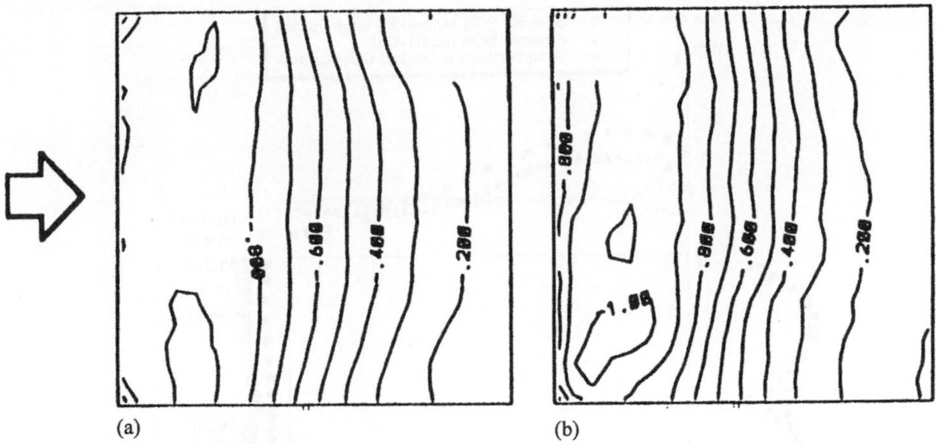

Fig. 7. Mean pressure coefficient on the roof. (a) Smagorinsky model. (b) Dynamic SGS model.

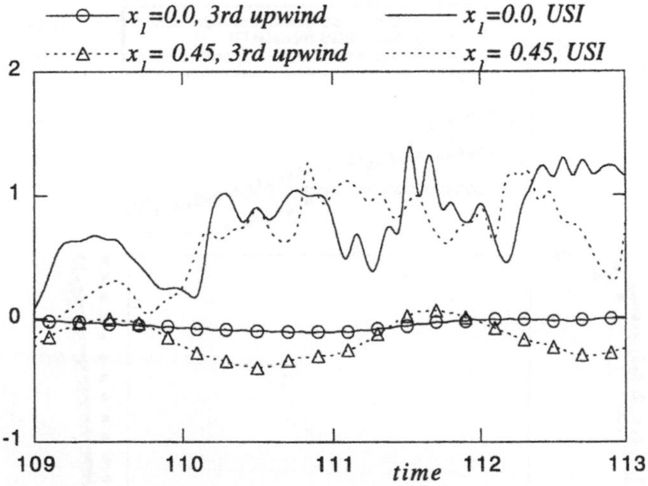

Fig. 8. Time history of streamwise velocity above the roof. ($x_1 = 0.0$, $+0.45$, $x_2 = 0.0$.)

of fluctuating pressure on the roof [12], but the difference in averaged pressure distribution was not so large as the difference seen in the AIJ Report. It is clear that the deviation in the AIJ Report was not caused by the difference of the adopted turbulent model.

4.2.4. Method of discretizing convection terms

Instead of the third-order upwind scheme Nozawa [11] tested the USI (upstream-shifted interpolated) method [13], whose numerical dissipation is smaller than that of the third-order upwind scheme. Fig. 8 shows the time history of streamwise velocity

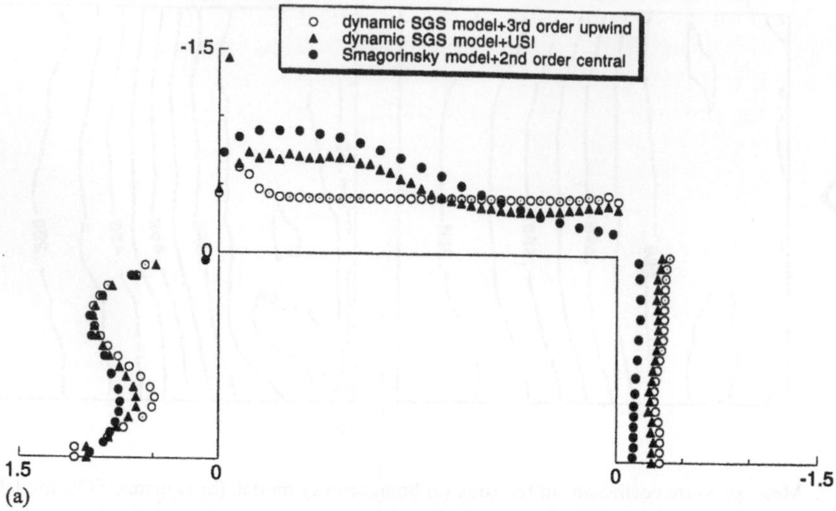

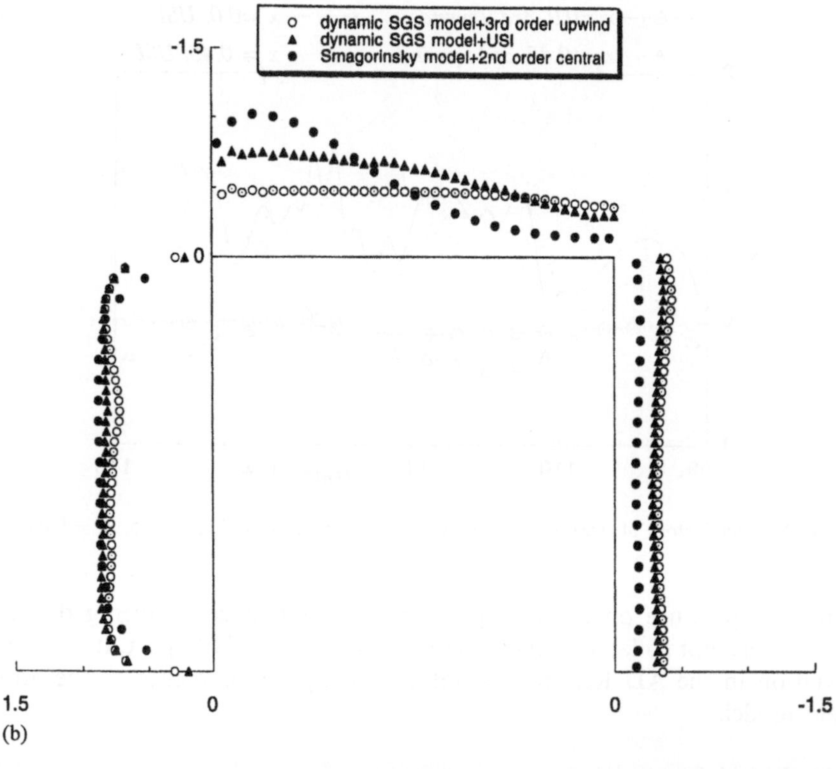

Fig. 9. Effects of numerical scheme on mean pressure coefficients on central section. (a) Vertical section ($x_2 = 0.0$). (b) Horizontal section ($x_3 = 0.5H$).

above the roof on the central section. In the case using the 3rd order upwind scheme, the streamwise velocity was almost steady or low-frequency fluctuation (freq. < 0.05) was observed. In the case using USI, high-frequency fluctuation was observed, and this enabled the pressure recovery on the roof (Fig. 9).

5. Conclusions

On the basis of results of the AIJ project, we have examined the appropriate CFD techniques for prediction of wind loading on buildings and structures from the view points of numerical accuracy and computational costs. For practical use we have proposed two methods: one is the improvement of the k–ε models as a simple model for the complex flows around a building and the other is the LES for unsteady problems.

Several subjects for further development of the CFD technique are clarified. In the case of the k–ε model, the modification based on the physical mechanism is so effective that the various kinds of flow patterns are successfully simulated. In the case of LES, it is found that the numerical scheme has an important role for the computed results.

References

[1] AIJ Working Group, Numerical prediction of wind loading on buildings and structures, AIJ, 1994.
[2] Proc. IWEF workshop on CFD for prediction of wind loading on buildings and structures, Yokohama, Japan, IWEF, 1995.
[3] Proc. IWEF Workshop on CWE/CFD for prediction of wind effects on structures, Colorado, USA, IWEF, 1996.
[4] B.E. Launder, D.B. Spalding, The numerical computation of turbulent flows, Comput. Meth. Appl. Mech. 3 (1974) 269–289.
[5] M. Kato, B.E. Launder, The modeling of turbulent flow around stationary and vibrating square cylinder, Proc. 9th Turbulent Shear Flows, 10–4, 1993 pp. 1–6.
[6] S. Kawamoto, An improved k–ε turbulence model for accurate prediction of pressure distribution, 7th CFD Symp, 1993 (in Japanese).
[7] S. Kawamoto, Improved turbulence models for estimation of wind loading, these Proceedings, J. Wind Eng. Ind. Aerodyn. 67&68 (1997) 589–599.
[8] J.S. Smagorinsky, General circulation experiments with the primitive equations; I. The basic experiment, Monthly Weather Rev. 91 (1963) 99–164.
[9] H.Werner, H. Wengle, Large-eddy simulation of turbulent flow over and around a cube in plate channel, Proc. 8th Symp. on Turbulent Shear Flows, 19–4, 1991 pp. 1–6.
[10] M. Germano, U. Piomelli, P. Moin, W.H. Cabot, A dynamic subgrid-scale eddy viscosity model, Phys. Fluids A 3 (1991) 1760–1765.
[11] K. Nozawa, Materials for WG of AIJ.
[12] S. Sakamoto, Flow field around 1 : 1 : 0.5 rectangular prism predicted by LES, Summaries of Tech. Papers of Annual Meeting, AIJ, 1995.
[13] T. Kajishima, Upstream-shifted interpolation method for numerical simulation of incompressible flows, Trans. Japan Soc. Mech. Eng. 60–578 (1994) 3319–3326.

Journal of Wind Engineering
and Industrial Aerodynamics 67&68 (1997) 687–696

The assessment of wind loads on roof overhang of low-rise buildings

Tore Wiik[a,*], Ernst W.M. Hansen[b]

[a] *Building Science, Narvik Institute of Technology, P.O. Box 385, N-8500 Narvik, Norway*
[b] *SINTEF, Thermal Energy and Hydropower, N-7034 Trondheim, Norway*

Abstract

Numerical simulations are compared with experimental mean values of characteristic pressure (C_p) values on a gable wall and on a roof overhang of a low-rise building. The inclined wind direction investigated is normal to the gable wall. Two-model designs with an overhang of 0.3 and 3.4 m are examined. The main issue in this paper is to investigate what influence a large roof overhang has on the C_p-values on the house, compared to an ordinary roof. Also from the engineering point of view, it is interesting to see how well a commercial CFD-code, with a basic turbulence model, can predict the C_p-values for a house. The numerical simulations are performed with a k–ε turbulence model.

Keywords: Low-rise building; Wind forces; Roof overhang; CFD simulations; Wind tunnel experiments

1. Introduction

In the assessment of the wind load on roofs and walls, the effects of a roof overhang are often ignored. Few studies concerning the effect of roof overhang on the flow pattern and wind loads of typical low-rise buildings have been made [1,2]. The Norwegian Code of Practice [3], regarding wind forces, is mainly based on wind tunnel experiments done by Jensen and Franck [4]. In their eminent work, they investigated wind forces on several cube designs and on free standing roofs, but they did not consider roof overhang in great detail. Most of the other Standards and Codes of Practice also provide limited guidance with regard to wind loads on roof overhangs.

The present investigation undertakes the study of the local, critical wind patterns around a roof overhang at the gable wall of a low-rise building pertaining to Nordic regions. Numerical simulations are performed and compared to experimental

* Corresponding author.

PII S0167-6105(97)00110-4

measurements from wind tunnel studies [5] in order to obtain data of the wind loads and flow pattern around a low-rise building. The present work demonstrates the difference in flow pattern and characteristic pressure values for a building with and without roof overhang at the gable wall for incidence wind normal to the gable wall.

2. Numerical method

The equations modelling the wind effects on buildings are partial differential equations which are further developed by numerical methods into algebraic equations by a finite difference method (finite volume method). The resulting numerical model solves the wind effects in detail in terms of control volumes (mesh cells). The mean flow effects taken into account in simulation are fluid dynamic effects at length scales larger than the control volume sizes. Turbulence modelling takes account of small-scale fluid dynamic effects in numerical simulations.

The SOLA algorithm (a SOLution Algorithm) is used to solve the governing equations. Each control volume is assigned local average values for the velocity components and pressure. These variables are located at staggered cell positions, which stabilized the finite difference approximations and the setting of boundary conditions. The velocity and pressure coupling is solved in an iterative procedure until a satisfied convergent result is obtained in each control volume. The SOLA algorithm is programmed in three-dimensions in the general-purpose flow code FLOW-3D [6] which was developed for the analysis of fluid dynamic and thermal phenomena. The numerical simulations performed have been achieved by FLOW-3D. FLOW-3D is a software package that consists of four separate programs, the preprocessor, the main processor, the post processor and the image display program, a CFD-code. Computational fluid dynamics (CFD) is a generic name of a wide range of numerical techniques that are used for obtaining solutions to the governing equations of a thermodynamic fluid with or without chemical reactions.

The flow modelling and the numerical simulations, by means of FLOW-3D, have been performed in three-dimensions and the Cartesian coordinate system is used. Complex geometrical regions are formulated with area and volume porosity functions, FAVOR (Fractional Area/Volume Obstacle Representation method). Along the walls the rigid, no-slip wall boundary conditions have been used together with a law-of-the-wall velocity profile and the kappa–epsilon (k–ε) turbulence model is used further away from the walls to model the flow-field conditions.

3. Experimental method

The experiments are carried out in the 1.2×1.5 m industrial aerodynamic wind tunnel of the University of Hertfordshire, UK. This is an open return wind tunnel with a working section of 4.7 m. The atmospheric boundary layer is simulated by means of a barrier at the entrance and boards with graded roughness elements on the floor of the wind tunnel. The wind profile representing the approach wind flow is shown in

Fig. 1. The model scale used in the wind tunnel experiments is 1 : 30 of a full-scale house. The flow characteristics are measured by a pitot-static probe and a standard hot-wire probe, the mean pressure on the house surface is measured with pitot probes, connected to a multi-manometer by plastic tubes. The C_p-values on the surfaces are measured for each 15° incidence wind angle. The C_p-value is defined as

$$C_p = \frac{P_{x,y,z} - P_{\text{ref}}}{0.5\rho u_{\text{ref}}^2}, \tag{1}$$

where $P_{x,y,z}$ is the total pressure at the surface, P_{ref} is the static pressure at the reference height and $0.5\rho u_{\text{ref}}^2$ is the dynamic pressure at the reference height. The reference point is chosen to be at 10 m corresponding full-scale height. The reason for this choice is that most of the meteorological weather data are given at this height.

4. Numerical simulations

The incident wind direction for this case is normal to the gable wall. This is not the angle that provides highest forces on the roof, but it simplifies the numerical simulation since only half of the house need to be simulated due to the symmetric condition in the y-direction.

The simulations are performed at full-scale condition, and the approaching velocity profile at the left boundary is defined as

$$U(z) = \frac{U_*}{\kappa} \ln\left(\frac{z}{z_0}\right), \tag{2}$$

where U_* is the friction velocity, κ is the von Kármán constant (~ 0.4), z the instant height and z_0 is the friction height.

4.1. Domain size

The domain size of the simulation is (75, 20, 30) m in the (x, y, z) direction, and the total number of nodes is 138 700, i.e. (73, 38, 50) nodes in the (x, y, z) direction. Fig. 2 shows a sketch of the domain in the x–z plane.

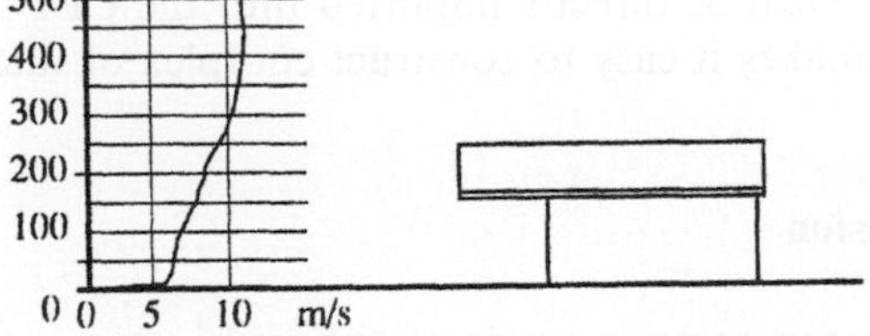

Fig. 1. The velocity profile and a model scale 1 : 30 configuration in the wind tunnel.

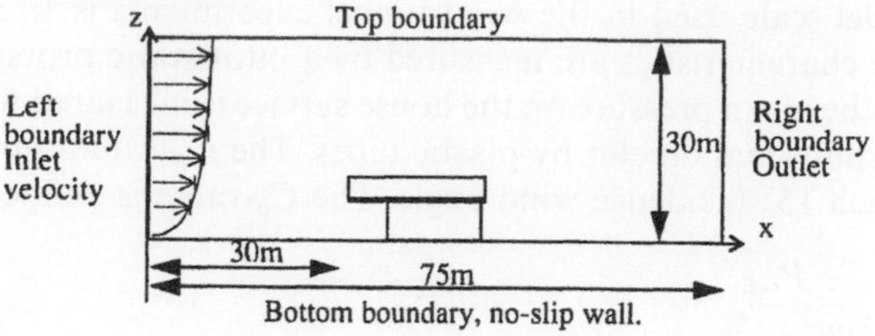

Fig. 2. Sketch of the simulation domain in x–z plane.

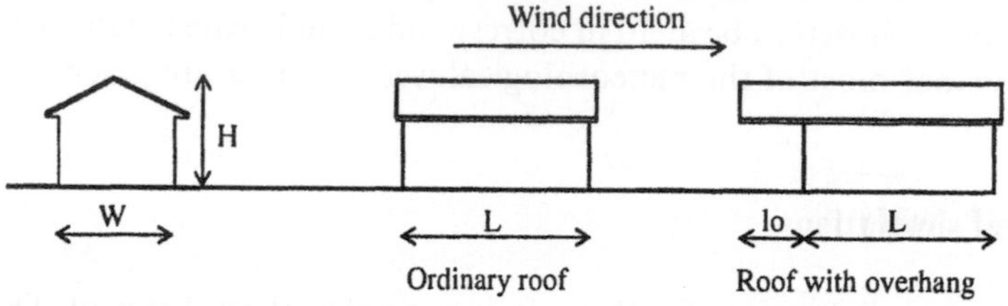

Fig. 3. Sketch of the house models with and without roof overhang at the gable wall.

4.2. Boundary condition

The inlet logarithmic velocity profile has $U_{max} = 13$ m/s at the top boundary. The friction velocity (U_*) is 0.736 m/s and friction height (z_0) is set to 0.025 m, in order to fit the measured velocity profile.

At the right boundary (outlet), the Neumann condition is assumed ($\partial u/\partial x = 0$ and $\partial w/\partial z = 0$), and along the bottom boundary, rigid no-slip wall condition is assumed. For the front and the back boundary (y-low and y-high), and for the top boundary the symmetric boundary conditions are applied.

4.3. Model sizes

The full-scale dimensions of the houses chosen are (11.0 m × 7.5 m × 7.5 m) (L, W, H), and the pitch of the roof is 30°. The long overhang at the gable wall (lo) is 3.4 m, eaves overhang is 0.5 m and "ordinary" overhang at gable wall is 0.3 m. Otherwise the design is closed as in Fig. 3.

The obstacles are created with a general CAD software [7], and exported to an STL-file format which can be directly imported into the CFD-software FLOW-3D. This saves time and makes it easy to construct complex obstacles.

5. Results and discussion

The experimental measurements are done with multi-manometers, and the readings of the fluid differences in this instrument will include some level of inaccuracy.

Especially small differences can be difficult to estimate correctly due to the resolution on the manometer scale. So although the readings of the manometer are done as accurately as possible, the experimental results cannot be taken as the exact answer regarding C_p-values.

The reader has to bear in mind that the models used in the experiments did not have the exact same probe location at the roof. This means that for the house with roof overhang, the C_p-plots do not start at the same distance from the gable edge as for the house with ordinary roof.

5.1. Comparisons of numerical and experimental results

At the gable wall of both houses, the results show good agreement between the simulations and experiments for both cases, see Figs. 4 and 5. The largest differences are at the corner of the houses, and in this region the numerical results seem to underestimate the C_p-values compared to the experimental results. The reason for this underestimation could be the same as explained for the numerical calculated C_p-values on the roof, see below.

In Fig. 6 the C_p-values at the underside of the overhang are compared, and also at this surface there is good agreement between simulated and experimental values. However, two of the probes need further explanation. The experimental value at the

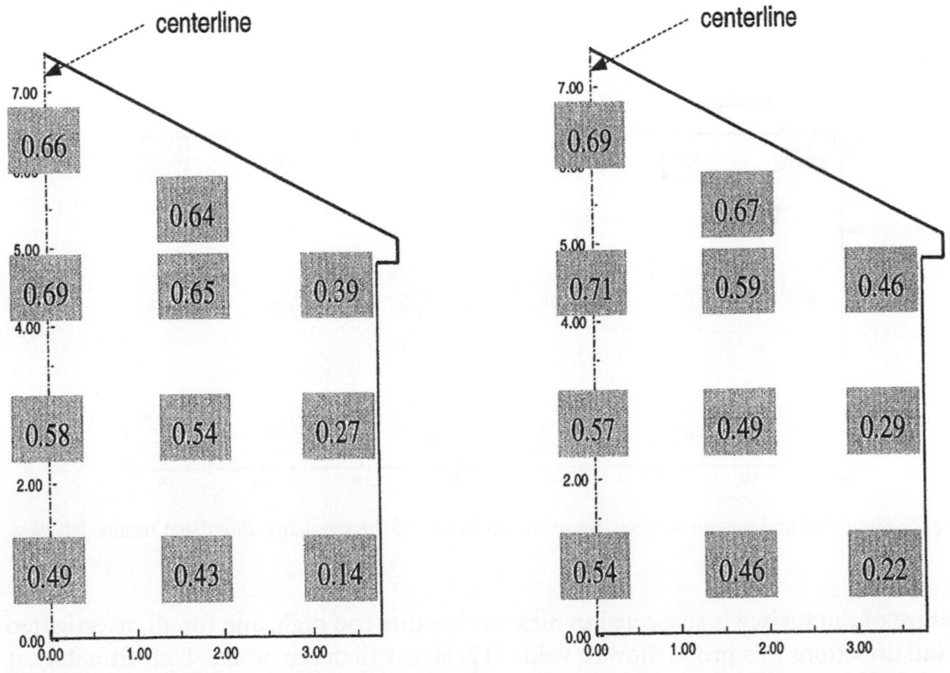

Fig. 4. Simulated and measured C_p-values on the gable wall, house with ordinary roof. Simulated results, left view.

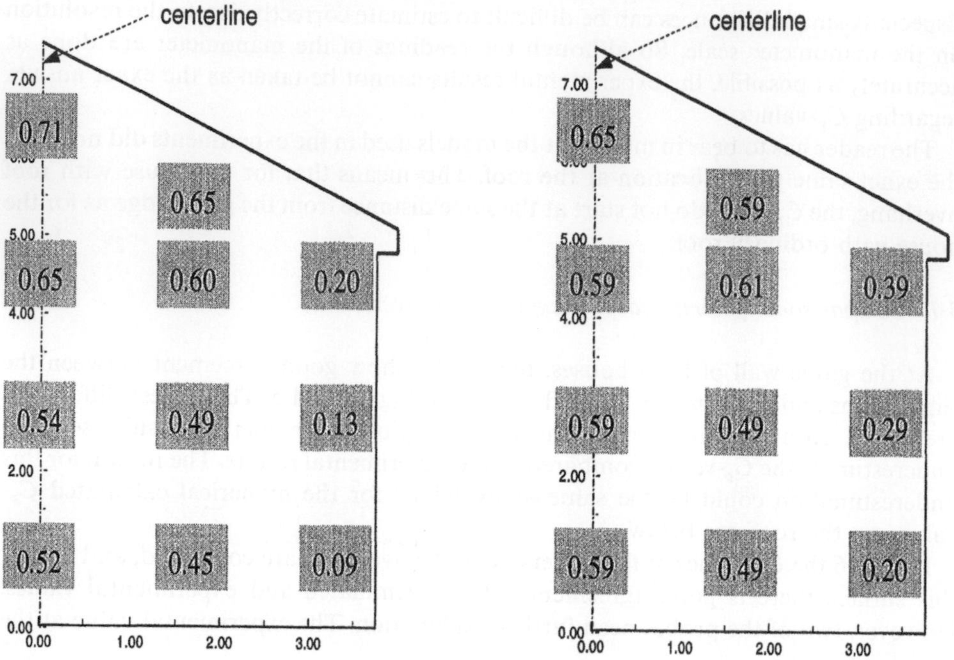

Fig. 5. Simulated and measured C_p-values on the gable wall, house with roof overhang. Simulated results, left view.

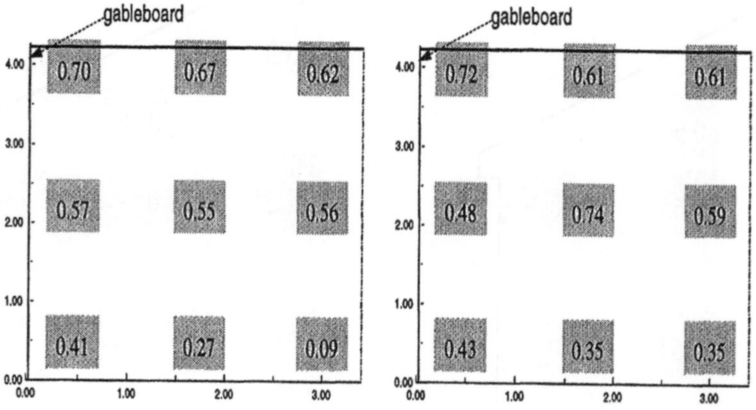

Fig. 6. Simulated and measured C_p-values at the underside of the overhang. Simulated results, left view.

centre of the surface is suspected to measure a value too high, and for all investigated wind directions this probe shows a value (0.2) larger than the others. Unfortunately, it has not been possible to check this probe later. The other probe location with variation is the probe closest to the gable wall and the eaves (probe at the bottom right corner in the plot). Also, here one possible explanation for the numerical results could be the overestimation of C_p-values near the edge in regions with suction.

At the topside of the roof, there are discrepancies between the experimental and numerical results for both houses, see Figs. 7 and 8. This was expected since earlier test cases with cubic obstacles have reported this phenomenon [8,9] when the k–ε turbulence model has been used. The reason for this is that the k–ε model, due to its isotropic eddy–viscosity concept [10], produces too much kinetic energy in the stagnation point on the windward side of the house. This prevents separation at the front edge of the roof, and the C_p-values will be overpredicted near the suction side of the edge.

Anyway, with these limitations in mind, the trend in C_p-values at the topside of the roofs is the same for both experimental and numerical results.

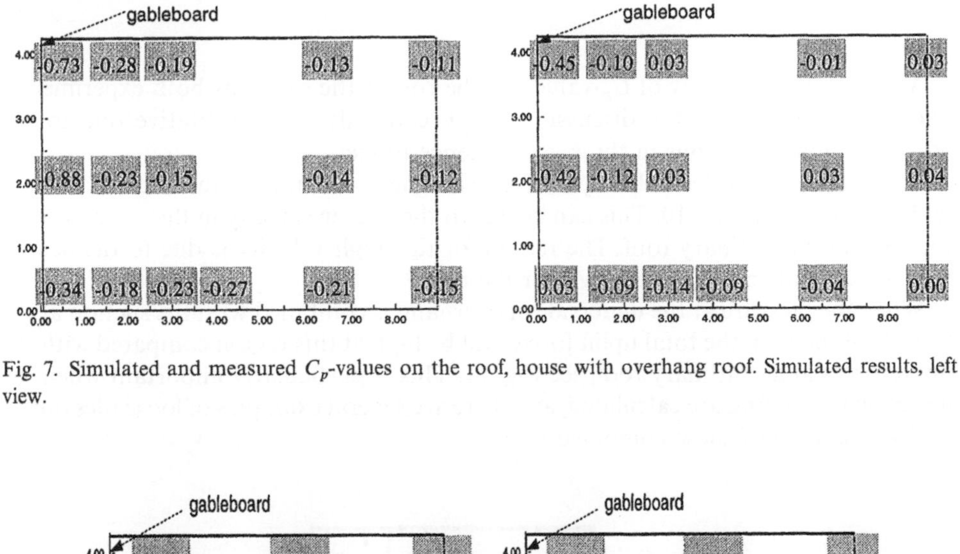

Fig. 7. Simulated and measured C_p-values on the roof, house with overhang roof. Simulated results, left view.

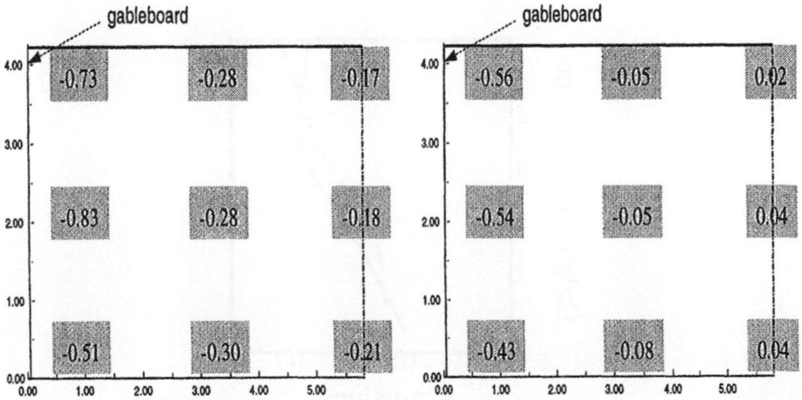

Fig. 8. Simulated and measured C_p-values on the roof, house with ordinary roof. Simulated results, left view.

5.2. Comparison of numerical calculated C_p-values on the two houses

5.2.1. Gable wall

The total forces on the gable wall are more or less the same for both houses, but the pressure distribution changes. In the case with ordinary roof, the traditional stagnation point appears at about $\frac{2}{3}$ of the height, and then the pressure reduces towards the roof. For the house with large roof overhang, the pressure has the stagnation point at the top of the wall. This means that the upper part of the wall with overhang will achieve higher wind forces than the wall with ordinary roof. Fig. 9 shows the differences in pressure distribution at the centreline of the gable wall for the two houses.

Since the Norwegian codification is based on wind tunnel experiments with models without overhang, which means even lower pressure at the top wall than shown here, the influence of an overhang must be important for the vertical stiffening of the gable wall.

5.2.2. Roof

Although the tendency of C_p-values on the roof is the same for both experimental and numerical results, the discussion below can only be quantitative due to the afore-mention limitations in the k–ε turbulence model.

The gable edge of the ordinary roof gains higher suction than the edge of the roof with overhang, see Fig. 10. This can be due to the higher velocity in the z-direction at the edge of the ordinary roof. The reason for this high velocity is due to the nearby wall which leads part of the flow over the roof.

When the pressure at the underside of overhang is combined with the suction at the topside of the roof, the total uplift forces will be high at this region compared with the uplift forces at an ordinary roof, see Fig. 11. This is particularly important when the forces on the roofing are calculated, and there are several examples of loose tiles due to underestimation of these combined forces.

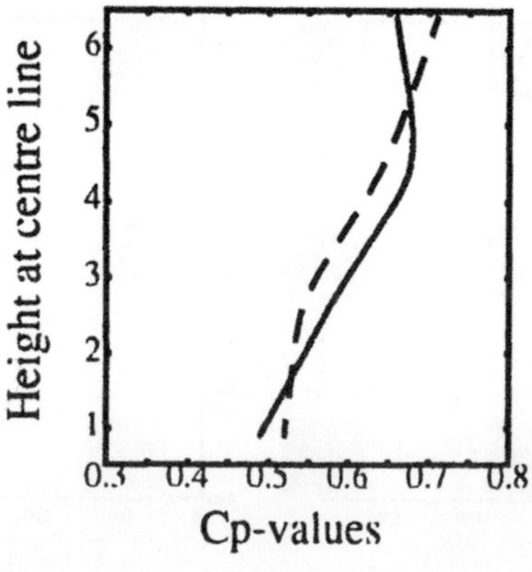

Fig. 9. C_p-values at the centre of gable wall: (—) case with ordinary roof; (– – –) roof with overhang.

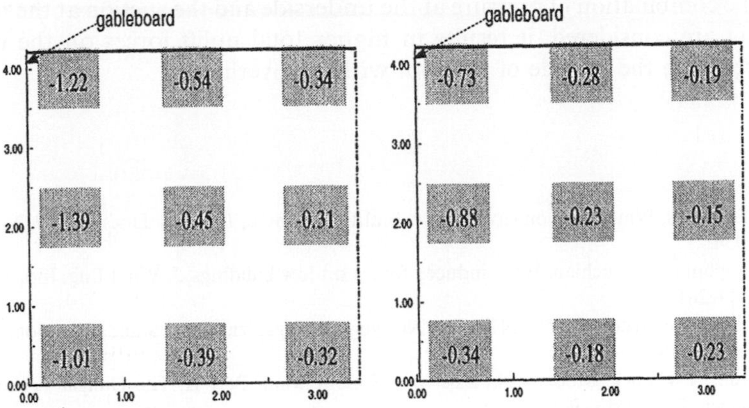

Fig. 10. Simulated C_p-values at the roof of the two different houses. C_p-values at the same locations from the gableboard, result for ordinary house at left.

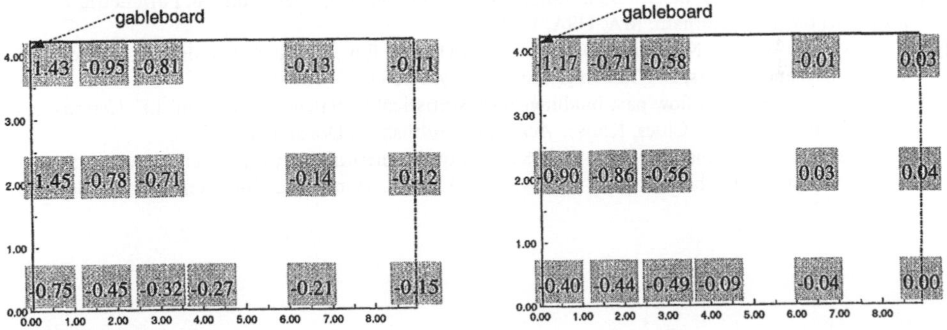

Fig. 11. Simulated and measured C_p-values of the total pressure on the roof for the house with overhang. Simulated results, left view.

6. Conclusions

- Numerical simulations with the basic k–ε turbulence model predicts well the pressure at the windward side of the houses, compared with experimental results.
- At edges with flow separation, the numerical simulations seem to underestimate the pressure at the windward side of the edge, and overestimate the negative pressure at the leeward side of the edge.
- The effect of a long roof overhang will change the pressure distribution on the low-standing wall. The pressure will increase at the upper part of the wall, compared with no overhang.
- At the gableboard edge, when only the upper part of the roof is considered, the uplift pressure on the overhang edge will be lower than for a roof with no overhang.

- When the combination of pressure at the underside and the suction at the topside of overhang are considered, it results in higher total uplift forces on the overhang compared with the topside of the roof without overhang.

References

[1] T. Stathopoulos, Wind loads on eaves of low buildings, ASCE, J. Struct. Division 107, No. 10 (1981) 1921–1934.

[2] T. Stathopoulos, H. Luchian, Wind induced forces on low buildings, J. Wind Eng. Ind. Aerodyn. 52 (1994) 249–261.

[3] Design loads for structures, NS-3479, The Norwegian Organization of Standardization, Oslo, Norway, Revision 1990.

[4] M. Jensen, N. Franck, Model-scale tests in turbulent wind, Part II, The Danish Technical Press, Copenhagen, Denmark, 1965.

[5] T. Wiik, Part of Dr. ingeniφr Thesis, currently in edition, Narvik Institute of Technology, Narvik, Norway, 1996.

[6] FLOW-3D, Computational modelling power for scientists and engineers, Users Manual, Flow Science, Los Alamos, USA, 1991.

[7] PTC Pro/ENGINEER, Basic design training guide, version 16.0, Users Manual, Parametric Technology Corporation, Waltham, MA, USA, 1996.

[8] F. Baetke, H. Werner, Numerical simulation of turbulent flow over surface mounted obstacles with sharp edges and corners, J. Wind Eng. Ind. Aerodyn. 35 (1990) 129–147.

[9] W. Rodi, Simulation of flow past buidlings with statistical turbulence models, in: J.E. Cermak et al. (Eds.), Wind Climate in Cities, Kluwer Academic Publishers, Dordrecht, 1995.

[10] S. Murakami, A. Mochida, Y. Hayashi, S. Sakamoto, Numerical study on velocity-pressure field and wind forces for bluff bodies by k–ε, ASM and LES, J. Wind Eng. Ind. Aerodyn. 41–44 (1992) 2841–2852.

ELSEVIER

Journal of Wind Engineering
and Industrial Aerodynamics 67&68 (1997) 697–708

JOURNAL OF wind engineering AND industrial aerodynamics

An efficient wind field simulation technique for bridges

W.W. Yang[1], T.Y.P. Chang[2,*], C.C. Chang[3]

Department of Civil and Structural Engineering, Hong Kong University of Science and Technology, Clear Water Bay, Kowloon, Hong Kong, China

Abstract

To perform a time-domain aerodynamic analysis for bridges, it is necessary to develop an efficient simulation scheme which generates wind velocity histories at locations along the bridge span with some given spectral density distributions. The spectral representation method is an unconditionally stable simulation approach. However, it becomes computationally prohibitive to simulate multidimensional processes. A modified spectral representation method is developed to generate the spanwise wind turbulence field for bridges. The method is based on the original spectral representation method under the assumptions that the bridge deck is on the same elevation and the wind field is homogeneous along the spanwise direction of the bridge. It is also assumed that the wind velocities are simulated at locations equally distributed along the bridge span. The current method requires neither Cholesky decomposition nor modal factorization, thus is more efficient than the original spectral representation method when a large number of wind processes are to be simulated. Numerical results show that the current method can save up to 50% of CPU time for one hundred wind velocity processes, when compared to the original spectral representation method using modal factorization with fifty modes.

Keywords: Wind turbulence; Stochastic simulation; Spectral representation; Bridge; Aerodynamic analysis

1. Introduction

Long-span bridges such as cable-stayed or suspension bridges are prone to wind-induced vibration. The aerodynamic performance of these bridges in strong wind regions has been a great concern to both the designers and the analysts [2,3]. The dynamic response of a bridge subjected to stochastic wind loads can be analyzed

* Corresponding author.

[1] Ph.D.

[2] Professor.

[3] Assistant Professor.

PII S0167-6105(97)00111-6

either in the frequency domain [10] or in the time domain [7]. For the frequency domain analysis, under the linear hypothesis, the response spectra of a bridge can be evaluated by numerically integrating the product of the bridge's transfer function and the wind load spectra. This assumption, however, is not valid for most of the long-span bridges, which are relatively flexible and behave nonlinearly due to either geometrical or aerodynamic effect. The nonlinear aerodynamic responses of long-span bridges can, however, be carried out in the time domain by using the step-by-step numerical integration techniques together with the Newton–Raphson iterative scheme. For these analyses, one will require wind load histories simulated at locations across the bridge span. As the wind velocities are correlated in both temporal and spatial domains, it is not trivial to accurately regenerate a large number of wind velocity histories which can match a set of given spectral density distributions.

There exist some simulation techniques which can regenerate wind turbulence based on given wind field spectra, e.g., the spectral representation method proposed by Shinozuka and his associates [11,12,14], and the auto-regressive and moving average (ARMA) methods by, among others, Iwatani [4], Samaras et al. [15], Li and Kareem [8], Mignolet and Spanos [9], Baran and Infield [1]. The spectral representation method is usually employed in conjunction with either the Cholesky decomposition [13] or the modal factorization [14] of a given spectral density matrix as part of the sample function generation procedure. The method can produce unconditionally stable results, nonetheless, it is computationally expensive due to the repetitive decomposition of the spectral density matrix when a large number of wind velocities are to be simulated. On the other hand, the ARMA modeling offers a computationally efficient scheme which does not require large computer storage. The ARMA algorithms, however, require extra attention on the stability of the discrete system associated with the algorithms.

In this study, a modified spectral representation method is developed specifically to generate the spanwise wind turbulence field for bridges. The method is based on the original spectral representation method under the assumptions that the bridge deck is on the same elevation and the wind field is homogeneous along the spanwise direction of the bridge. It is also assumed that the wind velocities are simulated at locations equally distributed along the bridge span. The advantage of the current method is that it does not require either Cholesky decomposition or modal factorization, thus is more efficient than the original spectral representation method when a large number of wind processes are to be simulated. The accuracy and the efficiency of this technique will be demonstrated by simulating the wind velocity field along the deck of a selective bridge.

2. Spectral representation method

The spectral representation method developed by Shinozuka and his associates [11,12] has been quite commonly used to digitally simulate a multivariate and multidimensional processes with a specified cross-spectral density matrix. This method is briefly outlined as follows for the benefit of further development.

Consider a set of m stationary Gaussian random processes $u_i^0(t)$ $(i = 1, 2, \ldots, m)$ with zero means and one-sided target cross-spectral density matrix $\boldsymbol{S}$, where the superscript 0 denotes the target function. It was shown that these multidimensional processes can be simulated by the following equation [12,14]:

$$u_i(t) = \sum_{l=1}^{m} \sum_{k=1}^{N} |H_{il}(\omega_k)| \sqrt{\Delta\omega_k} \cos[\omega_k t + \theta_{il}(\omega_k) + \phi_{lk}] \quad \text{for } i = 1, 2, \ldots, m, \qquad (1)$$

where $\Delta\omega_k$ is the frequency interval at frequency step k $(\omega_k = \omega_{k-1} + \Delta\omega_k)$; N is the number of frequency interval; ϕ_{lk} are the independent random phase angles identically distributed between 0 and 2π with uniform density of $1/2\pi$; and H_{il} is the (i,l) entry of the matrix $\boldsymbol{H}$ which can be determined by Cholesky decomposition [13] or modal factorization [14] on the cross-spectral density matrix $\boldsymbol{S}$. The relation between $\boldsymbol{S}$ and $\boldsymbol{H}$ matrices is

$$\boldsymbol{S}(\omega) = \boldsymbol{H}(\omega)\bar{\boldsymbol{H}}^{\mathrm{T}}(\omega). \qquad (2)$$

The angle θ_{il} is defined as

$$\theta_{il}(\omega) = \tan^{-1}\left[\frac{\mathscr{I}\{H_{il}(\omega)\}}{\mathscr{R}\{H_{il}(\omega)\}}\right] \qquad (3)$$

in which the super bar indicates the complex conjugate and superscript T denotes the matrix transposition; $\mathscr{I}\{\cdot\}$ and $\mathscr{R}\{\cdot\}$ indicate the imaginary and real components, respectively. It is noted from Eq. (1) that the most time-consuming portion of the simulation is at determining the $\boldsymbol{H}$ matrix for N number of frequencies. In the case that a large number of wind velocity processes are to be simulated, the modal factorization using only a limited number of modes may be used as suggested by Shinozuka et al. [14].

3. A modified spectral representation method for bridges

Consider m longitudinal wind turbulence processes, $u_i(t)$ $(i = 1, 2, \ldots, m)$, acting at m equally distributed locations along the spanwise direction of a bridge (see Fig. 1). The spectral density matrix for these m processes can be written as

$$\boldsymbol{S} = \begin{bmatrix} S_{11} & & & & \text{Symm} \\ S_{21} & S_{22} & & & \\ S_{31} & S_{32} & S_{33} & & \\ \vdots & \vdots & \vdots & \ddots & \\ S_{m1} & S_{m2} & S_{m3} & \cdots & S_{mm} \end{bmatrix}. \qquad (4)$$

The Kaimal wind spectrum [6] is used to present the longitudinal wind turbulence characteristics. For a homogeneous wind field along the bridge span, the one-sided auto-spectral density function of the longitudinal turbulence, $u_i(t)$ $(i = 1, 2, \ldots, m)$ can

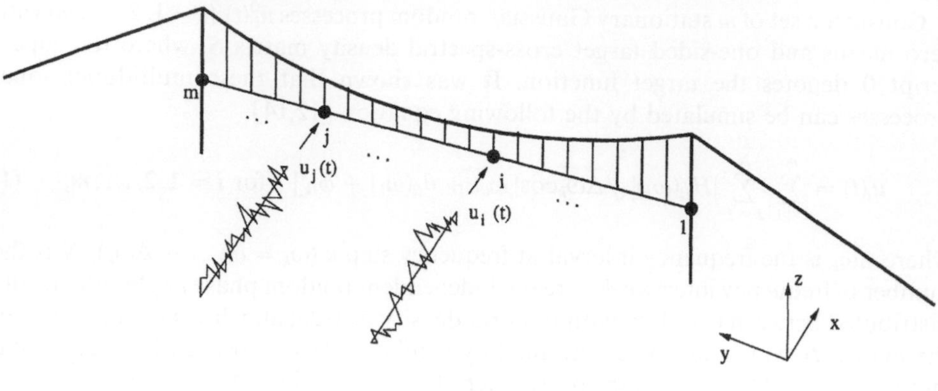

Fig. 1. Spanwise turbulent wind velocities components.

be stated as

$$S_{ii} = \frac{200 f u_*^2}{n(1+50f)^{5/3}} \quad \text{for } i = 1, 2, \ldots, m, \tag{5}$$

where, S_{ii} is the auto-spectrum of longitudinal wind turbulence $u_i(t)$ at location i; $f = nz_i/U_i$ is the non-dimensional Monin coordinate; U_i is the mean wind velocity at altitude z_i; n is the frequency in Hertz; u_* is the shear velocity of the wind flow.

The cross-spectral density function between locations i and j can be defined as [10]

$$S_{ij} = \sqrt{S_{ii}S_{jj}}\exp\left(-\frac{2n\sqrt{C_z^2\ (z_j - z_i)^2 + C_y^2(y_j - y_i)^2}}{U_j + U_i}\right), \tag{6}$$

where S_{ij} is the cross-spectrum between $u_i(t)$ and $u_j(t)$; (y_i, z_i) and (y_j, z_j) are the coordinates of locations i and j, respectively; C_y and C_z are the exponential decay coefficients. The suggested values of C_z and C_y for engineering application are 10 and 16, respectively [10].

Assume that the bridge deck is on a same elevation ($z_i = z_j$) and the wind velocity field is homogeneous along the bridge span ($U_i = U_j$), then

$$S_{ij} = S_{11}\exp\left(-\frac{n\ C_y\ |y_j - y_i|}{U_1}\right). \tag{7}$$

Under the assumption that m wind velocity processes are equally spaced along the bridge span, the cross-spectrum S_{ij} can further be written as

$$S_{ii} = S_{11} \quad \text{for } i = 1, 2, \ldots, m, \tag{8}$$

$$S_{ij} = S_{11}(\cos\alpha)^{j-i} \quad \text{for } j > i, \tag{9}$$

where,

$$\cos\alpha = S_{21}/S_{11}. \tag{10}$$

Substitution of Eqs. (8) and (9) into Eq. (4) yields

$$\boldsymbol{S} = S_{11}\begin{bmatrix} 1 & & & & \text{Symm} \\ \cos\alpha & 1 & & & \\ (\cos\alpha)^2 & \cos\alpha & 1 & & \\ \vdots & \vdots & \vdots & \ddots & \\ (\cos\alpha)^{m-1} & (\cos\alpha)^{m-2} & (\cos\alpha)^{m-3} & \cdots & 1 \end{bmatrix}. \tag{11}$$

A lower triangular real matrix $\boldsymbol{H}$ that satisfies Eq. (2) can be explicitly found as

$$\boldsymbol{H} = \sqrt{S_{11}}\begin{bmatrix} 1 & & & & 0 \\ \cos\alpha & \sin\alpha & & & \\ (\cos\alpha)^2 & \sin\alpha\cos\alpha & \sin\alpha & & \\ \vdots & \vdots & \vdots & \ddots & \\ (\cos\alpha)^{m-1} & \sin\alpha(\cos\alpha)^{m-2} & \sin\alpha(\cos\alpha)^{m-3} & \cdots & \sin\alpha \end{bmatrix}. \tag{12}$$

Substitution of Eq. (12) into Eq. (1) yields

$$u_i(t) = \sum_{k=1}^{N} \sqrt{2\Delta\omega_k S_{11}}\Big[(\cos\alpha)^{i-1}\cos(\omega_k t + \phi_{1k}) + \sin\alpha \sum_{l=2}^{i} (\cos\alpha)^{i-l}\cos(\omega_k t + \phi_{lk})\Big]. \tag{13}$$

When comparing Eq. (13) to Eq. (2), it is seen that Eq. (13) requires less computational effort. The Cholesky decomposition or the modal factorization is no longer needed since the $\boldsymbol{H}$ matrix can be explicitly derived under the current assumptions. To verify the current derivation, the mean, auto- and cross-covariance values of the simulated processes are examined. Based on Eq. (13), the ensemble average of the mean value of process $u_i(t)$ can be found as

$$E[u_i(t)] = \sum_{k=1}^{N} \sqrt{2\Delta\omega_k S_{11}}\,\tfrac{1}{2\pi}\Big[(\cos\alpha)^{i-1}\int_0^{2\pi} \cos(\omega_k t + \phi_1)d\phi_1 + \sin\alpha \sum_{l=2}^{i} (\cos\alpha)^{i-l}\int_0^{2\pi} \cos(\omega_k t + \phi_{lk})\,\mathrm{d}\phi_{lk}\Big] = 0. \tag{14}$$

The ensemble average of the auto-covariance function of process $u_i(t)$ is

$$E[u_i(t)u_i(t+\tau)] = \sum_{k=1}^{N} 2\Delta\omega_k S_{11} \tfrac{1}{2\pi}[(\cos\alpha)^{2i-2} \times \int_0^{2\pi} \cos(\omega_k t + \phi_{1k})\cos(\omega_k t + \omega_k\tau + \phi_{1k}) d\phi_{1k} + (\sin\alpha)^2 \times \sum_{l=2}^{i} (\cos\alpha)^{2i-2l} \int_0^{2\pi} \cos(\omega_k t + \phi_{lk})\cos(\omega_k t + \omega_k\tau + \phi_{lk})\, d\phi_{lk}]$$

$$= \sum_{k=1}^{N} S_{11}\Delta\omega_k[(\cos\alpha)^{2i-2} + (\sin\alpha)^2 \sum_{l=2}^{i} (\cos\alpha)^{2i-2k}]\cos(\omega_k\tau)$$

$$= \sum_{k=1}^{N} S_{11}\Delta\omega_k\cos(\omega_k\tau) \doteq \rho_{ii}(\tau). \tag{15}$$

The ensemble average of cross-covariance function between processes $u_i(t)$ and $u_j(t)(j > i)$ is

$$E[u_i(t)u_j(t+\tau)] = \sum_{k=1}^{N} 2\Delta\omega_k S_{11} \tfrac{1}{2\pi}[(\cos\alpha)^{i+j-2} \times \int_0^{2\pi} \cos(\omega_k t + \phi_{1k})\cos(\omega_k t + \omega_k\tau + \phi_{1k})\, d\phi_{1k} + (\sin\alpha)^2 \times \sum_{l=2}^{i} (\cos\alpha)^{i+j-2l} \int_0^{2\pi} \cos(\omega_k t + \phi_{lk})\cos(\omega_k t + \omega_k\tau + \phi_{lk})\, d\phi_{lk}]$$

$$= \sum_{k=1}^{N} S_{11}\Delta\omega_k[(\cos\alpha)^{i+j-2} + (\sin\alpha)^2 \sum_{l=2}^{i} (\cos\alpha)^{i+j-2l}]\cos(\omega_k\tau)$$

$$= \sum_{k=1}^{N} S_{ij}\Delta\omega_k\cos(\omega_k\tau) \doteq \rho_{ij}(\tau), \tag{16}$$

where $\rho_{ii}(\tau)$ and $\rho_{ij}(\tau)$ are target auto- and cross-covariance functions, respectively.

4. Simulation of the wind field for a bridge

The wind velocity field along the bridge span for a representative bridge is simulated using the proposed scheme. This bridge is a long-span cable-stayed bridge, which is located in a strong typhoon area of Southern China coast. The total length of the bridge is 673 m with a main span of 320 m and two symmetrical side spans of 176.5 m each as shown in Fig. 2. The bridge deck is 23 m above the average sea level. It is assumed that the mean wind speed at the deck level is 62 m/s and the shear velocity u_* of the wind flow is 2.77 m/s.

The auto-spectral density function is plotted in Fig. 3. It is seen that the turbulence energy content of the spectrum distributes mainly over the frequency range below 1 Hz. It is decided to use the cut-off frequency of 2 Hz which is broad enough to contain the wind turbulence energy as well as the first ten vibration modes of the bridge. The sampling time interval is 0.2 s and the number of time steps is 1000. It is

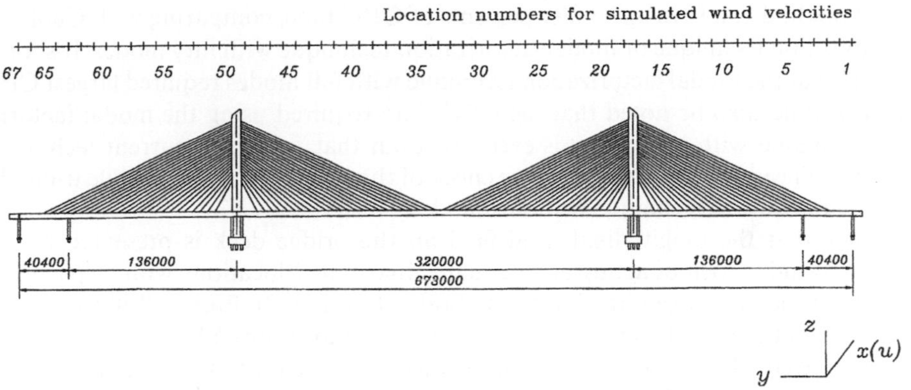

Fig. 2. Configuration of a representative bridge.

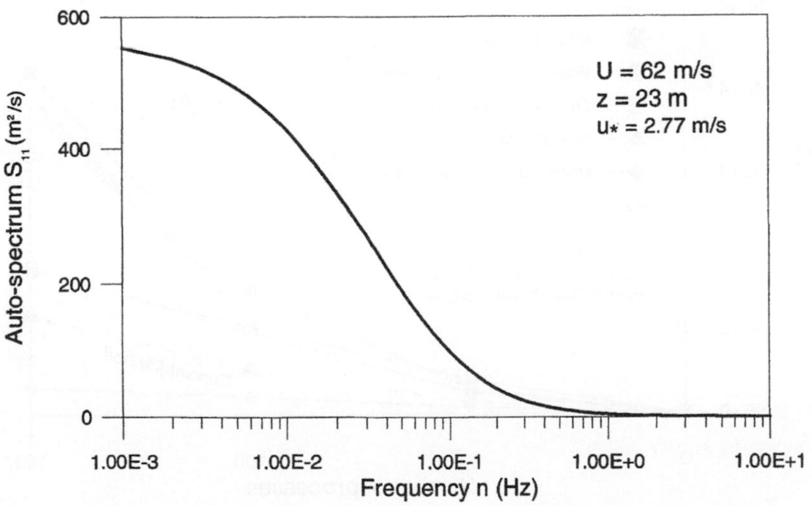

Fig. 3. Auto-spectrum of longitudinal turbulent wind velocities.

noted that the sampling frequency has to be at least twice as big as the highest frequency in the target spectrum in order to prevent the phenomenon of aliasing [5]. It is suggested by Jeffries et al. [5] that to avoid repetition of the simulated time series, the number of frequency steps must be equal to or lager than the number of sampling time steps. In the current study, the number of frequency is assumed to be 1000.

To demonstrate the efficiency of the proposed technique, simulation of turbulent wind velocity components along the bridge spanwise direction is carried out. Fig. 4 depicts the comparison of the CPU time (on SGI Power Indiogo 2) using current technique and the original spectral representation method with the Cholesky decomposition [13] and the modal factorization techniques [14]. It is seen that the current technique is quite efficient in comparison with the original spectral representation method. For example, to simulate one hundred wind velocity processes the current method can save up to fifty percents of CPU time, comparing with Cholesky decomposition technique or modal factorization technique with fifty modes. It should be noted that the modal factorization technique with full modes required largest CPU time. It should also be noted that the CPU time required using the modal factorization technique with ten modes is even less than that using the current technique. However, it needs an extra convergence check of the simulation results while using the modal factorization technique with few modes.

Assume that the longitudinal wind field on the bridge deck is presented by the turbulent wind velocities acting at 67 equally distributed locations with a spacing of 10 m along the spanwise direction of the bridge (see Fig. 2). Figs. 5–7 illustrate one realization of the turbulent wind velocity histories at locations 50 (left tower), 40 and 30, respectively. In order to study the statistic characteristics of simulated wind processes, the ensemble averages of auto- and cross-covariance functions are

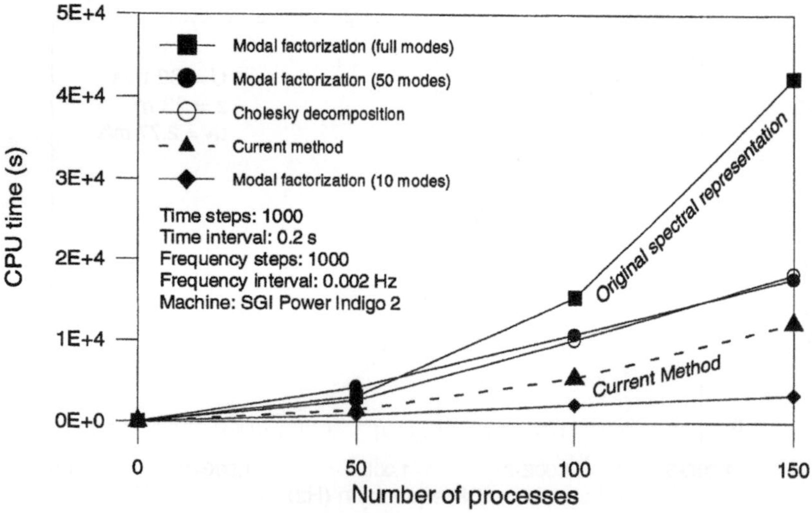

Fig. 4. Comparison of CPU time.

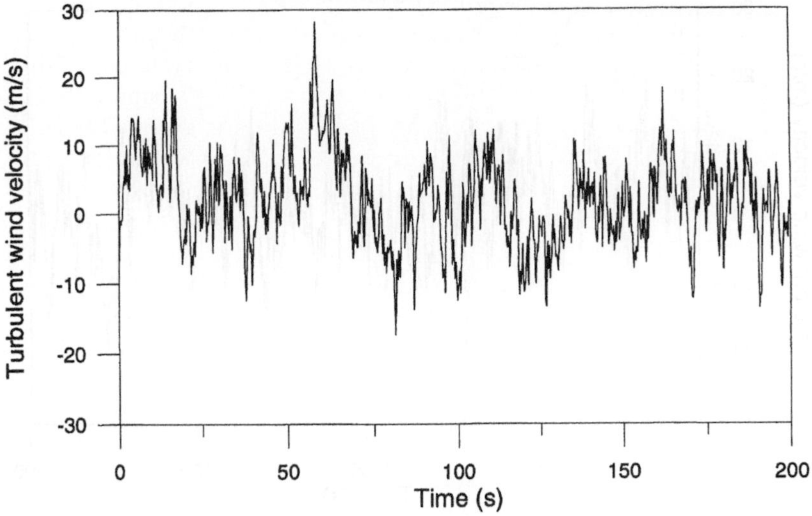

Fig. 5. Simulated turbulent wind velocity at location 50.

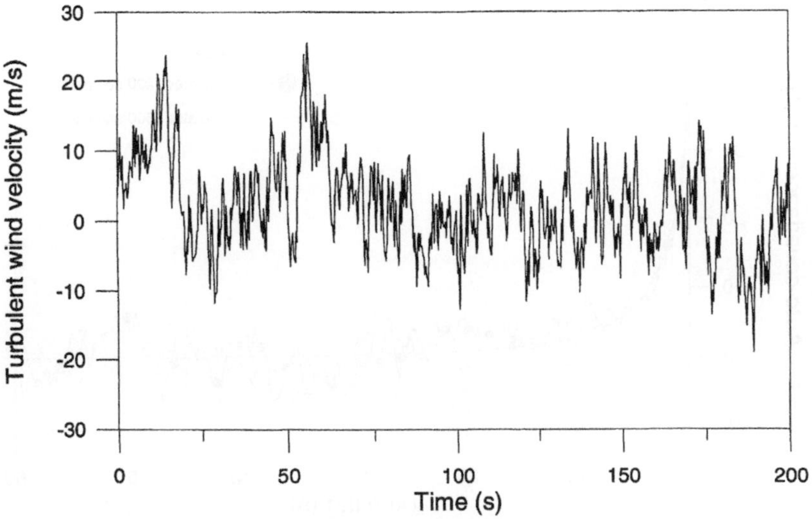

Fig. 6. Simulated turbulent wind velocity at location 40.

calculated using 200 and 500 samples, respectively. The comparison of the simulated and the target auto-covariance functions at locations 50 is presented in Fig. 8. It clearly shows that the simulated auto-covariance function converges quite well to the target auto-covariance function. Some simulated cross-covariance functions between

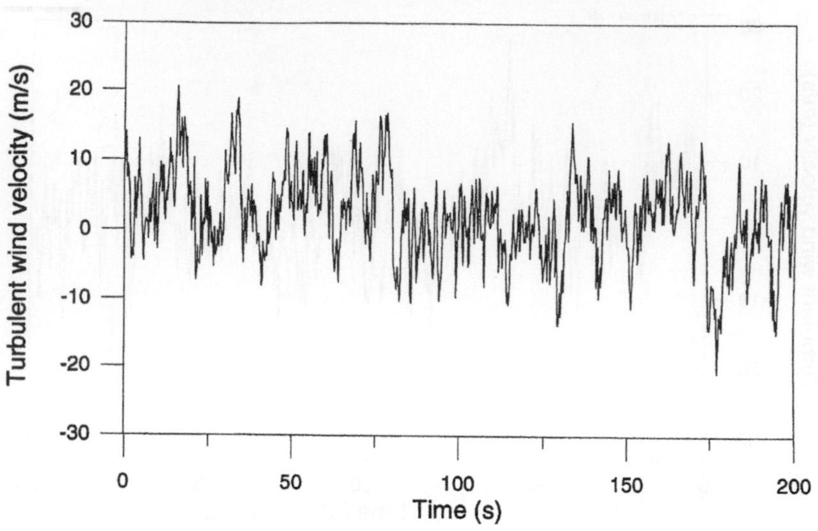

Fig. 7. Simulated turbulent wind velocity at location 30.

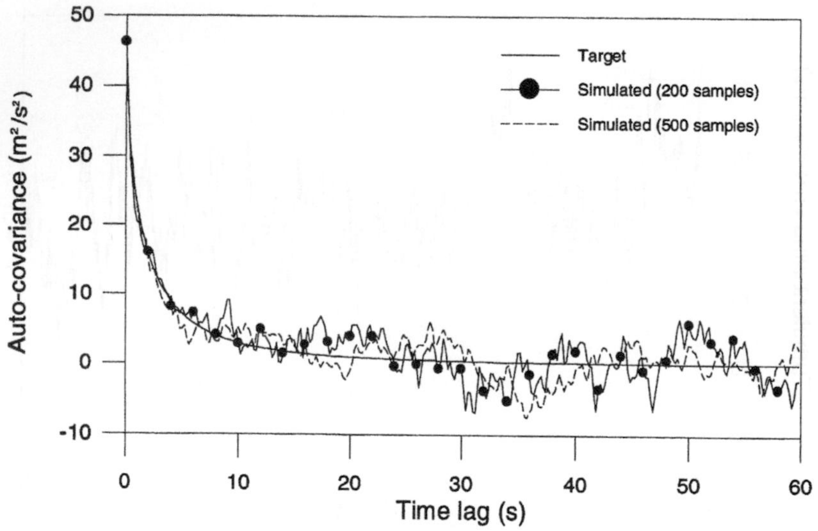

Fig. 8. Comparison of auto-covariance functions at location 50.

different processes at different locations are plotted and compared with the target functions in Figs. 9 and 10. For example, the simulated cross-correlation with $\Delta y = 50\,\text{m}$ (see Fig. 9) and the cross-correlation with $\Delta y = 250\,\text{m}$ (see Fig. 10) converge reasonably well to their respective target functions.

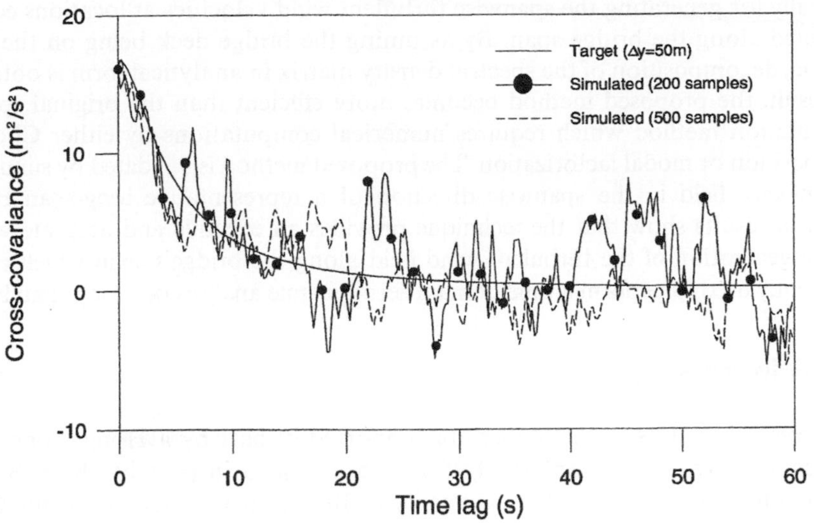

Fig. 9. Comparison of cross-covariance functions between locations 45 and 50.

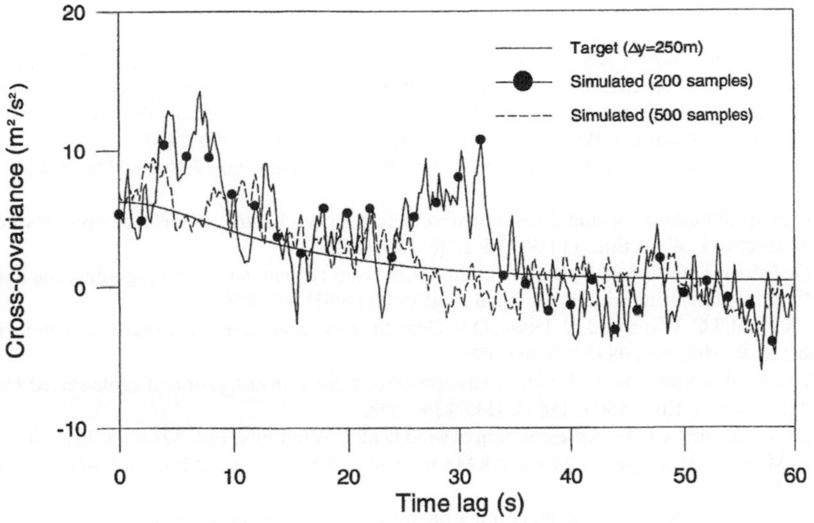

Fig. 10. Comparison of cross-covariance function between locations 25 and 50.

5. Conclusions

To perform a time-domain aerodynamic analysis of long-span bridges situated in a wind-prone area requires a digital regeneration of wind velocity field based on some given auto- and cross-spectral density functions. In this paper, an efficient simulation technique, which is based on the spectral representation method, is proposed to simulate the wind velocity field for bridges. The proposed method is developed specifically for generating the spanwise turbulent wind velocities at locations equally distributed along the bridge span. By assuming the bridge deck being on the same elevation, decomposition of the spectral density matrix in analytical form is obtained. As a result, the proposed method becomes more efficient than the original spectral representation method which requires numerical computations by either Cholesky decomposition or modal factorization. The proposed method is validated by simulating wind velocity field in the spanwise direction of a representative long-span bridge. Numerical results show that the technique provides an effective and accurate way of digital regeneration of the turbulent wind field along the bridge's span which can be conveniently used to perform a time-domain aerodynamic analysis of a long-span bridge.

Acknowledgements

The research results of this paper are supported in part by a Hong Kong UGC Research Infrastructure Grant RI94/95 EG.05 and also in part by the U.S. NSF under a subcontract through the University of Illinois at Chicago (grant no. CMS-9796091).

References

[1] A.J. Baran, D.G. Infield, Simulating atmospheric turbulence by synthetic realization of time series in relation to power spectra, J. Sound Vib. 180 (5) (1995) 627–635.
[2] M. Bocciolone, F. Cheli, A. Curami, A. Zasso, Wind measurements on the Humber bridge and numerical simulations, J. Wind Eng. Ind. Aerodyn. 41–44 (1992) 1393–1404.
[3] A.G. Davenport, The buffeting of a suspension bridge by storm winds, J. Struct. Div. ASCE 88 (1962) 233–268.
[4] Y. Iwatani, Simulation of multidimensional wind fluctuations having any arbitrary power spectra and cross-spectra, J. Wind Eng. 11 (1982) 5–18 (in Japanese).
[5] W.Q. Jeffries, D.G. Infield, J. Manwell, Limitations and recommendations regarding the Shinozuka method for simulating wind data, Wind Eng. 15 (3) (1991) 147–154.
[6] J.C. Kaimal, J.C. Wyngaard, Y. Izumi, O.R. Cote, Spectral characteristics of surface-layer turbulence, Quart. J. R. Met. Soc. 98 (1972) 563–589.
[7] I. Kovacs, H.S. Svensson, E. Jordet, Analytical aerodynamic investigation of cable-stayed Helgeland bridge, J. Struct. Eng. ASCE 118 (1) (1992) 147–168.
[8] Y. Li, A. Kareem, ARMA representation of wind field, J. Wind Eng. Ind. Aerodyn. 36 (1990) 415–427.
[9] M.P. Mignolet, P.D. Spanos, MA to ARMA modeling of wind, J. Wind Eng. Ind. Aerodyn. 36 (1990) 429–438.
[10] E. Simiu, R.H. Scanlan, Wind effects on structures, Wiley, New York, 1978.
[11] M. Shinozuka, Simulation of multivariate and multidimensional random processes, J. Acoust. Soc. Amer. 49 (1971) 357–368.
[12] M. Shinozuka, C.-M. Jan, Digital simulation of random processes and its applications, J. Sound Vib. 25 (1) (1972) 111–128.
[13] M. Shinozuka, Stochastic fields and their digital simulation, Stochastic methods in structural dynamics, G.I. Schueller, M. Shinozuka (Eds.), Martinus Nijhoff Publishers, Boston, 1987, pp. 93–133.
[14] M. Shinozuka, C.B. Yun, H. Seya, Stochastic methods in wind engineering, J. Wind Eng. Ind. Aerodyn. 36 (1990) 829–843.
[15] E. Samaras, M. Shinozuka, A. Tsurui, ARMA representation of random processes, J. Eng. Mech. ASCE 111 (3) (1985) 449–461.

Air Pollution

ELSEVIER

Journal of Wind Engineering
and Industrial Aerodynamics 67&68 (1997) 711–719

JOURNAL OF
wind engineering
AND
industrial
aerodynamics

A resistance approach to analysis of natural ventilation airflow networks

Richard M. Aynsley[1]

Australian Institute of Tropical Architecture, James Cook University, Townsville QLD 4811, Australia

Abstract

Many buildings in warm humid climates, particularly in tropical regions, rely for much of the time on natural ventilation from prevailing breezes for indoor thermal comfort. Much effort in recent years has been directed toward the use of computational fluid dynamics in evaluating airflow through buildings based on the solution of Navier–Stokes equations incorporating a turbulence model. This approach requires extensive data preparation and a reasonably powerful computer to yield results within an acceptable computation time for both numerical solution and simulated flow visualisation. Quantitative evaluation of natural ventilation through many low budget buildings in tropical regions is not evaluated due to a lack of suitable simple computer programs. What is needed are programs that can run on modest personal computers and be used quickly to compare the relative natural ventilation performance of alternative building layouts for prevailing breeze directions during the preliminary design stage. Smaller buildings are often designed for cross ventilation by prevailing breezes with flow entering a windward opening and exhausting through a leeward opening. Such flow through a limited number of openings in series can be calculated very quickly on a personal computer using an orifice flow approach based on estimates of pressure differences and discharge coefficients of openings. When buildings have external ventilating openings in a number of rooms and flow branches within the building, it is no longer possible to calculate directly the airflow in the various branches of the airflow network. Flow in such networks can be analysed iteratively on a personal computer by repetitive solution of simultaneous equations for flow rates in branches at nodes and conservation of mass flow through the network. The procedure described in the paper uses the *Hardy Cross* method of balancing flows at network nodes until errors throughout the network are acceptably small. Sources of data on wind pressure distributions over building walls and shielding influence of nearby buildings are provided together with a detailed description of a procedure for solving network airflows sufficient for readers to write their own computer code.

Keywords: Networks; Airflow; Natural; Ventilation; Resistance

[1] E-mail: richard.aynsley@jcu.edu.au.

PII S0167-6105(97)00112-8

1. Introduction

Significant advances have been made over the past decade in computational methods for estimating airflow through buildings. Recent efforts have focused mainly on computational fluid dynamics based on solution of the Navier–Stokes equations [1,2]. Three-dimensional modelling using this technique can provide detailed information on indoor airflows but requires costly software and considerable computation. Such a technique is unlikely to be used in the design of the numerous low-cost buildings built in developing countries in humid tropical regions where natural ventilation is often the only means of achieving indoor thermal comfort. The low-speed indoor airflow in lazy eddy regions away from the main air streams is usually of little value for indoor thermal comfort. Simpler computational methods that restrict airflow information to the main air stream through large openings are generally adequate for basic design purposes [3]. These methods make use of pressure differences and discharge coefficients for simple cross ventilation and network flow analysis where multiple inlets and outlets and internal flow branching occurs [4,5]. These computational techniques can be performed on modest personal computers using public-domain software developed at National Institute of Standards and Technology by Walton [6].

2. Shielding influence of adjacent buildings

Most wind pressure distribution data on building shapes is for buildings without nearby obstructions. In reality there often are buildings of a similar size within six building heights of the building of interest. The shielding effects of these adjacent buildings on wind pressure differences between windward and leeward walls of the shielded building have been studied by Lee et al. [7].

These studies suggest that for cuboid buildings of similar size arranged in a regular grid staggered grid patterns there are three distinct flow regimes, skimming, wake interference and isolated roughness regimes.

For cuboid buildings in a regular grid a skimming flow dominates when clear spacing between buildings is less than 1.4 building heights. This results in wind pressure differences between windward and leeward walls of the shielded building being reduced virtually to zero at very close spaces rising to approximately 25% of the corresponding pressure difference on an isolated building.

For cuboid buildings in a regular grid a wake interference flow dominates when clear spacing between buildings is between 1.4 and 2.6 building heights. This results in wind pressure differences between the windward and leeward walls of the shielded building being reduced from 25% to 50% of the corresponding pressure difference on an isolated building.

For cuboid buildings in a regular grid an isolated roughness flow regime dominates when clear spacing between buildings exceeds 2.6 building heights. This results in wind pressure differences between the windward and leeward walls of the shielded building being reduced from 50% of the corresponding pressure difference on an isolated building at a spacing of 2.6 building heights rising to around 100% when clear spacings approach six building heights.

For cuboid buildings arranged in a staggered grid a skimming flow dominates when clear spacing between buildings is less than 1.4 building heights. This results in wind pressure differences between the windward and leeward walls of the shielded building being reduced virtually to zero at very close spaces rising to approximately 12% of the corresponding pressure difference on an isolated building.

For cuboid buildings arranged in a staggered grid a wake interference flow dominates when clear spacing between buildings is between 1.4 and 2.6 building heights. This results in wind pressure differences between the windward and leeward walls of the shielded building being reduced from 12% to 33% of the corresponding pressure difference on an isolated building.

For cuboid buildings arranged in a staggered grid, an isolated roughness flow regime dominates when clear spacing between buildings exceeds 2.6 building heights. This results in wind pressure differences between the windward and leeward walls of the shielded building being reduced from 33% of the corresponding pressure difference on an isolated building at a spacing of 2.6 building heights rising to around 100% when clear spacings approach 7.5 building heights.

The limitless permutations and combinations of possible building shapes and spacings preclude the likelihood of definitive data being available for specific building shapes and shielding situations. Short of conducting a boundary layer wind tunnel study for a specific shielding situation, the data above for arrays of cuboids will give an indication of the type of effects to be expected.

3. Discharge coefficients

A simple means for estimating the volumetric turbulent flow rate or discharge through an opening in a pipe due to a nominated pressure difference (head loss) is to apply the Bernoulli equation to points along a streamline upstream and downstream of the opening. As this method cannot accommodate the complex fluid dynamics of separated flow downstream of the opening an empirical correction factor or discharge coefficient is used to obtain a realistic estimate [8].

The volumetric airflow rate, Q (m^3/s) through an opening with a free area A (m^2) and a discharge coefficient, C_d (dimensionless) and usually taken as 0.65 for sharp edged rectangular openings, as the head loss is H_L (Pa), is

$$Q = C_d A (H_L)^{0.5} \quad (m^3/s). \tag{1}$$

The mean velocity, V (m/s) through an opening with turbulent flow can be calculated by dividing the volumetric flow rate by the area of the opening A (m^2):

$$V = C_d (H_L)^{0.5} \quad (m/s). \tag{2}$$

Head loss, the difference between the upstream total pressure and the downstream static pressure, is well defined in the case of orifice flow in pipes. Specific locations are defined for pressure measurement upstream and downstream of the orifice plate. In the case of flow through wall openings in buildings the equivalent head loss would be between the total pressure (dynamic + static) at the windward opening and the static pressure near the wall beside the leeward opening. The pressure energy in the form of dynamic pressure in the air jet issuing from leeward opening does not contribute to the head loss between windward and leeward wall openings and is dissipated downstream from the building.

4. Resistance approach

Many of the airflows of interest to wind engineers are external flows in the Earth's turbulent boundary layers or around solid objects. This paper focuses on airflow through openings in building envelopes or through the interior of buildings. Discharge coefficients which reflect the discharge efficiency of openings are commonly used to estimate airflow through openings by wind engineers [9–11]. In simple situations, such as cross ventilation where there is only one windward opening and one leeward opening, this approach using discharge coefficients is satisfactory.

When more complex airflows with multiple inlet and outlet openings and internal flows occur through a network of alternate branching flow paths, electrical circuit analogies can be used. Analysis of complex airflow networks using computer software based on electrical circuit analogies such as Kirchhoff's first and second laws and Atkinson's equation is commonplace in mine ventilation engineering [4,12–15].

Atkinson's equation relates, H_L, the head (pressure) losses (Pa) in an airway proportionally to the square of the discharge, Q (m^3/s), through the airway with the constant of proportionality being the resistance, R ($N\,s^2/m^8$) of the airway:

$$H_L = RQ^2 \quad \text{(Pa)}. \tag{3}$$

Kirchhoff's first law for air circuits states the quantity of air leaving a junction must equal the quantity of air entering the junction. Kirchhoff's second law states that the sum of pressure drops around any closed path must be equal to zero. Pressure differences or head losses are analogous to voltage, electrical current is analogous to volumetric airflow rate and electrical resistance is analogous to airflow resistance [12]. This approach provides a useful framework when developing computer software packages for calculating airflow through complex networks.

Airflow resistance of wall openings can be expressed in terms of their discharge coefficients C_d, the mass density of air, ρ (usually 1.2 Kg/m^3), and the area of the opening, A (m^2).

$$R = (\rho/2)/(C_d^2 A^2) \quad (N\,s^2/m^8). \tag{4}$$

The volumetric turbulent airflow rate Q through an opening is

$$Q = (H_L/R)^{0.5} \quad (m^3/s). \tag{5}$$

The mean velocity, V (m/s) through an opening can be calculated by dividing the volumetric flow rate by the area of the opening A (m^2):

$$V = (H_L/R)^{0.5}/A \quad \text{(m/s)}. \tag{6}$$

When indoor airflow passes through a number of sequential openings with resistances $R_1, R_2, \ldots, R_n$, the equivalent resistance R_{eq} for all the openings in series can be calculated using the equation:

$$R_{eq} = R_1 + R_2 + \cdots + R_n \quad (\text{N s}^2/\text{m}^8). \tag{7}$$

When indoor airflow passes through a number of openings with resistances $R_1, R_2, \ldots, R_n$ in parallel, the equivalent resistance R_{eq} for all the openings can be calculated using the equation:

$$1/R_{eq} = 1/R_1 + 1/R_2 + \cdots + 1/R_n \quad (\text{m}^8/\text{N s}^2). \tag{8}$$

5. Estimating head loss from wind pressure distributions on external walls

The total pressure at a windward wall opening is difficult to define as flow is not contained as is the case in pipe flow for which discharge coefficients were developed. Static pressure at the leeward wall of buildings presents less of a problem. For practical reasons, estimates of head loss for natural ventilation usually are based on pressure distributions measured on isolated solid building models in boundary layer wind tunnels. There are also data on mean wind pressure coefficients of walls of high and low-rise rectangular buildings derived from numerous wind tunnel studies [16,17]. The term isolated indicates that the model on which the wind pressures on surfaces are measured has no other building models nearby, that is within approximately six building heights. Pressure coefficients on the surfaces of isolated solid models near the location of proposed inlet and outlet wall openings are used to estimate head loss.

$$H_L = 0.5\,\rho\,V^2(Cp_i - Cp_o) \quad \text{(Pa)} \tag{9}$$

where H_L is the estimate of head loss between inlet and outlet ventilation openings (Pa), ρ the mass density of air (1.2 Kg/m^3), V the mean approach wind speed at reference height associated with pressure coefficients, Cp_i the wind pressure coefficient on wall of a solid model near the location for the inlet opening, Cp_o the wind pressure coefficient on wall of a solid model near the location for the outlet opening.

Head at the inlet opening, H_i and outlet opening, H_o are

$$H_i = 0.5\,\rho\,V^2\,Cp_i \quad \text{(Pa)}, \tag{10}$$

$$H_o = 0.5\,\rho\,V^2\,Cp_o \quad \text{(Pa)}. \tag{11}$$

6. Complex network flows

Flows in networks with parallel branches can be analysed directly using Eqs. (1)–(8). When parallel branches overlap, or are interconnected, a complex network is created in which flows cannot be calculated directly and iterative approximation methods are employed.

A nomenclature has been established to describe complex networks. A complex network consists of *branches* which are segments of airflow between *nodes*. *Nodes* are locations where branch flows merge. Nodes where more than 2 branch flows merge are referred to as *junctions*. Complex networks are described by assigning unique numbers to each node. Branches are identified by the node numbers at each end. Flow direction in a branch is defined by using a strict order of node numbers for the branch [4]. Any node with only two branches are eliminated by converting the connected branches into a single equivalent branch with an equivalent resistance. Any sections of the network with parallel branches in parallel between the common nodes is converted to a single equivalent branch with an equivalent resistance. When this has been completed the network should conform to the following equation:

$$m_n = n_b - n_n + 1 \tag{12}$$

where m_n is the number of fundamental meshes in the network, n_b the number of branches in the network, n_n the number of nodes in the network.

A fundamental mesh consists of a unique series of interconnected branches completing a closed circuit which contain at least one of the nodes associated with a known head. One of these fundamental meshes will have the least resistance and incorporate the circuit between the two nodes for which the heads are known. To assist in the identification of this mesh it is useful to sort the branches in order of their resistances.

7. A procedure for solving complex network flows

Calculation of complex network flows is achieved by an iterative approach of successive approximations of network variables. Each iteration must conform with Kirchhoff's laws of conservation of mass and flow and balancing of heads. A variety of techniques can be used to accelerate convergence toward a solution with an acceptably small error.

One method used to perform this iteration is the Hardy Cross method of balancing flows. It is equivalent to Newton's method of tangents as it uses the derivative of an estimated flow to adjust the next iteration. In the case of natural ventilation, the knowns in the network are the estimated wind pressures at inlets and outlets and the resistance of branches. The unknowns in the network are the flow rates in each branch and the head losses throughout the network.

The Hardy Cross method keeps heads balanced in the network and balances flows by successive corrections. Initially unknown flows are assigned an arbitrary value.

Pressures at inlet and outlet nodes are then calculated from the velocity of the reference design wind speed and pressure coefficients at the inlet and outlet openings.

With the head loss established between inlet and outlet openings, flows are balanced at all junctions through the mesh with the least resistance followed by the other meshes in the network [4]. Flow rates in each branch are calculated using Eq. (5). These flows are summed at each junction assuming:

- flow into the junction is negative and
- flow out of the junction is positive.

Since Kirchhoff's first law requires the sum of flow at a junction to be zero, an equal and opposite flow equal to the sum of the flow at each junction is distributed between the branches at the junction using the following equation:

$$\Delta Q_i = -\frac{\sum_{j=1}^{n} R_j |Q_j|^2}{2\sum_{j=1}^{n} R_j Q_j} \quad (\mathrm{m^3/s}) \tag{13}$$

where ΔQ_i is the increment of flow required in each branch to balance total flow at n junctions in mesh i ($\mathrm{m^3/s}$), R_n the resistance of branch n ($\mathrm{N\,s^2/m^8}$), Q_n the current estimate of flow in branch n ($\mathrm{m^3/s}$).

The term $2R_j\,Q_j$ is the derivative of the term for total head loss along a branch, $R_j\,Q_j^2$. All terms between the symbols | | are absolute values. This requires a strict adherence to the sign of the flow at a junction when summing terms.

When summing the total head in a mesh, which includes the head loss due the wind pressure difference between inlet and outlet openings, the head loss due to this pressure difference must be deducted from the numerator in Eq. (13) to satisfy Kirchhoff's first law.

When flows have been balanced at a junction, flows at each end of each branch are summed. Since these net flows should also be zero, out of balance flow is balanced with an equal and opposite flow distributed equally between each end of the branch. There are exceptions to this procedure in the case of ends of branches which are inlets or outlets which are treated as infinite sources or sinks and the entire branch flow imbalance is absorbed.

When an iteration of balancing of flow in branches is completed, the net flows at each of the junctions is once again out of balance. The balancing of flows at junctions and in branches is repeated in the iterative process until the flow rate being distributed has reached a selected negligible amount. At this point of the computation, head losses for each branch are calculated using Eq. (3) and the analysis is complete [4].

There are a number of other methods used to perform iterations in solving network flows. Each method has different convergence characteristics for various types of networks [15]. Some networks, particularly those with both very high and low resistance branches can cause slow convergence toward a solution if balancing of flows at junctions is performed in a random fashion. To avoid slow convergence balancing of flows is performed by starting with junctions in the fundamental mesh with the lowest circuit resistance followed by junctions in the mesh with the next lower resistance. When matrix methods are used to solve the flows in networks, storage minimisation procedures such as skyline, or sparse matrix banding methods are often applied.

8. Conclusions

During the early design stage of low-cost naturally ventilated buildings in warm humid tropical regions, simple computer software for estimating natural ventilation in network flows can be extremely valuable for comparing the relative natural ventilation potential of alternative designs during the preliminary design stage. At this stage of building design, when many building features are in a state of flux, simple user friendly software is needed to accommodate numerous runs to allow exploration of design options when limited data is available.

The sophisticated three dimensional computational fluid dynamics software based on Navier–Stokes equations with turbulence modelling is often costly, requires users to have a good background in fluid dynamics and usually has tedious data entry procedures which are disincentives to their use.

Other simpler fluid modelling approaches such as orifice flow and network flow analysis do not offer as much detailed information on airflow outside the main air streams. These lazy eddy flows typically have mean velocities around 10% of the mean velocity in the main air stream and are of limited interest when the focus is on airflow for indoor thermal comfort. Average airflow due to wind and thermal forces in the main air stream near large wall or roof openings are adequate for these purposes.

A resistance approach to airflow in networks using electric circuit analogies can provide a useful framework when developing computer software packages for estimating airflow in complex indoor airflow networks. For such applications, computer software needs to be low cost, user friendly with simple data entry, fast in execution and suitable to run on commonly used personal computers. Public domain software such as CONTAM developed by Walton [6] at the National Institute of Standards and Technology in Gaithersburg contains AIRNET modules for solving complex network air flows.

More sophisticated fluids modelling software based on Navier–Stokes equations may be used for natural ventilation applications for estimating indoor thermal comfort in low cost buildings, but it is likely to be at a later design stage when the building design is approaching its final stage and the detailed configuration of the building is better established.

References

[1] J. Tsutsumi et al., Numerical simulation of cross-ventilation in a single-unit house, International Symposium on Room Air Convection and Ventilation Effectiveness, Tokyo, ASHRAE, 1993, pp. 447–451.

[2] S. Iwamoto et al., Numerical prediction of indoor airflow by cross-ventilation, international symposium on room air convection and ventilation effectiveness, Tokyo, ASHRAE, 1993, pp. 453–456.
[3] T.A. Reinhold (Ed.), Wind Tunnel Modeling for Civil Engineering Applications, Cambridge University Press, Cambridge, 1982, pp. 465–485.
[4] H. Cross, Analysis of flow in networks of conduits or conductors, Urbana. Univ. Ill. Bull. XXXIV (22) (1936) 29.
[5] E.A. Arens, N.S. Watanabe, A method for designing naturally cooled buildings using bin climate data, ASHRAE Trans. 92 (2B) (1986) 773–792.
[6] G.N. Walton, AIRNET – a computer program for building airflow network modeling, NISTIR 89-4072, National Institute of Standards and Technology, Gaithersburg, MD, 1989.
[7] ASHRAE Handbook of Fundamentals 1993, Ch. 23, Infiltration and Ventilation, ASHRAE, Tullie Circle, Atlanta, Ga.
[8] Ventilated Buildings Part 1: Ventilation model, Building Environ. 29 (4) (1994) 461–471.
[9] E.H. Mathews, P.G. Rousseau, A new integrated design tool for naturally ventilated buildings part 1: Ventilation model, Building Environ. 29 (4) (1994) 461–471.
[10] P.G. Rousseau, E.M. Mathews, A new integrated design tool for naturally ventilated buildings part 2: Integration and application, Building Environ. 29 (1994) 473–484.
[11] W.N. Shaw, Air currents and the laws of ventilation, Cambridge University Press, Cambridge, 1907, p. 94.
[12] H.L. Hartman (Ed.), Mine Ventilation and Air Conditioning, 2nd ed., Wiley-Interscience, New York, 1982, pp. 492–498.
[13] R.M. Aynsley, A resistance approach to estimating airflow through buildings with large openings due to wind, ASHRAE Trans. 94 (2) (1988) 1661–1669.
[14] G.N. Walton, Airflow network models for element-based building airflow modelling, ASHRAE Trans. 95 (2) (1989) 611–620.
[15] M.V. Swami, S. Chandra, Correlations for pressure distributions on buildings and calculation of natural ventilation, ASHRAE Trans. 90 (1) (1988) 243–266.
[16] ASHRAE handbook of fundamentals 1993, Ch. 14, Airflow Around Buildings, ASHRAE, Tullie Circle, Atlanta, Ga.
[17] B.E. Lee, M. Hussain, B. Soliman, Predicting natural ventilation forces upon low-rise buildings, ASHRAE J. (1980) 35–39.

ELSEVIER

Journal of Wind Engineering
and Industrial Aerodynamics 67&68 (1997) 721–732

JOURNAL OF wind engineering AND industrial aerodynamics

Numerical and wind tunnel simulation of gas dispersion around a rectangular building

D. Delaunay[a,*], D. Lakehal[b], C. Barré[a], C. Sacré[a]

[a] *Centre Scientifique et Technique du Bâtiment, 11 rue Henri Picherit, BP 82341, 44323 Nantes Cedex 3, France*
[b] *Institute for Hydromechanics, University of Karlsruhe, Kaiserstrasse 12, 76128 Karlsruhe, Germany*

Abstract

Some wind tunnel investigations of gas dispersion around a rectangular building placed in a simulated atmospheric boundary layer have been conducted. Numerical simulations of these experiments have been performed by solving the Reynolds-averaged Navier–Stokes equations, combined with a Reynolds-stress turbulence model, and two variants of the two-layer model due to Rodi. It appears that only the second moment closure correctly predict the recirculating zones on the faces. In this case, calculated values of gas concentrations on the building model faces agree generally well with measurements.

1. Introduction – the wind tunnel experiment

New developments in computational fluid dynamics now allow to foresee numerical estimations of wind flow and pollutant dispersion around buildings. Naturally, there is a need for confrontation with experimental results. Then, in the present study, numerical and wind tunnel simulations of the flow and diffusion over a building placed in an atmospheric boundary layer were conducted. The case of diffusion around a rectangular building with a perpendicular approach flow has been chosen, in order to allow computations with fine Cartesian grids and to minimize numerical diffusion.

The building is reproduced at a 1/125 scale in the CSTB atmospheric wind tunnel (8 m^2 working section). The model height H equals 0.144 m. The facades and the roof are smooth. The incident flow (longitudinal direction) is perpendicular to the largest side, which has a length of $3H$ (lateral direction). The length of the shorter side is H. A mixture of air and ethane is released out of six chimneys of 4 mm diameter and 96 mm height. The value of the ejection velocity W, as measured by a hot-wire anemometer is 2 m/s.

* Corresponding author. E-mail: delaunay@cstb.fr.

PII S0167-6105(97)00113-X

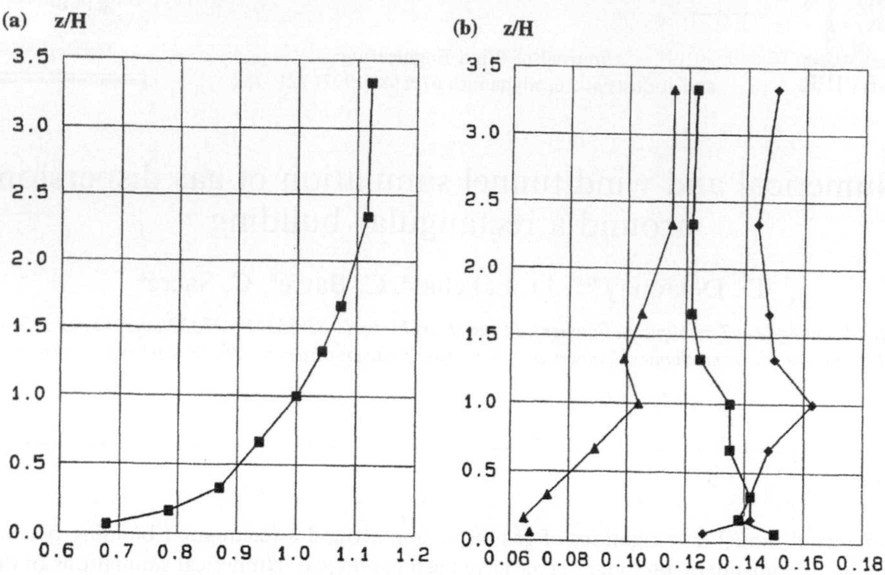

Fig. 1. Measured inlet profiles: (a) mean wind normalized by U (mean velocity at height H); (b) rms profiles normalized by U (◆: $\sqrt{\overline{u^2}}/U$; ■: $\sqrt{\overline{v^2}}/U$; Δ: $\sqrt{\overline{w^2}}/U$).

The mean and rms profiles of the incident velocity are measured 1 m upstream of the model (Fig. 1). Three values of the wind speed U at height H are tested: 1, 2 (reference case), or 4 m/s, corresponding to a W/U ratio of 0.5, 1 and 2, respectively, and to Reynolds numbers (based on the model height) ranging between 10 000 and 40 000.

Concentration measurements were performed by means of flame ionisation detectors [1], connected to 40 taps distributed on the façades and on the roof. The concentrations are averaged over 1 min. The precision of the concentration measurements is $\pm 10\%$.

Other configurations were simulated in the wind tunnel as well. They differ by their incident flow directions, building properties (aspect ratios, roof parapets), the presence of another nearby building, wall roughness, chimney heights. These cases are not presented here but they are described in Ref. [2] and could be used for validation of numerical models.

2. Numerical methods

2.1. Turbulence modelling

2.1.1. Reynolds-stress model

The numerical simulations were performed by solving the Reynolds-averaged Navier–Stokes equations, combined with the Reynolds-stress turbulence model of

Launder et al. (LRR model) [3]. In the Reynolds-stress transport equations, the pressure–strain correlations are estimated by the Isotropisation of Production Model [3], including the wall reflection term described by Gibson and Launder [4]. For reasons of coding, this wall reflection term was applied for ground but not for the building facades. We found that this approximation does not yield significative differences in this case. The diffusive transport is modelled by using the simple-gradient diffusion hypothesis of the Daly–Harlow model [5], and the viscous destruction is modelled by assuming local isotropy, solving the transport equation for the turbulent energy dissipation rate ε [3]. In the concentration equation, the turbulent fluxes are estimated by the simple-gradient diffusion hypothesis with a Schmidt number $S_c = 0.9$.

2.1.2. The two-layer turbulence model

The Rodi two-layer approach [6], which combines the standard k–ε model [7] in the outer region with a one-equation model in the viscosity-affected near-wall region has been tested for the reference case, as well as a variant of this model consisting to replace the standard k–ε model by the Kato–Launder model [8].

The k–ε model uses the eddy-viscosity concept which links the Reynolds stresses to the turbulent kinetic energy and the gradients of mean velocities via the Boussinesq approximation. The eddy viscosity is a function of k, the turbulent kinetic energy and ε, the dissipation rate of k. These two variables are solved by a transport equation. The isotropic eddy viscosity concept leads to an unrealistically high production of k in stagnation regions occurring in impinging flows.

The Kato–Launder model replaces the original production term in the k-equation by an expression which adds the vorticity invariant to the strain invariant so that this production term turns to zero in stagnation regions.

In the one-equation model, used in the near-wall region, the eddy viscosity is made proportional to a length scale prescribed algebraically and a velocity scale determined by solving the k-equation as in the standard model. The dissipation rate is related to k and a dissipation length scale which is also prescribed algebraically. A finer grid is therefore required very close to walls in order that the first point lies at a normalized distance to wall inferior to 5 (the normalized wall distance is defined as $y^+ = yu_*/\nu$, where y is the distance to the wall, u_* the friction velocity and ν the kinematic viscosity). A typical viscous-affected layer contains at least 15 points. The two zones are matched dynamically at a location where the viscous effects become negligible.

This approach was used in the region covering the close vicinity of the obstacle, as its walls are smooth. However, an additional treatment was applied to the floor in order to take into account its surface roughness. In here, the outer region was matched to the roughness layer at a location where the ratio of the turbulent to laminar viscosity is high enough.

2.2. Numerical resolution

The computations using the LRR Reynolds stress model were performed at CSTB, while the computations involving the two-layer model were carried out at the

Institute for Hydrodynamics of Karlsruhe. So there are two kinds of numerical resolution.

2.2.1. CSTB computations

The set of equations is solved using the finite-volume code PHOENICS [9]. The diffusion fluxes at the cell faces are estimated by a centered approximation while the third-order upwind boundedness SMART scheme [10] is adopted for the advective fluxes. The pressure/velocity linkage is solved via the SIMPLEST algorithm [11]. Scalars and velocity components are computed on staggered grids. A linear under-relaxation is applied to the pressure during the iterative resolution, while a local false time step method [12] is used for the other variables.

Since the experiment is symmetric in the vertical median section of the building model, the computations cover only half the domain. The $93 \times 57 \times 44$ (longitudinal/lateral/vertical) Cartesian grid represents a domain lying from $5H$ upstream to $14H$ downstream of the model. The height and the lateral size of the computation domain are, respectively, $6.5H$ and $6H$. In the vicinity of the walls, the mesh size equals 4 mm. The height of the cells in the vicinity of the roof is 3.2 mm. Just downstream of the leading edge of the building, the length of the cells is only 1 mm in order to catch the high gradients of velocity (Fig. 2). The circular chimneys are approximated by square cylinders having the equivalent section area.

2.2.2. Institute for hydrodynamics computations

The program FAST-3D [13], used in this case, is based on a finite-volume approach, with non-staggered, cell-centered grid arrangement, using the momentum

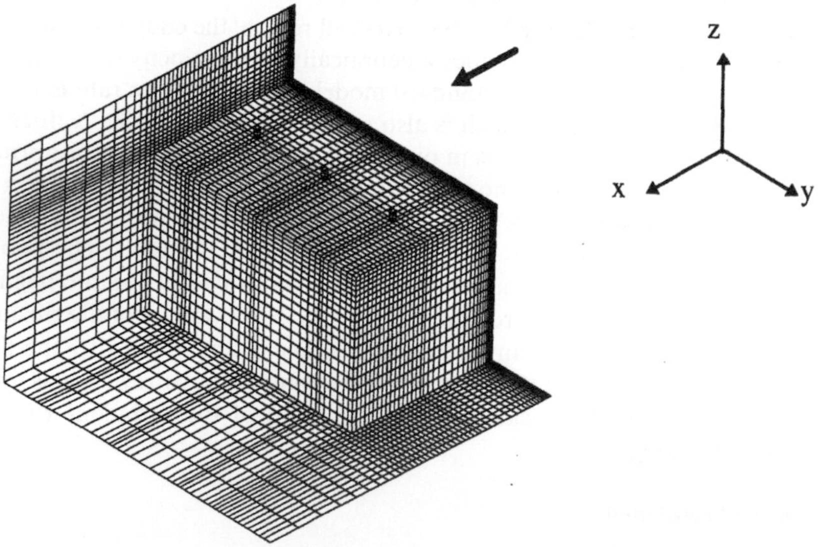

Fig. 2. The $97 \times 57 \times 44$ grid in the vicinity of the building model (LRR computations).

interpolation technique due to Rhie and Chow [14]. The pressure–velocity coupling is achieved with the SIMPLE algorithm. The diffusive fluxes are approximated by central differences while the HPLA second-order low-diffusive and oscillation-free scheme of Zhu [15] is applied for the convective part.

A fine mesh consisting of $146 \times 82 \times 88$ grid points (longitudinal/lateral/vertical) is applied to form a computational domain with an upstream length of $5.5H$, a downstream length of $10H$, a width of $9H$, and a height of $10H$. The grid covers half the domain taking into account the symmetry condition. The size of the smallest cell is $0.005H$.

2.3. Boundary conditions

The approach flow profile at the upstream boundary is specified to fit at best the wind tunnel measurements. The mean wind follows a logarithmic profile with roughness length equal to 0.03 mm with the velocity U at height H equal to 1, 2, or 4 m/s. The Reynolds stresses are constant in height and given by

$$\overline{u^2} = (0.14U)^2, \quad \overline{v^2} = (0.13U)^2, \quad \overline{w^2} = (0.095U)^2, \quad \overline{uw} = U_*^2, \quad \overline{uv} = \overline{vw} = 0,$$

where u, v, w are the turbulent fluctuations of the longitudinal, lateral and vertical velocities, and U_* the friction velocity deduced from the logarithmic profile. The rate of dissipation ε is computed following the boundary-layer approximation:

$$\varepsilon = U_*^3/\kappa z \quad \text{where } z \text{ is the height and } \kappa \text{ the von Karman constant.}$$

At the chimney outlets, the vertical velocity of 2 m/s is imposed as well as the corresponding mass flux. At walls and on the ground, the non-equilibrium smooth wall function of Launder and Spalding [7] is applied with the LRR model, and the laminar wall function is used with the two-layer model.

At the outlet sides and at the upper boundary, zero-gradient conditions are applied. At the symmetry plane, the lateral fluxes are set to zero, as well as the shear stresses involving the lateral fluctuations.

3. Results

3.1. The reference case – LRR model

The predicted wind field around the building in the vertical symmetry plane is shown in Fig. 3a. We can observe a well-developed vortex in front of the model, with a stagnation point at about half the model height. This vortex is due to the shear of the approach flow near the floor. The flow separates on the windward corner and does not reattach, inducing a reversed flow over the whole roof length. These computed features agree well with the experimental visualization. In the wake of the building, the recirculation zone extends up to a distance of about $5.5H$. Near the lateral edge of the roof, near the building side, the strength of the reversed flow decreases and on the sides the flow turns downward forming a counter-rotating vortex (Fig. 4).

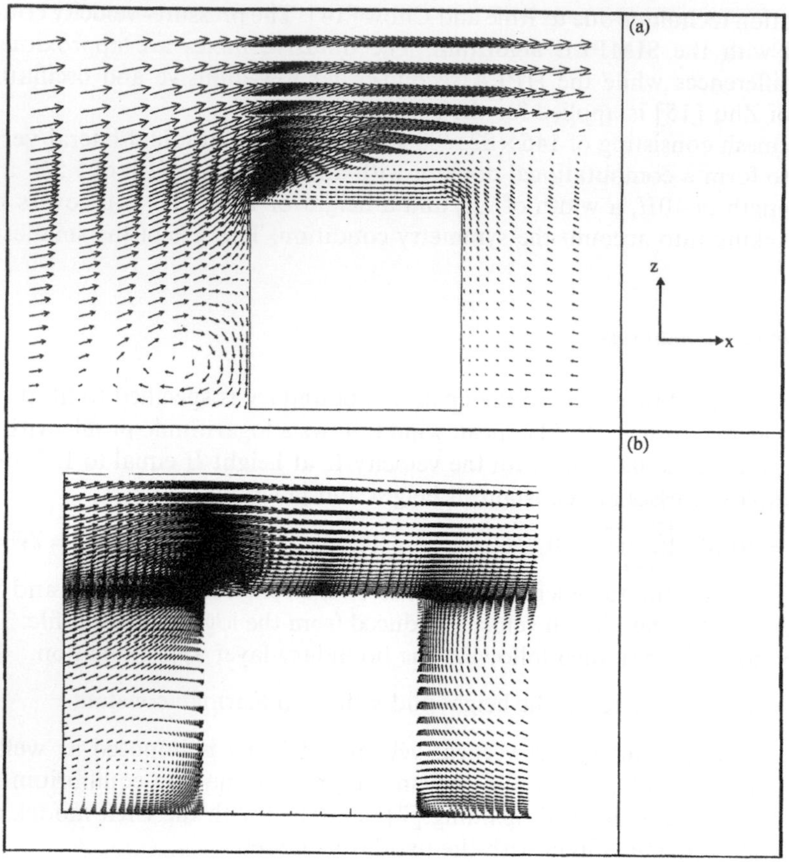

Fig. 3. Flow field in the vertical symmetry plane. (a) LRR model; (b) two layer model.

In horizontal planes, the flow separates on the windward corners, and no reattachment occurs, except in the upper part of the model. Horse-shoe vortices are created behind the building. Up to a height of about $0.75H$ there is a well-defined recirculating flow on the sides (Fig. 5a). On the roof, a deviation of the reversed flow towards the lateral edge is observed (Fig. 5b). These features have not been visualized in the experiment, but their reality can be checked by the observation of the concentration field on the building faces.

Fig. 6b (reference case) compares measured values of concentration at points on the faces and the isolines estimated by the computations. The homogeneity and the level of the concentration on the back face, are very well predicted. On the roof, both in computations and measurements, the recirculation leads to higher values upstream the chimneys than downstream. The computations slightly overestimate this effect. On the side faces, values are well reproduced except in the leeward top. In this region and also near the lateral edge of the roof, the concentrations are greatly overestimated.

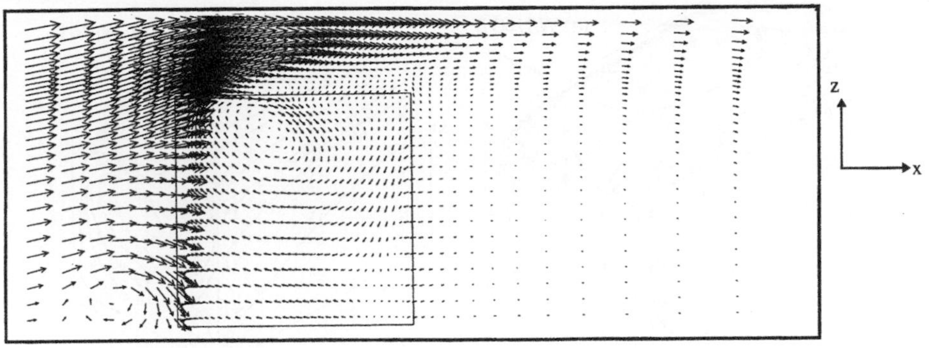

Fig. 4. Flow field in the vertical plane close to the side face (LRR model).

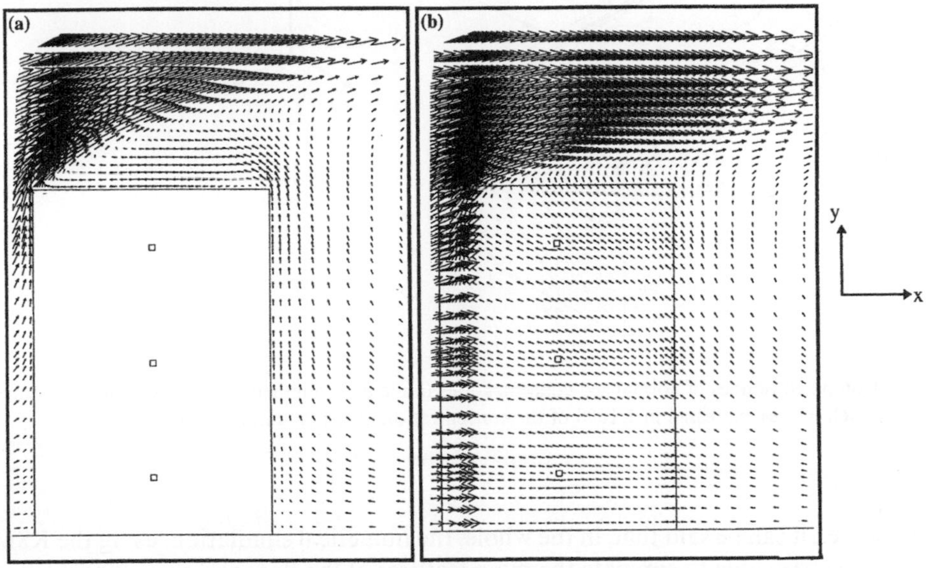

Fig. 5. Flow fields in horizontal planes at the model half-height: (a) and at the roof level (b) (LRR model).

The good agreement for the concentrations on the back face shows that the vortices behind the building are well computed. On the roof, the differences between measurements and computations can be due to either an overestimation of the reverse flow on the roof, either an underestimation of the vertical jet length at the outlet of the chimneys, as the mesh size could be too large to correctly estimate the gradients of velocity in this region. On the side faces, the recirculation zone seems to be correctly predicted on the lower part. The main departure from measurements lies on the upper part of the side face, because of an overestimation of the deviation effect of the roof reversed flow.

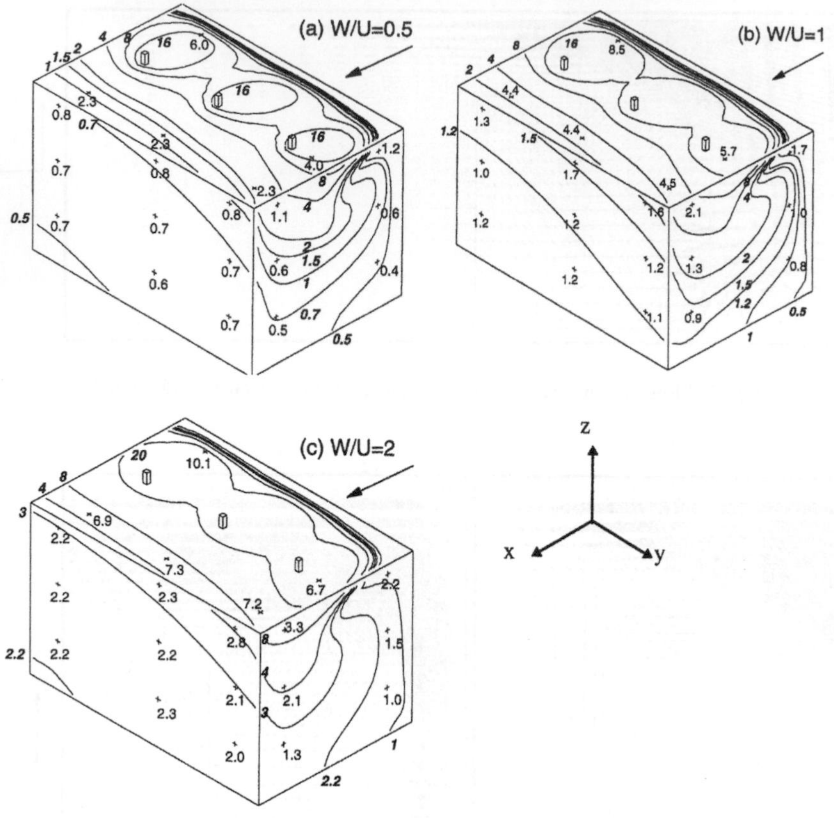

Fig. 6. Computed isolines of normalized concentrations (italic bold numbers) versus experimental values (LRR model) (values expressed in 1/1000 of the concentration at the chimney outlets).

However, it can be said that, in the whole, the numerical simulation, using the RSM model, has been able to simulate the main features of the flow and the concentration levels in the major parts of the building faces.

3.2. The reference case – the two-layer model

With the two-layer model (applied with the standard k–ε or the Kato–Launder model), a recirculation bubble is observed on the roof, but the reattachment occurs before half the roof length (Fig. 3b). The recirculating flow on the side face is not well estimated except at the bottom. It appears that the enhancement due to the use of near-wall one-equation models in the zones where production is dominated by shear stress is not observed in this case, as a consequence of the important turbulence intensity of the approach flow. The near-wall fine resolution allowed by the two-layer model is insignificant in this case. Furthermore, the use of the Kato–Launder model,

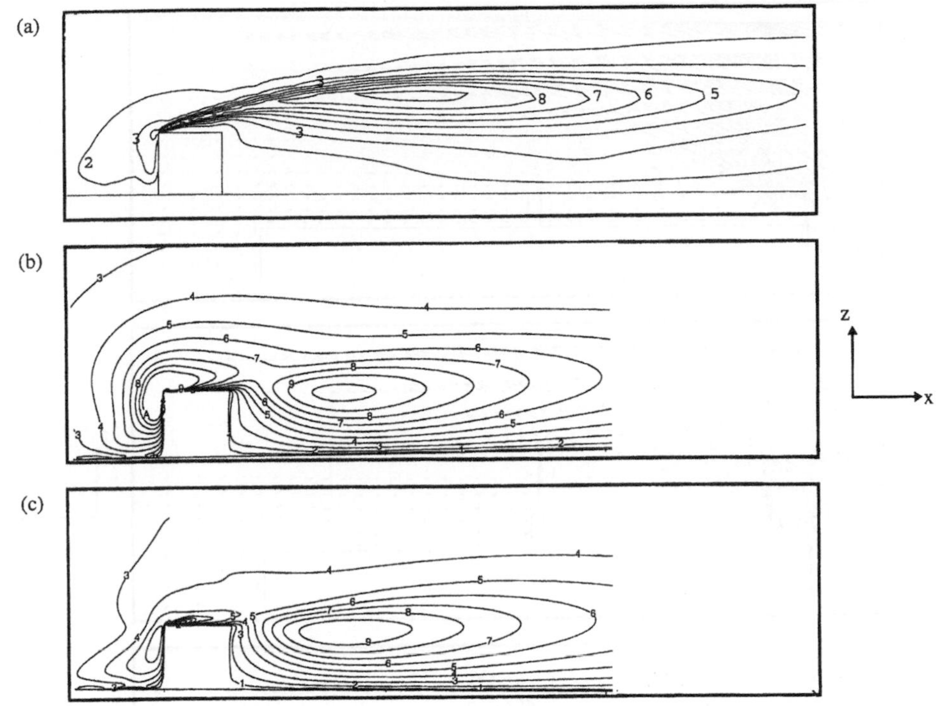

Fig. 7. k isolines in the vertical symmetry plane: (a) LRR model; (b) two-layer model – standard k–ε; (c) two-layer model – Kato–Launder (the values are multiplied by 20 and expressed in m^2/s^2).

does not show noticeable improvement: the results show that only a part of the k-production in stagnation point is eliminated (Fig. 7) so that consequence on the recirculating zone is still important. This implies that for atmospheric boundary-layer flows, characterized by a higher level of oncoming turbulence, this modification does not perform well as in channel flows.

3.3. *Effect of the ratio W/U*

Computations involving the LRR model, were also carried out with $W/U = 0.5$ and $W/U = 2$. Besides the dilution effect proportional to the flux of fresh air, these cases differ from the reference test case ($W/U = 1$) by the behaviour of the vertical jet at the outlet of the chimneys (Fig. 8). Hence, the convection and the diffusion of the pollutant immediately above the chimney is dependant on the ratio W/U. The overall distribution of pollutants on the building faces (Fig. 6) is similar to the reference case. However, due to the differences in the extension of the outlet jets, the effect of the roof recirculation on the diffusion is accentuated for $W/U = 0.5$ and reduced for $W/U = 2$.

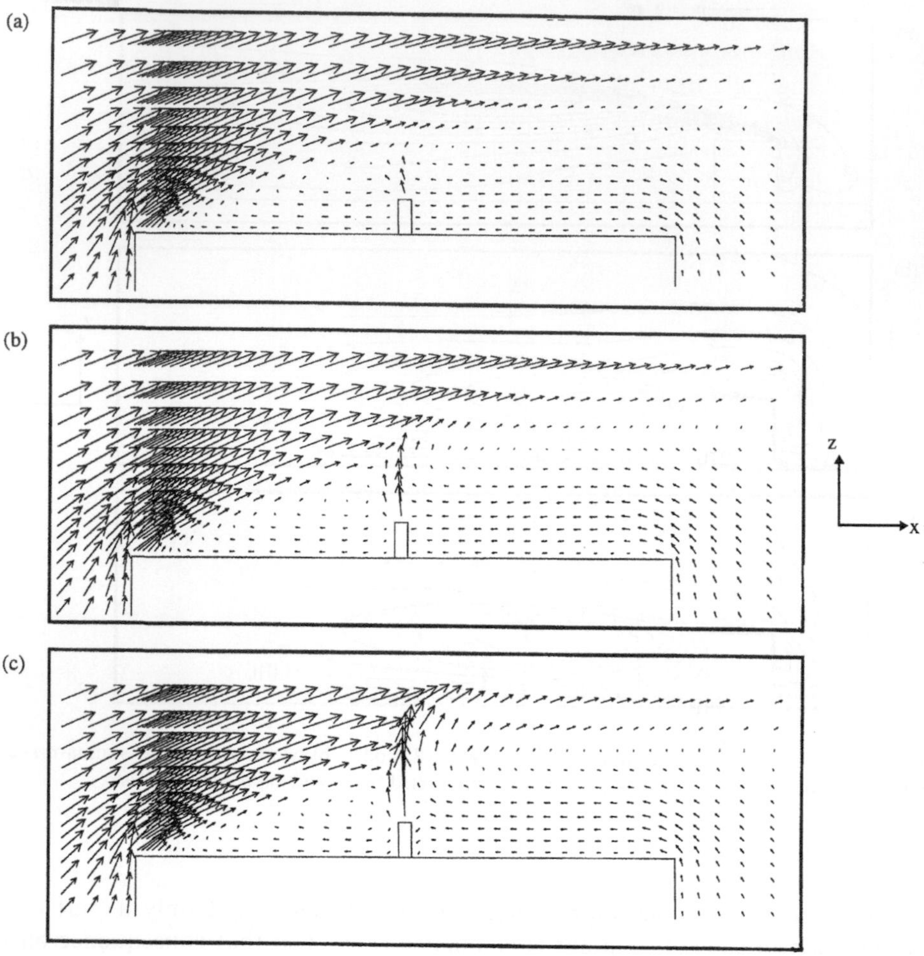

Fig. 8. Flow field in the vicinity of a chimney: (a) $W/U = 0.5$; (b) $W/U = 1.0$; (c) $W/U = 2.0$.

3.4. Discussion on numerical diffusion

Different tests have been conducted in order to evaluate the effects of numerical uncertainties, mainly the numerical diffusion:

Applying the first-order upwind scheme instead of the SMART scheme leads to a modification of the recirculating zone on the side face. The separation is well reproduced but, for instance, at half height, the reattachment occurs just before the leeward corner, so that the diffusion of the pollutant on the side face is considerably reduced. On the other faces, no major effects has been observed.

Grid refinements (mesh size divided by 2) at the front face have not change the computed flow field. In the same manner, modifying the mesh size near the side faces and in the wake have brought little modifications. Therefore, we can say that the numerical diffusion does not affect the computation of the flow around the building, when applying high-order schemes.

However, as pointed out before, numerical diffusion could affect the description of the jet at the chimney outlets. Probably, this could slightly change the distribution of concentrations on the roof but not on the lee and side faces. This point should be checked in the future.

4. Conclusion

These computations have shown that a second-order moment closure for turbulence modelling seems necessary to reproduce recirculating flows on the roof and on the side faces of a rectangular building inside a turbulent atmospheric boundary layer. The two-layer eddy-viscosity model failed to predict these features, due to the overproduction of turbulence upstream the obstacle.

On the global performance of the simulation with the LRR model, it can be said, despite some deficiencies which still have to be more precisely explained, that the flow past the building has been well predicted, leading to a good prediction of the pollutant concentrations on the faces.

References

[1] C. Solliec, G. Barnaud, R. Vilagines, Conception et performances d'une sonde rapide de mesure des concentrations gazeuses; application en soufflerie à couche limite turbulente, J. Wind. Eng. Ind. Aerodyn. 41–44 (1992) 2773–2783.

[2] C. Barré, D. Delaunay, Pollution des admissions d'air par l'air vicié extrait des bâtiments - approche expérimentale et numérique, CSTB Report EN-ECA 94.20L, 1994.

[3] B.E. Launder, G.J. Reece, W. Rodi, Progress in the development of a Reynolds-stress turbulence closure, J. Fluid Mech. 68 (1975) 537–566.

[4] M.M. Gibson, B.E. Launder, Ground effects on pressure fluctuations in the atmospheric boundary layer, J.Fluid Mech. 86 (1978) 491.

[5] B.J. Daly, F.H. Harlow, Transport equations of turbulence, Phys. Fluids 13 (1970) 2634.

[6] W. Rodi, Experience with two-layer models combining the k–ε model with a one-equation model near the walls, AIAA paper, AIAA-91-0216, 1991.

[7] B.E. Launder, D.B. Spalding, The numerical computation of turbulent flows, Comput. Meth. Appl. Mech. Eng. (3) (1974) 269–289.

[8] M. Kato, B.E. Launder, The modeling of turbulent flow around stationary and vibrating square cylinders, Proc. 9th Symp. on Turbulent Shear Flows, Kyoto, 1993.

[9] D.B. Spalding, A general purpose computer program for multi-dimensional one and two phase flow, Math. Comput. Simulation 8 (1981) 267–276.

[10] P. Gaskell, A. Lau, Curvature compensated convective transport; SMART, a new boundedness-preserving transport algorithm, Int. J. Numer. Methods Fluids 8 (1988) 1165–1181.

[11] D.B. Spalding, Mathematical modelling of fluid mechanics, heat transfer and mass transfer processes, Mech. Engrg. Report no. HTS/80/1, Imperial College, London, 1980.

[12] J.P. Van Doormal, G.D. Raithby, Enhancements of the SIMPLE method for predicting incompressible fluid flows, Numerical Heat Transfer 7 (1984) 147–163.

[13] S. Majumdar, W. Rodi, J. Zhu, Three-dimensional finite-volume method for incompressible flows with complex boundaries, J. Fluid Eng. 114 (1992) 496–503.
[14] C.M. Rhie, W.L. Chow, A numerical study of the the turbulent flow past an isolated airfoil with trailing edge separation, AIAA J. 21 (1983) 1225–1532.
[15] J. Zhu, A low-diffusive and oscillating free convective scheme, Appl. Numer. Methods 7 (1991) 225–232.

ELSEVIER

Journal of Wind Engineering
and Industrial Aerodynamics 67&68 (1997) 733–744

JOURNAL OF
wind engineering
AND
industrial
aerodynamics

Modelling particulate dispersion in the wake of a vehicle

Z.E. Hider[a,]*, S. Hibberd[a], C.J. Baker[b]

[a] *Department of Theoretical Mechanics, University of Nottingham, University Park, Nottingham NG7 2RD, UK*
[b] *Department of Civil Engineering, University of Nottingham, University Park, Nottingham NG7 2RD, UK*

Abstract

There is growing evidence to suggest that particulates within road vehicle exhaust emissions are a serious health risk. It is important to understand how these particles are dispersed in the wake of the vehicle, to enable roadside pollutant levels, etc. to be accurately predicted. In the past, the dispersion of gaseous exhaust pollutants has been studied in detail usually using a Gaussian Plume approach. In contrast, Eskridge and Hunt studied gaseous emissions within vehicle wakes using a boundary-layer approach to obtain a simple formulation for the turbulent velocity distribution of the air behind a vehicle in the direction of travel in the absence of wind effects. In subsequent work, they go on to derive a method for predicting gaseous pollutant dispersion behind a series of vehicles. In this present work, the wake formulation has been extended to give the velocity distribution in the cross-wake and vertical directions before moving on to investigate the dispersion of gaseous pollutants behind a single vehicle. The movement of particles within the wake is investigated by considering the mechanics of individual particle behaviour within the wake. This method makes use of the wake velocities and particle trajectories obtained by numerically solving a system of ordinary differential equations. Results are presented for particle trajectories for different particle sizes and different initial exhaust conditions together with concentration profiles for the case of solutes. This paper also investigates the changes in predicted wake behaviour of a single vehicle experiencing a crosswind as well as the effect of including a viscous sub-layer.

Keywords: Vehicle wakes; Vehicle pollution; Particulate dispersion; Solute dispersion; Wake velocities

1. Introduction

In recent years, there has been growing evidence to suggest that emissions from burning fossil fuels have a direct impact on many health and environmental issues. In most urban areas, the chief source of fossil fuel pollution is the road vehicle, which

* Corresponding author. E-mail: etxzh@thmech.nottingham.ac.uk.

PII S0167-6105(97)00114-1

emits both solute and particulate pollution, such as carbon monoxide, carbon dioxide, nitrous oxides, sulphur dioxide, benzene, carbon and unburnt hydrocarbons.

In the past, there have been various approaches to calculating solute pollution, for example, the Gaussian approach and the use of computational fluid dynamics. However, Eskridge and Hunt [1] developed an approximate method for the calculation of the velocity fields in the wake of a vehicle. Subsequently, their results are used in a companion paper [2] to calculate the concentration distribution of solute pollution behind a series of vehicles.

In Section 2, a theory for calculating the wake behaviour of a single vehicle is outlined, including the calculation of the velocity deficits in the downwake, cross-wake and vertical directions, using the principles first outlined by Eskridge and Hunt [1]. This paper goes on to investigate other situations including a vehicle experiencing a constant crosswind, and the effect of a viscous sub-layer beneath the vehicle wake.

In Section 3, the approximate governing equation for solute pollution is given and is derived from the conservation of species, or advection–diffusion equation. Computational results are given for the concentration distribution at different distances downstream of a single vehicle travelling at constant speed for the original case. The concentration distribution in terms of wake coordinates are also given for the crosswind case.

Since there is growing evidence to suggest some particulates are carcinogenic, it is necessary to understand their behaviour. Section 4 outlines the equations of motion for particles and then goes on to show the results obtained, to date, of the trajectories and landing positions of these particles in vehicle wakes.

2. Wake behaviour of a single vehicle

In 1979, Eskridge and Hunt developed a new method for investigating the behaviour of the wake of a vehicle [1]. Their paper outlines a method for calculating the velocity profile in the wake of a vehicle experiencing no prevailing wind before going on to suggest possible extensions to crosswinds. However, details and results for the calculation of the velocity profile were presented for the downwake direction only.

Before the dispersal of pollution from a single vehicle can be accurately modelled, the velocity field in the cross-wake and vertical directions is required. Further, the original results obtained by Eskridge and Hunt [1] need to be modified in order to consider the behaviour of the wake under other wind conditions.

2.1. *Vehicle experiencing no wind*

A single vehicle is assumed to be travelling at a constant speed U, on flat terrain, and experiencing no prevailing wind. We consider two coordinate systems, (X,Y,Z), fixed relative to the ground, and (x,y,z) fixed relative to the vehicle, as shown in Fig. 1. Thus, the vehicle is travelling at a speed U, in the $-x$-direction and experiences a velocity deficit in its wake.

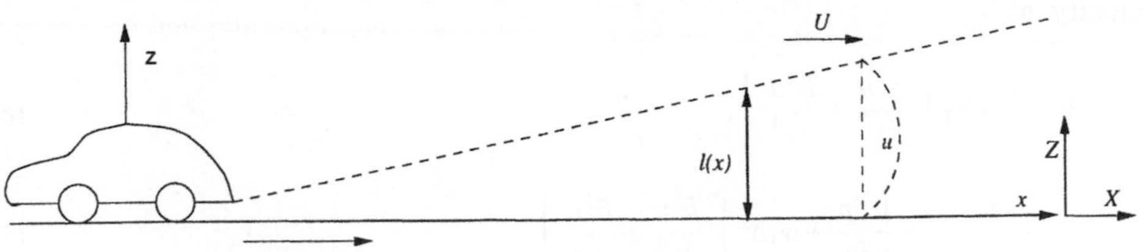

Fig. 1. Vehicle moving with constant velocity U in the $-x$-direction.

Immediately behind the vehicle, the wake is highly disordered. However, several vehicle heights downstream, the wake settles into a spatially homogeneous turbulent structure spreading in the cross-wake and vertical directions. The wake thickness $l(x)$ is assumed to be much smaller than the wake length L i.e. $l/L = \delta \ll 1$, giving rise to a boundary-layer problem with the derivatives in the x-direction being much smaller than the derivatives in the other directions.

The mean velocity is given, several heights downstream, as

$$\bar{\boldsymbol{u}} = (U + u, v, w)\,, \tag{1}$$

where $|u|, |v|, |w| \ll U$.

The governing Navier–Stokes equations are non-dimensionalized with $\varepsilon \ll 1$, using $\bar{u} = U(1 + \varepsilon u)$, $\bar{v} = U\delta\varepsilon v$, $\bar{w} = U\delta\varepsilon w$, $\bar{x} = Lx$, $\bar{y} = ly$, $\bar{z} = lz$, $\bar{\tau} = l\delta\varepsilon\tau$ and $\bar{p} = U\varepsilon p$. So to leading order, Navier–Stokes equations become

$$\frac{\partial u}{\partial x} + \frac{\partial v}{\partial y} + \frac{\partial w}{\partial z} = 0\,, \tag{2}$$

$$U\frac{\partial u}{\partial x} = \frac{\partial \tau_{xy}}{\partial y} + \frac{\partial \tau_{xz}}{\partial z}\,, \tag{3}$$

$$\delta^2 U\frac{\partial v}{\partial x} = -\frac{1}{\rho}\frac{\partial p}{\partial y} + \delta^2\frac{\partial \tau_{yy}}{\partial y} + \delta^2\frac{\partial \tau_{yz}}{\partial z}\,, \tag{4}$$

$$\delta^2 U\frac{\partial w}{\partial x} = -\frac{1}{\rho}\frac{\partial p}{\partial z} + \delta^2\frac{\partial \tau_{zy}}{\partial y} + \delta^2\frac{\partial \tau_{zz}}{\partial z}\,, \tag{5}$$

where τ_{ij} and p are the stress components and pressure terms, respectively.

The associated boundary conditions typically are, the no-slip condition on the ground, i.e. when $z = 0$ then $u = v = w = 0$, and at distances downstream of the car, in all directions, perturbation velocities vanish, i.e. when $z \to \infty$, $x \to \infty$, $y \to \pm\infty$ then $u = v = w = 0$.

The effect of turbulence is described by the inclusion of an eddy diffusivity coefficient, ν_T within the stress components. Eqs. (3)–(5) can then be re-written in terms of velocity, as

$$U\frac{\partial u}{\partial x} = \nu_T\left[\frac{\partial^2 u}{\partial y^2} + \frac{\partial^2 u}{\partial z^2}\right], \tag{6}$$

$$\delta^2 U\frac{\partial v}{\partial x} = -\frac{1}{\rho}\frac{\partial p}{\partial y} + \nu_T\delta^2\left[\frac{\partial^2 v}{\partial y^2} + \frac{\partial^2 v}{\partial z^2}\right], \tag{7}$$

$$\delta^2 U\frac{\partial w}{\partial x} = -\frac{1}{\rho}\frac{\partial p}{\partial z} + \nu_T\delta^2\left[\frac{\partial^2 w}{\partial y^2} + \frac{\partial^2 w}{\partial z^2}\right], \tag{8}$$

where $\nu_T = \gamma l(x)u_{\max}(x)$.

The downwake velocity deficit can be solved from Eq. (6) by seeking a similarity solution of the form, $u = -U|A|(x/h)^{-m}f(\zeta, \eta)$, where $\eta = y/l(x)$ and $\zeta = z/l(x)$. Hunt [3] outlines a method for calculating m, $l(x)$ and A using the component of couple, C', caused by viscous and pressure forces, and Eq. (6), to give $m = \frac{3}{4}$, $l(x) = \gamma A(x/h)^{1/4}$ and

$$A = \left[\frac{C'}{A_1\gamma^3 h^3 U^2\sqrt{32\pi}}\right]^{1/4}, \tag{9}$$

where h is the vehicle height, $\gamma = 0.4$ and $A_1 = 4.13$.

A resulting partial differential equation for f, in terms of the similarity variables, ζ and η, is obtained and can be solved, using the boundary conditions, to give

$$f(\zeta, \eta) = 0.824\zeta\exp\left[-(\zeta^2 + \eta^2)/8\right]. \tag{10}$$

The final solution for u, is

$$u = -U|A|\left(\frac{x}{h}\right)^{-3/4} 0.824\zeta\exp\left[-\frac{(\zeta^2 + \eta^2)}{8}\right]. \tag{11}$$

The cross-wake and vertical velocity deficits are found by seeking similarity solutions for v, w and p, of a similar form to u. These solutions are made to satisfy continuity and Eqs. (7) and (8) at $O(\delta^2)$. These give

$$v = -U\gamma A^2\left(\frac{x}{h}\right)^{-3/2} 0.412\zeta\eta\exp\left[-\frac{(\zeta^2 + \eta^2)}{8}\right], \tag{12}$$

$$w = -U\gamma A^2\left(\frac{x}{h}\right)^{-3/2} 0.206\zeta^2\exp\left[-\frac{(\zeta^2 + \eta^2)}{8}\right], \tag{13}$$

$$p = \rho U^2\gamma^2 A^3\left(\frac{x}{h}\right)^{-7/4} 0.412\zeta\exp\left[-\frac{(\zeta^2 + \eta^2)}{8}\right]. \tag{14}$$

2.2. Vehicle experiencing a crosswind

An extension of the previous case is considered where the vehicle now experiences a constant crosswind, κV, where κV is of the same order as the vehicle speed U. The mean velocity is re-written as $\bar{\boldsymbol{u}} = (U + u, \kappa V + v, w)$ in the (x, y, z) coordinate system. However, for ease of calculation, the coordinate system is transformed into a new (s, n, z) coordinate system where s is the coordinate along the centreline of the wake and n is perpendicular to the centreline.

The mean velocity, in this case, is then $\bar{\boldsymbol{q}} = (Q + q, r, w)$ where $\bar{q} = \bar{u}\cos\theta + \bar{v}\sin\theta, \bar{r} = -\bar{u}\sin\theta + \bar{v}\cos\theta$. θ is the angle the centreline of the wake makes with the x-axis and in the far wake as $|u|, |v| \to 0$, $\tan\theta = \kappa$.

Transforming the Navier–Stokes equations to the new (s, n, z) coordinate system, solutions for q, r, w, can be obtained in a similar way as for the previous case, to give

$$q = -Q|A|\left(\frac{s}{h}\right)^{-3/4} 0.824\zeta \exp\left[-\frac{(\zeta^2 + \eta^2)}{8}\right], \tag{15}$$

$$r = Q\gamma A^2\left(\frac{s}{h}\right)^{-3/2} 0.412\zeta\eta \exp\left[-\frac{(\zeta^2 + \eta^2)}{8}\right], \tag{16}$$

$$w = Q\gamma A^2\left(\frac{s}{h}\right)^{-3/2} 0.206\zeta^2 \exp\left[-\frac{(\zeta^2 + \eta^2)}{8}\right]. \tag{17}$$

These equations, can be seen to be the same as before, only in terms of the (s, n, z) coordinate system instead of the (x, y, z) coordinate system. This means in the original coordinate system the velocity deficit, if $\bar{u} = \bar{q}\cos\theta - \bar{r}\sin\theta$ and $\bar{v} = \bar{q}\sin\theta + \bar{r}\cos\theta$, the velocity deficits in the (x, y, z) coordinate system, when compared to the results for the original case, are smaller in the x-direction and larger in the y-direction while at the same time being dependent on the wind direction.

2.3. The effect of a viscous sublayer near the ground

Behind the vehicle, the wake experiences a very high shear effect near the ground which may not be accurately modelled by imposing a no-slip boundary condition. It is suggested, by Hocking [4], that near the ground there exists a highly stressed viscous sub-layer and that this may be incorporated into the problem by modifying the bottom boundary condition to include a slip coefficient $c_s(x)$ which is dependent on the downstream turbulence. The revised boundary condition then becomes

$$u = c_s(x)\frac{\partial u}{\partial z} \quad \text{on} \quad z = 0. \tag{18}$$

Eqs. (6)–(8) are now solved with the modified boundary condition. This is done by again seeking a similarity solution of the form $u = -U|A|(x/h)^{-m}f(\zeta, \eta)$ where η and ζ are the same scaled coordinates of y and z as before. In order to keep this work consistent with Section 2.1, put $m = 3/4$, $l(x) = \gamma Ah(x/h)^{1/4}$ and

$$A = \left[\frac{C'}{\Lambda_1 \gamma^3 h^3 U^2 \sqrt{32\pi}}\right]^{1/4}. \tag{19}$$

Again, $\Lambda_1 = 4.13$, $\gamma = 0.4$ and h is the vehicle height. Applying the above similarity solution to the boundary conditions gives $c_s(x) = K_s \gamma Ah(x/h)^{1/4}$, where K_s is a constant.

Now the partial differential equation for f, in terms of ζ and η, can be solved, with the new boundary conditions to obtain the following equation for u,

$$u = -U|A|\left(\frac{x}{h}\right)^{-3/4} 0.824 \Bigg\{ K_s \exp(-\eta^2/8) + \left[\zeta + K_s \frac{i\sqrt{2\pi}}{4} \zeta \operatorname{erf}\left(\frac{i\sqrt{2}}{4}\zeta\right)\right] \exp\left[\frac{-(\eta^2 + \zeta^2)}{8}\right]\Bigg\}. \tag{20}$$

As can be seen, if $K_s = 0$, this then reduces to the case outlined in Section 2.1. The same transformation as in Section 2.2 can be made to obtain the velocity q in the crosswind case

$$q = -Q|A|\left(\frac{s}{h}\right)^{-3/4} 0.824 \Bigg\{ K_s \exp(-\eta^2/8) + \left[\zeta + K_s \frac{i\sqrt{2\pi}}{4} \zeta \operatorname{erf}\left(\frac{i\sqrt{2}}{4}\zeta\right)\right] \exp\left[\frac{-(\eta^2 + \zeta^2)}{8}\right]\Bigg\}. \tag{21}$$

The most important result of this work is the effect this boundary condition has on the prediction of the effective centreline of the wake in the crosswind case. Solving the streamline equations gives

$$x = U\lambda + \frac{3.256QK_s|A|h}{U}\left(\frac{Q\lambda}{h}\right)^{1/4} + \cdots \quad \text{and} \quad y = \kappa U\lambda. \tag{22}$$

This can be plotted to give the centreline position for the original case ($K_s = 0$), compared with the case where a shear layer is imposed. This is shown in Fig. 2. From this figure, it can be seen, when the shear layer is imposed, the centreline is shifted. This is due to the slight curvature of the centreline near the vehicle for the shear-layer case.

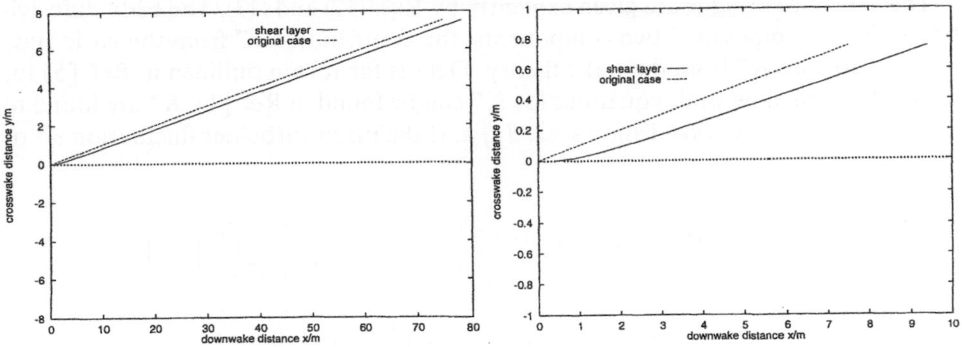

Fig. 2. Graphs of the centreline positions of the original case and the high shear-layer case.

3. The solute model

3.1. Vehicle experiencing no wind

To calculate the concentration distribution of solute pollutant behind a single vehicle the conservation of species equation is used. The form of this equation we use is given by Eskridge et al. [2]. This advection–diffusion equation, after a similar non-dimensionalization scaling and linearization as given in Section 2, becomes

$$\frac{\partial \chi}{\partial t} + u\frac{\partial \chi}{\partial x} + v\frac{\partial \chi}{\partial y} + w\frac{\partial \chi}{\partial z} = \frac{\partial}{\partial y}\left[K_y \frac{\partial \chi}{\partial y}\right] + \frac{\partial}{\partial z}\left[K_z \frac{\partial \chi}{\partial z}\right], \tag{23}$$

where χ is the concentration of solute per unit volume and K_y and K_z are the eddy diffusivities.

In the case of a continual release of solute, the pollutant spreads more in the cross-wake and vertical directions than in the downwake direction, so, $\partial\chi/\partial x \ll \partial\chi/\partial y, \partial\chi/\partial z$.

The boundary conditions are

$$\frac{\partial \chi}{\partial y} = 0 \quad \text{on} \quad y = \pm\, y_{\max}\,, \tag{24}$$

$$\frac{\partial \chi}{\partial z} = 0 \quad \text{on} \quad z = z_{\max} \quad \text{and} \quad \chi = 0 \quad \text{on} \quad z = 0\,. \tag{25}$$

These are not perfect physical restraints as bounded limits must be imposed in order to solve (23) numerically. The method of lines is used to obtain the concentration distribution. This requires discretization of the spatial derivatives, using central difference approximations, accompanied by using a Runge–Kutta–Mersen solver to integrate over time.

The velocities v and w are given explicitly by Eqs. (12) and (13). The eddy diffusivities are each composed of two components, the contribution K^s from the basic state atmosphere and K^w from the wake theory. Details for K^s are outlined in Ref. [5] for neutrally stable flow while equations for K^w can be found in Ref. [1]. K^w are found to be dependent on the wake length scale $l(x)$ and the mean turbulent fluctuation $\bar{v}'^2$ or $\bar{w}'^2$. Thus

$$K_y = \frac{4.4542\gamma U_w z}{\ln (H/z_0)} + 0.1192A^2Uh\left(\frac{h}{x}\right)^{1/2}\left[\left(1 - \frac{z^2}{4(\gamma Ah)^2}\left(\frac{h}{x}\right)^{1/2}\right)\right.$$

$$\left. + \frac{z^2y^2}{16(\gamma Ah)^4}\left(\frac{h}{x}\right)\right]^{1/4} \exp\left[-\frac{(y^2+z^2)}{16(\gamma Ah)^2}\left(\frac{h}{x}\right)^{1/2}\right], \tag{26}$$

$$K_z = \frac{\gamma^2 U_w z}{\ln (H/z_0)} + 0.0348A^2Uh\left(\frac{h}{x}\right)^{1/2}\left[\left(1 - \frac{z^2}{4(\gamma Ah)^2}\left(\frac{h}{x}\right)^{1/2}\right)\right.$$

$$\left. + \frac{z^2y^2}{16(\gamma Ah)^4}\left(\frac{h}{x}\right)\right]^{1/4} \exp\left[-\frac{(y^2+z^2)}{16(\gamma Ah)^2}\left(\frac{h}{x}\right)^{1/2}\right], \tag{27}$$

where U_w is the wind speed, H is the dispacement height and z_0 is the surface roughness height.

It is important to calculate the concentration distribution relative to a fixed point in space, to see how the concentration varies with time for a stationary observer. If the car is travelling at a speed U, then $x = Ut$ in the above equation. For illustration, an initial concentration distribution of 200 exp $[-50(y^2 + z^2)]$, equivalent to the diffusion of an instantaneous source released at $t = 0.005$, is taken. The results for the concentration distribution at three different distances behind the car are given in Fig. 3. These results are discussed in Section 5.

3.2. *Vehicle experiencing a constant crosswind*

As in Section 2.2 the transformation from the (x, y, z) to the (s, n, z) coordinate system must be used. After the same non-dimensionalization scaling and linearization as before, this gives the following conservation of species equation

$$\frac{\partial \chi}{\partial t} - q\frac{\partial \chi}{\partial s} + r\frac{\partial \chi}{\partial n} + w\frac{\partial \chi}{\partial z} = \frac{\partial}{\partial n}\left[K_n\frac{\partial \chi}{\partial n}\right] + \frac{\partial}{\partial z}\left[K_z\frac{\partial \chi}{\partial z}\right], \tag{28}$$

with the same boundary conditions as before. The method of lines is used in the same way as before to find a solution. K_n and K_z are the eddy diffusivities and have the same form as Eq. (27) and (28). As in the previous case, it can be assumed that the dominant spread of pollution is across the wake, therefore assuming that the contribution in the s-direction is negligible compared to the other directions. Also assume

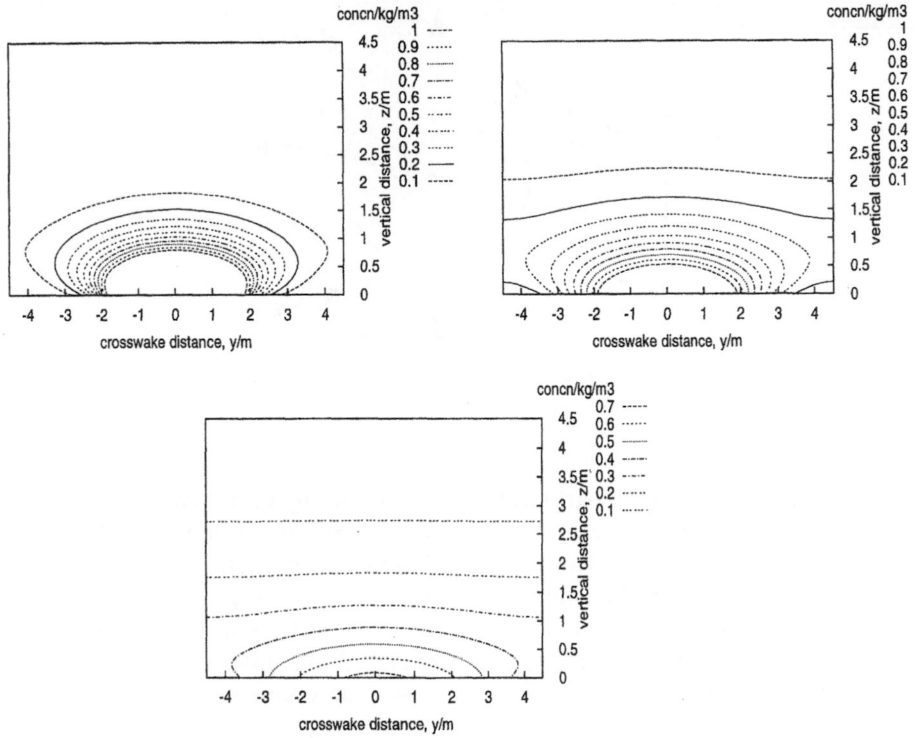

Fig. 3. Graphs of the concentration distribution for 3.5, 7.5 and 15 m downstream of a vehicle experiencing no wind.

$s = Qt$ to give the concentration distributions at different distances downwake of the vehicle. The results in this case for n,z are the same as the y,z distributions in the previous case (Fig. 3) and are discussed in Section 5.

4. The saltation of particulates

The concentration distribution of particulate pollution is different from that of solute pollution because particles possess a small but significant mass. This mass causes the particles to be dynamically active and affected by gravity as well as the local wake velocities. Because of this, it is important to understand the behaviour of particles before we can calculate the concentration distribution of particles in the wake of a vehicle. This section outlines a preliminary analysis about the behaviour of particles in vehicle wakes.

Consider a spherical particle, emitted from the exhaust of a car and assume the particle experiences forces generated by advection with the air flow and gravity only. This distribution of forces is shown in Fig. 4.

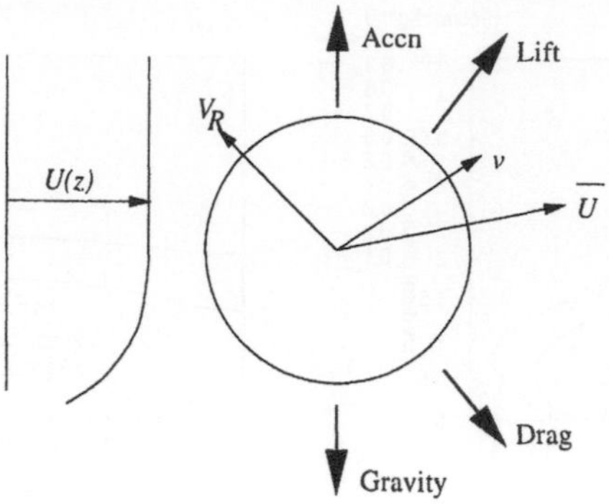

Fig. 4. Forces acting on a particle moving with velocity $\boldsymbol{v}$ through air moving with velocity $\bar{\boldsymbol{u}}$.

The relative velocity of the particle, to the surrounding air, is given as $\hat{\boldsymbol{V}}_{\mathrm{R}} = \hat{\boldsymbol{v}} - \bar{\boldsymbol{u}}$, where $\bar{\boldsymbol{u}} = (U + u, v, w)$ where u, v, w are the perturbation velocities calculated in Section 2. Also, $\hat{\boldsymbol{v}} = (v_1, v_2, v_3)$ and $\mathrm{d}\hat{\boldsymbol{x}}/\mathrm{d}\hat{t} = \hat{\boldsymbol{v}}$ and the particle is of radius a and density ρ_{p}. The fluid has kinematic viscosity ν and density ρ_a. Lift and inertial terms can be neglected because, as Nalpanis et al. [6] mention in their work, it has been shown from experimental work, there is nothing unusual about the trajectories of spinning particles. So the force equation becomes, when $C_{\mathrm{D_A}}$ is the drag coefficient of air,

$$\frac{\mathrm{d}\hat{\boldsymbol{v}}}{\mathrm{d}\hat{t}} = \frac{-3C_{\mathrm{D_A}}\rho_a}{8a\rho_{\mathrm{p}}}\hat{\boldsymbol{V}}_{\mathrm{R}}|\hat{\boldsymbol{V}}_{\mathrm{R}}| + \boldsymbol{g}\,. \tag{29}$$

This leads to a system of six first-order ODEs. To solve for the particle positions, the system can be solved using a Runge–Kutta–Mersen method.

Attention must be drawn to the coordinate system. As was mentioned in the previous section the (x, y, z) coordinate system moves with the car, so relative to a fixed point on the ground the position of a particle is taken, in this section, as $x_{\mathrm{A}} = -1.5 + x - Ut$.

For illustration purposes, the particles are initially assumed to have a uniform distribution across the wake, immediately behind the car. For different particle sizes, Fig. 5 shows the positions of the particles when they hit the ground and the centerline particle trajectory in each case.

In practical situations, this uniform distribution is not very realistic, consequently the particles were assumed to have an initial, bivariate distribution with the property that the particles are emitted from a greater height at the centre of the wake compared to the edges of the wake. Fig. 6 gives the landing positions of particles for the bivariate case compared to the uniform case. In both cases, the particles have the same radius of 10^{-5} m.

These results along with possible extensions of this work are discussed in Section 5.

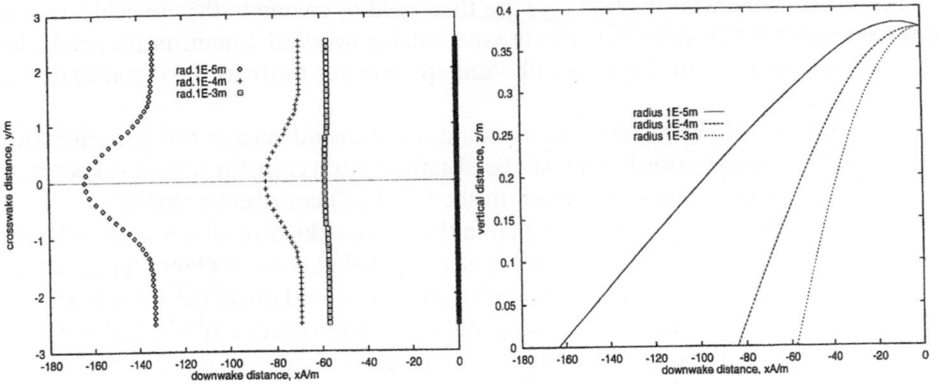

Fig. 5. Graphs of the final positions and centreline particle trajectories for particles of different sizes with an initial uniform distribution.

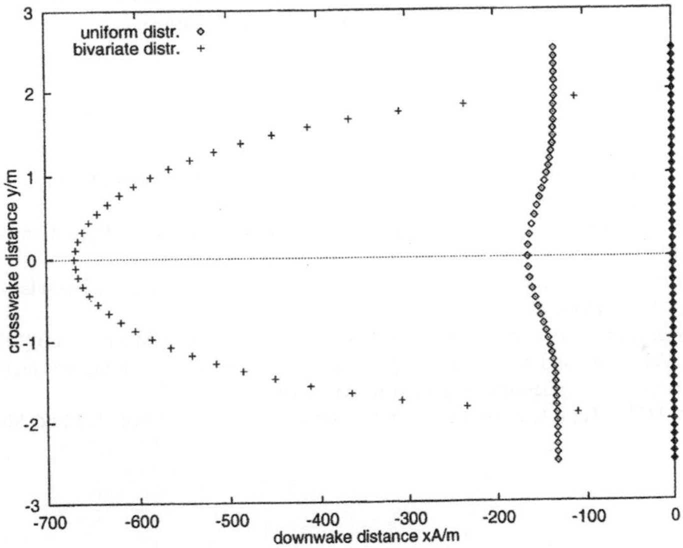

Fig. 6. Graph of the landing positions of particles for the bivariate distribution and the original uniform case. In both cases the particles have a radius of 10^{-5} m.

5. Discussion of results

As has been shown in this paper, it is necessary to fully understand the behaviour of the wake of a vehicle in order to be able to predict concentration and dispersion of the pollutants emitted from a vehicle. This paper outlines the results obtained for the velocity deficits in three different cases, the vehicle experiencing no wind, the vehicle experiencing crosswind and the effect of a viscous sublayer.

The velocity deficits in the first case are then used to calculate the concentration of solute pollutant in the wake of a vehicle experiencing no wind. Quantitative results for concentration dispersion show the pollutant spreads out more as the distance downstream increases.

The behaviour of the vehicle experiencing a crosswind means the concentration dispersion in the wake coordinates can be obtained but a transformation is necessary to obtain the concentration dispersion in the (X, Y, Z) coordinate system.

At the moment, only the velocity deficit in the downwake direction is known for the viscous sublayer case. It is noted that this velocity deficit contains extra terms, which are dependent on K_s, the slip constant coefficient. This could affect the velocity deficits in the other directions and the concentration distributions previously calculated. The most important result obtained from the rough surface case, to date, is the shift in centreline of the wake when the vehicle experiences a crosswind. This can be seen in Fig. 2.

In addition to solute pollution, vehicles also emit particulate pollution. This paper shows the behaviour of particles in the wake experiencing advective flow only.

This paper is part of an ongoing research project; future work is to compare experimental data with the analytical theory presented here.

References

[1] R.E. Eskridge, J.C.R. Hunt, Highway modelling Part I: prediction of velocity and turbulence fields in the wake of vehicles, J. Appl. Meteorol. 18 (4) (1979) 387–400.

[2] R.E. Eskridge, F.S. Binkowski, J.C.R. Hunt, T.L. Clark, K.L. Demerjian, Highway modelling Part II: advection of SF_6 tracer gas, J. Appl. Meteorol. 18 (4) (1979) 401–412.

[3] J.C.R. Hunt, A theory for the laminar wake of a two-dimensional body in a boundary layer, J. Fluid Mech. 49 (1) (1971) 159–178.

[4] L.M. Hocking, A moving fluid interface on a rough surface, J. Fluid Mech. 76 (4) (1976) 801–817.

[5] C.J. Baker, Wind flows and natural gas dispersion in the urban environment, Nottingham University Department of Civil Engineering, Report (FR95009), 1995.

[6] P. Nalpanis, J.C.R. Hunt, C.F. Barrett, Saltating particles over a flat bed, J. Fluid Mech. 251 (1993) 661–685.

ELSEVIER

Journal of Wind Engineering
and Industrial Aerodynamics 67&68 (1997) 745–755

JOURNAL OF
wind engineering
AND
industrial
aerodynamics

Concentration and flow distributions in the vicinity of U-shaped buildings: Wind-tunnel and computational data

Bernd M. Leitl[a,1], P. Kastner-Klein[b], M. Rau[b], R.N. Meroney[a]

[a] *Fluid Mechanics and Wind Engineering Program, Civil Engineering Department, Colorado State University, Fort Collins, CO 80523, USA*

[b] *Institut für Hydrologie und Wasserwirtschaft, Universität Karlsruhe, Kaiserstrasse 12, 76128 Karlsruhe, Germany*

Abstract

The flow and dispersion of gases emitted by point sources located near a U-shaped building were determined by the prognostic model FLUENT using the RNG k–ε turbulent closure approximation. Calculations are compared against wind-tunnel measurements about such a U-shaped building and several other prognostic and diagnostic numerical models. FLUENT gives a mixed image in terms of accuracy of predicted concentrations compared to the wind tunnel experiment. For identical boundary conditions higher as well as lower concentration values are calculated for different test cases. Ground level sources show higher discrepancies than situations where the tracer was emitted from the roof of the model building. A major error source was found to be the stationary solution procedure that was chosen for all simulations.

Keywords: Dispersion; Wind tunnel; Air pollution aerodynamics; Numerical simulation

1. Introduction

The flow patterns which develop around individual buildings govern the local distribution of pollution about the building and in its wake. The superposition and interaction of flow patterns associated with adjacent buildings govern the movement of pollutants in urban and industrial complexes. Sources are often located close to buildings, as in short stacks from thermal or chemical processes or ventilation shafts from parking garages.

[1] Lynen Research Fellow of the Alexander von Humboldt Foundation, Germany. Present address: Meteorology Institute, University of Hamburg, D-20148 Hamburg, Germany.

PII S0167-6105(97)00115-3

A number of prognostic and diagnostic numerical models have been developed to predict flow and dispersion within the arrangements of buildings, vegetation and vehicle corridors which constitute a city. Prognostic models solve the equations of motion with a selected closure model (mixing length, kinetic energy, Reynolds stress, or sub-grid scale assumptions; e.g. MISKAM [1]; FLUENT 1983 [11]; or LES [2]). Diagnostic models produce mass-consistent 3-D mean wind fields, using intelligent initialization procedures (often based on wind-tunnel model measurements) which analyze the building structure at each grid point and choose typical wind components for the stagnation separation and recirculation zones as a first approximation (e.g. ABC [3]; DASIM [4]; ASMUS [5].) In both cases a dispersion model solves the advection–diffusion equation based on the precalculated wind field and specified exchange coefficients.

Klein et al. [6] compared a number of such models against wind-tunnel measurements around a U-shaped building. The building shape was chosen because it can be represented by a number of alternative combinations of basic rectangular building components. Unfortunately, given such simple rectangular building elements, some diagnostic models produce different flow patterns for the same overall building envelope [7]. Klein et al. [6] found when they compared ABC, ASMUS, DASIM and MISKAM to the wind-tunnel measurements that there were no clear wind, source and/or building configurations where all models succeed or fail. For some wind orientation situations the numerical models produced similar maximum concentrations, but these differed from physical model measurements. In other wind directions even the differences between numerical calculations were huge. Their final conclusion was "it is impossible to recommend one of the models at present". They also noted that it would be necessary to compare local results from the wind field and turbulence parameterization in order to determine the reasons for the differences of the models between one another and the wind tunnel results.

2. Potential of RNG k–ε models

Murakami [2] has compared the calculated flow fields around a cube immersed in a turbulent boundary layer using various turbulence models (standard k–ε, ASM and LES). He concluded that the correspondence between experiment and simulations of the mean wind field "were fairly good, but serious and non-serious discrepancies existed in the pressure and turbulence statistics". He observed that the standard k–ε model overpredicts the value of k in regions of strong shear. This leads to excessive mixing and subsequent incorrect prediction of regions of separation and reattachment.

A renormalized-group-theory (RNG) version of the k–ε model has been proposed which only requires the existence of a large-Reynolds number region of turbulent cascading to generate a set of improved transport equations with no unspecified constants [8]. This model has led to improved predictions of flow and separation around bluff bodies like backward facing steps and cubical bluff bodies. In this paper the flow and dispersion produced by the prognostic model FLUENT using the RNG

k–ε turbulent closure is compared with the measurements from Klein et al. [6] for a U-shaped building.

3. Wind-tunnel measurements about the U-shaped building

The U-shaped building model was presumed to replicate at a scale of 1 : 200 a building with a planform base 40 m by 52 m with a 28 m square courtyard located along the 52 m face of the building. Wind-tunnel measurements of concentration were taken about buildings of heights 16, 28 and 40 m. Gas sources were presumed to lie on the roof of the base of the U, and 20 m up and downwind of this source at ground level (see Fig. 1). Wind approach angles of 0°, 45°, 90°, 120°, 135° and 180° were examined [9].

The experiments were performed in a boundary layer wind tunnel with a test-section about 10.5 m long, 2 m wide and 1 m high. A thick turbulent boundary layer was generated along the floor of the tunnel by a combination of vortex generators and roughness on the floor of the wind tunnel. The meteorological parameters of wind velocity profile, turbulence intensity and turbulence spectra were chosen to be similar in full scale and in the wind tunnel [10]. During the wind-tunnel experiments the power law profile of the approach flow had an exponent of 0.28 and the surface level turbulent intensity reached 33% for a wind speed u_{10} of 5 m/s. Approach wind and turbulence profiles are shown in Fig. 2. Emissions were simulated by releasing a tracer gas (SF_6), and concentrations were detected by analyzing samples with a leak detector. A total of 43 combinations of source location (3), building heights (3), and wind directions (6) were examined.

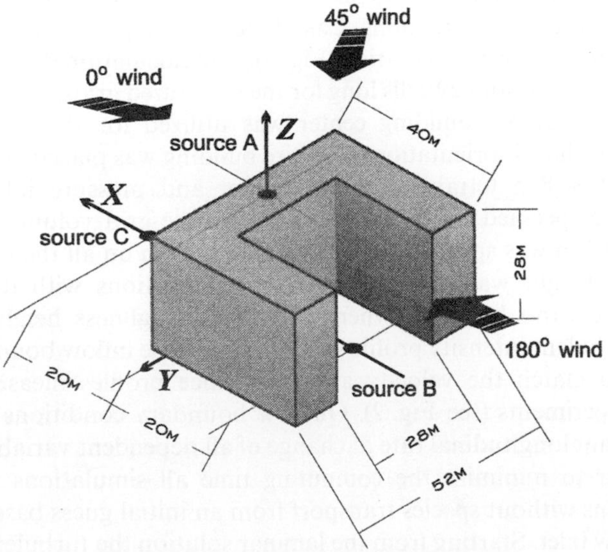

Fig. 1. Configuration of U-shaped building.

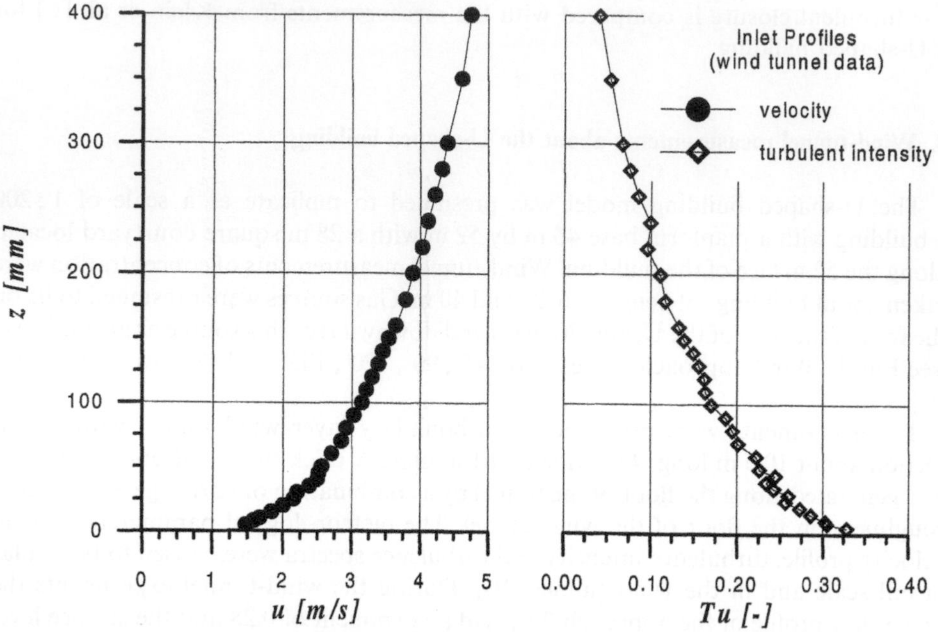

Fig. 2. Mean velocity and turbulent intensity profiles approaching U-shaped building.

4. Numerical model configuration

Version 4.3 of the FLUENT code as well as the new FLUENT/UNS Version 4.0 were used for numerical simulations. Calculations were performed with both structured and unstructured grid generation (Fig. 3). The calculation domain was a grid 94 cells high, 23 cells wide, and 29 cells long for the structured grids. A plane of symmetry through the longitudinal building center was utilized for the 0° and 180° wind orientations. For the 45° orientation the entire building was placed in an unstructured grid volume of 36 875 tetrahedral cells. Outlet and pressure inlet or symmetry boundaries were specified at the sides and top of the grid volume, while a surface roughness of 0.01 m was specified at the ground and 0 m on all the building surfaces. The roughness height was selected based on calculations with different uniform ground roughness (no building) where at 0.01 m roughness height no change in velocity and turbulent intensity profile was observed. The inflow boundary conditions were chosen to match the velocity and turbulence profiles measured during the wind-tunnel experiments (see Fig. 2). Outflow boundary conditions were chosen to maintain constant longitudinal rate of change of all dependent variables (i.e. constant slope). In order to minimize the computing time all simulations were started as laminar solutions without species transport from an initial guess based on the values given at the flow inlet. Starting from the laminar solution the turbulent flow field was calculated in the second step and the species transport was added to the solution

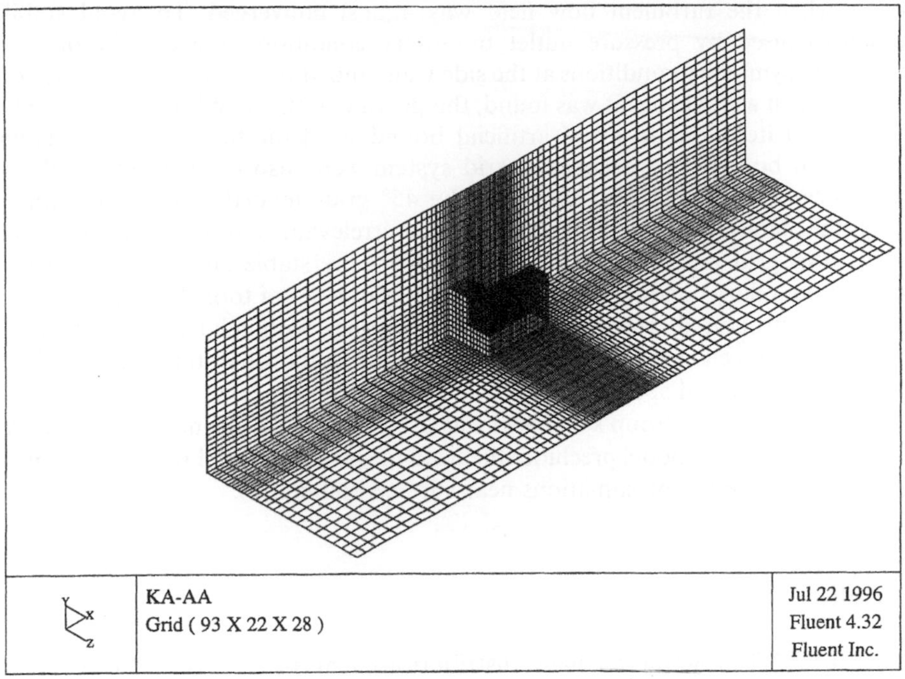

(a)

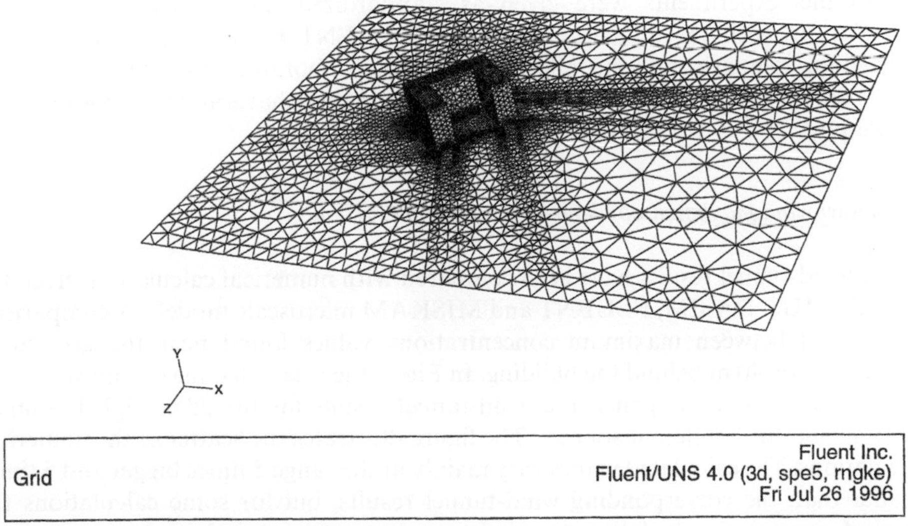

(b)

Fig. 3. (a) Typical structured grid used for 0°/180° wind direction, (b) unstructured boundary mesh used for tetrahedral mesh generation (45° setup).

process when the turbulent flow field was almost converged. To avoid stability problems caused by pressure outlet boundary conditions, the calculations were started with symmetry conditions at the side walls and at the top of the computational domain. When a stable result was found, the pressure outlet conditions were used for a final set of iterations to avoid artificial bounding of the flow. Several tests with a structured body-fitted rectangular grid system were also carried out for the 45° setup. Because of higher grid skewness the 45° grids tended to cause instabilities during the iteration process which produced irrelevant solutions. The optimized unstructured grid was found to be significantly more stable and robust even when changing boundary conditions. FLUENT provides a set of tools for generating and adapting unstructured triangular/tetrahedral meshes with relative ease. Thus, the total number of cells used in an unstructured grid could be optimized for solution accuracy, stability and speed.

The ReNormalized Group k–ε model was used to provide turbulent closure during the calculations. This model precludes the need to specify any calibration constants or to use wall function approximations near surface boundaries.

5. Results

The FLUENT program calculated distributions of mean velocity, turbulent energy, pressure and mean concentrations. Since the mean concentrations available from the wind-tunnel experiments were given as standardized normalized concentration, $K = Cu_{10}/Q$, with the dimensions $1/\mathrm{m}^2$ all FLUENT results were similarly normalized for comparison in the following diagrams. Laboratory and numerical model results are presented in the form of scatter diagrams, bar charts and iso-concentration profile plots.

5.1. Downwind maximum concentrations near the ground

The wind-tunnel results have been compared with numerical calculations from the ABC, ASMUS, DASIM, FLUENT and MISKAM microscale models. A comparison is provided between maximum concentrations values found near the ground in a cross-section 40 m behind the building. In Fig. 4 the calculated maximum values for all models are plotted against the wind-tunnel results for the 28 m high U-shaped building and the 3 different sources. The figure illustrates the scatter of the numerical calculations. The calculated values vary mainly in the range 5 times bigger and 5 times smaller than the corresponding wind-tunnel results, but for some calculations the ratio of numerical calculation to wind-tunnel experiment is less than one tenth. FLUENT calculations with the RNG turbulent model fall in the same range of accuracy except for source B and C at 0° wind direction and source C at 180° wind direction. FLUENT predicted significantly higher concentrations (up to a factor of 32), when a stationary vortex structure was near the 40 m profile. Otherwise the model DASIM produced the worst comparison range.

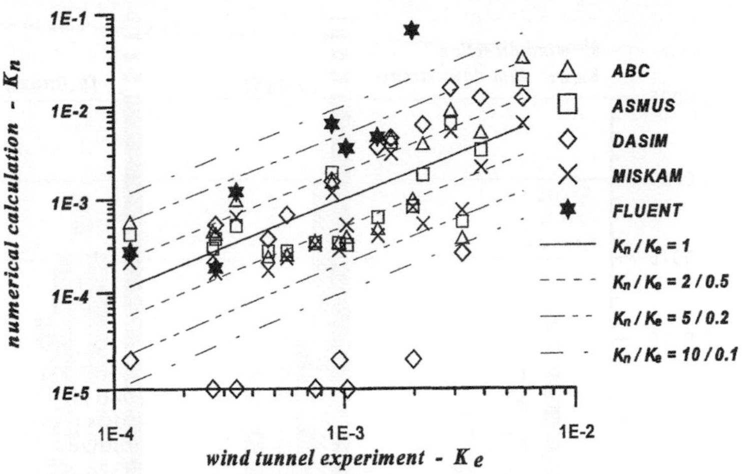

Fig. 4. Comparison between wind tunnel and numerical models using maximum concentrations near ground 40 m behind U-shaped building, height $H_b = 28$ m.

In Fig. 5 bar diagrams are shown, where calculated maximum concentrations at 40 m downstream of the building are compared with the corresponding wind-tunnel values. DASIM consistently underpredicts pollution for source A, located on the roof of the building. It also incorrectly predicts that almost no pollutants reach the ground; hence, the large discrepancies in the scatter diagram, Fig. 4. Concentrations calculated with FLUENT agree well with wind-tunnel results for the roof source situations (source A, 0°/180° wind direction) except at the 40 m downstream section. In a steady-state FLUENT simulation, stationary vortex structures are calculated at the edges of the building, and these structures are usually enriched with higher concentrations of tracer gas. In a wind tunnel as well as in nature, vortex shedding would lead to significantly higher mixing and lower local tracer concentrations. A lower lateral dispersion downstream of the building was observed for all FLUENT simulations. Once again the stationary simulation which does not lead to the characteristic vortex shedding at the building obviously caused this discrepancy, and a time dependent solution should give at least more intensive lateral mixing.

Large discrepancies were also found for most of the ground level release situations. Even if the roughness height was chosen to give similar inflow and outflow profiles for test calculations with a rough surface, it might be necessary to explicitly simulate sharp edged surface roughness elements. An artificial increase of turbulent intensity at the flow inlet as well as a significant increase of the surface roughness did not show a corresponding increase in turbulent mixing near the ground. Source B release at 180° wind direction was the only case were FLUENT predicted lower concentrations than measured in the wind tunnel.

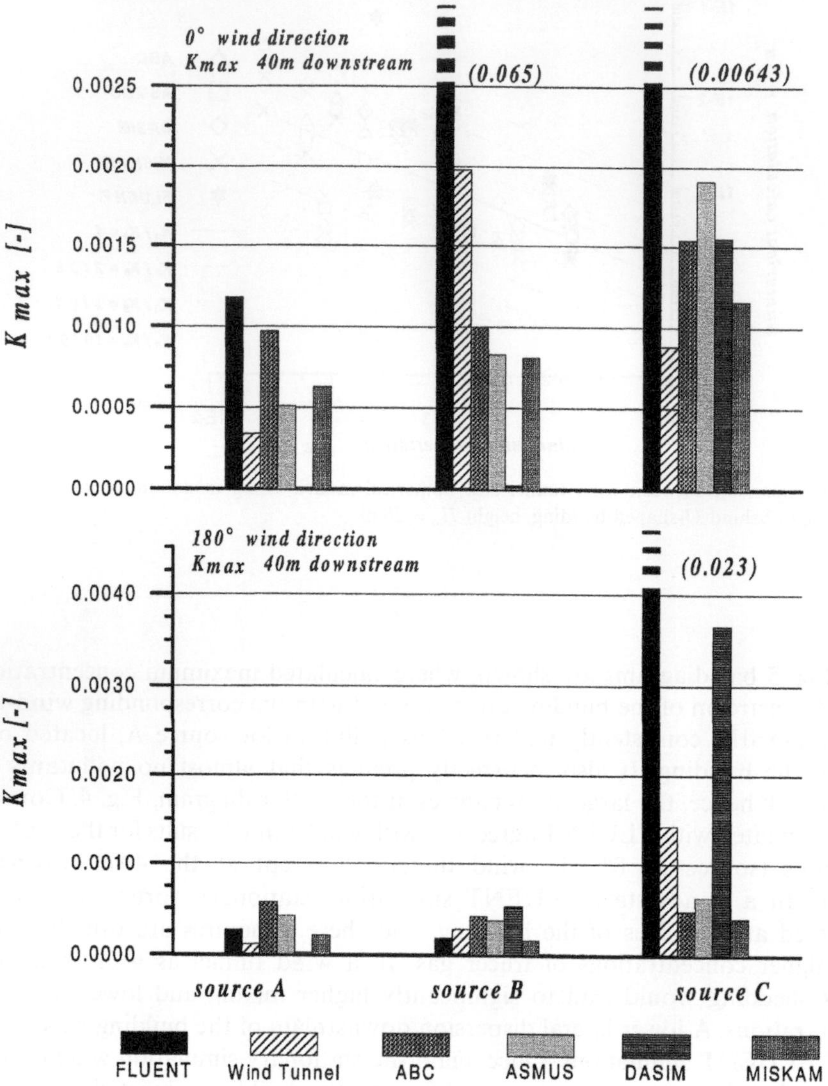

Fig. 5. Comparison between wind tunnel and numerical models using maximum normalized concentrations measured near ground 40 m behind U-shaped building, height $H_b = 28$ m.

5.2. Isolines along a symmetry axis

Isolines of concentration in a vertical plane through the building symmetry axis are shown for the case of source B and wind direction 0° in Fig. 6 (K values are multiplied by a factor of 1000). The figure demonstrates the differences between the calculated plumes. ASMUS, FLUENT and MISKAM show the best agreement with the

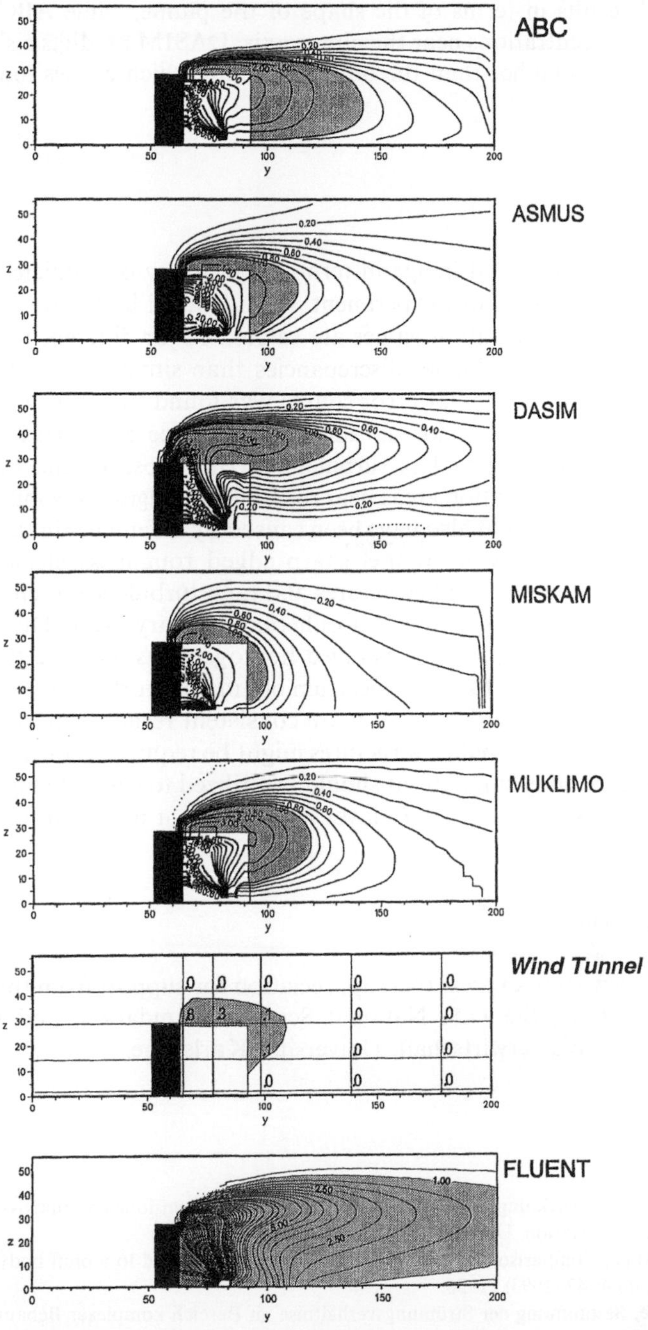

Fig. 6. Iso-concentration lines from wind tunnel and numerical models along longitudinal centerline cross-sections for the case of source B and wind direction 0°.

wind-tunnel results in terms of the shape of the plume, while ABC and FLUENT predict high concentrations near the plume axis. DASIM predicts a shift of the plume axis up to heights higher than the building height which causes lower ground level concentrations.

6. Conclusions

FLUENT gives a mixed image in terms of accuracy of predicted concentrations compared to the wind-tunnel experiment. For identical boundary conditions higher as well as lower concentration values are calculated for different test cases. Ground level sources show much higher discrepancies than situations where the tracer was emitted from the roof. A major error source was found to be the stationary solution procedure that was chosen for all simulations. Since no vortex shedding at the building edges is calculated less turbulent mixing close to the building leads to stationary high concentration areas near the building edges. Less mixing observed for ground level releases might also have been caused by differences in turbulent structure close to the wall. In a wind tunnel sharp edged roughness elements are used to simulate large scaled turbulent structures and high turbulence intensities close to the wall that meet conditions in the atmospheric boundary layer. During a stationary FLUENT simulation a spectral-averaged homogeneous-isotrope turbulence is presumed whereas in a wind tunnel a spectrum of different turbulent structures take part in turbulent mixing. In order to get more consistent results a more accurate way of simulating large-scale turbulent structures might be required. At least time-dependent solutions or large eddy simulations should be utilized to simulate high turbulent flow fields in an atmospheric boundary layer and turbulent mixing in urban areas.

Acknowledgements

The authors wish to express their appreciation for support from the Alexander von Humboldt Stiftung, the U.S. National Science Foundation, and the Institut für Hydrologie und Wasserwirtschaft, Universität Karlsruhe.

References

[1] J. Eichorn, Entwicklung und Anwendung eines dreidimensionalen mikroskaligen Stadtklima-Modelles, Dissertation, Universität Mainz, Germany (1989).

[2] S. Murakami, Comparison of various turbulence models applied to a bluff body, J. Wind Eng. Ind. Aerodyn. 46 & 47 (1993) 21–36.

[3] R. Röckle, Bestimmung der Strömungsverhältnise im Bereich komplexer Bebauungstrukturen, Dissertation, TH Darmstadt, Germany (1990).

[4] S. Blinda et al., Entwicklung und Verifizierung eines Rechenmodells zur Simulation der Ausbreitung von KFZ-Emissionen in Berich komplexer Gebäudekonfigurationen, Abschlußbericht MURL, NRW, Inst. F. Meteorogie, TH Darmstadt, Germany (1992).

[5] G. Gross et al., ASMUS – Ein numerisches Modell zur Berechnung der Strömung und der Schadstoffverteilung im Bereich einzelner Gebäude, Met. Rundschau (1994).
[6] P. Klein, M. Rau, R. Röckle, E.J. Plate, Concentration estimation around point sources located in the vicinity of U-shape buildings and in a built-up area, 2nd Int. Conf. Air Pollution, Barcelona, Spain, 27–29 September 1994.
[7] H. Panskus, Ein mikroskaliges diagnostisches Strömungs-und Ausbreitungsmodell für komplex bebautes Gelände – Evaluierung und Validation, Diplomarbeit im Fach Meteorologie, Universität Hamburg, Hamburg, Germany, (1995) 154 pp.
[8] V. Yakhot, S.A. Orzag, Renormalized group analysis of turbulence – I: basic theory, J. Sci. Comput. 1 (1986) 1–51.
[9] P. Klein, M. Rau, Z. Wang, E.J. Plate, Ermittlung des Strömungs- und Konzentrationsfeldes im Nahfeld typischer Gebäudekonfigurationen (Experimente), Paper 10, Statuskolloquium des PEF, 15–17 März 1994 und PEF-Abschlussbericht, Kernforschungszentrum Karlsruhe, Germany (1994a).
[10] E.J. Plate (1982) Wind tunnel modelling of wind effects in engineering, in: Engineering Meteorology, Ch. 13, Elsevier, Amsterdam, pp. 573–639.
[11] Fluent User's Guide, Vols. 1–4, Fluent Inc., Centerra Resource Park, Lebanon, New Hampshire, USA.

ELSEVIER

Journal of Wind Engineering
and Industrial Aerodynamics 67&68 (1997) 757–766

JOURNAL OF wind engineering AND industrial aerodynamics

Numerical evaluation of wind-induced dispersion of pollutants around a building

Y. Li, T. Stathopoulos*

Centre for Building Studies, Concordia University, 1455 de Maisonneuve Blvd, Montreal, Quebec, Canada, H3G 1M8

Abstract

Wind flow perturbations, recirculations and turbulence generated by buildings often dominate air pollutant distributions around buildings; this may be a great concern for engineers, architects and health professionals. The paper refers to an attempt to evaluate the air pollutant distribution around a building by solving the concentration equation based on the previously simulated wind flow field. The paper shows the application of the hybrid scheme for the evaluation of pollutant concentration around a rectangular building for two different sources under conditions of neutral atmospheric stratification. Results have been compared to the available experimental data from previous studies in boundary layer wind tunnels. Data agree well far downwind of the building but agree less satisfactorily close to the wall and within the wake zone.

Keywords: Wind; Building; Pollutant dispersion; Turbulence model

1. Introduction

Air pollutants disposed from or near a building can often have much higher than the allowable concentration at some points and if these points happen to be on the location of an open window or the intake of the ventilation system, the health of residents can severely be influenced. Sometimes this may even require the evacuation of a building.

The flowfield around a single building placed in a surface boundary layer is fully turbulent and very complex with separation and recirculation on each surface of the building: when this problem is extended to an arbitrary configuration of buildings and conditions of pollutant dispersion, this complexity is clearly compounded.

* Corresponding author. E-mail: statho@cbs-engr.concordia.ca.

PII S0167-6105(97)00116-5

Traditionally, the problem has been studied using models of buildings in wind-tunnels. The task of constructing a series of scale-model experiments to explore systematically the general "model space" of a collection of buildings is laborious because of the multiplicity of configurations that must be investigated. Furthermore, wind-tunnel experiments require resources of time and expertise which are often not directly available to architects and planners.

A reliable computer simulation of wind flow around buildings can make a contribution to this problem by facilitating the less time-consuming exploration of the model space. In principle, a computer wind-flow simulation can make wind-related design information accessible to an architect at every stage of the design process. In contrast to the increasing cost of performing experiments, this alternative becomes more and more encouraging by considering the relative cost of computation for a given algorithm and flow decrease by a factor of 10 every 8 years [1].

2. Review of numerical simulation

Although there have been several studies on the evaluation of dispersion of pollutants around buildings by physical simulation (e.g. Refs. [2–9]), very few results of numerical simulation of air pollution around buildings have appeared in the literature. Dawson et al. [10] computed the dispersion of a building rooftop release. Additional computations of transport and dispersion of plume over a 300 m conical hill were also carried out and results were compared with near ground-level field measurement data. The effect of false diffusion appeared to be more significant in ground-level regions where agreement between the calculated and measured concentrations was rather poor. Zhang [11] investigated the effects of approach flow shear, turbulence and atmospheric stability on the flow and pollutant diffusion around a cubic building, especially in the wake area. The study calculated the concentration field for rooftop releases. The simulation predicted the concentration field reasonably well both on the ground and in the wake of the building but underestimated the lateral diffusion (σ_Y) and overpredicted the vertical diffusion inside the wake. Both studies used the computer code TEMPEST [12] which is an Eulerian, finite-difference code designed to solve the time-dependent equations of motion, continuity, and energy conservation for turbulent flow in incompressible fluids. The solution technique in TEMPEST is similar to the Simplified Marker-And-Cell (SMAC) technique [13]; at each time step, the momentum equations are solved explicitly and pressure equations implicitly; temperature, turbulent kinetic energy, dissipation of kinetic energy, and scalar transport equations are solved using an implicit continuation procedure. The numerical simulation of diffusion field around a building complex by large eddy simulation has been carried out by Murakami et al. [14]. Their numerical results compared well with wind-tunnel experimental data. Selvam and Huber [15] provided a general review of the current status of numerical modelling of pollutant dispersion around buildings.

3. Present computational method

In the computations presented in this paper a steady state k–ε model of turbulence is used. The Navier–Stokes Equations with the k–ε model are as follows:

$$U_j \frac{\partial U_i}{\partial x_j} = -\frac{\partial P}{\partial x_i} + \frac{\partial}{\partial x_j}\left(\nu_t \frac{\partial U_i}{\partial x_j}\right) + \frac{\partial \nu_t}{\partial x_j}\frac{\partial U_j}{\partial x_i}, \quad (1)$$

$$\frac{\partial U_i}{\partial x_i} = 0, \quad (2)$$

$$\nu_t = C_\mu \frac{k^2}{\varepsilon}, \quad (3)$$

$$U_i \frac{\partial k}{\partial x_i} = \frac{\partial}{\partial x_i}\left(\frac{\nu_t \partial k}{\sigma_k \partial x_i}\right) + \nu_t\left(\frac{\partial U_i}{\partial x_j} + \frac{\partial U_j}{\partial x_i}\right)\frac{\partial U_i}{\partial x_j} - \varepsilon, \quad (4)$$

$$U_i \frac{\partial \varepsilon}{\partial x_i} = \frac{\partial}{\partial x_i}\left(\frac{\nu_t}{\sigma_\varepsilon}\frac{\partial \varepsilon}{\partial x_i}\right) + C_{1\varepsilon}\frac{\varepsilon}{k}\nu_t\left(\frac{\partial U_i}{\partial x_j} + \frac{\partial U_j}{\partial x_i}\right)\frac{\partial U_i}{\partial x_j} - C_{2\varepsilon}\frac{\varepsilon^2}{k}, \quad (5)$$

$$U_i \frac{\partial C}{\partial x_i} = \frac{\partial}{\partial x_i}\left(\frac{\nu_t}{\sigma_c}\frac{\partial C}{\partial x_i}\right) + S, \quad (6)$$

where

$$P = \frac{\overline{P}}{\rho} + \frac{2}{3}k, \quad (7)$$

is the augmented pressure and $\overline{P}$ is the mean pressure; U_i represents the mean velocity components along the x, y and z axes; k is the turbulent kinetic energy; ε is the isotropic dissipation of turbulence kinetic energy; ν_t is the eddy viscosity; ρ is the fluid density; S is the mean volume contaminant source generation rate and C is the concentration of pollutant; the constants assume the approximate empirical values of $C_\mu = 0.09$, $\sigma_k = 1.0$, $\sigma_\varepsilon = 1.3$, $C_{1\varepsilon} = 1.44$ and $C_{2\varepsilon} = 1.92$, $\sigma_c = 0.7$ (see Refs. [15–17]).

A flowfield was calculated by Turbulent Wind Simulation Technique (TWIST), which is a code developed in the Centre for Building Studies at Concordia University for the prediction of wind flow around buildings and wind-induced pressures on building surfaces. This code is a finite difference code based on the control volume concept using Semi-Implicit Method for Pressure-Linked Equations (SIMPLE) method. The turbulence is modelled by the k–ε model (Eq. (3)). The staggered rectangular grid system has been adopted in the present study. Special care is taken in arranging the computational grid nodes, particularly those near the building surfaces. In generating a grid system, the program first reads in the distance from the first gridline to its adjacent solid boundary; successive gridlines are then generated at the specific expanding factor defined for each region starting from that solid boundary. This grid generating system can ensure that regions where physical properties change fast have high-density grids, whereas the slow-change regions have coarse grids, thus providing the most economical use of computer resources.

Since the computational domain adopted is very large compared with the building model, the free boundary conditions are imposed at air-to-air boundaries as follows:

$$U(x, y, z) = U_g\left(\frac{z}{z_g}\right)^{\alpha}, \quad V(x, y, z) = 0, \quad W(x, y, z) = 0,$$

$$k(x, y, z) = k_0(z), \quad \varepsilon(x, y, z) = \varepsilon_0(z), \quad P(x, y, z) = P_0(z) = \frac{2}{3}k_0(z),$$

where U_g is the reference velocity at reference height z_g, α is the power-law velocity profile exponent; $k_0(z)$ and $\varepsilon_0(z)$ are obtained from simplified k and ε equations under conditions $V = 0$, $W = 0$, $\partial/\partial x = 0$, $\partial/\partial y = 0$.

At solid boundaries (ground, walls, roof), log-linear-law wall functions are used for velocities:

$$U_i = \frac{(\tau_i/\rho)\ln(EY^+)}{C_\mu^{1/4}k^{1/2}\kappa} \quad (Y^+ > 11.6) \quad \text{or } U_i = \frac{(\tau_i/\rho)Y^+}{C_\mu^{1/4}k^{1/2}} \quad (Y^+ \leqslant 11.6),$$

where U_i represents the component of tangential velocity along the x_i direction, τ_i is the corresponding shear stress on the wall, Y^+ is the dimensionless distance from the wall to the grid node expressed as $Y^+ = C_\mu^{1/4}k^{1/2}d/\nu$, in which d is the distance from the grid node to the wall. Zonal treatments are used for k and ε:

$$k_s = k_{es}\frac{d_s^2}{d_{es}^2} \quad \text{and} \quad \varepsilon_s = \frac{2\nu k_{es}}{d_{es}^2},$$

where subscripts es and s indicate "at the edge" and "within" the viscous sublayer, respectively. Zero-gradient conditions are applied for all the boundaries for concentration.

The results of present simulation are compared with measured concentrations from a wind-tunnel study [18] and with simulated results from Ref. [15]. In the experimental study, a rectangular-shaped building with its length equal to twice its height and width was selected. The long side of the building was oriented perpendicularly to the approaching wind. The model building height was $H_b = 25$ cm and the depth of the boundary layer δ was 1.8 m. The velocity profile U/U_r was found to fit a one-sixth power law, which is generally representative of neutral atmospheric flow over moderately smooth terrain [19]. The reference velocity, $U_r = 2.34$ m/s, was measured at $1.5H_b$ above the ground. The concentration measured in the experiment was used in nondimensional form $CU_rH_b^2/Q$, where Q is the emission rate. For this study, a 1 : 200 scale model of a 50 m high-prototype building would be an appropriate example. At this scale, the wind-tunnel boundary layer represents a 360 m deep atmospheric boundary layer.

The numerical study simulated the wind tunnel experiment. The computational domain was extended 417 cm (streamwise) × 280 cm (spanwise) × 144 cm (height), and was divided into 32 × 21 × 19 grid points, as shown in Fig. 1. The cases studied include a point source at the ground (position 1) and at a height $1.2H_b$ above the ground (position 2) at the center of the leeward building wall, as shown in Fig. 2. These cases

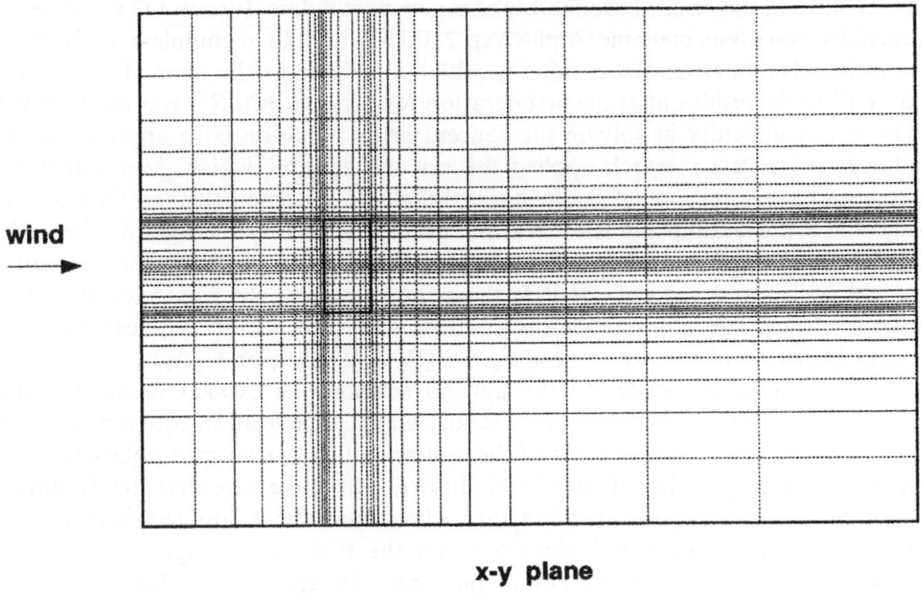

Fig. 1. Grid system.

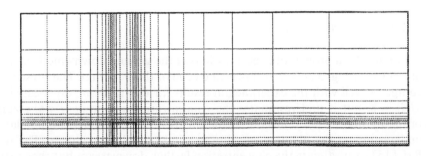

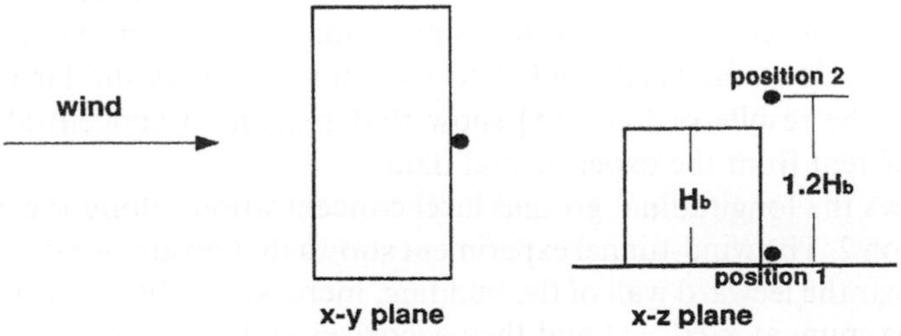

Fig. 2. Source position.

were selected on the basis of availability of experimental data. It took 132 iterations in 4 min (CPU) on a Vax machine (Alpha Axp-2100 Model 500) to complete the flowfield calculation. The final relative residue r_n, which is defined as the ratio of the current residue R^n to the residue after the first iteration R^1, i.e. $r_n = |R^n|/|R^1|$, was less than 0.1.

The major difficulty in solving the concentration equation is to approximate its nonlinear part. Past research applied the upwind scheme, which gives stable but inaccurate results because of the artificial diffusivity inherited therein. One way to reduce the artificial diffusion is to treat the nonlinear part by the hybrid scheme for advection term and a central-difference for diffusion term. Using the computed flowfield of U_i, and ν_t from TWIST, the concentration field was calculated with a hybrid scheme. The hybrid scheme is identical with the central-difference scheme for the grid Peclet number range $-2 \leqslant P_e \leqslant 2$ ($P_e = U_1\Delta x/\nu_t$ or $U_2\Delta y/\nu_t$ or $U_3\Delta z/\nu_t$); outside this range it reduces to the upwind scheme. An explicit first-order time marching method is used in solving the steady-state concentration equation until the root mean square of the variation of the concentrations at all grid points between successive time steps is less than 0.001%. In order to make sure that steady state is reached for this convergence criterion, the code is re-run with doubled time iteration steps. There was no large difference between the two results suggesting that the convergence criterion is satisfied. For a typical run with approximate 10 000 steps and time step dt = 0.0002, it takes 40 min CPU time on the Vax machine.

4. Results and discussion

Longitudinal profiles of concentrations at the ground downwind from source 1 are presented in Fig. 3 ($x = 0$ at downwind edge of the building). The present simulation compares well with the experimental results far downstream, especially farther than $x/H_b = 8$, but it underestimates the concentration in the near wake zone. This is probably due to the diffusion coefficient used in the numerical simulation, which underestimates the plume dispersion in wind-tunnel experiments. The results from Ref. [15] overestimate the concentration farther downstream but they seem to work better in the near wake zone.

Fig. 4 shows the comparison of present simulation results with previous simulation and wind tunnel experimental data for the vertical concentration profile at $3H_b$ downstream from the leeward building wall. Generally, though the present simulation somewhat underestimates the concentration, the trend of concentration variation with height is predicted well. The present simulation shows the maximum concentration approximately at the height of $0.7H_b$ and then decreases similar to the experimental data. The results of Ref. [15] show that maximum concentration near the ground is different from the experimental data.

Fig. 5 shows the longitudinal ground-level concentrations along the centerline for source position 2. The wind-tunnel experiment shows that ground-level concentration is very low near the leeward wall of the building, increases farther downstream until it reaches a maximum at $x/H_b = 3$ and then decreases and becomes relatively flat after $x/H_b = 7$. The present simulation predicts higher concentration near the leeward wall,

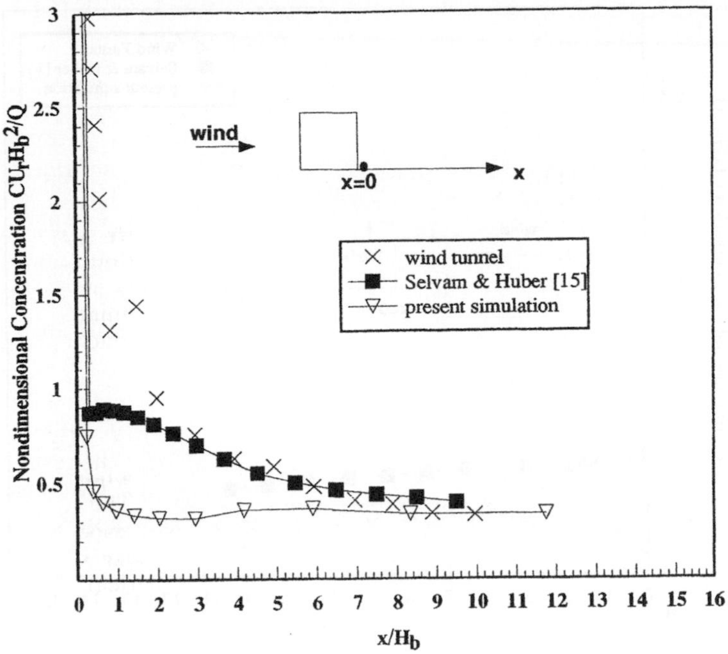

Fig. 3. Comparison of longitudinal ground-level concentration along the centerline for source position 1.

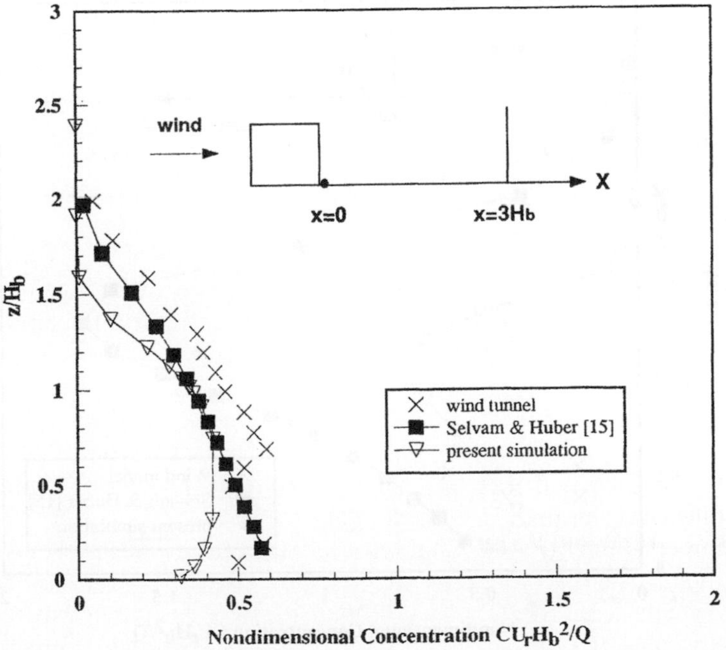

Fig. 4. Comparison of vertical profiles of concentrations along the centerline at $x/H_b = 3$ for source position 1.

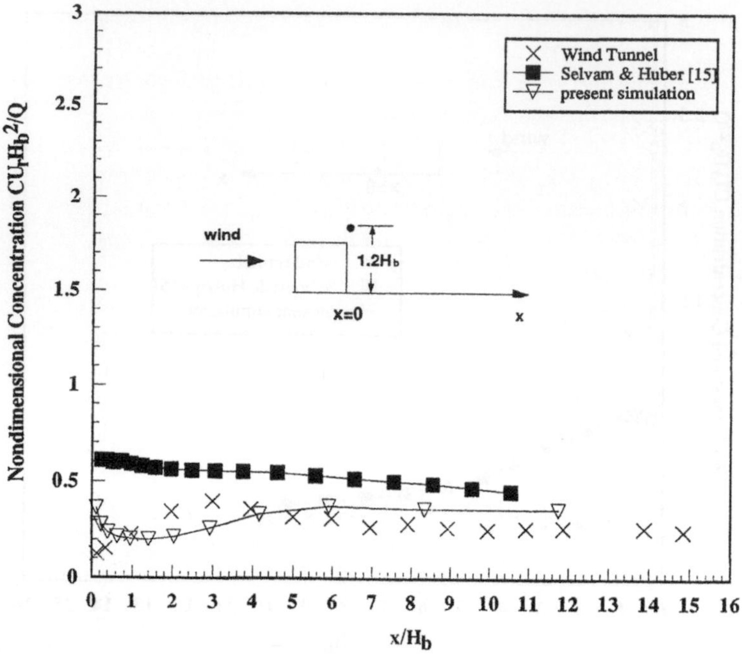

Fig. 5. Comparison of longitudinal ground-level concentration along the centerline for source position 2.

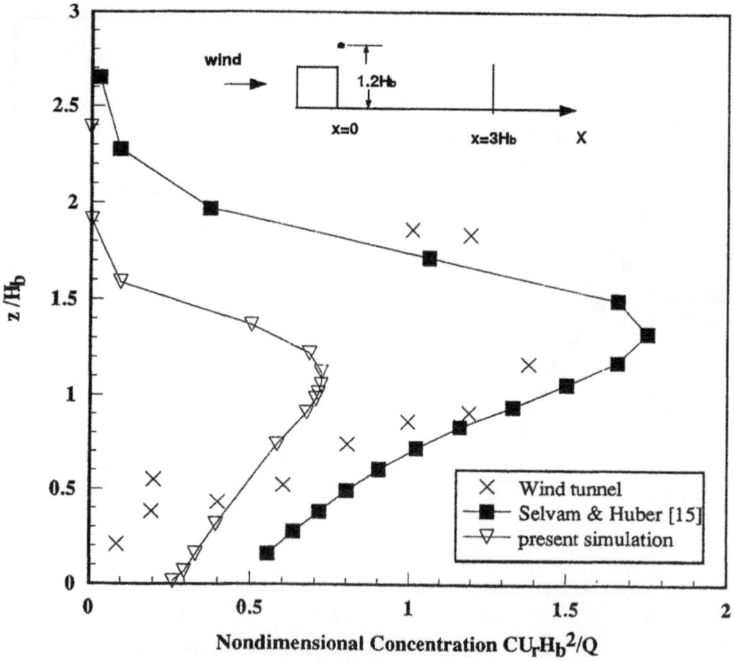

Fig. 6. Comparison of vertical profiles of concentrations along the centerline at $x/H_b = 3$ for source position 2.

decreases along the x-direction up to $x/H_b = 1$ and then it follows a similar trend with the experimental results. The prediction from Ref. [15] is rather poor; the simulated values are always higher than the experimental results and the ground-level concentration keeps decreasing farther downstream.

Fig. 6 shows the vertical profiles of concentration along the centerline at $3H_b$ for source position 2. The prediction is consistent with the experimental data at the lower part of the building height but it fails clearly at the higher part.

The prediction from Ref. [15] agrees much better with the experimental results at the higher part but not at the lower part.

Neither method seems to be completely satisfactory at the moment. Some improvements have been made in the numerical evaluation of wind flowfield near buildings – see, for instance, Ref. [20] – additional attempts may be made to improve the prediction of pollutant concentrations in the building environment. In particular, the application of a two-layer method combining with the k–ε model in external fully turbulent flow region and the Norris and Reynolds' one-equation model [21] in inner region is expected to be more successful because of the more accurate representation of simulated flow conditions.

5. Conclusions

The paper demonstrates the application of a numerical approach for the evaluation of wind-induced dispersion of pollutants in the vicinity of a building. A hybrid scheme has been used successfully in the numerical solution of the concentration equation after the general wind flow field has been simulated by the k–ε turbulence model. The comparison of the results with experimental data is satisfactory further downstream but improvements are necessary near the wall.

References

[1] D.R. Chapman, Computational aerodynamics development and outlook, AIAA J. 17 (12) (1979) 1293–1313.

[2] J. Halitsky, Gas diffusion near buildings, ASHRAE Trans. 69 (1963) 464–485.

[3] A.G. Robins, I.P. Castro, A wind Tunnel Investigation of Plume Dispersion in the Vicinity of a Surface Mounted Cube – II. The Concentration Field, Atmos. Environ. 11 (1977) 299–311.

[4] D.J. Wilson, R.E. Britter, Estimates of building surface concentrations from nearby point sources, Atmos. Environ. 16 (11) (1982) 2631–2646.

[5] Wen-whai Li, R.N. Meroney, Gas dispersion near a cubical model building. Part I. Mean concentration measurements, J. Wind Eng. Ind. Aerodyn. 12 (1983) 15–33.

[6] Y. Ogawa, S. Oikawa, K. Uehara, Field and wind tunnel study of the flow and diffusion around a model cube – II. nearfield and cube surface flow and concentration patterns, Atmos. Environ. 17 (6) (1983) 1161–1171.

[7] P.J. Saathoff, T. Stathopoulos, M. Dobrescu, Effects of model scale in estimating pollutant dispersion near buildings, J. Wind Eng. Ind. Aerodyn. 54–55 (1995) 549–559.

[8] B. Lamb, D. Cronn, Fume hood exhaust re-entry into a chemistry building, Am. Ind. Hyg. Assoc. J. 47 (2) (1986) 115–123.

[9] H.L. Higson, R.F. Griffiths, C.D. Jones, D.J. Hall, Concentration measurements around an isolated building: A comparison between wind tunnel and field data, Atmos. Environ. 28 (11) (1994) 1827–1836.
[10] P. Dawson, D.E. Stock, B. Lamb, The numerical simulation of airflow and dispersion in three-dimensional atmospheric recirculation zones, J. Appl. Meteorol. 30 (1991) 1005–1024.
[11] Y.Q. Zhang, Numerical simulation of flow and dispersion around buildings, Ph.D. Thesis, Department of Marine, Earth and Atmospheric Science, North Carolina State University, 1993.
[12] S.D. Trent, L.L. Eyler, TEMPEST, A Three-Dimensional Time-Dependent Computer Program for Hydrothermal Analysis, vol. 1: Numerical Methods and Input Instructions, PNL-4348 Pacific Northwest Laboratory, Battelle, WA, 1989.
[13] A.A. Amsden, F.H. Harlow, The SMAC method: a numerical technique for calculating incompressible flows, Report LA-4370, Los Alamos Scientific Laboratory, Los Alamos, NM, 1970.
[14] S. Murakami, A. Mochida, Y. Hayashi, Numerical simulation of velocity field and diffusion field in an urban area, Energy Buildings 15–16 (1991) 345–356.
[15] R.P. Selvam, A.H. Huber, Computer modeling of pollutant dispersion around buildings: current status, Proc. 9th Int. Conf. Wind Eng., New Delhi, India, 1995, pp. 596–605.
[16] B.E. Launder, D.B. Spalding, The numerical computation of turbulent flows, Comput. Methods Appl. Mech. Eng. 3 (1974) 269–289.
[17] D.B. Spalding, Concentration fluctuations in a round turbulent free jet, Chem. Eng. Sci. 26 (1971) 95–107.
[18] A.H. Huber, W.H. Snyder, R.E. Lawson, Jr., The effects of a squat building on short stack effluents: a wind-tunnel study, Report EPA-600/4-80-055, U.S. Environment Protection Agency, Research Triangle Park, NC, 1980.
[19] A.G. Davenport, The relationship of wind structure to wind loading, Paper 2. Proc. Conf. Wind Effects on Buildings and Structures, National Physics Laboratory, H.M.S.O., London, June 1965, pp. 54–102.
[20] Y.S. Zhou, T. Stathopoulos, Application of Two-Layer Methods for the Evaluation of Wind Effects on a Cubic Building, ASHRAE Trans. 102 (1), Paper AT-96-10-2, 1996.
[21] L.H. Norris, W.C. Reynolds, Turbulent channel flow with a moving wavy boundary, Report No. FM-10, Stanford University, Department of Mechanical Engineering, Stanford, CA, USA, 1975.

ELSEVIER

Journal of Wind Engineering
and Industrial Aerodynamics 67&68 (1997) 767–779

JOURNAL OF
wind engineering
AND
industrial
aerodynamics

Numerical study of atmospheric dispersion under unstably stratified atmosphere

C.H. Liu, D.Y.C. Leung*

Department of Mechanical Engineering, 7/F, Haking Wong Building, The University of Hong Kong, Pokfulam Road, Hong Kong, China

Abstract

Pollutant dispersion under unstably stratified atmosphere was investigated numerically using a second-order closure turbulent dispersion model. The effect of atmospheric stability, which affects the plume trajectory, was studied by carrying out two-dimensional calculations in the horizontal and vertical planes. It was found that the present numerical model can predict several non-Gaussian features of plume behaviour under unstably stratified atmosphere, such as the descent and the rise of a plume, both of which cannot be predicted by *k*-theory dispersion and conventional Gaussian models. The calculations agreed well with recent findings from experimental studies and another numerical model. Based on the above observations, the second-order closure dispersion model was found to be a promising technique for studying the plume behaviour under different atmospheric stabilities.

Keywords: Atmospheric dispersion modelling; Finite element method; Unstably stratified atmosphere

Nomenclature

A	turbulent closure constant = 0.75
b	turbulent closure constant = 0.125
c	pollutant concentration
$\overline{c'\theta'}$	covariance of concentration and temperature
c_z, c_y	crosswind integrated concentration in the horizontal and vertical planes, respectively
$\overline{c'\theta'_y}$	crosswind integrated covariance of concentration and temperature in the vertical plane

* Corresponding author. E-mail: ycleung@hkucc.hku.hk.

PII S0167-6105(97)00117-7

C	dimensionless crosswind integrated concentration $= (U_m z_i c)/Q$
E	turbulent kinetic energy $= (1/2)q^2$
g	gravity acceleration
$[J]$	Jacobian matrix
k	eddy diffusivity
$[K]$	stiffness matrix
$[M]$	mass matrix
$[N]$	interpolation function $= [N_1\ N_2\ N_3\ N_4]$
q	turbulent velocity $= (\overline{u'^2} + \overline{v'^2} + \overline{w'^2})^{1/2}$
Q	pollutant emission rate
s	turbulent closure constant = 1.8
t	time
U	mean wind speed along x-direction
U_m	mean wind speed in the convective boundary layer
U_s	wind speed at source height
$\overline{v'^2}, \overline{w'^2}$	crosswind speed variances
$\overline{v'c'}, \overline{w'c'}$	concentration fluxes
$\overline{v'c'_z}$	crosswind integrated flux in the horizontal plane
υ_c	turbulent closure constant = 0.3
$\overline{w'\theta'}$	vertical heat flux
$\overline{w'c'_y}$	crosswind integrated flux in the vertical plane
w_*	convective velocity scale
x, y, z	Cartesian coordinate in the streamwise, lateral and vertical directions, respectively
$\bar{x}$	downwind distance from the point source where boundary condition applies
(x_i, y_i)	global coordinate of node point i
$\{x\}$	global node point x coordinate in an element $= \{x_1\ x_2\ x_3\ x_4\}^T$
X	dimensionless streamwise distance $= (w_* x)/(z_i U_m)$
$\{y\}$	global node point y coordinate in an element $= \{y_1\ y_2\ y_3\ y_4\}^T$
Y	dimensionless lateral distance $= y/z_i$
z_i	height of convective boundary layer
Z	dimensionless vertical height $= z/z_i$
Δt	time interval
Λ	turbulent length scale
θ	potential temperature
(ξ_i, η_i)	local coordinate of node point i in an element
σ_y, σ_z	pollutant dispersion coefficients for horizontal and vertical planes, respectively
$\{\phi\}$	node value $= \{\phi_1\ \phi_2\ \phi_3\ \phi_4\}^T$

1. Introduction

In air pollution studies, atmospheric dispersion of passive contaminants in convective boundary layers attracts the interests of many researchers because of the complexity it involves and the important consequences it may bring to our environment. Convective boundary layer is usually created on a clear day with an unstable temperature gradient. Under this situation, pollutants diffuse rapidly but there is a good chance of high concentrations when sporadically reaching ground level. Laboratory experiments carried out by Willis and Deardorff [1] and Willis [2] showed that some of the plume behaviour cannot be described by conventional Gaussian models and *k*-theory dispersion models. In particular, the plume rise enhancement in unstably stratified atmosphere is difficult to be accurately obtained from these models. Furthermore, the fluctuation of wind direction is also a main factor affecting the dispersion of pollutants in the horizontal plane. Leung and Liu [3] found that assuming constant wind direction in the Gaussian models may overpredict the maximum concentrations by several times, which is in line with the findings of Hanna [4]. Empirically determined dispersion coefficients are used by the *k*-theory and Gaussian models to deal with the effect of wind fluctuations. This may introduce error since the dispersion coefficients are, in most cases, site specific.

Lewellen [5] derived a second-order closure dispersion model from an exact Reynolds stress equation to compute the spreading rate of a pollutant under neutral conditions, which was found to be very close to that predicted by Pasquill [6]. Furthermore, the model predicted several non-Gaussian features of pollutant dispersion caused by mechanisms such as wind shear variations in both magnitude and direction with altitude, and interaction with the inversion layer capping the boundary layer. Later, Sykes et al. [7] presented a system of transport equations for the second-order correlations of passive scalar fluxes and fluctuations, and demonstrated its ability to predict dispersion characteristics in a wind-tunnel environment.

Based on the results of these studies, it can be observed that although the second-order closure model requires solving a complicated set of non-linear, coupled partial differential equations, it does provide a very general framework within which the dispersion under complex flow fields can be studied more accurately. Another advantage in the application of the second-order closure model over Gaussian and *k*-theory dispersion models is the direct input of turbulent flow data into the model instead of the empirically determined eddy diffusivities or dispersion coefficients. The turbulence data can be obtained from either field measurements or atmospheric simulation models.

In this paper, the atmospheric dispersion of a hypothetical source will be studied in both horizontal and vertical planes under unstably stratified atmosphere.

2. Numerical model

The set of governing transport equations for the present model in a horizontal and a vertical plane is described in this section. The crosswind integrated mean scalar concentration c_z and flux $\overline{v'c'_z}$ in the horizontal plane can be given as

$$\frac{\partial}{\partial t}c_z + U\frac{\partial}{\partial x}c_z = -\frac{\partial}{\partial y}\overline{v'c'_z}, \tag{1}$$

$$\frac{\partial}{\partial t}\overline{v'c'_z} + U\frac{\partial}{\partial x}\overline{v'c'_z} = -\overline{v'^2}\frac{\partial}{\partial y}c_z + \upsilon_c\frac{\partial}{\partial y}\left(q\Lambda\frac{\partial}{\partial y}\overline{v'c'_z}\right) - \frac{Aq}{\Lambda}\overline{v'c'_z}. \tag{2}$$

Similarly, the crosswind integrated mean scalar concentration c_y and flux $\overline{w'c'_y}$ in the vertical plane can be given as

$$\frac{\partial}{\partial t}c_y + U\frac{\partial}{\partial x}c_y = -\frac{\partial}{\partial z}\overline{w'c'_y}, \tag{3}$$

$$\frac{\partial}{\partial t}\overline{w'c'_y} + U\frac{\partial}{\partial x}\overline{w'c'_y} = -\overline{w'^2}\frac{\partial}{\partial z}c_y + \frac{g}{\theta}\overline{c'\theta'_y} + \upsilon_c\frac{\partial}{\partial z}\left(q\Lambda\frac{\partial}{\partial z}\overline{w'c'_y}\right) - \frac{Aq}{\Lambda}\overline{w'c'_y}. \tag{4}$$

The crosswind integrated temperature concentration correlation $\overline{c'\theta'_y}$ is governed by the following equation:

$$\frac{\partial}{\partial t}\overline{c'\theta'_y} + U\frac{\partial}{\partial x}\overline{c'\theta'_y} = -\overline{w'c'_y}\frac{\partial T}{\partial z} - \overline{w'\theta'}\frac{\partial}{\partial z}c_y + \upsilon_c\frac{\partial}{\partial z}\left(q\Lambda\frac{\partial}{\partial z}\overline{c'\theta'_y}\right) - \left(\frac{2bsq}{\Lambda}\overline{c'\theta'_y}\right). \tag{5}$$

For the turbulent closure constants of A, υ_c, b and s those empirical values suggested by Sykes et al. [8] are used so as to close the system of equations in an unstably stratified atmosphere. To validate the model, two dimensional cross-wind transport equations are used. Several crosswind integrated terms were used in the present computation, which are defined as follows. The crosswind integrated concentration is

$$c_n = \int_{n_1}^{n_2} c(x, y, z)\,\mathrm{d}n \tag{6}$$

where $n_1 = 0$, $n_2 = z_i$ and $n = z$ in the horizontal plane and $n_1 = -\infty$, $n_2 = \infty$ and $n = y$ in the vertical plane. The crosswind integrated fluxes are

$$\overline{v'c'_z} = \int_0^{z_i} \overline{v'c'}(x, y, z)\,\mathrm{d}z, \tag{7}$$

and

$$\overline{w'c'_y} = \int_{-\infty}^{\infty} \overline{w'c'}(x, y, z)\,\mathrm{d}y \tag{8}$$

for the horizontal and vertical plane, respectively. The cross-wind integrated temperature concentration covariance in the vertical plane is

$$\overline{c'\theta'_y} = \int_{-\infty}^{\infty} \overline{c'\theta'}(x, y, z)\,\mathrm{d}y. \tag{9}$$

3. Boundary conditions

To avoid the singularity at $x = 0$ (i.e. the point source centre where $c = \infty$) a finite value of c at a particular location is assigned, which is used as one of the boundary conditions. The following Gaussian distributions for c were used to simplify the calculation:

$$c_z(\overline{x}, y) = \frac{Q}{\sqrt{2\pi}\sigma_y(\overline{x})U_s}\left[\exp\left(\frac{-y^2}{2\sigma_y^2(\overline{x})}\right)\right] \tag{10}$$

for the calculation in the horizontal plane, and

$$c_y(\overline{x}, z) = \frac{Q}{\sqrt{2\pi}\sigma_z(\overline{x})U_s}\left[\exp\left(\frac{-(z-z_s)^2}{2\sigma_z^2(\overline{x})}\right) + \exp\left(\frac{-(z+z_s)^2}{2\sigma_z^2(\overline{x})}\right)\right] \tag{11}$$

for the calculation in the vertical plane. Several values of $\overline{x}$ were tested and it was found that $\overline{x} = 0.06$ is sufficiently small to give convergent results. It should be noted that in the vertical plane calculation, the method of image has been applied to account for the reflection of pollutant from ground surface.

The following expressions suggested by Pai and Tsang [9] are used as dispersion coefficients at $\overline{x}$:

$$\sigma_y(\overline{x}) = (\overline{v'^2})^{1/2}\frac{\overline{x}}{U_s} \tag{12}$$

and

$$\sigma_z(\overline{x}) = (\overline{w'^2})^{1/2}\frac{\overline{x}}{U_s}. \tag{13}$$

The boundary conditions for the concentration flux at $\overline{x}$ are

$$\overline{v'c'_z} = -k\frac{\partial c_z}{\partial y} \tag{14}$$

and

$$\overline{w'c'_y} = -k\frac{\partial c_y}{\partial z} \tag{15}$$

for the calculation in the horizontal and in the vertical plane, respectively. The eddy diffusivity k for unstably stratified atmosphere is

$$k = 3 \times 0.1 \Lambda \sqrt{E}. \tag{16}$$

In calculating the turbulent length scale Λ, the following expression suggested by Sun and Chang [10] is used:

$$\Lambda = 0.25\left\{1.8z_i\left[1 - \exp\left(\frac{-4z}{z_i}\right) - 0.0003\exp\left(\frac{8z}{z_i}\right)\right]\right\}. \tag{17}$$

In the horizontal plane calculation, both mean scalar concentration gradient $\partial c_z/\partial y$ and turbulence flux gradient $\partial \overline{v'c_z'}/\partial y$ are set equal to zero at the centreline ($y = 0$) and at the edge of the computation domain ($y = y_{\max}$), respectively.

In the vertical plane calculation, $\partial c_y/\partial z$, $\overline{w'c_y'}$ and $\overline{c'\theta_y'}$ are set equal to zero at the bottom of the boundary while $\partial c_y/\partial z$, $\partial \overline{w'c_y'}/\partial z$ and $\partial \overline{c'\theta_y'}/\partial z$ are set equal to zero at the top of the boundary.

For all the calculations, a natural boundary condition is applied at the right most boundary. Only half of the domain was calculated in the horizontal plane to save computer memory and time of calculation. The other half will be the same because of symmetry.

4. Numerical method

Eqs. (1)–(5) were solved by a linear basis function finite element scheme. Four-node rectangular element and linear basis interpolation functions were used as shown in Fig. 1 where (ξ_i, η_i) represents the local coordinate of the node point i in the element.

It is assumed that the variables ϕ (i.e. c_z, c_y, $\overline{v'c_z'}$, $\overline{w'c_y'}$ and $\overline{c'\theta_y'}$) can be approximated by a linear basis function which, expressed in matrix notation, is:

$$\phi = [N]\{\phi\}. \tag{18}$$

Linear basis interpolation function N_i which is defined as

$$N_i = \tfrac{1}{4}(1 + \xi_i\xi)(1 + \eta_i\eta) \tag{19}$$

is used at each node point i.

To facilitate the numerical integration, isoparametric finite element method is used for the calculation of the stiffness and mass matrices. It can be carried out by assuming that the global coordinate (x, y) can be expressed by the interpolation function N_i and the global node point coordinate (x_i, y_i) as

$$x = [N]\{x\} \tag{20}$$

and

$$y = [N]\{y\}. \tag{21}$$

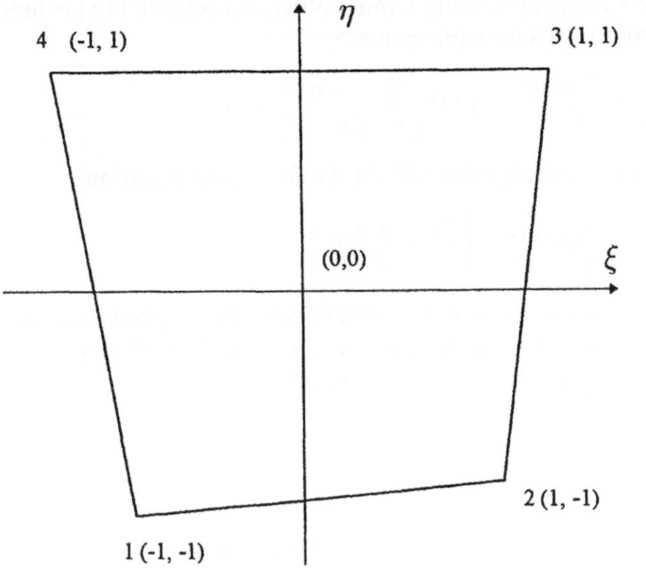

Fig. 1. 4-node rectangular linear basis element.

By differentiating Eq. (19) with respect to ξ and η, respectively, and rearranging in matrix form, the following expression is obtained:

$$\begin{bmatrix} \frac{\partial N_i}{\partial \xi} \\ \frac{\partial N_i}{\partial \eta} \end{bmatrix} = \begin{bmatrix} \frac{\partial x}{\partial \xi} & \frac{\partial y}{\partial \xi} \\ \frac{\partial x}{\partial \eta} & \frac{\partial y}{\partial \eta} \end{bmatrix} \begin{bmatrix} \frac{\partial N_i}{\partial x} \\ \frac{\partial N_i}{\partial y} \end{bmatrix}. \tag{22}$$

Let the 2×2 matrix be $[J]$, then the derivative of N_i with respect to x and y can be expressed as

$$\begin{bmatrix} \frac{\partial N_i}{\partial x} \\ \frac{\partial N_i}{\partial y} \end{bmatrix} = [J]^{-1} \begin{bmatrix} \frac{\partial N_i}{\partial \xi} \\ \frac{\partial N_i}{\partial \eta} \end{bmatrix}. \tag{23}$$

The derivatives of N_i are then expressed in term of local coordinate (ξ, η). This is done to simplify the calculation of the stiffness and mass matrices from analytical integration to numerical integration. The numerical integration was then carried out by 3×3 Gaussian Quadrature approximation [11].

The calculated mass and stiffness matrices were then substituted back into the weak Galerkin form of Eqs. (1)–(5) to obtain the following standard semi-discretized finite element equation

$$[K]\{\phi\} + [M]\left\{\frac{d\phi}{dt}\right\} = \{0\}. \tag{24}$$

For greater numerical stability Crank–Nicolson scheme [12] is used to integrate the semi-discretized Eq. (24) with time as

$$[K]\left\{\frac{\phi_i^{n+1}+\phi_i^n}{2}\right\}+[M]\left\{\frac{\phi_i^{n+1}-\phi_i^n}{\Delta t}\right\}=\{0\}. \tag{25}$$

Eq. (25) can be rearranged to obtain the following equation:

$$\left[\frac{M}{\Delta t}+\frac{K}{2}\right]\{\phi_i^{n+1}\}=\left[\frac{M}{\Delta t}-\frac{K}{2}\right]\{\phi_i^n\}. \tag{26}$$

After applying the initial and boundary conditions specified in Section 3, the system of Eqs. (1)–(5) was then solved by successive overrelaxation method. The value of node point i at time n, ϕ_i^n can then be obtained.

5. Results and discussion

5.1. *Concentration contour in the horizontal plane*

The contour of non-dimensional crosswind integrated pollutant concentration C obtained from the present numerical model is plotted in Fig. 2a. Fig. 2b shows the results of the water channel experiment conducted by Willis and Deardorff [1]. The water channel used in the experiment of Willis and Deardorff was heated at the bottom to make it convective in nature and create a highly turbulent environment for the diffusion of the passive pollutant. The streamwise velocity was constant in the vertical plane. These experimental conditions were simulated by the present numerical model.

It can be observed that the numerical results have similar trends in concentration distribution with experimental observations. However, the curvature of each contour shown in Fig. 2b is greater than that in Fig. 2a indicating that the spread of the pollutant in the water channel experiment is wider than that obtained by the present model. This difference is mainly due to the lack of background turbulence information and the use of a constant turbulent length scale Λ in the computation. The accuracy of the prediction can be improved if more detailed turbulence information is available.

5.2. *Concentration contour in the vertical plane*

Fig. 3a shows the contour plot of C predicted by the present model for the release of pollutant in the middle of the mixing layer. Fig. 3b and Fig. 3c show the numerical results of Pai and Tsang [9] and the experimental observation of Willis and Deardorff [1], respectively. The present model indicates that the plume centreline descends and impings on the ground with $C = 2.0$ at $X = 0.7$, which agree with both experimental observations ($C = 1.6$ and $X = 0.8$) [2] and the numerical results of Pai and Tsang ($C = 1.2$ at $X = 0.73$) [9]. There is also a minimum point at ground surface ($C = 0.4$ at $X = 2.0$) which also agrees with that of Pai and Tsang ($C = 0.8$ at $X = 2.0$) [9].

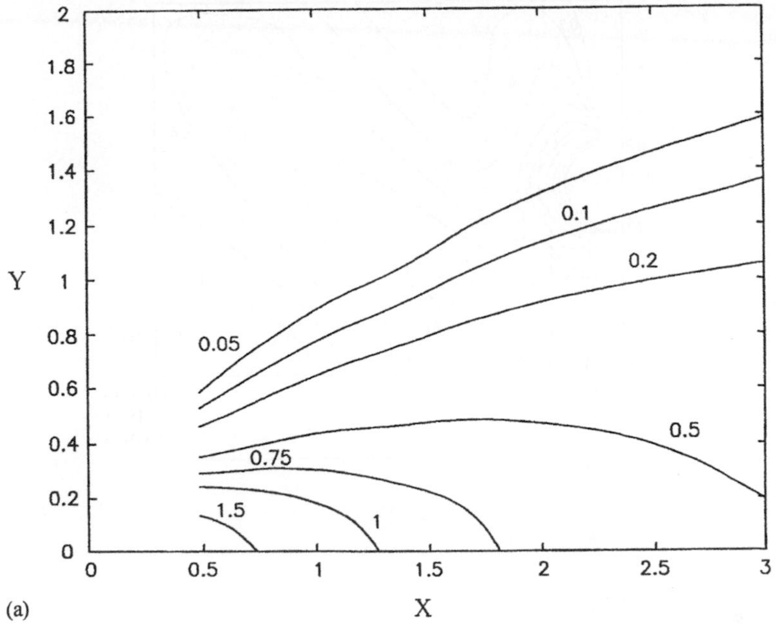

(a)

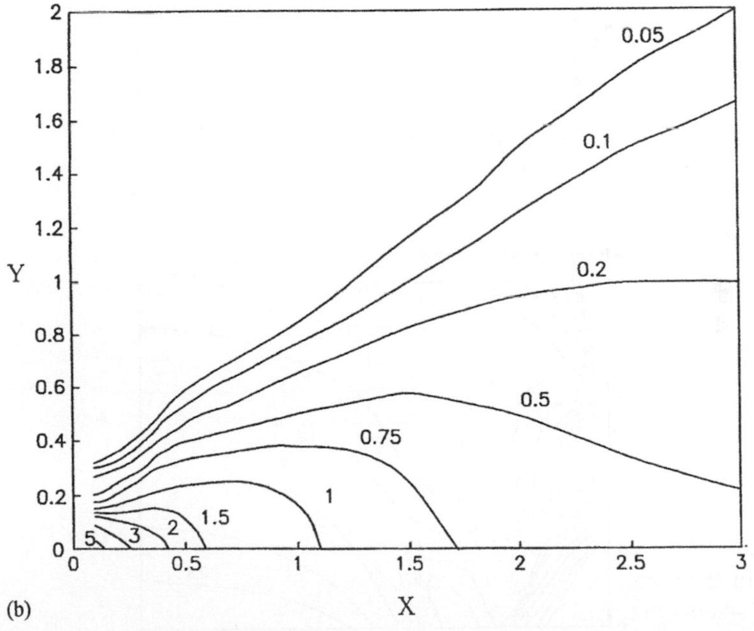

(b)

Fig. 2. Contour of non-dimensional crosswind integrated pollutant concentration in the horizontal plane: (a) present result; (b) experimental observation of Willis and Deardorff [1]. (Figures in the diagrams represent the magnitude of the z-integrated concentrations.)

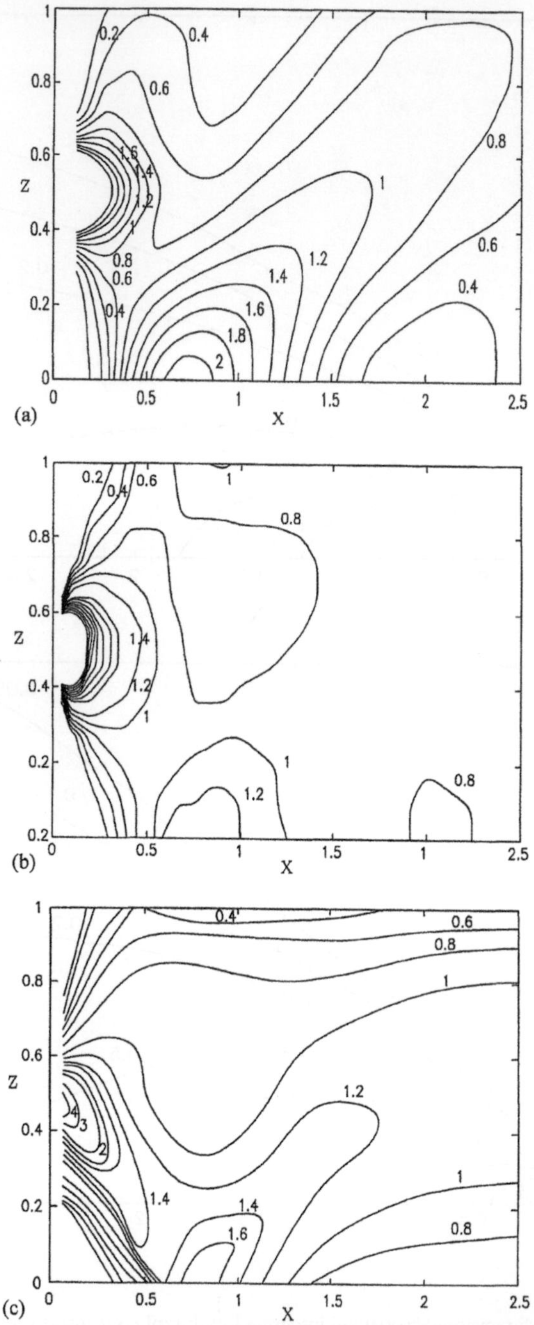

Fig. 3. Contour of non-dimensional crosswind integrated pollutant concentration in the vertical plane: (a) present results; (b) numerical result of Pai and Tsang [9]; (c) experimental observation of Willis and Deardorff [2]. (Figures in the diagrams represent the magnitude of the y-integrated concentrations.)

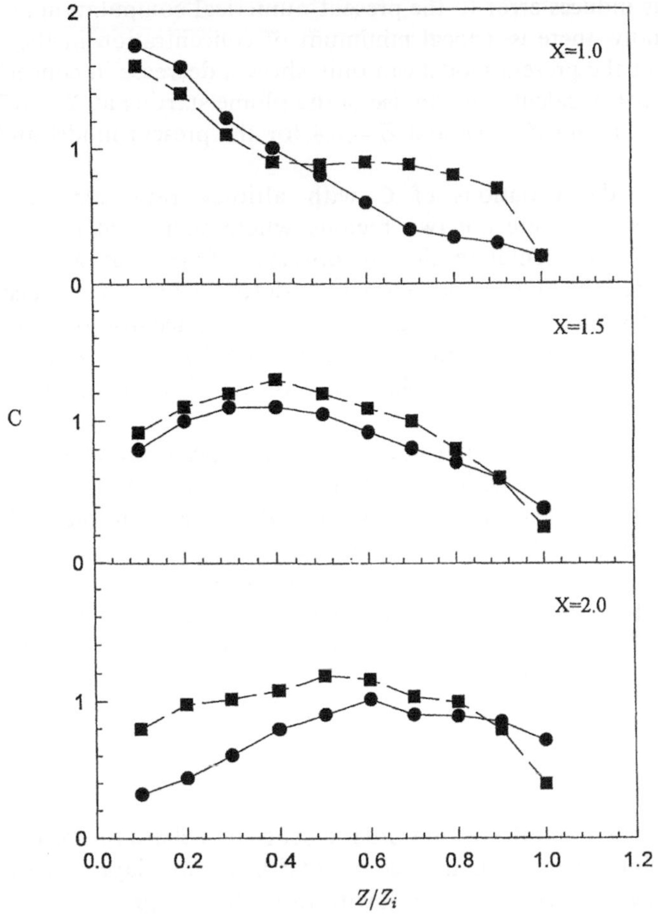

Fig. 4. Concentration at various altitudes for different vertical planes: (●) present result; (■) experimental result of Willis and Deardorff [2].

It can be observed that the agreement between the present numerical model with experimental observations is better than that with the numerical result of Pai and Tsang [9]. This is mainly due to the fact that Pai and Tsang [9] used the background turbulence from the Wangara experiment [9] which is different from that of the water channel experiment [13] currently adopted.

Several improvements could be made to the model for better comparison with experimental observations. Firstly, though the present numerical model can simulate the descent and impingement of the plume on the ground, the predicted trajectory of the plume is slightly different from that of the experimental observations. This is mainly due to the use of Gaussian distribution as the boundary conditions, causing the inability to model the effect of stability on the plume trajectory at the left most boundary. This induces error to the present numerical computation at $X = 0.9$ and $Z = 0.7$. Secondly, there is a local minimum of concentration in the experimental observation, but the present model can only show a decrease in concentration. This also induces error in calculating the rise of the plume starting at $X = 0.7$. The plume ceased to rise at about $Z = 0.8$ and $Z = 0.4$ for the present model and experiment respectively.

Fig. 4 shows the variations of C with altitude ratio z/z_i at three downstream distances X. There are two regions where significant deviations between numerical and experimental results are observed. One is at $X = 1.0$ and $0.6 < z/z_i < 0.9$ where the numerical model underpredicted the concentration by about 50%. Another is at $X = 2.0$ where the concentrations obtained from numerical model were underpredicted by 20% to 50% for $z/z_i < 0.5$. These underpredictions are due to the higher plume rise obtained from the present numerical model.

Although there were several deviations from the experimental results, generally the agreement between the numerical model and the experimental observation is quite good. In particular, the features of the descent and the rise of the plume, both of which cannot be well modelled by the k-theory and Gaussian dispersion models, can be predicted by the present model. The main source of error in the present study is the background turbulence data used in the dispersion model, which may not be accurate. If more detailed background turbulence data is available, the accuracy of the calculation can be greatly improved.

6. Conclusions

It is demonstrated that the second-order closure model is a promising technique to simulate the plume behaviour in a convective boundary layer. The results of the present numerical model compare well with those from a previous numerical model and with experimental data. The present model is therefore suitable to model those plume behaviours which cannot be predicted by k-theory or Gaussian dispersion models. The empirically determined dispersion coefficients are replaced by direct input of background turbulence data. Since model parameters and turbulence data have great effect on the plume behaviour, detailed input data of background turbulence and model parameters are necessary for accurate calculation of pollutant concentration. More works need to be done in order to quantify the error and make statistical comparisons with field data. The analysis would also be extended to three-dimensional cases to simulate a fully three-dimensional dispersion problem.

Acknowledgements

The authors wish to acknowledge the Hong Kong Research Grant Council for supporting this project.

References

[1] G.E. Willis, J.W. Deardorff, A laboratory model of diffusion into convective planetary boundary layer, Quart. J. R. Met. Soc. 102 (1976) 427–445.
[2] G.E. Willis, Laboratory modelling of dispersion in the convectively mixed layer, in: 5th Symp. on Turbulence Diffusion and Air Pollution, Amer. Meteorol. Soc., 1980, pp. 155–156.
[3] Y.C. Leung, C.H. Liu, Effect of wind fluctuations on pollutant dispersion modelling, in: Proc. POLMET 94, 1994, pp. 215–223.
[4] S.R. Hanna, Diurnal variation of horizontal wind direction fluctuations in complex terrain at geysers, Bound.-Layer Met. 18 (1991) 207–213.
[5] W.S. Lewellen, Use of invariant modelling, in: Handbook of Turbulence, 1977, pp. 237–280.
[6] F. Pasquill, Atmospheric Diffusion, Halstead Press, 2nd ed., Wiley, New York, 1974.
[7] R.I. Sykes, W.S. Lewellen, S.F. Parker, A turbulent-transport model for concentration fluctuations and flux, J. Fluid. Mech. 139 (1984) 193–218.
[8] R.I. Sykes, W.S. Lewellen, S.F. Parker, A Gaussian plume model of atmospheric dispersion based on second-order closure, J. Climate Appl. Met. 25 (1986) 322–331.
[9] P. Pai, T.H. Tsang, A finite element solution to turbulent diffusion in a convective boundary layer, Int. J. Numer. Methods Fluids 12 (1991) 179–195.
[10] W.Y. Sun, C.Z. Chang, Diffusion model for a convective layer, Part I: numerical simulation of convective boundary layer, J. Climate Appl. Met. 25 (1986) 1445–1453.
[11] A.H. Stroud, Gaussian Quadrature Formulas, Prentice-Hall, Englewood Cliffs, NJ, 1966.
[12] C.A.J. Fletcher, Computational Techniques for Fluid Dynamics, Springer, Berlin, 1991.
[13] G.E. Willis, J.W. Deardorff, A laboratory model for the unstable planetary boundary layer, J. Atmos. Sci. 31 (1974) 1297–1307.

ELSEVIER

Journal of Wind Engineering
and Industrial Aerodynamics 67&68 (1997) 781–791

JOURNAL OF
wind engineering
AND
industrial
aerodynamics

The use of the MERCURE CFD code to deal with an air pollution problem due to building wake effects

Dennis Moon[a,*], Armand Albergel[b], Florence Jasmin[b], Gerard Thibaut[c]

[a] SSESCO, 3490 Lexington Ave. N, Shoreview, MN 55126-8044, USA
[b] ARIA Technologies, 14-30 Rue de Mantes, 92700 Colombes, France
[c] Ville de Paris, Service Pollutions Atmospherique, Paris, France

Abstract

The technical services of the city of Paris were concerned with a very small-scale air pollution problem: the contamination of the air conditioning system of a public building from an outdoor source. In order to better understand which source was causing these episodes, a study was undertaken to evaluate the contribution of an underground parking lot exhaust system. The outlet of the system is located on the roof of the public building. The aim of the study was a quantitative estimation of the pollution transfer between the ventilation system and a fresh air inlet located about 50 m away. The MERCURE CFD code is described. Assumptions concerning meteorological choices and source term evaluation are discussed. Results of the study are described. Thermal effects of air conditioning units installed on the same roof are also simulated and discussed.

Keywords: Air pollution; CFD; Building wakes

1. Introduction

In 1993, a day-care facility in Paris had several apparent pollution episodes serious enough to cause evacuation. People inside the facility became ill, and staff reported smelling exhaust from internal combustion engines. Two possible sources were suspected: emissions from the exhaust vents of an underground traffic tunnel, and emissions from an emergency electric power generator. The exhaust sites for both of these potential sources are located on the top of the building of the day-care. A study was commissioned by the City of Paris to perform a set of numerical simulations to evaluate the possible contribution of the traffic tunnel exhaust vent. The goal of the study was to obtain a quantitative estimate of the degree to which

* Corresponding author. E-mail: dam@ssesco.com.

PII S0167-6105(97)00118-9

emissions from the tunnel are transferred to the fresh air inlet serving the day-care center. The CFD code MERCURE was chosen to simulate the transfer using meteorological conditions present on two of the actual episodes, as well as to look at a range of meteorological factors including wind speed and direction and thermal convective stability (dT/dz). A run was also made to determine the influence of a thermal plume from air conditioner units on the building on the pollution transfer, and to look at the effect of raising the elevation of the tunnel output vent.

1.1. Description of the site

Fig. 1 shows a plan view of the block of buildings. The tunnel vent and the air conditioning units are on the roof of the building 21.2 m above street level. The two courtyards are 3.6 m above street level and are connected by an open passageway. The fresh air intake for the day-care is located 1 m above the courtyard level in the courtyard to the left. The building dimensions are 106.4 m in the x-direction, and 88.6 m in the y-direction. Note that the y-axis is shifted by 12° from true north. It is clear that transfer would be expected to be an issue on days with winds from the north east. Also note that this would bring the plume directly over the air conditioning units. It was expected that the hot air rising from the units would act to elevate the plume and reduce the transfer to the fresh air intake.

The 3-D domain used for the simulations is shown in Fig. 2. The computational domain is 177 m long in the x-direction, 187 m long in the y-direction, and 200 m high. The grid is logically regular with cell counts of $53 \times 56 \times 32$ in the x,y,z directions, respectively. Grid cell size is approximately 1.5 m near the building. Above the building the vertical grid spacing gradually stretches to a maximum of approximately 38 m at the top of the domain.

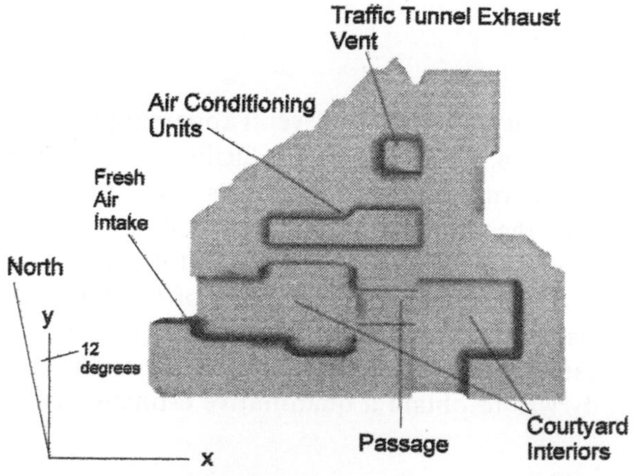

Fig. 1. Building layout.

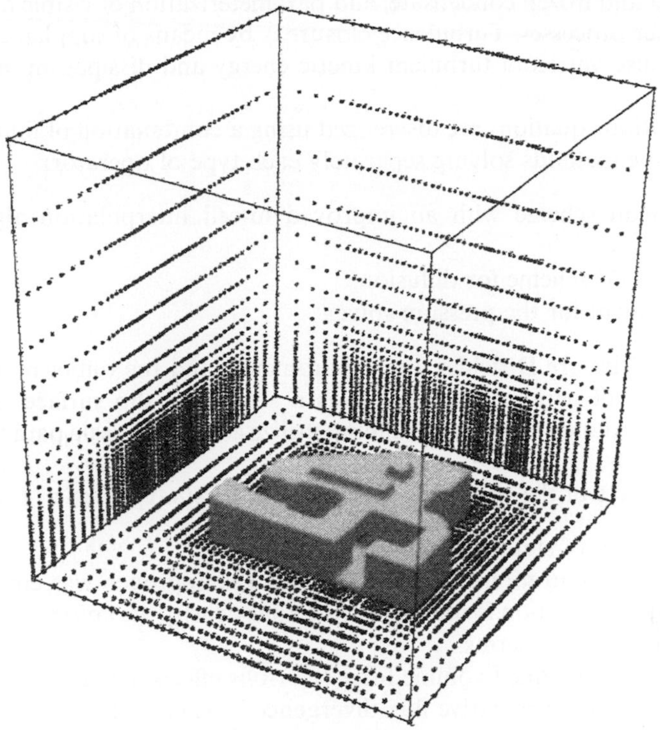

Fig. 2. Computational domain and grid definition.

2. The MERCURE model

The MERCURE model is the atmospheric adaptation of the CFD code Ensemble de Simulations Tridimensionnelles d'Ecoulements Turbulents known as ESTET, developed by the research department at Electricité de France. ESTET is commonly used for industrial CFD aplications at the Laboratoire National d'Hydraulique. The MERCURE code has undergone extensive validation, see Refs. [1,3]. Some relevant aspects of the code include:

- 3-D flow simulation,
- includes influence of terrain and obstacles,
- handles multiple fluids,
- non-hydrostatic formulation.

The system of equations are the classic Navier–Stokes equations with adaptations for multiple fluids and for passive scalar tracer variables. A conservation relation for thermodynamic energy is optionally solved. Solving the thermal energy equation implies that thermal buoyancy effects are included in the solution. Other optional features include conservation and conversion terms for water vapor and several species of liquid and frozen condensate, and parameterization of visible and infra-red radiative transfer processes. Turbulence closure is by means of supplementary equations for the conservation of turbulent kinetic energy and dissipation using the k–ε model.

The conservation equations are discretized using a combination of finite difference and finite-volume methods solving separately each type of operator:

- semi-Lagrangian scheme with an improved in-cell interpolation algorithm for advection,
- fully implicit ADI scheme for diffusion,
- conjugate gradient for the pressure solver.

The inlet boundaries are Dirichlet for all parameters and the outlet boundaries are zero gradient for all parameters. The top is a solid wall slip surface. The cells in contact with the ground surface and building utilize a sub-grid scale parameterization of the log wall laws.

Important aspects of the MERCURE setup *for this study* include:

- ideal gas equation of state,
- given the low concentrations involved, a single gas approach was employed,
- Boussinesq approximation is used, implying that density variations only affect the flow through buoyancy terms,
- gravity is the only retained volume force (Coriolis effects are ignored),
- thermal forcing due to radiative flux divergence is negligible,
- low-speed incompressible flow (Mach number $\ll 1$).

For a complete description of the MERCURE model see Ref. [2].

3. The simulations

Two types of MERCURE runs were made:

(1) The establishment of a steady-state flow field for a given set of meteorological conditions.

(2) The Eulerian calculation of the dispersion of the pollutant using the flow field generated in the first run.

The concentration variable in the MERCURE output was normalized by the concentration at the release area (C/C_0). Concentration for any of several exhaust gases is obtained by multiplying the normalized concentration value with the release concentration for the gas. While CO, NO_x, and HC were looked at in this study, CO was the primary consideration.

3.1. The source characteristics

The source concentrations were derived by looking at the traffic volume through the tunnel for the morning rush hour for different vehicle classes. The four exhaust vents (with areas ranging from 6.48 to 8.1 m^2) had two modes of operation: low-speed and high speed. Only the low-speed case was examined as it was expected that the effect of the higher-speed mode would be to elevate the plume and reduce the pollution transfer. Based on measurements the exit speed for the exhaust gases in the four vents ranged from 3.33 to 5.0 m/s for the low-speed case. The temperature of the exhaust gas was 20°C as specified by the City of Paris. In winter, the plume would tend to elevate from buoyancy effects and reduce the downward transport of the pollutant. If the estimated exhaust gas temperature were too high, this effect would be overestimated. In order to pin-down the range of variation due to this effect, simulations were performed with no buoyancy effects included.

3.2. Meteorological conditions

The important parameters here are the vertical profiles of wind speed and direction, and temperature. The temperature profile determines stability of the system with respect to convective overturning. It thus exerts a strong control over the vertical eddy structure and the amount of turbulent vertical transport. It therefore influences the development of the wind profile. In the situation considered here, it is doubtful that the thermal stability would have a strong effect on the pollution transfer to the air intake because of the small spatial scale; mechanically induced turbulence will dominate. The main effect here of the vertical temperature gradient is to modify the buoyantly induced plume rise.

Wind and temperature measurements from nearby meteorological stations were analyzed for representativeness, and profiles synthesized from them for the two episodes in question. The wind speed profile was related to the vertical temperature gradient using a power-law formulation recommended by the EPA: $u = u_{10\,\mathrm{m}}\,(z/10)^p$. The power p is equal to 0.15 for all cases except case 3 where it is set to 0.35 because of the slightly higher thermal stability. In addition to the two days studied, six other cases were formulated covering a range of other wind directions and stability values. Five of the cases used $\mathrm{d}T/\mathrm{d}z = -0.9°\mathrm{C}/100\ \mathrm{m}$ which is considered neutral stability. Case 6 looks at the effects of the heat released by the air conditioning units, and case 7 looks at the effect of raising the height of the exhaust vents. Case 8 used light winds of 1.25 m/s from the northeast as an expected worst-case condition. The cases are summarized in Table 1.

4. Results

We will begin by looking at the graphic results for two of the cases, and then summarize the overall results. Cases 6 and 8 are at two extremes in terms of the level of pollution transferred. Both cases use the same wind direction as case 1. In case 8 the wind speed of case 1 is cut in half and the thermal-energy equation is not solved so there are no thermal plume rise effects. We expect maximum pollution transfer in case 8. Case 6 includes thermal plume rise effects as well as the added buoyant boost from the air conditioner units. This case would be expected to have a lower transfer of

Table 1
Characteristics of study cases

Case	Wind direction (degrees) speed (m/s)	Surface temperature (°C)	dT/dz (°C per 100 m)	Thermal energy equation	Notes
1	45° (northeast) 2.5	10	− 0.9	Yes	10/27/93 episode
2	90° (east) 2.0	3.2	− 0.63	Yes	11/17/93 episode
3	0° (north) 2.0	20	0	No	
4	30° (northeast) 2.5	10	− 0.9	Yes	
5	60° (northeast) 2.5	10	− 0.9	Yes	
6	45° (northeast) 2.5	10	− 0.9	Yes	Effect of AC units
7	45° (northeast) 2.5	10	− 0.9	Yes	Effect of raising outlet
8	45° (northeast) 1.25	20	− 0.9	No	Worst-case

pollution to the air intake. Keep in mind that the concentration values in the MERCURE graphics are the fraction of the initial concentration. Concentration of 0.5 indicates a concentration of one-half of that in the release gas.

Figs. 3 and 4 show the flow pattern at the intake height, $Z = 4.6$ m, for cases 8 and 6, respectively. The flow direction is shown as 2-D streamlines (streamlines of the velocity vector projected onto the plane of the slice). The flow speed is shown with gray shading. Note the color bar showing the flow speed in case 6 as more than twice of that in case 8. Flow is from the northeast in both cases. The flow patterns are qualitatively quite similar but there are noticeable differences particularly in the downstream wake region. Also, notice that in case 8 the flow is continuous going from the left to the right courtyard through the passageway while in case 6 we have signs of flow blockage in the left courtyard.

Figs. 5 and 6 use gray shading to indicate concentration for cases 8 and 6, respectively. Note first that the levels in Fig. 5 are more than an order of magnitude greater than in Fig. 6. Significant concentrations SW of the building in Fig. 5 indicate that the plume is lower than in Fig. 6 and is intersecting the ground to a greater degree. Also notice that the pollutant clearly advects through the passageway in Fig. 5. If we look at the grid cell corresponding to the location of the intake the normalized concentration is 0.023 g/g, or 2.3% of the exhaust concentration.

Figs. 7 and 8 show 3-D isosurfaces of plume concentration for the two cases. The view is from the west. The outer transparent isosurface is at a value of 0.007 g/g, and the inner, opaque isosurface is at 0.1 g/g. In Fig. 7 the 0.007 g/g plume shows clearly an interaction with the building wake. On the downwind side of the building the

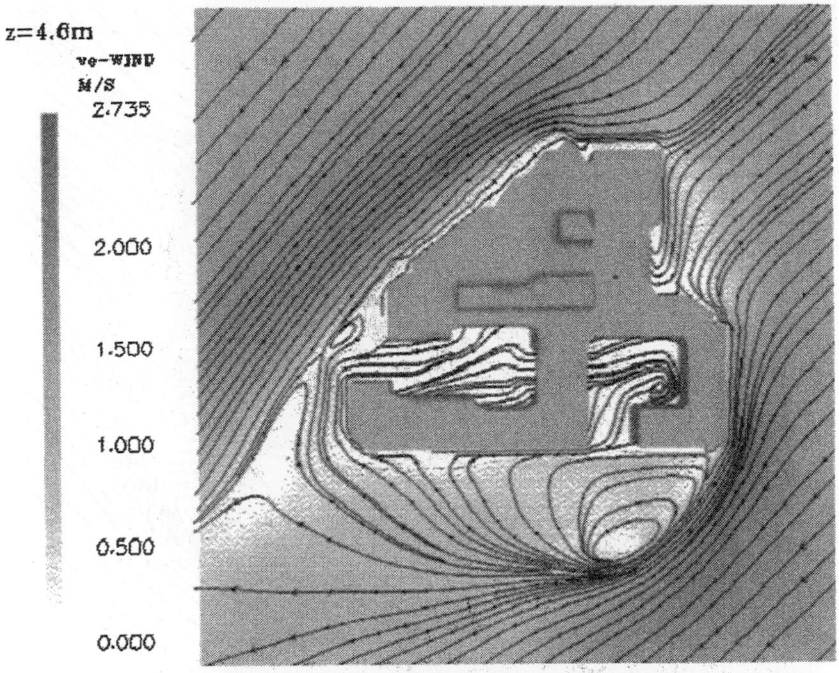

Fig. 3. Wind speed and direction at $Z = 4.6$ m for case 8.

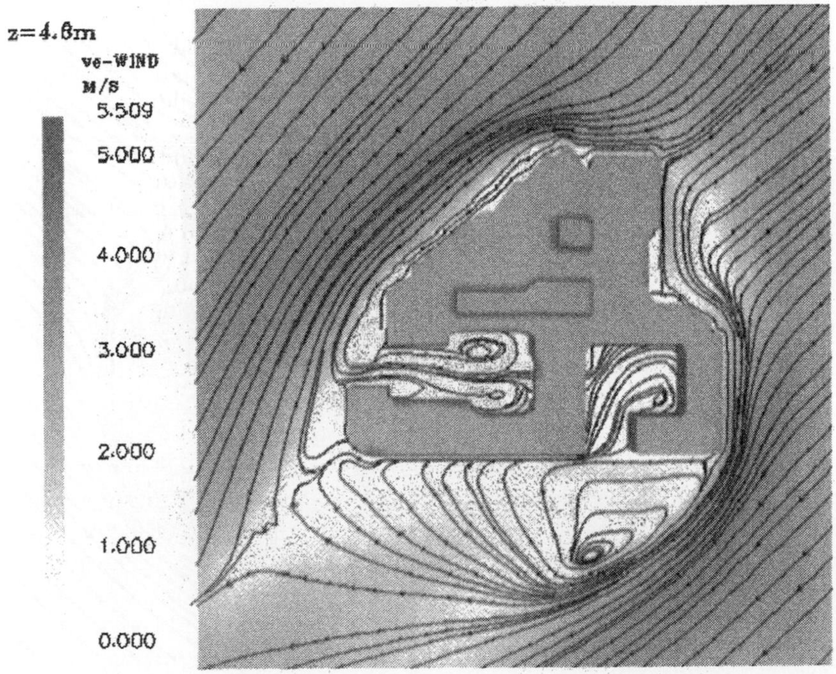

Fig. 4. Wind speed and direction at $Z = 4.6$ m for case 6.

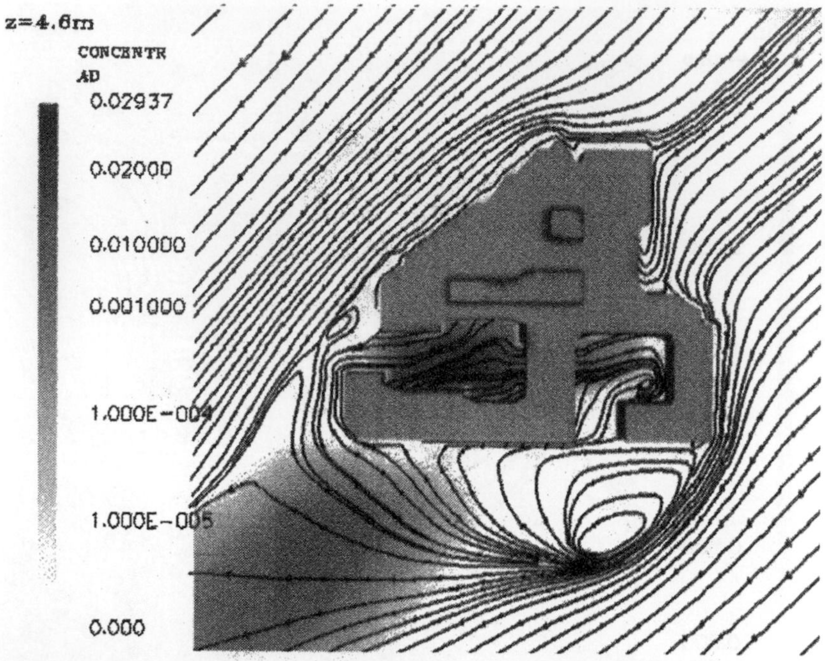

Fig. 5. Normalized pollution concentration and streamlines at $Z = 4.6$ m for case 8.

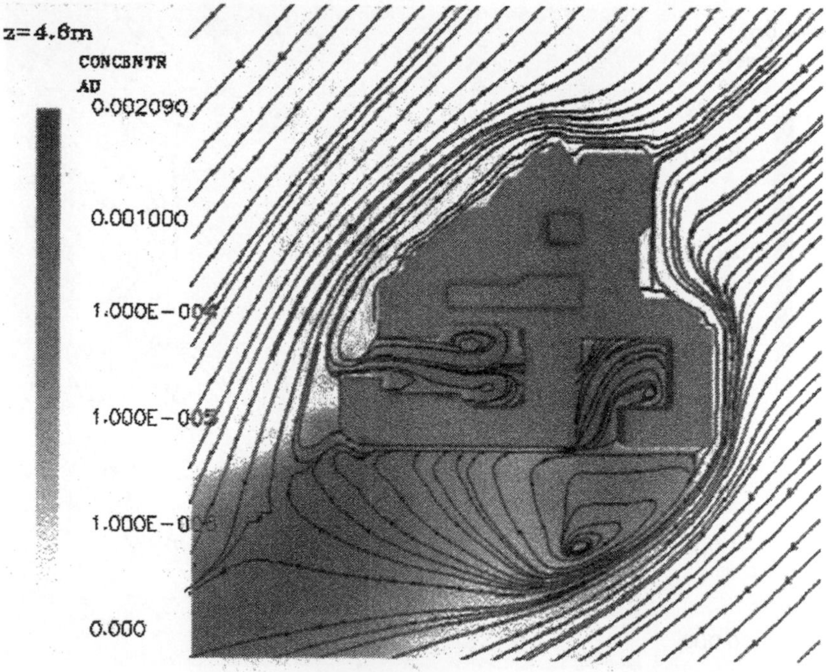

Fig. 6. Normalized pollution concentration and streamlines at $Z = 4.6$ m for case 6.

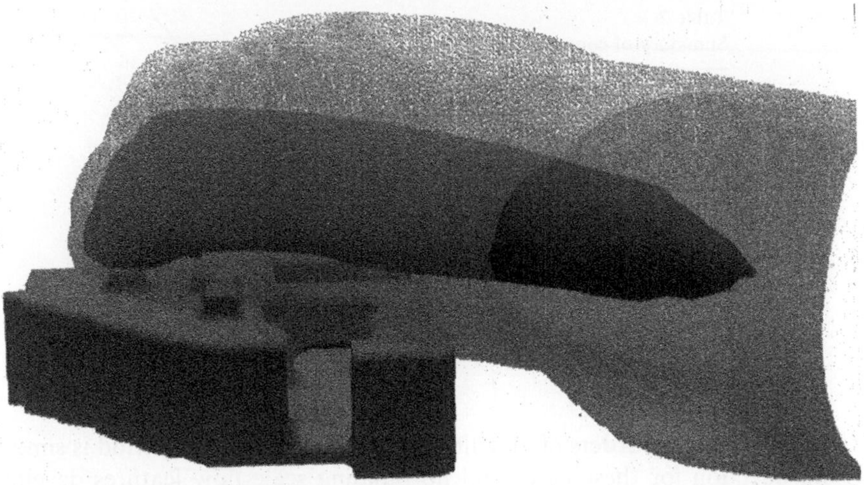

Fig. 7. Inner isosurface is for normalized concentration at a value of 0.1 g/g, and the outer isosurface is for concentration at a value of 0.007 g/g. For case 8.

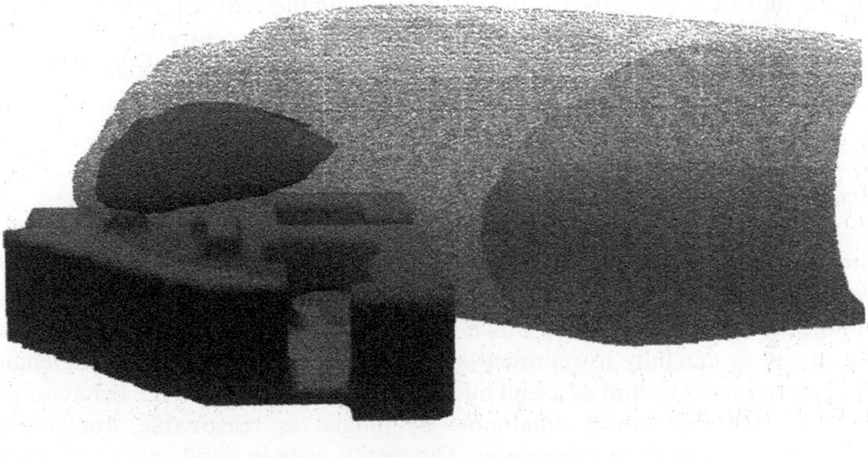

Fig. 8. As in Fig. 7 except for case 6.

pollutant is entrained into the wake. In addition, the plume is being pulled into with the west courtyard. In Fig. 8 we see no evidence of wake entrainment and the interaction with the courtyard is much weaker. Also, in Fig. 8 notice how the 0.1 g/g plume is disrupted close to its source by the hot-air plume from the air conditioner units. This raises the plume and greatly enhances the mixing process, accelerating the plume's dilution.

A summary of the resulting concentration at the fresh air intake is given for all six cases in Table 2. The concentrations at the air intake are lower than in case 1

Table 2
Summary of concentrations at the fresh air inlet for the cases

Case	Normalized concentration	Conc. divided by case 1
1	1.1×10^{-2}	1.0
2	7.2×10^{-8}	1.0×10^{-6}
3	1.1×10^{-12}	1.0×10^{-10}
4	1.1×10^{-2}	0.9
5	1.1×10^{-2}	0.9
6	1.4×10^{-3}	0.1
7	9.7×10^{-3}	0.9
8	2.3×10^{-2}	2.1

(as a reference) by many orders of magnitude for cases 2 and 3. The wind is simply in the wrong direction for these cases and no building scale flow features develop to facilitate the transfer. Cases 4 and 5 are similar to case 1 except that the wind directions are 30° and 60°, respectively. The concentrations at the air intake are within 10% for all three cases. Case 6 looked at the effects of the air conditioning units on case 1. The concentration at the air inlet is reduced by a factor of ten for the reasons described above. Case 7 looked at the effect of raising the vent height by 4.8 m for the traffic tunnel exhaust. This had the effect of lowering the concentration at the air inlet by 12% compared with case 1. Finally, case 8, a worst-case simulation, yielded a concentration just over twice of that for case 1.

5. Conclusions

The project described in this paper represents an engineering application where a model is used for guidance to the decision-making process. As in many such situations, there is little data with which to verify the results. Certainly the input data has been chosen carefully to represent actual conditions. Beyond that, the engineer must treat the model output as useful but imperfect guidance as to the behavior of the physical system. The model data must be judged as reasonable, applying past experimental and modeling experiences. The results were in good agreement with the expectations of the modeling team.

The following conclusions can be drawn:

- A transfer of pollution from the exhaust vent to the fresh air inlet is physically possible when the winds are from the northeast.
- The concentration at the air inlet due to the traffic tunnel emissions is in all cases less than 3% of the emitted concentration.
- The transfer is favored by weak winds and ambient temperatures close to the release temperature of 20°C.
- When running, the air conditioning units located on the same roof act to significantly reduce the concentration at the air inlet.

- Given an initial CO concentration of 8.5 mg/m^3 we arrive at a plume concentration of 0.2 mg/m^3 at the fresh air inlet for the worst case. The ambient morning rush hour CO concentration in the Halles district is around 2 mg/m^3. The traffic tunnel emissions do not appear to be a significant cause for the observed episodes in the day-care facility. Other sources of the pollution are being investigated.

References

[1] Y. Riou, Comparison between the MERCURE-GL code calculations, wind tunnel measurements and Thorney Island field trials, J. Hazardous Mater. 16 (1987) 247–265.

[2] B. Carissimo, E. Dupont, L. Musson-Genon, O. Marchand, Note de Principe du Code MERCURE Version 3.1, Electricité de France, Department Environnement, Groupe Meteorologie et Climat., 1995. Available from: EDF/DER Documentation Center, 6 Quai Watier, 78401 Chatou, France.

[3] E. Dupont, B. Carissimo, O. Marchand, E. Cayrol, L. Musson-Genon, Recueil de Fiches de Validation du Code MERCURE Version 3.1, Electricité de France, Department Environnement, Groupe Meteorologie et Climat., 1995. Available from: EDF/DER Documentation Center, 6 Quai Watier, 78401 Chatou, France.

Journal of Wind Engineering
and Industrial Aerodynamics 67&68 (1997) 793–804

A numerical study of a thermally stratified boundary layer under various stable conditions

Y. Ohya[a,*], H. Hashimoto[b,1], S. Ozono[a]

[a] *Research Institute for Applied Mechanics, Kyusyu University, Kasuga 816, Japan*
[b] *Interdisciplinary Graduate School of Engineering Sciences, Kyusyu University, Kasuga 816, Japan*

Abstract

A direct numerical simulation of a stratified boundary layer under various stable conditions is made by a finite-difference method without any turbulence model. The numerical simulation is very successful, showing variations of turbulent quantities due to thermal stratification. Namely, the intensities and fluxes of velocity and temperature fluctuations are significantly suppressed with increasing stability similar to the experimental results. Flow visualization of computational results is very useful to understand changes in the flow structure under stable stratification. However, compared with experimental results, the discrepancies in turbulence structure in the lowest part of the boundary layer (especially for $z/\delta < 0.2$, δ is the boundary layer thickness) are observed.

Keywords: Stably stratified flow; Turbulent boundary layer; Finite-difference method; Direct numerical simulation

1. Introduction

This paper describes a numerical study of a thermally stratified boundary layer under various stable conditions. The atmospheric boundary layer under stable stratification is difficult to describe and model, because the balance between mechanical generation of turbulence and damping by stability varies from case to case, creating stable boundary layers that range very widely from well mixed to nonturbulent [1]. We have been undertaking experimental studies of the buoyancy effects on the turbulence structure for a wide range of stability in stably stratified boundary layers

* Corresponding author. E-mail: ohya@riam.kyushu-u.ac.jp.
[1] Present address: Nippon Sanso Corporation, Tsukuba Laboratory, 10 Ohkubo, Tsukuba 300-33, Japan.

PII S0167-6105(97)00119-0

by using a specially designed wind tunnel [2]. From the experiment, we have obtained the following features: (1) stable stratification rapidly suppresses the fluctuations of velocity and temperature; (2) the momentum and heat fluxes are also significantly decreased with increasing stability; (3) the vertical profiles of turbulence quantities exhibit different behavior in three distinct stability regimes, i.e., the neutral flows, the stratified flows with weak stability and those with strong stability. However, the detailed flow structure in stratified boundary layers, which causes the various turbulence characteristics mentioned above, remains unclear. To investigate the stably stratified flows in more detail, we challenged to simulate a thermally stratified boundary layer numerically by using a finite-difference method. First, attention is focused on the possibilities of numerical simulation of stratified turbulent boundary layers. Next, we examine the stratification effects on the turbulence structure. Direct comparison with the experimental studies performed under similar conditions is made.

2. Numerical model and procedure

We consider a turbulent boundary layer developed on the ground as shown in Fig. 1. A small block of size $(0.15H \times 0.5H \times 0.055H)$ is placed near the entrance as a turbulence generator. The airflow with a high temperature enters into the calculation domain, the bottom of which is cooled at a low temperature and a stably stratified boundary layer is developed on a downstream ground behind the block. The calculation domain has a height of H in the z-direction, a streamwise length of $11H$ in the x-direction, and a spanwise length of $0.5H$ in the y-direction. A Cartesian grid system consists of horizontal grids with an equal size of $\Delta x = \Delta y = 5 \times 10^{-2}H$ in the x- and y-directions, and vertical non-uniform grids in the z-direction, concentrated toward the ground and of minimum size of $\Delta z = 4 \times 10^{-3}$ ($\Delta z^{+}(= zu^{*}/\nu) = 2$, u^{*} is the friction velocity), as shown in Fig. 2. The number of grid points in the x, y and z-directions are 221, 11, 41, respectively. The staggered arrangement for variables is employed. Under the Boussinesq approximation, the governing equations consist of the continuity equation, momentum equations, and an energy equation for three-dimensional incompressible stratified flow. Nondimensionalizing variables using U_{∞} (ambient airflow speed), $\theta_{\infty} - \theta_{s}$ (temperature difference between the

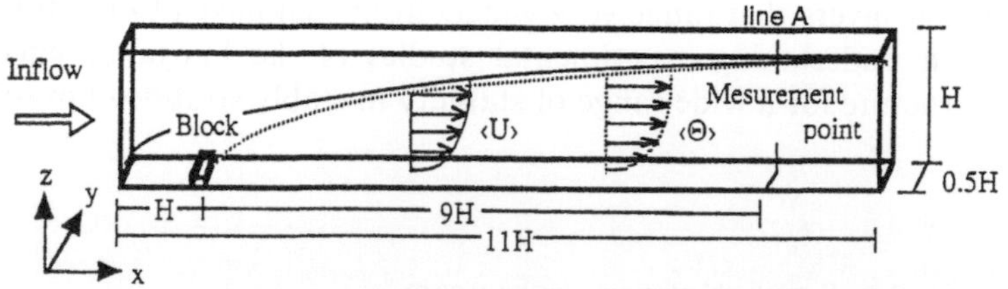

Fig. 1. Calculation domain.

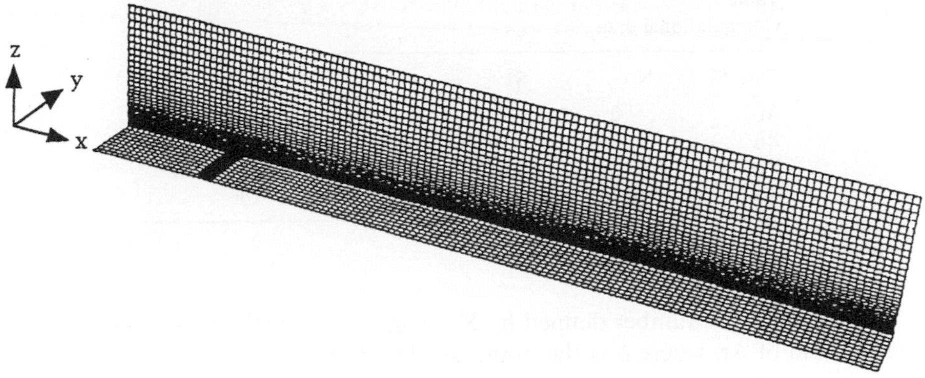

Fig. 2. Part of the computational grid.

ambient airflow and the bottom), and H, we have three dimensionless equations as follows:

$$\partial u_i/\partial x_i = 0 \tag{1}$$

$$\frac{\partial u_i}{\partial t} + \frac{\partial u_j u_i}{\partial x_j} = -\frac{\partial p}{\partial x_i} + \frac{1}{\mathrm{Re}}\frac{\partial^2 u_i}{\partial x_j \partial x_j} - \mathrm{Ar}\,\theta\delta_{i3} \tag{2}$$

$$\frac{\partial \theta}{\partial t} + \frac{\partial u_j \theta}{\partial x_j} = \frac{1}{\mathrm{Pr}\,\mathrm{Re}}\frac{\partial^2 \theta}{\partial x_j^2} \tag{3}$$

These equations include three dimensionless parameters, i.e., the Reynolds number $\mathrm{Re} = U_\infty H/\nu$, the Prandtl number Pr, and the Archimedes number Ar $(= -g\beta_0(\theta_\infty - \theta_s)H/U_\infty^2)$, where g is the acceleration due to gravity, β_0 is the volume expansion ratio at a reference temperature θ_0. The boundary conditions are as follows: for velocity, $U_\infty = 1$ (inflow), $\partial f/\partial x = 0$ (outflow), free slip (top and side walls), no slip (bottom and block surface), for temperature, $\theta = \theta_\infty = 0.5$ (inflow), $\theta = \theta_s = -0.5$ (bottom), $\partial\theta/\partial n = 0$ (other boundaries).

The numerical method used is based on a fractional step method [3] of incompressible stratified flow. For time integrations, the Adams–Bashforth method for convection terms and the Crank–Nicolson method for diffusion terms are employed. All of the spatial derivatives are approximated by the second-order central difference. For convection terms, Arakawa's form [4] is employed and Kajishima's discretization form [5] is used for its non-conservative terms. Taking the divergence of momentum equations, we obtain a Poisson equation for pressure which is solved by the SOR method. As for the momentum and energy equations, we have followed the Beam and Warming method [6] and have employed an approximated factorization method to use a Tri-Directional-Matrix-Algorithm. The initial condition is set with $U_\infty = 1$ and $\theta = 0$. The time step is 2×10^{-3}. $\mathrm{Re} = 10^4$ and $\mathrm{Pr} = 0.71$. We have calculated four cases of stability with $\mathrm{Ar} = -1.0 - 0$ as shown in Table 1. For convenience, we use

Table 1
Computational data

Case No.	N	S1	S2	S3
Ar	0	− 0.2	− 0.6	− 1.0
Ri_δ	0	0.09	− 0.24	0.36
Re_δ	5560	5590	5120	4680
Symbol	-●-	-Δ-	-□-	-○-

the bulk Richardson number defined by $Ri_\delta = g\beta_0(\theta_\infty - \theta_s)\delta/U_\infty^2$ as a stability parameter instead of Ar, where δ is the boundary layer thickness.

3. Results and discussions

3.1. Vertical profiles of mean and turbulent quantities

We have evaluated the turbulence quantities at a downstream position of $10H$ shown as the vertical line A in Fig. 1. After a fully developed boundary layer has been reached, the average values of 15 000 time steps are used. The vertical profiles of mean and turbulent quantities in Figs. 3–8 are normalized by the ambient velocity U_∞ and temperature difference $\Delta\theta$ ($= \theta_\infty - \theta_s$). These are shown for the normalized height z/δ. Also, all the numerical results shown in Figs. 3–8 are compared with the corresponding results obtained by a similar experiment [2] made in a thermally stratified wind tunnel (Table 2). The angular bracket $\langle\ \rangle$ in Figs. 3–8 means the time-average value and u', w' and θ' mean the fluctuating components.

Fig. 3 shows the vertical profiles of mean velocity of the u-component. Variations in the 'defect' profiles in the lowest part of $z/\delta < 0.2$ are due to changes in thermal stratification. For $z/\delta > 0.2$, all the profiles show almost the same distribution and follow the log-law profile. Compared with the profiles from the experiment, the changes in the velocity profiles are smaller over the whole boundary layer depth. The vertical profiles of mean temperature are shown in Fig. 4. As the stability increases from case S1 to S3, the "defect" portion in the profiles gradually increases.

Fig. 5 shows the vertical profiles of r.m.s. values of u-component fluctuations which are in good agreement with those from the experiment over the whole boundary layer depth. Accordingly, it should be noted that there are three characteristic profiles of u-component, i.e., a neutral flow ($Ri_\delta = 0$, case N), a stratified flow with weak stability of $Ri_\delta = 0.09$ (case S1), and stratified flows with strong stability of $Ri_\delta = 0.24$ and 0.36 (cases S2 and S3). Fig. 6 shows the vertical profiles of w-component r.m.s. fluctuations. Similar to the u-component in Fig. 5, we can see three different profiles of neutral flow and stratified flows with weak and strong stability. The intensities of the w-component fluctuation are reduced remarkably with increasing stability, because buoyancy forces in stratified flows extract energy directly from the vertical component w of the velocity. The intensities of u and θ fluctuations are also reduced with increasing

Table 2
Experimental data

Case No.	N	S1	S2	S3	S4
Ri_δ	0	0.12	0.20	0.39	0.47
Re_δ	127 000	109 000	74 100	46 000	42 000
Symbol	–●–	–△–	–◇–	–□–	–○–

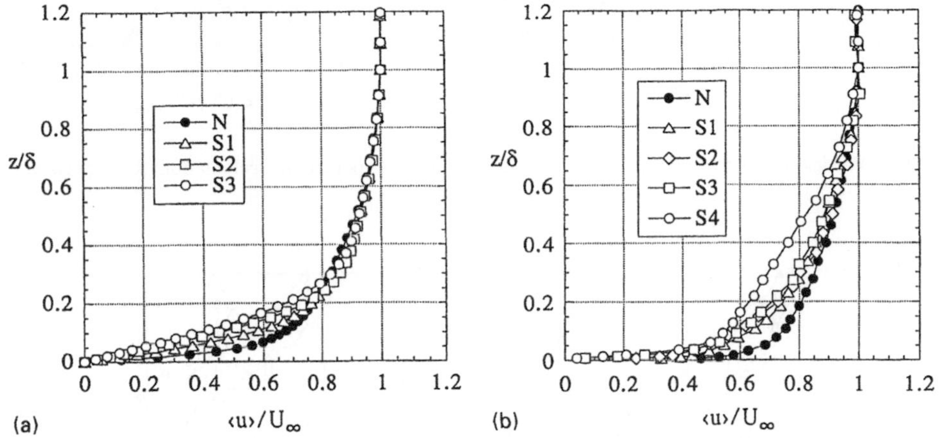

Fig. 3. Vertical profiles of the mean streamwise velocity: (a) calculation; (b) experiment.

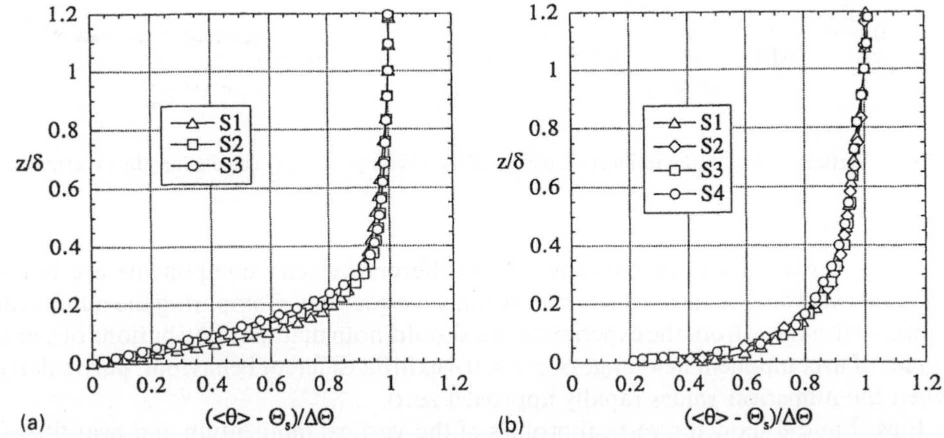

Fig. 4. Vertical profiles of the mean temperature: (a) calculation; (b) experiment.

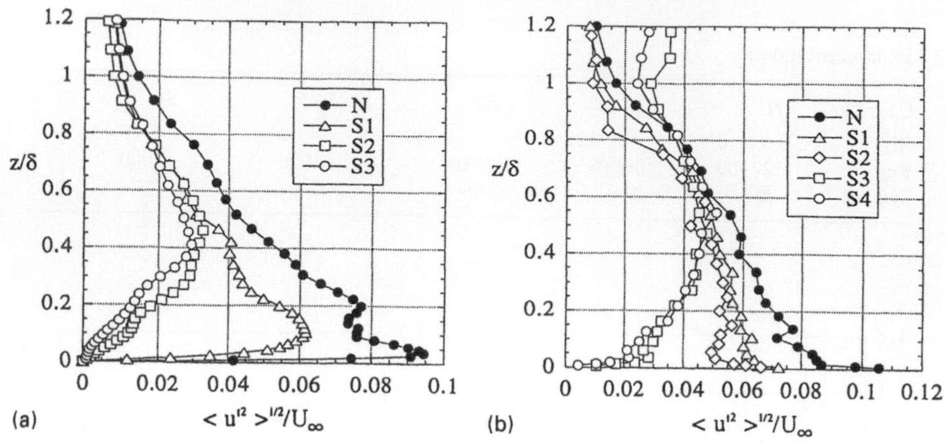

Fig. 5. Vertical profiles of the turbulent intensity of the *u*-component: (a) calculation; (b) experiment.

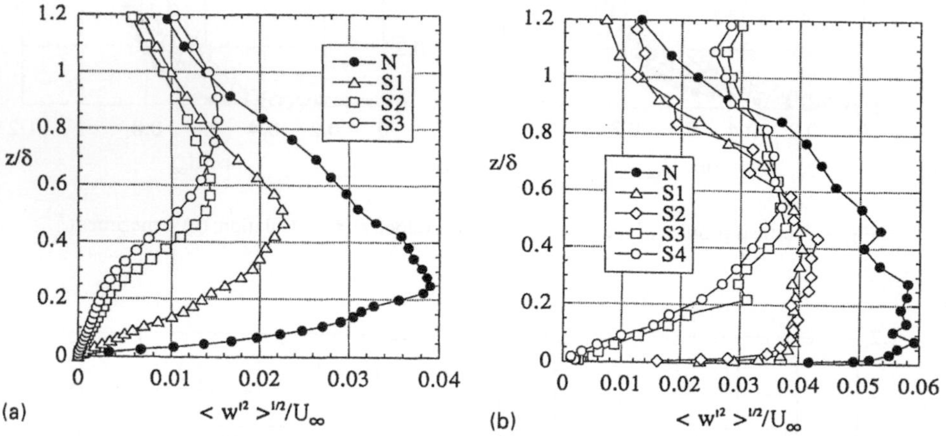

Fig. 6. Vertical profiles of the turbulent intensity of the *w*-component: (a) calculation; (b) experiment.

stability, because energies associated with different velocity components are redistributed according to their kinetic energy budget equations. Comparing the numerical profiles with those from the experiment, we should note that the distributions of r.m.s. values of w-component in a range of $z/\delta < 0.4$ exhibit different behaviour, particularly when the numerical values rapidly approach zero.

Figs. 7 and 8 show the vertical profiles of the vertical momentum and heat fluxes, respectively. In Fig. 7, the momentum fluxes $-\langle u'w' \rangle$ for stratified flows decrease remarkably compared with that of neutral flow and similar to the experimental result. The profiles of heat flux $-\langle w'\theta' \rangle$ for stratified flows with weak and strong stability

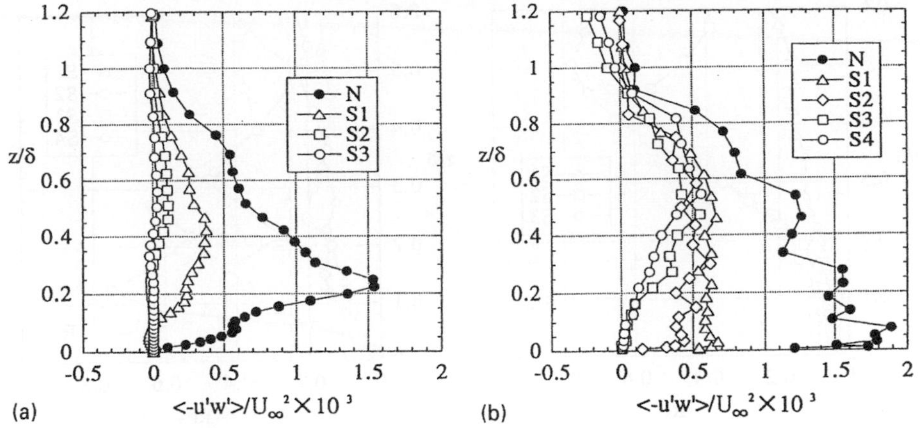

Fig. 7. Vertical profiles of the vertical turbulent momentum flux: (a) calculation; (b) experiment.

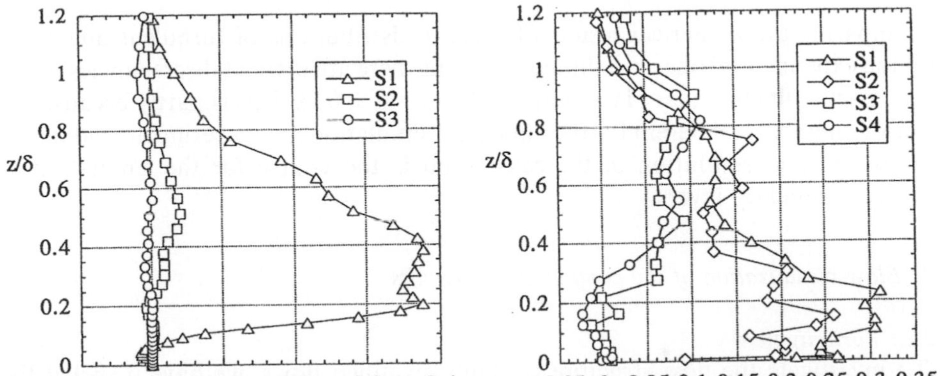

Fig. 8. Vertical profiles of the vertical turbulent heat flux: (a) calculation; (b) experiment.

are very different from each other as seen in Fig. 8, similar to the experimental result. For both Figs. 7 and 8, the distributions of fluxes in a range of $z/\delta < 0.2$ show a rapid decrease, different from those of the experiment.

Fig. 9 shows the vertical distributions of local gradient Richardson number Ri. Corresponding to the two groups of stratified flows with weak and strong stability as seen in Figs. 5–8, the profiles of these two groups are clearly different. However, compared with the experimental results, the distributions for stratified flows with strong stability are very different. This seems to be closely related to the difference in profiles of the mean velocity and temperature between numerical and experimental results as seen in Figs. 3 and 4.

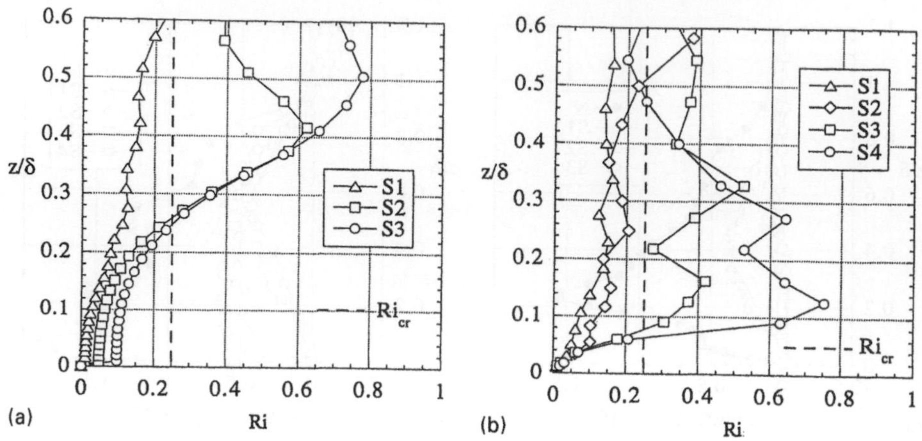

Fig. 9. Vertical profiles of the local gradient Richardson number: (a) calculation; (b) experiment.

Comparing the numerical results of vertical distributions of turbulent quantities with the experimental ones, we should note that there are distinct discrepancies in the lowest part of the boundary layer (especially for $z/\delta < 0.2$). This is partly because the Reynolds number is different for the numerical simulation and experiment, and partly because the grid resolution in the calculation is too coarse for the simulation of turbulent boundary layers.

3.2. *Flow visualization of the computational results*

3.2.1. *Vorticity field*

To investigate the flow structure in stably stratified flows, we have depicted the vorticity contours as shown in Figs. 10 and 11. Fig. 10 shows the vorticity field in the x–z section for a range of $x = (8\text{–}11)H$ at $y = 0.25H$ and Fig. 11 shows the vorticity field in the y–z section at $x = 10H$. In both Figs. 10 and 11, we can clearly see that vorticity diminishes as stability increases. For the stratified flows with strong stability in Fig. 10c, Fig. 10d, Fig. 11c, and Fig. 11d, horizontal lines in the vorticity contours, which mean the strong vertical shear of u-component, are dominant in the lowest part of the boundary layer.

3.2.2. *Particle movement*

To visualize the fluid motion, marker particles were successively released at upstream positions with a time interval of 0.05. Fig. 12 shows the front view of particle movement after 40 intervals released from the all grid points of the section at $x = 9H$. We can see a large difference in fluid motion especially in the lowest part of the boundary layer. As the stability increases, both vertical and horizontal motions are significantly suppressed and a viscous layer gradually develops. This tendency

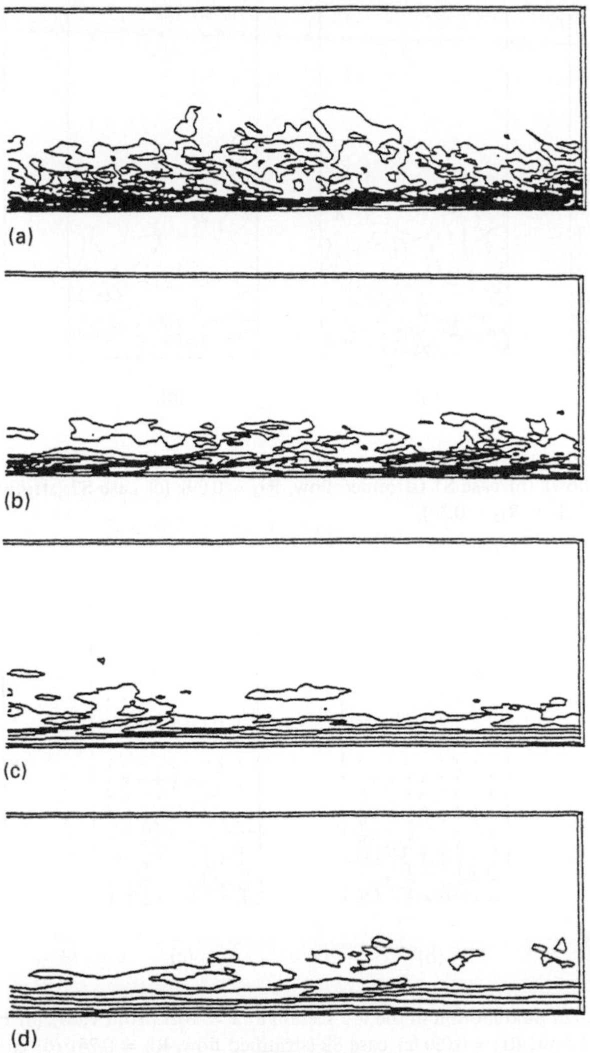

Fig. 10. Vorticity field in the x–z section for a range of $x = (8\text{–}11)H$ at $y = 0.25H$, $t = 60$, the interval of vorticity contours is $\Delta\omega = 2$: (a) case N (neutral flow); (b) case S1 (stratified flow, $\mathrm{Ri}_\delta = 0.09$); (c) case S2 (stratified flow, $\mathrm{Ri}_\delta = 0.24$); (d) case S3 (stratified flow, $\mathrm{Ri}_\delta = 0.36$).

corresponds to the rapid decrease of turbulent quantities in a range of $z/\delta < 0.2$ as seen in Figs. 5–8.

Fig. 13 shows the side and top views of particle movement after 80 intervals released from all grid points of a vertical line at $x = 7H$ and $y = 0.25H$. The side view displays the suppression of vertical motion with increasing stability. The spanwise motion is also suppressed as seen in the top views. One can find that

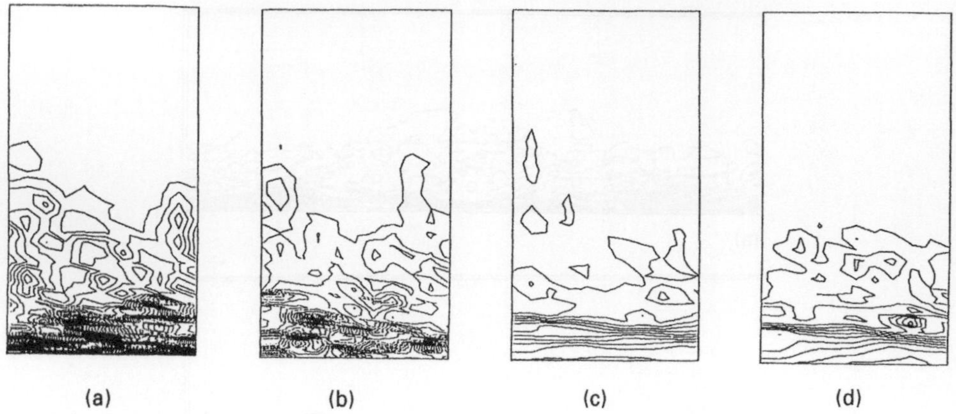

Fig. 11. Vorticity field in the y–z section at $x = 10H$, $t = 60$, the interval of vorticity contours is $\Delta\omega = 1$: (a) case N (neutral flow); (b) case S1 (stratified flow, $\mathrm{Ri}_\delta = 0.09$); (c) case S2 (stratified flow, $\mathrm{Ri}_\delta = 0.24$); (d) case S3 (stratified flow, $\mathrm{Ri}_\delta = 0.36$).

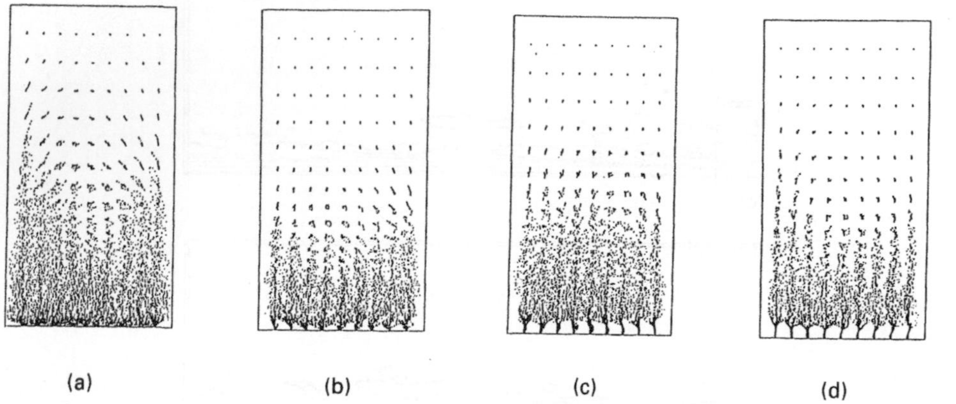

Fig. 12. Visualization of fluid motion in the y–z section at $x = 10H$ (front view): (a) case N (neutral flow); (b) case S1 (stratified flow, $\mathrm{Ri}_\delta = 0.09$) (c) case S2 (stratified flow, $\mathrm{Ri}_\delta = 0.24$); (d) case S3 (stratified flow, $\mathrm{Ri}_\delta = 0.36$).

a meandering motion in a horizontal plane occurs for the stratified flow with strongest stability as displayed in Fig. 13d.

4. Conclusions

A direct numerical simulation of a stratified boundary layer under various stable conditions has been made by a finite-difference method without any turbulence model. The results obtained are summarized as follows:

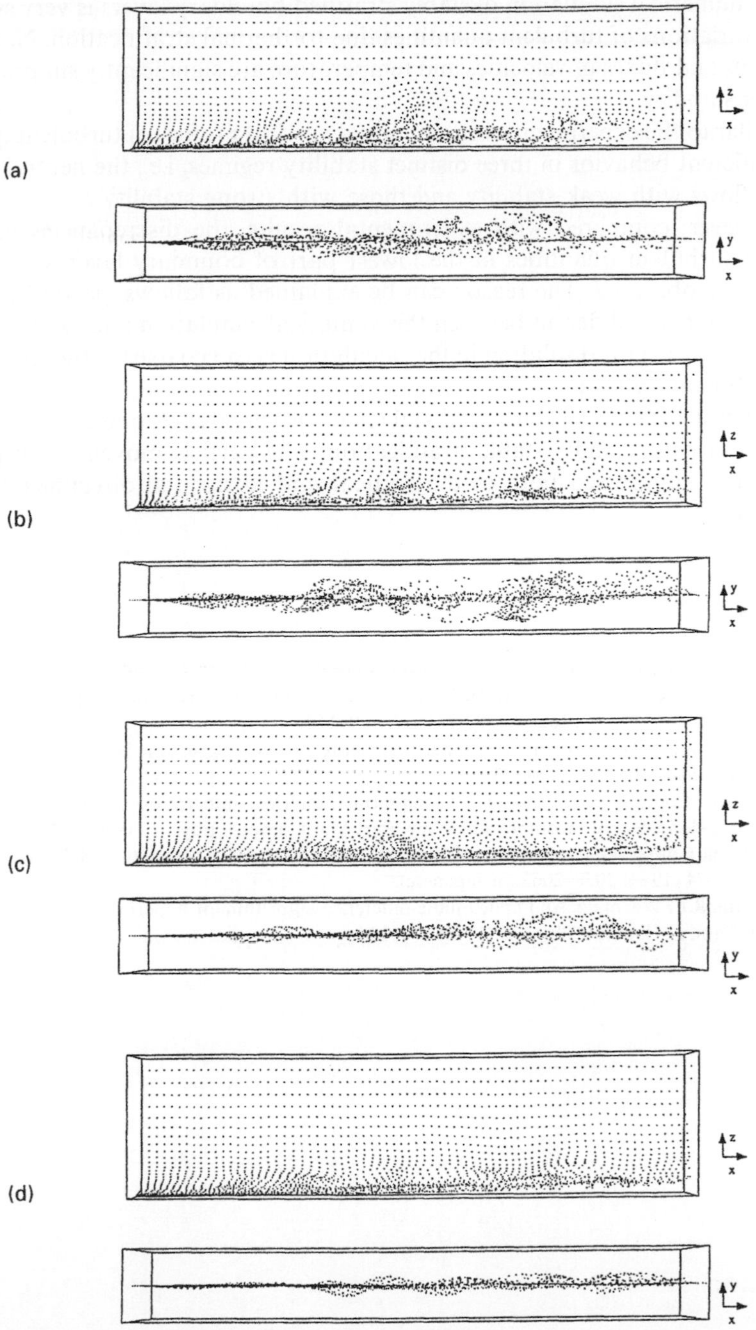

Fig. 13. Visualization of fluid motion in the x–z section (side view) and the x–y section (top view) in a range of $x = (7\text{–}11)H$: (a) case N (neutral flow); (b) case S1 (stratified flow, $Ri_\delta = 0.09$); (c) case S2 (stratified flow, $Ri_\delta = 0.24$); (d) case S3 (stratified flow, $Ri_\delta = 0.36$).

(1) The numerical simulation of stably stratified boundary layers is very successful, showing variations of turbulent quantities due to thermal stratification. Namely, the fluctuations and fluxes of velocities and temperature are significantly suppressed with increasing stability.

(2) Similar to the experiment, computational vertical profiles of turbulent quantities exhibit different behavior in three distinct stability regimes, i.e., the neutral flow, the stratified flows with weak stability and those with strong stability.

(3) However, compared with experimental results, the discrepancies in vertical profiles of turbulent quantities in the lowest part of boundary layer (especially for $z/\delta < 0.2$) are observed. The reason can be explained as follows: partly because the Reynolds number is different between the numerical simulation and experiment, and partly because the grid resolution in the calculation is too coarse for the simulation of turbulent boundary layers.

(4) Flow visualization of computational results is very useful to understand changes in the flow structure under stable stratification. The particle movement displays the fluid motions suppressed both in the vertical and horizontal directions by strong stratification.

References

[1] R.B. Stull, An Introduction to Boundary Layer Meteorology, Kluwer, Dordrecht, 1988.

[2] Y. Ohya, D.E. Neff, R.N. Meroney, Turbulence structure in a stratified boundary layer under stable conditions, Bound.-Layer Meteorol. 83-1 (1997) 139–161.

[3] J. Kim, P. Moin, Application of a fractional-step method to incompressible Navier–Stokes equations, J. Comput. Phys. 59 (1985) 308–323.

[4] A. Arakawa, Computational design for long-term numerical integration of the equation of fluid motion: two-dimensional incompressible flow, Part I, J. Comput. Phys. 1 (1966) 119–143.

[5] A. Kajishima, Conservation properties of finite difference method for convection, Bull. Japan Soc. Mec. Eng. B 60–574 (1994) 2058–2063 (in Japanese).

[6] R.M. Beam, R.F. Warming, An implicit finite-difference algorithm for hyperbolic systems in conservation-law form, J. Comput. Phys. 22 (1976) 87–110.

ELSEVIER

Journal of Wind Engineering
and Industrial Aerodynamics 67&68 (1997) 805–814

JOURNAL OF wind engineering AND industrial aerodynamics

Numerical simulation of pollutant dispersion around a building using FEM

R. Panneer Selvam[1]

BELL 4190, University of Arkansas, Fayetteville, AR 72701, USA

Abstract

Numerical simulation of pollutant dispersion around a building is reviewed. A computer model based on the k–ε turbulence model is used to compute flow around a building. Using this simulation flow-field pollutant concentrations around the building are computed using upwind and streamline upwind Petro–Galerkine (SUPG) finite element procedure. In this study the source is kept adjacent to the bottom (position 1) and top (position 2) of the leeward center wall. The computed concentrations using upwind and SUPG procedure are compared with wind-tunnel measurements. The SUPG procedure computes the trend of the concentration much closer to the wind-tunnel measurement than the upwind procedure.

Keywords: Computational fluid dynamics; Computational wind engineering; Pollutant transport; Turbulence; Buildings

1. Introduction

Concentrations of pollutants as emitted into the air are often at unhealthy levels. The dilution of these pollutants with surrounding air or the use of control devices to lower the emission quantity are often necessary in order to avoid entrainment of unhealthy air into buildings or meet ambient air-quality standards. This paper examines numerical methods for modeling the entrainment of pollutants emitted from short stacks or ground-level emissions into the complex flow around buildings. Computer models that are able to predict the concentrations of pollutants near buildings are essential tools for ensuring a healthy urban environment.

In order to model pollutant dispersion around a building, the flow around the building is modeled first and then the pollutant transport is predicted using the flow simulation. The computation of pollutant transport knowing the flow is done either

[1] E-mail: rps@engr.uark.edu.

PII S0167-6105(97)00120-7

using the Eulerian or Lagrangian framework. Lee and Leone [1] and Naslund et al. [2] have used the Lagrangian framework. In this approach the point sources can be modeled accurately and the numerical diffusion is less than for an Eulerian framework. Naslund et al. [2] used two stochastic models to study the pollutant transport around a building. For a ground-level point source located at the center of the leeward wall, the experimental plume widths are twice as large as the simulated widths. Hence, it is difficult to model diffusion near the ground in a complex flow using this approach. Also the Lagrangian approach requires more computational time and computer storage.

The Eulerian-grid based approach is easier to implement but a major problem is reducing the error involved in approximating the advection term. The usual upwind differencing procedure gives large numerical diffusion and the other existing procedures are also not completely satisfactory. This led to the continuing search for a better model.

There have been a number of numerical studies of pollutant dispersion near buildings and street canyons [1–8]. Except for the studies of Dawson et al. [3] and Selvam et al. [7,8], none of them verified their computed results with wind-tunnel or field measurements. Dawson et al. [3] computed the flow and pollutant transport for a single case using a coarse grid due to lack of computer time and storage. Selvam and Huber [7] computed the pollutant transport for the source kept at the ground and at roof level for one set of grids. The computed concentrations were compared with wind-tunnel measurements. The computer model using the k–ε model compared very well with wind-tunnel measurement when the source is kept at the ground level. Whereas when the source is kept at the roof level; the computed ground-level concentrations were about 33% larger than the measured ground-level concentration at the position of maximum concentration and also the shape is different from the measured value. This may be due to the error involved in approximating the advection term using the upwind procedure.

In this work, the streamline upwind Petro–Galerkine (SUPG) finite element procedure [9] which conserves mass and gives small numerical diffusion is used to approximate the transport equation. Next, to apply this procedure to compute the pollutant transport around building, the flow around the building is computed using the k–ε turbulence model. An eddy viscosity based on the k–ε turbulence model is used for calculating pollutant dispersion. The computed concentration levels at two different locations are compared with wind-tunnel model measurements.

2. Mathematical modelling

The flow around a building is a highly turbulent and complex flow field. Several methods are available to compute turbulent flow field. A short survey is given in Selvam and Huber [7]. In this work, the k–ε turbulence model which is widely used for computational wind-engineering applications is considered. It is simple to use and consumes less computer time compared to large eddy simulation. The three-dimensional equations for an incompressible fluid using a k–ε turbulence model in general tensor notation are as follows:

Continuity equation: $U_{i,i} = 0$. (1)

Momentum equation:

$$U_{i,t} + U_j U_{i,j} = -(p/\rho + 2k/3)_{,i} + [\nu_t(U_{i,j} + U_{j,i})]_{,j}. \quad (2)$$

Equation for k: $k_{,t} + U_j k_{,j} = (\nu_t k_{,j}/\sigma_k)_{,j} + P - \varepsilon$. (3)

Equation for ε: $\varepsilon_{,t} + U_j \varepsilon_{,j} = (\nu_t \varepsilon_{,j}/\sigma_\varepsilon)_{,j} + C_{\varepsilon 1} P\varepsilon/k - C_{\varepsilon 2}\varepsilon^2/k$. (4)

Equation for concentration: $C_{,t} + U_j C_{,j} = (\nu_t C_{,j}/\sigma_c)_{,j} + S$, (5)

where $\nu_t = C_\mu k^2/\varepsilon$ and $P = \nu_t(U_{i,j} + U_{j,i})U_{i,j}$.

Empirical constants: $C_\mu = 0.09$, $C_{\varepsilon 1} = 1.44$, $C_{\varepsilon 2} = 1.92$, $\sigma_k = 1.0$, $\sigma_\varepsilon = 1.3$, $\sigma_c = 0.7$, where U_i, p, and C are the mean velocity, pressure and concentration, respectively, k is the turbulent kinetic energy, ε is the dissipation rate, S is the mean volume contaminant source generation rate, ν_t is the eddy viscosity, P is the production term and ρ is the fluid density. The empirical constants used here are the standard values suggested by Launder and Spalding [10]. Here, a comma represents differentiation, t represents time and $i = 1, 2$ and 3, mean variables in the x-, y- and z-directions. First the flow field is computed using Eqs. (1)–(4). Using the flow field the pollutant transport is computed using Eq. (5).

To implement higher-order approximation of the convection term the following expression is used in Eq. (2) instead of $U_j U_{i,j}$:

$$U_j U_{i,j} - \frac{\theta}{2}(U_j U_k U_{i,j})_{,k}. \quad (6)$$

Depending upon the values of θ different procedures can be implemented. For the balance tensor diffusivity (BTD) scheme $\theta = \delta t$ is used; where δt is the time step used in the integration. For streamline upwind procedure suggested by Brooks and Hughes [9] $\theta = 1/\max(|U_1|/\mathrm{d}x, |U_2|/\mathrm{d}y, |U_3|/\mathrm{d}z)$. Here, d$x$, d$y$ and dz are the control volume length in the x-, y- and z-directions. In this computation a value of 0.4 is used for flow and 1.0 is used for pollutant transport.

3. Computer modelling of flow field

3.1. Numerical procedure

The equations are solved on a rectangular grid using control volume procedures, and the variables are stored on a node centered nonstaggered grid location as explained by Selvam [11]. The diffusion term in Eqs. (2)–(4) is approximated by the usual control-volume procedure. The convection term is approximated by upwind procedure as discussed in Ref. [12] or using Eq. (6).

The Navier–Stokes (NS) equations are solved using an implicit method similar to that of Choi and Moin [13] and applied for nonstaggered, node centered, control-volume procedure by Selvam in Ref. [11]. The four-step advancement scheme for Eqs. (1) and (2) is as follows:

Step 1: Solve for U_i from Eq. (2) using backward Euler for convection and diffusion.

Step 2: Get new velocities as $U_i^* = U_i + \delta t(p_{,i})$ where U_i is not specified.

Step 3: Solve for pressure from $(p_{,i})_{,i} = U_{i,i}^*/\delta t$

Step 4: Correct the velocity for incompressibility: $U_i = U_i^* - \delta t(p_{,i})$ where U_i is not specified.

Step 2 eliminates the checkerboard pressure field when using a nonstaggered grid. Implicit treatment of the convective and diffusive terms eliminates the numerical stability restrictions. In this work the time step is kept for CFL (Courant–Frederick–Lewis) number less than one.

3.2. *Solution procedure and convergence criterion*

Eq. (2) is solved using point iteration with an underrelaxation factor of 0.7 for each time step. Usually about 10–20 iterations are sufficient to reduce the absolute sum of the residue to be less than 0.01. The pressure equations take considerable computer time. In this work a preconditioned conjugate-gradient (PCG) procedure is used. The iteration is done until the absolute sum of the residue of the pressure equation reduces to 0.01 for each time step. The algorithm for the implementation of PCG on a finite difference grid was presented by the author [14] and the advantages of the procedure are discussed in Ref. [15]. For a $43 \times 35 \times 20$ grid to reach steady state, about 75 time steps are needed using a time step of 0.1. In a Sun-Sparcstation 20 the total computer time is about 45 min for 75 time steps.

3.3. *Solution region and boundary conditions*

The exterior boundaries on all sides are located about ten times the width of the building in that direction. The top boundary is situated above the ground at more than 15 times the height of the building. These dimensions were decided from computational experiments. On all exterior boundaries the undisturbed velocities, k and ε are prescribed. The undisturbed velocities are assumed to have a log-law profile. The expressions for undisturbed k and ε are $k = u_*^2/\sqrt{C_\mu}$ and $\varepsilon = u_*^3/(Kz)$; where u_* is the frictional velocity, K is the von Karman constant equal to 0.4 and z is the vertical distance from the ground. On the wall, the law of the wall condition is used as suggested by Launder and Spalding [10].

4. Computer modelling of pollutant dispersion

Pollutant dispersion is computed by solving Eq. (5). The major difficulty in solving the equation is to approximate the advection terms in an Eulerian-grid framework. Many different procedures have been used. The upwind (UW) procedure is a method commonly used and easy to implement. It produces a large amount of numerical diffusion. To reduce this error a control volume streamline (CVSU) procedure which conserves mass, computes positive concentration and gives small numerical diffusion is developed and applied for pollutant transport around buildings in Ref. [8]. The computed concentration along the centerline of the building is little higher than the wind-tunnel measurement when the pollutant is kept at the height of the building. Comparing to the upwind procedure the CVSU procedure computed a trend of the longitudinal ground-level concentration similar to the wind-tunnel measurement. In this work the SUPG procedure developed by Brooks and Hughes [9] is used to solve Eq. (5). The equation is approximated as a steady state and solved implicitly using point iteration. The details of the FEM implementation are given in Ref. [16]. The equations are assembled here element-by-element to reduce storage space. An under-relaxation factor of 0.4 is used. The iteration is done until the absolute sum of the residue is below 0.1 or 250 iterations.

4.1. *Incorporation of a source term*

The source term in Eq. (5) is modeled using the procedure discussed by Dawson et al. [3]. A point source term, S (g/s/m^3), in Eq. (5) is set equal to Q (g/s), the effluent rate divided by the emission cell volume.

4.2. *Boundary conditions*

Zero-gradient boundary conditions for the concentration equations are applied at the top and side boundaries, at the ground and at the building surfaces.

5. Comparison of computed concentrations with wind tunnel measurements

The performance of different numerical procedures are compared with measured concentrations from the wind-tunnel study of Huber et al. [17]. In this study measurements near a model block building of 25 cm in height (H_b), length perpendicular to the wind of 50 cm ($L/H_b = 2$) and a width of 25 cm along the direction of wind flow ($B/H_b = 1$) were taken for a range of source conditions. For our numerical computations, the building is modeled on a 1 : 200 scale. On this scale, the building is 50 m high and placed in a 360 m deep atmospheric boundary layer. The wind-tunnel velocity profile U/U_r was found to fit a one-sixth power law. The reference velocity $U_r = 2.35$ m/s was measured at $1.5H_b$ above the surface. The roughness length z_0 in the wind tunnel is scaled to be 0.1 m. A logarithmic upstream wind profile using the scaled z_0 and U_r was used in computations presented herein since it provides a better match to winds near the surface. The flow region is divided into $45 \times 42 \times 25$ grid points. The building is modeled into a $10 \times 10 \times 10$ grid.

The cases studied herein include a point source at the ground (position 1) and at 60 m ($1.2H_b$) above the ground (position 2), at the center of the leeward building wall

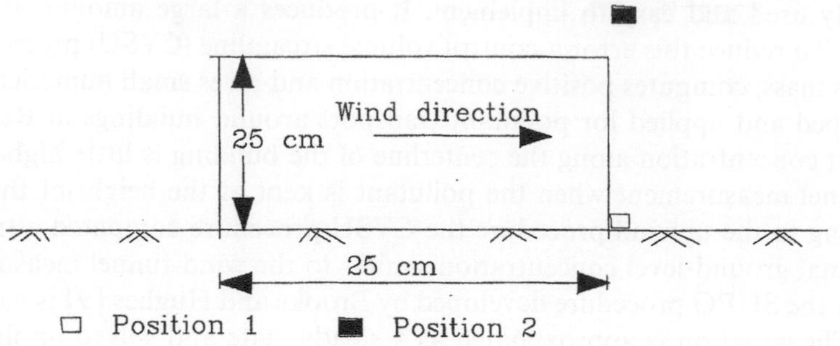

Fig. 1. Source position with respect to the building and wind flow (not to scale).

(see Fig. 1). The computed concentrations are compared with wind tunnel results as a nondimensional concentration $CU_rH_b^2/Q$.

Longitudinal profiles of concentrations at the ground and downwind from the source for positions 1 and 2 are presented in Fig. 2. The computed concentration using the upwind and SUPG procedure compare reasonably well with wind-tunnel measurements for position 1 in Fig. 2a. The SUPG procedure predicted well the wind-tunnel measurement close to the wall, and beyond $x/H_b = 2$ the computed values are about twice the wind-tunnel values. Whereas the upwind procedure under-predicted close to the wall and away from the wall it is much closer to the wind-tunnel measurement. For position 2 in Fig. 2b, the computed concentration using the upwind procedure decreases from 0.6–0.55 when x/H_b is 0–3. Whereas the wind-tunnel data is varying from 0.1 to 0.4 when x/H_b is 0–3 and then decreases to about 0.3. Hence, the computed values are higher than the wind-tunnel data using the upwind procedure. The computed concentrations using the SUPG procedure agree very well with the wind tunnel until $x/H_b = 3$ and then increase to 0.54. Hence, close to the building it is more accurate than away from the building.

The computed vertical concentration profile at three building heights downstream from the leeward building wall are compared with wind-tunnel data for positions 1 and 2 in Fig. 3. For position 1 in Fig. 3a, the upwind procedure is much closer to the wind tunnel measurement than the SUPG procedure. Whereas in Fig. 3b for position 2, the SUPG procedure is much closer to the wind-tunnel data than the upwind procedure. The upwind procedure overpredicts from 1.4–1.7 at $z/H_b = 1.5$.

The flow is very complex in the region. The flow is not completely verified with measurements using the k–ε turbulence model. Schmidt and Larsen [18] observed that the computed velocities are lower than the wind-tunnel measurement for different versions of the k–ε turbulence model for two-dimensional flow around a surface mounted obstacle. In addition to this the implementation of the law of the wall, numerical diffusion and dispersion in approximating the convection term and the grid resolution close to the wall complicates the final velocity field. The

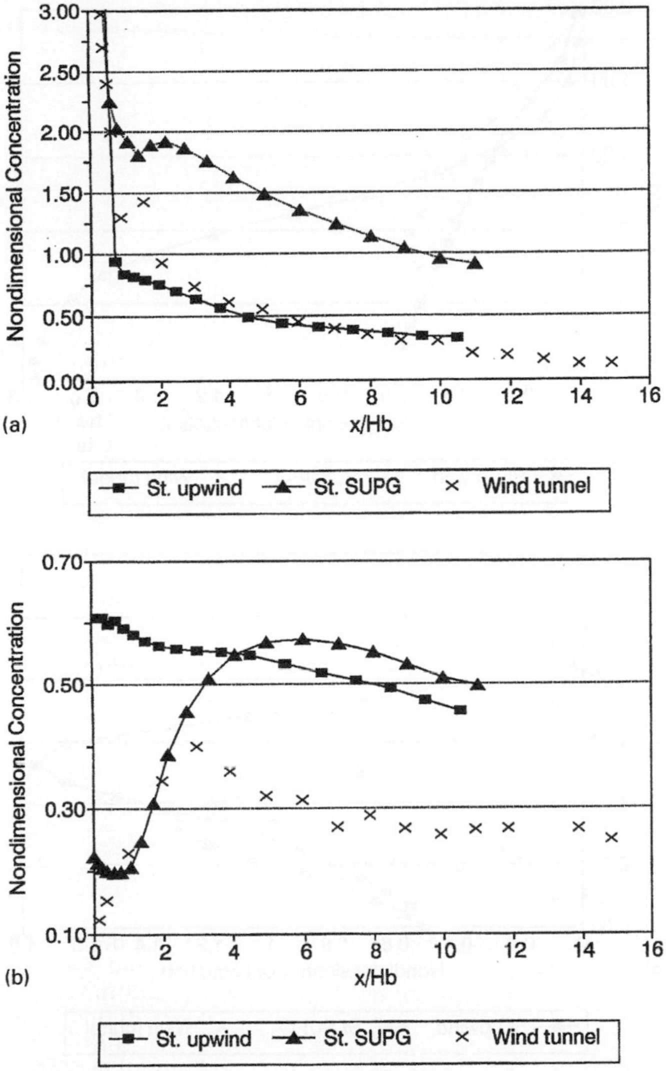

Fig. 2. Comparison of longitudinal ground-level concentrations along the centerline for source at (a) position 1 and (b) position 2.

SUPG procedure has less numerical diffusion and is much more accurate than the upwind procedure. The trend of numerical diffusion using the upwind procedure is clearly seen in the computed values. That is, close to the source the concentrations are lower than the wind-tunnel data and far away from the source higher than the wind-tunnel data. Further work is necessary to establish the computed flow using different turbulence models in complex flow regions. For this good wind-tunnel

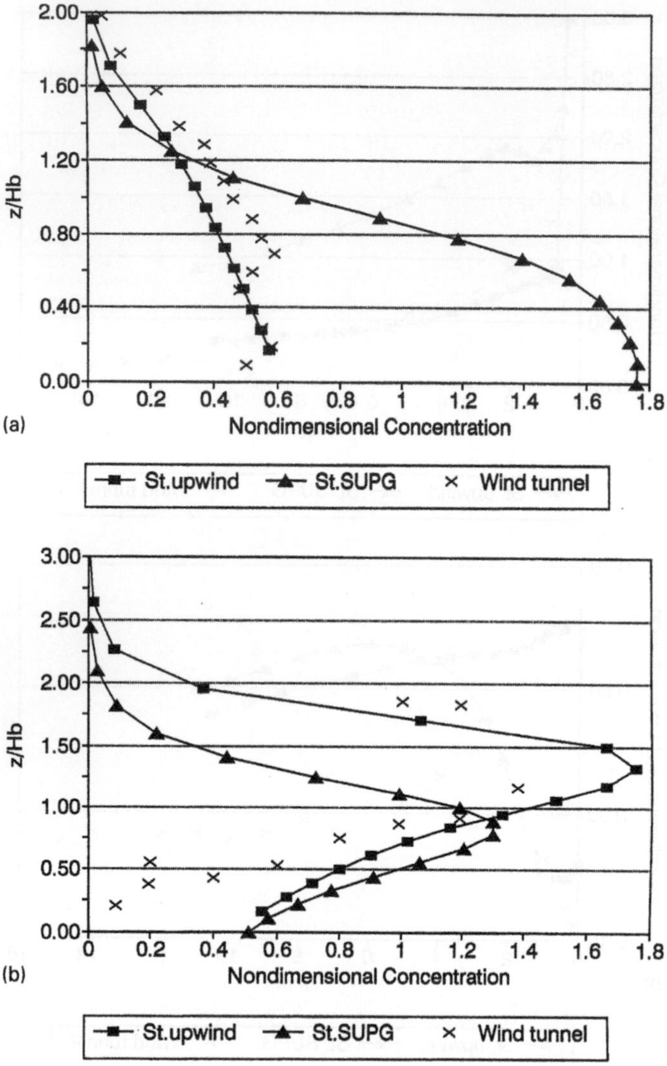

Fig. 3. Comparison of vertical profile of concentrations along the center line at $x/H_b = 3$ for source at (a) position 1 and (b) position 2.

data in the complex region is needed. Over all the computed results using the SUPG procedure are much closer to the wind-tunnel data. The error due to both procedures is not very large compared to the nondimensional point source release of 5875 kept at position 1 and 2. That is, the maximum concentrations reported in Figs. 2 and 3 are about one in 2000 of the concentration around the source.

6. Conclusions

A streamline upwind Petro–Galerkine (SUPG) procedure which conserves mass, and gives low numerical diffusion compared to upwind procedure is used to compute the pollutant transport equation. The pollutant dispersion around a building is modeled using an eddy viscosity based on the k–ε turbulence model. Both the upwind and SUPG procedures are used to approximate the advection term. The concentrations for a point source kept at the ground and at an elevated point (located at 1.2 times the building height above the ground) behind the center of the leeward wall are computed and compared with available wind-tunnel measurements. The SUPG procedure predicted a variation of concentration on the ground much closer to the wind tunnel data than the upwind procedure. Overall the error in the computed concentration is much less compared to the nondimensional point source release of 5875 kept at the source. Further research is underway to see the performance of this method with much finer grids and using large eddy simulation.

Acknowledgements

The author acknowledges the support provided by the Department of Energy Engineering, Technical University of Denmark and the Danish Technical Research Council under grant no: 5.26.16.31 for performing part of this work as a visiting professor.

References

[1] R.L. Lee, J.M. Leone Jr., Numerical modeling of turbulent dispersion around structures using a particle-in-cell method, 84th Meeting of the Air and Waste Management Association, Vancouver, Canada, 16–21 June 1991, p. 16.

[2] E. Naslund, H.C. Rodean ean, J.S. Nasstrom, A comparison between two stochastic diffusion models in a complex three-dimensional flow, Bound.-Layer Meteorol. 67 (1994) 369–384.

[3] P. Dawson, D.E. Stock, B. Lamb, The numerical simulation of airflow and dispersion in three-dimensional atmospheric recirculation zones, J. App. Meteorol. 30 (1991) 1005–1024.

[4] S. Murakami, A. Mochida, Y. Hayashi, Numerical simulation of velocity field and diffusion field in an urban area, Energy Buildings 15/16 (1990, 1991) 345–356.

[5] Y. Moriquchi, K. Uehera, Numerical and experimental simulation of vehicle exhaust gas dispersion for complex urban roadways and their surroundings, J. Wind Eng. Ind. Aerodyn. 46/47 (1993) 689–695.

[6] Y. Qin, S.C. Kot, Validation of computer modelling of vehicular exhaust dispersion near a tower block, Building Environ. 25 (1990) 125–131.

[7] R.P. Selvam, A.H. Huber, Computer modelling of pollutant dispersion around buildings: current status, in: Wind Engineering Retrospect and Prospect, vol. 2, Wiley, New Delhi, India, 1995, pp. 594–605.

[8] R.P. Selvam, K.S. Rao, A.H. Huber, Numerical simulation of pollutant dispersion around a building, Proc. 9th Joint Conf. on Air Pollution and Meterolgy with A&WMA, Atlanta, GA, 28 January–2 February 1996, pp. 329–332.

[9] A. Brooks, T.J.R. Hughes, Streamline upwind/Petrov–Galerkin formulations for convection dominated flow with particular emphasis on the incompressible Navier–Stokes equations, Comput. Meth. Appl. Mech. Eng. 32 (1982) 199–259.

[10] B.E. Launder, D.B. Spalding, The numerical computation of turbulent flows, Comput. Meth. Appl. Mech. Eng. 3 (1974) 269–289.
[11] R.P. Selvam, Computation of pressures on Texas Tech University building using large eddy simulation, these Proceedings, J. Wind Eng. Ind. Aerodyn. 67&68 (1997) 647–657.
[12] S.V. Patankar, Numerical Heat Transfer and Fluid Flow, Hemisphere, Washington, DC, 1980.
[13] H. Choi, P. Moin, Effects of the computational time step on numerical solutions of turbulent flow, J. Comput. Phys. 113 (1994) 1–4.
[14] R.P. Selvam, Computation of flow around Texas Tech building using k–ε and Kato–Launder k–ε turbulence model, Eng. Struct., accepted for publication.
[15] R.P. Selvam, Numerical simulation of flow and pressure around a building, ASHRAE Trans. 102, 1996, accepted for publication.
[16] R.P. Selvam, Finite element modelling of flow around a circular cylinder using LES, these Proceedings, J. Wind Eng. Ind. Aerodyn. 67&68 (1997) 129–139.
[17] A.H. Huber, W.H. Snyder, R.E. Lawson Jr., The effect of a squat building on short stack effulents: a wind-tunnel study, Report EPA-600/4-80-055, Research Triangle Park, NC, 1980.
[18] J.J. Schmidt, P.S. Larsen, Two-equation models for low-Reynolds number flows, ETMA Workshop, UMIST, Manchester, 14–17 November 1994, p. 8.

ELSEVIER

Journal of Wind Engineering
and Industrial Aerodynamics 67&68 (1997) 815–825

JOURNAL OF wind engineering AND industrial aerodynamics

Computational modelling in the prediction of building internal pressure gain functions

R.N. Sharma[a,]*, P.J. Richards[b]

[a] *Department of Technology, University of the South Pacific, PO Box 1168, Suva, Fiji*

[b] *Department of Mechanical Engineering, University of Auckland, Private Bag 92019, Auckland, New Zealand*

Abstract

This paper describes a methodology based on computational and analytical modelling techniques in the prediction of building internal pressure gain functions. It first involves computational modelling of the transient response following a sudden opening, which predicts both the Helmholtz frequency and the damping characteristics fairly accurately. The parameters derived by fitting the analytical model to the computed response are then used in the prediction of the gain function. Good agreement is obtained between measured and predicted internal pressure gain functions for the case of a cylinder and a 1 : 50 scale model of the TTU test building.

Keywords: Computational modelling; Internal pressure; Sudden opening; Helmholtz resonance; Gain function

1. Introduction

The safety of a building structure in strong winds depends as much on internal pressure as it does on external pressure. Two of the key issues concerning internal pressure are the level of overshoot in the transient response following a sudden dominant opening, and the subsequent steady-state (resonant) response to fluctuating external pressure in the presence of the opening. Various researchers, including Stathopoulos and Luchian [1], Vickery and Bloxham [2], and Yeatts and Mehta [3] have shown that the first of these is of little significance since the transients are lost amidst turbulence induced fluctuations and the overshoot if any is usually smaller than the peak internal pressure attained at steady-state. The steady-state response of internal pressure to wind turbulence in the presence of a dominant opening, which is a Helmholtz resonance type of response as first shown by Holmes [4], is therefore of

* Corresponding author.

PII S0167-6105(97)00121-9

greater significance in determining the ultimate design pressure. Despite the fact that in real life, the initial transient response is relatively unimportant, a proper understanding of the transient behaviour is however necessary in order to be able to predict correctly the steady-state (resonant) response to fluctuating external pressure. An understanding of the transient response results in two of the key parameters, firstly the natural or Helmholtz frequency and secondly the damping coefficient, required in the prediction of the resonant response.

For the limiting case of a rigid non-porous building with a dominant windward wall opening, a number of experimental and analytical studies in the past (see, e.g., Refs. [4–6]), have increased our understanding of internal pressure behaviour. Sharma and Richards [7] recently applied computational modelling for the first time in a study of the transient response behaviour of building internal pressure to sudden openings. It was shown that the computer model, based on the commercial package PHOENICS™, predicts the Helmholtz oscillation frequency as well as the damping characteristics fairly accurately. The study also established through computational flow visualisation, that formation of a vena-contracta does take place past the opening, despite the unsteady nature of the flow.

This paper describes an extension of the study discussed in Ref. [7] to utilise the transient response obtained from computational modelling in predicting the building internal pressure gain function which is important in the study of the steady-state response. The studies discussed in Ref. [7] and in this paper are part of a more comprehensive study on building internal pressure described in detail by Sharma [8].

2. Theoretical considerations

It can be shown [7,8] that the differential equation governing the dynamics of internal pressure in the presence of a dominant opening is of the form:

$$\frac{\rho_a L_e \forall_o}{\gamma c A_o p_a}\ddot{C}_{p_i} + C_L \frac{\rho_a q \forall_o^2}{2(\gamma A_o p_a)^2}|\dot{C}_{p_i}|\dot{C}_{p_i} + \frac{(\mu_{eff}/\Delta r)PL_e\forall_o}{\gamma c^2 A_o^2 p_a}\dot{C}_{p_i} + C_{p_i} = C_{p_e}. \tag{1}$$

The linear damping term arises as a result of apertrure wall shear stresses and is significant only at model-scale or for very small openings. The Helmholtz frequency f_{HH} is readily obtained from Eq. (1), and is dependent on the area of the opening A_o, the building volume $\forall_o$, and the effective length of the air slug that oscillates at the opening L_e:

$$f_{HH} = \frac{\omega_{HH}}{2\pi} = \frac{1}{2\pi}\sqrt{\frac{\gamma c A_o p_a}{\rho_a L_e \forall_o}}. \tag{2}$$

In these equations, ρ_a, p_a, γ, and μ_{eff} are the density, pressure, specific heat ratio, and viscosity of ambient air respectively, $\forall_o$ is the building internal volume, c and C_L are the discharge and loss coefficients for the opening of area A_o and circumferential perimeter P, L_e is the effective length of the air jet/slug at the opening (physical length

L_o), and Δr is a distance used in defining the wall shear stress in the aperture. Internal and external pressures are represented by the internal and external pressure coefficients $C_{p_i} = p_i/q$ and $C_{p_e} = p_e/q$ where $q = \frac{1}{2}\rho_a \overline{U}_h^2$ is the reference dynamic pressure based on the ridge-height velocity $\overline{U}_h$.

In order to determine the previously ill-defined parameters, namely the discharge coefficient c, the loss coefficient C_L, and the effective length of the air slug L_e, appropriate guidelines were developed. This was possible from computational and experimental modelling of sudden openings on a range of building and cavity models, as well as from spectral measurements in the wind tunnel. The details of this study are contained in Ref. [8], and only the results summarised hereafter are considered appropriate in the context of the present discussions. The guidelines involve classification of openings into thin and long openings according to the ratio of its physical length L_o to the effective radius $r_{eff} = \sqrt{A_o/\pi}$ as follows:

1. For *long openings* when $L_o/r_{eff} > 1.0$:

$$c = 1.00, \quad C_L = 1.50, \quad L_e = L_o + 1.73\sqrt{A_o/\pi}.$$

2. For *thin openings* when $L_o/r_{eff} < 1.0$:

$$c = 0.60, \quad C_L = 1.20.$$

(a) Window near the centre of the wall (i.e. away from side walls, floor, and the roof):

$$L_e = L_o + 1.39\sqrt{A_o/\pi}.$$

(b) Window near side wall or door near the floor:

$$L_e = L_o + (1.17\text{–}1.29)\sqrt{A_o/\pi}.$$

(c) Door at the corner of the wall (i.e. adjacent to the side wall and the floor):

$$L_e = L_o + 1.61\sqrt{A_o/\pi}.$$

It is important to note that the discharge coefficient c appears because flow contraction can take place past the opening thus reducing the inertia of the effective air jet/slug. Its presence therefore in the governing equations determines the inertia of the air slug and therefore the Helmholtz frequency, and does not in any way represent the losses through the opening. When a transient response of internal pressure to a step change in steady external pressure (i.e. that does not fluctuate) is obtained from computational modelling, the unknown parameters in Eq. (1) may be readily obtained by fitting the numerical solution to Eq. (1) to the computed response.

In order to obtain the gain function from Eq. (1), it must first be linearised. This is possible by equating the energy dissipated by the sum of the non-linear and linear damping terms to that dissipated by an equivalent linear damping term $c_{eq}\dot{C}_{p_i}$. If as assumed by Vickery and Bloxham [2], that internal pressure is of Gaussian distribution (this applies when excitation, in this case external pressure, can be approximated as being of Gaussian distribution), then an estimate of the equivalent damping term is determined by integration to yield:

$$c_{\mathrm{eq}} = \sqrt{8\pi}\frac{C_{\mathrm{L}} q \forall_{\mathrm{o}} f_{\mathrm{HH}} \tilde{C}_{p_i}}{\gamma(L_{\mathrm{e}}/c) A_{\mathrm{o}} p_{\mathrm{a}}} + \frac{(\mu_{\mathrm{eff}}/\Delta r)P}{\rho c A_{\mathrm{o}}}. \tag{3}$$

In this expression, $\tilde{C}_{p_i}$ is the root-mean-square value of internal pressure coefficient fluctuations, which may be determined from equations derived by Vickery and Bloxham [2]. The linearised equation:

$$\ddot{C}_{p_i} + c_{\mathrm{eq}}\dot{C}_{p_i} + \omega_{\mathrm{HH}}^2 C_{p_i} = \omega_{\mathrm{HH}}^2 C_{p_e} \tag{4}$$

is readily (Laplace) transformed to obtain an expression for the gain function of internal pressure over the external:

$$|\chi_{p_i - p_e}| = \frac{|C_{p_i}|}{|C_{p_e}|} = \left[\left(\frac{\omega_{\mathrm{HH}}^2 - \omega^2}{\omega_{\mathrm{HH}}^2}\right)^2 + \left(\frac{\omega c_{\mathrm{eq}}}{\omega_{\mathrm{HH}}^2}\right)^2\right]^{-1/2} \tag{5}$$

where $\omega = 2\pi f$ is the angular frequency. If a computational model can be made to predict both the Helmholtz frequency and the damping characteristics correctly, this approach may be used to obtain the coefficients of the terms in the differential equation, and subsequently the gain of internal pressure over the external.

3. Computational modelling of the sudden opening

Some of the advantages of computational modelling of the transient response to a sudden opening include the possibility of a three-dimensional solution, the ability to monitor both pressure and velocities internally and in the vicinity of the opening, and the capability of flow visualisation in the ensuing unsteady flow. With experimental modelling on the other hand, only pressures can be measured with ease, while measurement of velocities and visualisation of flow near the opening would be quite difficult, if not impossible.

Computational modelling of sudden openings in the present study is the same as that described in Refs. [7,8], and which is based on the commercial package PHOENICS™. The package uses a finite volume code to solve the conservation equations for mass, momentum, turbulence kinetic energy (TKE) per unit mass k and the rate of dissipation of TKE ε. The solution algorithm is a variant of SIMPLE (Semi-Implicit Method for Pressure Linked Equations), in which velocities are obtained by solving the momentum conservation equations using the most recent estimates of the pressure field that is corrected by the imbalances in the mass conservation equations. Other conservation equations are then solved and the procedure iterated until convergence.

Three types of sudden opening problems may be studied, that include either a boundary layer onset flow [9], a smooth onset flow, or no onset flow but with an initial internal–external pressure difference (hereafter referred to as static tests). It has previously [7,8] been shown that the static sudden opening test is sufficient in determining the unknown parameters. In the simulation of a sudden opening, the building is first modelled by blocking appropriate cells/cell faces, and then obtaining a steady solution corresponding to the conditions prior to the creation of the opening. In the case of static sudden opening tests, internal and external pressures are set to the desired values. A sudden opening is created by removing the cell/cell face blockages corresponding to the window or door opening and the transient solution advanced over many time steps. The number of sweeps at each time step is usually determined so as to obtain a balance between convergence requirements and the cost in terms of computation time. The time step size is typically of the order of 1/100th of the expected period of Helmholtz oscillations $1/f_{\mathrm{HH}}$, but may usually be increased considerably without compromising accuracy to any significant level. It should be noted that the time step size determines the suddenness of the creation of the opening, and may therefore also be varied to achieve various degrees of opening rapidity. This is a major advantage of computational modelling since in the experimental simulation of sudden openings, the time for the creation of an opening is usually not controllable.

Since the physics of the problem involves compressible flow, it is essential that the isentropic density formulation

$$\frac{p}{\rho^{\gamma}} = \frac{p_{\mathrm{a}}}{\rho_{\mathrm{a}}^{\gamma}} = \mathrm{const} \tag{6}$$

is invoked for the entire flow field. Note that it is the compressibility of the air that leads to the changes in internal pressure.

4. Experimental measurement of the internal pressure gain function

Sharma and Richards [10] have recently shown that the relevant forcing function for the internal pressure system is the external area-averaged pressure over the extent of the opening. The internal pressure gain function may therefore be determined experimentally from a measurement of the internal pressure spectrum $S_{p_{\mathrm{i}}}(f)$ and the spectrum of external area-averaged pressure $S_{p_{\mathrm{e}}}(f)$ from the relation for a linear system:

$$|\chi_{p_{\mathrm{i}}-p_{\mathrm{e}}}| = \sqrt{S_{p_{\mathrm{e}}}(f)/S_{p_{\mathrm{i}}}(f)}. \tag{7}$$

The external area-averaged pressure spectrum is measured without the opening, over an area covering the extent of the opening, while the internal pressure spectrum is measured in the presence of the opening. For the tests described later in this paper, measurements were conducted in a 1 : 50 scale wind tunnel boundary layer flow, simulating full-scale conditions at the Texas Tech. University (TTU) facility (i.e. full-scale equivalent roughness length $z_{0} = 29$ mm and 4 m height turbulence intensity $I_{u} = 0.18$). Internal pressure was measured using a restricted tubing having a flat response up to 200 Hz, while the external pressure with a 10 input manifold pneumatic averaging system having a flat response up to 300 Hz. The pressure signals were transmitted to Honeywell 163PC differential pressure transducers, and the electrical outputs were then digitised using a Keithley-Metrabyte DAS1602 analogue to digital converter on an IBM compatible personal computer.

5. Results for a cylindrical model

On the basis outlined earlier, the response of the cylinder illustrated in Fig. 1, having internal dimensions 247 mm long and 140 mm internal diameter; a 19 mm long and 25 mm diameter aperture, was measured in the wind tunnel turbulent boundary layer flow. The relevant parameters for this test were as follows:

$$A_o = 4.909 \times 10^{-4}\ \text{m}^2,\quad \forall_o = 3.803 \times 10^{-3}\ \text{m}^3,\quad L_o = 19\ \text{mm},\quad P = \pi(25\ \text{mm}),$$

$$f_{\text{HH}}(\text{measured}) = 97\ \text{Hz which gives } L_e/c = 0.04\ \text{m},$$

$$\tilde{C}_{p_i}(\text{measured}) = 0.237,\quad q = 27\ \text{Pa}.$$

The initial conditions from the wind tunnel test were duplicated for computational modelling of the corresponding sudden opening problem under static conditions. The transient response of internal pressure thus obtained is shown in Fig. 2. It exhibits an oscillation frequency of 97 Hz. Eq. (1) was fitted to the first three to four cycles of the computed response using $C_L = 1.5$, $c = 1.00$ and $\mu_{\text{eff}}/\Delta r = 0.16\ \text{kg m}^{-2}\,\text{s}^{-1}$. These parameters yield a value for the equivalent damping coefficient $c_{\text{eq}} = 32.6\ \text{s}^{-1}$. The fitted transient response of internal pressure also appears in Fig. 2.

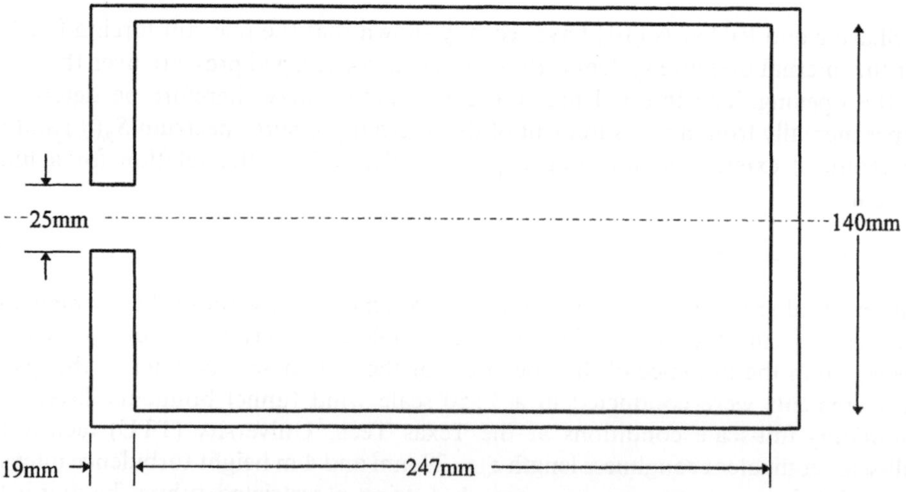

Fig. 1. Details of the cylindrical model.

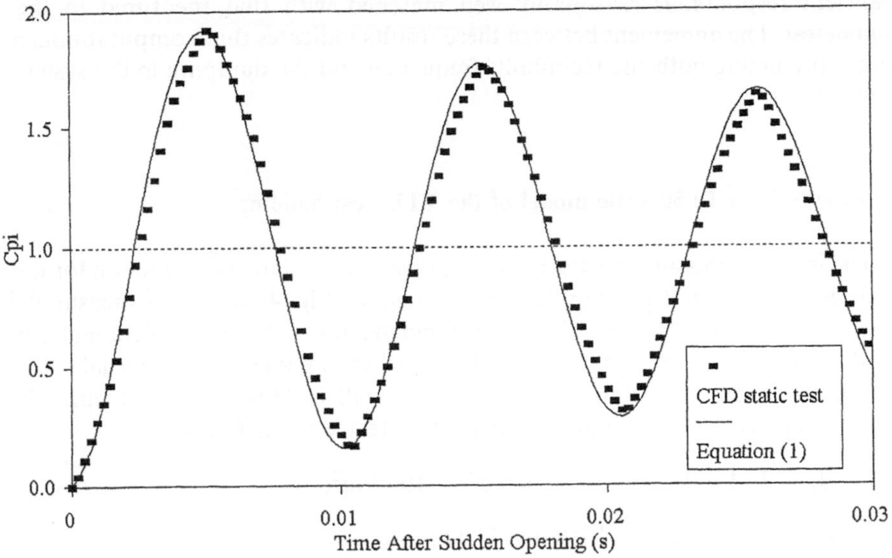

Fig. 2. Computational and analytic responses for the cylinder.

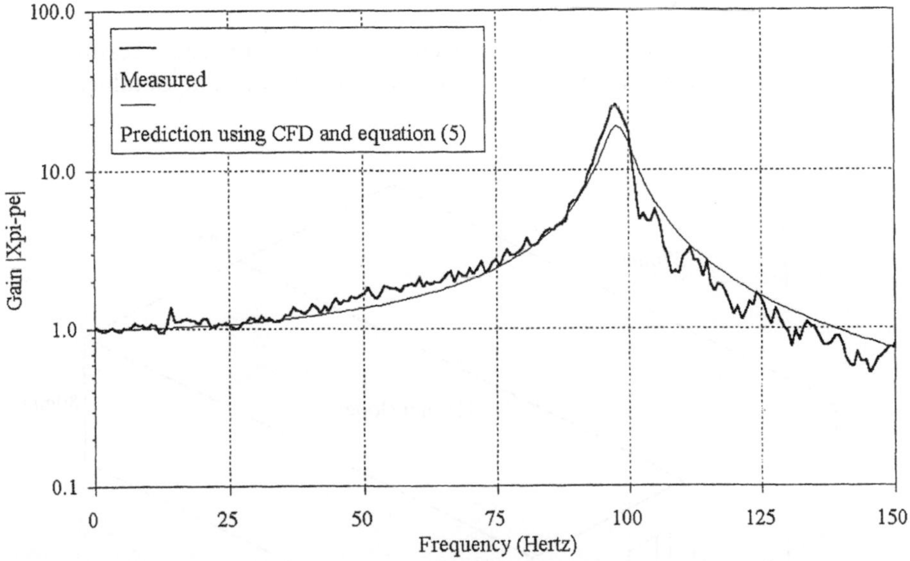

Fig. 3. Comparison of measured and predicted gain functions of internal pressure over the external for the cylinder.

Fig. 3 shows a comparison of the gain functions (square-root of admittance) obtained from the wind tunnel test and Eq. (5). The predicted gain function using computational fluid dynamics (CFD) modelling and the analytic model fitted to the transient response, is reasonably well matched with that measured in the wind tunnel test. The agreement between these results indicates that computational modelling is predicting both the Helmholtz frequency and the damping in the system fairly well.

6. Results for a 1 : 50 scale model of the TTU test building

The internal pressure to external area-averaged pressure gain function for the 1 : 50 scale model of the TTU building (illustrated in Fig. 4) was also measured in the boundary layer wind tunnel. The average dimensions of the nearly rectangular perspex model are as follows: external 276 mm × 184 mm × 80 mm; internal 264 mm × 172 mm × 67 mm. The door opening on the wall is 43 mm high, 18 mm wide and 6 mm deep. The relevant parameters for this test were as follows:

$$A_o = 7.74 \times 10^{-4}\ \text{m}^2, \quad \forall_o = 3.042 \times 10^{-3}\ \text{m}^3,$$

$$L_o = 6\ \text{mm}, \quad P = 2 \times (18\ \text{mm} + 43\ \text{mm}),$$

$$f_{HH}(\text{measured}) = 136\ \text{Hz which gives } L_e/c = 0.0405\ \text{m},$$

$$\tilde{C}_{p_i}(\text{measured}) = 0.21, \quad q = 30.5\ \text{Pa}.$$

Fig. 5 shows a comparison of the transient responses of internal pressure obtained from computational modelling ($q = 308$ Pa was used) and that obtained by fitting

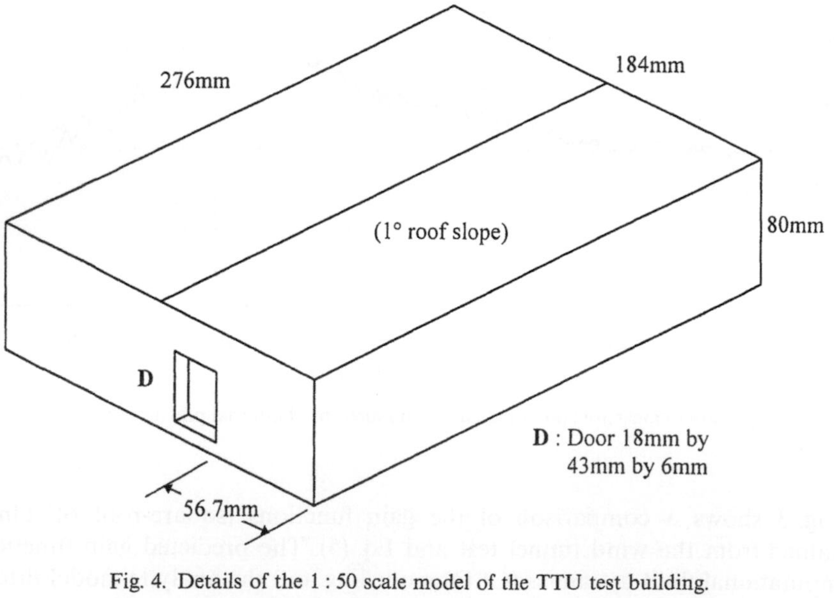

Fig. 4. Details of the 1 : 50 scale model of the TTU test building.

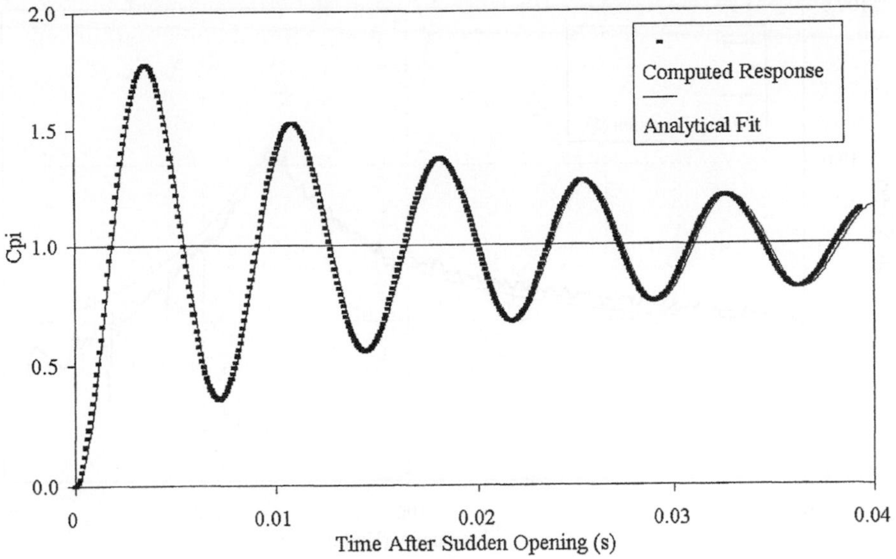

Fig. 5. Computational and analytic responses for the TTU model.

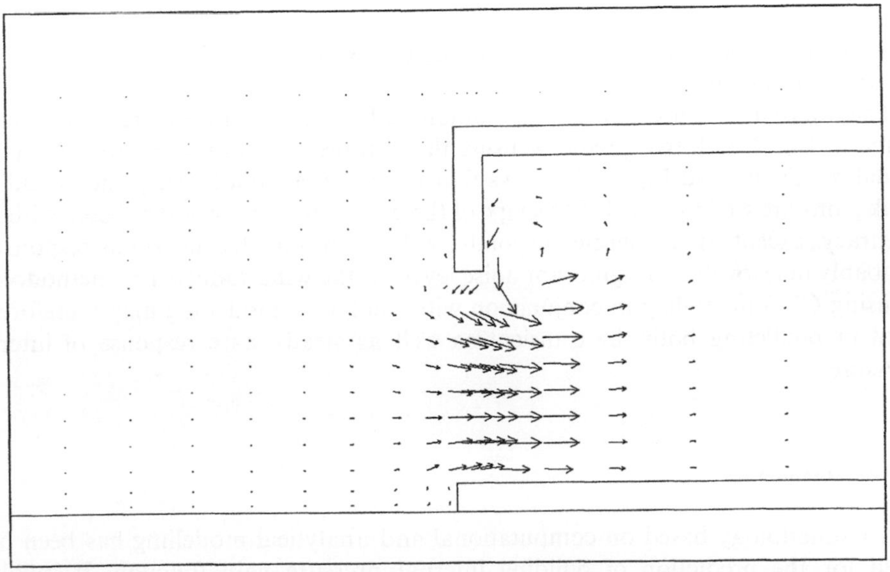

Fig. 6. Velocity vectors in the central longitudinal plane of the door showing the formation of a vena-contracta.

Eq. (1). Note that the oscillations have a frequency of 136 Hz, in exact agreement with that obtained from spectral measurements. Fig. 6 illustrates the formation of a vena-contracta in an in-flow section of the oscillating flow cycle. In order to fit Eq. (1) to the

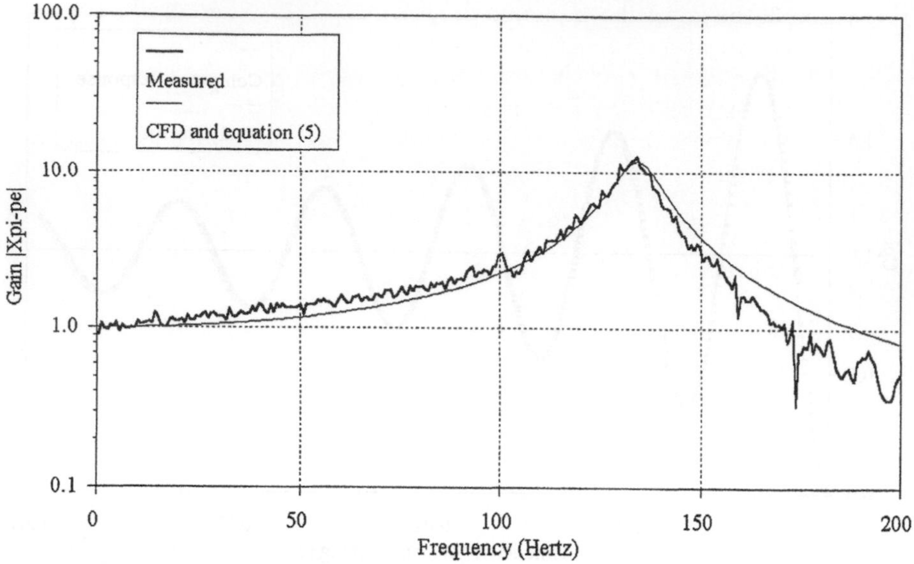

Fig. 7. Comparison of gain functions obtained from measurements and that predicted from computational modelling and Eq. (5).

computed response, the following parameters were required: $C_L = 1.20$, $c = 0.6$, $\mu_{eff}/\Delta r = 0.24\ \mathrm{kg\,m^{-2}\,s^{-1}}$.

The internal pressure gain function obtained from spectral measurements is compared in Fig. 7 with that predicted using the parameters established from computational modelling and Eq. (5). It shows that both the Helmholtz frequencies and the peak gains are well matched. The shape of the gain function is also predicted with fair accuracy, except at frequencies beyond 160 Hz – where the measured response is probably in error due to significant noise levels in the wind tunnel. The methodology of using CFD modelling in conjunction with analytical modelling may therefore be used in predicting both the transient as well as steady-state response of internal pressure.

7. Conclusions

A methodology based on computational and analytical modelling has been outlined for the prediction of building internal pressure gain function. It involves computational modelling of the transient response of building internal pressure to a sudden opening for the case of a step change in steady external pressure. Computational modelling in this manner predicts both the Helmholtz frequency as well as the damping characteristics fairly well. By fitting the analytical model to the computed response, the ill-defined parameters of the governing equation, such as the values for L_e, C_L and $\mu_{eff}/\Delta r$, are readily obtained. The gain function (Eq. (5)) obtained by linearisation of the governing equation utilises these parameters. It has been shown that for a cylindrical and a building model, good agreement is obtained between measured and predicted gain functions of internal pressure over the external area-averaged pressure. Where adequate computing resources are available, this methodology may prove to be extremely useful, particularly for complex opening and building geometry situations.

Acknowledgements

This study was funded through the University of the South Pacific (USP) under the New Zealand (NZ) Government–USP fellowship scheme and by the NZ Ministry of External Relations and Trade while the principal author was on training leave at the University of Auckland. The authors wish to thank both the University of the South Pacific and the New Zealand Government for the financial assistance.

References

[1] T. Stathopoulos, H.D. Luchian, Transient wind-induced internal pressures, J. Eng. Struct. Mech. Div. 115 (EM7) (1989) 1501–1513.
[2] B.J. Vickery, C. Bloxham, Internal pressure dynamics with a dominant opening, J. Wind Eng. Ind. Aerodyn. 41 (1992) 193–204.
[3] B.B. Yeatts, K.C. Mehta, (1993) Field study of internal pressures, in: Proc. 7th US Nat. Conf. on Wind Engineering, June 27–30, 1993, University of California, USA, pp. 889–897.
[4] J.D. Holmes, Mean and fluctuating internal pressures induced by wind, in: Proc. 5th Int. Conf. on Wind Engineering, Fort Collins, USA, July 1979, Pergamon, Oxford, 1980, pp. 435–450.
[5] H. Liu, P.J. Saathoff, Building internal pressure: sudden change, J. Eng. Mech. Div. 107 (EM2) (1981) 309–321.
[6] B.J. Vickery, Internal pressures and interactions with the building envelope, J. Wind Eng. Ind. Aerodyn. 53 (1995) 125–144.
[7] R.N. Sharma, P.J. Richards, Computational modelling of the transient response of building internal pressure to a sudden opening, in: Proc. 9th Int. Conf. on Wind Engineering, New Delhi, India, 9–13 January 1995, pp. 637–648, J. Wind Eng. Ind. Aerodyn., to be published.
[8] R.N. Sharma, The influence of internal pressure on wind loading under tropical cyclone conditions, Ph.D Thesis in Mechanical Engineering, University of Auckland, 1996.
[9] P.J. Richards, R.P. Hoxey, Computational and wind tunnel modelling of mean wind loads on the silsoe structures building, J. Wind Eng. Ind. Aerodyn. 43 (1992) 1641–1652.
[10] R.N. Sharma, P.J. Richards, Windward wall pressure admittance functions for low-rise buildings, 3rd Int. Colloq. on Bluff Body Aerodynamics and Applications, Virginia, USA, July 1996.

ELSEVIER

Journal of Wind Engineering
and Industrial Aerodynamics 67&68 (1997) 827–841

JOURNAL OF
wind engineering
AND
industrial
aerodynamics

CFD prediction of gaseous diffusion around a cubic model using a dynamic mixed SGS model based on composite grid technique

Y. Tominaga[a,*], S. Murakami[b], A. Mochida[a]

[a] *Niigata Institute of Technology, 1719, Fujihashi, Kashiwazaki-shi, Niigata 945-11, Japan*
[b] *I.I.S., University of Tokyo, 7-22-1, Roppongi, Minato-ku, Tokyo 106, Japan*

Abstract

Turbulent diffusion of gaseous contaminant near a cubic-shaped building model is predicted by LES using a Dynamic Mixed SGS (DM) model with the aid of the composite grid technique. The results obtained using a composite grid show better agreement with experiment than does the case using a single grid. Some discrepancies from the experimental results are observed in the results of the standard Smagorinsky model with a composite grid. These discrepancies are improved remarkably by using the DM model.

Keywords: Gaseous diffusion; LES; Dynamic mixed SGS model; Composite grid; Fortified solution algorithm

Nomenclature

x_i	three components of the spatial coordinate ($i = 1, 2, 3$: streamwise, lateral, vertical)
u_i	three components of the velocity vector
f	instantaneous value of a quantity
$\bar{f}$	grid-filtered value of f
$\langle f \rangle$	time averaged value of f
$\hat{\bar{f}}$	test-filtered value of f
c_g	gas concentration
$\langle c_{g0} \rangle$	standard gas concentration ($= q/\langle u_b \rangle / H_b^2$)
q	gas emission rate
ρ_s	gas density
ρ_a	ambient air density
x_n^+	distance from the wall ($x_n^+ = x_n u^* / \nu$)

* Corresponding author. E-mail: tominaga@abe.niit.ac.jp.

PII S0167-6105(97)00122-0

H_b height of model
u_b $\langle u_1 \rangle$ value at inflow of computational domain at height H_b

When values are made dimensionless, representative length scale H_b, velocity scale $\langle u_b \rangle$ and air density ρ_a are used.

1. Introduction

Prediction of contaminant dispersion around buildings is one of the most important subjects in the fields of wind engineering, air-conditioning engineering, etc. A number of investigations have been carried out for predicting turbulent diffusion around buildings using wind tunnel tests. However, there are several difficulties and limitations with wind tunnel tests in comprehending the very complicated turbulent diffusion processes around buildings located in atmospheric boundary layers.

Numerical methods for simulating flow and diffusion fields have developed rapidly as a new analysis tool for wind engineering. In particular, the development of dynamic type subgrid-scale (SGS) models for Large Eddy Simulation (LES) [1,2] has made it possible to analyze the various types of flowfields treated in wind engineering with certain accuracy. Furthermore, a notable feature of LES from the viewpoint of wind engineering is that it can reproduce time-dependent fluctuating flow and diffusion fields over a wide spectrum, while RANS (Reynolds averaged Navier–Stokes equation) models provide information on only ensemble averaged fields.

In recent years, the present authors have carried out computations of turbulent flow past a bluff body, such as a cube, two-dimensional square cylinder, etc., using various types of SGS models used in LES; i.e., the conventional standard Smagorinsky model (S model), Dynamic Smagorinsky model (DS model) [2,3] and Dynamic Mixed model (DM model) [4], etc. In our experience, results from dynamic type SGS models (DS and DM models) showed much better agreement with the experimental data in comparison with the static type of conventional S model. Particularly, the DM model provided more accurate results than did the DS model in many respects [1].

In this study, the DM model is applied to the analysis of turbulent diffusion of a gaseous contaminant near a cubical building model. Here, a composite grid technique supported by a fortified solution algorithm is adopted [5,6]. Computations based on the S model are also carried out for comparison. The accuracy of these computations is examined precisely by comparing with experimental data, including information on concentration fluctuation, $\langle c_g'^2 \rangle$, measured by the present authors [7,8].

2. Outline of wind tunnel experiment [7,8]

Fig. 1 illustrates the flow situation analyzed here. A cubic-shaped model, 0.2 m in height, is located in the turbulent boundary layer expressed as $\langle u_1(x_3) \rangle \propto x_3^{1/4}$. Here,

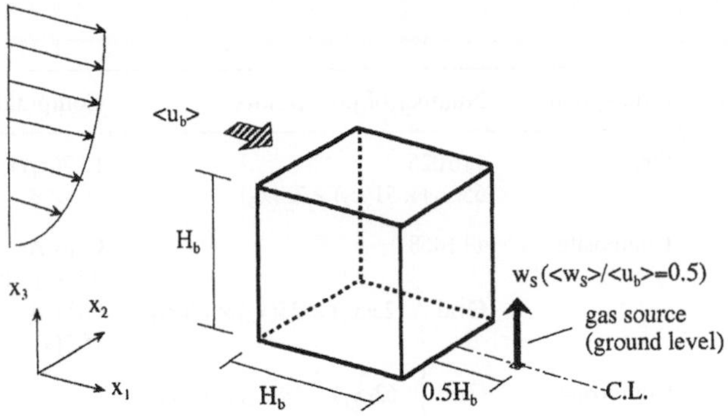

Fig. 1. Model flowfield.

x_3 is the vertical direction. The Reynolds number based on $\langle u_b \rangle$ and H_b is 5.7×10^3. A square-shaped gas source, the side length of which is $0.025H_b$, is set at ground level in the recirculation region behind the cube. Exit gas speed $\langle w_S \rangle$ is $0.5\langle u_b \rangle$.

The wind velocity is measured by a tandem-type hot wire anemometer which can monitor each component of an instantaneous velocity vector. The gas concentration is measured by high response FID (HFR300, Combustion Co. Ltd.) [9]. C_2H_4 is used for tracer gas. Thus, the density of discharged gas is almost the same as that of air (with no buoyancy effect).

3. Outline of computations

3.1. Computed cases

Table 1 lists the three cases compared: two different grid systems, and two different SGS models. A composite grid technique [5,6] is adopted in cases 2 and 3. Two types of SGS models (S model and DM model) are compared using the composite grid technique for cases 2 and 3. Computation using a usual rectangular structural grid (case 1) is also carried out and compared with the results of computations on the composite grid. In the computations conducted in this study, tracer gas is treated as a passive scalar ($\rho_s/\rho_a = 1.0$). The size and position of the gas source and exit gas speed (w_S) are set completely the same as those in the wind tunnel tests. Numerical methods and boundary conditions are summarized in Appendix A.

3.2. Grid system used

Fig. 2 illustrates the grid layout employed for cases 2 and 3. Grid B is used to cover the whole computational domain ($15.7(x_1) \times 9.7(x_2) \times 5.2(x_3)$) and grid A is applied to the region near the gas exit ($0.5(x_1) \times 0.5(x_2) \times 0.25(x_3)$). In order to avoid serious numerical instability caused by inappropriate connection at the connecting regions,

Table 1
Computed cases

Case	SGS model	Grid system	Number of grid points	Computational domain
1	S model	Single	116025 $65(x_1)\times 51(x_2)\times 35(x_3)$	$15.7(x_1)\times 9.7(x_2)\times 5.2(x_3)$
2	S model	Composite	112458 Grid A: $25(x_1)\times 25(x_2)\times 12(x_3)$	Grid A: $0.5(x_1)\times 0.5(x_2)\times 0.25(x_3)$ Grid B: $15.7(x_1)\times 9.7(x_2)\times 5.2(x_3)$
3	DM model	Composite	Grid B: $63(x_1)\times 49(x_2)\times 34(x_3)$	

S model: Smagorinsky model, DM model: Dynamic Mixed model.

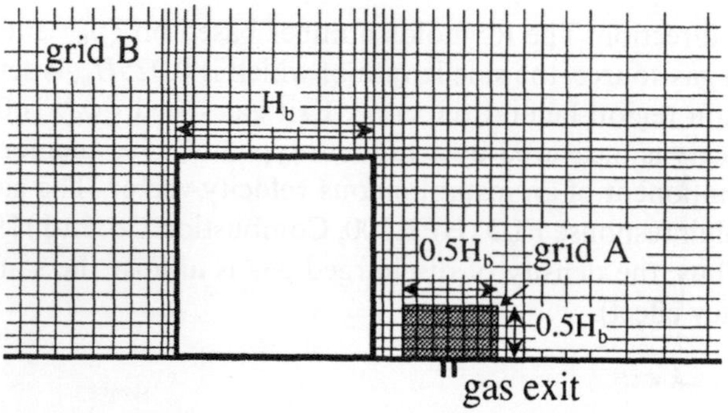

Fig. 2. Grid layout for cases 2 and 3 (composite grid; grid A overlaps onto grid B).

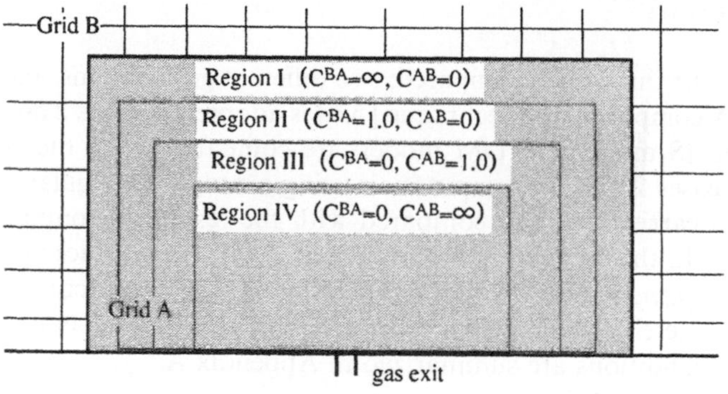

Fig. 3. Distributions of switching parameters.

the composite grid technique is supported by the fortified Navier–Stokes approach [5,6]. Details of the fortified Navier–Stokes approach are given in Appendix B. The distribution of the switching parameters and the connecting regions set in the region of grid A are indicated in Fig. 3.

The total number of grid points is 112458 ($63(x_1) \times 49(x_2) \times 34(x_3)$ for grid B and $25(x_1) \times 25(x_2) \times 12(x_3)$ for grid A) for cases 2 and 3. The smallest mesh size on grid A is $0.005H_b$ at the area just near the gas exit in these cases. In case 1, the usual single grid system is used. In this case, the total number of grid points is 116025 ($= 65(x_1) \times 51(x_2) \times 35(x_3)$), almost the same as for cases 2 and 3. The smallest mesh size in case 1 is $0.025H_b$ at the area near the gas exit.

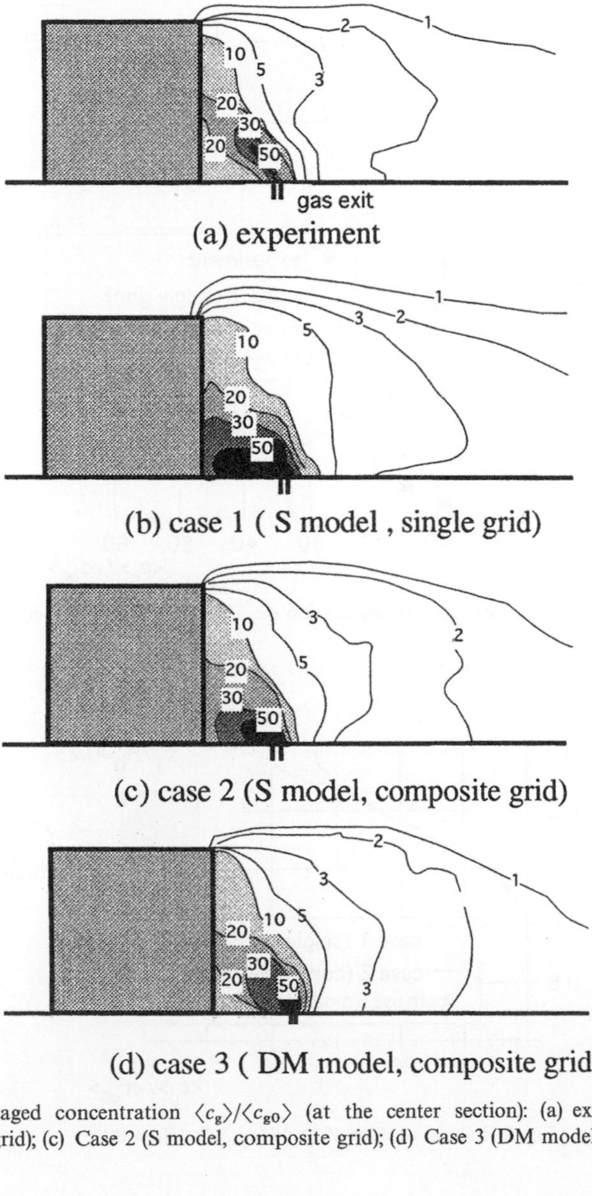

Fig. 4. Time-averaged concentration $\langle c_g \rangle / \langle c_{g0} \rangle$ (at the center section): (a) experiment; (b) Case 1 (S model, single grid); (c) Case 2 (S model, composite grid); (d) Case 3 (DM model, composite grid).

3.3. SGS models compared

Two types of SGS models, the static type of conventional Smagorinsky model (S model, cases 1 and 2 in Table 1) and the dynamic mixed SGS model (DM model, case 3) are compared. The details of the dynamic mixed SGS model used here are described in Appendix C.

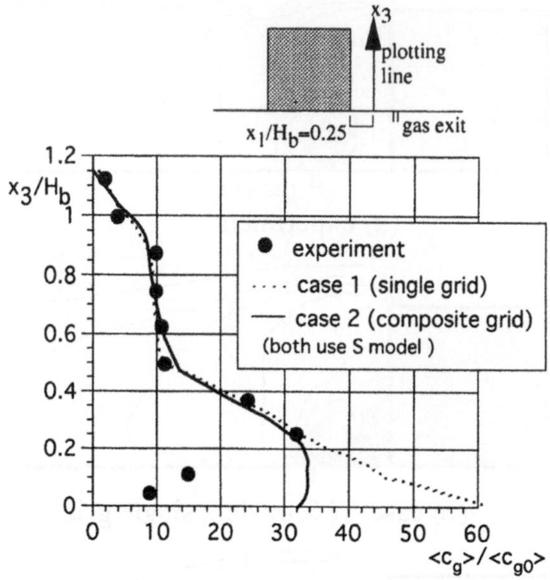

Fig. 5. Vertical profiles of time-averaged concentration $\langle c_g \rangle / \langle c_{g0} \rangle$ (comparison between cases 1 and 2).

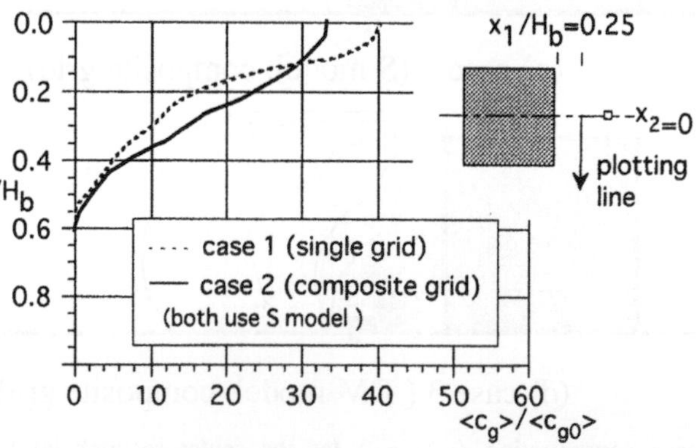

Fig. 6. Lateral profiles of $\langle c_g \rangle / \langle c_{g0} \rangle$ ($x_3/H_b = 0.15$).

4. Results and discussion

4.1. Influence of mesh spacing near gas exit: Comparison between results on single grid system (case 1) and composite grid system (case 2) with S model

Firstly, results from the S model using a single grid (case 1) and a composite grid (case 2) are compared. In Fig. 4, the distributions of time-averaged concentration $\langle c_g \rangle$ in the center section are shown. In the result of the experiment, tracer gas discharged from the ground is advected upstream by the reverse flow in the recirculation region behind the cube, and thus the value of $\langle c_g \rangle$ is relatively large in the area between the leeward face of the cube and the gas exit (Fig. 4a).

As shown in Fig. 4, $\langle c_g \rangle$ is overestimated in the region near the leeward face of the cube in the results of cases 1 and 2. However, the result given from case 2 (composite grid) shows better agreement with the experiment than does case 1 (single grid) (cf. Figs. 4a–4c). The better agreement for distribution of $\langle c_g \rangle$ in case 2 is produced by the fineness of mesh spacing in the vicinity of the gas exit in this case.

Fig. 5 shows the vertical profiles of $\langle c_g \rangle$ in cases 1 and 2. In the area near the leeward face of the cube, $\langle c_g \rangle$ given from the experiment decreases rapidly as the measuring point approaches ground level. This tendency is not reproduced at all in case 1 (single grid). In comparison with the result of case 1, the result of case 2 agrees well with experimental data. Fig. 6 compares lateral profiles of $\langle c_g \rangle$ behind the cube at $x_3/H_b = 0.15$. The value of $\langle c_g \rangle$ in case 2 becomes smaller than that in case 1 near the centerline. This is because the diffusion effect in the lateral direction is well reproduced in case 2 by using the finer mesh spacing near the gas exit.

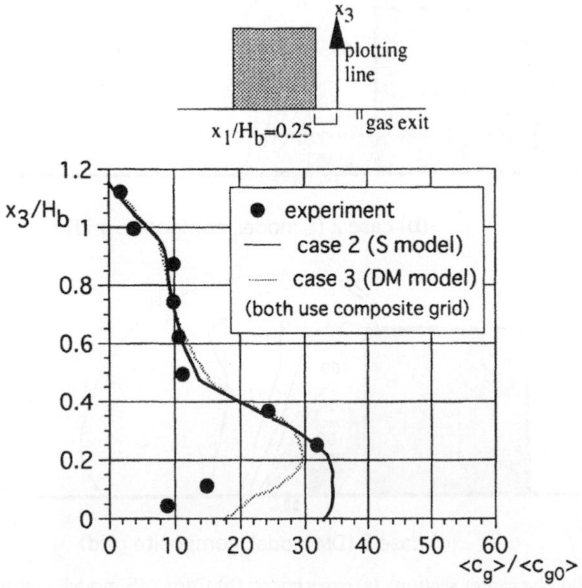

Fig. 7. Vertical profiles of time-averaged concentration $\langle c_g \rangle / \langle c_{g0} \rangle$ (comparison between cases 2 and 3).

4.2. Influence of SGS models: Comparison between results with S model (case 2) and DM model (case 3) on composite grid system

(1) *Distributions of time-averaged gas concentration* $\langle c_g \rangle$ (Fig. 4c, Fig. 4d and Fig. 7). The result of case 3 (DM model) shows rather good agreement with the experiment (Fig. 4a and Fig. 4d). The result of the experiment shows a maximum peak of $\langle c_g \rangle$ at $x_3/H_b \fallingdotseq 0.25$ just behind the cube, as is shown in Fig. 7. The result given by case 3 (DM model) reproduces well the decrease of $\langle c_g \rangle$ near the ground. On the other hand, the decrease of $\langle c_g \rangle$ near the ground is not well reproduced in case 2 (S model).

(2) *Distributions of concentration fluctuation* $\langle c_g'^2 \rangle$ (Figs. 8 and 9). Fig. 8 illustrates the distributions of $\langle c_g'^2 \rangle$. It can be seen that the patterns of the distributions of $\langle c_g'^2 \rangle$ are almost similar to those of $\langle c_g \rangle$ for each case. Fig. 9 indicates the vertical profiles of

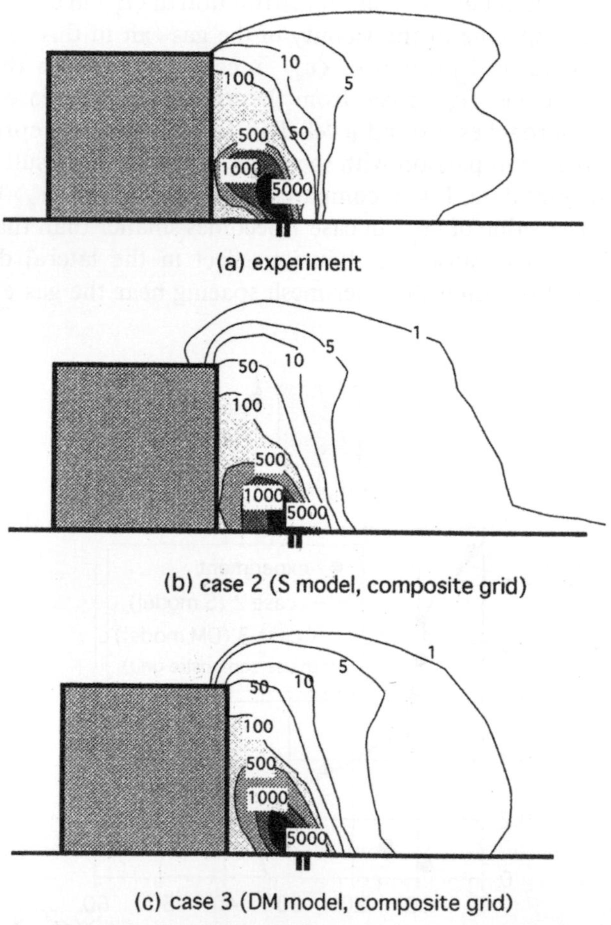

Fig. 8. $\langle c_g'^2 \rangle / \langle c_{g0} \rangle^2$ (at the center section): (a) experiment; (b) Case 2 (S model, composite grid); (c) Case 3 (DM model, composite grid).

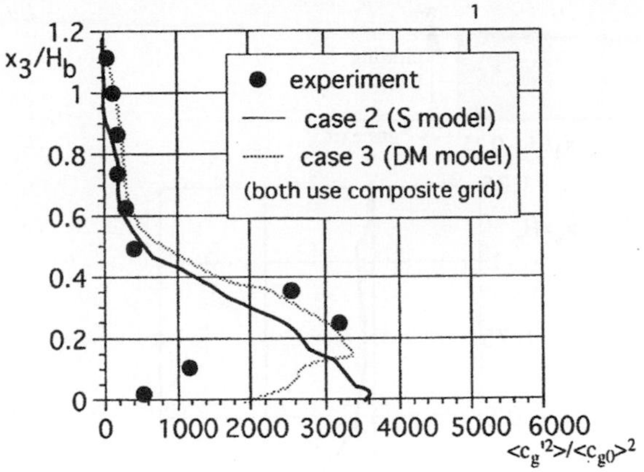

Fig. 9. Vertical profiles of concentration fluctuation $\langle c_g'^2\rangle/\langle c_{g0}\rangle^2$ (comparison between cases 2 and 3).

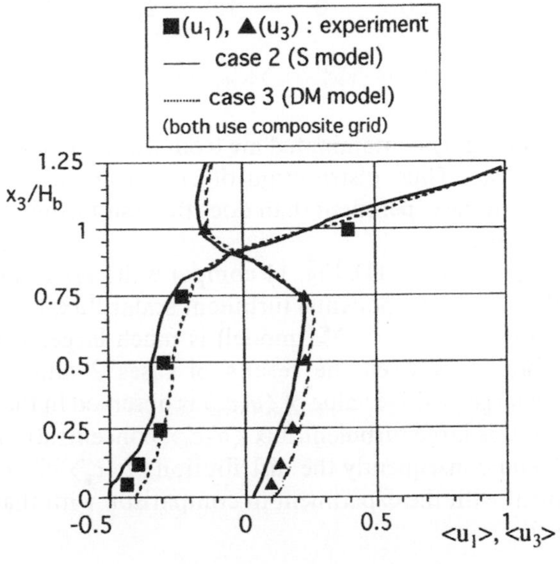

Fig. 10. Vertical profiles of $\langle u_1\rangle$ and $\langle u_3\rangle$.

$\langle c_g'^2\rangle$ behind the cube. Case 3 (DM model) shows much better agreement with the experiment in comparison with that of case 2 (S model).

(3) *Distributions of time-averaged velocities* $\langle u_1\rangle$ and $\langle u_3\rangle$ (Fig. 10). Fig. 10 shows a comparison of the vertical distributions of time averaged velocities $\langle u_1\rangle$ and $\langle u_3\rangle$ behind the cube in the center section. In this region, values from both cases agree rather well with those of the experiment. However, near the ground the S model overestimates the absolute value of $\langle u_1\rangle$ (in this region, $\langle u_1\rangle$ is negative) and underestimates the value of $\langle u_3\rangle$. With this decrease of $\langle u_3\rangle$ in this region in case 2

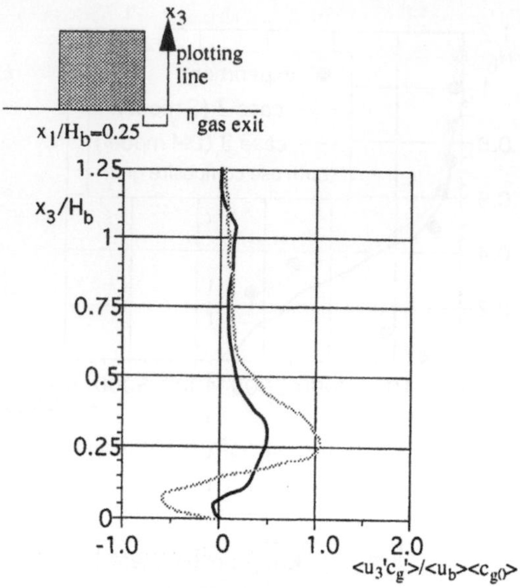

Fig. 11. Vertical profiles of $\langle u_3' c_g' \rangle / \langle u_b \rangle \langle c_{g0} \rangle$ (total turbulent scalar flux).

(S model), advection of highly contaminated air from the area near the ground to the upper area also decreases. Thus, distribution of $\langle c_g \rangle$ from case 2 (S model) corresponds more poorly with the experiment than does the result from case 3 (DM model), as is shown in Figs. 4 and 7.

(4) *Distributions of* $\langle u_3' c_g' \rangle$ (Fig. 11). Fig. 11 compares the vertical profiles of $\langle u_3' c_g' \rangle$, the total (Grid Scale + SGS components) turbulent scalar flux, behind the cube. The absolute value of $\langle u_3' c_g' \rangle$ in case 3 (DM model) is much larger than that in case 2 (S model). The difference between the results of cases 2 and 3 is significant at $x_3/H_b \fallingdotseq 0.25$, where a large positive value of $\langle u_3' c_g' \rangle$ is observed in the result of case 3 in Fig. 11. According to this large turbulent flux $\langle u_3' c_g' \rangle$, concentration near the ground decreases in case 3, and consequently the distribution of $\langle c_g \rangle$ given by case 3 shows much better agreement with the experiment in comparison with that given by case 2, as is shown in Fig. 7.

5. Conclusions

(1) The results using a composite grid show better agreement with the experiment than does the case using a single grid. This indicates that the fineness of mesh spacing near the gas exit is very important for reproducing the early process of gas diffusion.

(2) However, some discrepancies from the experimental results are observed in the distributions of $\langle c_g \rangle$ and $\langle c_g'^2 \rangle$ near the ground in the results of case 2, which employed the S model with a composite grid. These discrepancies are improved remarkably by using the DM model (case 3).

(3) The DM model reliably reproduces the distributions of mean concentration $\langle c_g \rangle$ and concentration fluctuation $\langle c_g'^2 \rangle$.

Appendix A. Numerical methods and boundary conditions

For the computations conducted here, a second-order centered difference scheme was adopted for the spatial derivatives. For time advancement, the second order Adams–Bashforth scheme was used for the convection terms and the Crank–Nicolson scheme for the diffusion terms. The interval for time advancement is 1.0×10^{-3} for all cases in a non-dimensional time scale based on $\langle u_b \rangle$ and H_b.

At the inflow boundary, the time history of $\bar{u}_i(t)$ in a fully developed channel flow predicted by LES was utilized. The profiles of $\langle \bar{u}_1(x_3) \rangle$ and $k(x_3)$ in the boundary layer of the channel given by LES correspond well to those of the experiment. The mean velocity profile at the inflow boundary is expressed as $\langle u_1(x_3) \rangle \propto x_3^{1/4}$ and the value of k at height H_b is 0.03. At the downstream boundary, zero gradient conditions are imposed. The normal gradients of the tangential velocity components and the normal velocity components were set to zero at the upper and side faces of the computational domain. For the boundary condition at the solid walls, Werner and Wengle's approach [10] was adopted, in which a linear or 1/7 power law distribution of the instantaneous velocity is assumed:

$$\frac{\bar{u}}{u^*} = x_n^+ \quad (x_n^+ \leqslant 11.81), \tag{A.1}$$

$$\frac{\bar{u}}{u^*} = 8.3\, x_n^{+\,1/7} \quad (x_n^+ > 11.81). \tag{A.2}$$

Appendix B. Fortified Navier–Stokes approach [5,6]

In this approach, Navier–Stokes equations and the transport equation of gas concentration are modified to include the forcing term. As an example, the modified Navier–Stokes equation for the streamwise velocity component for the fortification from grid B to grid A is given by

$$\frac{\partial U^{A}_{I,J,K}}{\partial t} + \mathrm{Cx}^{A}_{I,J,K} = -\,\mathrm{Px}^{A}_{I,J,K} + \mathrm{Dx}^{A}_{I,J,K} + \underline{C^{BA}_{I,J,K}(U^{B}_{(I,J,K)} - U^{A}_{I,J,K})}, \tag{B.1}$$

where, I, J and K denote spatial positions at x_1, x_2 and x_3 directions, respectively. $U^{A}_{I,J,K}$ indicates streamwise velocity U, defined at the nodal point (I,J,K) on grid A and $U^{B}_{(I,J,K)}$ is the interpolated value onto the nodal point (I,J,K) using velocities defined at the nodal points on grid B. Cx, Px and Dx in Eq. (B.1) are convection, pressure gradient and diffusion terms, respectively.

The underlined term on the right-hand side in Eq. (B.1) is the added forcing term. C^{BA} in the forcing term is a switching parameter. The superscript BA indicates the fortification from grid B to grid A.

The fortified Navier–Stokes equations for fortification from grid A to grid B can easily be derived in the same manner as that for Eq. (B.1). In this case, the switching parameter is expressed as C^{AB}. C^{BA} and C^{AB} are varied from 0 to infinity in the whole computational domain. When C^{BA} is 0, Eq. (B.1) becomes the usual Navier–Stokes equation for the streamwise velocity component, since the forcing term vanishes. When C^{BA} is infinite, Eq. (B.1) becomes equivalent $U^{A}_{I,J,K} = U^{B}_{(I,J,K)}$, since all other terms in Eq. (B.1) become negligibly small except for the forcing term. This means that the velocity value $U^{A}_{I,J,K}$ at the nodal point on grid A is completely replaced by the interpolated value $U^{B}_{(I,J,K)}$ given from grid B. Fortification from grid B to grid A is performed in this manner, when $C^{BA} = \infty$. When C^{BA} takes a finite value, $U^{A}_{I,J,K}$, the velocity defined at the nodal point on grid A, blends with $U^{B}_{(I,J,K)}$, the interpolated value of velocity from grid B. Thus, $U^{A}_{I,J,K}$ becomes the weighted average between velocity values on grids A and B. Through this blending, smooth fortification is attained in the connecting region.

Regions I–IV in Fig. 3 illustrate connecting regions set in the region of grid A. Here, regions II and III are the blending regions. Distribution of the switching parameters is also indicated in Fig. 3.

Appendix C. SGS models compared

In the S model, the anisotropic part of the SGS stress τ_{ij} is modeled as follows:

$$\tau^{a}_{ij} = -2C(f_{\mu}\bar{\Delta})^2|\bar{S}|\bar{S}_{ij}. \tag{C.1}$$

The superscript "a" denotes the anisotropic part of the tensor. C is the model coefficient, overbar denotes the grid-filtered value, $\bar{\Delta}$ is the width of the grid-filter and $\bar{S}_{ij}$ is the resolved-scale strain rate tensor,

$$\bar{S}_{ij} = \frac{1}{2}\left(\frac{\partial \bar{u}_i}{\partial x_j} + \frac{\partial \bar{u}_j}{\partial x_i}\right), \qquad |\bar{S}| = (2\bar{S}_{ij}\bar{S}_{ij})^{1/2}. \tag{C.2}$$

In the S model (cases 1 and 2), C is treated as a constant. The value of 0.0169 is selected in cases 1 and 2. This value corresponds to 0.13 of the so-called Smagorinsky constant C_S. $\bar{\Delta}$ is multiplied by the Van Driest type wall damping function f_{μ}, $1 - \exp(-x_n^+/25)$, in order to account for the near wall effect in the S model, while f_{μ} is not necessary in the dynamic SGS models.

In dynamic SGS models, model coefficient C is determined dynamically. Following Germano et al. [2], a test filter (denoted by a hat in this paper) is introduced to derive an expression for C. The test filter width is taken to be twice the grid filter width. Germano et al. defined the resolved turbulent stress as follows (Germano identity):

$$\mathcal{L}_{ij} = \widehat{\bar{u}_i\bar{u}_j} - \hat{\bar{u}}_i\hat{\bar{u}}_j\,. \tag{C.3}$$

$\mathscr{L}_{ij}$ can be related to the SGS stress τ_{ij} and the subtest-scale stress $T_{ij} = \widehat{\bar{u}_i\bar{u}_j} - \hat{\bar{u}}_i\hat{\bar{u}}_j$ by

$$\mathscr{L}_{ij} = T_{ij} - \hat{\tau}_{ij}. \tag{C.4}$$

The DM model [4] employs linear combination of the dynamic Smagorinsky model and the scale similarity model [11]. In the DM model, the anisotropic part of SGS stresses τ_{ij} and subtest-scale T_{ij} are expressed as:

$$\tau^{a}_{ij} = -2C\beta_{ij} + b_{ij}, \tag{C.5}$$

$$T^{a}_{ij} = -2C\alpha_{ij} + B_{ij}, \tag{C.6}$$

where

$$\alpha_{ij} = \hat{\bar{\Delta}}|\hat{\bar{S}}|\hat{\bar{S}}_{ij}, \qquad \beta_{ij} = \bar{\Delta}|\bar{S}|\bar{S}_{ij}, \tag{C.7}$$

$$B_{ij} = \widehat{\bar{\bar{u}}_i\bar{\bar{u}}_j} - \hat{\bar{\bar{u}}}_i\hat{\bar{\bar{u}}}_j, \qquad b_{ij} = \overline{\bar{u}_i\bar{u}_j} - \bar{\bar{u}}_i\bar{\bar{u}}_j. \tag{C.8}$$

The first and the second terms in the right-hand side of Eqs. (C.5) and (C.6) derived from the Smagorinsky model and the scale similarity model, respectively.

Substituting Eqs. (C.5) and (C.6) into Eq. (C.4) yields

$$\mathscr{L}^{a}_{ij} = -2C\alpha_{ij} + 2\widehat{C\beta}_{ij} + \gamma_{ij}, \tag{C.9}$$

where

$$\gamma_{ij} = B_{ij} - \hat{b}_{ij} = \widehat{\bar{\bar{u}}_i\bar{\bar{u}}_j} - \hat{\bar{\bar{u}}}_i\hat{\bar{\bar{u}}}_j, \tag{C.10}$$

$\mathscr{L}^{a}_{ij}$ indicates the anisotropic part of $\mathscr{L}_{ij}$.

Eq. (C.9) cannot be solved explicitly for C, since C appears inside a filtering operation in the right-hand side in Eq. (C.9). Previous authors [2–4] assumed $\widehat{C\beta}_{ij} = C\hat{\beta}_{ij}$ to avoid this problem. However, this assumption is obviously inconsistent with the concept of dynamic LES, in which C is treated as a function of space and time. In this study, a localization technique proposed by Piomelli et al. [12] is employed.

Piomelli et al. developed an approximate localization technique based on recasting the expression in Eq. (C.9) in the form

$$-2C\alpha_{ij} = \mathscr{L}^{a}_{ij} - \gamma_{ij} - 2\widehat{C^{*}\beta}_{ij}\,. \tag{C.11}$$

Here, C^* is the model coefficient inside the test filter, which is treated as the known value shown below. Eq. (C.11) is a tensor equation with six independent components. Hence, we need a procedure to specify the scalar value of C from the tensor equation. We use a least-squares method suggested by Lilly [3], in which the sum of the squares of the residual of Eq. (C.11) is minimized. Using this method, the expression for a scalar value of C is deduced from Eq. (C.11) [12]:

$$C = -\frac{1}{2}\frac{(\mathscr{L}^{a}_{ij} - \gamma_{ij} - 2\widehat{C^{*}\beta_{ij}})\alpha_{ij}}{\alpha^{2}_{kl}}. \tag{C.12}$$

Since C^* must be treated as a known value in order to solve Eq. (C.12), we assume $C^* = C^{n-1}$ for C^* at time step n following Piomelli et al. [12].

The dynamic SGS model can be extended easily to the modeling of SGS scaler flux $(h_j = \overline{u_j c_g} - \bar{u}_j \bar{c}_g)$.

In the S model, h_j is modeled as follows:

$$h_j = -\frac{\nu_{SGS}}{Sc_{SGS}} \frac{\partial \bar{c}_g}{\partial x_j} = -\frac{C\bar{\Delta}^2}{Sc_{SGS}} |\bar{S}| \frac{\partial \bar{c}_g}{\partial x_j}, \tag{C.13}$$

where Sc_{SGS} is the SGS Schmit number defined as follows:

$$Sc_{SGS} = \frac{\nu_{SGS}}{K_{SGS}}. \tag{C.14}$$

K_{SGS} is the SGS eddy diffusivity. In this study, Sc_{SGS} is treated as a constant equal to 0.5 [13] in the S model.

In the DM model, h_j is expressed by

$$h_j = -\frac{C\bar{\Delta}^2 |\bar{S}|}{Sc_{SGS}} \frac{\partial \bar{c}_g}{\partial x_j} + b_{jc}, \tag{C.15}$$

$$b_{jc} = \overline{\bar{u}_j \bar{c}_g} - \bar{\bar{u}}_j \bar{\bar{c}}_g \,. \tag{C.16}$$

b_{jc} corresponds to b_{ij} in Eq. (C.5), which derives from the scale similarity model. We employed the localization technique here, as well as the case for estimating the momentum flux:

$$\frac{1}{Sc_{SGS}} = -\frac{(\mathscr{L}_{jc} - \gamma_{jc} - (\widehat{\beta_{jc}}/Sc^*_{SGS}))\alpha_{jc}}{\alpha_{lc}^2}, \tag{C.17}$$

where

$$\gamma_{jc} = B_{jc} - \hat{b}_{jc} = \widehat{\bar{\bar{u}}_j \bar{\bar{c}}_g} - \hat{\bar{\bar{u}}}_j \widehat{\bar{\bar{c}}_g}\,, \tag{C.18}$$

$$\alpha_{jc} = C\hat{\bar{\Delta}}^2 |\hat{\bar{S}}| \frac{\partial \widehat{\bar{c}_g}}{\partial x_j}\,, \qquad \beta_{jc} = C\bar{\Delta}^2 |\bar{S}| \frac{\partial \bar{c}_g}{\partial x_j}\,. \tag{C.19}$$

In this study, the negative value of C and $1/Sc_{SGS}$ appeared even though this localization technique was used. Thereby we used the clipping technique, i.e., the coefficients C and $1/Sc_{SGS}$ were set equal to zero wherever they were estimated to be negative by Eqs. (C.12) and (C.17).

References

[1] S. Murakami, Current status and future trends in CWE, these Proceedings, J. Wind Eng. Ind. Aerodyn. 67&68 (1997) 3–34.

[2] M. Germano, U. Piomelli, P. Moin, W.H. Cabot, A dynamic subgrid-scale eddy viscosity model, Phys. Fluids A 3 (1991) 1760–1765.

[3] D.K. Lilly, A proposed modification of the Germano subgrid-scale closure method, Phys. Fluids A 4 (1992) 633–635.

[4] Y. Zang, R.L. Street, J.R. Koseff, A dynamic mixed subgrid-scale model and its application to turbulent recirculating flows, Phys. Fluids A 5 (12) (1993) 3186–3196.
[5] W.R.V. Delsem, J.L. Steger, The fortified Navier–Stokes approach, in: Workshop on Computational Fluid Dynamics, University of California, Davis, 1986.
[6] A. Mochida, S. Murakami, Y. Ishida, Numerical study on flowfield around structures with oblique wind angle based on composite grid system, in: 7th US National Conf. on Wind Engineering, 1993, pp. 463–472.
[7] Y. Tominaga, S. Murakami, A. Mochida, A. Shibuya, Y. Noguchi, Wind tunnel test on turbulent diffusion and concentration fluctuation of buoyant gas near building, in: 12th Japan National Symp. on Wind Eng., 1992, pp. 119–124 (in Japanese).
[8] S. Murakami, A. Mochida, Y. Tominaga, Numerical simulation of turbulent diffusion in cities, Wind Climate in Cities, Kluwer Academic Publishers, Dordrecht, 1995, pp. 681–701.
[9] N. Collings, A new technique for measuring HC concentration in real time in a running engine, International Congress and Exposition, Detroit, MI, 1988.
[10] H. Werner, H. Wengle, Large-eddy simulation of turbulent flow over and around a cube in a plate channel, in: 8th Symp. on Turbulent Shear Flows 19-4, 1991, pp. 155–165.
[11] J. Bardina, J.H. Ferziger, W.C. Reynolds, Improved subgrid-scale models for large-eddy simulation, AIAA paper-80, 1981.
[12] U. Piomelli, J. Liu, Large-eddy simulation of rotating channel flows using a localized dynamic model, Phys. Fluids 7 (4) (1995) 839–848.
[13] M. Antonopoulos-Domis, Large-eddy simulation of a passive scalar in isotropic turbulence, J. Fluid Mech. 104 (1981) 55–79.

ELSEVIER

Journal of Wind Engineering
and Industrial Aerodynamics 67&68 (1997) 843–857

JOURNAL OF
wind engineering
AND
industrial
aerodynamics

A vertical round jet issuing into an unsteady crossflow consisting of a mean current and a sinusoidal fluctuating component

L.P. Xia, K.M. Lam*

Department of Civil and Structural Engineering, The University of Hong Kong, Pokfulam Road, Hong Kong, China

Abstract

We applied a CFD code to investigate the dispersion of a round jet issuing into an unsteady ambient crossflowing stream which consists of coherent sinusoidal fluctuations imposed upon an otherwise steady mean flow. The computed flow patterns were compared with our laboratory observations where we employed an alternative approach to simulate the fluctuating crossflow. Both sets of results showed that coherent oscillations in the ambient crossflow organized the jet fluids into successive large-scale clouds or patches and that the formation of the patches was regular. As a result, the mean dispersion of the jet was enhanced. However, the coherent fluid patches could lead to very high instantaneous concentration levels. The CFD results also served to support the validity of our experimental technique.

Keywords: Jet in crossflow; Unsteady flow; Oscillating motion

1. Introduction

Many environmental problems are derivatives of the basic configuration of a jet in a crossflow, one wind-engineering example being the discharge of chimney emission into the atmosphere. Under certain situations, periodic and coherent wind velocity fluctuations can be found in the incoming wind field due to vortex shedding from upstream obstacles such as hills and tall buildings. The main objective of our study is to investigate how the additional sinusoidal coherent flow oscillations of the cross-wind modify the time-averaged and instantaneous dispersion pattern of emission from a vertical round jet.

The behavior of jet discharging into an unsteady ambient flow has received increased attention in the past decade. Sharp [1] carried out a qualitative study on the

* Corresponding author. E-mail: kmlam@hku.hk.

PII S0167-6105(97)00123-2

dispersion of a buoyant jet in deep water and shallow water waves. In the presence of shallow water waves and when the discharge nozzle was perpendicular to the wave direction, an interesting phenomenon, referred to as the "dumbbell effect", was observed and the jet developed into two different clouds of effluent due to the oscillating motion in the water body. Koole and Swan [2] reported experimental investigations on a two-dimensional horizontal non-buoyant jet discharging under progressive waves. They concluded that the oscillatory wave motion has a significant effect upon both the mean-velocity profiles and the magnitude of turbulent fluctuations. In an earlier study, we made use of a novel experimental technique and computer-aided flow visualizations to investigate the discharge of a turbulent round jet into a crossflow with a sinusoidal oscillating component [3]. The present work is aimed at numerical simulation of the same problem. We hope that the computational fluid dynamics results predict the same flow patterns as those obtained in our experiments and thus verify our experimental technique.

2. Experimental investigations

2.1. Experimental set-up

Fig. 1 shows a schematic diagram of the experimental set-up. The experiments were carried out in a 10 m long × 0.3 m wide × 0.45 m deep laboratory flume. A round turbulent jet was produced by discharging water through a circular nozzle of exit

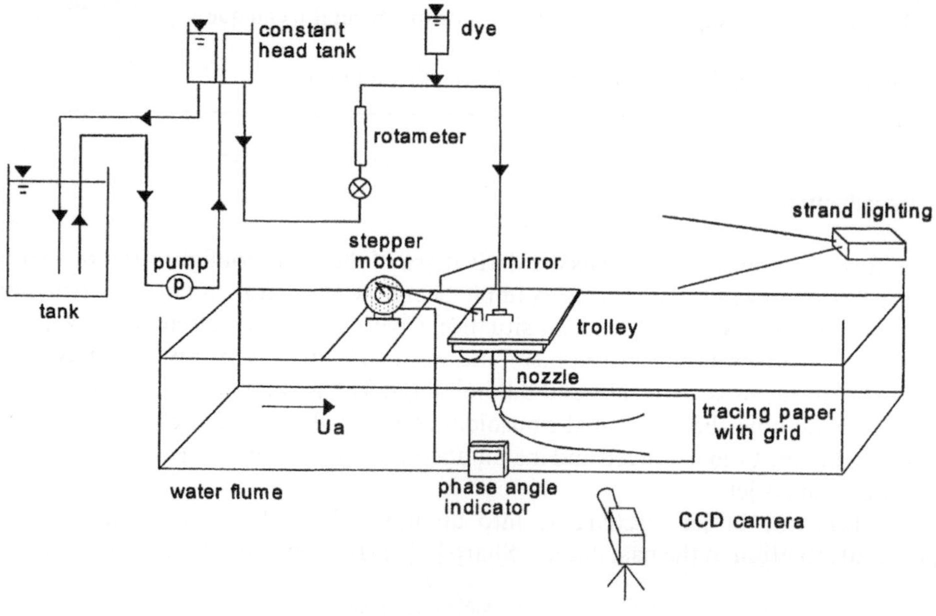

Fig. 1. Schematic diagram of experimental set-up.

diameter $D = 7.5$ mm at the end of a vertical pipe. For the sake of convenience, the discharge was vertically downwards with the nozzle exit about $5D$ below the water surface. The jet discharge was fed from a constant head tank with the jet exit velocity V_j adjusted and measured with a calibrated rotameter.

The flow problem we were looking at is the discharge of a vertical round jet into a crossflow which consists of a mean current and a sinusoidal oscillating component, that is $U(t) = U_a + u_p \sin(2\pi ft)$. It was difficult to produce this type of flow in the laboratory and we adopted an alternative approach. An unsteady ambient current was not produced in the flume; rather a steady current U_a was maintained and a pipe-nozzle assembly was mounted on a trolley and imposed a sinusoidal horizontal motion by means of a crank-yoke mechanism. The crank was driven by a stepper motor so that the frequency of oscillation f could be precisely adjusted. The stroke of oscillation A could also be adjusted and together with f, determined the amplitude of velocity oscillations in the crossflow $u_p = 2\pi fA$. A phase-angle indicator derived the phase angle of oscillation from the stepper motor driving signal and displayed it in a phase number representing multiples of $\frac{1}{20}$ cycle of a full oscillation.

The subsequent flow visualization results obtained at a particular phase were interpreted on a frame of reference as defined by the current jet nozzle position at that phase. Obviously, the frame of reference is oscillating and thus non-inertial. The uncertainty involved in the coordinate transformation depended on the ratio of the peak acceleration in the horizontal direction, $2\pi fu_p$ to the characteristic inertia force in the vertical direction which is related to V_j^2/D. The governing non-dimensional parameter was thus fu_pD/V_j^2 and it will be shown later that it has small values over our range of interest.

2.2. Image processing method

Spreading of the jet was analysed with a computer-aided flow-visualization technique using image processing. With a dye added into the jet discharge, the jet dispersion pattern was projected onto a tracing paper mounted along a side of the flume with a parallel light beam illuminated from the background by a 2500 W strand lighting. Shadowgraphs of the marked jet together with the display of the phase-angle indicator were video recorded with a CCD camera. The video images were grabbed into a 486 computer with a Data Translation DT3852 frame grabber. Each image frame was represented by a 640×480 pixel matrix of 8-bit grey level elements. Phase-locked averaging of jet dispersion was performed with an image processing procedure as follows: All the grabbed images underwent a contrast enhancement with the standard histogram sliding and stretching technique [4]. The grabbed image frames were sorted into groups of the same phase number, with an ensemble size of 10 frames for each phase. The averaged flow image at the particular phase was obtained by ensemble averaging of the brightness matrices of all frames at that phase.

2.3. Experimental parameters

Simple dimensional analysis suggests that the flow problem is governed by the following parameters: jet-to-current velocity ratio V_j/U_a; unsteadiness parameter of the crossflow u_p/U_a; and non-dimensional frequency of crossflow oscillation or the Strouhal number $St = fD/U_a$. We have carried out experiments under different combinations of these parameters. In all experiments, the jet exit velocity was kept at $V_j = 80$ cm/s, and the mean crossflow current was kept at $U_a = 10$ cm/s. The jet-to-current velocity ratio was thus fixed at $V_j/U_a = 8$. The unsteadiness parameter u_p/U_a varied from 0.25 to 1.5. For $u_p/U_a > 1.0$, the crossflow actually reversed its direction for some time in a cycle. The oscillation frequency, represented non-dimensionally by the Strouhal number $St = fD/U_a$, varied from 0.0375 to 0.12. The parameter fu_pD/V_j^2 governing the uncertainty of coordinate transformation thus had a maximum value of 0.0028 and the oscillating nozzle experimental technique was expected to model well the oscillating crossflow situation.

3. Mathematical model

3.1. The governing equations

The governing equations of continuous incompressible flow used in the model are conservation of mass

$$\frac{\partial \rho}{\partial t} + \frac{\partial(\rho u_i)}{\partial x_i} = 0,$$

and conservation of momentum

$$\frac{\partial u_i}{\partial t} + u_j\frac{\partial u_i}{\partial x_j} = -\frac{1}{\rho}\frac{\partial p}{\partial x_i} + \frac{1}{\rho}\frac{\partial \tau_{ij}}{\partial x_j} + g_i,$$

where ρ is the density of fluid, u_i are the time-averaged velocity components in the three coordinate directions, p is the static pressure, $\tau_{ij} = -\rho\overline{u_i'u_j'}$ is the Reynolds shear stress and g_i is the gravitational acceleration in the i-direction.

The k–ε model of Launder and Spalding [5] is adopted for turbulent closure:

$$\frac{\partial}{\partial t}(\rho k) + \frac{\partial}{\partial x_i}(\rho u_i k) = \frac{\partial}{\partial x_i}\left(\frac{\mu_t}{\sigma_k}\frac{\partial k}{\partial x_i}\right) + G_k - \rho\varepsilon,$$

$$\frac{\partial}{\partial t}(\rho\varepsilon) + \frac{\partial}{\partial x_i}(\rho u_i\varepsilon) = \frac{\partial}{\partial x_i}\left(\frac{\mu_t}{\sigma_\varepsilon}\frac{\partial \varepsilon}{\partial x_i}\right) + C_{1\varepsilon}\frac{\varepsilon}{k} - C_{2\varepsilon}\rho\frac{\varepsilon^2}{k},$$

where $\mu_t = \rho C_\mu k^2/\varepsilon$ is the turbulence, viscosity, σ_k and σ_ε are Prandtl numbers, $C_{1\varepsilon}$, $C_{2\varepsilon}$ and C_μ are empirical constants, G_k is generation rate of turbulent kinetic energy k,

$$G_k = \mu_t\left(\frac{\partial u_j}{\partial x_i} + \frac{\partial u_i}{\partial x_j}\right)\frac{\partial u_i}{\partial x_i}.$$

Standard values are used for the coefficients $C_{1\varepsilon}$, $C_{2\varepsilon}$, C_μ, σ_k and σ_ε, that is, $C_{1\varepsilon} = 1.44$, $C_{2\varepsilon} = 1.92$, $C_\mu = 0.09$, $\sigma_k = 1.0$, $\sigma_\varepsilon = 1.3$ [6].

3.2. Numerical solution procedure

The flow problem is computed with the FLUENT code which uses a control volume based technique [7] to solve the governing partial differential equations. In FLUENT the discrete control volumes are defined using a non-staggered grid storage scheme. The same control volume is employed for integration of all the conservation equations and all variables, that is, pressure, Cartesian velocities, Reynolds stress components and scalars, are stored at the centers of the control volumes. The cell-face values of the unknowns are calculated via the power-law differencing interpolation scheme. The set of finite-difference equations are solved using the SIMPLE iterative algorithm which starts from arbitrary initial conditions except at the boundaries [7].

Temporal discretization used in FLUENT is an unconditionally stable implicit scheme. The integration of the transient terms in the time-dependent equation for conservation of ϕ is straightforward, as described below:

$$\int_t^{t+\Delta t} \frac{\partial}{\partial t}(\rho\phi)\,\mathrm{d}t = (\rho\phi)^1 - (\rho\phi)^0.$$

For other terms, e.g., convection terms such as $\rho u\phi$, assuming that

$$\int_t^{t+\Delta t} (\rho u\phi)\,\mathrm{d}t = [f(\rho u\phi)^1 + (1-f)(\rho u\phi)^0]\,\Delta t$$

$f = 1$ is adopted in FLUENT, which is a fully implicit scheme. The conservation equations are solved iteratively with the SIMPLE algorithm at each time step.

3.3. Computational domain and boundary conditions

Fig. 2a shows the top view (xy plane) of the computation domain which is a 1.0 m × 0.3 m × 0.4 m rectangular box corresponding to a section of the laboratory flume. Since the flow problem is symmetrical about the plane at $y = 0$, only half of the physical domain, $y \geqslant 0$, was considered in the computations. The jet exit was positioned at the top face of the box which was treated as an impermeable wall except inside the jet injection hole. The main stream, which varied with time as $U(t) = U_a + u_p \sin(2\pi ft)$, was introduced from left to right. Consequently, these two faces were treated as velocity inlet and free outlet, respectively. The upstream boundary was placed 0.2 m before the jet injection point. The other two faces were naturally treated as wall boundaries.

A grid pattern 59 × 23 × 30 was used in the calculations. Structured body-fitted meshes were generated whereby in the x- and y-direction finer grids were employed near the jet inlet and coarser grids at the positions away from the jet, while uniform grids were employed in the z-direction. The grid-sizes ranged from 0.375 cm, which corresponded to the jet nozzle radius, to 3.5 cm. The computational grid is shown in Fig. 2b. The mesh size was selected from a compromise between mesh fineness and

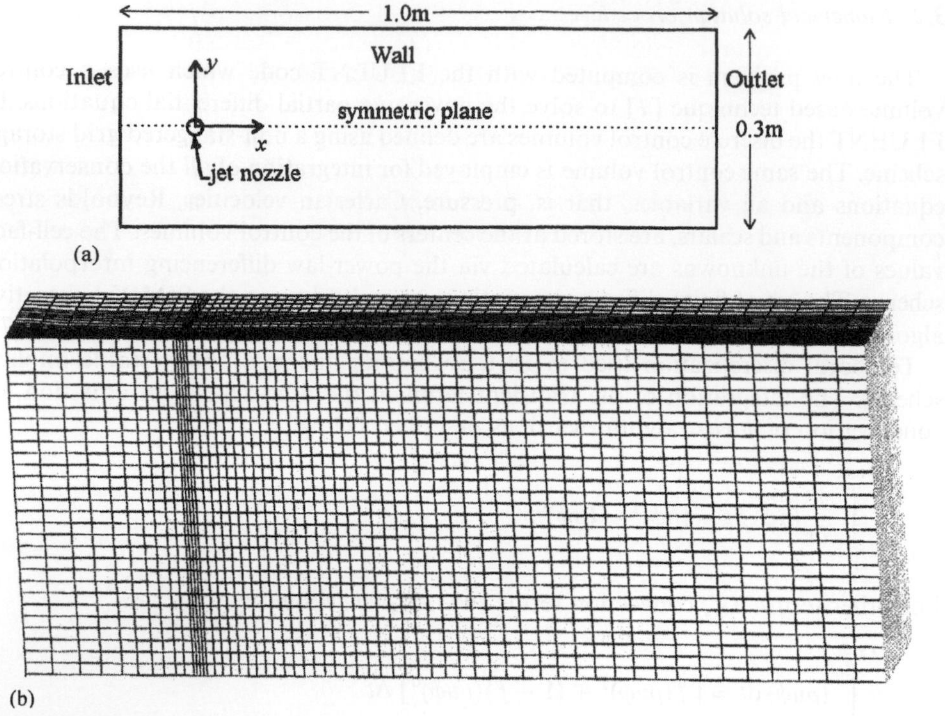

Fig. 2. (a) Top view of physical geometry. (b) Computational surface grid (59 × 23 × 30).

computation time. It should be noted that the flow to be computed was an unsteady one for which flow fields were required at a number of phases. For the steady-flow case of a jet into a non-time-varying crossflow, the chosen mesh size has been found to be fine enough. Six flow variables, namely, pressure, the three velocity components, kinetic energy and dissipation rate, were selected to be computed. Iterations were terminated when the sum of normalized residuals of the six variables became less than 10^{-3}, a value which was subjectively judged to be small enough. The time step Δt took the value of $0.005T$, where $T = 1/f$ was the time period of crossflow oscillation. Approximately 15–20 iterations were required for convergence at each Δt.

To simulate the unsteady crossflow, a time-dependent horizontal velocity along the x-direction $U(t) = U_a + u_p \sin(2\pi ft)$ was imposed on the inlet boundary of the computational domain. The y- and z-direction velocity components, v and w, at the velocity inlet boundary were set to zero for all the cases. At the jet inlet section, the z-direction velocity was equal to the jet discharging velocity $V_j = 80$ cm/s. The characteristic lengths of inlet boundaries were assigned values of their hydraulic radius. The evaluation of turbulence intensities for the inlet boundaries will be discussed later. Initial input was not required for the outlet boundary (right face of the domain) and wall boundaries (bottom, lateral and top faces of the domain except for jet injection).

We found that, with this CFD code, the numerical solutions of flow field were rather insensitive to the boundary conditions of turbulence intensities assumed within a reasonable range. Herein, we empirically assumed that the turbulence intensity in the oscillating crossflow had the value of 8% as the boundary condition. Actually, the turbulence intensity in the laboratory flume was about 5–8%. The jet exit boundary layer was reasonably inferred to be fully developed turbulent and the turbulence intensities corresponding to this kind of flow normally had values as great as 10% [8]. Therefore, we set the turbulence intensity at the jet exit as 10% in the computations.

4. Results and discussions

4.1. Experimental observations

The typical flow pattern can be observed in the flow case with $V_j/U_a = 8.0$, $u_p/U_a = 1.0$ and $St = 0.0375$. Fig. 3 shows the phase-averaged dye pattern over an oscillation cycle of the crossflow. These flow images were obtained from the image processing procedures as described in Section 2.2. The phase number at which each image was obtained could be seen on the phase indicator at the lower-left corner of the picture. To observe the effect of oscillating crossflow, the flow pattern should be interpreted with respect to the current nozzle position at that particular phase. The jet discharge fluid, marked by the dye, was observed to be organized into successive large-scale clouds or patches. The formation of these patches appeared to be caused by the acceleration and deceleration of the crossflow within the cycle. At phase number 0, that is $t = 0$, the crossflow speed was at its time-averaged mean value U_a and some jet fluids were being discharged from the nozzle into the crossflow with the momentum effect still dominating, that is, the jet had not been bent over by the crossflow. From phase number 0 to phase number 4, that is within the first quarter of a cycle period, the crossflow was accelerating with the jet discharging fluid still discharging in a near vertical direction. From phase number 6 to 14, that is from $t = 1/4T$ to $t = 3/4T$, the crossflow was decelerating. The discharged jet fluid, while still dominated by the vertical momentum, was not convected by the crossflow and accumulated to form a coherent patch. From phase number 16 to 18, the crossflow accelerated again and at the same time, the coherent patch of jet fluid was now far enough from the jet exit so that it behaved similar to the bent-over stage of a jet in steady crossflow. The coherent fluid patch was thus convecting with the accelerating crossflow and became clearly distinguishable as a bulge-shaped cloud patch protruding downwards from the bent-over location of the jet. The jet fluid upward of the bent-over location then underwent another cycle and eventually formed the succeeding fluid patch.

The time-averaged dispersion pattern of the jet into the fluctuating crossflow could be obtained from an average of the phase-locked patterns over all phases of a cycle. The pattern obtained is shown in Fig. 4. It should be noted that in obtaining Fig. 4 the phase-locked patterns have undergone a coordinate transformation so as to align the jet nozzle position to a fixed location. Compared with a round jet issuing into

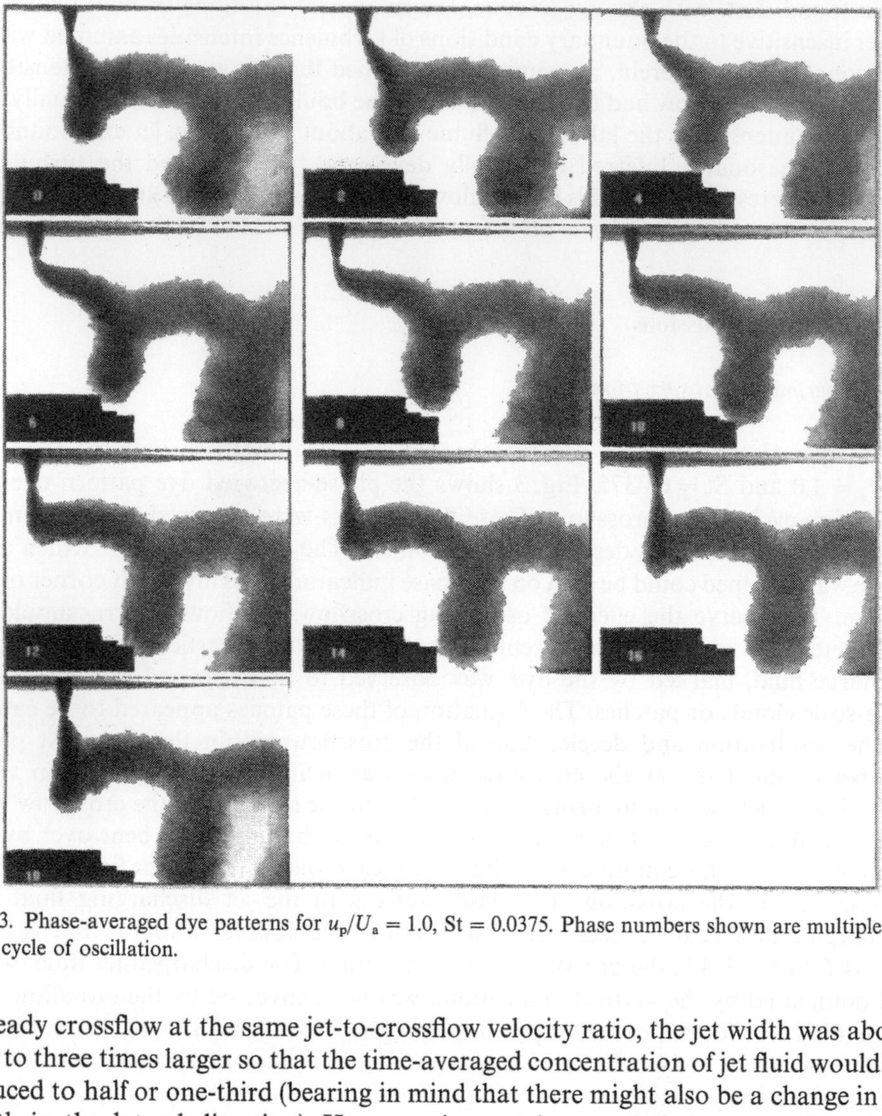

Fig. 3. Phase-averaged dye patterns for $u_p/U_a = 1.0$, St = 0.0375. Phase numbers shown are multiples of 1/20 cycle of oscillation.

a steady crossflow at the same jet-to-crossflow velocity ratio, the jet width was about two to three times larger so that the time-averaged concentration of jet fluid would be reduced to half or one-third (bearing in mind that there might also be a change in jet width in the lateral direction). However, it must be noted that at some locations passed by the coherent fluid patches, the peak concentration might be increased.

4.2. Computational results

The same flow case was computed with the FLUENT code. Fig. 5 shows the computed velocity fields on the $y = 0$ plane at four phases of a cycle. It is interesting to note that although the time-varying crossflow never reversed its direction, the jet fluid which initially issued vertically, did at some time discharge into the upstream direction. If we concentrate on a small fluid mass at jet exit in Fig. 5a, we can observe that

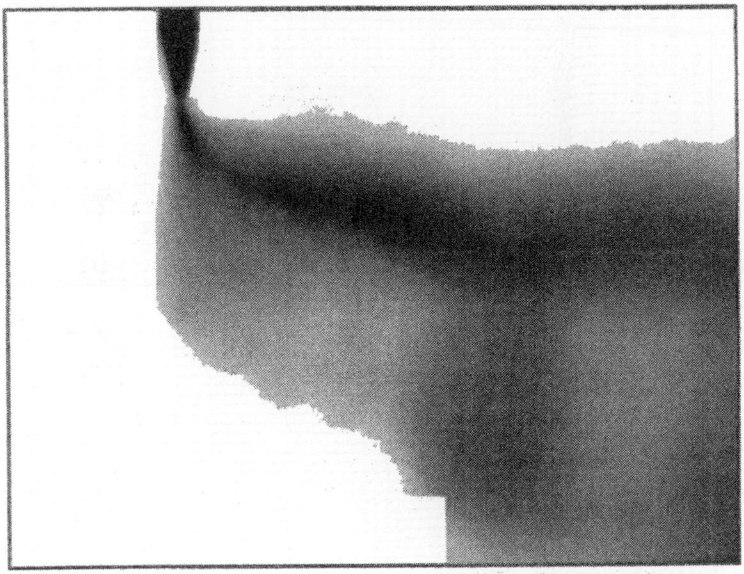

Fig. 4. Time-averaged dye pattern for $u_p/U_a = 1.0$, St = 0.0375.

as it discharged vertically downwards, it was not significantly convected downstream with the accelerating crossflow in Fig. 5b or upstream with the decelerating crossflow in Fig. 5c, and it came almost directly underneath the jet exit in Fig. 5d. It was, of course, now at a large distance away from the jet exit and, all the time, the fluid mass had been maintaining its vertical momentum until it reached the farthest vertical distance from the jet exit in Fig. 5a. Afterwards, with the loss of most of its vertical momentum, it was convected with the crossflow and developed into a coherent fluid patch in Fig. 5b.

It must be noted that the experimental observations were based on dye-patterns discharged from the jet exit so that they were streakline patterns and integrated in the z-direction. On the other hand, the computed velocity fields showed the changing streamline patterns and only on the $y = 0$ plane. This renders a direct comparison difficult between the CFD results and the experimental observations. However, both sets of results appear to support the proposed mechanism of patch formation. The most encouraging point is that the agreement between the two sets of results strengthens our belief that the vibrating nozzle method employed in the experiments could simulate the oscillating crossflow situation adequately.

Since the vertical velocity component originated from the vertically discharging jet fluids, we chose to present in Fig. 6 the contours of w-velocity magnitudes of the CFD results. The dye concentration contours from the experimental data in Fig. 3 are also shown for comparison.

When the frequency of oscillation was increased, say St = 0.0825, while the unsteadiness parameter remained at $u_p/U_a = 1.0$, organization of jet discharging fluids

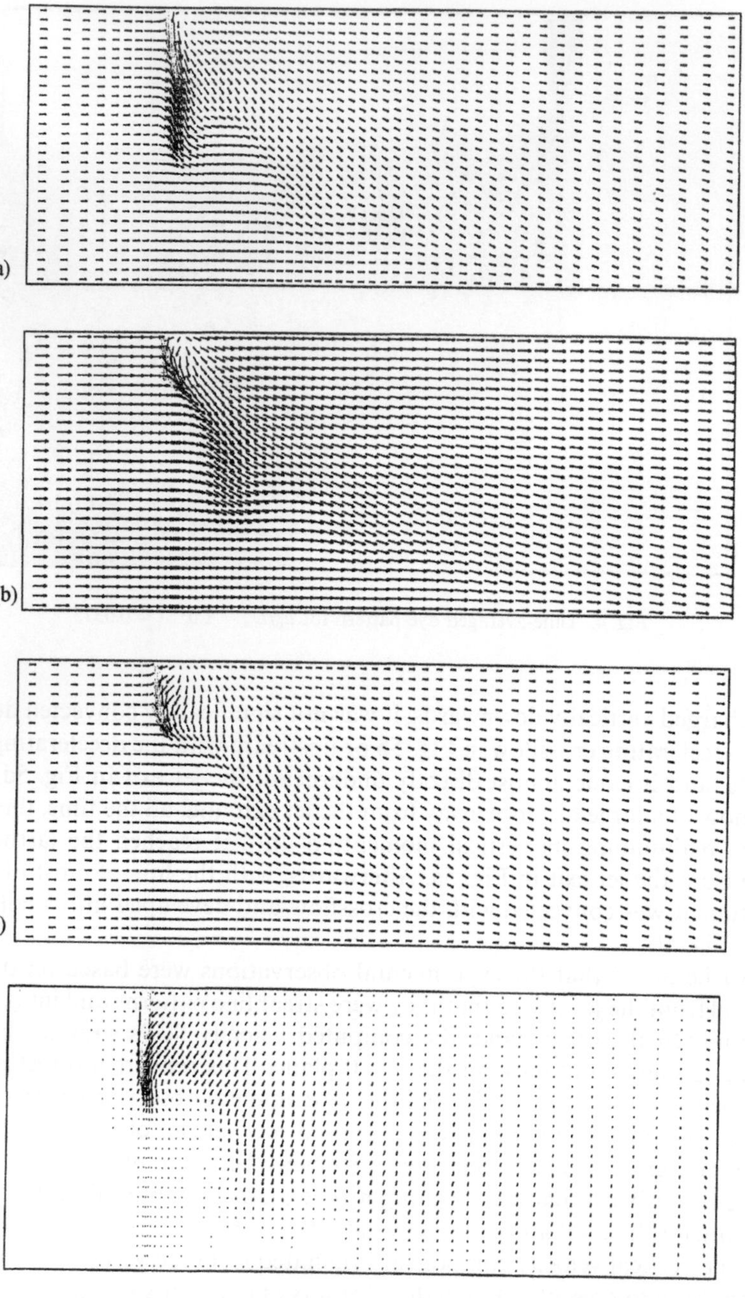

(a)
(b)
(c)
(d)

Fig. 5. Computed velocity vectors at the four phases of one cycle for $u_p/U_a = 1.0$, St = 0.0375. (a) $t = 0$; (b) $t = \frac{1}{4}T$; (c) $t = \frac{1}{2}T$; (d) $t = \frac{3}{4}T$.

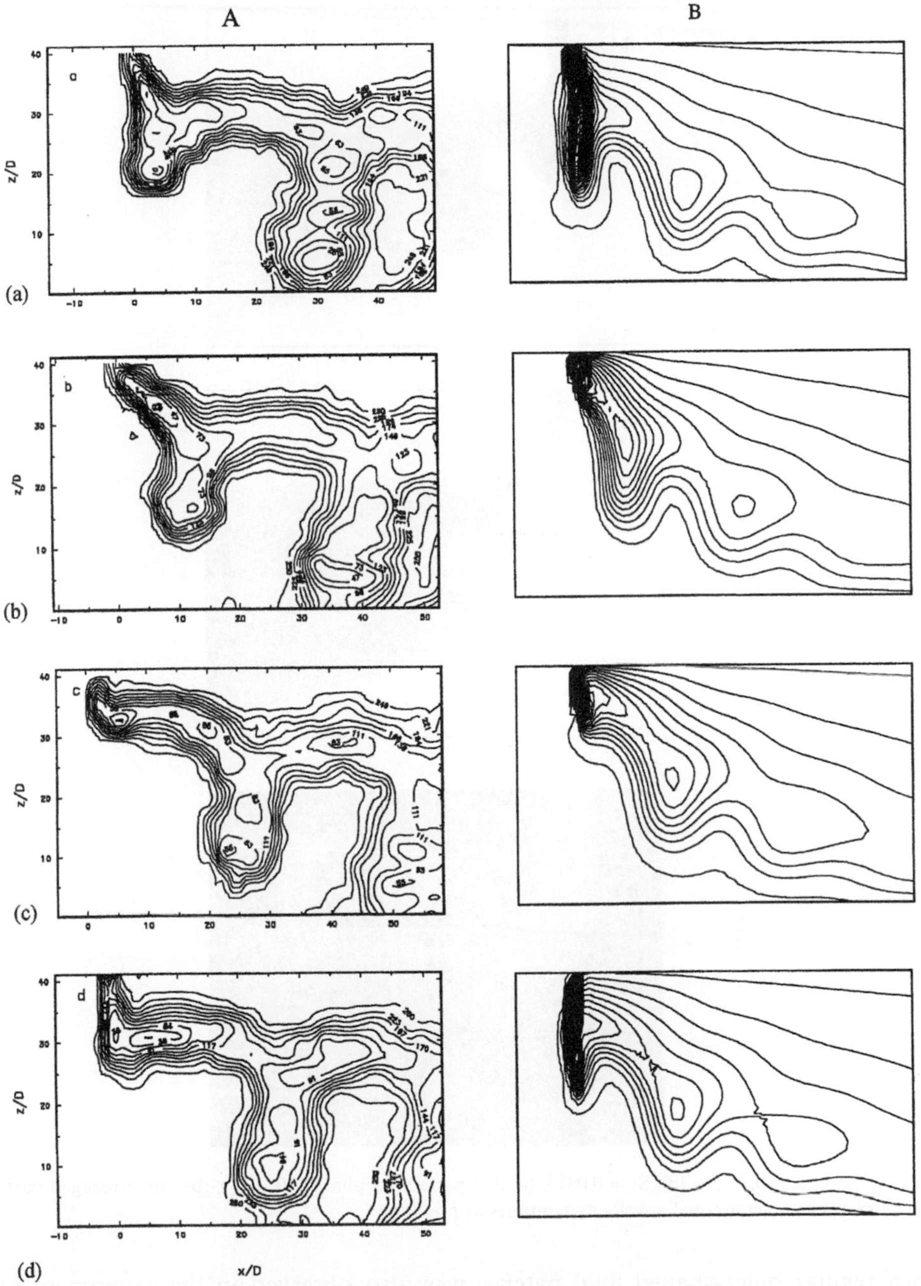

Fig. 6. (A) Experimental dye concentration contours; (B) Computed w-velocity contours at the four phases of one cycle for $u_p/U_a = 1.0$, St = 0.0375. (a) $t = 0$; (b) $t = \frac{1}{4}T$; (c) $t = \frac{1}{2}T$; (d) $t = \frac{3}{4}T$.

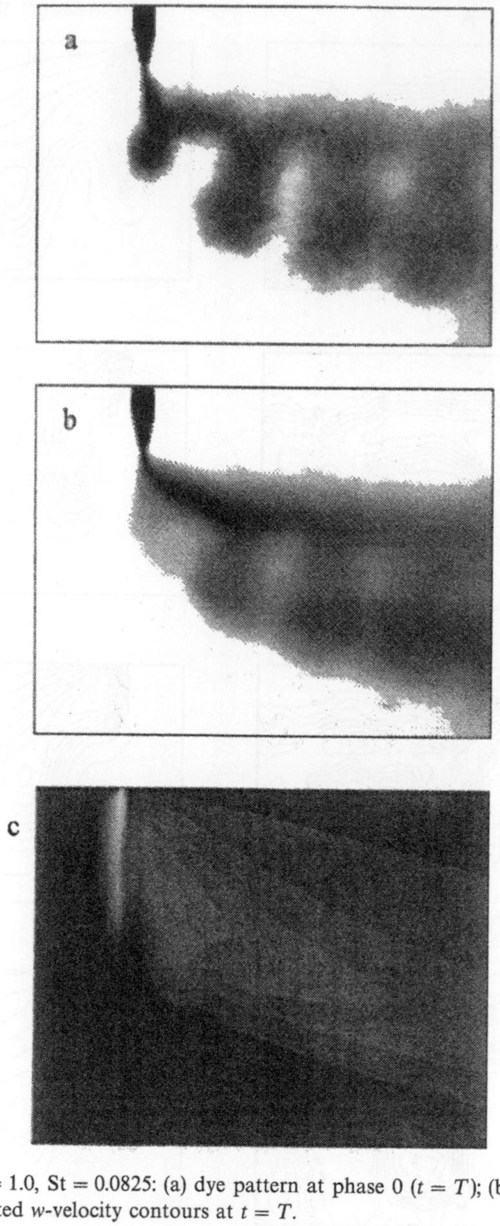

Fig. 7. Results at $u_p/U_a = 1.0$, St = 0.0825: (a) dye pattern at phase 0 ($t = T$); (b) time-averaged pattern over one cycle; (c) computed w-velocity contours at $t = T$.

into regular bulge-shaped fluid patches was also observed in the experiment and computation (Fig. 7). The formation is more frequent since one patch was formed every cycle of oscillation in the crossflow. The time-averaged width of the jet depended on how far the jet fluid issued vertically into the crossflow. This, in turn, depended on the maximum (and perhaps the minimum) jet-to-crossflow velocity ratio in the cycle,

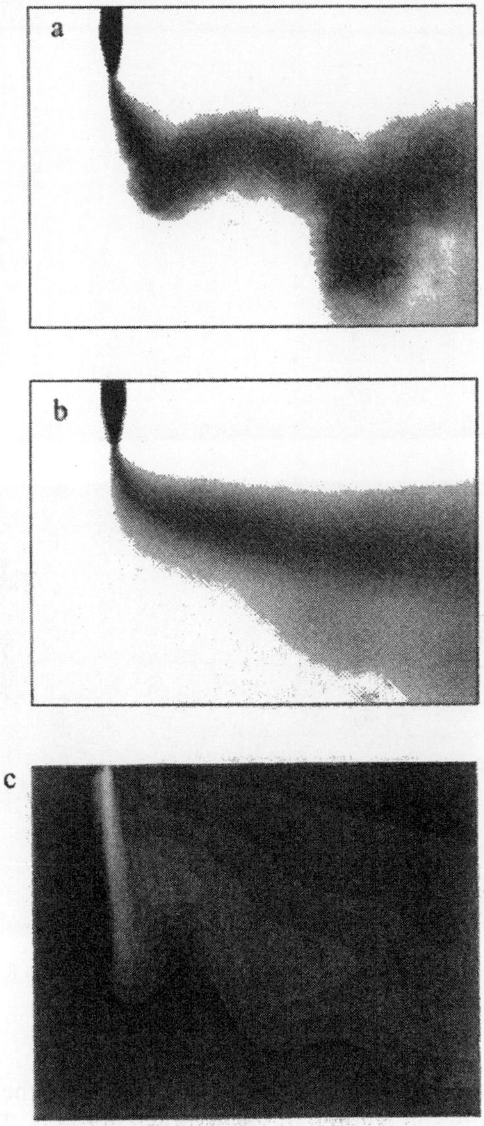

Fig. 8. Results at $u_p/U_a = 0.5$, St = 0.0375: (a) dye pattern at phase 0 ($t = T$); (b) time-averaged pattern over one cycle; (c) computed w-velocity contours at $t = T$.

which was determined by the unsteadiness parameter. Hence, it was observed that the time-averaged width in Fig. 7b was similar to that at St = 0.0375 in Fig. 4.

On the other hand, when the unsteadiness parameter was reduced, say $u_p/U_a = 0.5$, the peak jet-to-crossflow velocity ratio in the cycle was reduced. The jet was bent over to a larger extent and the coherent patches formed were less protruding vertically a

b

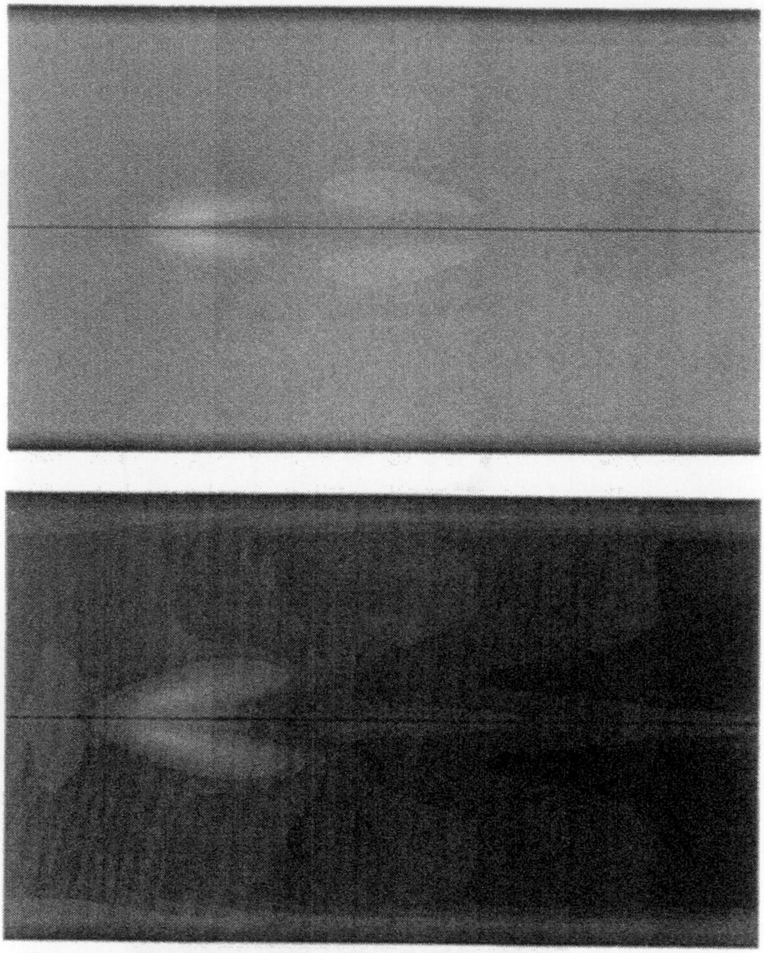

Fig. 9. Computed results at instant $t = T/4$ on $z/D = 27.6$ plane for $V_j/U_a = 8$, $u_p/U_a = 0.5$, St = 0.0375: (a) velocity magnitude contours; (b) turbulence kinetic-energy contours.

downwards from the connecting jet fluids on the inner side of the jet. This flow case as observed in experiments and computations is shown in Fig. 8. The time-averaged jet width in this case was widened to a smaller extent as compared with Fig. 4 or Fig. 7 where $u_p/U_a = 1.0$.

So far, the computational results presented are on the $y = 0$ plane. Now, we show the velocity magnitude contours on a horizontal plane in Fig. 9a. The flow case was $V_j/U_a = 8$, $u_p/U_a = 0.5$ and St = 0.0375 and the section was at $z/D = 27.6$ at instant $t = T/4$. This section intersected a fluid patch which revealed the kidney-shaped bifurcation pattern typical of a round jet in crossflow. Fig. 9b shows the turbulence kinetic energy contours which indicated extensive mixing and turbulence inside the "double-vortex" section.

5. Conclusions

The time-dependent flow of a round jet issuing into a fluctuating crossflow has been studied with a noble experimental technique. Sinusoidal oscillations imposed on an otherwise steady crossflow were simulated in the laboratory by vibrating the vertical jet–nozzle assembly in the direction of a mean crossflow. Phase-locked flow visualizations revealed that the jet discharging fluids were organized into regular fluid cloud patches which bulged towards the outer side of the bent-over jet. The prototype flow situation of jet in a fluctuating crossflow was studied with a CFD package. The computed flow patterns were in line with the experimental observations. This provided an indirect evidence that the experimental technique was valid. A mechanism of patch formation was proposed and some three-dimensional details of the flow were described based on the CFD results.

Acknowledgements

This investigation is supported by a research grant of The University of Hong Kong.

References

[1] J.J. Sharp, The effects of waves on buoyant jet, Proc. Inst. Civ. Eng. 81 (1986) 471–475.
[2] R. Koole, C. Swan, Measurements of a 2-D non-buoyant jet in a wave environment, Coastal Eng. 24 (1994) 151–169.
[3] K.M. Lam, L.P. Xia, Interaction of a round jet with an unsteady crossflow, in: Proc. 1st Int. Conf. on Flow Interaction, Hong Kong, September 1994, pp. 512–515.
[4] K.M. Lam, H.C. Chan, Investigation of turbulent jets issuing into a counter-flowing stream using digital image processing, Exp. Fluids 18 (1995) 210–212.
[5] B.E. Launder, D.B. Spalding, Lectures in Mathematical Models of Turbulence, Academic Press, London, 1972.
[6] B.E. Launder, D.B. Spalding, The numerical prediction of turbulent flows, Comput. Methods Appl. Mech. Eng. 3 (1974) 269–289.
[7] S.V. Patankar, Numerical Heat Transfer and Fluid flow, Hemisphere, Washington, DC, 1980.
[8] G.W. Govier, K. Aziz, The Flow of Complex Mixtures in Pipes, Krieger, New York, 1977.

Lab Methodology and Validation

ELSEVIER

Journal of Wind Engineering
and Industrial Aerodynamics 67&68 (1997) 861–872

JOURNAL OF wind engineering AND industrial aerodynamics

Turbulence control in multiple-fan wind tunnels

A. Nishi*, H. Kikugawa, Y. Matsuda, D. Tashiro

Department of Applied Physics, Faculty of Engineering, Miyazaki University, Miyazaki 889-21, Japan

Abstract

Many types of meteorological, environmental and boundary layer wind tunnels have been constructed to simulate atmospheric phenomena on buildings and structures. New types of two- and three-dimensional wind tunnels have been constructed. These tunnels have multiple-fans and oscillating blade rows to control the turbulence in the tunnels, and are controlled by computers. In this study a few turbulence parameter profiles were obtained in both longitudinal and vertical directions, and a trial to control the Reynolds stress was carried out.

Keywords: Wind tunnel; Multiple-fan wind tunnel; Turbulence control; Computer control

1. Introduction

Many types of wind tunnels have been built for meteorological, environmental and wind engineering purposes. In order to investigate the effects of strong wind, conventional boundary layer wind tunnels (BLWTs) have been used. Many spires and roughness blocks are set on the floor of the tunnel to produce the boundary layer velocity profiles and turbulence characteristics. In general, however, it is difficult to produce arbitrary turbulence characteristics in the conventional BLWTs.

Teunissen [1] has proposed a multiple-jet wind tunnel. Cermak and Cochran [2] and Kobayashi and Hatanaka [3] have noted new attempts to control the turbulence in the wind tunnel by using oscillating blades.

New types of two- and three-dimensional multiple-fan wind tunnels have been constructed and tested in our laboratory [4–6]. Each fan is controlled independently by a computer and many different turbulences are produced only by changing the computer program.

First, the configuration of two- and three-dimensional experimental wind tunnels is explained. Second, the turbulence control methods are introduced. Third, the profiles of turbulence parameters for both longitudinal and vertical directions obtained

* Corresponding author.

PII S0167-6105(97)00124-4

experimentally are discussed. Fourth, Reynolds stress, an important factor in a turbulent flow, produced by controlling the fans and oscillating blades is examined simultaneously.

2. Multiple-fan wind tunnels

Schematics of two- and three-dimensional experimental wind tunnels are shown in Fig. 1. The two-dimensional tunnel had 11 fans arranged vertically; the three-dimensional one had six rows of two-dimensional tunnels with a total of 66 fans. Each fan was connected to a computer through a driver. The test section dimensions (length, width and height) of each tunnel were (3.8, 0.18, 1.0 m) and (5.0, 1.0, 1.0 m), respectively. They had very short lengths compared to conventional BLWTs. The wind speed of each tunnel was 0–11 m/s, corresponding to the output of computer signals of 0–10 V. The fans installed in the two-dimensional tunnel were driven by high performance AC servo-motors. The response was sensitive to the computer output signals. On the other hand, conventional AC induction motors were used for the fans in the three-dimensional tunnel, so the response was inferior to that of the two-dimensional tunnel.

At the exit of all channels connected to each fan, a honeycomb core was installed to eliminate the secondary flow effect in the curved channels. To decrease the effect of the channel-wall boundary layer, cylindrical vortex generators 12 mm in diameter were inserted just after the honeycomb core. Each tunnel had a computer-controlled traverse gear system to measure the mean and turbulent velocities by hot-wire anemometers.

Oscillating blades were installed in the mid-section of the tunnel ($x/H = 2.2$ in the two-dimensional wind tunnel, where x is the down-flow length from the tunnel inlet and H the tunnel height). Each blade was driven by a stepping motor.

The control system consisted of three computers connected by a local-area network. The first computer controlled the fans, the second one controlled the blades and

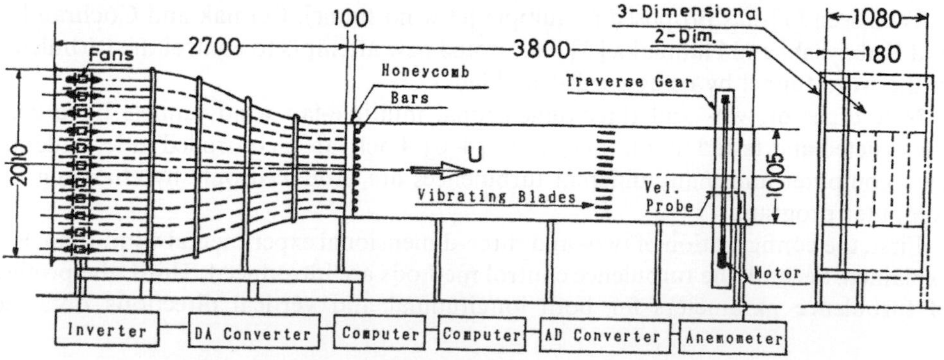

Fig. 1. Schematics of the two- and three-dimensional experimental wind tunnels.

the third one was used to control the traverse gear, as well as the measurement of velocity and the control of all system.

3. Flow in the two-dimensional wind tunnel

The instantaneous flow velocity was measured to observe the flow pattern in the two-dimensional wind tunnel. An example of a shear flow vector profile is shown in Fig. 2. Each velocity vector has a mean value of 5 s, which consists of 500 data points measured at an interval of 10 ms. In this figure, the flow at the inlet of the tunnel has small differences in velocity, however, at a down-flow section of $x/H = 3$ (about 3 m from the inlet), an almost smooth flow pattern is observed. This means that a shorter length of the tunnel than in conventional BLWTs is sufficient for this type of wind tunnel. The effect of the vortex generator bars was clearly seen by a comparison of two experiments with and without the vortex generator bars at the inlet of the tunnel.

4. Turbulence control methods

4.1. Power spectrum modification method

The power spectral density in the longitudinal velocity component has been investigated by a number of researchers, and a number of empirical equations have been proposed [7]. In these equations, the von Kármán equation is given in Eq. (1) for

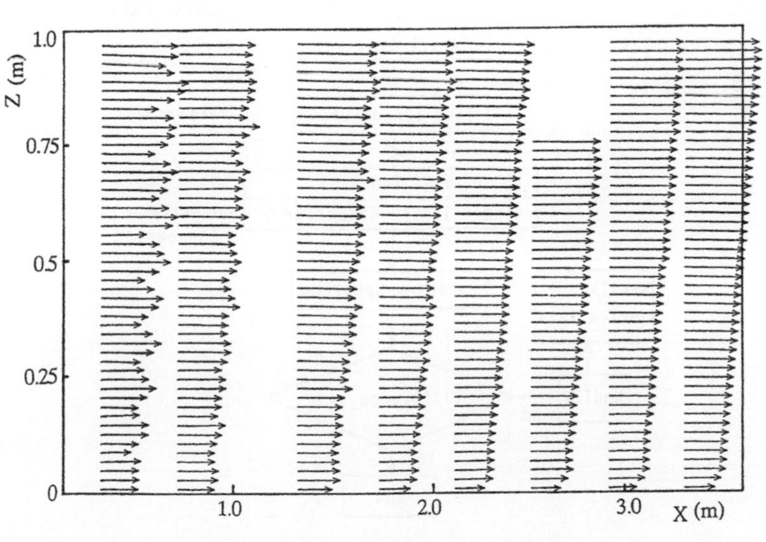

Fig. 2. Flow vector profiles for a shear flow.

the longitudinal flow component

$$S_u(f) = \frac{4I_u^2 L_u U}{\{1 + 70.7(fL_u/U)^2\}^{5/6}}, \tag{1}$$

where $S_u(f)$ is the power spectral density, I_u the turbulence intensity, L_u the turbulence scale, U the mean velocity and f the frequency of turbulence. It is convenient to use this equation when the turbulence parameters of U, I_u and L_u are given and has been used in the following analysis. The flow diagram of the iterative procedure to simulate the target turbulence is shown in Fig. 3. The target spectrum can be calculated by Eq. (1) for the assumed profiles of U, I_u and L_u. It was analyzed by making use of the inverse fast Fourier transform (inverse FFT) program, and a time historical velocity of turbulence is produced. This was given to the fan motor. The turbulent wind velocity was measured by a hot-wire anemometer at the test section ($x/H = 3.2$). These data were analyzed by the FFT method, and the measured spectrum was compared to the target one. The deviated part of the measured spectrum curve from the target one was modified and a new spectrum was estimated and given to the fan again. These procedures were repeated several times. This was called the "spectrum modification method".

The turbulent velocity spectrum in the vertical direction is given by

$$S_w(f) = \frac{4I_w^2 L_w U\{1 + 188.4(2fL_w/U)^2\}}{\{1 + 70.7(2fL_w/U)^2\}^{11/6}}. \tag{2}$$

This was also analyzed by the inverse FFT program and given to the oscillating blade motors.

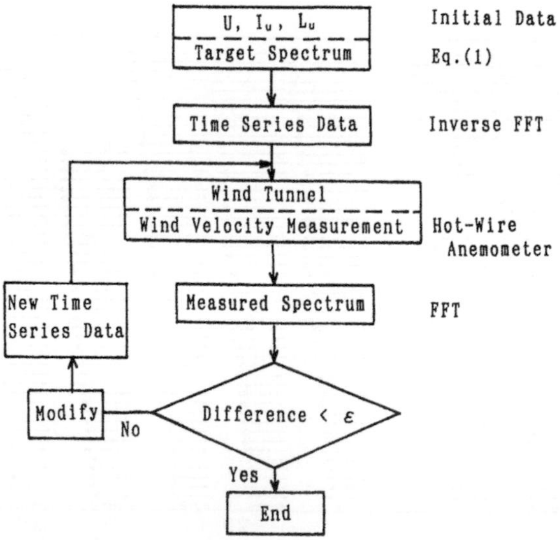

Fig. 3. Block diagram of turbulence control.

The required inclination angle of each blade is given by

$$\theta = \tan^{-1}\{cw'/(U + u')\}, \tag{3}$$

where θ is the wind velocity angle, c an adjustable coefficient and u' and w' the horizontal and vertical turbulence velocities, respectively. c is adjusted to obtain a higher correlation between the target and observed time-historical velocities.

4.2. Time-lag modification method

There is a certain time lag in the wind tunnel. When a fluctuating input signal was given to the fan motor, almost the same fluctuating velocity could be measured at the test section with a certain time lag. This time lag consisted of fans and flow channel. Therefore, if the input signal was modified depending on the time lag in advance, a modified turbulent velocity would be observed.

The input signal was modified by Eq. (4) for a small sampling interval of Δt,

$$y_{\mathrm{m}}(t + \Delta t) = \frac{y(t + \Delta t) - y(t)}{1 - \mathrm{e}^{-t/T}} + y(t), \tag{4}$$

where $y_{\mathrm{m}}(t + \Delta t)$ is the modified velocity, $\Delta y = y(t + \Delta t) - y(t)$, the increment of output within the interval of Δt, T the time constant of the time lag, when the first-order lag was assumed. This is called the "time-lag modification method", and is applied to the control of fan motor.

5. Profiles of turbulence parameters

5.1. Results of power spectrum modification method

The target profiles of the mean velocity, turbulence intensity and turbulence scale in the longitudinal velocity component were assumed by conventional equations, as follows:

$$U(z)/U_0 = (z/Z_0)^{\alpha}, \tag{5}$$

$$I_u(z)/I_{u0} = (z/Z_0)^{-\alpha - 0.05}, \tag{6}$$

$$L_u(z)/L_{u0} = (z/Z_0)^{0.5}, \tag{7}$$

where U_0 is the mean velocity at the upper point of boundary layer, Z_0 the height of the boundary layer, I_{u0}, L_{u0} the turbulence intensity and turbulence scale at the upper point of the boundary layer, respectively, and α the coefficient of power-law profile. In this case, $\alpha = 0.3$, $I_{u0} = 7\%$, $U_0 = 6$ m/s, $Z_0 = 0.8$ m and $L_{u0} = 1.63$ m were assumed.

The target profiles of turbulence intensity and scale in the vertical turbulence components were assumed as follows:

$$I_w(z) = I_u(z)/2, \tag{8}$$

$$L_w(z) = L_u(z)/4. \tag{9}$$

Examples of profiles obtained from these parameters are shown in Figs. 4 and 5 for longitudinal and vertical components, respectively. Each figure presents a result obtained by repeating the iterative methods four times. Good agreement is shown between the target and measured parameters. In Figs. 4 and 5, the comparisons of power spectra are shown for different heights. In addition, good agreements are presented for the target and measured spectra. The total sampling time of each point

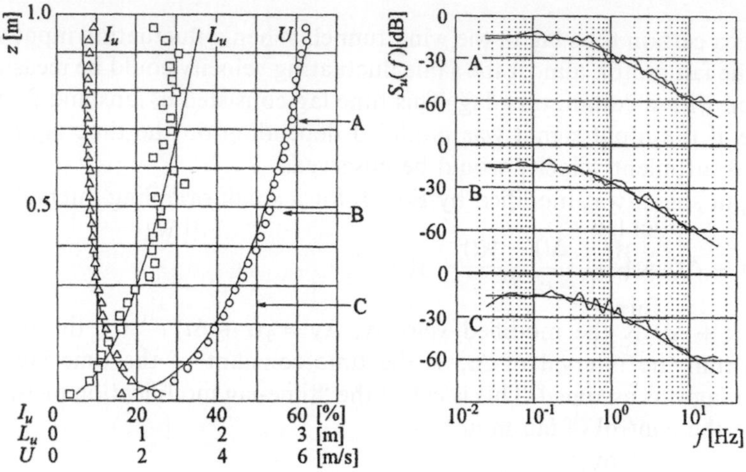

(a) Distributions of U-component.

(b) Spectra of U-component.

Fig. 4. Turbulence parameter profiles for longitudinal direction.

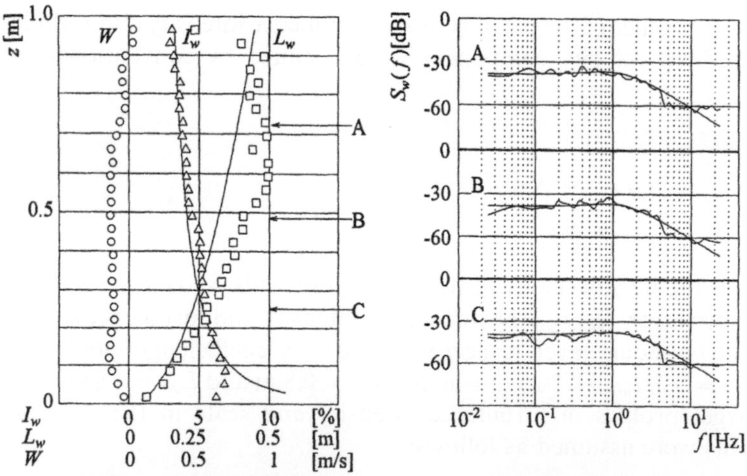

(a) Distributions of W-component.

(b) Spectra of W-component.

Fig. 5. Turbulence parameter profiles for vertical direction.

was about 40 s, which produced a total data number of 2048 measured in a sampling interval of 20 ms.

5.2. *Results of the time-lag modification method*

The time lag of the two-dimensional tunnel is shown in Fig. 6. Different time lags were observed for acceleration and deceleration of wind speed. These delays can be seen as a first-order lag with a small amount of dead time. It was caused by the inertias of the rotating parts of the fan and the air mass in the tunnel.

First, the uniform velocity profile was examined. Ten time modifications were examined continuously. The variations of turbulence parameters and the correlations between the target and measured time-historical velocities are shown in Fig. 7a and Fig. 7b, respectively. From these figures, it can be seen that one trial is enough to obtain the improved parameters. Moreover, it is the merit of this method that both the power spectrum and the phase angle are improved simultaneously, as shown in Fig. 8a and Fig. 8b. An example of the comparison of simulated time-historical velocities is given in Fig. 9, and about 96% of the correlation coefficient was obtained.

Second, the time-lag modification method was applied to the boundary layer velocity profile. An example is shown in Fig. 10 for one trial. Almost the same agreements were obtained between the target and measured profiles compared with

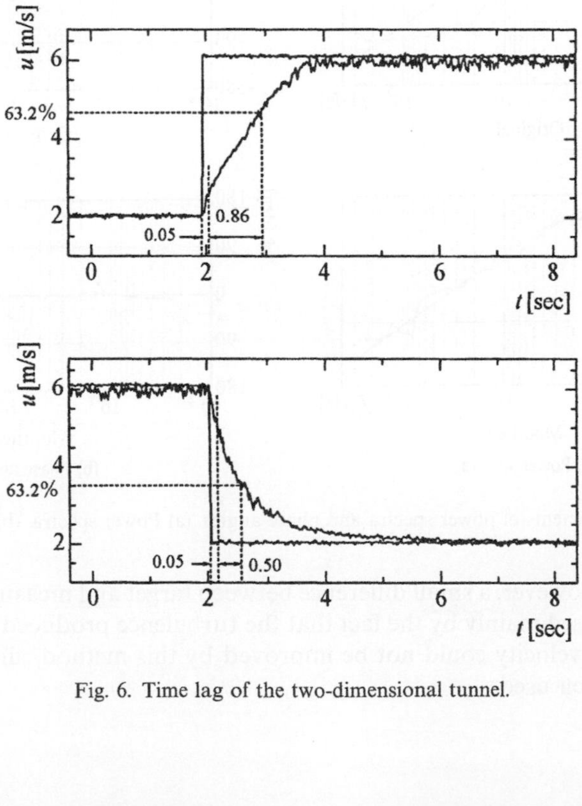

Fig. 6. Time lag of the two-dimensional tunnel.

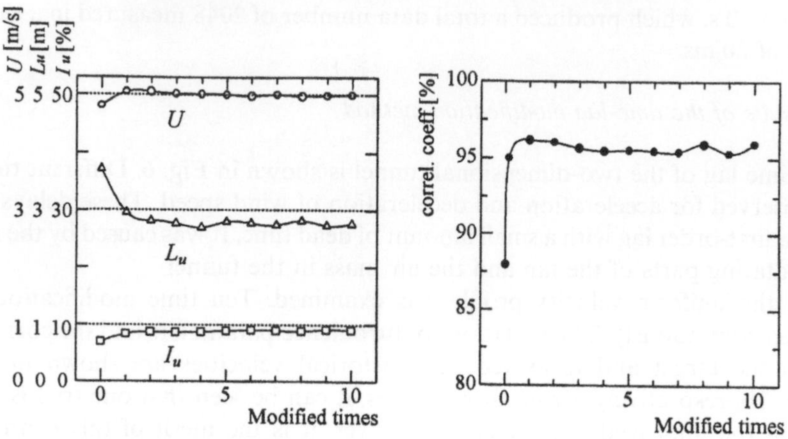

(a) Turbulence parameters. (b) Correlation coefficients.

Fig. 7. Changes of turbulence parameters and correlation coefficients for ten times modifications. (a) Turbulence parameters. (b) Correlation coefficients.

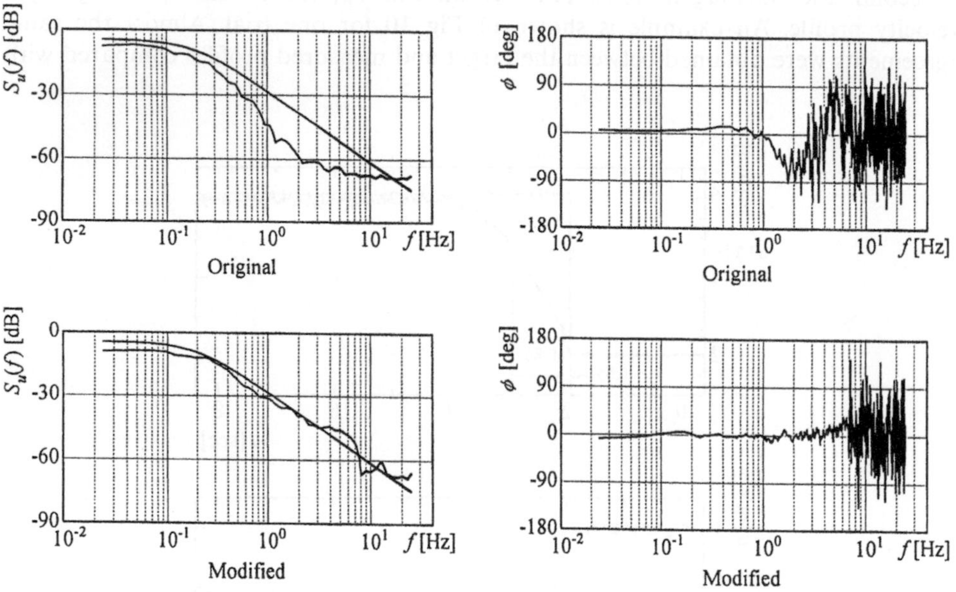

(a) Power spectra. (b) Phase angles.

Fig. 8. Improvements of power spectra and phase angles. (a) Power spectra. (b) Phase angles.

those in Fig. 4. However, a small difference between target and measured profiles near the floor was caused mainly by the fact that the turbulence produced in the tunnel by the strong shear velocity could not be improved by this method, since the feedback trials had not been used.

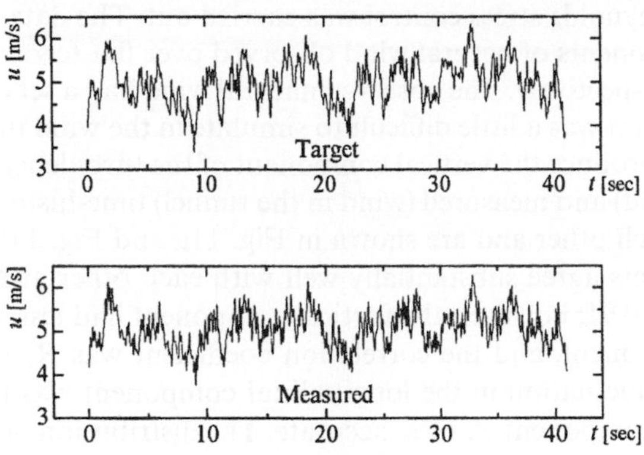

Fig. 9. Comparison of target and measured velocities.

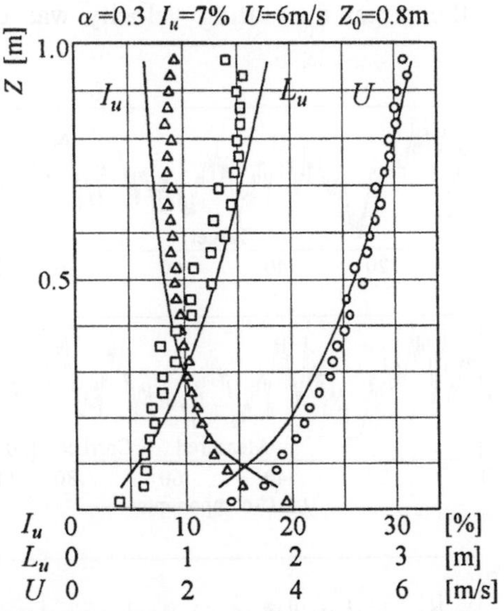

Fig. 10. Turbulence parameter profiles obtained by the time-lag modification method.

6. A trial of Reynolds stress control

Reynolds stress is the most important factor in turbulent wind. It is well known that natural wind has an almost constant Reynolds stress profiles in the surface layer [8]. Therefore, it is important to investigate the Reynolds stress profile in the wind tunnel boundary layer. In the two-dimensional wind tunnel, the longitudinal and vertical turbulences were produced by controlling the fans and airfoils, respectively. These controls were independent of each other. The Reynolds stress could be produced by the synchronized control of the phase angles of fans and blades.

A trial of the Reynolds stress control was carried out. The data of the longitudinal and vertical components of natural wind observed over flat terrain were given to the fans and blades, respectively. The observed natural wind had a set of larger turbulence parameters, so that it was a little difficult to simulate in the wind tunnel. In particular, it was difficult to produce the vertical component of the turbulence by the blades. The target (natural wind) and measured (wind in the tunnel) time-historical velocities were compared with each other and are shown in Fig. 11a and Fig. 11b. The longitudinal velocity components agree substantially well with each other and had a correlation coefficient of $R_l = 0.91$; however, the vertical component had less agreement than the longitudinal component, and the correlation coefficient was $R_v = 0.75$. In addition, the amplitude of fluctuation in the longitudinal component was in good agreement, while the vertical component was less accurate. The distribution of points on the u'–w' plane is given in Fig. 12a and Fig. 12b for the target and measured points, respectively. The Reynolds stress coefficients of each distribution were $\mathrm{Re}_t = \overline{uw}/(\sigma_u\sigma_w) = -0.271$ for the target and $\mathrm{Re}_m = -0.131$ for the measured distribution, respectively. Although the target fluctuating velocity was comparatively large in

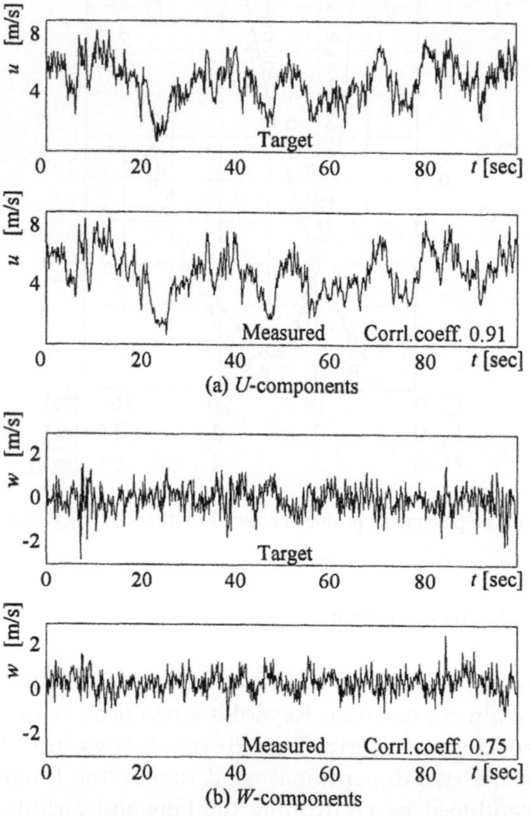

Fig. 11. Comparisons of target (natural wind) and measured velocities in the longitudinal and vertical directions.

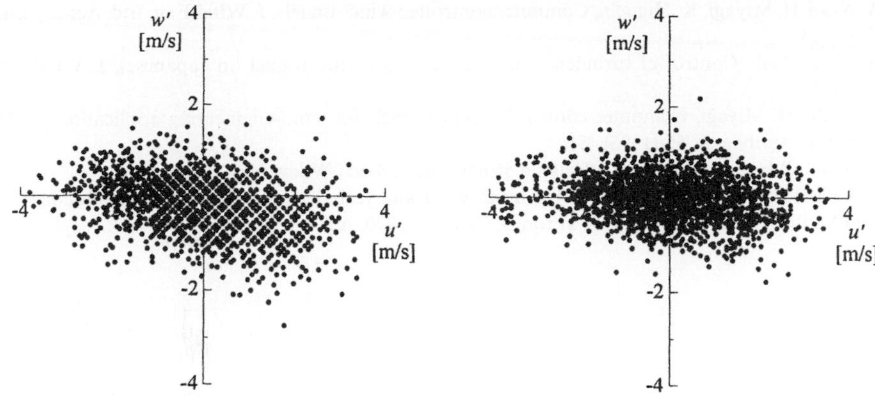

(a) Reynolds stress coeff. -0.271 (target)

(b) Reynolds stress coeff. -0.131 (measured)

Fig. 12. Comparison of target (natural wind) and measured distributions of turbulence velocities.

this case, it was shown that the required Reynolds stress could be produced by controlling vertical turbulence more precisely. The next step of investigation is to obtain the Reynolds stress profile in the wind tunnel boundary layer.

7. Conclusions

New types of multiple-fan wind tunnels have been constructed and tested in our laboratory. From these experiments the following conclusions are obtained:

1. A tunnel of shorter length is sufficient for these types of wind tunnels, compared to the conventional BLWTs. This is demonstrated by the figures of the flow vector profiles.
2. The profiles of turbulence parameters for longitudinal and vertical velocity components were obtained simultaneously by the spectrum modification method, and good agreements between the target and observed spectra were obtained.
3. A good agreement was also shown between the target and measured turbulence parameter profiles by using the time-lag modification method.
4. A trial of Reynolds stress control was carried out. It was clear that Reynolds stress can be simulated by controlling the phase angles of the fans and oscillating blades.

References

[1] H.W. Teunissen, Simulation of the planetary boundary layer in a multiple-jet wind tunnel, Atmos. Environ. 9 (1975) 145.

[2] J.E. Cermak, L.S. Cochran, Physical modeling of the atmospheric surface layer, J. Wind Eng. Ind. Aerodyn. 41/44 (1992) 935.

[3] H. Kobayashi, A. Hatanaka, Active generation of wind gust in a two-dimensional wind tunnel, J. Wind Eng. Ind. Aerodyn. 41/44 (1992) 959.

[4] A. Nishi, H. Miyagi, K. Higuchi, Computer controlled wind tunnels, J. Wind Eng. Ind. Aerodyn. 46/47 (1993) 837.
[5] A. Nishi et al., Control of turbulence in a multiple-fan wind tunnel (in Japanese), J. Wind Eng. 6 (1994) 1.
[6] A. Nishi, H. Miyagi, Computer controlled wind tunnels for wind-engineering applications, J. Wind Eng. Ind. Aerdyn. 54/55 (1995) 493.
[7] E. Simiu, R.H. Scanlan, Wind effects on structures, 2nd ed., Wiley, New York, 1986.
[8] H.W. Tewnissen, Characteristic of the mean wind and turbulence in the planetary boundary layer, UTIAS Review 32, University of Toronto, Canada, 1970, p. 22.

ELSEVIER

Journal of Wind Engineering
and Industrial Aerodynamics 67&68 (1997) 873–883

JOURNAL OF
wind engineering
AND
industrial
aerodynamics

Application of infrared thermography and a knowledge-based system to the evaluation of the pedestrian-level wind environment around buildings

Ryoji Sasaki[a,*,1], Yasushi Uematsu[b,2], Motohiko Yamada[b,3], Hiromu Saeki[b,4]

[a] *Technical Research Institute of Nishimatsu Construction Co., Ltd., 2570-4, Shimo-tsuruma, Yamato 242, Japan*
[b] *Department of Architecture, Urban Planning and Building Engineering, Tohoku University, Sendai 980-77, Japan*

Abstract

A convenient technique for evaluating the wind environment around buildings in wind tunnel tests has been developed. Infrared thermography and a knowledge-based system are applied to this technique. The evaluation system developed in this study can be used effectively at an early design stage of a tall building development, e.g., when the cross sections and/or the arrangement of buildings are examined from the view point of the pedestrian-level wind environment.

Keywords: Wind environment; Infrared thermography; Knowledge-based system; Pedestrian-level

1. Introduction

With the advent of high-rise buildings, the wind environment around them has become an important technological and social problem. To minimize the occurrence of unpleasant and dangerous wind conditions at the pedestrian level, an evaluation of the wind environment is an indispensable element in the design of these buildings. To date, the best evaluation still relies on wind speed measurements using anemometers in a wind tunnel. However, it generally takes a long time and a lot of effort to conduct the experiment and to process the data. Hence, it is hoped that a convenient method of

* Corresponding author. E-mail: sasakir@gw.ri.nishimatsu.co.jp.
[1] Research Engineer.
[2] Associate Professor.
[3] Professor.
[4] Graduate student.

PII S0167-6105(97)00125-6

wind tunnel experimentation can be developed to provide an overall indication of the wind conditions around buildings. A scour technique, or sand erosion technique has often been used for this purpose. Livesey et al. [1] have improved this technique using a digital image processor. However, the scour pattern depends on the experimental conditions, such as the choice of particles. Furthermore, the pattern may change with time, even if the wind speed is kept constant. These features result in difficulties in obtaining quantitatively reliable data with this technique.

Recently, we have developed a flow visualization technique using infrared thermography and have confirmed its validity [2–4]. This technique is based on the fact that the heat transfer from a heated body to the flow is closely related to the flow conditions near the body surface. Applying this technique, we have developed a convenient technique for evaluating the pedestrian-level wind environment around buildings; this yields a visual representation of the wind condition near the ground. In the experiments using this technique, a model is placed on a heated floor and immersed in a flow. The temperature distribution on the floor surface is measured by infrared thermography and is displayed as a thermal image. Some experimental results were published by Uematsu and Yamada [5] and Yamada et al. [6]. In these studies, we investigated the relationship between the surface temperature and the wind speed based on the experimental results. It was found that the relationship was represented by a simple function in the regions where wind speeds increase (referred to as 'higher-speed regions', hereafter), and an empirical formula was presented. On the other hand, it seemed difficult to derive a simple description for the relationship in the regions where wind speeds decrease. However, this is not a serious problem from the practical point of view. The most important thing is to predict where and how much the wind speeds increase around the buildings. Once the higher-speed regions are known, we can predict the wind speeds easily and almost in real time within an allowable margin of error in these regions.

In this study, we have applied a knowledge-based system to the prediction of higher-speed regions. A combination of the thermography technique and the knowledge-based system may give a more reasonable evaluation of the pedestrian-level wind environment. In the present paper, emphasis is on the description of the knowledge-based system which we are now developing. The thermography technique is outlined (regarding the details, see Ref. [6]).

2. Thermography technique

2.1. Experimental setup

The setup for the thermography tests is schematically illustrated in Fig. 1. A part of the wind tunnel floor is made of a 12 mm-thick acrylic plate and is warmed by hot water. A building model is placed on this part and immersed in the flow. In approximately 30 s, since the wind tunnel is run, we obtain a steady temperature distribution on the floor surface. The distribution is measured by an infrared camera (Nippon Avionics Co. Ltd., TVS-2000) mounted above the wind tunnel.

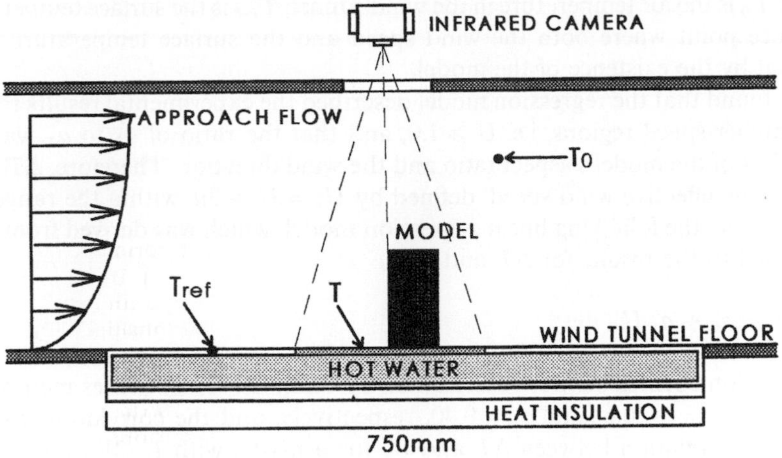

Fig. 1. Illustration of the experimental set-up.

2.2. Relationship between surface temperature and wind speed

The relationship between surface temperature and wind speed was investigated based on the experimental results of a thermography technique and a wind speed measurement made using a hot-wire anemometer. The experiments were made in a turbulent boundary layer with a power law exponent of $\alpha = 0.20$. Several rectangular prisms with square cross-sections were used as building models. Two wind angles were tested, i.e., $\theta = 0°$ and 45°, where $\theta = 0°$ represents a direction normal to one of the model's walls.

The surface temperature, or in other words, the local heat transfer from the floor to the flow, depends both on the mean and on the fluctuating wind speed. In this study the rms wind speed u' is used as an indication of the wind speed fluctuation. In order to examine the dependence of the surface temperature T on the mean and fluctuating wind speeds, the following multiple regression model was applied to the results,

$$\Delta T = a_0 + a_1 U + a_2 u', \tag{1}$$

where a_0, a_1 and a_2 are regression coefficients to be determined by using the least squares method; U and u' are the mean and rms wind speeds, respectively, which are both non-dimensionalized by the mean wind speed measured at the same location without model in position; and ΔT represents the normalized surface temperature defined as

$$\Delta T = \frac{T_{\text{ref}} - T}{T_{\text{ref}} - T_0} \tag{2}$$

in which T_0 is the air temperature in the wind tunnel; T_{ref} is the surface temperature at a reference point where both the wind speed and the surface temperature are not influenced by the existence of the model.

It was found that the regression model described the experimental results relatively well in higher-speed regions, i.e. $U \geqslant 1.0$, and that the ratio of a_2 to a_1 was about 3 regardless of the model's aspect ratio and the wind direction. Therefore, ΔT may be related to an 'effective wind speed' defined by $U_e = U + 3u'$ within the range where $U \geqslant 1.0$. Then, the following linear regression model, which was derived from Eq. (1), was applied to the results for ΔT and U_e,

$$\Delta T = a'_0 + a'_1 U_e. \tag{3}$$

The regression coefficients were determined by using the least squares method; these values were $a'_0 = -0.64$ and $a'_1 = 0.40$, respectively, and the correlation coefficient was 0.81. The relation between ΔT and U_e for a model with $H : W : D = 3 : 1 : 1$ is plotted in Fig. 2; the circle and cross represent the result for $U \geqslant 1.0$ and $U < 1.0$, respectively. It means that the coefficients are suitable within the higher-speed regions.

Since the effective wind speed U_e approximately corresponds to a gust wind speed in higher-speed regions [7], the distribution of U_e itself may give an indication of wind environment. However, the wind environment is often evaluated based on the mean wind speed U. Hence, we examine the relation between U and U_e. Some models were applied to the experimental results. It was found that the general behavior of U with U_e was expressed by the following equation regardless of the model's aspect ratio and the wind direction, as shown in Fig. 3:

$$U = -1.25 + 1.71\sqrt{U_e}. \tag{4}$$

The coefficients in the above equation were determined by using the least squares method; the correlation coefficient was 0.81.

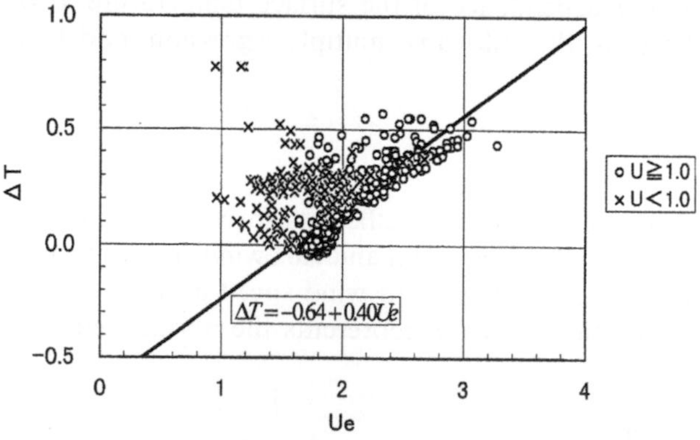

Fig. 2. Relation between ΔT and U_e ($H : W : D = 3 : 1 : 1$, $\theta = 0°$).

2.3. Prediction of wind speeds from the surface temperature

Using the formula obtained above, we can easily convert the thermal image into the mean wind speed distribution and display it as an image on a personal computer. The procedure is summarized in Fig. 4. The formula was obtained based on the results for isolated models. However, its application to two models was confirmed [8].

Fig. 5 shows the results for square buildings with an aspect ratio of 3. The prediction of U in the whole area is displayed in Fig. 5(a), while the results only in the higher-speed region are displayed in Fig. 5(b) after magnification. Comparing these results with those of a wind-speed measurement made by using a hot-wire anemometer, shown in Fig. 6, it is found that the thermography technique overestimates the

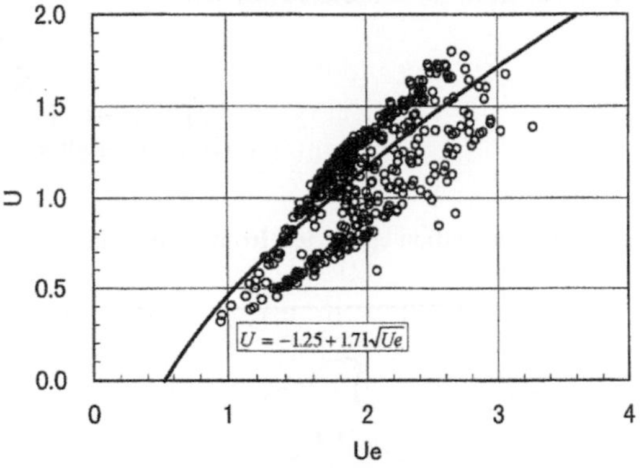

Fig. 3. Relation between U and U_e ($H:W:D = 3:1:1$, $\theta = 0°$).

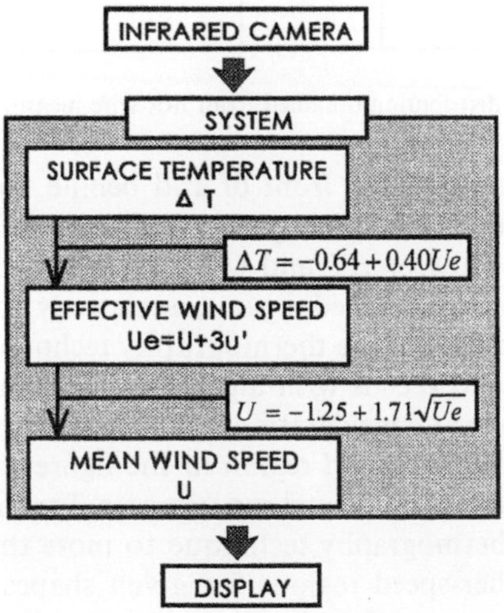

Fig. 4. Conversion of surface temperature into wind speed.

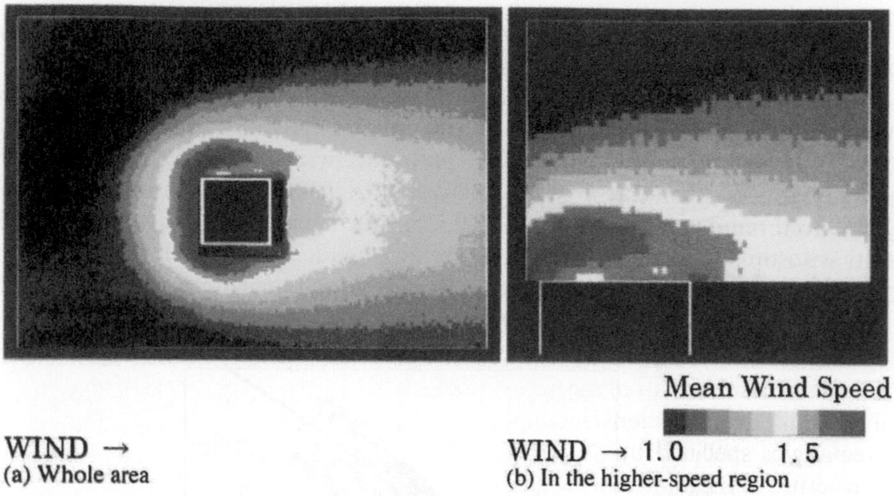

Fig. 5. Mean wind speed distribution converted from thermal image ($H:W:D=3:1:1$).

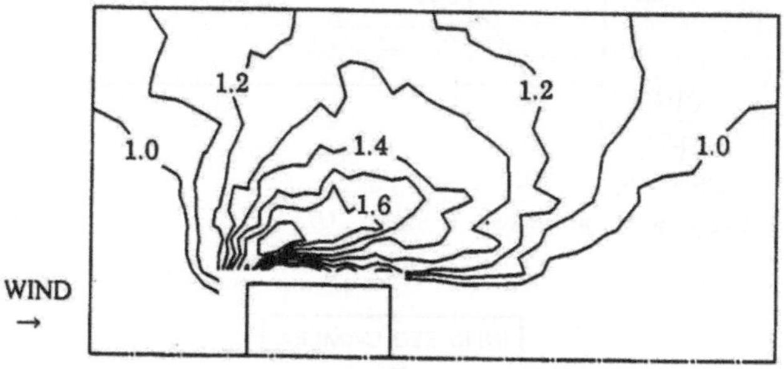

Fig. 6. Mean wind speed distribution obtained from hot-wire measurement ($H:W:D=3:1:1$).

mean wind speed in the regions in front of and behind the model, where the mean wind speed is low but the turbulence is great. This indicates that it may be difficult to express the relationship between ΔT and U by a unique formula for the whole area. On the other hand, in the higher-speed regions, a relatively good agreement is obtained. This is one of the advantages of the thermography technique. That is, this technique can predict the mean wind speeds with an allowable error in higher-speed regions, which is the most important consideration in the evaluation of wind environment.

The boundary of the higher-speed region in the figure was tentatively determined from the results of previous wind-tunnel experiments. The most important problem in the application of the thermography technique to more than two buildings may be how to predict the higher-speed regions for given shapes and arrangement of the buildings. Once the higher-speed regions are known, the thermography technique presents a reasonable estimation of the wind speed in these regions.

3. Knowledge-based system

3.1. Composition of the knowledge-based system

The knowledge-based system which infers the area of the higher-speed regions around buildings consists of a knowledge-base, a data-base, a user-interface and an inference engine as illustrated in Fig. 7. This system has been developed using an expert-system tool, 'Daisogen' (A.I. Soft, Japan). This system starts to predict the areas in reply to input data regarding the shapes and arrangement of buildings. As the first step of development, we focused on two rectangular buildings in various arrangements.

3.2. Data-base

The data-base consists of the experimental data for the higher-speed regions around isolated buildings. For constructing the data-base, the regions are simply represented by rectangles specified by $S_{x1} \times S_{y1}$ and $S_{x2} \times S_{y2}$ as shown in Fig. 8. Fig. 9 shows the structure of the data-base. It has a hierarchic frame structure, which is specified by the width-to-depth ratio (W/D), the aspect ratio (H/D) and the wind direction (θ). The

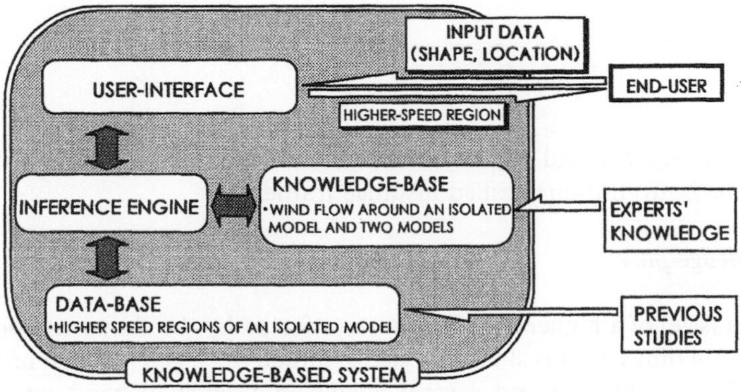

Fig. 7. Knowledge-based system of evaluating wind environment.

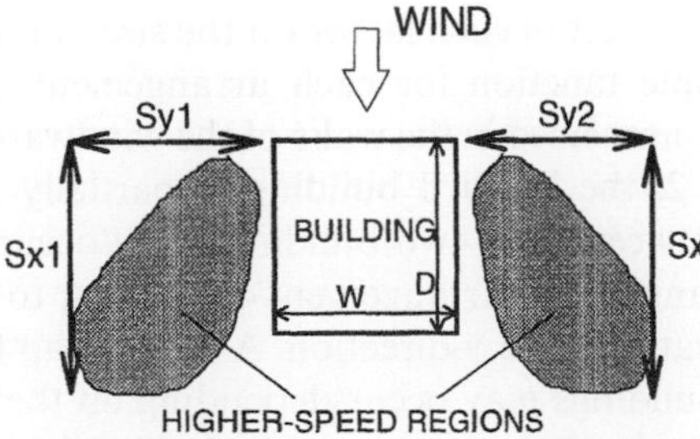

Fig. 8. Representation of higher-speed regions.

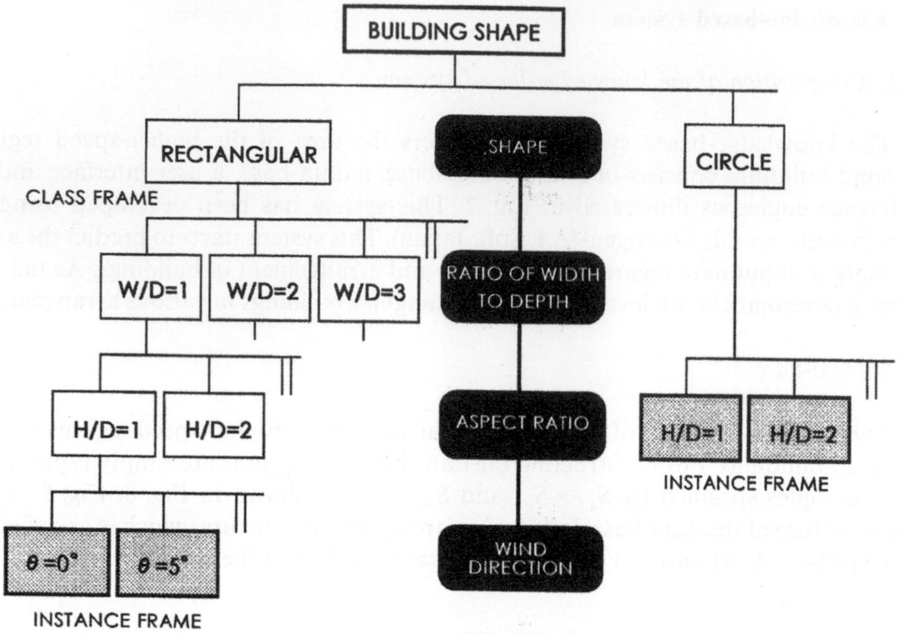

Fig. 9. Structure of data-base.

values of S_{x1}, S_{x2}, S_{y1} and S_{y2}, non-dimensionalized by the building's depth D, are stored in the bottom frames called instance frames.

3.3. Knowledge-base

The area (size) of a higher-speed region is affected by the shape and positioning of the nearby buildings. In this study, the size assumed to be given by the product of that for an isolated condition and several coefficients which depend on the size and arrangement of the two buildings. The knowledge-base consists of many heuristic rules regarding these coefficients. First the arrangement of two buildings was classified into four types shown in Fig. 10, because a detailed examination of the experimental results indicated that the effect of each factor on the size of higher-speed region could be described by a simple function for each arrangement. In Arrangement 1, the leeward building is fully immersed in the wake of the windward building. On the other hand, in Arrangement 2, the leeward building is partially immersed in the wake. A strong interaction between these two buildings may occur. Arrangement 3 represents a side-by-side arrangement. Arrangement 4 is similar to Arrangement 2, but the two buildings are separated in the y-direction. A strong gap flow or no flow interaction between the two buildings may occur depending on the distance. In the case of arrangement 2, for example, the parameters which describe the shapes and arrangement of the buildings are illustrated in Fig. 10. The wind speeds may increase in the four regions labeled I–IV, where the size of each higher-speed region in the x- and y-directions are determined by these parameters.

3.4. Inference of higher-speed regions

The procedure of inference is illustrated in Fig. 11. First the system requires the user to input the data of the shapes and arrangement of the two buildings, i.e., height

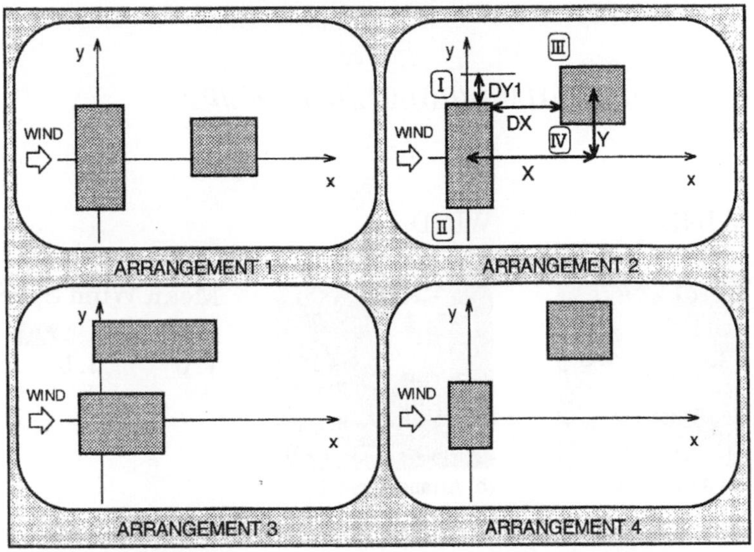

Fig. 10. Classification of the arrangement of two rectangular buildings.

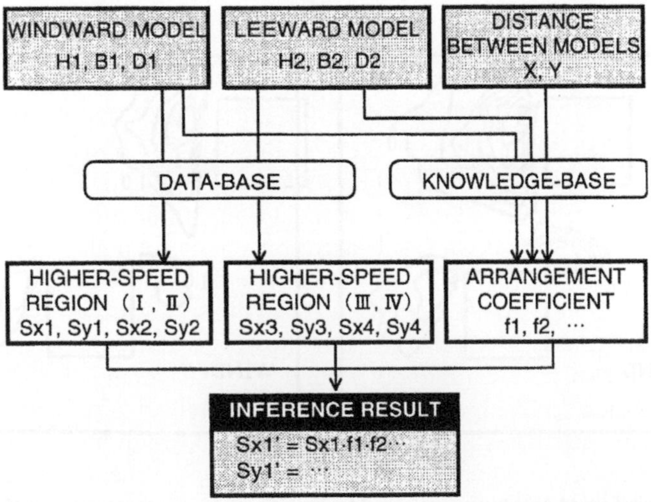

Fig. 11. Flowchart of inferring higher-speed region around two models.

(H_1, H_2), depth (D_1, D_2), width (W_1, W_2), and distance (X, Y) between these two buildings in the x- and y-directions. Based on these data, the system determines the size (S_{x1}, S_{y1}, etc.) of higher-speed regions for both buildings when they are isolated, using the data-base. The arrangement type, one from the four specified arrangements, is determined too. Then, the coefficients describing the effects of the nearby building are computed by using the knowledge-base. Finally, the sizes ($S_{x1'}$, $S_{y1'}$, etc.) of higher-speed regions for the two buildings are determined by multiplying those for the isolated condition and these coefficients. For example, in the case of Arrangement 2, the size $S_{x1'}$ is given by

$$S_{x1'} = S_{x1} f_1(D_{X1}/D_1) f_2(D_{Y1}/D_1) f_3(H_2/H_1) f_4(W_2/D_2). \tag{5}$$

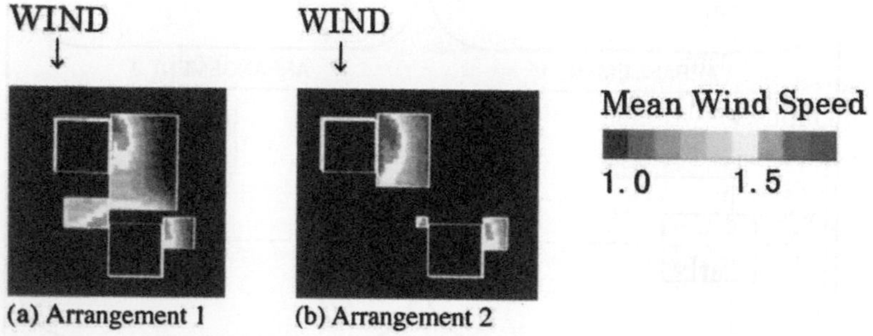

Fig. 12. Mean wind speed distribution converted from thermal image (windward $H:W:D = 2:1:1$, leeward $H:W:D = 1:1:1$): (a) Type-1, (b) Type-2.

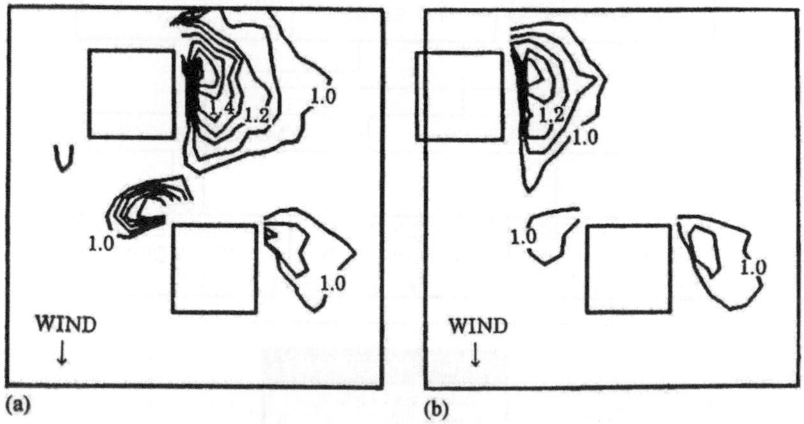

Fig. 13. Mean wind speed distribution measured by hot-wire anemometer (windward $H:W:D = 2:1:1$, leeward $H:W:D = 1:1:1$). (a) Arrangement 1; (b) arrangement 2.

3.5. Results of inference

Fig. 12 shows sample results for two square buildings in staggered arrangement; the aspect ratios of the windward and leeward buildings are 2 and 1, respectively. In the figures, the distribution of the mean wind speed, obtained from the thermography technique, is displayed as an image only in the higher-speed regions which are predicted by the knowledge-based system. For the purpose of comparison, the corresponding results of the hot-wire measurements are shown in Fig. 13. The relatively good agreement between these two results is obtained; this indicates that the technique developed in this study presents numerically reliable results as well as an overall representation of the pedestrian-level wind environment.

4. Concluding remarks

A convenient technique for predicting the pedestrian-level wind environment by using infrared thermography and a knowledge-based system has been developed. Using the empirical formula, which describes the relationship between the surface temperature and the mean wind speed, we can easily obtain the distribution of the mean wind speed from the temperature distribution. This thermography technique gives a reasonable estimation of the wind speeds in the regions where the mean wind speeds increase, i.e. $U \geqslant 1.0$. However, it overestimates the mean wind speed in the regions where mean wind speed is low but turbulence is great.

In order to correct this defect and improve the application of the thermography technique, we have incorporated a knowledge-based system into this technique. That is, the wind speeds are displayed only in or together with the higher-speed regions inferred by the knowledge-based system. Such a representation may result in a more reasonable evaluation of the wind environment. As the first step of development, we constructed a prototype of the knowledge-based system for two rectangular buildings. It was found that the system predicted the higher-speed regions around the two buildings within an allowable error. However, the application of this system is still limited to simple arrangements. Further investigations are planned for improving this technique.

References

[1] F. Livesey, D. Morrish, M. Mikitiuk, N. Isyumov, J. Wind Eng. Ind. Aerodyn. 44 (1992) 2265.

[2] M. Yamada, Y. Uematsu, T. Koshihara, J. Wind Eng. JAWE 35 (1988) 35 (in Japanese with English summary).

[3] M. Yamada, Y. Uematsu, T. Koshihara, J. Wind Eng. JAWE 36 (1988) 1 (in Japanese with English summary).

[4] Y. Uematsu, M. Yamada, M. Kikuchi, J. Wind Eng. JAWE 43 (1990) 9 (in Japanese with English summary).

[5] Y. Uematsu, M. Yamada, Proc. 1st. Int. Conf. Exp. Fluid Mech., Chengdu, China, 1992, p. 505.

[6] M. Yamada, Y. Uematsu, R. Sasaki, EECWE'94, Warsaw, Poland, 1994, p. 223.

[7] Y. Uematsu, M. Yamada, H. Higashiyama, T. Orimo, J. Wind Eng. Ind. Aerodyn. 44 (1992) 2289.

[8] H. Saeki, R. Sasaki, Y. Uematsu, M. Yamada, Summary Papers of AIJ Annual Meeting, (1996) B-1, 277 (in Japanese).

Journal of Wind Engineering
and Industrial Aerodynamics 67&68 (1997) 885–893

Some remarks on the validation of small-scale dispersion models with field and laboratory data

Michael Schatzmann*, Stilianos Rafailidis, Michel Pavageau

Zentrum für Meeres- und Klimaforschung, Meteorological Institute, University of Hamburg, Bundesstrasse 55, D-20146 Hamburg, Germany

Abstract

The objective of the paper is to contribute to the discussion on appropriate procedures for the evaluation of numerical models. This is done using as an example micro-scale atmospheric dispersion models as they are commonly applied for the prediction of local mean concentrations and higher percentile values. The paper starts with a description of problems associated with the direct comparison of numerical model results and measured data. After some remarks on numerical solutions to the problem, it concentrates on experimental strategies suitable for assuring the quality of validation data, and for quantifying systematic differences between simulated and experimental results.

Keywords: Model evaluation; Dispersion models; Local scale; Wind tunnel data; Field data

1. Introduction

The emergence of more and more sophisticated numerical models is paralleled by an increasing public awareness that many of these models have never been put through a procedure of careful evaluation, but nevertheless are used to assist in making decisions with profound political and economic consequences. Only recently have scientific organisations, like the German Engineering Society (VDI), or authorities like the Commission of the European Communities (CEC), responded to this unsatisfactory state of affairs by issuing some guidance on how a proper model evaluation should be made.

In accordance with the 'Guidelines for Model Developers' released by the CEC's Model Evaluation Group [1], the evaluation of a numerical model comprises

* Corresponding author.

PII S0167-6105(97)00126-8

basically three parts: a thorough model description, model verification and model validation.

The *model description* comprises a user-oriented and scientific assessment. It provides the user with sufficient information on the concepts followed during the model development, the application of the model, its performance objectives and inherent limitations. It contains a comprehensive description of the model content, a list of assumptions and parameterisations applied during model development and clear documentation on how the model must be implemented and applied. It should provide any information necessary to enable the user to judge whether the model meets his requirements or not.

During the *model verification* stage, the developer should prove that his code is producing output in accordance with the model specification, for example, that mass and momentum are conserved, etc. Verification is a substantial part of the model development phase and requires comprehensive detailed documentation (quality control).

Model validation, finally, deals with the comparison of results from the model with data sets. The term 'data sets' here comprises not only experimental data. It also includes results from simple test cases for which an analytical solution may be derived. Data-bases have to be selected so that they are suitable to test either the various parts or parameterisations of a given model, or the model as a whole. The particular choice of a certain data set needs to be justified, the uncertainties in the model and the data need to be estimated. Conclusions must be drawn and recommendations concerning the proper scope of model applicability must be given.

A more detailed evaluation strategy, which goes beyond this three-step procedure and which would be universally applicable to any model, is probably unachieveable. The models are simply too diverse with respect to scale (micro- or meso-scale), content (Gauss-models, 3-D hydrodynamic models) or field of application (pedestrian comfort or accidental release models). Even more critical is that certain strategies for evaluating the quality of a model can only in general be based on scientific principles (e.g. principle of falsification). Models which fail a validation test may be regarded as invalid whereas a model which passes a given test is not necessarily valid. Which particular tests and which model/database comparison should be done for a given model type can only be based on consensus built-up within the scientific community (see, e.g., Ref. [2], which contains suggestions for the evaluation of meso-scale models).

2. Problems associated with model validation

If we follow the definitions given in Section 1, the validation of a model is certainly the most difficult part of the evaluation process. To simply compare model results with measured data is often inappropriate since data generated in field or laboratory experiments and those from model simulations exhibit systematic differences. This is demonstrated in Fig. 1, for the example of a small area source which continuously discharges a passive tracer into a street canyon. Shown are the traces of concentration versus time (in excess above background) at the same receptor point and under

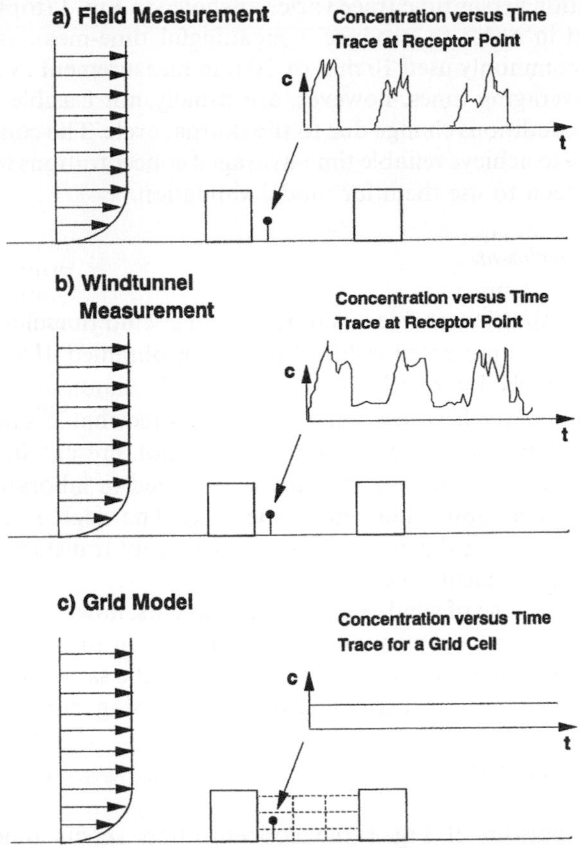

Fig. 1. Comparison of concentration versus time traces typical for field measurements (top), wind tunnel measurements (centre) and numerical model results (bottom) (excess concentrations only).

identical steady-state ambient conditions as they might be found in (a) a field experiment, (b) in a wind-tunnel experiment, or (c) in a numerical simulation with full turbulence parameterisation.

2.1. Field experiments

High-resolution field measurements provide usually highly intermittent signals, i.e. periods of near zero concentration are interspersed with non-zero fluctuating concentrations. It is to be expected that the intermittency of the signal depends largely on the turbulence structure within the canyon and the wind direction fluctuations. In general, the source dimensions relative to the cross-sectional area of the canyon and the distance between the source and the receptor point should also influence the intermittency (since there will be hardly any intermittency anymore when the concentration is uniformly dispersed over the whole cross section).

If the concentration versus time trace varies as shown in Fig. 1 (top), long averaging times are required in order to produce a meaningful time-mean-value. It is to be expected that the commonly used 10 min or 30 min measurement cycles are not long enough. Longer averaging times, however, are usually not feasible since the atmospheric boundary conditions change due to the diurnal cycle. The conclusion is that it appears impossible to achieve reliable time-averaged concentrations in field situations as described, and then to use them for model validation.

2.2. *Laboratory experiments*

When the same dispersion problem is modelled in a wind tunnel or water channel, the concentration signal presented in Fig. 1 (centre) is obtained. If all main similarity parameters were matched properly in the small-scale simulation, the time series should resemble that of the field test, but it would be somewhat less intermittent since the low frequency wind directional variations are not present in a ducted flow. Therefore, time mean concentration maxima determined in laboratory experiments are usually larger than those obtained in the field. The degree of overestimation depends again on the source dimensions, the source/receptor distance and the turbulence structure of the ambient flow.

An important advantage of wind-tunnel measurements, however (in comparison to field tests), is that the boundary conditions can be chosen to be appropriate to the problem being solved, and that numerous repetitions of the same case can be made in order to determine the inherent variability of the dispersing cloud characteristics.

2.3. *Numerical model results*

Finally, at the bottom of Fig. 1 the concentration versus time trace, as obtained from a common grid model, is displayed. Since the model considers turbulent fluctuations only in parameterised form, than with constant boundary conditions it delivers a stationary concentration value. In contrast to the point-by-point experimental data, this value represents not only a time-mean but also a space-mean concentration representative of the characteristics of the whole volume of the grid cell. In zones with large concentration gradients, small grids are required in order to obtain results independent of the actual spatial resolution. Since time-step size is linked to grid size, the demand for storage and computer time required for dispersion calculations within the canopy layer often becomes so large that even the most advanced supercomputers are unable to handle the problem anymore.

Validating conventional grid models entails a comparison of their results with field or laboratory data. In view of the remarks made above, it must be concluded that such a comparison often resembles the proverbial comparison of apples with oranges.

The problems associated with the validation of dispersion models are not at all new. Wilson [3] gives reason to believe that mean concentrations from field experiments can vary by more than a factor of 2 and peak values by more than a factor of

10 over an ensemble of identical releases. Hanna [4], came to the conclusion that, due to errors in the data base alone, a significant level of uncertainty exists in predicting mean concentrations even if the model has been set up in a very sophisticated way.

3. Numerical solutions to the problem

In order to overcome these problems, several approaches have been followed in the past. On the numerical side, models of different complexity were developed which allow the computation of concentration time series and therefore allow a direct comparison of the statistical properties of measured concentration versus time traces.

Most prominent among them are large eddy simulation models (LES) [5,6], in which at least the larger structures of atmospheric turbulence are directly simulated and only the small scale turbulence is parameterised. Then instantaneous puffs of a tracer are released and followed through the velocity field. The results are encouraging, above all, under unstable stability-conditions in which the eddies are large. The method is, however, computationally expensive. The generation of concentration time series in realistic environments, such as cities or industrial complexes, seems to still be out of reach due to a lack of sufficient computational power. The more common solution, turbulence kinetic energy closure models (TKE) employ empirical formulations for the turbulent transport properties. Since these are fitted to data which contain all the shortcomings described in Section 2, the quality of the model output is similarly limited.

Another approach, which can deliver concentration fluctuations, are the so-called particle models. There are a large number of Lagrangian models in use (see, e.g., Ref. [7]) for which the range of applicability can be extended to concentration fluctuation predictions when particle pairs correlated with each other are released [8]. Broadly equivalent to large eddy simulation in the field of particle models is the group of kinetic simulation particle models (KSP) [9]. Here, in utilizing information from power spectra of atmospheric boundary layer turbulence, large-scale eddy motions are superimposed on the mean flow which then influence the path of individual particles. The quality of Lagrangian as well as of KSP models depends largely on the quality of the meteorological pre-processor employed to calculate the mean and statistical properties of atmospheric turbulence. In general, these pre-processors take obstacles like buildings, trees, etc. into account only within the roughness length parameterisation. This restricts the range of applicability of this group of models to above-canopy-layer dispersion.

The same statement can be made for meandering plume models [10] which are usually based on Gaussian concentration profiles. They produce closed-form analytic solutions and are often applied in a regulatory context. A comprehensive review of these models is given in Ref. [3]. Like most other types of models discussed here, meandering plume models depend heavily on empirical input, i.e. accurate estimates of turbulent time or length scales which, according to literature, show a high degree of variability in existing measurements.

4. Experimental solutions to the problem

On the experimental side, strenuous efforts have been made to develop, first of all, concentration measurement instrumentation with sufficient resolution in space and time (see, e.g., Ref. [11]). Then, concepts were developed to derive meaningful statistical plume properties from concentration time series. These comprise mean and fluctuation intensities determined from conditional (zero concentration intervals excluded) and unconditional (zero concentration intervals included) concentration sampling which provides the basis for the determination of an intermittency factor [12] and other unambiguous properties of a time series.

Statistical analysis of time series from field experiments confirms the expectation that long series (or large numbers of repetitions, in case of instantaneous releases) are needed to obtain reliable mean concentrations and higher percentile values which could be used with confidence for model validation purposes. The fact that the atmosphere is not usually stationary enough to provide sufficiently long-time intervals with constant conditions, leads to the conclusion that the validation of micro-scale atmospheric dispersion models must be based predominantly on laboratory data. However, existing wind tunnel or water channel measurements are often poorly documented or of unknown quality. Therefore, in the following, some suggestions are made as to how the quality of future laboratory data should be assured, so that the potential of small-scale experiments be best utilised and data sets really suitable for model validation purposes are generated.

4.1. Quality assurance of laboratory data

The experimental program must be based on a thorough dimensional analysis which shows the complete set of similarity parameters influencing the dispersion and provides the basis for subsequent scaling-up of the model results to prototype or real scale. A detailed discussion is required showing which of the similarity parameters were matched and which had to be relaxed or distorted during the small-scale experiments. The consequences of any such parameter distortion need likewise be discussed.

The quality of the wind-tunnel boundary layer should be demonstrated through comparison of vertical profiles of mean velocity, turbulence intensities and cross-correlations (e.g. $u'w'$) with those of the atmosphere. In addition, turbulence power spectra taken at 2 or 3 different heights above ground should be compared with those of the atmosphere.

The measurement and data acquisition techniques must be defined properly and their appropriateness for the purpose proven (e.g. is sufficient frequency response provided, etc.). Selected measurements should be repeated frequently to determine the inherent variability between individual measurements, due to limitations of the instrumentation, non-uniformity in tunnel operation or other factors.

Finally, Reynolds-number independence of the data must be proven through appropriate tests such as the release of an inert tracer from a point source under varying wind velocities and the determination of the velocity range over which the concentration field is inversely proportional to the mean wind velocity (see Ref. [13]).

4.2. Utilisation of the data sets

Since laboratory simulations suffer from their own limitations, trends rather than absolute quantities should be compared with model results. Trends in the data can be found through experimental programs within which one particular similarity parameter is changed over the parameter range of interest whilst all other parameters are kept constant (see Ref. [14]). A good quality numerical model should reproduce the trends found in the laboratory experiments. Certain differences must be expected nevertheless, due to the different nature of the processes. For example, it is normal that the absolute centre-line concentrations of a plume are underpredicted by the numerical model, since in the wind tunnel plume low-frequency meandering is suppressed through the presence of the walls.

4.3. Analysing field tests in the laboratory

In many cases the wind tunnel may also assist by either helping analyse field observations more comprehensively, or by refining field studies to adapt them to the needs of particular model-validation exercises. A recent such example was the enhancement through wind tunnelling of field results form a monitoring station located in a heavy traffic canyon-like street in Hannover, Germany (Göttinger Straße). The field measurements comprised all information needed for model validation purposes. The source was defined through automatic monitoring of the traffic density which, by means of car-type specific emission factors and knowledge of the actual composition of the German vchicle fleet, could be used to obtain a good estimate of the source strength. The meteorological parameters were monitored both above roof level and inside the canyon. The pollutant measurements included several traffic-related substances, among them the sum of nitrogen oxides (NO_x) which may be used as a sort of passive tracer for the pollution dispersion. The background concentration outside the canyon was also continuously monitored.

At Hamburg University, a scale model of the site and its surroundings will be studied in a neutrally stratified boundary layer wind tunnel. The concentration fluctuations, at the corresponding measurement position as in the field, will be monitored with a fast-flame ionisation detector with a frequency response up to 200 Hz. The measurements will be repeated for 36 wind directions and several wind velocities. The source will be modelled as a uniformly and continuously discharging line source. The experimental program has the following objectives:

1. Investigation of the natural variability of 10 and 30 min time-averaged mean concentrations (the times correspond to field measurement conditions). This will

be achieved through repetition of the same measurement under constant source and boundary conditions at averaging periods scaled down to wind-tunnel time equivalents. Comparison of these results with those from quasi-infinite averaging will also be performed.
2. Investigation of the effect of vehicle-induced turbulence on concentration measurements. This will be done through a repeat of several of the measurements described before, but this time under the additional influence of moving traffic represented by small moving plates scaled in size and velocity according to the pertinent similarity laws.
3. Investigation of the effect of insufficient spatial resolution on numerical model results. Many models use Cartesian grids which limit their ability to resolve finely enough the obstacles and the flow field. This will be studied with a third series of measurements within which, instead of the detailed scale model, a model containing only rectangular blocks matched to an appropriately chosen Cartesian grid will be used. The results will be compared with those of the detailed model.

The above example is just one illustration of how to provide combined field and laboratory data banks for improved model validation. However, it should be clear from this example that there are ample opportunities for proper validation, particularly of models of micro-scale (=local) phenomena. This is less true at larger scales which, due to Reynolds-number effects, are usually beyond the scope of laboratory experiments, and for which the validation of a model must be based on field data alone.

Acknowledgements

The authors are grateful for financial support from Projekt Europäisches Forschungszentrum, Karlsruhe.

References

[1] Model Evaluation Group (MEG), Guidelines for model developers, Commission of the European Communities, DG XII, 1994.
[2] K.H. Schlünzen, On the validation of high-resolution atmospheric mesoscale models, these Proceedings, J. Wind Eng. Ind. Aerodyn. 67&68 (1997) 479–492.
[3] D.J. Wilson, Concentration fluctuations and averaging time in vapor clouds, Center for Chem. Process Safety of the Amer. Inst. of Chem. Eng., New York, 1995.
[4] S.R. Hanna, Uncertainties in air quality model predictions, Bound.-Layer Meteorol. 62 (1993) 3–20.
[5] H. Schmidt, U. Schumann, Coherent structures of the convective boundary layer derived from large-eddy simulations, J. Fluid Mech. 200 (1989) 511–562.
[6] A. Andren, A.R. Brown, J. Graf, P.J. Mason, C.-H. Moeng, F.T.M. Nieuwstadt, U. Schumann, Large-eddy simulation of a neutrally stratified boundary layer: A comparison of four computer codes, Q. J. R. Meterol. Soc. 120 (1994) 1457–1484.
[7] L. Janicke, W.J. Kost, R. Röckle, Modelling of motor vehicle emissions in a street system by combination of Lagrange models and surface wind field simulation in complex city structures, in: J.E. Cermak et al. (Eds.), Wind Climate in Cities, Kluwer Academic Publishers, Dordrecht, NL, 1995.

[8] D.J. Thompson, A stochastic model for the motion of particle pairs in isotropic high Reynolds-number turbulence and its application to the problem of concentration variance, J. Fluid Mech. 210 (1990) 113–153.
[9] R.J. Yamartino, D. Strimaitis, A. Graff, Evaluation of the kinematic simulation partical model using tracer experiments, in: Proc. 4th Workshop on Harmonisation within Atmos. Disp. Mod. for Regul. Purposes, Oostende, Belgium 1 (1996) 59–66.
[10] B.M. Bara, D.J. Wilson, B.W. Zelt, Concentration fluctuation profiles from a water channel simulation of ground-level-release, Atmos. Environ. A 26 (1992) 1053–1062.
[11] K.R. Mylne, P.J. Mason, Concentration fluctuation measurements in a dispersing plume at a range of up to 1000 m, Q. J. R. Meteorol. Soc. 117 (1991) 177–206.
[12] D.J. Wilson, A.G. Robins, J.E. Fackrell, Intermittency and conditionally-averaged concentration fluctuation statistics in plumes, Atmos. Environ. 19 (1985) 1053–1064.
[13] M. Schatzmann, W.H. Snyder, R.E. Lawson, Experiments with heavy gas jets in laminar and turbulent cross flows, Atmos. Environ. A 27 (1993) 1105–1116.
[14] R.N. Meroney, M. Pavageau, S. Rafailidis, M. Schatzmann, Study of line source characteristics for 2-D physical modelling of pollutant dispersion in street canyons, J. Wind Eng. Ind. Aerodyn. 62 (1996) 37–56.

New Computational Schemes

Journal of Wind Engineering
and Industrial Aerodynamics 67&68 (1997) 897–908

Turbulence closure model "constants" and the problems of "inactive" atmospheric turbulence

Marcel Bottema[a,b,1]

[a] *Laboratoire de Mécanique des Fluides, URA CNRS 1217 & SUB-MESO Groupement de Recherches CNRS, 1102, École Centrale de Nantes, B.P. 92101, F-44321, Nantes Cedex 3, France*

[b] *Department of Physical Geography, University of Groningen, Kerklaan 30, 9751 NN Haren, The Netherlands*

Abstract

Inactive turbulence is associated with waves and large eddies that are relatively ineffective in mixing. Many numerical models evaluate turbulent mixing using turbulent kinetic energy k, which may contain significant amounts of inactive turbulence (e.g., in real or simulated atmospheric boundary layers). Inactive turbulence as an unresolved phenomenon may yield case and even location-dependent model constants. Neglecting inactive turbulence (approach "A" and "B" in this paper) often gives unsatisfactory results, but accounting for inactive turbulence by tuning the model to the approach flow properties (approach "C") gives only partial improvement for isolated obstacle flows. Problems with location-dependent model constants occur especially in the range of 0–3 obstacle heights, the roughness sub-layer (RSL). Within the RSL, *measured* turbulence data suggest that k–ε and ASM model constants should be location dependent. Sensitivity tests suggest that the problem is sufficiently serious to consider an alternative approach by either zonal modelling or explicit modelling of inactive turbulence effects. Finally, an attempt is made to find the "hidden physics" and to explain the above model non-universality.

Keywords: Anisotropy; Roughness sub-layer; Universality; k–ε model; ASM; CFD

1. Introduction

A really universal turbulence model is a dream and will be a dream possibly for ever [1]. This is probably due to the fact that each turbulence model has its unresolved physics. For example, anisotropy is a well-known unresolved feature of the k–ε model with its scalar (isotropic) eddy viscosity [2]. This paper will discuss the part of

[1] Corresponding address: Department of Physical Geography, University of Groningen, Kerklaan 30, 9751 NN Haren, The Netherlands. E-mail: m.bottema@biol.rug.nl.

PII S0167-6105(97)00127-X

anisotropy which is related to "inactive turbulence"; a feature that is often neglected in computational wind engineering practice.

The concept of inactive turbulence originates from Townsend [3], who noticed that different boundary-layer profiles of turbulent variances did not collapse on one curve, as opposed to boundary-layer stress profiles ($u'w'(z)$). He explained this by assuming that the different boundary layers had various amounts of "inactive turbulence" (turbulence inactive in mixing). Inactive turbulence is generally related to gravity waves, or to low-frequency (large-scale) contributions caused by for example, "thermals" and upstream topography [4,5]. In the present paper, we will define and use an *inactive turbulence parameter* $c_k = k/u_*^2 = k/(-u'w')^{0.5}$; c_k^{-1} indicates "the mixing efficiency of turbulence" [6,7]. The relevance of inactive turbulence to several Reynolds-averaged Navier–Stokes (RANS) models becomes clear if we realise that most of these models use turbulent variances or turbulent kinetic energy k without specifying a spectral range, and without separating between, e.g. "active" turbulence and "inactive" wave action.

The aim of this paper is to show that inactive turbulence as an unresolved phenomenon may yield case-dependent and even location-dependent model constants, and significant uncertainties in computational results. Another aim is to try to identify the hidden feature (instead of just naming it "inactive turbulence") which should be a first step towards future improved modelling techniques.

Although "inactive turbulence" seems to be relevant to several turbulence closures, we will, for the sake of simplicity, focus on the k–ε model and some of its variants and alternatives. In Section 2, the equations of the k–ε model and related models will be tuned for the case of boundary layers; in this way it will be shown that the model constants depend on c_k. Section 3 will combine new experimental results of roughness sub-layer (RSL) research [8] with the model equations; it will be shown that location-dependent model constants would be needed to follow experimental trends. Finally, the implications of these model constant uncertainties will be further explored in Sections 4 and 5.

Tuning the model constants to the approach flow c_k is not the only possible approach. Modelling approaches discussed in this paper are:

(A) standard turbulence model with measured inflow turbulence profiles,
(B) as "A" but with synthetic inflow turbulence profiles (computation of current model); approach flow inactive turbulence is totally neglected,
(C) measured turbulence inflow profile together with tuned model constants; approach flow inactive turbulence is fully accounted for.

2. Tuning of model constants for boundary layers

It is important to have a good approach-flow simulation as obstacle flows are sensitive to approach-flow conditions (see, e.g., Refs. [9–11]). Approach-flow simulation requirements for the k–ε model and similar models are outlined below. Generally, the k–ε model is also tuned to decaying grid turbulence and homogeneous shear flow [12]; these tunings will be discussed.

Some measured relations [4] for the neutrally stratified surface layer are

$$U(z) = \frac{u_*}{\kappa} \ln\left(\frac{z - z_\mathrm{d}}{z_\mathrm{o}}\right), \tag{1a}$$

$$\nu_t = u_* \kappa (z - z_\mathrm{d}), \tag{1b}$$

$$-\frac{\overline{u'w'}}{u_*^2} = 1, \tag{1c}$$

$$\varepsilon = P_k = \frac{u_*^3}{\kappa z}, \tag{1d}$$

$$k \equiv \tfrac{1}{2}\sigma_{ii} = c_k u_*^2, \tag{1e}$$

$$\frac{\sigma_u}{u_*} = 2.4, \tag{1f}$$

$$\frac{\sigma_v}{u_*} = 1.9, \tag{1g}$$

$$\frac{\sigma_w}{u_*} = 1.25, \tag{1h}$$

where u_*, z_o, z_d, κ, ν_t, k and ε represent: friction velocity, roughness length, zero displacement height, the Von Karman constant (0.4), eddy viscosity, turbulent kinetic energy and the dissipation rate of k; P_k is the production rate of k. In an undisturbed, neutral surface layer, c_k is 5.5–6; with upstream topography, or in thermally unstable conditions c_k may well exceed 10; large c_k values correspond to large relative amounts of "inactive" turbulence.

We can tune the k–ε model for boundary-layer flows by inserting Eqs. (1b), (1d) and (1e) into the model relation for turbulent viscosity, Eq. (2a), which yields Eq. (2b) [7,12,13],

$$\nu_t = C_\mu \frac{k^2}{\varepsilon}, \tag{2a}$$

$$C_{\mu\mathrm{SL}} = \frac{u_*^4}{k^2} = c_k^{-2}. \tag{2b}$$

The resulting "surface layer value" ($C_{\mu\mathrm{SL}}$) of C_μ is 0.03. The standard C_μ value is 0.09 [14], corresponding to a rather low c_k value of 3.3. This low c_k value originates from Bradshaw et al. [15], who reported a large scatter in c_k due to inactive turbulence, and finally took a laboratory value at the lower end of the measured range in order to discard the influence of "inactive" turbulence.

The $C_{\varepsilon 2}$ constant can be evaluated for homogeneous, decaying-grid turbulence [12]; one finds $C_{\varepsilon 2} = 1.83$ [12] to 1.92 [14]. The relation between $C_{\varepsilon 1}$ and $C_{\varepsilon 2}$ can be determined for homogeneous shear flow, which leads to $C_{\varepsilon 1} \approx 1.44$ [14] (1.48 in Ref. [12]). Also, one can relate $C_{\varepsilon 1}$ to the critical Richardson number [12]. The Prandtl number for ε, σ_ε, is evaluated by imposing $\partial\varepsilon/\partial x = 0$. The dissipation equation [14,16]

$$\frac{\partial \varepsilon}{\partial t} + U_j \frac{\partial \varepsilon}{\partial x_j} = \frac{\partial}{\partial x_j}\left(\frac{\nu_t}{\sigma_\varepsilon}\frac{\partial \varepsilon}{\partial x_j}\right) + \frac{\varepsilon}{k}(C_{\varepsilon 1} P_k - C_{\varepsilon 2}\varepsilon) \tag{3a}$$

can be combined with Eqs. (1b), (1d) and (1e) which leads to [12,13]

$$\sigma_\varepsilon = \frac{\kappa^2}{(C_{\varepsilon 2} - C_{\varepsilon 1})\sqrt{C_\mu}}. \tag{3b}$$

The proposed σ_ε-values depend strongly on C_μ and vary from 1.3 [14] to 2.38 [12], but model results were quite insensitive to σ_ε [9]. No closure relation is known for σ_k. A $\sigma_k = 1.0$ is often used [12,14]. Chen and Kim [17] note that σ_k should be less than 1 (they use $\sigma_k = 0.75$) because boundary-layer edges for k extend wider than those for mean velocity. Also, realisability implies [17] that near the boundary-layer edges, ε should decrease quicker than k, which implies $\sigma_k < \sigma_\varepsilon$.

Before discussing the tuning of other turbulence models, it is interesting to note the better performance of tuned model predictions (approach "C") compared to especially approach "A". See Table 1 and Refs. [7,9,18,19]. Approach "A" yields even significant deviations (10% in U and 30% in k after 4 m of "wind tunnel fetch") for an empty domain test carried out before the simulations of Table 1.

A well-known k–ε model deficiency is the turbulence overproduction in front of an obstacle. Model tuning (case of Table 1, $C_{\mu\mathrm{SL}} = 0.03$ by Eq. (2b), for $c_k = 5.6$) yields at least mitigation of the effect: $k/k_o \approx 1.3$ near the stagnation point (k_o is the upstream value); k/k_o was similar for approach "B" and much higher ($k/k_o \approx 2$) for approach "A". k/k_o was also large when instead of σ_ε, $C_{\varepsilon 1}$ was adapted by Eq. (3b).

Next, we will briefly discuss the tuning of the three variants of the k–ε model. The first two variants [17,20] aim to reduce turbulence overproduction near stagnation points by an extra term in the dissipation equation [17], or by a modification in the Boussinesq relation [20]. For the boundary layer without obstacle (Eqs. (1a)–(1h))

Table 1

Pressure coefficients C_p (referred to undisturbed roof height wind speed) averaged over front and lee wall, for a wide obstacle with $W/H = 25$, $L/H = 0.5$, $H/z_o = 300$, normal flow, $c_k = 5.6$, blockage: 5% [7,9]

	$C_{p,\mathrm{front}}$	$C_{p,\mathrm{lee}}$	ΔC_p
Experiment	0.42	− 0.52	0.94
Approach "A": standard k–ε model (2-D)	0.53	− 0.89	1.42
Approach "C": k–ε model, $C_\mu = 0.03$ (2-D)	0.35	− 0.78	1.13
Approach "C": k–ε model, $C_\mu = 0.03$ (3-D)	0.38	− 0.58	0.96

both satisfy Eq. (2b), so that the "tuned" C_μ depends on the flow type. The same result is found for the non-linear k–ε model of Ref. [22]. Also its non-linear anisotropic term is proportional to his "C" $= 2C_\mu$ so that an increase of c_k, and thereby a decrease of C_μ, would yield less anisotropy, the opposite of what is expected.

Next, we consider the algebraic stress model (ASM) [16]

$$\overline{u_i u_j} = k\left[\frac{2}{3}\delta_{ij} + \frac{(1-\gamma)((P_{ij}/\varepsilon) - \frac{2}{3}\delta_{ij}(P_k/\varepsilon))}{c_1 + (P_k/\varepsilon) - 1}\right], \tag{4}$$

where γ, and c_1 are model constants (≈ 0.55 and 1.5–2.2 [16]). P_{ij} is a shear stress production term ($P_{ij} = -u_i' u_m' \partial U_j/\partial x_m - -u_j' u_m' \partial U_i/\partial x_m$); $P_k = P_{ii}$. Using the relations of Eqs. (1a)–(1h) and writing $w'^2/u'w'$ as c_w^2 (≈ 1.56) yields

$$\overline{u'w'} = -c_w^2 \frac{(1-\gamma)(P_k/\varepsilon)}{c_1 + (P_k/\varepsilon) - 1}, \tag{5a}$$

$$\frac{1}{c_k} = \text{“}C_\mu^{0.5}\text{”} = -c_w^2 \frac{(1-\gamma)}{c_1}. \tag{5b}$$

For the last step, P_k/ε was set to 1 (see Eq. (1d)) and $u'w'/k$ was interpreted as c_k^{-1} (or in k–ε model terms as "$C_\mu^{0.5}$"). Note that Eqs. (5a) and (5b) contain *two* "free" constants that need tuning. The dissipation equation constant C_ε [22] can be tuned by imposing $\partial\varepsilon/\partial x = 0$. The resulting equation is similar to Eqs. (3a) and (3b), but with $\sigma_\varepsilon = (c_k C_\varepsilon)^{-1}$, and $C_\mu =$ "C_μ". Although in many engineering flows, ASM performs better than the k–ε model, ASM seems not to be free of problems in (simulated or real) atmospheric flows. Firstly, ASM predicts $\sigma_v = \sigma_w$ which is not in accordance with experimental evidence (Eqs. (1g), (1h), [4]). Also, it can be shown that tuning to a larger c_k yields an unexpected *decrease* in predicted anisotropy.

Tuning of second-order models [2] is still more complicated. In the variances (σ_i^2), no distinction is made between "active" and "inactive" parts (scales), so that here, re-adjustments of model constants may be needed too.

Model tuning also affects wall functions [14] since, for the k–ε model, they use C_μ to convert k (by Eq. (2b)) to a $u_{*\mathrm{turb}}$ that is used by the wall function (which defines the wall stress τ_w as $\rho u_{*\mathrm{prof}} u_{*\mathrm{turb}}$; see Ref. [14]). The use of wall functions may cause grid resolution problems because the logarithmic profile used to evaluate $u_{*\mathrm{prof}}$ is only allowed (accurate) well above the roughness elements ($z_{\min} > 20z_0 + z_d$ [23]).

This is one reason why it is tempting to abandon wall functions altogether, and to go right down to the viscous sub-layer. For rough walls this implies explicit modelling of all roughness elements which generally is not feasible; in practice, one must often make a choice between no roughness and good grid resolution, or correct roughness with less grid resolution.

The perspective of advanced and less restrictive rough wall treatment is one reason for investigating turbulence properties in the roughness sub-layer (RSL), the rough wall equivalent of the buffer layer. Another reason for RSL research is the amount of buidings (probably most of them) within the RSL. However, turbulence modelling within the RSL is far from trivial, as will be shown below.

3. Turbulence modelling – the roughness sub-layer

The roughness sub-layer (RSL) is the layer between the logarithmic (surface) layer, and the viscous sub-layer on the roughness elements and the ground surface. Its depth, h_{RSL}, varies between 2 and 5 obstacle heights [8]; Ref. [24] suggests a typical $h_{\text{RSL}} = 2h$–$2.5h$; the link between h_{RSL} and the surface geometry has not been fully clarified yet.

In recent years, understanding of RSL-turbulence over vegetation, and to a lesser extent over buildings, has considerably increased [8]. A striking feature is the decrease of the variances and of the "inactive" turbulence: near the obstacle tops, $\sigma_u/u_{*\text{SL}}$, $\sigma_v/u_{*\text{SL}}$ and $\sigma_w/u_{*\text{SL}}$ (the subscript SL indicating the surface layer values) may decrease to 1.5, 1.3 and 1.0 [8]; notice that "$u_{*\text{RSL}}$" $((-u'w')^{0.5})$ remains close to $u_{*\text{SL}}$ [8]. This yields a c_k of 2.5 (surface layer value: 5.5–6); a similar value is given in Ref. [25]. These low c_k values suggest that *different C_μ values might even be needed within one computational domain* (see also below), thus indicating some unresolved phenomena.

The parameters that are needed for evaluating C_μ or its (ASM-) equivalent(s) are: k, ε, and ν_t (Eq. (2a)). Eddy viscosity ν_t is roughly constant and equal to $\kappa u_*(h_{\text{RSL}}\text{-}z_d)$ [8] instead of $\kappa u_*(z - z_d)$, but data in Refs. [25,26] suggest that the $\nu_{t,\text{RSL}}$ does not exceed 2.5 times the extrapolated surface layer value. k/u_*^2 or $k/(-u'w')$ is generally lower than the surface layer value, with a lower bound of about 2.5. The only (approximate) dissipation rate data that could be found [26] suggest that for $z/h < 1.5$, ε is a factor 2–2.5 lower than the extrapolated surface layer value of Eq. (1d); the ratio P_k/ε remains close to 1 without apparent trend. The above results can be summarised as

$$\frac{C_{\mu,\text{RSL}}}{C_{\mu,\text{SL}}} = \frac{R_{\nu_t} R_\varepsilon}{R_k^2}, \quad R_k = \frac{k_{\text{RSL}}}{k_{\text{SL}}}, \quad \text{etc.,}$$

$$R_{\nu_t} \approx \max\left(\frac{h_{\text{RSL}} - z_d}{z - z_d}, 2\right), \quad 0.5 \leqslant R_\varepsilon \leqslant 1, \quad 0.5 \leqslant R_k \leqslant 1. \tag{6}$$

The above results imply that well into the RSL, C_μ probably is about 4 times as high as the surface layer value, i.e. 0.12 instead of 0.03. The exact value is still uncertain due to the small number of data sets available, and because of experimental problems associated with RSL measurements. For the same reasons, no firm conclusions can yet be drawn on the RSL tuning of other model constants than C_μ.

The results of Eq. (6) also apply to the variants [17,20,21], provided that P_k/ε remains close to 1 (for Ref. [17]) and that U and $\partial U/\partial z$ remain the main flow terms (for Refs. [20,21]). The latter is expected for skimming flow, and at some distance of the roughness elements. Even for the "medium density" strips of Ref. [26], the horizontal average of UW is negligible compared to $u'w'$ for heights in excess of $1.1h$, suggesting that $\partial U/\partial z$ remains much larger than the other gradients.

For ASM, the same assumptions yield once more Eqs. (5a) and (5b). However, k_{RSL}/u_{*SL}^2 is much smaller than k_{SL}/u_{*SL}^2, and generally, $(-u'w')_{RSL}$ is close to u_{*SL}^2 [8]. Then for ASM too, surface layer model constants are different from RSL constants.

Apparently, the model "constants" do not only depend on the flow type, but also on the flow region within a computational domain. The importance of these non-universality's, and of the unresolved physics which seems to be their cause, will be discussed in the next section.

Finally, spatially variant model constants may also be needed for isolated obstacle flows [7,18,19], where a tuned $C_{\mu SL} = 0.03$ gave good results for near wake flows, but yielded too slow a flow recovery in the far wake. This is confirmed by results of approach "B" where both the ratio and the difference of k_{model} and $k_{meas.}$ were far from constant. Streamline curvature may partly explain this as accounting for it Ref. [27] would yield a locally increased C_μ in the (near) wake.

4. Validation tests for the roughness sub-layer

In the previous section, it was suggested that different model constant(s) (tunings) should apply for the surface layer and for the RSL. A zonal approach [28] does not yet seem feasible by lack of sufficiently detailed data, and because the *reasons* for the spatially variant C_μ are not yet fully understood. All that can be done now is to compare different simulations with each their different but spatially invariant C_μ value. Thereby, it may become clear whether an alternative (e.g., zonal) approach is justified by a sufficient increase in accuracy.

Early validation tests [7,18,19] for skimming flow over 6 quasi-two-dimensional street canyons with a street width over building height ratio of 2 show that wind speeds in (and mixing into) the streets are underestimated by a modified k–ε model with $C_{\mu SL} = 0.03$ (Fig. 1). The standard model with $C_\mu = 0.09$ performed better for this case, as opposed to the isolated obstacle case. Notice the quicker flow recovery in the wake for $C_\mu = 0.09$, and the "bumpy" upper profile, indicating increased drag by the buildings. No tests were carried out with RSL-tuned model constants. Also, the test case was not perfect because of the short fetch.

Some further tests were carried with the k–ε model CHENSI [29] of the Laboratoire de Mécanique des Fluides in Nantes. The test case consisted of two-dimensional square bars at 25% plan area density. The computational domain (including the buildings) was $4h \times 18h$ (h being the building height); grid resolution (a non-uniform 42×60 grid with periodic boundary conditions) was similar to that of other tests discussed in this paper. Three values of C_μ and σ_ε were tested: 0.03 and 1.86 (surface layer), 0.09 and 1.30 (standard [14]) and 0.12 and 1.00 (RSL).

Fig. 2 shows the results of the three variants, together with experimental data from [30,31]. There are significant differences in the mean velocity profiles (Fig. 2a); The $C_\mu = 0.03$ reflects a smaller flow penetration into the streets, and a smaller overall roughness; the opposite happens in the other two cases (without however a substantial change in the one-vortex flow regime). The sharpest local velocity gradients are found near roof height, and for the $C_\mu = 0.03$ case.

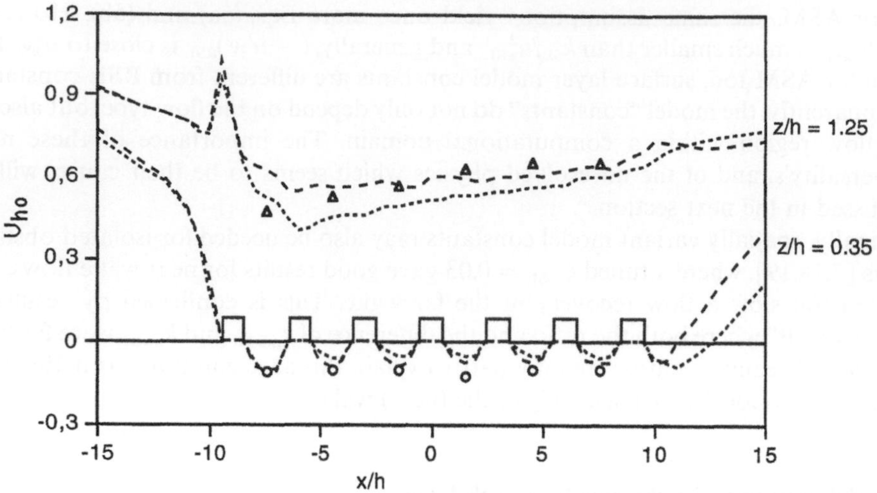

Fig. 1. Computed wind speeds and laser Doppler data (symbols) for wind flow over 7 square bars at 33% plan area density; after Refs. [7,18,19]; long dashed lines: $C_\mu = 0.09$; short dashed lines; $C_\mu = 0.03$; upper data: $z/h = 1.25$; lower data; $z/h = 0.35$. All data are normalised by the approach flow wind speed at roof height U_{h_o}; the approach flow $h/z_o = 8000$.

The $C_\mu = 0.03$ case is the only one for which k remains about constant down to roof height (Fig. 2b). This feature is not confirmed by turbulence data of Ref. [30], suggesting once again that the surface-layer tuned version ($C_\mu = 0.03$) is not appropriate for flow over large building groups.

For the mean profiles, comparison with data yields ambiguous results. The data of Ref. [31] suggest that even for this RSL flow case, $C_\mu = 0.03$ might be a good choice, whereas on the other hand, the data in Ref. [30] suggest optimum agreement for $C_\mu > 0.12$. The difference between these highly similar (fetch, Re-number, measuring technique, etc.) data sets is too large for measuring errors and totally unexpected; it will be discussed in the next section.

We must conclude that the differences between different model versions are significant. However, we cannot yet optimise the choice of C_μ because of unexpected differences between experimental data sets. It remains an open question whether such tuning is sufficient; other unresolved phenomena could, strictly speaking, still play a role (although streamline curvature effects, etc. might be weak for the present skimming flow cases).

A final remark concerns the relations of Eq. (6). Obviously, the model cannot reproduce the spatial variation of C_μ. Yet, the model results show the same overall trends for k, ε and ν_t, although for this particular geometry, most RSL-deviations occur for heights (well) below 1.5 obstacle heights. A final feature worth noting is the spatial average of UW, which for all heights is at least an order of magnitude smaller than (the approximated) $u'w'$, as expected.

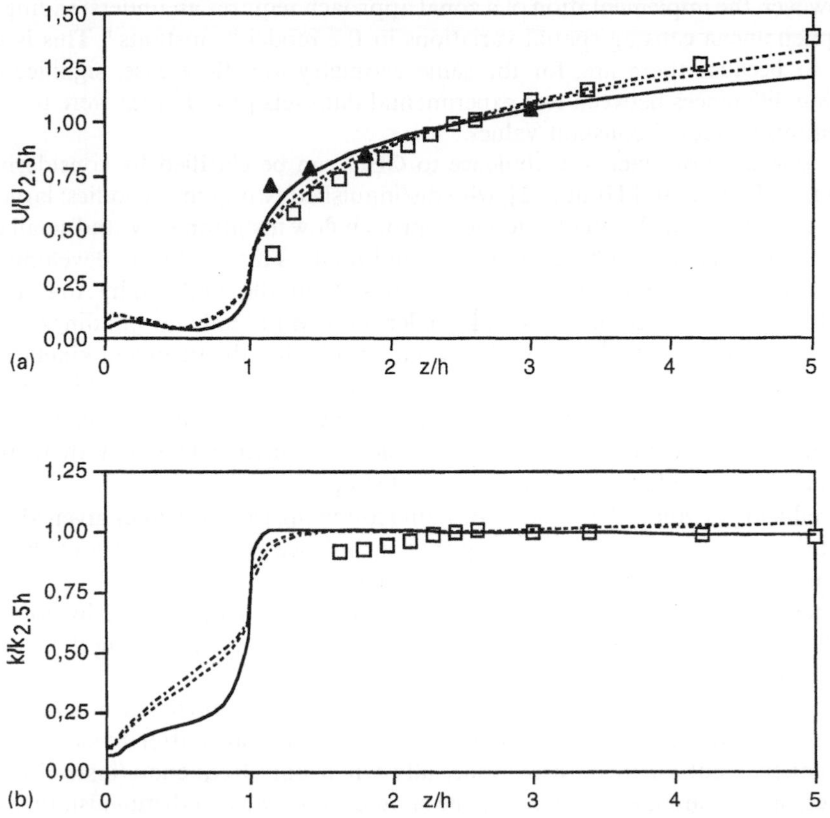

Fig. 2. Computed vertical wind and turbulence profiles (computed results are horizontally averaged) over two-dimensional square bars at 25% plan area density (infinite fetch). All variables are normalised by the values at 2.5 obstacle heights. Solid line: $C_\mu = 0.03$; dashed line: $C_\mu = 0.09$; dot-dashed line: $C_\mu = 0.12$. Experimental data: squares, [30]; triangles, [31], (a) mean wind speed $U(z)/U_{2.5h}$, (b) turbulent kinetic energy $k(z)/k_{2.5h}$.

5. Conclusions and discussion

The results of Section 2 suggest that in undisturbed boundary layers (more precisely, surface layers), the k–ε models and ASM have optimum model constants that depend on the inactive turbulence parameter $c_k = k/u_*^2$; therefore, the model constants are not universal. These results are relevant for isolated obstacle flows [7,18,19] as well.

The results of Section 3 suggest that the optimum value of model constants may even depend on the flow zone: optimum values of the "logarithmic" surface layer (SL) are different from those in the roughness sub-layer (RSL) below (and probably from those in wakes [7,18,19]). The tests of Section 4 (Figs. 1 and 2) with different, but spatially invariant, C_μ values suggest that the computed flow is sufficiently sensitive to C_μ to justify investigating the possibilities for an alternative (e.g., zonal) approach.

However, the implementation of a zonal approach requires an understanding of the flow phenomena causing spatial variations in the model "constants". This is all the more so because there are, for the same geometry and flow case, significant and puzzling differences between two experimental data sets [30,31] that were to be used for optimising model constant values.

The relevance of inactive turbulence to CFD can be clarified by considering the findings of Durbin and Hunt [32], who distinguished two types of eddies: large-scale eddies that are "seen" by an obstacle as approach flow instationarity, and small eddies that influence separation behaviour and wind loads. This led to the development of a small-scale turbulence parameter [33] based on the philosophy that there is a limited spectral range that has strong effects on wind loads on buildings.

The above ideas on "relevant scales" would imply that the turbulent kinetic energy k is only a correct measure of spectral density at the "relevant scales" if the overall shape of the spectrum remains unmodified. If the actual spectrum is different from the original tuning case due to a larger low-frequency ("inactive") content, there are two ways of accounting for the different spectral shapes:

- by reducing C_μ and other model constants following the guidelines given above, so that the extra "inactive" or low-frequency turbulence gives no extra contribution to turbulent mixing (approach "C" of Section 1);
- by neglecting inactive turbulence altogether [15]; this can be done by computing turbulent inflow values from the mean wind profile, instead of using the measured inflow turbulence values (approach "B").

For the "B"-approach, a first problem is related to the turbulence data used for validation: even if spectra are available, it is not a trivial matter to determine the low-frequency cut-off to be used, especially when only frequency (instead of wave number) spectra are measured. Also, one should find a way to distinguish turbulence from waves and general instationarity. These things are important because the experimental data discussed in Section 3 suggest that the fraction of inactive turbulence may depend on the flow region. Since the author used CFD mainly for applications for which turbulence data were needed (pedestrian comfort [7,18] and dispersion), he never fully tested approach "B".

There is also a fundamental difference between the "B" and "C" approach. Bradshaw neglects inactive turbulence altogether by reducing his k. In the present case, only the effect of inactive turbulence on turbulent mixing (Eqs. (2a) and (2b)) is considered, but not *on* the time scale $\tau_\varepsilon = k/\varepsilon$ which appears in the dissipation equation. If "inactive" turbulence is "real" – large-scale – turbulence, there appears to be no reason to modify this time scale. A preliminary calculation for the isolated building test case of Table 1 suggests that neglecting inactive turbulence altogether following Bradshaw [15] yields the following results:

- computed building pressures are almost the same as those of approach "C";
- it is difficult to convert "active" to "total" turbulence levels, because zones with large turbulence generation or destruction seem to have deviating (different) inactive turbulence levels (see last paragraph of Section 3).
- the recirculation zone length is 40% longer than for the "C"-approach, and downstream flow recovery is even slower than for "C". This is understandable

because – in the shear layer above the recirculation zone, and if $P_k > \varepsilon C_{\varepsilon 2}/C_{\varepsilon 1}$ – a reduced k implies an enhanced $U_j \partial\varepsilon/\partial x_j$ (Eq. (3a)) and thereby enhanced downstream dissipation.

Finally, the difference between the measurements [30,31] of Fig. 2 needs to be explained. In view of the similarity (Section 4) of the experiments, the only likely explanations, except for large experimental errors, can be

- even a fetch of $300h$ (the fetch used in Ref. [30]), h being the obstacle height, is not sufficient for equilibrium conditions up to a height of $5h$–$8h$
- the experiments have different amounts of inactive turbulence. This may well be so because the boundary layer depths, and thereby the range of turbulent scales, of the two experiments are very different: 3 inches ($24h$) above the obstacle tops in Ref. [30] and about 400 mm ($133h$) in Ref. [31]. Indeed, $k/u_{*\mathrm{prof}}^2$, where $u_{*\mathrm{prof}}$ is the wind profile derived u_*, differs significantly (about 3.5 and 6.5). Yet it is remarkable that the effect of inactive turbulence on the overall drag is so strong.

If the latter explanation is true, it implies that "inactive" turbulence is not as "inactive" as expected. Then we should explicitly model the effect of inactive turbulence, rather than to isolate and neglect it some way. This could be done with Large Eddy Simulation (the expensive one), or – if the geometry is such that there is one dominant scale – a dual-scale turbulence model. The latter type has already been proposed as a k–ε model with constants that depend on P_k/ε [34], like in ASM. What is needed here is a model that explicitly considers the turbulence to building size ratio L_{turb}/B, like is done experimentally by Durbin and Hunt [32].

Acknowledgements

This work was supported by CNRS, France, and by the European Commission under Human Capital and Mobility Grant no. ERBCHBGCT930493 (DISPURB).

References

[1] J.C. Rotta, Experience of second order turbulence flow closure models, Eurmech Coll. 180, Karlsr., FRG; Z. Flugwiss. Weltraumforsch. 10 (1986) 401–407.

[2] K. Hanjalic, Advanced turbulence closure models: a view of current status and future prospects, Int. J. Heat Fluid Flow 15 (3) (1994) 178–203.

[3] A.A. Townsend, Equilibrium layers and wall turbulence, J. Fluid Mech. 11 (1961) 97–120.

[4] H.A. Panofsky, J.A. Dutton, Atmospheric Turbulence, Wiley, New York, 1984.

[5] S. Zilitinkevitch, Non-local turbulent transport: pollution dispersion aspects of coherent structure of convective flows, in: Proc. 3rd Int. Conf. on Air Pollution, Porto Carras, Greece, part 1, WIT-Southampton, UK, 1995, pp. 53–60.

[6] M. Roth, Turbulence transfer relationships over an urban surface, part II: integral statistics, Q. J. R. Meteorol. Soc. 119 (1993) 1105–1120.

[7] M. Bottema, J.A. Wisse, J.A. Leene, Towards forecasting of wind comfort, J. Wind Eng. Ind. Aerodyn. 44 (1992) 2365–2376.

[8] M.R. Raupach, R.A. Antonia, S. Rajagopalan, Rough wall turbulent boundary layers, Appl. Mech. Rev. 44 (1991) 1–25.

[9] M. Bottema, J.A. Wisse, Effects of turntable roughness on low rise building pressures, in: Proc. 9th Int. Conf. on Wind Engineering, New Delhi, India, 1995, pp. 334–345.
[10] D.A. Paterson, Simulation past a cube in a turbulent boundary layer, J. Wind Eng. Ind. Aerodyn. 35 (1990) 149–176.
[11] M. Jensen, The model-law for phenomena in natural wind, vols. 2–4, Ingenioren, International ed., Copenhagen, 1958.
[12] P.G. Duynkerke, Application of the E–ε model to the neutral and stable atmospheric boundary layer, J. Atmos. Sci. 45 (1988) 865–880.
[13] P. Richards, R. Hoxey, Appropriate boundary conditions for computational wind engineering models using the k–ε turbulence model, J. Wind Eng. Ind. Aerodyn. 46 (1990) 145–153.
[14] B.E. Launder, D.B. Spalding, The numerical computation of turbulent flows, Comput. Meth. Appl. Mech. Eng. 3 (1974) 269–289.
[15] P. Bradshaw, D.H. Ferriss, N.P. Atwell, Calculation of boundary layer development using the turbulent energy equation, J. Fluid Mech. 28 (1967) 593–616.
[16] W. Rodi, Turbulence models for environmental problems, in: W. Kollmann (Ed.), Prediction Methods for Turbulent Flows, Hemisphere, Washington, DC, 1980, pp. 259–350.
[17] Y.-S. Chen, S.-W. Kim, Computation of turbulent flows using an extended k–ε turbulence closure model, NASA Contractor Report CR-179204, Marshall, Alabama, 1987.
[18] M. Bottema, Wind climate and urban geometry, Ph.D. Thesis, Department of Architecture & Building Technology, Eindhoven University of Technology, NL, 1992.
[19] M. Bottema, Numerical simulation of the lower urban boundary conditions with k–ε models and with ASM; reliability and (im)possibilities, Document SUB-MESO #22 (Report), Ecole Centrale de Nantes, 1995.
[20] M. Kato, B.E. Launder, The modelling of turbulent flow around stationary and vibrating square cylinders, in: Proc. 9th Symp. on Turbulent Shear Flows, Kyoto, Japan, 1993, pp. 10-4-1–10-4-6.
[21] C.G. Speziale, On non-linear k–l and k–ε models of turbulence, J. Fluid Mech. 178 (1987) 459–475.
[22] S. Murakami, Comparison of various turbulence models applied to a bluff body, J. Wind Eng. Ind. Aerodyn. 46&47 (1993) 21–36.
[23] J. Wieringa, Representative roughness parameters for homogeneous terrain, Bound.-Layer Meteorol. 63 (1993) 323–363.
[24] M.R. Raupach, A.S. Thom, I. Edwards, A wind-tunnel study of turbulent flow close to regularly arrayed rough surfaces, Bound.-Layer Meteorol. 18 (1980) 373–397.
[25] M. Rotach, Profiles of turbulence statistics in and above an urban street canyon, Atmos. Environ. 29 (13) (1995) 1473–1486.
[26] M.R. Raupach, P.A. Coppin, B.J. Legg, Experiments on scalar dispersion within a plant canopy, part 1: the turbulence structure, Bound.-Layer Meteorol. 35 (1986) 21–52.
[27] R.W. Benodekar, A.J.H. Goddard, A.D. Gosman, R.I. Issa, Numerical prediction of flows over surface mounted ribs, AIAA J. 23 (1985) 359–366.
[28] J.L. Ferziger, Approaches to turbulent flow computation: applications to flow over obstacles, J. Wind Eng. Ind. Aerodyn. 35 (1990) 195–212.
[29] J.-F. Sini, S. Anquetin, P.G. Mestayer, Pollutant dispersion and thermal effects in urban street canyons, Atmos. Environ. 30B (15) (1996) 2659–2677.
[30] R.A. Antonia, R.E. Luxton, The response of a turbulent boundary layer to a step change in surface roughness. Part 1: smooth to rough, J. Fluid Mech. 48 (1971) 721–761.
[31] P. Mulhearn, Turbulent flow over a periodic rough surface, Phys. Fluids 21–27 (1978) 1113–1115.
[32] P.A. Durbin, J.C.R. Hunt, On surface pressures beneath turbulent flow round bluff bodies, J. Fluid Mech. 100–101 (1980) 161–184.
[33] H.W. Tieleman, Simulation of surface winds for assessment of extreme wind loads, in: Proc. 9th Int. Conf. on Wind Engineering, New Delhi, India, 1995, pp. 1162–1169.
[34] S.-W. Kim, P. Chen, A multiple-time-scale turbulence model based on variable partitioning of turbulent kinetic energy spectrum, Proc. AIAA 26th Aerospace Sc. Meeting, Reno, Nevada, AIAA 88-0221, 1988, pp. 1–10.

Journal of Wind Engineering
and Industrial Aerodynamics 67&68 (1997) 909–922

Determination of plume capture by the building wake

Michel A. Brzoska[a], David Stock[b,*], Brian Lamb[c]

[a] *Department of Technology, Eastern Washington University, Cheney, WA, USA*
[b] *School of Mechanical and Materials Engineering, Washington State University, Pullman, WA 99164-2920, USA*
[c] *Civil and Environmental Engineering, Washington State University, Pullman, WA 99164, USA*

Abstract

Flow and dispersion about a cubical building were computed using a fourth-order accurate finite elements scheme. The time-averaged Navier–Stokes equations were closed with the standard k–ε turbulence model as well as with a k–ε turbulence model modified to allow production of turbulent kinetic energy to depend on the product of the strain rate and vorticity. The computed flow agreed with the wind tunnel measurements. Releases from a stack located at various positions within the recirculation zone behind the building were simulated. The effect of stack velocity on the concentration in the recirculation cavity was quantified by comparing the mass of pollutant in the recirculation cavity at very low stack exit velocities with the mass of pollutant in the recirculation cavity for higher stack exit velocities. The mass of pollutant in the recirculation zone decreased considerably at the higher stack velocities. The results of this work can be used to help develop and improve the modeling of pollutant transport in recirculation zones and wakes.

Keywords: Wind engineering; CFD application; k–ε turbulence model; Revised k–ε turbulence model; Turbulence model; Pollutant dispersion; Building; Building wake

1. Introduction

Considerable research has been directed toward quantifying the amount of pollutant that is entrained into the wake behind a building from a stack located either upwind of the building or on the roof of the building. This research is well justified since it is typical of many situations found in the field. However, in certain wind

* Corresponding author. E-mail: stock@wsu.edu.

PII S0167-6105(97)00128-1

conditions, the exhaust stack can discharge directly into the wake behind a building. The resulting pollutant concentration in the wake will depend on the location of the stack, the pollutant concentration in the exhaust gas, and the momentum of the exhaust gases. The purpose of our work was to quantify the capture of pollutants from stack releases in the building wake.

The EPA-recommended SCREEN cavity downwash model makes several assumptions in calculating plume downwash. One assumption is that the concentration is well mixed and uniform inside the recirculation cavity. This assumption of uniform mixing is, in many cases, an oversimplification. It applies to the case with the stack discharge outside the recirculation cavity: the pollutant diffuses through the shear layer and is then transported into the recirculation zone. This process of entraining a pollutant into the recirculation zone results in nearly uniform mixing in the recirculation zone [1,2]. A stack in the lee cavity recirculation zone with its exit at the height of the cavity may result in the pollutant being quickly carried to the ground by the downward curving shear layer bounding the cavity [3] and then some fraction of the pollutant being carried forward in the recirculation zone. In a wind tunnel study with pollutant sources in the building lee cavity region, Thompson [4] found that the concentration was not as uniform as expected and that the concentration in the cavity could vary by 20–100% depending on building shape and size. A distinction needs to be made between stacks discharging outside of a recirculation zone and ones discharging within the recirculation zone.

In the PRIME model, Schulman and Scire [2,5] determined the fraction of the plume captured from a source outside the recirculation zone. They calculate a recirculation cavity length using the method of Fackrell [6] and the cavity height using the method of Wilson and Britter [7]. They then calculate the fraction of the plume captured by the cavity as that portion of the plume that has diffused below the height of the recirculation zone. This method cannot be used for stack exhausts inside the wake recirculation zone.

A method is needed to quantify the fraction of a plume captured in the wake behind a building for the case of a discharge within the wake. The fraction captured will depend on the wind speed and profile, the building size and shape, and characteristics of the discharge. We restricted our consideration to one approach flow velocity profile and one building shape and size. In this case the fraction captured will depend on the stack location, the stack diameter, and discharge momentum. In most cases, the buoyancy of the plume will not be important in determining the flow within the recirculation zone and the stack exit velocity can be used instead of the stack momentum as one of the parameters. We will not attempt to non-dimensionalize the results since we do not know the controlling non-dimensional parameters. In the future when this problem is better understood, other researchers can convert our results to the appropriate form.

A method for estimating the fraction of pollutant captured by the lee recirculation region is presented. In this formulation, the effect of stack velocity on the concentration in the recirculation cavity is quantified by comparing the mass of pollutant in the recirculation zone at very low stack velocities (V_s) with the mass of pollutant in the recirculation zone at higher stack velocities. The plume was assumed to be completely captured in the recirculation at very low stack velocities. Thus, comparisons to the concentration at very low stack emission velocities can be used to quantify the capture fraction. As an initial trial of the method, we computed the flow and dispersion from a stack located in the wake of a cubical building using different stack locations and heights. Only neutral stability boundary layer and a non-buoyant exhaust plume were considered.

2. Numerical computations

A fourth-order accurate finite element code, FEAT [8], employing the steady, time-averaged Navier–Stokes equations and k–ε and strain/vorticity modified k–ε [9] turbulence models was used to predict airflow and pollutant concentrations over a cubical building (40 m). Flows normal to the cubical building (Fig. 1) were simulated and releases from stacks at several locations were made. The flow-field results from the k–ε model were compared to experimental results of Snyder and Lawson [10]. These comparisons are documented by Brzoska et al. [11]. Dispersion from a stack located in the building wake was simulated and the results used to calculate the fraction of the plume captured by the wake for a variety of stack locations, heights and exit velocities.

2.1. Review of numerical simulations of flow about buildings

A variety of studies have attempted to predict the three-dimensional flow over buildings and cubes [12–19]. Paterson and Apelt [17,18] used a k–ε model to predict the mean velocities and the mean pressures in the flows around a number of buildings. The predictions were compared with experimentally measured values from wind tunnel data [1]. The agreement between the computed wind velocities above and in the wake of the building with the measurements ranged from very good to poor. Pressures on the sides of the buildings were poorly predicted. Murakami and Mochida [12] compared the results of computations using upwind differencing and the k–ε model to wind tunnel data that included precise measurements of the turbulent kinetic energy around the building. Their predictions could reproduce the mean

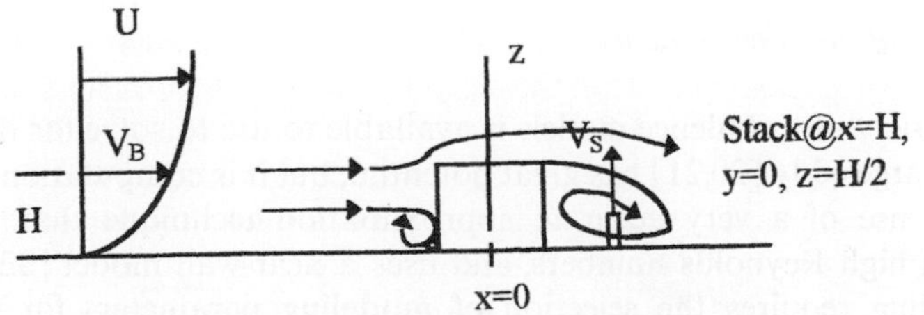

Fig. 1. Modeled building with stack.

velocity field and mean pressure field around the model accurately using a very fine mesh, but there existed significant differences in the distribution of the turbulent energy around the windward corner and in the wake. They suggested that further efforts including the modification of turbulence modeling should be made to improve the accuracy of the numerical simulation. Murakami et al. [20] calculated velocities near a cube employing large eddy simulation. Murakami [15] further contrasted solutions from the k–ε and algebraic second moment closure model methods to large eddy simulation. Murakami [15] pointed out shortcomings in predicting flow in the wakes of buildings using both the k–ε closure and the algebraic second-moment closure.

2.2. Selection of numerical method

A fourth-order-accurate, finite element numerical scheme, FEAT [8], was used to calculate the flow and dispersion about the cube. We have found that the finite element technique is better able to simulate steep gradients than finite difference methods, typically employing multi-node elements that can result in the use of coarser grids but still maintain higher-order accuracy. We have also found that for three-dimensional calculations, the finite element method seems to be faster than the finite difference method. However, it is well known that finite element models have convergence problems when the k–ε turbulence model is used. This problem was avoided in the FEAT code by using the q–f equations (where f is a measure of the frequency of large-scale turbulent eddies and $k = q^2$). Finite difference codes often avoid this problem by using upwind differencing in the k and ε equations. When certain assumptions are made, the q–f equation becomes equivalent to the k–ε equations. FEAT uses the Galerkin method and employs 20-node brick elements incorporating quadratic basis functions and eight internal nodes using linear basis functions for pressure terms. The code uses special wall elements that employ wall functions (law of the wall).

To reduce continuity errors, a penalty term is employed for pressure. Thus, there is a continuous part of the pressure and a piecewise constant part (Galerkin-PALM) providing an extra degree of freedom for pressure on each element. FEAT uses Newton–Raphson iteration to solve the close-coupled groups (velocity, pressure, temperature), turbulence kinetic energy, and turbulence dissipation and uses the frontal algorithm (Gaussian elimination) to solve the linearized system of equations for each iteration.

2.3. Selection of turbulence model

A wide variety of turbulence models is available to use to solve for the flow about a building. Large eddy [20,21] has great potential, but it is computationally intensive, requires the use of a very accurate approximation technique that can result in instability at high Reynolds numbers, and uses a near-wall model [22,23]. Second-order modeling requires the selection of modeling parameters for the pressure-strain terms and the resulting increase in uncertainty of solution as well as more computation time. The k–ε model is the most widely applied approach to engineering problems and is very promising in wind engineering. It does have two main problems: the use of first-order closure and the use of the law of the wall.

As a result of being a first-order closure scheme, k–ε fails to accurately model non-isotropic effects of flow such as streamline curvature. The turbulent kinetic energy production term is modeled by the strain-rate tensor, $\frac{1}{2}(\partial U_i/\partial x_j + \partial U_j/\partial x_i)$. When the flow undergoes rapid distortion (i.e., at the front corner of buildings), large values of $\partial U/\partial X$ are present; therefore, large values of turbulent kinetic energy are computed. However, we know that in rapid distortion the eddies are distorted by the flow sufficiently rapidly that the turbulence is not significantly modified by the change in strain rate; thus, the kinetic energy should not increase rapidly. Therefore, the use of the strain-rate tensor to model turbulent kinetic energy production results in an overprediction of the turbulent kinetic energy. As rapid distortion theory indicates, the eddies are modified by the stretching or compressing of the individual vortex elements. This has led modelers to use strain/vorticity to model the production of turbulent kinetic energy. The standard k–ε model gives large values of turbulent kinetic energy at the front corner of the building, which results in reduction or elimination of the recirculation zone on the top of the building due to excessive diffusion. In the recirculation zone behind the building, the turbulent kinetic energy is underestimated ($\partial V/\partial Y$ small), resulting in less diffusion with a subsequent increase in the recirculation cavity [15]. The strain/vorticity modified form of the k–ε model has the potential for correcting these shortcomings.

Near-wall modeling quite often involves the use of the law of the wall which may result in inaccurate velocities and turbulent kinetic energy values [24]. Computations can be extended to the wall by using a low Reynolds number form of the k–ε closure model or the k–ε model. In either case, the computation time will be increased by one-third and our experience shows little change in recirculation zone when these extensions are used. Use of the law of the wall typically slightly underpredicts flow separation. In this work, we used the strain/vorticity modified k–ε turbulence. Details of this model are given below.

2.4. Governing equations

The time-averaged Navier–Stokes equations for steady, incompressible turbulent flow were used with the gradient-transport relation for turbulent stresses. The eddy diffusivity, ν_t, was modeled as isotropic (ν_t is proportional to velocity scale times the length scale). The FEAT k–ε closure expresses the eddy diffusivity as a function of k (q^2) and l which is determined through q and f transport equations [25]. The eddy diffusivity is given by $\nu_t = \rho q \ell$, where ℓ is given by $\ell = C_\mu q/f$.

The turbulent concentration transport flux in the concentration equation is isotropic and given by

$$\overline{u_j'c'} = -D_t \partial C/\partial x_j, \qquad (1)$$

where the eddy mass diffusivity of scalar C is given by

$$D_t = \nu_t / \mathrm{Sc}_t. \tag{2}$$

The value of the turbulent Schmidt number, Sc_t, is assumed to be 0.8. It is assumed that the mean velocity has been previously calculated and the concentration does not affect the mean flow.

2.5. *Strain/vorticity modified k–ε turbulence model*

To improve the turbulence prediction near a stagnation point, Launder and Kato [9] investigated a strain/vorticity modified k–ε model. The production of turbulent kinetic energy (P_k) in the standard k–ε equation is given by

$$P_k = C_\mu \varepsilon S^2, \tag{3}$$

but in the Launder and Kato modified k–ε model, P_k is modified as

$$P_k = C_\mu \varepsilon S \Omega, \tag{4}$$

where

$$S = k/\varepsilon (0.5 S_{ij} S_{ij})^{0.5} \tag{5a}$$

and

$$\Omega = k/\varepsilon (0.5 \Omega_{ij} \Omega_{ij})^{0.5} \tag{5b}$$

with

$$S_{ij} = \partial U_i / \partial x_j + \partial U_j / \partial x_i \tag{6a}$$

and

$$\Omega_{ij} = \partial U_i / \partial x_j - \partial U_j / \partial x_i. \tag{6b}$$

S and Ω are equal in a simple shear flow, but at a stagnation point S becomes excessive as the $\partial U / \partial x$ term becomes quite large. In the modeled turbulent kinetic energy equation, the turbulence production term, $C_\mu \varepsilon S^2$, provides excessive production of turbulent kinetic energy near stagnation points as a result of the large normal Reynolds-stress term (i.e., $\partial U / \partial x$). In other words, the principal axes of Reynolds stress and the rate-of-strain tensor are aligned. This occurs in pure strain but not in flow with "mean vorticity" [26]. Since the flow is nearly irrotational at the stagnation point, this new term eliminates the excessive production term while maintaining the same production or turbulent kinetic energy in simple shear flow ($S \approx \Omega$ in simple shear flows).

Launder and Kato [9] utilized this strain/vorticity modified k–ε model for two-dimensional flow about a square cylinder. Their research showed a small improvement in lift coefficient with the use of the modified k–ε with less grid sensitivity than the standard k–ε model.

3. Numerical simulation

Two cases were studied numerically. In case I, the k–ε model was used and in case II the strain/vorticity modified k–ε model [9] was used. The flow field around the building was solved for each case and the mean velocity and turbulent kinetic energy compared with wind tunnel measurements. Stack releases at two locations downwind of the building and two stack heights for a range of stack exit velocities were then simulated. The results of these computations were then used to calculate the fraction of the plume captured by the recirculation zone.

3.1. Experimental study used for comparison

Snyder and Lawson [10] conducted wind tunnel experiments (scaled 200 : 1) using a 3.7 m wide, 2.1 m high, and 18.3 m long wind tunnel. A simulated neutral atmospheric boundary layer was generated for the flow approaching a 0.2 m cube. The boundary layer depth was 2 m and the roughness length was 1.0 mm. Measurements were taken of the three components of velocity and turbulence intensity in the vicinity of the building. The wind tunnel inlet velocity profiles used in the experiments varied slightly; thus, an average of these profiles was used in the numerical work as follows:

$$U = (u^*/\kappa)\ln((z - d)/z_0), \tag{7}$$

where $u^* = 0.25$, $\kappa = 0.4$, $z_0 = 1.2$ mm, and displacement thickness, $d = 1.2$ mm. In the outer region, the velocity profile approximated a power-law curve with an exponent of 0.16.

3.2. Boundary conditions and mesh

Wind tunnel conditions were approximated in the numerical simulations with a flow area 2 m wide, 2 m high, and 7 m long. At the inlet, velocity and turbulent kinetic energy values matched those of Snyder and Lawson [10]. Symmetry conditions were used at the center vertical plane allowing half of the cubic building to be modeled. Turbulent kinetic energy and turbulent dissipation were given normal gradients of zero on the symmetry planes. The maximum horizontal velocity, U, was fixed at 4.27 m/s. At the outlet, tangential gradients were set to zero. No slip conditions were used at the solid surfaces. The bottom center of the building was at x, y, and z = zero.

A mesh was created with five layers of nodes around the cubical building (approximately 8000 nodes for the flow normal to the cube and increased to 12 500 nodes for uncertainty analysis). The mesh near the cubical building is shown in Fig. 2.

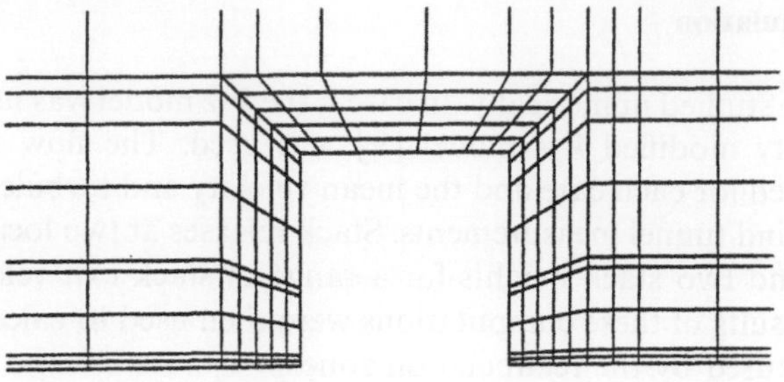

Fig. 2. Mesh near building.

Table 1
Recirculation length for two-dimensional backward-facing step

Reynolds number	Computed recirculation length ratio	Type of flow	Experimental recirculation length ratio	Reference
200	4.92	Laminar	5	Armaly et al. [28]
450	8.87	Laminar	8.5	Armaly et al. [28]
50 000	5.05	Turbulent	5.2	Moss and Baker [29]

3.3. *Uncertainty of numerical computations*

The FEAT code was validated on the two-dimensional backward-facing step for both laminar and turbulent conditions. FEAT compared favorably to the experiments and to other codes presented in the 1993 ASME CFD Triathlon [27] for laminar flow (Table 1). FEAT solutions to the laminar flow backward-facing step problem employed many fewer finite element nodes to obtain as accurate a solution as obtained by the other codes. The results for the turbulent flow case are as good or better than most codes using the k–ε turbulence closure and the law of the wall. Computing to the wall instead of using the law of the wall did not significantly change the results.

Richardson's extrapolation [30] and Roache's grid convergence index (GCI) [31] can be used to quantify truncation convergence error (i.e., grid refinement error). In Roache's method, a grid convergence index (GCI) is determined to allow different order-of-accuracy techniques to be compared. The order-of-accuracy can be determined through the computation of the change in results for three grid solutions to the same problem. For finite element solutions, an effective grid refinements ratio is calculated allowing the grid to be increased by a fraction. Because the three-dimensional mesh requires considerable computing time to be doubled, the mesh was increased from 8000 to 14 000 nodes and then to 12 500 nodes. The order-of-accuracy was determined to be 3 while the error in the computation of the recirculation zone length at the building centerline was 1% with GCI of 5% for the 8000 nodes model.

4. Discussion of results

4.1. Flow field

Velocity vector diagrams in a plane were made from the calculated flow at several locations. Fig. 3a and Fig. 3b show planar vector fields at $y = 0$ and $z = 0.1$ (one-half building height, $H/2$), respectively. These results are compared to the data of Snyder and Lawson [10] in another report [11]. Generally, the experimental and computed velocity fields are quite similar except that the recirculation region is extended by approximately 20% in the k–ε model's computed field.

The vector plots show the recirculation area in front of and in the building lee very clearly. The difference between the k–ε and the strain/vorticity modified k–ε model is too small to allow comparison in vector plots, although the lee recirculation cavity of the strain/vorticity modified k–ε model was approximately 2.2% smaller than the

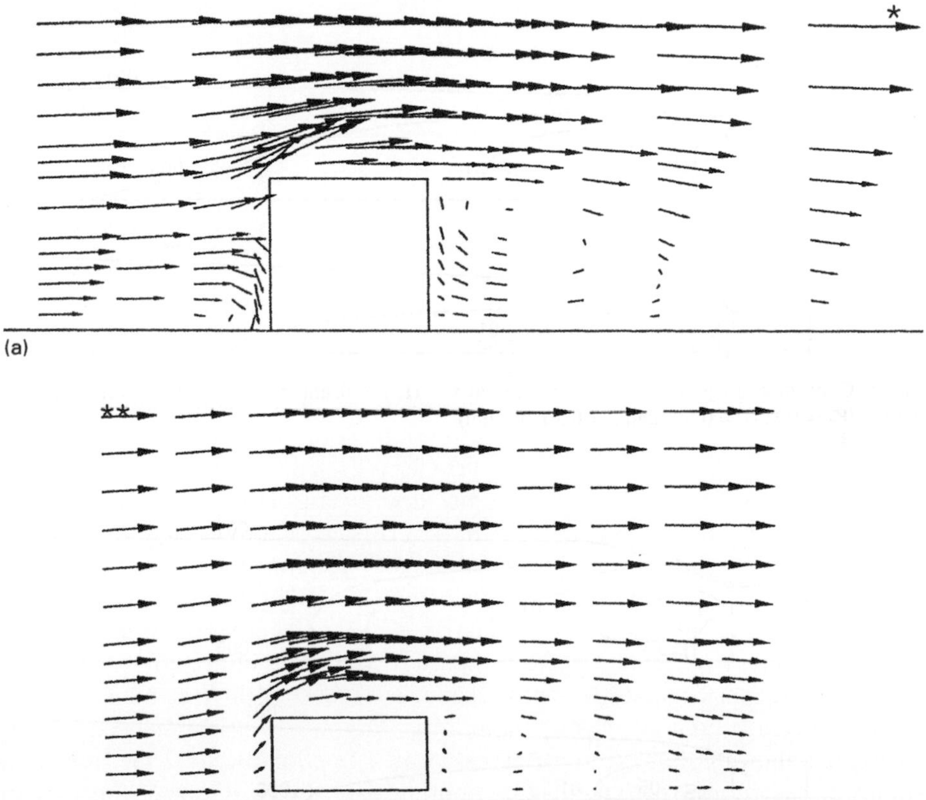

(a)

(b)

Fig. 3. (a) Velocity vector diagram at $y = 0.0$ for flow normal to building (scale: vector* = 3.8 m/s). (b) Velocity vector diagram at $z = 0.5H$ for flow normal to building (scale: vector** = 2.8 m/s).

cavity size of the k–ε model. The strain/vorticity modified k–ε model compares more favorably to the cavity size measured in Snyder and Lawson [10]. For a rectangular building, the standard k–ε results become progressively worse as the building becomes wider.

4.2. Plume capture

Non-buoyant releases were made from two different locations and for stack heights equal to the building height and one-half the building height. In all cases the pollutant release was 0.00045 g/s and, the effective stack exit area was decreased as the stack velocity increased. The case simulated is representative of a powerplant with a fixed exhaust flow rate, but a variable exit diameter on the stack. For the case simulated the concentration of pollutant in the exhaust was held constant.

Figs. 4 and 5 show the effects of the stack exit velocity for a release at $x = H, y = 0$ and $z = H/2$ with the stack velocity/building top velocity $(V_s/V_B) = 0.33$ in Fig. 4, $V_s/V_B = 6.0$ in Fig. 5, and the concentration in g/m^3. The effect of the higher stack

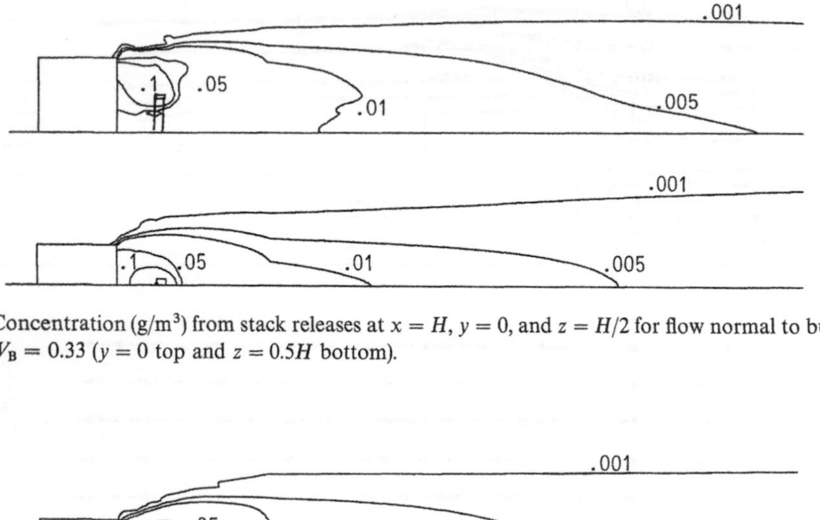

Fig. 4. Concentration (g/m^3) from stack releases at $x = H$, $y = 0$, and $z = H/2$ for flow normal to building with $V_s/V_B = 0.33$ ($y = 0$ top and $z = 0.5H$ bottom).

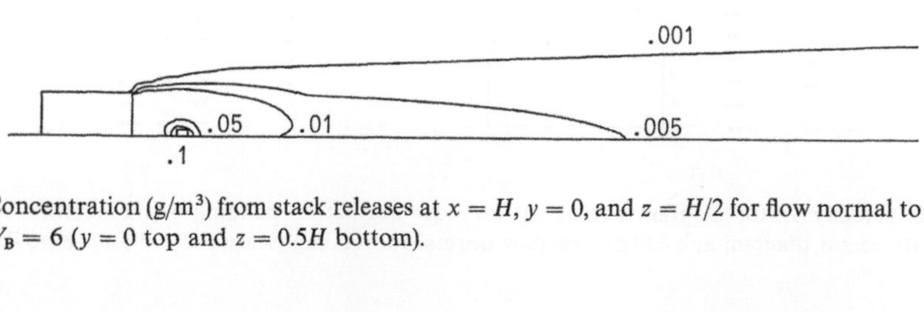

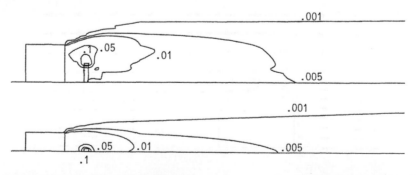

Fig. 5. Concentration (g/m^3) from stack releases at $x = H$, $y = 0$, and $z = H/2$ for flow normal to building with $V_s/V_B = 6$ ($y = 0$ top and $z = 0.5H$ bottom).

velocity is evident with the pollutant dispersing further up and out of the recirculation cavity, and lower concentrations in the lower recirculation zone region.

The effect of the stack exit velocity on the amount of pollutant in the recirculation zone was quantified by comparing the mass of pollutant in the recirculation zone at a given stack velocity to the mass in the recirculation zone for a stack with very low exit velocity. The first task was to identify which computational cells are in the recirculation zone. We defined the recirculation zone as being composed of the cells behind the building and having a negative longitudinal mean velocity. This is the region that will carry the pollutant toward the building and cause it to recirculate. The size of the recirculation zone changed very little as the stack velocity increased. It was relatively easy to calculate the mass of pollutant in the recirculation zone by summing the product of the cell volume and concentration for all the cells with a negative longitudinal mean velocity. For the reference case we used a stack discharge velocity of 0.5 m/s and assumed all the discharge was captured in the recirculation zone.

Fig. 6 shows the recirculation zone capture fraction for both the k–ε and strain/vorticity modified k–ε model for a stack located a building height behind the building and discharging at half the height of the building. The difference is quite small with the modified k–ε model indicating a larger amount of pollutant captured. More variation in the results is expected with a wider building. The fraction captured decreases in an exponential manner as the stack velocity increases.

Fig. 7 shows the fraction captured for a stack at the same location as that shown in Fig. 6 except in this case the stack is as tall as the building. There is almost no difference in the fraction captured even though the stack is twice as high. Fig. 8 shows the fraction captured for a stack that is two building heights behind the building and discharging at a height of half the building height. Again we found that the fraction captured is almost identical to the other two cases.

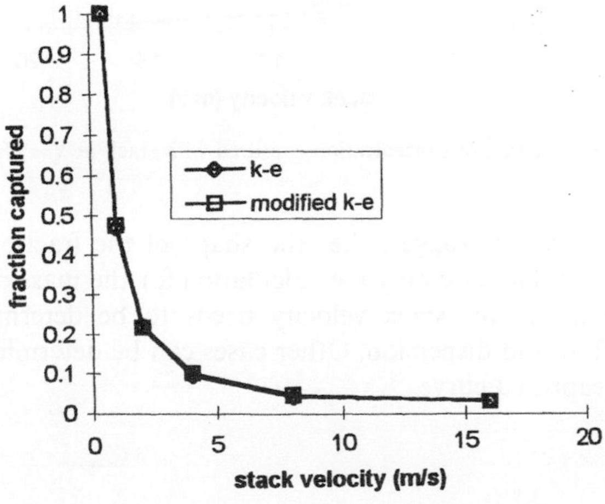

Fig. 6. Recirculation zone fraction of concentration captured with stack at $x = H$, $y = 0$, $z = H/2$, and $V_B = 3.1$ m/s.

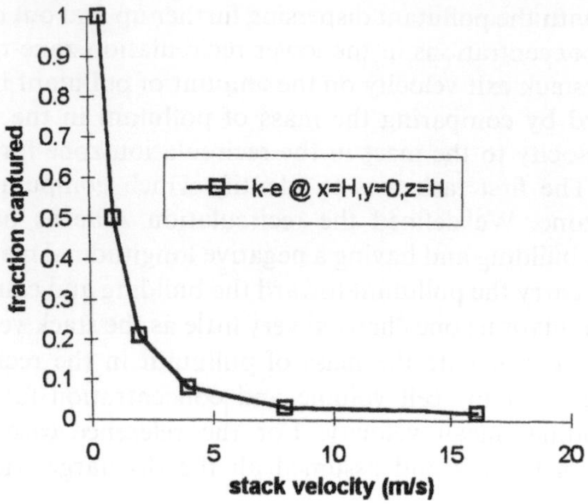

Fig. 7. Recirculation zone fraction of concentration captured with stack at $x = H$, $y = 0$, $z = H$, and $V_B = 3.1$ m/s.

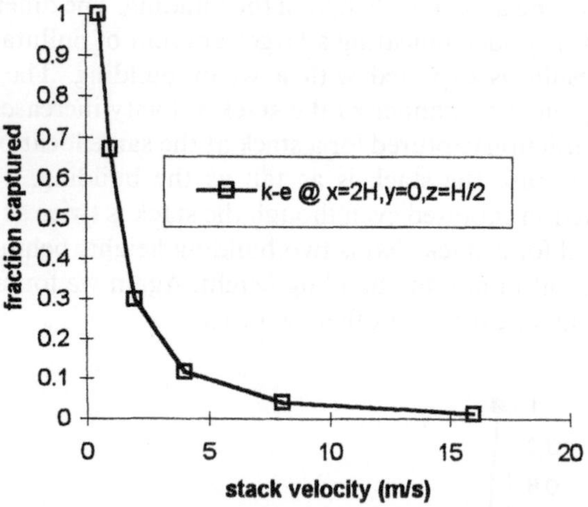

Fig. 8. Recirculation zone fraction of concentration captured with stack at $x = 2H$, $y = 0$, $z = H/2$, and $V_B = 3.1$ m/s.

These preliminary results suggest that the shape of the fraction captured curve might be universal. In that case only the calculation for the mass of pollutant in the recirculation zone for a low stack velocity needs to be determined from a full simulation of the flow and dispersion. Other cases can be determined from a nearly universal fraction captured curve.

5. Conclusions

Numerical simulation of the flow and dispersion of a plume from a stack located in the recirculation zone behind a cubical building was used to quantify the fraction of the plume that was captured by the recirculation zone. The fraction captured was found to decrease rapidly as the stack exit velocity was increased. For the limited number of locations studied all showed the same decrease in the fraction of the plume captured as the exit velocity increased. More work is needed to determine the effect of the location of the stack on the fraction captured and the effect of building shape and orientation.

Acknowledgements

We acknowledge Charles Hakkarinen, Project Technical Manager, Electric Power Research Institute and Lloyd Schulman, Earth Tech, for their support; William Snyder and Robert Lawson, EPA, for providing wind tunnel data; and Richard Smith, Stephen Hickmott, Mike Rabbit, Tony Hutton, and Mike Reeks of Nuclear Electric for the use of FEAT.

References

[1] A.G. Robins, I.P. Castro, A wind tunnel investigation of plume dispersion in the vicinity of a surface mounted cube – II, the concentration field, Atmos. Environ. 11 (1977) 299–311.
[2] L.L. Schulman, J.S. Scire, The development of the plume rise model enhancements (PRIME): The EPRI plume rise and downwash modeling project, in: Proc. 9th Joint Conf. on Applications of Air Pollution Meteorology with AWMA, Atlanta, GA, 1996, pp. 307–310.
[3] S.R. Hanna, G.A. Briggs, R.P. Hosker, Handbook on Atmospheric Diffusion, Technical Information Center US Department of Energy, Ch. 3, DE82002045(DOE/TIC-11223), 1982.
[4] R.S. Thompson, Building amplification factors for sources near buildings: a wind-tunnel study, Atmos. Environ. A 27 (15) (1993) 2313–2325.
[5] L.L. Schulman, J.S. Scire, Building downwash screening modeling for the downwind recirculation cavity, J. Air Waste Management Assoc. 43 (1993) 1122–1127.
[6] J.E. Fackrell, Parameters characterizing dispersion in the near wake of buildings, J. Wind Eng. Ind. Aerodyn. 16 (1984) 97–118.
[7] D.J. Wilson, R.E. Britter, Estimates of building surfaces concentrations from nearby point sources, Atmos. Environ. 16 (1982) 2631–2646.
[8] Nuclear Electric, FEAT User Guide, Nuclear Electric plc, Barnettway, Barnwood, Gloucester, UK GL47RS, 1996.
[9] B.E. Launder, M. Kato, Modelling flow-induced oscillations in turbulent flow around a square cylinder, FED-Vol. 157 (1993) 189–199.
[10] W.H. Snyder, R.E. Lawson, Wind-tunnel measurements of flow fields in the vicinity of buildings, in: Proc. 8th AMS Conf. on Appl. Air Pollution Meteorology, with AWMA, 23–28 January, 1993, Nashville, TN.
[11] M.A. Brzoska, D.E. Stock, B. Lamb, Three-dimensional finite element numerical simulations of airflow and dispersion around buildings, in: Proc. 9th Joint Conf. on Applications of Air Pollution Meteorology, Atlanta, GA, 1996, pp. 322–328.
[12] S. Murakami, A. Mochida, 3-D numerical simulation of airflow around a cubic model by means of the k–ε model, J. Wind Eng. Ind. Aerodyn. 31 (1988) 283–303.
[13] S. Murakami, A. Mochida, Three-dimensional numerical simulation of turbulent flow around buildings using the k–ε turbulence model, Building Environ. 24 (1) (1989) 51–64.
[14] S. Murakami, A. Mochida, Numerical simulations of air flow around surface-mounted square rib by means of ASM and k–ε EVM, in: Proc. 9th Structures Conf., ASCE, Indianapolis 1991, pp. 639–642.

[15] S. Murakami, Comparison of various turbulence models applied to a bluff body, J. Wind Eng. Ind. Aerodyn. 46&47 (1993) 21–36.
[16] P.J. Dawson, D.E. Stock, B. Lamb, The numerical simulation of airflow and dispersion in three-dimensional atmospheric recirculation zones, J. Appl. Meteor. 30 (1991) 1005–1024.
[17] D.A. Paterson, C.J. Apelt, Computation of wind flows over three-dimensional buildings, J. Wind Eng. Ind. Aerodyn. 24 (1986) 193–213.
[18] D.A. Paterson, C.J. Apelt, Simulation of flow past a cube in a turbulent boundary layer, J. Wind Eng. Ind. Aerodyn. 35 (1990) 149–176.
[19] A. Guenther, B. Lamb, D. Stock, Three-dimensional numerical simulation of plume downwash with a k–ε turbulence model, J. Appl. Meteor. 29 (7) (1990) 633–643.
[20] S. Murakami, A. Mochida, K. Hibi, Three-dimensional numerical simulation of air flow around a cubic model by means of large eddy simulation, J. Wind Eng. Ind. Aerodyn. 25 (1987) 291–305.
[21] S. Sakamoto, S. Murakami, A. Mochida, Numerical study of flow past 2D square cylinder by large eddy simulation: comparison between 2d and 3d computation, J. Wind Eng. Ind. Aerodyn. 50 (1993) 6–68.
[22] M.A. Leschziner, Computational modelling of complex turbulent flow expectations, reality and prospects, J. Wind Eng. Ind. Aerodyn. 46&47 (1993) 37–51.
[23] W.C. Reynolds, The potential and limitation of direct and large eddy simulations, Wither turbulence? Turbulence at the crossroads, in: J.L. Lumley (Ed.), Lecture Notes in Physics, vol. 357, Springer, Berlin, 1990, pp. 313–343.
[24] V. Haroutunian, M.S. Engelman, Two-equation simulations of turbulent flows: a commentary on physical and numerical aspects, American Society of Mechanical Engineers FED vol. 171, Advances in Finite Element Analysis in Fluid Dynamics, 1993, pp. 95–105.
[25] R.M. Smith, A practical method of two-equation turbulence modeling using finite elements, Int. J. Numer. Meth. Fluids 4 (1984) 321–336.
[26] W.C. Reynolds, Computation of turbulent flows, Ann. Rev. Fluid Mech. 8 (1976) 183–208.
[27] ASME, The CFD Triathlon – Three laminar flow simulations by commercial CFD Codes, in: Proc. of The Fluids Engineering Conference, Washington, DC, 20–24 June, 1993, FED vol. 160, pp. 1–69.
[28] B.F. Armaly, F. Durst, J.C.F. Pereira, B. Schonung, Experimental and theoretical investigation of backward-facing step flow, J. Fluid Mech. 127 (1983) 473–496.
[29] W.D. Moss, S. Baker, Re-circulating flows associated with two-dimensional steps, The Aeronautical Quart. 31 (1980) 151–172.
[30] L.F. Richardson, The approximate arithmetical solution by finite differences of physical problems involving differential equations, with an application to the stresses in a masonry dam, Trans. Roy. Soc. London Ser. A. 210 (1910) 307–357.
[31] P.J. Roache, A method for uniform reporting of grid refinements studies, ASME Fed summer meeting, (1993) pp. 1–12.

ELSEVIER

Journal of Wind Engineering
and Industrial Aerodynamics 67&68 (1997) 923–933

JOURNAL OF wind engineering AND industrial aerodynamics

Simulating the dynamics of spray droplets in the atmosphere using ballistic and random-walk models combined

M.L. Mokeba, D.W. Salt, B.E. Lee*, M.G. Ford

University of Portsmouth, Portsmouth, UK

Abstract

This paper presents a simulation model based on earlier work that combines both ballistic and random-walk models to describe the three-dimensional dynamics of spray droplets released in a specified direction from ground-based appliances in various weather conditions. The velocity of spray droplets is considered as a weighted sum of their ballistic and random-walk velocities scaled by a factor $(1 - \beta)$ and β, respectively, where β is defined as the ratio of the sedimentation velocity and the relative velocity between the spray droplets and the ambient wind speed. The contribution of the random-walk model to the initial velocity is seen to be negligible at first, but increases progressively, though not proportionally, as β increases. As soon as the spray droplets attain their sedimentation velocities, $\beta = 1$, the random-walk velocity component predominates and β plays no further part in the calculations. The predicted effects close to the sprayer of the drop size, wind velocity and direction, evaporation on the transport process have been evaluated and combined to provide an analysis of spray drift.

Keywords: Spray droplets; Dynamics; Ground-based appliances; Simulation; Wind; Spray drift

1. Introduction

Agrochemical sprays are widely used to protect crops from the ravages of weeds, pests and diseases. When these sprays are released in air, from ground-based appliances in the form of droplets with some initial velocity, their trajectories will inevitably be controlled by the prevailing atmospheric conditions. Their motion will be influenced by frictional forces due to wind motion, itself dependent on the roughness of upstream terrain, together with external body forces, notably gravity, which will act to accelerate or decelerate the droplets. The net force acting on the droplets in any given direction will thus be the vectorial sum of the external forces and the frictional forces, where the latter are dependent on the instantaneous velocity of the spray droplets.

* Corresponding author. E-mail: blee@civl.port.ac.uk.

PII S0167-6105(97)00129-3

When the spray droplets have decelerated sufficiently to attain their sedimentation velocities, their motion will be dominated solely by the fluctuating effects of the wind rather than on their initial velocity. The interplay between these separate processes in the atmosphere can be predicted and analysed accurately using relatively simple computer models.

However, the mechanisms which determine the behaviour of sprays under field conditions are poorly understood despite considerable effort by scientists and engineers over the years to investigate the physical processes involved in the transport of spray droplets in air from sprayer to the crop surface [1–5].

There are very strong motives for wishing to understand and simulate the dynamics of agricultural spray droplets in the atmosphere. Approximately (25–35%) of the potential harvest of the world's food crops is lost to plant diseases and pests and more is lost by bad post-harvest storage conditions [6]. Therefore, the need to increase productivity and maximise efficiency has necessitated the intensive use of chemicals in the form of sprays. As an example of this large-scale application taken from studies in China alone, it is estimated that over 100 million farmers regularly apply crop protectants [3].

Unfortunately, there are serious problems associated with this excessive use of potentially hazardous substances. Firstly, due to the stochastic nature of the wind, spray drift to off-target sites downwind can occur with serious contamination to air, soil, water and this may result in adverse effects to both wildlife and humans. Secondly, there are numerous problems of inefficient pest control, under-treated plant interiors, chemically scorched outer leaves; these effects are due in part to the large scale of application, and the difficulty of presenting the sprays to its target species.

Because of these very diverse practical problems, an understanding of the mechanisms involved is necessary if spraying is to be undertaken properly and the benefits associated with its utilisation maximised. This will help in a variety of application; not only in agriculture, forestry, biology, but also in addressing long-term environmental issues such as the global balance of chemicals in the atmosphere.

Since field experiments are prohibitively expensive and very difficult to carry out because of the many meteorological and spraying parameters involved, little scientific understanding of these problems has been achieved to date. Simulation has become increasingly popular in recent years as an alternative means of addressing some of these problems and has a number of advantages [2,4,5]. Using models, experiments can be conducted in a fraction of the time and cost required by field studies and the parameters involved combined in a way that might not otherwise be possible. When dealing with complex chains of events, simulation models allow a statistical representation of the likely outcome to be gained. Additionally, they might also help to optimise the elements of a spraying programme perhaps leading to a reduction of waste and environmental contamination [4].

Nevertheless, it must be stressed that despite these advantages, models have limitations. Due to the many complexities involved when trying to simulate the dynamics of spray droplets in air, certain features have to be removed and simplifying assumptions made. For this reason alone, simulation models do not "predict" the dynamics of spray droplets in the colloquial sense of the word. Thus, predictions based on simulation should always be verified by comparison with experimental observations and not considered as separate entities. They may, however, broaden our understanding of the consequences of the decisions made while formulating the model. These decisions must always be presented in a form which can be grasped with little explanation and cast in terms which are straightforward to apply. Therefore, fully effective simulation models must always be able to strike a balance between not being too complex to use in a wider sense and not being so simple that they are unable to simulate the physical processes involved in a realistic way.

The simulation models presented by the authors mentioned earlier have realised this objective. The model developed in this study is based on earlier work, the novel element here being in the consideration of the effects of turbulence. Miller and Hadfield [2] considered the effects of turbulence only when the spray droplets have attained their sedimentation velocities (i.e. the terminal velocity of the droplets). This simplifying assumption can be unrealistic since although the effects of turbulence will be irrelevant close to the sprayer, its effects will start to become noticeable before the sedimentation velocities of the droplets are attained. It is therefore necessary to ensure that these apparent fluctuations of droplet velocity close to their terminal velocity are not ignored. As a first approximation, the velocity of the spray droplets have been considered as weighted sums of their ballistic and random-walk velocities scaled by a factor $(1 - \beta)$ and β, respectively, where β is the ratio of the sedimentation velocity and the relative velocity between the droplets and the surrounding airstream. This ratio β was formulated like this so that, as the relative velocity between the spray droplets and the ambient wind speed decreases due to viscous forces resulting from wind shear, the value of β increases. This will ensure that, the contribution to the velocity of the spray droplets provided by the random-walk velocity component will increase likewise and attains a maximum when the sedimentation velocity is reached. At this point, $\beta = 1$, the contribution to the velocity of the spray droplets from the ballistic velocity component is nill. There after, the random-walk velocity component of the spray droplets will predominate their dynamics and β plays no further part in the calculations.

To terminate this foreward, it is worth mentioning that with the exception of Williamson and Treadgill [5], most earlier simulation models have considered the motion of spray droplets in two-dimensions only, by assuming that the crosswind diffusion is insignificant. The current model considers the full three-dimensional dynamics of the spray droplets.

2. Combined model

The effects of air turbulence on the motion of spray droplets, though inconsequentially close to the sprayer as a result of air entrainment effects, will start to become significant as the sedimentation velocity of the spray droplets is approached due to the continuous loss of their initial momentum to the ambient wind speed. The size of this momentum flux lost is directly proportional to the instantaneous velocity of the spray droplets and is continuously driven downwards by turbulence.

A model was therefore developed to account for the combined effects of turbulence and momentum loss by considering the velocities of spray droplets as a weighted sum of their ballistic and random-walk velocities. The ballistic velocities of spray droplets was calculated using Newton's second law of motion. This law can be expressed in component form in the lateral, horizontal and vertical directions as follows:

$$mA_{px} = F_D \cos(G_1), \tag{1}$$

$$mA_{py} = F_D \cos(G_2), \tag{2}$$

$$mA_{pz} = mg - F_D \cos(G_3), \tag{3}$$

where $\{F_D = \frac{1}{2}C_D \rho A_r |V_r|$ is the drag force acting on the spray droplets; with $|V_r|$ $(= |V_{as} + V_{ent} - V_p|)$ as the magnitude of the relative velocity, V_{as} is the velocity of the airstream defined in Section 2.1, V_{ent} is the velocity of entrained air defined in Section 2.2, V_p is the velocity of the droplet, C_D is the drag coefficient which is a function of the Reynolds number Re $(= |V_r| d/\nu_{is})$, ν_{is} $(= 1.5 \times 10^{-5}\ \mathrm{m^2\ s^{-1}})$ is the kinematic viscosity of air, ρ $(= 870\ \mathrm{kg\ m^{-3}})$ is the density of the droplet, A_r $(= \pi d^2/4)$ is the area of the droplet, d is the diameter of the droplet, g $(= -9.81\ \mathrm{m\ s^{-2}})$ is the acceleration due to gravity, A_{px}, A_{py} and A_{pz} are the accelerations of the spray droplet in the x-, y- and z-axis, respectively, m $(= \rho\pi d^3/6)$ is the mass of the droplets, $\cos(G_1)$, $\cos(G_2)$ and $\cos(G_3)$ are the direction cosines of the drag force in the lateral, horizontal and vertical axes, respectively.

Integrating Eqs. (1)–(3) with initial conditions, the ballistic velocities in the component axes; V_{px}, V_{py} and V_{pz}were obtained as follows:

$$V_{px} = V_0 \cos(A) + \int_0^t A_{px}, \tag{4}$$

$$V_{py} = V_0 \cos(B) + \int_0^t A_{py}, \tag{5}$$

$$V_{pz} = V_0 \cos(C) + \int_0^t A_{pz}, \tag{6}$$

where V_0 $(= 10\ \mathrm{m\ s^{-1}})$ is the initial velocity of the spray droplets, $\cos(A)$, $\cos(B)$ and $\cos(C)$ are the direction cosines of V_0 in the lateral, horizontal and vertical axes, respectively.

The random-walk velocity component of the spray droplets at the $(i+1)$th time step in the lateral axis was obtained using the Markov process as follows:

$$V_{rx_{i+1}} = \alpha V_{rx_i} + \chi_{i+1}\sigma_x(1 - \alpha^2)^{1/2}. \tag{7}$$

The horizontal and vertical component of the random-walk velocity was shown by Thompson and Ley [4] to be

$$V_{\mathrm{ry}_{i+1}} = \alpha V_{\mathrm{ry}_i} + (1-\alpha)V_{ay} + \gamma\sigma_y(1-\alpha^2)^{1/2}, \tag{8}$$

$$V_{\mathrm{rz}_{i+1}} = \alpha(V_{\mathrm{rz}_i} + V_{s_i}) + \eta_{i+1}\sigma_z(1-\alpha^2)^{1/2} - V_{s_{i+1}}, \tag{9}$$

where $\alpha = \exp(-h/\tau_L)$ is the correlation of the component velocities over the time step h, provided that $h < \tau_L$, $\tau_L = [ku^*(z-d_0)/\sigma_z^2][1-16(z-d_0)/L]^{1/2}$ is the Lagrangian scale of turbulence, with z, u^*, d_0 and L defined in Section 2.1; η_{i+1} and γ are correlated random variables drawn from a standard normal distribution such that $\gamma = 0.5r\eta_{i+1} + (1-1.25r^2)^{1/2}\varepsilon_{i+1}$, with $r = (u_*^2/\sigma_x\sigma_z)$, σ_x, σ_y and σ_z as the size of the turbulent fluctuating velocities in the x- , y- and z-axis also defined in Section 2.1. The random numbers ε_{i+1} and χ_{i+1} are different from η_{i+1} and γ but are also drawn from a standard normal distribution. V_s is termed the sedimentation velocity and is given by Ref. [4] as follows:

$$V_s = 4.47\times10^{-3}d - 0.191 \quad \text{for } d > 100\ \mu\text{m}, \tag{10}$$

and

$$V_s = 3.2\times10^{-5}d^2 - 6.4\times10^{-8}d^3 \quad \text{for } d < 100\ \mu\text{m}. \tag{11}$$

Hence, the velocities of the combined model at the (i + 1)th time step with the ballistic and random-walk velocities scaled by a factor $(1-\beta)$ and β, respectively, are as follows:

$$V_{\mathrm{cx}_{i+1}} = (1-\beta)V_{\mathrm{px}_{i+1}} + \beta V_{\mathrm{rx}_{i+1}}, \tag{12}$$

$$V_{\mathrm{cy}_{i+1}} = (1-\beta)V_{\mathrm{py}_{i+1}} + \beta V_{\mathrm{ry}_{i+1}}, \tag{13}$$

$$V_{\mathrm{cz}_{i+1}} = (1-\beta)V_{\mathrm{pz}_{i+1}} + \beta V_{\mathrm{rz}_{i+1}}, \tag{14}$$

where β is defined mathematically as (V_s/V_r) provided that $V_r \geqslant V_s$.

The position of the droplets after n time steps of integration released in the (x, y, z) plane from a source at an origin $(0, 0, l)$ is then

$$(x,y,z) = \left(\sum_{i=1}^{n} V_{\mathrm{cx}_i}h, \sum_{i=1}^{n} V_{\mathrm{cy}_i}h, l - \sum_{i=1}^{n} V_{\mathrm{cz}_i}h\right), \tag{15}$$

where V_{cx_i}, V_{cy_i} and V_{cz_i} are the velocities of the combined model in the x- , y- and z-axis after n time steps of integration and l is the release height of droplets above the ground.

2.1. *Wind profile model*

Within the lower atmosphere (i.e. the first few hundred metres above the earth's surface), the variation of wind speed with height is dominated by the frictional drag exerted on the flow by the underlying terrain. Drag retards the horizontal wind velocity close to the ground. In neutral weather conditions, the logarithmic wind velocity profile can be expressed as follows:

$$V_{as} = \left(\frac{u^*}{k}\right)\ln\left(\frac{z - d_0}{z_0}\right), \tag{16}$$

where u^* is the friction velocity, z_0 (= 0.1CH) is the roughness length, k (= 0.4) is the von Karman constant, z is the height above the ground and d_0 (= 0.63CH) is the zero-plane displacement with CH (= 0.2 m) as the crop height. Eq. (16) was used in the simulation model to represent wind conditions in the region above the crop.

In unstable weather conditions, Eq. (16) was modified by Ref. [4] to take into account the rapid upward convection of heat as follows:

$$V_{as} = \left(\frac{u^*}{k}\right)[f(\xi) - f(\xi_0)], \tag{17}$$

where

$$\xi = \frac{(z - d_0)}{L}, \quad \xi_0 = \frac{z_0}{L}$$

and

$$f(\xi) = 2\tan^{-1}[(1 - 16\xi)^{1/4}] - \ln\left[\left|\frac{(1 + 16\xi)^{1/4}}{(1 - 16\xi)^{1/4}}\right|\right]. \tag{18}$$

The symbol L denotes a scaling length termed the Monin–Obukhov length [7].

For unstable weather condition, L is numerically small and negative (about − 20 m) while for near-neutral condition, the value of L is about − 1000 m.

The size of the turbulent fluctuations is represented in the horizontal, lateral and vertical directions by Panofsky et al. [7] for unstable weather conditions as follows:

$$\sigma_x \approx \sigma_y = u^*\left(12 - \frac{z}{2L}\right)^{1/3}, \tag{19}$$

$$\sigma_z = 1.3u^*\left(1 - \frac{3z}{L}\right)^{1/3}, \quad 0 < \frac{z}{L} < -6. \tag{20}$$

The boundary layer height sufficiently accurate for specifying the turbulence statistics in a random-walk model close to the surface layer is approximately 1000 m [4].

For neutral weather conditions, these turbulent fluctuations are given by Panofsky and Dutton [8] in the respective directions as follows:

$$\sigma_x \approx \sigma_y = 2.3u^*, \tag{21}$$

$$\sigma_z = 1.3u^*. \tag{22}$$

2.2. *The effects of air entrainment*

The effects of air entrainment for fan jet nozzles was accounted for in the combined model by including the velocity of entrained air along the axis of the jet V_{ent}, proposed by Briffa and Dombrowski [9] as

$$V_{\text{ent}} = V_0 \left(\frac{L_c}{h_n} \right)^{\delta^2/2p_1}, \tag{23}$$

where L_c (= 0.025 m) is the spray sheet coherent length, V_0 is defined in Section 2, h_n is the distance below the nozzle, δ is a constant with a value of 0.4 and p_1 is a constant defined as the width of the spray fan at right angles to the spray sheet at a given distance from the nozzle [2]. The characteristic ratio ($\delta^2/2p_1$) was also suggested by Miller and Hadfield [2] to have a value of 0.95.

2.3. Evaporation model

The theory proposed by Ranz and Marshall [10] for describing the change in diameter with time for an evaporating drop formed the basis of a model of droplet evaporation. This may be expressed as

$$\frac{\mathrm{d}d}{\mathrm{d}t} = -2\frac{M_{\mathrm{v}}}{M_{\mathrm{a}}}\frac{D_{\mathrm{v}}}{d}\frac{\rho_{\mathrm{a}}}{\rho}\frac{\Delta P}{P_{\mathrm{a}}}(2 + (0.6)\mathrm{S}_{\mathrm{c}}^{1/3}\mathrm{Re}^{1/3}), \tag{24}$$

where M_{v} (= 106) is the molecular weight of diffusing vapour, M_{a} (= 29) is the molecular weight of air, D_{v} ($= 7 \times 10^{-6}\ \mathrm{m^2\,s^{-1}}$) is the diffusion coefficient of vapour in air, ρ_{a} ($= 1.2\ \mathrm{kg\,m^{-3}}$) is the density of air, ΔP (= 790 Pa) is the vapour pressure difference, P_{a} ($= 10^5$ Pa) is the partial pressure of air, atmospheric temperature is 20°C and $\mathrm{S_c}$ ($= V_{\mathrm{is}}/D_{\mathrm{v}}$) is the Schmidt number.

2.4. Methods employed in programming the model

The four equations Eqs. (12)–(14) and (24) comprise a model which simulates the dynamics of evaporating spray droplets. The equations were programmed in QBASIC on an Elonex computer (PC-466/VL). In order to obtain the velocity components V_{px}, V_{py} and V_{pz} in Eqs. (12)–(14) and hence their respective positions together with the diameter of the droplet, the ballistic accelerations A_{px}, A_{py} and A_{pz} in the lateral, horizontal and vertical axes, respectively, as given in Eqs. (4)–(6) together with Eq. (24) were integrated using a numerical integration technique known as the Runge–Kutta algorithm [11]. In this algorithm, the integration interval is divided into subintervals, the derivatives are found at each of the subintervals, and a weighted average is taken of these multiplied by the integration interval in order to determine the increment to add to the dependent variable. Since numerical integration is bound to give errors, the selection of the time step is crucial if such errors are to be minimised. The time step of integration must always be selected within the framework of two conflicting requirements [1]. It must be small enough to preserve accuracy and to prevent numerical instability occurring but yet not too small to increase both the number of steps required for integration and the overall rounding error to such an extent that no benefit accrues. Since it is difficult to give stability bounds for non-linear differential equations used in this model, trial and error method was used and a time step of 2 ms was found convenient. The computed results were plotted using the graphics package GRAFTOOL that is capable of generating complex 3-D graphs using full vector based graphics with advanced hidden-line removal. execution of the program was terminated when the spray droplets hit the top of the crop canopy or when the droplet diameter reaches a minimum.

3. Results and discussion

A preliminary evaluation of the model and its properties have been undertaken. Fig. 1 gives an example of the simulated trajectories of five spray droplets using the combined model with the diameters selected from normal distribution of mean 150 μm and standard deviation 30 μm. The spray droplets were projected downwards from a 110° flat fan nozzle with an initial velocity of 10 m/s towards the top of a crop canopy situated at 0.5 m below the droplet discharge point. The simulated wind velocity was 2 m/s which is considered appropriate for moderate sunshine in neutral weather conditions. The relative humidity was 50% at an atmospheric temperature of 20°C. Trajectory angles were also sampled from a normal distribution with a mean of zero and a standard deviation of 0.4 times the nozzle angle [2]. It can be discerned

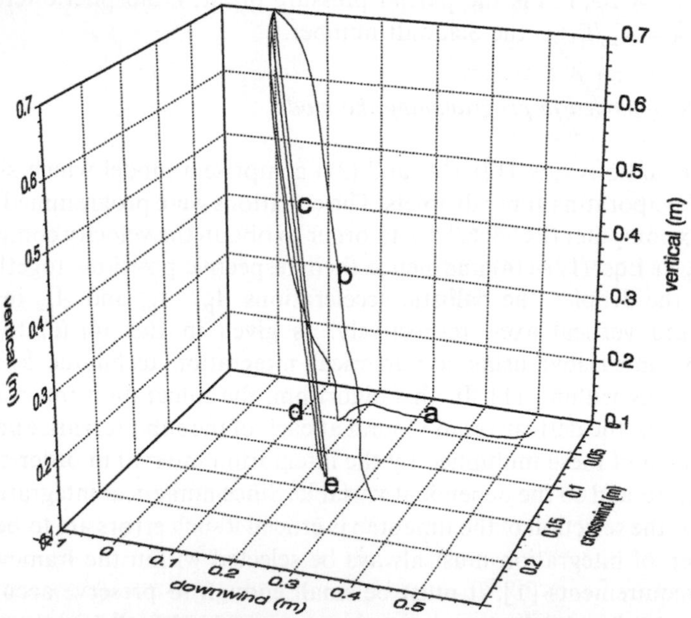

Fig. 1. Droplet trajectory in neutral weather conditons. (a) 60 μm drop size; (b) 90 μm drop size; (c) 120 μm drop size; (d) 150 μm drop size; (e) 180 μm drop size.

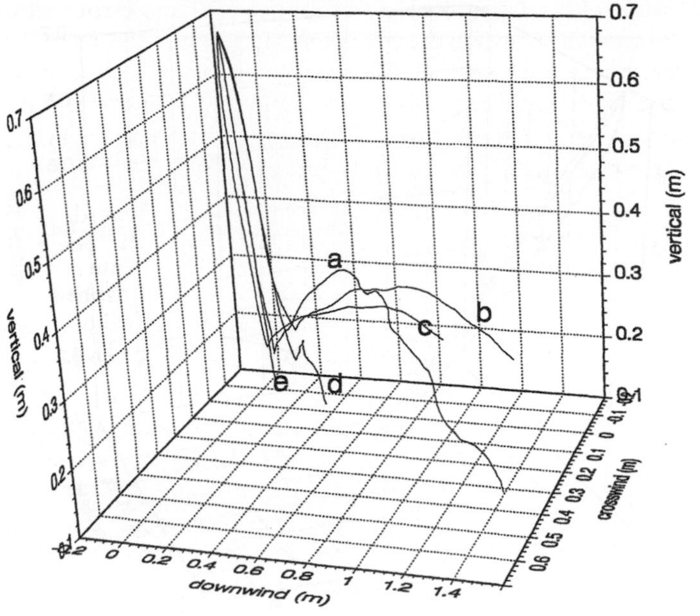

Fig. 2. Droplet trajectory in unstable weather conditons for the same droplets as in Fig. 1.

from the graph that, the effects of this moderate wind has little effect on the larger droplet trajectories but has an effect only on the small 60 μm droplet.

Fig. 2 shows a repeat of the above set of data when the simulated wind velocity was 5 m/s which is considered appropriate for strong sunshine in unstable weather conditions. It can be observed that, increasing the wind speed increases spray drift for all the droplet trajectories with more pronounced effects on the smaller droplets.

Fig. 3 shows the effects of evaporation on the same set of data given in Fig. 1. Owing to the reduction in size of the spray droplets as they approach the canopy top due to evaporation effects, they become more susceptible to the stochastic nature of the wind, especially the 60 and 90 μm spray droplets. However, for the large 120, 150 and 180 μm droplets, evaporation seems to have no effects on their trajectories. This insignificance in the effects of evaporation is attributed to the short time and distance scale that it takes the droplets to reach the canopy top.

Fig. 4 shows a 3-D histogram of the simulation of spray drift from twelve nozzles mounted on a boom in the crosswind direction with each nozzle having a different Normal distribution of drop sizes ranging from 10 to 300 μm using the combined model. The boom of length 6 m consist of six 80° nozzles and six 110° nozzles with an interval of 0.5 m between each nozzle. The wind velocity was 5 m/s at atmospheric temperatures of 20°C. The drift from all the twelve nozzles was simulated by calculating the drift from a single nozzle in each class category and then displacing it in the cross wind direction of the line of nozzles mounted on a boom by a distance equal to that of the nozzle spacing to obtain the mass flux field due to the complete set of

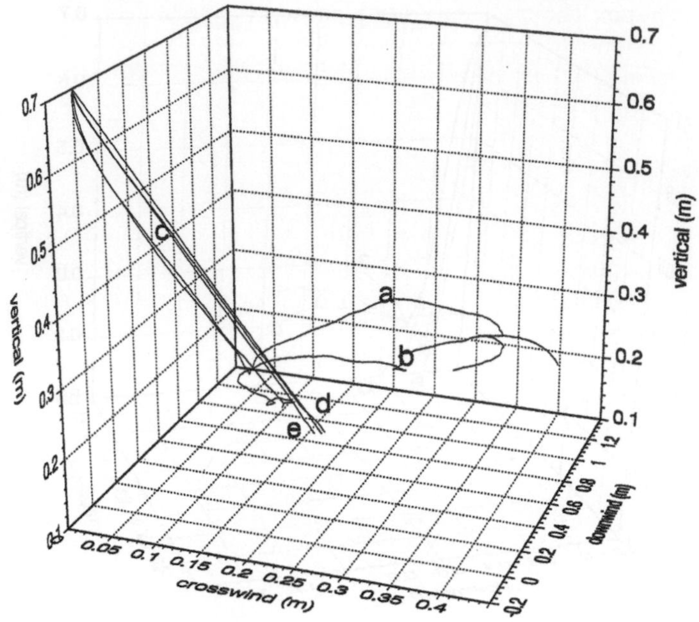

Fig. 3. Effect of evaporation on droplet trajectory in neutral weather conditons for the same droplets as in Fig. 1.

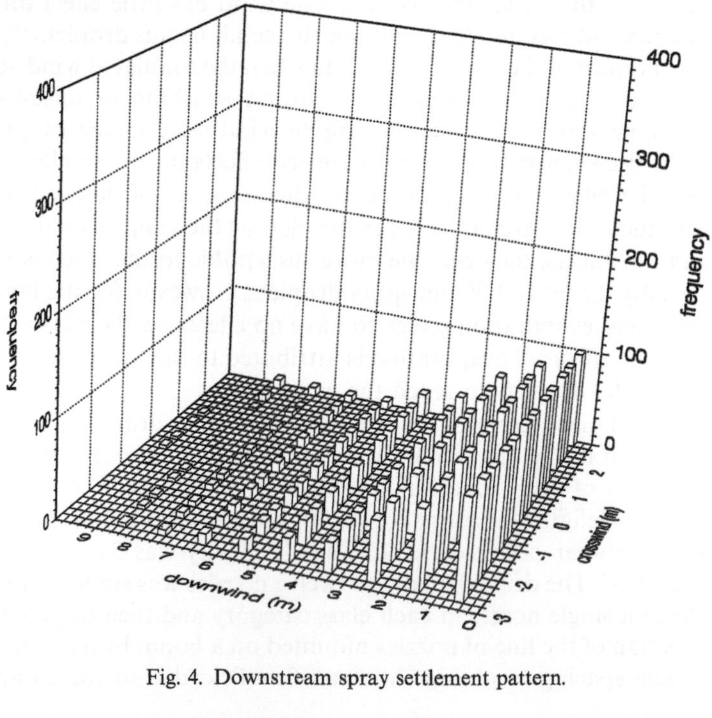

Fig. 4. Downstream spray settlement pattern.

nozzles on the boom. The spray deposition pattern is seen to decrease downwind from the point of ejection with larger spray droplets depositing closer to the source than smaller ones owing to their large inertia.

4. Conclusion

A simulation model that combines both ballistic and random-walk models to analyse ground deposits close to the sprayer in three-dimensions has been developed from earlier work. The effects of wind speed, drop size, evaporation and spray drift on the transfer process from sprayer to the crop surface in two weather conditions have been elucidated and appropriate inference drawn. It is planned that, in the near future, experiments will be conducted in a wind tunnel to validate the assumptions made in the model.

Acknowledgements

The authors are grateful to the referees for their careful comments.

References

[1] J.A. Marchant, Calculation of spray droplet trajectory in a moving airstream, J. Agric. Eng. Res. 22 (1977) 93–96.
[2] P.C.H. Miller, D.J. Hadfield, A simulation of the spray drift from hydraulic nozzles, J. Agric. Eng. Res. 42 (1988) 135–147.
[3] G.W. Schaefer, K. Allsopp, Spray droplet behaviour above and within a canopy, in: Proc. 10th Int. Congr. on Plant Protection, vol. 3, 1983, pp. 1057–1065.
[4] N. Thompson, A.J. Ley, Estimating spray drift using random-walk model of evaporating drops, J. Agric. Eng. Res. 28 (1983) 419–435.
[5] R.E. Williamson, E.D. Threadgill, A simulation for the dynamics of evaporating spray droplets in agricultural spraying, Trans. Am. Soc. Agric. Eng. 17 (1974) 254–261.
[6] D.H. Bache, D.R. Johnson, Microclimate and Spray Dispersion, Ellis Horwood, Chichester, West Sussex, 1992, p. 239.
[7] H.A. Panofsky, H. Tennekes, D.H. Lenshow, J.C. Wyngaard, The characteristics of turbulent velocity components in the surface layer under convective conditions, Boundary-layer Meteorol. 11 (1977) 355–361.
[8] H.A. Panofsky, J.A. Dutton, Atmospheric Turbulence, Wiley Interscience, New York, 1984, p. 397.
[9] F.E.J. Briffa, N. Dombrowski, Entrainment of air into a liquid spray, Am. Inst. Chem. Eng. 12 (4) (1966) 708–717.
[10] W.E. Ranz, W.R. Marshall, Evaporation from drops, Chem. Eng. Prog. 48 (1952) 173–180.
[11] R.L. Lafara, Computer methods for Science and Engineering, Hayden Book Company, New York, 1973, p. 326.

APPENDIX

Abstracts of papers presented but not published

ELSEVIER

Journal of Wind Engineering
and Industrial Aerodynamics 67&68 (1997) 937

JOURNAL OF wind engineering AND industrial aerodynamics

Wind-induced response of double articulated towers

Suhail Ahmad, N. Islam, N.A. Siddiqui

Department of Applied Mechanics, I.I.T., New Delhi 110 016, India

Abstract

Double hinged articulated off-shore towers are highly flexible against rotation at hinges and derive stability by means of their inherent large buoyancy. Their displacement response is mainly governed by rigid body mode of vibration at very low frequency, Emil Simiu's fluctuating wind velocity spectrum exclusively meant for compliant off-shore structures has been employed. This spectrum has high energy content in low-frequency region and, therefore, wind induces significant dynamic response of such towers. Hydrodynamic response due to surrounding waves becomes complex in the presence of wind. This paper deals with the investigation of structural behaviour due to random wind and wave environment. The Monte Carlo simulation technique is used to model this environment. Iterative time-domain solution procedure is adopted to take care of time-dependent parameters and nonlinearities. Stochastic response is characterised by Power Spectral Density Functions (PSDF) for various parametric combinations. Studies of wind effects are found to be imperative for double hinged articulated towers to serve and survive in the hostile off-shore environment, dominated by high wind velocities.

0167-6105/97/$17.00
PII S0167-6105(97)00134-7

ELSEVIER

Journal of Wind Engineering
and Industrial Aerodynamics 67&68 (1997) 938,939

JOURNAL OF wind engineering AND industrial aerodynamics

Forecasting roof snow accumulation using numerical models

A. Baskaran[1], A. Kashef

Building Performance Laboratory, Institute for Research in Construction, National Research Council Canada, Ottawa, Ont., Canada K1A 0R6

Abstract

To forecast snow accumulation rate on large roofs, the present study used the Computational Fluid Dynamics (CFD) tools and techniques. A numerical model is formulated using the 3D Navier–Stokes Equations (NSE) together with the standard k–ε turbulence model. For the model development, the control volume technique is employed to discretize the differential equations into difference form and the developed model has been used to simulate the air flow predictions over variety of roof configurations. To assess the snow-drift condition for each configuration, the velocity components are extracted from the simulated air flow results and the local wind speeds are calculated at different locations. The threshold values of snow particles are calculated for three different types of snow fall: light, medium, and heavy. A comparison of the calculated wind speed with the threshold value determines whether a snow-drift process will take place. Comparing the local wind speed and the threshold value for each type of snow, the forecast for the possibility of snow accumulation is presented by "Yes" or "No". This has been performed for a flat roof building configuration of about 112 m long, 24 m wide and 12 m high. For this unusual building aspect ratios (5 and 10 respectively for length/width and length/height), local wind environmental conditions are evaluated and snow accumulations are forecasted.

The model has been extended not only to include complex roof configurations but also to quantify the snow accumulation on roofs for a given ground snow fall. Modeled configurations are: a three level slope roof (15°, 5° and 5°) and a 13° slope roof with two levels. Both roofs envelop a building of about 63 m long, 57 m wide and 13 m high. Architectural features such as parapets can significantly modify the local wind flow conditions over a roof. Effect of small architectural features of about 1 m height on these large roofs are also predicted based on the developed model. Combining the present investigation with the theories developed based on experimental work, a general procedure to compute snow accumulation has been developed and it

[1] E-mail: bas.baskaran@nrc.ca.

PII S0167-6105(97)00134-7

involves the following steps:

1. calculation of the wind environmental conditions to obtain the friction velocity at each nodal point,
2. determination of snow transport rate,
3. estimation of erosion or deposition rates

The simplified procedure is used to quantify snow accumulations for variety of configurations derived by modifying the three level roof geometry. Efforts are also made to quantify the validity of model predictions through comparisons with field observation. The above evaluations demonstrate that the CFD techniques can be effectively applied during the early stage of the design process as a predictive tool in forecasting snow accumulation on various roof configurations.

Journal of Wind Engineering
and Industrial Aerodynamics 67&68 (1997) 940

Simulation of unsteady flow around oscillating bluff bodies using a hybrid Eulerian Lagrangian scheme

Pavit S. Brar[a], Omar M. Knio[b], Robert H. Scanlan[a]

[a] *Department of Civil Engineering, The Johns Hopkins University, Baltimore, MD-21218, USA*
[b] *Department of Mechanical Engineering, The Johns Hopkins University, Baltimore, MD-21218, USA*

Abstract

Simulation of 2D, moderate-Reynolds-number flows around oscillating bluff bodies is performed using a multi-domain multi-method technique. The latter is based on decomposition of the computational domain into two regions: an Eulerian region surrounding the bluff body, and a Lagrangian region in the remainder of the flow. Within the Eulerian region, a second-order finite-difference discretization of the vorticity transport equation and the streamfunction Poisson equation is used. The resulting discrete equations are integrated using an alternating direction implicit (ADI) scheme. Meanwhile, a vortex element technique is used within the Lagrangian region. Vorticity is discretized into Lagrangian vortex elements of circular overlapping cores. The vortex elements are advected along particle trajectories, and their vorticity changes according to local diffusion fluxes which are computed using a conservative, deterministic particle exchange algorithm. Solutions within Eulerian and Lagrangian subdomains are joined along common boundaries using a coupling scheme which ensures continuity of the velocity and vorticity fluxes. The present construction combines the accuracy and flexibility of finite-difference methods with the efficiency of Lagrangian particle techniques. Implementation of the numerical scheme is discussed in light of results for rectangular bluff-body sections. Application of the simulations to compute unsteady aeroelastic forces and moments and to extract flutter derivatives is also discussed.

PII S0167-6105(97)00134-7

ELSEVIER

Journal of Wind Engineering
and Industrial Aerodynamics 67&68 (1997) 941

JOURNAL OF wind engineering AND industrial aerodynamics

Complex terrain surface wind field modeling

Edmond D.H. Cheng, Jie Shang

Department of Civil Engineering, University of Hawaii at Manoa, Honolulu, Hawaii 96822, USA

Abstract

The study of surface wind field is very important because it provides the primary input for decision making in air quality control, structural design improvement, wind energy development, and many other applications. The numerical model used in the study is a three-dimensional mass-consistent kinematic flow model. The major elements considered in the surface wind field modeling are topography and surface roughness.

Topographical features such as mountains, valleys, hills, canyons, and cliffs can cause drastic changes in local wind speed and direction. Spatial variability of the atmospheric boundary layer flow increases with the complexity of the terrain. Yet, traditionally complex terrains are mostly interpolated by ridge amplification, valley separation and surface roughness. It is, therefore, difficult to model the surface wind field with the desired degree of accuracy in an area of complex topography. Clearly, a digital elevation model (to describe the terrain) and a detailed land use and land cover digital model are the most effective means of achieving better accuracy in complex terrain surface wind field modeling. However, the handling of the massive amount of terrain and surface roughness data is a task by itself. A computerized Geographic Information Systems (GIS) is used in this study to accomplish the task of handling data.

The advantages of integrating surface wind field modeling with GIS are numerous. It allows one to utilize the GIS database, which has a higher degree of accuracy in data interpretation, and can significantly reduce the required time for data preparation. The unique features of GIS in manipulating, editing, and graphically viewing large input and output data array are essential in modeling technique. Further, the integrated surface wind field model can be applied to other areas as long as the required GIS database is available.

The purpose of this study is to demonstrate that surface wind field with a complex terrain can be successfully modeled by a GIS assisted solution of the kinematic flow model. Wind data at the Honolulu International Airport will be used as an input to the boundary layer flow model. Wind data at this station has been well researched and long-term (100-year) hourly wind data has been simulated by means of a stochastic model [Cheng and Chiu, ASCE J. Struct. Eng. 111(1), Paper no. 19429 (1985); in: Proc. the 9th Internat. Conf. on Wind Eng., New Delhi, India, 9–13 January, 1995]. Comparison of results will be discussed.

PII S0167-6105(97)00134-7

ELSEVIER

Journal of Wind Engineering
and Industrial Aerodynamics 67&68 (1997) 942

JOURNAL OF wind engineering AND industrial aerodynamics

Computer simulation of the dynamic response of wind-loaded large bridges under limit load conditions

Imre Kovács[a], Karlheinz Beyer[a], Narimichi Oba[b]

[a] *Büro für Baudynamik GmbH, Bessemerstrasse 2, 70435 Stuttgart, Germany*
[b] *Obayashi Corporation, Hongo 2-2-9, Bunkyo-ku, Tokyo 113, Japan*

Abstract

Gusty wind often represents the dimensioning case for large bridges, for which it is difficult to predict realistic limit loads. The spectral-analytical method, normally applied for checking the behaviour of such bridges under wind loads, uses a number of approximations (linearizations, simplified flow/body interaction). The simulation method presented here represents a more realistic approach, taking account both of the complex interaction between flow and vibrating body and of the non-linear stiffness of the load-carrying system. The simulation is performed under η_0-fold wind loads. The load-carrying safety is checked through the statistical evaluation of simulation results.

PII S0167-6105(97)00134-7

ELSEVIER

Journal of Wind Engineering
and Industrial Aerodynamics 67&68 (1997) 943

JOURNAL OF
wind engineering
AND
industrial
aerodynamics

Wind speed distribution over the bay by numerical solution of 3D Navier–Stokes equation

Kiyonori Kushioka, Toru Saito, Akihiro Honda

Fluid Dynamics & Heat Transfer Lab., Nagasaki Research & Development Center, Mitsubishi Heavy Industries, Ltd. 1-1, Akunoura-machi, Nagasaki 850-91, Japan

Abstract

Prediction of wind speed at the point where structures are constructed is important for their design against wind. Wind tunnel test has been used for prediction of wind speed over terrain. However, measurement of wind velocity at many points using wind tunnel needs much time and labor to determine the spatially velocity distribution. Otherwise, the velocity distribution can be obtained by numerical simulation, easily. The authors applied the conventional method of numerical flow simulation (SIMPLE method for Navier–Stokes and Algebraic Stress Model) to the flow over complex terrain. The result of simulation was compared with the measurement of flow velocity in wind tunnel test. As the result, calculated wind speed corresponds to that of experiment except for the region near the surface. The wind speed near the surface is improved by considering roughness.

PII S0167-6105(97)00134-7

ELSEVIER

Journal of Wind Engineering
and Industrial Aerodynamics 67&68 (1997) 944

JOURNAL OF wind engineering AND industrial aerodynamics

Numerical simulation of flow over two successive buildings of square cross-section

S.K. Maharana[a], A.K. Ghosh[b]

[a] *Indira Gandhi Institute of Technology, Talcher, 759146 Orissa, India*
[b] *Aerospace Engineering, I.I.T., Kharagpur, 721302 West Bengal, India*

Abstract

The present paper deals with 3D simulation of the wind flow pattern around two successive buildings of square cross-section. One of the buildings is cubic and the other is a rectangular prism of height twice that of the cube (H). The spacing between the buildings has been chosen as H and $2H$ and the wind speeds 9 and 18 m/s. In addition, two possible configurations of the building with respect to air flow have been considered. The numerical modelling is based on the SIMPLE algorithm and velocity components and pressures have been obtained as output. The results are presented as velocity contour plots and pressure coefficients. The results show that the wind flow pattern around the buildings is mainly influenced by wind speed and spacing.

PII S0167-6105(97)00134-7

ELSEVIER

Journal of Wind Engineering
and Industrial Aerodynamics 67&68 (1997) 945, 946

JOURNAL OF wind engineering AND industrial aerodynamics

Wind induced vibrations of bell-towers: An experimental procedure to find vibration frequencies and test simulation based on wind engineering

Alfonso Nappi[a], Francesco Pedrielli[b], Michela Pian[a]

[a] *Department of Civil Engineering, University of Trieste, Piazzale Europa 1, 34127 Trieste, Italy*
[b] *Department of Physics, University of Ferrara, Via Paradiso 12, 44100 Ferrara, Italy*

Abstract

The knowledge of some vibration frequencies usually represents a basic information in Structural Dynamics. It often happens that a reliable computation of these frequencies is not possible, because mass and/or stiffness distributions are not available with the required accuracy. This is the case of historical monuments, particularly in Europe, where the mechanical properties of old buildings can hardly be determined and damage often affects their structural behaviour. With structures of this kind, it may be very important to find a reasonable number of vibration frequencies by using an adequate test method. Next, by means of appropriate parameter estimation techniques, optimal stiffness parameters can be determined, which allow one to study the structural behaviour. In addition, a periodical experimental check of vibration frequencies can be used as a monitoring technique. Indeed, some interesting information can be obtained about the state of the monument, since progressive damage may severely affect the values of its frequencies.

This paper deals with a simple low-cost experimental method already applied to a few slender structures (chimneys and towers). It is based on the use of a point light source located at the top of the structure. At the bottom a telescope is focused on the light and transmits a beam that hits a silicon plate. When the structure is deformed, the light source tends to move. Thus, also the beam is shifted along the silicon plate and displacements of the light source can be measured. Next, a number of records can be processed. Usually, we consider samples that consist of 256 displacements measured at intervals of 0.02 s. For each sample the Fast Fourier Transform is applied and a spectrum is obtained. Noise may highly affect the spectra, but a simple filtering technique can be used. If n spectra are available and m frequencies f_i are selected for each spectrum ($i = 1, \dots, m$), it is possible to compute an *average* amplitude $A(f_i)$ by using the relationship $A(f_i) = (1/n) \sum_{j=1,n} a_j(f_i)$, where $a_j(f_i)$ denotes the amplitude related to f_i according to the jth spectrum. Eventually, the m amplitudes $A(f_i)$ provide a new spectrum, where a number of peaks can be noted. Each peak corresponds to

PII S0167-6105(97)00134-7

a vibration frequency of the structure, since the *averaging* procedure tends to cross out the effects due to noise, as *n* increases.

The efficiency of the above technique is quite good in the presence of fairly large amplitudes, which tend to reduce the influence of measuring errors. Thus, very good results based upon a limited number of samples can be obtained by using a vibrodine. This device, however, cannot be used with confidence when historical monuments are considered (particularly in the presence of severe damage). Therefore, natural excitation (first of all wind excitation) should be exploited. In this case vibration amplitudes are usually low. Thus, for a better understanding of the experimental procedure, of its performance and of possible limitations, a numerical approach was developed in order to simulate and check the test method. A bell tower was discretised and its vibration frequencies were computed. Next, a convenient numerical model developed in the context of *Wind Engineering* was considered in order to define suitable load histories and to simulate wind action upon the structure. Namely, load histories were introduced by considering a suitable mathematical model which properly describes the stochastic features of wind velocities at different levels and takes into account the interaction between wind and structure. In this way it was possible to simulate the acquisition of any number of sample records (with and without the superposition of measuring errors of any kind). Eventually, by means of the numerical simulation discussed in the paper the basic features of the experimental method could be determined with a satisfactory level of confidence.

ELSEVIER

Journal of Wind Engineering
and Industrial Aerodynamics 67&68 (1997) 947,948

JOURNAL OF wind engineering AND industrial aerodynamics

Prediction of large-scale wind field over complex terrain by finite element method

Takashi Nomura

Department of Civil Engineering, Nihon University, 1-8, Kanda-Surugadai, Chiyoda-ku, Tokyo 101, Japan

Abstract

In order to determine appropriate design wind velocity for civil engineering structures, it is desired to introduce the influence of complex terrain which surrounds the objective structures. For the prediction of such wind fields by means of numerical flow analysis, it is necessary to deal with a large-scale computational domain, tens of kilometers horizontally and several kilometers vertically. This domain covers the Ekman layer where the Coriolis force plays an important role. On the other hand, the structure itself is not so large and the surface layer wind field should also be taken into account. Since 3D analysis is mandatory and it is difficult to cover the large-scale computational domain of the Ekman layer with a fine mesh which can resolve the surface layer, it seems to couple computations of two scale domains: the small-scale one with sufficient fine resolution for the surface layer around the structure; large-scale domain with rough resolution for the surface layer but fine enough for the Ekman layer. The latter large domain computation is expected to give the boundary condition for the former small domain one.

The present work is an attempt of predicting wind fields in the large domain of computation. In order to take into account the Coriolis effect in the Ekman layer, we have employed the Ekman potential model of Yamada et al. [J. Hydraulic Coastal Environ. Eng. JSCE No. 503/II-29 (1994) 49–58]. This model is based on the analytical solution of the wind field in the Ekman potential layer, which expresses a typical velocity profile called the Ekman spiral; exponential decay of velocity magnitude from the geostrophic wind to zero on the terrain surface in association with rotation of the velocity direction. The fundamental velocity potential which represents this velocity profile is modified by superimposing an additional velocity potential which accounts for the irregular boundary geometry due to complex topography. Unlike the Fourier analysis of the original work of Yamada, the boundary value problem to determine the overall velocity potential field is solved by the finite element method which is more straightforward in utilizing distributed digital data of terrain heights.

Fig. 1 shows an example of the computed wind field over an existing mountainous area in Japan. The computational domain covers 8 km × 8 km square area

PII S0167-6105(97)00134-7

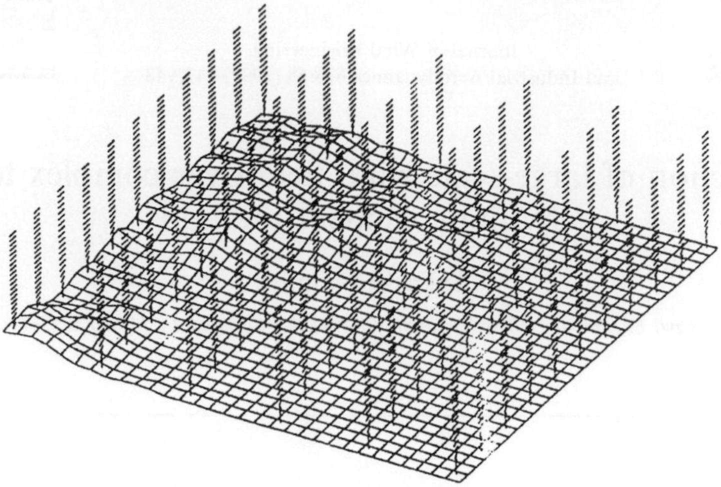

Fig. 1. Predicted wind field.

and 2 km height. The above features of the wind field in the Ekman layer can be observed.

ELSEVIER

Journal of Wind Engineering
and Industrial Aerodynamics 67&68 (1997) 949

JOURNAL OF wind engineering AND industrial aerodynamics

A boundary element-finite difference formulation for wind flow simulation

Okey U. Onyemelukwe

Department of Civil Engineering, University of Central Florida, Orlando, FL 32816, USA

Abstract

This study utilizes a numerical technique combining the Finite-Difference Method (FDM) and the Boundary Element Method (BEM), for computer simulation of wind flow around bluff bodies. The numerical technique developed is used for solving the governing unsteady flow equations of motion for bluff body flow.

A FDM formulation often involves determining the divergence of the Navier–Stokes equations, while enforcing continuity conditions. This mathematical operation results in a Poisson type pressure equation, which is solved along with the momentum equations (in case of 2D, two momentum equations, one for the along-wind direction and the other for the direction normal to it), using Dirichlet and Neuman type boundary conditions. A major drawback of the FDM approach lies in the enormous amount of computational effort and time required for the pressure equation to converge during each time step of the simulation.

To solve this pressure equation a domain-independent numerical technique such as the BEM is favored over the domain-dependent finite-element method (FEM). The BEM which involves discretization of only the boundary, results in the use of smaller sized matrices. The BEM also leads to a direct determination of the wall surface pressures, thus, eliminating the need for interpolation as is the case with FDM.

The technique is then used to simulate the flow around a circular cylinder, a bluff shape for which measured data and numerically computed data are available. The results from this study are compared with the existing data, and the computational efficiency of this FDM–BEM hybrid technique is also evaluated, in comparison with the techniques used by other researchers.

PII S0167-6105(97)00134-7

ELSEVIER

Journal of Wind Engineering
and Industrial Aerodynamics 67&68 (1997) 950

JOURNAL OF
wind engineering
AND
industrial
aerodynamics

Navier–Stokes' solutions for a pair of circular cylinders

Sharad Purohit[1], Girish Chafle, Neeraj Khurana, Sampath Narayanan

Centre for Development of Advanced Computing, Pune, India

Abstract

The unsteady, compressible Navier–Stokes equations, in mass averaged variables, are numerically solved for flow around two right circular cylinders placed one after the other at one and half times the diameter apart in the direction of incoming flow. The vortical flow in the near wake region is investigated with emphasis on surface pressure oscillations. The computations are performed on a parallel machine at the National Param Supercomputing Facility (NPSF) using the explicit finite difference scheme. For high Reynolds number, the time-dependent calculation is carried out for 82×64 grid points in the computational domain to achieve a steady-state periodic solution. The entire flowfield is analysed and comparison with available experimental data is reported.

[1]E-mail: purohit@parcom.ernet.in.

PII S0167-6105(97)00134-7

ELSEVIER

Journal of Wind Engineering
and Industrial Aerodynamics 67&68 (1997) 951

JOURNAL OF wind engineering AND industrial aerodynamics

Effect of wind environment on the response of buildings in Oman

M. Qamaruddin*, Salim K. Al-Oraimi

Department of Civil Engineering, Sultan Qaboos University, P.O. Box 33, Al-Khod 123, Oman

Abstract

An investigation has been carried out to study the effect of various wind parameters on the response of multistory buildings subjected to wind induced forces. This study is concerned with the wind response of the buildings located at six building sites in Sohar, Thumrait and Saiq regions of Oman. A typical plan of a multistory framed building has been chosen for the present investigation. The wind and the other related data appropriate to these regions have been used in the analysis. It turns out from the present study that the response of the building located at a site in a country terrain of a region has higher values than the response obtained for the same building if constructed in a town terrain of the same region with other parameter combinations remaining invariant. It is recommended that a database for the wind parameters, based on the available wind and other related data for different regions of Oman, should be developed for the design of buildings subjected to wind induced forces.

* Corresponding author.

PII S0167-6105(97)00134-7

ELSEVIER

Journal of Wind Engineering
and Industrial Aerodynamics 67&68 (1997) 952

JOURNAL OF wind engineering AND industrial aerodynamics

Multigrid methods for computational wind engineering

R. Panneer Selvam[1]

BELL 4190, University of Arkansas, Fayetteville, AR 72701, USA

Abstract

A review of existing solution procedures to solve steady and unsteady Navier–Stokes equations (NS) using multigrid methods and their relevance to computational wind engineering is discussed. Even though multigrid procedures for steady NS equations are robust; they are not preferable for many wind engineering applications due to their unsteady nature. Hence methods that are available for the solution of unsteady NS equations and when to prefer which procedure will be discussed.

In CWE when time-dependent phenomena are investigated using large eddy simulation (LES), the most time consuming part is the solution of pressure equations. The pressure equation is solved thousands of times for many of the practical problems. Here flow around a circular cylinder and/or flow over a cube is considered for illustration to compare the performance of the multigrid (MG), preconditioned conjugate gradient (PCG) and MG-conjugate gradient (MGCG) solution procedures for the pressure equations. The different smoothers considered for MG and MGCG are Gauss–Siedel (GS), Incomplete Cholesky factorization (ILU), and line GS (LGS). It is found from this study that MGCG using ILU smoother is the most robust for the optimum storage. The pressure equations could be solved in about 5-V cycles or about 10 work units (WU). A WU is the amount of computational work required to perform one GS iteration. If more storage is available, one form PCG is the fastest. MG using ILU smoother is preferable when one wants to use less storage. For general applications and for easy programming PCG using ILU as the preconditioner is preferable.

[1]E-mail: rps@engr.uark.edu.

PII S0167-6105(97)00134-7

ELSEVIER

Journal of Wind Engineering
and Industrial Aerodynamics 67&68 (1997) 953,954

JOURNAL OF wind engineering AND industrial aerodynamics

Parallel performance of a three-dimensional grid-free vortex method for wind engineering simulations

George Turkiyyah[a,*], Dorothy Reed[a], Cecile Viozat[a], Calvin Lin[b]

[a] *Department of Civil Engineering, Box 352700, University of Washington, Seattle, WA 98195, USA*
[b] *Department of Computer Science and Engineering, University of Washington, Seattle, WA 98195, USA*

Abstract

Vortex methods are Lagrangian particle-based numerical simulation schemes especially appropriate for fluid flow characterized by high Reynolds numbers and complex geometries. When coupled with fast solvers for computing vortex interactions, they become ideally suited for wind engineering simulations which tend to be characterized by large Reynolds numbers and irregular geometries. Fast vortex methods have several advantages over grid-based methods: they do not suffer from numerical diffusion; they are simpler to implement particularly as the geometry becomes more complex; and they can exploit the architecture of distributed-memory computers more effectively. In this paper, initial results are presented for a parallel fast vortex method for the simulation of the atmospheric boundary layer wind flow around a bluff body under open terrain conditions.

The basic computational element of vortex methods is a smoothed Lagrangian particle (vortex blob) whose position and the amount of vorticity it carries are updated at each time interval. The locations and strengths of these computational elements determine the vorticity, and therefore the velocity at any point in the domain of computation through a Bio-Savart law. The main computational bottleneck of vortex methods is the computation of pairwise interactions among vortex blobs which must be performed at each time step in order to compute the velocity field. A direct method that evaluates all interactions requires $O(N^2)$ work and is prohibitively expensive. "Fast" vortex methods, that require an essentially linear amount of work, allow a dramatic reduction in the computational cost. Our simulation uses two N-body solvers for computing vortex interactions: one solver computes vortex interactions in a thin region around the ground and building boundaries (a "numerical" boundary layer) while another handles the interactions in the exterior region of the flow. This latter solver represents by far the most computationally expensive part of the computation.

* Corresponding author. E-mail: george@ce.washington.edu.

PII S0167-6105(97)00134-7

Parallel solutions to this component of the simulations are particularly important because meaningful 3D problems may typically require hundreds of thousands of vortices and cannot be solved on high performance workstations. Computational savings in the fast solver are obtained by combining large numbers of particles into a small set of discrete values (rings) whose effect approximates the effect of the particle cluster. Fast solvers construct and evaluate these approximations in a hierarchical fashion much like multigrid solvers. The main parallel data structure consists of a hierarchy of distributed grids that store these ring approximations and transfer information between adjacent levels. We will describe the details of the data structures and the inter-level and intra-level communications required during computation. Our 3D prototype implementation achieves a speedup of more than 11 on a 16-processor Paragon even for relatively small problems.

ELSEVIER

Journal of Wind Engineering
and Industrial Aerodynamics 67&68 (1997) 955,956

JOURNAL OF wind engineering AND industrial aerodynamics

A multigrid method and high-order discretisation for turbulent flow over a building

Nigel Wright

Department of Civil Engineering, University of Nottingham, Nottingham NG7 2RD, UK

Abstract

In pursuit of more insight into the effects of wind on buildings experimental, theoretical and more recently computational approaches have been taken. The latter is limited both by the computing power available and the level of sophistication of models for the physical phenomena – particularly turbulence. In recent years there have been significant advances in both of these that make realisation of accurate numerical solutions feasible. At the University of Nottingham expertise has been developed both in experimental/analytical techniques for wind engineering and in computational fluid dynamics. The latter has included the authors work on multigrid methods and high-order discretisations for incompressible flows and the development of adaptive, unstructured grid techniques for channel flows. The work to be presented here combines these two in order to develop and rigorously validate a computational approach to flows around buildings. This includes the following features:

- A multi block domain structure which allows for irregular geometries and for grid refinement in localised areas. This allows us to set up a finer grid around the structure itself whilst having a coarser one throughout the rest of the domain.
- Algebraic multigrid for efficient solutions. Especially with the refinement mentioned above these problems require large computational grids. An algebraic multigrid method can accelerate the solution of these dramatically.
- k–ε and Reynolds stress models for turbulence with the possibility of extending to Large Eddy Simulation. It has been shown that k-models are not adequate in these cases and that Reynolds stress models offer better predictions.
- A high-order bounded discretisation, CCCT, that overcomes the numerical diffusion of upwind schemes without the under/over shoots of other high-order schemes that are so problematic in turbulence modelling. It has been shown elsewhere that such an approach is essential if the benefits of Reynolds stress models are not to be largely lost through numerical diffusion.

These features were selected because of their relevance to the problems to be considered and are implemented through a CFD package, CFDS-FLOW3D, either by use of advanced features or by developing our own additional modules. Given that

PII S0167-6105(97)00134-7

the underlying data structure and problem definition is dealt with by the package we have been able to develop a complex model quickly and with some confidence due to the extensive validation already carried out for the package. This model is now easily adaptable for more complex geometries without recoding and holds open the possibility of a useful design tool which can take a design and predict flow patterns with the minimum of effort.

We have tested the technique on the problem of a cube in flow with different angles of incidence. There has been extensive work on this situation at Nottingham and data is readily available. In addition to the flow features observed with a 90° angle of incidence, this situation leads to a delta wing vortex that is difficult to predict. The results demonstrate the effectiveness of our approach. The Reynolds stress model gives better results than k–ε, but there is a clear need for more detail particularly given the anisotropy of this situation.

This initial work is to be scaled up and extended to more complicated structures which we hope to present at the conference. Combination of this with LES may be too demanding for present computers, but we believe that we have a validated approach that will be able to take the work further with the minimum of redevelopment as computer power becomes available.

Discussions

ELSEVIER

Journal of Wind Engineering
and Industrial Aerodynamics 67&68 (1997) 959–961

JOURNAL OF
wind engineering
AND
industrial
aerodynamics

DISCUSSIONS OF BLUFF BODY AERODYNAMICS

Discrete vortex model of flow over a square cylinder
Authors: D. Bergstrom, J. Wang

George Turkiyyah: Your simulation presumes a *known* location for the separation point. This is not necessary. If you enforce the traditional no-slip boundary condition by introducing vortex elements on the boundary, you do not need to decide in advance the separation location.

Response: This is true. However, one might argue that an advantage of DVM is that it *avoids* modelling the boundary layer in detail, something perhaps better done with pressure-velocity formulations. For a square cylinder, e.g. bluff body, we know approximately where separation occurs. Therefore the method we used may have the advantage of being simpler for some flow configurations.

Camarero: Is the DVM extensible to 3D?

Response: Yes it is, perhaps not precisely this methodology, but vortex methods are successfully used for simulating 3D flows.

Calculation of the flow past a surface-mounted cube with two-layer turbulence models
Authors: D. Lakehal, W. Rodi

Leighton Cochran: Experimental mean-velocity data in the separated regions of your plots seem to be consistently larger than the CFD results. If these data were collected by a single hot film then the inability of a film to distinguish direction may be contributing to the higher mean. A single film will yield a mean of the absolute values rather than the mean of a highly-fluctuating, reversing flow. The use of cross-films in these highly turbulent environments to discern direction is not very successful either. Would you comment on this as a partial reason for the mismatch in mean-velocity data?

Response: None.

0167-6105/97/$17.00
PII S0167-6105(97)00136-0

Unsteady aerodynamic force prediction on a square cylinder using k–ε turbulence models
Author: S. Lee

Ferziger: Lyn and Rodiś flow was reported to be periodic but it is not actually periodic. Does this affect the validity of your predictions?

Response: I agree with you in that the experimental flow is not periodic as you commented. But if the signal obtained in the experiment has outstanding frequency where the power is at least orders of magnitude higher than the background (or purely turbulent/stochastic) signals, I think the conventional k–ε turbulence model can still be used to predict unsteady flows.

I also agree with you in that this argument cannot be extended to general turbulent flows without any outstanding signal periodicity.

Allan Larsen: Are your calculations 2D or 3D?

Response: 2D.

Jens Honoré Walther: I have conducted simulations on the flow past a circular cylinder at Reynolds number of 3000, and found agreement within 10% of experimental data, using RNG and low Reynolds, number model.

Do you have any comment on today's keynote presentation regarding unsteady flow using k–ε models?

Response: I think that, in these flow cases, the time scales are so different that the turbulence model is capable of picking up the low transient flow while modelling the high frequency turbulence through "equilibrium" k–ε turbulence models.

Finite element modelling of flow around a circular cylinder using LES
Author: R.P. Selvam

Eddy Willemser: The aerodynamic coefficients of a circular cylinder in the supercritical Reynolds number region strongly depend on the surface roughness. Can you cope with this effect in your numerical simulation?

Response: I am modelling for smooth cylinder. This is done by (i) using a smooth surface; (ii) using a proper roughness value in the implementation of the law of the wall.

Lakehal: It sounds that, without talking about turbulence modelling, the finite element method and the control volume approach can produce for the same purpose considerable deviations between results?

Response: For the given grid, as I compared the performance of different procedures for the bench mark problem it seems the characteristic FEM is better than CV procedures.

Development of a parallel code to simulate skewed flow over a bluff body
Authors: T. Thomas, J. Williams

Ted Stathopoulus: How long does it take you for a typical run?

Response: The present simulation using 239 blocks mapped to 128 modes of a CRAY T3D took about 3 s per time step, and about 24 h for the run.

Two-dimensional discrete vortex method for application to bluff body aerodynamics
Authors: J.H. Walther, A. Larsen

Pavit S. Brar: In deriving aerodynamic derivatives, did you simulate flow for different reduced velocity values?

Response: Yes! Typically aerodynamic derivatives will be calculated for reduced wind speeds, V/fB, ranging between 4 and 16 typically, i.e. $V/fB = 4,6,8,\ldots,16$.

B. Bienkiewicz: Did you introduce perturbation to promote flow unsteadiness?

Response: None.

Numerical simulation of flow around rectangular prisms
Authors: D.-H. Yu, A. Kareem

R. Panneer Selvam: For $\text{Re} = 10^5$, the boundary layer depth around the cylinder is approximately $\approx\sqrt{\text{Re}} \leqslant 0.01$. But your grid spacing (fine) $1/20 = 0.05$ is very coarse. So without the law of the wall, modelling grid the flow is not correct. You should run a much finer of and compare the results.

Response: Thank you for your comments. We are looking into further grid refinement and addition of a wall function. However, at this stage, due to the lack of appropriate wall function for the flow shield around prisms we are not very optimistic regarding its influence in the flow field.

Sangsan Lee: If 2D and 3D LES results are similar (even without physical representation in 2D LES), is it ok to say we do not need LES at all?

At $\text{Re} = 10^5$ where you did not use wall functions, what is Δy^+ adjacent to the wall?

Response: None.

ELSEVIER

Journal of Wind Engineering
and Industrial Aerodynamics 67&68 (1997) 962–965

JOURNAL OF
wind engineering
AND
industrial
aerodynamics

DISCUSSIONS OF BRIDGE AERODYNAMICS

A fluid mechanicians view of wind engineering: Large eddy simulation of flow past a cubic obstacle

Authors: K. Shah, J.H. Ferziger

R. Panneer Selvam: In your work, you are using central difference. You refined the grid to reduce oscillations. Did you do any study to know how much the numerical dispersion affects the turbulence or changes the turbulence that you want?

Response: We refined the grid until oscillations disappeared (< 0.5%). This is a necessary but not sufficient criterion for accuracy. We have also looked at some spectra and found that they are sufficiently resolved. I do not believe that there is any justification for using upwind methods in LES (except at shocks); you will get smooth but quantitatively incorrect results.

S. Kato: (1) Do you think it is possible to develop a subgrid model for the coarse grid systems? In the wind engineering field, we always suffer from insufficient mesh dividing systems. (2) What is the difference between the CSC and LES with the coarse grid system?

Response: (1) I believe that it will be possible to do LES on coarse grids. The danger is that the flow may be periodic when it should not be and that is why I think it may be necessary to introduce a random force. (2) They are essentially the same thing. The objective of CSC is to capture the large coherent structures and as little else as necessary.

Jens Honoré Walther: Back scatter seems to be the problem of the dynamic LES. Do you believe, that the mesh can be refined sufficiently to remove backscatter without going to direct numerical simulation?

Response: No.

Roger Pielke, Sr.: You clearly demonstrated the value of the LES approach, in contrast to the Reynold's Averaging Method, to obtain estimates of extreme values

0167-6105/97/$17.00 © 1997 Published by Elsevier Science B.V. All rights reserved.
PII S0167-6105(97)00137-2

(e.g. of pressure, skin stress, etc.). This information should be of considerable value for wind engineers. I would like your comments on this capability of the LES method.

Response: I agree that LES can produce information about the pressure fluctuations. In fact, we are currently producing the PDF of the pressure fluctuations and will try to fit it to the standard statistical models.

Ian Castro: Many people are using unsteady RANS to calculate periodic vortex shedding flows. What is your view of this procedure?

Response: I do not understand how RANS can be used for unsteady flows. The pressumption of RANS is that all unsteadiness is modeled so if an unsteady flow is predicted this premise is rotated. There is also a danger that a periodic flow will be predicted when the actual flow is not periodic.

Fatigue strength design for vortex-induced oscillation and buffetting of a bridge
Authors: M. Hosomi, H. Kobayashi, Y. Nitta

N. Shiraishi: The method presented is the fatigue examination. Have you applied your method for the actual fatigue damaged case of structural members under action of wind such as damaged cases of hangers of arch and others?

Response: We have not applied for actual cases. But, we are going to apply for a girder of a cable stayed bridge. In this study, we checked fatigue strengths of a girder member.

Numerical simulation of flow around a box girder of a long span suspension bridge
Author: S. Kuroda

Allan Larsen: Re is correct (assumed Re $\neq 3.10^5$). What is the CPU time required for your computations? On which computer were they run?

Response: None.

Pavit S. Brar: How do you plan to include body motion for dynamic tests?

Response: In the case there is only one moving body. What I must do is only moving the whole grid rigidly according to the body motion. And the terms related to the grid speed added to the inviscid flux terms.

In the case there is relative motion between the bodies, the minor grid is slid on the main grid (when I use the overlaid grid system).

Aeroelastic analysis of bridge girder sections based on discrete vortex simulations
Authors: A. Larsen, J.H. Walther

W.W. Yang: (1) How do you consider the cross-correlation of the wind turbulence field spanwise? (2) Is the geometrical nonlinearity of the structure included in the model?

Response: (1) Calculations made are 2D and for a steady onset flow, hence there is no effect of upwind turbulence or its lateral correlation. The effect of turbulence on the bridge structure is calculated by standard buffeting routines for which coefficients calculated by the code are introduced. (2) No. the model only considers a 2D cross-section of the bridge deck.

Hiromichi Shirato: (1) How many vortex elements are defined on the body surface? (2) When the free vortex is getting closer to the body surface, the induced wind velocity is sometimes overestimated by the Biot–Savart equation. Could you give us some idea of special treatment for that undesirable phenomenon in your simulation?

Response: (1) The number of vortex panels on a section will typically vary between 100 and 200. At the end of simulations 30.000–60.000 vortices will typically be present in the flow. (2) The Biot-Savart equation is modified by introducing a Gaussian core in which the velocity decreases to 0 at the center of the vortex.

Pavit S. Brar: I noted the discrepancy in St number between your results and experimental results for the Gibraltar bridge section. Discrete vortex methods are known to have some problems in predicting generation of vorticity in the boundary layer. Do you think this could be a factor in the Strouhal number discrepancy found above?

Response: For all "mono-box" sections investigated St have been in very good agreement with experiments, hence the vortex method is not expected to be inaccurate with respect to prediction of St. At the present stage discrepancy is not clear but computations as well as model experiments will be reviewed. It is noted that the predicted St is almost exactly 1/2x the measured St indicating perhaps that the model frequency which locked on to the vortex shedding frequency was the 2nd structural harmonic.

Jae Seok Lee: Your approach is very good to evaluate the aeroelastic behavior of a bridge in a practical point of view.

Have you ever performed dynamic analysis of the 3D bridge structural model using the unsteady wind forces obtained by discrete vortex simulation?

If you have not performed yet, the structural dynamic analysis of 3D model using unsteady wind forces is highly recommended, because it gives more understanding of the structural response of the bridge under unsteady wind forces.

Response: I have not tried yet. I agree with your comment.

Prediction of vortex-induced wind loading on long-span bridges
Authors: S. Lee, J. Lee, J. Kim

Allan Larsen: Did your calculations include effects of structural motion? Can your code handle moving boundaries and thus accommodate non-linear effects?

Response: None.

Karlmanz Beyer: The assumption of all forces in phase along the bridge deck may lead to either safe or unsafe results, depending on length relations and stiffness distributions.

A realistic assumption of coherence should be included.

Response: In-phase forcing along the bridge span is the first possible choice, when forces on the 2–D deck model are available.

Even though the in-phase forcing is not the extreme choice as commented, it may serve as a good starting point to buildup further ideas.

Kangpyo Cho: Did you check the fundamental frequency using spectral analysis from wind tunnel test result? Your first peak of displacement is due to resonance effect. What is the meaning of the second peak?

Response: We checked the fundamental frequency of the structure by comparing with the other researchers' models for the same bridge used in structural analysis and in wind tunnel test.

The second resonance seems to be from the sub-harmonic resonance, whose frequency is twice the fundamental frequency.

An analysis of vortex induced oscillation of long span bridge
Author: S. Lee

N. Shiraishi: In connection with the Seohae Bridge, torsional oscillation seems to tend to occur because of cross-sectional shape. Have you any kind of investigation on vortex-induced oscillations of torsional modes?

Response: We computed torsional moments on the bridge, and they are then applied in the structural analysis. Whenever the vertical displacements are small, the torsional motion does not give any undesirably large oscillations.

ELSEVIER

Journal of Wind Engineering
and Industrial Aerodynamics 67&68 (1997) 966–968

JOURNAL OF wind engineering AND industrial aerodynamics

DISCUSSIONS OF VEHICLE AERODYNAMICS AND DISPERSION

Numerical simulation of air flow in an urban area with regularly aligned blocks
Authors: P. He, T. Katayama, T. Hayashi, J. Tsutsumi, J. Tanimoto, I. Hosooka

P.J. Richards: Your grid appears to be very coarse. Do you think your grid is fine enough to capture important flow features?

Response: The grid scale was decided from our experience. We think that it is appropriate to get only average wind speed. However, it may be not fine enough to discuss the characteristics of turbulence. We are going to try finer grids.

Car exhaust dispersion in a street canyon. Numerical critique of a wind tunnel experiment
Authors: B. Leitl, R.N. Meroney

Nigel Wright: Which of the available discretizations did you use?

Response: As far as I know, there is no chance to change the form of the discretization used. If your question is about the interpolation scheme we have used, we carried out calculations with first order as well as power law and second order upwind scheme. The differences we found were small in terms of the variation of the concentration results but significant in terms of the stability of the solution. All final results were found with second order upwind interpolation.

C.J. Baker: A comment that may cause depression! I have been involved in making both wind tunnel and full scale measurements of street canyon dispersion. The agreement between the two sets of measurements was very poor, and perhaps should act as a caution to our usual assumption that wind tunnel tests are a good representation of reality.

Response: None.

Marcel Bottema: You mentioned measuring problems in street canyons.

Is it perhaps an idea to do an indirect validation of in-canyon flows by considering the development of the internal boundary layer over the obstacle group?

0167-6105/97/$17.00
PII S0167-6105(97)00138-4

Response: As long as we are not quite sure that a numerical simulation represents the "real" flow field I would not push the interpretation of results that far. But you are right – it might be a way that should be considered – an indirect validation of in canyon flow.

Shuzo Murakami: The velocity distribution given from CFD is rather similar to the result of wind tunnel test. However the concentration distribution is rather greatly different from the experiment. It seems curious because the pollutant is connected by the velocity.

Response: I think the main discrepancy is caused by the stationary solution procedure we have chosen. Even if the mean flow structure is predicted quite well we just do not simulate non-stationary vortex shedding or things like a "wash out" of the whole street canyon – and those are effects that dominate the spreading/dispersion driving the wind tunnel experiment.

Sangsan Lee: Some of bluff body flow features can only be predicted accurately by unsteady computations. Why don't you try unsteady instead of steady computations?

Response: I would rather simulate the unsteady situation too but to get a sufficient number of unsteady solutions for averaging our work station needs quite a while for the calculations. But I think we will try an unsteady solution as soon as possible to get an idea about what happens!

An investigation of the ventilation of a day-old chick transport vehicle
Authors: A. Quinn, C. Baker

P.J. Richards: I know that when dealing with flow through porous fences, and when some of my students have looked at flow through porous media, it has been necessary to add sink terms in the momentum, turbulent kinetic energy and dissipation equations in order to correctly model the flow. Have you used anything of this type in your model?

Response: Terms for the momentum sink were included in my model but sink terms in the k and ε equations were not included, but these are clearly necessary.

The effect of wind profile and twist on downwind sail performance
Author: P.J. Richards

R. Hoxey: In the case of attached flow is a k–ε model sufficient to model the boundary layer development on the sail? (Necessary for the prediction of lift, drag, etc.)

Response: In this model the forces primarily result from the pressure. Skin friction is only significant in affecting separation points. Qualitatively the separation occurring in the model appears to be reasonable. At the present time I have two final year students looking at what happens to the flow in our wind tunnel so that we can compare this with the CFD flow patterns.

R. Camarero: Is the grid a stack of 2D horizontal girds? Are the "vertical" grid surfaces forced to go through the top vertex of each sail?

Response: In mapping the grid onto the sail surfaces most of the distortion is in the horizontal plane, but there is some vertical movement of grid modes as well.

In creating the sail surfaces a vertical rectangle of cell faces in the initial Cartesian grid is mapped onto a curved surface as specified by a sail making program. As a result all the "verticals" are forced to go through the top vertex of the sail.

ELSEVIER

Journal of Wind Engineering
and Industrial Aerodynamics 67&68 (1997) 969

JOURNAL OF wind engineering AND industrial aerodynamics

DISCUSSIONS OF STRUCTURAL RESPONSE

Numerical study on the suppression of the vortex-induced vibration of a circular cylinder by acoustic excitation
Author: S. Hiejima, T. Nomura, K. Kimura, Y. Fujino

Allan Larsen: (1) What is the practical application of acoustic suppression vortex shedding? (2) Acoustic excitation leads to an increase of the vortex shedding frequency. Will acoustic excitation also lead to a decrease in the fluctuating lift force?

Response: None.

PII S0167-6105(97)00139-6

ELSEVIER

Journal of Wind Engineering
and Industrial Aerodynamics 67&68 (1997) 970–973

JOURNAL OF
wind engineering
AND
industrial
aerodynamics

DISCUSSIONS OF TERRAIN AERODYNAMICS

Use of meterological models in computational wind engineering
Authors: R. Pielke, M. Nicholls

Kishor C. Mehta: An excellent presentation; I agree that different disciplines need to work together to solve problems of windstorm hazard. From the point of view of design for occupant protection and maximum probable loss of insured property loss we need to know highest wind speed expected. From the atmospheric science models based on physics of the atmosphere is it possible to estimate maximum probable wind speeds at ground surface in hurricanes and tornadoes?

Response: Thank you for your comment.
I feel that the model provides a valuable procedure to obtain realistic estimates of maximum potential winds just above the ground during hurricanes and tornadoes. Since they are based on the fundamental principles of atmospheric flow, the models can provide realistically constrained results.

Sharud Purohit: Is the local meterological model influenced by the global model (from other part of the world)?

Response: Yes, but they give mathematically the same, and it is only a question of incorporating in the basic equations.

Flow and dispersion over hills: Comparison between numerical predictions and experimental data
Authors: D. Apsley, I.P. Castro

Fernando Bocon: Were different grid orientations used to simulate different wind orientations?

Response: Yes, the grid was aligned with the upwind flow direction in every case.

Heinke Schlünzen: How did you calculate the horizontal exchange coefficient? Does it account for stratification?

Response: We used a standard isotropic diffusivity tensor. An anisotropic alternative would be much more complex to code in curvilinear coordinates and, in any case,

PII S0167-6105(97)00140-2

there is a theoretical argument against gradient–diffusion hypotheses (even anisotropic ones) since the eddies responsible for plume spread may be considerably larger than the plume dimensions.

Marcel Bottema: You used wind tunnel measured turbulence inflow conditions which generally contain a significant amount of large scale or inactive turbulence. However, inactive turbulence is not considered in the development of the k–ε model (see Dradshaweal).

How did you manage to get good results?

– Did you tune your radar to the approach flow?

– Does the limited length scale k–ε model account in some way for inactive turbulence?

Response: We did not consider inactive motions directly. But we were careful to match our inlet profiles of velocity and turbulence with properly applied surface conditions (in the application of wall log-law), so that a calculation of the boundary layer in the *absence* of the hill led to only the slow and expected boundary layer growth. The limited length scale model may have some implicit link with inactive motions, and was certainly found to be necessary if good agreement with field measurements of the ABL was required (see our forthcoming papers in Boundary Layer Meteorology).

Cellular convection embedded in the convective planetary boundary layer surface layer

Authors: D.S. DeCroix, Y.-L. Lin, D. Schowalter

Fernando Bocon: Your first point above ground was placed at 5 m height. Did you try to place this point closer to the ground to see if this can cause differences in the results? (The large gradients are near the ground surface.)

Response: This will be addressed in subsequent nested-grid simulations. However, the implicit assumption in using Monin–Obukhov similarity as the surface boundary condition is that the first grid point from the wall is within the surface layer of the planetary boundary layer being simulated. 5 m is certainly within this layer for convective PBLs.

Melville Nicholls: What was the strength of the initial uniform wind and how did the wind field evolve in time?

Response: The environmental uniform wind was 0.1 m/s, since a non-zero value was required. The resultant velocity profile is one typical of a turbulent boundary layer due to the mixing in the convective PBL. However, this effect should be more fully explored since we are trying to simulate free convection.

Flow separation and hydraulic transitions over hills modelled by the Reynolds equations
Authors: K. Eidsvik, T. Utnes

Ian Castro: (1) The conclusion that k–ε might not be too bad for lee-wave dominated flows is consistent with our numerical work.

(2) There is strong experimental and numerical evidence that these can be strong, essentially laminar down-slope winds (jets), with a strongly mixed region (in the breaking wave).

So I am not convinced that k–ε predictions of such a low-turbulence region is a result of inadequacies in k–ε model.

Response: (1) Ok.

(2) Our experience on stably stratified wall comes from self generated turbidity flows. Here the difference between a dynamic Reynolds stress model and a k–ε-model is important for the "vertical" sediment flux and therefore for the whole flux. This difference, may not be important for the lee slope jet behind a hill.

Application of k–ε model to the stable ABL: Pollution in complex terrain
Authors: A. Huser, P. Nilsen, H. Skåtun

Yuji Ohya: (1) What kind of stability parameter do you use in your calculation and how do you define it?

(2) For a stably stratified flow with strong stability, the eddy diffusivity tends to become zero in the lowest part of the atmospheric boundary layer. In your calculation, a certain value of the eddy diffusivity was seen even in the lower part of the boundary layer. What do you think about it?

Response: (1) The parameter in the Drammen model stability parameter is defined based on the measurement at one location 1 and 10 m above ground. This is very sparce and results may suggest that a stronger stability should be applied.

(2) We are assuming a logarithmic temperature profile. This gives finite value of temperature near the surface. Under more stable conditions we should specify a linear temperature profile which will give reduced values or zero.

Wind field in a typhoon boundary layer
Authors: Y. Meng, M. Matsui, K. Hibi

Kishor C. Mehta: The numerical study looks good. Wind load implications of winds in typhoon/hurricane are significant. Is there any field data measurements in typhoons available that will allow validation and improvement in wind field?

Response: There are few field data available to investigate the vertical characteristics of TBL (typhoon boundary layer). We adopted, therefore, a numerical study in which the code was validated by a few field measurements.

We are going to simulate a lot of time histories of typhoon wind speeds and directions for several meteorological stations. It will validate our proposed model indirectly.

Statistical-dynamical downscaling of wind climatologies
Author: H. Mengelkamp

David DeCroix: Regarding the discrepancy between the model results of wind field and direction and the "other" (not NWS) observations, could the difference be due to the lack of a long time-series of observed data? Is the difference due to biases or errors in the completed statistics for the "non-NWS" stations?

Response: Different time periods of observations and simulations may be only one reason for the discrepancy. Influence of the observations by close obstacles or iteration may also be a reason. Because the measurement devices were moved already the latter cannot be investigated. Also inspection of data from synoptic stations for a longer time period did not give an answer.

CFD analysis of mesoscale climate in the greater Tokyo area
Authors: A. Mochida, S. Murakami, T. Ojima, S. Kim, R. Ooka, H. Sugiyama

Heinke Schlünzen: Do you use only one land-use characteristic per grid-cell? What influence on the model results do you expect when allowing several different land-uses per grid cell and taking their effect on the fluxes within one grid-cell into account?

Response: In our calculations, the present situation of land-use in Japan was given by utilizing the numerical data-base for land-use compiled by the National Land Agency of Japan. This data-base provides land-use information for each 100 m × 100 m mesh, while horizontal grid spacings are 8 km × 8 km. So, surface parameters were imposed at each computational mesh, i.e., artificial heating were given by averaging the 6400 values assigned at each 100 m × 100 m mesh, based on the parameterizations shown in Table 2.

A numerical study of stably stratified flows over a two-dimensional hill Part I. Free slip condition on the ground
Authors: T. Uchida, Y. Ohya

Heinke Schlünzen: Is the effect of channel height (H) on model results investigated and what is its influence on the results? Have you considered reflective effects on top and end wall boundary conditions on stratified flow wave behavior?

Response: We think that the effect of the domain depth (H) is very important because the stratification parameter $K(= nHmo)$ contains the domain depth (H).

The flow is very strongly dependent on K as you can see in the numerical results presented in our paper.

ELSEVIER

Journal of Wind Engineering
and Industrial Aerodynamics 67&68 (1997) 974–980

JOURNAL OF
wind engineering
AND
industrial
aerodynamics

DISCUSSIONS OF BUILDING AERODYNAMICS

Computational wind engineering: Past achievements and future challenges
Author: T. Stathopoulos

Ferziger: (1) We showed that you can get excellent agreement between carefully controlled experiments and carefully done simulations. We therefore need to assess the reasons for the variability in field data, experiments and computations.

(2) Numerical error estimation is not difficult and should be required of every author.

Response: (1) The variability of field data is due to the varying atmospheric conditions, as well as to issues related to instrumentation; the latter is also a source of error in the experiment, although lack of adequate representation of field conditions is probably the main reason. Regarding the variability of computational results, one suspects modeling inadequacies, solution errors, invalid assumptions, etc.

(2) I totally agree with this comment. Errors must be estimated when this is possible.

Jack Cermak: Comparisons of mean pressures by CWE and experiment are in fair agreement but RMS pressures are not. What is the prospect for obtaining pressure statistics (pdfs, peak values, etc.) by CWE? These data are essential for wind-engineering application.

Response: It is always more difficult to obtain fluctuating pressures in comparison with mean pressures and this is also true in the case of physical simulation. However, our objective is to compile pdfs, peaks, etc. since these are absolutely necessary for design purposes. LES modeling technique appears promising in this regard.

R. Hoxey: It would be more accurate to predict RMS and peak values from mean pressure coefficients using quasi-steady methods than to compute these within CFD.

Response: This is debatable, particularly for peak values since, in many cases, their probability distribution is far from Gaussion. As you know, also in the case of physical simulation we always measure peak and RMS values instead of calculating them.

0167-6105/97/$17.00
PII S0167-6105(97)00141-4

R. Hoxey: Is the poor comparison of peak and RMS values between full-scale and CFD related to inability of CFD boundary conditions to represent the disturbances in the atmosphere associated with turbulence outside the boundary layer generated turbulence.

Response: I believe that it is more than that; for instance, building generated turbulence is not modelled properly in most cases. Different length and time scales involved in the process of wind-building interaction may not be represented well by the generalized set of equations used. Direct numerical simulation would help solve this issue but, as you know, computational technology has not reached this level as yet.

Numerical considerations for simulations of flow and dispersion around buildings
Authors: I. Cowan, I. Castro, A. Robins

Ferziger: (1) Results should not depend on the method so long as sufficient grid is used. (2) Errors due to numerical methods can be estimated and estimation should be required.

Response: (1) This is, of course, true but the differencing schemes often used have grid convergence rates which are much too slow (with reducing mesh size) to allow many "industrial" users to perform proper grid independency tests. (2) Agreed on both counts.

R. Panneer Selvam: I agree with your comment about solving the equations accurately. Without that we may not have proper solutions.

It is clearly known that the methods are highly diffusive, but then why this test is done?

Response: Some, but not all, numerical methods are highly diffusive. This is one, but not the only one of the features of the different solutions which led to the variability – different groups opted to use different numerical schemes.

A numerical study of wind flow around the TTU building and the roof corner vortex
Authors: J. He, C. Song

Camarero: What are the advantages of using the weakly compressible formulation instead of the incompressible formulation?

Response: Our recent bench mark comparison indicated that the weekly compressible flow approach is 100 times faster than a typical incompressible flow approach for time-dependent flow computation.

This is made possible because the outer solution of the weakly compressible flow equation is incompressible flow, and it is independent of the assumed sound speed. So we can accelerate the solution by assuming small sound speed.

S. Lee: Since you observed little differences in the mean flow field between the medium and the refined grid with so much change in the rims pressure, did you check the change in the trajectories of the corner vortices between the medium and the refined grid?

Response: No change in the time averaged trajectories of the corner vortices occurred for different grid systems. We did not check with the time dependence of the trajectory but I doubt whether we can find any change.

Ted Stathopoulos: Your finest mesh might be sufficient for the model-scale building. Have you made any comparisons with wind-tunnel results?

Response: Yes, we compared with the wind tunnel results also. As indicated in the paper, no change in time averaged quantities simulated with a medium size mesh, and the fine mesh. However, the fine mesh system gives much more turbulent fluctuations than the medium size mesh system. Therefore, we believe that the small scale turbulence requires a very fine mesh to resolve.

R. Hoxey: Did you have a log-law inlet boundary layer, and was it in stability through the solution domain?

Response: We used a power law velocity distribution at the inlet. I do not understand the second part of the question.

R. Panneer Selvam: I assume you do not have inflow turbulence. If you have turbulence at the inflow; how much will it affect the pressures and flow field on the corner vortex?

Response: Although we input turbulence at the upstream boundary our grids are too coarse to transport the small scale turbulence relevant to the corner vortex instability. So as far as the corner vortex turbulence is concerned, it is the same as not putting the free stream turbulence.

Chained analysis of wind tunnel test and CFD on cross ventilation of large-scale market building

Authors: S. Kato, S. Murakami, T. Takahashi, T. Gyobu

Bogusz Bienkiewicz: In my view, hybrid studies in which wind-tunnel / CFD results are coupled, as you have illustrated in your case study, have a great potential for wider application in wind engineering.

To add another example, let me mention our experience in combining wind-tunnel data-external measure on a roof with a numerical model, to determine net wind uplift on loose-laid roofing systems.

Response: Thank you for your comment.

As stated in our paper, the balance of accuracy between CFD and wind tunnel tests is quite important for rational analysis.

Improved turbulence models for estimation of wind loading
Author: S. Kawamoto

Shuzo Murakami: (1) How about the stability of the k–ε–ϕ model compared to the standard k–ε model?

(2) How much is the CPU time increase of the k–ε–ϕ model compared to the standard k–ε model?

Response: (1) Stability of the k–ε–ϕ model is about the same as the standard k–ε model.

(2) It depends on the problems. A steady method can be applied to the standard k–ε model when the inflow turbulence intensity is large because the standard k–ε model leads to steady solutions. But an unsteady time marching method must be applied to the k–ε–ϕ model. Therefore we need much more CPU time for the k–ε–ϕ model. But the increase of the CPU time is caused not by the turbulence model but the time marching method. Therefore, the CPU time is about the same if the same time marching method is applied.

P.J. Richards: Do you see your model housing any advantages or disadvantages compared with the MMK model?

Response: The advantage of my model is the high accuracy for the strongly accelerated flow field, oblique impinging flow field, adverse pressure gradient flow field and the helical flow field. And the disadvantage is the complicated modeling, coding, and the adjustment of the coefficients.

Bob Meroney: Algebraic stress models introduced in 1970s also adjusted for anisotropy and replaced gradient transport assumption. These did *not* require additional transport equations only algebric expressions i.e. of (k–ε). Why have these been abandoned now for additional transport expressions for ϕ?

Response: The anisotropy effect of the ASM (Algebraic Stress Model) is not enough for complicated flow fields, and the anisotropy effect of the DSM (Differential Stress Model) also is not enough because of the difficulty of modeling the pressure–strain correlation term including wall reflection. I intended to introduce the anisotropy effect in a much simpler way in the k–ε–ϕ model. Moreover the drawback of the standard

k–ε model is not only the imperfection of the anisotropy effect but also other imperfections. Though I call the variable ϕ anisotropy parameter, the variable ϕ connects not only the anisotropy effect but also all drawbacks of the standard k–ε model, e.g. the imperfection of the definition of the vortex viscosity and the imperfection of the production term of the turbulence energy dissipation rate ε.

On the application of Thompson's random flight. model to the prediction of particle dispersion within a ventilated airspace
Author: A. Reynolds

Burns: I noticed that your numerical results far from the source were predicting almost zero concentrations, while your measurements indicated significant concentrations near the walls/floor/ceiling. Could this possibly be due to surface contamination?

Response: Since I have written the paper, I have noticed that there are significant transients in the flow, which tends to "smear" out the concentrations.

Full-scale measurements and computational predictions of wind loads on free-standing walls
Authors: A. Robertson, R. Hoxey, P. Richards, W. Ferguson

M. Bottema: During my Ph.D. work I found that for oblique flow corner stream wind speeds keep increasing with relative building width W/H, in accordance with your corner pressure results.
Have you got any explanation for this phenonenon?

Response: The pressure on the front face of the wall is largely independent of length. However the vortex generated on the back of the wall is clearly sensitive to length with no apparent maximum value within the experimental range of results I reported i.e. $L/H = 13$. This applies for oblique flow $\simeq 40°$ to normal. I cannot explain why or how the increasing load will continue with increasing length.

Heinke Schlünzen: What is the accuracy of the measurements?

Response: The pressure coefficients are derived from 10 min mean measurements of surface pressure and wind dynamic pressure. The number of repeat measurements gives an estimated accuracy of 0.01 in C_P around 1%.

Full-scale measurement on free study walls
Author: R.P. Hoxey

A. Baskaran: How were the front to back pressures measured?
What is the reference pressured used in field and CFD model?

Response: All pressures were measured relative to the upstream static pressure sensed by a probe which was calibrated and is insensitive to the horizontal flow direction. The pressures from the 30 tapping points were recorded simultaneously at 5 Hz and differenced to give the overall load.

Mean pressures were integrated in the CFD in a similar manner.

Computation of pressures on Texas Tech University Building using large eddy simulation
Author: R.P. Selvam

Ted Stathopoulos: (1) Since it is well known that wind speed near the ground is non-Gaussian, what was your rationale in using a Gaussian process for turbulence simulation?

(2) Your numerically simulated peak suctions on the roof do not follow the pattern expected, i.e. the highest values are not near the roof edge. How do you justify this?

Response: (1) I tried three different methods of generating turbulence for input: one of them is a Gaussian distribution. This was used by Song and He. So I try to see the performance. The unique part in this work is the use of field measured TTU turbulence data.

(2) I used only 7 equally spaced points on the roof. This is very coarse. If we use a much finer grid, we can have a much more reasonable pressure distribution.

Akashi Mochida: Values of peak pressures are affected by the length of data sampling period. Do you think that the length of data sampling period in your computation is sufficiently long to obtain reliable peak values?

Response: You may be correct that for longer time we may have a much higher peak value. At the same time from our experiments we find that the peak pressures occur when the max and min velocity occur in a short time. More work is needed to learn more about it. By this we can compute the peak pressures with less CPU.

S. Lee: As it was pointed out in the paper that the peak pressure on the building surface corresponds closely to the inflow velocity fluctuations it cannot be over emphasized that the inflow turbulence boundary condition is to be realistic (or physical) to get meaningful computational results. Just a comment.

Response: Thank you for the support and understanding.

Allan Larsen: Have you mapped out the flow field (instantaneous) around the building?

Response: Not this time. I will try to do in the future. Prof. Murakami plotted the instantaneous velocity for this kind of work in many of his papers.

Application of computational techniques for studies of wind pressure coefficients around the odd-geometrical building
Authors: S.-H. Suh, H.-W. Roh, H.-R. Kim, K.-Y. Lee, K.-S. Kim

A. Mochida: Did you examine the accuracy of your CFD code based RNG-ε model through the comparison of CFD results with measured data?

Response: None.

Numerical prediction of wind loading on buildings and structures activities of AIJ cooperative project on CFD
authors: T. Tamura, H. Kawai, S. Kawamoto, K. Nozawa, S. Sakamoto, T. Ohkuma

Jae Seok Lee: Your extensive work on prediction of unsteady wind loading on structures would be useful guidelines for other's further computational efforts to evaluate wind loadings on bridge and buildings.

Could you give me rough ideas on grid size and computer time of your work? I think the computational methods would be practical in case all computations could be done within practically allowable time and costs.

Do you have any plans to continue this type of cooperative project on CFD?

Response: I cannot say exactly because grid size and computer time varies case by case. Rough ideas on that would be given later.

Works on improved turbulence model and LES would be the focus of the ongoing cooperative project on CFD.

The assessment of wind loads on roof overhang of low-rise buildings
Authors: T. Wiik, E. Hansen

Marcel Bottema: Your inflow turbulence levels seem to be extremely low. What kind of inflow turbulence conditions did you use?

Response: Your observation is correct. For this simulations I did not model the turbulence level at inlet boundary. Still my main velocity profile was the same through an "empty tunnel".

ELSEVIER
Journal of Wind Engineering and Industrial Aerodynamics 67&68 (1997) 981–983
JOURNAL OF wind engineering AND industrial aerodynamics

DISCUSSIONS OF AIR POLLUTION

A resistance approach to analysis of natural ventilation airflow networks
Author: R. Aynsley

Burns: I noticed that the dimensions of e.g. (8) might not be correct.

Response: I will have to check on this.

Modelling particulate dispersion in the wake of a vehicle
Authors: Z.E. Hider, S. Hibberd, C. Baker

D.Y.C. Leung: (1) What particle sizes have been considered? (Particulate emitted from vehicles may be as small as 0.1 μm or as large as 10 μm.)

(2) Have you considered the ground reflection effect?

(3) How can you apply your results in a real situation with the influence of recontainment of dust on the road?

Response: (1) 10 μm to 1 mm.

(2) Yes – ground boundary conditions were imposed.

(3) The results will be applied to the real situation.

Castro: Vehicle wakes contain strong axial vorticity (typically, in the *mean*, counter-rotating vortex pairs).

The similarity approaches used do not embody such phenomena.

To what extent do you think this axial vorticity will control or at least significantly effect the dispersion process?

Response: I agree that vehicle wakes do contain strong axial vorticity: However, this vorticity is most dominant near the vehicle, to within 2 or 3 car lengths. After this the effect became less important. The work I have been doing is concerned with the far wake case, i.e. at distances greater than 3–5 car lengths.

Lakehal: (1) Have you used a Lagrangian approach?

(2) Do you think that if you take into account the effect of turbulence on the particle movement will improve your predictions?

Response: (1) The approach described in the paper is a Lagrangian one.

(2) The effect of turbulence will be incorporated into the model in the near future.

PII S0167-6105(97)00142-6

Marcel Bottema: For wide obstacles in oblique flow the wake is laterally displaced whereas right behind the obstacle there is downwash and increased wind speeds (like for in a crosswind)

Did you take this into account, or will you so?

Response: Primarily the work we have been considering is at several vehicle lengths downstream of the car. The situation you are describing is an effect that occurs near the back of the car. Although, some of these features can be observed for the case where the vehicle experiences a high shear layer effect.

Concentration and flow distributions in the vicinity of U-shaped buildings: Wind tunnel and computational data
Authors: B. Leitl, P. Kastner-Klein, M. Rau, R. Meroney

Ron Petersen: Did you make your wind tunnel versus model comparisons at full scale or model scale? Would tunnel wall effects be an issue?

Response: The comparisons with the FLUENT-calculations were done on model-scale. The calculations with the models used in Germany were based on full-scale. Wall effects on dispersion of gases released from point sources placed in the center of the wind-tunnel are ineligible. Furthermore the big differences between the various model results could not be explained by these effects as well as the fact that one model shows over-prediction for wind direction *A* and under prediction for wind-direction *B*.

Numerical evaluation of wind-induced dispersion of pollutants around a building
Authors: Y. Li, T. Stathopoulos

Yoshihide Tominaga: Do you compare the velocity field given from computation with that given from experiment?

In that comparison, is good agreement shown?

Response: The velocity fields were not compared with experimental results for the two cases simulated because those fields were not given in the experiments. Comparison between experimental and numerical results for other cases can be seen in the literature from the paper authored by Dr. T. Stathopoulos and Dr. Baskaran.

A numerical study of a thermally stratified boundary layer under various stable conditions
Authors: Y. Ohya, H. Hashimoto, S. Ozono

Shuzo Murakami: Is some type of turbulence model used? If it is direct numerical simulation, is the mesh discretization fine enough?

Response: We have used a finite-difference method without any turbulence model, i.e., a direct simulation was employed.

In the present direct simulation, the grid resolution was not enough, because a discrepancy in the behaviour of fluid motion in the lowest part of the boundary layer between calculations and experiments was observed.

Ian Castro: Your calculations were at a Re substantially below that of the experiments. One might expect the effects of increasing Richardson on the turbulence structure to be more significant at low Reynolds numbers.

Might this be the major reason for the differences you found between your experiments and calculations?

Response: For the differences in turbulence characteristics between the calculation and experiment, the reason can be explained partly because the Reynolds number is different and partly because the grid resolution in the calculation is too coarse for the simulation of turbulent boundary layers.

CFD prediction of gaseous diffusion around a cubic model using a dynamic mixed SGS model based on composite grid technique
Authors: Y. Tominaga, S. Murakami, A. Mochida

R.P. Selvam: (1) Did you compare with experiment transport problems where the source is kept at the roof level? This is a much more difficult problem than the one on the ground.

(2) Do you have any idea how much numerical diffusion you have using central difference for convection?

Response: (1) Not yet.

(2) In this computation, the numerical diffusion is small because mesh independence had been checked. It is important to check the mesh independence.

A vertical round jet issuing into an unsteady crossflow consisting of a mean current and a sinusoidal fluctuating component
Authors: L.P. Xia, K. Lam

Fernando Bocon: How many grid cells were used to represent the jet nozzle?

Response: Eight or so.

Journal of Wind Engineering
and Industrial Aerodynamics 67&68 (1997) 984, 985

DISCUSSIONS OF LAB METHODOLOGY AND VALIDATION

Turbulence control in multiple-fan wind tunnels
Authors: A. Nishi, H. Kikugawa, Y. Matsuda, D. Tashiro

Bob Meroney: Can you use your feedback control approach to simulate non-stationary wind episodes like thunderstorm winds on downbursts?

Response: It depends on both the fan performance and the time-lag of the wind tunnel. If the sensitive fans are used and the wind tunnel time constant is small, the sharp and large fluctuations can be reproduced by feedback control method. So that, a large number of sensitive small fans should be used and shorter tunnel length is advantageous.

Application of infrared thermography and a knowledge-based system to the evaluation of the pedestrian-level wind environment around buildings
Authors: R. Sasaki, Y. Uematsu, M. Yamada, H. Saeki

Marcel Bottema: You corrected your results with an expert. It seems to me that your infrared thermography results can be applied directly as a measure of wind chill potential. (Actually you measure heat exchange), and that thus it is complementary to conventional methods for evaluating mechanical wind effects on people.
Could you comment on this?

Response: Our purpose in this study is to establish the method of evaluating mean wind speed using infrared thermography, because mean wind speed is usually used to evaluate wind environment. As I presented, infrared thermography tends to overestimate the mean wind speed in the regions in front of and behind the model. So, we used the knowledge-based system to correct it.

On the other hand, we are very interested in the original results of infrared thermography. It may indicate the wind effect on people, as you said. Therefore, we have been developing another method of evaluating wind environment based on human sense.

PII S0167-6105(97)00143-8

Some remarks on the validation of small-scale dispersion models with field and laboratory data
Authors: M. Schatzmann, S. Rafailidis, M. Pavageau

Bob Meroney: Is it appropriate to arrange a laboratory facility comparison exercise to validate the use of different wind/water equipment and measurement techniques?

Response: Yes, of course it is. In fact, the German Wind Engineering Society has already adopted this idea. All German wind tunnel facilities do presently the same set of experiments. The result will probably be available early next year.

Lakehal: Amongst the numerical approaches that you pointed out as eventual more accurate models vis-a-vis the steady-state one, you have mentioned the Lagrangian approach. From a numerical point of view, these techniques are not able to identify the effect of dispersed phase on the transporting phase's turbulence.

Response: I mentioned the Lagrangian models just as an example. When pairs of particles are released which are correlated to each other, they can (amongst other models) also produce time series. I see the need for numerical models which produce time series. Then, the model time series and the time series from the experiment can be treated statistically, and the statistical properties can then be compared with each other.

ELSEVIER

Journal of Wind Engineering
and Industrial Aerodynamics 67&68 (1997) 986, 987

JOURNAL OF
wind engineering
AND
industrial
aerodynamics

DISCUSSIONS OF NEW COMPUTATIONAL SCHEMES

Turbulence closure model "constants" and the problems of "inactive" atmospheric turbulence
Author: M. Bottema

Castro: When you adjusted c_μ for your sensitivity tests, did you also make the appropriate adjustment to δ_ε (which is a function of c_μ)?

Response: In the atmosphere surface layer, a c_μ modification should indeed be accompanied by a δ_ε modification for streamwise equilibrium. In ε $(\partial \in / \partial \in = 0)$, and so I did. It is not clear what should be done in the roughness sub-layer (RSL) but previous tests for isolated obstacle flows suggest that these flows are rather insensitive to δ_ε-modifications.

Determination of plume capture by the building wake
Authors: M. Brzoska, D. Stock, B. Lamb

Camarero: In your grid adaption procedure, what is the error estimation procedure?

Response: We used Roache's grid convergence index (Roache "A method for uniform reporting of grid refinement studies", American Society of Mechanism Engineers FED Summer Meeting, 1993, pp. 1–12).

Simulating the dynamics of spray droplets in the atmosphere using Ballistic and random-walk models combined
Authors: M. Mokeba, D. Salt, B. Lee, M. Ford

R. Hoxey: Are you modeling the estimated downflow and also the resulting upward movement of air?

Response: We are modeling estimated downflow but not the resulting upward circulation further away from the spray zone.

PII S0167-6105(97)00144-X

Allan Larsen: Do you plan on experimental verification of the spray model?

Response: Yes. We plan to have a cooperative program with Silsoe Research Station in the UK who will carry out tests in a specially constructed wind tunnel for pesticide and insecticide spraying.

A.C. Allar: Does the model include the effect and evaporative cooling on the buoyancy.

Response: Yes, it does.

Chris Balsi: What is the displacement height used in the log law profile, and what is the near crop turbulence intensity?

Response: (1) There is no assumption of a displacement length in the trajectory model being used at present. The next step in the development of the model is to assume that the crop surface is oscillating with a motion determined by the structural dynamic characteristics of the crop itself. Thus the vertical motion of the ground plane negates the requirement for a displacement length. It should also be noted that most particles will have their motion dominated by the turbulence random walk model when close to the surface.

(2) The turbulence intensity at all points is determined from the formulation by Pamofsky, Ref. [xx] in the paper.

ELSEVIER

Journal of Wind Engineering
and Industrial Aerodynamics 67&68 (1997) 988–991

JOURNAL OF
wind engineering
AND
industrial
aerodynamics

DISCUSSIONS OF THE APPENDIX

Forecasting roof snow accumulation using numerical models
Authors: A. Baskaran, A. Kashef

A.K. Ghosh: Snow accumulation for particular slope angle was studied. How does the snow accumulation vary with slope angle of the roof?

Response: Yes; snow accumulation depends on roof slope. Equally it depends on the wind angle. We studied three different roof slopes for three directions. This created a matrix of 9 cases.

Jon peterka: Do you have a quantitative validation for your method?

Response: Yes, for the simulated mean wind velocity. We do not have experimental data that are similar. The modelled case, for snow accumulation.

Tore Wiik: Do you calculate the snow transpirtation (erosion/deposition) on the ground? If so, what kind of equation do you use for this movement?

Response: Yes, we calculate snow accumulation around the building on the ground. We used the same equation as computed on the building roof.

Tore Wiik: Have you considered the snow accumulation over time? Would not the accumulation over time change the flow pattern near the surfaces?

Response: Yes, I agree with you that snow accumulation will change with time. However, our study is focused on steady flow conditions.

Simulation of unsteady flow around oscillating bluff bodies using a hybrid Eulerian Lagrangian scheme
Authors: P.S. Brar, O.M. Knio, R.H. Scanlan

Allan Larsen: How do you determine the size of the Eulerian mesh region? At large structural motions vortices from the Lagrangian region may wash into the Eulerian region.

PII S0167-6105(97)00135-9

Response: We have not looked into the question of determining the size of the Eulerian region yet but the essential idea is to capture the near-wall viscous region by an Eulerian scheme.

As described in my presentation, influx of vorticity into the Eulerian region is taken care of by a triangle shaped cloud (TSC) scheme so large structural motions do not cause any extra problem.

Computer simulation of the dynamic response of wind-loaded large bridges under limit load conditions
Authors: I. Kovacs, K. Beyer, N. Oba

Kichiro Kimura: Have you included the aerodynamic admittance in the formulation of buffeting forces?

Response: The aerodynamic admittance is directly included through the consideration of the coherences. These functions however vary along the longitudinal axis.

Numerical simulation of flow over two successive buildings of square cross section
Authors: S.K. Maharana, A.K. Ghosh

P.J. Richards: (1) What form of inlet boundary profiles for velocity ε did you use. (2) Jensen showed about 30 years ago that wind tunnel tests conducted in uniform flow generated unrealistic results. Why are you now making the same errors in your computer model?

Response: (1) Uniform velocity profile. (2) We have not considered boundary layer profile as input for this presentation. A realistic boundary layer profile will certainly be incorporated in the next phase of our work.

J. Tsutsumi: I cannot understand the distribution of C_p. Please explain the nearly zero value on the windward wall and positive value on the roof.

Response: The possible explanation for this trend is that the incoming wind is parallel with a uniform velocity distribution, and a boundary layer profile for approach flow has not been considered in the analysis.

Wind induced vibration of bell-towers: An experimental procedure to find vibration frequencies and test simulation based on wind engineering
Authors: A. Nappi, F. Pedrielli, M. Pian

Allan Larsen: Did you measure mode shapes in addition to frequencies?

Response: None.

Navier–Stokes' solutions for a pair of circular cylinders
Authors: S. Purohit, G. Chafle, N. Khurana, S. Narayanan

Camarero: What is the geometric representation of the obstacles. I.e. does your grid generation package accept IGES or NURBS standard?

Response: Circular cylinder. The grid generation package has a library of different curves.

R. Panneer Selvam: What is the smallest spacing of the grid in the radial direction. What is the total number of grid points used.

Response: The number of grid points was 82×64 with a stretched grid.

Castro: You did not employ a turbulence model. What was your flow Reynolds number?

Response: I still need to know about an appropriate Reynolds number for the typical flow situations. The Reynolds number in this case was one million.

Multigrid methods for computational wind engineering
Author: R.P. Selvam

Nigel Wright: (1) How do you decide on convergence, as MG will perform better as the convergence criterion is reduced? (2) How do you coarsen the grid? How would this be possible in more complex geometries?

Response: (1) Yes. I agree with you. The convergence is faster for MG when the convergence criterion is reduced. (2) I am coarsening the grid in the usual way suggested by Brandt. For complex geometries, this is not applicable. But methods by Lohner et al. using an unstructured grid from one level to another level can be used.

A multigrid method and high-order discretisation for turbulent flow over a building
Author: N. Wright

Jens Honoré Walther: Did you use upwind in the turbulence equation?

Response: Yes and no. The default is upwind (commercial code), but tried higher order upwind. This caused stability problems which were overcome by pseudo-transient approach.

T. Wiik: What was the cell aspect ratio in your simulations?

Response: Generally near 1, but in some cases near the inlet and outlet up to 10.

Castro: You concluded that the k–ε turbulence model is used, it is not worth using higher order discretisation scheme.

I would argue that whatever turbulence model is used, the best possible discretisation and grid should be used. Otherwise, no useful conclusions can be reached from comparison of the results with other solutions or experimental data.

Response: The main conclusion is that both improved discretisation and turbulence model are required. High order discretisation does improve the k–ε solution, so I would agree with your interpretation.

ELSEVIER

Journal of Wind Engineering
and Industrial Aerodynamics 67 & 68 (1997) 993–998

JOURNAL OF wind engineering AND industrial aerodynamics

Author index to volumes 67 & 68 (1997)

Elsevier Science B.V.